Pharmaceutical Organic Chemistry

Pharmaceutical Organic Chemistry

Dr. V. Alagarsamy
M. Pharm, PhD, FIC, DOMH.
Professor and Principal,
MNR College of Pharmacy,
Sangareddy - 502 294,
Gr. Hyderabad, TS, India.

PharmaMed Press
An imprint of Pharma Book Syndicate
A Unit of BSP Books Pvt. Ltd.
4-4-309/316, Giriraj Lane,
Sultan Bazar, Hyderabad - 500 095.

Pharmaceutical Organic Chemistry by ***Dr. V. Alagarsamy***

Published by:

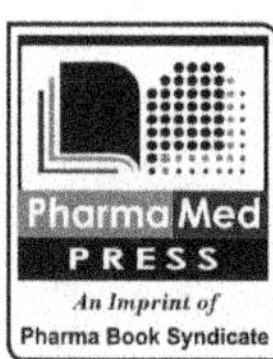

PharmaMed Press

An imprint of Pharma Book Syndicate

A Unit of BSP Books Pvt. Ltd.

4-4-309/316, Giriraj Lane, Sultan Bazar, Hyderabad - 500 095.
Phone: 040-23445600, 23445688; Fax: 91+40-23445611
E-mail: info@pharmamedpress.com
www.pharmamedpress.com/pharmamedpress.net

ISBN: 978-93-89974-37-9

Dedication

My First & Best Teacher - Beloved Mother

Gave me not only Respiration but also Inspiration

Shrimathi. V. KAMUTHAI

The inspirational words of my mother ringing in my ears are.......

"The two most important days in your life are the day you are born and the day you find out why you are born. In your life either write something worth reading or do something worth writing. You should dedicate yourself to your profession and do your best that should answer why you are born."

Whatever the mind of man can conceive and believe, it can achieve. Challenges are what makes life interesting and overcoming them is what makes life meaningful. In order to succeed in your life, your desire for success should be greater than your fear of failure. It does not matter how slowly you go as long as you do not stop.

We become what we think about; hence your thoughts should always be of high-quality. Definiteness of purpose is the starting point of all achievement. If you believe you can do it you are halfway there. Whatever you can do, or dream you can, begin it. Boldness has genius, power and magic in it. It will give you the required strength. "Nothing is Impossible in the world, because the word impossible itself says, I'm possible!"

When everything seems to be going against you, remember that the aircraft takes off against the wind, not with it. That is why it is reaching the desired destination. If you compromise for others, you cannot reach your destination, instead you will reach others destination.

Life shrinks or expands in proportion to one's courage. Limitations live only in our minds. But if we use our imaginations, our possibilities become limitless. Hence you have enough opportunities in the world if you have courage. Do what you can, where you are, with what you have. You should have a dream in your life and to achieve it if you have good idea and clear plan with full devotion, it will fetch you sure success even if the God says impossible. Because the hard work has its own power, it will never fail. You should not be a product of your circumstances. You should be a product of your decisions.

Preface

Pharmaceutical organic chemistry is the main branch of organic chemistry deals with the study of preparation, structure and reactions of organic compounds. As it deals with all the chemical reactions related to life, study of Pharmaceutical organic chemistry is important. Application of Organic chemistry in the development of pharmaceuticals, resulted in evolving Pharmaceutical organic chemistry. Hence studying Organic chemistry and applying this knowledge in Pharmaceutical substances is called as Pharmaceutical organic chemistry. Organic chemistry forms the basis of biochemistry, in which various aspects of health and diseases are studied. The biochemical knowledge is very important for the practice of nutritional, medical and related life sciences. In addition Organic chemistry paved way for the development of medicinal chemistry, Pharmaceutical organic chemistry, bioinformatics, biotechnology, gene therapy, Pharmacology, pathology, chemical engineering, dental science and so on. Organic substances play such a vital role in our daily life that all of us should know about organic chemistry in order to understand the manner how it influence our life process.

In the given conditions, a specific compound is inert or reactive, and if it is reactive, how will it react? Such knowledge will be used to design the structure of a substance that will have a specifically desired property. We will then know what substance to be used for making the parts for various instruments? Which drug to be used for a specific disease? Many of careers such as Doctors, Engineers, Pharmacists, Veterinarians, Dentists, Pharmacologists and Chemists make use of the knowledge of this fundamental subject. Hence the study of this basic Organic chemistry subject will make you to become successful professionals.

The field of organic chemistry is incredibly vast subject. Why is organic chemistry considered to be so difficult? in which "How to start" ? "What to study" ? and "How to study and remember the chemical reactions"? Is this either an antiquated misconception, or is absolutely true ? ? ?

The book has been designed to meet the needs of the intended readers with reference to the above questions.

Students should start learning organic chemistry by understanding only, not through mechanical memorization like "a poem learnt by rote in childhood". Organic chemistry is not a difficult subject, and once you understand, it will become an enjoyable subject and you blast your way by proposing your own way of reaction one after another.

This book is a product of my vision to design the best book on Pharmaceutical organic chemistry, which deals with the origin of organic chemistry, the concise description of structure of atom & organic molecules and their related properties, the nature of organic reactions & their mechanisms, nomenclature of organic compounds, clear classification of various organic compounds, preparation of each class of organic compounds by various routes, its chemical structure, physical properties and chemical reactions with the mechanism in a simplified manner and drugs derived from each class along with their applications in medicine. Swathing the entire features of Pharmaceutical organic chemistry, first of its kind, is the unique feature of this book. It facilitates the students to understand the subject more easily and make the subject interest.

As the students entering graduate course, understanding of Pharmaceutical organic chemistry is always been a difficult task especially the various chemical reactions with their mechanisms of different kind of organic compounds. Hence my efforts have been devoted to authoring a book by equipping with the challenging requirements of the subject for the new generations of the teachers all over India and is easy to read for students who are not necessarily of Pharmacy program, but mainly for the students of first time reading organic reactions, their mechanisms and their applications in science, Pharmacy and medicine.

Methodical description of each chapter, enriching chemical and pharmaceutical background of the medicinally important organic compounds, proceeding each group of organic compounds in a systematic way in easy-to-understand style in the larger interest of the students without any difficulty and making the reader acquainted thoroughly with chemistry of organic compounds is the unique feature of this book.

In preparing this text book, I have tried to enrich the importance of organic compounds in medicine and pharmacy, so that the anticipated audience of this book will feel the importance of organic compounds and made the book a comprehensive. The students of Pharmacy graduates of our country faced with the scarcity of books to serve their needs. Few of the authors dealt well about the basics of organic chemistry, but the reactions of organic compounds are not presented in a easy to understand manner which students felt difficult to learn. Some of the books of organic chemistry fail to give the chemical structures for all reactions, hence the students are unable to understand the reactions clearly. Hence the content of this book is made as a humble attempt to cater the needs of academicians belonging to all Indian Universities by incorporating the chemical structure of all reactions and enriching basic principles of each organic class of compounds.

The book has covered the entire Pharmaceutical organic chemistry, starting from origin of organic chemistry to advanced topics like stereochemistry and heterocyclic compounds and it is divided into 31 chapters. **Chapter 1 to 7** deals with the basics of the organic chemistry, wherein the fundamentals like origin and development of organic chemistry, structure of organic molecules and their related properties are described. Classification and nomenclature of organic compounds and general terms used are also presented in a systematic way, which is easy to understood and able to reproduce well in examinations.

Chapter 8 to 25 deals with Aliphatic and Aromatic compounds which are further divided into different chapters and each chapter is dealt with Introduction, Importance of each class of compounds, Nomenclature, General methods of preparation, General physical and chemical properties and Pharmaceutically important organic compounds of each chapter are described in a easy to understand manner. Summary of methods of preparation and chemical reactions in a flow chart manner presented is unique and helps student to remember for exam which is the first of its kind.

Chapter 26 to 28, Isomerism and detailed description of optical and geometrical isomerism is presented in a simplified manner.

Chapter 29 to 30, Heterocyclic rings of various types and their utility in pharmaceutical chemistry is described well. The medicinal compounds derived from each heterocycles also exemplified which makes the reader inspiring.

In **Chapter 31,** Important reactions and reagents used in organic chemistry and some of the special reactions are described with their mechanism and applications in Pharmaceutical organic chemistry.

To inspire the readers and make them attracted, interesting facts about great scientists and Organic compounds and their discovery etc are given under each chapters.

In order to help students to remember well and reproduce well in exams, the style and presentation are followed simple and similar pattern in all chapters. We have also used some special abbreviations like **"CAD"** (which helps to remember cold Alkaline dilute).

To help the students to learn and magnetize the attention we have used color in equations and diagrams. We have done it this thoughtfully and purposefully and not just to make the book attractive. These small inputs helps a lot for the fresher's who enter into degree program.

We hope that this special volume will be a good source of information and reference for not only to graduates and post-graduate students but also for basic and applied researchers in this field. Moreover it will also be of interest to a wide range of scientists who are involving in the Pharmaceutical organic chemistry related research. I welcome suggestions and constructive criticism from all corners of scientific community.

V. Alagarsamy
drvalagarsamy@gmail.com

What is the best way to study organic chemistry?

- Develop the desire to study organic chemistry
- Read the basic points thoroughly
- Read the chemical structures and correlate with the basics studied
- Be through with the nomenclature
- Always prepare your own notes by applying the above knowledge
- Read each and every step of the reaction and mechanism thoroughly
- Practice daily
- Make use of the lab time
- Discuss in a group and clear the doubts spontaneously
- Break large tasks in smaller ones

Acknowledgement

A warm response to my earlier Books on the "Text Book of Medicinal Chemistry", "Pharmaceutical Chemistry of Natural Products" (Published by Elsevier), Pharmaceutical Inorganic Chemistry (Published by Pharma Book Syndiacte) prompted me to write this Book on "Pharmaceutical Organic Chemistry", covering all the topics suggested by the Pharmacy Council of India (PCI) for graduate students. I dedicate this book to the many hundreds of budding graduates and pharmacy students, whom I have taught over the years and my Teachers who have encouraged me to convert my class notes into the text book in order to reach into wide range of academic community.

It is my pleasure to place on record my heartfelt thanks to everyone who have made this book possible, especially my beloved teachers of 10+2 class to Ph.D., who made me to read the Organic Chemistry in a simplified manner while the other students felt difficulty in reading and reproducing in the Exams.

I am immensely grateful to Prof. K. Chinnaswamy, Dr. B. Suresh (President, Pharmacy Council of India) and Dr. R.K. Goyal, Dr. Rajani Giridhar, Dr. M.R. Yadav, Dr. C.J. Shishoo and Dr. U.S. Pathak for their constant support, inspiration and initiation extended to me to author this book.

I gratefully acknowledge the constant and continuous encouragement and moral support extended by Shri M.N. Raju, Chairman, and Mr. M. Ravi Varma, Vice Chairman, MNR Educational Trust, Hyderabad, for my entire academic and research pursuit, is a great motivation for me in taking up this new challenge.

I thank Prof. R. Shyam Sunder and Dr. Kavita Waghray Faculty of technology, Osmania University, Hyderabad for their constant support in all my academic activities.

I express my sincere appreciation to my students, research scholars and friends especially, Dr. V. Raja Solomon (Postdoctoral Research Associate, Laurentian University, Canada), Dr. G. Saravanan, Mrs. M.T. Sulthana, Mr. B. Narendhar, Mrs. K. Lahari and Dr. P. Subhash Chandra Bose for their support in making this book.

I thank Mr. Anil Shah, Managing Director, PharmaMed Press for recognising and inviting me to write this book. The friendly interaction experienced with the Pharma Book Syndicate, Mr. Naresh (Production Manager) and his team offered a cordial support, which always made me furnish my inputs to make this Book one of the Best. Getting such a cooperative and energetic team encourage the author to continue their writing always. I thank them whole heartedly for accepting all views while designing the book and helping me reach this target.

I also express my sincere thanks to my father-in-law (Shri. V. Sundara Rajan) and mother-in-law (Ms. S. Chinnammal) for their kind encouragement and moral support throughout my Career. My father-in-law born in a small village, studied upto 10th class only, had joined as a Police and came up to the level of Police Inspector is a great motivation for me.

In all my academic efforts the stimulation I gain from my father (Mr. P. R. Veerachamy), mother (Ms. V. Kamuthai), sister, brothers and wife A. Sathyabhama to reach this goal is like the nature provides sunlight for photosynthesis for healthy maintenance of humans, and the patience and cooperation extended by my children, A. Dharshini Aishwarya (B.Pharm) and A. Dharshini Abhinaya made me think of the goal without any diversion. To express my thankfulness, I pray the Almighty to bless my children with teachers like those I got in my life so that they too are inspired by their teachers and dedicate to the field of Pharmacy and, in turn, serve for the mankind.

V. Alagarsamy

drvalagarsamy@gmail.com

WhatsApp: +91 7674893936

Contents

1

Introduction to Organic Chemistry

Organic chemistry is the major branch of chemistry which deals with the scientific study of preparation, structure, properties, composition and reactions of carbon containing compounds. In organic chemistry, not only hydrocarbons are studied but also compounds in which carbon is bonded with any other atoms like oxygen, halogens, nitrogen, phosphorus and sulfur *etc.* Almost all organic compounds contain at least one carbon hydrogen bond (C-H) in it.

Origin of Organic Chemistry

The history of organic chemistry can be traced back to ancient times when the tribes extracted chemicals from plants and animals to treat their ailments. But they did not name it as organic chemistry. The records what they kept are useful in the development of modern organic chemistry.

In 1700s, the French noble man and chemist, *Antonie-Laurent De Lavoisier* showed that nearly all substances of plant origin are composed of carbon, hydrogen and oxygen. Organic compounds of animal's origin also consist of similar three elements but frequently contain nitrogen, phosphorous or sulphur either alone or in combination.

The German scientist *John Jakob Berzelius*, was the first scientist who defined organic chemistry as a branch of modern science in early 1800's. He had classified the chemical compounds into organic and inorganic compounds with the following definition.

Organic compounds: Compounds derived from living organisms were believed to contain an immeasurable vital force and the essence of life called organic compounds.

Inorganic compounds: Compounds derived from minerals and those lacking of that vital force were called inorganic compounds.

The Vital Force Theory "Essence of Life"

According to the *Berzelius* discovery, the chemists could create a life in the laboratory; they assumed that they could not create any compound with a vital force. This vital force theory put forwarded by *Berzelius* had no scientific background but held the field till 1828. A German chemist *Friedrich Wohler* synthesized an organic compound named urea by heating ammonium cyanate, a compound that did not come from a living organism.

   **JOHN-JAKOB BERZELIUS (1779-1848)**   

A German scientist not only coined the term "inorganic and organic" but also invented the system of chemical symbols used even today. He discovered or purified the elements silicon, cerium, lithium, titanium, thorium and zirconium. He published the first list of accurate atomic weights and proposed the idea that atoms carry an electric charge.

For the first time an "organic compound" is obtained from something other than living system without the aid of any kind of vital force.

The synthesis of urea was indicated by the leading chemists of the day. The vital force concept (essence of life) did not die quickly. It was only after the synthesis of acetic acid by *Kolbe* in 1845 and the synthesis of methane in 1856 the belief in the essence of life theory was abandoned. Thus, we see that the vital force theory died only a lingering death.

$$NH_4^+OCN^- \xrightarrow{\text{Heat}} H_2N-\overset{\overset{\displaystyle O}{\|}}{C}-NH_2$$

Ammonium cyanate → Urea (Well known compound isolated earlier from human urine by Roulle in 1780)

 **FRIEDRICH-WOHLER (1800-1882)**   

A German chemist discovered the fact that two different chemicals could have the same molecular formula and also developed methods for purification of aluminium and beryllium "the most expensive metals on earth".

Definition of Organic Chemistry

From the beginning of synthesis of urea, methane and acetic acid in the laboratory, chemists needed a new definition for organic compounds.

Organic compounds are now defined as "compounds that contain carbon". Then the next question arises "Why is an entire branch of organic chemistry devoted to the study of carbon containing compounds? The so called organic compounds have been found to be mainly compounds of carbon and hydrogen (called hydrocarbons) and their derivatives. Finally, organic chemistry is defined today as "The chemistry of hydrocarbons and their derivatives". There is no sharp line of demarcation between organic and inorganic chemistry, the difference between **organic and inorganic** compounds is still being retained as a matter of convenience than of principle.

Reasons for Treating Organic Chemistry as a Separate Branch of Chemistry

The following number of characteristics of organic compounds justifies the treatment of organic chemistry as a separate branch of chemistry.

(i) Large number or Catenation (making bond with all other atoms easily) property of carbon: Compared with the compounds of other elements, the number of organic compounds is very large. About more than one million organic compounds are already known and more being added to the list everyday by chemists all over the world. Why are so many organic compounds present? What makes carbon so special? The answer is the position of carbon in the periodic table. In periodic table, carbon is present in the centre of the second row of elements. The atoms to the left of carbon have a tendency to give up electrons, whereas the atoms to the right of carbon have a tendency to accept electrons (Fig. 1.1).

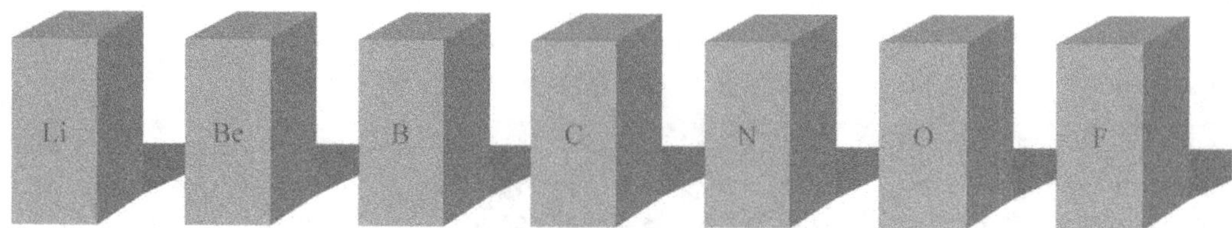

Figure 1.1 Shows the position of carbon atom in periodic table (second row elements).

Since carbon is in the middle, it neither readily gives up nor readily accepts electrons, instead it shares the electrons. Carbon not only can share the electrons with different kind of atoms, it can also share electrons with other carbon atoms which allow it to form the vast variety of chain like molecules. Because carbon can share electrons with atoms other than carbon also adds to the range of stable carbon containing compounds. Consequently it also forms millions of more stable compounds with various ranges of chemical properties by sharing its electrons.

This peculiar property of carbon is known as *Catenation* (ability of carbon atom to make a bond with all atoms). It is not possible to add all these huge carbon compounds in the chapter on carbon in any textbook and hence organic compounds need to be studied separately.

(ii) Complexity of molecules: Some types of organic molecules have more complex structures and possess high molecular weight than inorganic molecules. For example, molecular weight of proteins ranges from several thousands to million.

(iii) Isomerism: Representation of more than one organic compound for the same molecular formula is called as isomers and this phenomenon is known as isomerism. For example, the molecular formula $C_4H_{10}O$ stands for three types of ether and four types of alcohol.

Different type of ethers

$C_2H_5 - O - C_2H_5$
Diethyl ether

$CH_3 - O - CH_2 - CH_2 - CH_3$
Methyl-*n*-propyl ether

$CH_3 - O - CH(CH_3)_2$
Methyl *iso*-propyl ether

Different type of alcohols

$CH_3 - CH_2 - CH_2 - CH_2 - OH$
n-Butanol

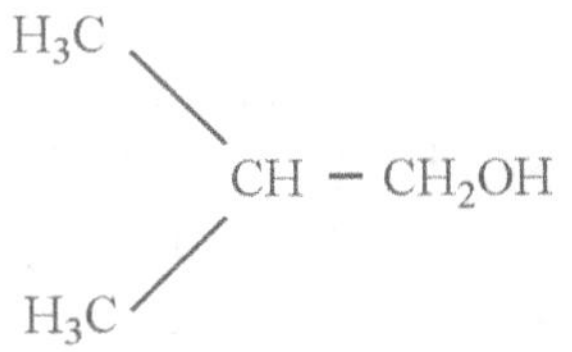

$(H_3C)_2CH - CH_2OH$ ⟶ Primary alcohol
iso-Butyl alcohol

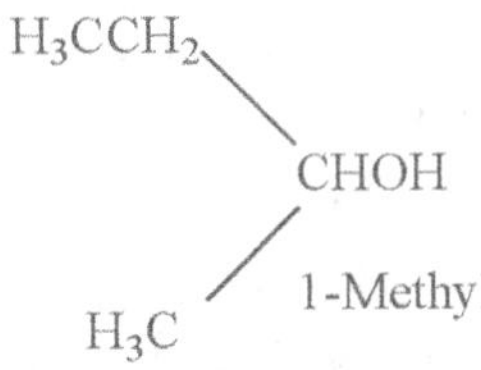

$(H_3CCH_2)(H_3C)CHOH$ ⟶ Secondary alcohol
1-Methyl propanol

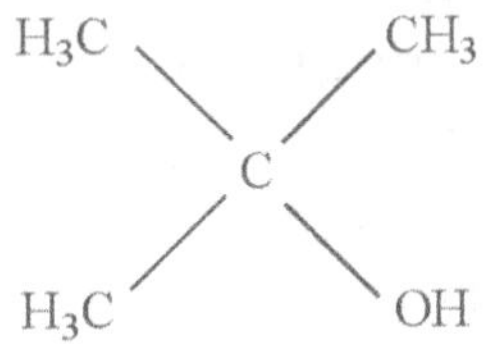

$(H_3C)_2C(CH_3)OH$ ⟶ Tertiary alcohol
(tertiary Butanol)

All the above isomers possess different properties due to different arrangement of an atom. In contrast, in inorganic chemistry, one molecular formula stands for only one compound. For example H_2SO_4 indicates only sulphuric acid and nothing else (only a few inorganic complex compounds also show isomerism).

(iv) Polymerism: When the molecular formula of one organic compound is a simple multiple of the other (*i.e.*, both of them have the same empirical formula), they exhibit polymerism.

For example: benzene exhibits polymerism with acetylene (Benzene is a polymer of acetylene)

C_2H_2	C_6H_6
Acetylene	Benzene

A large number of organic compounds exhibit polymerism but a very few inorganic compounds show polymerism.

(v) Non-ionic character of organic compounds: Organic compounds have covalent bonds and therefore do not get ionize when dissolved in water.

For example: Chloroform ($CHCl_3$) is insoluble in water and does not give white precipitation with sliver nitrate solution (a characteristic property of inorganic chloride salts).

On the other hand, the inorganic compounds usually ionize in water and therefore mainly all inorganic reactions undergo ionic reactions. For example; all chloride salts react with silver nitrate solution and produces white precipitate which is a characteristic reaction of chloride ions.

$$KCl \rightleftharpoons K^+ + Cl^-$$

$$AgNO_3 \rightleftharpoons Ag^+ + NO_3^-$$

$$Ag^+ + Cl^- \rightleftharpoons AgCl\downarrow \text{ (White precipitate)}$$

(vi) Comparative instability of organic compounds: The covalent bond nature of organic compounds makes them more liable to heat and light. On heating the covalent bond, it easily undergoes decomposition. Hence all of the organic molecules possess low melting point and boiling point.

But all the inorganic compounds have ionic bonds which cannot be decomposed at low temperature. More amount of energy is needed for breaking the bond. Hence all inorganic compounds have high melting and boiling point.

(vii) Solubility: Organic compounds, unlike inorganic compounds are mostly soluble in organic solvents such as alcohol, ether, acetone, benzene *etc.* and they are sparingly soluble (formation of weak hydrogen bond) or insoluble in water (lack of formation of hydrogen bond with water).

(viii) Homologues series: Organic compounds are classified into families or groups called as homologues series. Different members belonging to a homologues series are known as homologues and they are characterized by the presence of a particular characteristic group so can be indicated by a general formula. The common difference between the consecutive members in the homologues series is CH_2. These members possess similar chemical properties and regular gradation in their physical properties. They are prepared by general methods of preparation as described in the individual chapters.

This property of organic compounds reduced the studies of over a million of organic compounds to few homologues series. Similar properties of different homologues are the properties of the functional group in each one of them.

(ix) Action of heat: Most of the organic compounds burns in air to give carbon dioxide and water and some of them decompose on heating to leave a black residue (ash or charcoal). But generally inorganic compounds are stable towards heat.

(x) Structure: The arrangement of atoms or structure in most of the organic molecules is well established. But it is more difficult to determine the structure of inorganic compounds. A comparatively very few of them have been completely worked out.

(xi) Origin: Most of the organic compounds are obtained from animal or vegetable kingdom in comparison to the inorganic compounds which are obtained from mineral origin.

(xii) Rates of organic reactions: Due to the reversibility and invariability accompanied by side reactions, most of the reactions between organic compounds are slow. It indicates the activity of molecules rather than ions. They rarely proceed to completion and their yield is generally low.

Rise of Organic Chemistry

After the various discoveries of *Wohler* (1882), *Kolbe* (1840) and *Berthelot* (1850's), chemists found that it was not the vital force which imparted uniqueness to organic chemistry rather a simple fact that organic compounds were all compounds of carbon. The perfection in technique of combustion analysis of carbon and oxygen gave new dimension to organic chemistry. Hence, the first time accurate formulae were available for a good number of fairly complicated organic compounds. Various theories were advanced in order to describe the complexities of substitution, isomerism *etc.*

Frankland, in 1852 discovered the concept of valence. In 1858 *Kekule* & *Cooper* proposed the tetravalency concept of carbon and its ability to link with each other. In 1858, *Cannizaro* and *Avogadro* discovered a hypothesis for the determination of accurate molecular weight. Chemists started thinking about molecular structure and chemical bond related terms. *Kekule* introduced the idea about a bond between atoms.

Nowadays, organic chemistry has matured as a major scientific discipline. Over 95 % the known chemical compounds are organic in nature and roughly half of the present-day chemists are organic chemists. Rapid and fast growth in the field of organic chemistry is due to the fact that organic chemical industry contributes a major role in the world economy and organic compounds are literally the "**stuff of life**".

Need for Studying Organic Chemistry

Organic chemistry is an essential aspect in biological and medical field. All the living organisms are composed of organic substances. Evolution of life postulated that life started from a single organic compound, nucleotide which polymerises or joined together to form a building blocks of life known as DNA.

In most of the chemical industries and pharmaceutical companies in and around us, we encounter organic reactions. Maximum number of life saving drugs are organic compounds only, hence any field e.g. chemical engineering, food and flavoring industry, electrical engineering (liquid crystals displays for digital watches) is surrounded by basics of organic chemistry only. Doctor or Pharmacist are surrounded by organic compounds throughout their life because maximum numbers of life saving drugs are organic compounds only.

For the discipline of biochemistry, molecular biology or another branch of life sciences, a good background of organic chemistry is essential. Specialization in any other branch of chemistry needs good knowledge of organic chemistry.

Apart from these, study of organic chemistry extremely stimulate intellectual pursuit and promotes logical thinking. Organic compounds are vital for the sustenance of all life. Many medical disorders or diseases are due to disruption of organic molecules in the body. Enzymes, proteins, carbohydrates, lipids and catalysts present in our body are organic substances which play a major role in our day to day life.

Classification of Organic Compounds

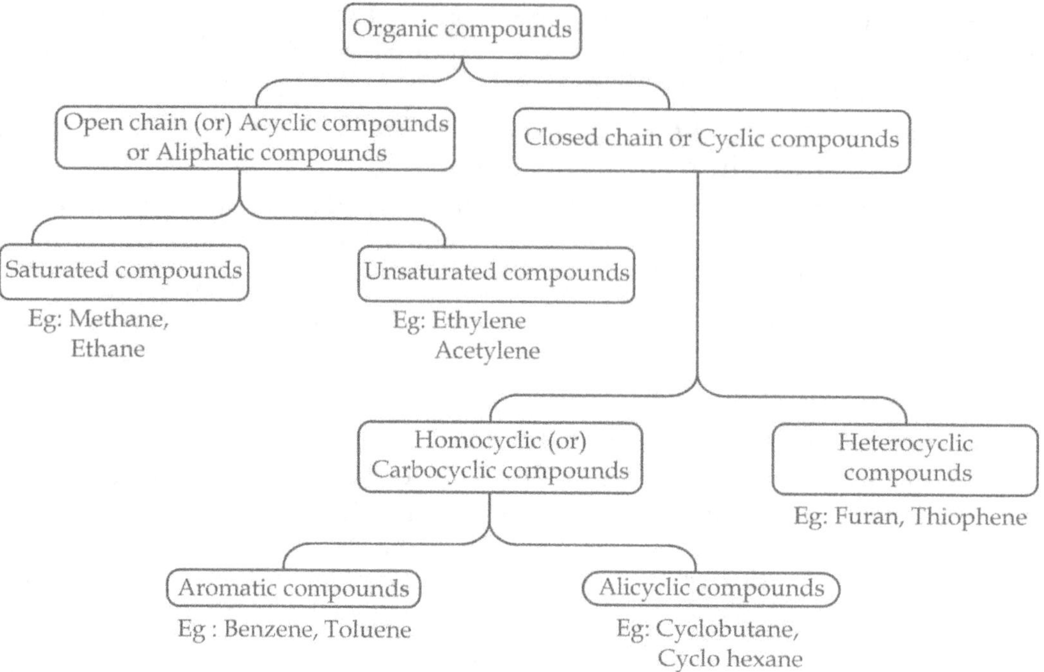

Organic compounds are categorized into two major groups such as

1. Open chain or acyclic compounds or aliphatic compounds.
2. Cyclic compounds or closed chain compounds.

1. ***Open chain or acyclic compounds:*** Carbon compounds having open chain of carbon atoms whether branched or unbranched are called as open chain compounds. The name aliphatic was derived from the Greek word aleipher (meaning fat) as the earliest known compounds of this type were obtained from fats.

CH_4	Methane
$CH_3 - CH_3$	Ethane
$CH_3 - CH_2 - CH_3$	Propane
$CH_3 - CH_2 - CH_2Br$	*n*-Propyl bromide
$CH_3 - CH_2 - OH$	Ethanol

Saturated compounds: A hydrocarbon (compounds that contains only carbon and hydrogen) contains only one carbon-carbon (C-C) single bond are called as saturated compounds (Four valency of carbon atom is saturated with different atoms or groups).

CH_4	Methane
$CH_3 - CH_3$	Ethane

Unsaturated compounds: A hydrocarbon having carbon-carbon double bond (C=C) or carbon-carbon triple bond ($C \equiv C$) are called as unsaturated compounds.

$CH_2 = CH_2$	Ethylene
$CH \equiv CH$	Acetylene
$CH_2 = CH - CH = CH_2$	1,3-Butadiene

2. ***Cyclic compounds or closed chain compounds:*** Organic compounds with a closed chain of atoms are called as closed chain or cyclic compounds.

Polycyclic compounds: Molecules containing two or more ring system is known as polycyclic compounds.

Homocyclic or carbocyclic compounds: The cyclic compounds entirely composed of carbon atom are called as homocyclic or carbocyclic compounds.

These are further sub divided in to two important groups.

Aromatic compounds: Compounds containing a ring structure of six carbon atom of benzene are known as benzenoid or aromatic compounds as shown in following examples.

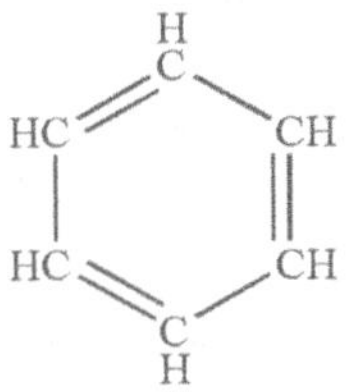

Benzene 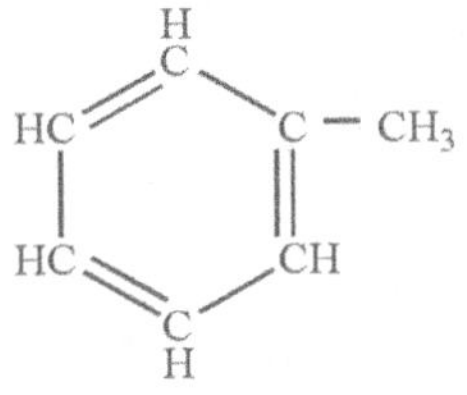 Toluene 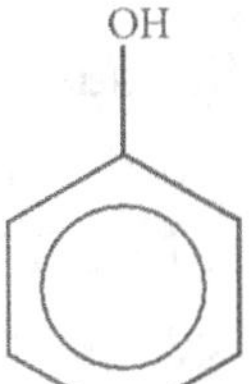 Phenol

Most of the organic compounds occur in plants and many of them have fragrant odour, so they are named as aromatic compounds (Greek *aroma* = sweet smell).

Alicyclic compounds or aliphatic cyclic compounds: Some of the homocyclic compounds contain a ring structure but behave like aliphatic compounds, hence, they are named alicyclic or cyclic aliphatic compounds. It includes the important compounds of polymethylenes or their derivatives as shown below.

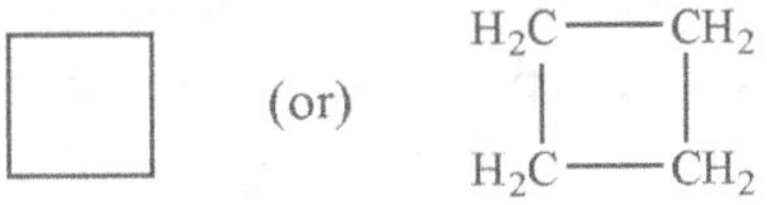

(or)

Cyclobutane

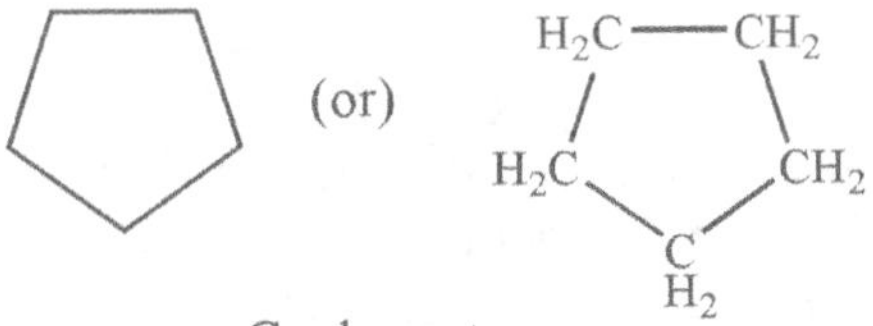

(or)

Cyclopentane

(or)

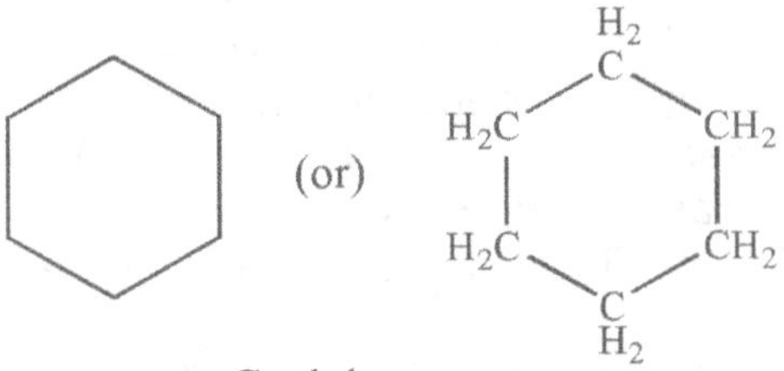

Cyclohexane

Heterocyclic compounds: Cyclic compounds in which one or more carbon atoms are replaced by other atoms are called as heterocyclic compounds. The non carbon atoms present in the ring are called as hetero atoms. Examples are nitrogen, oxygen and sulphur. Some of the examples for heterocyclic compounds are shown below.

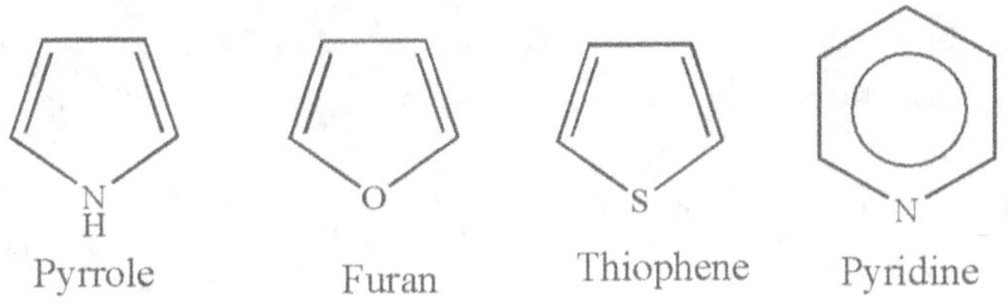

Pyrrole Furan Thiophene Pyridine

Organic Chemistry in the Service of Mankind

All molecules such as proteins, amino acids, enzymes, lipids, DNA & RNA which make life possible contain carbon atoms and the great majority of the chemical reactions that occur within body or living system are organic reactions. We depend on organic compounds that occur in nature for our day to day life needs as given below:

Foods: Starch, sucrose, glucose, fats, vitamins, proteins *etc.*

Household and commercial articles: Soap, ointments, oils, cosmetics, flavoring agents, paints, varnishes *etc.*

Drugs and disinfectants: Aspirin, sulphonamides, penicillin, chloroform, dettol, lysol *etc.*

Poisons: Strychnine, opium and different kind of insecticides.

Perfumes: Camphor, ionone and vanillin.

Dyes: Indigo, malachite green, congored *etc.*

Explosives: Trinitrotoluene, nitroglycerin, dynamite and picric acid *etc.*

War gases: Lewisite, mustard gas and chloropicrin.

Hence, organic compounds penetrate into each and every phase of our life. We look and count on an organic chemist for the manufacture of the above compounds and for the development of various chemical industries.

Various sources of organic compounds: As mentioned earlier, organic compounds are mainly obtained from the natural sources and synthesized in laboratory.

Plants and animal sources: By using suitable method of isolation, various types of organic compounds are obtained directly from plant and animal sources. For example: carbohydrates (sugars, starch, cellulose), proteins (silk, wool, food proteins, casein), alkaloids (morphine, quinine, strychnine), fats and oils (soyabean oil, cotton seed oil, sunflower oil, butter), hormones, vitamins, resins, perfumes and flavouring agents.

1. **Natural gas and petroleum:** These are used as fuels as well as the major source for organic compounds. Through various types of synthetic organic reaction, natural gas and fuels are used as a source for the preparation of hundreds of useful organic substances such as explosives, solvents, synthetic rubber and plastics.
2. **Coal:** It is another major source of lot of organic compounds. Pyrolysis or the destructive distillation of coal produces coke and coal tar. Nearly more than 200 varieties of organic compounds are isolated from coal tar. The coal tar produces the starting materials for the manufacture of thousands of useful aromatic compounds such as drugs, dyes, perfumes and others.
3. **Synthesis:** By using synthetic strategies and reactions, the simple organic compounds obtained from petroleum and coal tar have been converted in to thousands of useful materials.

   Many examples can be cited of synthetic organic compounds replacing those obtained from natural sources but it need volumes. In several cases, the synthetic compounds obtained are more superior to natural compounds. We will study the basic principles of organic chemistry which will help us to learn complex organic compounds and their chemistry in further stages.

Probable Questions

1. Define vital force theory. Why was it abandoned?
2. What is meant by isomerism and polymerism? Write the various isomers of $C_4H_{10}O$.
3. Write the importance of organic chemistry in medicines or medical field.
4. Write a brief note on the rise and development of organic chemistry.
5. Define organic chemistry, and how will you justify organic compounds belongs to separate branch of chemistry?
6. Write a short note on organic chemistry in the service of mankind.

2

Nomenclature of Organic Compounds

Introduction

In early days, scientists named the compounds based on historic background. For example, "wood spirit" was named so because it was obtained from distillation of wood. Later it was named as methanol based on Greek words (methu = wine and hale = wood). Similarly, the name "acetic acid" was derived from vinegar (Latin; acetum = vinegar), because the acetic acid is the major constituent of vinegar. These are called as common names or trivial names.

As the number of compounds were discovered more, naming by history becomes difficult. Hence systematic naming becomes important. The systematization of names was carried out by the International congress of leading chemistry held in Geneva, 1892.

Rational system of nomenclature was formed and it is called as Geneva system of nomenclature. Slight revision and improvements were carried out time to time. One such being held at Liege (Belgium) 1930 by International Union of Chemistry and it is called as IUC system of nomenclature. The IUC was later modified by the International Union of Pure and Applied Chemistry in 1958 and it is called as IUPAC system of nomenclature.

To name the organic compound according to IUPAC nomenclature a set of rules were framed and all the compounds are named accordingly. However, even today some of the common names are used for organic compounds. Hence the chemists should also be aware of the common names apart from IUPAC nomenclature.

Non-systemic nomenclature of organic compounds like common name, trivial name *etc* are described in individual chapters.

IUPAC System of Nomenclature

IUPAC nomenclature is been used now a days to name organic compounds. However, some of the simple compounds are named by trivial names. Earlier names have been continued even today but complex organic compound can be given using IUPAC nomenclature only. Various rules are followed for naming compounds by IUPAC system:

Rule 1: Longest chain rule: "In the given organic compound longest possible chain of carbon atoms is selected and the compound is named as a derivative of this alkane."

For example, the compound given below have five carbons in horizontal line and six carbons in the longest chain hence we should select it as a hexane derivative only.

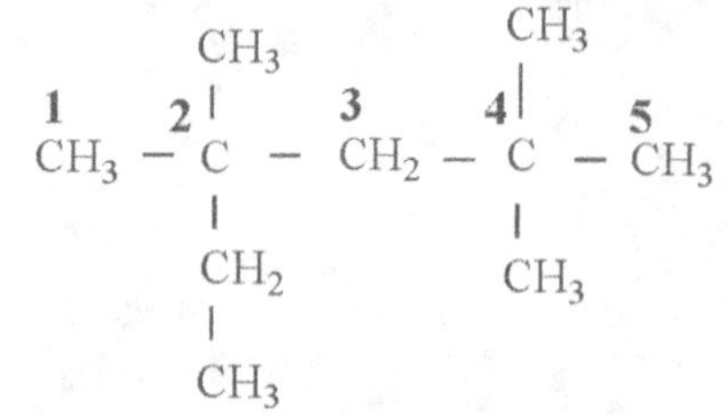

```
         CH3           CH3
1       2|      3     4|     5
CH3  –  C  –  CH2  –  C  –  CH3    Pentane derivative
         |             |
        CH2           CH3                (Wrong)
         |
        CH3

        CH3
       4|            CH3
CH3  –  C  –  CH2 –  C – CH3       Hexane derivative
        |     3     2|    1
      5 CH2          |                  (Correct)
        |           CH3
      6 CH3
```

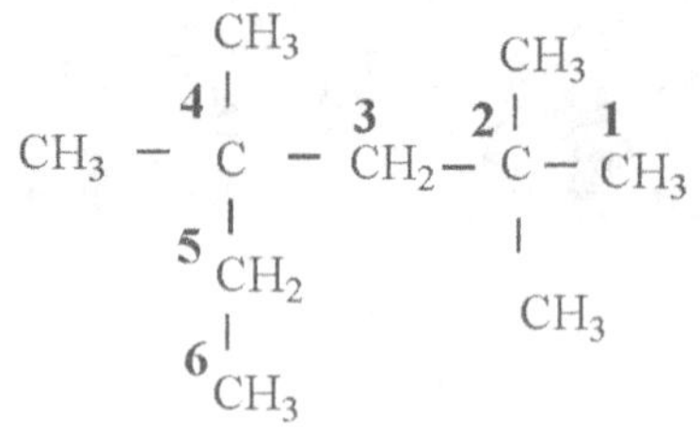

Rule 2: Lowest number for substituents rule: After selecting the longest chain, the numbering should be given from one end to the other end. While giving the number, the substituents should be given lowest possible number.

For example, the compound given below is named in two ways.

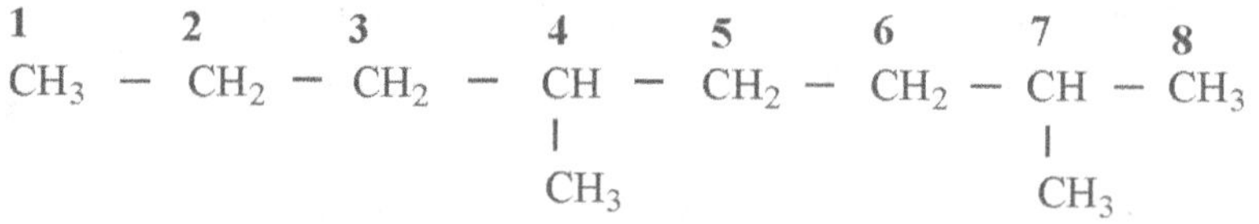

```
 1       2       3       4       5       6       7      8
CH3  –  CH2  –  CH2  –  CH  –  CH2  –  CH2  –  CH  –  CH3
                         |                      |
                        CH3                    CH3
```

4,7-Dimethyl octane (Wrong)

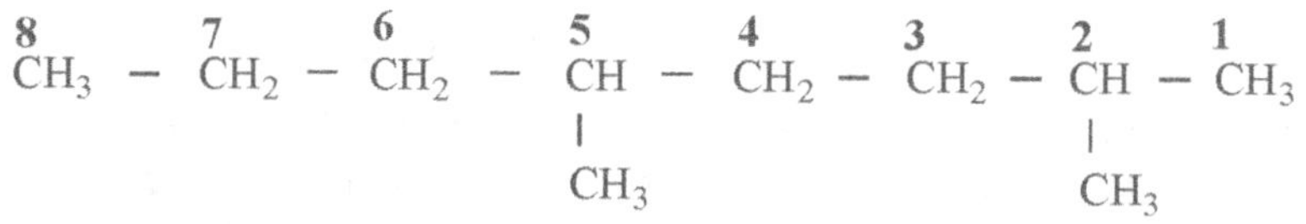

```
 8       7       6       5       4       3       2      1
CH3  –  CH2  –  CH2  –  CH  –  CH2  –  CH2  –  CH  –  CH3
                         |                      |
                        CH3                    CH3
```

2,5-Dimethyl octane (Correct)

In first case, naming 4,7-dimethyl octane is not correct because 2,5-di methyl octane have lowest numbers for the substituents.

If different alkyl groups are in equivalent positions in relation to the end of the chain, preference is given to the end where the radical has fewer carbon atoms (methyl, ethyl, *etc*).

In the following example, the first case of naming is correct because methyl group is given preference over ethyl group.

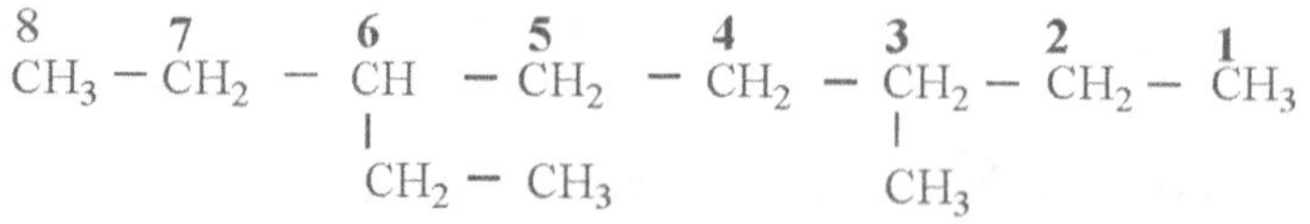

```
 8     7       6      5       4       3      2      1
CH3 – CH2  –  CH  – CH2  –  CH2  –  CH2 – CH2 – CH3
               |                     |
              CH2 – CH3             CH3
```

6-Ethyl-3-methyl octane (Correct)

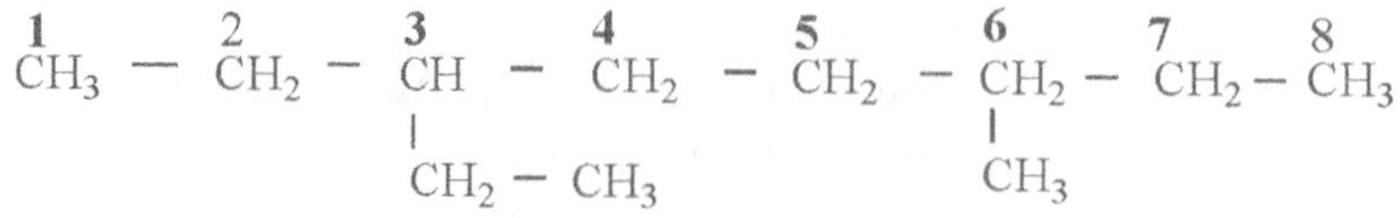

```
 1       2      3       4       5       6      7      8
CH3  –  CH2 –  CH  –  CH2  –  CH2  –  CH2 – CH2 – CH3
                |                      |
               CH2 – CH3              CH3
```

3-Ethyl-6-methyloctance (Wrong)

If identical radicals are at equal distance in the chain then the numbering starts from the end where it is more branched.

In the following example, the first way of naming is correct where the branching end is given preference.

$$\overset{1}{CH_3} - \underset{\underset{CH_3}{|}}{\overset{2}{CH}} - \underset{\underset{CH_3}{|}}{\overset{3}{CH}} - \overset{4}{CH_2} - \overset{5}{CH_2} - \underset{\underset{CH_3}{|}}{\overset{6}{CH}} - \overset{7}{CH_3}$$

2,3,6-Trimethyl heptane (Correct)

$$\overset{7}{CH_3} - \underset{\underset{CH_3}{|}}{\overset{6}{CH}} - \underset{\underset{CH_3}{|}}{\overset{5}{CH}} - \overset{4}{CH_2} - \overset{3}{CH_2} - \underset{\underset{CH_3}{|}}{\overset{2}{CH}} - \overset{1}{CH_3}$$

2,5,6-Trimethyl heptane (Wrong)

If two sets of numbers are possible for the given chain, then order of prefix in the name will decide the numbering (alphabetical order of the substituents).

For example, the given compound can be named as 1-bromo-4-chloro butane or 1-chloro-4-bromo butane. As the prefix bromo is first, the first name is correct.

$$\underset{\underset{Br}{|}}{\overset{1}{CH_2}} - \overset{2}{CH_2} - \overset{3}{CH_2} - \underset{\underset{Cl}{|}}{\overset{4}{CH_2}}$$

1-Bromo-4-chloro butane (Correct)

$$\underset{\underset{Br}{|}}{\overset{4}{CH_2}} - \overset{3}{CH_2} - \overset{2}{CH_2} - \underset{\underset{Cl}{|}}{\overset{1}{CH_2}}$$

1-Chloro-4-bromo butane (Wrong)

If chains of equal length are competing for selection as the parent chain in a branched alkane, the preference goes to the chain carrying more branches.

For example, in the given organic compound first way of naming *i.e.*, 3-ethyl-2,6-dimethyl heptane is correct where as 5-isopropyl-2-methyl heptane is wrong.

$$\overset{1}{CH_3} - \underset{\underset{CH_3}{|}}{\overset{2}{CH}} - \underset{\underset{\underset{CH_3}{|}}{CH_2}}{\overset{3}{CH}} - \overset{4}{CH_2} - \overset{5}{CH_2} - \underset{\underset{CH_3}{|}}{\overset{6}{CH}} - \overset{7}{CH_3}$$

3-Ethyl-2,6-dimethyl heptane (Correct)

$$CH_3 - \underset{\underset{CH_3}{|}}{CH} - \underset{\underset{\underset{7\,CH_3}{|}}{6\,CH_2}}{\overset{5}{CH}} - \overset{4}{CH_2} - \overset{3}{CH_2} - \underset{\underset{CH_3}{|}}{\overset{2}{CH}} - \overset{1}{CH_3}$$

5-Isopropyl-2-methyl heptane (Wrong)

Rule 3: Arrangement of prefixes: When there is more than one group attached in the chain, they should be arranged alphabetically. If same group is presented in two or three places of chain then the prefix di or tri *etc* are used.

For example, the given organic compound is named as 5-ethyl-2,3-dimethyl heptane.

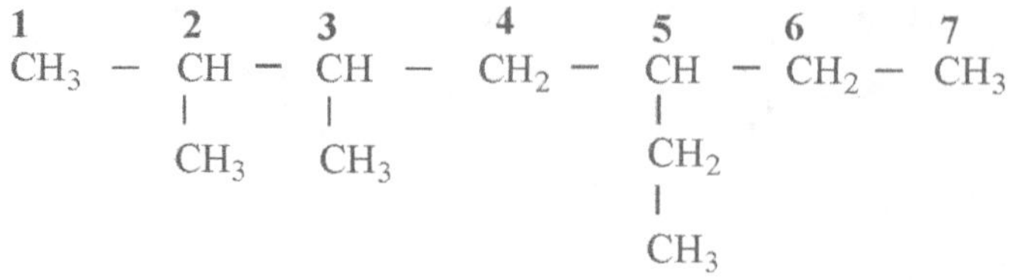

$$\begin{array}{ccccccccccccc} 1 & & 2 & & 3 & & 4 & & 5 & & 6 & & 7 \\ CH_3 & - & CH & - & CH & - & CH_2 & - & CH & - & CH_2 & - & CH_3 \\ & & | & & | & & & & | & & & & \\ & & CH_3 & & CH_3 & & & & CH_2 & & & & \\ & & & & & & & & | & & & & \\ & & & & & & & & CH_3 & & & & \end{array}$$

5-Ethyl-2,3-dimethyl heptane (correct)

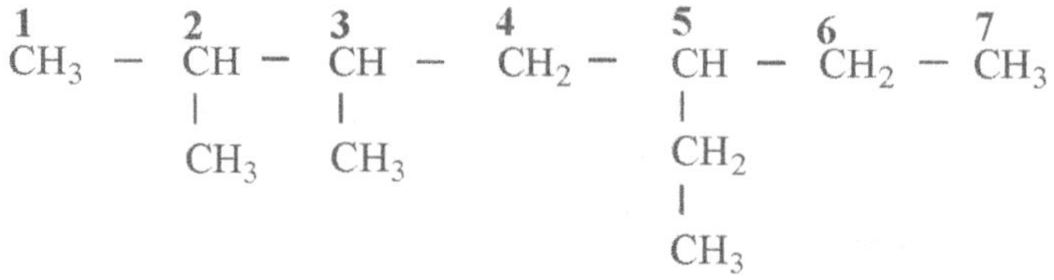

$$\begin{array}{ccccccccccccc} 1 & & 2 & & 3 & & 4 & & 5 & & 6 & & 7 \\ CH_3 & - & CH & - & CH & - & CH_2 & - & CH & - & CH_2 & - & CH_3 \\ & & | & & | & & & & | & & & & \\ & & CH_3 & & CH_3 & & & & CH_2 & & & & \\ & & & & & & & & | & & & & \\ & & & & & & & & CH_3 & & & & \end{array}$$

2,3-Dimethyl-5-ethyl heptane (wrong)

Rule 4: Lowest number for functional group: When the functional group is present in the chain, it should be given first preference even if it violates lowest number rule 2. Double bond or triple bond also considered as functional groups.

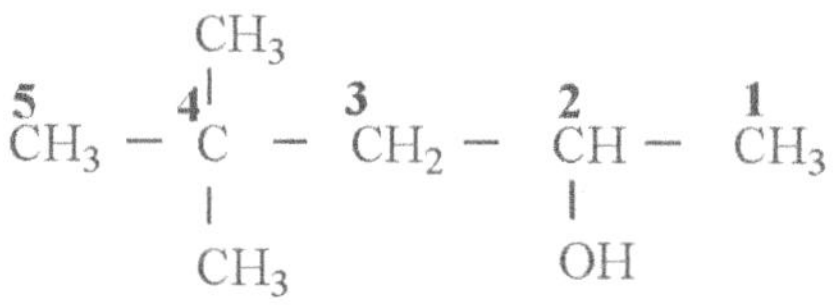

$$\begin{array}{ccccccccc} & & CH_3 & & & & & & \\ 5 & & 4| & & 3 & & 2 & & 1 \\ CH_3 & - & C & - & CH_2 & - & CH & - & CH_3 \\ & & | & & & & | & & \\ & & CH_3 & & & & OH & & \end{array}$$

4,4-Dimethyl-2-pentanol (Correct)

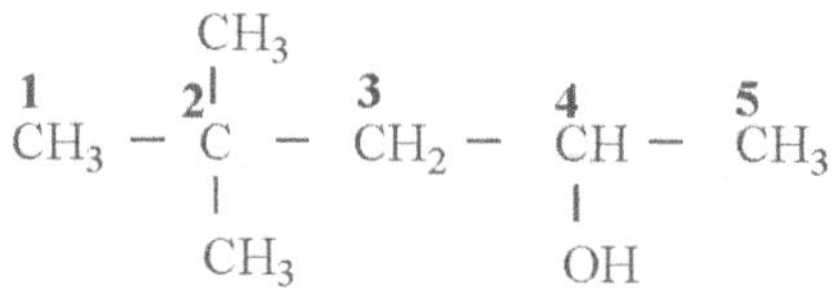

$$\begin{array}{ccccccccc} & & CH_3 & & & & & & \\ 1 & & 2| & & 3 & & 4 & & 5 \\ CH_3 & - & C & - & CH_2 & - & CH & - & CH_3 \\ & & | & & & & | & & \\ & & CH_3 & & & & OH & & \end{array}$$

2,2-Dimethyl-4-pentanol (Wrong)

The order of preference of numbering is as follows.

(i) To the principal functional group of a compound.
(ii) To the double or triple bond.
(iii) To the substituent atoms or groups.

When more than one functional group present in the compound, then the order of preference is as follows.

1. Carboxylic acids
2. Carboxylic acid derivatives
3. Aldehydes
4. Nitriles
5. Ketones

6. Alcohols
7. Amines
8. Ethers
9. Olefins
10. Acetylenes

Systematic name for allyl alcohol is

$$\overset{3}{CH_2} = \overset{2}{CH} - \overset{1}{CH_2} - OH$$

2-Propene-1-ol

For nomenclature purpose the following functional groups are considered as substituents not as functional groups (halo, nitroso and azo as they do not have ending).

When there is more than one functional group in the compound, one is principal functional group and the other is secondary functional group. The prefixes and suffixes used for various functional groups are depicted in the following Table 2.1.

Table 2.1 Groups cited only as prefixes.

S. No.	Functional Group	*Prefix* name
1.	– F	Fluoro
2.	– Br	Bromo
3.	– Cl	Chloro
4.	– ClO	Chlorosyl
5.	– ClO_2	Chloryl
6.	– ClO_3	Perchloryl
7.	– I	Iodo
8.	$=N_2$	Diazo
9.	– N_2	Azido
10.	– NC	Carbylamino
11.	– NO	Nitroso
12.	– NO_2	Nitro
13.	– N(O)OH	*aci*-Nitro
14.	– OR	Alkyl or aryl-oxy
15.	– SR	Alkyl-thio
16.	$-OC\equiv N$	Cyanato
17.	– OOH	Hydroperoxy
18.	– OR	Alkyl-oxy
19.	– OOR	Alkyl-dioxy
20.	$-S-C\equiv N$	Thiocyanato
21.	– SR	Alkyl-thio
22.	– S(O)R	Alkyl-sulphinyl
23.	– SO_2R	Alkyl-sulphonyl
24.	– SSR	Alkyl-dithio
25.	$-N=C=O$	Carbonylamino
26.	$-N=C=S$	Thiocarbonylamino

Table 2.2 Groups cited as prefixes or suffixes.

S. No.	Functional Group	*Prefix* name	*Suffix* Name
1.	$-COOH$	Carboxy	Carboxylic acid
2.	$-SO_2OH$	Sulpho	– sulphonic acid
3.	$-COX$	---	– oyl(-yl) halide
4.	$-CONH_2$	Carbamoyl	– carboxamide or amide
5.	$-C\equiv N$	Cyano	– nitrile
6.	$-\overset{\oplus}{N}\equiv\overset{\ominus}{C}$	Cyano	– Isonitrile
7.	$-CHO$	Formyl	– al
8.	$>C=O$	Oxo	– one
9.	$>C=S$	Thioxo	– thione
10.	$-OH$	Hydroxy	– ol
11.	$-SH$	Mercapto	– thiol
12.	$-NH_2$	Amino	– amine
13.	$=NH$	Imino	– imine

Rule 5: Writing names for compounds containing more than one functional group: Whenever more than one functional group are present in the given compound then the ending is suitably modified. Carbon - carbon multiple bonds and second functional groups are combined in endings or the important functional group is considered as substituent.

Some of the examples are shown below.

$$\overset{1}{CH_2}(OH)-\overset{2}{CH}=\overset{3}{CH_2}$$

2-Propene-1-ol

$$\overset{1}{CH_2}=\overset{2}{CH}-\overset{3}{CH}=\overset{4}{CH_2}$$

1,3-Butadiene

Name of some compounds containing two or more functional groups are shown in Table 2.3.

Table 2.3 Nomenclature of some simple polyfunctional compounds.

S. No.	Functional Groups	Generic name	Specific examples	
			Structure	Name
1.	Two hydroxyl	Alkanediol	CH_2OH-CH_2OH	Ethane-1,2-diol
2.	Three hydroxyl	Alkanetriol	$CH_2OH-CH(OH)-CH_2OH$	1,2,3-Propane triol
3.	Two double bonds	Alkadienes	$CH_2=CH-CH=CH_2$	1,3-Butadiene

Table 2.3 *Contd...*

S. No.	Functional Groups	Generic name	Specific examples	
			Structure	Name
4.	Two acids	Alkanedioic acid	COOH | $(CH_2)_3$ | COOH	Pentanedioic acid
5.	Two aldehydes	Alkanedial	CHO | $(CH_2)_3$ | CHO	Pentanedial
6.	Two ketones	Alkanedione	CH_3 | $(CO)_2$ | CH_3	2,3-Butanedione
7.	Three acids	Tricarboxylic acid	CH_2COOH | CHCOOH | CH_2COOH	1,2,3-Propane tricarboxylic acid

Table 2.4 Nomenclature of some unsaturated compounds of Simple Functions.

S. No.	Unsaturation and Functional Groups	Generic name	Specific examples	
			Structure	Name
1.	Double bond and acid	Alkenoic acid	$CH_2{=}CH{-}COOH$	Propenoic acid
2.	Triple bond and aldehyde	Alkynal	$CH{\equiv}C{-}CHO$	Propynal
3.	Double bond and ketone	Alkenone	$CH_2{=}CH{-}C({=}O){-}CH_3$	3-Butene-2-one
4.	Two double bonds and alcohol	Alkadienol	$(CH_2{=}CH{-}CH_2)_2\,CHOH$	1,6-Hepatadien-4-ol
5.	Double bond and two hydroxyls	Alkenediol	$HOCH_2{-}CH{=}CH{-}CH_2OH$	2-Butene-1,4-diol

Table 2.5 Nomenclature of some compounds of complex Functions.

S .No.	Complex functional Groups	Generic name	Specific examples	
			Structure	Name
1.	Keto acid	Oxoalkanoic acid	$CH_3{-}C({=}O){-}CH_2{-}CH_2{-}COOH$	4-Oxopentanoic acid
2.	Hydroxy acid	Hydroxy alkanoic acid	$HOCH_2{-}COOH$	Hydroxy ethanoic acid
3.	Amino acid	Amino alkanoic acid	$H_2N(CH_2)_5COOH$	6-Amino hexanoic acid
4.	Cyano acid	Cyano alkanoic acid	$N{\equiv}C{-}CH_2{-}COOH$	Cyano ethanoic acid
5.	Aminoketone	Aminoalkanone	$H_2N(CH_2)_2COCH_3$	4-Amino-2-butanone
6.	Alkoxy alcohol	Alkoxy alkanol	$CH_3O{-}CH_2{-}CH_2OH$	2-Methoxy ethanol

Rule 6: Treatment of "like things alike": All groups of one kind which occurs in a single molecule should be given the same treatment as far as possible.

For example, in the given example carboxylic acid is the main functional group, the parent compound should include two or three functional groups as possible.

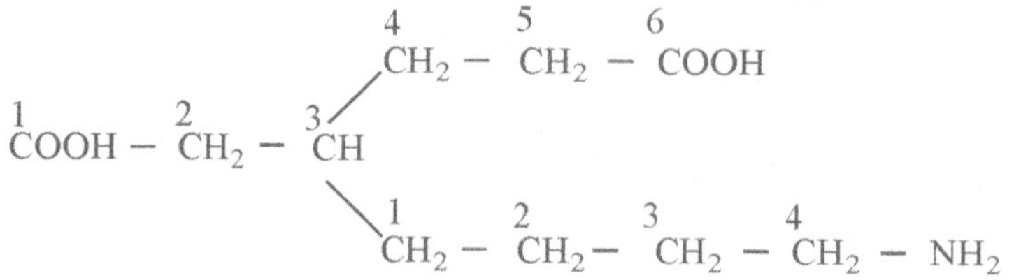

```
                        4       5        6
                        CH2 – CH2 – COOH
  1         2        3 /
 COOH – CH2 – CH
                       \ 1       2        3        4
                        CH2 – CH2– CH2 – CH2 – NH2
```

3-(4-aminobutyl)hexane-1,6-dioic acid

(Preferred)

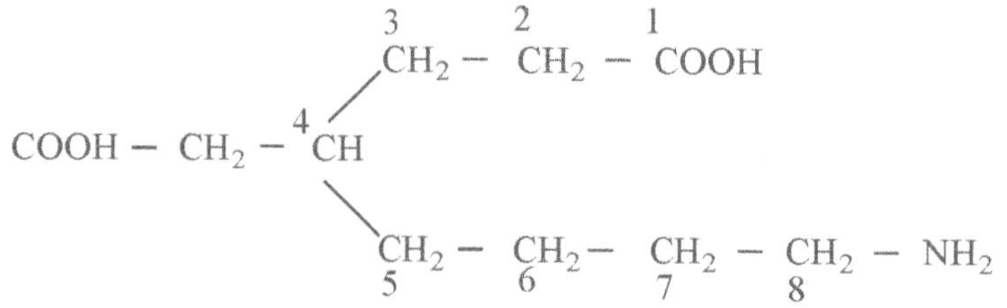

```
                        3       2        1
                        CH2 – CH2 – COOH
                     4 /
 COOH – CH2 – CH
                       \
                        CH2 – CH2– CH2 – CH2 – NH2
                        5       6        7        8
```

8-Amino-4-(carboxyl methyl) octanoic acid

(Not preferred)

Rule 7: Functional groups and the selected chain: Maximum number of functional groups must be included in the carbon chain even if it violates longest chain rule (Rule 1), as shown in the following example.

```
                          2       3       4       5
 CH3 – CH2 – CH2 – CH – CH2 – CH2 – CH3
                          |
                         1CH2 – OH
```

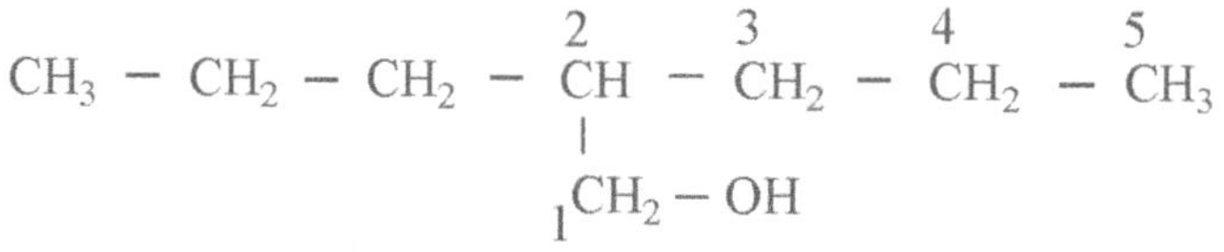

2-Propyl-1-pentanol

When there is a side chain with side chain, the latter is numbered and the name of the complex is considered to start with the first letter of its complete name, as shown in the following example.

```
                                CH3
                               1|    2      3
                          CH3 – C – CH2 – CH3
 9       8       7       6      5|    4      3      2      1
 CH3 – CH2 – CH2 – CH2 – C – CH2 – CH2 – CH2 – CH3
                                 |1   2      3
                                CH2 – CH – CH3
                                       |
                                      CH3
```

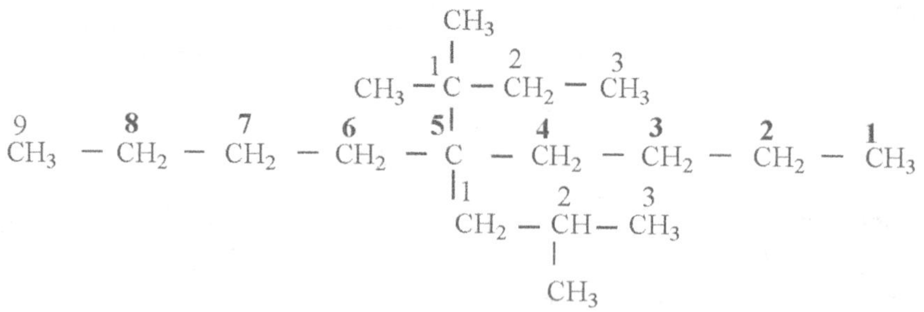

5-(1,1- Dimethylpropyl)-5-(2-methylpropyl) nonane

In addition to these rules, following points mentioned are also useful in writing IUPAC name of compound.

Steps involved in writing IUPAC name of the compound

Step 1: Locate the longest chain containing principal functional group and as many as secondary functional group and carbon-carbon multiple bonds.

Step 2: Select the root word corresponding to the chain length. For example Hex for six carbon atom chain.

Step 3: Number the longest chain selected from the end near to the principal functional group.

Step 4: Based on the carbon-carbon bonds $C - C$, $C = C$, $C \equiv C$ attach the suffix -ane, -ene or yne respectively to the root word of carbon chain.

Step 5: Add suitable prefixes and suffixes with numerals to indicate the number and position of each side chain, substituent, or functional group.

Example:

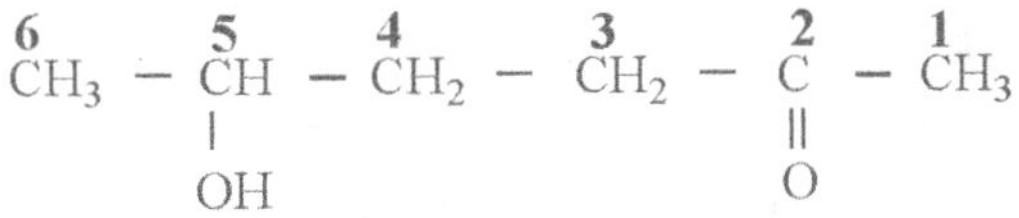

$$\overset{6}{CH_3} - \underset{|}{\overset{5}{CH}}\!\!\!\!\underset{OH}{} - \overset{4}{CH_2} - \overset{3}{CH_2} - \underset{\underset{O}{||}}{\overset{2}{C}} - \overset{1}{CH_3}$$

5-Hydroxy-2-hexanone

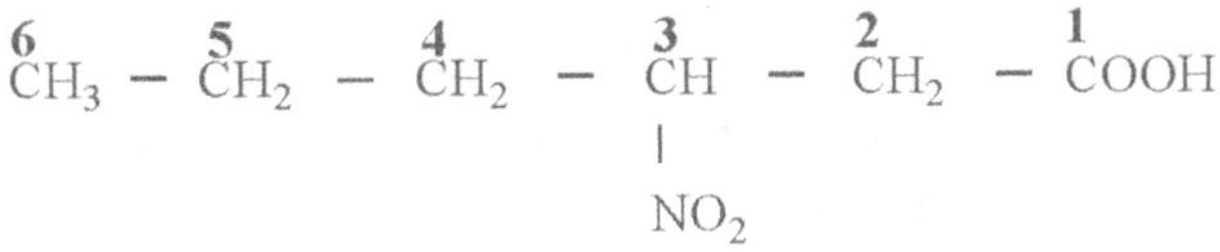

$$\overset{6}{CH_3} - \overset{5}{CH_2} - \overset{4}{CH_2} - \underset{\underset{NO_2}{|}}{\overset{3}{CH}} - \overset{2}{CH_2} - \overset{1}{COOH}$$

3-Nitrohexanoic acid

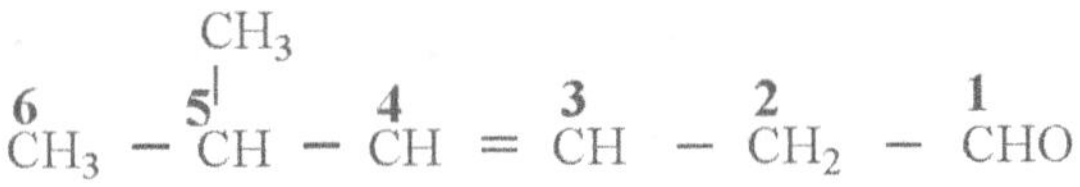

$$\overset{6}{CH_3} - \overset{\overset{CH_3}{|}}{\overset{5}{CH}} - \overset{4}{CH} = \overset{3}{CH} - \overset{2}{CH_2} - \overset{1}{CHO}$$

5-Methyl-3-hexen-1-al

$$\overset{5}{CH_2} = \overset{4}{CH} - \overset{3}{CH_2} - \overset{\overset{O}{||}}{\overset{2}{C}} - \overset{1}{CH_3}$$

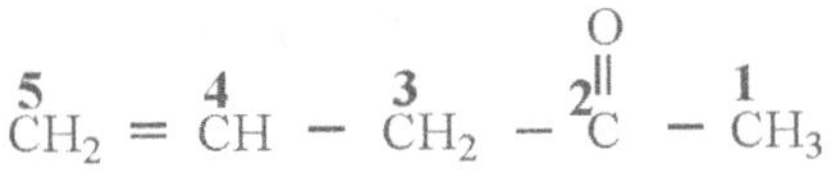

4-Pentene-2-one

(or)

Pent-4-en-2-one

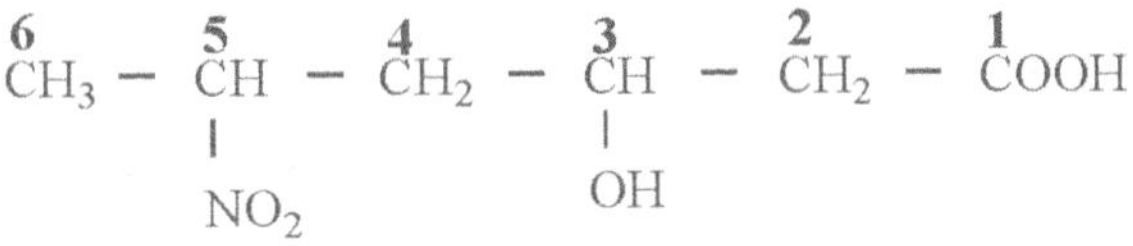

$$\overset{6}{CH_3} - \underset{\underset{NO_2}{|}}{\overset{5}{CH}} - \overset{4}{CH_2} - \underset{\underset{OH}{|}}{\overset{3}{CH}} - \overset{2}{CH_2} - \overset{1}{COOH}$$

3-Hydroxy-5-nitro hexanoic acid

Notes:

1. **Position of numerals used in the enumeration of substituents:** Numerals representing location of unsaturation or functional groups are placed before the name stem as

2-Pentene	not pentene-2
1-Chloro-2-pentene	not 1-chloropentene-2
1-hexene-3-yne	Not hexenyne-3.

2. **Writing names:** The names of radical replacing hydrogen atom in compound are carried out. For example, Chlorotoluene (chlorine replaced "H" atom of toluene).
   Elision of vowels: To avoid ambiguity vowels, whether pronounced or silent are generally retained in systematic naming. This results in using of double vowels, e,g, cyclooctane.

However it has been accepted to elide following vowel a,e and o in the following circumstances.

(i) When preceding the suffix name of a functional group, example
Propanol not propaneol
Hexamine not hexane amine.

(ii) Naming Alkenes and Alkynes on the same compound-example Pentenyne not penteneyne.

3. **Punctuation marks:** Most commonly used punctuation in naming organic compounds are hyphens, commas and enclosing brackets.

(i) ***Hyphens:***

(a) Used to connect numbers and letters serving as a locants.
For example: 2-Chloropropanone 1-Bromo-3-chlorobutane.

(b) Used to connect the prefixes like *cis, trans* (configurational prefixes) or structural prefixes (*sec, tert, neo*) with the compound name.
For example: *Cis*-2-butene, tert–butyl alcohol
The prefixes *cis, trans, iso, neo, tert etc* should be in italics.

(ii) ***Commas:*** Used to separate individual members of a series of locants.
Example: 1,1,2-Trichloro propane.

(iii) ***Enclosing brackets:*** Parenthesis ( ) and square [ ] are used as demarcation symbols when the locants are related to complete names. Example: 4-amino-N-(hydroxyl ethyl) butyramide.

Writing the structural formula from the given IUPAC name: To write the chemical strcture of a given compound from the IUPAC name, the following steps are to be adopted.

(i) ***Locate the parent alkane:*** From the name write the number of carbon atoms of the alkane in a straight chain and number them from any one of the end.

(ii) ***Locate the suffix:*** Locating suffix gives information about chain length, nature of functional group along with the positions.

(iii) ***Locate the groups / substituents:*** As mentioned in prefix locates the groups position in the chain.

(iv) ***Add hydrogen atoms if required to satisfy:*** Four valencies of each carbon atoms to get the formula.

Thus, for writing the structural formulae of 3-ethyl-2,5-dimethyl-1,4-octadiene.

(i) Parent alkane is octane. Write eight carbon atoms in a straight chain and number it.

```
8   7   6   5   4   3   2   1
C − C − C − C − C − C − C − C
```

(ii) The suffix diene indicates two double bonds in 1 and 4 points.

```
8   7   6   5   4   3   2   1
C − C − C − C = C − C − C = C
```

(iii) To locate the groups mentioned in prefix we attach ethyl group on 3rd C and methyl groups to 2nd C and 5th C to get the desired compounds.

```
8   7   6   5   4   3   2   1
C − C − C − C = C − C − C = C
            |       |   |
           CH3      |  CH3
                  CH2CH3
```

(iv) Finally to satisfy valencies hydrogen atoms are added.

```
 8     7     6    5  4    3    2  1
CH3—CH2—CH2—C=CH—CH—C=CH2
            |     |   |
           CH3    |  CH3
                CH2CH3
```

3-Ethyl-2,5-dimethyl-1,4-octadiene

Probable Questions

1. Mention the functional group and general molecular formula for each of the following homologues series.
   (a) Alkanes (b) Alkenes
   (c) Alkynes (d) Alcohols
   (e) Ethers (f) Aldehydes
   (g) Ketones (h) Amines
   (i) Esters (j) Carboxylic acids
   (k) Acid amides (l) Acid chlorides
   (m) Alkyl cyanides (n) Nitroalkanes and isocyanides.
2. Write the molecular formula of the following compounds.
   (a) Methane (b) Ethane
   (c) Ethylene (d) Acetylene
   (e) Chloroform (f) Methane thiol
   (g) Propyl chloride (h) Acetic anhydride
   (i) Ethylidene chloride (j) Ethanol
   (k) Ethyl acetate (l) Aceto acetic ester
   (m) Methyl propionate (n) Acetonitrile
   (o) Oxalic acid
3. What do you understand about IUPAC nomenclature system? and write the various rules and steps involved in it.
4. Write the structure and name of the following compounds according to IUPAC rules.
   (a) Ethane (b) Butene
   (c) Ethanol (d) Formic acid
   (e) Acetyl chloride (f) Propionamide
   (g) Citric acid (h) Methyl propionate
   (i) Ethylene glycol
5. Write down the formulae of the following compounds.
   (a) Cyclo hexane (b) 3- ethyl -1,1-dimethyl cyclo hexanone
   (c) Cyclobutene (d) 4-bromo – 4 – methyl cyclohexanone
   (e) 3-Methylcyclohexene (e) 4-Methyl-2-cyclopenten-1-one
6. Mention the IUPAC names of the various compounds represented by the molecular formula $C_4H_{10}O$.
7. Write the IUPAC name for the following compounds.
   (a) $(CH_3)_2C = CH_2$ (b) $(CH_3)_2 - C = CH_2$
   (c) $CH_3 - CH = CH - CH_3$ (d) $CH_3 - CH = C = CH_2$
   (e) $CH_3\text{-}CH_2\text{-}O\text{-}CH_2\text{-}CH_2\text{-}CH_3$
8. Write the structure of the following compounds.
   (a) 1,5-Hexadiene (b) 4-Methyl-2-pentyne
   (c) 1-Methyl-2-penten-1-ol (d) 2-Chloro butanoyl chloride.

9. Write the IUPAC nomenclature following compounds.

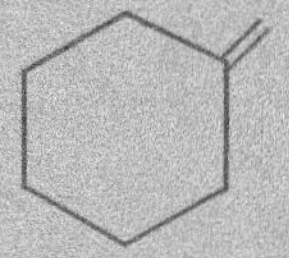

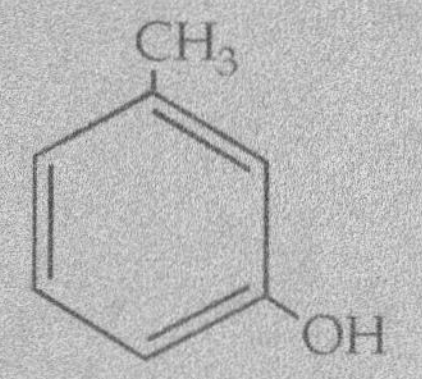

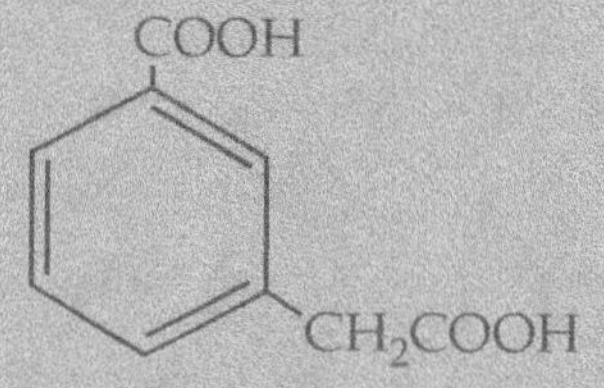

CH_3 OH   COOH CH_2COOH

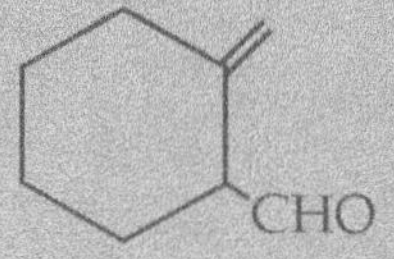

CHO $C_6H_5CH{=}CHCOOH$

3

Structure of Organic Molecules and their Relative Properties

Introduction to Atom/Molecule

The wide diversity of chemical behavior of different compounds is due to the difference in the internal structure of constituting atoms of these compounds.

We know that the atom is made up of certain fundamental particles like electrons, protons, neutrons and subatomic particles like mesons and positrons. The protons and neutrons are present in the nucleus while electrons are present in the extra nuclear part which is divided into different orbitals. Electrons are the basic constituent of all atoms.

Wave nature of electrons and wave equations: *Louis de Broglie*, a French scientist in 1924 suggested that the electrons in an atom possess both particle and wave like properties. The properties of electrons in atoms can be better described as waves rather than particles.

There are two types of waves. One is travelling waves and another is standing waves. Example for travelling waves is sound waves and standing waves are waves found inside an organ pipe.

An electron in an atomic orbital is like a stationary, bound such as vibration: a standing wave. The three-dimensional standing wave features of an orbital are easily explained by using a guitar string as one-dimensional analogy.

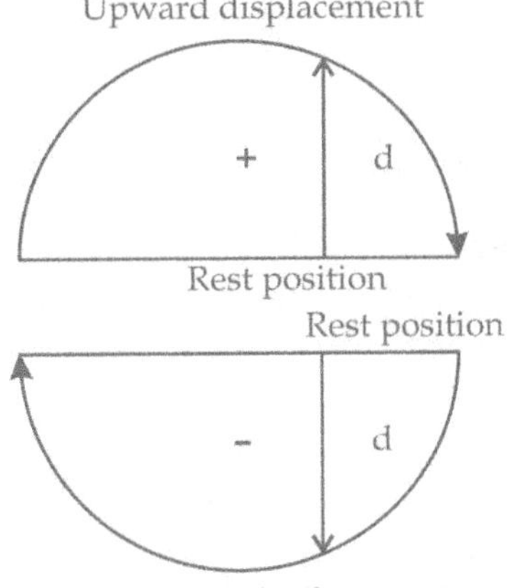

If we pluck a guitar string at its middle it produce standing waves. This vibration has spread into the upward and downward direction equally in a second (with equal time). If we draw an instantaneous diagram of the wave form, it shows the string displaced in a smooth curve either upward or downward, depending on the exact instant of the picture.

Now, *1s* orbital of an atom is considered a string on a guitar (except 3D direction). The orbital is described by its wave equation ψ. Wave equation is a mathematical description of the shape of the wave as it vibrates. All of the waves are positive sign for a brief instant and then it is negative sign. The density of an electron at any point is given by ψ^2. The *1s* orbital is spherical in shape and symmetrical, so it is indicated by a circle with the nucleus in centre and with a plus and minus sign to represent the instantaneous sign of the wave function (Fig. 3.1).

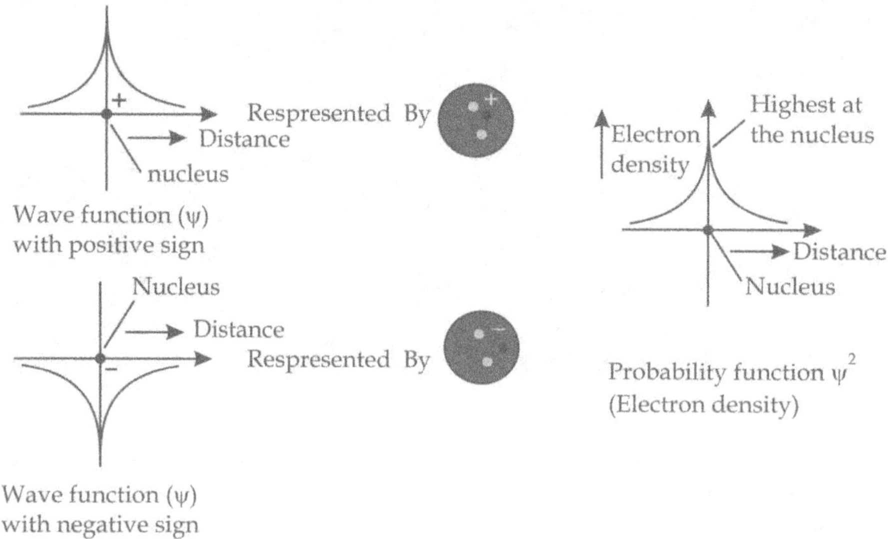

Figure 3.1 Similarity of the 1*s* orbital with fundamental vibration of guitar string. The wave function square provides electron density. The spherically symmetrical orbital is represented by a circle with a nucleus.

When we gently place our finger at the centre of a guitar string while plucking the string from moving, the position at the midpoint is always zero so this point is node. The string vibrates in two halves at the mid point of string and the two halves are vibrating in opposite side. *i.e.,* upward and downward displacement (the two halves of the string is out of phase). Fig. 3.2 shows the first harmonic of the guitar string.

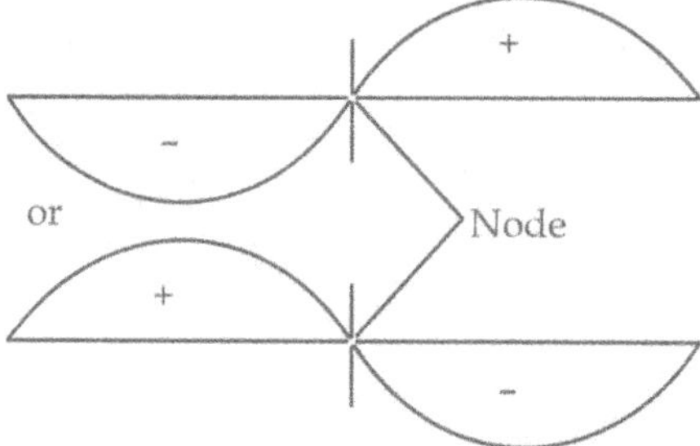

Figure 3.2 The first harmonic of the guitar resembles 2*p*-orbital.

The two lobes of 2*p* orbital are separated by a nodal plane. The two lobes are out of phase with each other. When one lobe has plus sign, the other lobe has minus sign as shown in Fig. 3.2 (a)

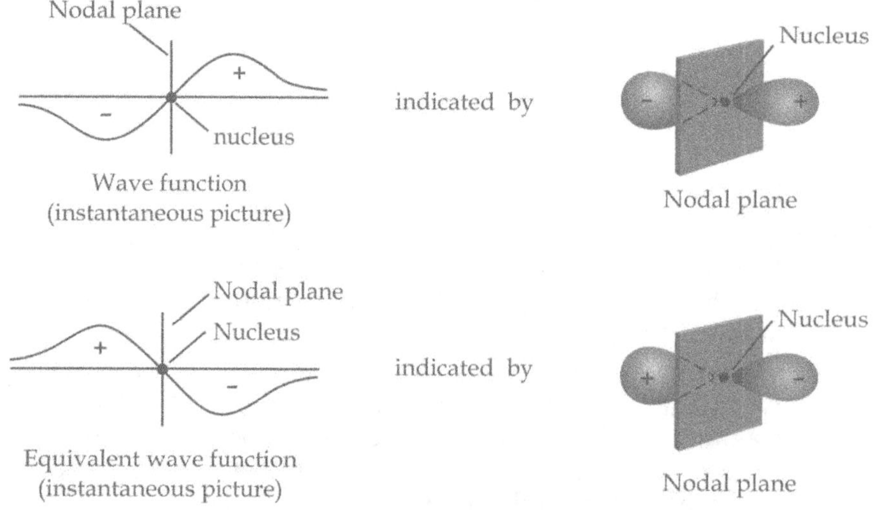

Figure 3.2(a) The two lobes of 2*p* orbital are separated by a nodal plane.

The wave nature of the electron is indicated by following equation.

Wave equation or Schrodinger wave equation

$$\frac{\delta^2\psi}{\delta x^2}+\frac{\delta^2\psi}{\delta y^2}+\frac{\delta^2\psi}{\delta z^2}+\frac{8\pi^2 m}{h^2}(E-P.E)\psi=0$$

Where, x, y, z = Three space coordinates.

ψ (psi) = Wave equation of the electron.

m = Mass of electron.

E = Total energy.

P.E. = Potential energy.

h = Planck's constant.

The above equation indicates the state of an electron in terms of its mass, total energy, its potential energy relative to the nucleus, Planck's constant and a quantity, ψ (psi) known as the wave function of the electron. Hence if we know the all other terms in Schrodinger equation, we can calculate ψ.

Quantum Mechanics

Quantum mechanics uses mathematical equation that explains the wave motion of electrons. According to this theory, the wave motion of guitar string is used to describe the motion of electron around the nucleus. In 1926, *Erwin-Schrodinger* proposed a wave equation of an electron. Wave equations possess a series of solution that are called as wave functions. Solving this wave equation for a given electron tells us the volume of space around the nucleus where the electron is likely to be present and is known as orbital.

Atomic Orbitals involved in Organic Molecules

Electrons move forth and back about the nucleus hence it would not be visible as a particle but it gives the appearance of diffused spherical **cloud** due to its motion. This is called as **electron cloud**. This concept can be explained by referring ceiling fan. At rest or slow movement, blades of fan occupy only a part of space but at high speed it occupies entire circular space. Similarly electrons occupy entire space around the nucleus but at some time it is found somewhere in space (Fig. 3.3 & 3.3(a)).

An orbital is a region in space where there is a maximum probability of finding the moving electron. **Hence atomic orbital can also be defined as "one electron wave function denoted by ψ, is the probability of finding an electron at that point".** Therefore it is a mathematical symbol or representation of electron density distribution which has all the properties associated with waves. It has a numerical value which may be positive or negative (corresponding to the crest or trough of the wave respectively) or zero (which corresponds to a node which is a region where a crest and trough meet).

Fan

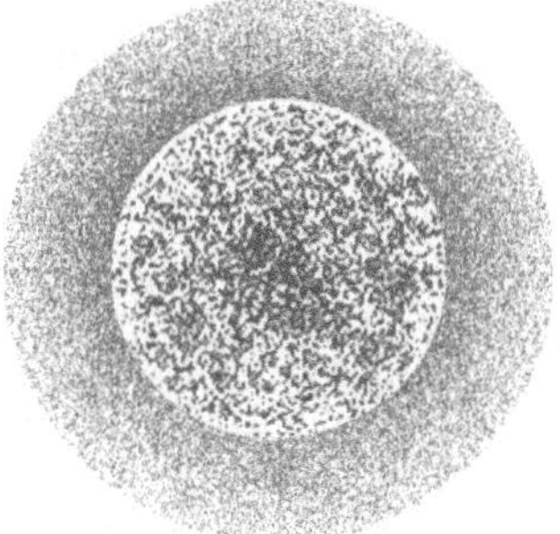

Cloud

Figure 3.3 Only a portion of the circle occupied by a fan, but whole circular space is occupied when it rotates rapidly. Formation of a cloud when electron moves briskly in a shell is also similar to this.

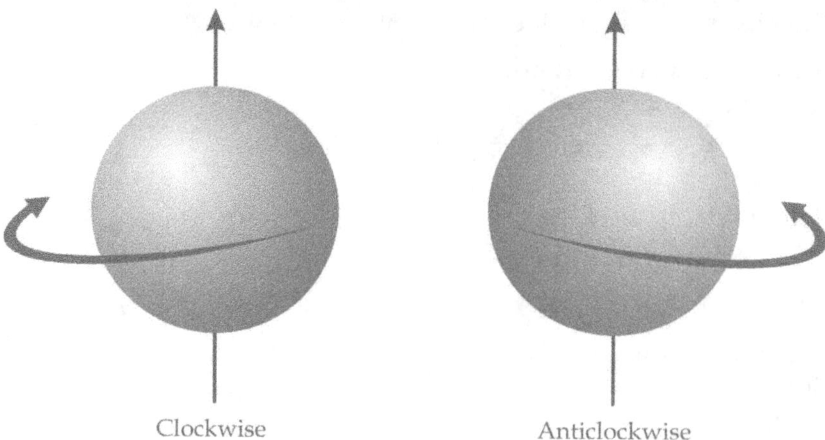

Figure 3.3(a) Spinning of an electron around its own axis in clockwise and anticlockwise direction.

Shells, Sub-shells and Orbitals

The nucleus of an atom remains intact in most of the organic reactions and the outermost electrons only take part in chemical reactions. From the electronic structure of atoms, it is understood that atoms of elements heavier than hydrogen contain more than one electron which occupy definite orbitals. These orbitals further divided into several shells depending upon the principal quantum number "*n*". The shell corresponding to a principal quantum number **1** is also called the **K** shell and the shell corresponding to principal quantum number **2** is also called the **L** shell. Likewise, the shells corresponding to higher principal quantum number **3, 4, 5, 6** and **7** are termed **M, N, O, P** and **Q** respectively.

Quantum Numbers

Principal quantum number (*n*, refers to the size of orbital): It denotes energy and distance of the electron from the nucleus. It cannot have value equal to zero which shows that electrons are not in nucleus as proved by *Rutherford*.

Azimuthal or subsidary quantum number (*l*, refers to the shape of the orbital): It is the geometric shape and the direction in which the electron is most likely to be found in an atom. It may have whole number values from 0 to (n-1). The area corresponding to the azimuthal quantum number is called subshells. The sub shells with respect to ***l*** = 0, 1, 2, and 3 are also designated by letters ***s, p, d*** and ***f*** respectively. The solid number of subshells in a shell is equal to its principal quantum number. For example, 1 or K shell has only one sub shell ***l*** (*1s*), 2 or L shell has two subshells (*2s* and *2p*) and 3 or M shell has three subshells (*3s*, *3p* and *3d*). Similarly, four subshells in the 4 or N shells are the *4s*, *4p*, *4d* and *4f*. The term $2p^4$ means that there are four electrons in the *p* shell of the second or L shell.

The arrangement of shells, subshells and orbitals in an atom and the distribution of electrons are shown in Fig 3.4 and 3.4 a.

Magnetic quantum number (*m*, refers to the orientation of the orbital): It is the behavior of the electron in the magnetic field. It can have all integral values from – l to 0 to +l. Thus for *1s* electrons, since *l* = *0*; it can have only one value 0 which means that s orbital will not change in the magnetic field. When ***l*** = 1 *i.e.*, for *p* orbital, ***m*** can have values –1, 0 and 1 which means that the ***p***-orbital will orient itself in three different direction in the magnetic field represented by three axes. Such, orbitals are termed as p_x, p_y and p_z orbitals and are represented as shown in Fig. 3.5a. Similarly, the *d*- orbitals (*l* = 2) will give rise to five values of the magnetic quantum number (–2, –1, 0, 1, 2) and therefore ***d*** orbitals will split into five different orbitals while *f* orbital will split into seven different *f* orbitals corresponding to seven values of magnetic quantum number (***m*** = -3, -2 -1, 0, 1, 2 and 3) when placed in the magnetic field.

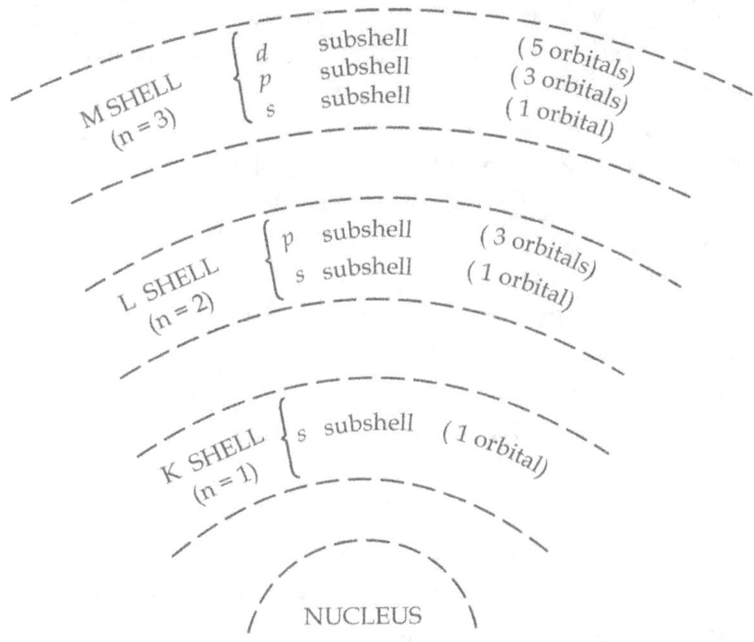

Figure 3.4 Arrangement of shells and subshells in an atom.

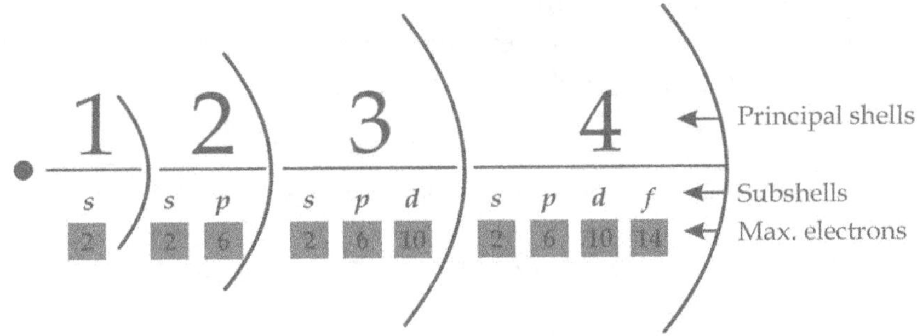

Figure 3.4(a) Distribution of electrons in shells and subshells.

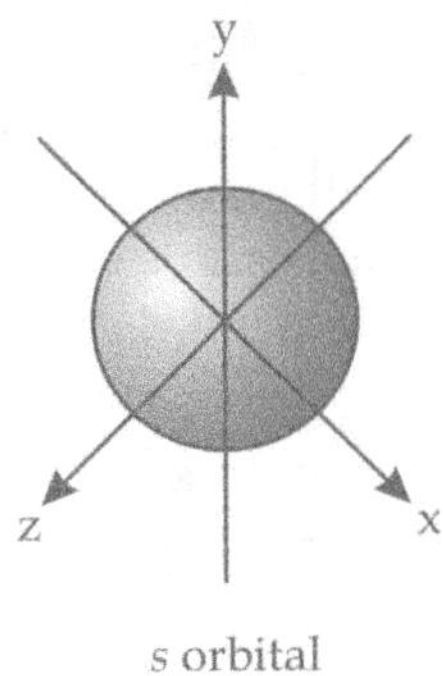

Figure 3.5 s orbital-Geometrical representation.

Spin quantum number (s, refers the direction of the spin of the electron about its own axis): It is a behavior of the electrons in the electrical field. It can be either + ½ or – ½ corresponding to each magnetic quantum number, as difference between the spin quantum numbers is always unity. Thus each of the *p*-orbitals (p_x, p_y and p_z) will have two electrons corresponding to spin quantum number of either + ½ or – ½. The *p*-sub shell therefore has six electrons (2 × 3 *p*-orbitals), *d*-orbitals ten electrons (2 × 5 *d*-orbitals) and *f*-orbitals fourteen electrons (2 × 7 *f*-orbitals).

So, the four quantum numbers give the following information:

1. **n (Principal quantum number)** defines the shell and determines the size of the orbital along with the energy of orbital. There are n subshells in the nth shell.
2. **l (Azimuthal quantum number)** identifies the subshell and determines the shape of the orbital. There are $(2l + 1)$ orbitals present in each type of a subshell that is, one *s* orbital ($l = 0$), three *p* orbitals ($l = 1$) and five *d*-orbitals ($l = 2$) per sub shell. The l also determines the energy of the orbital in a multi-electron atom to some extent.
3. **m (Magnetic quantum number)** indicates the orientation of the orbital. For a given value of l, m_l has $(2l + 1)$ values, the same as the number of orbitals per subshell *i.e.,* the number of orbitals is equal to the number of ways in which they are oriented.
4. **s (Spin quantum number)** indicates the orientation of the spin of the electron about its own axis.

Shapes of Atomic Orbitals

The directional properties of atomic orbitals play a major role in the determination of molecular shape. Hence in the formation of covalent bond only *s* and *p* orbitals are involved.

Shape of *s* orbital: According to wave mechanical calculations, the *s* orbital is spherical in shape (Fig. 3.5(a)) and symmetrical about the nucleus which is indicated by (+) sign. All atomic *s* orbitals possess spherical shape but only differ in their size *i.e.,* size of the orbital increases with higher shell number. Thus *1s* < *2s* < *3s*. Hence *1s* orbital is surrounded by the larger *2s* orbitals which in turn surrounded by still larger *3s* orbital and so on. There is a space between every two *s* adjacent orbitals (between *1s* and *2s*) where the probability of finding an electron is zero (a spherical nodal space).

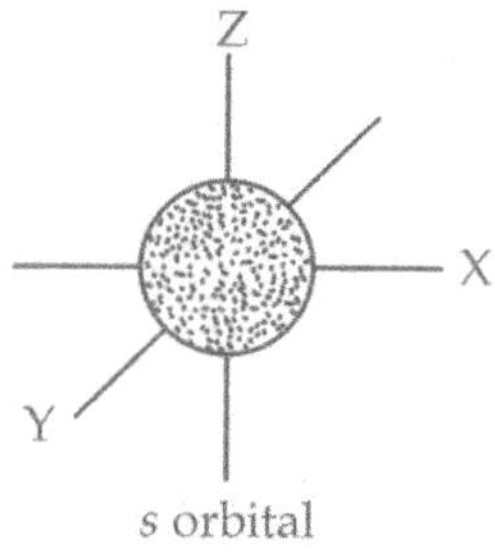

Figure 3.5(a)

Shape of *p* orbital: The *p* orbital is not spherical and symmetrical (Fig 3.5(b)) but it consists of two lobes to form a dumb-bell shaped structure. The three orbitals are identical in shape and energy but differ in their orientation about the x, y and z axes. The three orbitals are represented as p_x, p_y and p_z depending upon its orientation along the coordinate axes. The two lobes of the *p* orbitals do not touch with each other at the nucleus and are concentric with the nucleus which is called as **node.** In this node where the probability of finding an electron is zero and the plane passing through this point is termed as **nodal plane** that separates the two lobes of each dumb-bell.

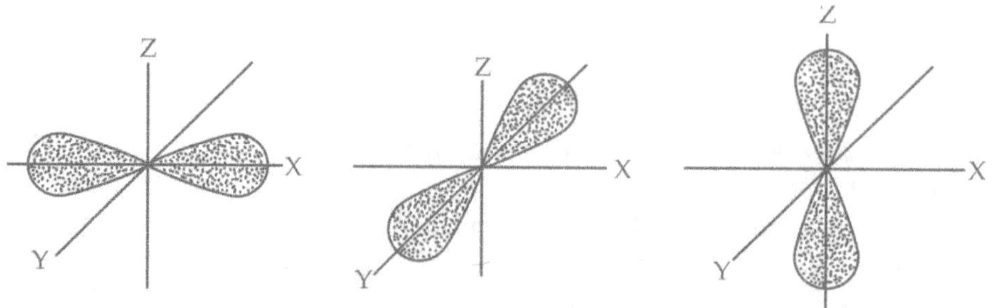

Figure 3.5(b) *p* orbitals-Geometrical representation.

Rules for Distribution of Electrons into various Shells, Subshells and Orbitals

Based on the spectroscopic, magnetic, experimental and theoretical considerations, various rules for distribution of electrons into different shells, subshells and orbitals can be summarized as follows:

1. The maximum number of electrons in any shell is $2n^2$ where n is the principal quantum number of the shell. Hence the maximum number of electrons in the first shell is 2, in the second shell is 8; in the third shell is 18; and the fourth shell is 32.
2. The electrons always enter into the available orbital of the lowest energy. The new electron enters the orbital where $(n+1)$ is minimum. Whenever $(n+1)$ has the same value for two or more orbitals, the new electron enters the orbital where n is minimum. This is known as the *Aufbau principle* (Electrons are filled in the orbitals with increasing order of energy). Few orbitals arranged in the order of increasing energy are as follows:

   1s, 2s, 2p, 3s, 3p, 4s, 3d, 4p, 5s, 4d, 5p, 6s, 4f, 5d, 6p, 7s, 5f, 6d, 7p...

   The order may be remembered by the direction of the arrows which gives the order of filling of orbitals that is starting from right top to bottom left (Fig. 3.6)

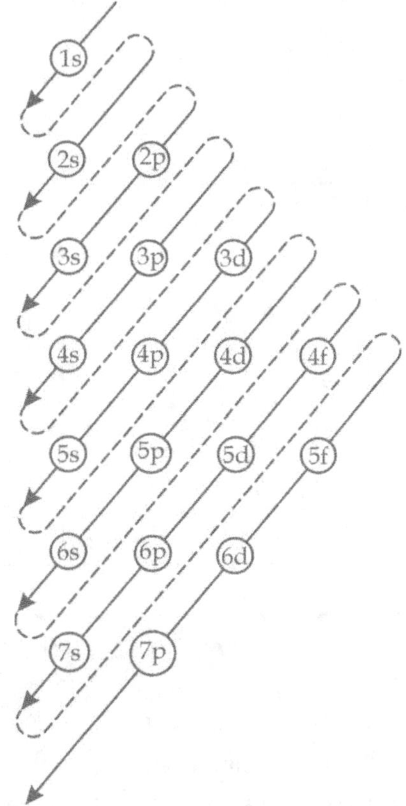

Figure 3.6 Order of filling of orbitals.

3. **Pauli exclusion principle:** Only two electrons may exist in the same orbital and these electrons must have opposite spin.
4. As long as empty orbitals are available electrons will never pair, this is called as **Hund's rule of maximum multiplicity**. That is pairing of electrons begins only with the introduction of second electron in the *s*-orbital, the fourth electron in *p*-orbital, the sixth electron in the *d*-orbital, and the eigth electron in the *f*-orbital. For example, two *p*-electrons in carbon are unpaired and three *p* electrons in nitrogen are also unpaired. Pairing begins with the introduction of fourth electron in *p* orbitals as in oxygen.

5. **Stability of subshells:** Completely full or exactly half-full, subshells are more stable. For example, nitrogen, with its *p*-orbitals half-full, is more stable than both of its neighbours, carbon and oxygen.
6. To explain the above points, electronic configuration of orbital elements from hydrogen (atomic number = 1) to neon (atomic number = 10) is given below in Table 3.1 & 3.1(a).

Table 3.1 & 3.1(a) Electron configuration of first ten elements.

Element	At. No.	Electron configuration				
H	1	$1s^1$				
He	2	$1s^2$				
Li	3	$1s^2$	$2s^1$			
Be	4	$1s^2$	$2s^2$			
B	5	$1s^2$	$2s^2$	$2p_x^1$		
C	6	$1s^2$	$2s^2$	$2p_x^1$	$2p_y^1$	
N	7	$1s^2$	$2s^2$	$2p_x^1$	$2p_y^1$	$2p_z^1$
O	8	$1s^2$	$2s^2$	$2p_x^2$	$2p_y^1$	$2p_z^1$
F	9	$1s^2$	$2s^2$	$2p_x^2$	$2p_y^2$	$2p_z^1$
Ne	10	$1s^2$	$2s^2$	$2p_x^2$	$2p_y^2$	$2p_z^2$

Symbol	At. No.	Electron configuration	1s	2s	$2p_x 2p_y 2p_z$
Be	4	$1s^2 2s^2$	↑↓	↑↓	
B	5	$1s^2 2s^2 2p_x^1$	↑↓	↑↓	↑ | |
C	6	$1s^2 2s^2 2p_x^1 2p_y^1$	↑↓	↑↓	↑ | ↑ |
N	7	$1s^2 2s^2 2p_x^1 2p_y^1 2p_z^1$	↑↓	↑↓	↑ | ↑ | ↑
O	8	$1s^2 2s^2 2p_x^2 2p_y^1 2p_z^1$	↑↓	↑↓	↑↓ | ↑ | ↑
F	9	$1s^2 2s^2 2p_x^2 2p_y^2 2p_z^1$	↑↓	↑↓	↑↓ | ↑↓ | ↑
Ne	10	$1s^2 2s^2 2p_x^2 2p_y^2 2p_z^2$	↑↓	↑↓	↑↓ | ↑↓ | ↑↓

Probable Questions

1. Define atom and write a brief note on atomic number and mass number of an atom.
2. Mention the wave equation and write its significance.
3. Explain in detail about the wave nature of electron and its properties.
4. What is meant by Schrodinger's equation? and write its uses.
5. Define electron cloud and atomic orbital.
6. Write the concept of shells, subshells and their properties.
7. Explain the four kinds of quantum numbers and their significance.
8. Write a brief note on the shape of various orbitals.
9. Define and explain the following terms.
   (a) Hund's rule.
   (b) Pauli exclusion principle.
   (c) Aufbau principle.

10. How many electrons are present in valence shell of following elements?
    (a) Chlorine.
    (b) Fluorine.
    (c) Carbon.
    (d) Magnesium.
    (e) Oxygen.
    (f) Nitrogen.
11. Write the ground state electronic configuration of the following elements. Number in bracket indicates atomic number of each element.
    (a) Carbon (6).
    (b) Copper (29).
    (c) Zinc (30).
    (d) Sodium (11).
12. Write down the maximum number of electrons in a given atom that can occupy each of the following.
    (a) The *L* shell.
    (b) The n = 3 shell.
    (c) The *2p* orbitals.
    (d) The *3d* orbitals.
    (e) The *4f* orbitals.
13. Fill any one pair of electrons that are missing from the following line structures.
    (a) $C_2H_5-O-C_2H_5$.
    (b) CH_3-NH-H.
    (c) C_2H_5Br.
    (d) CH_3SH.
14. Explain the various rules for filling of electrons in atomic orbitals.
15. Explain different shapes of various orbitals.

4

Bonds in Organic Compounds

Introduction

Chemical reactions occur by the formation of bond between the atoms or breaking of bond between the atoms and leads to formation of stable products. Hence, examining the nature of various kind of bonds between atoms is an important aspect of organic chemistry.

Why atoms want to make bonds? The answer for this question is the *Lewis-Langmuir* concept of stable electronic configuration or octet rule. According to this concept, *G.N. Lewis* proposed that "an atom attains most stability when it has filled shell or the outermost shell of an atom has eight electrons and no electrons in highest energy state. For attaining this inactive condition or maximum stability, an atom gives or accepts or shares the electron with other atoms".

The Nature of Bond between Atoms

During chemical reactions, bonds between the atoms of a molecule in a reaction are broken and new bonds between atoms are formed which leads to the formation of products. At this point of study of organic chemistry we need to examine the nature of different bonds formed between atoms.

The *Lewis-Langmuir* concept of stable electronic configuration: There are eight electrons in the outer most orbital of all noble gases except helium (2) electrons in all in first shell. These gases are practically do not involve in any reaction. This condition of maximum stability and inactivity is related to their electronic arrangement.

Stable electronic configuration: The electronic configuration or arrangement in which eight electrons are present in the outermost orbital or valency shell is called stable configuration (octet rule). If the outermost orbital is the first one (as in helium), the number of electron for stable configuration is two. Hence the compound without stable electronic configuration involve in the chemical reactions. From this *Lewis and Langmuir* stated that in all cases "the elements tend to enter into chemical reactions in such a way to acquire stable configuration". We can say that elements react with each other to attain a stable electronic configuration.

Types of bonds: It is based on the nature of an atom which combines with other atom to attain stable electronic configuration. It can be achieved in three different ways which leads to the formation of three kinds of bonds.

1. **Ionic bond or Electrovalent bond** – This bond is formed due to the electrostatic attraction between two oppositely charged stable ions.
2. **Covalent bond –** Formed by mutual sharing of electrons between two atoms.
3. **Co-ordinate covalent bond or Dative bond –** A bond formed by donating a pair of electrons to another atom.

1. **Ionic bond or Electrovalent bond:** This bond is formed due to the electrostatic attraction between two oppositely charged stable ions formed by the complete transfer of electrons between one atom to another. For example, Formation of NaCl, KCl, and AlF_3.

   Formation of NaCl: Let us consider the formation of sodium chloride molecule.

   Electronic configuration of sodium atom is $\boldsymbol{1s^2\,2s^2\,2p^6\,3s^1}$.

   The number of electrons present in the outer most orbital of sodium is 1, it needs to loose one electron for attaining inert gas electronic configuration of neon. According to the definition of ionic bond, the ions of an atom are formed during the bond formation, so sodium atom loses its one electron from the outer most orbital *i.e.*, $3s^1$ which leads to the formation of sodium ion with the electronic configuration of $\boldsymbol{1s^2 2s^2 2p^6}$.

$$\underset{\substack{(2,8,1)\\ \text{Sodium atom}}}{\text{Na}} \xrightarrow{\text{energy}} \underset{\substack{(2,8)\ \longrightarrow\ \text{Neon electronic configuration}\\ \text{Sodium ion}}}{\mathbf{Na^+}} + \mathbf{e^-}$$

   Now, the electronic configuration of chlorine atom is $\boldsymbol{1s^2 2s^2 2p^6 3s^2 3p^5}$. Chlorine atom is having one electron short or less to attain its stable, inert gas electronic configuration of Argon. Hence, chlorine atom is ready to accept one electron which leads to the formation of chloride anion with electronic configuration of $1s^2 2s^2 2p^6 3s^2 3p^6$.

$$\underset{\substack{(2,8,7)\\ \text{(Chlorine atom)}}}{\text{Cl}} + e^- \xrightarrow{\text{energy}} \underset{\substack{(2,8,8)\ \longrightarrow\ \text{Argon electronic configuration}\\ \text{(Chloride ion)}}}{\text{Cl}^-}$$

   Hence, in case of sodium chloride, the outer most electron (valence electron) of sodium and chlorine atom are held together by strong electrostatic attraction of their opposites charges *via* the formation of two ionic species such as Na^+, Cl^- (inert gas electronic configuration)

$$\underset{\substack{(2,8,1)\\ \text{Sodium atom}}}{\text{Na}^{\times}} + \underset{\substack{(2,8,7)\\ \text{Chlorine atom}}}{\cdot\ddot{\underset{\cdot\cdot}{\text{Cl}}}:} \longrightarrow \underset{\substack{2,8\ (2,8,8)\\ \text{Sodium chloride}}}{\text{Na}\ {}^{\times}_{\cdot}\ddot{\underset{\cdot\cdot}{\text{Cl}}}:}$$

Ions are formed in this so it is called as ionic bond. Transfer of electrons between the atoms alters the valency so it is called as electrovalent bond.

The formation of ionic compound mainly depends upon the ease of formation of anions and cations from the neutral atoms. This is influenced by two important factors. They are as follows.

(a) **Ionisation Energy (I):** Minimum amount of energy needed to remove an electron from a gaseous state of an atom in its ground state to form a gaseous ion.

$$M \longrightarrow M^+ + e^- \quad \Delta H^\circ = 1$$

Smaller the value of I, greater will be the ease of formation of cation from it.

(b) **Electron Affinity:** Electron affinity is defined as the "amount of energy released when an electron is added to a neutral atom or molecule in gaseous state to form a negative ion. *i.e.*, energy of the reaction".

$$X_{(g)}^+ + e^- \longrightarrow X_{(g)}^- \quad \Delta H^\circ = -E$$

Higher the value of 'E' greater will be the ease of formation of anion from it. In periodic table, element at the far left (Li, Na, Mg) have relatively less ionisation energy and are called electropositive elements. Similarly, elements at the far right (F, Cl, Br) have relatively high electron affinities are called as electronegative elements. So, the ionic bond is formed between the electropositive elements present at the far left of the periodic table and electronegative element present at the far right of the periodic table.

2. **Covalent Bond:** Consider the elements present in the middle part of the periodic table. More amount of energy is required to gain or lose electrons to form ions of these atoms attaining octet electronic configuration.
   Consequently, there is no possibility of forming bond by transfer of electrons between the atoms.

$$\cdot\dot{\underset{\cdot}{C}}\cdot \longrightarrow C^{4+} + 4e^- \quad I = 6195 \text{ Kj/mole}$$

$$\text{x}\overset{\text{x}}{B}\text{x} \longrightarrow B^{3+} + 3e^- \quad I = 3641 \text{ Kj/mole}$$

Hence this type of atoms forms a stable molecule by mutual sharing of their one or more electrons pair in such a way that each atom attains their nearest inert gas stable electronic configuration. Hence,

"Mutual sharing of electrons between two atoms leads to the formation of covalent bond".

For example: Formation of hydrogen molecule from two hydrogen atoms.

Each hydrogen atom has only one electron in its valency shell so it is unstable but both hydrogen atoms equally share their valency electrons to form a pair of electrons in such a way that they attain their stable inert gas electronic configuration of helium.

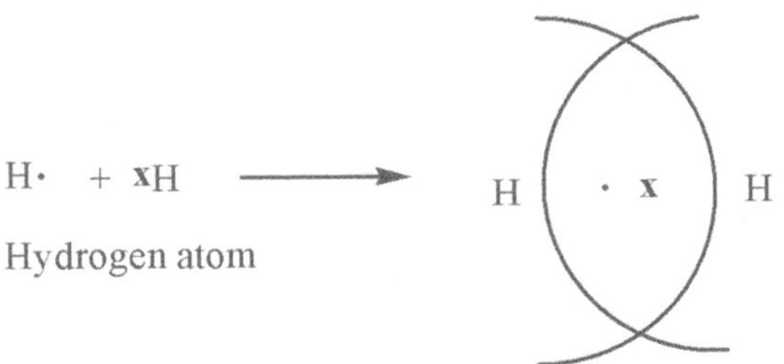

Hydrogen molecule
(Helium electronic configuration)

The covalent bond is denoted by a solid line (-).

Types of covalent bond: There are three types of covalent bonds *i.e.,* **single, double and triple bonds.**

Single Bond: This covalent bond is formed by mutual sharing of one electron pair between two atoms. For Example:

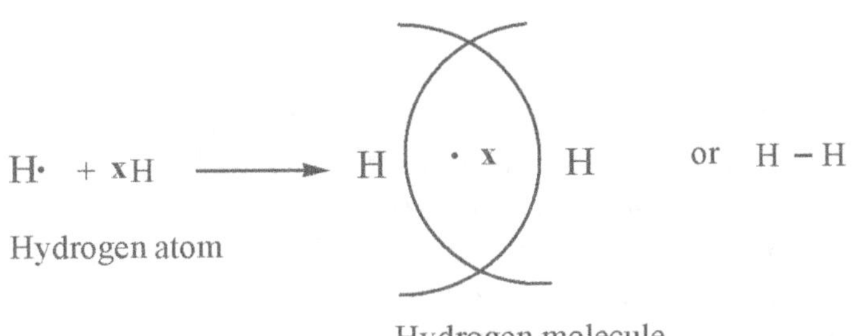

Hydrogen molecule

Double Bond: This type of covalent bond is formed by mutual sharing of two electron pair between two atoms. Example: Formation of oxygen, ethylene molecule *etc.*

$\ddot{O}:$ or $O = O$

Oxygen atom

Oxygen molecule

$H \overset{\times\cdot}{\underset{\times}{:}} C :: C \overset{\times\cdot}{\underset{\times}{:}} H$

$CH_2 = CH_2$

Ethylene

Triple Bond: This type covalent bond is formed by the mutual sharing of three electron pairs between two atoms. ***Example:*** Formation of nitrogen molecule and acetylene molecule *etc.*

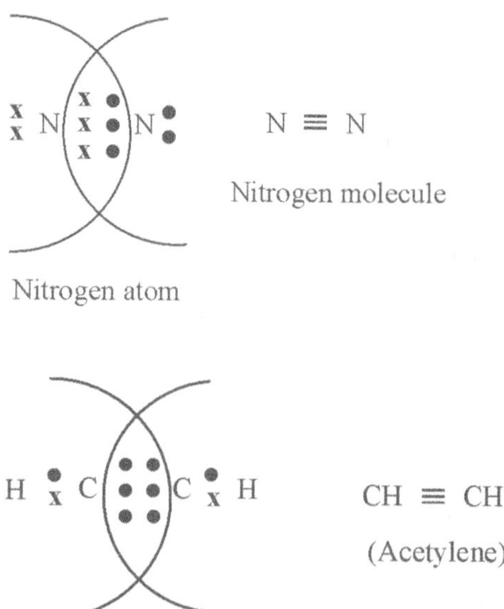

3. **Co-ordinate covalent bond or Dative bond:** (By one side or single sided sharing of electrons). **A bond formed by donating a pair of electrons to another atom is called dative bond.** Let us consider the formation of ozone molecule. Ozone molecule contains three oxygen atoms. The pictorial representation of the formation of O_3 molecule is as follows.

$O::O + \ddot{O}: \longrightarrow O::O:\ddot{O}:$ (OR) $O = O \rightarrow O$

Oxygen molecule Oxygen atom Ozone molecule

Ozone molecule is formed by the formation of bond between one O_2 molecule and one oxygen atom. In oxygen molecule the octet electronic configuration is completed by mutual sharing of two electron pairs between two oxygen atoms. If one new oxygen atom with two electrons deficient in nature, shares one electron pair belongs to any one of the oxygen atom of the oxygen molecule and it attains its stable inert gas electronic configuration. Now the bond formation occurs by sharing of electron pair between two atoms in which the electron pair comes from any one of the single atom even though it is shared by two atoms. This type of bond is known as **co-ordinate covalent bond or dative covalent or co-ordinate bond.** This bond is indicated by arrow. The head of the arrow indicates the atom which accepts electron (electron acceptor).

For example: Formation of ammonium ion between ammonia molecule and hydrogen ion.

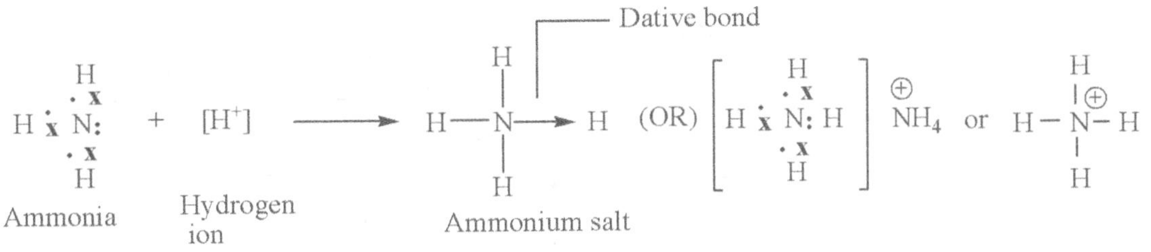

Ammonia molecule possess inert gas electronic configuration. It donates a lone pair of electrons to a hydrogen atom and form ammonium ion.

Atom that donates an electron pair is known as **donor.**

Atom that accepts an electron pair is known as **acceptor.**

Due to the transfer of electron pair between the donor and acceptor atom, dative bond is formed. The donor attains a unit positive charge and the acceptor atom attains a unit negative charge.

Example: Representation of nitric acid.

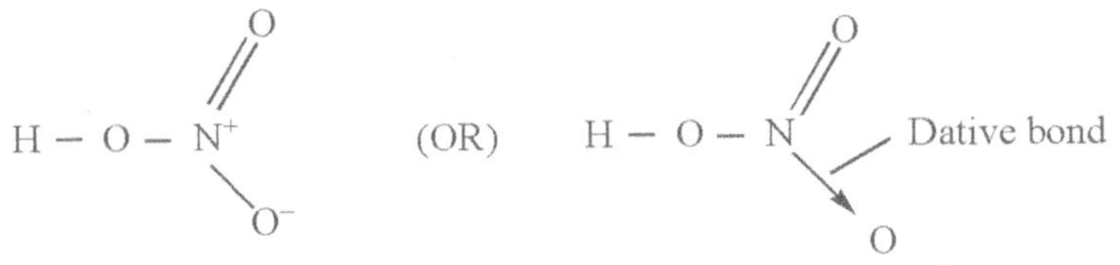

Electronic Theory of Valency

The number of electrons present in the outer most shell or orbital of an atom is known as valency. If the valence electrons are removed then the remaining portion of the atom is called as "kernel of the atom" (core electrons are not participated in bond formation). A chemical bond is formed between two elements involving either transfer of valence electrons to the other elements or sharing of valence electrons between two atoms to attain inert gas electronic configuration. Hence the chemical properties of an element depends upon its electronic configuration.

An electron lending atom is known as electropositive and an electron borrowing atom is known as electronegative.

Characteristic Property of Covalent Bond

Bond length: A bond is formed between two atoms by overlapping of atomic orbitals to produce molecular orbitals. This overlapping of two atomic orbitals at certain distance between the bonded atoms is called as bond length.

The distance between the center of the two orbitals is called as bond length or bond distance. The unit of bond length is Angstrom ($1A^{\circ} = 10^{-8}$ cm). The bond length between two given atoms of different molecule of the same series is essentially constant for most bonds. The bond length value lies between 1° to 2 A°.

Bond length is measured by X-ray diffraction, electron diffraction and spectroscopic methods. Some common bond length of the covalent bond is listed below.

Covalent bond	Distance in A° unit
C—C	1.54
C—N	1.47
C—O	1.43
C—F	1.42
C—Cl	1.77
C—Br	1.91
C—I	2.13
C—H	1.12
C=C	1.40
C=O	1.20
C=N	1.28
C≡C	1.21
C≡N	1.15
Cl—Cl	1.99
N—N	1.09
H—H	0.75
H—O	0.97
H—N	1.03
H—S	1.35
H—F	0.92
H—Cl	1.27
H—Br	1.41
H—I	1.61

Bond Angle

It is defined as "an angle between the directions of two covalent bonds". It does not have a fixed value because the atoms in a molecule are in constant motion with respect to each other. But, the average bond angle can be defined and calculated. Basically, bond angle largely depends upon the nature of bonds concerned. As "*s*" orbital is spherical and symmetrical, it does not project in any direction and it can overlap with each other orbitals equally in all directions. But three "*p*" orbitals perpendicular to one another naturally, hence their overlapping will be more in these directions. So those atoms (example: oxygen, nitrogen *etc*) using "*p*" orbitals would be expected to form bonds with inter bond angle of 90°. For example, expected bond angle in water molecule is shown in Figure 4.1.

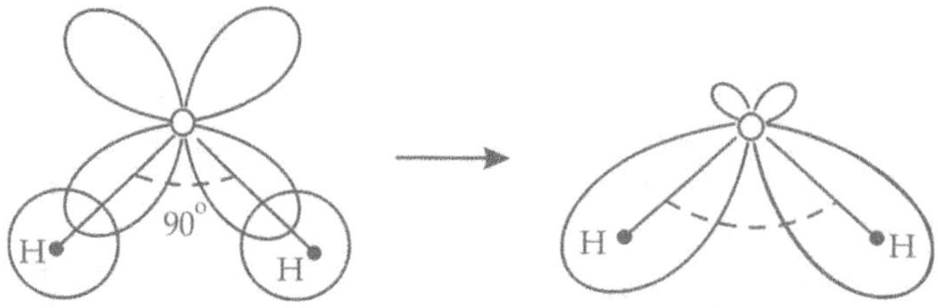

Figure 4.1 Bond angle in water molecule.

Water molecule is formed by the overlapping of the two "s" orbitals of the hydrogen atoms with one "2p" orbital of the oxygen atom. So, the bond angle should be 90°. But the original measured value between H-O-H bond is 104° 31′. This difference in value from theoretical value is attributed to the following two factors.

1. Repulsion between the atoms or groups attached to the central atom.
2. Hybridization of bonding orbitals.

They are described below in detail.

1. **Repulsion between the atoms or groups attached to the central atom:** In water molecule, two hydrogen atoms are attached to the central oxygen atom. Due to the high electronegativity of oxygen atom, the electron pair is somewhat shifted near to the central oxygen. Hence oxygen atom attains a partial negative charge and two hydrogen atoms attain a partial positive charge. The positive charge on the hydrogen atoms causes repulsion among them and will increases the bond angle to 104° 31′.
2. **Hybridization of bonding orbitals:** Hybridization of bonding orbitals in a molecule also plays a important role for determining the bond angle.

In hybridization the "s" character of the hybrid orbitals or hybrid bonds increases the bond angle.

Hybridized bond types and their respective bond angles are depicted in the following Table 4.1.

Table 4.1 Types of hybridization and their respective bond angles.

Bond type or type of hybridization	Bond angle
sp^3	109° 28'
sp^2	120°
sp	180°

Table 4.2 Bond angle in some of the important organic molecules.

S. No.	Name of the molecule	Bond angle	Value of bond angle
1.	H_2O	H-O-H	104° 31'
2.	NH_3	H-N-H	107° 3'
3.	H_2S	H-S-H	92° 2'
4.	CH_3OH	H-O-CH_3	108°
5.	CH_3SH	H-S-CH_3	96° 5'

The bond angle values can be determined from measured values of bond lengths, dipole moments, electron or neutron diffraction measurements and X-rays.

Bond energy: There are two kinds of bonds energies in organic compounds.

(i) Bond dissociation energy.
(ii) Bond energy.

(i) **Bond dissociation energy (D):** It is denoted by letter D. It is the amount of energy required to break a particular bond in a poly atomic molecule in a gaseous state into neutral fragments or free radicals in their gaseous state.

$$\underset{\text{Molecule}}{A - B\ (g)} \longrightarrow \underset{\text{Free radicals}}{A\cdot(g) + B\cdot(g)}$$

(ii) **Bond energy (E):** Consider the bond dissociation of methane molecule.

```
     H
     |
 H — C — H
     |
     H
```

In methane, all four C-H bonds are equivalent but the energy required to break the first C-H bond is not same as that of energy required to break the second bond and so on. It indicates that the energy for breaking the individual four C-H bonds in methane is different; hence the dissociation energy of a bond depends upon the nature of the rest of the molecule also. **Bond energy is defined as the average value of all the bonds dissociation energy of that molecule.** The unit of bond energy is KJ / mole or Kcal / mole.

Some of the bond energy values are shown below. For example: Bond dissociation energy of individual bond in methane molecule is shown here.

$$CH_4 \longrightarrow CH_3^{\cdot} + H\cdot \quad D(CH_3-H) = 104 \text{ Kcal/mole}$$

$$CH_3 \longrightarrow CH_2^{\cdot} + H\cdot \quad D(CH_2-H) = 105 \text{ Kcal/mole}$$

$$CH_2 \longrightarrow CH^{\cdot} + H\cdot \quad D(CH-H) = 108 \text{ Kcal/mole}$$

$$CH \longrightarrow C^{\cdot} + H^{\cdot} \quad D(C-H) = 81 \text{ Kcal/mole}$$

Hence the carbon hydrogen bond energy in methane is 99 Kcal/mole is a single average value.

$$CH_4 \longrightarrow C + 4H\cdot \quad \Delta H = 398 \text{ Kcal/mole}$$

$$E(C-H) = 398/4 = 99 \text{ Kcal/mole}$$

Bond Breaking

Breaking of bond between two atoms takes place by two ways.

1. Homolysis.
2. Heterolysis.

1. **Homolysis:** Each atom or group in a molecule takes one electron of the shared electron pair is called homolysis or homolytic fission. In this process two free radicals are formed.

$$\underset{\text{Molecule}}{A : B} \longrightarrow \underset{\text{Free radical}}{A\cdot + B\cdot}$$

This type of fission mainly occurs in substitution reactions called as free-radical substitution reaction.

2. **Heterolysis:** Both electrons in a bond go to anyone of the fragment which leads to the formation of an anion and cation.

$$\underset{\text{Molecule}}{A : B} \longrightarrow \underset{\text{Anion}}{\overline{A:}} + \underset{\text{Cation}}{B^+}$$

Bond Order

Number of covalent bonds in the molecule is known as bond order. The bond order for H_2 is 1, O_2 is 2 and N_2 is 3. For understanding the stability of the molecules, it can be generally correlated that with increase in bond order, bond enthalpy decreases and bond length decreases.

Polarity of Bond and Dipole Moments

A covalent bond is formed between two atoms by mutual sharing of electrons between them (equal sharing of electrons). It is again classified into two types whether the electron pair is shared unequally or equally between the atoms.

(a) Polar bond.

(b) Non-polar bond.

For example, the bond between hydrogen atoms in H_2 molecule and chloride atoms in Cl_2 molecule are known as non-polar because the two electrons between the atoms are equally shared.

$$H \overset{\bullet}{\times} H \qquad :\ddot{\underset{\bullet\bullet}{Cl}} \overset{\bullet}{\times} \underset{\times\times}{\overset{\times\times}{Cl}}\,\substack{\times \\ \times}$$

In case of hydrogen fluoride, the electron pair is unequally shared between hydrogen and fluoride and the electron cloud is denser in one atom than the other. In hydrogen fluoride molecule, fluoride atom is more electronegative than hydrogen and it pulls the bonded electrons towards itself. Therefore, the hydrogen atom appears partial positive charge and fluorine atom appears partial negative charge. These types of bond are known as polar bond.

Note: Electronegativity: The ability of atoms to attract the electrons.

Polarity of a bond is indicated by using symbols δ^+ and δ^- (delta plus and delta minus)

$$\overset{\delta^+}{H} - \overset{\delta^-}{F} \qquad \underset{\delta^+}{H} - \overset{\delta^-}{O} - \underset{\delta^+}{H} \qquad \underset{\delta^+}{H} - \overset{\delta^-}{N}(-\underset{\delta^+}{H}) - \underset{\delta^+}{H}$$

The strength of bond depends upon the electronegativity of the atom involved in the bond. Greater the difference between the electronegativity of the bonded atoms more polar is the bond. In the upper right hand side of the periodic table higher electronegative elements are present. The order of electronegativity of the elements is as follows

$$F > O > Cl > N > Br > CH$$

According to Pauli's scales the electronegativity value lies between 0-4 and the electronegativity between carbon and hydrogen atom is 2.1 as a standard.

Types of Orbital Overlapping and their Orbital Diagrams

Various types of overlapping of orbitals are discussed here.

1. **s-s overlapping:** Hydrogen atom has only one electron which is present in its *s* orbital. In the formation of a hydrogen molecule, spherical *s* orbital of one atom overlaps with that the other *s* orbitals of hydrogen atom to form a single covalent bond.

   For hydrogen molecule the system is most stable when the distance between the nuclei is 0.74 A° (1 A° = 10^{-8} cm). In this distance, the stabilizing effect of overlapping is exactly balanced by repulsive forces between similarly charged nuclei. This distance is called as bond length. The resulting hydrogen molecule contains less energy than the hydrogen atoms from which it is formed. So, we can state that H-H bond has a length of 0.74 A° and strength of 453.1 KJ.

   The bonding orbital is sausage shaped with its long axis lying along the line joining the nuclei as shown in Fig. 4.2. It is symmetrical about the long axis.

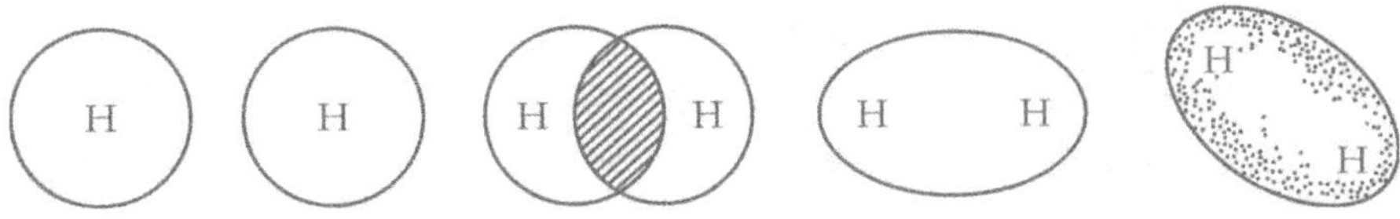

Figure 4.2 Overlap of s-s orbitals.

2. ***s-p* overlapping:** The electronic configuration of fluorine atom is $1s^2$, $2s^2$, $2px^2$, $2py^2$, $2pz^1$ having one of the electrons ($2p_z$) unpaired. Hence the molecule of hydrogen (H-F) is formed by the overlapping of *1s* orbital of hydrogen with $2p_z$ orbital of fluoride (*s-p* overlapping) as in Fig. 4.3.

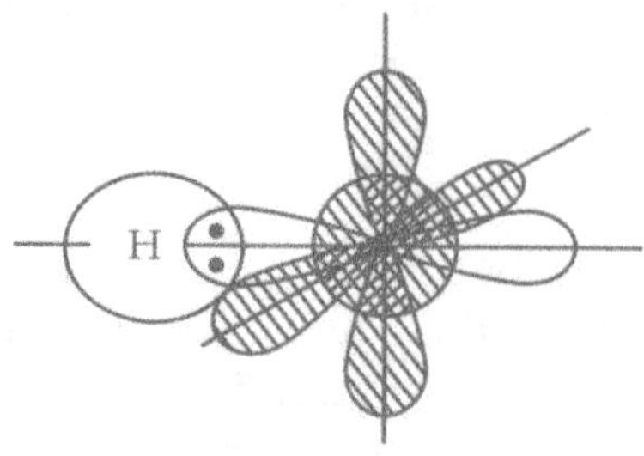

Figure 4.3 Overlap of s-p orbitals.

3. ***p-p* overlapping:** The electronic configuration of chlorine is $1s^2$, $2s^2$, $2p^6$, $3s^2$, $3px^2$, $3py^2$, $3pz^1$ with one electron ($3p_z$) being unpaired. The molecule of chlorine is produced by the overlapping of the $3p_z$ orbitals of two atoms (Fig. 4.4)

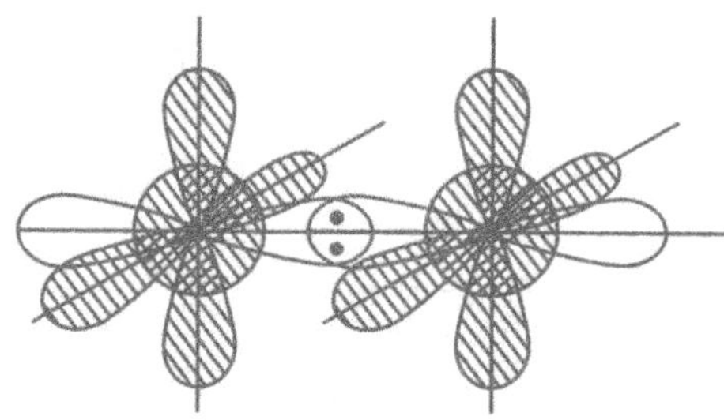

Figure 4.4 Overlap of p-p orbitals.

Formation of F-F and Br-Br molecules also explained in a same manner.

The structure of water (H_2O) and ammonia (NH_3) molecules can be described as follows. Based on Hund's rule of maximum multiplicity the electronic arrangement of nitrogen and oxygen atoms can be represented as shown below.

Nitrogen: $1s^2$, $2s^2$, $2px^1$, $2py^1$, $2pz^1$

Oxygen: $1s^2$, $2s^2$, $2px^2$, $2py^1$, $2pz^1$

Nitrogen thus has three unpaired electrons while oxygen has two. Hence nitrogen exhibits a valency of three and oxygen has a valency of two.

When the nitrogen atom combines with three hydrogen atoms to form a molecule of NH_3 the three N-H bonds must be at right angles to one another. Similarly, the two O-H bonds in the formation of water should be at right angles to each other as shown diagrammatically in Figs. 4.5 and 4.6.

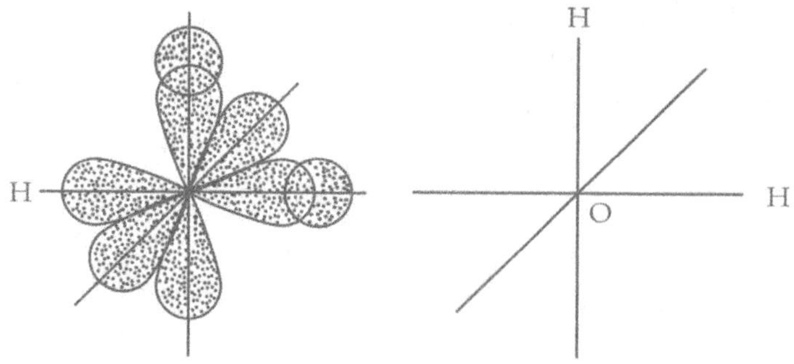

Figure 4.5 Representation of directed bonds in water molecule.

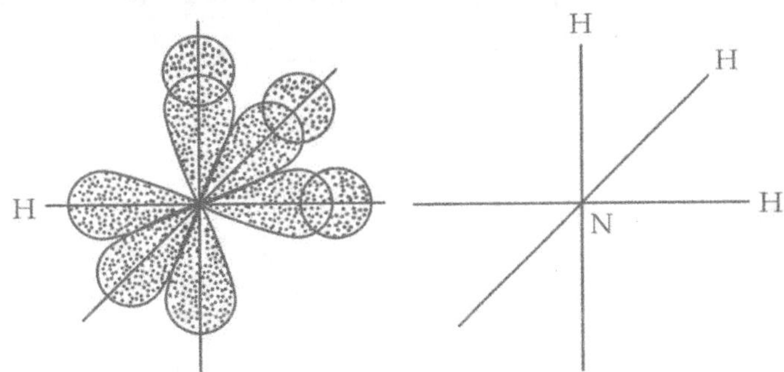

Figure 4.6 Representation of directed bonds in ammonia molecule.

However, the actual angle between O-H bonds in H_2O is 105° instead of 90° and bond angle between N-H bonds in NH_3 is 108° instead of 90°. This abnormality is explained by the concept of hybridization.

Hybridization

Linus Pauling in 1931 introduced the concept of hybridization to explain the characteristic geometrical shapes of polyatomic molecules like CH_4, NH_3 and H_2O *etc.* According to him, **the atomic orbitals combine to form new set of equivalent orbitals known as hybrid orbitals. Unlike pure orbitals, the hybrid orbitals are used in bond formation. The phenomenon is known as hybridization. Thus, hybridization is defined as the process of intermixing of the orbitals of slightly different energies so as to redistribute their energies, resulting in the formation of new set of orbitals or equivalent energies and shape.** For example, when one 2*s* and three 2*p* orbitals of carbon hybridizes, there is the formation of four new sp^3 hybrid orbitals.

Salient features of hybridization: The main features of hybridization are enumerated as follows.

1. The number of hybrid orbitals is equal to the number of the atomic orbitals that get hybridized.
2. The hybridized orbitals are always equivalent in energy and shape.
3. The hybrid orbitals are more effective in forming stable bonds than the pure atomic orbitals.
4. The hybrid orbitals are placed in space in some preferred direction to have minimum repulsion between electron pairs and thus get stable arrangement. Hence, the type of hybridization indicates the geometry of the molecules.

Conditions for hybridization:

1. The orbitals present in the valence shell of the atom are hybridized.
2. The orbitals undergoing hybridization should have almost equal energy.
3. Promotion of electron is not essential condition prior to hybridization.
4. It is not necessary that only half-filled orbitals only participate in hybridization. In some cases, even filled orbitals of valence shell also take part in hybridization.

Theory of Hybridization for the Formation of Covalent Bond (Sidwick – Powell Theory)

The electronic configuration of carbon atom is $1s^2$, $2s^2$, $2p_x^{\,1}$, $2p_y^{\,1}$, $2p_z^{\,0}$ (Ground state). There are only two unpaired electrons and hence carbon should have a valency of 2.

In few compounds carbon is bivalent in nature but in most cases, carbon is tetravalent. The tetra valency of carbon atom is explained on the following basis that one of the 2*s* electron is promoted to $2p_z$ which will now get higher energy and less stability. In this state the electronic configuration of carbon atom is $1s^1$, $2s^1$, $2p_x1$, $2p_y1$, $2p_z1$. This state of the carbon atom is known as excited state. Now in the excited state, there are four unpaired electrons and hence carbon atom will form four bonds by overlapping with four different kinds of orbitals (Fig. 4.7).

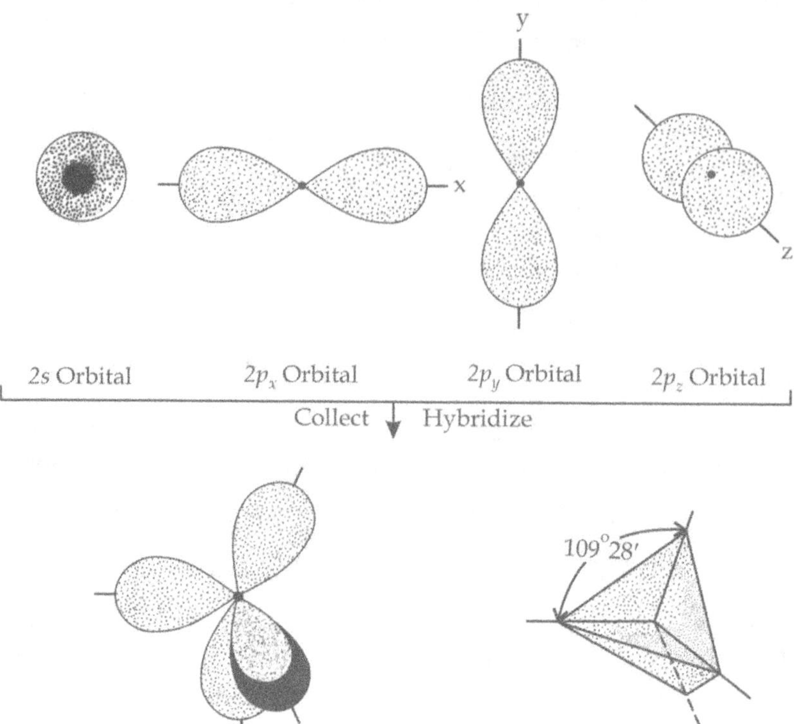

Figure 4.7 Hybridization of atomic orbitals in carbon atom (Tetrahedral or sp^3 hybridization).

Actually, the excitation of electrons from 2*s* orbitals to $2p_z$ requires considerable amount of energy and takes place in two ways:

1. Unpairing of electrons in the 2*s* orbitals
2. Promotion of one paired electron from 2*s* orbital to empty $2p_z$ orbital

Source of Energy required for Hybridization

The energy required for hybridization is obtained through the formation of bonds which are stronger with the hybrid orbitals than with the pure orbitals. More energy is released with the former than with the latter one when a chemical bond is formed.

The energy needed for this process is equivalent to the energy difference between $2p_z$ and 2*s* orbitals. Similarly, the energy needed for this excitation of electrons and consequent increase in stability is obtained by the formation of four covalent bonds. The energy supplied is more than that required for the excitation or promotion of electrons. Hence it is understood that the process of hybridization occurs only whenever there is a bond formation. All the steps in this process are schematically represented below in Fig. 4.8

Types of hybridization: There are mainly three types of hybridization in organic chemistry, they are as follows

1. sp^3 Hybridization or tetrahedral hybridization.
2. sp^2 Hybridization or trigonal hybridization.
3. *sp* Hybridization or diagonal hybridization.

sp^3 Hybridization or tetrahedral hybridization: This type of hybridization involves mixing or merging of one *s* and three *p* orbitals to give four new equivalent sp^3 hybridized orbitals.

Generally, this kind of overlapping will form two different types of bonds.

One involving 2*s* electrons which will not have any oriental direction and second involving three 2*p* electrons which will be present at right angle to each other (90°). Again, three dumb-bell shaped *p* electron clouds would form stronger bonds than the one spherically symmetrical *s* electron cloud because dumb-bell shaped orbitals can overlap more effectively than the spherical orbitals. Moreover the energy of *2s* orbital is less than the energy of the *2p* orbitals. So, the carbon atom forms one bond (uses *2s* orbital) of different strength than the other three *2p* orbitals. However, in reality the four C-H bonds in methane are exactly equivalent and the angle is of 109° 28′. To explain this, it is assumed that the four orbitals *2s*, $2p_x$, $2p_y$ and $2p_z$ can be mixed or hybridized with each other to form four new identical and symmetrically directed orbitals (each one with one electron) and inclined to each oher at 109° 28′. Each of the sp^3 orbitals has one electron. It has 25 % *s*-character and 75 % *p*-character becuse it is formed from one *s* and three *p* orbitals.

In methane molecule, each of the four hybridized orbitals of the carbon atom overlaps (head to head or end to end) with the *1s* orbital of the hydrogen atom to fom four carbons – hydrogen sigma (σ) bonds as shown in Figs. 8.8 and 8.9 The four sp^3 orbitals have the same energy, arrange themselves in space in such a way that they are far away from each other *i.e.*, ointed towards the four corner of the normal tetrahedron. As the four equivalent orbitals are arrangd tetrahedrally it is also termed as tetrahedral hybridization. The hybridized bonds are stronger than the original s or *p* electron bonds.

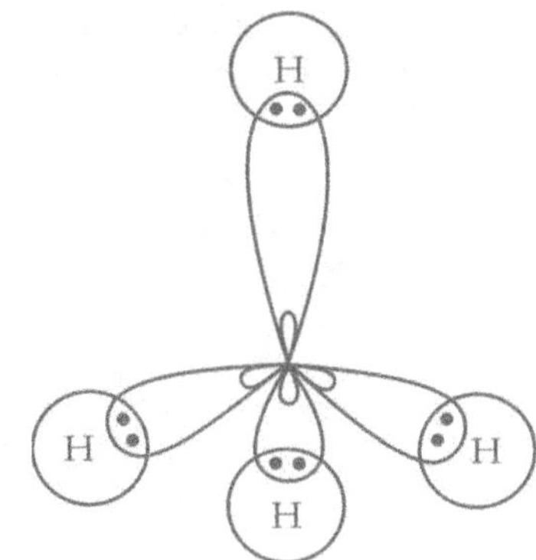

Figure 4.8 Orbital picture of methane.

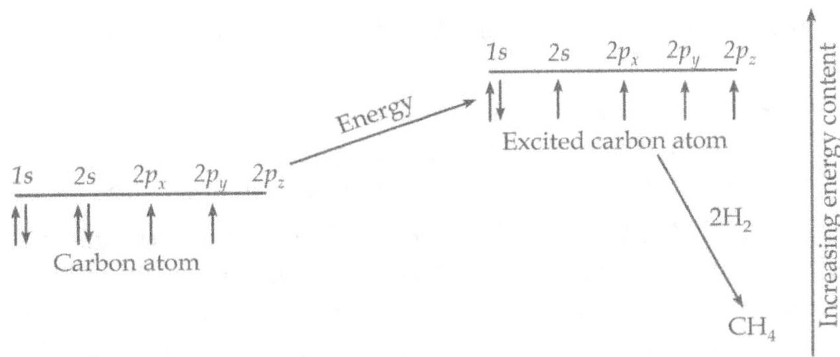

Figure 4.9 Formation of four stable bonds in carbon.

sp^2 Hybridization or trigonal hybridization: This type of hybridization involves mixing or merging of one *s* and two *p* pure orbitals to give three new equivalent sp^2 hybridized orbitals.

This type of hybridization leads to the formation of C=C, C=O type bonds. For example: Formation of ethylene molecule.

In ethylene molecule (C_2H_4), three orbitals (one *s* orbital and two *p* orbitals) of each carbon atom unite (hybridize) *i.e.*, combine to form identical orbitals, each called as sp^2 hybrid orbital. The three orbitals differ only in direction. They lie in the same plane and are mutually inclined at 120°. Hence the electronic configuration of the carbon atom becomes $C\ (sp^2)^1$, $C\ (sp^2)^1$, $C\ (sp^2)^1$, $p_z^{\ 1}$ or $1s^2\ 2(sp^2)^1\ 2(sp^2)^1 2(sp^2)^1 2p_z^{\ 1}$ *i.e.*, its four valence electrons are placed in each of the hybridized and unhybridized orbitals.

In ethylene molecule, two carbon atoms are joined to each other by a sigma bond formed by head to head overlap of two sp^2 orbitals (one from each carbon). The four C-H bonds are sigma bonds which result from the overlap of sp^2 orbital of carbon with 1*s* orbital of hydrogen as shown in Fig. 4.10.

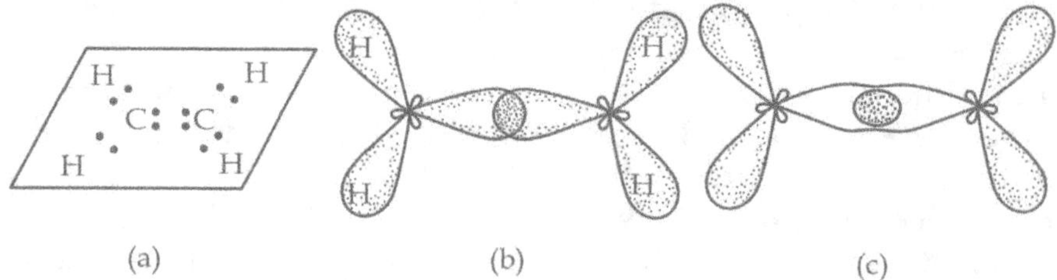

Figure 4.10 Sigma bond in the formation ethylene (a) Electronic formula (b) Formation of five sigma bonds by s-sp^2 overlap (in case of four C-H bonds and one C-C bond) (c) Structure of molecular orbital.

Each one of these five bonds has a symmetrical distribution of the electron cloud linked with it and called as sigma bond. After formation of sigma bonds each carbon atom has one unhybridised *p*-orbital which is perpendicular to the plane of the C-H and C-C bonds. The second bond (π bond) between the carbon atoms can be formed by the sideways overlap of these *p* orbitals as shown in **Fig. 4.11** wherein sigma bonds have been represented by straight lines.

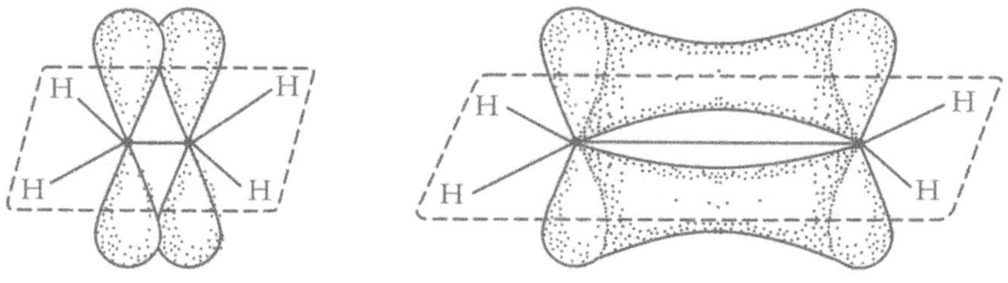

Figure 4.11 π bond in ethylene.

So in ethylene molecule there are four C- H sigma bonds and one σ and one π bond between two carbon atoms are present.

The π electron cloud of ethylene molecule is not delocalized directly between the carbon atoms, and concentrated above and below the plane of carbon and hydrogen atoms. The orbital arrangement of ethylene molecule is like a **"hot dog"**. Hence the double bond is a mixture of sigma and pi bond.

***sp* Hybridization or Diagonal hybridization:** This type of hybridization occurs by mixing or merging of one *s* and one *p* pure orbital to give two new equivalents collinear or *sp* hybridized orbitals. This type of hybridization leads to the formation of carbon-carbon triple bond. Example: Formation of acetylene molecule.

In acetylene molecule, two orbitals one *s* orbital and one *p* orbital of each carbon atom hybridize to form two identical orbitals called as *sp* hybridized orbital. The *sp* orbitals are linear and the electronic configuration of carbon atom becomes as shown below (Fig. 4.12).

$$C\ (sp)^1,\ (sp)^1,\ px^1,\ py^1$$

S-orbital

2*p* Orbital

Two *sp*-hybrid atomic orbitals with common origin

Figure 4.12 sp hybridisation or diagonal hybridisation.

Among these orbitals, one of the two *sp* hybrid orbitals of each carbon atom overlap with each other to form C-C sigma bond and the remaining one *sp* hybrid orbital of each carbon atom overlaps with the two *1s* orbitals of hydrogen atoms to form two C-H sigma bonds (Fig. 4.13).

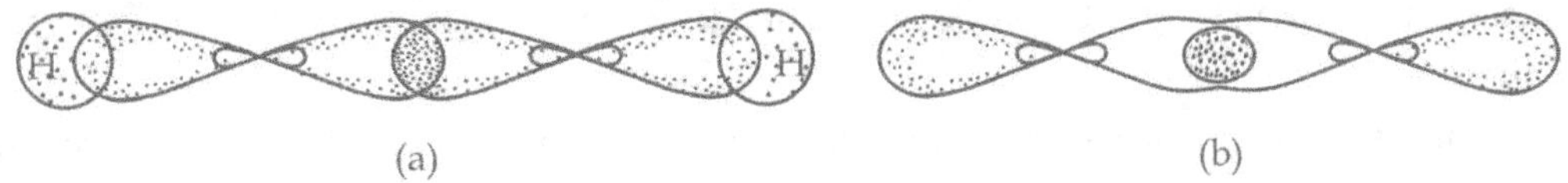

Figure 4.13 Sigma bond formation in the acetylene a) Formation of three sigma bonds by s-sp overlap (in case of two C-H bonds and one C-C bond) b) Resultant molecular orbital.

Each carbon atom possesses two unhybridised p (p_y *and* p_z) orbitals. After formation of sigma bond, these two p orbitals are perpendicular to one another and also perpendicular to the sigma bond between them. Each pair of these p orbitals forms a pi (π) bond by sideways overlap and the sigma bonds are represented by straight lines. In acetylene molecule, the triple bond is present between the two carbon atoms which are made up of one sigma (σ) and two pi (π) bonds. A crude model of the orbital arrangement of acetylene molecule is keeping like two bananas above and below a flat surface such as a piece of glass or card board (Fig. 4.14).

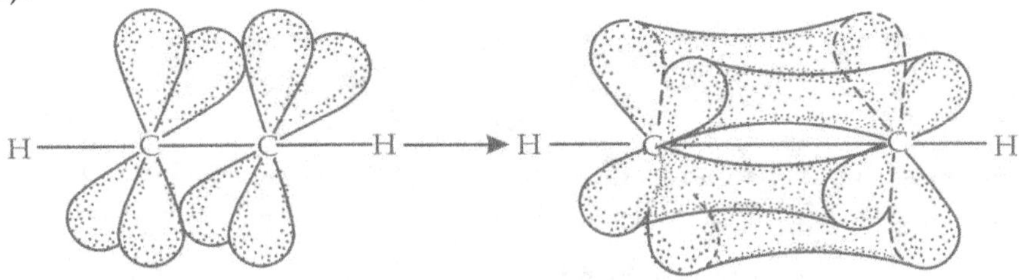

Figure 4.14 Two π bonds in acetylene

Effectiveness of Overlap

The strength of a covalent bond depends upon the degree of overlap of the two orbitals forming a bond or localized molecular orbital. The strength of the resulting covalent bond is predicted by considering the following points.

1. Bonding between two *s* orbitals is weak especially when two orbitals are at different energy levels.
2. The more the *s* character of a hybrid orbital, the weaker the bond strength..
3. A strong sigma bond is formed by the end to end overlap of two p orbitals at the same energy level. The bond formed is weaker when it is formed by the orbitals with different principal quantum numbers and the bond energy approaches to zero as the distance between the two levels increases. For example, C-Br bond in bromobenzene is much weaker than C-F bond in fluorobenzene.
4. Shorter the bond, greater the compression between atomic nuclei and stronger the bond.

Molecular Orbital Theory

It was developed by *F. Hund* and *R.S. Mulliken* in 1932. The salient features of this theory are:

1. The electrons in a molecule are present in the molecular orbitals as the electrons of atoms are present in the atomic orbitals.
2. The combination of atomic orbitals of comparable energies and proper symmetry leads to the formation of molecular orbitals.
3. Electron in an atomic orbital is influenced by one nucleus whereas in a molecular orbital it is influenced by two or more nuclei based on the number of atoms in the molecule. Therefore an atomic orbital is monocentric while a molecular orbital is polycentric.
4. The number of molecular orbital formed as per the number of combining atomic orbitals. For example, When two atomic orbitals combine, two molecular orbitals are formed. In this one is called as bonding molecular orbital while the other is called anti bonding molecular orbital.

5. The bonding molecular orbital has lower energy and hence greater stability with respect to antibonding molecular orbital.
6. As in the atom electron probability distribution around a nucleus is given by an atomic orbital, the electron probability distribution in a molecule is given by a molecular orbital.
7. Like atomic orbitals the molecular orbitals are filled in as per *Aufbau* principle by following the *Pauli* exclusion principle and the *Hund's* rule of maximum multiplicity.

In the valence bond theory, a bond is formed between two atoms when they are coming close enough close to ensure maximum overlap of the atomic orbitals. One electron from each bonded atom loses its identity and moves in the outer atomic orbitals of both bonded atoms.

According to MO theory, the molecular orbitals are formed by the coalescence of the individual atomic orbitals when the atoms to be bonded come together. Atomic orbitals are denoted by the letters *s*, *p* and *d* similarly molecular orbitals are denoted by the Greek letter sigma (σ), pi (π) and delta (δ).

Probable Questions

1. Why the atoms want to attain stable electronic configuration? Or why atoms are involved in bond formation?
2. Define the doublet and octet rule.
3. Write the *Lewis-Langmuir* concept of electronic configuration.
4. Explain the various types of bonds between the atoms.
5. Define ionic bond and explain the formation of this bond by considering NaCl molecule as an example.
6. Define ionisation energy and explain how it influences the formation of ionic bond.
7. Define electron affinity and explain how it influences the formation of ionic bond.
8. Explain the term covalent bond and the mechanism involved in the formation of covalent bond.
9. Write in detail about the various types of covalent bond with examples.
10. Define co-ordinate covalent bond and explain the formation of it with example.
11. Explain in detail about electronic theory of valency.
12. Describe various characteristic properties of covalent bond.
13. Define the following terms.
    (a) Bond angle.
    (b) Bond length.
    (c) Bond order.
14. Describe the factors responsible for the deviation of original bond angle from its theoretical bond angle.
15. Write a short note on types of hybridisation and their respective bond angles.
16. Explain the concept of bond dissociation energy and bond energy with appropriate examples.
17. Define homolysis with one example.
18. Define heterolysis with one example.
19. Explain the concept of polarity with appropriate examples.
20. Write a short note on polarity of atoms.
21. Define non-polar bond.
22. Mention the number of σ and π bonds present in the following compounds.
    (a) $CH_3CH_2NH_2$
    (b) $CH_3CH = CH_2$
    (c) $CH_2 = CH = CH_2$

23. Mention the type of bonding orbitals in each of the following and predict the geometrical configuration for each.
    (a) CH_4
    (b) NH_3
    (c) H_2O
24. How do bond length and bond strength differs in the following.
    (a) C- C
    (b) C = C
    (c) C- Cl
25. Explain the different types of orbitals overlapping.
26. Define hybridisation with example and write various types of it.
27. Write in detail about the salient features of hybridisation.
28. Describe the various conditions for hybridisation.
29. Write in detail about *Sidwick – Powell theory.*
30. Explain in detail about the energy requirement for hybridisation.
31. Define sp^3 hybridisation and write its salient features with neat orbital picture.
32. Define sp^2 hybridisation and write its salient features with neat orbital picture.
33. Define *sp* hybridisation and write its salient features with neat orbital picture.
34. Describe the orbital structure of methane.
35. Describe the orbital structure of ethane.
36. Describe the orbital structure of ethylene.
37. Describe the orbital structure of acetylene.

5

Factors Influencing Chemical Reaction or Electron Displacements in Molecules

Introduction

Ideal covalent bond is electrically neutral. It develops a polarity by the displacement of their bonding electrons. The electronic displacement in a molecule is due to certain effects, some of which are permanent effect (inductive effect and mesomeric effect) and other effects are temporary (electromeric and inductomeric effects). These effects are classified into two types.

1. Polarisation effects.
2. Polarisability effects.

Polarisation effects: Inductive effect and mesomeric effect permanently operates in the real molecule and this effect is polarization effect.

Polarisability effects: Electromeric and inductomeric effects depend on attacking reagent and as soon as the attacking reagent is removed, the electronic displacement disappears and this effect is known as polarisability effects. These effects involving displacement of electrons in a substrate is known as electronic displacement effects.

The following are the various factors which influences the chemical reaction.

I. Inductive effect.
II. Mesomeric and Resonance effect.
III. Electromeric effect.
IV. Resonance.

Inductive Effect

It is defined as "The permanent displacement of electrons producing polarity in a covalent bond due to the presence of high electronegativity of one atom than another atom". It is denoted by the symbol $\rightarrow$, the arrow pointed towards the high electronegative atom. Let us consider the following example.

$$\rightarrow C_4 \rightarrow C_3 \rightarrow C_2 \rightarrow \overset{\delta^+}{C_1} \rightarrow \overset{\delta^-}{Cl}$$

(or)

$$\overset{\delta\delta\delta\delta^+}{CH_3} \rightarrow \overset{\delta\delta\delta^+}{CH_2} \rightarrow \overset{\delta\delta^+}{CH_2} \rightarrow \overset{\delta^+}{CH_2} \rightarrow \overset{\delta^-}{Cl}$$

In the above example, the chlorine atom is highly electronegative than carbon atom, so the chlorine atom pulls the bonded electrons towards itself. Hence the electrons are displaced from the middle of the bond somewhat nearer to the chlorine atom. Due to this effect the chlorine atom acquires a small negative charge (δ^-) and the carbon atom C_1 attains a small positive charge (δ^+). Now C_1 attracts the electrons between C_1 and C_2 and attains small positive charge but it will be lesser than C_1. Similarly, C_3 will acquire smaller positive charge than C_2, this effect is produced throughout the carbon chain in a molecule. But the intensity decreases as the distance from the C-Cl decreases. For better understanding of inductive effect, the C-H bond is considered as a standard one.

Types of Inductive Effect: There are two types of inductive effects, they are as follows.

1. +I effect.
2. –I effect.

Let us discuss these effects in detail

1. **+ I effect (Electron releasing group):** Any group or atom that attracts the electrons less strongly than hydrogen is said to have + I effect, they are called as electron releasing or electron repelling group.

   Example: For + I effect groups:

   $$C_6H_5O^- > COO^- > R_3C > CHR_2 > CH_2R > CH_3 > H$$

   Electron releasing group (+I groups) increases the stability or stabilize the carbonium ions by partially neutralizing positive charge on carbon. Hence, the tertiary carbonium ion is more stable than that of secondary which is more stable than primary carbonium ion. So, the order of stability of carbonium ion is as follows.

   The presence of an electron withdrawing groups (-I effect) tends to intensify the positive charge of the carbonium ion and makes the carbocation less stable.

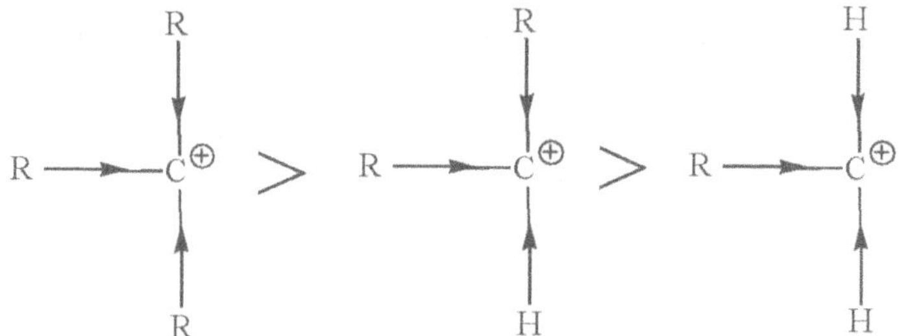

Tertiary carbonium ion Secondary carbonium ion Primary carbonium ion

   By dispersal of the charges, stability of the charged system is increased. Hence any factor that disperses the positive charge of the electron deficient carbon and distributes it in rest of the molecule stabilizes carbocation. Inductive effect is the main factor which disperses the positive charge of the alkyl carbocation.

   In carbocation the positively charged carbon atom is present in the center of the ion. So the electron density of the sigma bonds is shifted towards the positive carbon atom.

2. **-I effect (Electron attracting group):** Any group or atom that attracts the electrons more strongly than hydrogen is said to have -I effect, they are called as electron attracting or electron withdrawing groups.

   Example: For -I effect groups (Electron attracting or electro withdrawing groups).

   $N(CH_3)_3 > NH_3 > NO_2 > CN > COOH > F > Cl > Br > I > OAr > COOR > OR > OH > C_6H_5 > H$

Mesomeric and Resonance Effect

It is defined as "The permanent effect in which π electrons are transferred from a multiple bond to a single covalent bond or from an atom with lone pair of electrons to the adjacent single covalent bond". The mesomeric effect mainly occurs in unsaturated compounds and especially in conjugated system *via* their π orbitals. Let us consider the carbonyl compound.

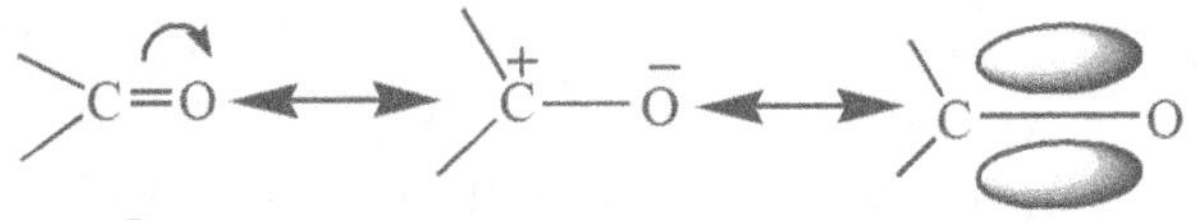

$$>C{=}O \longleftrightarrow >\overset{+}{C}{-}\overset{-}{O} \longleftrightarrow >C{-}O$$

I II III

The oxygen atom is highly electronegative atom which attracts the π electrons of the double bond, so they displace towards the oxygen atom which leads to the above resonance structure. If the carbonyl group is attached with C=C type of system, the polarity is transmitted further *via* the electrons.

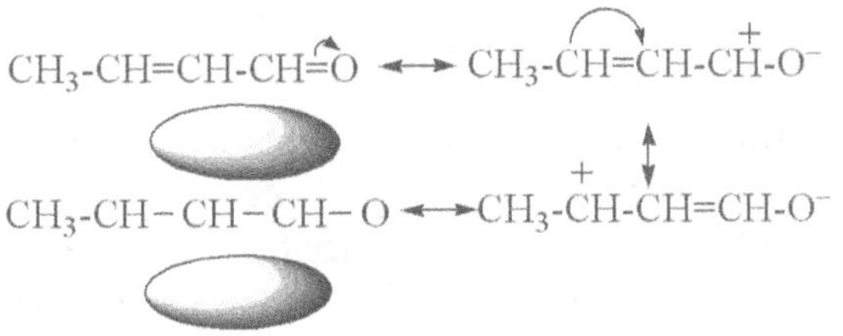

$$CH_3\text{-}CH{=}CH\text{-}CH{=}O \longleftrightarrow CH_3\text{-}CH{=}CH\text{-}\overset{+}{C}H\text{-}O^-$$

$$CH_3\text{-}CH{-}CH{-}CH{-}O \longleftrightarrow CH_3\text{-}\overset{+}{C}H\text{-}CH{=}CH\text{-}O^-$$

It is also produced by the presence of an atom containing at least one lone pair of electrons in conjugation with a conjugated system.

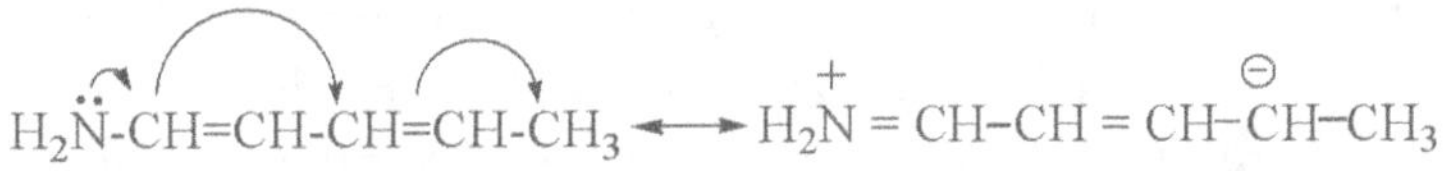

$$H_2\ddot{N}\text{-}CH{=}CH\text{-}CH{=}CH\text{-}CH_3 \longleftrightarrow H_2\overset{+}{N}{=}CH{-}CH{=}CH{-}\overset{\ominus}{C}H{-}CH_3$$

Like inductive effect, it is a permanent effect and denoted by a letter M and symbol of curved arrow (↷).

Types:

1. **+ M effect:** It is shown by groups having lone pair of electrons which are capable of exhibiting conjugation with an attached group.

   - OR, - OH, - C≡N, - NR_2, and - SR

2. **- M effect:** It is shown by electronegative atoms like nitrogen or oxygen which behaves as an electron "sink". Example,

   $-NO_2$, >C=O, $-SO_3H$

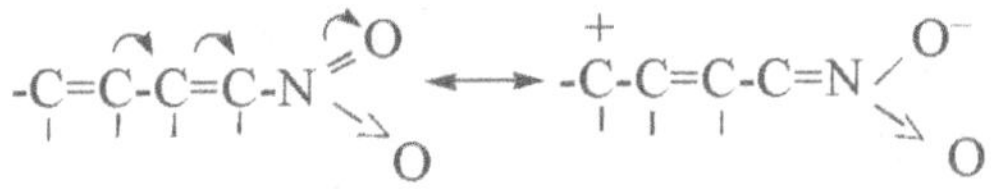

$$\text{-}C{=}C\text{-}C{=}C\text{-}N(=O)\text{-}O \longleftrightarrow \text{-}\overset{+}{C}\text{-}C{=}C\text{-}C{=}N(O^-)\text{-}O$$

Mesomeric effect also occurs in aromatic ring where the + M effect group or –M effect group is directly attached to the phenyl group so the lone pair of electrons is conjugated with the benzene nucleus and produces +M effect.

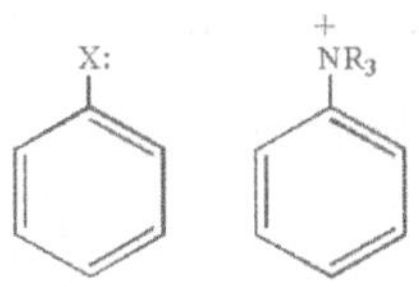

X: $\overset{+}{N}R_3$

+M effect -M effect

Electromeric Effect

In some of the chemical reactions, the two or three shared electron pairs between two atoms (joined by double or triple bond) are completely transferred from one of the atoms to another atom in the presence of an attacking reagent.

$$>C=C< \xrightarrow[\text{reagent}]{\text{Polar}} >\overset{+}{C}-\overset{..}{\overset{-}{C}}<$$

We know already that, a multiple bond is made up of a σ bond and π bond. The electrons of the π bond are loosely held and easily polarisable. Hence, when a charged reagent approaches a double bond or triple bond, the electrons of π bond are completely polarised or transferred to any one of the constituent atom. These transfers of electrons is indicated by a curved arrow (↷), starting from the original position of the electron pair and ending to the new position of the electron pair.

Due to the complete transfer of electrons, it leads to the development of full positive and negative charges on the carbon atoms.

The electromeric effect may be defined as "**temporary effect in which a shared pair of electrons is completely transferred from a double or a triple bond to another atom joined from the bond in the presence of polar attacking reagent**". As soon as the attacking reagent is removed, original electronic condition is restored. This effect is indicated by the letter "E".

$$CH_2=CH_2 \xrightarrow[\text{reagent}]{\text{Polar}} \overset{+}{C}H_2-\overset{-}{C}H_2$$

$$\overset{+}{C}H_2-\overset{-}{C}H_2 \xrightarrow[\text{polar reagent}]{\text{Removal of}} CH_2=CH_2$$

Examples:

1. When the multiple bonds are present between two similar atoms, the electromeric shift takes place in any direction. Example in ethylene molecule, the electromeric shift in both the directions will give the similar structure.

$$CH_2=CH_2 \xrightarrow[\text{reagent}]{\text{Polar}} H_2\overset{+}{C}-\overset{..}{\overset{-}{C}}H_2$$

$$CH_2=CH_2 \xrightarrow[\text{reagent}]{\text{Polar}} H_2\overset{..}{\overset{-}{C}}-\overset{+}{C}H_2$$

2. If the multiple bonds are present between two dissimilar atoms, the direction of electromeric shift is determined by the direction of inductive effects exhibited in the molecule. For example, in propylene molecule, the first reaction is preferred over the second reaction and the electromeric effect will occur in two ways (depending upon the direction of inductive effect).

$$CH_3 \rightarrow CH=CH_2 \longrightarrow CH_3-\overset{+}{C}H-\overset{..}{\overset{-}{C}}H_2 \quad \text{Favoured by inductive effect, so possible}$$

$$CH_3 \rightarrow CH=CH_2 \longrightarrow CH_3-\overset{..}{\overset{-}{C}}H-\overset{+}{C}H_4 \quad \text{Opposed by inductive effect, so not possible.}$$

3. In some cases, the electromeric effect overcomes the inductive effect. Example: Vinyl bromide.

$CH_2 = CH - Br \longrightarrow \overset{+}{C}H_2 - \overset{-}{\ddot{C}}H - Br$ Electrometric effect operating in the same direction of inductive effect.

$CH_2 = CH - Br \longrightarrow H_2\overset{-}{\ddot{C}} - \overset{+}{C}H - Br$ Electrometric effect operating in the opposite direction of inductive effect. (In this case it is more effective than above)

4. In carbonyl compounds, due to the presence of highly electronegative oxygen atom, initially small positive charge exists on carbon atom and small negative charge exists on oxygen atom. When a polar reagent approaches the carbonyl group, the π electrons are completely transferred to oxygen atom. Due to this the later one attains full negative charge and carbon atom attains full positive charge.

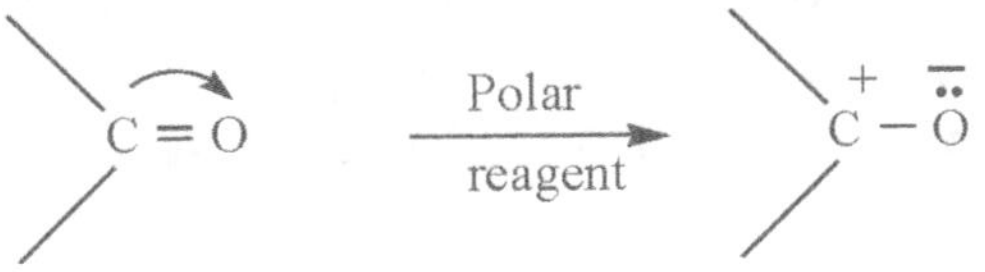

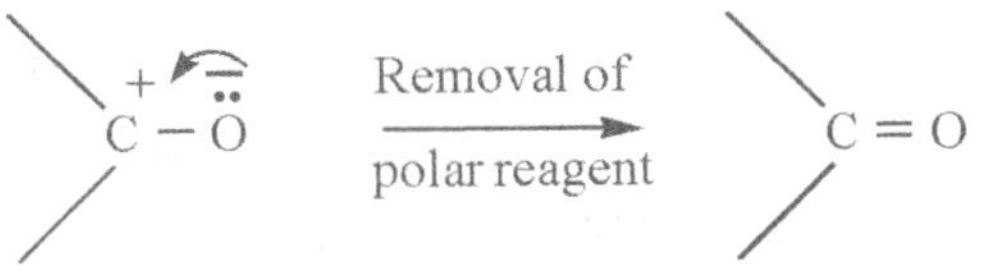

$>C=O \xrightarrow[\text{reagent}]{\text{Polar}} >\overset{+}{C}-\overset{-}{\ddot{O}}$

$>\overset{+}{C}-\overset{-}{\ddot{O}} \xrightarrow[\text{polar reagent}]{\text{Removal of}} >C=O$

Like inductive effect, electromeric effect is also classified into two types.

1. (+) E effect.
2. (-) E effect.

Let us discuss in detail.

1. **(+) E effect:** Displacement or transfer of electron pair away from the group or atom.

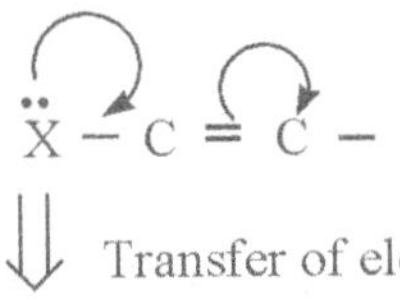

$\ddot{X} - C = C -$

⇓ Transfer of electrons away from X

2. **(-) E effect:** Displacement or transfer of electron pair towards the group or atom.

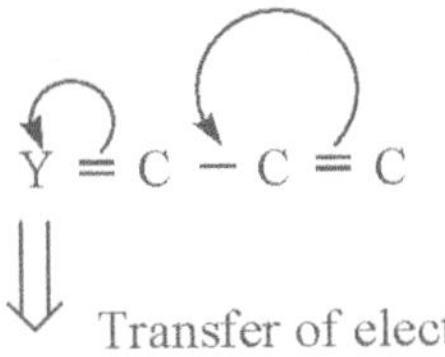

$Y = C - C = C$

⇓ Transfer of electrons towards group Y

Resonance

Occasionally a single electronic picture or a single structural formula drawn for a particular molecule does not satisfactorily explain all properties of the molecule. For example the structure of CO_2 molecule is represented by three ways.

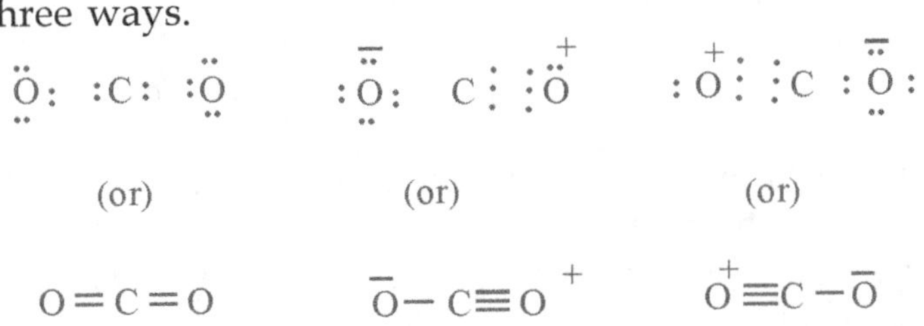

$\ddot{O}::C::\ddot{O}$ (or) $O=C=O$

$:\overset{-}{\ddot{O}}:C:::\overset{+}{O}$ (or) $\overset{-}{O}-C\equiv\overset{+}{O}$

$:\overset{+}{O}:::C:\overset{-}{\ddot{O}}:$ (or) $\overset{+}{O}\equiv C-\overset{-}{O}$

The calculated heat of formation of CO_2 for one formula is 1464 Kcal/mole and the O-O distance should be 2.40 A°. It indicates that none of the three structure of CO_2 molecule satisfactorily explains the properties of CO_2. This leads to the idea that these type of compounds exists in a state which is roughly a combination of two or more electronic structures, each one of which makes some contribution to its properties. The true structure of the compound is an intermediate between two or more valence structure.

Resonance may be defined as "**The phenomenon in which two or more structures, involving identical position of atoms can be written for a particular compound**" or

"**When several structures are assumed to represent the true structure of a molecule, none of them can be said to represent it completely it is known as resonance hybrid and the phenomenon is known as resonance**".

The above definition indicates that, the resonance structure is not an impression of the molecule but the molecule resonates from one structure to another structure due to the back and forth jumping of electrons pair from one bond to another bond. Hence, the compound or a molecule possess only one electronic structure that can't be described physically.

The concept of resonance is explained by the following crude analogy.

(a) Mule is a hybrid of the Mare and the Jack ass. It possesses or inherited with the characteristics of both parents. For example: It can run fast like a Mare and can carry load like Jack ass.

(b) When we look at a Mule, we don't see Jack ass at one glance and we don't see Mare at one glance. We always look at Mule. Like this, a resonance hybrid does not oscillate between its canonical structure and these properties are fixed which indicate the properties of original structure.

Chief conditions for occurrence of resonance:

1. Each canonical structure must have the same number of unpaired electrons.
2. The energy content of all the canonical structures is nearly same.
3. The arrangement of atoms must be identical in every formula.

Criteria of Resonance:

1. ***Heat of formation of a resonance hybrid is abnormally high:*** For example, the observed heat of formation of carbon dioxide is 1590 KJ/mole which is greater than the calculated value of 1464 KJ/mole. The difference between the two energy level is 126 Kcal/mole is known as resonance energy. It is defined as "The amount of energy lost when a molecule undergoes resonance or the energy difference between the resonance hybrid and the most stable canonical structure of a particular molecule".
2. ***Stability:*** Greater the value of resonance energy, smaller the internal energy of the hybrid and greater its stability [That is why a system always stabilizes itself as a result of resonance].
3. ***Shortening of bond length:*** The resonance hybrid always possesses shorter bond length than that of any of its canonical structures.

Many of the organic compounds exhibit resonance. For example: aromatic compound "benzene" exhibits the concept of resonance and the structure of benzene is exhibiting any one of the alternative representations.

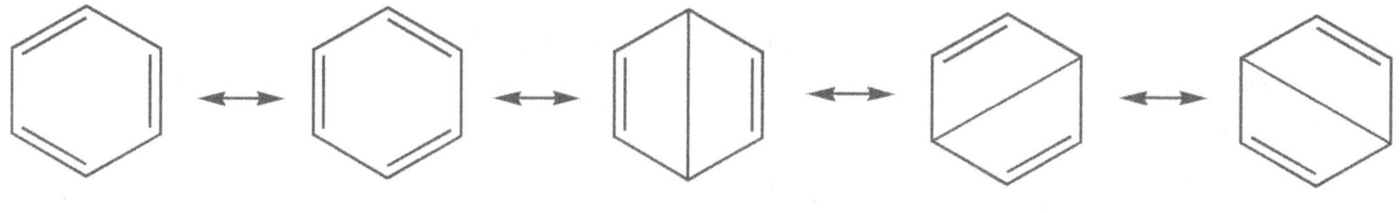

The energy of benzene is less than 150 Kcal / mole than the energy of anyone of its contributing structures. It indicates that the structure of benzene is an intermediate between any one of these structures, so that the benzene ring is not formed by alternate single and double bonds but made up of six equivalent bonds.

A resonance hybrid denoted by keeping double headed arrows between various canonical structures.

$$O = C = O \longleftrightarrow O^{-} - C \equiv O^{+} \longleftrightarrow O^{+} \equiv C - O^{-}$$

Hyperconjugation

Hyperconjugation is a special type of resonance or delocalisation of electrons involving σ bond orbital.

According to the heat of hydrogenation of alkene molecules "stability of alkenes not only depends on conjugation but also by the presence of alkyl group". It indicates that "greater the number of alkyl groups attached to the doubly bonded carbon atoms, the more stable is the alkene" (Saytzeff's rule).

Example 1: Stability of propylene is more than ethylene.

$CH_3 - CH = CH_2$	$CH_2 = CH_2$
More stable due to the presence of alkyl (methyl) - CH_3 group	Less stable due to the absence of alkyl group

Example 2: Stability of 1,3-butadiene (High stability): 1,3-butadiene (conjugated diene) is more stable than that of isolated dienes. It has been attributed to the delocalization of π electrons, *via* the overlapping of the "*p*" orbital of the π bond and a σ bond orbital of the alkyl group. In butadiene, the four carbon atoms possess an unhybridised "*p*" orbital. Each "*p*" orbital is perpendicular to the plane of σ bonds. So all the four "*p*" orbitals overlap with each other to form a larger π molecular orbital. Each pair of electrons is attracted by four carbons. **Delocalization of this type involving a σ bond orbital is known as hyper conjugation.**

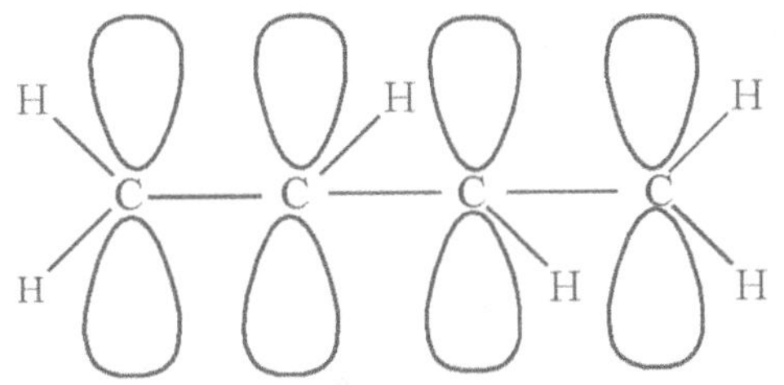

H H H C C C C H H H H

Probable Questions

1. What is meant by polarization effect and polarisability effects? Explain with examples.
2. Mention the various factors which influence a chemical reaction.
3. Define inductive effect and write its types with examples.
4. Define mesomeric effect and explain its effects in a chemical reaction.
5. How can you differentiate inductive effect and mesomeric effect?
6. Define electromeric effect and explain the effect of electromeric effect in a chemical reaction.
7. What is meant by resonance? and explain the basic concept behind resonance.
8. Mention the important requirements of resonance.
9. Describe the various criteria of resonance.
10. What is meant by hyperconjugation?

6

Organic Reactions and Mechanism

Introduction

In an organic reaction, the organic molecule reacts with an appropriate attacking reagent with the formation of one or more intermediates and finally products.

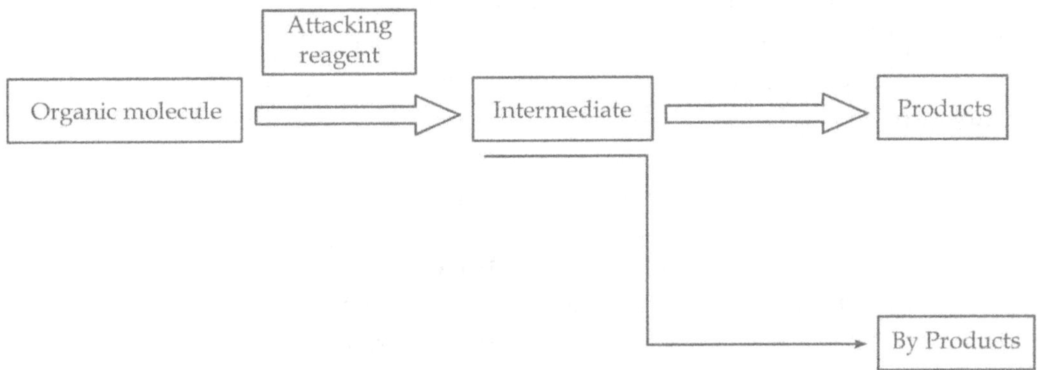

In the above reaction, the covalent bond between two atoms (carbon - carbon or carbon - other atoms) is broken and new bonds between two atoms (carbon - carbon or carbon - other atoms) are formed.

The study of sequential account of each step describing details of electron movement, energetics during bond cleavage and bond formation and the kinetic of a reaction (*i.e.* the rate at which reactants are converted into products) is called as **reaction mechanism**.

The knowledge of reaction mechanism helps us to understand the reactivity of organic compounds and their synthesis.

Some of the principles which explain the organic reactions are described below.

Bond Breaking (Homolysis and Heterolysis)

Breaking of bond between two atoms can takes place by two ways.

1. Homolysis
2. Heterolysis.

1. **Homolysis:** Each atom or group in a molecule takes one electron of the shared electron pair is called homolysis or homolytic fission. In this process two free radicals are formed.

$$\underset{\text{Molecule}}{A : B} \longrightarrow \underset{\text{Free radical}}{A\cdot + B\cdot}$$

This type of fission mainly occurs in substitution reactions called as free-radical substitution reactions.

2. **Heterolysis:** Both electrons in a bond go to anyone of the fragment which leads to the formation of an anion and cation.

$$\underset{\text{Molecule}}{A : B} \longrightarrow \underset{\text{anion}}{A\!:^{-}} + \underset{\text{cation}}{B^{+}}$$

(Both electrons go to A)

Organic Reagents (Nucleophiles and Electrophiles)

They are classified into two types.

1. Nucleophiles
2. Electrophiles

1. **Nucleophiles (Nucleus loving species or Electron rich species):** Organic reagent that can donate an electron pair in a chemical reaction is known as nucleophile. It attacks the positive portion (region with low electron density) of the substrate molecule. These are negatively charged species (carbanion also) or neutral species with free electron pairs. These are denoted by the symbol Nu^-.

   Types: Nucleophiles are further classified in to two types.

   (i) Negative nucleophiles: These species carry an excess electron pair and negative charge. They attack on the positively charged reactants to gives neutral products. Examples for negative nucleophiles are given below.

$$H\ddot{O}:^{\ominus},\ :\ddot{X}:^{\ominus},\ R\ddot{O}:^{\ominus},\ :\ddot{N}H_2,\ :\bar{C}\equiv N,\ :CR_3,\ COO^{\ominus}$$

   (ii) Neutral nucleophiles: Species which are rich in electrons (due to the presence of lone pair of electrons) and are neutrally charged. They attack on the positively charged reactants to give positively charged products. Example for neutral nucleophiles is given below.

$$H-\ddot{O}-H,\ R-\ddot{O}-H,\ R-\ddot{O}-R,\ R-\ddot{S}-H,\ \ddot{N}H_3,\ R\ddot{N}H_2$$

2. **Electrophiles (Electron loving species or Electron deficient species):** Organic reagents that can accept a pair of electron in a chemical reaction is known as electrophile. It attacks the negative portion (region with high electron density) of the substrate molecule. They are positively charged species (carbonium ion also) or neutral species with electron deficient centers, these are denoted by the symbol E^+.

   Types: Electrophiles are of two types.

   (i) Positive electrophiles: These species carry positive charge. They attack on the negatively charged reactants and gives neutral products. Example for positive electrophiles are given below.

$$H^+,\ Br^+,\ Cl^+,\ NH_4^+,\ H_3O^+,\ CR^{3+},\ Ar-N=\overset{\oplus}{N}$$

   (ii) Neutral electrophiles: These species are deficient in electrons and neutrally charged. They attack on the negatively charged reactants to gives negatively charged products. Example for neutral electrophiles are given below.

$$BF_3,\ AlCl_3,\ ZnCl_2,\ :CCl_2$$

   The distinction between electrophiles and nucleophiles are given in the following table.

Table 6.1 The difference between electrophiles and nucleophiles

S. No.	Nucleophiles	Electrophiles
1.	Electron rich species hence they are anions.	Electron deficient species hence they are cations.
2.	Donate an electron pair.	Accept an electron pair.
3.	Carry negative charge.	Carry positive charge.
4.	They have lone pair of electrons which are strongly attached to the atomic nucleus.	Having empty "*p*" orbital to accommodate lone pair of electrons from nucleophiles.
5.	They can increase their covalency by one unit.	Having the ability to form alternative or extra bond with nucleophile.

Types of Organic Reactions

Organic reactions are classified into four types according to the manner of reaction how it proceeds.

1. Substitution reactions.
2. Addition reactions.
3. Elimination reactions.
4. Rearrangement reactions.

These reactions are described in detail as follows.

1. **Substitution reactions:** Reactions which involves direct substitution or replacement of an atom or group of atoms by some other atom or group of atoms are known as substitution reactions.

$$\underset{\text{Methane}}{CH_4} + Cl_2 \longrightarrow \underset{\text{Methyl chloride}}{CH_3Cl} + HCl$$

In the above reaction, one hydrogen atom of the methane molecule is substituted by one chlorine atom. Substitution reactions are further classified in to three types.

(i) Free radical substitution reactions.
(ii) Nucleophilic substitution reactions.
(iii) Electrophilic substitution reactions.

(i) **Free radical substitution reactions:** Substitution reactions initiated by free radicals is called as free radical substitution reactions. Example, halogenation of alkanes (chlorination of methane).

$$\underset{\text{Methane}}{CH_4} + Cl_2 \longrightarrow \underset{\text{Methyl chloride}}{CH_3Cl} + HCl$$

Methane reacts with chlorine in the presence of heat or UV light to give chlorinated methane. This reaction is initiated by the formation of chlorine free radical.

(ii) **Nucleophilic substitution reactions:** Substitution reactions initiated by the attack of nucleophile is called as nucleophilic substitution reactions. Hydrolysis of alkyl halides by base (NaOH) is an example for this reaction.

$$\underset{\text{Methyl chloride}}{CH_3{-}Cl} + OH^- \longrightarrow \underset{\text{Methanol}}{CH_3{-}OH} + Cl^-$$

In this reaction the nucleophile (OH^-) replaces the chlorine atom. There are two types of Nucleophilic substitution reactions:

(a) S_N1 Reactions.

(b) S_N2 Reactions.

(a) S_N1 Reactions (Unimolecular nucleophilic substitution reactions): The rate of the substitution reaction depends only on the concentration of alkyl halide hence it called as **S_N1 Reactions (Unimolecular nucleophilic substitution reactions).**

$$(H_3C)_3C - Br + OH^- \longrightarrow (CH_3)_3COH + Br^-$$

t-Butyl bromide → *t*-butanol

(b) S_N2 Reactions (Bimolecular nucleophilic substitution reactions): The rate of the substitution reaction depends upon the concentration of alkyl halide as well the nucleophile hence it is called as **S_N2 Reactions (Bimolecular nucleophilic substitution reaction).**

$$CH_3{-}Cl + OH^- \longrightarrow CH_3{-}OH + Cl^-$$

Methyl chloride → Methanol

(iii) Electrophilic substitution reactions: Substitution reactions which are initiated by the attack of electrophile is called as electrophilic substitution reactions. Example, chlorination of benzene.

$$C_6H_6 + Cl_2 \xrightarrow{FeCl_3} C_6H_5Cl + HCl$$

Benzene → Chlorobenzene

2. Addition reactions: Organic reactions in which an atom or group of atoms are simply added to the unsaturated compounds (double or triple bonded compounds) without the elimination of another atom or group. In this reaction one π bond is broken and two new σ bonds are formed. Addition reactions convert the unsaturated compounds into saturated compounds.

Example: Bromination of alkenes. Ethylene reacts with bromine to give dibromo ethane. In this reaction bromine atoms are added to the double bonded ethylene molecule.

$$CH_2 = CH_2 + Br_2 \longrightarrow Br - CH_2 - CH_2 - Br$$

Ethene → 1,2-Dibromo ethane

There are two types of addition reactions:

(i) Electrophilic addition.

(ii) Nucleophilic addition.

(i) Electrophilic addition: Addition reactions which are initiated by the attack of electrophile, is called as electrophilic addition (If the given alkene is unsymmetrical, addition follows Markovnikov's rule).

$$CH_2 = CH_2 + HBr \longrightarrow CH_3CH_2Br$$

Ethene → Bromoethane

(ii) Nucleophilic addition: Addition reactions which are initiated by the attack of nucleophile, is called as nucleophilic addition. Example, reaction between acetaldehyde and HCN.

$$H_3C-\overset{O}{\overset{\|}{C}}-H + HCN \rightleftharpoons H_3C-\underset{CN}{\underset{|}{\overset{OH}{\overset{|}{C}}}}-H$$

Acetaldehyde → Acetaldehyde cyanohydrin

3. **Elimination reactions:** Organic reactions in which two atoms or group of atoms are eliminated from the two adjacent atoms in a reactant molecule is called as elimination reaction. During this reaction, two σ bonds are utilized and one new π bond is formed. Elimination reaction convert saturated compound in to unsaturated compound. Example, dehydration of ethanol.

$$\underset{\text{Ethanol}}{CH_3CH_2OH} \xrightarrow[430\ K]{H_2SO_4} \underset{\text{Ethene}}{CH_2=CH_2} + H_2O$$

Two type of elimination reactions are there as mentioned below.

(i) E_2 Elimination.

(ii) E_1 Elimination.

(i) E_2 Elimination: In this reaction, the rate of elimination depends on the concentration of alkyl halide as well as the base. Hence it is called as E_2 elimination. E indicates elimination and 2 indicates bimolecular.

$$\underset{\text{Ethyl bromide}}{CH_3-CH_2-Br} + OH^- \longrightarrow \underset{\text{Ethylene}}{CH_2=CH_2} + H_2O + Br^-$$

(ii) E_1 Elimination: In this reaction, the rate of elimination reaction depends upon the concentration of alkyl halide only. Hence it is called as E_1 elimination. E indicates elimination and 1 indicates unimolecular.

$$\underset{\text{Tertiary butyl bromide}}{H_3C-\underset{CH_3}{\underset{|}{\overset{CH_3}{\overset{|}{C}}}}-Br} + H_2O \longrightarrow \underset{\text{2-Methyl-propene}}{CH_3-\overset{CH_3}{\overset{|}{C}}=CH_2} + \underset{\text{Hydronium ion}}{H_3O^+} + Br^-$$

4. **Rearrangement reactions:** Shifting of an atom or group within the molecule with the formation of new molecule is called as rearrangement reaction.

$$\underset{n\text{-Butane}}{CH_3CH_2CH_2CH_3} \xrightarrow{AlCl_3} \underset{iso\text{-butane}}{CH_3-\overset{CH_3}{\overset{|}{C}H}-CH_3}$$

Types of rearrangements:

(i) Rearrangements to electron deficient carbon.

(a) Pinacol-Pinacolone rearrangement.

(b) Wanger-Meerwin rearrangement.

(c) Benzilic acid rearrangement.

(d) Wolf rearrangement.

(e) Allylic rearrangement.

(f) Sommelet-Hauser rearrangement.

(ii) Rearrangements to electron deficient nitrogen.
- (a) Hofmann rearrangement.
- (b) Curtius rearrangement.
- (c) Schmidt rearrangement.
- (d) Lossen rearrangement.
- (e) Beckmann rearrangement.
- (f) Neber rearrangement.

(iii) Rearrangements to electron deficient oxygen.
- (a) Baeyer-Villiger reaction.
- (b) Cumene hydroperoxide rearrangement.
- (c) Dakin reaction.

(iv) Rearrangements to electron rich atom (Electronic rearrangements).
- (a) Stevens rearrangement.
- (b) Wittig rearrangement.
- (c) Favorskii rearrangement.

(v) Rearrangements to free radical species (Free radical rearrangements).

(vi) Aromatic rearrangements.
1. Intermolecular aromatic rearrangements.
   - (a) Orton rearrangement.
   - (b) Hofmann – Martius rearrangement.
2. Intramolecular aromatic rearrangements.
   - (a) Claisen rearrangement.
   - (b) Benzidine rearrangement.
3. Mixed type of aromatic rearrangements.
   - (a) Fries rearrangement.

Reaction Mechanisms

Mechanism of the above reactions are described as follows.

1. Substitution reactions:
   - (i) Nucleophilic substitution: Described under, chapter no- 14.
   - (ii) Electrophilic substitution: Described under, chapter no. 20 – Benzene.
   - (iii) Free radical substitution: Described under, chapter no- 8 - Alkanes.
2. Elimination reactions:
   E_1 and E_2 Elimination: Described under, chapter no- 14.
3. Addition reactions:
   - (i) Electrophilic addition: Described under, chapter no- 9 - Alkenes.
   - (ii) Nucleophilic addition: Described under, chapter no- 15 – Aldehydes and ketones.

Reactive Intermediates or Reaction Intermediates

Homolytic and heterolytic bond fission leads to the formation of active radicals or ions which are known as reactive intermediates or reaction intermediates. The important reaction intermediates are as follows.

- I. Carbonium ions or Carbocations.
- II. Carbanions.
- III. Carbon free radicals.
- IV. Carbenes.

V. Nitrenes or imidogens.

VI. Vinylamines or enamines or α, β - unsaturated amines.

Let us discuss these intermediates in detail.

I. Carbonium ions or Carbocations

Carbonium ions or Carbocations are positively charged species containing a carbon atom [carbon atom with positive charge] having only six electrons with three bond (the carbon atom lacking of one pair of electron in its valency shell). These are produced by heterolytic bond fission.

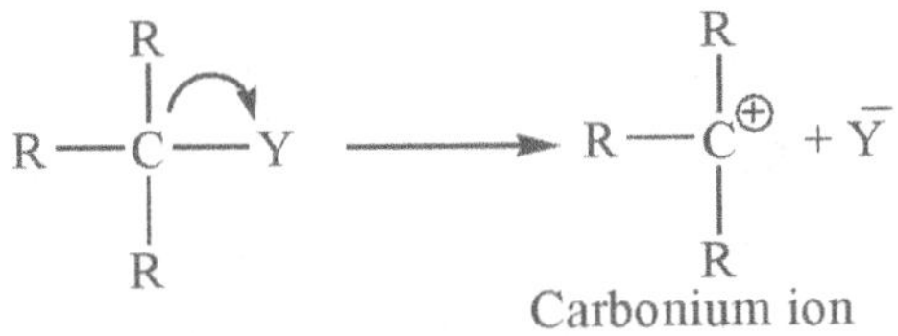

$$R_3C\text{—}Y \longrightarrow R_3C^{\oplus} + \bar{Y}$$

Carbonium ion

Where R = alkyl group / aryl group

Nomenclature: Carbonium ions are named according to two conventions.

1. According to carbonium ion nomenclature, the positive carbon is known as carbonium ion and the names of the attached groups are adjoined as a prefix.
2. According to cation nomenclature, the name of the corresponding parent group is used. The cation system is more frequently used. For Example:

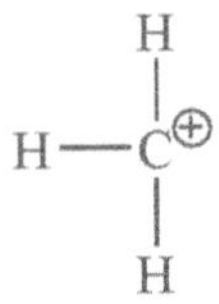

$$H\text{—}CH_2^{\oplus} \qquad CH_3\text{—}CH_2^{\oplus}$$

Methyl cation (Methyl carbonium ion) — Ethyl cation (Ethyl carbonium ion)

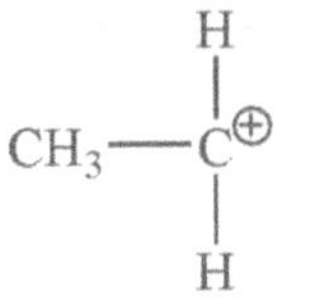

Types: Carbonium ions are classified into primary, secondary and tertiary according to the nature of carbon atom bearing positive charge.

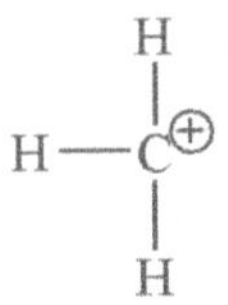

$$H\text{—}CH_2^{\oplus} \qquad CH_3\text{—}CH_2^{\oplus}$$

Methyl cation (Methyl carbonium ion) — Ethyl cation (Ethyl carbonium ion)

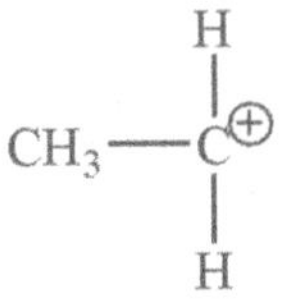

Primary carbocation

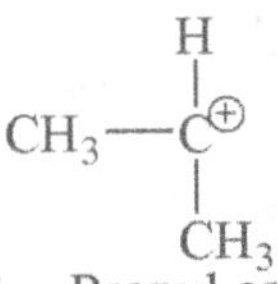

$$CH_3\text{—}CH^{\oplus}\text{—}CH_3 \qquad CH_3\text{—}C^{\oplus}(CH_3)_2$$

iso-Propyl cation (Dimethyl carbonium ion) **Secondary carbocation**

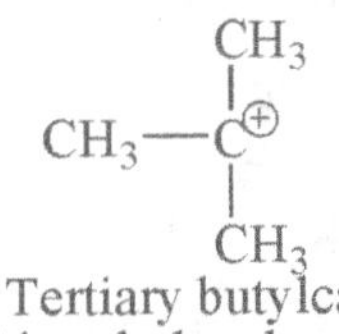

Tertiary butylcation (Trimethyl carbonium ion) **Tertiary carbocation**

Generation of carbonium ions: Carbonium ions are formed widely and are of great importance in several chemical reactions. They are formed by any one of the following methods.

(i) Direct ionization of alkyl or aryl halides: Several kinds of alkyl and aryl halides produce carbonium ions in the presence of high polar medium.

$$(CH_3)_3C{-}Cl \xrightarrow[\text{medium}]{\text{High polar}} (CH_3)_3C^{\oplus} + Cl^-$$

tert-Butyl chloride → *tert*-Butyl cation + Chloride anion

$$CH_2{=}CH{-}CH_2Cl \xrightarrow[\text{medium}]{\text{High polar}} CH_2{=}CH{-}\overset{+}{C}H_2 + Cl^-$$

Allyl chloride → Allyl cation + Chloride anion

$$(C_6H_5)_3C{-}Cl \xrightarrow[\text{medium}]{\text{High polar}} (C_6H_5)_3C^{\oplus} + Cl^-$$

Triphenyl methyl chloride (White solid) → Triphenyl methyl cation (Orange solution) + Chloride anion

(ii) By the protonation of alkyl or acyl halides: Alkyl or acyl halides react with a proton donating solvent or Lewis acid to give carbonium ion.

$$R{-}X + H^+ \rightleftharpoons R{-}\overset{\oplus}{X}{-}H \rightleftharpoons \overset{\oplus}{R} + H{-}X$$

Carbonium ion

$$R{-}Cl + AlCl_3 \rightleftharpoons R{-}\overset{\oplus}{Cl}{-}\overset{\ominus}{Al}Cl_3 \rightleftharpoons \overset{\oplus}{R} + \overset{\ominus}{Al}Cl_4$$

Carbonium ion

$$R{-}\overset{O}{\overset{\|}{C}}{-}Cl + AlCl_3 \rightleftharpoons R{-}\overset{O}{\overset{\|}{C}}{-}\overset{\oplus}{Cl}{-}\overset{\ominus}{Al}Cl_3 \rightleftharpoons R{-}\overset{O}{\overset{\|}{\underset{\oplus}{C}}}H + \overset{\ominus}{Al}Cl_4$$

Acyl Carbonium ion

It is explained by the following facts. Ethyl chloride is non-conductor but its solution in aluminum chloride is electrolysed during which aluminum is deposited at the anode. It clearly indicates that some charged species are produced during the reaction.

Carbonium ion as an intermediate: In the following reaction, carbonium ions are formed as an intermediate.

1. **Substitution reactions:** S_N1 reactions and electrophilic aromatic substitution reactions.
2. **Addition reaction**: Addition reactions of alkenes.
3. **Elimination reactions:** E_1 reaction.
4. **Rearrangements**: Allylic rearrangement, Pinacol-pinacolone rearrangement, Hoffmann rearrangement, Lossen and Curtius reaction, Beckmann's rearrangement, Wager-Meerwin rearrangement, Neopentyl rearrangement, rearrangement of peroxides and Baeyer-Villeger oxidation.

II. Carbanions

Definition: It is defined as "a negatively charged species containing a carbon atom with three bonds and an unshared pair of electrons".

$$-\overset{|}{\underset{|}{C}}-Y \longrightarrow -\overset{|}{\underset{|}{C}}:^{-} + Y^{+}$$

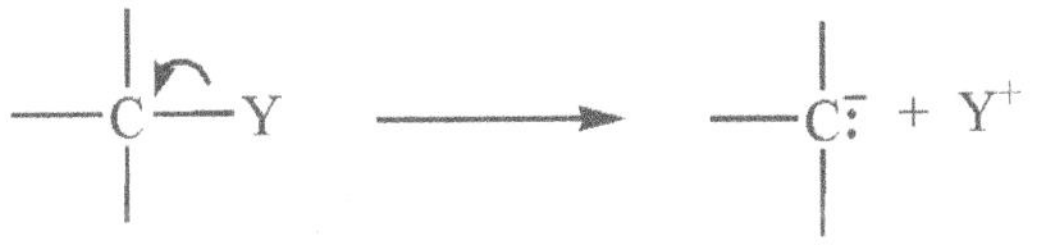

Where y = less electronegative atom than carbon. Carbanions can also be defined as "negatively charged carbon atoms contain three bonds and an unshared pair of electrons.

$$R-CH_2-X + :B \longrightarrow R-\overset{\ominus}{C}HX + B-H$$

Structure of carbanions: The carbon atom of the carbanion is sp^3 hybridized. Three sp^3 hybrid orbitals form covalent bond with other three groups or atoms while the fourth sp^3 hybrid orbital possess unshared pair of electrons, hence carbanion possess pyramidal shape.

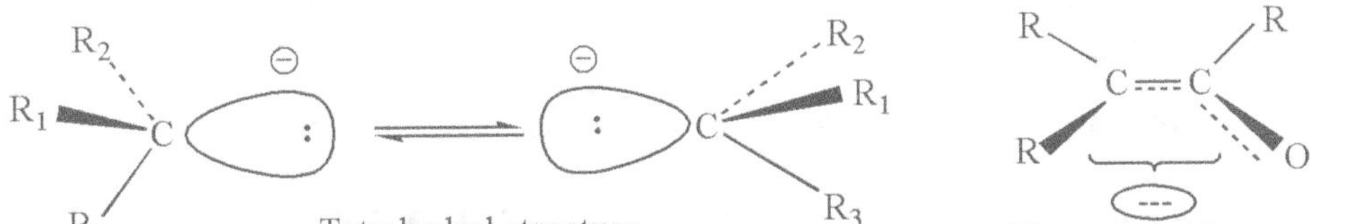

Tetrahedral structure
(Spontaneous inversion)

Planar hybrid structure

The above structures concluded that "the carbon atom of an unconjugated carbanion is in sp^3 hybridised state with a pyramidal shape while the carbon atom of a conjugated carbanion is in sp^2 hybridised state with a planar structure.

Type of carbanions: According to their existence of half life time, carbanions are classified into two types.

1. Stable carbanions.
2. Transient carbanions.

Reaction involving carbanions as an intermediate

1. **Substitution reaction:** Alkylation of β-keto esters, malonic ester, β-dicarbonyl compounds, Wurtz reaction, Riemer Tiemann reaction, decarboxylation of carboxylic acid, haloform reaction and halogenation of ketones.
2. **Addition reaction:** Aldol condensation, Benzoin condensation, Claisen-ester condensation, Darzen's glycidic ester reaction, Kolbe's reaction, Michael addition, Mannich reaction, Perkin reaction and Knovenegal condensation reaction *etc.*
3. **Rearrangement reaction:** Wittig reaction.

III. Carbon Free Radicals

Free radicals are defined as "neutral species which are having odd or unpaired electrons". These are electrically neutral and produced by homolytic fission.

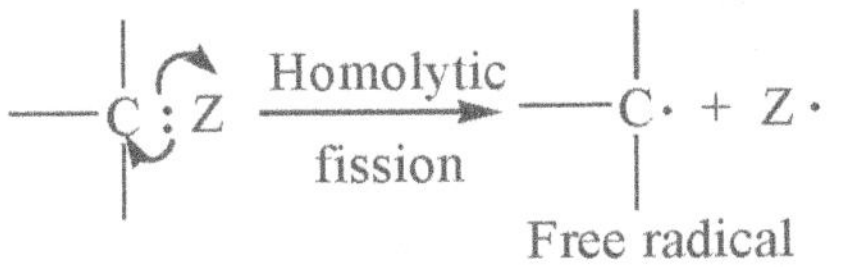

$$-\overset{|}{\underset{|}{C}}:Z \xrightarrow[\text{fission}]{\text{Homolytic}} -\overset{|}{\underset{|}{C}}\cdot + Z\cdot$$

Free radical

Here both carbon and Z possess similar electronegativities.

Structure of free radical:

sp^2 hybridised carbon

The carbon atom of the carbon free radical possesses sp^2 hybrid orbitals to form three sigma bonds. A half-filled p orbital is present above and below the plane of sigma bonds. Hence carbon free radicals are neutral in charge and contain one odd electron or unpaired electron. Due to the presence of odd electrons, these are highly reactive species. They collide with other free radicals or with other molecules to give larger free radicals.

Nomenclature: Carbon free radicals are named after their parent alkyl group by adding the word free radical in suffix. Example, methyl free radical ($\dot{C}H_3$).

Stability of free radicals: Free radicals are obtained by homolytic fission of alkanes. [When alkane dissociates, energy has to be supplied to increase its potential energy]. It indicates that smaller the value of bond dissociation energy, greater the ease of formation and less will be the potential energy which will increase the stability.

Table 6.2 Bond dissociation energy of some molecules.

Bond	Dissociation energy (Kcal/mole)
H - H	104
$CH_3 - H$	102
$CH_3CH_2 - H$	98
$(CH_3)_2CH$-H	94
$(CH_3)_3C$-H	90
$CH_3 - CH_3$	84
$CH_3 - OH$	90
$CH_3 - Cl$	82
$CH_3 - Br$	67
$C_6H_5CH_2$-H	78
C_6H_5-H	102
HO - H	120
F - H	135
F – F	37
Cl - H	103
Cl - Cl	58

For example, from the above data (Table 6.2), it is observed that dissociation energy & potential energy decreases when we pass from methyl ($-CH_3$) to -*t*-butyl radical. The decrease in potential energy of radical increases the stability of the radical from CH_3 to $(CH_3)_3C$. The order of stability of free radicals is as follows.

3° radical > 2° radical > 1° radical > methyl radical.

$(CH_3)_3C > (CH_3)_2CH > CH_3CH_2 > CH_3$

Generation of free radical:

(a) Photochemical fission: Molecules absorb electromagnetic radiation which causes bond fission to give free radicals. For example: Photochemical fission of chlorine, bromine, peroxides, acetone *etc.*

(b) Oxidation or reduction reactions: Oxidation or reduction of inorganic ions and metals give free radicals. For example:

(i) Kolbe's electrolytic synthesis of hydrocarbons: Electrolysis of aqueous solution of alkali salt of fatty acid yields alkyl free radical.

(ii) Oxidation of ferrous ion with hydrogen peroxide in the presence of Fenton's reagent.

(c) Thermal fission: An organic compound undergoes pyrolysis to yield free radicals.

For example:

(i) Heating of lead tetra ethyl undergoes decomposition to give free radicals.

(ii) Organic peroxide undergoes decomposition at quite low temperature (ranging between room temperature and 370 K) to yield free radicals.

(iii) Azomethane undergoes decomposition at 570 K to give methyl radical.

Reactions of free radicals: Free radicals undergo the following type of reactions.

(i) Combination: Two free radicals dimerise to yield respective products.

(ii) Abstraction: Free radicals abstract an atom from saturated organic compounds.

(iii) Addition to multiple bonds: Free radicals add to the double bond or triple bond to yield respective radicals.

(iv) Fragmentation: Some type of free radicals themselves undergoes fission leads to the formation of new radical and non-radical fragment.

IV. Carbenes

Carbenes are defined as "neutrally charged carbon atom with two bonds and two electrons". Hence, they require a pair of electrons to complete their outer most orbital, hence they behave as electrophiles.

If both the electrons are present in one orbital (opposite spins or antiparallel spins) it is called as singlet state with no magnetic field. Conversely, if the two electrons are present in different orbitals (parallel spins) it is known as triplet state.

$$\overset{\downarrow\uparrow}{—C—} \qquad \overset{\uparrow}{\underset{\uparrow}{—C—}}$$

Singlet carbene Triplet state carbene

Generation of carbenes: They are generated by following methods.

1. **Photolysis or pyrolysis of aliphatic diazo compounds or ketones:**

i) $\underset{\text{Diazomethane}}{CH_2N_2} \longrightarrow \underset{\text{Carbene}}{:CH_2 + N_2}$

ii) $\underset{\text{Acyldiazo methane}}{R—\overset{O}{\overset{\|}{C}}—CHN_2} \longrightarrow \underset{\text{Acyl carbene}}{R—\overset{O}{\overset{\|}{C}}—CH: + N_2}$

iii) $\underset{\text{Diazoethylacetate}}{N_2CH—COOC_2H_5} \longrightarrow \underset{\text{Carbethoxy carbene}}{:CH—COOC_2H_5 + N_2}$

iv) $\underset{\text{Ketene}}{CH_2CO} \longrightarrow \underset{\text{Carbene}}{:CH_2 + CO}$

v) $R_2C{=}C{=}O \longrightarrow :CR_2 + CO$

2. **Reaction between a base with suitable poly halogen compound:**

$$\underset{\textit{tert}\text{-butoxide ion}}{(CH_3)_3CO^-} + \underset{\text{Chloroform}}{CHCl_3} \rightleftharpoons \underset{\textit{tert}\text{-butanol}}{(CH_3)_3COH} + :\bar{C}Cl_3$$

$$:\bar{C}Cl_3 \longrightarrow \underset{\text{Dichlorocarbene}}{:CCl_2} + Cl^-$$

$$\underset{\text{Dichloromethane}}{CH_2Cl_2} + LiR \xrightarrow[(-RH)]{} [LiCHCl_2] \xrightarrow[-LiCl]{} \underset{\text{Chlorocarbene}}{:CHCl}$$

Reactions of carbenes: Carbenes undergoes various important reactions and they are as follows.

(a) Addition to olefins.
(b) Ring expansion reactions.
(c) Insertion to C-H bond. These are described as follows.

Reaction involving carbenes as an intermediate: Following reactions are the examples of carbenes as an intermediate. Carbylamine reaction, Reimer–Tiemann reaction, Wittig reaction, synthesis of spiro compounds and synthesis of alkanes.

V. Nitrenes or Imidogens

Nitrenes are defined as "electron deficient species in which nitrogen has a sextet of electrons".

$$R—\ddot{\underset{\cdot\cdot}{N}}$$

1. **Generation of nitrenes:** Photolysis of N_3H yields nitrenes.

$$\underset{\text{Hydrazoic acid}}{N_3H} \xrightarrow{h\nu} N_2 + \underset{\text{Nitrene}}{H\ddot{\underset{\cdot\cdot}{N}}}$$

VI. Vinylamines or Enamines or α,β-Unsaturated Amines

Enamines are obtained by the condensation of a carbonyl compound having at least one α-hydrogen atom with a secondary amine (Enamines can be obtained from primary amines are not stable because they are immediately converted into imines).

$$\underset{\text{Carbonyl compound with } \alpha \text{ - hydrogen}}{>CH—\overset{|}{C}=O} + HN< \xrightarrow{-H_2O} \underset{\text{Enamines (}\alpha,\beta\text{-Unsaturated amines)}}{>\underset{\beta}{C}=\underset{\alpha}{\overset{|}{C}}—N<}$$

Structure: Enamines are represented by the following resonating structures.

$$\underset{(A)}{>C=\overset{|}{C}—\ddot{N}<} \longleftrightarrow \underset{(B)}{>\underset{\beta}{\overset{\ominus}{\ddot{C}}}—\underset{\alpha}{\overset{|}{C}}=\overset{\oplus}{N}<}$$

In the above structure (B) the β-carbon atom is electron rich, so it acts as a nucleophilic centre and attacked by an electrophile. Hence enamines are important intermediates in organic syntheses.

Probable Questions

1. Define homolysis with examples.
2. Define heterolysis with examples.
3. What is meant by nucleophiles? and write various types with suitable examples.
4. What is meant by electrophiles? and write various types with suitable examples.
5. Define organic reaction and write the sequence of organic reaction.
6. Mention the various types of organic reaction with appropriate examples.
7. Define substitution reaction with examples.
8. What is meant by free radical substitution reaction? and which type of organic compound undergoes this reaction and why?
9. Describe in brief about nucleophilic substitution reaction and its types.
10. Write in detail about S_N1 and S_N2 reactions with examples.
11. Describe briefly about electrophilic aromatic substitution reaction and its types.
12. Define addition reaction and write its types with examples.
13. What is meant by elimination reaction? and write its two types.
14. Write a short note on rearrangement reaction and its types.
15. Define reaction intermediates with examples.
16. Write in detail about formation of carbonium ions.
17. Write in detail about formation of carbanions.
18. Write in detail about formation of free radicals.
19. Write in detail about formation of carbenes.
20. Write in detail about formation of nitrenes.
21. Write in detail about formation of enamines.

7

General Terms Used in Organic Chemistry

Absolute configuration: The three dimensional arrangement of a chiral compound. It is indicated by R or S.

Acetylation: It is the chemical reaction in which acetyl group is added to the molecule with the formation of a product. Usually carboxylic acid derivatives like chlorides or anhydrides are used for acetylation, which are more reactive than carboxylic acids.

$$\underset{\text{Alcohol}}{R-OH} + \underset{\text{Acetyl chloride}}{CH_3COCl} \xrightarrow[-HCl]{} \underset{\text{Ester}}{R-O-CO-CH_3}$$

$$\underset{\text{Amine}}{R-NH_2} + \underset{\text{Acetyl chloride}}{CH_3COCl} \xrightarrow[-HCl]{} \underset{\text{Amide}}{R-NH-CO-CH_3}$$

Acid chloride: Replacement of hydroxyl group of the carboxyl group (– COOH) by chloro(–Cl) group is called as acid chloride.

$$R-\overset{\overset{\displaystyle O}{\|}}{C}-Cl$$

Acid chloride

Acid anhydride: Removal of water molecule from two molecules of carboxylic acid gives acid anhydrides.

$$R-\overset{\overset{\displaystyle O}{\|}}{C}-O-\overset{\overset{\displaystyle O}{\|}}{C}-R$$

Acid anhydride

Acid catalysed reaction: A chemical reaction catalysed by acid.

Activating group: Substituents which increases the reactivity. Example, in benzene ring, electron- donating substituents activate benzene ring towards electrophilic attack and electron- attracting substituents activate benzene ring towards nucleophilic attack.

Activation energy: The minimum amount of energy must be provided by a collision for a reaction to occur is known as activation energy. It is denoted by E_{act}.

Acyl group: A carbonyl group attached to an alkyl or aryl group ($-RCO$, $-ArCO$).

1,2-Addition or direct addition: Addition occurs at 1-and 2-position in a diene molecule is known as 1,2-addition or direct addition.

$$CH_2 = CH - CH = CH_2 + Cl_2 \longrightarrow \underset{Cl}{CH_2} - \underset{Cl}{CH} - CH = CH_2$$

1,3-Butadiene (1 mole) + (1 mole) → 1,2-Dichloro-3-butene (1,2-addition product)

1,4-Addition or conjugate addition: Addition occurs at 1-and 4-position in a diene molecule is known as 1,4- addition or conjugate addition.

$$CH_2 = CH - CH = CH_2 + Cl_2 \longrightarrow \underset{Cl}{CH_2} - CH = CH - \underset{Cl}{CH_2}$$

1,3-Butadiene (1 mole) + (1 mole) → 1,4-Dichloro-2-butene (1,4-addition product)

Alcohols : These are hydroxy (-OH) derivative of hydrocarbons in which one or more hydrogens are replaced by hydroxyl groups. Example, methanaol (CH_3OH) and ethanol (C_2H_5OH).

Alkylation: It is the chemical reaction in which alkyl groups are added to the molecule to form a alkylated product. Alkyl halides are used to carry out this reaction.

$$R-NH_2 + R'X \xrightarrow{-HX} R-NH-R'$$

Primary amine → Secondary amine (Alkylated Product)

In case of methylation, dimethyl sulphate or diazomethane are also used for alkylation.

Aldehydes and ketones: These are carbonyl compounds possessing carbonyl (-C=O) group in their structure. Example for aldehyde is acetaldehyde (CH_3CHO) and for ketone is acetone (CH_3COCH_3).

Aldol condensation: Condensation of two molecules of same or different carbonyl compounds possessing α–hydrogen atoms in the presence of base is known as aldol condensation. Aldehydes undergo this reaction to yield the product known as aldols.

Alcoholysis: Chemical reaction in which a bond is cleaved using alcohol with the addition of alkoxide group in the product is called as alcoholysis. It is similar to that of hydrolysis.

$$(H_3C)_3C - Cl + C_2H_5OH \longrightarrow (H_3C)_3C - O - C_2H_5$$

tert-Butyl chloride + Ethanol → Ether (*tert*-Butyl ethyl ether)

Alicyclic compounds or aliphatic cyclic compounds: Some of the homocyclic compounds containing a ring structure but behaves like aliphatic compounds are referred as alicyclic compounds. Example is cycolobutane.

□ (or) H_2C-CH_2 / H_2C-CH_2 (four-membered ring)

Cyclobutane

Alkanes: Alkanes are the simple aliphatic organic saturated hydrocarbons. Carbon and hydrogen are connected to each other by a single covalent (sigma) bond. They are called as "saturated hydrocarbons" (four valence of carbon is saturated) and also called as "paraffins" (less affinity towards any reagents). Example methane (CH_4) and ethane (C_2H_6).

Alkyl groups: An alkyl group is obtained by removing one hydrogen atom from alkane. Their name is derived from the alkane by removing **-ane** (from concerned alkane) and adding **-yl** to the corresponding alkane.

$$\underset{\substack{\text{Alkane}\\(C_nH_{2n+2})}}{RH} \xrightarrow{-H} \underset{\substack{\text{Alkyl}\\(C_nH_{2n+1})}}{-R}$$

$$\underset{\text{Methane}}{CH_4} \xrightarrow{-H} \underset{\text{Methyl}}{-CH_3}$$

Alkenes: The unsaturated hydrocarbons which contain at least one carbon-carbon double bond [C=C] are known as alkenes. They are also named olefins, because of the fact that lower members form oily products upon reaction with halogens and halogen acids. Examples are given below.

Ethylene $CH_2{=}\ CH_2$ Propylene $CH_2{=}\ CH-CH_3$

Alkenyl groups: The monovalent groups formed by removing hydrogen from alkenes are called as alkenyl group. For example

$$\underset{\text{Ethenyl}}{CH_2{=}CH-} \qquad \underset{\text{2-Propenyl}}{\overset{3}{C}H_2{=}\overset{2}{C}H-\overset{1}{C}H_2-}$$

Alkylation: A chemical reaction in which an alkyl group is added to the substrate.

Alkyl halides: Compounds in which halogen atoms are attached to the carbon atom. Example methylchloride (CH_3Cl) and ethyl chloride (C_2H_5Cl).

Alkyl nitrites: They are esters of nitrous acid having formula R-O-N=O. They are considered as isomers of nitrates. Example ethyl nitrite (CH_3-CH_2-O-N=O).

Alkyl cyanides: The alkyl derivatives of cyanide forms are known as alkyl cyanides or alkyl nitriles. Example methyl cyanide (CH_3CN).

Alkyl isocyanides: The alkyl derivatives of isocyanide forms are known as alkyl isonitriles or alkyl isocyanides. Example methyl isocyanide (CH_3NC)

Alkynes: The unsaturated hydrocarbons, which contain at least one carbon-carbon triple bond ($CH\equiv C-$) are called as alkynes. Example,Acetylene ($CH\equiv CH$) .

Allyl group: $CH_2 = CH - CH_2$-

Allylic carbon: A sp^3 carbon attached to the vinylic carbon.

Allylic cation: Molecule with a positively charged carbon atom in allylic carbon.

Allylic substitution: Substitution occurs in allylic carbon is known as allylic substitution.

$$\underset{\text{Propylene}}{CH_2=CH-CH_3} + Cl_2 \xrightarrow{\text{High temperature}} \underset{\text{Chloro propylene}}{CH_2=CH-CH_2Cl} + HCl$$

Alpha (α) carbon: A carbon atom adjacent to a functional group.

Amide: Replacement of hydroxyl group of the carboxyl group (- COOH) by amino group (-NH_2) is called as amide or acid amide.

$$\underset{\text{Acid amide/Amide}}{R-\overset{\overset{O}{\|}}{C}-NH_2}$$

Amines: These are alkyl or aryl derivatives of ammonia (NH_3) in which one or more of the hydrogen atoms of the NH_3 are replaced by alkyl or aryl groups. Example is Methyl amine (CH_3 NH_2).

Amino acid: Compound possessing amino and carboxylic acid groups.

Ammonolysis: Chemical reaction in which a bond is cleaved and amino group ($-NH_2$) is added to the molecule is called as ammonolysis. This reaction is used in the synthesis of amines.

$$\underset{\text{Alkyl halide}}{R-X} + NH_3 \longrightarrow \underset{\text{Amine}}{R-NH_2} + HX$$

Angstrom: Unit of length, 10^{-8} cm = 1 angstrom.

Angle strain: The strain produced in a molecule due to the result of its bond angles being distorted from their normal values.

Annulation reaction: A ring forming reaction.

Antarafacial bond formation: Formation of two new σ bonds on opposite sides of the π bond.

Anti Markovnikov's rule: The negative part of the addendum adds on to the unsymmetrical carbon containing more number of hydrogen atoms.

$$\underset{\text{1-Propene}}{CH_3-CH=CH_2} \xrightarrow[\text{Peroxide}]{\text{HBr}} \underset{\text{1-Bromopropane}}{CH_3-CH_2-CH_2Br}$$

Annulenes: These are conjugated monocyclic hydrocarbons with alternate single and double bonds.

Araliphatic amines: Amino group attached to the side chain of the aromatic ring is known as araliphatic amines. Example is benzylamine.

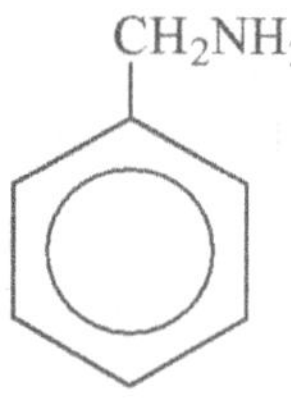

CH_2NH_2

Benzylamine

Aromatic aldehydes: These are compounds in which -CHO group is directly attached to aryl ring. Example, benzaldehyde and salicylaldehyde.

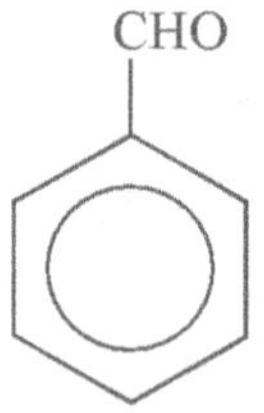

CHO | CHO, OH

Benzaldehyde 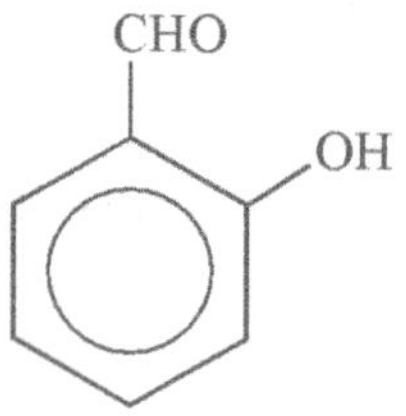 | Salicylaldehyde

Aromatic amines: These are compounds in which the amino group is attached to a aryl ring. Example, aniline.

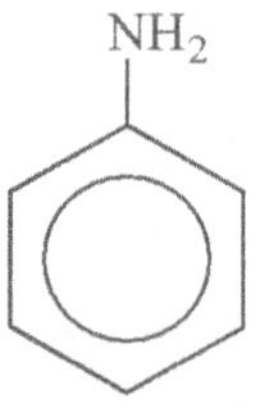

NH_2

Aniline

Aromatic carboxylic acids: Compounds in which the carboxylic acid group directly attached to a aryl ring. Example: benzoic acid.

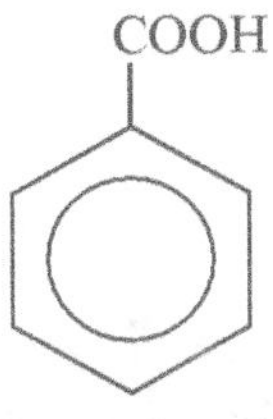

COOH

Benzoic acid

Aromatisation: Alkanes containing six or more carbons when heated under pressure in presence of catalyst yields cyclised and dehydrogenated product called aromatic compounds and the process of formation of aromatic compound is known as aromatization.

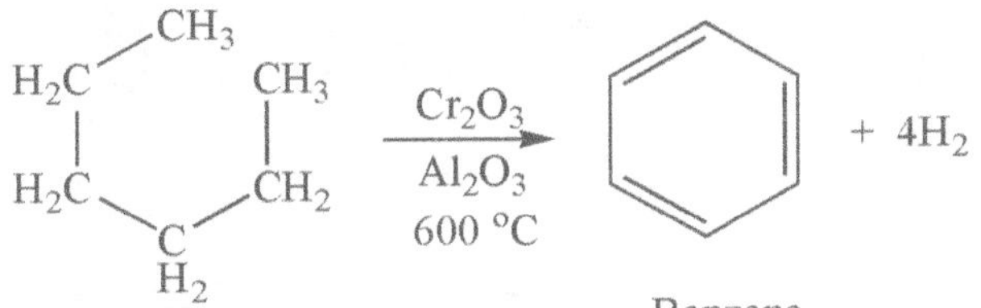

$$\text{n-Hexane} \xrightarrow[\text{600 °C}]{Cr_2O_3 \ / \ Al_2O_3} \text{Benzene} + 4H_2$$

n-Hexane Benzene

Aromatic compounds: Compounds containing a ring structure of carbon atoms alone or along with hetero atoms possessing continuous delocalized π electrons are known as aromatic compounds. Example, Benzene & phenol.

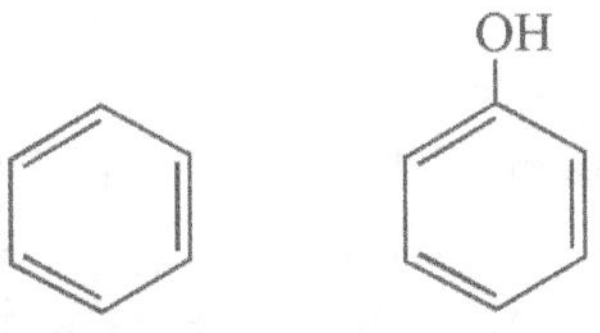

OH

Benzene Phenol

Aromatic nitro compounds: These are substitution products of aryl compounds in which one or more hydrogen atoms are substituted with nitro groups. Example, nitrobenzene.

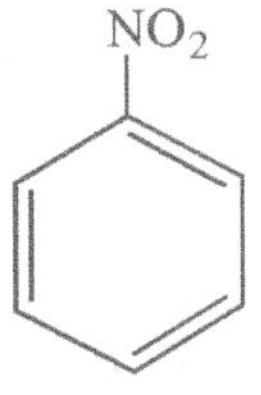

NO_2

Nitrobenzene

Aromatic suphonic acids: In aromatic suphonic acids one or more hydrogen atoms of aryl ring is replaced with sulphonic acid group ($-SO_3H$ / $-SO_2OH$). Example benzene sulphonic acid.

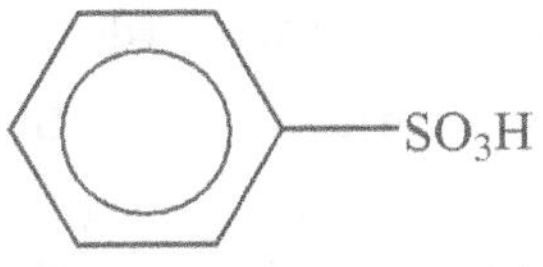

SO_3H

Benzene Sulphonic acid

Aryl group: Benzene or substituted benzene or any ring with aromaticity in which one hydrogen is removed. Example phenyl group — C_6H_5 or

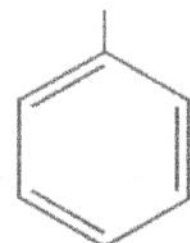

Aryl halides: These are derivatives of aryl compounds in which the hydrogen atom of the ring has been replaced by halogen atom X (Cl, Br, F, I). Example Chlorobenzene C_6H_5Cl.

Asymmetrical ether: Ether with two different substituents attached to the oxygen atom. Example $CH_3OC_2H_5$ Ethyl methyl ether.

Atomic number: The number of electrons or number of protons present in the neutral atom.

Axial hydrogens: Six hydrogen atoms are vertical and alternative above and below the rings (perpendicular to the plain) are called as axial hydrogens.

Axial bond

Back side attack: Nucleophile or electrophile attacks the carbon atom from the side opposite to that of the leaving group.

$$CH_3Br + OH^- \longrightarrow OH-CH_3 + Br^-$$

Methyl bromide   Nucleophile   Methanol

Basicity: It describes the ability of a compound to share its lone pair of electrons with a proton.

Bayer's Strain Theory: Any deviation of bond angles from the normal tetrahedral value would impose a condition of internal strain on the ring is known as angle strain. It is explained as Bayer's strain theory.

Bimolecular reaction: The rate of the reaction depends upon the concentration of the two reactants involving in that reaction.

Birch reduction: Partial reduction of benzene to 1,4-cyclo hexadiene using sodium or lithium in liquid ammonia.

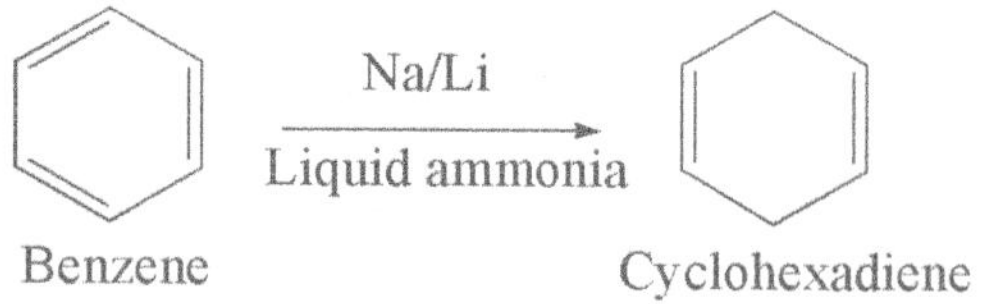

Bond angle: It is defined as an angle between the directions of two covalent bonds.

Boat conformation: One of the conformer of cyclohexane resembles like a boat.

Boiling point: Boiling point is defined as the temperature at which a liquid boil and convert in to vapours.

Bond dissociation energy: It is the amount of energy required to break a particular bond into neutral fragments or free radicals in their gaseous state in a poly atomic molecule.

Bond energy: It is defined as the average value of all the bonds dissociation energy values of that molecule. The unit of bond energy is KJ/mole or Kcal/mole.

Bond length: The distance between the center of the two orbitals of atoms is called as bond length or bond distance. The unit of bond length is Angstrom (A°) (1A° = 10^{-8} cm).

Cannizaro reaction: Aldehydes with the lack of α-hydrogen atom when heated with concentrated NaOH undergoes self reduction and oxidation to form primary alcohol and a salt of a carboxylic acid.

$$\underset{\text{Formaldehyde}}{2HCHO} + NaOH \longrightarrow \underset{\text{Methanol}}{CH_3OH} + \underset{\text{Sodium formate}}{HCOONa}$$

Carbanions: A negatively charged species containing a carbon atom with three bonds and an unshared pair of electrons.

Carbenes: Neutrally charged carbon atom with two bonds and two electrons.

Carbonium ions or Carbocations: Carbonium ions or Carbocations are positively charged species containing a carbon atom [carbon atom with positive charge] having only six electrons with three bond (the carbon atom lacking of one pair of electron in its valency shell).

Carbonyl carbon: The carbon atom of a carbonyl group.

Carbonyl group: A carbon atom doubly bonded to the oxygen atom (-C=O).

Carboxylation: Chemical reaction in which carboxy group (-COO) is added to the molecule with the formation of a product is called as carboxylation.

OH
COOH
NaOH
CO_2
High atm.
pressure
Phenol Salicylic acid

Carboxylic acid: These are the carboxyl derivatives of hydrocarbons. In hydrocarbons one or more hydrogen atoms are replaced with one or more carboxyl (–COOH) groups.

$$\underset{\text{Methane}}{CH_4} \xrightarrow[\text{by one - COOH group}]{\text{Replacement of one H}} \underset{\text{Acetic acid}}{CH_3COOH}$$

$$\underset{\text{Methane}}{CH_4} \xrightarrow[\text{by two - COOH group}]{\text{Replacement of two H}} CH_2(COOH)_2$$

COOH
CH_2
COOH
Malonic acid

Carbylamine reaction: Reaction between primary amines with chloroform and KOH gives isocyanides with the offensive smell of carbylamines.

$$\underset{\text{Methylamine}}{CH_3NH_2} + \underset{\text{Chloroform}}{CHCl_3} + 3KOH \longrightarrow \underset{\text{Methylisocyanide}}{CH_3N{\equiv}C} + 3KCl + 3H_2O$$

Catalytic hydrogenation: Addition of hydrogen atom or atoms to the double or triple bonded compounds in the presence of metal catalyst.

Chiral carbon: A carbon atom with four different substituents.

Chiral center: A tetrahedral atom with four different substituents.

Clemmensen reduction: Carbonyl compounds, especially ketones can be reduced to alkanes using Zn/Hg and conc. HCl. This reaction is known as Clemmensen reduction.

$$\underset{\text{Acetone}}{CH_3-\underset{\underset{O}{\|}}{C}-CH_3} \xrightarrow[HCl]{Zn\text{-}Hg} \underset{\text{Propane}}{CH_3-CH_2-CH_3}$$

Common name: Name given for compound in a simple manner without following any rules or system. Non systematic nomenclature of the compounds.

Complete racemisation: Formation of pair of enantiomers in equal amounts.

Condensation: It is a chemical reaction in which two molecules combine to form a product with the elimination of a small molecules like water, ammonia, methanol *etc.* It is similar to that of dehydration reaction. Condensation reaction is of two types *i.e.,* intermolecular and intramolecular condensation.

$$\underset{\text{Aldehyde}}{R-\overset{\overset{O}{\|}}{C}-H} + \underset{\text{Hydrazine}}{NH_2-NH_2} \xrightarrow[-H_2O]{\text{Intermolecular condensation}} \underset{\text{Hydrazone}}{R-\overset{\overset{N-NH_2}{\|}}{C}-H}$$

$$\underset{\text{Aldehyde}}{R-\overset{\overset{O}{\|}}{C}-H} + \underset{\text{Phenyl hydrazine}}{C_6H_5NHNH_2} \xrightarrow[-H_2O]{\text{Intermolecular condensation}} \underset{\text{Phenyl hydrazone}}{R-\overset{\overset{N-NH-C_6H_5}{\|}}{C}-H}$$

$$\text{Phthalic acid} \xrightarrow[-H_2O]{\text{Intramolecular condensation}} \text{Phthalic anhydride}$$

Conformations: Three dimensional arrangement of a molecule that can change as a result of rotation of sigma bond.

Conformational isomers: The different conformations of a molecule.

Conjugated dienes: Double bonds in the diene are separated by one single bond is known as conjugated dienes. Example 1,3-butadiene $CH_2=CH-CH=CH_2$.

Covalent bond: Chemical bond formed between the atoms by a mutual sharing of electrons.

Cyclisation: Conversation of a open chain compound into a cyclic product is called as cyclisation.

$$\underset{\text{1,6-Dibromo hexane}}{BrCH_2-CH_2-CH_2-CH_2-CH_2-CH_2Br} \xrightarrow[-2NaBr]{2Na} \underset{\text{Cyclohexane}}{(CH_2)_6}$$

Cycloaddition reaction: A reaction in which two π bond containing compounds reacts to forms a cyclic compound.

[4+2] Cycloaddition reaction: A cycloaddition reaction in which 4π electrons are coming from one substrate and 2π electrons are coming from another one substrate.

Cycloalkanes/cyclo paraffins/polymethylenes/alicyclic compounds: These are saturated, closed chain hydrocarbons in which the carbon atoms are connected by single covalent bonds lead to the formation of rings. Examples are shown below.

△ Cyclo propane   □ Cyclo butane

Deactivating group or substituent: A substituent decreases the reactivity of the ring.

Decarboxylation: Chemical reaction in which carboxyl group is removed with the formation of a product is called as decarboxylation. Catalyst used is sodalime (NaOH/CaO).

$$\underset{\text{Carboxylic acid}}{R-COOH} \xrightarrow[-CO_2]{NaOH/CaO} \underset{\text{Alkane}}{R-H}$$

Dehydration: It is a chemical reaction in which a molecule of water is eliminated by the condensation of two molecules or from a same molecule. It is reverse of hydration reaction. It is a type of condensation reaction.

$$\underset{\text{Alcohol}}{R-OH} + \underset{\text{Alcohol}}{R'-OH} \xrightarrow{-H_2O} \underset{\text{Ether}}{R-O-R'}$$

Though it is a simple chemical reaction used in many organic synthesis, some of them are listed below.

1. Synthesis of nitriles from amides.

$$\underset{\text{Amide}}{R-\overset{\overset{\displaystyle O}{\|}}{C}-NH_2} \xrightarrow{-H_2O} \underset{\text{Nitrile}}{R-C\equiv N}$$

2. Synthesis of alkenes from alcohols

$$\underset{\text{Alcohol}}{R-CH_2-CH_2OH} \xrightarrow{-H_2O} \underset{\text{Alkene}}{R-CH=CH_2}$$

Dehydrogenation: Chemical reaction in which a molecule of hydrogen is removed with the formation of a new product is called as dehydrogenation.

$$\underset{\text{Alkane}}{R-CH_2-CH_3} \longrightarrow \underset{\text{Alkene}}{R-CH=CH_2} + H_2$$

1. Usually this reaction is used to convert alkanes to alkenes which in turn is used for the synthesis of many organic compounds like alcohols, aldehydes *etc.*

$$\underset{\text{Alkane}}{R-CH_2-CH_3} \xrightarrow{-H_2} \underset{\text{Alkene}}{R-CH=CH_2} \xrightarrow{HOH} \underset{\text{Alcohol}}{R-\overset{\overset{\displaystyle OH}{|}}{C}H-CH_3}$$

2. In food industry this reaction is used to convert saturated fats into unsaturated fats.

Dehalogenation: It is the chemical reaction, in which halogen is removed from the molecule in presence of a reducing agents like nickel, zinc *etc* results in the formation of a product.

$$\underset{\text{1,2-Dihalo alkane}}{R-CHX-CH_2-X} \xrightarrow[-ZnX_2]{Zn} \underset{\text{Alkene}}{R-CH=CH_2}$$

This reaction is used in the organic chemistry to convert halo compounds to alkene/alkanes. It can be used to convert toxic organic halides to less hazardous products.

Dehydrohalogenation: Chemical reaction in which hydrogen and a halide is eliminated with the formation of a product is called dehydrohalogenation. It is a type of elimination reaction.

$$\underset{\text{Alkyl halide}}{R-CH_2-CH_2-X} \xrightarrow{-HX} \underset{\text{Alkene}}{R-CH=CH_2}$$

It is widely used in organic compound synthesis and drug synthesis.

Demercuration: Removal of mercury atom from a compound is called as demercuration. Example organo mercury compound upon reduction with sodium borohydride ($NaBH_4$) gives an alcohol.

$$\underset{\text{Organomercury compound}}{CH_3-CH(OH)-CH_2-HgO-C(=O)-CH_3} \xrightarrow{NaBH_4} \underset{\text{2-Propanol}}{CH_3-CH(OH)-CH_3}$$

Denatured alcohol: Ethanol with denaturants such as methanol or benzene.

Deoxygenation: Removal of oxygen from a reactant.

Desulphonation (Removal of -SO_3H group) : Example, Benzene sulphonic acid is heated with dilute HCl or H_2SO_4 at 420-470 K whereby the -SO_3H group is replaced by hydrogen atom to form the parent aromatic hydrocarbon.

$$\underset{\text{Benzene sulphonic acid}}{C_6H_5-SO_3H} \xrightleftharpoons[420\text{-}470\ K]{H_2O\ \&\ H_2SO_4} \underset{\text{Benzene}}{C_6H_6} + H_2SO_4$$

Dextrorotatory: An enantiomer which rotates the plane-polarised light in clockwise direction.

Diastereomers: A configurational stereoisomers which are not enantiomers.

Diazonium salts: These are important class of organic aromatic compounds which possess the functional group $-\overset{+}{N}\equiv N$ (Diazonium ion), which is directly attached to a benzene ring.

$$C_6H_5-N\equiv N^{+}Cl^{-}$$ Benzene diazonium chloride

Diazotization: The formation of a diazonium salt by the reaction between aromatic primary amine and sodium nitrite solution in the presence of acid in ice cold condition is known as diazotization reaction. Aryl diazonium salts are mainly prepared by diazotization reaction.

$$\underset{\text{Aniline}}{Ar-NH_2} + \underset{\text{Sodium nitrite}}{NaNO_2} + 2HCl \xrightarrow{0\text{-}5\ ^\circ C} \underset{\text{Aryl diazonium chloride}}{Ar-N\equiv N^+Cl^-} + 2H_2O + NaCl$$

Dicarboxylic acids: It contains two carboxyl groups in their structure; each one is present at the end of a saturated hydrocarbon.

$$HO-\overset{O}{\overset{\|}{C}}-(CH_2)n-\overset{O}{\overset{\|}{C}}-OH \quad \text{(or)} \quad HOOC(CH_2)nCOOH$$

Where n = 0,1,2,3.......

Dihydric alcohol: Alcohol containing two hydroxyl groups are called as dihydric alcohol, the two -OH groups are attached to different carbon atoms. Example, ethylene glycol $OHCH_2CH_2OH$.

Dieckmann condensation: An intramolecular Claisen condensation.

Diel's Alder reaction: Conjugated diene (Diene) reacts with a compound containing a carbon-carbon double bond (Dienophile) to form a cyclic molecule (Adduct). This reaction is called as Diel's Alder reaction.

Diene + Dienophile → Adduct

Dienes or alkadienes: Hydrocarbons with two carbon-carbon double bonds are called as dienes or alkadienes. Example 1,3-butadiene $CH_2=CH-CH=CH_2$.

β-diketone: A ketone with a second carbonyl group at β – position.

Dimer: A molecule formed by the joining of two identical molecules.

Dipole-dipole interaction: Interaction between the dipole of one molecule with dipole of another molecule.

Dipole moment: Dipole moment of a molecule (μ) is equal to the magnitude of the charge (e) multiplies by the distance (d) between the centers of two charges.

$$\mu = e \times d$$

Double Bond: This covalent bond is formed by mutual sharing of two electron pair between two atoms. Example, formation of oxygen, ethylene molecule *etc.*

Dow process: Conversion of chlorobenzene to phenol by the action of NaOH and HCl at high temperature and pressure.

Chlorobenzene (Cl) $\xrightarrow[\text{300 °C/150 atm, -HCl}]{NaOH}$ Sodium phenoxide ($\bar{O}\overset{+}{Na}$) $\xrightarrow[-NaCl]{HCl/H_2O}$ Phenol (OH)

Dye: Dye is a coloured substance, which is capable of colouring a solid support like cloth or making the solution colour. The colour of dye is due to the presence of a chromophore and fixing property to the basic or acidic auxochromic groups like -OH, $-NH_2$, $-NR_2$ *etc.*

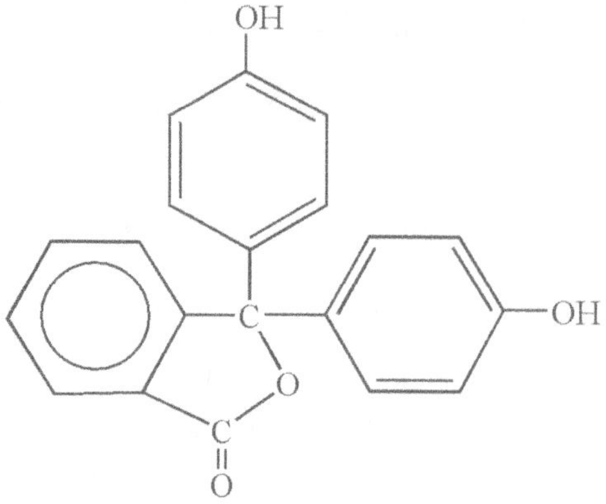

Phenolphthalein
(Used as acid-base indicator)

***E* isomer:** The isomer with high priority groups on opposite sides of the double bond.

Electrocyclic addition: A type of pericyclic reaction in which a π bond in the reaction is destroyed leads to a formation of a cyclic compound with a new σ bond.

Electrophiles (Electron loving species or Electron deficient species): Organic reagents that can accept an electron pair in a chemical reaction is known as electrophile. These are denoted by the symbol E^+.

Empirical formula: The relative numbers of different kinds of atoms in a molecule.

Enamines: These are obtained by the condensation of a carbonyl compound having at least one α-hydrogen atom with a secondary amine.

Enantiomerically pure: Compound containing only one isomer.

Enantiomers: Non superimposable mirror images of the isomers.

Enantioselective reaction: A reaction that forms more amount of one enantiomer than another one.

Endergonic reaction: A reaction with a positive ΔG^o value.

Endothermic reaction: Chemical reactions take place with the consumption of energy (positive ΔH^o) are called endothermic reaction.

Enolization: Keto-enol tautomerism in which ketone is converted into enol.

Entropy: Determination of the freedom of motion in a system.

Epimerization: An optically active compound contains two or more chiral carbon atoms and the configuration of only one of the carbon atom changes in a reaction, this process is called as epimerization.

Epoxidation: Formation of epoxides.

Equatorial bond: A bond of the chair conformer of cyclohexane that just out from the ring in approximately the same plane that contains the chair.

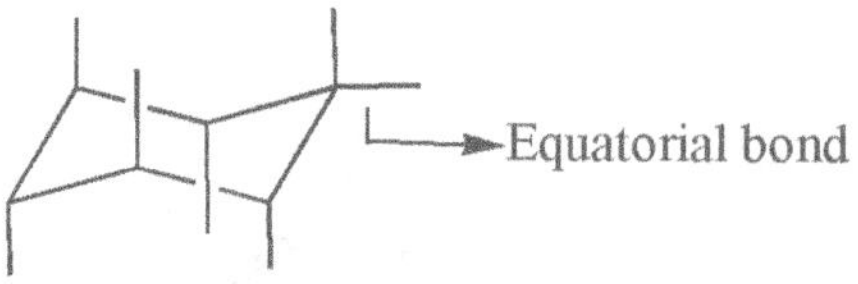

Equatorial bond

E_1 reaction: A first order elimination reaction.

E_2 reaction: A second order elimination reaction

Esters: Compounds obtained by replacing -OH group of the carboxyl (- COOH) group by alkoxy (-OR) group are called as esters.

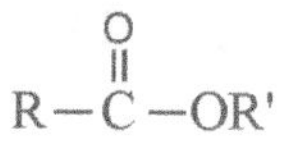

$$R-\overset{\overset{O}{\|}}{C}-OR'$$

Ester

Esterification: When a carboxylic acid or acid chlorides or acid anhydrides react with an alcohol in presence of concentrated sulphuric acid, an ester is formed. This reaction is called esterification.

$$\underset{\text{Carboxylic acid}}{R-\overset{O}{\overset{\|}{C}}-OH} + \underset{\text{Alcohol}}{R'-OH} \xrightarrow{\text{Conc. } H_2SO_4} \underset{\text{Ester}}{R-\overset{O}{\overset{\|}{C}}-OR'} + H_2O$$

Actually, ester is a derivative of carboxylic acid in which -OH group is replaced by an -OR group.

Esterification is a reversible reaction. In order to prevent the backward reaction and improve the ester yield, excess alcohol and use of dehydrating agents like conc. sulphuric acid (removes the water and favors the forward reaction) are preferred.

Evaporation: It is a purification method for the removal of volatile impurities from the solution.

Exergonic reaction: A reaction with a negative ΔG^o value.

Exhaustive methylation: Reaction between amine and excess amount of methyl iodide to give quaternary ammonium iodide.

Exothermic reaction: In a chemical reaction energy is liberated in the form of heat (negative ΔH^o) is called as exothermic reaction.

Free radicals: Neutral species which are having odd or unpaired electrons. These are electrically neutral and produced by homolytic fission.

Free radical initiator: The compounds which can initiate a free radical substitution reaction is called as free radical initiator or free radical catalysts. Example, UV light, Benzoyl peroxides, Peroxybenzoic acid, Azobisisobutyro nitrile (AIBN).

Free radical inhibitors or free radical trap: Compounds which inhibit the free radical reactions are known as free radical inhibitors or free radical trap. Example, Iodine

Friedel - Craft's acylation: An electrophilic aromatic substitution reaction that substitutes acyl group in place of hydrogen of benzene ring.

$$\underset{\text{Benzene}}{C_6H_6} \; \underset{\text{Acyl cloride}}{R-\overset{O}{\overset{\|}{C}}-Cl} \xrightarrow{AlCl_3} \underset{\text{Acylbenzene}}{C_6H_5-\overset{O}{\overset{\|}{C}}=R} + HCl$$

Friedel - Craft's alkylation: An electrophilic aromatic substitution reaction that substitutes alkyl group in place of hydrogen of a benzene ring.

$$\underset{\text{Benzene}}{C_6H_6} + \underset{\text{Alkyl cloride}}{RCl} \xrightarrow{AlCl_3} \underset{\text{Alkyl benzene}}{C_6H_5-R} + HCl$$

Fries rearrangement: Phenol when reacted with acetic anhydride in NaOH yields phenyl acetate. The ester upon reaction with $AlCl_3$ the acyl group migrates into *o* or *p* position to yield *o*-and *p*-hydroxy acetophenone.

$$\underset{\text{Phenol}}{C_6H_5OH} \xrightarrow[\text{NaOH}, \; -CH_3COOH]{(CH_3CO)_2O} \underset{\text{Phenylacetate}}{C_6H_5OCOCH_3} \xrightarrow[\Delta]{AlCl_3} \underset{o\text{-Hydroxyacetophenone}}{o\text{-}HOC_6H_4COCH_3} + \underset{p\text{-Hydroxyacetophenone}}{p\text{-}HOC_6H_4COCH_3}$$

First order rate constant: The rate constant of first order reaction.

First order reaction: The rate of the reaction depends upon the concentration of any one of the reactants.

Fischer esterification reaction: Synthesis of ester by the reaction between carboxylic acid and alcohol in the presence of acid catalyst.

Fischer projection: Represents the spatial arrangement of groups bonded to a chiral center. The chiral center is the point of intersection of two perpendicular lines. The horizontal lines indicates bonds that project out of the plane of the paper toward the viewer and the vertical lines indicates bonds that point back from the plane of the paper away from the viewer.

Functional group: The reactive or active part of the molecule responsible for chemical property of the molecule.

Gabriel phthalimide method: Conversion of alkyl halide into a primary amine by using phthalimide as a starting material.

Gattermann reaction: Conversion of phenol in to salicylaldehyde by the action of HCN and HCl in presence of $AlCl_3$ catalyst.

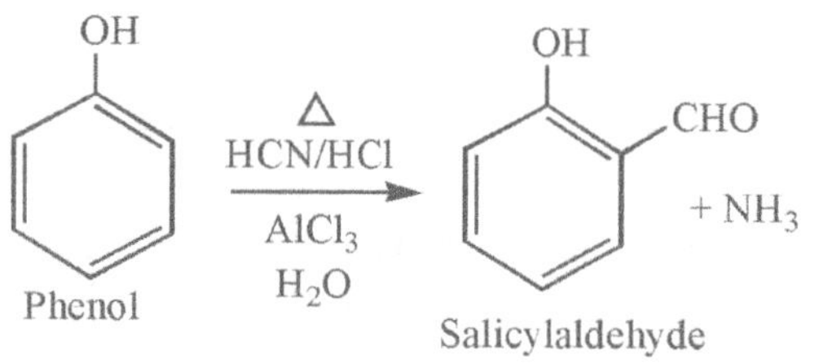

$$\text{Phenol} \xrightarrow[\text{AlCl}_3,\ \text{H}_2\text{O}]{\Delta,\ \text{HCN/HCl}} \text{Salicylaldehyde} + NH_3$$

Phenol

Salicylaldehyde

Gem-**diol:** A compound with two-OH groups on the same carbon.

Geminal dihalides or Gem dihalides: In alkyl halides two halogen atoms are attached to the same carbon atom.

$$CH_3-CH(Br)_2$$

Ethylidene dibromide or dibromo ethane

Grignard reagent or organometallic compound: Organic compounds in which a metal is directly attached to a carbon atom are known as organometallic compound.

$$C_2H_5-Mg-I$$

Ethyl magnesium iodide

Haloform reaction: Acetone reacts with halogens in the presence of excess base (*i.e.,* in hypohalite solutions) to yield trihalo substituted ketone wherein three hydrogen atoms of the methyl group are replaced by halogen atoms. This reaction is known as haloform reaction.

$$H_3C-\overset{O}{\overset{\|}{C}}-CH_3 + 3Br_2 + 4NaOH \longrightarrow H_3C-\overset{O}{\overset{\|}{C}}-\bar{O}Na^+ + CHBr_3 + 3H_2O + 3NaBr$$

Acetone Sodium acetate Bromoform

Halogenation of Benzene: Benzene reacts with halogens like Br_2 or Cl_2 in the presence of Lewis acid catalyst ($FeBr_3$ or $FeCl_3$) to give bromobenzene or chlorobenzene respectively. Example

$$\text{Benzene} + Br_2 \xrightarrow{FeBr_3} \text{Bromobenzene (Br)}$$

Halohydrin: Organic compound with halogen atom and hydroxyl group on adjacent carbon atoms.

Heat of combustion: The amount of heat released when a carbon containing compound is burnt completely with oxygen.

Heat of hydrogenation: The amount of heat released during hydrogenation reaction.

Heteroatom: Atoms other than carbon and hydrogen.

Heterolysis: During the cleavage of bond both electrons in a bond go to anyone of the fragment which leads to the formation of an anion and cation.

$$\underset{\text{Molecule}}{A : B} \longrightarrow \underset{\text{anion}}{A\!:^{-}} + \underset{\text{cation}}{B^{+}}$$

(Both electrons go to A)

Heterocyclic compounds: Cyclic compounds in which one or more carbon atoms are replaced by other atoms like O, S, N *etc* are called as heterocyclic compounds. Examples are given below.

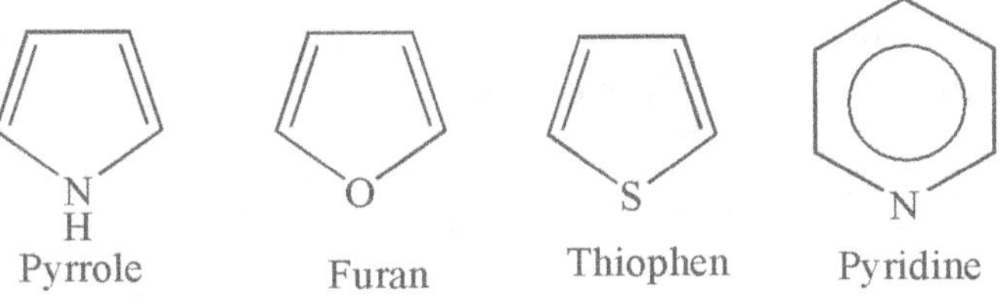

Pyrrole Furan Thiophen Pyridine

Hinsberg method: Method used for the separation of primary, secondary and tertiary amines.

Hoffman's degradation of amides: Primary amides react with chlorine or bromine in the presence of strong alkali to give primary amines.

$$\underset{\text{Amide}}{R-\overset{\overset{O}{\|}}{C}-\ddot{N}H_2} + X_2 + 4NaOH \longrightarrow \underset{\text{Primary amine}}{R-\ddot{N}H_2} + 2NaX + Na_2CO_3 + 2H_2O$$

$(X_2 = Cl_2/Br_2)$

Hofmann elimination or anti elimination: The hydrogen atom is eliminated from the β -carbon atom which is attached to more hydrogen atoms.

Homocyclic or carbocyclic compounds: The cyclic compounds entirely composed of carbon atoms are called as homocyclic or carbocyclic compounds.

Homologue: A member of homologues series.

Homologues series: The chemical properties of a particular group of organic compounds depend upon the nature of the functional group present in the molecules. The rest of the part of molecule has very little effect on chemical properties, which differs from each compound by $-CH_2$.

A group or a series of organic compounds each containing one characteristic group or functional group and differs by -CH_2 constitutes homologues series of organic compounds. They are classified in to a number of homologues series.

CH_3OH	Methanol
CH_3CH_2OH	Ethanol
$CH_3CH_2CH_2OH$	*n*-Propanol
$CH_3CH_2CH_2CH_2OH$	*n*-Butanol

Homolysis: Each atom or group in a molecule split with one electron of the shared electron pair is called homolysis or homolytic fission. In this process two free radicals are formed.

$$A : B \longrightarrow A\cdot + B\cdot$$

Molecule → Free radical

Huckel's rule: It states that for a compound to be aromatic it must contain (4n+2) π electrons, where n is an integer. It also indicates that aromatic compound must contain odd number of pairs of electrons.

Hund's rule of maximum multiplicity: As long as empty orbitals are available electrons will never pair, this is called as Hund's rule of maximum multiplicity.

Hunsdiecker or borodin Hunsdiecker reaction: The silver salts of carboxylic acids undergo decomposition reaction by bromine or chlorine to give alkyl halides.

$$CH_3-COO-Ag + Br_2 \longrightarrow CH_3-Br + CO_2 + AgBr$$

Silver acetate → Methyl bromide

Hybrid orbital: An orbital obtained after hybridization.

Hybridization: It is the process of intermixing of the orbitals of slightly different energies so as to redistribute their energies, resulting in the formation of new set of orbitals of equivalent energies and shape.

***sp*³ Hybridization or tetrahedral hybridization:** This type of hybridization involves mixing or merging of one *s* and three *p* orbitals to give four new equivalent sp^3 hybridized orbitals. Example: Hybridization in methane molecule .

***sp*² Hybridization or trigonal hybridization:** This type of hybridization involves mixing or merging of one *s* and two *p* pure orbitals to give three new equivalent sp^2 hybridized orbitals. Example: Hybridization in ethylene molecule.

***sp* Hybridization or Diagonal hybridization:** This type of hybridization occurs by mixing or merging of one *s* and one *p* pure orbital to give two new equivalents collinear or *sp* hybridized orbitals. Example: Hybridization in acetylene molecule.

Hydrated: Water has been added to a molecule.

Hydration: Chemical reaction in which a water molecule is added (*i.e.*, -OH group and -H atom) to form a new product is called as hydration. This reaction is opposite to that of dehydration in which a molecule of water is eliminated and forms product. As it is an addition reaction, usually unsaturated compounds are converted to saturated compounds.

Hydration is a common reaction in organic chemistry and used in the formation of many products as shown below.

1. Hydration of ethylene to ethanol

$$CH_2=CH_2 \xrightarrow{HOH} CH_3-CH_2-OH$$

Ethylene → Ethanol

Hydride ion: Negatively charged hydrogen.

1,2-Hydride shift: The movement of hydride ion from one carbon to adjacent carbon atom.

Hydrogenation: It is chemical reaction in which a molecule of hydrogen is added to another molecule to form a product. This reaction is carried out usually in presence of catalysts such as platinum, zinc and nickel *etc.*

Applications:

1. Hydrogenation is used to convert unsaturated compounds (alkenes and alkynes, dienes) into saturated compound (alkanes).
2. Hydrogenation of nitriles yields primary amines.

$$\underset{\text{Nitrile}}{R-C\equiv N} \xrightarrow{(H)} \underset{\text{Primary amine}}{R-CH_2-NH_2}$$

3. Hydrogenation is used in food industry to convert unsaturated fatty acids into saturated fatty acids.

Hydrogen bonding: A weak bond (5 kcal/mole) formed between a hydrogen atom and non-bonding electrons of the electronegative atoms like oxygen, nitrogen or sulphur *etc.*

Hydrolysis: A chemical reaction in which the bond of a compound is broken by using water is called hydrolysis. The term hydrolysis is derived from Greek words *i.e.*, Hydro means water; lysis means breaking.

$$\underset{\text{Ester}}{R-\overset{\overset{O}{\|}}{C}-OR'} + H-OH \longrightarrow \underset{\text{Carboxylic acid}}{R-\overset{\overset{O}{\|}}{C}-OH} + \underset{\text{Alcohol}}{R'-OH}$$

Hydrolysis and condensation are opposite type of reaction. In the example given below the forward reaction is a hydrolysis in which ester is broken into carboxylic acid and alcohol, while in the backward reaction the carboxylic acid and alcohol combines to form an ester.

$$\underset{\text{Ester}}{R-\overset{\overset{O}{\|}}{C}-O\ R'} + H-OH \underset{\text{Condensation}}{\overset{\text{Hydrolysis}}{\rightleftharpoons}} \underset{\text{Carboxylic acid}}{R-\overset{\overset{O}{\|}}{C}-OH} + \underset{\text{Alcohol}}{R'-OH}$$

Ester and acid hydrolysis is the common examples of hydrolysis. It can be catalysed by acid or base.

Applications:

Apart from ester or amide hydrolysis some of the hydrolysis reactions applicable today are as follows.

1. Base catalysed hydrolysis of triglycerides to fatty acids and soap are used.
2. By means of hydrolysis the carbohydrates are digested into monosaccharides *i.e.*, Polysaccharides are broken into simple monosaccharides.

Hyperconjugation: Delocalization of electrons due to the overlap of carbon-carbon or carbon-hydrogen σ bonds with an empty *p* orbital.

Inductive effect: The permanent displacement of electrons producing polarity in a covalent bond due to the presence of high electronegative atom in a molecule. It is denoted by the symbol $\longrightarrow$, the arrow pointed towards the high electronegative atom.

Ionic bond or Electrovalent bond: Chemical bond formed between the atoms by transfer of electrons from one atom to another atom.

Ionisation energy (I): Minimum amount of energy needed to remove an electron from a gaseous state of an atom in its ground state to form a gaseous ion.

Intramolecular forces: Any force that hold or keep the atoms together making up a molecule or compound. Example: Repulsive and attractive forces.

Intermolecular forces: Neutral molecules are held together by electrostatic forces. This electrostatic force constitutes the attraction between positive charge and negative charge. Example Dipole- dipole interaction.

Intermediate: A compound formed in the middle of the chemical reaction and it cannot be isolated.

Intermolecular hydrogen bonding: Hydrogen bonding between the molecules.

Intermolecular reaction: A chemical reaction takes place between two molecules.

Intramolecular reaction: A chemical reaction takes place within the molecule.

Intramolecular hydrogen bonding: Hydrogen bonding within the molecule.

Imine: It is a functional group in which carbon linked to nitrogen by a double bond. The nitrogen atom attached to a hydrogen (H) or an organic group ($R_2C = NR$) or ($R_2C = NH$).

Incipient carbocation: Carbocation complexed with a catalyst in a reaction.

Inversion of configuration: Inverting the configuration of carbon inside out like an umbrella in a windstorm so that the resulting compound has an inverted configuration to that of the reactant.

Isolated dienes: Double bonds in the diene are separated by more than one single bond carbon. For Example:

For example: $CH_2 = CH - CH_2 - CH = CH_2$

1,4-Pentadiene

Isomers: Non-identical compounds with same molecular formula.

Isomerisation: Conversion of one isomer in to another one isomer of a compound is called as isomerisation.

$$\underset{\text{n-Butene}}{CH_3 - CH_2 - CH = CH_2} \xrightarrow[Al_2(SO_4)_3]{\Delta} \underset{iso\text{-Butene}}{CH_3 - \underset{\large CH_3}{\underset{|}{C}} = CH_2}$$

Isotopes: Atoms with same number of protons but different number of neutrons.

IUPAC nomenclature: Systematic nomenclature of chemical compounds.

Keto-enol tautomerism: Interconversion of ketone and enol form of compound.

Kekule's structure: A model which represents the bonds between atoms as lines.

Kolbe's reaction: Phenol upon reaction with CO_2 in presence of NaOH at high atmospheric pressure followed by hydrolysis yields salicylic acid. This reaction is used to introduce -COOH group in *ortho* position. It is otherwise known as Kolbe's-Schmidt reaction.

$$\underset{\text{Phenol}}{C_6H_5OH} \xrightarrow[-H_2O]{NaOH} \underset{\text{Sodium phenoxide}}{C_6H_5ONa} \xrightarrow[125\ ^\circ C/6atm]{CO_2} \underset{\text{Sodium salicylate}}{o\text{-}HO\text{-}C_6H_4\text{-}COONa} \xrightarrow{H^+/H_2O} \underset{\text{Salicylic acid}}{o\text{-}HO\text{-}C_6H_4\text{-}COOH}$$

Kolbes synthesis: Electrolysis of a concentrated aqueous solution of the sodium or potassium salt of carboxylic acid yields alkanes. This reaction is as called Kolbe's electrolytic reaction.

$$\underset{\text{Sodium carboxylate}}{2R-COONa}$$

At anode

$$\begin{matrix} RCOO^- \\ RCOO^- \end{matrix} \longrightarrow \underset{\text{Alkane}}{R-R} + 2CO_2$$

At cathode

$$2Na^+ \xrightarrow{2HOH} 2NaOH + H_2\uparrow$$

Lactam: A cyclic amide.

Example

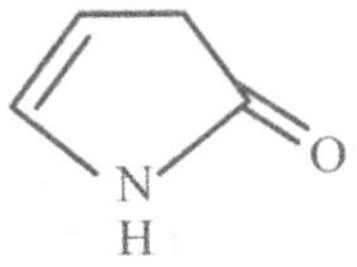

Lactone: A cyclic ester.

Example

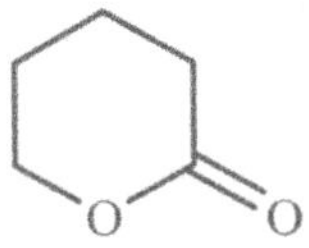

Leaving group: A group that is displaced in a substitution reaction.

London forces: Induced dipole - dipole interactions.

Lone - pair of electrons: Valence electrons not involving in bonding.

Lucas test: A chemical test used to find out whether the alcohol is primary or secondary or tertiary.

Magnetic quantum number (*m*, refers to the orientation of the orbital): It is the behavior of the electron in the magnetic field. It can have all integral values from – l to 0 to +l.

Mass number: The sum of the number of protons and number of neutrons in an atom.

Markovnikov's rule: The negative part of the addendum adds on to the unsymmetrical carbon containing less number of hydrogen atoms.

$$\underset{\text{1-Propene}}{CH_3 - CH = CH_2} \xrightarrow{HBr} \underset{\text{2-Bromopropane}}{CH_3 - \overset{\overset{Br}{|}}{CH} - CH_3}$$

Mechanism of a reaction: A detailed description of the step -by -step process by which the reactants are converted in to products.

Meerwin-Pondorf-Verley Reduction: Reduction reaction of aldehydes and ketones to the corresponding alcohols by treating with aluminium isopropoxide in isopropanol solution.

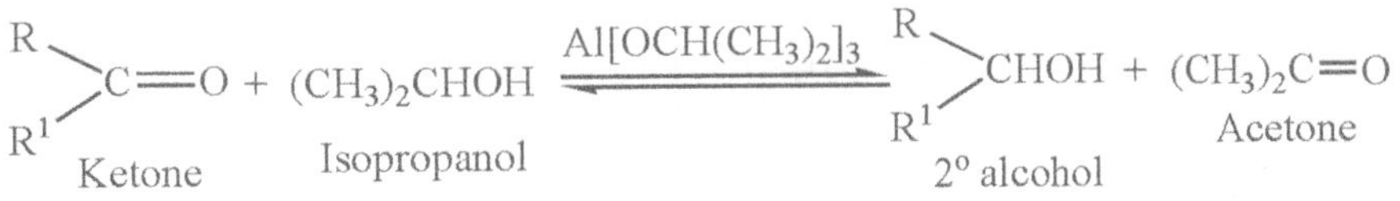

$$\underset{\text{Ketone}}{RR^1C{=}O} + \underset{\text{Isopropanol}}{(CH_3)_2CHOH} \xrightleftharpoons{Al[OCH(CH_3)_2]_3} \underset{2^\circ\text{ alcohol}}{RR^1CHOH} + \underset{\text{Acetone}}{(CH_3)_2C{=}O}$$

Mercaptan: A sulphur analogue of an alcohol molecule (RSH).

Mercuration of benzene: Reaction between benzene and mercuric acetate gives phenyl mercuric acetate. It is called as mercuration.

Melting point: The point or temperature at which a crystalline solid substance is converted in to liquid.

Meso compound: A compound with chiral center and a plane of symmetry.

Mesomeric and resonance effect: The permanent effect in which π electrons are transferred from a multiple bond to a single covalent bond or from an atom with lone pair of electrons to the adjacent single covalent bond. It is denoted by a letter M and symbol of curved arrow (↷).

***meta*-director:** A substituent which directs the incoming substituent to the *meta* position with respect to the existing substituent.

Nitration: Substitution of hydrogen atom of substrate with nitro (-NO_2) group is called as nitration.

$$\underset{\text{Methane}}{CH_4} + HNO_3 \xrightarrow[\text{Pressure}]{400\text{-}500°C} \underset{\text{Nitromethane}}{CH_3NO_2} + H_2O$$

For nitration of benzene conc. HNO_3 acid and conc. H_2SO_4 (nitrating mixture) is used to give nitrobenzene.

$$\underset{\text{Benzene}}{C_6H_6} + HNO_3 \xrightarrow{\text{Con. } H_2SO_4} \underset{\text{Nitrobenzene}}{C_6H_5NO_2}$$

Nucleophiles (Nucleus loving species or Electron rich species): Organic reagent that can donate an electron pair in a chemical reaction is known as nucleophile. It attacks the positive portion (region with low electron density) of the substrate molecule. These are negatively charged species (carbanion also) or neutral species with free electron pairs. These are denoted by the symbol Nu^-.

Octet rule: The electronic configuration or arrangement in which eight electrons are present in the outermost orbital or valency shell is called stable configuration or octet rule.

Organic chemistry: The chemistry of hydrocarbons and their derivatives.

Organic compounds: Compounds that contain carbon.

Open chain compounds: Carbon compounds having open chain of carbon atoms whether branched or unbranched are called as open chain compounds. Example pentane (CH_3CH_2 CH_2 CH_2CH_3).

Optical isomers: Stereoisomers which contains chiral centres.

Optically active: Compounds rotates the plane polarised light.

Optically inactive: Compounds does not rotate plane polarised light.

Organic synthesis: Synthesis of a compound containing carbon.

Orientation: The process of finding the relative position of various groups attached to the nucleus in an unknown derivative.

***ortho/ para*-directing substituents:** Substituents direct the incoming substituents to the *ortho/para* position with respect to already existing substituent in a ring.

Osazone: A compound obtained by the treatment of an aldose or ketose with excess amount of phenyl hydrazine.

s-s overlapping: Overlapping of *s* orbital of one atom with *s* orbital of another atom. Example: Formation of hydrogen molecule.

***s-p* overlapping:** Overlapping of *s* orbital of one atom with *p* orbital of another atom. Example: Formation of hydrogen fluoride molecule.

***p-p* overlapping:** Overlapping of *p* orbital of one atom with *p* orbital of another atom. Example: Formation of fluorine molecule.

Oxidation: Loss of electrons or addition of oxygen.

Oxidation-reduction (Redox) reaction: Reduction and oxidation takes place in a reaction simultaneously.

Ozonide: A five membered compound obtained in ozonolysis reaction after rearrangement of molonozide.

Ozonolysis: The formation of an unstable ozonide upon reaction of ozone with alkene and breaking it into different products with the help of reagents is called ozonolysis.

$$\underset{\text{Alkene}}{RR'C=CH-R''} + \underset{\text{Ozone}}{O_3} \longrightarrow \underset{\text{Ozonide}}{RR'C(-O-)CH-R''\ (\text{with } O-O \text{ bridge})}$$

Ozonide $\xrightarrow{H_2O_2}$ $RR'C=O$ (Ketone) + $R''COOH$ (Carboxylic acid)

Ozonide $\xrightarrow{H_2/Ni}$ $RR'C=O$ (Ketone) + $R''CHO$ (Aldehyde)

Ozonide $\xrightarrow{LiAlH_4}$ $RR'CH-OH$ (2° Alcohol) + $R''CH_2OH$ (1° Alcohol)

Paraffin: Another name for alkanes.

Parent hydrocarbon: A longest continuous carbon chain in an organic compound.

Partial racemization: Formation of a pair of enantiomers in unequal amounts.

Pauli exclusion principle: Only two electrons may exist in the same orbital and these electrons must have opposite spin.

Pericyclic reactions: A concerted reaction that occurs as the result of a cyclic rearrangement of electrons.

Peroxy acid: A carboxylic acid carbon with -OOOH group. Example perbenzoic acid (C_6H_5COOOH)

Perkin reaction: Aromatic aldehydes when heated with acid anhydrides in the presence of sodium salt of the corresponding carboxylic acid yields an α,β– unsaturated acid, this reaction in known as Perkin reaction.

$$\underset{\text{Benzaldehyde}}{C_6H_5CHO} + \underset{\text{Acetic anhydride}}{(CH_3CO)_2O} \xrightarrow{CH_3COONa} \underset{\text{Cinnamic acid}}{C_6H_5-CH=CH-COOH}$$

Phenols: Phenols are compounds containing hydroxy (-OH) group attached directly into an aromatic ring.

Phenyl group: Group obtained by removal of one hydrogen from benzene is called phenyl group. Example,

[benzene ring with one bond] or [benzene ring with one bond]

Phenyl

Photochemical reaction: A chemical reaction takes place when the reactant molecules absorb light.

***Pi* (π) bond:** A bond formed by the side wise overlap of *p* orbitals.

***Pi* (π) complex:** A complex obtained by the interaction between electrophile and a π electron cloud.

Pinacol -Pinacolone rearrangement: Fully substituted 1,2 -diols (Pinacol) reacts with mineral acid to form ketone (pinacolone), this is called as Pinacol–Pinacolone rearrangement (dehydration followed by rearrangement occurs).

$$CH_3-\underset{OH}{\overset{CH_3}{C}}-\underset{OH}{\overset{CH_3}{C}}-CH_3 \xrightarrow[-H_2O]{H^+} CH_3-\underset{O}{\overset{}{C}}-\underset{CH_3}{\overset{CH_3}{C}}-CH_3$$

2,3-Dimethyl-2,3-butanediol (Pinacol) → 3,3-Dimethyl-2-butanone (Pinacolone)

Plane of symmetry: An imaginary plane which bisects a molecule in to mirror images.

Polar bond: Bond with two charges.

Polarimeter: An instrument used to determine the rotation of plane polarized light.

Polarisability: An indication of the ease with which the electron cloud of an atom can be distorted.

Polarised light: Light that oscillates in only one plane.

Polar reaction: Reaction between electrophile and nucleophile.

Polymerization: Chemical reaction in which a molecule or monomer is converted into polymolecule or polymer is called as polymerization.

$$\underset{\text{Ethylene}}{CH_2 = CH_2} \xrightarrow{\text{Polymerisation}} \underset{\text{Polyethylene}}{\left(CH_2 - CH_2 \right)_n}$$

If the polymer is made up of same type of monomers then it is called homopolymer, if different then heteropolymer.

Primary alcohol: An alcohol in which the - OH group is attached with primary carbon atom.

Primary alkyl halide: An alkyl halide in which the halogen atom is attached with primary carbon atom.

Primary amine: An amine in which one alkyl or aryl group is attached with amine nitrogen atom.

Principal quantum number (*n*, refers to the size of orbital): It denotes energy and distance of the electron from the nucleus.

Pyrolysis and Cracking: Pyrolysis is a Greek word means breaking of a compound by heat (Pyro-fire, lysis-breaking). The phenomenon of pyrolysis of alkanes is called cracking.

$$\underset{\text{Propene}}{CH_3-CH=CH_2} + H_2 \xleftarrow[\text{C-H linkage}]{\text{Breaking of}} \underset{\text{Propane}}{CH_3CH_2CH_3} \xrightarrow[\text{C-C linkage}]{\text{Breaking of}} \underset{\text{Methane}}{CH_4} + \underset{\text{Ethylene}}{CH_2=CH_2}$$

Qualitative analysis: Identification of the chemical constituents or various atoms present in a mixture of chemical substances is called as qualitative analysis.

Quantitative analysis: It is used to find out the purity of chemical substances or to determine the amount of active chemical constituent present in the given substance or dosage form.

Quantum mechanics: It uses mathematical equation that explains the wave motion of electrons.

Quantum numbers: Numbers arising from the quantum mechanical treatment of an atom that describes the properties of the electrons in the atom.

Quaternary ammonium salt: A quaternary ammonium ion and anion.

Racemic mixture: A mixture of equal amount of pair of enantiomers.

Racemisation: The phenomenon of converting optically active compound into optically inactive racemic mixture is called as racemisation. For example when (+) tartaric acid in water is heated, it is completely transformed into optically inactive mixture of the racemic and mesotartaric acid.

Radical or Group or Functional group: Generally, an organic compound is composed of two or more parts, each of which is called a group or radical. Each group shows its typical characteristic properties

$CH_3 - CH_2 - OH$ ⟶ Ethanol

$CH_3 - CH_2-$ ⟶ Ethyl radical

$- OH$ ⟶ Hydroxyl group or Functional group

Functional group is the reactive portion of an organic compound which determines the chemical properties of organic compounds to a great extent. For example, identical chemical properties are shown by all alcohols (Methanol, Ethanol and Propanol).

Rancidification: Oxidation of fats or oils when exposed to atmospheric moisture and oxidation takes place due to the action of aerobic bacteria with the formation of aldehydes, ketones and acids which results in the formation of bad odor and taste, This process is called as rancidification. It is two types, *i.e.*, hydrolytic and oxidative rancidification.

Rate determining step: The step in a chemical reaction that has the transition state with the highest energy. It determines the rate of reaction.

Reimer–Tiemann reaction: Reaction of phenol with chloroform in aqueous NaOH solution, followed by acid hydrolysis yields salicylaldehyde. If CCl_4 is used in place of chloroform, salicylic acid is obtained. Reimer -Tiemann reaction is used to introduce -CHO or -COOH group *ortho* to -OH group of phenol.

Phenol (OH) $\xrightarrow[-HCl]{CHCl_3/\ NaOH}$ Salicylaldehyde (OH, CHO)

Reduction: Addition of hydrogen or gaining of electrons.

Reformatsky reaction: It involves the reaction of aldehydes and ketones with α- bromoester in the presence of zinc metal to give β–hydroxy ester after hydrolysis.

Regioselective reaction: A reaction that leads to the preferential formation of one constitutional isomer than other.

Resolution of racemic mixture: Separation of racemic mixture in to an individual enantiomers.

Resonance: The phenomenon in which two or more structures, involving identical position of atoms can be written for a particular compound.

Resonance energy: The extra stability associated with the compound as a result of its delocalized electrons.

Resonance hybrid: The actual structure of a molecule with delocalized electrons and it is indicated by more than one structure.

Retrosynthesis: Working backward from the target molecule to the starting material.

Ring flip (interconversion of chair- chair conformation): Due to the rotation of carbon carbon single bond in cyclohexane one chair form is converted in to another one chair form. The axial hydrogens in one becomes equatorial in another one.

Robinson annulation: A Michael reaction followed by an intramolecular aldol condensation.

Rosenmund's reduction: Acid chlorides react with hydrogen in the presence of palladium poisoned with $BaSO_4$ to give aldehyde.

$$H_3C-\overset{O}{\overset{\|}{C}}-Cl + H_2 \xrightarrow[BaSO_4]{Pd} H_3C-\overset{O}{\overset{\|}{C}}-H + HCl$$

Acetyl chloride ⟶ Acetaldehyde

Sabatier-Senderens reaction: Hydrogenation of unsaturated hydrocarbons in the presence of nickel at 570 K yields alkanes. This reaction is called Sabatier-Senderens reaction.

$$\underset{\text{Ethylene}}{H_2C = CH_4} + H_2 \xrightarrow[570\text{ K}]{Ni} \underset{\text{Ethane}}{CH_3 - CH_3}$$

Salting out: It is a method of purification used to precipitate proteins by reducing their solubility in high ionic strength solvent. It is also known as salt induced precipitation.

Sandmeyer's reaction: The reaction of aryl diazonium salt with a cuprous salt and the concerned halogen acid is known as Sandmeyer's reaction.

$$\underset{\text{Aryl diazonium chloride}}{Ar - N^+ \equiv NCl^-} \xrightarrow[X = Cl,\ Br,\ CN]{CuX/HX} Ar - X + N_2\uparrow$$

Saponification: Alkaline hydrolysis of fats and oils yields soap or candles and glycerol. This reaction is called as saponification.

$$\underset{\text{Fat or oil}}{\begin{matrix} CH_2OCOR \\ | \\ CHOCOR \\ | \\ CH_2OCOR \end{matrix}} + 3NaOH \longrightarrow \underset{\text{Glycerol}}{\begin{matrix} CH_2OH \\ | \\ CHOH \\ | \\ CH_2OH \end{matrix}} + \underset{\text{Soap}}{3RCOONa}$$

Saturated compounds: Hydrocarbons containing only one carbon-carbon (C-C) single bond are called as saturated compounds.

Saturated hydrocarbon: A hydrocarbon that is completely saturated with hydrogen.

Sawhorse projection: In this formula the molecule is viewed at the carbon - carbon bond from an oblique angle. It is used to determine the enantiomeric and diastereomeric relationship of a molecule.

H H H H H H

S-configuration: Relative priorities to the four groups bonded to a chirality centre to be given first. If the lowest priority groups are on vertical axis in a Fischer projection, an arrow drawn from the highest priority group to the next -highest priority group goes in counter clockwise direction.

Sigmatropic rearrangement: A reaction in which a σ bond is broken in the starting compound, a new σ bond is formed in the product and the π bonds gets rearranged.

Single bond: This covalent bond is formed by mutual sharing of one electron pair between two atoms.

S_N1: Nucleophilic substitution reaction which follows first order kinetics.

S_N2: Nucleophilic substitution reaction which follows second order kinetics.

S_NAr reaction: A nucleophilic aromatic substitution reaction.

Solvation: An interaction between the solvent and solute molecule.

Stereochemistry: The field of chemistry deals with the 3D structure of the molecules (spatial arrangement of atoms in a molecule).

Stereoisomers: Isomers showing stereochemical property.

Stereoselective reaction: A reaction preferential γ forms one isomer more than another isomer.

Stereospecific reaction: A reaction in which the reactant can exists as stereoisomers and each stereoisomeric reactant leads to a different stereoisomeric product or set of products.

Steric effects: Effects due to the group that occupy the particular volume of space.

Steric hindrance: It refers to bulk groups at the site of reaction hinders the reactants to approach each other.

Steric strain: The repulsion between the electron cloud of one atom and the electron cloud of another atom.

Structural isomers: Molecules with same molecular formula but differ in the structure (arrangement of atoms).

Sublimation: The process of direct conversion of solid into vapor is called as sublimation. This process is used for the separation of substances which sublimates on heating from non-volatile impurities and *vice versa.*

Substituted acids: These are the derivatives of carboxylic acids obtained by replacing one or more hydrogen atoms of the alkyl radical by different atoms or groups such as – Cl, - Br, $-NH_2$, -OH *etc.*

CH_3-COOH	$CH_2Cl-COOH$	$CH_2Br-COOH$	CH_2NH_2-COOH	$CH_2OH-COOH$
Acetic acid	Chloroacetic acid	Bromoacetic acid	Aminoacetic acid	Hydroxyacetic acid

Substitution reaction: Organic reactions which involve direct substitution or replacement of an atom or group of atoms by some other atom or group of atom are known as substitution reactions.

$$\underset{\text{Methane}}{CH_4} + Cl_2 \longrightarrow \underset{\text{Methyl chloride}}{CH_3Cl} + HCl$$

In the above reaction, one hydrogen atom of the methane molecule is substituted by one chlorine atom.

Sulphonation: Substitution of hydrogen atom of alkyl or aryl compounds with $-SO_3H$ group is called as sulphonation.

$$\underset{\text{Methane}}{H_3C-H} + \underset{\text{(Fuming) Sulphuric acid}}{OH\text{-}SO_3H} \longrightarrow \underset{\text{Methanesulphonic acid}}{H_3C-SO_3H} + H_2O$$

Tautomers: Isomers that differs in the position of double bond and a hydrogen.

Terminal alkyne: Alkynes with triple bond at the end of the carbon skeleton.

Termination step: When two free radicals combined and form the product.

Tetrahedral intermediate: An intermediate formed during the nucleophilic acyl substitution reaction.

Thiols or thioalcohols or mercaptans: The sulphur analogues of alcohols are called thiols or mercaptans or alkyl hydrogen sulfides. Example methanethiol (CH_3SH).

Thioethers or sulfides: The sulfur analogues of ethers are called thioethers or sulfides.

$$\underset{\text{Dimethyl sulphide}}{H_3C-S-CH_3}$$

Torsional strain: It is the strain induced by the repulsion of bonding electrons of one substituent with the bonding electrons of nearby substituent.

Trihydric alcohols: Compounds containing three hydroxyl groups are called trihydric alcohols or triols.

$$\begin{array}{l} CH_2OH \\ | \\ CHOH \\ | \\ CH_2OH \end{array}$$

Glycerol

Trihalogen derivatives: In this type three halogen atoms are attached to a molecule.

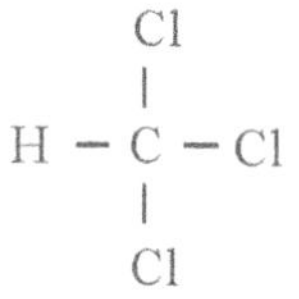

$$\begin{array}{c} Cl \\ | \\ H - C - Cl \\ | \\ Cl \end{array}$$

Chloroform or Trichloromethane

Triple Bond: This covalent bond is formed by the mutual sharing of three electron pairs between two atoms.

Unsaturated alcohols: Monohydric alcohols containing double bonds or triple bonds are known as unsaturated alcohols.

$CH_2 = CH - OH$ Vinyl alcohol

$CH_2 = CH - CH_2 - OH$ Allyl alcohol

Unsaturated compounds: A hydrocarbon having carbon-carbon double bond (C=C) or carbon-carbon triple bond ($C \equiv C$) are called as unsaturated compounds.

$CH_2 = CH_2$ Ethylene

$CH \equiv CH$ Acetylene

α,β-unsaturated carbonyl compounds: Compounds containing both carbon-carbon double bond and a carbon-oxygen double bond are known as α,β-unsaturated carbonyl compounds.

$$CH_2 = CH - \overset{\overset{\displaystyle O}{\|}}{C} - H$$

Acrolein

Van der Waals forces: It is a weak attractive force between the molecules which are produced by the dipole- dipole interaction. It holds the molecules in liquid state.

Valence electrons: Number of electrons present in the outer most orbitals of an atom.

Victor- Meyer's method: It is a method used for the determination of molecular mass of volatile substances.

Vicinal dihalides or Vic-dihalides: Both halogen atoms are attached to the adjacent carbon atoms.

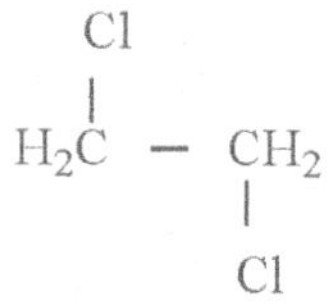

$$\begin{array}{ccc} Cl & & \\ | & & \\ H_2C & - & CH_2 \\ & & | \\ & & Cl \end{array}$$

Ethylene dichloride

Vicinal diol: A compound with two-OH groups attached to a adjacent carbon atom. Example, ethylene glycol

Vinyl group: $CH_2 = CH$-

Vinylic carbon: A carbon atom in a carbon -carbon double bond.

Wacker process (From alkenes): Reaction between alkenes with palladium chloride and cupric chloride to give aldehydes and ketones.

$$\underset{\text{Ethylene}}{CH_2{=}CH_2} + PdCl_2 + H_2O \xrightarrow{CuCl_2} \underset{\text{Acetaldehyde}}{CH_3CHO} + Pd + 2HCl$$

Wave equation: An equation explains the behaviour of each electron in an atom or molecule.

Wave function: A series of solutions of a wave equation.

Williamson's ether synthesis: Sodium or potassium alkoxide is heated with alkyl halide to give ether. This reaction is called as Williamson's ether synthesis.

$$\underset{\text{Sodium ethoxide}}{C_2H_5{-}ONa} + \underset{\text{Ethyl bromide}}{C_2H_5Br} \xrightarrow[\text{-NaBr}]{\Delta} \underset{\text{Diethyl ether}}{C_2H_5{-}O{-}C_2H_5}$$

Willgerodt reaction: The conversion of aryl ketones into an amide by using yellow ammonium polysulphide is called Willgerdot reaction.

$$\underset{\text{Acetophenone}}{C_6H_5\overset{O}{\overset{\|}{C}}CH_3} \xrightarrow{(NH_4)_2Sx} \underset{\substack{\text{Phenyl acetamide}\\(50\%\text{ yield})}}{C_6H_5CH_2CONH_2} + \underset{\substack{\text{Ammonium}\\\text{phenyl acetate}\\(13.5\%\text{ yield})}}{C_6H_5CH_2COONH_4} + \underset{\text{Ethyl benzene}}{C_6H_5CH_2CH_3}$$

Woodward-Hofmann rules: A series of selection rules in pericyclic reactions.

Wolf-Kishner reduction: Aldehydes and ketones are first converted into their derivatives such as hydrazones, semicarbazones, azines (by treatment with hydrazines, semicarbazides *etc*) which upon further reaction with a strong base such as potassium hydroxide or potassium *tert*- butoxide in dimethyl sulphoxide (DMSO) solvent yield alkanes.

$$\underset{\text{Cyclohexyl methyl ketone}}{C_6H_{11}{-}\overset{O}{\overset{\|}{C}}{-}CH_3} \xrightarrow{NH_2NH_2} \underset{\text{Cyclohexyl methyl ketone hydrazone}}{C_6H_{11}{-}\overset{NNH_2}{\overset{\|}{C}}{-}CH_3}$$

$$\xrightarrow{\frac{KOH}{DMSO}} \underset{\text{Ethyl cylohexane}}{C_6H_{11}{-}CH_2CH_3} + N_2\uparrow$$

Wurtz-Fittig Reaction: Conversion of aryl halides in to aryl homologues by using alkyl halides and metallic sodium in dry ether.

$$\underset{\text{Phenyl bromide}}{C_6H_5{-}Br} + 2Na + \underset{\text{Methyl iodide}}{CH_3I} \xrightarrow{\text{Dry ether}} \underset{\text{Toluene}}{C_6H_5{-}CH_3} + NaBr + NaI$$

Wurtz Synthesis: Reaction between a solution of alkyl halide (especially alkyl bromide or alkyl iodide) and metallic sodium yields alkanes.

$$\underset{\text{Methyl iodide}}{CH_3I} + 2Na + \underset{\text{Methyl iodide}}{CH_3I} \xrightarrow{\text{Dry ether}} \underset{\text{Ethane}}{CH_3—CH_3} + \underset{\text{Sodium iodide}}{2NaI}$$

Z-conformer: A geometrical isomer in which the higher priority groups on the same side of the molecule.

Zeisel method: A method used for determination of number of methoxy groups in an organic compound.

Zwitter ion: A compound in which both positive and negative charges are present.

8

Alkanes

Introduction

Alkanes are the simple aliphatic organic saturated hydrocarbons, having the general formula C_nH_{2n+2} where n = 1, 2, 3 *etc*. Carbon and hydrogen are connected to each other by a single covalent (sigma) bond. They do not contain π bond in their structure and the 4 valency of carbon atom is saturated by **carbon-hydrogen** (C-H) and carbon–carbon (C-C) bonds and hence they are called as "**saturated hydrocarbons**". As alkanes contain strong C-C and C-H bonds, they do not react with variety of reagents at normal conditions therefore they are also called as **paraffins** (The term derived from Latin words parum= little and affins = affinity *i.e.*, Less affinity towards any reagent).

```
    H            H   H           H   H   H
    |            |   |           |   |   |
H — C — H    H — C — C — H   H — C — C — C — H
    |            |   |           |   |   |
    H            H   H           H   H   H
 Methane         Ethane           Propane
```

All members of alkanes differ from each other by their molecular formula by the unit of -CH_2- and are called as members of "homologues series". Removal of one hydrogen from alkane yields alkyl group (–R). General formula for alkyl group is C_nH_{2n+1}.

Occurrence: The main source of alkanes is petroleum or natural gas, both of them found in underground deposits. Natural gas contains about 80% methane, 10% ethane and 10% of higher alkanes.

Importance of Alkanes

Hydrocarbons having long chain alkanes of carbon atoms (C_{25}-C_{35}) are important in nature as components of waxes. Natural hydrocarbons exist as terpenes from which bioactive compounds are derived, example Limonene, Pinene and Phellanderene *etc*.

Functional derivatives of hydrocarbons have biological importance. Various functional groups attached with alkyl group elicit different kind of pharmacological activities. Some of the drugs which possess alkyl group and their therapeutic uses in the treatment of various diseases are shown in Table 8.1.

Table 8.1 Drugs containing alkyl groups and their therapeutic uses.

Uses	Drug Name	Structure
Sedative and hypnotic	Pentobarbital	5-ethyl-5-(1-methylbutyl)barbituric acid: ring O=C(NH)₂ with C=O groups; substituents C_2H_5 and $CH(CH_3)CH_2CH_2CH_3$
	Amobarbital	Barbituric acid ring; substituents C_2H_5 and $CH_2CH_2CH(CH_3)_2$
	Chlorbutanol	$CH_3-C(CH_3)(OH)-CCl_3$
General anaesthetic	Thiopentone sodium	Na^+S^- thiobarbiturate ring; substituents C_2H_5 and $CH(CH_3)-CH_2-CH_2-CH_3$
Anticancer drug	Mechlorethamine	$H_3C-N(CH_2CH_2Cl)_2$

Nomenclature of Alkanes

Alkanes can be named by three systems.

1. **Common name system or Trivial system:** This system of naming is based on branching in alkanes. The parent name of isomeric alkanes is same and the prefix distinguishes the each isomer based on branching.
   (i) The prefix *'n'* (means normal) used for alkanes in which all carbon atoms is in continuous chain. Example: *n*-Butane, *n*-Pentane.

$$CH_3-CH_2-CH_2-CH_3 \quad \text{n-Butane} \qquad CH_3-CH_2-CH_2-CH_2-CH_3 \quad \text{n-Pentane}$$

   (ii) The prefix *'iso'* used for alkanes in which one methyl group branching is present. Example: *iso*-Butane, *iso*-Pentane.

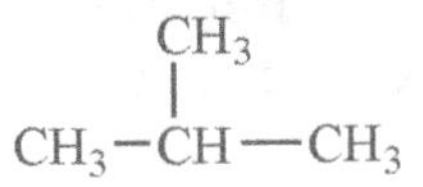

$$CH_3-\underset{}{\overset{CH_3}{\overset{|}{C}H}}-CH_3$$

iso-Butane

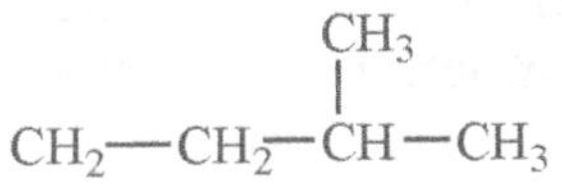

$$CH_2-CH_2-\overset{CH_3}{\overset{|}{C}H}-CH_3$$

iso-Pentane

(iii) The prefix *'neo'* used for alkanes in which two methyl groups are branched in same carbon atom e.g. *'neo'*-Hexane

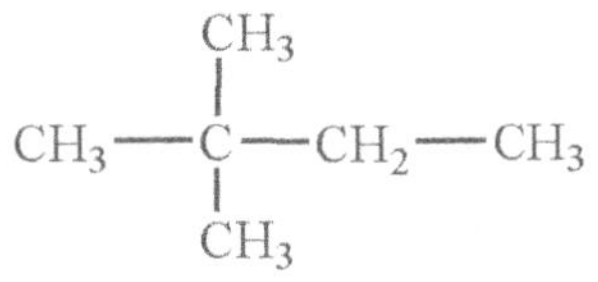

$$CH_3-\underset{CH_3}{\underset{|}{\overset{CH_3}{\overset{|}{C}}}}-CH_2-CH_3$$

neo-Hexane

The first four alkanes are known by their common names (which are derived from history). Fifth member onwards are derived from Latin or Greek numerals which indicate the number of carbon atoms in a molecule.

Alkyl groups: An alkyl group is obtained by removing one hydrogen atom from alkane. Their name is derived from the alkane by removing **-ane** from concerned alkane and adding **-yl** (Alkane-ane +yl = Alkyl) to the corresponding alkane.

$$\underset{\substack{\text{Alkane} \\ (C_nH_{2n+2})}}{RH} \xrightarrow{-H} \underset{\substack{\text{Alkyl} \\ (C_nH_{2n+1})}}{-R}$$

$$\underset{\text{Methane}}{CH_4} \xrightarrow{-H} \underset{\text{Methyl}}{-CH_3}$$

Types of alkyl groups: Based on the structural formula, alkyl groups are classified into primary, secondary and tertiary carbon as shown in Table 8.2.

Table 8.2 Structural Formula of Different Types of Alkyl Groups.

Primary (1°)	Secondary (2°)	Tertiary (3°)	Quarternary (4°)
Directly linked with one another or no other carbon atoms.	Directly linked with two other carbon atoms.	Directly linked with three other carbon atoms.	Directly linked with four other carbon atoms.

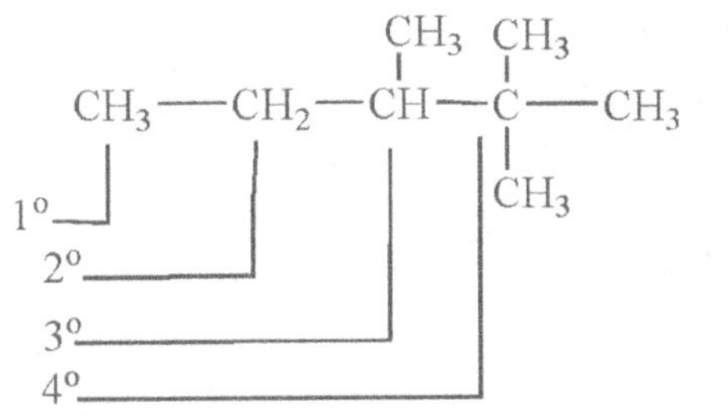

$$CH_3-CH_2-\overset{CH_3}{\overset{|}{C}H}-\underset{CH_3}{\underset{|}{\overset{CH_3}{\overset{|}{C}}}}-CH_3$$

1° (first CH_3), 2° (CH_2), 3° (CH), 4° (C)

2. **Substituted methane system:** The most highly branched carbon atom is considered as the parent and the alkyl groups attached to this carbon atom are named in alphabetical order.

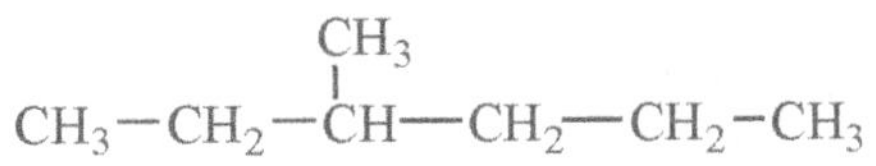

$$CH_3-CH_2-\overset{CH_3}{\overset{|}{C}H}-CH_2-CH_2-CH_3$$

Ethyl methyl propyl methane

3. **IUPAC name system:** To overcome the limitation of earlier methods, IUPAC system of naming organic compounds was initiated. In this system some of the common names also retained. The steps involved in the IUPAC nomenclature are described herewith examples.

Step 1. Select the longest chain: The longest continuous carbon chain is selected to give the parent name.

$$\overset{1}{CH_3}-\overset{2}{\underset{\underset{CH_3}{|}}{CH}}-\overset{3}{CH_2}-\overset{4}{CH_3}$$

The longest continuous chain has four carbon atoms *i.e.,* Butane

Step 2. Number the longest chain: The carbon atoms in the longest chain are numbered. The numbering should be given in such a way that the carbon atom containing substituent will get lowest number.

$$\overset{1}{CH_3}-\overset{2}{\underset{\underset{CH_3}{|}}{CH}}-\overset{3}{CH_2}-\overset{4}{CH_3} \qquad \overset{4}{CH_3}-\overset{3}{\underset{\underset{CH_3}{|}}{CH}}-\overset{2}{CH_2}-\overset{1}{CH_3}$$

(Correct) (Wrong)

The branched carbon gets lower number in first case, hence it is correct.

Step 3. Locate and name the substituents: Mark the position and name of each substituent in the carbon chain.

$$\overset{1}{CH_3}-\overset{2}{\underset{\underset{CH_3}{|}}{CH}}-\overset{3}{CH_2}-\overset{4}{CH_3}$$

2-Methylbutane

Attached group is methyl and attached at 2nd position of the carbon chain.

Step 4. Name the compound by combining longest chain and substituents:

(i) The position and the substituents are combined with the longest chain and the name is written as one word.

$$\overset{1}{CH_3}-\overset{2}{\underset{\underset{CH_3}{|}}{CH}}-\overset{3}{CH_2}-\overset{4}{CH_3}$$

2-Methylbutane

Longest chain is butane and the methyl group is attached at 2nd position. Hence the IUPAC name of the above compound is 2-methylbutane.

(ii) Indicate the number of groups present and their position:

$$\overset{1}{CH_3}-\overset{2}{\underset{\underset{CH_3}{|}}{CH}}-\overset{3}{\underset{\underset{CH_3}{|}}{CH}}-\overset{4}{CH_2}-\overset{5}{CH_3} \qquad \overset{4}{CH_3}-\overset{3}{CH_2}-\overset{\overset{CH_3}{|}}{\underset{\underset{CH_3}{|}}{\overset{2}{C}}}-\overset{1}{CH_3}$$

2,3-Dimethylpentane 2,2-Dimethylbutane

(a) If same substituent is present more than once then it is indicated by adding prefix **-di, -tri, -tetra** *etc.*

(b) If more than one type of substituents is present, then arrange the substituents in alphabetical order.

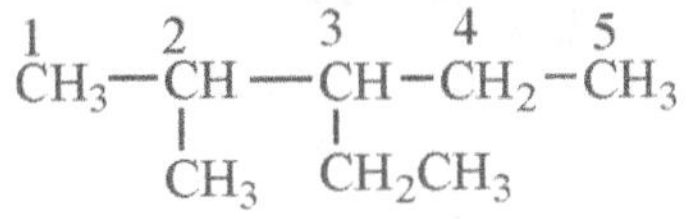

$$\overset{1}{CH_3}-\overset{2}{CH}(CH_3)-\overset{3}{CH}(CH_2CH_3)-\overset{4}{CH_2}-\overset{5}{CH_3}$$

3-Ethyl-2-methylpentane

Trivial names of the following branched alkanes have been recognized in 1957 IUPAC rules. Common names and IUPAC name of first 10 members are shown in Table 14.3.

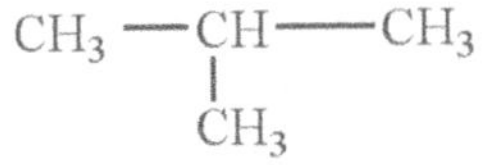

$$CH_3-CH(CH_3)-CH_3$$

iso-Butane

$$CH_3-CH_2-HC(CH_3)-CH_3$$

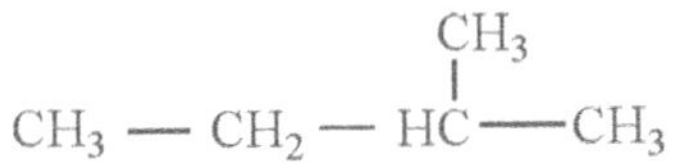

iso-Pentane

Table 8.3 Common name and IUPAC Name of Alkanes.

No. of Carbons	Mol. Formula	Structural Formula	Common name	IUPAC name
1.	CH_4	CH_4	Methane	*n*-Methane
2.	C_2H_6	CH_3CH_3	Ethane	*n*-Ethane
3.	C_3H_8	$CH_3CH_2CH_3$	Propane	*n*-Propane
4.	C_4H_{10}	$CH_3CH_2CH_2CH_3$	Butane	*n*-Butane
5.	C_5H_{12}	$CH_3CH_2CH_2CH_2CH_3$	Pentane	*n*-Pentane
6.	C_6H_{14}	$CH_3CH_2CH_2CH_2CH_2CH_3$	Hexane	*n*-Hexane
7.	C_7H_{16}	$CH_3CH_2CH_2CH_2CH_2CH_2CH_3$	Heptane	*n*-Heptane
8.	C_8H_{18}	$CH_3CH_2CH_2CH_2CH_2CH_2CH_2CH_3$	Octane	*n*-Octane
9.	C_9H_{20}	$CH_3CH_2CH_2CH_2CH_2CH_2CH_2CH_2CH_3$	Nonane	*n*-Nonane
10.	$C_{10}H_{22}$	$CH_3CH_2CH_2CH_2CH_2CH_2CH_2CH_2CH_2CH_3$	Decane	*n*-Decane

Isomerism in alkanes: *n*-Alkanes when heated they are converted into their branched chain isomers is called as isomerisation.

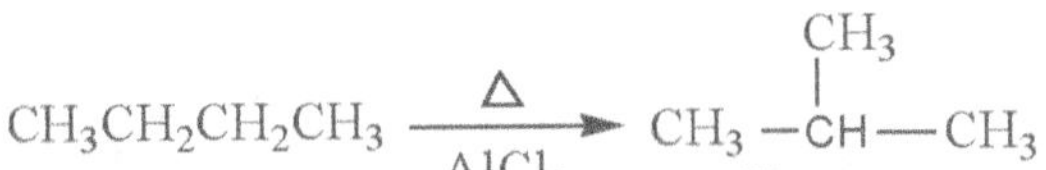

$$\underset{n\text{-Butane}}{CH_3CH_2CH_2CH_3} \xrightarrow[AlCl_3]{\Delta} \underset{iso\text{-Butane}}{CH_3-CH(CH_3)-CH_3}$$

As there is higher possibility of different arrangements of carbon atoms in compounds containing more than three carbons, they exhibit wide structural isomerism especially of chain isomerism. First three members of hydrocarbon series *i.e.*, methane, ethane and propane do not exhibit any isomerism. Butane exhibits **2** isomers, Pentane **3** isomers, Hexane **5** isomers, Heptane **9** isomers, Octane **18** isomers, and Decane **75** isomers.

Butane (C_4H_{10}):

$$CH_3-CH_2-CH_3$$

n-Butane

$$CH_3-CH(CH_3)-CH_3$$

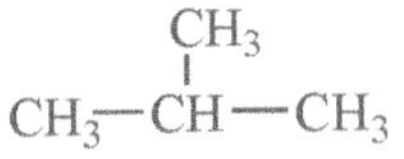

iso-Butane

(2-Methylpropane)

Pentane (C_5H_{12}):

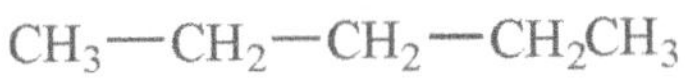

$$CH_3-CH_2-CH_2-CH_2CH_3$$

n-Pentane

$$CH_3-CH_2-CH(CH_3)-CH_3$$

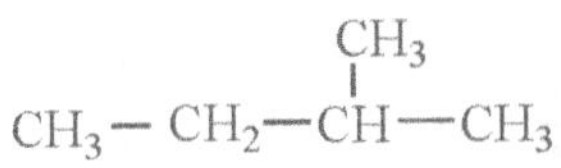

iso-Pentane

(2-Methylbutane)

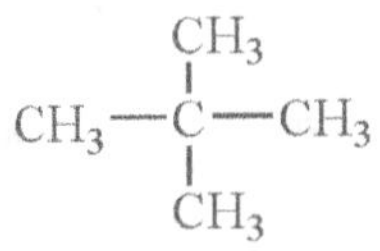

$$CH_3-C(CH_3)_2-CH_3$$

neo-Pentane

(2,2-Dimethylpropane)

Hexane (C_6H_{14}):

$$CH_3-CH_2-\underset{}{\overset{CH_3}{\overset{|}{C}H}}-CH_2-CH_3$$

3-Methyl pentane

$$CH_3-\overset{CH_3}{\overset{|}{C}H}-\overset{CH_3}{\overset{|}{C}H}-CH_3$$

2,3-Dimethyl butane

$$CH_3-\underset{CH_3}{\underset{|}{\overset{CH_3}{\overset{|}{C}}}}-CH_2-CH_3$$

2,2-Dimethylbutane
(*neo*-Hexane)

$$CH_3-CH_2-CH_2-\overset{CH_3}{\overset{|}{C}H}-CH_3$$

2-Methyl pentane
(*iso*-Hexane)

$$CH_3-CH_2-CH_2-CH_2-CH_2-CH_3$$

n-Hexane

General Methods of Preparation

The characteristic feature of homologues series is that its homologues can be prepared by adopting similar methods of preparation, hence the following general methods of preparation are described with a view to apply these methods for preparing all homologues members of alkanes.

1. **Decarboxylation of Carboxylic Acids:** Alkanes are prepared by heating a mixture of anhydrous sodium salt of carboxylic acid with soda lime (NaOH + CaO). During this reaction, a molecule of carbon dioxide is eliminated with the formation of alkane with one carbon less than the original carboxylic acid. The process of elimination of carbon dioxide from carboxylic acid (-COOH) is called as **"Decarboxylation".**

$$\underset{\text{Sodium carboxylate}}{RCOONa} + NaOH \xrightarrow[\Delta]{CaO} \underset{\text{Alkane}}{R\text{-}H} + Na_2CO_3$$

$$\underset{\text{Sodium acetate}}{CH_3COONa} + NaOH \xrightarrow[\Delta]{CaO} \underset{\text{Methane}}{CH_4} + Na_2CO_3$$

2. **Reduction of Alkenes or Alkynes:** Hydrogenation of unsaturated hydrocarbons in the presence of Nickel at 570 K yields alkanes. This reaction is called **Sabatier-Senderens** reaction. Examples for other hydrogenating agents are platinum and palladium.

i) $$\underset{\text{Alkene}}{C_nH_{2n}} + H_2 \xrightarrow[570\ K]{Ni} \underset{\text{Alkane}}{C_nH_{2n+2}}$$

$$\underset{\text{Alkene}}{R-CH=CH_2} + H_2 \xrightarrow[570\ K]{Ni} \underset{\text{Alkane}}{R-CH_2-CH_3}$$

$$\underset{\text{Ethylene}}{CH_2=CH_2} + H_2 \xrightarrow[570\ K]{Ni} \underset{\text{Ethane}}{CH_3-CH_3}$$

ii) $$\underset{\text{Alkyne}}{C_nH_{2n-2}} + 2H_2 \xrightarrow[570\ K]{Ni} \underset{\text{Alkane}}{C_nH_{2n+2}}$$

$$\underset{\text{Alkyne}}{R-C\equiv CH} + 2H_2 \xrightarrow[570\ K]{Ni} \underset{\text{Alkane}}{R-CH_2-CH_3}$$

$$\underset{\text{Acetylene}}{CH\equiv CH} + 2H_2 \xrightarrow[570\ K]{Ni} \underset{\text{Ethane}}{CH_3-CH_3}$$

3. **Reduction of Alkyl Halides:** Reduction of alkyl halides with nascent hydrogen yields alkanes. Common reducing agents used are: Zn/HCl, Zn/CH_3COOH, Zn/Cu, Ni, Pt, $LiAlH_4$, H_2 (gas), C_2H_5OH/Na, HI/P.

$$\underset{\text{Alkyl halide}}{R{-}X} + 2[H] \longrightarrow \underset{\text{Alkane}}{R{-}H} + HX$$

$$\underset{\text{Methyl iodide}}{CH_3{-}I} + 2[H] \longrightarrow \underset{\text{Methane}}{CH_4} + HI$$

4. **Hydrolysis of Grignard Reagents:** Upon hydrolysis, Grignard reagent (alkyl magnesium halide) yields concerned alkanes.

$$\underset{\text{Grignard reagent}}{RMgX} + HOH \longrightarrow \underset{\text{Alkane}}{RH} + Mg\begin{matrix} X \\ OH \end{matrix}$$

$$\underset{\text{Methyl magnesium iodide}}{CH_3MgI} + HOH \longrightarrow \underset{\text{Methane}}{CH_4} + Mg\begin{matrix} I \\ OH \end{matrix}$$

5. **Reduction of Alcohols, Aldehydes, Ketones or Carboxylic acids:** Reduction of alcohols, aldehydes, ketones or carboxylic acids with excess of hydroiodic acid at high temperature in a sealed tube in presence of red phosphorous yields alkanes.

$$\underset{\text{Alcohol}}{ROH} + 2HI \xrightarrow[420\ K]{\text{Red P}} \underset{\text{Alkane}}{RH} + I_2 + H_2O$$

$$\underset{\text{Aldehyde}}{R{-}\overset{\displaystyle O}{\overset{\|}{C}}{-}H} + 4HI \xrightarrow[420\ K]{\text{Red P}} \underset{\text{Alkane}}{R{-}CH_3} + 2I_2 + H_2O$$

$$\underset{\text{Ketone}}{R{-}\overset{\displaystyle O}{\overset{\|}{C}}{-}R} + 4HI \xrightarrow[420\ K]{\text{Red P}} \underset{\text{Alkane}}{R{-}CH_2{-}R'} + 2I_2 + H_2O$$

$$\underset{\text{Carboxylic acid}}{R{-}\overset{\displaystyle O}{\overset{\|}{C}}{-}OH} + 6HI \xrightarrow[420\ K]{\text{Red P}} \underset{\text{Alkane}}{R{-}CH_3} + 3I_2 + 2H_2O$$

Carbonyl compounds, especially ketones can be reduced with Zn/Hg and conc. HCl. This reaction is known as **Clemmensen reduction.**

$$\underset{\text{Ketone}}{R{-}\overset{\displaystyle O}{\overset{\|}{C}}{-}R'} \xrightarrow[HCl]{\text{Zn-Hg}} \underset{\text{Alkane}}{R{-}CH_2{-}R'}$$

$$\underset{\text{Acetone}}{CH_3{-}\overset{\displaystyle O}{\overset{\|}{C}}{-}CH_3} \xrightarrow[HCl]{\text{Zn-Hg}} \underset{\text{Propane}}{CH_3{-}CH_2{-}CH_3}$$

6. **Kolbes Synthesis:** Electrolysis of a concentrated aqueous solution of the sodium or potassium salt of carboxylic acid yields alkanes. This reaction is as called **Kolbe's electrolytic reaction.**

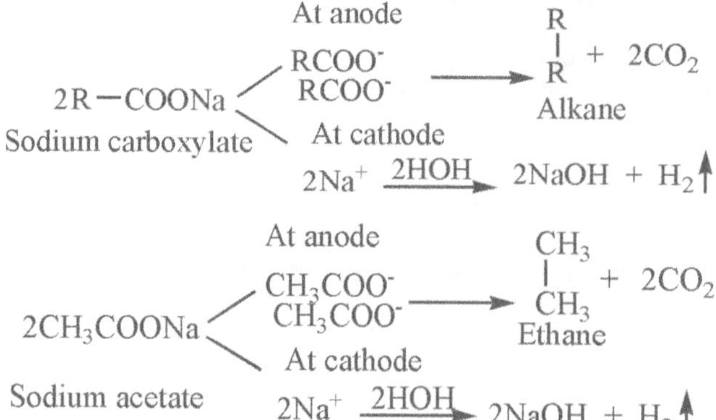

7. **Wurtz Synthesis:** Reaction between a solution of alkyl halide (especially alkyl bromide or alkyl iodide) and metallic sodium in the presence of dry ether yields alkanes. In this method alkanes containing double the number of carbon atoms of alkyl halides are obtained. This method is commonly used for the synthesis of symmetrical alkanes.

$$\underset{\text{Alkyl halide}}{RX + 2Na + RX} \xrightarrow{\text{Dry ether}} \underset{\text{Alkane}}{R—R} + 2NaX$$

$$\underset{\text{Methyl iodide}}{CH_3I} + 2Na + \underset{\text{Methyl iodide}}{CH_3I} \xrightarrow{\text{Dry ether}} \underset{\text{Ethane}}{CH_3—CH_3} + \underset{\text{Sodium iodide}}{2NaI}$$

Limitations:

(i) Methane cannot be prepared by this method.

(ii) Tertiary alkyl halides fail to react.

(iii) Instead of one alkyl halide if we use mixture of two different alkyl halide like methyl iodide and ethyl iodide a mixture of three alkanes such as ethane, propane and butane is obtained.

$$\underset{\text{Methyliodide}}{CH_3I} + 2Na + \underset{\text{Methyliodide}}{CH_3I} \longrightarrow \underset{\text{Ethane}}{CH_3\text{-}CH_3} + 2NaI$$

$$\underset{\text{Methyliodide}}{CH_3CH_2I} + 2Na + CH_3CH_2I \longrightarrow \underset{\text{Butane}}{CH_3CH_2CH_2CH_3} + CH_3CH_3 + CH_3CH_2CH_3 + 2NaI$$

$$\underset{\text{Ethyliodide}}{CH_3CH_2I} + 2Na + CH_3CH_2I \longrightarrow \underset{\text{Butane}}{CH_3CH_2CH_2CH_3} + 2NaI$$

$$\underset{\text{Methyliodide}}{CH_3I} + 2Na + \underset{\text{Ethyliodide}}{CH_3CH_2I} \longrightarrow \underset{\text{Propane}}{CH_3CH_2CH_3} + 2NaI$$

Separations of these alkanes are difficult because of their close boiling point. Hence this method is not suitable for preparing unsymmetrical alkanes.

8. **Corey-house Synthesis:** Eliminating the limitations of Wurtz synthesis and coupling of two alkyl halides are carried out by **Corey-house Synthesis.**

$$\underset{\text{Ethyl bromide}}{CH_3CH_2Br} + \underset{\text{Lithium dimethyl cuprate}}{Li(CH_3)_2Cu} \longrightarrow \underset{\text{Propane}}{CH_3CH_2CH_3} + \underset{\text{Methyl lithium}}{CH_3Li} + CuBr$$

The lithium dimethyl cuprate needed for this reaction is obtained by the following method.

$$\underset{\text{Methyl bromide}}{CH_3Br + 2Li} \longrightarrow \underset{\text{Methyllithium}}{CH_3Li} + LiBr$$

$$\underset{\text{Methyllithium}}{2CH_3Li} + CuI \longrightarrow \underset{\text{Lithium dimethyl cuprate}}{Li(CH_3)_2Cu} + LiI$$

9. **Synthesis of Long chain Alkanes:** They are prepared by coupling alkyl boranes in presence of silver nitrate containing sodium hydroxide at 300 K. The alkyl borane needed for this reaction is prepared from alkenes.

$$\underset{\text{1-Pentene}}{6CH_3(CH_2)_2CH{=}CH_2} \xrightarrow{2B_2H_6} \underset{\text{Pentyl borane}}{[2CH_3(CH_2)_3CH_2]_3B}$$

$$\xrightarrow{NaOH \mid AgNO_3} \underset{\text{Decane}}{3CH_3(CH_2)_3CH_2{-}CH_2(CH_2)_3CH_3}$$

Summary of General Methods of Preparation of Alkanes

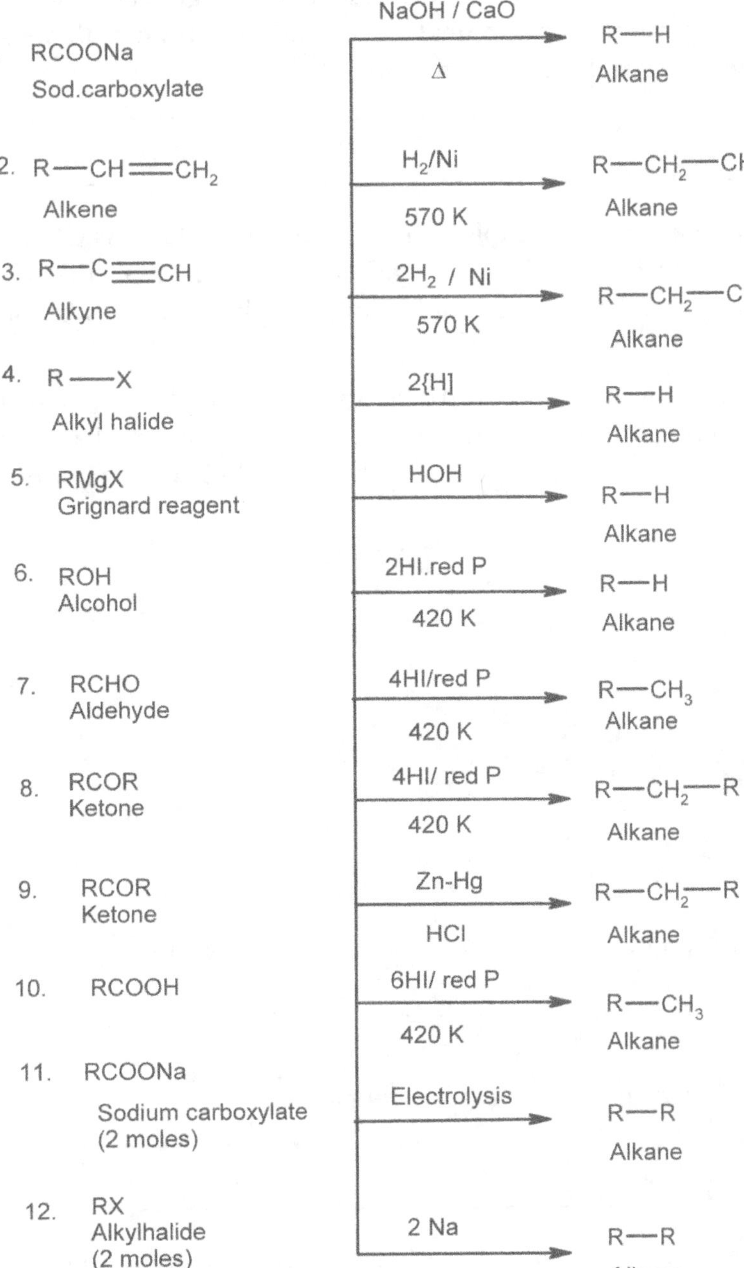

Physical Properties

First four members of alkanes (Methane, Ethane, Propane, and Butane) are colorless, odorless, gases and the members C_5 to C_{17} are colorless, odorless liquids and the further members are waxy solids. As alkanes are non-polar in nature, they are soluble in non-polar solvents like ether, benzene and insoluble in polar solvents like water. As the molecular weight increases their solubility decreases, other physical constants like density, melting point, boiling point *etc.* increase as the molecular weight increases, in the normal alkanes boiling point increases by 20 – 30 °C for each carbon added to the chain.

In case of isomeric alkanes, the normal or straight chain alkanes have higher boiling point and melting point than branched chain isomers. Branching lowers the alkane's boiling point because the intermolecular attraction is diminished due to compact nature and hence lesser energy required for converting the liquid to vapor phase.

The melting point also increased with increase in molecular weight. Alkanes with even number of carbon atoms have higher melting point than the next lower and next higher alkanes having odd number of carbon atoms, due to greater intermolecular attraction.

Structure

Methane is the first member of alkane and the molecular formula of methane is CH_4. All the four hydrogen atoms linked to carbon by covalent bond. All C-H bonds are covalent (σ) in nature. Bonding occurred between carbon and hydrogen due to the overlapping of sp^3 hybrid orbitals of carbon and s-orbital of hydrogen

The configuration of ground state is tetrahedran and in which all four hydrogen are located at four corners of the tetrahedran. The σ bond is very strong, because in the tetrahedral arrangement most effective overlapping of orbitals takes place. The bond angle in methane is 109.5° and the bond length of C-H is 1.09 A° as shown in Figure 8.1.

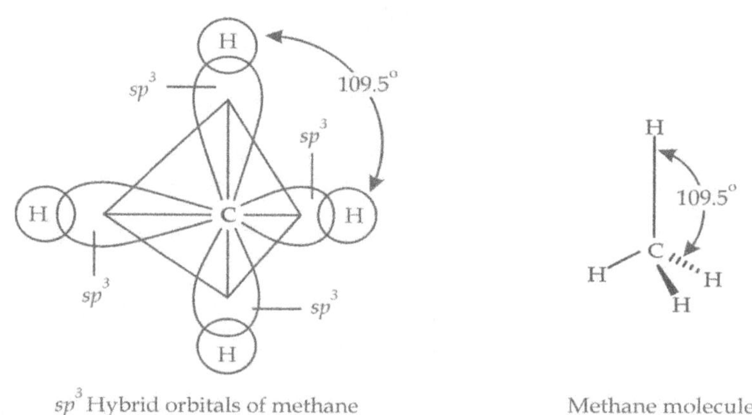

Figure 8.1 Structure of methane.

sp^3 Hybridisation in alkanes: Refer chapter No. 4 (Covalent bond).

Chemical Properties

The electronegativity of carbon (2.6) and hydrogen (2.1) do not differ considerably, hence the bonding electrons in C – H bond are shared equally and the bond is weakly non-polar. The C – C bond is completely non-polar; therefore alkanes are quite inert towards acids, alkalis and oxidizing agents *etc.* at room temperature. However alkanes undergo two types of reactions.

(I) Substitution reactions.

(II) Thermal and catalytic reactions.

These reactions take place due to the formation of **highly reactive free-radicals.**

(I) Substitution reactions:

1. Halogenation: Reaction of alkanes with halogens are described below:

(i) Chlorination: Alkanes when react with chlorine in presence of UV light, heat or catalyst (halogen carriers, Iron fillings) yields a mixture of products depending upon the amount of chlorine used.

$$\underset{\text{Methane}}{CH_4} + Cl_2 \longrightarrow \underset{\text{Methyl chloride}}{CH_3\text{-}Cl} + HCl$$

$$\underset{\text{Methyl chloride}}{CH_3\text{-}Cl} + Cl_2 \longrightarrow \underset{\text{Dichloromethane}}{CH_2Cl_2} + HCl$$

$$\underset{\text{Dichloro methane}}{CH_2Cl_2} + Cl_2 \longrightarrow \underset{\text{Trichloromethane or Chloroform}}{CHCl_3} + HCl$$

$$\underset{\text{Chloroform}}{CHCl_3} + Cl_2 \longrightarrow \underset{\text{Tetrachloromethane or Carbon tetrachloride}}{CCl_4} + HCl$$

In this reaction chloride atom replaces hydrogen atom of alkane and forms alkyl chloride, this is called ***Substitution reaction*** and the products obtained are called as ***Substitution products.*** In the above reaction, if chlorine used in excess amount, tetrachloro derivative is the main product, whereas if methane concentration is more, methyl chloride is the major product. In higher alkanes all possible isomeric monochloro derivatives are obtained and the order of substitution is **tertiary hydrogen > secondary hydrogen > primary hydrogen.** The order of reactivity of halogens is **Fluorination > Chlorination > Bromination > Iodination.**

(ii) Bromination: Bromination of alkanes is also similar but not so vigorous like chlorination.

$$\underset{\text{Methane}}{CH_4} + Br_2 \longrightarrow \underset{\text{Methyl bromide}}{CH_3Br + HBr}$$

(iii) Iodination: Iodination of alkanes is a reversible reaction.

$$\underset{\text{Methane}}{CH_4} + I_2 \rightleftharpoons \underset{\text{Methyl iodide}}{CH_3I + HI}$$

The hydroiodic acid formed as a byproduct reduces alkyl halide (because it is a strong reducing agent) to alkane. Hence iodination must be carried out in presence of oxidising agents like nitric acid which removes the HI formed during reaction.

$$\underset{\text{Hydroiodic acid}}{5HI} + HIO_3 \longrightarrow \underset{\text{Iodine}}{3I_2} + 3H_2O$$

(iv) Fluorination: Direct fluorination of alkanes is explosive; hence it can be carried out by diluting fluorine with nitrogen.

$$\underset{\text{Methane}}{CH_4} + F_2 \xrightarrow{N_2} \underset{\text{Methyl fluoride}}{CH_3F} + HF$$

Mechanism of chlorination: Chlorination of methane in UV light or heat takes place through free radical formation. The steps involved are described below:

(a) Chain initiation: Chlorine molecule yields chlorine free radical by homolytic fission.

$$\underset{\text{Chlorine}}{Cl{:}Cl} \xrightarrow[\Delta]{\text{UV light}} \underset{\text{Chlorine free radical}}{Cl^{\cdot} + Cl^{\cdot}}$$

(b) Chain propagation:

(a) Chlorine free radical attacks methane and yield methyl free radical.

$$Cl^{\cdot} + \underset{\text{Methane}}{\mathbf{H}{:}CH_3} \longrightarrow \underset{\text{Methyl free radical}}{CH_3^{\cdot}} + H{:}Cl$$

(b) Methyl free radical further attacks chlorine molecule to form chlorine free radical and methyl chloride.

$$CH_3^{\cdot} + Cl : Cl \longrightarrow \underset{\text{Methyl chloride}}{CH_3{:}Cl} + Cl^{\cdot}$$

Steps (a) and (b) are repeated and free radicals are propagated.

(c) Chain termination: When two free radicals combine, a product is formed and the chain reaction comes to an end.

$$Cl^{\cdot} + CH_3^{\cdot} \longrightarrow \underset{\text{Methyl chloride}}{CH_3{-}Cl}$$

$$Cl^{\cdot} + Cl^{\cdot} \longrightarrow \underset{\text{Chlorine}}{Cl{-}Cl}$$

$$CH_3^{\cdot} + CH_3^{\cdot} \longrightarrow \underset{\text{Ethane}}{CH_3{-}CH_3}$$

A chlorine free radical can also attack methyl chloride and forms chloro methylene free radical and further it reacts with chlorine molecule to yield dichloromethane and chlorine free radical.

$$Cl^{\cdot} + H : CH_2{-}Cl \longrightarrow Cl : H + \underset{\text{Chloromethylene radical}}{{}^{\cdot}CH_2{-}Cl}$$

$$Cl{-}CH_2^{\cdot} + Cl : Cl \longrightarrow \underset{\text{Dichloromethane}}{Cl{-}CH_2{-}Cl} + Cl^{\cdot}$$

In the same manner, trichloromethane (chloroform) and tetra chloromethane (carbon tetrachloride) are obtained upon further chain reaction.

Free radical initiator: The compounds which can initiate a free radical substitution reaction is called as free radical initiator or free radical catalysts example UV light. Any compound that easily decomposes into free radicals acts as an initiator. For example, Benzoyl peroxides, Peroxybenzoic acid and Azobisisobutyro nitrile (AIBN) are acts as free radical initiators.

Free radical inhibitors: Compounds which inhibit the free radical reactions are known as free radical inhibitors or free radical trap example iodine.

Order of abstraction of hydrogen: The free radical substitution reaction includes abstraction of proton. The abstraction of proton takes place on the basis of nature of hydrogen atom. The hydrogen atoms present in the organic compounds are classified into primary hydrogen ($-CH_3$), secondary hydrogen (attached to secondary carbon), tertiary hydrogen (attached to tertiary carbon), allylic and benzylic hydrogens (carbon adjacent to a double bond and carbon adjacent to an aromatic ring respectively). All these different hydrogen atoms are not abstracted easily and the abstraction of hydrogen occurs selectively. Let us consider the following example, chlorination of propane, to give two substituted products *i.e.,* 1-chloropropane and 2-chloropropane.

$$\underset{\text{Propane}}{CH_3CH_2CH_3} + Cl_2 \xrightarrow{h\upsilon} \underset{\substack{\text{1-Chloropropane} \\ (45\ \%)}}{CH_3CH_2CH_2Cl} + \underset{\substack{\text{2-Chloropropane} \\ (55\ \%)}}{CH_3\overset{\overset{\large Cl}{|}}{C}HCH_3}$$

1. Propane molecule contains six primary hydrogen atoms and two secondary hydrogen atoms. The ratio of primary and secondary hydrogen atom is 6/2 = 3/1.
2. So, if all hydrogen atoms undergo abstraction then 1-chloropropane is formed in more amount than 2-chloropropane (3:1 ratio). But, here 2-chloropropane predominates which indicates that secondary hydrogen gets abstracted faster than primary hydrogen.

Reactivity - selectivity principle: By considering the reaction of bromine and chlorine with alkanes, bromine is found to be more reactive and more selective because a chlorine free radical forms primary, secondary and tertiary alkyl radical in equal ease, but bromine radical forms only tertiary alkyl radical (easily formed one) and also bromine is relatively less reactive. In contrast, chlorine radical is highly reactive, but less selective. Hence the reactivity-selectivity theory states that, "greater the reactivity of the species, the less selective it will be".

INTERESTING FACT

Food preservatives: Free radical inhibitors that are present in the food are called as preservatives or antioxidants. They preserve the food by inhibiting unwanted radical reactions. Vitamin E is a naturally occurring antioxidant or preservative present in vegetable oil. BHA (butylated hydroxy anisole) and BHT (butylated hydroxy toluene) are synthetic preservatives that are added to many of the packaged food item. Vitamin C and Vitamin E are two radical inhibitors present in our biological systems.

OH, $C(CH_3)_3$, OCH_3 — Butylated hydroxy anisole (BHA)

OH, $C(CH_3)_3$, OCH_3 — Butylated hydroxy anisole (BHA)

OH, $(H_3C)_3C$, $C(CH_3)_3$, OCH_3 — Butylated hydroxy toluene (BHT)

2. **Nitration:** Substitution of hydrogen atom of alkanes with nitro ($-NO_2$) group is called as nitration. The reaction between alkanes and nitric acid takes place in vapour phase at 400 – 500 °C yields nitro alkanes. As the reaction takes place at high temperature, C – C fission takes place which results in the formation of mixture of nitroalkanes.

$$\underset{\text{Methane}}{CH_4} + HNO_3 \xrightarrow[\text{Pressure}]{400\text{-}500^\circ C} \underset{\text{Nitromethane}}{CH_3NO_2} + H_2O$$

$$\underset{\text{Propane}}{CH_3CH_2CH_3} + HNO_3 \xrightarrow[\text{Pressure}]{400\text{-}500^\circ C} \underset{\text{1-Nitropropane}}{CH_3CH_2CH_2NO_2} + \underset{\text{2-Nitropropane}}{H_3C-\underset{\underset{\large NO_2}{|}}{C}H-CH_3} + \underset{\text{Nitroethane}}{CH_3CH_2NO_2} + \underset{\text{Nitromethane}}{CH_3NO_2}$$

Mechanism:

Step 1: Chain initiation: Thermal splitting of propane yields propyl free radical.

$$CH_3\text{-}CH_2\text{-}CH_2{\cdot\cdot}H \xrightarrow{\Delta} CH_3CH_2\dot{C}H_2 + \dot{H}$$

Propane — Propyl free radical

Step 2: Chain propagation: Propyl free radical attacks nitric acid to yields nitropropane.

$$CH_3CH_2\dot{C}H_2 + OH{\cdot\cdot}NO_2 \longrightarrow CH_3CH_2CH_2NO_2 + \dot{O}H$$

Nitropropane — Hydroxy free radical

$$CH_3CH_2CH_2\text{—}H + \cdot OH \longrightarrow CH_3CH_2\dot{C}H_2 + H_2O$$

Propane — Propyl free radical

Step 3: Chain termination:

$$CH_3CH_2\dot{C}H_2 + CH_3CH_2\dot{C}H_2 \longrightarrow CH_3(CH_2)_4CH_3$$

Propyl free radical — Hexane

3. **Sulphonation:** Substitution of hydrogen atom of alkanes with $-SO_3H$ group is called as sulphonation. The reaction does not take place at room temperature and lower alkanes up to pentane cannot be sulphonated. From hexane, the alkanes react with oleum [Fuming sulphuric acid and sulphur trioxide] and form alkane sulphonic acid.

$$R{-}H + OH\text{-}SO_3H \longrightarrow R{-}SO_3H + H_2O$$

Alkane (Fuming Sulphuric acid) Alkane sulphonic acid

4. **Oxidation (Combustion reactions):**
   (i) **Combustion:** Oxidation of alkanes in presence of excess of oxygen forms carbon dioxide and water. If it is oxidised with insufficient oxygen it forms carbon monoxide and carbon.

$$\underset{\text{Methane}}{CH_4} \xrightarrow{2O_2} CO_2 + 2H_2O$$

$$\underset{\text{Methane}}{2CH_4} \xrightarrow{3O_2} 2CO + 4H_2O$$

$$\underset{\text{Methane}}{CH_4} \xrightarrow{O_2} C + 2H_2O$$

   (ii) **Reaction with oxidising agents**: Tertiary alkanes are oxidised to tertiary alcohol by alkaline potassium permanganate.

$$\underset{\textit{iso}\text{-Butane}}{(CH_3)_3CH} + [O] \xrightarrow[\text{KMno}_4]{\text{Alkaline}} \underset{\textit{tert}\text{- Butyl acohol}}{(CH_3)_3COH}$$

   (iii) **Catalytic oxidation**: Oxidation with oxygen in copper tubes yield alcohol.

$$\underset{\text{Methane}}{2CH_4} + O_2 \xrightarrow[\text{100 atm, 410 K}]{\text{Cu tube}} \underset{\text{Methanol}}{2CH_3OH}$$

Oxidation using molybdenum or manganese acetate yields aldehydes and carboxylic acids respectively.

$$\underset{\text{Methane}}{CH_4} + O_2 \xrightarrow{\text{Molybdenum}} \underset{\text{Formaldehyde}}{HCHO} + H_2O$$

$$\underset{\text{Higher alkane}}{2R{-}CH_3} + 3O_2 \xrightarrow{(CH_3COO)_2Mn} \underset{\text{Carboxylic acid}}{2RCOOH} + 2H_2O$$

(iv) **Slow combustion:** At high pressure and low temperature, alkanes undergo gradual oxidation with oxygen.

$$\underset{\text{Methane}}{CH_4} \xrightarrow{(O)} \underset{\text{Methanol}}{CH_3OH} \xrightarrow{(O)} \underset{\text{Methanediol}}{CH_2(OH)_2} \xrightarrow{-H_2O} \underset{\text{Formaldehyde}}{HCHO} \xrightarrow{(O)} \underset{\text{Formic acid}}{HCOOH}$$

$$HCOOH \xrightarrow{(O)} \underset{\text{Carbon dioxide}}{CO_2} + H_2O$$

5. **Thermal Decomposition or Pyrolysis or Cracking:** Pyrolysis is a Greek word means breaking of a compound by heat (Pyro-fire, lysis-breaking). The phenomenon of pyrolysis of alkanes is called cracking. Upon heating alkanes at high temperature (500 °C - 600 °C) in the absence of air, pyrolysis or thermal decomposition occurs and large alkane molecules are broken into smaller alkanes, alkenes and hydrogen with the cleavage of C-C bond and C-H bond in alkanes.

Mechanism: It follows free radical mechanism.

$$\underset{\text{Propene}}{CH_3{-}CH{=}CH_2} + H_2 \xleftarrow[\text{C-H linkage}]{\text{Breaking of}} \underset{\text{Propane}}{CH_3CH_2CH_3} \xrightarrow[\text{C-C linkage}]{\text{Breaking of}} \underset{\text{Methane}}{CH_4} + \underset{\text{Ethylene}}{CH_2{=}CH_2}$$

Mechanism:

$$\underset{\text{Propane}}{CH_3CH_2CH_2{-}H} \longrightarrow \underset{\text{Propyl free radical}}{CH_3CH_2\dot{C}H_2} + H^{\bullet} \longrightarrow \underset{\text{Propene}}{CH_3{-}CH{=}CH_2} + H_2$$

$$CH_3CH_2CH_2{-}CH_3 \longrightarrow \underset{\text{Ethyl free radical}}{CH_3\dot{C}H_2} + \underset{\text{Methyl free radical}}{\dot{C}H_3} \xrightarrow{\text{Dispropotionation}} \underset{\text{Ethylene}}{CH_2{=}CH_2} + CH_4$$

Importance of cracking: It is mainly used in petroleum industry because it forms smaller and volatile hydrocarbon from higher less volatile hydrocarbon.

6. **Aromatisation:** Alkanes containing six or more carbons when heated under pressure in presence of catalyst yields cyclised and dehydrogenated product called aromatic compounds and the process of formation of aromatic compounds in this reaction is called aromatisation. It is a good method for converting aliphatic compound into aromatic.

$$\underset{\text{n-Hexane}}{CH_3CH_2CH_2CH_2CH_2CH_3} \xrightarrow[\text{600 °C}]{Cr_2O_3 / Al_2O_3} \underset{\text{Benzene}}{C_6H_6} + 4H_2$$

Summary of Chemical Reactions

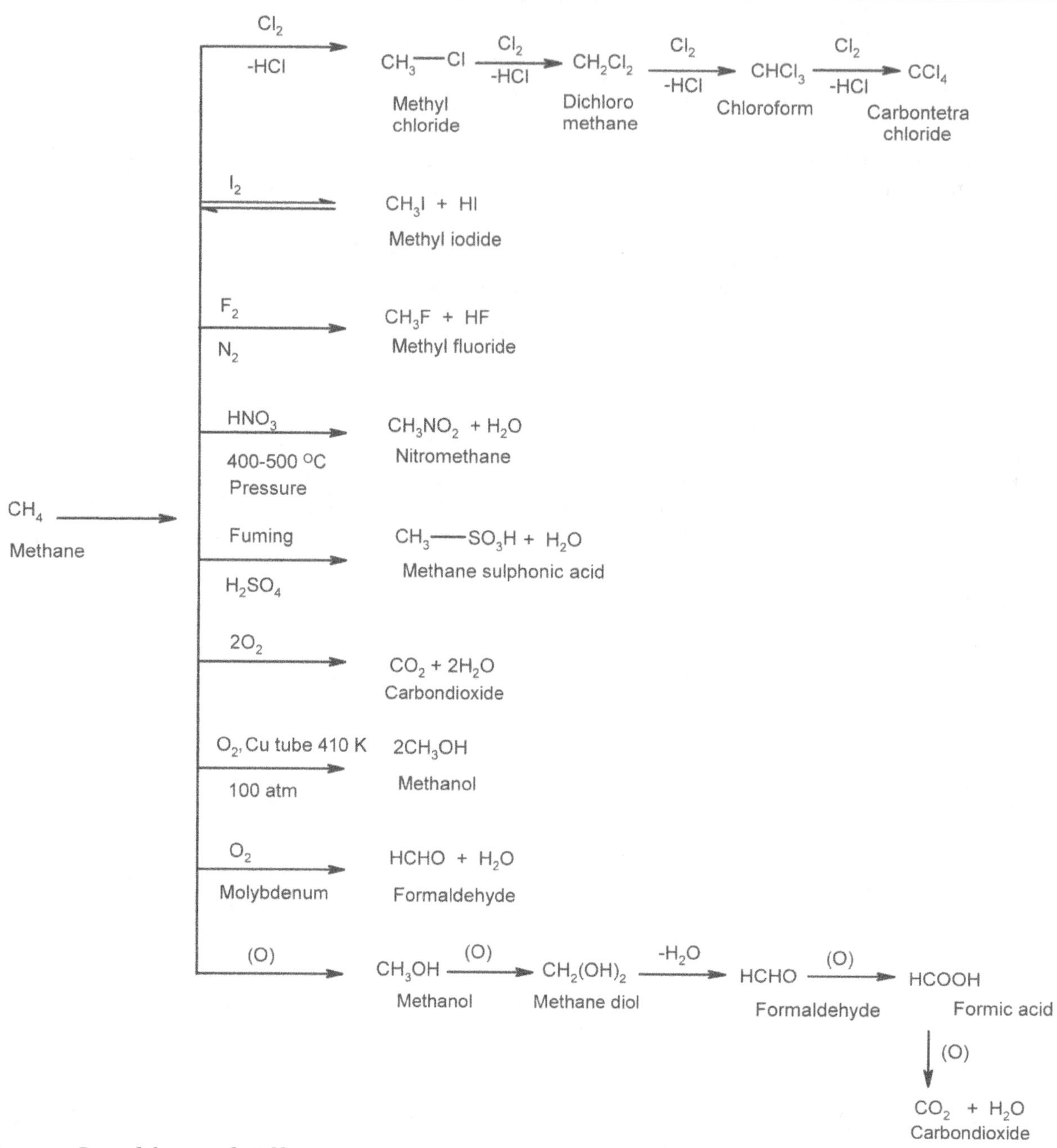

Pyrolysis or Cracking of Alkanes

$$\underset{\text{Propane}}{CH_3CH_2CH_3} \xrightarrow[\text{Breaking C-H bond}]{H_2} \underset{\text{Propene}}{CH_3-CH=CH_2} \xrightarrow[\text{Breaking C-C bond}]{} \underset{\text{Methane}}{CH_4} + \underset{\text{Ethylene}}{CH_2=CH_2}$$

$$\underset{\text{n-Hexane}}{C_6H_{14}} \xrightarrow[Al_2O_3,\ 600^\circ C]{Cr_2O_3} \underset{\text{Benzene}}{C_6H_6} + 4H_2$$

Uses of paraffins:

1. Liquid paraffin is commonly used to treat eczema, dry skin and constipation.
2. Paraffins are generally used as fuel for rockets and jet engines.
3. Liquid paraffin is used as a lubricant in machinery.
4. Paraffins are also used as a coolant for electrical systems, as a hydraulic fluid and as a solvent for greases and insecticides.
5. Paraffin wax is used as an adhesive and water proofing agent.

Probable Questions

1. Define alkanes and mention its importance?
2. Write a detailed note on nomenclature of alkanes.
3. Write a short note on isomerism exhibited by alkanes with suitable examples.
4. Enlist various general methods of preparation of alkanes and explain any five in detail.
5. Explain Sabatier-Senderens reactions & Clemmensen reduction.
6. How are alkanes synthesized by Kolbes & Wurtz method of synthesis?
7. What is Corey-house synthesis?
8. Describe the general structure & physical properties of alkanes.
9. Explain the various chemical properties of alkanes.
10. Write a note on substitution reactions of alkanes.
11. Write short notes on thermal and catalytic reactions of alkanes.
12. Write the halogenation reactions of alkanes with suitable example and explain its reaction mechanism.
13. Define free radical and explain free radical reactions.
14. Write a note on combustion reactions in alkanes.
15. How alkanes undergo sulphonation & nitration reactions?
16. Define pyrolysis and describes pyrolysis reactions with the mechanism.
17. Write a note on conformations of alkanes.
18. Write the uses of paraffins.
19. Define aromatization reactions.
20. Write a note on cracking.
21. How alkanes are synthesized from sodium carboxylate? Explain with mechanism.
22. Write the reactions involved in the synthesis of alkanes from alkyl halide.
23. How alkanes are synthesized from unsaturated compounds?
24. How long chain alkanes are synthesized?
25. Explain the term saturated hydrocarbons.

9

Alkenes or Olefins

Introduction

The unsaturated hydrocarbons which contain at least one carbon-carbon double bond [C=C] are known as alkenes. They are also named olefins, because of the fact that lower members form oily products upon reaction with halogens and halogen acids. They have the general formula C_nH_{2n} where, n = 1, 2, 3.... In contrast to alkanes (general formula C_nH_{2n+2}) they have two hydrogens atoms less and hence the tetravalency of carbon is satisfied by presence of carbon-carbon double bond [C=C].

First member of alkenes is ethylene. Alkenes occur free in nature but large quantities of alkenes are obtained during cracking of petroleum. The carbon-carbon double bond between adjacent carbon atoms in alkenes is made up of one σ bond and one π bond. Because of the presence of more exposed π electrons, alkenes are more reactive than alkanes which have only σ bonds.

Importance of Alkenes

One of the major commercial chemical used in the synthesis of organic compounds in alkene is ethylene. Polyethylene, a famous plastic is made up of ethylene monomers. In medicinal chemistry, alkenyl groups attached with various functional groups elicit different kind of pharmacological activities based on the core structure to which it is attached. Some of the drugs which possess alkenyl group and their therapeutic uses in the treatment of various diseases are shown in Table 9.1.

Table 9.1 Drugs containing alkenyl groups and their therapeutic uses.

Category	Drug	Structure
General anaesthetic	Trichloro ethylene	Cl—C(Cl)=C(H)—Cl
Adrenergic blocker	Oxprenalol	$H_2C=HC-H_2CO-C_6H_4-OCH_2CH(OH)-CH_2NHCH(CH_3)_2$

Table 9.1 *Contd...*

Category	Drug	Structure
Antiarrhythmic agent	Quinidine	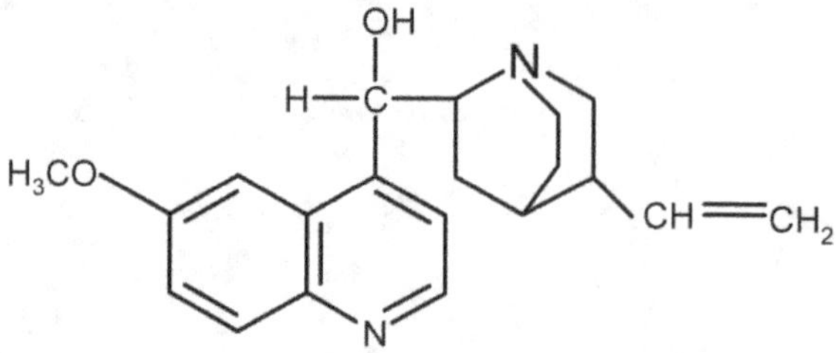
Antihelminthic agent	Pyrantel Pamoate	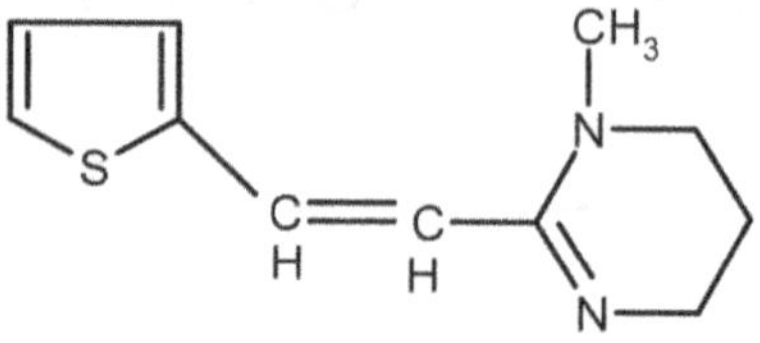
Anticancer agent	Tamoxifen	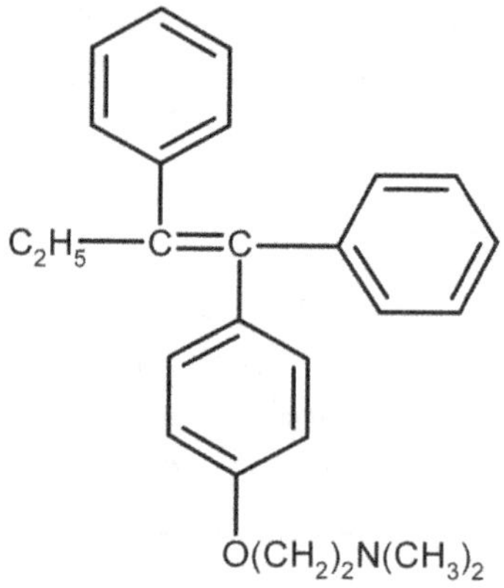
Antihyperlipidemic	Fluvastatin	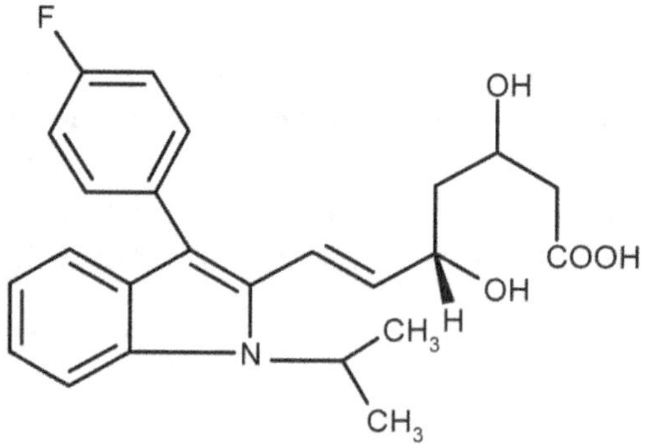
Antihypertensive agent	Ethacrynic acid	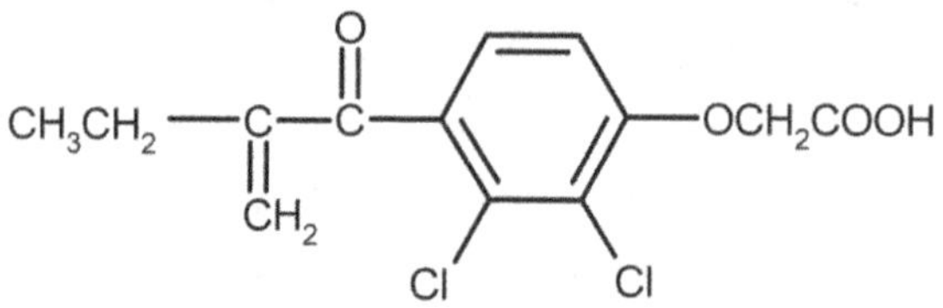
Antifungal agent	Terbinafine	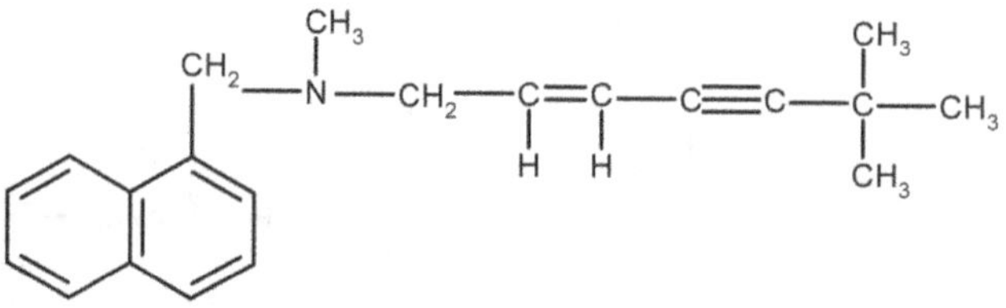

Nomenclature

1. **Common Name System:** This method is used to name only first four members by replacing the suffix ***-ane*** from alkane with **- *ylene***. The symbols α and β used to indicate the position of double bonds. α indicates the double bond in first carbon and β in second carbon. Some scientists still use another Greek letter (Δ) to indicate the position of double bond. The number placed at the right hand corner of the symbol (Δ) indicates the position of double bond

$CH_2 = CH_2$	Ethylene
$CH_2{=}CH{-}CH_3$	Propylene
$CH_3{-}C(CH_3){=}CH_2$	*iso*-Butylene
$CH_2{=}CH{-}CH_2{-}CH_3$	α -Butylene
$CH_3{-}CH{=}CH{-}CH_3$	β-Butylene
$CH_3{-}CH_2{-}CH{=}CH{-}CH_2{-}CH_3$	Δ^3- Hexene

2. **Another Common Method:** In this method alkenes are named as derivatives of ethylene. Dialkyl derivatives exhibit isomerism. The isomers are named as symmetrical or asymmetrical. If the alkyl groups attached to two different carbon atoms, then it is called as symmetrical (sym or s). If they are attached with same carbon atom they are called as asymmetrical or unsymmetrical (*unsym* or *as*).

$CH_3{-}CH{=}CH{-}CH_3$	$(CH_3)_2{-}C{=}CH_2$
s-Dimethyl ethylene	*as*-Dimethyl ethylene

3. **IUPAC System:** Named by adopting the following steps.
   (i) The name of the hydrocarbon is based on the parent alkane having longest chain in which one double bond is present then the suffix - ***ane*** of alkane is replaced with ***-ene***.

   $CH_3 - CH_2 - CH = CH_2$ Butene

   (ii) The hydrocarbon chain is numbered from the end near the double bond and the position of double bond is indicated by mentioning the number of that carbon atom.

   $\overset{5}{CH_3} - \overset{4}{CH_2} - \overset{3}{CH} = \overset{2}{CH} - \overset{1}{CH_3}$ 2- Pentene

   (iii) Name of the parent hydrocarbon is chosen and the position of double bond is placed before that. Then the location and name of the substituent are prefixed.

   $\underset{1}{CH_3} - \underset{2}{CH} = \underset{3}{CH} - \underset{4}{CH}(CH_3) - \underset{5}{CH_3}$

   4-Methyl-2-pentene

   $\underset{1}{CH_3} - \underset{2}{CH_2}(CH_3) - \underset{3}{CH} = \underset{4}{CH} - \underset{5}{CH}(CH_3) - \underset{6}{CH_3}$

   2,5-Dimethyl-3-hexene

(iv) When more than one double bond is present then the suffix **-ene** of alkene is replaced with **-diene** or **- triene** for two or three bonds respectively along with the position of the double bonds.

$$\underset{1}{CH_3} - \underset{2}{CH} = \underset{3}{CH} - \underset{4}{CH} = \underset{5}{CH} - \underset{6}{CH_3}$$

2,4-Hexadiene

$$\underset{1}{CH_2} = \underset{2}{C} = \underset{3}{CH} - \underset{4}{CH} = \underset{5}{CH} - \underset{6}{CH_3}$$

1,2,4-Hexatriene

Table 9.2 Common names and IUPAC names of various alkenes.

Molecular formula	Respective alkane	Common name	IUPAC name
C_nH_{2n}	C_nH_{2n+2}	Alkylene	Alkene
$:CH_2$	CH_4, Methane	Methylene	Methene
$CH_2{=}CH_2$	C_2H_6, Ethane	Ethylene	Ethene
$CH_3{-}CH{=}CH_2$	C_3H_8, Propane	Propylene	Propene
$CH_3{-}CH_2{-}CH{=}CH_2$	C_4H_{10}, Butane	1-Butylene	1-Butene
$CH_3{-}CH{=}CH{-}CH_3$	C_4H_{10}, Butane	2-Butylene	2-Butene

Alkenyl groups: The monovalent groups formed by removing hydrogen from alkene are called as alkenyl group. For example

$CH_2 = CH-$ Ethenyl $\quad CH_2 = CH - CH_2-$ 2-Propenyl

Isomerism

Isomerism in alkenes is due to change of position of double bond and the attached groups. First two members do not exhibit isomerism. Butene exhibits **3** isomeric forms *i.e.,* two positional isomers *i.e.* (I, II (due to change in position of double bond) and one chain isomer III (change of position of groups).

$CH_3 - CH_2 - CH = CH_2$ 1-Butene (I) $\quad CH_3 - CH = CH - CH_3$ 2-Butene (II)

$$\underset{1}{CH_3} - \underset{2}{\overset{CH_3}{\overset{|}{C}}} = \underset{3}{CH_2}$$

2-Methylpropene (III)

Geometrical isomerism: Refer Chapter-28.

General Methods of Preparation

$$-\underset{X}{\underset{|}{\overset{|}{C}}} - \underset{Y}{\underset{|}{\overset{|}{C}}} - \longrightarrow -\overset{|}{C} = \overset{|}{C} - \; + \; XY$$

1. **Dehydration of alcohols:** Alcohols upon reaction with sulphuric acid lose a molecule of water and form an alkene.

$$\underset{\text{Alcohol}}{R-CH_2-CH_2-OH} \xrightarrow[180^{o}C]{H_2SO_4} \underset{\text{Alkene}}{R-CH=CH_2} + H_2O$$

$$\underset{\text{Ethanol}}{CH_3-CH_2-OH} \xrightarrow[180^{o}C]{H_2SO_4} \underset{\text{Ehylene}}{CH_2=CH_2} + H_2O$$

In this reaction, sulphuric acid acts as a dehydrating agent. Other dehydrating agents can be used are glacial phosphoric acid, phosphorus pentoxide and alumina. The order of dehydration reaction of alcohols is 3° alcohol > 2° alcohol > 1° alcohol.

Unsymmetrical alcohols yield mixture of alkenes upon dehydration as shown below.

$$\underset{\text{2-Butanol}}{CH_3-CH_2-\overset{OH}{\overset{|}{CH}}-CH_3} \xrightarrow[180\ ^{o}C]{H_2SO_4} \underset{\text{1-Butene (20\%)}}{CH_3-CH_2-CH=CH_2}$$

$$\underset{\text{2-Butanol}}{CH_3-CH_2-\overset{OH}{\overset{|}{CH}}-CH_3} \xrightarrow[H_2SO_4]{180\ ^{o}C} \underset{\text{2-Butene (80\%)}}{CH_3-CH=CH-CH_3} + H_2O$$

According to Saytzeff rule "when two alkenes are possible to obtain in the reaction, the alkene with more branched one predominates". Hence in the above example 2-Butene is the major product.

2. **Dehydrohalogenation of alkyl halides:** Reaction of alkyl halide with alcoholic potash yields alkene with the elimination of hydrogen halide.

$$\underset{\text{Alkyl halide}}{R-CH_2-CH_2-X} \xrightarrow{\text{Alcoholic KOH}} \underset{\text{Alkene}}{R-CH=CH_2} + HX$$

$$\underset{\text{Bromopropane}}{CH_3-CH_2-CH_2-Br} \xrightarrow{\text{Alcoholic KOH}} \underset{\text{Propylene}}{CH_3-CH=CH_2} + HBr$$

The order of reactivity of alkyl halides are 3° > 2° > 1°. In unsymmetrical alkyl halides, the reaction takes place as per Saytzeff's rule.

$$\underset{\text{2-Bromobutane}}{CH_3-\underset{Br}{\underset{|}{CH}}-CH_2-CH_3} \xrightarrow{\text{Alcoholic KOH}} \underset{\text{2-Butene (80 \%)}}{CH_3-CH=CH-CH_3} + HBr$$

$$\underset{\text{2-Bromobutane}}{CH_3-\underset{Br}{\underset{|}{CH}}-CH_2-CH_3} \xrightarrow{\text{Alcoholic KOH}} \underset{\text{1-Butene (20 \%)}}{CH_2=CH-CH_2-CH_3} + HBr$$

When unsymmetrical alkyl derivatives are used, on account of greater predictability of the position of double bond formed, dehydrohalogenation method is better than dehydration of alcohols.

Note: E_1 and E_2 reactions, kinetics, order of reactivity of alkyl halides, rearrangement of carbocations, saytizeff's orientation and evidences, E_1 versus E_2 reactions, factors affecting E_1 and E_2 reactions - Refer Chapter No: 14.

3. **Dehalogenation of vicinal dihalides:** When vicinal dihalides (two halogen atoms on adjacent carbon atom) are treated with zinc dust in the presence of alcohols, alkenes are obtained with the removal of halogen molecule as zinc halide.

$$\underset{\textit{vic}\text{-Dihalide}}{R-\underset{X}{\underset{|}{CH}}-\underset{X}{\underset{|}{CH_2}}} \xrightarrow[\text{Alcohol}\ \Delta]{\text{Zn dust}} \underset{\text{Alkene}}{R-CH=CH_2} + \underset{\text{Zinc halide}}{ZnX_2}$$

$$\underset{\text{1,2-Dibromopropane}}{CH_3-\underset{Br}{\underset{|}{CH}}-\underset{Br}{\underset{|}{CH_2}}} \xrightarrow[\text{Alcohol}\ \Delta]{\text{Zn dust}} \underset{\text{Propylene}}{CH_3-CH=CH_2} + \underset{\text{Zinc bromide}}{ZnBr_2}$$

4. **Hydrogenation of alkynes:** Alkynes upon controlled hydrogenation by using Lindlar's catalyst (Pd poisoned with $CaCO_3$ + quinoline) yields alkenes.

$$\underset{\text{Alkynes}}{C_nH_{2n-2}} + H_2 \xrightarrow[\text{Quinoline}]{Pd/CaCO_3} \underset{\text{Alkenes}}{C_nH_{2n}}$$

$$\underset{\text{Alkyne}}{R-C\equiv CH} + H_2 \xrightarrow[\text{Quinoline}]{Pd/CaCO_3} \underset{\text{Alkene}}{R-CH=CH_2}$$

$$\underset{\text{Propyne}}{CH_3-C\equiv CH} \xrightarrow[\text{Quinoline}]{Pd/CaCO_3} \underset{\text{Propylene}}{CH_3-CH=CH}$$

5. **By heating quaternary ammonium hydroxide:** Heating quaternary ammonium hydroxide yields alkene.

$$\underset{\text{Tetraethyl ammonium hydroxide}}{(C_2H_5)_4N^+OH^-} \xrightarrow{\Delta} \underset{\text{Ethylene}}{CH_2=CH_2} + \underset{\text{Triethylamine}}{(C_2H_5)_3N} + H_2O$$

When the alkyl group possesses more than two carbon atoms in the quaternary ammonium hydroxide, then the alkene with less substituted one predominates This is called **Hoffmann rule of elimination.** (In contrast to Saytzeff rule).

$$\underset{\text{N-(2-Butyl)-N,N,N-trimethyl ammonium hydroxide}}{CH_3-CH_2-\underset{N^+(CH_3)_3\ OH^-}{\underset{|}{CH}}-CH_3} \xrightarrow{\Delta} \underset{\text{2-Butene (5 \%)}}{CH_3-CH=CH-CH_3} + (CH_3)_3N + H_2O$$

$$\xrightarrow{\Delta} \underset{\text{1-Butene (95 \%)}}{CH_3-CH_2-CH=CH_2} + \underset{\text{Trimethylamine}}{(CH_3)_3N} + H_2O$$

6. **By cracking of alkanes:** Cracking or pyrolysis of alkanes at 500 – 800 °C in the absence of air yields lower alkenes.

$$\underset{\text{Alkane}}{R-CH_2-CH_3} \xrightarrow{500\text{-}800\,^{\circ}C} \underset{\text{Alkene}}{CH_2=CH_2} + \underset{\text{Lower alkane}}{R-H}$$

$$\underset{\text{Propane}}{CH_3-CH_2-CH_3} \xrightarrow{500\text{-}800\,^{\circ}C} \underset{\text{Ethylene}}{CH_2=CH_2} + \underset{\text{Methane}}{CH_4}$$

Summary of Methods of Preparation

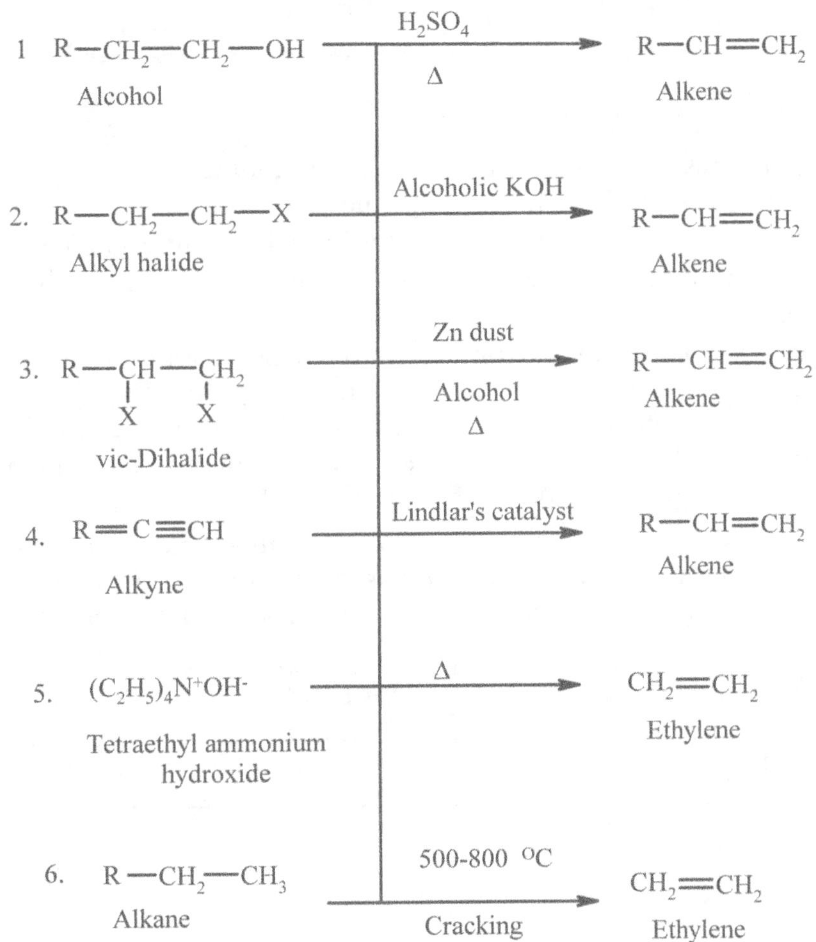

Structure

Ethylene is the first member of alkene and the molecular formula is C_2H_4. Its structure is represented in Fig. 9.1. The two carbon atoms of ethylene are bonded by σ and π bond. The σ bond is formed by overlapping of one sp^2 hybrid orbital of each carbon and the π bond is formed by the parallel overlapping of 2*p* orbital of each carbon. The four hydrogen atoms linked to carbon by σ bond, formed by overlapping of sp^2orbitals of carbons and *s* orbital of hydrogen.

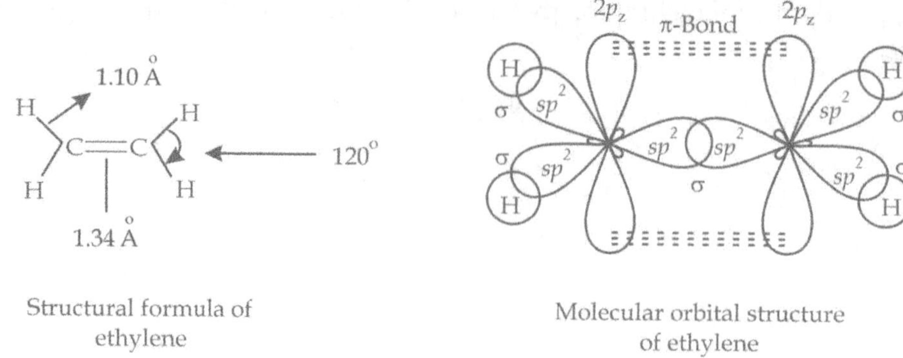

Figure 9.1 Structure of ethylene.

sp^2 hybridisation in alkenes: Refer Chapter No: 4, Covalent bond.

Physical Properties

First three members (C_2-C_4) are gases at room temperature, next members (C_5-C_{17}) are liquids, more than C_{18} are solids. All are colorless, odorless but ethylene has pleasant odor.

1. All alkenes and its homologous are insoluble in water but soluble in organic solvents.
2. When molecular weight of homologues series of alkenes increases, their boiling point (B.P), melting point (M.P) also increases.
3. Alkenes exhibits *cis-trans* isomerism due to the restricted rotation about double bond. Both the isomers possess different physical properties like boiling point, melting point, density. The boiling point of *cis*-2-butene is 4 °C but *trans*-2-butene is 1 °C. In 2-butene the alkyl group attached to the any one of the doubly bonded carbon atom donates electrons to sp^2 carbon which induces polarity in the molecule (sp^3 carbon gets positive charge, sp^2 carbon gets partial negative charge).

In a *cis* isomer, the two C_{sp2}-C_{sp3} bond dipoles reinforce each other and produces a small dipole thus shows polarity, but in *trans* isomer the polarity created gets cancel and the net dipole moment is zero . So *cis* isomer is slightly polar than *trans* isomer which leads to its higher boiling point and melting point than *trans* isomer.

sp^2 Hybridisation in alkenes: Refer Chapter No: 4, Covalent bond.

Chemical Properties of Alkenes

Small dipole *cis*-isomer

No dipole *trans*-isomer

Alkenes undergo different type of chemical reactions as described below.

(I) Electrophilic addition reaction:

Mechanism of electrophilic addition reaction: Alkenes contain a strong σ bond and weak π bond, the π bond made up of π electrons; the π electrons of alkenes are less firmly bound to the carbon nuclei and are more exposed. The carbon-carbon double bond is electron rich and it can supply electrons to electrophilic reagent (electrophile, which is electron deficient).

In addition reactions of alkenes, the π bond is broken and it is replaced by two new σ bonds.

Addition reaction

Alkene (π bond) + E – Nu → Addition product (σ bonds)

Addition to carbon - carbon double bond involves breaking of π bond. The breaking takes place by homolytic or heterolytic fission.

Charged species ← Heterolytic fission — Ethylene — Homolytic fission → Radicals

In polar solvents addition proceeds by ionic mechanism while in the presence of light or free radicals addition takes place by free radical mechanism.

Mechanism: Electrophilic addition takes place in two steps.

Step 1: The reagent (E-Nu) ionized in presence of alkenes and give positive ion (Electrophile) and a negative ion (Nucleophile). Thus, the formed electrophiles initiate the addition reaction, this step is slow and rate determining step, hence these reactions are called as electrophilic addition reactions.

$$E - Nu \longrightarrow \underset{\text{Electrophile}}{E^+} + \underset{\text{Nucleophile}}{:Nu^-}$$

The electrophile (E^+) attacks the carbon-carbon double bond and forms a new covalent bond with one carbon and the other carbon becomes positively charged with the formation of carbonium ion.

Alkene + E^+ → Carbonium ion

Step 2: The nucleophile attacks the carbonium ion and form addition product. The nucleophile attacks the carbonium ion (carbocation) from the side opposite to the C-E bond; hence it is called ***trans* addition.**

Carbonium ion + $:Nu^-$ → Addition product

Common Electrophiles: H^+, Br^+, Cl^+

Common Nucleophiles: Br^-, Cl^-, OH^- *etc*

Some of the common electrophilic addition reactions of alkenes are described below:

1. **Addition of hydrogen:** Alkenes combine with hydrogen in presence of catalyst (Pd supported Ni [**Sabatier- Senderens reduction**] and Raney Nickel and yields alkanes.

$$R-CH=CH_2 + H_2 \xrightarrow[\Delta]{Ni} R-CH_2-CH_3$$

Alkene → Alkane

$$CH_3-CH=CH_2 + H_2 \xrightarrow[\Delta]{Ni} CH_3-CH_2-CH_3$$

Propylene → Propane

It is a *cis* addition and the order of addition is

$$CH_2=CH_2 > R-CH=CH_2 > R-CH=CH-R > R_2C=CHR$$

Mechanism: The mechanism of addition of hydrogen does not take place *via* ionic or polar mechanism. It follows adsorption phenomenon due to the high affinity of hydrogen gas to some types of metals like Na, Co, Ni *etc*. When the hydrogen gets adsorbed on the surface of the metals the alkenes also adsorbed by overlapping its *p* orbitals with vacant orbitals of metal, the rearrangement of electrons occur due to the weakening of H-H bond to form metal-H bond. Addition of hydrogen to alkenes is *cis* - addition because the alkene and hydrogen molecules present side by side on the catalyst surface during the time of reaction (Fig.9.2).

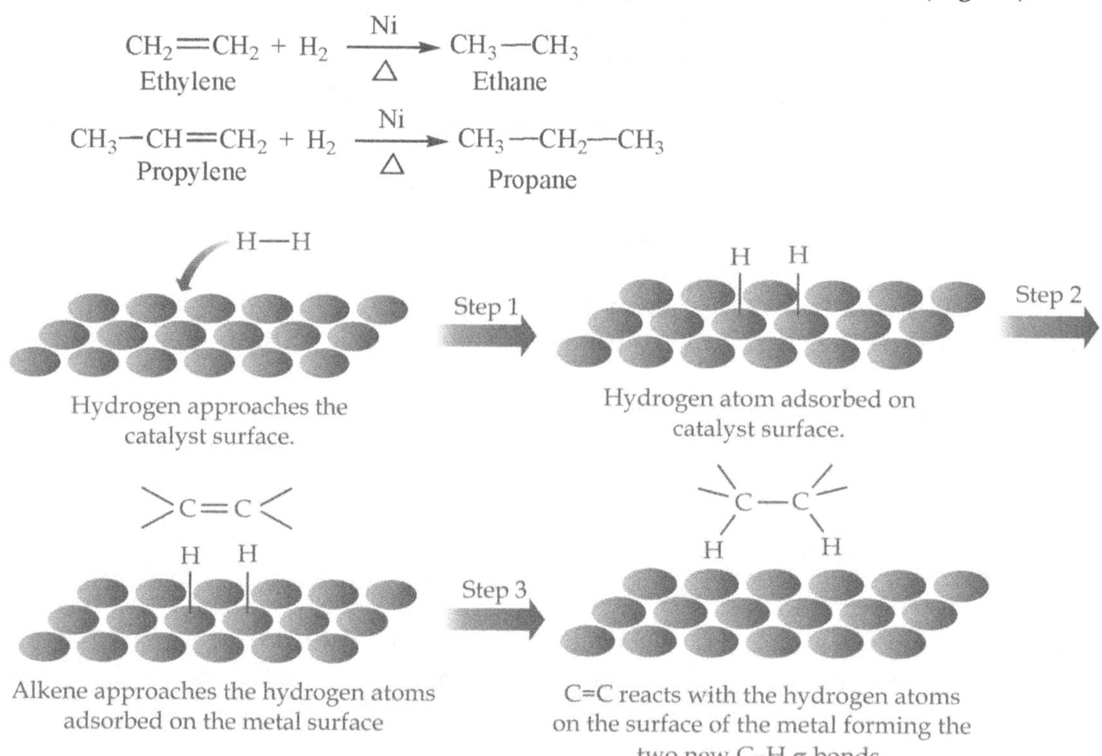

Figure 9.2 Hydrogenation of alkenes.

Stabilities of alkenes (Heat of hydrogenation): Stability of alkenes depend on their heat of hydrogenation when the heat of hydergenation increases, the stability of alkenes decreases and vice versa. Heat of hydrogenation is defined as "the energy difference between the starting alkene compounds and the resulting product of the alkene". It is determined from the amount of heat released during the hydrogenation. Let us consider the following reaction.

$CH_3-CH_2-CH=CH_2$ (1-Butene)
$CH_3-CH=CH-CH_3$ (*cis*-2-Butene)
$CH_3-CH=CH-CH_3$ (*trans*-2-Butene)

$$\xrightarrow{H_2/Pt} CH_3-CH_2-CH_2-CH_3$$

Butane

1-Butene ⟶ $\Delta H = 30.3$ Kcal/mole } Differ 1.7 Kcal/mole

cis-2-Butene ⟶ $\Delta H = 28.6$ Kcal/mole

trans-2-Butene ⟶ $\Delta H = 27.6$ Kcal/mole } 1 Kcal/mole

It indicates that heat of hydrogenation (ΔH) value increases, stability decreases. Hence *trans*-butene is more stable than *cis* and 1-butene. The order of relative stability of alkenes is as follows.

$$\underset{trans\text{-2-Butene}}{CH_3-CH{=}CH-CH_3} > \underset{cis\text{-2-Butene}}{CH_3-CH{=}CH-CH_3} > \underset{\text{1-Butene}}{CH_3-CH_2-CH{=}CH_2}$$

Glossary:

1. Alkyl substitution of the double bonded carbon atom increases the stability of alkenes. The order of stability of alkenes is as follows

$$CH_2=CH_2 < RCH=CH_2 < R_2C=CHR < R_2C=CR_2$$

2. *trans* isomers are more stable due to less steric repulsion between the groups as they are present on opposite side as compared to *cis* - in which bulky groups are present on same side.
3. Due to delocalization of π electrons in conjugated dienes, they are more stable than isolated dienes **(Refer chapter 10 - Alkadienes).**

2. **Addition of Halogen:** Alkenes react with halogens in the presence of inert solvents and form dihalides. The order of reactivity of halogens is $Cl_2 > Br_2 > I_2$.

$$\underset{\text{Alkene}}{R-CH=CH_2} + Br_2 \longrightarrow \underset{\text{1, 2-Dibromoalkane}}{R-CH(Br)-CH_2(Br)}$$

$$\underset{\text{Propene}}{CH_3-CH=CH_2} + Br_2 \longrightarrow \underset{\text{1, 2-Dibromopropane}}{CH_3-CH(Br)-CH_2(Br)}$$

$$\underset{\text{Alkene}}{R-CH=CH_2} + Cl_2 \longrightarrow \underset{\text{1, 2-Dichloroalkane}}{R-CH(Cl)-CH_2(Cl)}$$

$$\underset{\text{Propene}}{CH_3-CH=CH_2} + Cl_2 \longrightarrow \underset{\text{1, 2-Dichloropropane}}{CH_3-CH(Cl)-CH_2(Cl)}$$

The relative reactivity of halogens to olefins is as follows

$$CH_2=CH_2 < R_2C=CH_2 < RCH=CHR < R_2C=CHR < R_2C=CR_2$$

Note: Alkenes decolourises bromine in CCl_4 without evolution of HBr. This is used as an identification test and to differentiate alkenes from alkanes.

Mechanism: The addition of halogen to an alkene is an ***anti*** addition. It follows three steps *via* the formation of halonium ion intermediate.

***Step* 1:** The π electrons of alkene attack the halogen and gives halonium ion (three membered ring with positive charge in halogen atom) and halide ion.

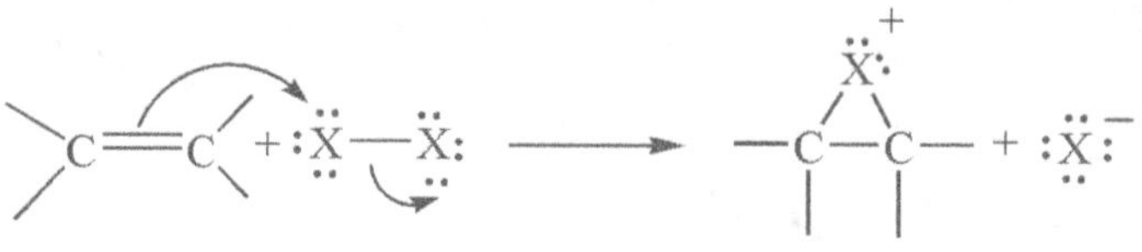

Halonium ion

***Step* 2:** The halide ion attacks the halonium ion from the back side to give *vic*-dihalides.

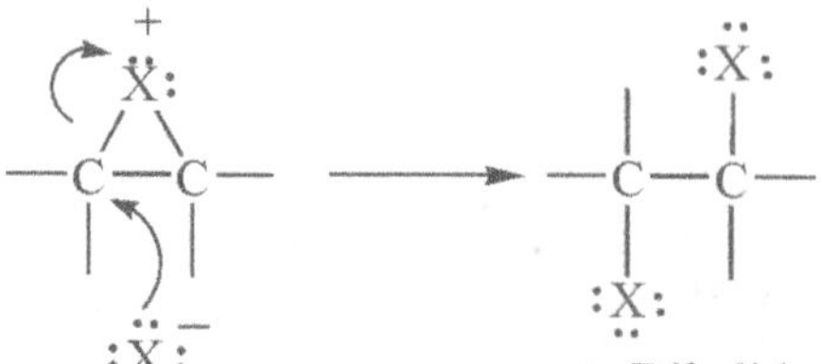

vic-Dihalide

For example: Bromination of Alkenes:

$$CH_2{=}CH_2 + Br_2 \longrightarrow BrCH_2{-}CH_2Br$$

Ethylene → 1,2-Dibromoethane

***Step* 1:** The π electrons of ethylene attack the bromine and gives bromonium ion and bromide ion.

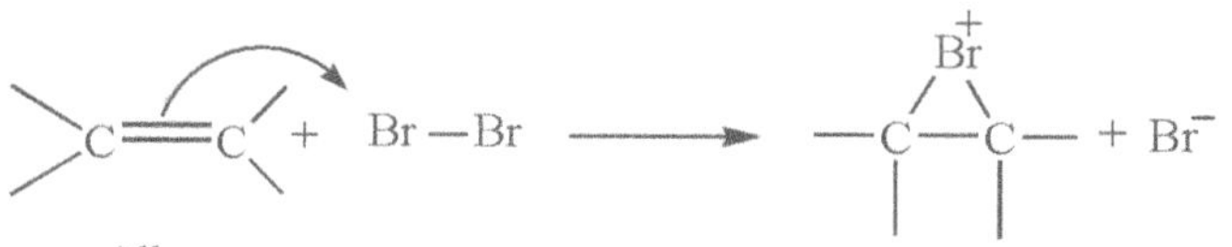

Alkene → Bromonium ion $+ Br^-$

***Step* 2:** The bromide ion attacks the bromonium ion from the back side to give *vic*-dibromide.

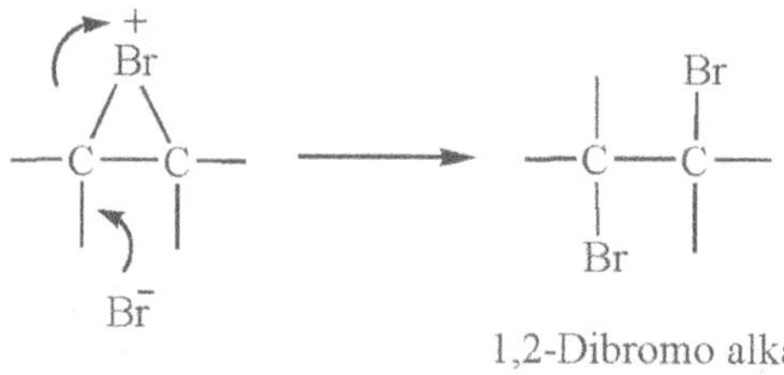

1,2-Dibromo alkane

3. **Markovnikov's Rule (Addition of Halogen acids):** Alkenes react with halogen acids and form alkyl halides and the order of reactivity is

$$HI > HBr > HCl.$$

$$CH_2 = CH_2 \xrightarrow{HBr} CH_3 - CH_2Br$$

Ethylene → Ethyl bromide

When alkene is symmetrical example ethylene, the product formed is same no matter which carbon reacts with hydrogen or bromide.

$$CH_2 = CH_2 + HBr \longrightarrow CH_3 - CH_2Br$$

Ethylene → Ethyl bromide

$$CH_2 = CH_2 + HBr \longrightarrow BrCH_2 - CH_3$$

Ethylene → Ethyl bromide

If the alkene is unsymmetrical, two alternatives are possible for reaction with HBr.

$$CH_3-CH=CH_2 \xrightarrow{HBr} CH_3-\underset{Br}{CH}-CH_3$$

Propene → 2-Bromopropane

$$CH_3-CH=CH_2 \xrightarrow{HBr} CH_3-CH_2-CH_2Br$$

Propene → 1-Bromopropane

In practice, 2-bromopropane is found to be the main product. *Vladimir Markovnikov's* studied many of these kinds of reactions and formulated the rule called Markovnikov's rule. It states that "**The negative part of the addendum adds on to the unsymmetrical carbon containing less number of hydrogen atoms".**

Formation of 2-bromopropane from propylene confirms the above rule. The mechanism of this reaction is as follows.

***Step* 1:** Attacking of H-Br protonates the alkene to gives carbocation and bromide ion. In this reaction two types of carbocations are possible.

$$CH_3-CH=CH_2 + HBr \longrightarrow CH_3-CH_2-\overset{\oplus}{C}H_2 + Br^-$$

Propene → 1° Carbocation (less stable)

$$CH_3-CH=CH_2 + HBr \longrightarrow CH_3-\overset{\oplus}{C}H-CH_3 + Br^-$$

Propene → 2° Carbocation (More stable)

***Step* 2:** Primary carbocation formed is less stable, so the bromide ion attacks the carbocation which is more stable (2^0 carbocation). So, the product 2-bromo propane predominates.

$$CH_3-\overset{+}{C}H-CH_3 + Br^- \longrightarrow CH_3-\underset{Br}{CH}-CH_3$$

2° Carbocation → 2-Bromopropane

Peroxide effect or Kharasch-peroxide effect or Anti Markovnikov's rule (or) Free radical addition reactions of alkenes.

In 1933, Kharasch discovered that "the addition of HBr to unsymmetrical alkenes in presence of air or peroxides takes place opposite to that of Markovnikov's rule". This is called as Kharasch-peroxide effect or Peroxide effect. When propylene reacts with HBr in presence of peroxide, 1-bromopropane is obtained as a major product.

$$CH_3-CH=CH_2 + HBr \xrightarrow{Peroxide} CH_3-CH_2-CH_2Br$$

Propene → 1-Bromopropane (Anti-Markovnikov's product)

$$CH_3-CH=CH_2 + HBr \xrightarrow{No\ peroxide} CH_3-\underset{Br}{CH}-CH_3$$

Propene → 2-Bromopropane (Markovnikov's product)

Mechanism of this reaction is as follows

***Step* 1:** Homolysis of peroxide to yield alkoxy free radicals.

$$R-O-O-R \xrightarrow{\Delta} 2RO^{\bullet}$$

Peroxide → Alkoxy free radical

***Step* 2:** Formation of bromine free radical:

The alkoxy free radical attacks HBr to give alcohol and bromine free radical.

$$RO^{\bullet} + H-Br \longrightarrow R-OH + Br^{\bullet}$$

Alcohol

***Step* 3:** Bromine free radical attacks the propylene molecule to yield 1° and 2° propyl free radical.

$$CH_3-CH=CH_2 + Br^{\bullet} \longrightarrow CH_3-CH(Br)-\dot{C}H_2$$

1° Freeradical (less stable)

$$CH_3-CH=CH_2 + Br^{\bullet} \longrightarrow CH_4-\dot{C}H-CH_2Br$$

2° Freeradical (More stable)

***Step* 4:** H-Br molecule attacks the more stable 2° propyl free radical to give *n*-propyl bromide.

$$CH_3-\dot{C}H-CH_3Br + H:Br \longrightarrow CH_3-CH_2-CH_2Br + Br^{\bullet}$$

2° propyl Free radical → 1-Bromo propane

Note:

- Only H-Br molecule gives anti Markovnikov's product.
- HCl and HI do not give anti Markovnikov's product because H-Cl bond is stronger than H-Br bond, therefore it cannot be cleaved by alkoxy free radicals.
- H-I bond is weaker than H-Br bond, therefore it is easily cleaved by alkoxy free radicals. But the iodide atoms so produced immediately combine with each other to give iodine molecule instead of attacking the alkene.

4. **Addition of Sulphuric acid:** Alkenes react with H_2SO_4 and forms alkyl hydrogen sulphate, which upon further reaction with water yields alcohol. The addition takes place as per Markovnikov's rule.

$$CH_3-CH=CH_2 + H^+HSO_4^- \longrightarrow CH_3-CH(HSO_4)-CH_3$$

Propene + (H_2SO_4) Sulphuric acid → *iso*-Propyl hydrogensulphate

$$CH_3-CH(HSO_4)-CH_3 + HOH \xrightarrow{\Delta} CH_3-CH(OH)-CH_3 + H_2SO_4$$

iso-Propyl hydrogen sulphate → *iso*-Propanol

This method of formation of alcohol from alkene by the addition of water molecule in presence of H_2SO_4 is called as **hydration** and is used in the preparation of alcohols.

Mechanism:

***Step* 1:** Formation of electrophile.

$$H-OSO_3H \longrightarrow H^+ + OSO_3\bar{H}$$

***Step* 2:** Attacking of H^+ to alkene produce a more stable carbocation.

$$\underset{\text{1- Propene}}{CH_3-CH=CH_2} + H^+ \longrightarrow \underset{2^\circ\text{ Propyl carbocation}}{CH_3-\overset{+}{C}H-CH_3}$$

***Step* 3:** Attack of nucleophile to yield the product.

$$\underset{2^\circ\text{ Carbocation}}{CH_3-\overset{+}{C}H-CH_3} + OSO_3\bar{H} \longrightarrow \underset{\textit{iso}\text{-Propyl hydrogen sulphate}}{CH_3-\underset{}{\overset{OSO_3H}{\overset{|}{C}H}}-CH_3} \xrightarrow{\text{H-OH}} \underset{\textit{iso}\text{-Propyl alcohol}}{CH_3-\overset{OH}{\overset{|}{C}H}-CH_3} + H_2SO_4$$

5. **Addition of hyphohalous acid:** Alkenes react with hyphohalous acid (HOX) and yields halohydrin. Addition takes place as per Markovnikov's rule. For example, on addition of HOCl, Cl carries positive charge and OH carries negative.

$$\underset{\text{Propene}}{CH_3-CH=CH_2} + \underset{\text{Hypochlorous acid}}{HO^- - Cl^+} \longrightarrow \underset{\text{Propylene chlorohydrin}}{CH_3-\underset{OH}{\underset{|}{C}H}-\underset{Cl}{\underset{|}{C}H_2}}$$

6. **Hydroboration:** Alkenes react with diboranes $[(BH_3)_2]$ and yield trialkyl boranes, the positive part is H and negative part is B.

$$\underset{\text{Propylene}}{3CH_3-CH=CH_2} + \underset{\text{Diborane}}{(BH_3)_2} \longrightarrow \underset{\text{Tri propylborane}}{2(CH_3-CH_2-CH_2)_3B}$$

Trialkyl boranes upon oxidation yields primary alcohol

$$\underset{\text{Tri Propylborane}}{(CH_3-CH_2-CH_2)_3B} + 3H_2O_2 \xrightarrow{OH^-} \underset{\substack{\text{1-Propanol}\\(1^\circ\text{Alcohol})}}{3CH_3-CH_2-CH_2OH} + \underset{\text{Boric acid}}{H_3BO_3}$$

The overall reaction takes place as anti Markovnikov's addition.

$$\underset{\text{Propylene}}{CH_3-CH=CH_2} \xrightarrow[\text{(ii) } 3H_2O_2/\,OH^-]{\text{(i) } (BH_3)_2} \underset{\text{1-Propanol}}{CH_3-CH_2-CH_2-OH} + \underset{\text{Boric acid}}{H_3BO_3}$$

7. **Hydration of olefins:** Alkenes can be hydrated to alcohols upon reaction with H_2SO_4 followed by hydrolysis as described in reaction 4. However they can also be catalytically hydrated in dilute acid solutions. For example, *iso*-butylene is hydrated to *t*-butyl alcohol in dilute acid solutions.

$$\underset{iso\text{-Butylene}}{(CH_3)_2-C=CH_2} + H_2O \xrightarrow{H^+} \underset{\text{Tertiary butyl alcohol}}{(CH_3)_3C-OH}$$

8. **Addition of oxygen:** Lower alkenes upon reaction with oxygen in the presence of catalyst (Ag) yields alkene-oxides or epoxides.

$$2\ \underset{\text{Ethylene}}{CH_2=CH2} + O_2 \xrightarrow{Ag} 2\ \underset{\text{Ethylene oxide}}{CH_2-CH2\ (\text{bridged by } O)}$$

These epoxides can also be obtained upon reaction of alkenes with salts of peracids.

$$\underset{\text{Alkene}}{R-CH=CH-R'} + \underset{\text{Sodiumperbenzoate}}{C_6H_5COOONa} \longrightarrow \underset{\text{Epoxide}}{R-CH-CH-R'\ (\text{bridged by } O)} + \underset{\text{Sodium benzoate}}{C_6H_5COONa}$$

9. **Alkylation:** Some of the alkanes add to alkenes and yields higher alkanes.

$$\underset{\text{Alkane}}{RH} + \underset{\text{Alkene}}{>C=C<} \longrightarrow \underset{\text{Higher alkane}}{R-\overset{|}{\underset{|}{C}}-\overset{|}{\underset{|}{C}}-H}$$

$$\underset{iso\text{-Butane}}{(CH_3)_3CH} + \underset{iso\text{-Butylene}}{CH_2=C(CH_3)_2} \longrightarrow \underset{neo\text{-Octane}}{(CH_3)_3C-CH_2-CH(CH_3)_2}$$

10. **Addition of nitrosyl chloride or bromide:** Alkenes react with nitrosyl chloride or bromide and yields alkene nitrosochloride or bromide.

$$\underset{\text{Propylene}}{CH_3-CH=CH_2} + \underset{\text{Nitrosobromide}}{NO-Br} \longrightarrow \underset{\text{Propylene nitrosobromide}}{CH_3-\underset{Br}{\underset{|}{CH}}-\underset{NO}{\underset{|}{CH_2}}}$$

11. **Addition of ozone:** Alkenes on reaction with ozone in organic solvents give an ozonide.

$$\underset{\text{Alkene}}{RR'C=CH-CH_2R''} + \underset{\text{Ozone}}{O_3} \xrightarrow{\text{Organic solvent}} \underset{\text{Ozonide}}{RR'\underset{\underset{O}{|}}{C}-O-\underset{\underset{O}{|}}{CH}-CH_2R''\ (O-O)}$$

$$\underset{\text{Ethylene}}{CH_2=CH_2} + \underset{\text{Ozone}}{O_3} \xrightarrow{\text{Organic solvent}} \underset{\text{Ethylene ozonide}}{\underset{\underset{O}{|}}{CH_2}-O-\underset{\underset{O}{|}}{CH_2}\ (O-O)}$$

The ozonide on reaction with reducing or oxidizing agents yields products accordingly. From the nature of the product obtained, the position of double bond in alkene can be located.

Ozonolysis: The formation of unstable ozonide upon reaction of ozone with alkene and breaking it into different products with the help of reagents is called ozonolysis.

$$\underset{\text{Alkene}}{RR'C=CH-CH_2R''} + \underset{\text{Ozone}}{O_3} \longrightarrow \underset{\text{Ozonide}}{RR'C(-O-)CH-CH_2R''\ (\text{O}-\text{O bridge})}$$

Ozonide $\xrightarrow{H_2O_2}$ $\underset{\text{Ketone}}{RR'C=O}$ + $\underset{\text{Carboxylic acid}}{R''COOH}$

Ozonide $\xrightarrow{H_2/Ni}$ $\underset{\text{Ketone}}{RR'C=O}$ + $\underset{\text{Aldehyde}}{R''CHO}$

Ozonide $\xrightarrow{LiAlH_4}$ $\underset{2^{\circ}\text{ Alcohol}}{RR'CH-OH}$ + $\underset{1^{\circ}\text{ Alcohol}}{R''CH_2OH}$

12. **Addition of Carbenes:** Carbenes are added to alkenes to yield a substituted cyclopropane. This reaction is stereospecific in nature. The *cis* alkene gives *cis*-cyclopropane and the *trans* alkene give *trans*-cyclopropane derivative.

$$\underset{cis\text{-Alkene}}{RHC=CHR} + \underset{\text{Carbene}}{:CH_2} \longrightarrow \underset{\text{(meso)}}{cis\text{-Dialkyl cyclopropane}}$$

$$\underset{trans\text{-Alkene}}{RHC=CHR} + :CH_2 \longrightarrow \underset{\text{(Racemic mixture)}}{trans\text{-Dialkyl cyclopropane}}$$

Note: For more details about carbenes, refer chapter No: 6.

13. **Oxymercuration and demercuration:** Alkenes react with mercuric acetate and water molecule to give an organomercury compound. This reaction is known as oxymercuration of alkenes.

$$\underset{\text{Propene}}{CH_3-CH=CH_2} + \underset{\text{Mercuric acetate}}{(CH_3COO)_2Hg} \xrightarrow{H_2O} \underset{\text{Organomercury compound}}{CH_3-\overset{OH_2}{\overset{|}{C}}H-\underset{HgO-\underset{\overset{||}{O}}{C}-CH_3}{\underset{|}{C}H_2}} + CH_3COOH$$

The organo mercury compound obtained in the above reaction further undergoes reduction with sodium borohydride ($NaBH_4$) to give an alcohol. This reaction is known as demercuration reaction.

$$\underset{\text{Organomercury compound}}{CH_3-H\overset{OH}{\overset{|}{C}}-\underset{HgO-\underset{\overset{||}{O}}{C}-CH_3}{\underset{|}{C}H_2}} \xrightarrow{NaBH_4} \underset{\text{2-Propanol}}{CH_3-H\overset{OH}{\overset{|}{C}}-CH_3}$$

The addition follows Markovnikov's rule.

Mechanism: It involves three steps.

***Step* 1:** Electrophilic addition of mercuric acetate to the alkene molecule to yield mercurinium ion.

$$\underset{\text{Alkene}}{R-CH=CH_2} + \underset{\text{Mercuric acetate}}{\underset{AcO}{\underset{|}{Hg}}-OAc} \xrightarrow[-OAc]{} \underset{\text{Mercurinium ion}}{R-HC\overset{\overset{+}{Hg}-OAc}{\diagup\diagdown}CH_2}$$

***Step* 2:** Attack of water molecule:

$$\underset{\text{Mercurinium ion}}{R-HC\overset{Hg-OAc}{\diagup\diagdown}CH_2} + H_2\ddot{O}: \longrightarrow R-\underset{\underset{H\quad H}{\overset{|}{O^+}}}{CH}-\overset{HgOAc}{\overset{|}{C}H_2}$$

***Step* 3:** Abstractions of proton by acetate ion.

$$R-\underset{\underset{H\quad H\quad :\ddot{O}Ac}{\overset{|}{O^+}}}{CH}-\overset{HgOAc}{\overset{|}{C}H_2} \xrightarrow[-AcOH]{} \underset{\text{Organo mercury compound}}{R-\underset{HO}{\underset{|}{CH}}-\overset{HgOAc}{\overset{|}{C}H_2}}$$

(II) Oxidation: Alkenes are oxidized into different products based on the oxidizing agents used.

(i) **With cold, alkaline and dilute potassium permanganate solution:** Alkenes upon reaction with Bayer's reagent (Cold alkaline dilute (CAD) $KMnO_4$) undergo hydroxylation and yield dihydroxy compound called glycols.

$$\underset{\text{Ethylene}}{CH_2 = CH_2} + MnO_4^- \longrightarrow \begin{matrix} CH_2 - O \\ | \\ CH_2 - O \end{matrix} \!\!> Mn <\!\! \begin{matrix} O^- \\ \\ O^- \end{matrix} \xrightarrow{OH^-} \underset{\textit{cis-}\text{ glycol}}{\begin{matrix} CH_2 - OH \\ | \\ CH_2 - OH \end{matrix}}$$

During this reaction the purple colour of potassium permanganate is decolourised, hence it is used as one of the identification test for alkenes called **Bayer's test.**

(ii) **Oxidation with hot $KMnO_4$ solution:** Upon reaction with hot $KMnO_4$ solution the double bond of alkenes is splitted to yield ketones and or carboxylic acids.

$$\underset{\text{Propene}}{CH_3 - CH = CH_2} \xrightarrow[\text{(O)}]{\text{Hot } KMnO_4} \underset{\text{Carboxylic acid}}{CH_3 - \overset{\overset{O}{\|}}{C} - OH} + CO_2 + H_2O$$

$$\underset{\text{2-Methyl-2-butene}}{CH_3 - \overset{\overset{CH_3}{|}}{C} = C - CH_3} \xrightarrow[\text{(O)}]{\text{Hot } KMnO_4} \underset{\text{Acetone}}{CH_3 - \overset{\overset{CH_3}{|}}{C} = O} + \underset{\text{Acetic acid}}{CH_3COOH}$$

(iii) **Oxidation with osmium tetroxide & hydrogen peroxide in glacial acetic acid:** Alkenes upon reaction with osmium tetroxide yields *cis* glycol.

$$\underset{\text{Alkene}}{\begin{matrix} R - CH \\ \| \\ R - CH \end{matrix}} + \underset{\text{Osmium tetroxide}}{OsO_4} \longrightarrow \begin{matrix} R - CH - O \\ | \\ R - CH - O \end{matrix} \!\!> Os <\!\! \begin{matrix} =O \\ \\ =O \end{matrix} \xrightarrow{NaHSO_4} \underset{\textit{cis}\text{-Glycol}}{\begin{matrix} R - CH - OH \\ | \\ R - CH - OH \end{matrix}}$$

(iv) **Oxidation of alkenes with peracids:** Alkenes upon reaction with peracids form epoxide which upon hydrolysis yields *trans* glycol.

$$\underset{\text{Alkene}}{R - CH = CH - R'} \xrightarrow{\text{Peracid}} \underset{\text{Epoxide}}{R - \underset{\diagdown \; O \; \diagup}{CH - CH} - R'} \xrightarrow[H_2O]{\text{Dilute acid}} \underset{\textit{trans}\text{-Glycol}}{R\underset{\underset{OH}{|}}{CH} - \overset{\overset{OH}{|}}{CH} - R'}$$

(v) **Reaction with periodic acid or lead tetraacetate:** Alkenes reacts with lead tetraacetate and initially form glycols which undergoes further oxidation and yields aldehydes or ketones depending upon starting alkenes. The oxidation product helps us to locate the position of double bonds.

$$\underset{\text{Alkene}}{\begin{matrix} R_2 - CH \\ \| \\ CH_2 \end{matrix}} \xrightarrow{Pb(COOCH_3)_4} \underset{\text{Glycol}}{\begin{matrix} R_2 - CH - OH \\ | \\ H - CH - OH \end{matrix}} \xrightarrow{HIO_4} \underset{\text{Ketone}}{\begin{matrix} R \\ & C = O \\ R \end{matrix}} + \underset{\text{Aldehyde}}{HCHO}$$

(vi) **Reaction with atmospheric oxygen:** Alkenes undergo slow oxidation by atmospheric oxygen and forms gums and waxes due to cross linking. Hence the presence of unsaturated compounds in lubricants is undesirable for use.

(vii) **Oxidation with ozone:** Described under ozonolysis.

(III) Substitution Reaction (Allylic Substitution): Alkyl group of alkenes undergo substitution reaction at high temperature. The reaction takes place by free radical mechanism. For example propylene reacts with chlorine and yield chloropropylene or allyl chloride. **Substitution occurs in allylic carbon, hence known as allylic substitution.**

$$\underset{\text{Propylene}}{CH_2=CH-CH_3} + Cl_2 \xrightarrow{\text{High temperature}} \underset{\text{Chloro propylene}}{CH_2=CH-CH_2Cl} + HCl$$

With branched chain olefins substitution takes place easily at room temperature. The order of reactivity is Allyl > 3° > 2° > 1° > CH_3 > Vinyl.

$$\underset{\textit{iso}\text{-Butene}}{(CH_3)_2C=CH_2} \xrightarrow{Cl_2} \underset{\substack{\text{3- Chloro-2-methyl}\\\text{-1-propylene}\\\text{(Main product)}}}{(ClCH_2)(CH_3)C=CH_2} + \underset{\substack{\text{1,2-Dichloro-2-methyl}\\\text{propane (Traces of addition}\\\text{product)}}}{CH_4-C(Cl)(CH_3)-CH_2Cl}$$

Mechanism: It follows free radical mechanism.

Step 1: Chain initiation step: Homolytic cleavage or homolysis of Cl_2 molecule leads to the formation of chlorine free radicals.

$$Cl-Cl \longrightarrow 2Cl^{\bullet}$$

Step 2: Chain propagation step: Chlorine free radicals attack propylene to give allyl free radical and HCl.

$$\dot{Cl} + H-CH_2-CH=CH_2 \longrightarrow HCl + \underset{\text{Allyl free radical}}{\dot{C}H_2-CH=CH_2}$$

The allyl free radical attacks the chlorine molecule to give allylchloride and chlorine free radical.

$$\underset{\text{Allyl free radical}}{\dot{C}H_2-CH=CH_2} + Cl-Cl \longrightarrow \underset{\text{Allyl chloride}}{Cl-CH_2-CH=CH_2} + \dot{Cl}$$

These two steps repeated again and again.

Step 3: Chain termination step: The two free radicals combine to give molecules.

$$Cl^{\bullet} + Cl^{\bullet} \longrightarrow Cl-Cl$$

$$\dot{Cl} + \dot{C}H_2—CH{=}CH_2 \longrightarrow Cl—CH_2—CH{=}CH_2$$

Allyl free radical → Allyl chloride

$$2\,\dot{C}H_2—CH{=}CH_2 \longrightarrow CH_2{=}CH-CH_2CH_2-CH{=}CH_2$$

Allyl freeradical → 1,5-Hexadiene

(IV) Isomerisation: Alkenes undergo isomerisation when heated with catalysts like aluminium sulphate. Isomerisation takes place due to shifting of double bond or a methyl group or both.

(i) Shifting of double bond

$$CH_3-CH_2-CH_2-CH{=}CH_2 \xrightarrow[Al_2(SO_4)_3]{\Delta} CH_3-CH_2-CH{=}CH-CH_3$$

1-Pentene → 2-Pentene

(ii) Shifting of methyl group:

$$CH_3-CH_2-CH{=}CH_2 \xrightarrow[Al_2(SO_4)_3]{\Delta} CH_3-\underset{\displaystyle CH_3}{\underset{|}{C}}{=}CH_2$$

1-Butene → *iso*-Butene

(V) Polymerisation: Simple alkenes undergo polymerisation in the presence of catalysts like HF, H_2SO_4 at high temperature and pressure yields polymer. **This process of forming large molecule from small molecule is called polymerisation.**

$$CH_2{=}CH_2 \xrightarrow[Pressure]{HF/\ High\ temp} +CH_2-CH_2+_n$$

Ethylene → Polyethylene

$$CH_3-CH{=}CH_2 \xrightarrow[Pressure]{H_2SO_4/\ High\ temp} +CH_2-CH_2-CH_2+_n$$

Propylene → Polypropylene

(VI) Combustion: Alkenes burn with luminous flame and give carbon dioxide and water with the liberation of heat. It forms explosive mixture with air or oxygen.

$$CH_2{=}CH_2 + 3O_2 \longrightarrow 2CO_2 + 2H_2O + Heat$$

Ethylene

Summary of Reactions of Alkenes

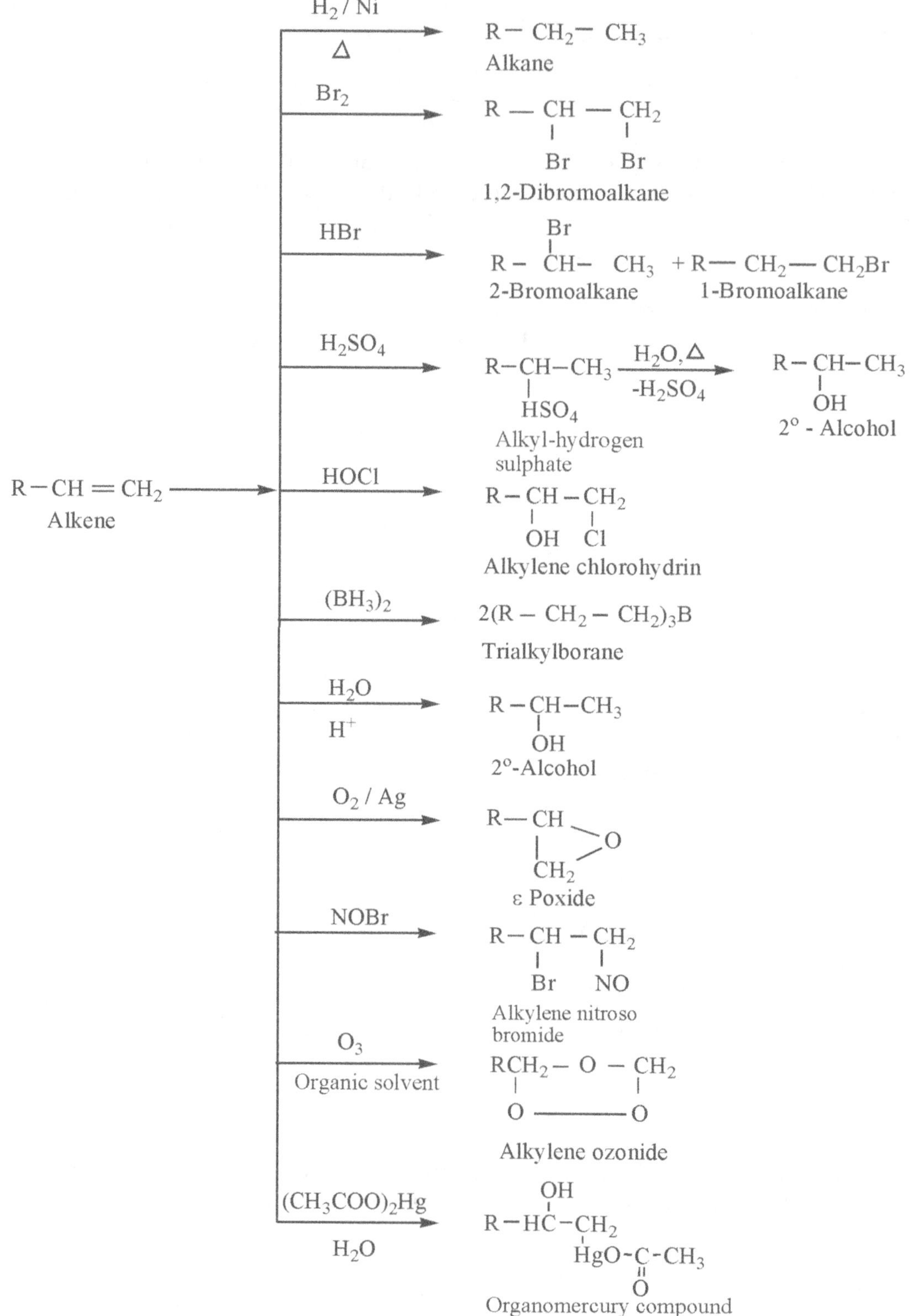

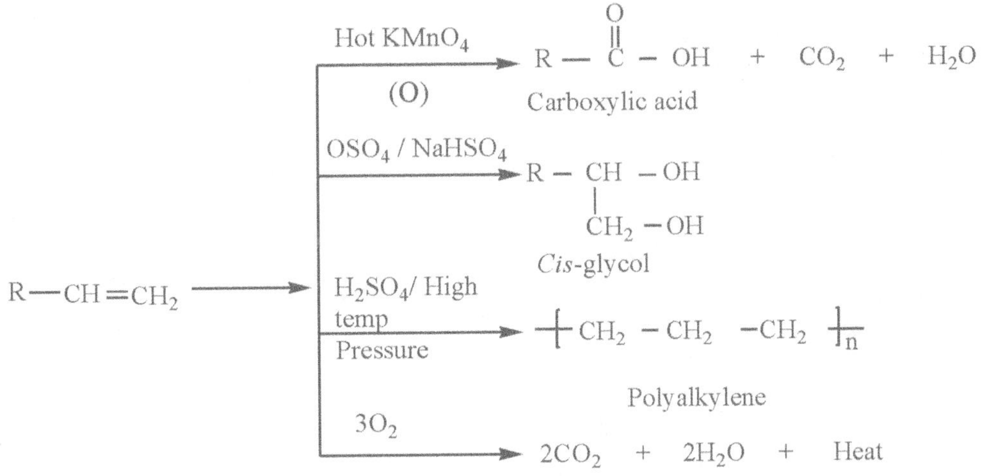

Probable Questions

1. Define alkenes and mention its importance.
2. Write a detailed note on nomenclature of alkenes.
3. Write a short note on isomerism exhibited by alkenes.
4. Enlist various general methods of preparation of alkenes and explain any four methods in detail.
5. How alkenes are synthesized from alkanes & alkynes?
6. Write the synthesis of alkenes from vicinal dihalides.
7. Describe the general structure & physical properties of alkenes.
8. Explain the various chemical properties of alkenes.
9. Write a note on electrophilic addition reactions of alkenes with mechanism of reaction.
10. Explain about Sabatier- Senderens reduction of alkenes.
11. Describe in detail about Markovnikov's rule & anti Markovnikov's rule.
12. What is Kharash peroxide effect?
13. Write a note on Saytzeff's rule.
14. Write short notes on ozonolysis of alkenes.
15. Write the various oxidation reactions of alkenes with suitable example.
16. Write the substitution reactions of alkenes with suitable example and explain its reaction mechanism.
17. How alkenes undergo polymerization & isomerisation reaction?

10

Alkadienes or Dienes or Diolefins

Introduction

Hydrocarbons with two carbon-carbon double bonds are called as dienes or alkadienes. The general formula is C_nH_{2n-2} (similar to alkynes). For example:

$$CH_2 = CH - CH = CH_2$$

1,3-Butadiene

$$CH_2 = CH - CH_2 - CH = CH_2$$

1,4-Pentadiene

Importance of Alkadienes

Many organic compounds of natural origin have conjugated double bonds. For example, natural rubber is a polymer of isoprene, a conjugated diene. Synthetic rubber called neoprenes are made from butadiene, chloroprene *i.e.* 2-chlorobuta-1,3-diene. Some naturally occurring compounds containing diene and triene system are shown in table 10.1. Some drugs which possess alkadienyl group show therapeutic activity. For example, Divinyl ether (CH_2= CH-O-CH=CH_2), is a general anesthetic.

Table 10.1 Natural Compounds containing diene and triene system.

Natural Compound	Source	Structure
α-Cadinene (diene)	Citronella oil	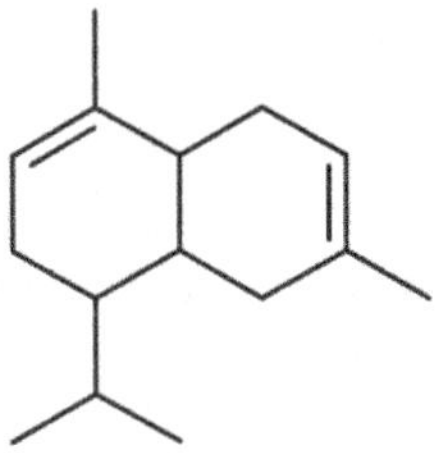
β-Selinene (diene)	Celery oil	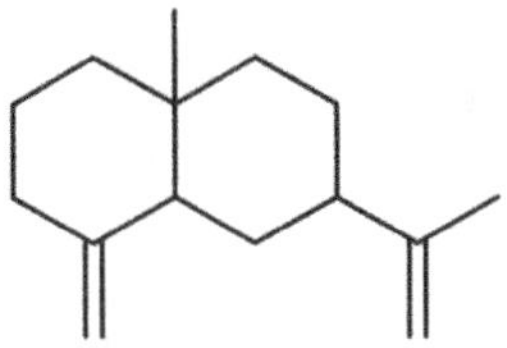

Table 10.1 *Contd...*

Natural Compound	Source	Structure
Zingiberene (triene)	Ginger oil	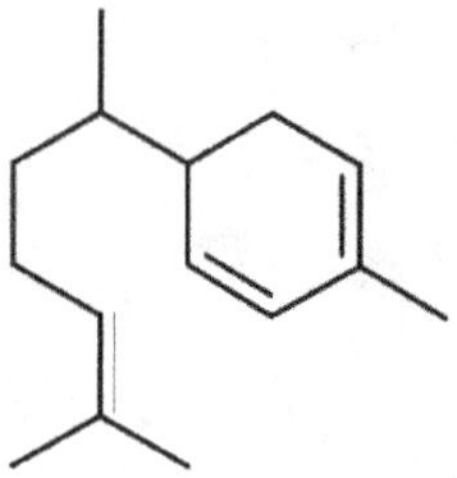
β -Carotene (polyene)	Carrot	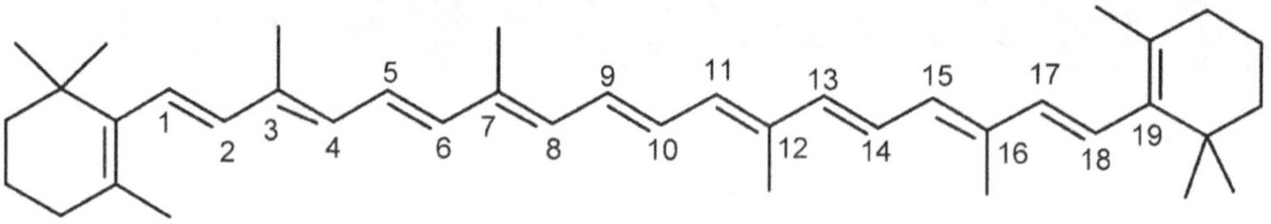

Classification: Dienes are classified into three types according to the position of double bonds:

1. **Isolated dienes:** The double bonds are separated by more than one single bond.
   For example:

   $CH_2 = CH - CH_2 - CH = CH_2$
   1,4-Pentadiene

2. **Conjugated dienes:** The double bonds are separated by one single bond.
   For example:

   $CH_2 = CH - CH = CH_2$
   1,3-Butadiene

3. **Cumulated dienes:** The double bonds are adjacent to each other. Cumulated dienes are also called allenes.
   For example:

   $CH_2 = C = CH_2$
   1,2-Propadiene

   $CH_3 - CH = C = CH_2$
   1,2-Butadiene

Nomenclature

The IUPAC system of naming dienes is similar to alkenes. Based on the carbon chain length the suffix –<u>**ene**</u> of alkene is replaced by –<u>**adiene**</u>. The position of two double bonds is indicated by giving least number to one of the double bond and the other get next minimum number. For example

$\overset{4}{C}H_2{=}\overset{3}{C}H{-}\overset{2}{C}H{=}\overset{1}{C}H_2$
1,3-Butadiene

$\overset{5}{C}H_2{=}\overset{4}{C}H{-}\overset{3}{C}H_2{-}\overset{2}{C}H{=}\overset{1}{C}H_2$
1,4-Pentadiene

Configurational isomers of dienes: There are four configurational isomers possible with 1-chloro-2,4-heptadiene because each one of double bonds have either the *E* or the *Z* configuration. Hence there are *E-E, Z-Z, E-Z* and *Z-E* isomers.

(2*Z*,4*Z*)-1-Chloro-2,4-heptadiene

(2*Z*,4*E*)-1-Chloro-2,4-heptadiene

(2*E*, 4*Z*)-1-Chloro-2,4-heptadiene

(2*E*, 4*E*)-1-Chloro-2,4-heptadiene

 **INTERESTING FACT**   

Some naturally available compounds containing dienes or trienes are as follows.

β-Carotene (Present in carrot)

(Polyene)

β - Carotene, a colored pigment present widely in carrot, responsible for its orange color. Various derivatives of carotene provide many of the colors that we see in variety of vegetables, fruits and autumn leaves.

Butadiene

Preparation:

Alkadienes are prepared by methods similar to those for the preparation of alkenes.

1. **Dehydrogenation of alkanes or alkenes:** Vapours of alkanes or alkenes on passing over heated Cr_2O_3 supported alumina catalyst at 800-900 K yields 1,3-butadiene.

$$\underset{n\text{-Butane}}{CH_3CH_2CH_2CH_3} \xrightarrow[800\text{-}900\ K]{Cr_2O_3/Al_2O_3} \underset{1,3\text{-Butadiene}}{CH_2{=}CH{-}CH{=}CH_2} + 2H_2$$

$$\underset{1\text{-Butene}}{CH_2{=}CH{-}CH_2CH_3} \xrightarrow[800\text{-}900\ K]{Cr_2O_3/Al_2O_3} \underset{1,3\text{-Butadiene}}{CH_2{=}CH{-}CH{=}CH_2}$$

2. **From 1,3-butane diol:** By passing a mixture of 1,3-butanediol and steam (4:1 ratio) over trisodium phosphate and H_3PO_4 (20%) at 540 K, 1,3-butadiene is obtained.

$$\underset{1,3\text{-Butanediol}}{CH_3{-}\overset{OH}{\overset{|}{C}H}{-}CH_2CH_2OH} \xrightarrow[540\ K]{Na_3PO_4/H_3PO_4} \underset{1,3\text{-Butadiene}}{CH_2{=}CH{-}CH{=}CH_2} + 2H_2O$$

3. **From ethanol:** Mixture of ethanol and acetaldehyde vapours are passed over heated catalyst (silica gel + 2% tantalum oxide) to yield 1,3-butadiene.

$$\underset{Ethanol}{C_2H_5OH} + \underset{Acetaldehyde}{CH_3CHO} \xrightarrow[Ta_2O_5]{Silica\ gel} \underset{1,3\text{-Butadiene}}{CH_2{=}CH{-}CH{=}CH_2} + 2H_2O$$

4. **From acetylene:** Acetylenes when treated with a mixture of cuprous chloride and ammonium chloride dimerizes to form vinyl acetylene which on reduction with Lindlar's catalyst yields 1,3-butadiene. Acetylene upon reaction with formaldehyde followed by dehydration also yields 1,3-butadiene.

$$\underset{Acetylene}{2CH{\equiv}CH} \xrightarrow[NH_4Cl]{CuCl} CH{\equiv}C{-}CH{=}CH_2 \xrightarrow[Pd/H_2SO_4]{H_2} \underset{1,3\text{-Butadiene}}{CH_2{=}CH{-}CH{=}CH_2}$$

$$\underset{Acetylene}{CH{\equiv}CH} + \underset{Formaldehyde}{2HCHO} \longrightarrow CH_2OH{-}C{\equiv}C{-}CH_2OH \xrightarrow{O_2} \underset{1,3\text{-Butadiene}}{CH_2{=}CH{-}CH{=}CH_2}$$

Summary of Methods of Preparation

1. $CH_3CH_2CH_2CH_3$ (n-Butane) $\xrightarrow[800–900\ K]{Cr_2O_3/Al_2O_3}$ $CH_2=CH-CH=CH_2$ (1, 3-Butadiene)

2. $CH_2=CH-CH_2CH_3$ (1-Butene) $\xrightarrow[800–900\ K]{Cr_2O_3/Al_2O_3}$ $CH_2=CH-CH=CH_2$ (1, 3 Butadiene)

3. $CH_3CH(OH)CH_2CH_2OH$ (1, 3-Butanediol) $\xrightarrow[540\ K]{Na_3PO_4/H_3PO_4}$ $CH_2=CH-CH=CH_2$ (1, 3-Butadiene)

4. C_2H_5OH (Ethanol) $\xrightarrow[Silica\ gel/Ta_2O_5]{CH_3CHO}$ $CH_2=CH-CH=CH_2$ (1, 3-Butadiene)

5. $CH\equiv CH$ Acetylene (2 moles) $\xrightarrow[H_2/Pd/H_2SO_4]{CuCl/NH_4Cl}$ $CH_2=CH-CH=CH_2$ (1, 3-Butadiene)

6. $CH\equiv CH$ (Acetylene) $\xrightarrow[-2H_2O]{2HCHO}$ $CH_2=CH-CH=CH_2$ (1, 3-Butadiene)

Relative Stability of Dienes: Unlike alkenes, the relative stability of substituted alkene molecules is determined by their relative heat of hydrogenation (–ΔH°). It indicates that the least stable alkene has the greatest heat of hydrogenation value *i.e.,* it releases maximum heat because it contains more energy to start with. Let us consider the following example.

$$CH_3-CH=C=CH-CH_3 + 2H_2 \xrightarrow{Pt} CH_3CH_2CH_2CH_2CH_3$$

2,3-Pentadiene → Pentane

[Δ H° = – 70.5 kcal/mol]
(– 295 KJ/mol)

$$CH_2=CH-CH_2-CH=CH_2 + 2H_2 \xrightarrow{Pt} CH_3CH_2CH_2CH_2CH_3$$

1,4-Pentadiene → Pentane

[Δ H° = – 60.2 kcal/mol]
(– 252 KJ/mol)

$$CH_2=CH-CH=CH-CH_3 + 2H_2 \xrightarrow{Pt} CH_3CH_2CH_2CH_2CH_3$$

1,3-Pentadiene → Pentane

[Δ H° = – 54.1 kcal/mol]
(– 226 KJ/mol)

The heat of hydrogenation of 2,3-pentadiene (a cumulative diene) is greater than the heat of hydrogenation of 1,4-pentadiene (an isolated diene) which in turn is greater than the heat of hydrogenation of 1,3-pentadiene (a conjugated diene).

From the value of heat of hydrogenation of the above dienes, it is understood that the conjugated dienes are more stable than isolated dienes which are more stable than cumulated dienes. So the order of relative stability of dienes is as follows.

Conjugated dienes > isolated dienes > cumulated dienes.

Stability of Conjugated Dienes

Two factors are responsible for the stability of conjugated dienes *i.e.,*

1. Hybridization in conjugated dienes.
2. Concept of Resonance.

1. **Hybridization in conjugated dienes:** Hybridization of orbitals leads to the formation of carbon-carbon single bonds. In 1,3-butadiene molecule, the C-C single bond is formed by the overlap of one sp^2 orbital with another sp^2 orbital. But in 1,4-pentadiene (isolated diene) the carbon-carbon single bond is formed from the overlap of an sp^3 orbital with an sp^2 orbital.

   Let us consider the structure of 1,3-butadiene. It has four carbon atoms, which are sp^2 hybridised. These are overlap with each other and overlap with the *s* orbitals of the hydrogen atoms to provide C-C and C-H σ bonds. Apart from this, the each carbon atoms possess an unhybridised "*p*" orbital which are perpendicular to the σ bonds. Hence, the *p* orbital of C-2 overlaps with the *p* orbitals of C-1 and C-3; and the p orbital of the C-3 overlaps with the *p* orbitals of C-2 and C-4 leads to the formation of large π molecular orbital (each pair of π electrons are attracted by four carbon atoms). The overlap of the *p* orbitals of C-2 and C-3 takes place in both directions, leads to the p electrons to be spread in large surface area known as delocalisation (formation of delocalized π Molecular orbitals), which is responsible for the greater stability of conjugated dienes when compared with isolated and cumulated dienes.

Single covalent bond formed by sp^2- sp^2overlap

$$CH_2 = CH - CH = CH_2$$

1,3-Butadiene (Conjugated diene)

$$CH_2 = CH - CH_2 - CH = CH_2$$

1,4-Pentadiene (Isolated diene)

Single bonds formed by sp^3- sp^2 overlap

   According to hybridization or molecular orbital theory, the strength and length of a bond depends upon how closer the electrons on the bonding orbitals are to the nucleus. *i.e.,* if the bond length is shorter, a bond formed by sp^2-sp^2 overlap is stronger than a bond formed by sp^3-sp^2 overlap ("*s*" character of a sp^2 orbital is 33.3 % and sp^3 orbital is only 25 %). Hence a conjugated diene possesses one stronger single bond than an isolated diene. As a result, the conjugated diene is more stable.

2. **Concept of resonance:** According to resonance, the π electrons in each of the double bonds present in the isolated diene are localized between two carbons but the four π electrons in a conjugated diene are delocalized over four carbons. So it increases stability.

$$\overset{-}{C}H_2 - CH = CH - \overset{+}{C}H_2 \longleftrightarrow CH_2 = CH - CH = CH_2$$

Conjugated dienes

$$\updownarrow$$

Resonating structures $\overset{+}{C}H_2 - CH = CH - \overset{-}{C}H_2$

CH₂ ⋯ CH ⋯ CH ⋯ CH₂

Resonance hybrid

Resonance hybrid indicates that 1,3–butadiene is not a pure single bond but has partial double-bond character also. So both electron delocalization and orbital formation account for the fact that a bond formed by sp^2-sp^3 overlap is shorter and stronger than a bond formed by sp^3-sp^3 overlap.

Reactivity consideration of dienes: Like alkenes and alkynes, dienes act as nucleophiles due to the electron density of their π bonds. Alkadienes react with electrophiles and undergo electrophilic addition reactions.

The reactivity of the compounds is characteristic feature of its functional group. If the compounds possess two or more functional groups which are sufficiently separated from each other then the compound shows specific reaction of the individual functional groups. If they are close enough to allow electron delocalization, then one functional group affects the reactivity of other functional group. Therefore, the isolated dienes undergo the same chemical reactions as that of alkenes whereas the conjugated dienes show little different reactions due to the influence of one double bond over the other.

Electrophilic addition reactions of isolated dienes: The isolated dienes undergo electrophilic addition reaction as that of alkenes. Isolated dienes react with excess amount of electrophilic reagent, two independent addition reactions will take place each following Markovnikov's rule as shown in example.

Mechanism: The mechanism follows the electrophilic addition reactions of alkenes. In this reaction, the electrophile (H⁺) first adds to the electron rich double bond to produce more stable carbocation to give addition product. Due to the presence of an excess amount of reagent, both double bonds undergo addition.

$$\underset{\text{1,5-Hexadiene}}{CH_2=CH-CH_2-CH_2-CH=CH_2} + H-\ddot{B}r: \longrightarrow CH_3-\overset{+}{C}H-CH_2-CH_2-CH=CH_2$$

$$\downarrow :\ddot{B}r:^-$$

$$CH_3-\underset{|}{\overset{Br}{C}}H-CH_2-CH_2-CH=CH_2 + H-Br$$

$$\downarrow$$

$$CH_3-\overset{Br}{\overset{|}{C}}H-CH_2-CH_2-\overset{+}{C}H-CH_3 + :\ddot{B}r:^-$$

$$\downarrow$$

$$CH_3-\overset{Br}{\overset{|}{C}}H-CH_2-CH_2-\overset{Br}{\overset{|}{C}}H-CH_3$$

2,5- Dibromohexane

If electrophile is present only in required amount that can be added to only one double bond then it will prefer the more reactive double bond. For example, the reaction of 2-methyl-1,5-hexadiene with HCl.

If the electrophile is added to the left-hand side of the double bond, it gives secondary carbocation. Similarly, if the electrophile is added to the right-hand side of the double bond it leads to the formation of tertiary carbocation. In transition state, the secondary carbocation rearranges to tertiary (3°) carbocation which is more stable. So, in the presence of enough HCl (no excess amount), the major product obtained will be 5-chloro-5-methyl-1-hexene.

$$CH_2{=}CH{-}CH_2{-}CH_2{-}\overset{CH_3}{\overset{|}{C}}{=}CH_2 + HCl \longrightarrow CH_2{=}CH{-}CH_2CH_2\underset{Cl}{\underset{|}{\overset{CH_3}{\overset{|}{C}}}}{-}CH_3$$

2-Methyl-1,5-hexadiene (1 mole) | (1 mole) | 5-Chloro-5-methyl-1-hexene (Major product)

Electrophilic addition reactions of conjugated dienes: A conjugated diene such as 1,3-butadiene reacts with a limited amount of electrophilic reagent, the addition takes place at only one of the double bonds that leads to the formation of two kinds of addition products.

***1,2-Addition or direct addition*:** Addition occurs at 1-and 2-position is known as 1,2-addition or direct addition.

***1,4-Addition or conjugate addition*:** Addition occurs at 1-and 4-position is known as 1,4- addition or conjugate addition.

$$CH_2{=}CH{-}CH{=}CH_2 + Cl_2 \longrightarrow \overset{Cl}{\overset{|}{CH_2}}{-}\overset{Cl}{\overset{|}{CH}}{-}CH{=}CH_2$$

1,3-Butadiene (1 mole) | (1 mole) | 3,4-Dichloro-1-butene (1,2-Addition product)

+

$$\overset{Cl}{\overset{|}{CH_2}}{-}CH{=}CH{-}\overset{Cl}{\overset{|}{CH_2}}$$

1,4-Dichloro-2-butene (1,4-Addition product)

$$CH_2{=}CH{-}CH{=}CH_2 + HBr \longrightarrow CH_3{-}\overset{Br}{\overset{|}{CH}}{-}CH{=}CH_2$$

1,3-Butadiene (1 mole) | (1 mole) | 3-Bromo-1-butene (1,2-Addition product)

+

$$CH_3{-}CH{=}CH{-}\overset{Br}{\overset{|}{CH_2}}$$

1-Bromo-2-butene (1,4-Addition product)

Note:

1. In 1,2-addition the reagents add to double bonds like alkenes.
2. In 1,4-addition, the reagent did not add to the adjacent carbons but also the double bond changed its position. In the product, double bond is present in between 2 & 3 position, whereas in the reactant the single bond is present in 2 & 3 positions.

Mechanism:

$$CH_2{=}CH-CH{=}CH_2 + H-\ddot{Br}: \longrightarrow CH_3-\overset{+}{C}H-CH{=}CH_2 + Br^-$$

$$CH_3-\overset{+}{C}H-CH{=}CH_2 \longleftrightarrow CH_3-CH{=}CH-\overset{+}{C}H_2$$

$$:\ddot{Br}:^-$$

$$CH_3-\underset{Br}{\underset{|}{CH}}-CH{=}CH_2 \qquad CH_3-CH{=}CH-\underset{Br}{\underset{|}{CH_2}}$$

3-Bromo-1-butene (1,2-Addition product) | 1-Bromo-2-butene (1,4-Addition product)

Reason for formation of both 1,2- & 1,4- addition products in the presence of limited number of reagents:

1. In the first step of the mechanism, the electrophile (H^+) adds to the C-1 which leads to the formation of allylic carbocation (positive charge on the carbon atom which is next to a double bonded carbon).
2. 1,3-butadiene molecule is a symmetrical one, addition of H^+ to C-1 is the same to the addition of H^+ to C-4.
3. The electrophile does not add to the C-2 or C-3 because this addition leads to the formation of primary carbocation which is less stable than allylic cation.
4. The allylic carbocation possesses two contributing resonance structures.

$$CH_3-\overset{+}{C}H-CH{=}CH_2 \longleftrightarrow CH_3-CH{=}CH-\overset{+}{C}H_2$$

Two Resonance contributors

$$CH_3-CH \;\text{-----}\; CH \;\text{-----}\; CH_2$$

In the above structures, the positive charge on the carbocation is not localized on C-2, but it is shared by both C-2 and C-4 as a result of delocalization of π electrons. Consequently, in the second step of the reaction, the bromide ion attacks either C-2 carbon to yield 1,2-addition product where as if it attacks C-4 carbon it yield 1,4-addition (conjugate addition) product.

Note: In all electrophilic addition reactions of dienes, the addition of electrophile is first step, the proton adds to one of the sp^2 carbons at the end of the conjugated system. This is the only way to obtain a carbocation which is stabilized by resonance. If the electrophile is added to the any other internal sp^2 carbons, the carbocation obtained is not stabilized by resonance.

Comparison of addition in isolated dienes Vs conjugated dienes: The carbocation formed by the addition of electrophile to an isolated diene is not stabilized by resonance. So, the positive charge is localized on one carbon atom and only 1,2-addition occurs. In contrast, in conjugated dienes, the positive charge is spread over between two carbons which is resonance stabilized and 1,2-addition and 1,4-addition takes place.

Example for addition to isolated diene:

$$CH_2 = CHCH_2CH_2CH = CH_2 \xrightarrow{H^+} CH_3 - \overset{+}{C}HCH_2CH_2CH = CH_2$$

1,5-Hexadiene

(Positive charge is accumulated in one carbon atom)

Example for addition to conjugated dienes:

If the diene is unsymmetrical, the major product obtained depends upon the addition of electrophile to whichever terminal sp^2 carbon, results in formation of the more stable carbocation. For example, addition reaction between 2-methyl-1,3-butadiene with HBr. In this reaction, the proton preferentially adds to C-1 because the addition to C-1 produces the positive charge on the carbon atom shared by a tertiary allylic and a primary allylic carbon (more stable carbocation). If the proton adds to the C-4 carbon atom, the positive charge on the carbon atom is shared by a secondary allylic and a primary allylic carbocation (less stable carbocation). So, the major product obtained is 3-bromo-3-methyl-1-butene and a minor amount of 1-bromo-3-methyl-2-butene is also formed.

$$CH_2 = C(CH_3) - CH = CH_2 + HBr \longrightarrow CH_3 - C(CH_3)(Br) - CH = CH_2 + CH_3C(CH_3) = CHCH_2Br$$

2-Methyl-1,3-butadiene; 3-Bromo-3-methyl-1-butene; 1-Bromo-3-methyl-2-butene

Carbocation formed by the attack or addition of H^+ to C - 1

$$CH_3 - \overset{\oplus}{C}(CH_3) - CH = CH_2 \longleftrightarrow CH_3 - C(CH_3) = CH - \overset{\oplus}{C}H_2$$

3° Allylic cation; 1° Allylic cation

$$CH_2 = C(CH_3) - \overset{\oplus}{C}H - CH_3 \longleftrightarrow {}^{+}CH_2 - C(CH_3) = CHCH_3$$

2° Allylic carbocation; 1° Allylic carbocation

Thermodynamic *Vs* kinetic control of reactions: When a conjugated diene undergoes electrophilic addition reaction to yield 1,2-addition and 1,4-addition products then the predominance of one product over the other is determined by the following two factors.

1. **Structure of the reactant**
2. **Temperature** at which the reaction is carried out.

Kinetic product and kinetically controlled reaction: When a reaction gives more than one product, the product formed most rapidly is called as kinetic product and the reactions that give the kinetic product as the major product are known as kinetically controlled reaction.

Thermodynamic product and thermodynamically controlled reaction: When a reaction gives more than one product, the most stable product is known as thermodynamic product and the reaction that give thermodynamics products as the major product are known as thermodynamically controlled products.

In maximum number of organic reactions, the kinetic and thermodynamic products are one and same. But the electrophilic addition of 1,3-butadiene is an example of the reaction in which the kinetic and the thermodynamics product are not the same.

$$\underset{\text{1,3-Butadiene}}{CH_2 = CH - CH = CH_2} + HBr \longrightarrow \underset{\substack{\text{3-Bromo-1-butene} \\ \text{(1,2-Addition product)}}}{CH_3 - \overset{\overset{Br}{|}}{C}H - CH = CH_2} + \underset{\substack{\text{1-Bromo-2-butene} \\ \text{(1,4-Addition product)}}}{CH_3 - CH = CH - CH_2 - Br}$$

Now let us determine which product in the preceding reaction is more stable *i.e.* thermodynamic product. According to Saytzeff rule, the stability of alkene depends upon the number of alkyl groups attached to the sp^2 carbon atom. Greater the number of alkyl groups, more stable is the alkene. So, in the above reaction, in 1,2-addition product contains one alkyl group but the 1,4-addition products contains two alkyl groups, so the 1,4-addition product is the major product. 1,3-butadiene reacts with one mole of HBr, to yield two products of different stabilities *i.e.,* the 1,2-addition product contain one alkyl group bonded to the sp^2 carbons, while the 1,4-addition product contain two alkyl groups bonded to sp^2 carbons. So the 1,4-addition product is more stable than 1,2-product. Therefore 1,4-addition product is thermodynamic product.

The formation of kinetic product is explained as follows:

(a) In the addition reaction, the first step is the addition of a proton to C-1. This is the same reaction whether the product is 1,2 or 1,4–product.

(b) In the second step, the bromide ion attacks either C-2 or C-4. To determine which of the product is formed faster, we must check the relative stabilities of the transition states for the second step of the reaction.

$$\underset{\text{1,3-Butadiene}}{CH_2 = CH - CH = CH_2} + HBr \longrightarrow CH_3 - \overset{+}{C}HCH = CH_2 \longleftrightarrow CH_3CH = CH\overset{+}{C}H_2$$

$$\xrightarrow{Br^-} \left[CH_3 - \overset{d^+}{C}HCH = CH_2 \ \cdots \ \overset{d^-}{:\ddot{Br}:} \right]^+ \qquad \xrightarrow{Br^-} \left[CH_3CH = CH - \overset{d^+}{C}H_2 \ \cdots \ d^- :\ddot{Br}: \right]^+$$

Transition state for the formation of the 1,2-product

Transition state for the formation of the 1,4-product

$$\downarrow \qquad\qquad \downarrow$$

$$\underset{\substack{\text{3-Bromo-1-butene} \\ \text{(1,2-Addition product)} \\ \text{(Kinetic product)}}}{CH_3 - \overset{\overset{Br}{|}}{C}H - CH = CH_2} \qquad \underset{\substack{\text{1-Bromo-2-butene} \\ \text{(1,4-Addition product)} \\ \text{(Thermodynamic product)}}}{CH_3 - CH = CH - CH_2 - Br}$$

The reactant for the second step is an allylic carbocation which is resonance stabilized. If the bromide ion approaches the allylic carbocation, the positive charge becomes concentrated on the carbon on which the bromide ion attacks. This indicates that the transition state for the formation of 1,2-product is the secondary allylic carbocation, where as the transition state for the formation of 1,4-product is the primary allylic carbon. A secondary allylic carbocation is more stable than a primary allylic cation. So the 1,2–product is formed faster than 1,4-product, and 1,2–product is the kinetic product. In any reaction, the kinetic and thermodynamic products are not the same and which product predominates depends upon the reaction conditions in which it is carried out. If the reaction is carried out under low temperature conditions, the reaction is irreversible, the major product obtained is kinetic product, For example, the addition reaction of 1,3-butadiene with HBr at -80 °C, the major product is the 1,2-product.

$$\underset{\text{1,3-Butadiene}}{CH_2 = CH - CH = CH_2} + HBr \xrightarrow{-80\ ^\circ C} \underset{\substack{\text{3-Bromo-1-butene} \\ \text{(1,2-Product)} \\ \text{80\% Yield}}}{CH_3 - \overset{\overset{Br}{|}}{C}HCH = CH_2} + \underset{\substack{\text{1-Bromo-2-butene} \\ \text{(1,4-Product)} \\ \text{20\% Yield}}}{CH_3CH_2 = CHCH_2BrH}$$

At -80 °C, there is enough energy for the reactions to overcome the energy barrier for the first step of the reaction to form carbocation and there is enough energy for the carbocation to form two addition products. But there is not enough energy for the reverse reaction to occur. The energy barrier for the formation of 1,4-addition products is higher than the 1,2-addition product, so 1,2–product predominates.

On the other hand, if the reaction is carried out at higher temperature or vigorous conditions, the reaction is reversible one, so the major product obtained is thermodynamic product. For example, addition reaction of 1,3-butadiene with HBr at 45 °C, the major product obtained is the 1,4–product.

$$\underset{\text{1,3-Butadiene}}{CH_2 = CH - CH = CH_2} + HBr \xrightarrow{45\ ^\circ C} \underset{\substack{\text{3-Bromo-1-butene} \\ \text{(1,2-Product)} \\ \text{15 \% Yield}}}{CH_3 - \overset{\overset{Br}{|}}{C}HCH = CH_2} + \underset{\substack{\text{1-Bromo-2-butene} \\ \text{(1,4-Product)} \\ \text{85 \% Yield}}}{CH_3CH = CHCH_2Br}$$

At 45 °C, sufficient energy is present for the reactant to get rid of energy barrier to form the product and sufficient energy for one or more of the products to go back to the intermediate carbocation.

So when the reaction is irreversible under the conditions used in this experiment, it is said to be under kinetic control and the relative amount of the product obtained depends upon the rate at which they are formed.

When the reaction is reversible under thermodynamic control, the relative amount of the product obtained depends upon their stabilities.

Diels - Alder reaction or addition of a dienophile to a conjugated diene: Reactions that form carbon-carbon bonds are very important in organic chemistry or synthetic chemistry because through these reactions a small carbon skeleton is converted into larger carbon skeleton. The Diels - Alder reaction is an important reaction, because it forms two carbon-carbon bonds in a way that forms a cyclic molecule.

In Diels Alder reaction, a conjugated diene reacts with a compound containing a carbon-carbon double bond and form a cyclic molecule. The carbon-carbon double bond compound is known as **dienophile** because it loves a **diene**.

$$CH_2 = CH - CH = CH_2 + CH_2 = CH - \overset{\overset{O}{\|}}{C} - CH_3 \longrightarrow \text{(cyclohexene)}-\overset{\overset{O}{\|}}{C} - CH_3$$

1,3-Butadiene (Diene) — 3-Oxo-1-butene (Dienophile) — (Diels-Alder adduct)

Electron withdrawing group attached to the dienophile increases the reactivity of the reaction. The electron withdrawing groups withdraw the electrons from the double bond and initiate the reaction by placing a partial positive charge on a carbon atom of the dienophile.

$$CH_2 = CH - \overset{\overset{O}{\|}}{C} - CH_3 \longleftrightarrow \overset{+}{C}H_2 - CH = \overset{\overset{O^-}{|}}{C} - CH_3$$

Resonance contributors of the dienophile

$$\overset{d^+}{CH_2} \overset{---}{-} CH \text{ ------ } \overset{\overset{d^-}{O}}{\overset{\vdots\|}{C}} - CH_3$$

Resonance hybrid

The partially positive carbon of the dienophile acts as an electrophile that is attracted by electrons from C-1 of the conjugated diene. The other sp^2 carbon of the dienophile is the nucleophile that attaches to the C-4 of the diene.

$$\underset{\text{Diene}}{\overset{4}{CH_2}=\overset{3}{CH}-\overset{2}{CH}=\overset{1}{CH_2}} + \underset{\text{Dienophile}}{CH_2 = CH - \overset{\overset{O}{\|}}{C} - CH_3} \xrightarrow{30\,^{o}C} \text{4-Acetyl-1-cyclohexene}$$

4-Acetyl-1-cyclohexene (Diel's-Alder adduct)

Diel's -Alder reaction is a simple 1,4-addition reaction of conjugated diene. But the main, difference between Diel's–alder and 1,4-addition reaction is that the Diel's Alder reaction is a concerted reaction and addition occurs in single step.

The Diel's –Alder reaction is an example for pericyclic reaction, a reaction takes place in one step by a cyclic shift of electrons. It is otherwise known as cyclo addition reaction in which two reactants add together to form a cyclic product. It is a type of addition because six π electrons are involved in the cyclic transition state, four electrons comes from conjugated diene and two electrons comes from dienophile.

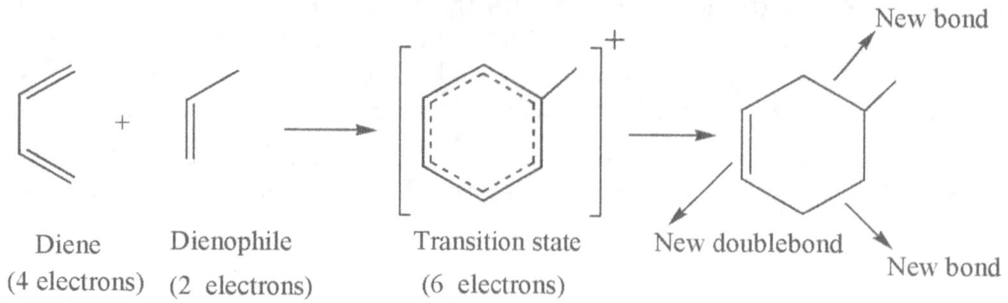

New bond

Diene (4 electrons) + Dienophile (2 electrons) → [Transition state (6 electrons)] → New doublebond, New bond

Stereo chemistry of Diel's - Alder reaction: If a Diel's-Alder reaction produces a chiral center in the products, identical amount of the R and S enantiomers are formed. So, the product will be a racemic mixture.

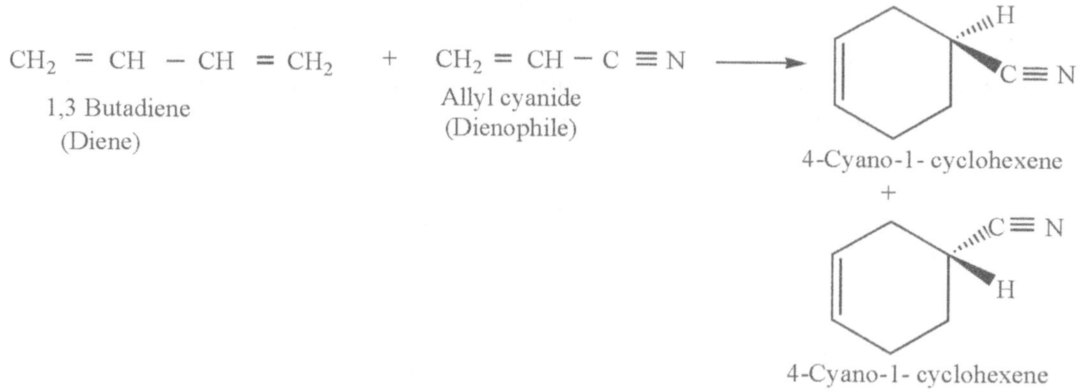

$CH_2 = CH - CH = CH_2$ (1,3 Butadiene (Diene)) + $CH_2 = CH - C \equiv N$ (Allyl cyanide (Dienophile)) → 4-Cyano-1- cyclohexene + 4-Cyano-1- cyclohexene

Diel's - Alder reaction is stereospecific because the configuration of reactants is maintained during the reaction. If the substituents of the dienophile in reaction are *cis*, they will be in *cis* in the product; if the substituents of the dienophile are *trans*, they will be in the trans form in the product also.

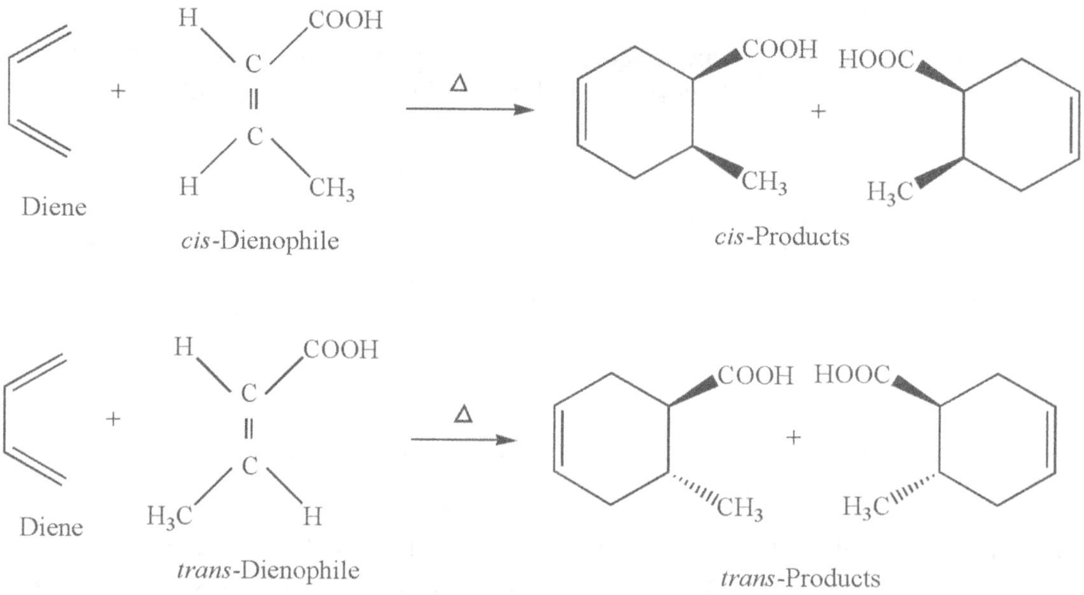

Diene + *cis*-Dienophile (H, COOH, H, CH_3) —Δ→ *cis*-Products (COOH, CH_3 + HOOC, H_3C)

Diene + *trans*-Dienophile (H, COOH, H_3C, H) —Δ→ *trans*-Products (COOH, CH_3 + HOOC, H_3C)

Each of above reaction gives products with new two chiral centers. So, each product has four stereoisomers.

Applications:

1. A wide variety of cyclic compounds can be obtained.

$$\text{1,3-Butadiene (Diene)} + \text{Furan 2,5-dione (Dienophile)} \xrightarrow{20^\circ C} \text{3a,4,7,7a-Tetrahydro isobenzofuran 1,3-dione}$$

1,3-Butadiene (Diene) — Furan 2,5-dione (Dienophile) — 3a,4,7,7a-Tetrahydro isobenzofuran 1,3-dione

2. Carbon-carbon triple bonds can also be used as dienophiles in the preparation of compounds with two double bonds.

$$\text{1,3-Butadiene (Diene)} + H-C \equiv C-COOCH_3 \xrightarrow{\Delta} \text{Methyl cyclohexa-1,4-diene carboxylate}$$

1,3-Butadiene (Diene) — Methyl propionate — Methyl cyclohexa-1,4-diene carboxylate

Summary of Chemical Reactions

$CH_2 = CH - CH = CH_2$ (1, 3-Butadiene) →

- Cl_2: $CH_2(Cl)-CH(Cl)-CH=CH_2 + CH_2(Cl)-CH=CH-CH_2(Cl)$
  3,4-Dichloro-1-butene (1,2-addition product) + 1,4-Dichloro-2-butene (1,4-addition product)
- HBr: $CH_3-CH(Br)-CH=CH_2 + CH_3-CH=CH-CH_2(Br)$
  3-Bromo-1-butene (1,2-addition product) + 1-Bromo-2-butene (1,4-addition product)
- $CH_2=CH-C(=O)-CH_3$: cyclohexenyl$-C(=O)-CH_3$
  Diel's-Alder adduct (or) 4-Acetyl-1-cyclohexene

Theory of Resonance

Allyl radical as a resonance hybrid:

The structure of allyl radical is represented by the following hybrid of the two resonance structures A & B.

$$CH_2=CH-\dot{C}H_2 \text{ (A)} \longleftrightarrow \dot{C}H_2-CH=CH_2 \text{ (B)}$$

It indicates that the actual structure lies in between the above two structures but does not corresponds to either A or B. Both are equivalent and posses equal stability. According to resonance theory the allyl radical does not contain one carbon - carbon single bond and carbon - carbon double bond (like A & B) but it contains two identical bonds each one is an intermediate between a single bond and a double bond. This type of bond is known as hybrid bond which is described as one -and -a-half bond. It shows that the allyl radical posses one-half single bond character and one-half double bond character. The odd electrons in the allyl radical are equally distributed over both terminal carbon atoms. It is indicated by the resonance hybrid C.

$$\left[\underset{A}{CH_2{=}CH{-}\dot{C}H_2},\ \underset{B}{\dot{C}H_2{-}CH{=}CH_2}\right] \equiv \underbrace{CH_2 \cdots CH \cdots CH_2}_{C}$$

Due to the hybrid character, allyl radical easily undergoes allylic rearrangement.

Stability of the allyl radical: According to the theory of resonance, as a resonance hybrid, the allyl radical is more stable due to less energy than the corresponding either of the contributing structures. The additional stability of the molecule is known **resonance energy**. The resonance energy of the allyl radical is determined by comparing the actual value of hybrid structure with resonance energy of hypothetic at structure A & B. But we can't perform experimentally. Hence, it can be determined by comparing the resonance energy of two reactions, such as dissociation reaction of propane to yield *n*-propyl radical and dissociation reaction of propylene to yield allyl radical.

$$\underset{\text{Propane}}{CH_3CH_2CH_3} \longrightarrow \underset{n\text{-Propyl free radical}}{CH_3CH_2\dot{C}H_2} + \dot{H} \quad \Delta H = +98 \text{ kcal}$$

$$\underset{\text{Propylene}}{CH_3\text{-}CH{=}CH_2} \longrightarrow \underset{\text{Allyl radical}}{\dot{C}H_2\text{-}CH{=}CH_2} + \dot{H} \quad \Delta H = +\ 88 \text{ kcal}$$

The above equation indicates that the dissociation energy difference between propyl and the allyl free radical is 10 kcal (98 - 88) than the energy difference between propane and propyl free radical. This lower dissociation energy is responsible for the stability of allyl free radical.

Orbital picture of allyl radical: How does the allyl radical attains its stability by resonance? It is explained as follows.

Let us consider the bond orbitals in the allyl radical.

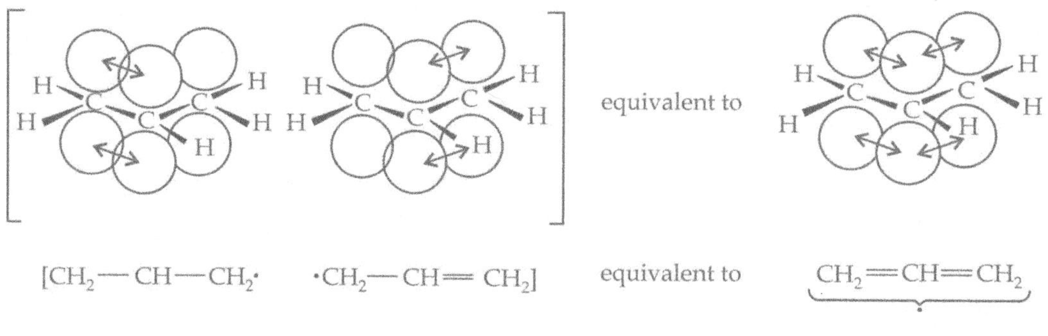

Figure 10.1 Orbital picture of allyl radical. The *p* orbital of centre carbon atom overlaps with the *p* orbital of two other carbon atoms leads to delocalisation.

According to the above orbital picture, each carbon atom in the allyl radical is bonded to three other carbon atoms by using sp^2 orbital. Overlap of these sp^2 orbital with each other and "s" orbital of five hydrogen atoms forms the molecular structure which is shown in the figure 12.1. The all bond angle is 120^0. All the carbon atoms possess "*p*" orbital which are having two equal lobes. One lobe is lying above the plane and the other is lying below the plane of a bonds. These "*p*" orbital are accommodated by a single electron. Similar to that ethylene molecule the *p* orbital of one carbon atom can overlap with the *p* orbital of adjacent carbon atom forms the electron pair and bond to the formed. But in allyl radical , the overlap of "*p*" orbital is not a limited one *i.e.*, the *p* orbital of the central carbon atom overlaps equally with the p orbital of both carbon atom to which it is attached. This overlapping of *p* orbital lead to the formation of two continues π electron clouds; one is lying above the plane and another one is lying below the plane. This ability of π electrons involving in several bonds is called as delocalisation of electrons which leads to the allyl radical is more stable with stronger bonds. Due to this reason, the word delocalisation energy is used frequently instead of resonance energy.

Resonance Stabilisation of Allyl Radicals-Hyper Conjugation

Allyl Cation as a Resonance Hybrid

The classical structure of allyl cation is as follows.

$$CH_2{=}CH_2{-}\overset{+}{C}H_2$$

The allyl cation can be represented by the following resonance structures.

$$\left[\underset{I}{CH_2{=}CH{-}\overset{+}{C}H_2},\ \underset{II}{\overset{+}{C}H_2{-}CH{=}CH_2}\right] \equiv \underbrace{CH_2{\cdots}CH{\cdots}CH_2}_{\oplus\ III}$$

According to resonance theory, the structures I & II reveals the cation structure, and the resonance hybrid is III. Both of the structures I & II are exactly equivalent and possess similar stability and they produce equal contributions to the hybrid. Like allyl radical, allyl cation also does not possess one carbon -carbon single bond and carbon-carbon double bond, but it contains two equivalent or identical bonds and each of the bond is an intermediate between a single and double bond. The positive charge is not accumulated in one carbon atom but it is spread over the both. The molecular orbital picture of the two resonating structures and the resonance hybrid is shown in the following Fig. 10.2. In either of these resonating structures, the electron deficient carbon atoms (carbocation) possess an empty "*p*" orbital. Overlap of this "*p*" orbital with π cloud of the double bond leads to delocalisation of π electrons. Each of these two electrons is used to hold together all these carbon nuclei.

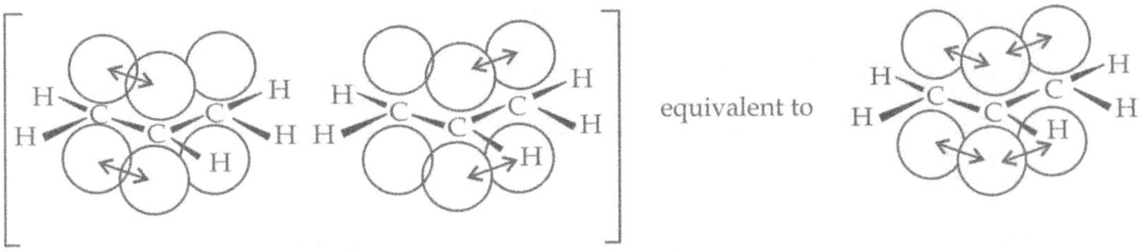

equivalent to

$[CH_2{=}CH{-}CH_2+\ ,\ +CH_2{-}CH{=}CH_2]$ equivalent to $\underbrace{CH_2{\cdots}CH{\cdots}CH_2}_{\oplus}$

Figure 10.2 Resonance stabilisation of allyl cation by delocalisation of electrons.

It explained that the two contributing structure I & II are exactly equivalent and resonance gives considerable stabilization of the cation. Hence the order of stability of carbocation is as follows.

$$3^o > 2^o > 1^o > CH_3^+$$

Nucleophilic Substitution in Allylic Substrates: S_N1 Reactivity and Allylic Rearrangement

1. **Allylic substrates reacts faster:** Allylic substrate undergoes nucleophilic substitution reaction. What are the predictions we can expect? Let us consider the allylic substrate undergoes S_N1 reaction. For example, solvolysis of allyl chloride or reaction between allyl alcohol and HBr.

$$\underset{\text{Allyl chloride}}{CH_2=CH\text{-}CH_2\text{-}Cl} + C_2H_5OH \longrightarrow \underset{\text{Allyl ethyl ether}}{CH_2=CH-CH_2-O-C_2H_5} + HCl$$

$$\underset{\text{Allyl alcohol}}{CH_2=CH\text{-}CH_2\text{-}OH} + HBr \longrightarrow \underset{\text{Allyl bromide}}{CH_2=CH\text{-}CH_2Br} + H_2O$$

The first step of the mechanism of this reaction is formation of allyl carbocation intermediate by the heterolysis of allyl chloride or allyl alcohol. This step is slow and rate determining step in the reaction. The rate of the reaction purely depends upon the nature of allylic carbocation formed.

$$\underset{\text{Allyl substrate}}{CH_2=CH\text{-}CH_2\text{-}X} \longrightarrow \underset{\text{Allyl cation}}{CH_2=CH\text{-}\overset{+}{C}H_2 + \overset{\oplus}{X^-}} \xrightarrow{Nu:^-} \underset{\text{Allyl product}}{CH_2=CH\text{-}CH_2\text{-}Nu + X^-}$$

In the previous sessions, we concluded that the stability of the allyl cation is similar to the secondary carbocation. It indicates that "allyl substrate will react as fast in S_N1 reaction like secondary substrate". In nucleophilic substitution reactions, the allyl substrate react million times as fast as its saturated analogues *i.e.*, allyl cation reacts much faster reacts than *n*-propyl cation. Hence the reactivity of allyl cation is as follows.

$$\text{Allyl} > 3^o > 2^o > 1^o > \overset{+}{C}H_3$$

The presence of alkyl substituent at either terminal of the allylic system further increases the stability.

2. **Allylic substrates shows allylic rearrangement in S_N1 reaction:** The second step of the SN[1] reaction is the attack of nucleophile to the carbocation. In allylic substrates the incoming nucleophile attack the either terminal carbon atoms of allylic cation, and give two different products if the structure permits.

$$\underset{\text{Allyl cation}}{RCH \overset{Nu}{=\!=} \underset{\oplus}{C}H =\!= CH_2} \begin{cases} \xrightarrow{a} \underset{|}{\overset{Nu}{R}}CH\text{-}CH=CH_2 \\ \xrightarrow{b} RCH=CH\text{-}CH_2Nu \end{cases}$$

Consider the following examples: Nucleophilic substitution reaction of isomeric allyl chlorides.

1-Chloro-2-butene and 3-chloro-1-butene reacts with sodium ethoxide in presence of ethanol and gives ethyl ether. The sequence of the reactions are as follows.

$$\underset{\text{1-Chloro-2-butene}}{CH_3\text{-}CH=CH\text{-}\underset{\substack{| \\ Cl}}{CH_2}} + C_2H_5ONa \xrightarrow{C_2H_5OH} \underset{\text{1-Ethoxy-2-butene}}{CH_3\text{-}CH=CH\text{-}\underset{\substack{| \\ OC_2H_5}}{CH_2}} + NaCl$$

(Only one product is formed)

$$\underset{\text{3-Chloro-1-butene}}{CH_3\text{-}\underset{\underset{Cl}{|}}{CH}\text{-}CH{=}CH_2} + C_2H_5ONa \xrightarrow{C_2H_5OH} \underset{\substack{\text{3-Ethoxy-1-butene}\\ \text{(Only one product is formed)}}}{CH_3\text{-}\underset{\underset{OC_2H_5}{|}}{CH}\text{-}CH{=}CH_2} + NaCl$$

In the presence of concentrated sodium ethoxide solution in ethanol, the two isomeric chlorides react to give only one product. This reaction follows S_N2 mechanism.

If the same reaction occurs by heating the both allylic substrates with ethanol in the absence of concentrated base solution, the reaction mixture contains both kind of ethers (two types of products *i.e.*, 1-ethoxy-2-butene and 3-ethoxy-1-butene). It indicates that, in solvolysis condition the above reaction follows S_N1 mechanism and if the structure permits allylic rearrangement takes place.

Nucleophilic Substitution in Allylic Substrates

S_N2 Reaction

We have already discussed the behaviours of allylic substrate in previous section, in SN^1 reaction the allylic substrates behave like secondary substrates. But in S_N2 reaction, allylic substrate is reactive as such as primary substrates. In a sense, the allyl group has the best of both words: a capacity for charge dispersal comparable to that of 2° groups, but without the bulkiness that would find direct nucleophilic attack.

Nucleophilic substitution in vinylic substrate - vinylic cations:

$$\underset{\text{Vinylic substrate}}{-\overset{|}{C}{=}\overset{|}{C}-L} + \overset{\ominus}{Nu:} \longrightarrow -\overset{|}{C}{=}\overset{|}{C}-Nu + \overset{\ominus}{L:}$$

Generally vinylic substrate is very much less reactive towards nucleophilic substitution reaction than their saturated analogues and they do not give white precipitate with $AgNO_3$ solution. The main reason for the low reactivity of vinylic substrate is the presence of unusual strong carbon-halogen bond. The main step in the nucleophilic substitution reaction (both S_N1 & S_N2) is the heterolysis of carbon–halogen bond. The heterolytic bond dissociation energy of vinyl chloride is 207 kcal, as compared with 191 kcal for ethyl chloride. The difference of 16 to 18 kcal more energy is needed to break the carbon-halogen bond in vinyl chloride. Hence the very less reactivity of vinylic substrates in nucleophillic substitution reaction is the presence of unusual strong carbon - halogen bond.

Vinylic cations easily undergoes S_N1 reaction if it fulfill the following conditions.

1. If an extremely good leaving group (super leaving group) is attached to the vinylic carbon.
2. If the vinylic group contains electron releasing substituent.

Free Radical Addition Reaction of Conjugated Dienes

Conjugated dienes undergo free radical addition reaction *via* the formation of free radicals. Like other free radical reactions it also follows the three step mechanism such as chain initiation step, chain propagation step and chain termination step. Chain initiation and chain propagation steps are similar to that of other radical addition reactions of alkenes. But the chain propagation step involves two steps.

General mechanism: It involves three steps.

1. **Chain initiation step:** Formation of peroxide free radical upon exposure to UV light.

$$R_2O_2 \xrightarrow{\text{Uv light}} 2\ RO\bullet$$

2. **Chain propagation step:**

   (i) **First propagation step:** Formation of resonance stabilized carbon free radical. It involves the initial attack of free radical at the site of carbon atom which produces more stable carbon free radical.

$$-C{=}C-C{=}C- \; + \; RO^{\bullet} \longrightarrow -C{=}C-\dot{C}-C-OR \longleftrightarrow -\dot{C}-C{=}C-C-OR$$

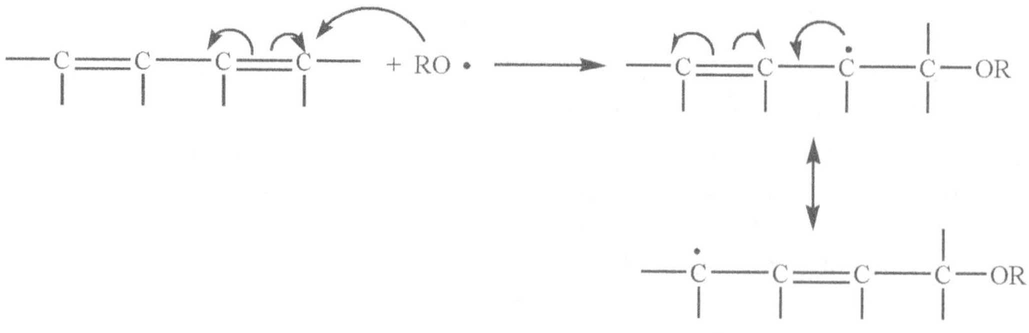

   (ii) **Second propagation step:** The more stable carbon free radical abstract a free radical from reagent and gives product.

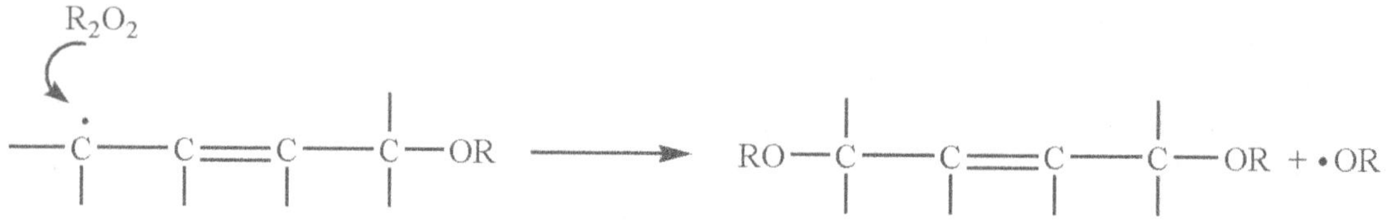

$$-\dot{C}-C{=}C-C-OR \xrightarrow{R_2O_2} RO-C-C{=}C-C-OR + {}^{\bullet}OR$$

1,4-Addition product

$$-C{=}C-\dot{C}-C-OR \xrightarrow{R_2O_2} -C{=}C-C(OR)-C-OR + {}^{\bullet}OR$$

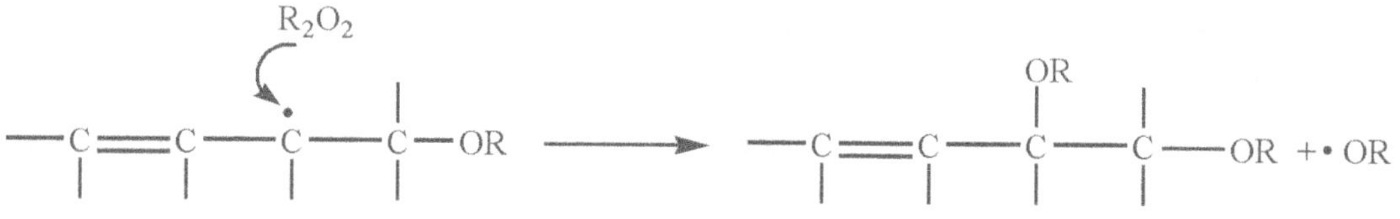

1,2-Addition product

3. **Chain termination step:** Two free radicals combine to form a molecule and terminate the reaction.

$$RO^{\bullet} + RO^{\bullet} \longrightarrow R_2O_2$$

Example 1: Polymerization reaction of butadiene.

Butadiene polymerizes in the presence of peroxide and yield polybutadiene or buna rubber. This reaction follows free radical addition mechanism.

$$n\,CH_2{=}CH{-}CH{=}CH_2 \longrightarrow \left[CH_2{-}CH{=}CH{-}CH_2 \right]_n$$

1,3-Butadiene → Polybutadiene

Mechanism: Chain initiation and chain termination steps are similar to that of general mechanism described above.

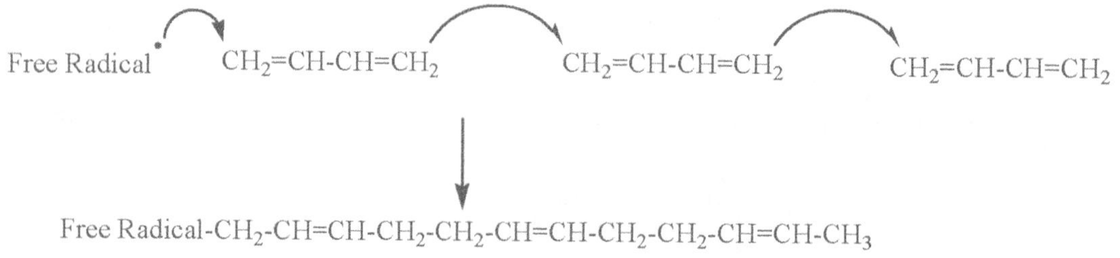

Free Radical$^{\bullet}$ $CH_2{=}CH{-}CH{=}CH_2$ $CH_2{=}CH{-}CH{=}CH_2$ $CH_2{=}CH{-}CH{=}CH_2$

↓

Free Radical-$CH_2{-}CH{=}CH{-}CH_2{-}CH_2{-}CH{=}CH{-}CH_2{-}CH_2{-}CH{=}CH{-}CH_3$

Note: In the above reaction, the initially formed butadiene free radical attacks the second molecule of butadiene and gives another free radical likewise and polymerises to yield buna rubber.

Example 2: Free radical addition of HBr to 1,3-butadiene.

$$CH_2{=}CH{-}CH{=}CH_2 + HBr \longrightarrow \underset{\text{1-Bromo-2-butene}}{\overset{Br}{\overset{|}{C}}H_2{-}CH{=}CH{-}CH_3} + Br^{\bullet}$$

Mechanism:

1. Chain initiation step:

$$H{-}Br \longrightarrow H^{\bullet} + Br^{\bullet}$$

2. Chain propagation step:

$$Br^{\bullet} + CH_2{=}CH{-}CH{=}CH_2 \longrightarrow \left[Br{-}CH_2{-}\dot{C}H{-}CH{=}CH_2 \longleftrightarrow Br{-}CH_2{-}CH{=}CH{-}\dot{C}H_2 \right]$$

$$Br{-}CH_2{-}\dot{C}H{\cdots}CH{\cdots}\dot{C}H_2 + H{\bullet\bullet}Br \longrightarrow Br{-}CH_2{-}CH{=}CH{-}CH_3 + Br^{\bullet}$$

3. Chain termination step:

$$Br^{\bullet} + H^{\bullet} \longrightarrow HBr$$

Probable Questions

1. What are alkadienes? Explain the importance of alkadienes.
2. Define and classify dienes.
3. Write a detailed note on nomenclature of alkadienes.
4. Write a short note on configurational isomers of dienes.
5. Explain various general methods of preparation of alkadienes.
6. Write a detailed note on relative stability of dienes.
7. Explain hybridization in conjugated dienes.
8. Explain the concept of resonance in dienes.
9. Explain various chemical properties of dienes.
10. Write a note on electrophilic addition reactions of alkadienes with its reaction mechanism.
11. What is 1,2-addition or direct addition & 1,4-addition or conjugate addition.
12. Compare addition reactions of isolated dienes & conjugated dienes.
13. Write about factors which determine the predominance of one product over the other in 1,2-addition & 1,4-addition reactions.
14. What is Diel's-Alder reaction? Explain the mechanism of reaction of Diels – Alder reaction.

15. Write a note on stereochemistry & applications of Diel's-Alder reaction.
16. Explain free radical addition reactions of conjugated dienes.
17. Write a short note on allylic rearrangement.
18. Explain indetail about the free radical addition reactions of conjugated dienes.

11

Cycloalkanes

Introduction

Most of the organic compounds are cyclic in nature. Antibiotics being used to treat infectious diseases have cyclic ring as a part of core structure, the carbohydrates that we eat to get energy are cyclic and the bases of nucleic acid DNA and RNA are cyclic in nature.

Cycloalkanes or cycloparaffins or polymethylenes are saturated, closed chain hydrocarbons in which the carbon atoms connected by single covalent bonds lead to the formation of rings, hence they are called as carbocyclic or homocyclic compounds. These are otherwise known as **alicyclic** compounds, the suffix **cyclic** indicates the presence of a **ring** and the prefix **ali** indicates their similarity to **aliphatic** compounds. The general formula for this series is C_nH_{2n}. Line angle formula is used to draw the structures of cycloalkanes. The general formula has fewer hydrogen atoms less than that of the formula for acyclic alkanes, because there is no end in the closed chain and no hydrogen atoms required to cap off the ends of the chain.

The first member of this series is cyclopropane. The four and five membered cycloalkanes are present in terpenes, the most important classes of alicyclic compounds. The hydrocarbons of petroleum have five and six membered cycloalkanes known as naphthenic acids as a major constituent.

   **INTERESTING FACT**   

Cyclopropane was previously used as general anesthetic since its vapour induces sleepiness and loss of consciousness. After inhalation, it easily reaches the systemic circulation due to its non-polar nature. It fastly leaves from the blood and enters into the CNS which leads to anesthesia. Due to its highly flammable nature (produce explosions when mixed with air) it is no longer used.

Importance of Cycloalkanes

Cycloalkanes are important organic compounds. Many biologically active compounds have cycloalkyl ring structure which when attached with various functional groups elicit different kind of pharmacological activity. Some of the drugs which possess cycloalkyl group and their therapeutic uses in the treatment of various diseases are shown in Table 11.1.

Table 11.1 Drugs containing cycloalkyl groups and their therapeutic uses.

Category	Drug	Structure
General anaesthetic	Cyclo propane	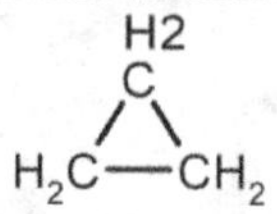 H2 C H_2C—CH_2
Antidepressant	Tranylcypromine	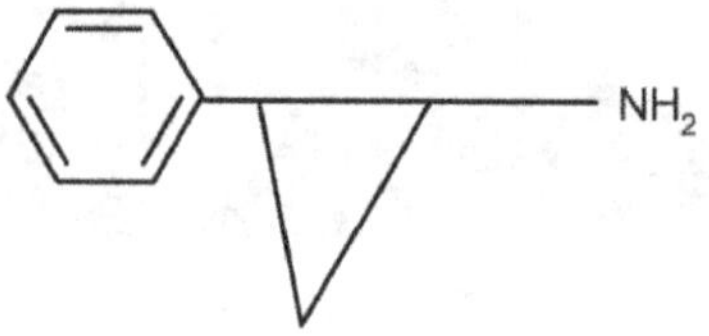 NH_2
Narcotic analgesic	Cyclazocine	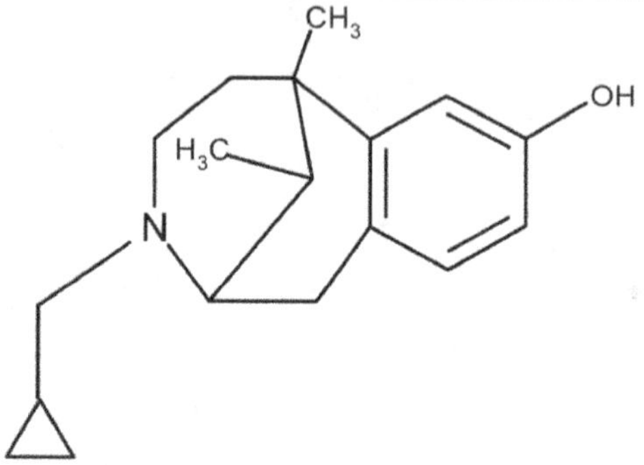 CH_3 OH H_3C N
	Ketazocine	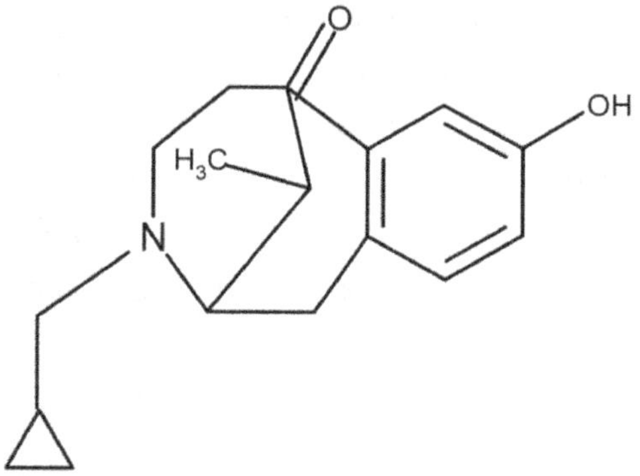 O OH H_3C N
Anticholinergic drug	Scopalamine (Hyoscine)	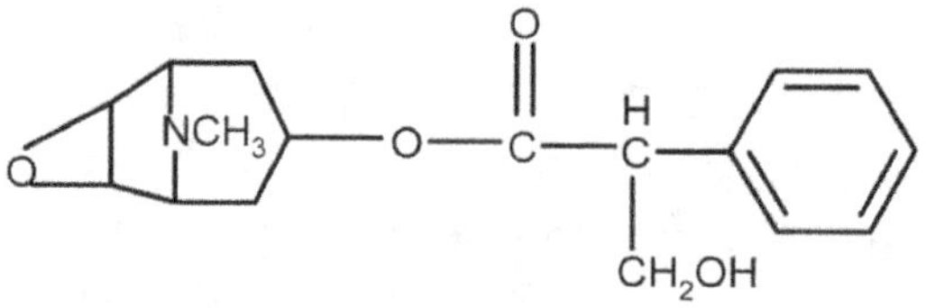 O NCH_3 O—C—C H CH_2OH

Nomenclature

IUPAC System: Named by adopting the following steps

1. The name of unsubstituted cycloalkanes are derived by adding the prefix – **cyclo** to the corresponding alkane possessing same number of carbon atoms, For example

$CH_3—CH_2—CH_3$
Propane

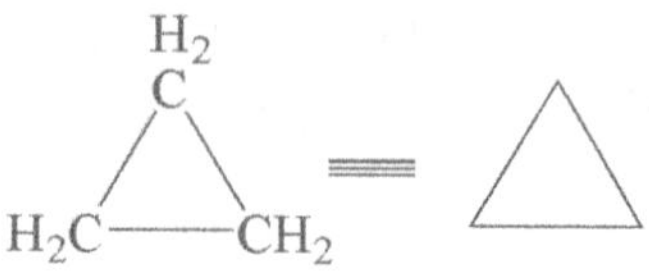

H_2 C
H_2C—CH_2 ═
Cyclopropane

2. The substituents on the ring are numbered by providing least number to the substituents and their positions are indicated in name. For example,

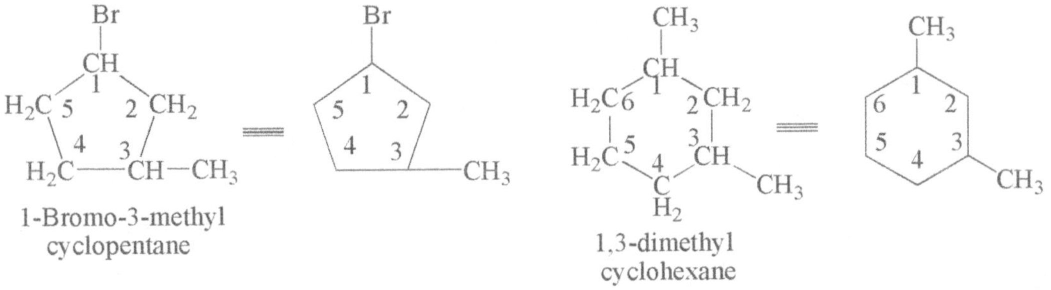

1-Bromo-3-methyl cyclopentane

1,3-dimethyl cyclohexane

General Methods of Preparation

1. **From dihalides or dihalogen derivatives of alkanes (extension of Wurtz reaction):** The dihalogen derivatives of the alkanes (terminal dihalides) are treated with sodium metal or zinc metal to yield corresponding cycloalkanes. For example: 1,3-dibromo propane reacts with Zn or Na to give cyclopropane. This method was discovered by Freund in the 1882.

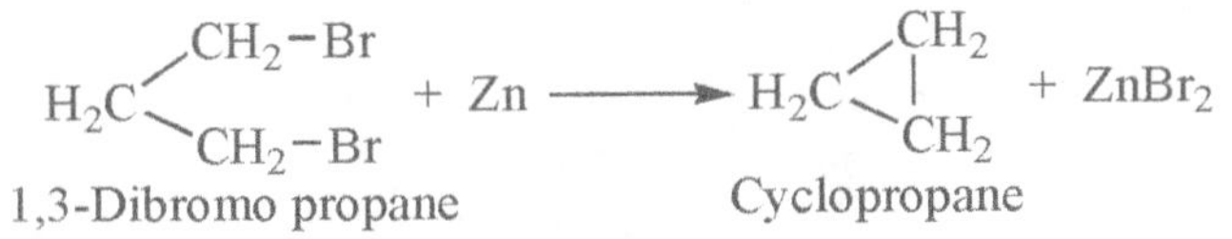

$$H_2C(CH_2-Br)_2 + Zn \longrightarrow (CH_2)_3 + ZnBr_2$$

1,3-Dibromo propane → Cyclopropane

2. **Dieckmann reaction (1901) or from esters of dicarboxylic acids or intramolecular Claisen condensation:** This reaction involves the reaction between the esters of dicarboxylic acids (adipic acid or suberic acid) with sodium or sodium ethoxide to give β-ketoester which upon hydrolysis yields cyclic ketones. This on reduction with Zn(Hg)/HCl gives the respective cycloalkanes.

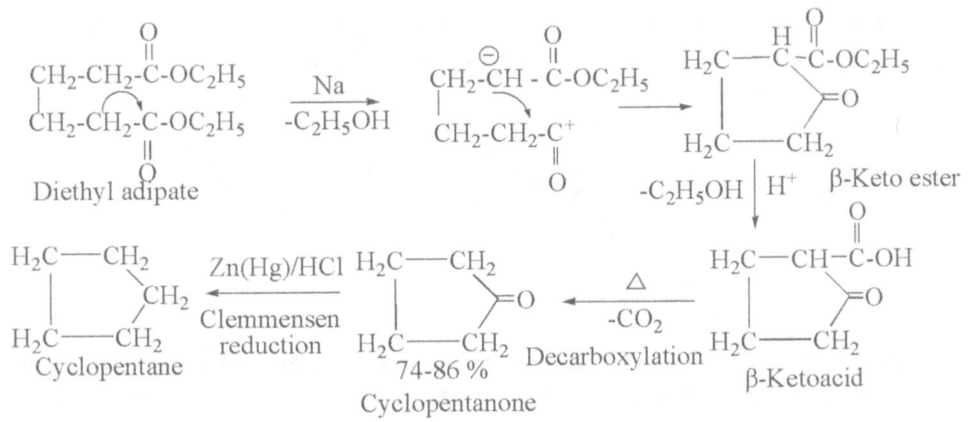

Diethyl adipate $\xrightarrow[-C_2H_5OH]{Na}$ β-Keto ester $\xrightarrow[-C_2H_5OH]{H^+}$ β-Ketoacid $\xrightarrow[\text{Decarboxylation}]{\Delta,\ -CO_2}$ Cyclopentanone (74-86 %) $\xrightarrow[\text{Clemmensen reduction}]{Zn(Hg)/HCl}$ Cyclopentane

3. **From calcium or barium salts of dicarboxylic acids:** When the barium or calcium salt of dicarboxylic acid is distilled, a cyclic ketone is obtained. For example, barium or calcium salt of adipic acid gives cyclopentanone. This method is useful for the synthesis of five, six and seven membered cycloketones which on further reduction with Zn(Hg)/HCl yields corresponding cycloalkanes.

Calcium adipate $\xrightarrow[-CaCO_3]{\Delta}$ Cyclopentanone $\xrightarrow[\text{Clemmensen reduction}]{Zn(Hg)/HCl}$ Cyclopentane

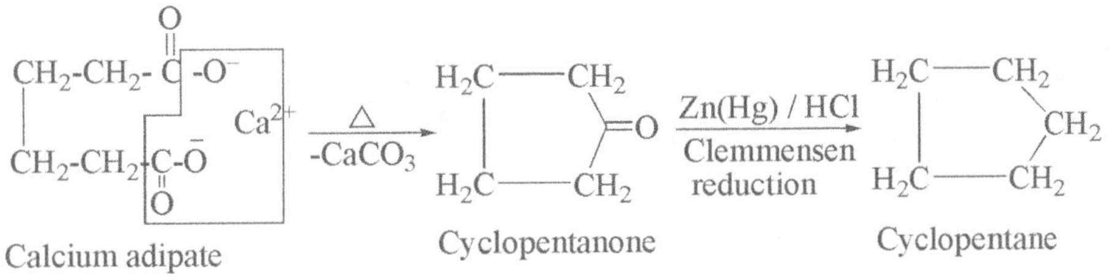

Alternatively the reduction is also effected by any one of the following methods.

Cyclopentanone $\xrightarrow[C_2H_5OH]{Na}$ (Bouveault reduction) Cyclopentanol $\xrightarrow[0\ °C,\ -H_2O]{HI}$ Cyclopentyl iodide $\xrightarrow{Zn/HCl \mid H^+}$ Cyclopentane

4. **From alkenes (Simmons - Smith reaction):** Alkenes on treating with methylene iodide (CH_2I_2) in the presence of zinc-copper couple and ether give cyclohexane.

$$H_3C{-}CH{=}CH_2 + CH_2I_2 \xrightarrow[Ether]{Zn/Cu} \text{Methyl cyclopropane} + ZnI_2$$

Propene → Methyl cyclopropane

$$CH_3{-}CH{=}CH{-}CH_3 + CH_2I_2 \xrightarrow[Ether]{Zn/Cu} \text{1,2-Dimethylcyclopropane} + ZnI_2$$

2-Butene → 1,2-Dimethylcyclopropane

With respect to alkene, the above said reaction is stereospecific. If the substituents present in the alkene are *cis* then the product obtained is also *cis* cyclo alkane only.

$$\textit{cis}\text{-2-Butene} + CH_2I_2 \xrightarrow[Ether]{Zn/Cu} \textit{cis}\ \text{1,2-Dimethyl cyclopropane} + ZnI_2$$

5. **From benzene and its derivatives:** Six membered alicyclic compounds are prepared from benzene and its derivatives by catalytic reduction using nickel catalyst under pressure.

$$\text{Benzene} + 3H_2 \xrightarrow[200\ °C]{Ni} \text{Cyclohexane}$$

$$\text{Phenol} + 3H_2 \xrightarrow[200\ °C]{Ni} \text{Cyclohexanol}$$

6. **Preparation of cyclobutane:** Cyclobutane is also prepared by malonic ester synthesis.

$$\underset{\text{1,3-Dibromopropane}}{CH_2Br-CH_2-CH_2Br} + \underset{\text{Diethyl malonate}}{H_2C(COOC_2H_5)_2} \xrightarrow[\text{-2C}_2\text{H}_5\text{OH, -2NaBr}]{2NaOC_2H_5} \text{cyclobutane-1,1-}(COOC_2H_5)_2$$

$$\xrightarrow[\text{-2C}_2\text{H}_5\text{OH}]{\text{Hydrolysis}} \text{cyclobutane-1,1-}(COOH)_2 \xrightarrow[-CO_2]{\Delta} \text{cyclobutane-}COOH \xrightarrow[\text{-2HCl}]{\text{i)PCl}_5\text{ ii)NH}_3} \text{cyclobutane-}CONH_2$$

$$\xrightarrow[-CO]{Br_2/KOH} \text{cyclobutane-}NH_2 \xrightarrow[\text{AgOH, - 2HI, -AgI}]{3CH_3I} \text{cyclobutane-}\overset{+}{N}(CH_3)_3OH^- \xrightarrow[\text{- N(CH}_3\text{), - H}_2\text{O}]{\text{Heat}} \underset{\text{Cyclobutene}}{C_4H_6} \xrightarrow{H_2/Ni} \underset{\text{Cyclobutane}}{\square}$$

Physical Properties

Most of the physical properties of cycloalkanes are similar to that of the physical properties of the non-cyclic, open chain alkanes. Cyclopropane and cyclobutane, the first two members of this series are gases at room temperature, next three are liquids and higher members are solid in nature. They are non-polar, insoluble in water but soluble in methanol and ether.

The boiling point and melting point of cycloalkanes depends upon their molecular weight. As the molecular weight increases, boiling point and melting point also increases. The physical properties of some common cycloalkanes are listed in Table 11.2.

Table 11.2 Physical properties of common cycloalkanes.

Name of the cyclo alkanes	Molecular formula	Boiling point (°C)	Melting point (°C)	Density
Cyclopropane	C_3H_6	– 33	– 128	0.72
Cyclobutane	C_4H_8	– 12	– 50	0.75
Cyclopentane	C_5H_{10}	49	– 94	0.75
Cyclohexane	C_6H_{12}	81	7	0.78
Cycloheptane	C_7H_{14}	118	12	0.81
Cyclooctane	C_8H_{16}	148	14	0.83

Chemical Properties

The chemical properties of cycloalkanes are similar to alkanes in many chemical aspects but the lower members such as cyclopropane and cyclobutane forms addition products by ring opening reactions. The chemical reactions of cycloalkanes are explained in two major headings.

I. Ring opening reactions.
II. Substitution reactions.

1. **Ring opening reactions:**
   (i) **Addition of halogens:** Cycloalkanes react with chlorine or bromine in the presence of carbon tetrachloride or carbon tetrabromide in dark to form halogenated cycloalkanes. But cyclopropane

reacts with chlorine in the presence of CCl_4 in dark to give a mixture of products such as 1,1, 1,2 and 1,3- dichloropropanes.

$$\underset{\text{Cyclopropane}}{(CH_2)_3} + Cl_2 \xrightarrow[\text{in dark}]{CCl_4} \underset{\text{1,1-Dichloropropane}}{CH_3\text{-}CH_2\text{-}CHCl_2} + \underset{\text{1,2-Dichloropropane}}{CH_3\text{-}CHCl\text{-}CH_2\text{-}Cl} + \underset{\text{1,3-Dichloropropane}}{ClCH_2\text{-}CH_2\text{-}CH_2\text{-}Cl}$$

(ii) Reaction with hydrogen iodide (HI): Generally concentrated HI does not react with cycloalkanes except with lower members. Cyclopropane, cyclobutane reacts with HI at room temperature to yield 1-iodopropane or *n*-propyl iodide and *n*-butyl iodide respectively.

$$\underset{\text{Cyclopropane}}{(CH_2)_3} + HI \xrightarrow{\text{Room temperature}} \underset{\text{1-Iodopropane}}{CH_3\text{-}CH_2\text{-}CH_2I}$$

$$\underset{\text{Cyclobutane}}{(CH_2)_4} + HI \xrightarrow{\text{Room temperature}} \underset{n\text{-Butyl iodide}}{CH_3\text{-}CH_2\text{-}CH_2\text{-}CH_2I}$$

(iii) Reaction with hydrogen bromide: Cyclopropane reacts with HBr at room temperature to yield *n*-bromo propane.

$$\underset{\text{Cyclopropane}}{(CH_2)_3} + HBr \longrightarrow \underset{n\text{-Bromopropane}}{CH_3\text{-}CH_2\text{-}CH_2Br}$$

(iv) Reaction with hydrogen: Cyclopropane and cyclobutane reacts with hydrogen in the presence of nickel catalyst at elevated temperature (80 °C for cyclopropane and 200 °C for cyclobutane) to yield *n*-propane and *n*-butane respectively.

$$\underset{\text{Cyclopropane}}{(CH_2)_3} + H_2 \xrightarrow[\text{at 80 °C}]{Ni} \underset{n\text{-Propane}}{CH_3\text{-}CH_2\text{-}CH_3}$$

$$\underset{\text{Cyclobutane}}{(CH_2)_4} + HI \xrightarrow[\text{at 200 °C}]{Ni} \underset{n\text{-Butane}}{CH_3\text{-}CH_2\text{-}CH_2\text{-}CH_3}$$

(v) Reaction with sulphuric acid: Cyclopropane reacts with sulphuric acid to yield propane sulphonic acid. Higher cycloalkanes do not give this reaction.

$$\underset{\text{Cyclopropane}}{(CH_2)_3} + H_2SO_4 \longrightarrow \underset{\text{Propane sulphonic acid}}{CH_3\text{-}CH_2\text{-}CH_2\text{-}OSO_3H}$$

(vi) Oxidation: Cyclohexane undergoes oxidation with hot potassium permanganate solution to give dicarboxylic acids such as adipic acid.

$$\text{Cyclohexane} + 5(O) \xrightarrow[\text{KMnO}_4/\text{OH}^-]{\text{Hot}} \underset{\text{Adipic acid}}{\begin{array}{l} CH_2\text{-}CH_2\text{-}COOH \\ | \\ CH_2\text{-}CH_2\text{-}COOH \end{array}} + H_2O$$

Substituted cycloalkanes react with unsymmetrical reagents (HBr) and results in the breaking of C-C bond in the ring with the formation of 2-substituted alkane as the major product.

$$\underset{\text{Methyl cyclopropane}}{CH_3\text{-}CH(CH_2\text{-}CH_2)} \xrightarrow{H^+} \underset{\text{Butyl carbocation intermediate}}{CH_3\text{-}\overset{+}{C}H\text{-}CH_2\text{-}CH_3} \xrightarrow{X^-} \underset{\text{2-Substituted butane}}{CH_3\text{-}CH(X)\text{-}CH_2\text{-}CH_3}$$

II. Substitution reactions:

(i) Reaction with halogens: Cycloalkanes react with chlorine and bromine in the presence of scattered light or UV light to give substituted products.

$$\underset{\text{Cyclopropane}}{C_3H_6} + Cl_2 \xrightarrow{\text{UV light}} \underset{\text{Chloro cyclopropane}}{C_3H_5Cl} + HCl$$

$$\underset{\text{Cyclohexane}}{C_6H_{12}} + Cl_2 \xrightarrow{\text{UV light}} \underset{\text{Chlorocyclohexane}}{C_6H_{11}\text{-}Cl} + HCl$$

Summary of Chemical Reactions

Cyclopropane (H_2C–CH_2–CH_2 ring) →

- $\xrightarrow{Cl_2, CCl_4 \text{ in dark}}$ $CH_3CH_2CHCl_2 + CH_3CH(Cl)CH_2Cl + CH_2(Cl)CH_2CH_2\text{-}Cl$ (1,1-Dichloropropane, 1,2-Dichloropropane, 1,3-Dichloropropane)
- $\xrightarrow{HI}$ $CH_3\text{-}CH_2\text{-}CH_2I$ (1-Iodopropane)
- $\xrightarrow{HBr}$ $CH_3\text{-}CH_2\text{-}CH_2Br$ (*n*-Bromopropane)
- $\xrightarrow{Ni, H_2 \text{ at } 80\,°C}$ $CH_3\text{-}CH_2\text{-}CH_3$ (*n*-Propane)
- $\xrightarrow{H_2SO_4}$ $CH_3\text{-}CH_2\text{-}CH_2\text{-}OSO_3H$ (Propane sulphonic acid)
- $\xrightarrow{Cl_2 \text{ UV light}}$ C_3H_5Cl (CH–Cl, H_2C, CH_2 ring) + HCl (Chlorocyclopropane)

Bayer's Strain Theory or Stability of Cycloalkanes or Ring Strain

Early researchers found that the cyclic compounds present in nature generally had five or six membered ring structures. Three or four membered rings are less frequently present in nature. The above observation suggested that compounds with three and four membered rings were not stable when compared with five and six membered ring compounds. In the late nineteenth century, the German chemist Adolf Van Bayer first attempted to explain the instability of three and four membered rings and he proposed that the instability is due to angle strain. The theory says that the normal angle between any pair of bonds of carbon atom is 109° 28′ or the angle between the centre of regular tetrahedran is 109° 28′ which lies between the values of angles in a regular pentagon (108°) and a regular hexagon (120°). **Bayer postulated that any deviation of bond angles from the normal tetrahedral value would impose a condition of internal strain on the ring it is called as Bayer's strain theory.** He determined that all cycloalkanes were planar with an angle of a regular triangle 60°.

But bond angle in a cyclopropane are compressed from the desired tetrahedral angle of 109° 28′ to 60° (Bayer's strain theory based on the mechanical concept of valency). The deviation of bond angle from the desired bond angle leads to the strain, known as **angle strain.**

This indicates that the normal tetrahedral angle of 109° 28′ is compressed to 60° and then each of the two bonds involved is pulled which is calculated by ½ (109° 28′-60°) = 24.44. This value represents the angle strain or the deviation through which each bond bends from the normal tetrahedral direction. Some of the cycloalkanes and their respective bond angle and angle strain values are depicted in the following Table 11.3.

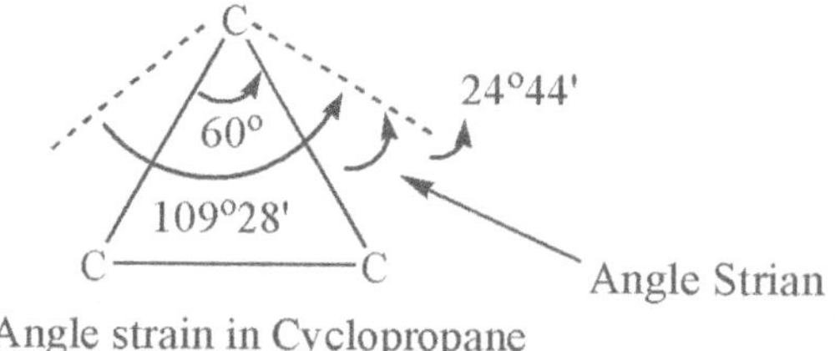

Angle strain in Cyclopropane

Table 11.3 Angle strain present in different cycloalkanes.

S. No.	Name of the cycloalkane	Bond angle	Angle strain
1.	Cyclopropane	60°	24° 44′
2.	Cyclobutane	90°	9° 44′
3.	Cyclopentane	108°	0° 44′
4.	Cyclohexane	120°	– 5° 16′

From the above table, it is learnt that the angle strain is maximum in the case of cyclopropane and minimum for cyclopentane. According to Bayer's principle cyclopropane has high strain so it is unstable and easily undergoes the ring opening reactions at room temperature.

In case of cyclobutane, the angle strain is less than that of cyclopropane, so it possesses more stability. *i.e.* it undergoes ring opening reactions at drastic conditions or elevated temperature.

Considering the minimum angle strain of cyclopentane, it possesses least strain and it has greater stability. It indicates that cyclopentane does not undergo ring opening reactions. Considering cyclohexane, the angle strain is higher than that of cyclopentane. The angle strain value increases along with the increasing number of carbon atoms in the ring. Depending upon the Bayer's postulations, cyclohexane and higher cycloalkanes are unstable and more reactive. In contrary to this, cyclohexanes and higher cycloalkanes are quite stable, do not give ring opening reactions but they undergo substitution reactions. The above facts indicate that Bayer's strain theory explains the exceptional or relative stability of first three members or lower cycloalkanes.

Molecular Orbital Theory of Cycloalkanes

The relative stability of cycloalkanes is explained by molecular orbital theory. Formation of a covalent bond involves the overlapping of orbitals of the two atoms which are involved in the bond formation. It indicates that greater the overlap, stronger the bond.

Consider the angle strain in cyclopropane. It can be explained on the basis of orbitals overlap leads to a formation of covalent bond. Generally, σ bonds are formed by the overlap of two sp^3 orbitals through their axis which point directly towards each other. In cyclopropane overlapping orbitals do not point directly towards each other. So the orbital overlap is less effective in cyclopropane when compared with normal alkanes. The above overlap of orbitals which produces the C-C bond is weaker than that normal C-C bond in alkanes which leads to angle strain and the shapes of the C-C bond in cyclopropane is like a banana, so it is called as **banana bond**.

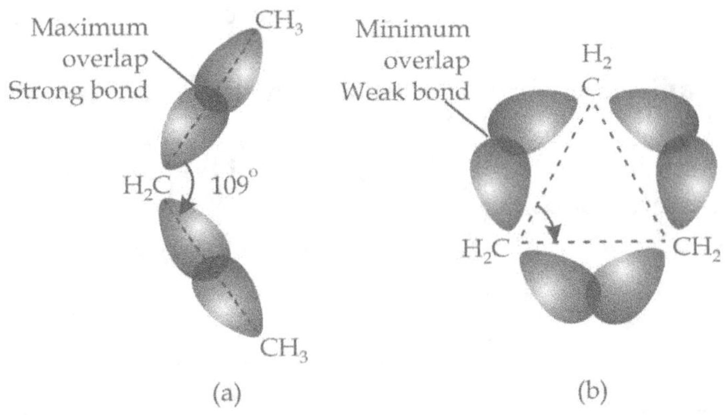

Figure 11.1 sp^3 orbitals overlap in a) Propane b) Cyclopropane. Maximum overlap occurs in propane.

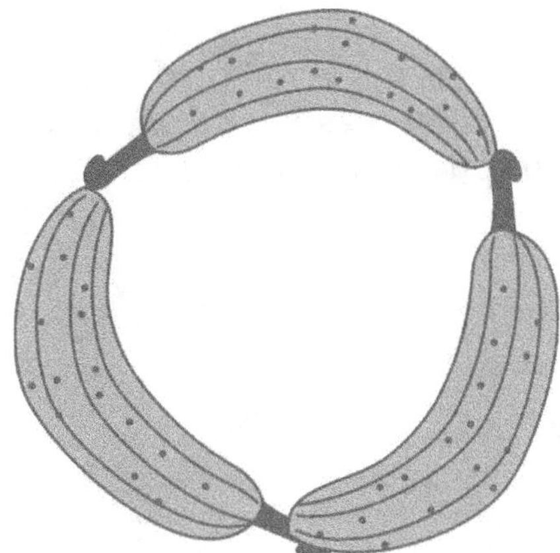

Figure 11.1(a) Banana bonds or bent bonds in cyclopropane formed by the overlap of C-C orbitals.

Along with angle strain, the cyclopropane possesses torsional strain since all the adjacent carbon hydrogen (C-H) bonds are eclipsed.

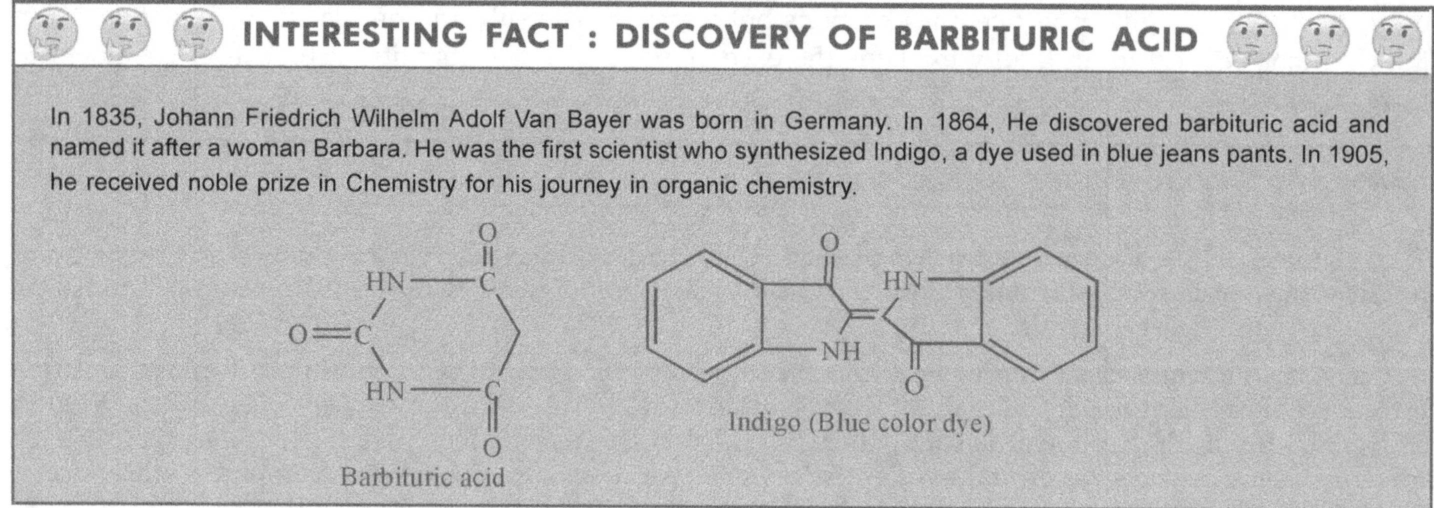

INTERESTING FACT : DISCOVERY OF BARBITURIC ACID

In 1835, Johann Friedrich Wilhelm Adolf Van Bayer was born in Germany. In 1864, He discovered barbituric acid and named it after a woman Barbara. He was the first scientist who synthesized Indigo, a dye used in blue jeans pants. In 1905, he received noble prize in Chemistry for his journey in organic chemistry.

Type of Strains

All of the cyclic compounds are not planar except cyclopropane. Cyclic compounds twist and bend to get a structure which is used to minimize the three different kind of strain that destabilizes the ring system. They are as follows:

1. **Angle strain:** It is the strain induced in a molecule when the C-C bond angle is different from the normal tetrahedral bond angle such as 109° 28′.
2. **Torsional strain:** It is the strain induced by the repulsion of bonding electrons of one substituent with the bonding electrons of nearby substituent.
3. **Steric strain:** It is caused by the atoms or group of atoms approaching too close to each other.

Sache-Mohr Theory

In 1918 Sache and Mohr explained about the stability of cyclohexane and higher members. According to this theory, cyclohexane and higher members are free from the strain if all the ring carbons are not forced into one plane which was postulated by Bayer's. If the cyclohexane ring is assumed a **folded or puckered** condition, the normal tetrahedral bond angle value 109° 28′ is maintained which leads to the relieving of strain within a ring.

For example, cyclohexane exhibits non-planar puckered conformation free from strain totally. It exhibits in two forms. Due to their shape these are known as boat form and chair form. This type of non-planar strain free ring in which the carbon atom possesses normal tetrahedral angle of 109°28′ is possible for higher cycloalkanes also (Fig. 11.2).

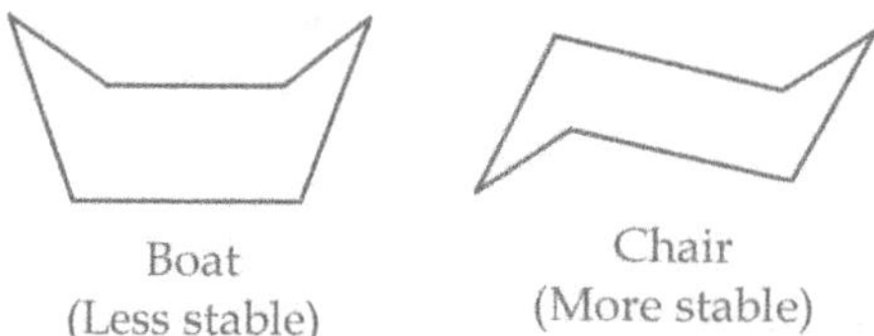

Figure 11.2 Boat and chair conformation of cyclohexane.

Structure of Two Forms of Cyclohexane

Under ordinary conditions, cyclohexane exists in chair form and boat form. Types of hydrogen atoms present in the chair form of cyclohexane are axial hydrogens and equatorial hydrogens.

Axial Hydrogens

Six hydrogen atoms are vertical and alternative above and below the rings (perpendicular to the plane) are called as axial hydrogens.

Axial bond

Equatorial bond

Equatorial Hydrogens

The remaining hydrogen atoms in cyclohexane are present slightly above or below the plane of cyclohexane ring are called as equatorial hydrogens.

Conformations of Cycloalkanes

Most common naturally available cyclic compounds contain six member rings because such rings exist in almost completely strain free conformation. This is called as chair conformation.

In the chair conformation, the bond angle is 111° which are near or closer to tetrahedral bond angle 109°28′ and all the adjacent carbon atoms are staggered which increases the stability.

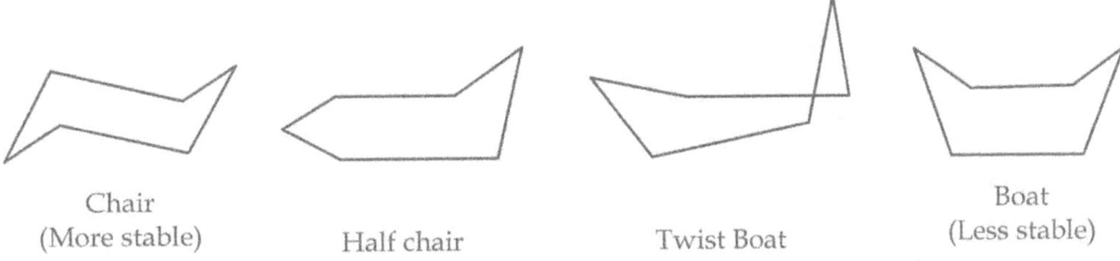

Figure 11.3 Different conformations of cyclohexane.

Drawing Chair Form of Cyclohexane Ring

The chair conformer of cyclohexane is important and more stable one. It should be drawn by as follows.

1. Draw two parallel lines of the same length, slanted upward position. Both lines should start at the same height.

2. Connect the tops of lines with a V; the left hand of the V should be slightly longer than the right hand side. Connect the bottoms of lines with an inverted V; the lines of the both V should be parallel. It forms the complete frame work of six membered ring system.

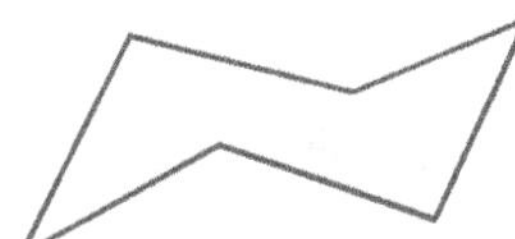

3. Each carbon possesses axial and equatorial hydrogens. The axial bonds are vertical and alternate above and below the ring. The one on the uppermost carbon is up and the next are down, the next is up and so on.

Axial bond

4. The equatorial bonds point towards out from the ring. If the axial bond is up, the equatorial bond on the same carbon is below the plane perpendicular to the bottom of the axial bond. If the axial bond is down, the equatorial bond on the same carbon is above the plane perpendicular to the bottom of the axial bond.

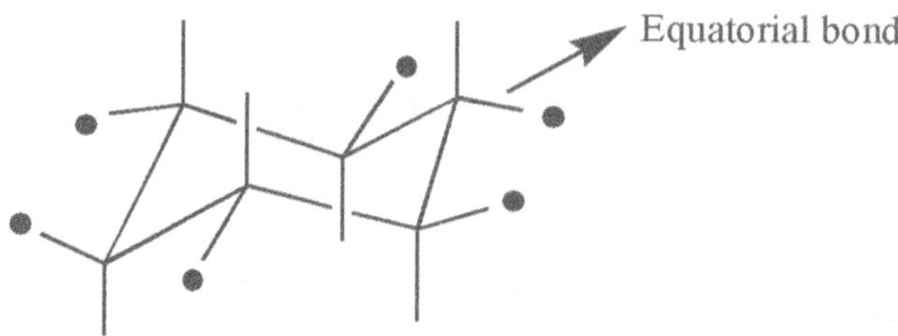

5. Observe that each equatorial bond is parallel to a ring bond two carbons over. The cyclohexane ring is always viewed on edge. The lower bonds of the ring are in front and the upper bonds of the ring are in back.

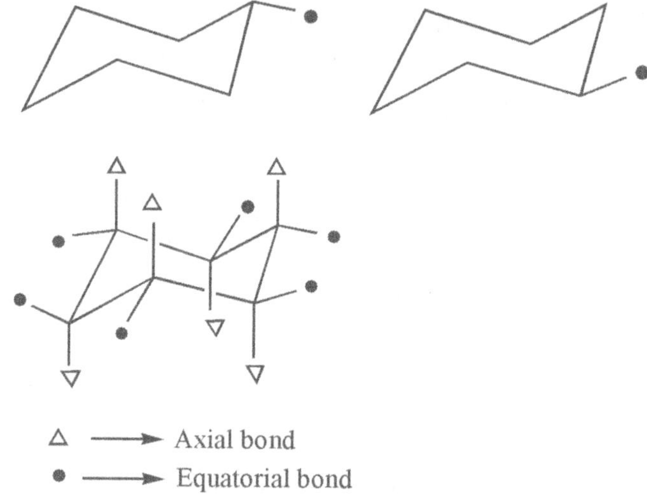

Due to the presence of rotation about carbon-carbon single covalent bond, cyclohexane interconverts between two stable chair conformations. This interconversion is called as ring flip (Fig 11.4).

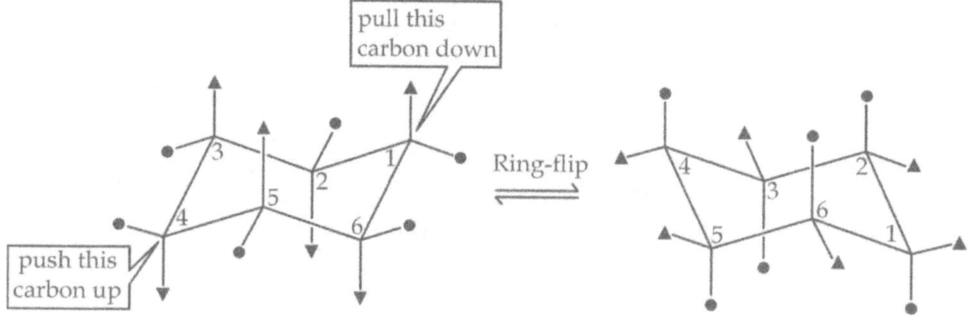

Figure 11.4 Ring flip in cyclohexane.

Cyclohexane can also be presented in boat conformation that is shown in figure 11.5.

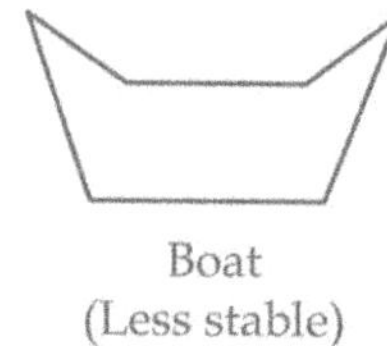

Boat
(Less stable)

Figure 11.5 Boat conformation of cyclohexane.

It is also free from angle strain, but it is less stable since some of the bonds are eclipsed which leads to torsional strain. It is also destabilized by the hydrogen atoms which are positioned in bow and stem position of the boat (flagpole hydrogen) which induces steric strain.

Interconversion of Conformations of Cyclohexane

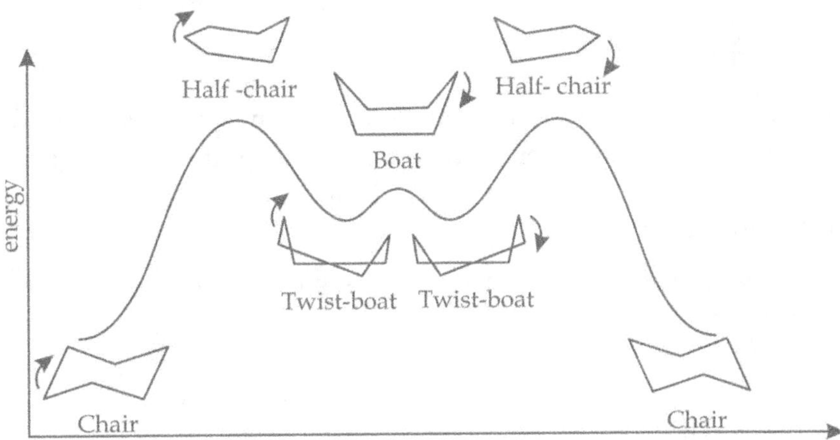

Figure 11.6 Various conformations of cyclohexane and their relative energies.

The inter conversion of one conformer to another conformer of cyclohexane is shown in graph.

1. To convert from the boat conformer to one of chair conformer, the top most carbon atoms of the boat conformer must be pulled down, it is converted to the bottom most carbon.
2. **Twist board** or **skew boat** conformer: It is obtained by just slightly pulling down the top most carbon. This conformer is more stable than that of boat conformer since the flag pole hydrogens in the boat conformer is far apart which leads to slight reduction in the steric strain.
3. **Half chair conformer:** When the top most carbon is pulled down to the point from where it is in the same plane as the sides of the boat then it leads to unstable half chair conformer.
4. Pulling the top most carbon down further produces chair conformer.
5. The energy barrier of cyclohexane molecule when it undergoes interconversion is shown in the Figure 11.6.

The energy barrier value of cyclohexane during the interconversion is calculated 10.8 kcal/mol which shows that cyclohexane undergoes 10^5 ring flips per second at room temperature (the two chair forms are in equilibrium). The chair conformers of cyclohexane are more stable and more number of chair conformers is exhibited than that of other conformers.

INTERESTING FACT

Some of the highly strained cycloalkanes are synthesized by researchers. They are cyclo (1.1.0) butane, cubane, prismane. Octanitro cyclooctane is most powerful explosive.

Conformations of Mono Substituted Cyclohexanes

In previous section we discussed about the two equivalent chair conformers of cyclohexane. Unlike cyclohexane, the two chair conformers of mono substituted cyclohexane are not equivalent. In the substituted cyclohexane, the methyl group is present in an equatorial position in one conformer and in an axial position in another conformer.

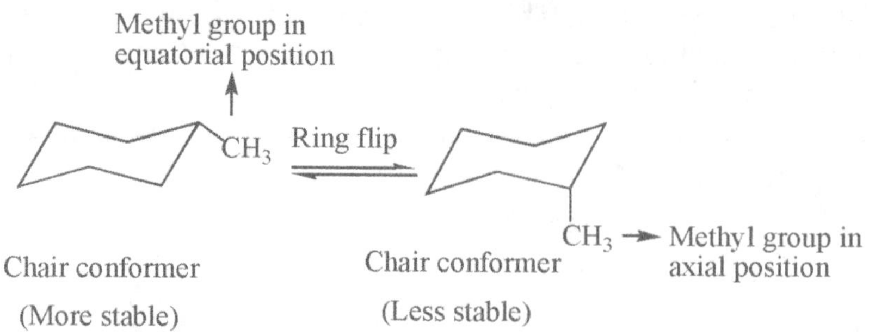

Methyl group in equatorial position

CH_3 Ring flip

CH_3 → Methyl group in axial position

Chair conformer (More stable)

Chair conformer (Less stable)

The chair conformer with methyl group in equatorial position is more stable because, a substitution has more space, therefore, lesser steric interactions when it is in an equatorial position. When the methyl group is in an equatorial position, it is anti to the C-3 and C-5 carbons. So, the methyl group extends into space away from the rest of the molecules.

In contrast, the methyl group is at axial position, it is gauche to the C-3 and C-5 carbons. These two positions are depicted in the following figure.

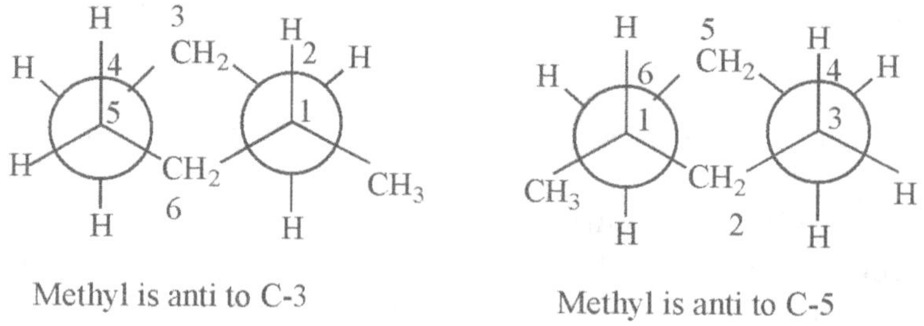

Methyl is anti to C-3

Methyl is anti to C-5

An equatorial substituent on the C-1 carbon is anti to the C-3 and C-5 carbons.

Methyl is gauche to C-3

Methyl is gauche to C-5

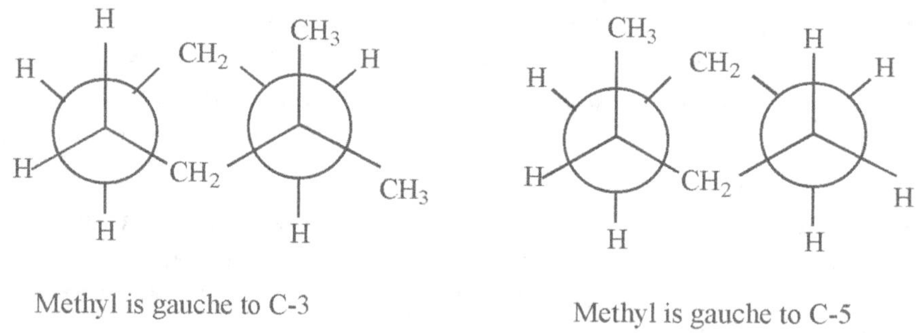

Due to this, the three axial bonds on the same side of the ring are parallel to each other; therefore, any axial substituent will be closer to axial substituents on the other two carbon atoms. Because the interacting substituents are on 1,3-positions relative to each other. Unfavourable steric interactions are known as 1,3-diaxial interactions.

Conformations of Disubstituted Cyclohexanes

If the cyclohexane possesses two substituents, then both substituents are to be considered while determining which of the two chair conformers is a more stable one. Let us consider 1,4-dimethyl cyclohexane. It exhibits geometrical isomerism (*cis-trans* isomers). One isomer is *cis* –isomer (both methyl groups are present in the same side of the cyclohexane ring) and the second isomer is the *trans*-isomer in which the two methyl groups are present in the opposite side of the cyclohexane ring.

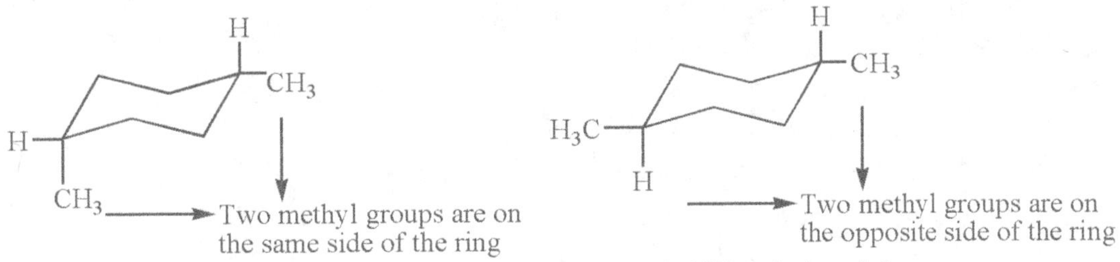

H
CH₃ → Two methyl groups are on the same side of the ring

cis-1,4-Dimethyl cyclohexane

trans-1,4-Dimethyl cyclohexane → Two methyl groups are on the opposite side of the ring

First of all we shall determine which of the two chair conformers of *cis*-1,4-dimethyl cyclohexane is more stable. For finding out this the following two structures are considered.

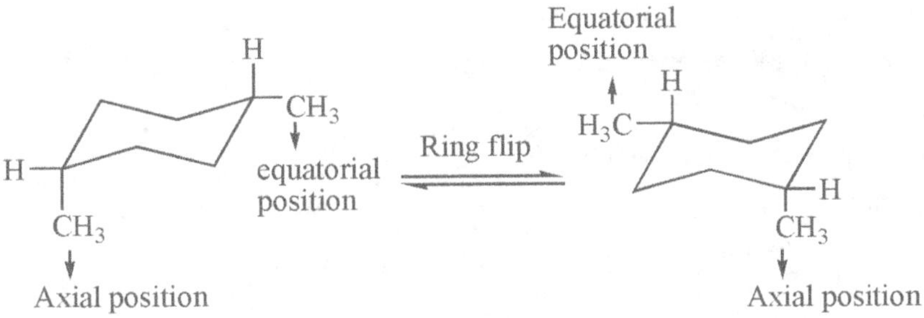

CH_3 equatorial position; CH_3 Axial position ⇌ (Ring flip) H_3C Equatorial position; CH_3 Axial position

In the above structures, both conformers are having one methyl group in equatorial position and other methyl group in axial position. So, both chair conformers of *cis*-1,4-dimethyl cyclohexane are more stable.

Coulson and Moffit's Theory of Maximum Overlapping of Carbon Orbitals

According to valency bond theory, the reactivity of double bond is explained by the presence of π electrons. When four identical atoms or groups are attached to the carbon atom, the four valences of carbon atom are equivalent and the bond angle or valency angle is 109° 28. When the four atoms or groups are different, then the four valences are not equivalent and the bond is not 109° 28′ (four corners of irregular tetrahedran).

According to Bayer's strain theory, the bond angle of cyclopropane is an equilateral triangle of 60°. This value is impossible as the carbon valency angles can never be less than 90°. But the angles are more than 90° when hybridisation occurs.

According to Coulson, the smallest carbon valency angle which one can reasonably expect to have is 10°. So, he proposed that in cyclopropane a partial overlap of *sp* hybrid orbitals occurs, so the C-C-C bond is decreased slightly (104°) and the H-C-H angle (120°) is opened from the normal tetrahedral angle. Infact the orbitals of carbon atom overlap at an angle to give bent bonds. Like cyclopropane, cyclobutane also possess bent bonds but the decrease in overlap is less than cyclopropane. Hence cyclobutane is more stable and less reactive than cyclopropane.

Probable Questions

1. What are cycloalkanes? Explain the importance of cycloalkanes.
2. Write a detailed note on nomenclature of cycloalkanes.
3. List out various general methods of preparation of cycloalkanes and explain any four methods in detail.
4. Explain Dieckmann reaction.
5. How cycloalkanes are synthesized by intra molecular Claisen condensation?
6. Write a note on Simmons – Smith reaction.

7. Write in detail about the relative stability of cycloalkanes
8. Enumerate the chemical reactions of cycloalkanes.
9. What is mean by Bater's strain theory and explain it.
10. Explain in detail about the various conformations of cycloalkanes.
11. How are cycloalkanes synthesized from benzene & its derivatives?
12. Describe physical properties of cycloalkanes.
13. Explain various chemical properties of cycloalkanes.
14. Write a note on ring opening reactions of cycloalkanes
15. Write short notes on substitution reactions of cycloalkanes.
16. Explain Bayer's strain theory.
17. Write a note on stability of cycloalkanes.
18. What is ring strain?
19. Explain in detail about molecular orbital theory of cycloalkanes.
20. Write a note on Sache-Mohr theory.
21. Write short notes on conformations of cycloalkanes.
22. Explain in detail about conformations of mono substituted cyclohexanes.
23. Write conformations of disubstituted cyclohexanes in detail.
24. What is Coulson and Moffit's theory? Explain.
25. Write a note on cyclopropane.
26. Write the chemical structure, synthesis & chemical reactions of cyclobutane.

12

Alcohols

Introduction

Alcohols are hydroxy (-OH) derivative of saturated hydrocarbons in which one or more hydrogens are replaced by hydroxyl groups. The general formula for a simple alicyclic alcohol is $C_nH_{2n+1}OH$. If the hydrocarbon contains one hydroxyl group, then it is called monohydric alcohols, if it contains two, three or more it is called as dihydric, trihydric or polyhydric alcohols respectively. If the carbon atom is attached with more than one -OH group it becomes unstable, hence hydroxyl groups are always attached to different carbon atoms.

CH_3CH_2OH	CH_2OH–CH_2OH	CH_2OH–$CHOH$–CH_2OH
Monohydric alcohol (Ethanol)	Dihydric alcohol (Ethylene glycol)	Trihydric alcohol (Glycerol)

Importance of Alcohols

Alcohols either exist free or as esters in nature. Esters of monohydric alcohols example methyl salicylate (oil of wintergreen) is used as fragrance and counter irritant. Esters of long chain alcohols like cetyl alcohol ($C_{16}H_{33}OH$) are important waxes. Terpene alcohols like geraniol, menthol are the important constituents of volatile oils. The dihydric and trihydric alcohols like panaxadiol, glycerol and polyhydric alcohols like erythritol exist in nature as important chemical constituents.

Alcohols find wide applications in our day to day life. Many of the natural and synthetic drugs also have the alcoholic function in it and elicit various pharmacological activity based on the core structure to which it is attached. Some of the drugs which possess alcohol group and their therapeutic applications in the treatment of various diseases are shown in Table 12.1.

Table 12.1 Drugs containing alcohol function and their therapeutic uses.

Category	Drug	Structure
Sedative and hypnotic	Amylene hydrate	CH_3–C(CH_3)(OH)–C_2H_5

Table 12.1 *Contd...*

Category	Drug	Structure
Sedative and hypnotic	Chlorbutanol	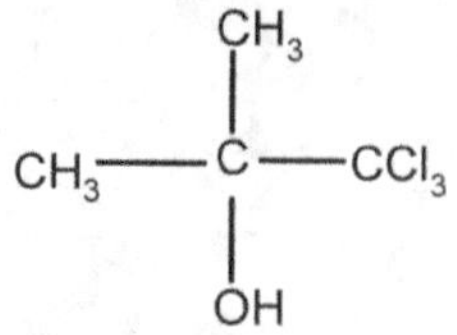 $CH_3-C(CH_3)(OH)-CCl_3$
	Chloral hydrate	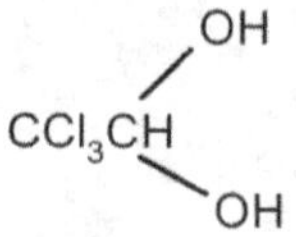 $CCl_3CH(OH)_2$
	Methyl pentynol	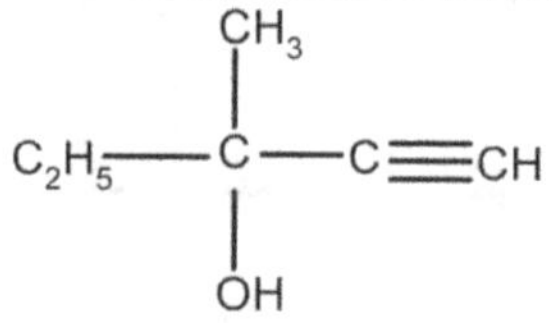 $C_2H_5-C(CH_3)(OH)-C\equiv CH$
General anaesthetic	Tribromo ethanol	Br_3CCH_2OH
Narcotic analgesic	Morphine	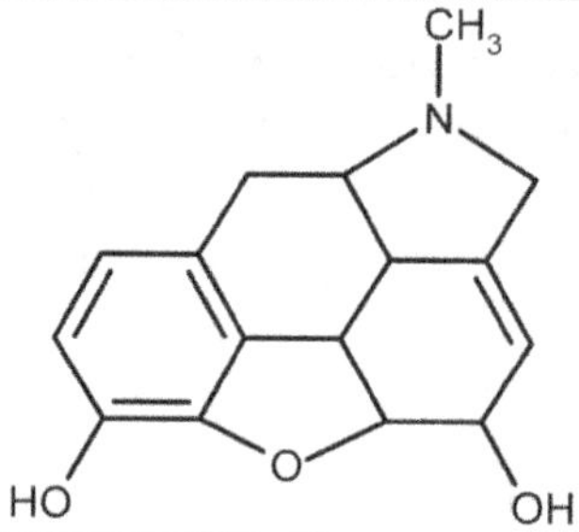
Antihyperlipidemic drug	Fluvastatin	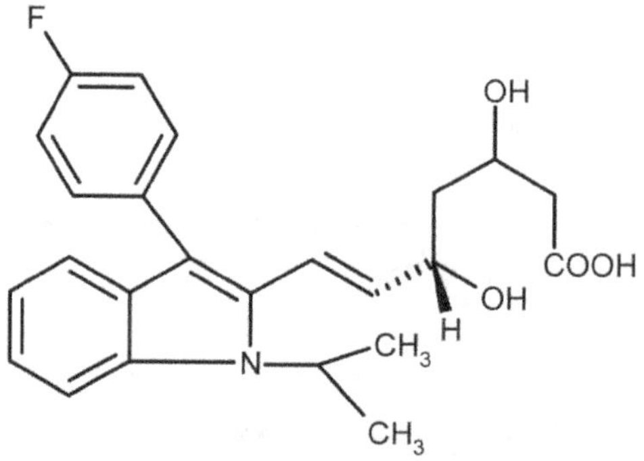
Tranquillizer	Haloperidol	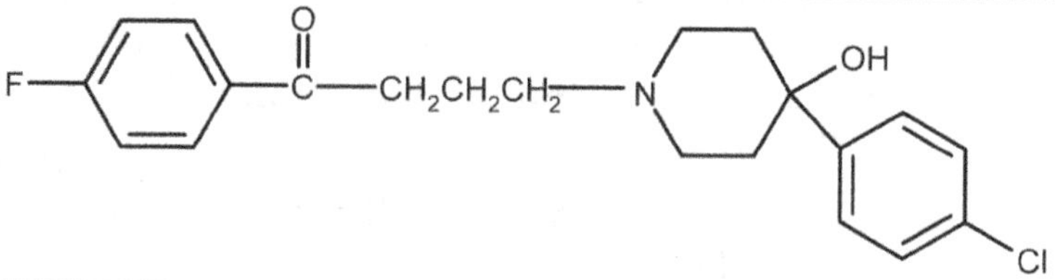
Adrenergic drug	Adrenaline	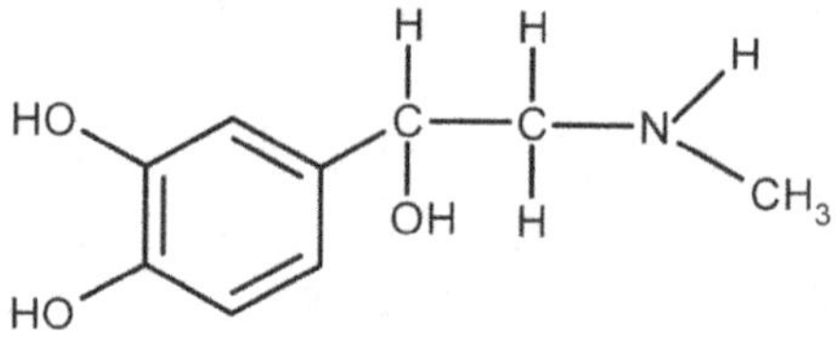
	Noradrenaline	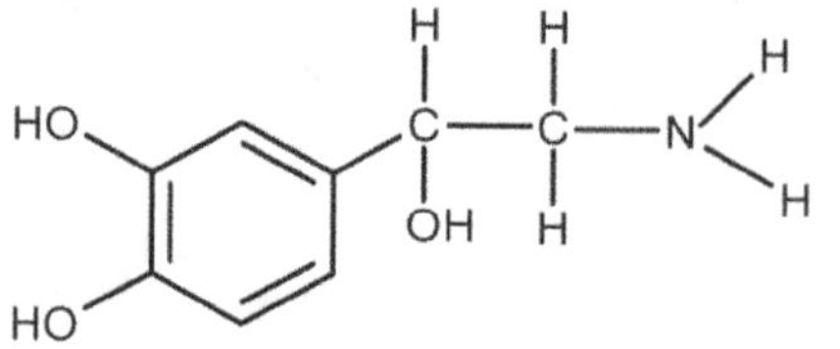

Monohydric Alcohols

Introduction

They are hydroxy saturated hydrocarbons represented by R-OH, formed by replacement of one H atom of alkane by -OH group.

$$\underset{\text{Alkane}}{\text{R-H}} \xrightarrow[\text{+OH}]{\text{-H}} \underset{\text{Alcohol}}{\text{R-OH}}$$

Classification

Monohydric alcohols are further classified in two ways.

1. **On the basis of number of carbon atoms directly linked with -OH group attached carbon:** They are classified as primary, secondary and tertiary alcohols. If the hydroxy group containing carbon atom is directly linked with one carbon atom, then it is called primary alcohol. Similarly if it is linked with two or three carbon atoms then it is called as secondary or tertiary alcohol respectively.

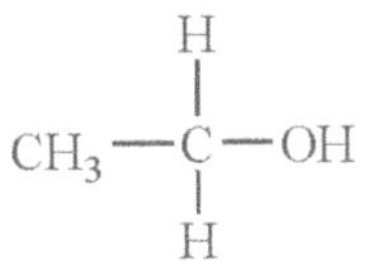

CH_3-CH_2-OH (C bearing two H) — Primary (1°) alcohol

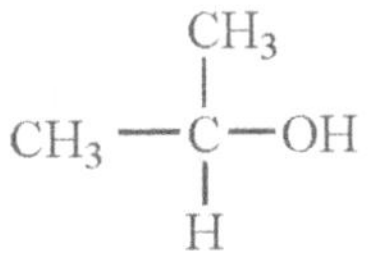

$CH_3-CH(CH_3)-OH$ (C bearing CH$_3$ and H) — Secondary (2°) alcohol

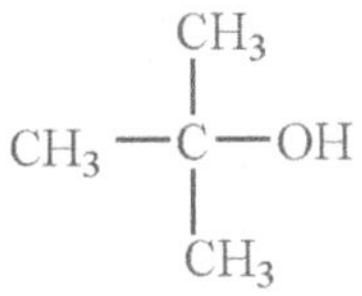

$CH_3-C(CH_3)_2-OH$ — Tertiary (3°) alcohol

2. **On the basis of hybridisation of the carbon atom to which -OH group is attached.**
   They are classified as follows.
   (i) ***Compounds possessing C sp³-OH bond*:** In this, the -OH group is attached to a *sp*^{3} hybridised carbon atom of an alkyl group. These include the following alcohols.
      (a) **All Primary, Secondary and Tertiary Alcohols:** Structures shown above.
      (b) **Allylic alcohols:** In this -OH group is attached to a *sp*^{3} hybridized carbon next to the carbon-carbon double bond (C=C) *i.e.*, an allylic carbon.

      $CH_2=CH-CH_2-OH$

      Allyl alcohol

      (c) **Benzylic alcohols:** In this -OH group is attached to a *sp*^{3} hybridized carbon atom next to aromatic ring.

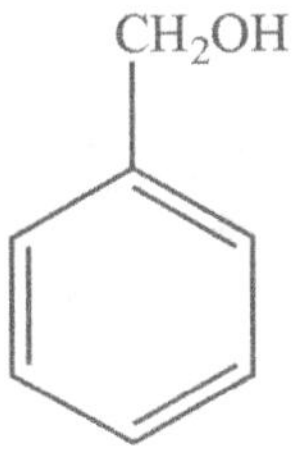

      $C_6H_5-CH_2OH$

      Benzyl alcohol

   (ii) ***Compounds possessing C sp²-OH bond*:** In this the -OH group bonded to a carbon- carbon double bond (C=C), *i.e.*, vinylic carbon or to an aryl carbon (Phenol).

$CH_2{=}CH{-}OH$

Vinyl alcohol

OH

Phenol

Nomenclature

Alcohols are named by three different methods as described below.

1. **Common or trivial name system:** Alcohols are named as alkyl alcohols *i.e.,* the prefix alkyl indicates the name of the alkyl group which is followed by word alcohol. For example, -C_2H_5-OH (ethyl alcohol). In the higher alcohols, the nature of alcohol *i.e.,* 1°, 2°, 3° is indicated as a prefix to the name of alcohol.

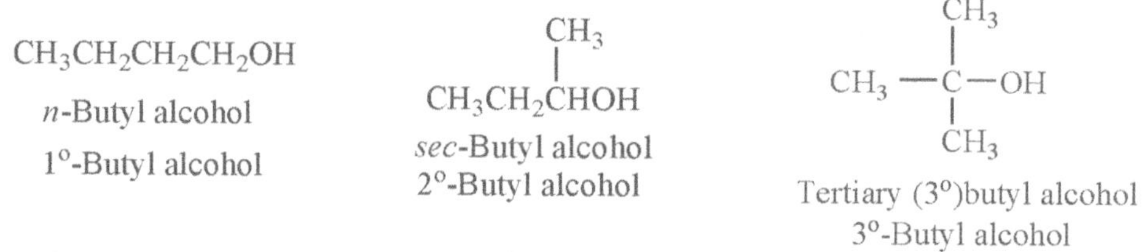

2. **Carbinol name system:** In this method, all alcohols are considered as derivatives of carbinol (methanol). The alkyl groups attached are indicated in alphabetical order.

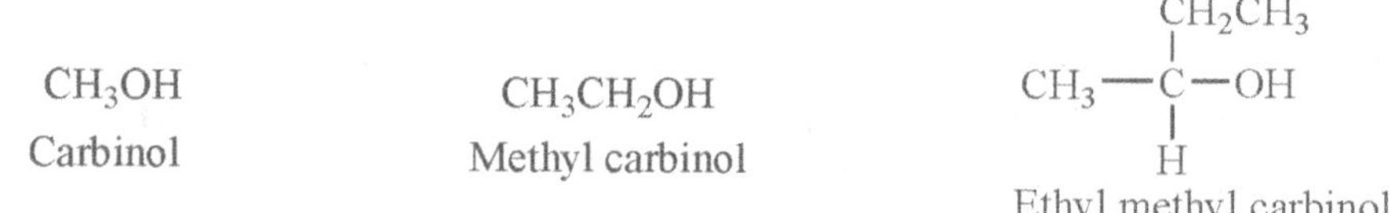

3. **IUPAC name system:** In this system alcohols are named as derivatives of alkanes, the suffix **-e** is replaced with **-ol**. *i.e.,* alkanol. The rules are
   (i) The longest carbon chain containing -OH is taken as parent chain.
   (ii) Name is derived by replacing the suffix **–e** of the corresponding alkane with **-ol.**
   (iii) Number the carbon chain in such a way the –OH group should get lowest possible number.
   (iv) Indicate the other substitutes, or multiple bonds by numbers.

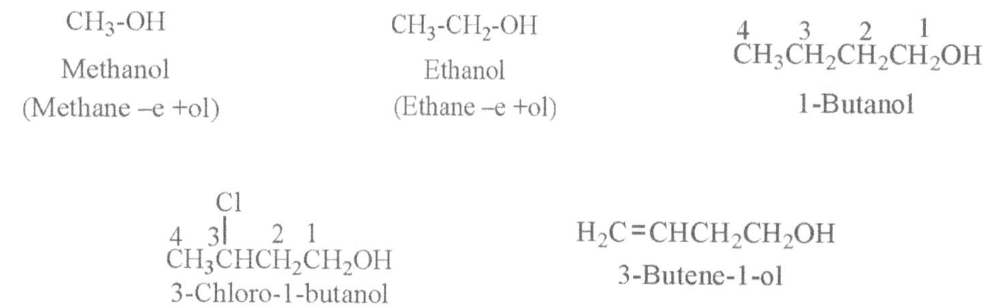

Common names and IUPAC names of some of the alcohols are indicated in the Table 12.2.

Table 12.2 Common name and IUPAC name of some alcohols.

Compound	Common name	IUPAC name
CH_3—OH	Methyl alcohol	Methanol
CH_3—CH_2—CH_2—OH	*n*-Propyl alcohol	Propan-1-ol
CH_3—CH(OH)—CH_3	*iso*-Propyl alcohol	Propan-2-ol
CH_3—CH_2—CH_2—CH_2—OH	*n*-Butyl alcohol	Butan-1-ol
CH_3—CH(OH)—CH_2—CH_3	*sec*-Butyl alcohol	Butan-2-ol
CH_3—CH(CH_3)—CH_2—OH	*iso*-Butyl alcohol	2-Methylpropan-1-ol
CH_3—C(CH_3)(CH_3)—OH	*tert*-Butyl alcohol	2-Methylpropan-2-ol
CH_2(OH)—CH(OH)—CH_2(OH)	Glycerol	Propane-1,2,3-triol

General Methods of Preparation

1. **Hydrolysis of alkyl halides:** The hydrolysis of alkyl halides with aqueous alkali (NaOH) or silver oxide suspended in water ($Ag_2O + H_2O \rightarrow 2AgOH$) yield alcohols.

$$\underset{\text{Alkyl halides}}{\text{R-X} + \text{NaOH}} \xrightarrow{H_2O} \underset{\text{Alcohol}}{\text{R-OH}} + \text{NaX}$$

$$\underset{\text{(Moist silver oxide)}}{\text{R-X} + \text{AgOH}} \longrightarrow \underset{\text{Alcohol}}{\text{R-OH}} + \text{AgX}$$

$$\underset{\text{Ethyl bromide}}{CH_3CH_2Br} + \text{NaOH} \longrightarrow \underset{\text{Ethanol}}{CH_3CH_2OH} + \text{NaBr}$$

$$\underset{\text{Ethyl bromide}}{CH_3CH_2Br} + \text{AgOH} \longrightarrow \underset{\text{Ethanol}}{CH_3CH_2OH} + \text{AgBr}$$

Mechanism: It follows S_N^2 mechanism which follows the formation of pentavalent transition state.

H Ö + H–C–Br (Methyl bromide) → [HO······C······Br]
Transition state
↓ $-\ddot{Br}$
HO–C(H_3)
Methanol

2. **Hydration of olefins:** Alkenes upon reaction with water in presence of sulphuric acid yields alcohol. As water is added to alkene this reaction is called as hydration of alkenes. Initially sulphuric acid reacts with olefins and form alkyl hydrogen sulphate, which upon hydrolysis yield alcohol. The reaction takes place as per Markovnikov's rule in unsymmetrical alkenes.

$$\underset{\text{Ethylene}}{CH_2{=}CH_2} \xrightarrow{H_2SO_4} \underset{\text{Ethyl hydrogen sulphate}}{CH_3CH_2HSO_4} \xrightarrow{H_2O} \underset{\text{Ethanol}}{CH_3CH_2OH} + H_2SO_4$$

Water molecule can be added to olefins in two indirect ways also, as described below.

1. Oxymercuration and demercuration.
2. Hydroboration-oxidation.

1. **Oxymercuration and demercuration:** Alkenes react with mercuric acetate in presence of water to give organomercury product. This reaction is known as **oxymercuration of alkenes.**

$$\underset{\text{Propene}}{CH_3-CH{=}CH_2} \xrightarrow{(CH_3COO)_2Hg} \underset{\text{Alkyl mercuric product}}{CH_3-CH(OH)-CH_2-HgO-C(=O)-CH_3}$$

The organomercury compound obtained by the oxymercuration reaction further undergoes reduction with sodium borohydride ($NaBH_4$) to give an alcohol. This reaction is known as **demercuration** reaction.

$$CH_3-\underset{}{\overset{OH}{\overset{|}{C}}}H-\underset{\underset{\underset{\underset{O}{\|}}{HgO\text{-}C-CH_3}}{|}}{CH_2}\xrightarrow{NaBH_4} CH_3-\overset{OH}{\overset{|}{C}}H-CH_3$$

Organomercury compound 2-Propanol

3. **Hydrolysis of esters:** Hydrolysis of esters in the presence of alkali yields alcohol with the formation of sodium salt of carboxylic acids as a by product. This method is of industrial importance, since some of the alcohol occurs as esters.

$$\underset{\text{Ester}}{R\text{-}COO\text{-}R'} + \underset{\text{(Dilute)}}{NaOH} \longrightarrow \underset{\text{Sodium salt of carboxylic acid}}{RCOONa} + \underset{\text{Alcohol}}{R'OH}$$

$$\underset{\text{Ethyl acetate}}{CH_3COOC_2H_5} + NaOH \longrightarrow \underset{\text{Sodium acetate}}{CH_3COONa} + \underset{\text{Ethanol}}{C_2H_5OH}$$

4. **Reduction of aldehydes, ketones and carboxylic acids:** Reduction of aldehydes & ketones by using the catalysts like H_2/Ni or $LiAlH_4$ yields primary and secondary alcohols respectively. Carboxylic acids can be reduced to alcohol using $LiAlH_4$.

$$\underset{\text{Aldehyde}}{R\text{-}\overset{O}{\overset{\|}{C}}\text{-}H} \xrightarrow{LiAlH_4} \underset{1^\circ\text{Alcohol}}{R\text{-}CH_2OH}$$

$$\underset{\text{Acetaldehyde}}{CH_3\text{-}\overset{O}{\overset{\|}{C}}\text{-}H} \xrightarrow{LiAlH_4} \underset{\text{Ethanol}}{CH_3CH_2OH}$$

$$\underset{\text{Ketone}}{R\text{-}\overset{O}{\overset{\|}{C}}\text{-}R'} \xrightarrow{LiAlH_4} \underset{2^\circ\text{Alcohol}}{R\text{-}\overset{OH}{\overset{|}{C}}H\text{-}R'}$$

$$\underset{\text{Acetone}}{CH_3\text{-}\overset{O}{\overset{\|}{C}}\text{-}CH_3} \xrightarrow{LiAlH_4} \underset{\text{2-Propanol}}{CH_3\overset{OH}{\overset{|}{C}}HCH_3}$$

$$RCOOH \xrightarrow{LiAlH_4} \underset{1^\circ\text{Alcohol}}{R\text{-}CH_2OH}$$

$$\underset{\text{Acetic acid}}{CH_3COOH} \xrightarrow{LiAlH_4} \underset{\text{Ethanol}}{CH_3\text{-}CH_2OH}$$

Note: $LiAlH_4$ do not reduce the olefinic double bond, hence it can be used in preparing unsaturated alcohols. The catalyst, excess of sodium and ethanol ($Na\text{-}C_2H_5OH$) can also be used as a reducing agent in the above reaction, then it is called **Bouveault – Blanc reduction.**

5. **By the action of nitrous acid on aliphatic primary amines:** Nitrous acid on reaction with aliphatic primary amines, yields primary alcohol.

$$\underset{\text{Alkylamine}}{R\text{-}NH_2} \xrightarrow{HONO} \underset{\text{Alcohol}}{R\text{-}OH} + N_2 + H_2O$$

$$\underset{\text{Ethylamine}}{CH_3\text{-}CH_2\text{-}NH_2} \xrightarrow{HONO} \underset{\text{Ethanol}}{CH_3\text{-}CH_2\text{-}OH} + N_2 + H_2O$$

6. **Hydroboration of alkenes:** Three molecules of alkenes react with borane and yield trialkyl boranes, the B-H bond in diborane is polar due to the difference in electronegativity between boron and hydrogen. Boron is less electronegative than hydrogen So, H bears a partial negative charge and boron bears partial positive.

$$3\underset{\text{Propylene}}{CH_3-CH=CH_2} + BH_3 \longrightarrow \underset{\text{Tripropylborane}}{(CH_3-CH_2-CH_2)_3B}$$

Trialkyl boranes upon oxidation yields primary alcohol

$$\underset{\text{Tripropylborane}}{(CH_3-CH_2-CH_2)_3B} + 3H_2O_2 \xrightarrow[OH^-]{} 3\underset{\substack{\text{1-Propanol} \\ (1^\circ\text{Alcohol})}}{CH_3-CH_2-CH_2OH} + H_3BO_3$$

The overall reaction takes place as anti-Markovnikov's addition as-OH group gets attached to less substituted carbon.

$$\underset{\text{Propylene}}{CH_3-CH=CH_2} \xrightarrow[\text{(ii) } H_2O_2/OH^-]{\text{(i) } (BH_3)_2} \underset{\text{1-Propanol}}{CH_3-CH_2-CH_2-OH}$$

7. **By the reaction between Grignard reagent and carbonyl compounds:** Reaction of Grignard reagent with carbonyl compounds (aldehyde & ketone) followed by hydrolysis yields alcohols. By this method all three (1°, 2°, 3°) alcohols can be prepared by changing the carbonyl compound.

   (i) Preparation of primary alcohol from formaldehyde:

$$\underset{\text{Formaldehyde}}{H\text{-}\overset{\overset{O}{\|}}{C}\text{-}H} + RMgX \longrightarrow H-\underset{\underset{R}{|}}{\overset{\overset{OMgX}{|}}{C}}-H \xrightarrow[\text{-MgXOH}]{HOH/H^+} \underset{1^\circ\text{Alcohol}}{H-\underset{\underset{R}{|}}{\overset{\overset{OH}{|}}{C}}-H}$$

   (ii) Preparation of secondary alcohol from aldehyde other than formaldehyde:

$$\underset{\text{Acetaldehyde}}{CH_3\text{-}\overset{\overset{O}{\|}}{C}\text{-}H} + RMgX \longrightarrow CH_3-\underset{\underset{R}{|}}{\overset{\overset{OMgX}{|}}{C}}-H \xrightarrow[\text{-MgXOH}]{HOH/H^+} \underset{2^\circ\text{Alcohol}}{CH_3-\underset{\underset{R}{|}}{\overset{\overset{OH}{|}}{C}}-H}$$

(iii) Preparation of tertiary alcohol from ketone:

$$\underset{\text{Ketone}}{R\text{-}\overset{\overset{O}{\|}}{C}\text{-}R^1} + R^2MgX \longrightarrow R-\underset{R^2}{\overset{OMgX}{\overset{|}{\underset{|}{C}}}}-R^1 \xrightarrow[\text{-MgXOH}]{HOH/H^+} \underset{3^\circ\text{Alcohol}}{R-\underset{R^2}{\overset{OH}{\overset{|}{\underset{|}{C}}}}-R^1}$$

8. **Fermentation of carbohydrates:** Fermentation of some of the carbohydrates and sugars under the influence of appropriate microorganism yields alcohol.

$$\underset{\text{Glucose}}{C_6H_{12}O_6} \xrightarrow[\text{Fermentation}]{\text{Yeast}} \underset{\text{Ethanol}}{2CH_3\text{-}CH_2\text{-}OH} + 2CO_2\uparrow$$

9. **Oxo process:** When carbon monoxide, hydrogen and alkanes are passed under pressure at 400-420 K over a catalyst (cobalt, thoria *etc*), followed by heating, mixture of alcohols are obtained.
10. **Reduction of acid chlorides and esters:** Acid chlorides and esters undergo reduction with $LiAlH_4$ to yields 1° alcohol.

$$\underset{\text{Acetyl Chloride}}{CH_3\overset{\overset{O}{\|}}{C}Cl} \xrightarrow[2)\ H_2O]{1)\ LiAlH_4} \underset{\text{Ethanol}}{CH_3CH_2OH} + HCl$$

$$\underset{\text{Ester}}{R-\overset{\overset{O}{\|}}{C}-OR'} \xrightarrow{LiAlH_4} \underset{\text{Primary alkoxide}}{RCH_2O^-Li^+ + R'-OLi} \xrightarrow{H_3O^+} \underset{\text{Primary alcohol}}{RCH_2OH + R'OH}$$

$$\underset{\text{Methyl Propanoate}}{CH_3CH_2\overset{\overset{O}{\|}}{C}-OCH_3} \xrightarrow{LiAlH_4} CH_3CH_2CH_2O^-Li + Li-OCH_3 \xrightarrow{H_3O^+} \underset{\text{Propanol}}{CH_3CH_2CH_2OH} + \underset{\text{Methanol}}{CH_3OH}$$

11. **Reduction of epoxides:** Epoxides are reduced with $LiAlH_4$ to yield alcohol by ring opening reaction.

$$\underset{\text{Epoxide}}{-\underset{\diagdown O \diagup}{\overset{|}{C}-\overset{|}{C}}-} \xrightarrow[2)\ H_2O]{1)\ LiAlH_4} \underset{\text{Alcohol}}{-\underset{OH}{\underset{|}{\overset{|}{C}}}-\underset{OH}{\underset{|}{\overset{|}{C}}}-}$$

Summary of Methods of Preparation

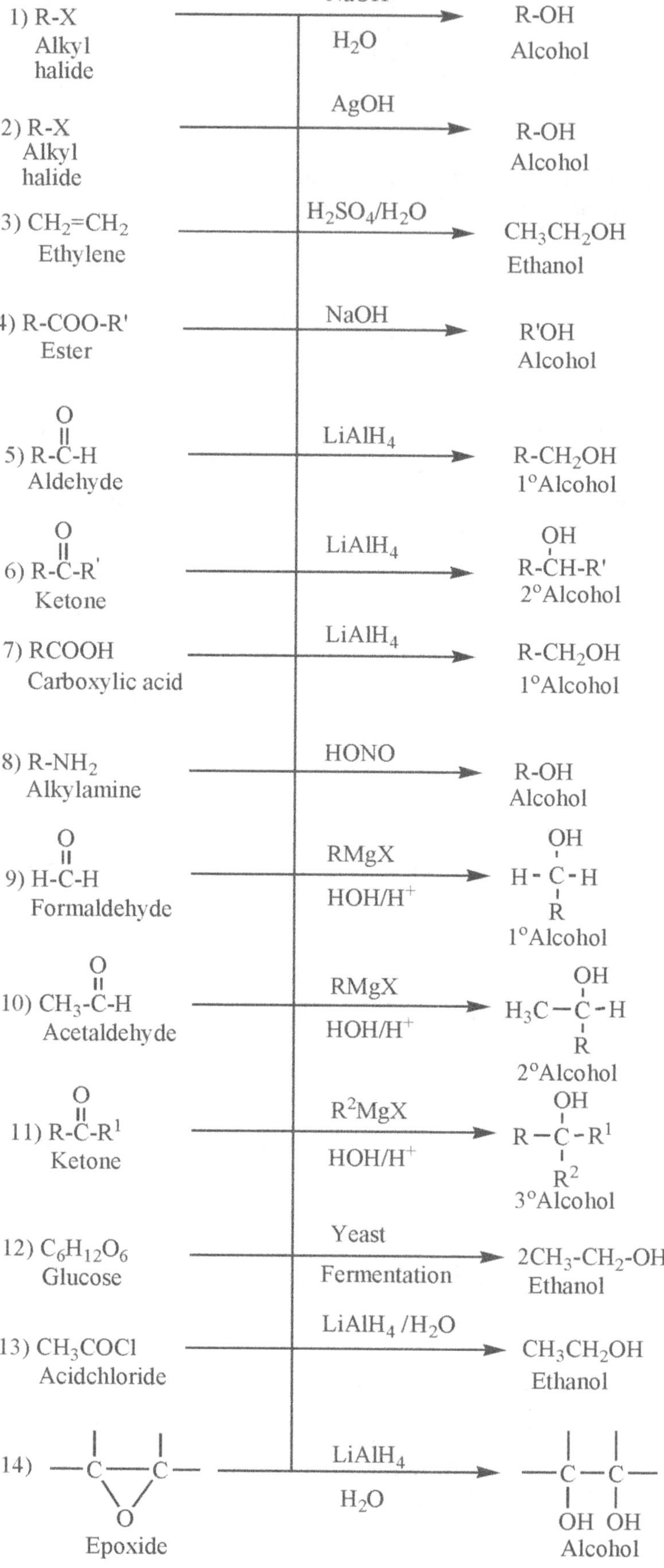

Physical Properties

(i) Lower alcohols are colorless mobile liquids, followed by oily liquids up to $C_{11}H_{23}OH$. Higher than C_{11} members are waxy solids.

(ii) Boiling point of alcohols increase with increase in carbon numbers.

(iii) In a group of isomeric alcohols, primary alcohol has higher boiling point than 2° and 3° alcohol. *i.e.,* branching leads to decrease in boiling point. The order of boiling point of alcohols are 1° > 2° > 3°.

(iv) Boiling point of alcohol is higher than those of hydrocarbons of same molecular mass due to the formation of hydrogen bonding between alcohol and water molecules.

(v) Lower members of alcohols are soluble in water and organic solvents, but the solubility decreases upon increasing carbon chain. The solubility is due to the oxygen atom of –OH group which forms hydrogen bond with water molecules. In lower alcohols the –OH group which forms a large part of molecule, but in higher alcohols the hydrocarbon character increases which diminishes the role of –OH group and hence the solubility decreases. Thus the order of solubility in water can be written as follows.

Methanol > ethanol > propanol > butanol > pentanol > hexanol.........

(vi) Lower member of alcohols have pleasant smell and burning taste but higher alcohols are odourless and tasteless. Monohydric alcohols are lighter than water while dihydric and trihydric alcohols are more viscous.

Chemical Properties

Structure

In alcohols, the –OH group is attached to a carbon by a sigma bond formed by the overlap of sp^3 hybridized orbital of oxygen.

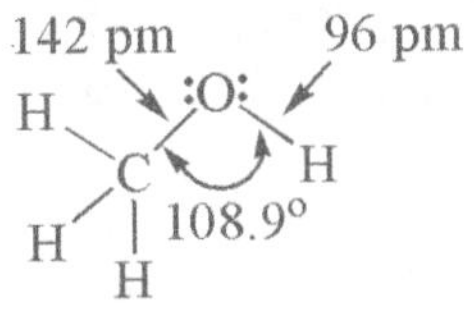

142 pm 96 pm
H :O: H
C
H | 108.9°
H

Methanol

The bond angle (108.9°) is slightly less than the tetrahedral angle (109° 28′), due to the repulsion between unshared pair of electrons of oxygen.

Monohydric alcohols have inherited chemical properties of alkanes and water, because alcohols can be considered as derivatives of alkanes in which one hydrogen atom is replaced by –OH group; they can also be considered as derivatives of water in which one –H atom is replaced by alkyl group as shown below.

$$\underset{\text{Methane}}{CH_4} \xrightarrow[\text{of H by OH}]{\text{Replacement}} \underset{\text{Methanol}}{CH_3OH} \xleftarrow[\text{of H by -}CH_3]{\text{Replacement}} \underset{\text{Water}}{\text{H-OH}}$$

Alcohols are polar ionic compounds, hence reacts with polar ionic reagents. As the 'O' atom of alcohol is electron rich and have unshared pair of electrons, the C-O bond and O-H bonds are polar and this makes the alcohols more reactive. Let us discuss the reactions of alcohols in detail.

Chemical reactions of alcohols are categorized as follows:

1. Reactions involving replacement of hydrogen of the hydroxyl group (O-H cleavage).
2. Reactions involving replacement of hydroxyl group (R-OH cleavage).

3. Reactions involving both alkyl group and hydroxyl group.
4. Chemical tests for alcohol.
5. Distinction between 1°, 2°, 3° alcohols (R-C-OH cleavage).

1. **Reactions involving replacement of hydrogen of hydroxyl group.**

   (i) **Action of alkali metals:** Alcohols react with sodium or potassium to form alkoxides with the liberation of hydrogen. The order of ease of formation of alkoxides are 1° > 2° > 3° alcohol.

$$\underset{\text{Alcohol}}{2R\text{-}OH} + 2Na \longrightarrow \underset{\text{Sodium alkoxide}}{2RONa} + H_2$$

$$\underset{\text{Ethanol}}{2CH_3CH_2OH} + 2Na \longrightarrow \underset{\text{Sodium ethoxide}}{2CH_3CH_2ONa} + H_2$$

This shows that alcohols are acidic in nature. The order of acidity is 1° > 2° > 3° alcohols. The alkoxides formed (RO^-) are highly electronegative, which readily decomposes in water and forms alcohols.

$$\underset{\text{Alkoxide}}{RONa} + \underset{\text{Water}}{H\text{-}OH} \rightleftharpoons \underset{\text{Alcohol}}{ROH} + NaOH$$

Acidity of alcohols: Due to the presence of polar O-H bond, alcohols are acidic in nature. Electron donating groups such as $-CH_3$, $-C_2H_5$ *etc.* increases electron density on oxygen atom and decreases the polarity of O-H bond which results in decrease of acid strength. The order of decreasing acidity is shown below.

$$\underset{\substack{\text{Primary}\\1^\circ}}{R \rightarrow CH_2OH} \qquad \underset{\substack{\text{Secondary}\\2^\circ}}{R_2CH\text{—}OH} \qquad \underset{\substack{\text{Tertiary}\\3^\circ}}{R_3C\text{—}OH}$$

   (ii) **Action of organic acids:** Alcohols react with organic acids and forms esters. This reaction is called **esterification.** This reaction is carried out in the presence of dehydrating agent such as conc.H_2SO_4. The order of reactivity of alcohols is 1° > 2° > 3° alcohols.

$$\underset{\text{Carboxylic acid}}{RCOOH} + \underset{\text{Alcohol}}{R'OH} \xrightarrow{\text{Conc.}H_2SO_4} \underset{\text{Ester}}{RCOOR'} + H_2O$$

$$\underset{\text{Acetic acid}}{CH_3COOH} + \underset{\text{Ethanol}}{C_2H_5OH} \xrightarrow{\text{Conc.}H_2SO_4} \underset{\text{Ethylacetate}}{CH_3COOC_2H_5} + H_2O$$

Mechanism:

Step 1: Protonation of the carbonyl portion of carboxylic acid.

$$R-\overset{O}{\overset{\|}{C}}-OH + H^+ \rightleftharpoons R-\overset{OH}{\overset{|}{\underset{+}{C}}}-OH$$

Step 2: Attack of nucleophile.

$$R-\overset{OH}{\overset{|}{\underset{+}{C}}}-OH + H\ddot{O}-R' \rightleftharpoons R-\overset{OH}{\overset{|}{\underset{OH}{\underset{|}{C}}}}-\overset{H}{\overset{|}{\underset{+}{O}}}-R'$$

Step **3:** Transfer of hydrogen ion (Hydride ion transfer)

$$R-\underset{\underset{OH}{|}}{\overset{\overset{:\ddot{O}H}{|}}{C}}-\overset{+}{O}(H)-R' \rightleftharpoons R-\underset{\underset{OH}{|}}{\overset{\overset{\overset{+}{O}H_2}{|}}{C}}-O-R'$$

Step **4:** Removal of water molecule and proton to give an ester.

$$R-\underset{\underset{O-H}{|}}{\overset{\overset{\overset{+}{O}H_2}{|}}{C}}-O-R' \rightleftharpoons \underset{\text{Ester}}{R-\underset{\underset{O}{||}}{C}-OR'} + H_2O + H^+$$

(iii) Acetylation: Alcohols when reacts with acetyl chloride or acetic anhydride, the –H atom of –OH group is replaced by acetyl group. This reaction is called **acetylation**. In general, alcohols react with any acid chlorides and anhydrides and gives acetylated products.

$$\underset{\text{Acetylchloride}}{CH_3COCl} + \underset{\text{Ethanol}}{C_2H_5OH} \longrightarrow \underset{\text{Ethylacetate}}{CH_3COOC_2H_5} + HCl$$

$$\underset{\text{Acetic anhydride}}{(CH_3CO)_2O} + \underset{\text{Ethanol}}{C_2H_5OH} \longrightarrow \underset{\text{Ethylacetate}}{CH_3COOC_2H_5} + \underset{\text{Acetic acid}}{CH_3COOH}$$

(iv) Reaction of Grignard reagent: Grignard reagent reacts with alcohol to form hydrocarbons.

$$\underset{\text{Alcohol}}{ROH} + \underset{\text{Grignard reagent}}{C_2H_5MgI} \longrightarrow \underset{\text{Ethane}}{C_2H_6} + Mg\begin{matrix} OR \\ I \end{matrix}$$

2. Reactions involving replacement of hydroxyl group

(i) Reaction of halogen acids: Reaction of alcohols with halogen acids yields alkyl halides. The order of reactivity of halogen acids HI > HBr > HCl and the order of reactivity of alcohols are 3° > 2° > 1° alcohols.

$$\underset{\text{Alcohol}}{R\text{-}OH} + \underset{\text{Halogen acid}}{HX} \longrightarrow \underset{\text{Alkyl halide}}{R\text{-}X} + H_2O$$

$$\underset{\text{Ethanol}}{CH_3CH_2OH} + \underset{\text{Hydrogen chloride}}{HCl} \longrightarrow \underset{\text{Ethyl chloride}}{CH_3CH_2Cl} + H_2O$$

Mechanism: It follows S_N1 and S_N2 (Nucleophilic substitution mechanism)

Step **1:** Protonation of OH group by the attack of H-Cl.

$$CH_3-CH_2-\ddot{O}H + H-Cl \longrightarrow CH_3-CH_2-\overset{+}{\ddot{O}}H_2 + \bar{C}l$$

(Good leaving group)

Step **2:** Attack of nucleophile (Cl^-) to yields alkyl chloride and water.

$$CH_3-CH_2-\overset{+}{\ddot{O}}H_2 + \bar{C}l \longrightarrow \underset{\text{Ethyl chloride}}{CH_3CH_2Cl} + H_2O$$

(ii) Action of phosphorous halides: Alcohols upon reaction with phosphorous halides (PCl_5, PCl_3, PBr_3, PI_3) yields alkyl halides.

$$\underset{\text{Alcohol}}{R\text{-}OH} + PCl_5 \longrightarrow \underset{\text{Alkyl chloride}}{R\text{-}Cl} + POCl_3 + HCl$$

$$\underset{\text{Ethanol}}{CH_3CH_2OH} + \underset{\text{Phosphorous pentacholoride}}{PCl_5} \longrightarrow \underset{\text{Ethyl chloride}}{CH_3CH_2Cl} + POCl_3 + HCl$$

(iii) Action with tosyl chloride (TsCl): Alcohols react with tosyl chloride (*p* - Toluene sulphonyl chloride, TsCl) in the presence of pyridine and yields tosylates.

$$\underset{\text{Ethanol}}{CH_3CH_2OH} + \underset{\text{Tosyl chloride}}{Cl\text{-}SO_2\text{-}C_6H_4\text{-}CH_3} \xrightarrow{\text{Pyridine}} \underset{\text{Tosylate}}{CH_3CH_2\text{-}O\text{-}SO_2\text{-}C_6H_4\text{-}CH_3} + \underset{\text{Pyridinium chloide}}{C_5H_5\overset{+}{N}\text{-}H + \overset{-}{Cl}}$$

(iv) Reaction of thionyl chloride: Alcohols upon reaction with thionyl chloride gives alkyl chlorides.

$$\underset{\text{Alcohol}}{R\text{-}OH} + SOCl_2 \longrightarrow RCl + SO_2 + HCl$$

$$\underset{\text{Ethanol}}{CH_3CH_2\text{-}OH} + SOCl_2 \longrightarrow \underset{\text{Ethyl Chloride}}{CH_3CH_2Cl} + SO_2 + HCl$$

(v) Reduction: Alcohols undergo reduction to form alkanes upon reaction with HI and red phosphorous (HI + P).

$$\underset{\text{Alcohol}}{R\text{-}OH} \xrightarrow{2HI + P} \underset{\text{Alkane}}{R\text{-}H} + I_2 + H_2O$$

$$\underset{\text{Ethanol}}{CH_3CH_2OH} \xrightarrow{2HI + P} \underset{\text{Ethane}}{CH_3CH_3} + I_2 + H_2O$$

3. Reactions involving both alkyl group and hydroxyl group

(i) Dehydration: Alcohols can be dehydrated to olefins by simple heating or using alumina or sulphuric acid. As the water is removed in this reaction it is called as **dehydration**. The order of reaction is 3° > 2° > 1° alcohol.

$$\underset{\text{Alcohol}}{RCH_2CH_2OH} \xrightarrow[\text{or alumina / } H_2SO_4]{\Delta} \underset{\text{Alkene}}{RCH{=}CH_2} + H_2O$$

$$\underset{\text{Ethanol}}{CH_3CH_2OH} \xrightarrow[\text{or alumina / } H_2SO_4]{\Delta} \underset{\text{Ethylene}}{CH_2{=}CH_2} + H_2O$$

Primary alcohols are dehydrated by using conc.H_2SO_4 where as dilute H_2SO_4 is used with secondary and tertiary alcohols.

Reaction of alcohol with Sulphuric acid.

Alcohol when reacted with sulphuric acid forms ethyl hydrogen sulphate

$$\underset{\text{Ethanol}}{CH_3CH_2\text{-}OH} + H_2SO_4 \longrightarrow \underset{\text{Ethyl hydrogen sulphate}}{CH_3CH_2HSO_4} + H_2O$$

Thus the formed ethyl hydrogen sulphate yields variety of products under different conditions upon reaction with different reagents.

(a) Ethyl hydrogen sulphate upon distillation under reduced pressure yield Diethylsulphate.

$$\underset{\text{Ethyl hydrogen sulphate}}{2CH_3CH_2HSO_4} \xrightarrow[\text{under pressure}]{\text{Distillation}} \underset{\text{Diethyl sulphate}}{(CH_3CH_2)_2SO_4} + H_2SO_4$$

(b) Ethyl hydrogen sulphate reacts with excess of ethanol and yield diethyl ether.

$$\underset{\text{Ethyl hydrogen sulphate}}{CH_3CH_2HSO_4} + \underset{\text{Ethanol}}{CH_3CH_2OH} \longrightarrow \underset{\text{Diethyl ether}}{CH_3CH_2OCH_2CH_3} + H_2SO_4$$

(c) Ethyl hydrogen sulphate reacts with excess of sulphuric acid and yield ethylene.

$$CH_3CH_2HSO_4 \xrightarrow[H_2SO_4]{\text{Excess of}} CH_2{=}CH_2 + H_2SO_4$$

Dehydration using $POCl_3$: During dehydration with strong acids, some of the organic compounds easily decompose. An alternate method uses phosphorous oxychlorides ($POCl_3$) and pyridine (in place of H_2SO_4).

$$\underset{\text{Cyclohexanol}}{C_6H_{11}OH} + POCl_3 \xrightarrow{POCl_3\text{ in Pyridine}} \underset{\text{Cyclohexene}}{C_6H_{10}} + H_2O$$

(ii) **Oxidation:** (Refer Distinction between primary, secondary and tertiary alcohol)

(iii) **Action of hot reduced copper:** The products obtained depend upon the class of alcohol (1° or 2° or 3° alcohol) (Refer distinction between 1° or 2° or 3° alcohol).

(iv) **Action of chlorine:** Chlorine oxidizes alcohol to aldehyde and also chlorinates the product obtained. Hence in this reaction chlorine acts as an oxidizing agent and chlorinating agent.

$$\underset{\text{Ethanol}}{CH_3CH_2\text{-}OH} \xrightarrow[\text{(O)}\ \ -2HCl]{Cl_2} \underset{\text{Acetaldehyde}}{CH_3CHO} \xrightarrow{Cl_2} \underset{\text{Chloral}}{CCl_3CHO} + 3HCl$$

4. **Chemical tests for alcohols:** The following chemical tests are used for the identification of alcohol.

   (i) **Sodium metal test:** When sodium metal is added to alcohol, hydrogen gas is evolved as bubbles. Evolution of hydrogen gas indicates the presence of alcohol.

$$\underset{\text{Alcohol}}{2R\text{-}OH} + 2Na \longrightarrow \underset{\text{Sodium alkoxide}}{2RONa} + H_2\uparrow$$

(ii) **Phosphorous pentachloride test:** When PCl_5 is added to alcohol, hydrogen chloride is evolved. Evolution of HCl indicates the presence of alcohol.

$$\underset{\text{Alcohol}}{\text{R-OH}} + PCl_5 \longrightarrow \underset{\text{Alkylchloride}}{\text{R-Cl}} + POCl_3 + HCl\uparrow$$

(iii) **Acylation test:** When acetyl chloride or benzoyl chloride is added to alcohol, HCl is liberated and leads to the formation of ester as oily layer.

$$\underset{\text{Alcohol}}{\text{R-OH}} + \underset{\text{Acetyl chloride}}{CH_3COCl} \longrightarrow \underset{\text{Acetyl ester}}{ROCOCH_3} + HCl\uparrow$$

$$\underset{\text{Alcohol}}{\text{R-OH}} + \underset{\text{Benzoyl chloride}}{C_6H_5COCl} \longrightarrow \underset{\text{Benzoyl ester}}{ROCOC_6H_5} + HCl\uparrow$$

5. **Distinction between primary, secondary and tertiary alcohols:** The three classes of alcohols can be distinguished by the following methods.

(i) **Oxidation method:** In this method alcohols are oxidised by using acidified $KMnO_4$ or alkaline $KMnO_4$ or dilute nitric acid and from the product obtained the nature of alcohol is identified.

(a) **Primary alcohol** on oxidation gives aldehyde and on further oxidation gives carboxylic acid. The carboxylic acid and the aldehyde will have same number of carbon atoms as in the parent alcohol.

$$\underset{\substack{\text{Ethanol}\\(2C)}}{CH_3CH_2OH} \xrightarrow{(O)} \underset{\substack{\text{Acetaldehyde}\\(2C)}}{CH_3CHO} \xrightarrow{(O)} \underset{\substack{\text{Acetic acid}\\(2C)}}{CH_3COOH}$$

(b) **Secondary alcohol** on oxidation gives ketone which upon further oxidation yields carboxylic acid. The carboxylic acid will have one carbon less than ketone while the ketone have same number of carbons as in alcohol

$$\underset{\textit{iso}\text{-propyl alcohol (3C)}}{(CH_3)_2CH\text{-}OH} \xrightarrow[-H_2O]{(O)} \underset{\text{Acetone (3C)}}{(CH_3)_2C\text{=}O} \xrightarrow{(O)} \underset{\text{Acetic acid (2C)}}{CH_3COOH} + HCOOH$$

(c) **Tertiary alcohol** upon oxidation gives ketone which upon further oxidation gives carboxylic acid; both the products have lesser number of carbons than alcohol.

$$\underset{\text{t-Butyl alcohol (4C)}}{(CH_3)_3C\text{-}OH} \xrightarrow[-HCOOH]{(O)} \underset{\text{Acetone (3C)}}{(CH_3)_2C\text{=}O} \xrightarrow{(O)} \underset{\text{Acetic acid (2C)}}{CH_3COOH} + HCOOH$$

(ii) **Dehydrogenation of alcohols by the action of hot reduced copper:** All the three classes of alcohols behave differently with hot reduced copper at 570 K.

(a) **Primary alcohol yields aldehyde** with the elimination of hydrogen.

$$\underset{\text{Ethanol}}{CH_3CH_2OH} \xrightarrow[570\text{ K}]{Cu} \underset{\text{Acetaldehyde}}{CH_3CHO} + H_2\uparrow$$

(b) Secondary alcohol yields ketone with the elimination of hydrogen.

$$(H_3C)_2CH\text{-}OH \xrightarrow[570K]{Cu} (H_3C)_2C{=}O + H_2\uparrow$$

iso-Propyl alcohol → Acetone

(c) Tertiary alcohol yields olefin with the elimination of water

$$(CH_3)_3C\text{-}OH \xrightarrow[570\ K]{Cu} (CH_3)_2C{=}CH_2 + H_2O$$

t-Butyl alcohol → *iso*-Butylene

(iii) Victor Meyers method: In this method, the given alcohol is first converted into alkyl iodide with phosphorus and iodine ($P+I_2$), which is then treated with silver nitrate to form nitroalkanes. Thus the formed nitro alkanes are reacted with nitrous acid (obtained by treating sodium nitrite and hydrochloric acid), and finally it is made alkaline with sodium hydroxide. The nitrous acid reacts differently with the different nitroalkanes from which the three classes of alcohols are distinguished as shown below.

$CH_3\text{-}CH_2\text{-}OH$ (Primary)	$(CH_3)_2\text{-}CH\text{-}OH$ (Secondary)	$(CH_3)_3\text{-}COH$ (Tertiary)
↓ $P + I_2$	↓ $P + I_2$	↓ $P + I_2$
$CH_3\text{-}CH_2\text{-}I$ Alkyl halide	$(CH_3)_2\text{-}CH\text{-}I$ Alkyl halide	$(CH_3)_3\text{-}C\text{-}I$ Alkyl halide
-AgI ↓ $AgNO_2$	-AgI ↓ $AgNO_2$	-AgI ↓ $AgNO_2$
$CH_3\text{-}CH_2\text{-}NO_2$ Nitroalkane	$(CH_3)_2\text{-}CH\text{-}NO_2$ Nitroalkane	$(CH_3)_3\text{-}C\text{-}NO_2$ Nitroalkane
$-H_2O$ ↓ O=N-OH	$-H_2O$ ↓ HO-N=O	↓ HONO
$CH_3\text{-}C(=N\text{-}OH)\text{-}NO_2$ Nitrolic acid	$(CH_3)_2\text{-}C(N{=}O)\text{-}NO_2$ Pseudonitrolic acid (Blue colour)	No reaction
↓ NaOH	↓ NaOH	
Blood red colour	No reaction	

(iv) Reaction of alcohol with hydrochloric acid and zinc chloride (Lucas test): All the different class of alcohols (1°, 2°, 3°) react at different rate with Lucas reagent (conc. HCl and anhydrous $ZnCl_2$) and forms alkyl chlorides as cloudiness or precipitate. From the rate of reaction and product obtained, the three classes of alcohols are distinguished.

$$CH_3\text{-}CH_2\text{-}OH \xrightarrow[HCl + ZnCl_2]{\text{Lucas reagent}} \textbf{No reaction}$$

Ethanol (1° Alcohol)

$$(CH_3)_2\text{-}CH\text{-}OH \xrightarrow[HCl + ZnCl_2]{\text{Lucas reagent}} (CH_3)_2\text{-}CH\text{-}Cl$$ **(Cloudiness appears in 5 minutes)**

iso-Propanol (2° Alcohol) → *iso*-Propyl chloride

$$\underset{\substack{\text{t-Butylalcohol} \\ (3^\circ \text{ Alcohol})}}{(CH_3)_3\text{-C-OH}} \xrightarrow[HCl + ZnCl_2]{\text{Lucas reagent}} \underset{\text{t-Butyl chloride}}{(CH_3)_3\text{-C-Cl}}$$ **(Cloudiness appears immediately)**

Summary of Chemical Reactions

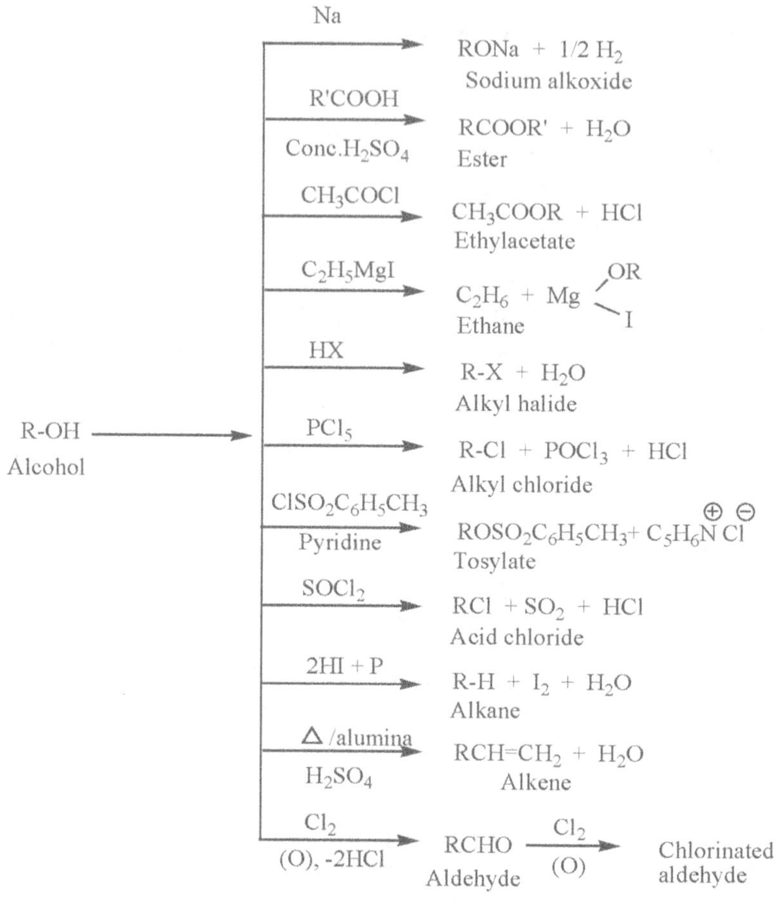

INTERESTING FACT

Grain alcohol and wood alcohol: Ethanol present in alcoholic beverages is obtained by fermentation of glucose from grapes and from grains such as corn, rye and wheat, so it is known as grain alcohol. Cooking the grain in the presence of malt (sprouted barley) converts maximum amount of starch into glucose. Yeast is added to convert glucose into ethanol and CO_2.

$$\underset{\text{Glucose}}{C_6H_{12}O_6} \xrightarrow{\text{Yeast}} \underset{\text{Ethanol}}{2CH_3CH_2OH} + 2CO_2$$

The variety of beverages produced (white or red wine, beer, scotch, champagne, bourbon) depends upon the plant species being fermented and whether the CO_2 formed is allowed to escape or adding or not adding of other substances and how the beverage is purified (by sedimentation for wines, by distillation for scotch and brandy).

The tax on liquor would make prohibitively expensive laboratory reagent, since it is used for wide variety of commercial processes. Laboratory alcohol is not taxed, but it is very carefully regulated by federal government to make sure it is not used for the illegal preparation of alcoholic beverages.

Denatured alcohol: It contains ethanol, made to undrinkable by adding of a denaturant such as methanol or benzene. It is also not taxed, but the presence of other substances makes it unusable for other purposes. Methanol is also called as wood alcohol because it is obtained from distillation of wood in the absence of air or oxygen. It is very toxic in nature. A small amount of ingestion of methanol leads to blindness. The antidote for methanol poisoning is antabuse (Disulfiram).

Dihydric Alcohols (or) Diols

Introduction

Alcohol containing two hydroxyl groups are called as dihydric alcohol, the two -OH groups are attached to different carbon atoms, they are also called as glycols as they are sweet in taste (in Greek word glycols means sweet).

Glycols are classified as α, β, γ glycols according to the relative position of two hydroxyl groups. 1,2-glycols are called as α-glycols, 1,3 as β-glycols, 1,4-glycols as γ and so on. Ethylene glycol is the simple and most important dihydric alcohol.

Nomenclature

1. **Common name method:** The α-glycols are named based on the olefins from which they are prepared by direct hydroxylation.
2. **IUPAC method:** They are considered as derivative of alkanes and the suffix diol is added,and called

$$OH\text{-}CH_2\text{-}CH_2\text{-}OH$$

Ethylene glycol

as alkane diols. As in alkane and alcohol, the longest chain containing both the-OH group is preferred and the position of two hydroxyl groups is indicated by numbers before adding **di** to the suffix **–ol.** For example:

$$OH\text{-}CH_2\text{-}CH_2\text{-}CH_2\text{-}OH$$

Propane-1,3-diol

$$CH_3\text{-}\underset{}{\overset{OH}{\overset{|}{C}H}}\text{-}\overset{OH}{\overset{|}{C}H_2}$$

Propane-1,2-diol

$$OH\text{-}CH_2\text{-}(CH_2)_2\text{-}CH_2\text{-}OH$$

Butane-1,4-diol

Trihydric Alcohols (or) Triols

Introduction

Compounds containing three hydroxyl groups are called trihydric alcohols or triols. The important trihydric alcohol is glycerine or glycerol which is present in all animal or vegetable oils and fats as glycerides. Glyceryl esters of fatty acids are known as glycerides.

$$\begin{array}{c} CH_2OH \\ | \\ CHOH \\ | \\ CH_2OH \end{array}$$

Glycerol

IUPAC name for glycerol is given as derivative of alkanes with the three hydroxyl groups at 1, 2, 3 positions hence named 1,2,3-tri hydroxy propane (1,2,3-Propane triol). All tri hydric alcohols are named in similar manner.

Pharmaceutically Important Alcohols

Nitroglycerin

It is otherwise known as glyceryl trinitrate or glyceryl nitrate.

$$\begin{array}{l} CH_2ONO_2 \\ | \\ CHONO_2 \\ | \\ CH_2ONO_2 \end{array}$$

Preparation:

Nitration of glycerine with nitric acid and anhydrous calcium nitrate.

$$\text{Glycerol} \xrightarrow{HNO_3\ .\ Ca(NO_3)_2} \text{Nitroglycerine}$$

Nitration of glycerin with nitric acid and sulfuric acid.

$$\text{Glycerol} \xrightarrow[-H_2O]{H_2SO_4\ .\ HNO_3} \text{Nitroglycerine}$$

Nitration of glycerin with nitric acid and phosphorous pentoxide.

$$\text{Glycerol} \xrightarrow{HNO_3, P_2O_5} \text{Nitroglycerine}$$

Nitration of glycerin with nitric acid.

$$\text{Glycerol} \xrightarrow{HNO_3} \text{Nitroglycerine}$$

Physical properties: It is a colorless liquid, oily in nature with sweet burning taste. Upon cooling, it forms crystals. It melts at 13.2 °C, insoluble in water, soluble in alcohol and ether. Vapors are poisonous which cause headache and loss of consciousness.

Chemical properties: Nitroglycerin is highly shock sensitive.

$$\underset{\text{Nitroglycerin}}{4C_3H_5(ONO_2)_3} \xrightarrow{\text{Shock}} 6\underset{(g)}{N_2} + 12\underset{(g)}{CO_2} + 10H_2O + \underset{(g)}{O_2}$$

The sudden liberation of the large volumes of gases in a space occupied initially by the liquid substance causes explosion.

Uses:

1. It is used as medicine for the treatment of heart disease "angina pectoris".
2. It is also used for asthma and blood pressure by dilating the blood vessel.
3. It is used for the preparation of a powerful explosive e.g. dynamite, gun cotton, cordite and blasting gelatin.

Unsaturated Alcohols

Monohydric alcohols containing double bonds are known as unsaturated alcohols. These are classified into two types depending upon the attachment of -OH group to a carbon which is singly or doubly linked to another carbon. The simplest member of these alcohols is as follows.

$$CH_2 = CH - OH \qquad CH_2 = CH - CH_2 - OH$$

Vinyl alcohol   Allyl alcohol

Vinyl Alcohol

$$CH_2 = CH - OH$$

Ethenol

This compound exhibits keto-enol tautomerism. Due to this effect, the vinyl alcohol obtained from synthetic way is immediately converted into acetaldehyde.

$$H-\underset{\underset{H}{|}}{C}=\overset{\overset{OH}{|}}{C}-H \longrightarrow H_3C-\overset{\overset{O}{||}}{C}-H$$

Vinyl alcohol (Enol form)   Acetaldehyde (Keto form)

But the derivatives of vinyl alcohol such as vinyl acetate and vinyl bromide are quite stable.

Allyl Alcohol

$$CH_2 = CH - CH_2OH \quad \text{2-propene -1-ol}$$

It is an important unsaturated alcohol.

Table 12.3 Alcohols of Pharmaceutical importance.

Compound name & Structure	Uses	Qualitative tests
Benzyl alcohol CH_2OH attached to benzene ring ($C_6H_5CH_2OH$)	1. Used as an antiseptic. 2. Used in the preparation of benzyl benzoate which is used to treat asthma and whooping cough.	It is identified by infrared absorption spectrophotometry.
Cetostearyl Alcohol $C_{34}H_{72}O_2$	1. It is used in the preparation of emulsifying wax. 2. It is one of the ingredient of simple ointment and paraffin ointment.	It is identified by infrared absorption spectrophotometry.
Chlorbutol or Chlorbutanol $(CH_3)_2C(CCl_3)$-OH	1. It is used as an antibacterial agent and it also has fungicidal property. 2. The 0.5% concentration of chlorbutanol is used as a preservative in eye drops, injections, creams and mouthwash. 3. It also used as sedative & hypnotic.	1. To a small amount of freshly prepared sample solution add 1ml of 1 M sodium hydroxide solution and then add 2ml of iodine solution. Yellow precipitate of iodoform is produced. 2. Heat a small quantity of sample with 2 ml of 10 M sodium hydroxide solution and 1 ml of pyridine on a water bath and shake well. The separated pyridine layer becomes red colour. 3. Warm gently about little quantity of sample with 5 ml of ammonical silver nitrate solution, Black precipitate is obtained.
Ethanol CH_3CH_2OH	1. It is used for the manufacturing of alcoholic beverages, drugs, flavoring extracts and perfumes. 2. In hospitals, it is used as an antiseptic. 3. It is also used as an industrial solvent and used for the manufacturing of acetaldehyde, ethyl chloride and acetic acid etc. 4. Acts as a preservative for biological specimens. 5. It is used as a low freezing and mobile liquid in thermometers. 6. It is used as an antifreezing agent in automobiles.	1. Mix small quantity of sample solution in a small beaker with 1ml of potassium permanganate solution and 0.25 ml of dilute sulphuric acid and cover the beaker immediately with a filter paper moistened with a solution of freshly prepared sodium nitroprusside solution and a pinch of piperazine hydrate. An intense blue color is produced on the filter paper and the colour fades after a few minutes. 2. To the sample solution add 1ml of 1M sodium hydroxide solution followed by slow addition of 2ml of iodine solution. The smell of iodoform develops and a yellow precipitate is produced.
Glycerol CH_2OH | CHOH | CH_2OH Trihydric alcohol	1. It is used as humectants. 2. It is used in soaps, hand lotions, tooth pastes and bakery products and used in the production of plastics, surface coatings. 3. It is used for the manufacturing of explosives such as dynamite etc.	1. It is identified by the infrared absorption spectrophotometry. 2. Mix a small amount of sample solution with 0.5 ml of nitric acid and superimpose 0.5 ml of potassium dichromate solution. A blue ring develops at the interface of two liquids. Allow to stand for 10 minutes, the blue colour does not diffuse into the lower layer. 3. Heat a small amount of sample solution with 2 gm of potassium hydrogen sulphate in an evaporating dish. Irritant vapours are evolved which blacken filter paper moistened with alkaline potassium mercuric-iodide solution.

Table 12.3 Contd...

Propylene Glycol $CH_2(OH)—CH(OH)—CH_3$ 1,2-Propylene glycol $CH_2(OH)—CH_2—CH_2—OH$ 1,3-Propylene glycol	1. Commonly it is used as a drug solubiliser in topical, oral and injections. 2. It is used as a pharmaceutical additive. 3. It is also used as a stabilizer for vitamins and as a water miscible co - solvent.	1. To a small amount of ice cooled solution of sample, add 5ml of a cooled mixture of 10 ml of water and 90 ml of sulphuric acid. Heat for 10 minutes on a water-bath at 70 °C, cool and add 0.2 ml of a 3% w/v solution of ninhydrin in a solution of sodium metabisulphite. Violet colour slowly appears. 2. Heat a small amount of sample solution with pinch of boric acid. A pleasant odor develops. 3. Add a small amount of sample solution to 0.5 gm of potassium bisulphate and heat gently, a fruity odour develops.
Methanol CH_3OH	1. It is used as a raw material for the synthesis of formaldehyde. 2. It is used as an antifreeze for automobile radiators and used as motor fuel (20 % mixture of methanol and gasoline). 3. It is also used for denaturing of ethanol and acts as a solvent for paints and varnishes.	Transfer a small amount of sample solution in a fusion tube. Introduce heated copper wire for 2-3 times. To this solution add Schiff's reagent. Violet or purple color is formed.

Table 12.4 Qualitative tests for alcohols.

S. No.	Procedure	Observation	Inference
1.	To the small quantity of the sample, add a piece of sodium metal and heat it in a water bath.	Brisk effervescence is produced.	Indicates the presence of alcohol.
	$2R—OH + 2Na \longrightarrow 2R—ONa + H_2\uparrow$ (Brisk effervescence) Alcohol → Sodium alkoxide		
2.	**Ceric ammonium nitrate test:** To the small amount of sample solution, add 2-3 drops of ceric ammonium nitrate solution.	Red color appears.	Indicates the presence of alcohol.
	$R\text{-}OH + (NH_4)_2Ce(NO_3)_6 \longrightarrow (NH_4)_2Ce(OR)(NO_3)_5 + HNO_3$ Alcohol, Cerric ammonium nitrate → (Red color)		
3.	**Esterification test:** To the small quantity of the sample, add equal quantity of acetic acid and 1 part of conc. H_2SO_4. Warm the mixture for 2 minutes. Cool and pour the mixture in to a beaker containing aqueous sodium carbonate solution and observe the smell.	Sweet fruity smell is observed.	Indicates the presence of alcohol.
	$R—OH + HOOC\text{-}R' \xrightarrow{H_2SO_4} RCOOR' + H_2O$ Alcohol + Acid → Ester (Sweet fruity smell)		

Table 12.4 Contd...

S. No.	Procedure	Observation	Inference
4.	**Xanthate test:** To the small amount of sample, add sodium hydroxide pellets, dissolve it and cool. To this add 5 drops of carbon disulphide reagent and mix it well.	Yellow color precipitate is obtained.	Indicates the presence of alcohol.
	$R\text{-}OH + NaOH \longrightarrow RONa + H_2O$ $RONa + CS_2 \longrightarrow$ (RO)(NaS)C=S Thiocarbamate (Yellow colour)		
5.	**Oxidation test:** Transfer small amount of sample solution in a fusion tube. Introduce heated copper wire for 2-3 times. Divide it into two parts. To one part of solution add Schiff's reagent. To the second part, add 2-3 drops of dinitrophenyl hydrazine (DPH).	Appearance of violet or purple color. Odour of bitter almond oil is observed.	Indicates the presence of methanol and ethanol. Indicates the presence of benzyl alcohol.
6.	**Difference between 1°, 2° and 3°alcohols:** To the sample solution, add 2ml of Lucas reagent (anhydrous $ZnCl_2$ + concentrated HCl). Shake it well and keep it aside for 5 minutes.	Two phases of liquids separate immediately.	Indicates the presence of 3° alcohols.
		Cloudy solution appears after 5 minutes.	Indicates the presence of 2° alcohols.
		Clear solution is observed.	Indicates the presence of 1°alcohols.
7.	**Reaction with Jones reagent:** To the small amount of sample, add one ml of acetone and one drop of Jones reagent, mix it well.	Green or bluish green precipitate or emulsion is formed.	Indicates the presence of alcohol.
	$\underset{\text{Alcohol}}{RCH_2OH} \xrightarrow{(O)} \underset{\text{Aldehyde}}{RCHO} \xrightarrow{(O)} \underset{\text{Carboxylic acid}}{RCOOH}$		

Probable Questions

1. What are alcohols? Mention their importance.
2. Define & classify monohydric alcohols.
3. Write a detailed note on nomenclature of alcohols.
4. List out various general methods of preparation of alcohols and explain any five methods in detail.
5. Explain S_N2 reaction with reaction mechanism.
6. How alcohols are synthesized from olefins? & explain its reaction mechanism.
7. What is oxymercuration of alkenes?
8. Explain Bouveault – Blanc reduction.
9. How various alcohols are synthesized from Grignard reagent?
10. Explain the synthesis of alcohols from acid chlorides and esters by reduction reaction.

11. Describe the general structure & physical properties of alcohols.
12. Explain the various chemical properties of alcohols.
13. Write about chemical reactions involving replacement of hydrogen of the hydroxyl group.
14. Explain about reactions involving replacement of hydroxyl group.
15. Write a note on reactions involving both alkyl group and hydroxyl group.
16. Write short notes on chemical tests used for identification of alcohol.
17. How will you distinguish 1°, 2° & 3° alcohols?
18. Write a note on acidity of alcohols.
19. Explain esterification reaction with reaction mechanism.
20. Write a note on Victor Meyer's method of distinguishing alcohols.
21. What is Lucas test?
22. Write the chemical structure, synthesis, physical properties, chemical properties & uses of nitroglycerin.
23. What is unsaturated alcohol? Give examples.
24. Write about vinyl alcohol.
25. Write the chemical structure, synthesis, physical properties, chemical properties & uses of allylalcohol.
26. Write the chemical structure and uses of chlorobutanol, cetosteryl alcohol, benzyl alcohol, & propylene glycol.
27. Write a note on qualitative tests for alcohols.
28. Explain the pharmaceutical importance of alcohols.
29. Write a note on the application of Lucas test.

13

Halogen Derivatives or Alkyl Halides

Introduction

Alkyl halides are purely synthetic compounds and are rarely present in the nature. Chemically they are highly reactive species and serves as the starting material for very large number of aliphatic compounds. These are analogues to inorganic metallic halides such as NaCl, NaBr, *etc.*, but differ from them by its covalent bond. These are prepared by the substitution reaction between alkanes and halogens to give different derivatives in which one or more hydrogen atoms are replaced with one or more halogen atoms.

Importance of Alkyl Halides

Alkyl halides when attached with various chemical structures elicit different kind of pharmacological activities based on the core structure to which it is attached. Some of the drugs which possess alkyl halides and their therapeutic uses in the treatment of various diseases are shown in Table 13.1.

Table 13.1 Drugs containing alkyl halides and their therapeutic uses.

Category	Dug name	Structure
General anaesthetic	Ethyl chloride	CH_3CH_2Cl
	Chloroform	$CHCl_3$
	Halothane	$CF_3CHClBr$
Adrenergic blockers	Phenoxy benzamine	$C_6H_5-CH_2-N(CH_2CH_2Cl)-CH(CH_3)-CH_2O-C_6H_5$
	Dibenamine	$C_6H_5-CH_2-N(CH_2CH_2Cl)-CH_2-C_6H_5$

Classification

(A) Depending upon the number of halogen atoms present in the molecules alkyl halides are classified into following types.

1. ***Monohalogen derivatives:*** These are having one halogen atom in the molecules. The general formula is $C_nH_{2n+1}X$. These are generally denoted by RX (alkyl halide), where X= Cl, Br, I and F.

CH_3CH_2Cl — Ethyl chloride

$CH_3-CH(Br)-CH_3$ — *iso*-Propyl bromide

$CH_3-C(CH_3)_2-Cl$ — *tertiary*-butyl chloride

2. ***Dihalogen derivatives:*** In these two halogen atoms are attached in the molecule. They are further divided into two types.

   (i) Germinal dihalides or Gem dihalides: When both halogen atoms are attached to the same carbon atom, they are known as geminal or gem dihalides. These are otherwise called as alkylidene halides or alkylene dihalides because of loss of two hydrogen atoms takes place from the same carbon atom.

   CH_3-CHBr_2 ⟶ Ethylidene dibromide or 2,2-Dibromo ethane

   $CH_3-CCl_2-CH_3$ ⟶ Isopropylidene dichloride or 2, 2 - Dichloro propane

   (ii) Vicinal dihalides or *Vic*-dihalides: Both halogen atoms are attached to the adjacent carbon atoms. These are named as the dihalides of alkene from which the alkylene dihalides are prepared by addition reactions.

   $CH_2(Cl)-CH_2(Cl)$ ⟶ Ethylene dichloride

   $CH_3-C(CH_3)(Br)-CH_2-Br$ ⟶ *iso*-Butyl dibromide

3. ***Trihalogen derivatives:*** In this type three halogen atoms are attached to a molecule.

$$H-\underset{Cl}{\overset{Cl}{\underset{|}{\overset{|}{C}}}}-Cl \longrightarrow \text{Chloroform or Trichloromethane}$$

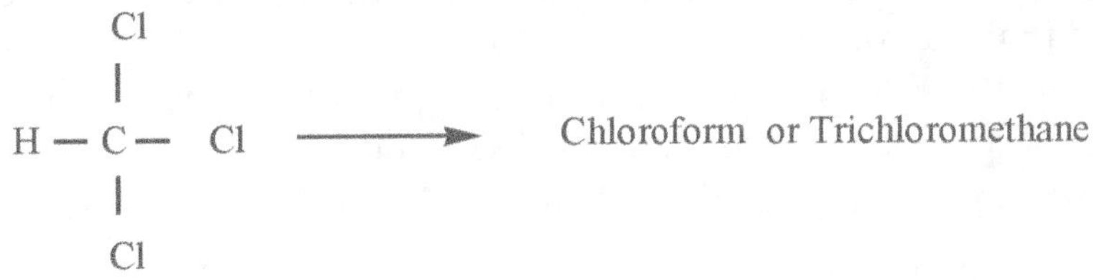

4. ***Tetra halogen derivatives:*** In this type four halogen atoms are attached to a molecule.

$$Cl-\underset{Cl}{\overset{Cl}{\underset{|}{\overset{|}{C}}}}-Cl \longrightarrow \text{Carbon tetrachloride}$$

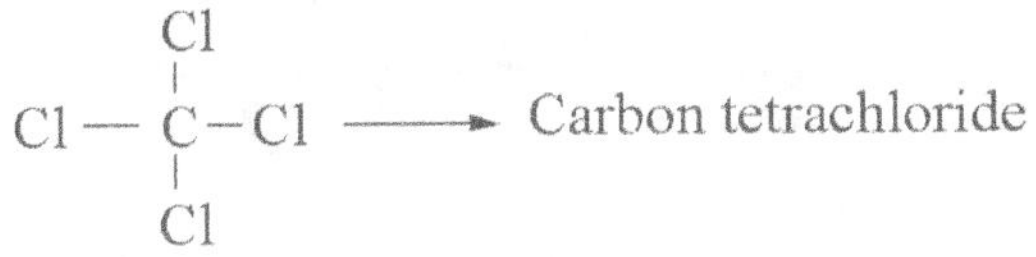

(B) In alkyl halides, the halogen atom is attached to the sp^3 carbon atom. Depending upon the nature of carbon atoms in which the halogen atoms are attached in alkyl halides, they are classified into primary, secondary and tertiary alkyl halides.

1. **Primary alkyl halides**

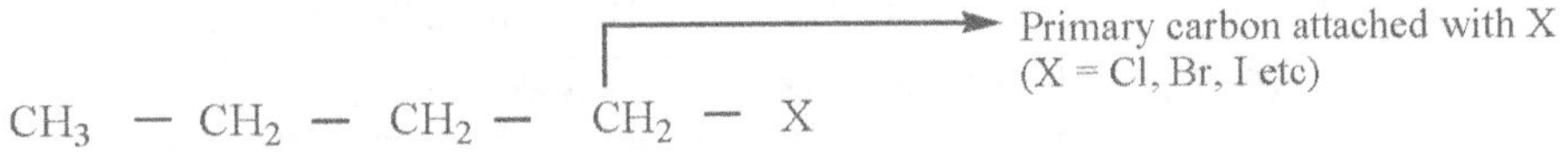

$$CH_3-CH_2-CH_2-CH_2-X$$

Primary carbon attached with X ($X = Cl, Br, I$ etc)

2. **Secondary alkyl halides**

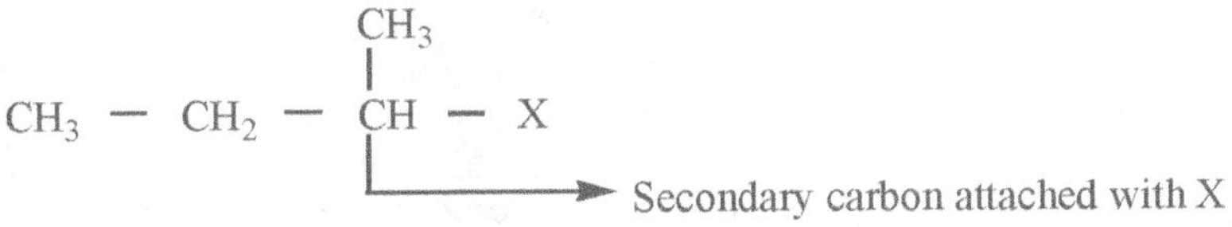

$$CH_3-CH_2-\overset{CH_3}{\overset{|}{CH}}-X$$

Secondary carbon attached with X

3. **Tertiary alkyl halides**

$$CH_3-\underset{CH_3}{\overset{X}{\underset{|}{\overset{|}{C}}}}-CH_3$$

Tertiary carbon attached with X

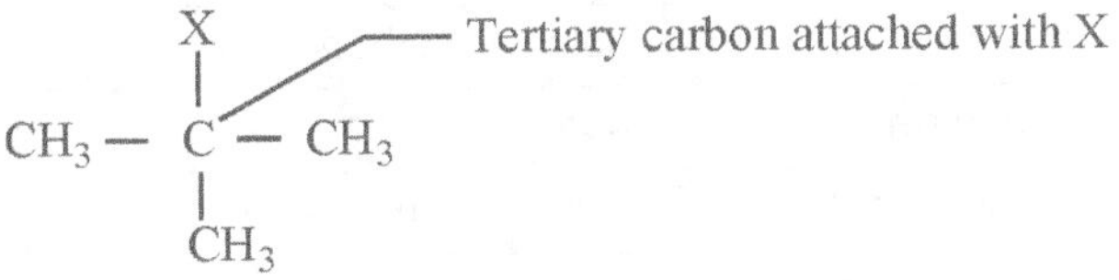

Nomenclature

They can be named by two ways.

1. ***Common name system:*** In this method alkyl group attached to halogen is mentioned first followed by the halide. The name of alkyl halide is mentioned as two words.

$R-X$	Alkyl halide
CH_3-CH_2-Cl	Ethyl chloride

2. ***IUPAC name system:*** The following rules are followed.
   (i) Longest continuous chain containing halogen is selected and the compound is considered as the derivative of that alkane.
   (ii) The carbon having the halogen atom should be given the least number and this number is used to indicate the position of halogen in alkyl halide.
   (iii) If any other substituents are present, indicate their position while writing the name of alkyl halide.

$$\underset{3}{CH_3}-\underset{2}{CH_2}-\underset{1}{CH_2}-Br$$

1-Bromo propane

$$\underset{5}{CH_3}-\underset{4}{CH_2}-\underset{H}{\overset{CH_3}{\underset{3}{C}}}-\underset{H}{\overset{Br}{\underset{2}{C}}}-\underset{1}{CH_3}$$

2-Bromo-3-methylpentane

Table 13.2 Common name and IUPAC name of some alkyl halides.

Structure	Common name	IUPAC name
$CH_3CH_2CH(Cl)CH_3$	*sec*-Butyl chloride	2-Chlorobutane
$(CH_3)_3CCH_2Br$	*neo*-Pentyl bromide	1-Bromo-2,2-dimethylpropane
$(CH_3)_3CBr$	*tert*-Butyl bromide	2-Bromo-2-methylpropane
$CH_2{=}CHCl$	Vinyl chloride	Chloro ethene
$CH_2{=}CHCH_2Br$	Allyl bromide	3-Bromopropene
(benzene ring with Cl and CH_3 ortho)	*o*-Chlorotoluene	1-Chloro-2-methylbenzene or 2-Chlorotoluene
(benzene ring with CH_2Cl)	Benzyl chloride	Chlorophenylmethane or Chloromethyl benzene
CH_2Cl_2	Methylene chloride	Dichloro methane
$CHCl_3$	Chloroform	Trichloro methane
$CHBr_3$	Bromoform	Tribromo methane
CCl_4	Carbon tetrachloride	Tetrachloro methane
$CH_3CH_2CH_2F$	*n*-Propyl fluoride	1-Fluoropropane

General Methods of Preparation

1. **From alkanes:** Alkanes undergo direct halogenation with Cl_2 or Br_2 in the presence of light or heat to yield respective alkyl halides with poly halogenated products. The monosubstituted products are obtained by controlling the ratio of chlorine to alkane.

$$\underset{\text{Methane}}{CH_4} + Cl_2 \xrightarrow[\text{or heat}]{\text{UV light}} \underset{\text{Methyl chloride}}{CH_3Cl} + HCl$$

The ease of halogenation is tertiary hydrogen atom > secondary hydrogen atom > primary hydrogen atom. Disadvantage: This method is not used for the laboratory synthesis of alkyl halides because separation of different halogenated products is a difficult process.

2. **From olefins or alkenes:** Alkenes undergo addition reaction with halogen acids (HCl, HBr *etc*) to give respective alkyl halides. The mechanism of addition reaction follows Markovnikov's rule or Anti-Markovnikov's rule (Refer alkenes for details).

$$\underset{\text{Ethylene}}{CH_2 = CH_2} + \underset{\text{Halogen acid}}{HX} \longrightarrow \underset{\text{Alkyl halide}}{CH_3 - CH_2 - X}$$

$$\underset{\text{Ethylene}}{CH_2 = CH_2} + HBr \longrightarrow \underset{\text{Ethyl bromide}}{CH_3 - CH_2 - Br}$$

Example for Markovnikov's addition

$$\underset{\text{Unsymmetrical alkene}}{R - CH = CH_2 + HX} \longrightarrow \underset{\text{Markovnikov's product}}{R - \underset{|}{\overset{X}{CH}} - CH_3}$$

$$\underset{\text{1-Propene}}{CH_3 - CH = CH_2} + HBr \longrightarrow \underset{\text{2-Bromo propane}}{CH_3 - \overset{Br}{\overset{|}{CH}} - CH_3}$$

Example for anti Markovnikov's addition

$$\underset{\text{Unsymmetrical alkene}}{R - CH = CH_2} + HX \xrightarrow{\text{Peroxide}} \underset{\text{Anti-Markovnikov's product}}{R - CH_2 - CH_2 - X}$$

$$\underset{\text{1-Propene}}{CH_3 - CH = CH_2} + HBr \xrightarrow{R-O-O-R'} \underset{\text{1-Bromopropane}}{CH_3 - CH_2 - CH_2 - Br}$$

3. **From alcohols or action of HX with alcohols:** It is the most widely used method for the synthesis of alkyl halides. In this the "-OH" group of alcohol is replaced by "X" atoms by the following ways.

   (a) Reaction of hydrogen chloride or hydrogen bromide with alcohol in the presence of zinc chloride or sulphuric acid.

$$\underset{\text{Alcohol}}{R-OH} + H-Cl \xrightarrow{ZnCl_2} \underset{\text{Alkyl chloride}}{R - Cl} + H_2O$$

$$\underset{\text{Ethanol}}{CH_3 - CH_2 - OH} + H - Cl \xrightarrow{ZnCl_2} \underset{\text{Ethyl chloride}}{CH_3 - CH_2 - Cl} + H_2O$$

$$R\text{—}OH + H\text{—}Br \xrightarrow{H_2SO_4} R\text{—}Br + H_2O$$

Alcohol → Alkyl bromide

$$CH_3\text{—}CH_2\text{—}OH + H\text{—}Br \xrightarrow{H_2SO_4} CH_3\text{—}CH_2\text{—}Br + H_2O$$

Ethanol → Ethyl bromide

Use of sulphuric acid is avoided when 2° and 3° alcohols are used, since it leads to dehydration of alcohols to olefins. In that case, alcohols are heated with potassium bromide and sulphuric acid.

$$ROH + KBr + H_2SO_4 \longrightarrow R\text{—}Br + KHSO_4 + H_2O$$

Alcohol → Alkyl bromide

Excess amount of constant boiling hydroiodic acid (57%) is used for the preparation of alkyl iodides. But good yields of alkyl iodides are obtained by heating alcohol with potassium or sodium iodide with 95% phosphoric acid.

(b) *Reaction of alcohols with thionyl chloride (Darzen's procedure):* Alcohols react with thionyl chloride in the presence of pyridine (1 mole) to give alkyl halides. Pyridine is used to absorb the HCl formed during the reaction.

$$R\text{—}OH + SOCl_2 \xrightarrow[C_5H_5N\ (Reflux)]{1\ mole} R\text{—}Cl + SO_2 + HCl$$

Alcohol + Thionyl chloride → Alkyl chloride

$$CH_3\text{—}CH_2\text{—}OH + SOCl_2 \xrightarrow[C_5H_5N\ (Reflux)]{1\ mole} CH_3\text{—}CH_2\text{—}Cl + SO_2 + HCl$$

Ethanol → Ethyl chloride

(c) *Reaction of alcohols with phosphorous halides:* Alcohols react with phosphorus pentahalide (phosphorus pentachloride) or phosphorus trihalide (phosphorus tribromide or triiodide) to give corresponding alkyl halides.

$$R\text{—}OH + PX_5 \longrightarrow R\text{—}X + POX_3 + HX$$

Alcohol + Phosphorous penta halide → Alkyl halide

$$H_3C\text{—}CH_2\text{—}OH + PCl_5 \longrightarrow CH_3\text{—}CH_2\text{—}Cl + POCl_3 + HCl$$

Ethanol + Phosphorous pentachloride → Ethyl chloride + Phosphorous oxy chloride

$$3R\text{—}OH + PX_3 \longrightarrow 3R\text{—}X + H_3PO_3$$

Alcohol + Phosphorous trihalide → Alkyl halide + Phosphoric acid

$$3CH_3\text{—}CH_2\text{—}OH + PBr_3 \longrightarrow 3CH_3\text{—}CH_2\text{—}Br + H_3PO_3$$

Ethanol + Phosphorous tribromide → Ethyl bromide + Phosphoric acid

$$3CH_3\text{—}CH_2\text{—}OH + PI_3 \longrightarrow 3CH_3\text{—}CH_2\text{—}I + H_3PO_3$$

Ethanol + Phosphorous triiodide → Ethyl iodide + Phosphoric acid

Note:

(a) The phosphorous tribromide or triodide needed for this reaction is produced insitu *i.e.*, bromine or iodine is added to a mixture of red phosphorous and alcohol followed by warming.

$$3Br_2 + 2P \xrightarrow[\Delta]{\text{Alcohol}} \underset{\text{Phosphorous tribromide}}{2PBr_3}$$

$$3I_2 + 2P \xrightarrow[\Delta]{\text{Alcohol}} \underset{\text{Phosphorous triiodide}}{2PI_3}$$

(b) This method is suitable for preparing primary alkyl halides with very good yield but less suitable for secondary and tertiary alkyl halides.

4. **By displacement reaction of one halogen atom by another (halogen atoms exchange reaction):** This method is most affordable for the preparation of alkyl iodides. By this method alkyl chloride or bromide in acetone or methanol is heated with sodium iodide.

$$\underset{\text{Alkyl chloride}}{R-Cl} + \underset{\text{Sodium iodide}}{Na-I} \xrightarrow[\Delta]{\text{Acetone/ methanol}} \underset{\text{Alkyl iodide}}{RI} + NaCl$$

$$\underset{\text{Ethyl chloride}}{CH_3-CH_2-Cl} + \underset{\text{Sodium iodide}}{Na-I} \xrightarrow[\Delta]{\text{Acetone/ methanol}} \underset{\text{Ethyl iodide}}{CH_3-CH_2-I} + NaCl$$

Alkyl fluorides are obtained by treating alkyl bromide or chloride with inorganic fluorides.

$$\underset{\text{Methyl chloride}}{2CH_3-Cl} + \underset{\text{Mercurous fluoride}}{Hg_2F_2} \longrightarrow \underset{\text{Methyl fluoride}}{2CH_3-F} + \underset{\text{Mercurous chloride}}{Hg_2Cl_2}$$

5. **Hunsdiecker or borodin Hunsdiecker reaction or decomposition of silver salts of carboxylic acids:** In this method, the silver salts of carboxylic acids undergo decomposition reaction by bromine or chlorine to give alkyl halides.

$$\underset{\text{Silver carboxylate}}{R-COO-Ag} + Br_2 \longrightarrow \underset{\text{Alkyl bromide}}{RBr} + CO_2 + \underset{\text{Silver bromide}}{AgBr}$$

$$\underset{\text{Silver acetate}}{CH_3-COO-Ag} + Br_2 \longrightarrow \underset{\text{Methyl bromide}}{CH_3-Br} + CO_2 + AgBr$$

The degree of yield of alkyl halide is primary > secondary > tertiary. Generally bromine is used because chlorine gives lesser yield.

Mechanism: The main mechanism of this reaction is formation of acyl hypohalite which upon further decomposition yields free radicals.

$$RCOO^{-}Ag^{+} \xrightarrow[CCl_4]{Br_2} RCOOBr + AgBr$$

$$RCOO-Br \xrightarrow{\Delta} RCOO^{\cdot} + Br^{\cdot}$$

$$RCOO^{\cdot} \longrightarrow R^{\cdot} + CO_2$$

$$R^{\cdot} + Br^{\cdot} \longrightarrow R-Br$$

Higher yields of alkyl bromides are obtained by refluxing bromine and carbon tetrachloride solution in the presence of excess mercuric oxide.

$$\underset{\text{Carboxylic acid}}{2RCOOH} + 2Br_2 + \underset{\text{Mercuric oxide}}{HgO} \longrightarrow \underset{\text{Alkyl bromide}}{2RBr} + 2CO_2\uparrow + HgBr_2 + H_2O$$

$$\underset{\text{Acetic acid}}{2CH_3COOH} + 2Br_2 + \underset{\text{Mercuric oxide}}{HgO} \longrightarrow \underset{\text{Methyl bromide}}{2CH_3Br} + 2CO_2\uparrow + HgBr_2 + H_2O$$

Summary of Methods of Preparation

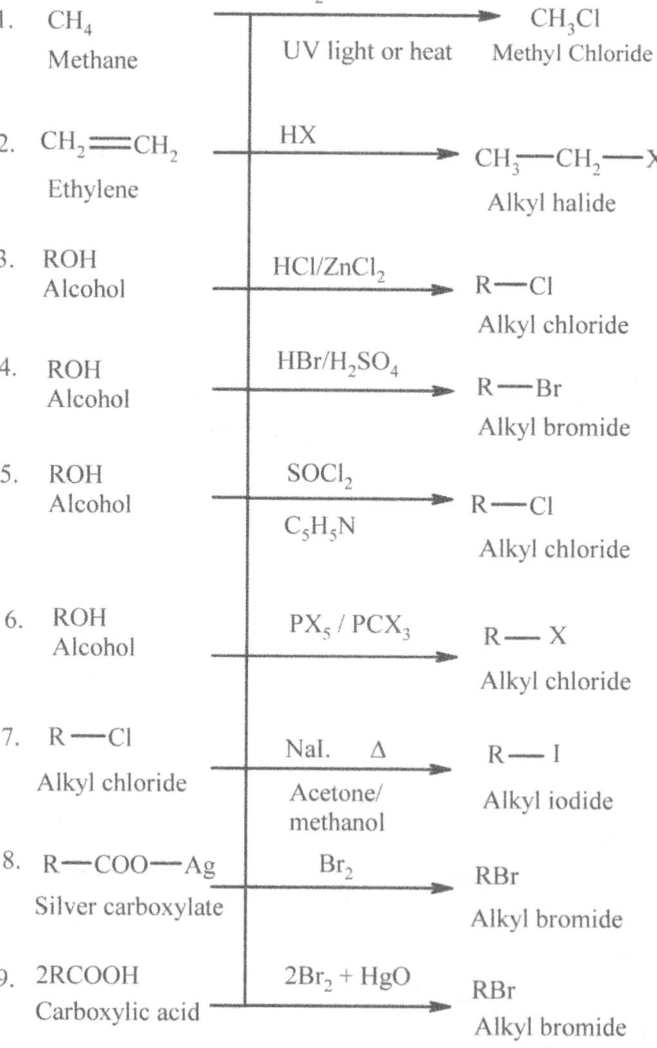

Structure

The first member of alkyl halide is methyl halide and the molecular formula is CH_3X. The carbon atom is in sp^3 hybridised state and the halogen atom has half-filled p orbital in the valence shell. The C-X bond is formed by co-axial overlap of sp^3 orbital of carbon and half-filled p-orbital of halogen with the formation of σ bond. The C-H bond is formed by the co-axial overlap of sp^3 orbital of carbon and s orbitals of hydrogen forming σ bond. The overall shape is tetrahedral.

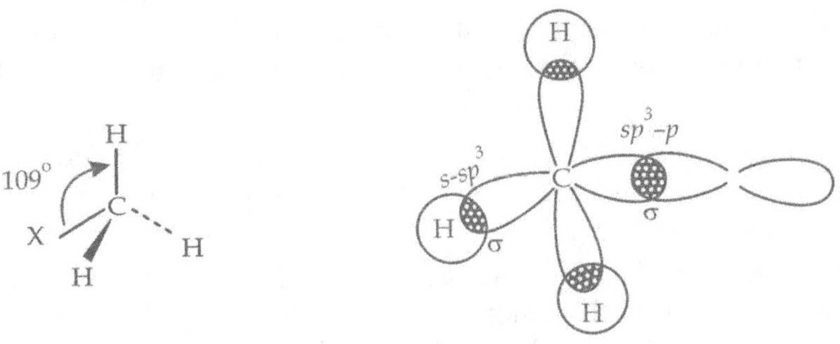

Orbital structure of Methylhalide

Figure 13.1 Structure of alkyl halide.

Physical Properties

1. The lower members of alkyl halides such as methyl chloride, methyl fluoride and ethyl chloride are gases at room temperature.
2. Upto C_{18} are colourless, sweet smelling liquids.
3. Higher alkyl halides greater than C_{18} are colorless, odorless solids.
4. These are insoluble in water (no hydrogen bond with water) but soluble in organic solvents.
5. Alkyl bromides and alkyl iodides are heavier than water.
6. They burn with green edged flame.
7. Alkyl iodides are easily decomposed by light and liberation of iodine takes place which is responsible for darkening of alkyl iodides upon standing. The reaction is due to the instability or unstable nature of C-I bond.
8. The order of boiling points and densities of alkyl halide is iodide > bromide > chloride > fluoride. Boiling point and density of alkyl halides are listed in the Table 13.3.

Table 13.3 Physical properties of some alkyl halides.

S. No.	Molecular Formula	Molecular weight	Boiling point (°C)	Density or w/ml
1.	CH_3F	34	– 78	0.88
2.	CH_3Cl	50.5	– 24	0.92
3.	CH_3Br	95	4	1.68
4.	CH_3I	142	42	2.28
5.	CH_2Cl_2	85	40	1.34
6.	$CHCl_3$	119	61	1.5
7.	CCl_4	154	77	1.6

For isomeric alkyl halides, the order of boiling point is

Primary > secondary > tertiary

Factors influencing the boiling point of alkyl halides

Two types of intermolecular forces influence the boiling point of alkyl halides. They are

1. London forces (strongest intermolecular force).
2. Dipole-dipole attraction (attractive forces which arise from C-X bond).

1. **London forces:** It is a surface attraction force due to temporary dipole effects. When surface area of a molecule is more, London force also more and boiling point also increases. The surface area of one alkyl halide varies from halogen atoms, which depends upon the surface area of halogen atoms. Alkyl halides with large surface area possess larger London forces leads to higher boiling points.

   For example, compare the boiling point of *n*-butyl fluoride (33 °C) and *n*-butane (0 °C). Generally, the surface areas (Vander walls radii) of alkyl halides are nearly similar to that of corresponding alkane. So the London forces are similar.

2. **Dipole-dipole attraction:** Consider about the dipole moment, alkyl halides have high dipole moment than alkane. The total attractive forces in the alkyl fluoride are slightly greater which leads to higher boiling point. So *n*-butyl fluoride boiling point is 33 °C, but the alkane, *n*-butane has the boiling point 0 °C.

   Alkyl chlorides possess higher boiling point due to more surface area of chlorine atom. Similarly alkyl bromide and alkyl iodides also possess higher boiling points. Alkyl halides with spherical shape and more branched chains possess lower boiling points than that of straight chain due to their small surface area.

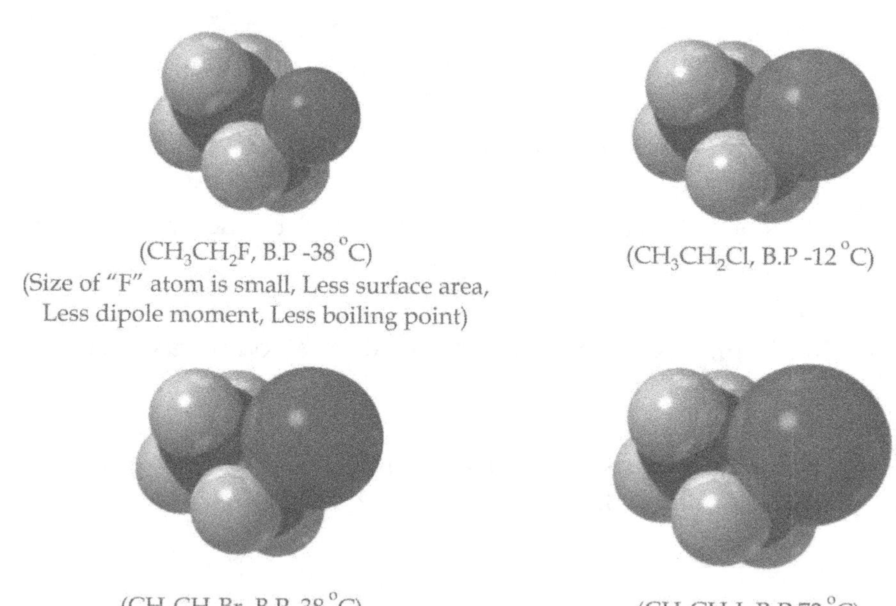

Figure 13.2 Boiling points of alkyl halides increases with increase in surface area.

Chemical Properties

Alkyl halides undergo two major types of chemical reaction (1 & 2). They are

1. **Nucleophilic substitution reactions.**
2. **Elimination reactions.**

3. Miscellaneous reactions

All alkyl halides are highly reactive in nature. The order of reactivity depends upon the nature of halogen atom.

The order of reactivity on the basis of halogen atom is as follows.

iodide > bromide > chloride

The order of reactivity on the basis of nature of alkyl group is as follows.

$$3^o > 2^o > 1^o > CH_3$$

The order of reactivity of 1°alkyl halide is

$$CH_3X > CH_3CH_2X > C_2H_5CH_2X$$

The order of reactivity is explained by the nature of C-X bonds. Due to the presence of highly electronegative halogen atom, the bond between alkyl group and halogen atom is polar and possess dipole.

Due to this, carbon atoms get partial positive charge and the halogen atom gets partial negative charge.

$$-\overset{\delta+}{C} - \overset{\delta-}{X}$$

Due to the formation of carbocation, which is easily attacked by an electron rich species such as nucleophiles, the halide ion is removed or eliminated as a leaving group. The leaving halide ion is substituted by attacking nuclophile, which leads to nucleophilic substitution reactions of alkyl halides. This reaction is one of the important reactions of alkyl halides which lead to the formation of various kinds of functional derivatives.

$$\underset{\text{Alkyl halide}}{R - X} + \underset{\text{Nucleophile}}{\overset{\ominus}{Nu:}} \longrightarrow R - Nu + \bar{X}:$$

1. **Nucleophilic substitution reactions:** Two types of nucleophilic substitution reactions take place in alkyl halides.

S_N1 Reaction: $\underset{\text{Alkyl halide}}{R - X} \xrightarrow{-\bar{X}} R^+ \xrightarrow{\bar{Nu}} R - Nu$

S_N2 Reaction: $\underset{\text{Alkyl halide}}{R - X} + \underset{\text{Nucleophile}}{\bar{Nu}:} \longrightarrow [Nu.....R.....X] \longrightarrow Nu - R + X^-$

For More Details Refer Chapter-14

2. **Elimination Reactions:** The positive charge of the carbon or the carbocation of an alkyl halide is propagated to the neighboring carbon atoms by inductive effect. When a strong base approaches this carbon, it has the tendency to lose a proton from the β-carbon atom. Such kind of reaction involves the elimination of halogen atom along with a proton from the β carbon. As a result alkenes are formed as a product.

   This reaction is called as elimination reaction. There are two types of elimination reactions and similar to nucleophilic substitution reaction, elimination reaction also follows two type of mechanism as described below.

Mechanism: E_1 reaction

$$R-\underset{H}{\overset{H}{C}}-\underset{X}{\overset{H}{C}}-H \xrightarrow{\text{Slow}} R-\underset{H}{\overset{H}{C}}-\overset{H}{\underset{+}{C}}-H + X^-$$

Alkyl halide

for detailed description about E_1 and E_2 reactions refer Chapter- 14

$$\overline{B}: \text{(Base)} \quad R-\underset{H}{\overset{H}{C}}-\overset{H}{\underset{+}{C}}-H \xrightarrow{\text{Fast}} (R)(H)C=C(H)(H) + B:H$$

Alkene

E_2 reaction

$$\overline{B}: \quad -\overset{H}{C}-\underset{X}{C}- \xrightarrow{\text{Slow}} \left[B\text{----}H \; -C \overset{\cdot\cdot}{=} C \; X \right] \xrightarrow{\text{Fast}} >C=C< + B:H + X^-$$

Alkyl halide — Alkene

The order of reactivity of alkyl halides depending upon the nature of alkyl groups is 3° > 2° > 1°

Reasons

1. The alkyl groups have electron releasing properties. As the number of alkyl groups attached to the carbon atom of alkyl halides increase, the electron density of that carbon atom also increases which leads to repulsion of electron pair towards the X atom of C-X bond.

$$\underset{1^\circ}{R \rightarrow CH_2 \rightarrow X} < \underset{R\;2^\circ}{R \rightarrow \overset{}{CH} \twoheadrightarrow X} < \underset{R\;3^\circ}{R \rightarrow \overset{R}{C} \rightarrow \rightarrow X}$$

The above fact indicates that the halogen atom is eliminated from the alkyl halide as an X^- ion. The order of elimination of halide ion is

3° > 2° > 1°

The primary alkyl halides undergo chemical reaction either by S_N2 or E_2 mechanism which involves the formation of an intermediate transition state.

The tertiary alkyl halides react by either S_N1 or E_1 mechanism by the formation of carbocation intermediate.

The secondary alkyl halides react by either one or both mechanisms depending upon the nature of alkyl halide and nature of reagents. **For more details refer chapter -14.**

Nucleophilic substitution reactions: Some of the important nucleophilic substitution reactions of alkyl halides are as follows.

(i) Williamson's ether synthesis (Reaction with sodium alkoxide or dry silver oxide): Alkyl halides react with sodium alkoxide or dry silver oxide to gives ether.

$$\underset{\text{Alkyl halide}}{RX} + \underset{\text{Sodium ethoxide}}{R'ONa} \xrightarrow{-NaX} \underset{\text{Ether}}{R-O-R'} + NaX$$

$$\underset{\text{Ethyl iodide}}{C_2H_5I} + \underset{\text{Sodium ethoxide}}{C_2H_5ONa} \longrightarrow \underset{\text{Diethyl ether}}{C_2H_5-O-C_2H_5} + NaI$$

$$\underset{\text{Ethyl iodide}}{2C_2H_5I} + Ag_2O \longrightarrow \underset{\text{Diethyl ether}}{C_2H_5-O-C_2H_5} + 2AgI$$

The sodium alkoxide needed for this reaction is obtained by dissolving metallic sodium in excess of alcohol.

$$\underset{\text{Ethanol}}{2CH_3CH_2OH} + 2Na \longrightarrow \underset{\text{Sodium ethoxide}}{2CH_3CH_2O^-Na^+} + H_2\uparrow$$

The reaction of alkyl halide with sodium ethoxide or silver oxide is an example of S_N2 reaction *via* the formation of transition state intermediate.

$$\underset{\text{Ethoxide ion}}{CH_3-CH_2-\ddot{O}:} + CH_3-I \longrightarrow \underset{\text{Transition state}}{\left[CH_3-CH_2-\overset{\delta^-}{O}\cdots\overset{\delta^+}{CH_3}\cdots\overset{\delta^-}{I}\right]}$$

$$\downarrow$$

$$\underset{\text{Ethyl methyl ether}}{CH_3-CH_2-O-CH_3} + I^-$$

(ii) Hydrolysis: Alkyl halides undergo hydrolysis very fast with boiling aqueous alkali or silver oxide suspended in boiling water.

$$\underset{\text{Alkyl halide}}{R-X} + \underset{\text{Water}}{HOH} \xrightarrow{\text{Slow}} \underset{\text{Alcohol}}{R-OH} + HX$$

$$\underset{\text{Alkyl halide}}{R-X} + \underset{\text{Potassium hydroxide (aqueous)}}{KOH} \xrightarrow{\text{Fast}} \underset{\text{Alcohol}}{R-OH} + KX$$

$$\underset{\text{Methyl iodide}}{CH_3I} + KOH \xrightarrow[\Delta]{H_2O} \underset{\text{Methanol}}{CH_3-OH} + KI$$

$$\underset{\text{Ethyl bromide}}{CH_3CH_2Br} + KOH \xrightarrow[\Delta]{H_2O} \underset{\text{Ethanol}}{CH_3CH_2-OH} + KBr$$

Mechanism: In the above said reaction, OH^- acts as a nucleophile.

$$:\overline{O}H + CH_3 - CH_2 - Br \xrightarrow{S_N^2} CH_3 - CH_2 - OH + \ddot{\overline{B}}r$$

Ethyl bromide

(iii) Reaction with sodio-alkynides (synthesis of higher alkynes): Higher alkynes are prepared by treating alkyl halides with sodio alkynides. Primary alkyl halides predominantly give this reaction.

$$CH_3 - I + Na^+C^- \equiv CH \longrightarrow CH_3 - C \equiv CH + NaI$$

Methyl iodide; Sodium alkynides (Sodium acetylide); Propyne

(iv) Reaction with ammonia: Alkyl halides when heated with alcoholic solution of ammonia in a sealed tube, a mixture of different type of amines are formed.

$$CH_3CH_2Br + NH_3 \xrightarrow{\Delta} CH_3 - CH_2 - NH_2 + HBr$$

Ethyl bromide; Ammonia; Ethylamine (Primary amine)

$$CH_3 - CH_2 - NH_2 + CH_3 - CH_2 - Br \xrightarrow{\Delta} (CH_3 - CH_2)_2NH + HBr$$

Ethylamine (Primary amine); Ethyl bromide; Diethyl amine (Secondary amine)

$$(CH_3 - CH_2)_2NH + CH_3 - CH_2 - Br \xrightarrow{\Delta} (CH_3 - CH_2)_3 - N + HBr$$

Diethyl amine (Secondary amine); Ethyl bromide; Triethyl amine (Tertiary amine)

$$(CH_3 - CH_2)_3 - N + CH_3 - CH_2 - Br \longrightarrow (CH_3CH_2)_4N^+Br^-$$

Triethyl amine (Tertiary amine); Ethyl bromide; Tetra ethyl quarternary amine

(v) Reaction with potassium or sodium cyanide: Alkyl halides when heated with potassium or sodium cyanide in the presence of aqueous ethanol yield alkyl cyanides or nitriles.

$$R - X + KCN \xrightarrow[\text{Aqueous ethanol}]{\Delta} RCN + KX$$

Alkylhalide; Alkyl cyanide

$$CH_3 - Br + KCN \xrightarrow[\text{Aqueous ethanol}]{\Delta} CH_3 - CN + KBr$$

Methyl bromide; Methyl cyanide

$$CH_3 - CH_2 - Br + KCN \xrightarrow[\text{Aqueous ethanol}]{\Delta} CH_3 - CH_2 - CN + KBr$$

Ethyl bromide; Ethyl cyanide

In the above reaction the CN^- (cyanide ion) acts as a nucleophile.

Mechanism

Step 1: Formation of nucleophile (CN^-).

$$K - CN \xrightarrow[\text{Ethanol}]{\text{Aqueous}} \overset{+}{K} + \overset{-}{\ddot{C}N}$$

Step 2: Nucleophile attacks ethyl bromide to give ethyl cyanide.

$$CH_3 - CH_2 - Br + CN^- \longrightarrow CH_3 - CH_2 - CN + \bar{Br}$$

Alkyl cyanides or alkyl nitriles are useful synthetic reagents to prepare a variety of derivatives such as carboxylic acid, primary amines and amides under different conditions.

$$\underset{\text{Ethyl cyanide}}{CH_3 - CH_2 - CN} \xrightarrow{2H_2O} \underset{\text{Propanoic acid}}{CH_3 - CH_2 - COOH} + NH_3$$

$$\underset{\text{Ethyl cyanide}}{CH_3 - CH_2 - CN} + 4[H] \xrightarrow[\text{Reduction}]{LiAlH_4} \underset{\text{Propylamine}}{CH_3 - CH_2 - CH_2 - NH_2}$$

$$\underset{\text{Alkyl cyanide}}{R - C \equiv N} + H_2O \xrightarrow[H_2O_2]{\text{Alkaline}} \underset{\text{Amide}}{RCONH_2}$$

$$\underset{\text{Methyl cyanide}}{CH_3 - C \equiv N} + H_2O \xrightarrow[H_2O_2]{\text{Alkaline}} \underset{\text{Acetamide}}{CH_3CONH_2}$$

(vi) Reaction with silver cyanide: Alkyl halides when heated with aqueous ethanolic silver cyanide, alkyl isocyanide and small amounts of isomeric cyanides are obtained.

$$\underset{\text{Ethyl iodide}}{CH_3 - CH_2 - I} + \underset{\text{(aqueous, ethanolic silver cyanide)}}{AgCN} \longrightarrow \underset{\text{Ethyl isocyanide}}{CH_3CH_2NC} + AgI$$

Silver cyanide is predominantly a covalent compound. In this case only N atom is free to donate electron pair. When an alkyl halide is treated with silver cyanide, the alkyl carbocation is driven to the nitrogen and the attack mostly occurs through the N- atom of the cyanide group to yield alkyl isocyanide.

$$\underset{\text{Silver cyanide}}{Ag - C \equiv N} + \underset{\text{Alkyl iodide}}{R - I} \longrightarrow [Ag - C \equiv N - R]^+ I^- \xrightarrow{-AgI} \underset{\text{Alkyl isocyanide}}{R - N \equiv C}$$

(vii) **Reaction with silver salt of carboxylic acid:** Alkyl halides when heated with ethanolic solution of silver salt of carboxylic acid yields ester.

$$\underset{\text{Silver salt of acid}}{R-COOAg} + X-R' \longrightarrow \underset{\text{Ester}}{RCOOR'} + AgX$$

$$\underset{\text{Silver acetate}}{CH_3-COOAg} + \underset{\text{Ethyl iodide}}{I-C_2H_5} \longrightarrow \underset{\text{Ethyl acetate}}{CH_3-COO-C_2H_5} + AgI$$

(viii) **Reaction with silver nitrite:** Alkyl halides when heated with the ethanolic solution of silver nitrite yields nitro alkanes. Some amount of alkyl nitrite is also formed in the reaction mixture.

$$\underset{\text{Alkyl halide}}{R-X} + \underset{\text{Silver nitrite}}{Ag-O-N=O} \longrightarrow \underset{\text{Nitro alkane}}{R-NO_2} + AgX$$

$$\underset{\text{Ethyl iodide}}{C_2H_5-I} + \underset{\text{Silver nitrite}}{Ag-O-N=O} \longrightarrow \underset{\text{Nitro ethane}}{C_2H_5-NO_2} + AgI$$

(ix) **Reaction with potassium nitrite:** Alkyl halides when heated with potassium nitrite in an aqueous ethanolic solution yields alkyl nitrite as a main product along with some nitro alkanes.

$$\underset{\text{Alkyl halide}}{R-X} + \underset{\text{Potassium nitrite}}{K-O-N=O} \longrightarrow \underset{\text{Alkyl nitrite}}{R-O-N=O} + KX$$

$$\underset{\text{Ethyl Iodide}}{C_2H_5-I} + \underset{\text{Potassium nitrite}}{K-O-N=O} \longrightarrow \underset{\text{Ethyl nitrite}}{C_2H_5-O-N=O} + KI$$

(x) **Reaction with mercaptides:** Alkyl halides when heated with an alcoholic solution of a mercaptide (metallic derivative of thio alcohol) or potassium sulphide gives thioethers.

$$\underset{\text{Alkyl halide}}{R-X} + \underset{\text{Sodium alkyl mercaptide}}{Na-S-R'} \xrightarrow[\text{Alcohol}]{\Delta} \underset{\text{Thio ether}}{R-S-R'} + NaX$$

$$\underset{\text{Ethyl iodide}}{C_2H_5-I} + \underset{\text{Sodium methyl mercaptide}}{Na-S-CH_3} \xrightarrow[\text{Alcohol}]{\Delta} \underset{\text{Ethyl methyl thio ether}}{C_2H_5-S-CH_3} + NaI$$

(xi) **Reaction with sodium or potassium hydrogen sulphide:** Alkyl halides when react with aqueous solution of alcoholic sodium or potassium hydrogen sulphide gives thioalcohols.

$$\underset{\text{Alkyl halide}}{R-X} + \underset{\text{Sodium hydrogen sulphide}}{Na-SH} \xrightarrow{\text{Ethanol}} \underset{\text{Thio alcohol}}{R-SH} + NaX$$

$$\underset{\text{Ethyl iodide}}{CH_3-CH_2-I} + \underset{\text{Sodium hydrogen sulphide}}{Na-SH} \xrightarrow{\text{Ethanol}} \underset{\text{Ethane thiol}}{CH_3-CH_2-SH} + NaX$$

Some common nucleophilic substitution reactions and their products are given below and summarized in Table 13.4.

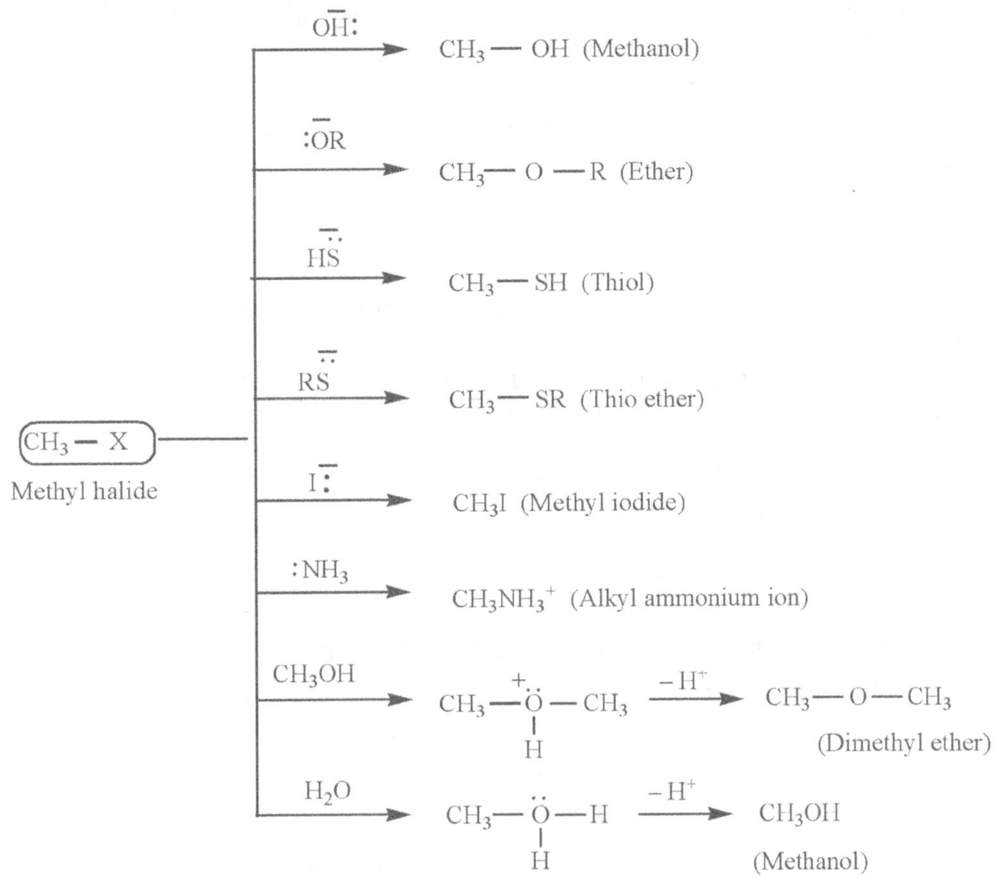

Table 13.4 Nucleophiles and their nucleophilic substitution products.

Reagent	Nucleophile	Substitution Product R-Nu	Class of main Product
NaOH (KOH)	$H\bar{O}$	ROH	Alcohol
H_2O	H_2O	ROH	Alcohol
NaOR′	$R'O^-$	ROR′	Ether
NaI	I	R-I	Alkyl iodide
NH_3	NH_3	RNH_2	1° amine
$R'NH_2$	RNH_2	RNHR′	2° amine
R′R″NH	R′R″NH	RNR′R″	3° amine
KCN	—CN⁻	RCN	Nitrile (cyanide)
AgCN	Ag-CN:	RNC (isocyanide)	Isonitrile
KNO_2	O=N-O	R-O-N=O	Alkyl nitrite
$AgNO_2$	Ag —Ö—N=O	$R-NO_2$	Nitroalkane
R′COOAg	$R'COO^-$	R′COOR	Ester
$LiAlH_4$	H	RH	Alkane
R'^-M^+	R'^-	RR′	Higher Alkane

If a nucleophile possesses negative charge then the atom which donates the electron pair becomes neutral in the reaction. If the nucleophile is neutral charged, then the atom which donates the electron pair becomes positive charge in the product which further undergoes proton transfer reaction to yield a neutral product.

Elimination reactions (Dehydrohalogenation): When alkyl halides with β-hydrogen atoms are boiled with alcoholic potassium hydroxide solution, formation of olefins or alkenes takes place. This reaction involves the elimination of HX from the alkyl halide, so it is called as dehydrohalogenation.

Note: Described in detail under chapter 14. (under elimination reactions,E1 & E2 mechanisms).

3. **Miscellaneous chemical reactions:**

(i) **Wurtz reaction:** Alkyl halides (bromide or iodide) when heated with sodium metal in the presence of an ether gives alkanes.

$$\underset{\text{Alkyl halide}}{RX} + R'X + 2Na \longrightarrow \underset{\text{Alkane}}{R-R'} + 2NaX$$

The product obtained possesses double the number of carbon atoms as are there in the alkyl halide.

(ii) **Reaction with magnesium (preparation of Grignard reagent):** Dry alkyl halides react with dry magnesium powder in dry ether or dioxan to yield Grignard reagents.

$$\underset{\text{Dry alkyl halide}}{R-X} + \underset{\text{Dry magnesium powder}}{Mg} \xrightarrow[\text{Dioxan}]{\text{Dry ether/}} \underset{\text{Grignard reagent}}{R-Mg-X}$$

(iii) **Reduction:** Alkyl halides undergo reduction with zinc/copper couple, sodium and ethanol or Sn/HCl to give corresponding alkanes. The reaction takes place by the electron transfer from the metal to alkyl halide.

$$\underset{\text{Alkyl halide}}{R-X} + H_2 \xrightarrow[/C_2H_5OH/\ Sn/HCl]{\text{Zn-Cu/Na}} \underset{\text{Alkane}}{R-H} + HX$$

$$\underset{\text{1-Iodo propane}}{CH_3-CH_2-CH_2-I} + 2[H] \xrightarrow{\text{Zn-Cu}} \underset{\text{Propane}}{CH_3-CH_2-CH_3} + HI$$

$$\underset{\text{2-Bromopropane}}{CH_3-\overset{\overset{\large Br}{|}}{CH}-CH_3} + 2[H] \xrightarrow[\text{Ether}]{LiAlH_4} \underset{\text{Propane}}{CH_3-CH_2-CH_3} + HBr$$

Mechanism:

$$Zn \longrightarrow Zn^{2+} + 2e^-$$

$$R-X \longrightarrow X^+ + R\bar{:}$$

$$R\bar{:} + C_2H_5OH \longrightarrow R-H + C_2H_5O^-$$

(iv) **Friedel Crafts alkylation:** Alkyl benzenes are obtained when alkyl halides are treated with benzene in the presence of $AlCl_3$

$$\underset{\text{Ethyl bromide}}{CH_3CH_2Br} + \underset{\text{Benzene}}{C_6H_6} \xrightarrow{AlCl_3} \underset{\text{Ethyl benzene}}{C_6H_5CH_2CH_3} + HBr$$

(v) Action of heat:

(a) Alkyl halides when heated at a temperature above 570 K yield olefins *via* the loss of halogen acid molecule.

$$\underset{\text{2-Iodopropane}}{CH_3-\underset{|}{\overset{}{CH}}(I)-\underset{|}{CH_2}(H)} \xrightarrow[\Delta]{> 570K} \underset{\text{1-Propene}}{CH_3-CH=CH_2} + HI$$

The ease of removal of halogen acid is iodide >bromide >chloride.

The order of reactivity of alkyl halide is Tertiary > secondary > primary.

(b) Alkyl halides when heated below 570 K or at lower temperature in the presence of $AlCl_3$, undergo rearrangement reaction

$$\underset{\text{1-Bromo butane}}{CH_3-CH_2-CH_2-CH_2-Br} \xrightarrow[AlCl_3]{570\ K} \underset{\text{2-Bromo butane}}{CH_3-CH_2-\underset{Br}{\underset{|}{CH_2}}-CH_3}$$

Mechanism

$$\underset{\text{1-Bromo butane}}{CH_3-CH_2-CH_2-CH_2-Br} \xrightarrow{Br-} \underset{\text{Primary carbocation (Less stable)}}{CH_3CH_2\overset{+}{C}H_2CH_2}$$

Hydride shift — Rearrangement

$$\underset{\text{Secondary carbocation (More stable)}}{CH_3CH_2\overset{+}{C}HCH_3} \xrightarrow{Br^-} \underset{\text{2-Bromobutane}}{CH_3-CH_2-\overset{Br}{\overset{|}{CH}}-CH_3}$$

If there is no hydrogen atom present on the carbon atom which is attached to carbon atom bearing the halogen i.e. C-X bond then alkyl group migrates. For example, *neo* pentyl chloride gives 2-chloro-2-methyl butane on heating.

$$\underset{\textit{neo-}\text{ pentyl chloride}}{(H_3C)_2\underset{CH_3}{\underset{|}{C}}-CH_2-Cl} \xrightarrow[AlCl_3]{570\ K} \underset{\text{2-Chloro-2-methyl butane}}{(H_3C)_2\overset{Cl}{\overset{|}{C}}-CH_2-CH_3}$$

(vi) Reaction with lithium: Alkyl halides when treated with lithium in the presence of dry ether yields alkyl lithium which are similar to that of Grignard reagents with increased activity.

$$\underset{\text{Alkyl halide}}{R-X} + 2Li \xrightarrow{\text{Dry ether}} \underset{\text{Alkyl lithium}}{R-Li} + LiX$$

$$\underset{\text{Ethyl bromide}}{CH_3-CH_2-Br} + 2Li \xrightarrow{\text{Dry ether}} \underset{\text{Ethyl lithium}}{CH_3-CH_2-Li} + LiBr$$

(vii) Halogenation: Alkyl halides on reaction with halogen (Cl_2 or Br_2) in the presence of UV light or elevated temperature yield polyhalogenated derivatives.

$$\underset{\text{Methyl chloride}}{CH_3Cl} + Cl_2 \xrightarrow[\text{Light / Heat}]{\text{UV}} \underset{\text{Methylene chloride}}{CH_2Cl_2} + HCl$$

$$\underset{\text{Methylene chloride}}{CH_2Cl_2} + Cl_2 \xrightarrow[\text{Light / Heat}]{\text{UV}} \underset{\text{Chloroform}}{CHCl_3} + HCl$$

$$\underset{\text{Chloroform}}{CHCl_3} + Cl_2 \xrightarrow[\text{Light / Heat}]{\text{UV}} \underset{\text{Carbon tetra chloride}}{CCl_4} + HCl$$

The above reaction takes place by free radical substitution mechanism (Refer part 2 chapter 14 alkanes).

Summary of Chemical Reactions

Nucleophilic Substitution Reactions

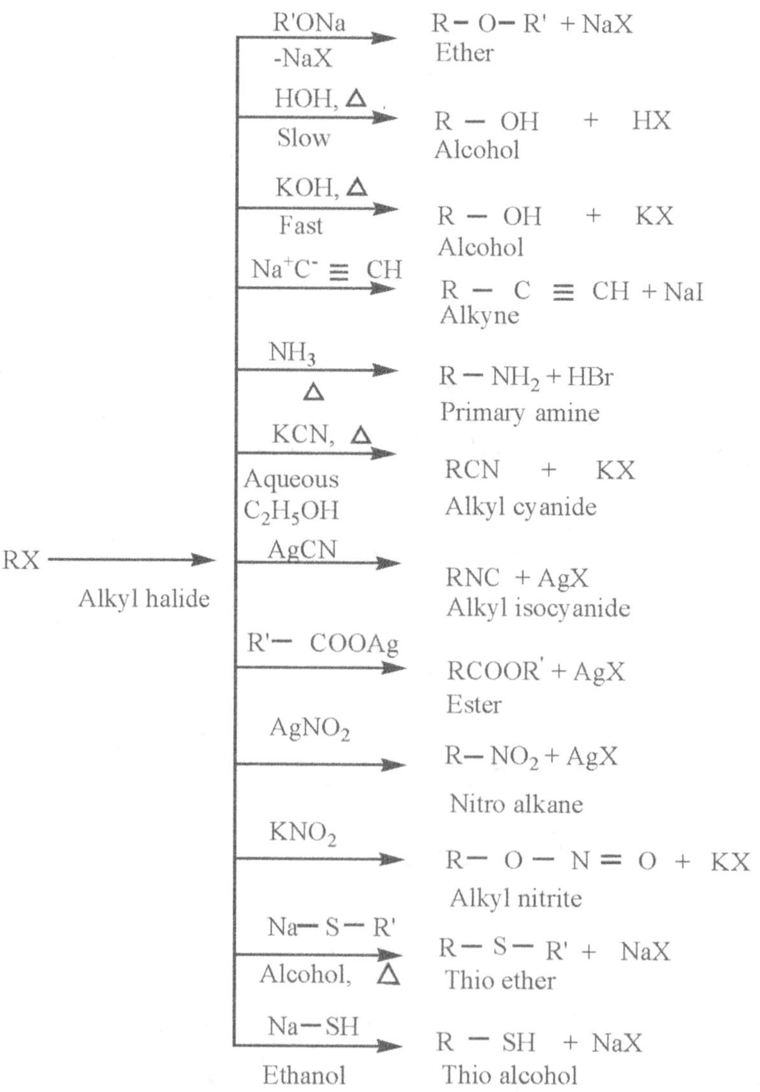

Miscellaneous reactions:

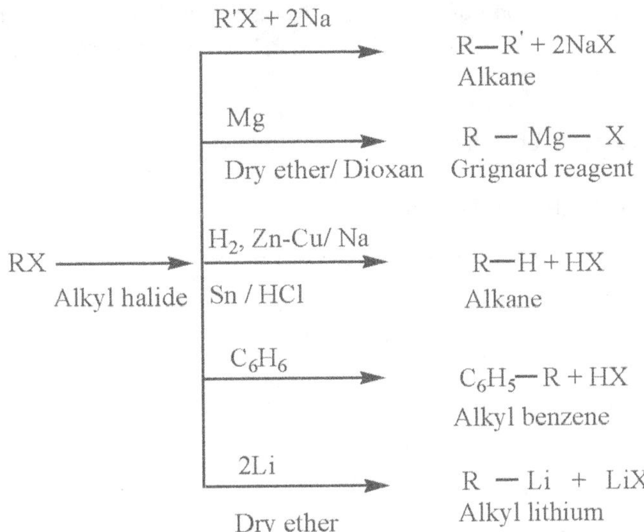

Table 13.5 Alkyl halides of Pharmaceutical importance.

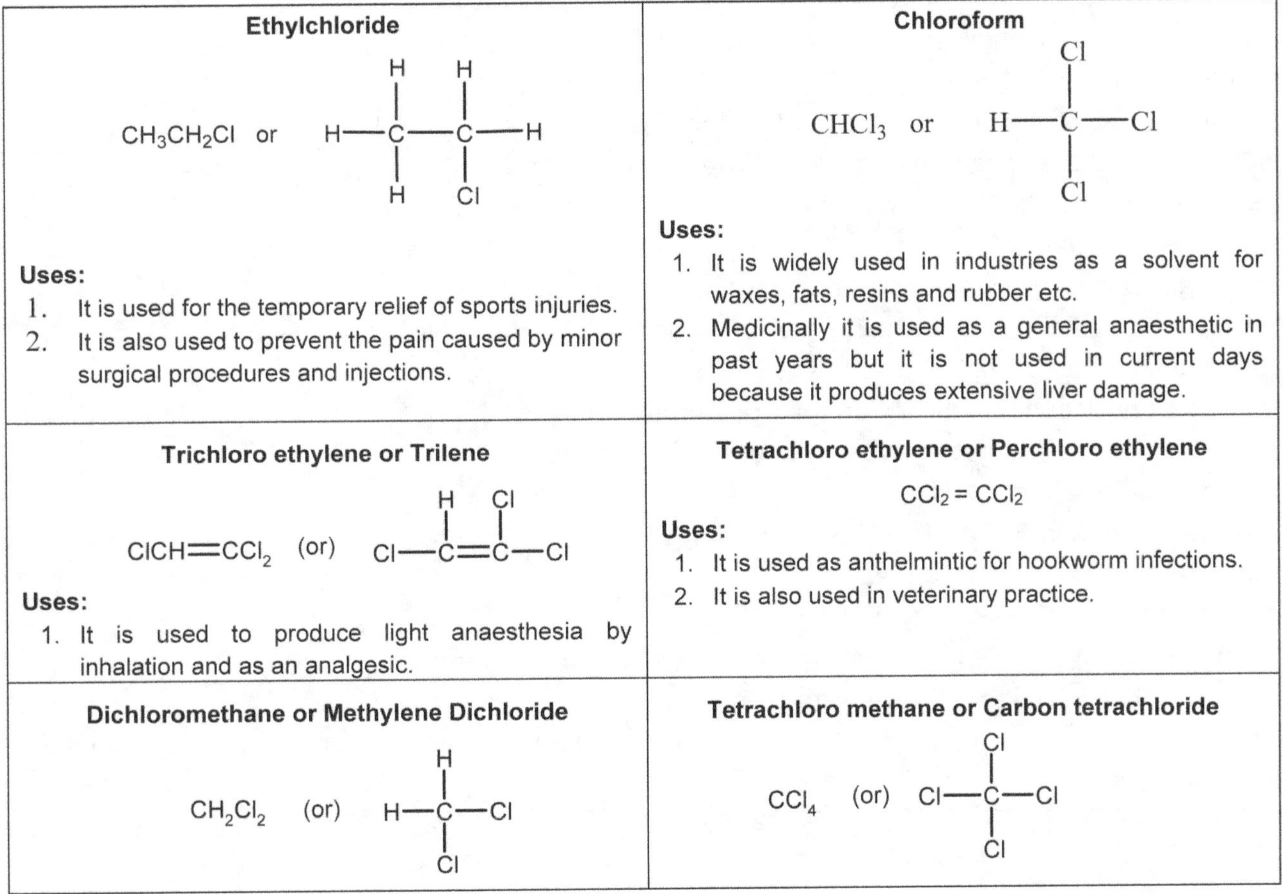

Ethylchloride	Chloroform
CH_3CH_2Cl or H—C(H)(H)—C(H)(Cl)—H **Uses:** 1. It is used for the temporary relief of sports injuries. 2. It is also used to prevent the pain caused by minor surgical procedures and injections.	$CHCl_3$ or H—C(Cl)(Cl)—Cl **Uses:** 1. It is widely used in industries as a solvent for waxes, fats, resins and rubber etc. 2. Medicinally it is used as a general anaesthetic in past years but it is not used in current days because it produces extensive liver damage.
Trichloro ethylene or Trilene	**Tetrachloro ethylene or Perchloro ethylene**
$ClCH{=}CCl_2$ (or) Cl—C(H)=C(Cl)—Cl **Uses:** 1. It is used to produce light anaesthesia by inhalation and as an analgesic.	$CCl_2 = CCl_2$ **Uses:** 1. It is used as anthelmintic for hookworm infections. 2. It is also used in veterinary practice.
Dichloromethane or Methylene Dichloride	**Tetrachloro methane or Carbon tetrachloride**
CH_2Cl_2 (or) H—C(H)(Cl)—Cl	CCl_4 (or) Cl—C(Cl)(Cl)—Cl

Table 13.5 *Contd...*

<table>
<tr>
<td>Uses:
1. It dissolves wide range of organic compounds, hence used as a solvent.
2. In food industry it is used to prepare decaffeinated coffee and tea.
3. It is also used to prepare extracts of hops and other flavourings.
4. Due to its volatile property, it is used as an aerosol spray propellant.</td>
<td>Uses:
1. It is used as propellant in aerosol spray.
2. It is also used as refrigerant.</td>
</tr>
<tr>
<td>Iodoform or Triiodomethane
CHI_3 (or) H—C(I)(I)—I
Uses:
1. It is used as an antiseptic.</td>
<td></td>
</tr>
</table>

Probable Questions

1. What are alkyl halides? Mention its importance.
2. Define & classify alkyl halides.
3. What is geminal dihalides & vicinal dihalides?
4. Write a detailed note on nomenclature of alkyl halides.
5. List out general methods of preparation of alkyl halides and explain any three methods in detail?
6. Explain Markovnikov's addition & anti Markovnikov's addition.
7. Write a note on Darzen's procedure.
8. Explain about Borodin Hunsdiecker reaction with reaction mechanism.
9. Describe in detail about structure & physical properties of alkyl halides.
10. Write a note on factors influencing the boiling point of alkyl halides.
11. Write about nucleophilic substitution reactions of alkyl halides.
12. Explain the elimination reactions of alkyl halides.
13. What is S_N1 & S_N2 reaction?
14. Write short notes on E_1 & E_2 reaction.
15. Write a note on Wurtz reaction.
16. What is Friedel Crafts alkylation reaction?
17. Describe in detail about various chemical properties of alkyl halides.
18. Compare substitution reaction with elimination reaction.
19. Write the chemical structure and uses of ethyl chloride, chloroform, trichloroethylene, tetrachloroethylene, dichloromethane, tetrachloromethane and iodoform.
20. Write the chemical structure, synthesis, physical properties & uses of methyl iodide.

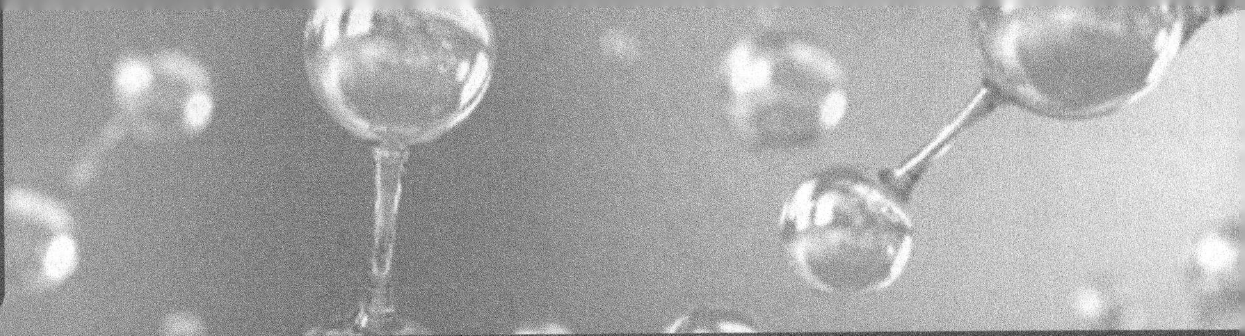

14

Nucleophilic Substitution and Elimination Reactions

Introduction

Alkyl halides easily undergo different kind of reactions and are regarded as useful synthetic tools for synthesis of compounds belonging to different functional groups. In nucleophilic substitution reactions, leaving group is a halogen which is replaced with nucleophile. The nucleophile replaces a leaving group of a carbon atom by donating or using its lone pair of electrons to form a new covalent bond. The chemical reaction is described as follows.

$$\underset{\substack{\downarrow\ \ \downarrow \\ \text{Alkyl group}\ \ \text{Leaving group}}}{R-L} + \underset{\text{Nucleophile}}{:Nu} \xrightarrow{\text{Solvent}} \underset{\text{Product}}{R-Nu} + :L$$

Components of Nucleophilic Substitution Reaction

Each and every nucleophilic substitution reaction has three components:

1. Substrate (alkyl group and leaving group).
2. Nucleophile.
3. Solvent.

Mechanisms

Nucleophilic substitution reactions take place by two mechanisms.

(I) S_N2 mechanism (Bimolecular nucleophilic substitution reaction).

(II) S_N1 mechanism (Unimolecular nucleophilic substitution reaction).

The S_N2 Mechanism

The rate of reaction depends upon the concentration of both alkyl halide and hydroxide ion or base, hence it is called as bimolecular nucleophilic substitution reaction. Consider the following reaction, methyl bromide (CH_3Br) reacts with sodium hydroxide (NaOH) to give methanol and leaving group (bromide ion).

$$\underset{\text{Methyl bromide}}{CH_3-Br} + \overset{\ominus}{:OH} \xrightarrow{\text{Solvent}} \underset{\text{Methanol}}{CH_3-OH} + \ddot{Br}$$

Factors influencing (S_N1 and S_N2) reaction - The course of a particular nucleophilic reaction is influenced by following factors:

(i) Structure of alkyl halide.
(ii) Structure and reactivity of the nucleophile.
(iii) The solvent in which the reaction is carried out.
(iv) Concentration of the nucleophile.

In the above reaction if the concentration of methyl bromide is doubled, the rate of reaction also increases. Similarly, if the concentration of hydroxide ion increases, the rate of reaction also doubles and if the concentration of both the reagents increases, the rate of the reaction quadruples.

$$\text{Rate} \propto [\text{alkyl halide}]\,[\text{nucleophile}]$$
$$\text{Rate} = K\,[\text{alkyl halide}]\,[\text{nucleophile}]$$

The above reaction is second order reaction since rate of the reaction depends upon the concentration of both the reagents.

Mechanism of the reaction is described below.

Consider the hydrolysis of methyl bromide with aqueous KOH.

When the hydroxide ion reacts or collides with methyl bromide, it attacks the carbon atom of alkyl halide (methyl bromide) molecule from the side opposite to that of leaving group *i.e.,* back side attack.

When the collisional energy is sufficient, a C-OH bond starts to form, and C-Br bond starts to break. Bond formation and bond breaking takes place simultaneously. This reaction takes place *via* the formation of transition state.

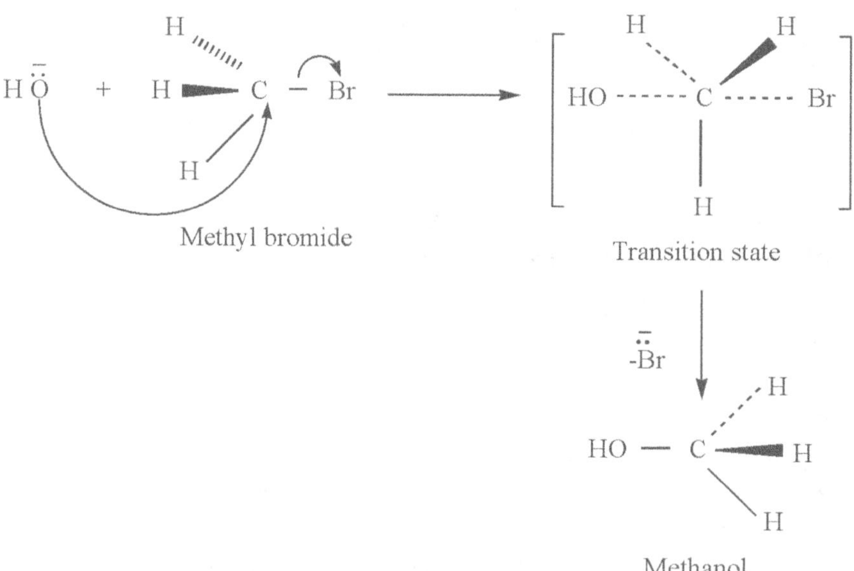

Transition state

1. In the transition state, the carbon atom is in pentavalent state *i.e.,* bonded to five atoms. The three substituents are linked to the carbon atom by complete covalent bonds, the hydroxide ion and leaving group are linked by partial bonds.
2. The transition state changes to product when C-OH bond is completely formed and the C-Br bond is completely cleaved.
3. During this transition state, the leaving group attains a negative charge because it removes the shared electrons from the carbon.
4. Similarly, the hydroxide ion attains diminished negative charge because it has begun to share its electron pair with carbon atom.

5. The -OH and -Br are present as far apart as possible, the three hydrogen atoms and the carbon present in one plane and the bond angle is 120°.
6. So the C-H bonds are arranged like a spoke of a wheel, while C-OH and C-Br are lying along the axle. The energy needed to break the C-Br is more than compensated by the energy released during the formation of C-OH bond.

   The mechanism of S_N2 reaction was proposed by Hughes and Ingold. He postulated that the mechanism of this reaction is concerted reaction (the reaction takes place in one step and no intermediates are produced). It is also called as **direct displacement reaction,** since the nucleophile displaces the leaving group in a single step.

Reason for back side attack

1. During the reaction, the back-side attack of carbon occurs because the orbital of the nucleophile which contains its non-bonding electrons interacts with the σ* empty molecular orbital belonging to the C-Br bond.
2. The largest lobe of this orbital is present on the side of the carbon atom directed away from the C-Br bond.
3. Similarly, the best overlap of the orbitals takes place in the back-side attack only.

Stereochemistry of S_N2 Reaction

During the reaction, the back-side attack of carbon occurs because the orbital of the nucleophile which contains its non-bonding electrons interacts with the σ* empty molecular orbital belonging to the C-Br bond. The mechanism of S_N2 reaction is good example for **Walden inversion** (reaction sequence in which the asymmetric carbon atom undergoes inversion of its configuration).

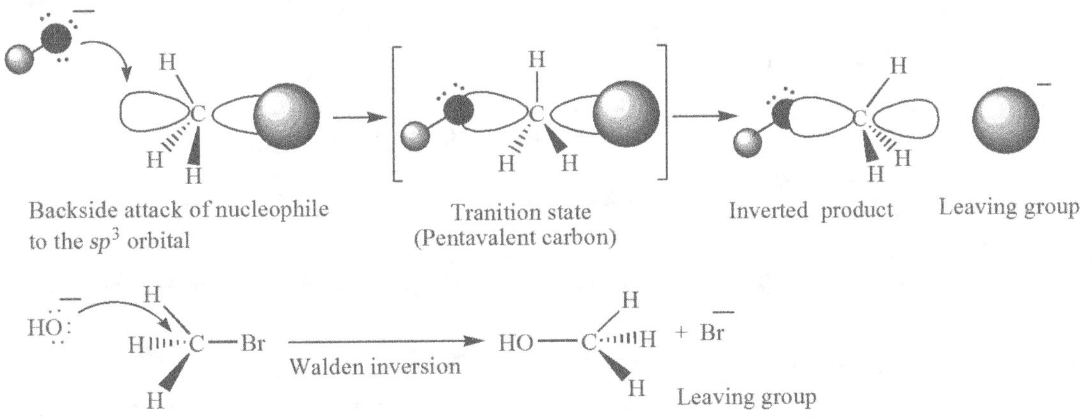

Figure 14.1 Back side attack of nucleophile in S_N2 reaction which inverts the tetrahedral arrangement of carbon atom, like the wind inverts umbrella.

The back-side attack of carbon atom is proved by explaining the following appropriate examples.

1. Hydrolysis of *cis*-1-bromo-3-methyl cyclopentane by potassium hydroxide solution. Theoretically it gives both *cis* and *trans*-alcohols due to the front and back side attack respectively.

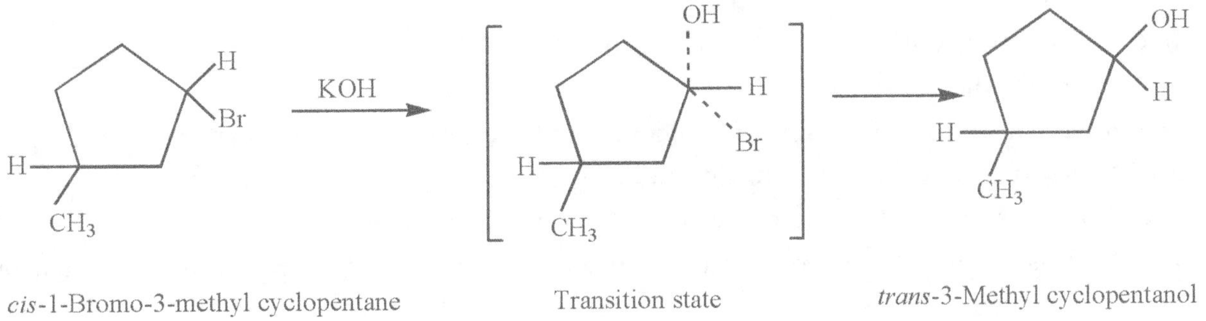

but practically, the *trans*-cyclopentanol predominates which indicates that the nucleophile attacks the carbon atom from the back side, *i.e.*, opposite to that of the leaving group.

2. S_N2 is a good example for stereospecific reaction in which different stereoisomers react with nucleophiles to give different kind of stereoisomers as the product.
3. In S_N2 reaction, after the back-side attack of nucleophile to the carbon atom, as soon as the nucleophile gets attached to the carbon atom, the three substrate bonds are forced apart until they attain planar spoke arrangement of transition state. As soon as the leaving group is removed, the three substrate bonds again changed to a tetrahedral arrangement with opposite configuration, like a turning made out of a umbrella in gale (strong wind). So the S_N2 reaction leads to **Walden inversion.**

Table 14.1 Various kind of S_N2 reactions of alkyl halides.

$$Nu^{-} + R-X \longrightarrow Nu-R + X^{-}$$

Nucleophile	Product	Class of product
$R-X + :\ddot{I}:^{-}$ ⟶	$R-\ddot{I}:$	Alkyl iodide
$R-X + :\ddot{O}H^{-}$ ⟶	$R-\ddot{O}H$	Alcohol
$R-X + :\bar{N}H_3$ ⟶	$R-NH_2$	Amine
$R-X + :\bar{C}\equiv C-R'$ ⟶	$R-C\equiv C-R'$	Alkyne
$R-X + :\ddot{O}R'^{-}$ ⟶	$R-\ddot{O}R$	Ether
$R-X + R'-CO\ddot{O}:^{-}$ ⟶	$R'-COO-R$	Ester
$R-X + :\ddot{S}H^{-}$ ⟶	$R-\ddot{S}H$	Thiol (Mercaptan)
$R-X + :\ddot{S}R^{-}$ ⟶	$R-\ddot{S}R'$	Thioether (Sulfide)
$R-X + \bar{:}C=N:$ ⟶	$R-C\equiv N:$	Nitrile

Factors Affecting S_N2 Reactions

The following factors influence the rate of S_N2 reactions.

1. Strength of nucleophile.
2. Steric effect on nucleophile.
3. Effect of solvents.
4. Effect of leaving group on the substrate.
5. Steric effect on the substrate.

1. **Strength of nucleophile:** Atoms or molecules with lone pair of electrons are sometimes known as bases and sometimes known as nucleophiles.

   Base: A base shares its lone pair of electrons with a proton.

   Basicity: It is defined as the equilibrium constant for abstracting a proton. It is denoted by Ka.

Nucleophile: A nucleophile shares its lone pair of electrons to attack an electron deficient atom (carbocation) other than a proton.

Nucleophilicity: It is a measure of the ability of nucleophile to attack electron deficient atom or group. It is measured by rate constant (K).

The nature of the nucleophile present in the substitution reaction strongly affects the rate of S_N2 reaction. A strong nucleophile is much more effective than that of weak nucleophile. For example,, consider methanol (CH_3OH) and methoxide ion (CH_3O^-) which can easily share their lone pair of electrons to the electron deficient carbon atom. But methoxide ion reacts about 1 million times faster than that of methanol. It indicates that the species with negative charge acts as a stronger nucleophile than that of neutral species.

In methoxide ion, the lone pair of electrons is easily available for bonding with electron deficient atom. In transition state, the negative charge in the methoxide ion is shared by the oxygen of methoxide ion and by the leaving group such as the halide ion.

But in case of methanol, there is no negative charge; in transition state it shows a partial negative charge on the leaving group and partial positive charge on the oxygen atom of the methanol. The structural illustration of the reaction is as follows.

$CH_3-\ddot{O}:^- + CH_3-Br \longrightarrow [CH_3-\ddot{O}\cdots CH_3 \cdots Br]^{\ddagger} \longrightarrow CH_3-O-CH_3 + Br^-$

Methoxide ion

Stronger nucleophile

Transition state (Lower E_a)

$CH_3-\ddot{O}H + CH_3-Br \longrightarrow [CH_3-\overset{\delta+}{\ddot{O}}(H)\cdots CH_3 \cdots \overset{\delta-}{Br}]^{\ddagger} \longrightarrow CH_3-\overset{+}{O}(H)-CH_3 + Br^-$

Methanol

Conjugate acid
Weaker nucleophile

Transition state (Higher E_a)

The above example shows that a base is always a strong nucleophile than its conjugate acid. The order of nucleophilicity is shown in Table 14.2.

Table 14.2 Some common nucleophiles with decreasing order of nucleophilicity in water and alcohols.

Strong nucleophiles	$(CH_3CH_2)_3P$: -SH -I- $-(CH_3CH_2)NH,-$ C≡N $(CH_3CH_2)_3NH$-O^-, CH_3-O^-
Moderate nucleophiles	Br^-, :NH_3, CH_3-S- CH_3,Cl^-
Weak nucleophiles	CH_3-COO^-, F^-, CH_3-OH, H_2O

In periodic table, the atoms present in the second row possess approximately the same size. If hydrogen atoms are attached to the second row elements, the relative acidity of the compounds is as follows.

$$CH_4 < NH_3 < H_2O < HF$$

Increasing acidity

Similarly the relative basic strength of conjugate base is as follows

$$CH_3 < NH_2 < OH < F$$

From all the above facts, it is concluded that the strong nucleophiles favours the S_N2 reaction while the weaker nucleophile does not favour the S_N2 reaction.

Nucleophilicity increases with increase in size and polarizability. In periodic table, down the column, the size of the atoms increases, electrons are present far apart from the nucleus (electrons are loosely held), the atom is polarisible. When the electrons present in the atom moves fastly towards positive charge it leads to strong bond in transition state (The increased mobility of its electrons enhances the atoms ability to begin to form a bond at a relatively long distance).

For example, Compare the polarizability effect of iodide and fluoride on a methyl halide.

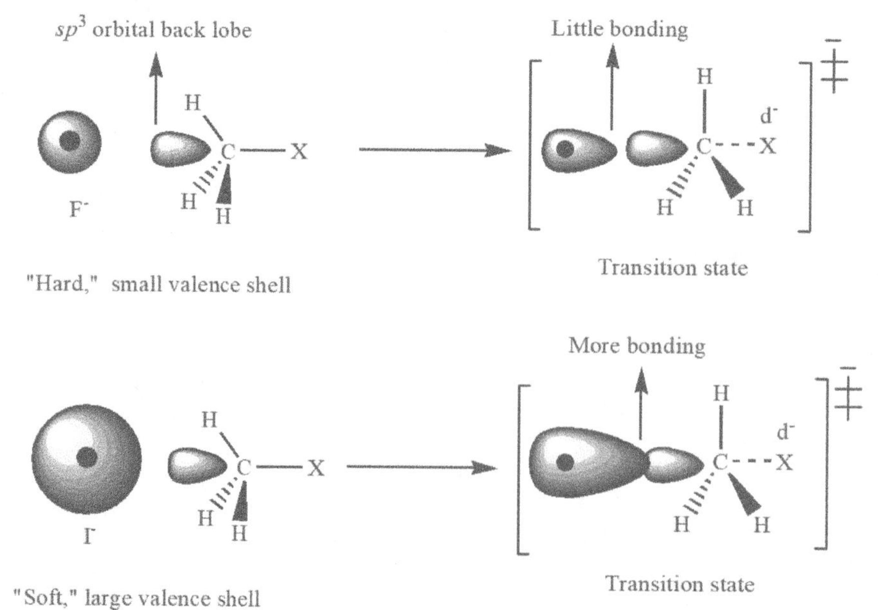

Figure 14.2 Polarisability effect of iodide and fluoride on methyl halide.

The outermost shell or the valency shell of the fluoride ion is second shell (L Shell). The electrons present in this shell are held tightly and close to the nucleus. So the fluoride ion possesses low polarizability and the fluorine atom is unable to form a bond with carbon atom until the atoms are close together.

But in iodide, the valency shell is fifth shell. The electrons present in the shell are loosely held and at a distance (far away) from the nucleus which makes iodide ion as a highly polarized (soft) nucleophile. These electrons forms a bond with carbon atom from further away in the transition state.

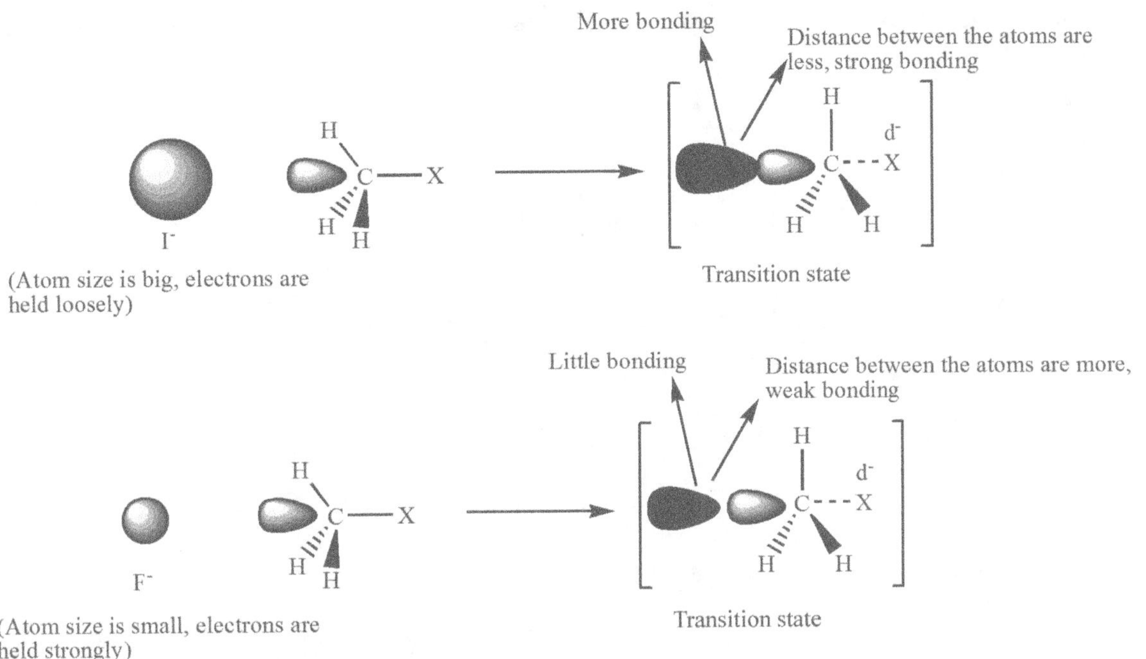

Figure 14.2a Effect of atom size in nucleophilic substitution reaction.

2. **Steric effects on nucleophile:** For an ion or a molecule to serve as a nucleophile, it must get close to attack the electron deficient atom. But presence of bulky groups in the nucleophile prevents this close approach and slows down the rate of reaction.
   For example, compare the nucleophilicity of *t*-butoxide and ethoxide.

$$CH_3-\overset{\displaystyle CH_3}{\underset{\displaystyle CH_3}{\overset{|}{\underset{|}{C}}}}-\ddot{\underset{..}{O}}{}^{-} \qquad\qquad CH_3-CH_2-\ddot{\underset{..}{O}}{:}^{-}$$

t-Butoxide (steric hindrance due to bulky groups), stronger base, but weaker nucleophile

Ethoxide ion (No steric hindrance), weaker base, but stronger nucleophile

In the above said examples, *t*-butoxide ion strongly abstracts the proton than the ethoxide ion, so it acts as a strong base, but the three methyl groups present in the *t*-butoxide ion hinders the close approach to a carbon atom, so it is a weaker nucleophile. In ethoxide ion no bulky groups are present and hence no steric hindrance present to approach a carbon atom which makes ethoxide ion as a stronger nucleophile.

Steric hindrance: When bulky groups interfere with a reaction by virtue of their size, is known as steric hindrance. [If we need a species to acts as a base, we use a bulky group reagent such as *t*-butoxide ion and if we need a species to acts as a nucleophile, we use a less hindered reagent such as ethoxide ion].

3. **Effects of solvents on nucleophilicity:** When a base or nucleophile or negatively charged species are placed in a protic solvent, the ion gets solvated. The solvent molecules arrange themselves, so their partially positive charged hydrogen atoms are pointing towards the negative species. This type of interaction between the ion and the dipole of the protic solvent is known as **ion-dipole interaction forces** and these forces should be broken before entering into the reaction. Strong bases interact strongly with protic solvents and weak bases interact weakly with protic solvents. It indicates that the breaking of iodide and solvent molecule is easy than that of breaking of bond between fluoride ion and solvent molecule. So iodide ion acts as a better nucleophile than that of fluoride ion in a protic solvent. (Solvent with acidic protons, in the form of N-H or O-H group forms hydrogen bond with nucleophiles).

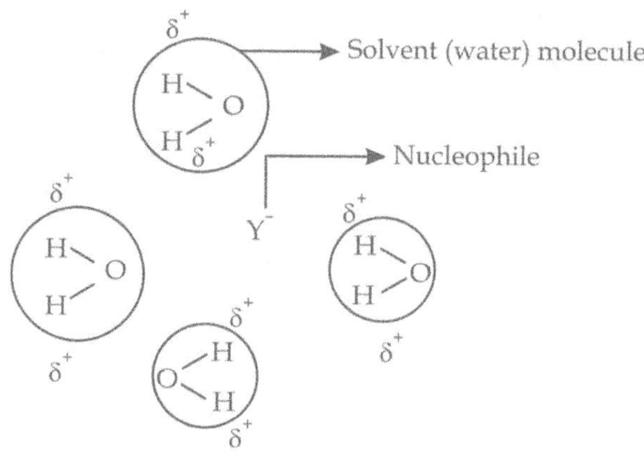

Figure 14.3 Ion dipole interactions between a nucleophile and water molecule in substitution reaction.

Small anions are solvated more strongly than that of large anions in a protic solvent because the solvent approaches the small anion very closely than that of large anion.

Sometimes energy is needed to "strip off" some of the solvent molecule (breaking of some hydrogen bonds that are used to stabilize the solvated anion). If the anion is small, more amount of energy is needed to strip off the solvent which leads to decrease in their nucleophilicity. It is reverse to the polarizability.

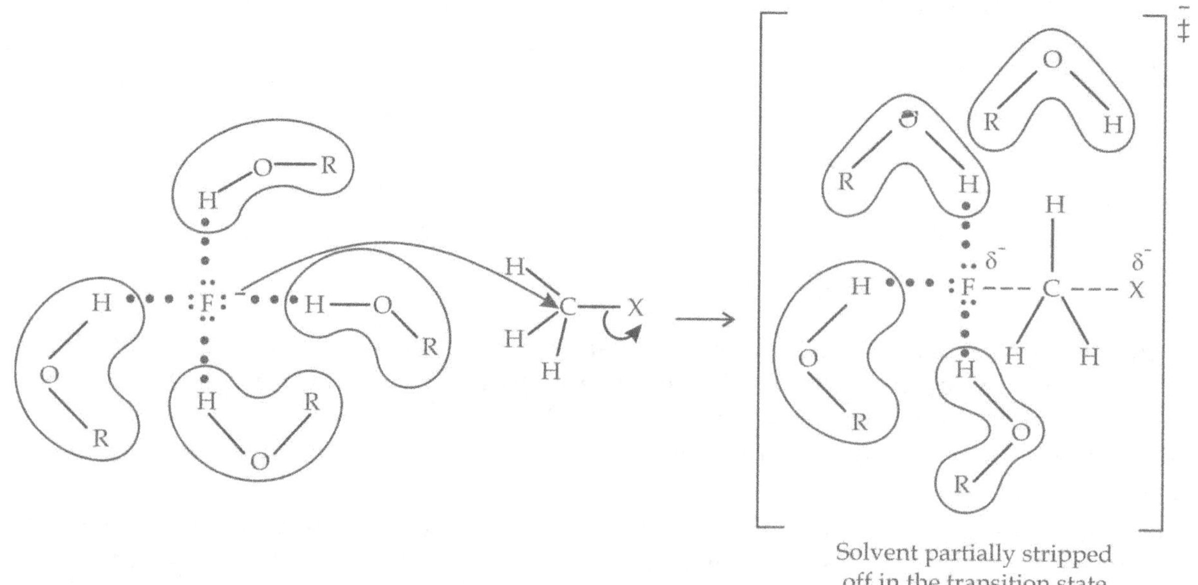

Figure 14.3a Protic solvents effect on nucleophilicity.

When atomic number increases, polarisability increases. When atomic number increases, solvation energy of protic solvent decreases.

Aprotic solvents: Solvents without O-H or N-H groups are known as aprotic solvents. These groups enhance the nucleophilicity of anions. An anion (nucleophile) is more reactive in aprotic solvents because it is not strongly solvated. Energy is not needed for strip off the solvent molecule.

Table 14.3 Relative nucleophilicity towards methyl iodide in methanol.

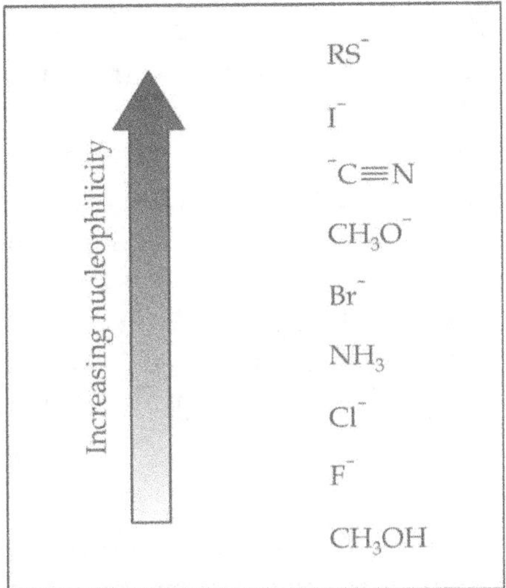

Ions which are insoluble in more polar solvents and soluble in polar aprotic solvents, possess strong dipole moments to increase the solubility (They don't have N-H or O-H groups to form a hydrogen bonds).

Examples: For polar aprotic solvents such as DMF (Dimethyl formamide), DMSO (Dimethyl sulfoxide), acetone, acetonitrile, sometimes specific solvating reagents are added to the reaction to increase the solubility without affecting the nucleophilicity of a nucleophile such as crown ethers.

The 18-crown ether-6 solvates potassium ions. If the potassium salt of a nucleophile is used, crown ether solvates the potassium ion which causes nucleophile to be dragged into solution.

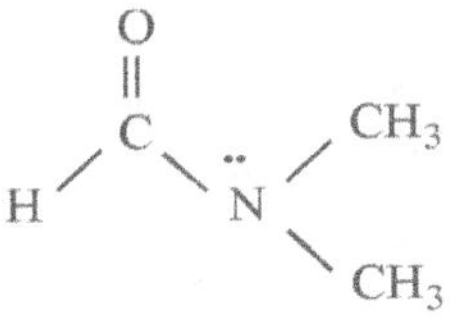

$H-C(=O)-\ddot{N}(CH_3)_2$

DMF

$H_3C-S(=O)-CH_3$

DMSO

$CH_3-C(=O)-CH_3$

Acetone

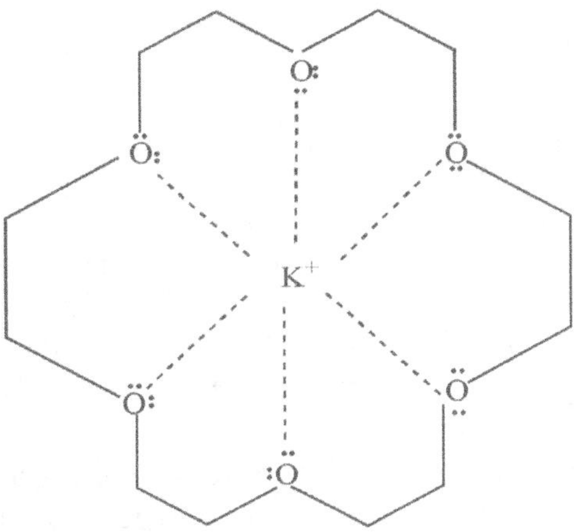

Crown ether

Reactivity of the substrate in S_N2 reaction: In S_N2 reaction, the structure of the substrate is equally important as that of the nucleophile. Commonly, alkyl halides are used as a substrate. A good substrate molecule needed for S_N2 reaction possess electrophilic carbon atom with a good leaving group, and it should not be sterically hindered as the reaction requires approaching from the nucleophile.

4. **Effect of leaving group on the substrate:** The main function of the leaving group in a S_N2 reaction is
   (i) Polarization of C-X bond which makes the carbon atom more acidic.
   (ii) Leaves along with its electron pair which is bonded to the electrophilic carbon atom.
   To fulfill the above said criteria, the leaving group should be
   (a) Highly electronegative or possessing electron withdrawing property to make the carbon atom electrophilic.
   (b) It should be stable (not like a strong base) after leaving the substrate.
   (c) Highly polarisable, to stabilize the transition state.
   Leaving group must be electronegative to form a partial positive charge on the carbon atom to make it electrophilic.
   It also stabilizes the transition state. The halogen atoms are highly electronegative in nature, so in S_N2 reaction, the alkyl halides generally serves as a substrate molecule. Other atoms oxygen, sulfur and nitrogen also form strong polarisable bond with carbon atom.
   The order of reactivity is

$$C < X \ (X = \text{Halogens}),\ C < X,\ C < N,\ C < S$$

The halogen atom or the leaving group must be more stable after leaving from the carbon atom along with its shared pair of electrons. Leaving of the substrate from transition state, increases the energy of the reaction and slows down the rate of the reaction. It also favors the equilibrium towards the reactants.

Bond forming — Bond breaking

$$\text{Nu}^{-}\!: \ + \ \rangle C - \ddot{X} \longrightarrow [\text{Nu} \text{----} C \text{----} X] \longrightarrow \text{Nu} - C\langle \ + \ :\ddot{X}^{-}$$

Transition state

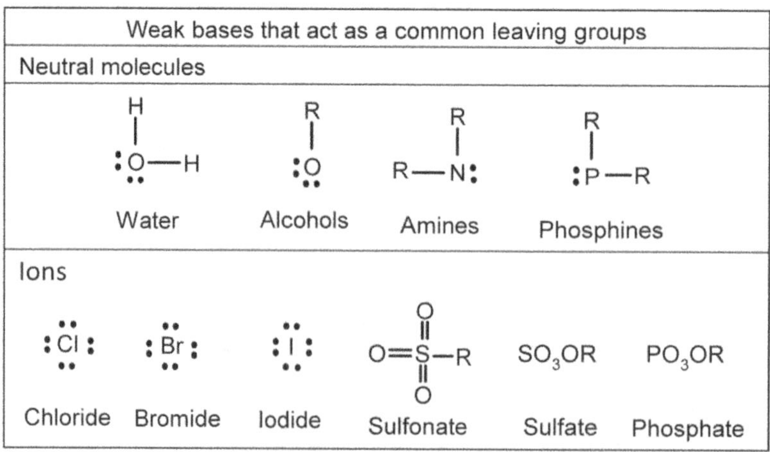

Weak bases that act as a common leaving groups					
Neutral molecules					
H–O–H	R–O–H	R_2N–R	R_2P–R		
Water	Alcohols	Amines	Phosphines		
Ions					
:Cl:	:Br:	:I:	O=S(=O)–R	SO_3OR	PO_3OR
Chloride	Bromide	Iodide	Sulfonate	Sulfate	Phosphate

Weak bases those act as a good leaving groups are conjugate bases of strong acids. Example HCl, HI and HBr. Alkoxide ions, hydroxide ions and other strong bases are poor leaving groups.

Finally a good leaving group should be polarisable, it should maintain the partial bonding with the carbon atom in transition state which reduces the activation energy. According to this concept, in S_N2 reaction, iodide ion acts as a good nucleophile and good leaving group. In contrast, fluoride ion is a small hard ion, so it acts as a poor nucleophile (protic solvents) and poor leaving group.

5. **Steric effect on the substrate:** Various kind of alkyl halides undergo S_N2 reaction at different rates. The structure of substrate is most important factor for its reactivity towards S_N2 reaction. The order of relative rates of S_N2 displacement of alkyl halides are as follows.

$$CH_3X > 3^o > 2^o > 1^o$$

Table 14.4 Steric effect of alkyl halide in substitution reaction.

S. No.	Halide class	Example	Relative rate
1.	Methyl	CH_3Br	> 1000
2.	Primary (1°)	CH_3- CH_2-Br	50
3.	Secondary (2°)	$(CH_3)_2$- CH-Br	1
4.	Tertiary (3°)	$(CH_3)_3$- C-Br	< 0.001
5.	*n*-Butyl (1°)	CH_3 CH_2CH_2- CH_2Br	20
6.	*iso*- Butyl (1°)	$(CH_3)_2$- CH CH_2 – Br	2
7.	*neo*- pentyl	$(CH_3)_3$ C- CH_2-Br	0.0005

The order of relative rate is explained as follows. In the above table, all of the slow reacting compounds possess one unique common property *i.e.,* the back side of the electrophilic carbon atom is crowded with bulky groups. For example, tertiary halides are more hindered than that of secondary halides which in turn are more crowded than primary alkyl halides. It shows that rather than the electronic effect, the reactivity of alkyl halides is affected by the presence of bulky alkyl groups in the substrate (steric hindrance).

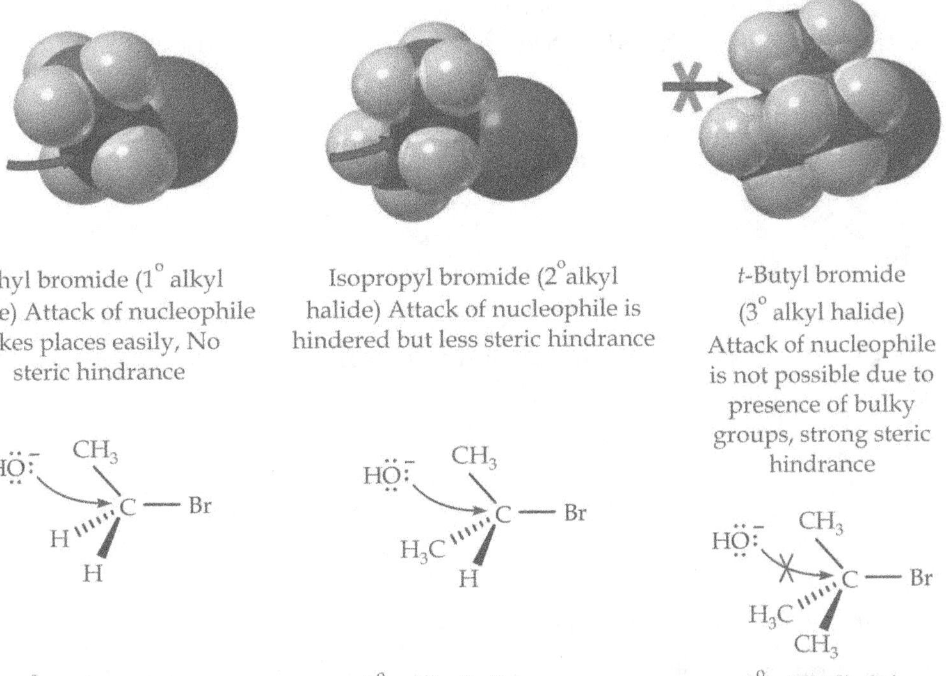

Figure 14.4 S_N2 reaction in primary, secondary and tertiary alkyl halides.

Figure 14.4 shows (the S_N2 reaction between hydroxide ion and ethyl bromide (1°), isopropyl bromide (2°) and *t*-butyl bromide (3°). The hydroxide ion easily attack the ethyl bromide, in case of isopropyl bromide, the attack is hindered but still possible. In contrast, the attack of -OH ion to the *t*-butyl bromide is impossible due to presence of three methyl groups (steric hindrance).

First Order Nucleophilic Substitution Reaction (S_N1) or Unimolecular Nucleophilic Substitution Reaction

The term unimolecular indicates that only one molecule is involved in the transition state of rate limiting step.

Let us consider the following chemical reaction.

$$(H_3C)_3C - Br \;+\; CH_3 - OH \xrightarrow{\text{Boil}} (H_3C)_3C - O - CH_3 \;+\; HBr$$

t-Butyl bromide   Methanol   Methyl-*t*-butylether

In the above reaction, *t*-butyl bromide is treated with boiling alcohol to give methyl-*t*-butylether. In this reaction the solvent molecule acts as a nucleophile, which is known as solvolysis (solvo indicates solvent, lysis means cleavage). This solvolysis process is a substitution reaction because the halide ion (bromide ion) is replaced by $-OCH_3$ ion (Methoxide ion). This reaction does not take place by S_N2 mechanism, since S_N2 reaction needs strong nucleophile and the substrate should be free from steric hindrance.

But in the above said reaction, *t*-butyl bromide is a hindered alkyl halide and methanol acts as weak nucleophile. An interesting fact about this reaction is, rate of reaction does not depend upon concentration of methanol (nucleophile) but only depends upon the concentration of *t*-butyl bromide (alkyl halide).

Rate of reaction α [alkyl halide]

Rate of reaction = K [alkyl halide]

The overall rate of reaction follows first order kinetics [follows first order kinetics with respect to the concentration of alkyl halide and follows zero order kinetics with respect to the concentration of alcohol (nucleophile)].

Mechanism:

The S_N1 mechanism takes place as follows.

Step 1: Ionization of *t*-butyl bromide leads to the formation of carbocation. This step is slow and so is the rate limiting step.

$$(H_3C)_3C - Br \xrightarrow{\text{Slow}} (CH_3)_3C^{+} + Br^{-}$$

t-Butyl bromide   Carbocation

Step 2: Nucleophile attacks on the carbocation which has the sextet of electrons.

$$(CH_3)_3C^{\oplus} + \ddot{O}(H) - CH_3 \xrightarrow{\text{Fast}} (CH_3)_3 - C - \overset{+}{\ddot{O}}(H) - CH_3$$

Carbocation   Methanol

Step 3: Loss of proton to solvent

$$(H_3C)_3C - \overset{+}{\ddot{O}}(H) - CH_3 + CH_3 - \ddot{O}H \overset{\text{Fast}}{\rightleftharpoons} (H_3C)_3C - \ddot{O} - CH_3 + CH_3 - \overset{\oplus}{\ddot{O}}(H) - H$$

Methyl-*t*-butyl ether

Note:

$$CH_3-\overset{\oplus}{\underset{|}{\underset{H}{O}}}-H + Br^- \longrightarrow \underset{\text{Methanol}}{CH_3OH} + HBr$$

The S_N1 reaction is a multistep process. The first step is the slow ionization of alkyl halide which leads to the formation of carbocation and the second step is the fast attack on the carbocation by nucleophile. The carbocation is a strong electrophile and reacts with both weak and strong nucleophiles.

If the nucleophile is NaOH, then the reaction takes place as follows.

$$\underset{t\text{-Butyl bromide}}{(CH_3)_3-C-Br} + :\ddot{O}\bar{H} \longrightarrow \underset{t\text{-Butyl alcohol}}{(H_3C)_3-\bar{C}OH} + Br^-$$

Step 1: Formation of carbocation.

$$\underset{t\text{-Butyl bromide}}{(H_3C)_3C-Br} \xrightarrow{\text{Slow}} \underset{t\text{-Butyl carbocation}}{(CH_3)_3C^{\oplus}} + Br^-$$

Step 2: Nucleophile attack on the carbocation.

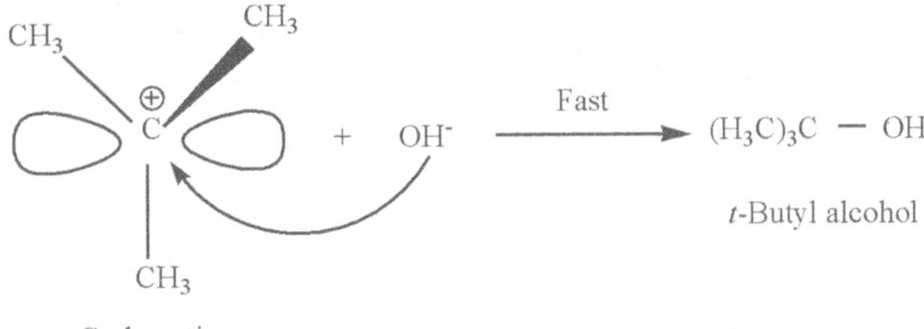

The energy diagram of the S_N1 reaction shows that the rate of reaction does not depend upon the concentration of the nucleophile. First step in the reaction is endothermic, so the activation energy determines the overall reaction. The second step is exothermic with low energy transition state.

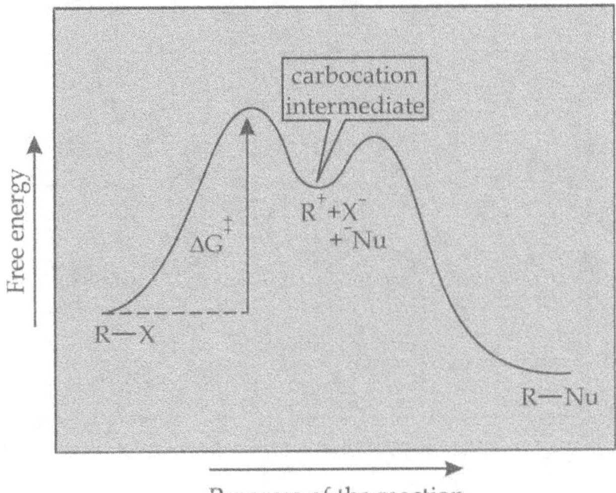

Graph 14.1 Reaction coordinate diagram of S_N1.

Stereochemistry of S_N1 Reaction

The first step in the S_N1 mechanism is the formation of carbocation. The carbon atom (central) is sp^2 hybridised. The three substituents attached to the carbon are present in one plane and the bond angle is 120°. The empty orbital of the central carbon atom is present perpendicular to the plane of sp^2 orbital. So the nucleophile attacks the two sides of the empty *p* orbital to form two different products. If the starting alkyl halide is optically active, the carbocation produces equal amounts of two isomers (one with retention of configuration and another one with inversion of configuration). So the product obtained is racemic. It indicates that the S_N1 reaction involves the **racemization of the optically active substrate.**

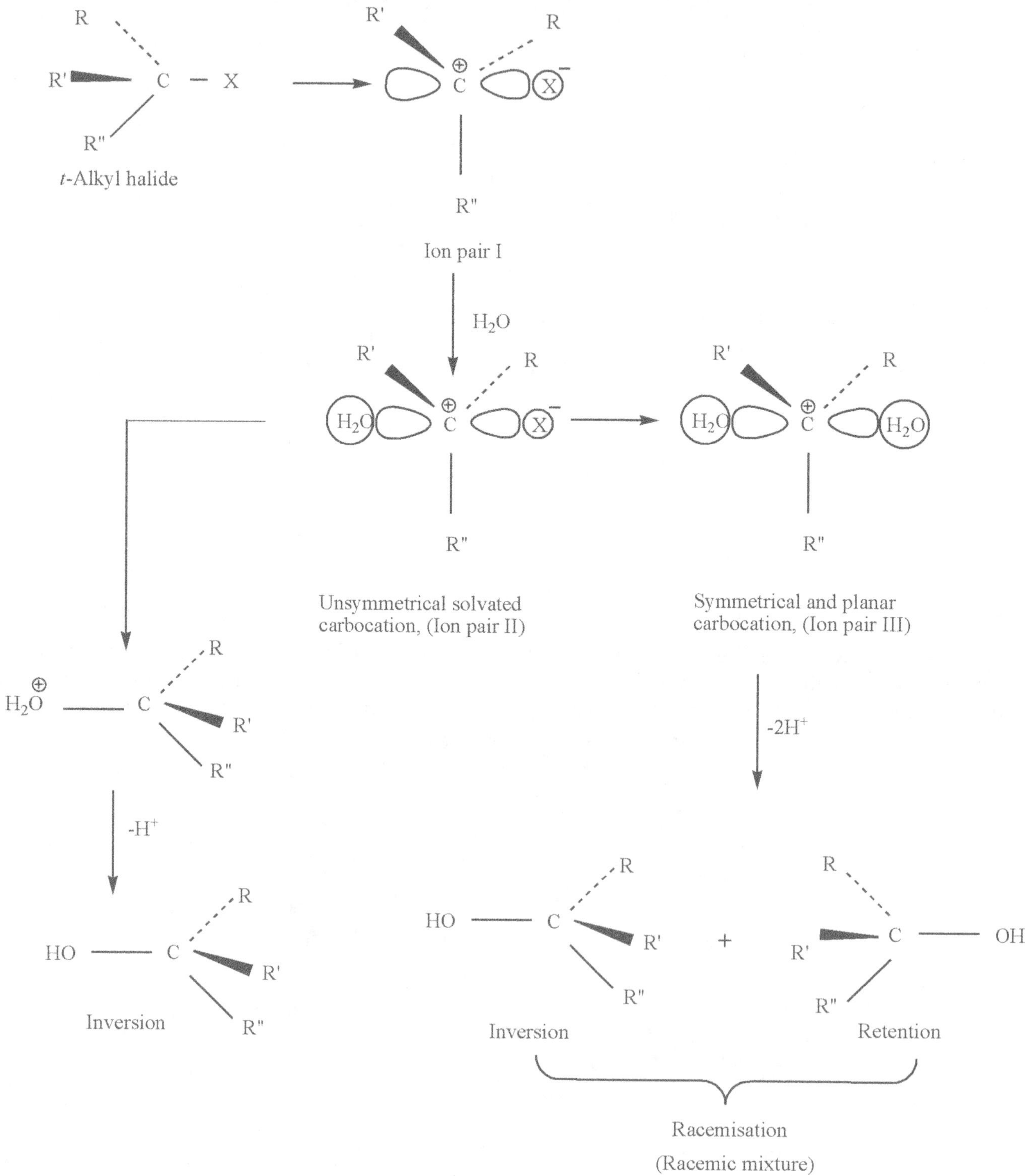

Racemization of this S_N1 reaction is only partial and not complete. During the formation of carbocation from the alkyl halide, the distance between carbon and the leaving group (halogen) increases until the covalent bond is broken. The two opposite charged ions are formed, but not immediately and completely, which leads to the formation of ion pair I. It is explained in the above chemical equation.

1. First step involves the formation of oppositely charged ions, initially the two ions are close together and leads to formation of ion pair I.
2. The leaving group or the halide ion is still attached to the charged carbon atom and prevents the front side attack of nucleophile, so the nucleophile attacks only from the back side of the carbon atom which gives unsymmetrical solvated carbocation ion pair II that may yield product with inverted configuration.
3. The carbocation I has longer stability, the leaving group is displaced with another one solvent molecule to yield symmetrical, planar solvated carbocation ion pair III, which is attacked by the nucleophile on both sides to yield racemic mixture of products, one with inverted configuration and the other with retention of configuration.
4. The rate of reaction is only determined by the stability of carbocation I. The more stable carbocation gives more racemized product due to symmetrical solvation, the less stable carbocation gives less racemic mixture due to unsymmetrical solvation. So it is concluded that S_N1 reaction proceeds with some net inversion.

Factors affecting S_N1 Mechanism

1. Substituent effect.
2. Effect of leaving group on S_N1 reaction.
3. Solvent effects.
4. Nature of nucleophile.

1. **Substituent effect:** The rate limiting step in S_N1 reaction is step 1, ionization of alkyl halide to form carbocation, which is an endothermic process. According to stereochemistry, it is concluded that rate of reaction depends upon the stability of carbocation. The alkyl substituents attached to the carbocation stabilizes it by two ways.
   (i) Inductive effect (donating electrons to the carbocation).
   (ii) Hyperconjugation (overlap of filled orbital with the unfilled *p* orbital of the carbocation).
   Carbocations with more substituents are more stable. The order of carbocation stability is as follows.

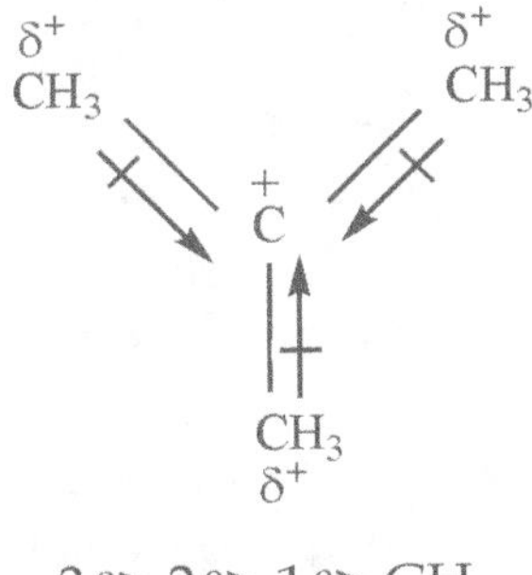

$3° > 2° > 1° > CH_3$

Figure 14.5 Substituents effect on S_N1.

The above said order is opposite to that of S_N2 reaction. In S_N2 reaction the alkyl group blocks or hinders the attack of the nucleophile by steric hindrance, but in S_N1, alkyl groups increases the stability of carbocation, thus favors S_N1.

Resonance effect of carbocation also favors the S_N1 mechanism. For example:

Allyl halide, a primary halide easily undergoes S_N1 reaction when compared with secondary halide, because the carbocation formed is stabilized by resonance and the positive charge is equally spread over the two carbon atoms. For example,

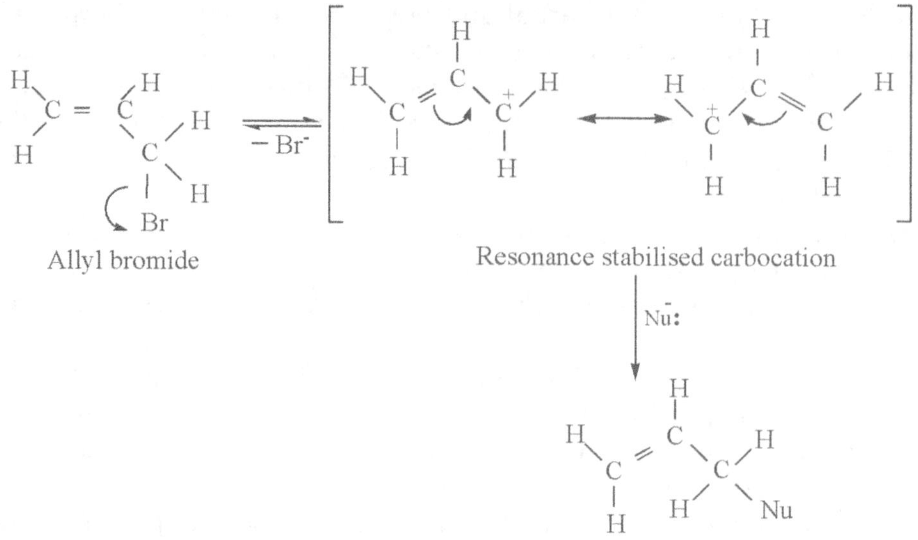

$$H_2C=CH-CH_2-Br \underset{-Br^-}{\rightleftharpoons} [H_2C=CH-\overset{+}{C}H_2 \longleftrightarrow H_2\overset{+}{C}-CH=CH_2] \xrightarrow{\bar{Nu}:} H_2C=CH-CH_2-Nu$$

Allyl bromide — Resonance stabilised carbocation

2. **Effect of leaving group on S_N1 reaction:** Two factors affecting the rate of reaction.
   (i) The ease with which the leaving group dissociates from the carbon atom.
   (ii) Stability of carbonium ion or carbocation formed.

   If the carbocation formed is more substituted, more will be its ease of formation from alkyl halide. But one question arises, what happen if series of alkyl halides with different leaving group dissociates to give same carbocation. In this case, the answer is similar to that of S_N2 reaction. The weaker the base, less tightly it is attached to the carbon atom and easier to break the carbon-halogen bond. So, the relative reactivity of alkyl halides in S_N1 reaction is

   $$RI > RBr > RCl > RF$$

   Benzylic halides & Alkylic halides: Both halides undergo S_N2 reaction unless they are 3°. (3° halides are not reactive due to steric hindrance).

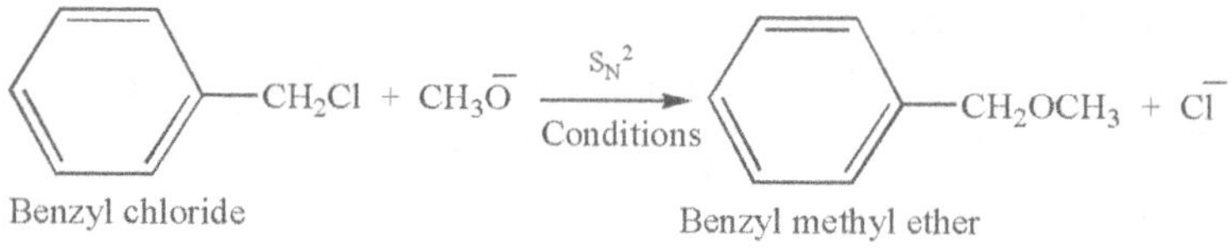

$$C_6H_5-CH_2Cl + CH_3\bar{O} \xrightarrow[\text{Conditions}]{S_N2} C_6H_5-CH_2OCH_3 + Cl^-$$

Benzyl chloride — Benzyl methyl ether

$$CH_3-CH=CHCH_2Br + H\bar{O} \xrightarrow{S_N2} CH_3-CH=CHCH_2OH + Br^-$$

2-Buten-1-bromide — 2-Buten-1-ol

They also undergo S_N1 reactions because they form stable carbocations which are stabilized by resonance.

Vinyl halides and aryl halides do not undergo S_N2 and S_N1 reaction. They do not give S_N2 reaction because if the nucleophile approaches the carbon from the side opposite to that of leaving group (back side attack), it is repelled by the π electron cloud of the double bond or the aromatic ring.

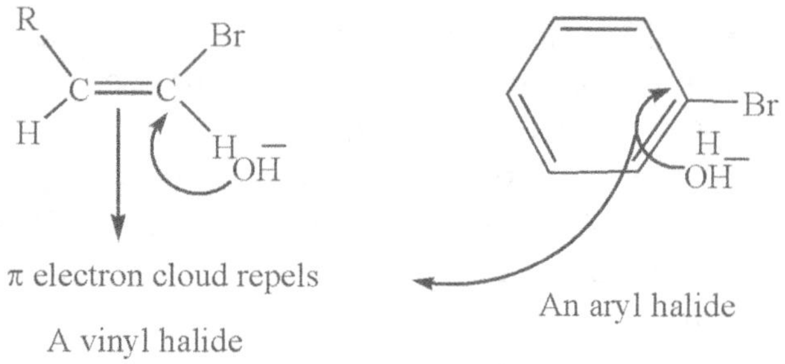

R(H)C=C(Br)H ← $H\bar{O}$; C_6H_5Br ← $H\bar{O}$

π electron cloud repels

A vinyl halide — An aryl halide

Why vinyl and aryl halides do not undergo S_N1 reaction. It is explained by two factors.

(i) Vinyl and aryl carbocations are more unstable because of *sp* carbons which are more resistant of becoming positively charged (*sp* carbon having positive charge).

(ii) sp^2 carbon atom forms stronger bonds than sp^3 carbons. Hence, it is harder to break the C-X bond when the halogen is bonded to sp^2 carbon.

$$R-CH=CHCl \not\longrightarrow R-CH=\overset{+}{C}H + \overline{C}l$$

Vinyl chloride — Too unstable to be formed (*sp* hybridised)

$$C_6H_5-Br \not\longrightarrow C_6H_5^{+} + Br^-$$

Aryl bromide — Too unstable to be formed

3. **Solvent effects:** Polar solvents favor S_N1 reaction. The ionization of substrate takes place in transition state. Polar solvents solvate these ions and promotes S_N1 mechanism. Ionization takes place by formation and separation of positive and negative charges. So it needs highly polar solvents for solvating the ions. The ionization takes place in highly polar solvents like water, alcohol *etc.*

$$R-\ddot{X}: \xrightarrow{\text{Ionisation}} R^+ + :\ddot{X}:^-$$

Alkyl halide

Solvated ions

Table 14.5 Effect of polarity of solvent on the rate of reaction.

The effect of polarity of solvent on the rate of reaction of *t*-butyl bromide in S_N1 reaction	
Solvent	**Rate of reaction**
100% water	1200
80% H_2O + 20% Ethanol	400
50% H_2O + 50% Ethanol	60
20% H_2O + 80% Ethanol	10
100% Ethanol	1

4. **Nature of the nucleophile:** In S_N1 reaction, the rate of the reaction only depends upon the concentration of alkyl halide. The nucleophile participates in the reaction after the rate determining step *i.e.,* after formation of carbocation, so the reactivity of nucleophile does not influence the rate of S_N1 mechanism.

Carbocation rearrangement in S_N1 reaction

S_N1 reaction takes place *via* the formation of carbocation and hence depends upon the stability of carbocation. If the carbocation formed is less stable, it undergoes rearrangement to a more stable carbocation.

In S_N2 reaction, no such type of rearrangement occurs because there is no formation of carbocation. The product obtained from the S_N1 reaction of 2-bromo-3-methyl butane differs from the S_N2 reaction. When the reaction takes place by S_N1 mechanism in appropriate reaction conditions, it leads to the formation of secondary carbocation which undergoes 1,2-hydride shift and rearranges to a more stable tertiary carbocation.

$$CH_3-CH(CH_3)-CH(Br)-CH_3 \xrightarrow[-Br^-]{S_N1} CH_3-CH(CH_3)-\overset{\oplus}{C}H-CH_3 \xrightarrow{\text{1,2 hydride shift}} CH_3-\overset{\oplus}{C}(CH_3)-CH_2CH_3$$

2-Bromo-3-methyl butane; 2° Carbocation (Less stable); 3° Carbocation

$$CH_3-\overset{\oplus}{C}(CH_3)-CH_2CH_3 \xrightarrow{H_2O} CH_3-C(CH_3)(\overset{+}{O}H_2)-CH_2CH_3 \underset{H^+}{\overset{-H^+}{\rightleftharpoons}} CH_3-C(CH_3)(OH)-CH_2CH_3$$

2-Methyl-2-butanol

$$CH_3-CH(CH_3)-CH(Br)-CH_3 \xrightarrow{S_N2} CH_3-CH(CH_3)-CH(OH)-CH_3$$

3-Methyl-2-butanol

Competition between S_N2 and S_N1

In earlier topics, we discussed about the reactivity of alkyl halides towards S_N2 and S_N1 reactions. When an alkyl halide undergoes S_N1 and S_N2 reactions, both reactions occur simultaneously and which one predominates depends upon the reaction conditions. Therefore, we have some experimental control over which reaction takes place.

What type of reaction conditions favor S_N2 reaction and what type of reaction condition favor S_N1 reaction?

S_N2 reaction leads to the formation of one single substituted product but the S_N1 reaction yields two substituted products if the starting alkyl halide possess chirality center. S_N1 reaction is also complicated one due to the rearrangement of carbocation intermediate. Hence, S_N2 reaction is synthetic chemist's friend but S_N1 reaction is synthetic chemist's nightmare. The conditions which determines whether the given chemical reaction follows S_N2 reaction or S_N1 reactions are as follows:

1. Concentration of nucleophile.
2. Reactivity of nucleophile.
3. Effect of solvent in which the reaction is carried out.

   For understanding this concept, we must find out the rate law of the two reactions.

   Rate law for S_N2 reaction:

   Rate = K_2 [alkyl halide] [nucleophile]

   Rate law for S_N1 reaction

   Rate = K_1 [alkyl halide]

Rate law for the reaction of alkyl halide which undergoes both reactions is sum of the rate law of the two reactions.

$$Rate = K_2\ [alkyl\ halide]\ [nucleophile] + K_1\ [alkyl\ halide]$$

The above equation of two rate laws indicate that an increase in the concentration of nucleophile increases the rate of S_N2 reaction but no effect on rate of S_N1 reaction. So when both reactions take place simultaneously, increasing the concentration of nucleophile, increases the S_N2 reaction. On other hand, decreasing the concentration of nucleophile decreases the S_N2 reaction. In all chemical reactions, the rate determining step is the slow reaction only. In S_N2 reaction, the rate determining step (slow step) is attack of nucleophile on the alkyl halide. Thus increasing the concentration of nucleophile, increases the rate constant because more reactive nucleophile is able to replace the leaving group better. Similarly, the rate determining step in S_N1 is the formation of carbocation or carbonium ion, which further reacts with any nucleophile in the second step (fast step) to give substituted products. It indicates that any change in the concentration of nucleophile does not affect the rate of S_N1 reaction. Hence, a good nucleophile and high concentration favors S_N2 reaction but decreases the rate of completing S_N2 reaction rate.

Comparison between S_N1 and S_N2 Reactions

1. **Effect of substrate:** The structure of the substrate is very important factor for determining the type of mechanism of substitution reaction. Primary halide and CH_3X does not easily undergo ionization and S_N1 reaction because primary and methyl carbocation possess high energy. They are relatively unhindered, hence they undergo S_N2 reaction. 3° are more sterically hindered, so easily undergo S_N1 reaction, by the formation of carbocations.

   S_N1 3° > 2°

   S_N2 CH_3X >1° > 2° > 3°

Table 14.6 Comparison of S_N2 Mechanism and S_N1 Mechanism.

S. No.	S_N2 Mechanism	S_N1 Mechanism
1	Bimolecular reaction.	Unimolecular reaction.
2	It takes place in one step.	It takes place in two steps.
3	It follows second order kinetics.	It follows first order kinetics.
4	Nucleophile attacks the carbon atom of the alkyl halide from back side.	Nucleophile attacks the carbon atom of the alkyl halide from both front and back side but back side attack predominates.
5	Complete inversion of configuration occurs (Walden inversion).	Partial racemization occurs if the starting material is optically active.
6	No rearranged product is obtained.	Formation of rearranged product is possible.
7	Order of reactivity of alkyl halides Methyl > 1° > 2° > 3° halides.	Order of reactivity of alkyl halides 3° > 2° > 1° > methyl halides.
8	Favored by strong and high concentration of nucleophile.	Favored by weak and low concentration of nucleophile.
9	Favored by low polar solvents.	Favored by high polar solvents.
10	Rate of the reaction is determined by steric factors.	Rate of the reaction is determined by electronic factors.
11	No catalyst is used.	Catalyzed by Lewis acids such as $ZnCl_2$, $AlCl_3$ etc.

Elimination Reactions (Reactions at sp^3 Hybridised Carbon)

Introduction

Elimination is defined as the loss of two atoms or groups from the substrate which leads to formation of π bond. Alkyl halides undergo elimination reaction. During this reaction, the halogen atom is eliminated from the carbon atom to which it is attached, and a proton is eliminated from the adjacent carbon atom. So, a double bond is formed between the two carbon atoms from which the atoms are eliminated. It is depicted in the following chemical reaction.

$$\underset{\text{Alkyl halide}}{CH_3CH_2CH_2X} + Y^- \longrightarrow \underset{\text{Alkene}}{CH_3 - CH = CH_2} + HY + X^-$$

Depending upon the reaction conditions and reagents used, the elimination reaction is classified into two types.

(I) E_1 Mechanism (First order/ Unimolecular).

(II) E_2 Mechanism (Second order/ Bimolecular) [or] **β**-elimination.

E_2 Mechanism [or] β–Elimination

E_2 stands for elimination bimolecular reaction. Elimination also takes place by second order kinetics because the rate of the reaction depends on the concentration of alkyl halide as well as the base. For example, consider the reaction between ethyl bromide and hydroxide ion.

$$\underset{\text{Ethyl bromide}}{CH_3 - CH_2 - Br} + OH^- \longrightarrow \underset{\text{Ethylene}}{CH_2 = CH_2} + H_2O + Br^-$$

The rate of reaction $R \propto [CH_3CH_2Br]\,[OH^-]$

$$R = K\,[CH_3CH_2Br]\,[OH^-]$$

The rate law indicates that the ethyl bromide and hydroxide ion are involved in the transition state of the rate determining step of the reaction. According to the above rate equation, E_2 reaction is a concerted reaction. The halide ion and the proton are removed in the same step and no formation of carbocation intermediate takes place

Mechanism:

$$H\ddot{\underset{\cdot\cdot}{O}}{:}^- \quad \underset{\underset{}{|}}{H}\!-\!CH_2 - \underset{\underset{Br}{|}}{CH_2} \longrightarrow \underset{\text{Ethylene}}{CH_2 = CH_2} + H_2O + Br^-$$

1. In an E_2 reaction, a base abstract a proton from the carbon atom which is adjacent to the carbon atom bonded with halogen. When the proton is removed, the electrons shared by that hydrogen atom with the carbon are moved towards the carbon atom which bears the halide atom.

2. Once these electrons move, the halogen atom leaves along with its shared electrons from carbon atom.
3. The electrons which are bonded to the reactant forms the π bond in the product.
4. Removal of proton (hydrogen) and a halogen is called **dehydrohalogenation.**
5. The carbon to which the halogen is attached or bonded is called **α**-carbon. A carbon which is adjacent to the **α**-carbon is known as **β**-carbon.
6. In E_2 mechanism, the elimination is initiated by the removal of hydrogen atom from the **α**-carbon, so it is called as **β-elimination.**
7. The hydrogen atom and the halogen atom is removed from the adjacent carbon atom so it is called as **1, 2-elimination.**

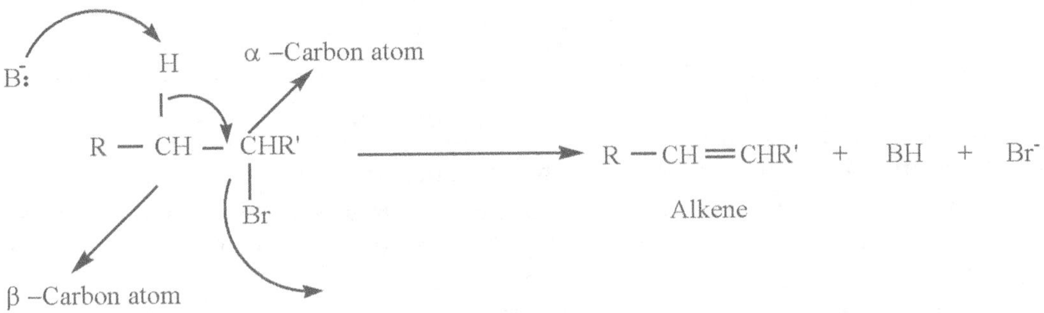

Equation No:1

β -Carbon atom

$$\underset{\text{2-Bromopropane}}{CH_3CH(Br)CH_3} + \underset{\text{Methoxide ion}}{CH_3O^-} \longrightarrow \underset{\text{1-Propene}}{CH_3-CH=CH_2} + CH_3OH + Br^-$$

Equation No:2

$$\underset{\text{2-Bromobutane}}{CH_3-CH(Br)-CH_2-CH_3} + \underset{\text{Methoxide ion}}{CH_3O^-} \longrightarrow \underset{\text{2-Butene (80 \%), Mixture of E and Z}}{CH_3-CH=CH-CH_3}$$

$$\underset{\text{1-Butene (20 \%)}}{CH_2=CHCH_2CH_3} + CH_3OH + Br^-$$

β - Carbon atom

8. In equation 1, presence of two equal or identical **α**-carbon atoms from which a proton can be removed equally and easily from any one side. The product obtained is propene. But in equation 2, 2-bromobutane has two structurally different **β** carbon from which a proton can be removed, so, when 2-bromobutane reacts with a base, two elimination products are formed i.e. 2-butene and 1-butene. What are the factors that dictate which of the two elimination products will be formed in the greater yield? To answer this question, we must determine which of the alkenes is formed more easily-that is, which is formed faster. The reaction coordinate diagram of E_2 elimination of 2-bromobutane and methoxide ion is shown in the Graph 14.2.

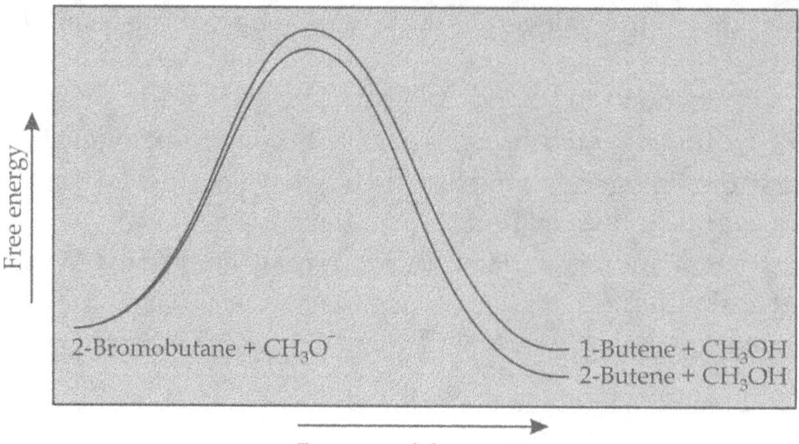

Progress of the reaction

Graph 14.2 Reaction coordinate diagram of E_2 reaction of 2- bromo butane and methoxide ion.

9. Stability of alkene depends upon the number of alkyl substituents attached to the sp^2 carbon atom. The greater the number of alkyl group, greater the stability of alkene (Zaitsev's rule)
10. In elimination reaction (equation no 2), two methyl groups are attached to the sp^2 carbon in 2 butene but in 1-butene, only one substituent (ethyl) is attached to sp^2 carbon. So 2-butene is more stable than that of 1-butene.
11. **Transition state:** In the transition state, the C-H and C-Br bonds are partially broken and the double bond is partially formed, giving alkene like structure. The one leading to 2-butene is more stable than the one leading to 1-butene. Thus the 2-butene is more stable than 1-butene.
12. The difference in the rate of formation of the two alkenes is not very great. Consequently, both products are formed, but the more stable of the two alkenes will be the major product of the reaction. So, the E_2 mechanism is regioselective since one of the constitutional isomer is formed more than that of another one.
13. Other examples of E_2 mechanism is the reaction between 2-bromo-2-methyl butane and hydroxide ion to give mixture of 2-methyl-2-butene (70 %) and 2-methyl-1-butene (30%).

OCH_3 ⋮ H ⋮ $CH_3 - CH \text{----} CHCH_3$ (with Br ⋮ on CH)

Transition state leading to 2-butene (More stable)

OCH_3 ⋮ H ⋮ $H_2C \text{----} CH - CH_2 - CH_3$ (with Br ⋮ on CH)

Transition state leading to 1-butene (Less stable)

$$CH_3-\underset{Br}{\overset{CH_3}{C}}-CH_2CH_3 + OH^- \longrightarrow CH_3-\overset{CH_3}{C}=CH-CH_3 + CH_2=\overset{CH_3}{C}-CH_2-CH_3 + H_2O + Br^-$$

2-Bromo-2-methyl butane → 2-Methyl-2-butene (70 %) + 2-Methyl-1-butene (30 %)

2-Methyl-2-butene is more stable (more alkyl substituted alkene), so it is obtained in high yield or as the major product.

Zaitsev's rule:

In 19th century, Russian chemist Alexander M. Zaitsev discovered a shortcut method to predict the more substituted alkene product. He proposed that "the more substituted alkene product is obtained when a proton is removed from the **β**–carbon that is bonded to the fewest hydrogens". In elimination reaction, the major product is the alkene which is more substituted because it is the more easily formed. This is called as Zaitsev's rule For example: 2-Chloropentane reacts with hydroxide ion to yield 2-pentene (67 %) and 1-pentene (33 %).

$$CH_3-CH_2-CH_2-\underset{\displaystyle |}{\overset{\displaystyle Cl}{CH}}-CH_3 + HO^- \longrightarrow CH_3CH_2CH=CH-CH_3$$

2- Chloro pentane

2-Pentene (67 %)

(Mixture of *E* and *Z*)

+

$$CH_3-CH_2-CH_2-CH=CH_2$$

1-Pentene (33 %)

According to Zaitsev's rule, the more substituted alkene will be the one formed by the removal of a proton from the one **β**-carbon that is bonded to two hydrogens rather than from the **β**-carbon that is bonded to three hydrogens.

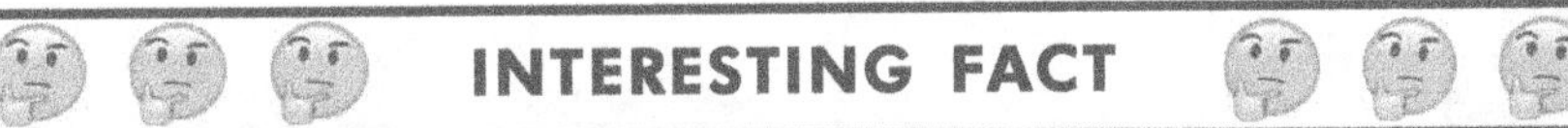

Alexander M. Zaitsev (1841-1910) was born in Kazan, Russia. In German transliteration is sometimes used for his family name.

$$CH_3-CH_2-\overset{\displaystyle CH_3}{\underset{\displaystyle Br}{C}}-CH_3 > \underset{\displaystyle Br}{CH_3CH_2CHCH_3} > CH_3CH_2CH_2CH_2Br$$

$$\downarrow \qquad\qquad \downarrow \qquad\qquad \downarrow$$

$${}^{1}CH_3-CH=\overset{\displaystyle {}^{3}CH_3}{C}-\underset{2}{CH_3} \qquad CH_3-CH=CH-CH_3 \qquad CH_3CH_2CH=CH_2$$

3-Alkyl substituents | 2-Alkyl substituents | 1-Alkyl substituent

Relative reactivities of alkyl halides in E_2 reaction are as follows.

3° alkyl halide > 2° alkyl halide > 1° alkyl halide.

In most of the elimination reactions, the highly or more substituted alkene is the major product because it is easily formed. But some of the elimination reactions are exceptional from this concept. For Example E_2 elimination reaction between 2-bromo-2-methylbutane and tertiary butoxide ion. In this reaction the base is sterically bulky, so it will preferentially eliminate the most accessible hydrogen.

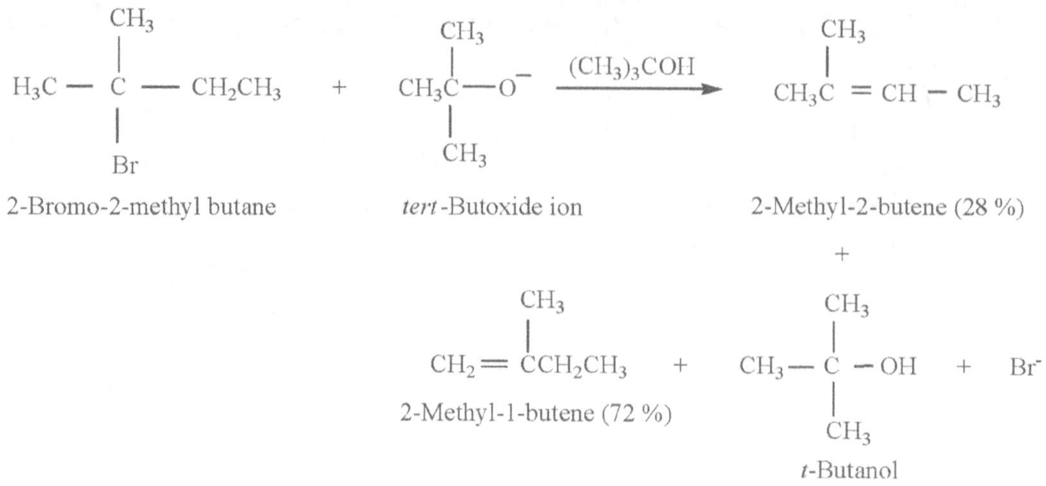

$$H_3C-\underset{Br}{\overset{CH_3}{C}}-CH_2CH_3 + CH_3\underset{CH_3}{\overset{CH_3}{C}}-O^- \xrightarrow{(CH_3)_3COH} CH_3\overset{CH_3}{C}=CH-CH_3$$

2-Bromo-2-methyl butane; *tert*-Butoxide ion; 2-Methyl-2-butene (28 %)

$$+ \; CH_2=\overset{CH_3}{C}CH_2CH_3 + CH_3-\underset{CH_3}{\overset{CH_3}{C}}-OH + Br^-$$

2-Methyl-1-butene (72 %); *t*-Butanol

In the above reaction, bulky bases such as *tert*-butoxide tend to give higher yields of the less substituted double bond isomers, a characteristic that has been attributed to steric hindrance. The effect of steric properties of the base on the formation of products is shown in the reaction and summarized in Table 14.7.

Table 14.7 Effect of steric properties of the base on the formation of products in an E_2 reaction.

Type of base used	More substituted product %	Less substituted product %
$CH_3CH_2O^-$	79 %	21 %
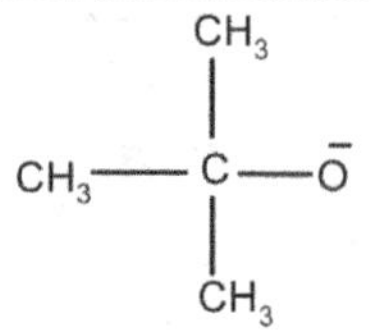 $CH_3-\underset{CH_3}{\overset{CH_3}{C}}-O^-$	27 %	73 %
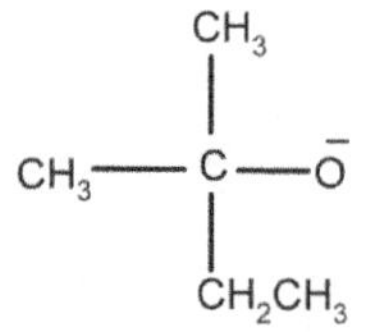 $CH_3-\underset{CH_2CH_3}{\overset{CH_3}{C}}-O^-$	19 %	81 %
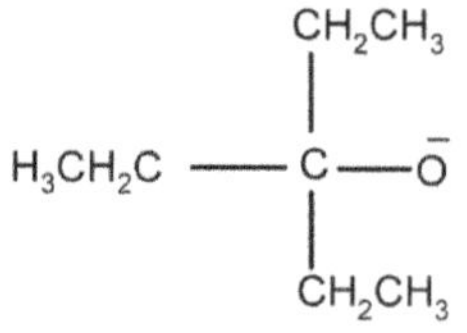 $H_3CH_2C-\underset{CH_2CH_3}{\overset{CH_2CH_3}{C}}-O^-$	8 %	92 %

$$CH_3-\underset{CH_3}{\overset{Br}{C}}-CH_2-CH_3 + RO^- \longrightarrow CH_3-\underset{CH_3}{C}=CHCH_3 + CH_2=\underset{CH_3}{C}-CH_2CH_3$$

2-Bromo-2-methylbutane; Base; 2-Methyl-2-butene (More substituted product); 2-Methyl-1-butene (Less substituted product)

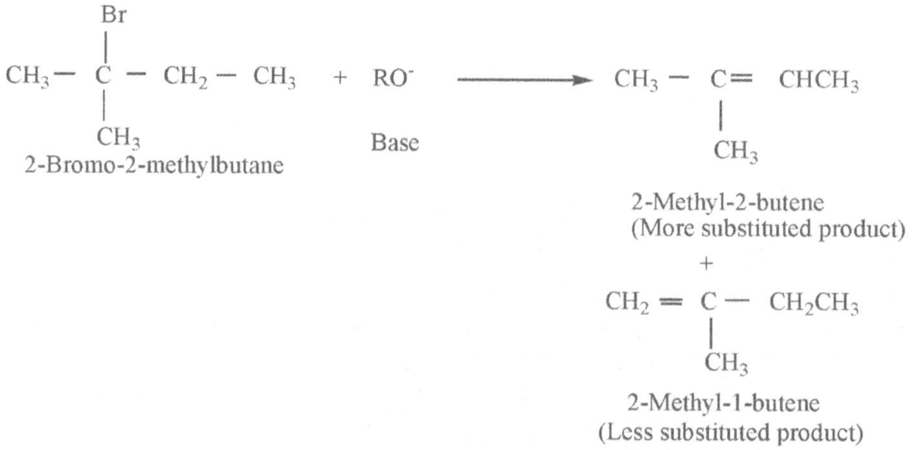

In E_2 dehydrohalogenation of alkyl chlorides, alkyl bromides and alkyl iodides the major product obtained is the high or more substituted alkenes (follows Zaitsev's rule) but for E_2 dehydrohalogenation of alkyl fluorides, the major product is the less substituted alkene.

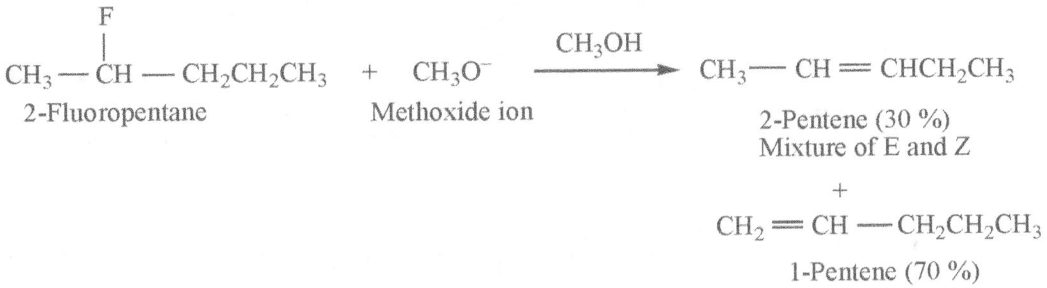

$$\underset{\text{2-Fluoropentane}}{CH_3-\overset{\overset{F}{|}}{C}H-CH_2CH_2CH_3} + \underset{\text{Methoxide ion}}{CH_3O^-} \xrightarrow{CH_3OH} \underset{\substack{\text{2-Pentene (30 \%)}\\ \text{Mixture of E and Z}}}{CH_3-CH=CHCH_2CH_3} + \underset{\text{1-Pentene (70 \%)}}{CH_2=CH-CH_2CH_2CH_3}$$

Explanation

1. If a hydrogen atom and halogen (chlorine, bromine or iodide) is eliminated from an alkyl halide, the halogen atom first starts to move from alkyl halide as soon as the base begins to remove the hydrogen atom. So, there is no development of negative charge on the carbon which is losing proton. (Because the halogen atom starts to leave with its bonding electron as soon as the proton begins to leave). So, the transition state resembles like that of alkene structure rather than that of carbanion.
2. In case of fluorine, it is a strongest base and a poorest leaving group. When a base begins to remove a proton from the alkyl halide; fluorine atom has less tendency to leave the alkyl halide. A negative charge is developed on the carbon atom from which the hydrogen atom is removed which leads to a formation of carbanion rather than alkene structure.

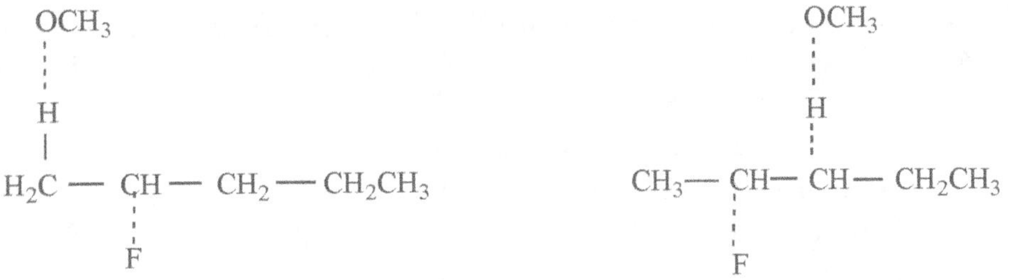

$$\underset{\substack{\text{Transition state leading to}\\ \text{1-pentene (More stable)}}}{\overset{\overset{OCH_3}{\vdots}}{\overset{H}{\overset{|}{H_2C}}}-\underset{\underset{F}{\vdots}}{C}H-CH_2-CH_2CH_3} \qquad \underset{\substack{\text{Transition state leading to}\\ \text{2-pentene (Less stable)}}}{CH_3-\underset{\underset{F}{\vdots}}{C}H-\overset{\overset{OCH_3}{\vdots}}{\overset{H}{\overset{\vdots}{C}}}H-CH_2CH_3}$$

Relative stability of carbanions: Carbanions are negatively charged, so they are destabilized by alkyl groups. Methyl carbanions are most stable and tertiary carbanions are least stable.

$$\underset{\substack{\text{Tertiary}\\ \text{carbanion}}}{R-\underset{\underset{R''}{|}}{\overset{\overset{R'}{|}}{C}}{:}^-} < \underset{\substack{\text{Secondary}\\ \text{carbanion}}}{R-\underset{\underset{H}{|}}{\overset{\overset{R'}{|}}{C}}{:}^-} < \underset{\substack{\text{Primary}\\ \text{carbanion}}}{R-\underset{\underset{H}{|}}{\overset{\overset{H}{|}}{C}}{:}^-} < \underset{\substack{\text{Methyl}\\ \text{carbanion}}}{H-\underset{\underset{H}{|}}{\overset{\overset{H}{|}}{C}}{:}^-}$$

Increasing stability →

In case of 2-fluoropentane, in the transition state, the negative charge is developed on the primary carbon atom (more stable carbanion), but in 2-pentene structure, the negative charge is developed on the secondary carbanion which is less stable. So the product obtained is opposite to that of Zaitsev's rule.

Table 14.8 Products obtained from E_2 elimination of methoxide ion and 2 halo alkanes.

			More substituted product	**Less substituted product**
$CH_3CH(X)CH_2CH_2CH_3$ + CH_3O^- ⟶			$CH_3CH{=}CHCH_2CH_2CH_3$ + **2-Hexene (mixture of E and Z)**	$CH_2{=}CHCH_2CH_2CH_2CH_3$ **1-Hexene**
Leaving group	**Conjugate acid**	**pK$_a$**		
X = I	HI	–10	81%	19%
X = Br	HBr	–9	72%	28%
X = Cl	HCl	–7	67%	33%
X = F	HF	3.2	30%	70%

The above table indicates that as the basicity of halide ion increases, the leaving ability decreases which leads to decrease in the formation of more substituted alkene product. The order of basicity of halogen atom is as follows

F > Br > Cl > I

Zaitsev's rule cannot be applicable for the determination of major product in the elimination reaction of conjugated dienes and benzene ring and also in any one of the following conditions.

(a) If a base is large (more basicity).
(b) If the alkyl halide is an alkyl fluoride.
(c) If the alkyl halides contain one or more double bonds.

Stereochemistry of E_2 Reaction

E_2 elimination is a concerted reaction because the two groups (one is hydrogen, another is halogen atom) are eliminated in the same step. So the bonds to the eliminated group (HX) must be in the same plane because the sp^3 orbital of the carbon bonded to H and sp^3 orbital of the carbon bonded to X become overlapping p orbitals in the alkene product.

In the transition state, the orbitals must overlap. This overlap is optimal if the orbitals are parallel (they must be in same plane). There are two possibilities in which the C-H and C-X bonds are in the same plane. They can be parallel to one another either on the same side of the molecule (*Syn*-periplanar) or opposite side of the molecule (*anti*-periplanar).

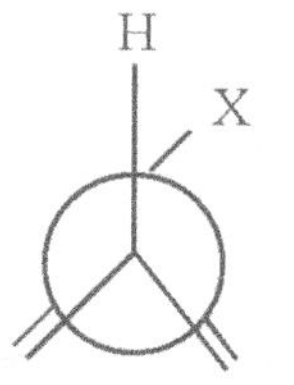

H X

An eclipsed conformer
(substituents are syn-periplanar)

H X

A staggered conformer
(substituents are anti-periplanar)

***Syn*-elimination:** Elimination reaction involves the removal of two substituents from the same side of the C-C bond. ***Anti*-elimination:** Elimination reaction involves the removal of two substituents from the opposite side of the C-C bond.

In both cases, eliminations occur but anti-elimination predominates to syn elimination.

Reason for fast anti-elimination

1. *Anti*-elimination requires staggered conformation but *syn*-elimination requires eclipsed conformation of the molecule.
2. Staggered conformation is more stable when compared with eclipsed conformation. In transition state, the elimination occurs in staggered conformation only.
3. It is also explained by drawing the sawhorse projections for the two conformers.

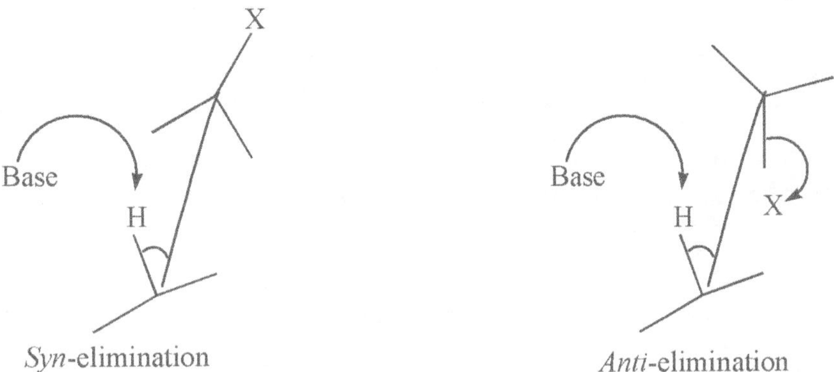

Syn-elimination   *Anti*-elimination

4. In syn-elimination, the electrons of the leaving atom moves to the front side of the carbon bonded to X.
5. But in anti-elimination, back side of the electrons of the leaving hydrogen atom move to the carbon bonded to X. So anti-elimination prevents the repulsion of electrons between base and halide ion.
6. According to the previous section, E_2 mechanism is regioselective one because one constitutional isomer is obtained more than the other.
7. Similarly, E_2 mechanism is stereoselective because one of the stereoisomer is formed more than that other. For example, 2-pentene obtained as the major product from the elimination reaction of 2-bromopentene exists as a pair of streoisomers, (E)-2-pentene is formed more than that of (Z)-2-pentene.

H, CH_2CH_3, C = C, H_3C, H

(E)-2-Pentene (41 %)

H, H, C = C, H_3C, CH_2CH_3

(Z)-2-Pentene

Key points:

1. If the two hydrogen atoms are present in the carbon atom from which a hydrogen atom is to be removed, both *E* and Z products will be formed since there are two conformers in which the groups to be eliminated are *anti*.
2. The alkene with the bulkiest groups on the opposite side of the double bond is formed more than that of same side groups (depends upon the stability).
3. Energy diagram or reaction coordinate diagram for E_2 reaction of 2-bromopentane with methoxide ion, is as follows.

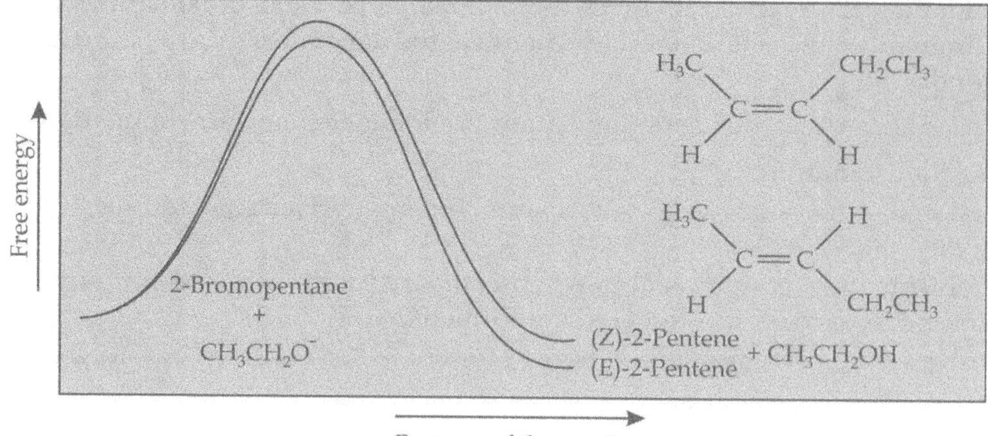

Graph 14.3 Reaction coordinate diagram of E_2 reaction of 2- bromo pentane and ethoxide ion.

E_1 Reaction or Unimolecular Elimination Reaction

In this reaction, the rate of elimination reaction depends only upon the concentration of alkyl halide . E indicates elimination and 1 indicates unimolecular. For example, reaction between *tert*-butyl bromide and water to yield 2-methyl propene.

Mechanism: The E_1 mechanism involves two steps.

$$\underset{\textit{tert}\text{ Butyl bromide}}{H_3C-\underset{CH_3}{\overset{CH_3}{\overset{|}{\underset{|}{C}}}}-Br} + H_2O \longrightarrow \underset{\text{2-Methyl-1-propene}}{CH_3-\overset{CH_3}{\overset{|}{C}}=CH_2} + \underset{\text{Hydronium ion}}{H_3O^+} + Br^-$$

Step 1: Formation of carbocation. Heterolytic dissociation of alkyl halide leads to the formation of carbocation.

$$\underset{\text{Tertiary butyl bromide}}{H_3C-\underset{CH_3}{\overset{CH_3}{\overset{|}{\underset{|}{C}}}}-Br} \underset{}{\overset{\text{Slow}}{\rightleftharpoons}} \underset{\text{Carbocation}}{CH_3-\underset{CH_3}{\overset{CH_3}{\overset{|}{\underset{|}{C^+}}}}} + Br^-$$

Step 2: Abstraction of proton from a carbon adjacent to the carbocation.

$$H_3C-\underset{\underset{\underset{H}{|}}{CH_2}}{\overset{CH_3}{\overset{|}{\underset{|}{C^+}}}} \quad H\ddot{O}H \xrightarrow{\text{Fast}} \underset{\text{2-Methyl-1-propene}}{H_3C-\underset{CH_2}{\overset{CH_3}{\overset{|}{\underset{\|}{C}}}}} + \underset{\text{Hydronium ion}}{H_3O^+}$$

In the above reaction, first step is the rate determining step (slow rate). If the concentration of the base is increased it influences the second step only and do not have any effect on the rate of reaction. Elimination involves the formation of two products in which more substituted alkene predominates.

For example, consider the following equation like elimination reaction of 2-chloro-2-methylbutane.

$$CH_3CH_2-\overset{\displaystyle CH_3}{\underset{\displaystyle Cl}{C}}-CH_3 + H_2O \longrightarrow CH_3CH=\overset{\displaystyle CH_3}{C}CH_3 + CH_3CH_2\overset{\displaystyle CH_3}{C}=CH_2 + H_3O^+ + Cl^-$$

2-Chloro-2-methyl butane → 2-Methyl-2-butene (Major product) + 2-Methyl-1-butene (Minor product)

Graph 14.4 shows the reaction co-ordinate diagram. Reaction coordinate diagram shows the E_1 reaction of 2-chloro-2-methylbutane. The major product is the more substituted alkene because of the greater stability of the transition state.

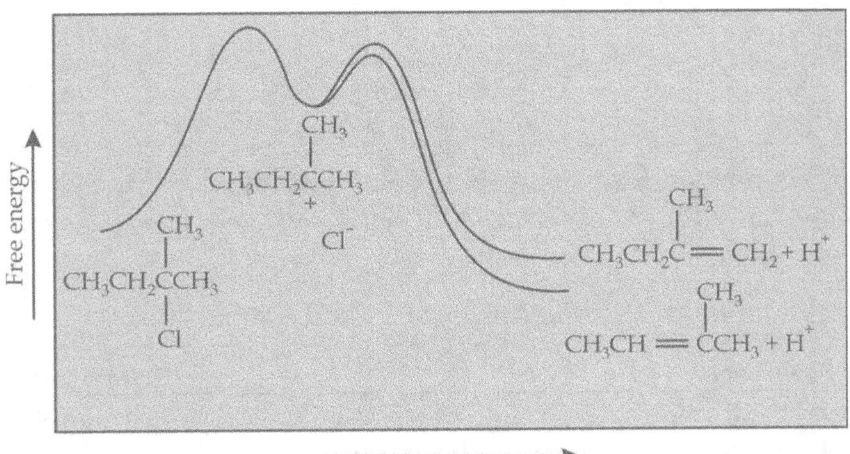

Graph 14.4 Reaction coordinate diagram of E_1 reaction of 2-chloro-2-methylbutane.

The rate of an E_1 reaction depends upon the ease of leaving group and stability of carbocation. Relative reactivities of alkyl halides = Relative stabilities of carbocation.

3° Benzylic >3° allylic >2° benzylic>1° benzylic ≈ 1° allylic≈ 2° >1 °>vinylic

Stereochemistry of E_1 Reaction

We have already discussed the mechanism of E_1 elimination in the previous topics. The first step of the reaction is the elimination of leaving group and in the second step of the reaction a proton is removed, following the Zaitsev's rule to form the more stable alkene. The carbocation formed in the first step of the reaction is planar. Hence the electron from the leaving proton can move towards to the positively charged carbon atom from either side. So both *syn* and *anti* elimination can takes place. It indicates that both syn and anti eliminations occurs in E_1, produces both *E* and Z products regardless of the β carbon from which the proton eliminated is bonded to one or two hydrogens. The major product obtained is one alkene with the bulkiest groups on the opposite side of the double bond since it is the more stable one.

$CH_3CH_2CHBr(CH_3)$ $\xrightarrow{-Br}$ $CH_3-CH_2-\overset{+}{C}H(CH_3)$ $\xrightarrow{-H^+}$ (E)-2-Butene + (Z)-2-Butene

β-carbon with two hydrogens

(*E*)-2-Butene (Major product) (*Z*)-2-Butene (Minor product)

$CH_3CH_2CH(CH_3)-C(CH_3)(Cl)-CH_2CH_3$ $\xrightarrow{-Cl}$ $CH_3CH_2-CH(CH_3)-\overset{+}{C}(CH_3)CH_2CH_3$ $\xrightarrow{-H^+}$ (*E*)-3,4-Dimethyl-3-hexene + (*Z*)-3,4-Dimethyl-3-hexene

β-carbon with one hydrogens

(*E*)-3,4-Dimethyl-3-hexene (Major product) (*Z*)-3,4-Dimethyl-3-hexene (Minor product)

But just recall that E_2 elimination forms both E and Z products only if the β carbon from which the proton is eliminated is bonded with two hydrogens. If the β carbon is attached with only one hydrogen only one product is obtained.

Table 14.9 Comparison of E_2 and E_1 reactions.

S. No.	E_2 Reaction	E_1 Reaction
1	Primary alkyl halides undergo E_2 only.	No reaction since difficulty in the formation of primary carbocation.
2	Secondary alkyl halides undergo E_2 reaction.	Undergoes E_1 also.
3	Tertiary alkyl halides undergo E_2 reaction.	Undergoes E_1 also.
4	Favored by high concentration of base.	Favored by weak base.
5	Takes place in aprotic polar solvents. example: DMSO.	Takes place in protic solvents. example: HCHO or ROH.

Competition between Substitution and Elimination Reactions

1. **Reaction conditions:** Whether the reaction condition favors (S_N2/ E_2) or (S_N1/ E_1) reaction. The condition which favors the S_N2 reaction also favors the E_2 reaction and the condition which favors the S_N1 also favors the E_1 reaction.
2. Primary alkyl halides undergo S_N2 mechanisms because (a) 1° substrates have little steric hindrance to nucleophilic attack and (b) 1° carbocations are relatively unstable, so it never undergoes (S_N1/ E_1) mechanism.
3. If the alkyl halide is secondary or tertiary, either it undergoes S_N2/ E_2 or S_N1/ E_1 depends upon the reaction conditions.
4. **Nature of the nucleophile:** S_N2/ E_2 reactions are favored by a high concentration of strong base or good nucleophile. Similarly, the S_N1/ E_1 reactions are favored by weak nucleophile. Both the above reactions also depend upon the nature of the solvents used.

5. After finding out the type of reaction whether S_N2/ E_2 or S_N1/ E_1, the obtained amount of substituted and elimination reaction products can be estimated and it depends upon whether the alkyl halide is primary, secondary or tertiary and nature of the nucleophile.

Conditions for S_N2/ E_2

The main conditions for the S_N2/ E_2 reactions are high concentration of alkyl halide and strong base or nucleophile.

The negatively charged base or nucleophile attacks th β carbon atom of the alkyl halide from the back side of the carbon atom which leads to substitution reaction. Similarly the base removes the proton or β-hydrogen from the carbocation intermediately which leads to elimination.

$$CH_3-CH_2-Br \text{ (Ethyl broide)} + OH^- \xrightarrow{\text{Substitution}} CH_3-CH_2-OH \text{ (Ethanol)} + Br^-$$

$$H_2C(H)-CH_2-Br + OH^- \xrightarrow{\text{Elimination}} CH_2=CH_2 \text{ (Ethylene)} + H_2O + Br^-$$

Table 14.10 Relative reactivities of alkyl halides.

S_N2 reaction	1° > 2°> 3°
E_2 reaction	3° > 2° >1°
S_N1 reaction	3° > 2°> 1°
E_1 reaction	1° >2° > 3°

If the primary alkyl halide or the nucleophile is sterically hindered, the elimination will predominate. The 2° alkyl halide gives both reactions but the amount of products obtained depends upon the base strength.

The 3° alkyl halides give elimination products only because they are least reactive in S_N2 reaction due to steric hindrance.

As temperature is increased, the relative amount of elimination products will increase relative to substitution products.

conditions for S_N1/ E_1

When condition favor S_N1/ E_1 reactions, the alkyl halides dissociates to form a carbocation, which will either combine with the nucleophile to form the substitution product or loses a proton to give a elimination product.

$$-\underset{H}{C}-C-Br \xrightarrow{-Br^-} -\underset{H}{C}-C^+ \begin{cases} \xrightarrow[H_2O]{\text{Substitution}} -\underset{H}{C}-C-OH \text{ (Alcohol)} \\ \xrightarrow[H_2O]{\text{Elimination}} >C=C< \text{ (Alkene)} + H_3O^{\oplus} \end{cases}$$

Alkyl halides possess same reactivity. All the alkyl halides which undergo S_N1/ E_1 reactions will give both substitution and elimination reactions. Substitution is favored at lower temperature and elimination is favored at higher temperature. But primary alkyl halides do not undergo S_N1/ E_1 reactions since primary carbocations are very unstable.

Elimination Reactions from Cyclic Compounds

E_2 Elimination from Cyclic Compounds

Elimination from cyclic compounds is similar to that of elimination of open - chain compounds. The main condition for E_2 elimination of cyclic compounds is that the two groups eliminated from a six membered ring must be present in the axial positons (anti periplanar geometry).

If the two groups are eliminated from the equatorial position, no possibility for E_2 reaction.

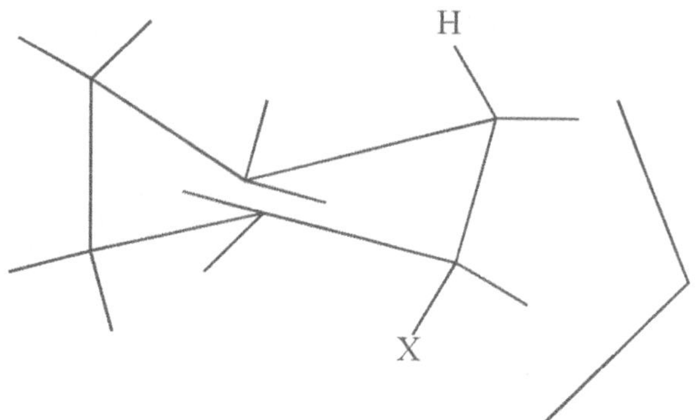

Both groups to be eliminated must be present in axial position.

The more stable conformer of chloro cyclohexane does not undergo an E_2 reaction because the chlorine atom is present in an equatorial position. But the less stable conformer of chlorocyclohexane easily undergoes elimination reaction because the chloro substituent is present in axial position. The rate of elimination reaction depends upon the stability of the conformer.

Equatorial
Keq
More stable
(Axial)
Less stable
K'
OH^-
E_2 conditions
+ $\overline{Cl}$
HO^-
E_2 conditions
No reaction

E_1 Elimination from Cyclic Compounds

Substituted cyclohexane undergoes E_1 reaction when the hydrogen and halogen atom are not in axial position as the reaction takes place in two steps.

Step 1: Formation of carbocation.

CH_3OH
E_1 conditions
+ Cl^-

Step 2: Abstraction of proton

H, CH_3, +, CH_3 ⟶ CH_3, CH_3 + HCl

+ Cl^-

Before determining the kind of elimination product, we must determine the rearrangement of carbocation either 1,2-hydride shift or 1,2-methyl shift.

Consecutive E_2 Elimination Reactions:

Geminal dihalides undergo two consecutive dehydrohalogenation reactions to give alkene with two double bonds as shown in following example,

$$\underset{\text{3,5-Dichloro-2,6-dimethyl heptane}}{CH_3\overset{CH_3}{C}H\underset{Cl}{C}HCH_2\underset{Cl}{C}H\overset{CH_3}{C}HCH_3} \xrightarrow{OH^-} \underset{\text{5-Chloro-2,6-dimethyl-2-heptene}}{H_3C\overset{CH_3}{C} = CHCH_2\underset{Cl}{C}H\overset{CH_3}{C}HCH_3} \xrightarrow{OH^-} \underset{\text{2,6-Dimethyl-2,4-heptadiene}}{H_3C\overset{CH_3}{C} = CHCH = CH\overset{CH_3}{C}H - CH_3}$$

Evidence of E_2 Mechanism: The following factors are the evidences for E_2 mechanism.

1. Follows 2nd order kinetics or not?
2. Rearrangement of products takes place or not.
3. Involves large H_2 isotope effect.
4. Involves no replacement of hydrogen atom.
5. The large element effect.

1. **Follows 2nd order kinetics or not:** The rate of reaction depends upon the concentration of both alkyl halides and nucleophile. So it follows 2nd order kinetics.
2. **Absence of rearrangements:** The reaction is concerted one *via* the formation of transition state, so there is no rearrangement.
3. **Involves large H_2 isotope effect:**

   Isotopes: Atoms with different mass number, electronic configuration and different chemical properties.

   Isotopic effect: The difference in the rate of reaction due to the presence of different isotope of an element in a reaction is called as isotope effect.

   Primary Isotope Effect: Isotope effect due to the breaking of a bond to the isotopic atom is called as primary isotope effect.

   For example, the bond proteium is very easily broken than that of deutrium.

   The presence or absence of isotopic effect of a particular reaction shows some important conditions of the reactions.

Consider the following examples.

$$C_6H_5CH_2CH_2Br + C_2H_5O^- \xrightarrow{K^H} C_6H_5CH{=}CH_2 + C_2H_5OH + Br^-$$

Phenyl ethyl bromide   Ethoxide ion

$$C_6H_5CD_2CH_2Br + C_2H_5O^- \xrightarrow{K^D} C_6H_5CH{=}CH_2 + C_2H_5{-}OD + Br^-$$

Deutriated phenyl ethyl bromide   Ethoxide ion

Compare the rate of reaction of the above two equations

$$K^H / K^D = 7$$

It indicates that compound containing proteium reacts 7 times faster than that of deuterium containing compound. It indicates that the only bond between hydrogen and β-carbon atom is broken, that is rate determining step in this reaction.

4. **Involves no replacement or absence of hydrogen atom:** The elimination reaction involves two steps.
   (i) Formation of carbocation: The alkyl halide molecules loses its proton to the base to give carbanion.

$$-\overset{X}{\underset{}{C}}-\overset{}{\underset{H}{C}}- \longrightarrow -\overset{X}{C}-\ddot{C}^{-}- + H_2O$$

Alkyl halide   :OH   Carbanion

   (ii) The carbanion loses its halide to form the product.

$$-\overset{X}{C}-\ddot{C}- \longrightarrow X^- + -C{=}C-$$

Consider the reaction between phenyl ethyl bromide and labeled ethanol (deutriated ethanol). What happened if the carbanion is formed again and again. This carbanion regains the proton from the solvent ethanol and it produces the starting material again. But, in the above said equation, the ethanol is replaced with deutriated ethanol, so the carbanion accept or regains the deuterium instead of proton.

The reaction is allowed to continue until 50 % of the substrate is converted into alkene. The unconsumed or unreacted 2-phenyl ethyl bromide is recovered. Mass spectrum of the compound showed that there is no deuterium. The similar reaction is carried out with other compounds, the similar results are obtained. It indicates that E_2 elimination does not involve the replacement or exchange of hydrogen atom.

5. **The element effect:** The strength of the carbon-halogen bonds is shown by their heterolytic bond dissociation energies. The sequence of heterolytic bond dissociation energy of alkyl halide is as follows.

R-F > R-Cl > R-Br > R-I

In substitution reaction the rate of bond broken between C-X is as follows

R-I > R- Br > R- Cl > R-F

The elimination reaction also follows the same order

$$R\text{-}I > R\text{-}Br > R\text{-}Cl > R\text{-}F$$

Alkyl bromides react 40 to 60 times as fast as chlorides, alkyl iodides reacts 25000 times as fast as the chlorides. It indicates that the rate at which the carbon-halogen cleaves does not affect the overall rate of elimination.

Table 14.11 Summary of the products for Substitution / Elimination Reactions.

S. No.	Name of alkyl halide	S_N2 vs E_2	S_N1 vs E_1
1.	1° Alkyl halide	Primarily undergoes substitution, if any steric hinderance is possible, elimination is favored.	Does not undergo S_N1/E_1 reaction.
2.	2° Alkyl halide	Both substitution and elimination take place. At high temperature, with strong and bulky base, elimination predominates.	Both substitution and elimination take place. At high temperature, with strong and bulkier the base, elimination predominates.
3.	3° Alkyl halide	Only elimination occurs.	Both substitution and elimination occur. At higher temperature, elimination predominates.

Table 14.12 Stereochemistry of Substitution and Elimination Reaction.

Mechanism	Products
S_N1	Both isomers are formed (Enantiomers).
E_1	Both E and Z stereoisomers are produced (More yield is the stereoisomer with the bulkiest groups on the opposite sides of the double bond)
S_N2	Only inverted products are obtained.
E_2	Both E and Z stereoisomers are formed. If the β-carbon of the alkyl halide has only one hydrogen, one stereoisomer is formed. Its configuration depends upon the configuration of the reactant.

Phase Transfer Catalysis

Phase Transfer Catalyst

A phase transfer catalyst is a compound which catalyses a reaction by transferring a reagent (usually an inorganic ion) into the phase in which it is needed. Quaternary ammonium compounds, crown ethers are the examples for phase transfer catalysts.

BITREX: Useful Bitter Tasted Compound

It is a quaternary ammonium compound which has most bitter taste but it is non toxic. It is applied to the back of the animals to keep them from biting one another. It is also applied to children's fingers to persuade them to stop sucking their thumbs or biting their finger nails and it is mixed to the toxic substances to keep them away from the accidental ingestion.

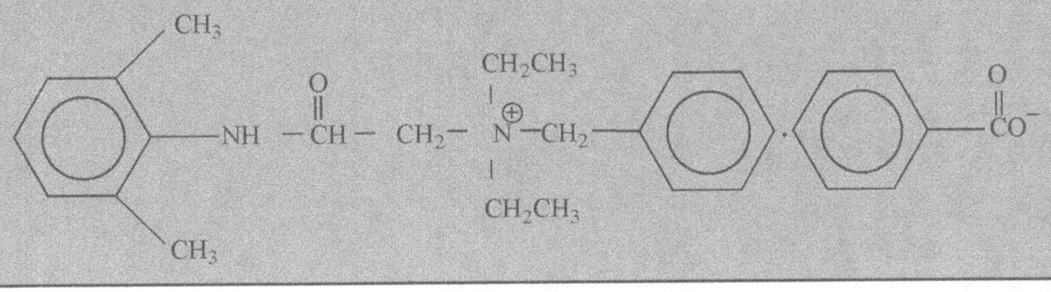

In laboratories, the main problems faced by the organic chemists are about the solvent system whether it dissolves all the reagents or not. For example: the reaction between 1-bromohexane and sodium cyanide.

$$\underset{\text{1-Bromohexane}}{CH_3CH_2CH_2CH_2CH_2CH_2Br} + \bar{C}\equiv N \longrightarrow \underset{\text{1-Cyanohexane}}{CH_3CH_2CH_2CH_2CH_2CH_2C\equiv N} + Br^-$$

If we mix the aqueous solution of sodium cyanide with organic solution of 1-bromohexane, the two solutions will forms two layers due to their immiscibility. Then how the cyanide ion reacts with an alkyl halide? If long time heating of this mixture also does not initiate the reaction.

But the two compounds will be able to react with each other when the catalytic amount of phase transfer catalyst is added to the reaction mixture. Example: Quaternary ammonium compounds such as tetrabutyl ammonium hydrogen sulphate *etc.*

$$\underset{\text{1-Bromohexane}}{CH_3CH_2CH_2CH_2CH_2CH_2Br} + \bar{C}\equiv N \xrightarrow{R_4N^+HSO_4^- \text{(PTC)}} \underset{\text{1-Cyanohexane}}{CH_3CH_2CH_2CH_2CH_2CH_2C\equiv N} + Br^-$$

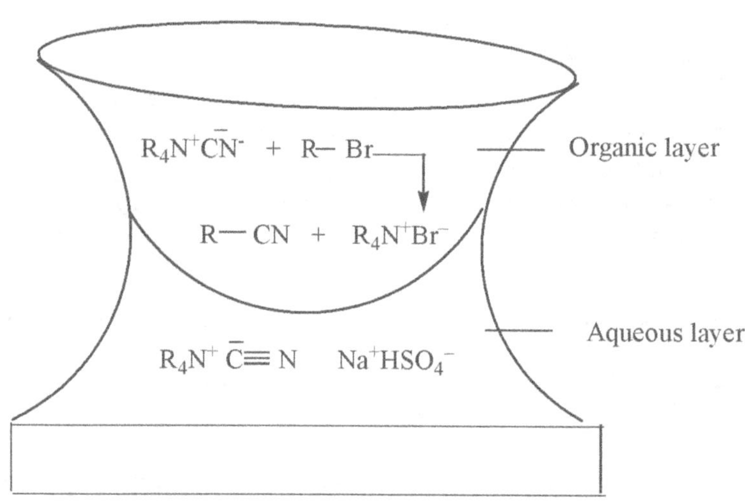

$$H_3CH_2CH_2CH_2C-\overset{+}{N}(CH_2CH_2CH_2CH_3)_2-CH_2CH_2CH_2CH_3 \quad HSO_4^-$$

Tetra butylammonium bisulphate

$$C_6H_5-CH_2\overset{+}{N}(CH_2CH_3)(CH_3)-CH_2CH_3 \quad HSO_4^-$$

Benzyl triethyl ammonium bisulphate

The reaction initiated in the presence of PTC is explained as follows.

1. The quaternary ammonium compound is soluble in organic solvent due to the presence of nonpolar alkyl group and it is also soluble in water due to the presence of positive charge.
2. It indicates that it acts as a mediator between two immiscible solvents.
3. The quaternary ammonium compound enters the organic phase, it must carry a counter ion along with it in order to balance its positive charge. The counter is either bisulphate ion or cyanide ion.
4. Once it enters into the organic layer, it reacts with alkyl halide (The hydrogen or bisulphate ion does not react in organic layer because it is weak base as well as poor nucleophile).
5. The quaternary ion will enter back into aqueous layer. It carries a counterion, either HSO_4^- or bromide ion.
6. So, the reaction takes place continuously with the PTC shuffling back and forth between the two phases. So, PTC is successfully used in wide variety organic compounds.

Probable Questions

1. Define nucleophilic substitution reaction and explain about various components of nucleophilic substitution reaction.
2. What is bimolecular nucleophilic substitution reaction? Explain in detail with suitable example along with reaction mechanism.
3. Write a short note on stereochemistry of S_N2 reaction.
4. Write a note on factors influencing the rate of S_N2 reactions.
5. What is unimolecular nucleophilic substitution reaction? Explain in detail with suitable example & reaction mechanism?
6. Write a short note on stereochemistry of S_N1 reaction.
7. Write a note on factors influencing the rate of S_N1 reactions.
8. Write a detailed note on factors influencing S_N1 and S_N2 reaction.
9. Write a note on competition between S_N1 and S_N2 reactions.
10. Compare S_N1 and S_N2 reactions.
11. What are aprotic solvents?
12. Explain carbocation rearrangement in S_N1 reaction.
13. Define and classify elimination reactions.
14. Write a note on β-elimination.
15. What is Zaitsev's rule?
16. Explain the relative reactivities of alkyl halides in E_2 reaction.
17. Write in detail about relative stability of carbanions.
18. What is E_2 reaction? Explain in detail with suitable example & reaction mechanism.
19. Write a short note on stereochemistry of E_2 reaction.
20. What is E_1 reaction? Explain in detail with suitable example & reaction mechanism.
21. Write a short note on stereochemistry of E_1 reaction.
22. Compare E_1 and E_2 reactions.
23. Write a note on competition between E_1 & E_2 reaction.
24. Compare nucleophillic substitution reaction with elimination reaction.
25. Write a note on competition between nucleophilic substitution reaction with elimination reaction.

26. Write a note on elimination reactions of cyclic compounds in detail.
27. Write in detail about consecutive E_2 elimination reactions.
28. Explain E_1 reactions of cyclic compounds.
29. Explain E_2 reactions of cyclic compounds.
30. Explain the evidence of E_2 mechanism in consecutive E_2 elimination reactions.
31. What is phase transfer catalyst? Explain with suitable example.
32. Explain in detail about the conditions required for S_N2/E_2 reactions.
33. Write a note on conditions required for S_N1/E_1 reactions.

15

Aldehydes and Ketones

Introduction

Aldehydes and ketones are carbonyl compounds which possess carbonyl (C=O) group in their structure. If the carbonyl group is attached to a primary carbon atom it is called an aldehyde (except formaldehyde which is attached to hydrogen) and if it is attached to a secondary carbon atom, it is called ketone. Both the aldehydes and ketones possess the general formula $C_nH_{2n}O$. The functional group of aldehyde is CHO which is present in the end of chain while the functional group of ketone is C=O which is not present in the end of chain but is joined to a carbon atom on either side. The simplest carbonyl compound is an aldehyde *i.e.* formaldehyde in which a carbonyl group is bonded to two hydrogen atoms. In all other aldehydes, the carbonyl group is attached to hydrogen and to an alkyl or aryl group.

$$\underset{\text{Aldehyde}}{R-\overset{\overset{O}{\|}}{C}-H} \qquad \underset{\text{Formaldehyde}}{H-\overset{\overset{O}{\|}}{C}-H} \qquad \underset{\text{Ketone}}{R-\overset{\overset{O}{\|}}{C}-R'}$$

Ketones and aldehydes are similar in structure and physical properties. But they differ in their chemical properties especially with nucleophiles and oxidizing agents. Aldehydes are more reactive than ketones. Some common types of carbonyl compounds are depicted in Table 15.1.

Table 15.1 Common carbonyl compounds.

S. No.	Chemical Class	General Formula
1.	Aldehydes	R-CHO
2.	Ketones	R-COR'
3.	Carboxylic acids	R-COOH
4.	Esters	$RCOOR^1$
5.	Acid chlorides	RCOCl
6.	Amides	$RCONH_2$

Importance of Aldehydes and Ketones

Carbonyl compounds are utmost importance in organic chemistry, biochemistry and biology. Apart from their uses as solvents and reagents, they are one of the constituents of flavoring agents, fabrics and plastics. Naturally occurring carbonyl compounds are carbohydrates, proteins and nucleic acids. They are constituents of drugs and pharmaceutical aids. They are wide spread in plant and animal kingdom. They play an important role in biochemical processes of life. Some of them are used as flavouring agents as mentioned in following example.

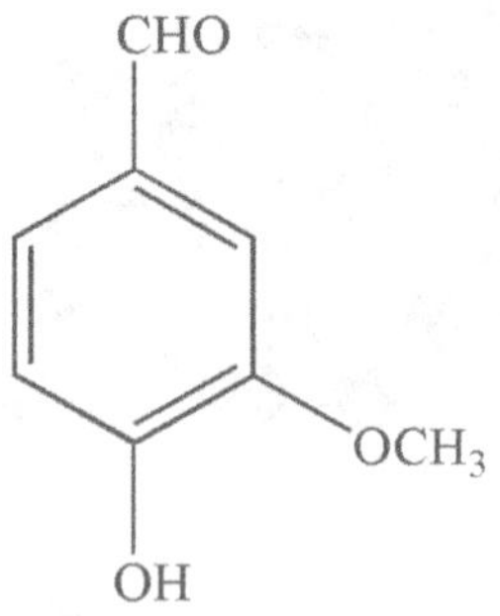

CHO OCH3 OH
Vanillin

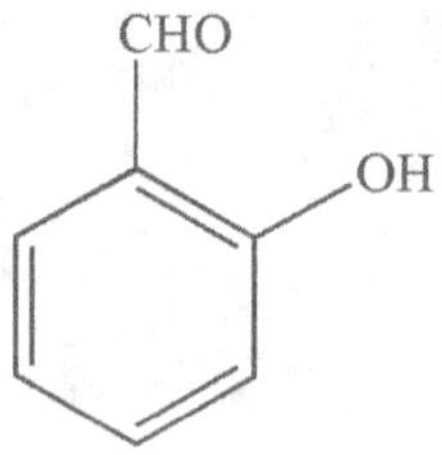

CHO OH
Salicylaldehyde

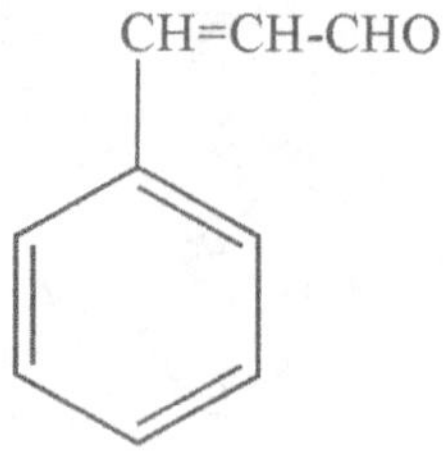

CH=CH-CHO
Cinnamaldehyde

Aldehydes and ketones when attached with various chemical structures elicit different kind of pharmacological activities based on the core structure to which it is attached. Some of the drugs which possess aldehydes and ketones and their therapeutic uses in the treatment of various diseases are shown in Table 15.2.

Table 15.2 Drugs containing aldehyde and ketone group and their therapeutic uses.

Category	Drug	Structure
Sedative and hypnotic	Paraldehyde	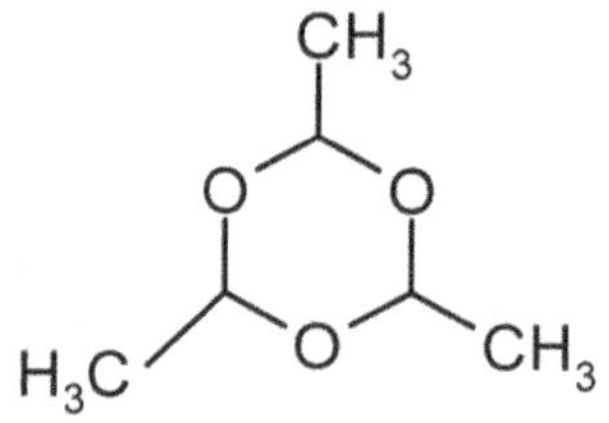 CH_3, O, O, O, H_3C, CH_3
	Sodium-5,5'-diethylbarbiturate	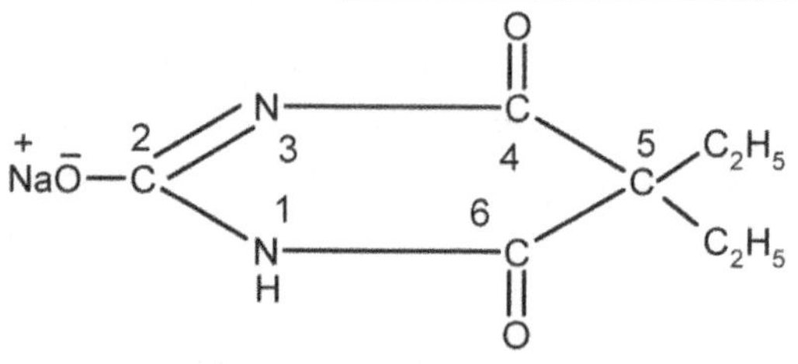 $\overset{+}{Na}\bar{O}$—C, N, H, O, C, C_2H_5, C_2H_5
Anticonvulsant	Ethosuximide	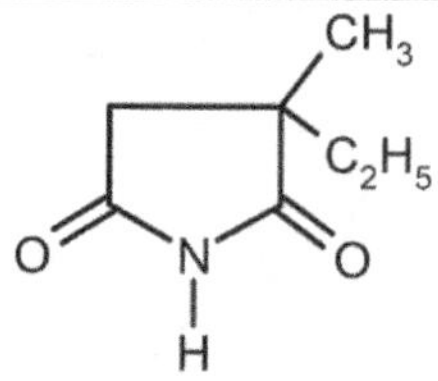 CH_3, C_2H_5, O, N, H, O
Cholinergic drug	Pilocarpine	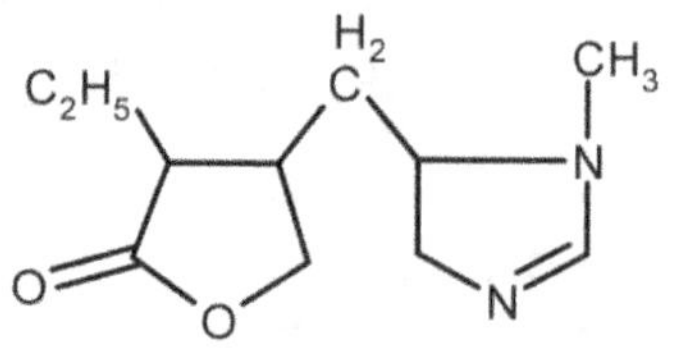 C_2H_5, H_2 C, CH_3, N, N, O, O

Industrial Importance of Aldehydes and Ketones

1. In chemical industries and laboratories, the aldehydes and ketones are used as solvents, reagents and starting materials for the synthesis of other products.
2. Formalin solution is used to preserve biological specimens.
3. Formaldehyde is used to prepare bakelite, urea-formaldehyde, phenol-formaldehyde, glues and other polymeric products.
4. Acetaldehyde is used as a starting material for the synthesis of acetic acid, polymers and drugs.
5. The most important commercial ketone is acetone. Both acetone and 2-butanone are used as common industrial solvents.

Nomenclature of Aldehydes

They can be named by two ways.

1. **Common name system:** Based on the corresponding carboxylic acid obtained from aldehydes, the names are given. The name of aldehyde is derived by replacing **–ic acid** of the acid with **–aldehyde.** For example, the name form**aldehyde** is derived from formic acid. The substituents present in the chain are indicated by the Greek letters α, β, γ *etc.* The carbon atom next to aldehyde group is called as α-carbon and next one is β and so on.

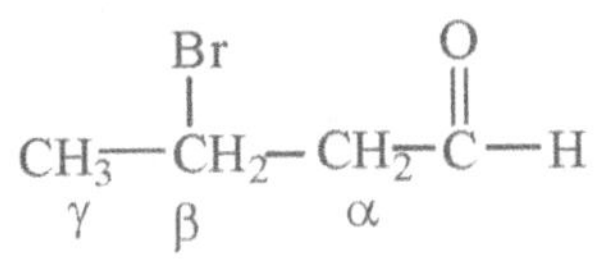

$$\underset{\gamma}{CH_3}-\underset{\beta}{\overset{\overset{Br}{|}}{CH}}-\underset{\alpha}{CH_2}-\overset{\overset{O}{||}}{C}-H$$

β–Bromo butyraldehyde

2. **IUPAC name system:** It is derived by replacing the **–e** of alkane with **–al.** The aldehyde function is always present in the end. When numbering is given to the chain, the carbonyl carbon gets number 1.

CH_3CHO
Ethanal (Ethane-e+al)

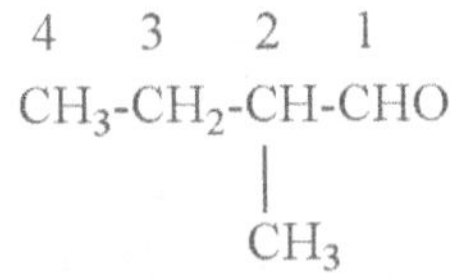

$$\overset{4}{CH_3}-\overset{3}{CH_2}-\overset{2}{\underset{\underset{CH_3}{|}}{CH}}-\overset{1}{CHO}$$

2-Methyl butanal

Nomenclature of Ketones

They can be named by two ways.

1. **Common name system:** In this method, they are named by adding the word ketone along with the name of two alkyl groups attached to the carbonyl group. When same alkyl groups are present then it is named as dialkyl ketone example, dimethyl ketone (acetone), otherwise the name of individual alkyl group is mentioned and then the word ketone is to be added in suffix. For example

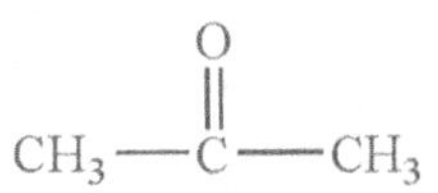

$$CH_3-\overset{\overset{O}{||}}{C}-CH_3$$

Dimethyl ketone

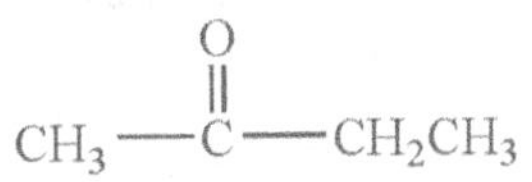

$$CH_3-\overset{\overset{O}{||}}{C}-CH_2CH_3$$

Ethyl methyl ketone

2. **IUPAC name system:** In this method, ketones are named by replacing **-e** of corresponding alkane with **-one.** While numbering the carbon chain, carbonyl group should be given the lowest possible number and this number is used to indicate the position of carbonyl group in the carbon chain as in the following example. Some common name and IUPAC name of aldehydes and ketones are depicted in the Table 15.3.

$$\overset{5}{CH_3}-\overset{4}{CH_2}-\overset{3}{CH_2}-\overset{2}{\underset{\underset{O}{\|}}{C}}-\overset{1}{CH_3}$$

2-Pentanone

Table 15.3 Common name and IUPAC name of some aldehydes and ketones.

Structure	Common name	IUPAC name
HCHO	Formaldehyde	Methanal
CH_3CHO	Acetaldehyde	Ethanal
$CH_3-CH(CH_3)-CHO$	*iso*-Butyraldehyde	2-Methyl propanal
H_3C–cyclohexane–CHO	γ-Methyl cyclohexane carbaldehyde	3-Methyl cyclohexane carbaldehyde
$CH_3CH(OCH_3)CHO$	α-Methoxy propionaldehyde	2-Methoxy propanal
$CH_3CH_2CH_2CH_2CHO$	Valeraldehyde	Pentanal
$CH_2{=}CHCHO$	Acrolein	Prop-2-enal
benzene with CHO, CHO (ortho)	Phthalaldehyde	Benzene-1,2-dicarbaldehyde
benzene with CHO, Br (meta)	*m*-Bromo benzaldehyde	3-Bromo benzene carbaldehyde or 3-Bromo benzaldehyde
$CH_3COCH_2CH_2CH_3$	Methyl *n*-propyl ketone	Pentan-2-one
$(CH_3)_2CHCOCH(CH_3)_2$	Diisopropyl ketone	2,4-Dimethylpentan-3-one
cyclohexanone with O, CH_3	α -Methylcyclohexanone	2-Methylcyclohexanone
$(CH_3)_2C{=}CHCOCH_3$	Mesityl oxide	4-Methylpent-3-en-2-one

General Methods of Preparation of Aldehydes and Ketones

Aldehydes and ketones are prepared from alcohols by two ways *i.e.,* Oxidation of alcohols and Dehydrogenation of alcohols. They are also prepared by various other methods.

1. **Oxidation of primary alcohols:** Primary alcohols undergo oxidation with acid dichromates or potassium dichromate or potassium permanganate and yield aldehydes.

$$\underset{1^\circ \text{ Alcohol}}{RCH_2OH} + [O] \xrightarrow[H^+]{K_2Cr_2O_7 / KMnO_4 / H_2CrO_4} \underset{\text{Aldehyde}}{RCHO} + H_2O$$

$$\underset{\text{Ethanol}}{CH_3CH_2OH} + [O] \xrightarrow[H^+]{K_2Cr_2O_7 / KMnO_4 / H_2CrO_4} \underset{\text{Acetaldehyde}}{CH_3CHO} + H_2O$$

During this reaction the aldehyde obtained is removed immediately from the reaction mixture otherwise the aldehyde formed further undergoes oxidation to give respective carboxylic acids.

$$\underset{\text{Ethanol}}{CH_3CH_2OH} + [O] \xrightarrow[H^+]{K_2Cr_2O_7 / KMnO_4 / H_2CrO_4} \underset{\text{Acetaldehyde}}{CH_3CHO} \xrightarrow[\text{oxidation}]{\text{Further}} \underset{\text{Acetic acid}}{CH_3COOH}$$

The oxidation of primary alcohol is easily stopped at the aldehyde stage if pyridinium chlorochromate (PCC) is used as an oxidizing agent and the reaction is carried out in anhydrous solvent such as dichloromethane.

$$\underset{\text{Ethanol}}{CH_3CH_2OH} + [O] \xrightarrow[CH_2Cl_2]{PCC} \underset{\text{Acetaldehyde}}{CH_3CHO} + H_2O$$

Oxidation of secondary alcohols: Secondary alcohols undergo oxidation to give ketones.

$$\underset{2^\circ \text{ Alcohol}}{(R)(R')CHOH} + [O] \xrightarrow[KMnO_4]{H_2CrO_4/K_2Cr_2O_7} \underset{\text{Ketone}}{(R)(R')C{=}O} + H_2O$$

$$\underset{\text{2-Propanol}}{CH_3CH(OH)CH_3} + [O] \xrightarrow[KMnO_4]{H_2CrO_4/K_2Cr_2O_7} \underset{\text{Acetone}}{CH_3C({=}O)CH_3} + H_2O$$

This method gives high yield of ketones and stops further oxidation.

2. **Catalytic dehydrogenation of alcohols (Removal of two hydrogen atoms):** Alcohol vapours are passed over heated copper or copper oxide at 570 K to yield aldehydes and ketones. It is not a suitable method for laboratory synthesis.

$$\underset{1^\circ \text{ Alcohol}}{RCH_2OH} \xrightarrow{\text{Cu, 570 K}} \underset{\text{Aldehyde}}{RCHO} + H_2$$

$$\underset{\text{Ethanol}}{CH_3CH_2OH} \xrightarrow{\text{Cu, 570 K}} \underset{\text{Acetaldehyde}}{CH_3CHO} + H_2$$

$$\underset{2^\circ \text{ Alcohol}}{RR'CHOH} \xrightarrow{\text{Cu, 570 K}} \underset{\text{Ketone}}{RR'C{=}O} + H_2$$

$$\underset{\text{2-Propanol}}{CH_3CH(OH)CH_3} \xrightarrow{\text{Cu, 570 K}} \underset{\text{Acetone}}{CH_3C(=O)CH_3} + H_2$$

  **INTERESTING FACT/USES OF OXIDATION OF ALCOHOLS**   

Determination of Blood alcohol content

When alcohol passes through the blood, the equilibrium is established between the alcohol content of one's blood and breath. So, if the concentration of any one is known, the other is easily determined.

By using this test, the law enforcement agencies find out approximate concentration of alcohol in person's blood which is based on the oxidation of breath ethanol by sodium dichromate. A more accurate determination is done by using breath analyzer.

Here a small volume of breath is bubbled through an acidic solution of sodium dichromate. The concentration of chromic ion is determined by spectrophotometer. This method is accurate one but time consuming. The main principle is the alcoholised breath undergoes oxidation; the red color of the chromate ion gets converted to green color. Chromate is reduced and alcohol gets oxidized to aldehyde. In this method, a sealed glass tube is used in which the oxidizing agent is impregnated to an inert material. One end is connected to the mouth and another end to a balloon like bag. The person blows into the tube and oxidation takes place where the red color of the chromate ion changes to to green color.

$$\underset{\text{Ethanol}}{CH_3CH_2OH} + \underset{\text{(Red colour)}}{Cr_2O_7^{2-}} \xrightarrow[Na_2Cr_2O_7]{H^+} \underset{\text{Acetaldehyde}}{CH_3CHO} + \underset{\text{Green}}{Cr^{3+}}$$

Another one type of breath analyzer is a small fuel cell in which oxidation of alcohol takes place by oxygen present in the air when blowing the alcohol mixed breath. The oxidation of alcohol generates an electrical current which is proportional to the amount of alcohol present in the blood.

3. **Ozonolysis of alkenes:** Alkenes reacts with ozone (O_3) at low temperature to form ozonide. The double bond in alkene gets cleaved and the carbon atoms that were doubly bonded to each other make themselves doubly bonded to oxygen. The ozonide subsequently cleaves into aldehydes and ketones using different reagents. This kind of oxidation is known as **Ozonolysis**.

$$\underset{\text{Alkene}}{>C{=}C<} \xrightarrow[\text{2) Work up}]{\text{1) 3 - 78 °C}} \underset{\text{Aldehydes and ketones}}{H{-}C{=}O + >C{=}O}$$

Alkene and ozone undergoes concerted cycloaddition reaction. In single step, the oxygen atoms are added to the two sp^2 carbon atoms. The product obtained from ozonolysis is unstable molozonide (One mole of ozone is added to the alkene) which rearranges to give more stable ozonide.

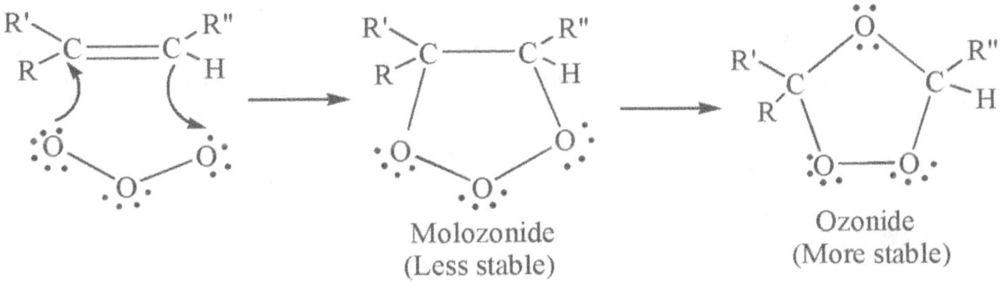

Molozonide
(Less stable)

Ozonide
(More stable)

Ozonides are not isolated since they are explosive in nature (solution state). These are easily cleaved in the presence of reducing agents like zinc or dimethyl sulphide to yield aldehydes or ketones. (Aldehyde is formed if at least one of the substituents present in the sp^2 carbon atom is hydrogen). Ketone is obtained if the sp^2 carbon atom of the alkene is bonded to two carbon containing substituents *i.e.,* alkyl groups).

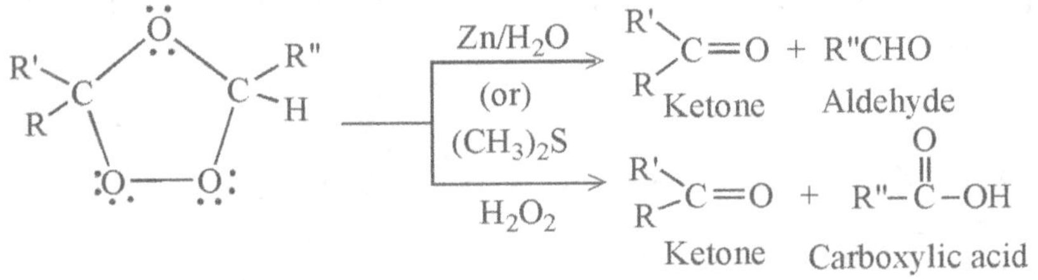

Zn/H_2O (or) $(CH_3)_2S$ → R'(R)C=O + R"CHO
Ketone Aldehyde

H_2O_2 → R'(R)C=O + R"–C(=O)–OH
Ketone Carboxylic acid

Cleaving or breaking of ozonides in the presence of reducing agents is referred to as "working up the ozonide under reducing conditions". Here the product obtained is aldehyde and ketone.

Cleaving or breaking of ozonides in the presence of hydrogen peroxide (oxidizing agent) is referred as "working up the ozonide under oxidizing conditions". Here the products obtained are ketone and carboxylic acids.

Ozonolysis is not a suitable method for the preparation of carbonyl compounds *i.e.,* aldehydes and ketone, because mixture of products are obtained. Only one type of carbonyl compounds are formed when the alkene is symmetrical.

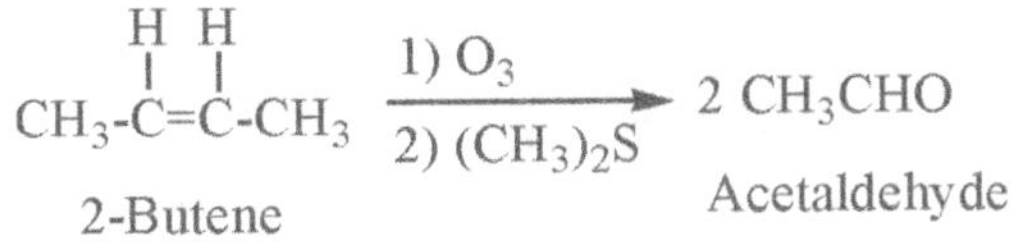

$$CH_3\text{-}CH\text{=}CH\text{-}CH_3 \xrightarrow[2)\ (CH_3)_2S]{1)\ O_3} 2\ CH_3CHO$$

2-Butene → Acetaldehyde

One of most important use of ozonolysis is the determination of position of double bonds in alkenes. For example

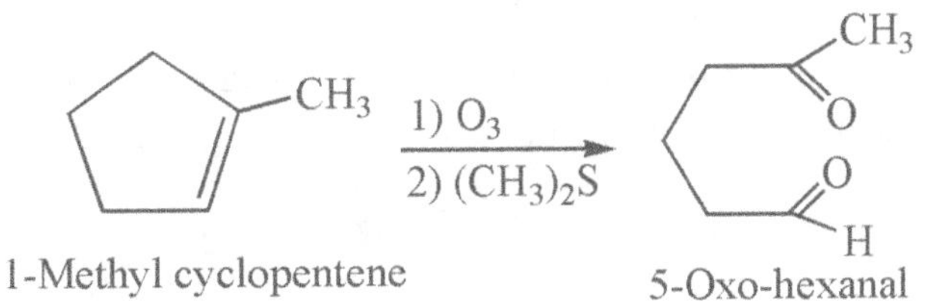

1-Methyl cyclopentene $\xrightarrow[2)\ (CH_3)_2S]{1)\ O_3}$ 5-Oxo-hexanal

4. **Hydration of alkynes:** Alkynes yield carbonyl compounds *via* two kinds of reactions.
   (a) Mercuric ion catalyzed hydration (Addition of water to alkynes).
   (b) By hydroboration - oxidation.
   (a) **Mercuric ion catalyzed hydration:** Alkynes reacts with water in the presence of a mixture of mercuric sulphate in aqueous sulphuric acid to give unstable enol intermediate which undergoes rearrangement to yield carbonyl compounds. The reaction follows Markovnikov's orientation

$$\underset{\text{Acetylene}}{HC\equiv CH} + H\text{-}OH \xrightarrow[H_2SO_4]{HgSO_4} \underset{\text{Unstable enol}}{\left[H-\overset{H}{\overset{|}{C}}=\overset{H}{\overset{|}{C}}-OH\right]} \longrightarrow \underset{\text{Acetaldehyde}}{H-\overset{H}{\overset{|}{\underset{H}{\underset{|}{C}}}}-CHO}$$

$$\underset{\text{1-Propyne}}{CH_3C\equiv CH} + H\text{-}OH \xrightarrow[H_2SO_4]{HgSO_4} \underset{\text{Unstable enol}}{\left[CH_3-\overset{OH}{\overset{|}{C}}=\overset{H}{\overset{|}{C}}-H\right]} \longrightarrow \underset{\text{Acetone}}{CH_3-\overset{O}{\overset{||}{C}}-CH_3}$$

5. **Heating calcium salts of fatty acids:** Calcium salts of fatty acids are heated to yield aldehydes. For example

$$\underset{\text{Calcium acetate}}{(CH_3COO)_2Ca^{2+}} + \underset{\text{Calcium formate}}{Ca(OOCH)_2} \xrightarrow{\Delta} \underset{\text{Acetaldehyde}}{2\,CH_3CHO} + 2\,CaCO_3$$

$$\underset{\text{Calcium acetate}}{(CH_3COO)_2Ca} \xrightarrow[300\,^\circ C]{\Delta} \underset{\text{Acetone}}{CH_3\text{-}\overset{O}{\overset{||}{C}}\text{-}CH_3} + CaCO_3$$

The yield of the reaction is poor. Barium, manganese and thorium salts of fatty acids give better yield than calcium salts.

6. **Catalytic degradation of fatty acids:** The vapors of fatty acids are passed over manganese oxide or thorium oxide at 570 K to give aldehyde or ketones.

$$\underset{\text{Acetic acid}}{CH_3COOH} + \underset{\text{Formic acid}}{HCOOH} \xrightarrow[570\,K]{MnO} \underset{\text{Acetaldehyde}}{CH_3CHO} + CO_2 + H_2O$$

$$\underset{\text{Acetic acid}}{2\,CH_3COOH} \xrightarrow[570\,K]{MnO} \underset{\text{Acetone}}{CH_3\text{-}\overset{O}{\overset{||}{C}}\text{-}CH_3} + CO_2 + H_2O$$

7. **From acid chlorides (Rosenmund's reduction):** Reduction of acid chlorides in the presence of a catalyst palladium suspended in barium sulphate gives aldehydes.

$$\underset{\text{Acetyl chloride}}{CH_3COCl} + H_2 \xrightarrow[\text{Poisoned}]{Pd/BaSO_4} \underset{\text{Acetaldehyde}}{CH_3CHO} + HCl$$

The aldehyde obtained can be further reduced to primary alcohol but the final product of this reaction is not an alcohol. As the barium sulphate used here acts as a poison for the palladium catalyst and inhibits the further reduction of aldehyde to an alcohol. Actually the barium sulphate is poisoned with small quantity of sulphur and quinoline.

8. **Stephen's method (From alkyl cyanides):** Alkyl cyanides or alkyl nitriles are reduced with stannous chloride and hydrochloric acid in the presence of ether (solvent for alkyl nitriles) to yield iminochloride which on further hydrolysis yields aldehyde. Ketones are not prepared by using this method.

$$\underset{\text{Alkyl nitrile}}{R-C\equiv N} \xrightarrow[HCl]{SnCl_2} \underset{\text{Iminochloride}}{RCH=NH.HCl} \xrightarrow{H_2O} \underset{\text{Aldehyde}}{RCHO} + NH_4Cl$$

9. **From Grignard reagent:** Grignard reagent reacts with hydrogen cyanide and alkyl cyanides to give aldehydes and ketones, respectively. During this reaction, addition products are formed which upon hydrolysis gives carbonyl compounds.

$$\underset{\text{Hydrogen cyanide}}{H-C\equiv N} + \underset{\text{Alkyl magnesium iodide}}{Mg\begin{smallmatrix}R\\I\end{smallmatrix}} \longrightarrow \underset{\text{Addition product}}{R-CH=N-MgI} \xrightarrow{2H_2O} \underset{\text{Aldehyde}}{R-CH=O} + Mg\begin{smallmatrix}OH\\I\end{smallmatrix} + NH_3$$

$$\underset{\text{Methyl cyanide}}{H_3C-C\equiv N} + \underset{\text{Methyl magnesium iodide}}{Mg\begin{smallmatrix}CH_3\\I\end{smallmatrix}} \longrightarrow (H_3C)_2C=N-MgI \xrightarrow{2H_2O} \underset{\text{Acetone}}{(H_3C)_2C=O} + Mg\begin{smallmatrix}OH\\I\end{smallmatrix} + NH_3$$

10. **Ketones are prepared from sodium salt of acetoacetic ester by ketonic hydrolysis:** Acetoacetic ester reacts with sodium ethoxide to gives sodium ethyl acetoacetate, which upon further reaction with alkyl halides give alkyl acetoacetic ester. The latter gives corresponding ketones by ketonic hydrolysis with acid. Aldehydes are not prepared by this method.

$$\underset{\text{Ethylacetoacetate}}{H_3C-\overset{O}{\overset{\|}{C}}-\overset{H_2}{C}-\overset{O}{\overset{\|}{C}}-OCH_2CH_3} \xrightarrow[-C_2H_5OH]{\overset{-}{C_2H_5O}\overset{+}{Na}} \underset{\text{Sodium salt of ethylacetoacetate}}{H_3C-\overset{O}{\overset{\|}{C}}-\overset{-}{\underset{Na^+}{\overset{H}{C}}}-\overset{O}{\overset{\|}{C}}-OCH_2CH_3} \xrightarrow[-NaX]{RX}$$

$$\underset{\text{Alkyl derivative of acetoacetic ester}}{H_3C-\overset{O}{\overset{\|}{C}}-\underset{H}{\overset{R}{C}}-\overset{O}{\overset{\|}{C}}-OCH_2CH_3} \xrightarrow[\text{Heat}]{\text{Dil.HCl}} \underset{\text{Methyl/alkyl ketone}}{H_3C-\overset{O}{\overset{\|}{C}}-CH_2R} + CO_2 + \underset{\text{Ethanol}}{CH_3CH_2OH}$$

11. **From Gem - dihalides:** An organic compound in which the two halogen atoms are attached to the same carbon is called as Gem dihalide which undergoes alkaline hydrolysis to give aldehydes.

$$\underset{\text{1,1-Dibromoethane}}{H_3C-\underset{H}{\overset{Br}{C}}-Br} \xrightarrow[\substack{2H_2O\\-2HBr}]{NaOH} \underset{\text{(Less stable)}}{H_3C-\underset{H}{\overset{OH}{C}}-OH} \xrightarrow{-H_2O} \underset{\text{Acetaldehyde}}{CH_3CHO}$$

Aldehydes are prepared from the gem dihalides in which the two halogen atoms are attached to the terminal carbon atom. Ketones are also prepared by similar process but in the gem-dihalides, the two halogen atoms are attached to a non - terminal carbon atom.

$$H_3C-\underset{Br}{\overset{Br}{C}}-CH_3 \xrightarrow[\text{-2HBr}]{\text{NaOH, }2H_2O} H_3C-\underset{OH}{\overset{OH}{C}}-CH_3 \xrightarrow{-H_2O} CH_3\overset{O}{\overset{\|}{C}}CH_3$$

2,2-Dibromopropane (Less stable) Acetone

This method is not much used since aldehydes are affected by alkalis.

12. **From 1,2-glycols or oxidation of 1,2-glycols:** Glycols are oxidised with lead tetra acetate and periodic acid to yield aldehydes and ketones.

$$\begin{array}{l} R-\overset{H}{C}-OH \\ R-\underset{H}{C}-OH \end{array} \xrightarrow[HIO_4]{(O)} 2RCHO + HIO_3 + H_2O$$

1,2-Glycol Aldehyde

$$\begin{array}{l} R_1, R'_1 > C-OH \\ R_2, R'_2 > C-OH \end{array} \xrightarrow{HIO_4} \begin{array}{c} R_1 \\ R'_1 \end{array}\!\!>C=O + \begin{array}{c} R_2 \\ R'_2 \end{array}\!\!>C=O + HIO_3 + H_2O$$

1,2-Diols Ketone

13. **Wacker process (From alkenes):** Alkenes react with acidified aqueous solution of palladium chloride and cupric chloride to give aldehydes and ketones.

$$CH_2{=}CH_2 + PdCl_2 + H_2O \xrightarrow{CuCl_2} CH_3CHO + Pd + 2HCl$$

Ethylene Acetaldehyde

$$CH_3CH{=}CH_2 + PdCl_2 + H_2O \xrightarrow{CuCl_2} CH_3COCH_3 + Pd + 2HCl$$

Propylene Acetone

The $PdCl_2$ is regenerated at the end of the reaction by the reaction between palladium and HCl in the presence of cupric chloride.

$$Pd + 2HCl \xrightarrow{CuCl_2} PdCl_2 + H_2$$

14. **Synthesis of aldehydes by oxo process:** Alkenes react with hydrogen and carbon monoxide in the presence of a catalyst cobalt carbonyl to give aldehydes. This process is known as oxo process which is the one of the important method for industrial preparation of aldehydes.

$$RCH = CH_2 + CO + H_2 \xrightarrow{[CO(Co)_4]_2} RCH_2\text{-}\underset{CHO}{CH_2}$$

Alkenes Aldehyde

The addition follows anti Markovnikov's rule. Ketones are not prepared by this method.

15. **Synthesis of phenyl aldehydes and ketones:** Done by Friedel-Crafts acylation:

Summary of Methods of Preparation

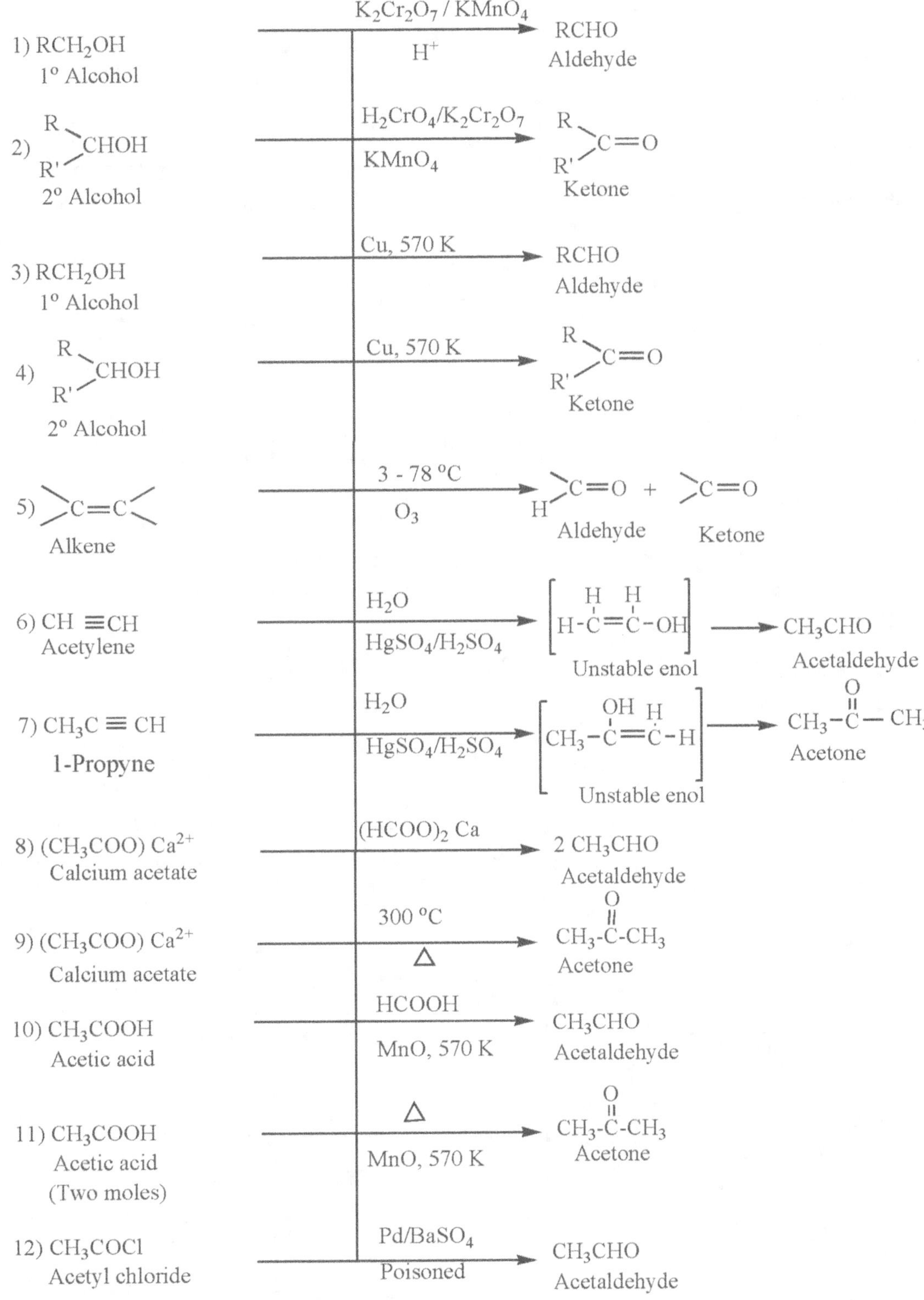

13) $R-\overset{O}{\overset{\|}{C}}-OH$
Carboxylic acid

14) RCOCl
Acid chloride (2 moles)

15) $R-C\equiv N$
Alkyl nitrile

16) $H-C\equiv N$
Hydrogen cyanide

17) $CH_3-C\equiv N$
Methyl cyanide

18) $H_3C-\overset{O}{\overset{\|}{C}}-\overset{H}{C}-\overset{O}{\overset{\|}{C}}-OCH_2CH_3$ Na^+
Sodium salt of ethyl aceto acetate

19 $H_3C-\overset{Br}{\overset{|}{C}}(H)-Br$
1,1-Dibromoethane

20 $H_3C-C(Br)_2-CH_3$
2,2-Dibromopropane

21 $R-CH(OH)-CH(OH)-R$
1,2-Glycol

22 $R_1R'_1C(OH)-C(OH)R_2R'_2$
1,2-Diols

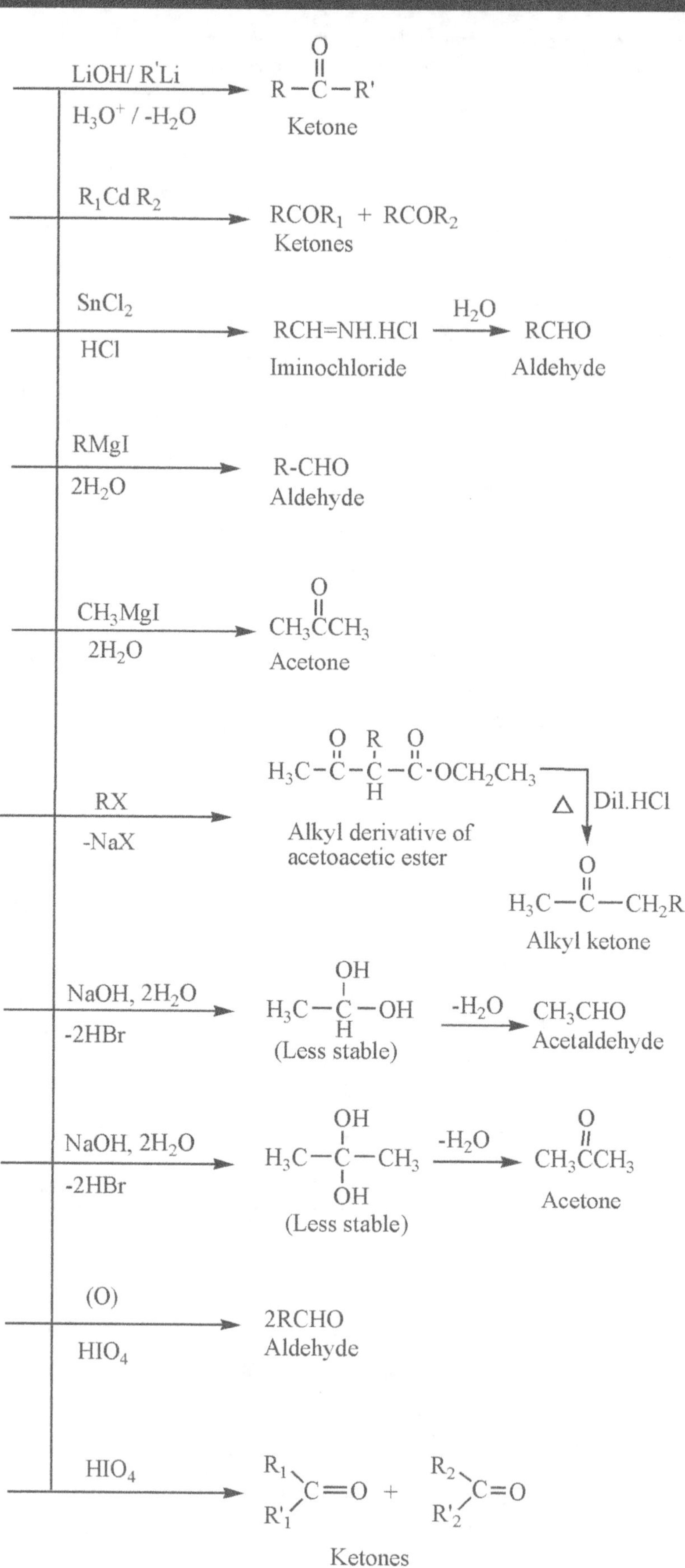

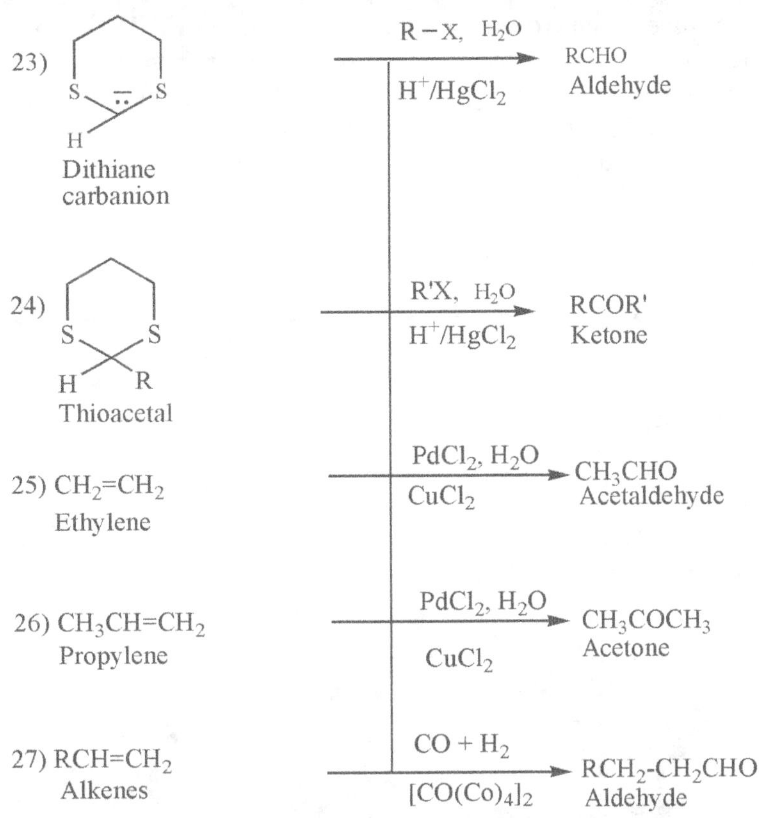

Physical Properties

1. Formaldehyde is a gas (B.P. 21 °C) at room temperature. Acetaldehyde and other low molecular weight aldehydes and ketones are colourless liquids.
2. Lower molecular weight aldehydes have unpleasant pungent smell whereas the ketones have pleasant and sweet odours.
3. Aldehydes and ketones are less dense than water.
4. **Boiling point of aldehydes and ketones:** The carbonyl group present in the aldehydes and ketones make them highly polar and hence they possess high boiling point than non polar compounds of comparable molecular weight. They are also having lower boiling point than comparable alcohol and carboxylic acids because they themselves are not capable of forming intermolecular hydrogen bonding. For example: *n*-butaraldehyde (B.P. 76 °C) compared to *n*-butyl alcohol (B.P 118 °C)
5. **Solubility:** Lower molecular weight aldehydes and ketones are soluble in water due to hydrogen bonding. Aldehydes and ketones with more than five carbon atoms are less soluble in water and are soluble in organic solvents.

   Structure

   Carbonyl groups in aldehydes and ketones have double bond between carbon and oxygen. One of these bonds is σ bond which is formed by overlap of sp^2 orbital of carbon while another bond is π bond which is formed by overlap of unhybridised p orbital of carbon and oxygen. The remaining two sp^2 orbitals of carbon form two σ bonds with other groups of atoms attached to it. The three sp^2 orbitals of carbon are inclined to each other at angle of 120° as shown in figure.

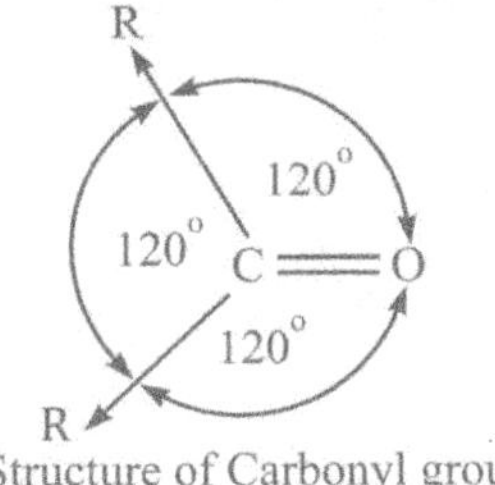

Structure of Carbonyl group

The oxygen atom has two unpaired electrons. The average C-X bond distance is 1.22 A°, which is shorter than typical C-O bond (1.41). As the oxygen is highly electronegative, the electron density of both σ and π bonds of (C-O) is shifted to oxygen. Hence, the carbonyl group is polarized and the carbon atom gets partial positive charge and the oxygen atom gets partial negative charge. This polarity is confirmed by its dipole moment value 2.7 D.

$$>\overset{\delta^+}{C}=\overset{\delta^-}{O}$$

Chemical Properties

Aldehydes and ketones undergo following type of chemical reactions,

I. Nucleophilic addition reactions.
II. Addition - Elimination reactions.
III. Oxidation of aldehydes and ketones.
IV. Reduction reactions of aldehydes and ketones.
V. Condensation reactions.

Butanedione

$$CH_3-\overset{O}{\overset{\|}{C}}-\overset{O}{\overset{\|}{C}}-CH_3$$

You may be able to smell butanedione by sniffing your armpits or someone's unwashed feet, because it is a contributor to the odor of fermenting perspiration. Fresh sweat is almost odorless, but the action of the bacterium *Streptococcus albus*, which is present on the skin, increases its acidity and makes an inviting feast for other bacteria; they, in turn, excrete pungent compounds including butanendione.

The carbonyl group of an aldehyde and ketone is highly polar. Due to inductive effect, partial positive charged carbon and partially negative charged oxygen atom is formed. So the partially positive charged carbon is attacked by nucleophiles and partially negative charged oxygen atom is attacked by electrophiles as shown below.

$$>\overset{\delta^+}{C}=\overset{\delta^-}{O}$$

O (δ^-) → Susceptible by the attack of electrophile and acids

C (δ^+) → Susceptible by the attack of nucleophile and bases

I. Nucleophilic addition reactions:

Relative reactivity of carbonyl compounds: The carbonyl group (C=O) present in the carbonyl compounds is polar because oxygen atom is more electronegative than carbon. So the oxygen atom pulls the electrons towards itself which produces partial positive and partial negative charges on the C=O bond. Due to the presence of the partial positive charges on the carbonyl carbon, it becomes a site for nucleophilic attack.

$$-\underset{|}{C}=\ddot{O} \longleftrightarrow -\overset{+}{\underset{|}{C}}-O^- \equiv -\underset{|}{\overset{\delta^+}{C}}=\overset{\delta^-}{O}$$

The alkyl group present in the ketone is electron releasing when compared to hydrogen group present in the aldehyde molecule, so an aldehyde molecule possess a greater partial positive charge on its carbonyl carbon than a ketone molecule which results in more reactivity of an aldehyde than a ketone molecule towards nucleophilic attack. The order of relative reactivities of aldehyde and ketones is as follows.

$H-CO-H$ > $H_3C-CO-H$ > $R-CO-R'$

Formaldehyde Acetaldehyde Ketone

The greater reactivity of aldehydes is also due to the contribution of steric factors. As the hydrogen atom attached to the carbonyl group of an aldehyde molecule is smaller than that of the alkyl group present in the carbonyl carbon of a ketone (steric hindrance) therefore the carbonyl carbon of an aldehyde easily accepts the nucleophile than the carbonyl carbon of a ketone.
Similarly, ketones with higher number of alkyl groups are less reactive than ketones with smaller number alkyl group (due to steric hindrance). The order of relative reactivities of ketones is

$H_3C-CO-CH_3$ > $H_3C-CO-CH(CH_3)CH_3$ > $H_3C(CH_3)CH-CO-CH(CH_3)CH_3$

Acetone Isopropylmethyl ketone Diisopropyl ketone (or) 2,4-Dimethyl-3-pentanone

Comparison of relative reactivities of aldehydes and ketones with carboxylic acid and their derivatives.

Relative reactivities of carbonyl compounds towards nucleophiles is as follows
Acyl halide > acid anhydride > aldehyde > ketone > ester > carboxylic acid > amide

$$RCOX > RCOOCOR > RCHO > RCOR > RCOOR > RCOOH > RCONR_2$$

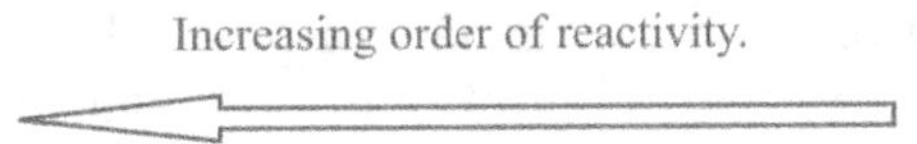
Increasing order of reactivity.

In the above scale, the aldehydes and ketones are in the right middle. *i.e.* They are less reactive than acyl halides and anhydrides but more reactive than esters, carboxylic acids and amides.
Except the aldehydes and ketones, all other carbonyl compounds such as RCOX, RCOOH, RCOOR, RCOOCOR, $RCONR_2$ possess a lone pair of electrons on an atom which is attached to the carbonyl group. These electrons are shared with the carbonyl carbon by resonance, which makes it less electron deficient. So, the reactivity of these carbonyl compounds depends upon their basicity of "Y" as indicated below.
When the Y^- is the weaker base, the reactivity of carbonyl group is more because the weak bases are better able to withdraw the electrons by inductive effect from the carbon but they are less able to donate electrons by resonance.

$R-C(=O)-\ddot{Y}$ ⟷ $R-C(-O^-)=Y^+$

So, in acyl halides and acid anhydrides the Y is weaker base than aldehydes and ketones, hence aldehydes and ketones are less reactive than acid halides and acid anhydrides. Aldehydes and ketones are more reactive than carboxylic acids, esters and amides because the nature of "Y" in the latter compounds is relatively strong base.

Reasons for nucleophilic addition reactions of aldehydes and ketones but nucleophilic acyl substitution reactions for carboxylic acid derivatives.

$$R-\overset{O}{\overset{\|}{C}}-R' + HZ \longrightarrow R-\overset{OH}{\overset{|}{\underset{Z}{\underset{|}{C}}}}-R'$$

Ketone Addition product

The carbonyl carbon of aldehyde or ketone is attached to H atom or alkyl group respectively which are too strong base and cannot be easily eliminated under normal conditions so it cannot be substituted by another group. Consequently, aldehydes and ketones react with nucleophiles to give addition products not substitution products. The nucleophile adds to the aldehyde or ketone to yield an alkoxide ion. During this reaction, the planar sp^2 hybridized carbon atom of the carbonyl group changes to a tetrahedral sp^3 hybridized state in the product.

$$\ddot{Nu} + R_2C^{\delta+}=O^{\delta-} \rightleftharpoons R_2C(Nu)-\ddot{\underset{..}{O}}:^- \overset{H^+}{\rightleftharpoons} R_2C(Nu)-\underset{..}{O}:-H$$

Alkoxide Addition product

Acidity of α-hydrogens: The hydrogen atoms attached to a carbon atom adjacent to the functional group is called α-hydrogens. The α-hydrogen bonded to a carbon adjacent to a carbonyl carbon is so much acidic than hydrogen bonded to other sp^3 hybridised carbons. The reason is that the base obtained after removal of proton from the α-carbon of the carbonyl group is more stable when compared with the base obtained from other sp^3 hybridised carbon and the acid strength is determined by the stability of the conjugate base which is obtained when the acid loses its proton.

When a proton is removed from a carbon atom adjacent to the carbonyl group, two factors are responsible to increase the stability of the base. First one is, the electrons left behind when the proton is removed are delocalized and this delocalization increases the stability of the base. The electrons are delocalized on to oxygen, an atom that is better able to accommodate the electrons because it is more electronegative than carbon.

$$R-\overset{H}{\overset{|}{C}H}-\overset{O}{\overset{\|}{C}}-R \overset{-H^+}{\rightleftharpoons} R-\overset{..}{\overset{-}{C}H}-\overset{O}{\overset{\|}{C}}-R \longleftrightarrow R-CH=\overset{O^-}{\overset{|}{C}}-R$$

(Resonance contributors)

But if a proton is removed from ethane, the electrons left behind are present on a carbon atom only, and the resulting carbanion is unstable and hence difficult to form. So, the pka value of its conjugate acid is very high.

$$CH_3-CH_3 \rightleftharpoons CH_3\ddot{C}H_2 + H^+$$

Carbanion

(Unstable)

Why aldehydes and ketones are more acidic than esters? The electrons left behind when an α-hydrogen is removed from an ester is not delocalized on to the carbonyl oxygen like aldehydes and ketones. Hence esters are less acidic than aldehydes and ketones.

$$R-\ddot{C}H^{-}-C(=\ddot{O})-\ddot{O}R \longleftrightarrow R-\ddot{C}H-C(-\ddot{O}:)=\overset{+}{\ddot{O}}R$$

Delocalisation of non bonding electrons on oxygen

(Resonance contributors)

Delocalisation electrons on the α–carbon

$$R-CH=C(-\ddot{O}:^{-})-\ddot{O}R$$

If the α-carbon is between two carbonyl groups, the acidity of α-hydrogen is greater. Let us consider the following examples, an α-hydrogen of ethyl 3-oxobutyrate and 2,4-pentanedione

$$CH_3-\overset{O}{\overset{\|}{C}}-CH_2-\overset{O}{\overset{\|}{C}}-OC_2H_5$$

pka = 10.7
(-CH_2 group flanked between one ketonic carbonyl group and ester carbonyl group)

$$CH_3-\overset{O}{\overset{\|}{C}}-CH_2-\overset{O}{\overset{\|}{C}}-CH_3$$

pka = 8.9
(-CH_2 group is flanked between two ketonic carbonyl group)

The α-hydrogen present in the above examples are more acidic because the electrons left behind when the proton is removed can be delocalized on to two oxygen atoms. α-Diketones have lower pka values than α-ketoesters because electrons are more readily delocalized on to ketone carbonyl groups than they are on to ester carbonyl groups.

$$CH_3-C(=\ddot{O})-CH=C(-\ddot{O}^{-})-CH_3 \longleftrightarrow CH_3-C(=\ddot{O})-\ddot{C}H^{-}-C(=\ddot{O})-CH_3 \longleftrightarrow CH_3-C(-\ddot{O}^{-})=CH-C(=\ddot{O})-CH_3$$

General mechanism for Nucleophilic addition reaction: It takes place in two steps.

***Step 1:* Addition of the nucleophile:** The nucleophile attacks the carbonyl carbon leads to a formation of new σ bond. Once the σ bond is formed between the Nu (nucleophile) and carbonyl carbon, the bond between the carbon and oxygen is broken, the oxygen gets its bonded electrons and attains negative charge.

$$-C=O + \ddot{N}u^{-} \longrightarrow Nu-C-\ddot{O}:^{-}$$

Step 2: **Protonation:** The electrophile or proton attacks the negatively charged oxygen to yield the addition product.

$$Nu-\overset{|}{\underset{|}{C}}-\ddot{O}:^{-} + H^{+} \longrightarrow Nu-\overset{|}{\underset{|}{C}}-\ddot{O}-H$$

Addition product

The above said addition reaction is catalysed by acids or bases. According to the nature of the catalyst used, the addition reaction is classified into two types.

(A) Base catalysed addition (Strong nucleophile): Strong bases change the weak nucleophile to strong by removal of its proton.

Step 1:

(i) Conversion of weak nucleophile to strong nucleophile by strong base.

$$\underset{\text{Weak nucleophile}}{Nu-H} + B: \longrightarrow \underset{\text{Strong nucleophile}}{\ddot{N}u^{-}} + BH$$

(ii) Attack of strong nucleophile to the carbonyl carbon.

$$>C=O + \underset{\text{Strong nucleophile}}{\ddot{N}u^{-}} \longrightarrow Nu-\overset{|}{\underset{|}{C}}-\ddot{O}:^{\ominus}$$

Step 2: Protonation.

$$Nu-\overset{|}{\underset{|}{C}}-\ddot{O}:^{\ominus} \xrightarrow{H^{+}} Nu-\overset{|}{\underset{|}{C}}-\ddot{O}-H$$

(B) Acid catalysed addition (Acidic conditions or Weak nucleophile):

Step 1: Protonation

The proton from the acid attacks the carbonyl oxygen to yield a protonated carbonyl group.

$$>C=\ddot{O}: + H\text{-}Nu \longrightarrow \left[>C=\overset{+}{O}-H \longleftrightarrow >\overset{+}{C}-\ddot{O}-H\right] + Nu^{-}$$

Protonated carbonyl group

Step 2: Addition of Nucleophile

$$\underset{\text{Protonated carbonyl group}}{>C=\overset{+}{O}-H} + \ddot{N}u^{-} \longrightarrow \underset{\text{Addition product}}{Nu-\overset{|}{\underset{|}{C}}-O-H}$$

The product obtained is the same for both base and acid catalysed addition.

Types of Nucleophilic addition:

(a) Addition of carbon nucleophiles.

(b) Addition of oxygen nucleophiles.

(a) Addition of carbon nucleophiles: The addition of carbon nucleophiles to the carbonyl compound is an example of an organic reaction which involves the formation of new carbon-

(i) Addition of hydrogen cyanide.

(ii) Addition of Grignard reagents.

(iii) Addition of hydride ions.

(iv) Addition of acetylide ions.

(i) Addition of hydrogen cyanide: Hydrogen cyanide adds to aldehydes and ketones to forms cyanohydrins. The product obtained has one carbon atom more than that of starting aldehyde or ketone.

$$\underset{\text{Aldehyde}}{R-\overset{O}{\overset{\|}{C}}-H} + HCN \longrightarrow \underset{\text{Cyanohydrin}}{R-\overset{OH}{\overset{|}{\underset{CN}{\underset{|}{C}}}}-H}$$

$$\underset{\text{Acetaldehyde}}{H_3C-\overset{O}{\overset{\|}{C}}-H} + HCN \rightleftharpoons \underset{\text{Acetaldehyde cyanohydrin}}{H_3C-\overset{OH}{\overset{|}{\underset{CN}{\underset{|}{C}}}}-H}$$

$$\underset{\text{Ketone}}{R-\overset{O}{\overset{\|}{C}}-R'} + HCN \longrightarrow \underset{\text{Ketone cyanohydrin}}{R-\overset{OH}{\overset{|}{\underset{CN}{\underset{|}{C}}}}-R'}$$

$$\underset{\text{Acetone}}{H_3C-\overset{O}{\overset{\|}{C}}-CH_3} + HCN \rightleftharpoons \underset{\text{Acetone cyanohydrin}}{H_3C-\overset{OH}{\overset{|}{\underset{CN}{\underset{|}{C}}}}-CH_3}$$

(ii) Addition of Grignard reagent: Grignard reagents on addition to the carbonyl compounds give new carbon- carbon bonds which leads to variety of products. The Grignard reagent needed for this reaction is prepared by the reaction between alkyhalide and magnesium turnings in diethyl ether.

$$\underset{\text{Aldehyde}}{R-\overset{O}{\overset{\|}{C}}-H} + \underset{\text{Grignard reagent}}{RMgX} \longrightarrow \left[R-\overset{OMgX}{\overset{|}{\underset{R}{\underset{|}{C}}}}-H\right] \xrightarrow{H^+/H_2O} \underset{\text{Alcohol}}{R-\overset{OH}{\overset{|}{\underset{R}{\underset{|}{C}}}}-H} + Mg\begin{smallmatrix}X\\OH\end{smallmatrix}$$

$$\underset{\text{Ethyl bromide}}{CH_3CH_2Br} \xrightarrow[Et_2O]{Mg} \underset{\text{Ethyl magnesium bromide}}{CH_3CH_2MgBr}$$

Mechanism:

Step 1: Formation of an alkoxide ion

Carbonyl carbon of aldehyde or ketone reacts with Grignard regents to give an alkoxide ion which is complexed with magnesium ion.

$$\underset{\text{Aldehyde}}{R-\overset{\overset{\ominus}{O}}{\overset{\|}{\underset{\oplus}{C}}}-H} + \overset{-\ +}{RMgX} \longrightarrow \underset{\text{Alkoxide ion complexed with Mg}}{\left[R-\overset{OMgX}{\overset{|}{\underset{R}{\underset{|}{C}}}}-H\right]}$$

Step 2: Acid hydrolysis of this complex gives alcohol.

$$R-\overset{OMgX}{\overset{|}{\underset{R}{\underset{|}{C}}}}-H \xrightarrow[\text{Hydrolysis}]{H^+/H_2O} \underset{\text{Alcohol}}{R-\overset{OH}{\overset{|}{\underset{R}{\underset{|}{C}}}}-H} + Mg\langle^{OH}_{X}$$

Application: Based on aldehyde or ketone used various alochols are prepared as shown below.

(a) Formaldehyde gives primary alcohols and other aldehydes gives 2° alcohols while ketones give 3° alcohols.

$$\underset{\text{Formaldehyde}}{H-\overset{O}{\overset{\|}{C}}-H} + \underset{\text{Methyl magnesium iodide}}{CH_3MgI} \longrightarrow \underset{\text{Alkoxide ion complexed with magnesium}}{\left[H-\overset{OMgI}{\overset{|}{\underset{CH_3}{\underset{|}{C}}}}-H\right]} \xrightarrow[\text{Hydrolysis}]{H^+/H_2O} \underset{\text{Ethanol}}{H-\overset{OH}{\overset{|}{\underset{CH_3}{\underset{|}{C}}}}-H} + Mg\langle^{OH}_{I}$$

$$\underset{\text{Acetaldehyde}}{H_3C-\overset{O}{\overset{\|}{C}}-H} + CH_3MgI \longrightarrow \left[H_3C-\overset{OMgI}{\overset{|}{\underset{CH_3}{\underset{|}{C}}}}-H\right] \xrightarrow[\text{Hydrolysis}]{H^+/H_2O} \underset{\text{2-Propanol}}{H_3C-\overset{OH}{\overset{|}{\underset{CH_3}{\underset{|}{C}}}}-H} + Mg\langle^{OH}_{I}$$

$$\underset{\text{Acetone}}{H_3C-\overset{O}{\overset{\|}{C}}-CH_3} + CH_3MgI \longrightarrow \left[H_3C-\overset{OMgI}{\overset{|}{\underset{CH_3}{\underset{|}{C}}}}-CH_3\right] \xrightarrow[\text{Hydrolysis}]{H^+/H_2O} \underset{\text{3°-Butanol}}{H_3C-\overset{OH}{\overset{|}{\underset{CH_3}{\underset{|}{C}}}}-CH_3} + Mg\langle^{OH}_{I}$$

(iii) ***Addition of hydride ion:*** Aldehydes and ketones react with hydride ions to give an alkoxide ion. Subsequent protonation of the later by weak acid or proton gives an alcohol. The overall reaction is the addition of H_2 to the carbonyl group.

$$\underset{\text{Aldehyde}}{R-\overset{O}{\overset{\|}{C}}-H} \xrightarrow{H^{-}} \underset{\text{Alkoxide ion}}{R-\overset{O^-}{\overset{|}{\underset{H}{\underset{|}{C}}}}-H} \underset{}{\overset{H^+}{\rightleftharpoons}} \underset{\text{1° Alcohol}}{R-\overset{OH}{\overset{|}{\underset{H}{\underset{|}{C}}}}-H}$$

$$R-\overset{O}{\overset{\|}{C}}-R' \xrightarrow{H^-} R-\overset{O^-}{\overset{|}{\underset{H}{\underset{|}{C}}}}-R' \underset{}{\overset{H^+}{\rightleftharpoons}} R-\overset{OH}{\overset{|}{\underset{H}{\underset{|}{C}}}}-R'$$

Ketone Alkoxide ion 2° Alcohol

$$H_3C-\overset{O}{\overset{\|}{C}}-H \xrightarrow{H^-} H_3C-\overset{O^-}{\overset{|}{\underset{H}{\underset{|}{C}}}}-H \overset{H^+}{\rightleftharpoons} H_3C-\overset{OH}{\overset{|}{\underset{H}{\underset{|}{C}}}}-H$$

Acetaldehyde Alkoxide ion Ethanol

$$H_3C-\overset{O}{\overset{\|}{C}}-CH_3 \xrightarrow{H^-} H_3C-\overset{O^-}{\overset{|}{\underset{H}{\underset{|}{C}}}}-CH_3 \overset{H^+}{\rightleftharpoons} H_3C-\overset{OH}{\overset{|}{\underset{H}{\underset{|}{C}}}}-CH_3$$

Acetone Isopropanol

(iv) ***Addition of acetylide ion:*** Acetylide ions are added to the aldehydes and ketones to give alkoxide ion which upon further protonation with acid gives the product.

$$CH_3CH_2\overset{O}{\overset{\|}{C}}H + CH_3C\equiv\bar{C}: \longrightarrow CH_3CH_2\overset{O^\ominus}{\overset{|}{\underset{H}{\underset{|}{C}}}}C\equiv CCH_3 \xrightarrow{H^+} CH_3CH_2\overset{OH}{\overset{|}{C}}HC\equiv CCH_3$$

Propanaldehyde Alkoxide ion 4-Hydroxy-2-hexyne

Note: Acetylide ion required for this reaction is prepared by the reaction between alkynes (acetylene) and sodamide in the presence of ammonia.

$$CH_3C\equiv CH \xrightarrow[NH_3]{NaNH_2} CH_3C\equiv\bar{C}:$$

Propyne Acetylide ion

(b) Addition of oxygen Nucleophiles:

(i) Addition of water.
(ii) Addition of alcohols.
(iii) Addition of bisulphite.

(i) **Addition of water:** Water adds to aldehydes and ketones to give hydrate or 1,1-diols or gem-diols. The reaction is reversible one and the equilibrium preferably exists on left side *i.e.,* carbonyl carbon side.

$$R-\overset{O}{\overset{\|}{C}}-H + H_2O \overset{H^+}{\rightleftharpoons} R-\overset{OH}{\overset{|}{\underset{OH}{\underset{|}{C}}}}-H$$

Aldehyde Hydrate or 1,1-Diol or Gem-diol

$$R-\overset{O}{\overset{\|}{C}}-R + H_2O \overset{H^+}{\rightleftharpoons} R-\overset{OH}{\overset{|}{\underset{OH}{\underset{|}{C}}}}-R'$$

Ketone Hydrate or diol

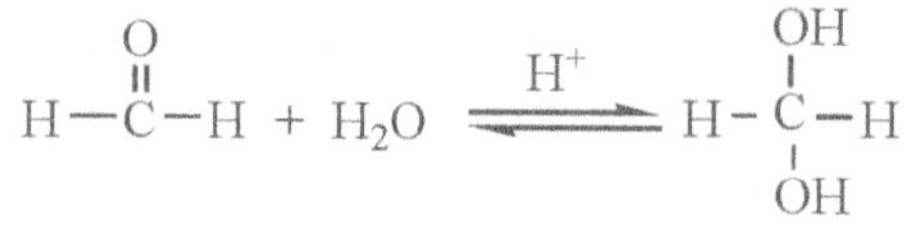

$$H-\overset{\overset{O}{\|}}{C}-H + H_2O \underset{}{\overset{H^+}{\rightleftharpoons}} H-\underset{\underset{OH}{|}}{\overset{\overset{OH}{|}}{C}}-H$$

Formaldehyde — Formalin

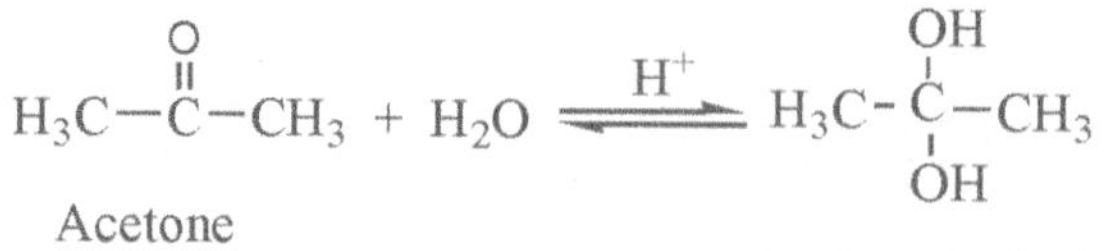

$$H_3C-\overset{\overset{O}{\|}}{C}-CH_3 + H_2O \overset{H^+}{\rightleftharpoons} H_3C-\underset{\underset{OH}{|}}{\overset{\overset{OH}{|}}{C}}-CH_3$$

Acetone — Acetone hydrate (2,2-Propanediol)

Mechanism: The reaction takes place in the presence of acidic or basic condition.

In acidic condition:

Step 1: Protonation.

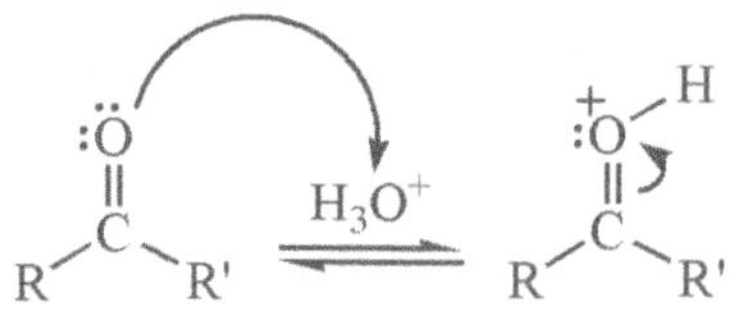

$$R-\overset{\overset{:\ddot{O}}{\|}}{C}-R' \overset{H_3O^+}{\rightleftharpoons} R-\overset{\overset{:\overset{+}{O}-H}{\|}}{C}-R'$$

Step 2: Addition of water molecule.

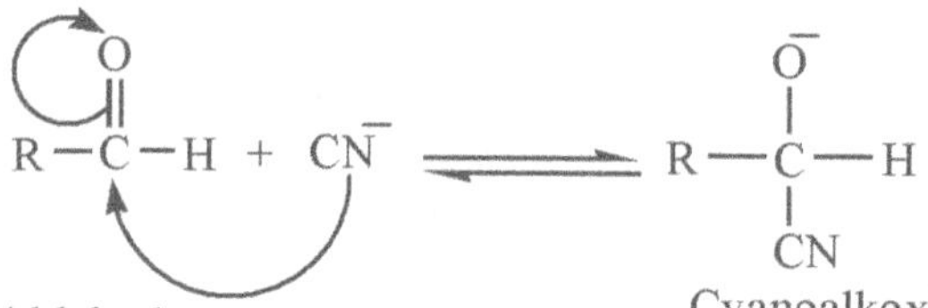

$$R-\overset{\overset{O}{\|}}{C}-H + \overline{CN} \rightleftharpoons R-\underset{\underset{CN}{|}}{\overset{\overset{\overline{O}}{|}}{C}}-H$$

Aldehyde — Cyanoalkoxide ion

Step 3: Deprotonation.

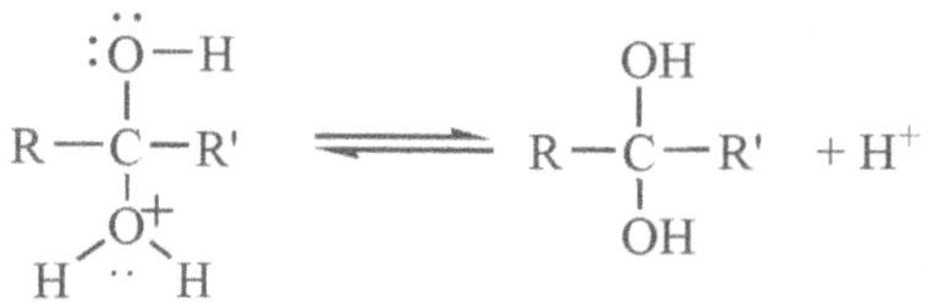

$$R-\underset{\underset{H\ \overset{+}{\ddot{O}}\ H}{|}}{\overset{\overset{:\ddot{O}-H}{|}}{C}}-R' \rightleftharpoons R-\underset{\underset{OH}{|}}{\overset{\overset{OH}{|}}{C}}-R' + H^+$$

1,1-Diol

In basic condition:

Step 1: Addition of hydroxide ion.

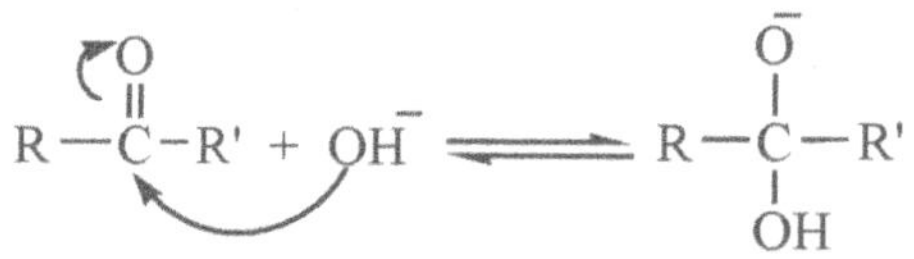

$$R-\overset{\overset{O}{\|}}{C}-R' + \overline{OH} \rightleftharpoons R-\underset{\underset{OH}{|}}{\overset{\overset{\overline{O}}{|}}{C}}-R'$$

1,1-Diol

Step 2: Protonation.

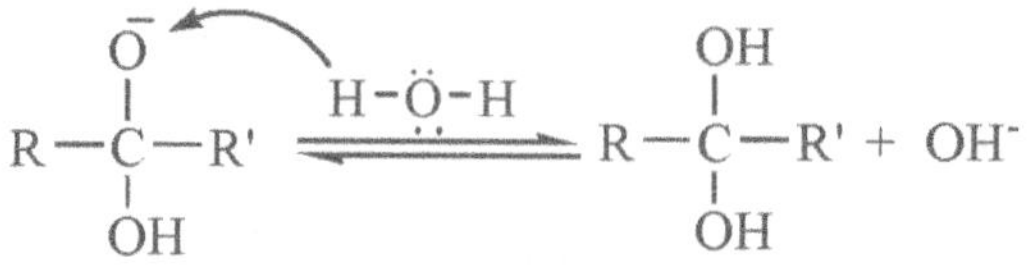

$$R-\underset{\underset{OH}{|}}{\overset{\overset{\overline{O}}{|}}{C}}-R' \overset{H-\ddot{O}-H}{\rightleftharpoons} R-\underset{\underset{OH}{|}}{\overset{\overset{OH}{|}}{C}}-R' + OH^-$$

Diol

Note: In aqueous solutions, the reaction depends upon the following

Nature of the carbonyl group: The extent to which the aldehyde or ketone is hydrated depends upon the nature of aldehyde and ketone only. For example- At equilibrium condition, only 0.2 % of acetone is hydrated, but 99.9 % formaldehyde gets hydrated. The difference in the equilibrium constant for this reaction depends upon the relative stabilities of carbonyl compound and the hydrate.

According to this concept, the alkyl groups present in the carbonyl carbon increases its stability but decreases the stability of hydrate. So, the alkyl groups present in the carbonyl compounds shifts the equilibrium towards left. So, formaldehyde is more reactive than acetone.

$$\underset{\text{Formaldehyde}}{H-\overset{\overset{O}{\|}}{C}-H} \qquad \underset{\text{Acetaldehyde}}{H_3C-\overset{\overset{O}{\|}}{C}-H} \qquad \underset{\text{Acetone}}{H_3C-\overset{\overset{O}{\|}}{C}-CH_3}$$

Stability increases (stability of aldehydes and ketones) →

$$\underset{\text{Formalin}}{H-\overset{OH}{\underset{OH}{C}}-H} \qquad \underset{\text{Acetaldehyde hydrate}}{H_3C-\overset{OH}{\underset{OH}{C}}-H} \qquad \underset{\text{Acetone hydrate}}{H_3C-\overset{OH}{\underset{OH}{C}}-CH_3}$$

Stability decreases (stability of hydrates) →

 **INTERESTING FACT**  

Preservation of Biological specimens:

A 37% aqueous solution of formaldehyde is used as formalin solution which is commonly used for preserving biological specimens.

In our body, chloral is converted to trichloro ethanol that is responsible for the drugs sleep inducing effect. It acts as a sedative but produces lethal effect.

(ii) Addition of alcohols: Aldehydes and ketones react with alcohols to give addition products known as hemiacetal and hemiketal respectively. If one molecule of alcohol is added to an aldehyde, the product is called hemiacetal. If two molecules of alcohol are added to an aldehyde, it is called acetal. Similarly, the product obtained from the addition of alcohol to the ketone molecule is known as hemiketal and ketal.

The addition of alcohol to aldehydes and ketones is always a reversible reaction. The equilibrium always lies in the left-hand side of the equation *i.e.,* carbonyl carbon side. Alcohol is also a weak nucleophile like water, so the rate of the reaction is increased by acid catalyst.

$$\underset{\text{Aldehyde}}{R-\overset{\overset{O}{\|}}{C}-H} + \underset{\text{Alcohol}}{R'OH} \overset{H^+}{\rightleftharpoons} \underset{\text{Hemiacetal}}{R-\overset{OR'}{\underset{OH}{C}}-H} \overset{R''OH,\ H^+}{\rightleftharpoons} \underset{\text{Acetal (carbon with OR' or OR'')}}{R-\overset{OR'}{\underset{OR''}{C}}-H} + H_2O$$

$$\underset{\text{Acetaldehyde}}{H_3C-\overset{\overset{O}{\|}}{C}-H} + \underset{\text{Methanol}}{CH_3OH} \overset{H^+}{\rightleftharpoons} \underset{\text{Hemiacetal}}{H_3C-\overset{OCH_3}{\underset{OH}{C}}-H} \underset{H^+}{\overset{CH_3OH}{\rightleftharpoons}} \underset{\text{Acetal}}{H_3C-\overset{OCH_3}{\underset{OCH_3}{C}}-H} + H_2O$$

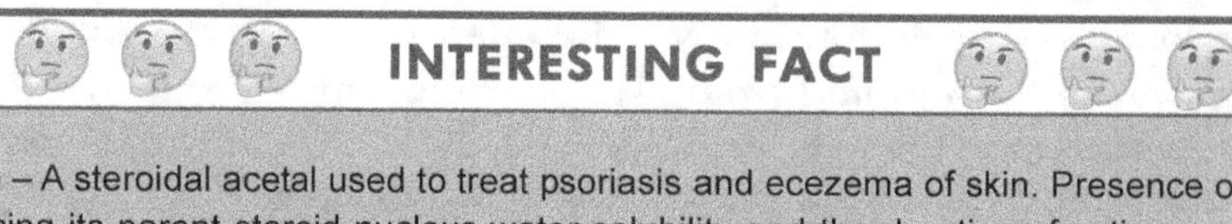

Flucinolone acetonide – A steroidal acetal used to treat psoriasis and ecezema of skin. Presence of acetal group increases its potency by enhancing its parent steroid nucleus water solubility and the duration of action.

Note: Carbohydrates and sugars are having cyclic acetal structures. For example, Gluose – Cyclic hemiacetal form, Lactose- disaccharide that has one acetal and one hemiacetal form.

Cyclic hemiacetal form

Glucose

Acetal form

Hemiacetal

Lactose

(iii) Addition of bisulphite: Solid sodium bisulphite reacts with aldehydes and ketone to give addition products.

For Aldehydes:

$$R-\overset{O}{\overset{\|}{C}}-H \xrightarrow[H_2O]{Na^+HSO_3^-} R-\overset{OH}{\overset{|}{\underset{H}{\underset{|}{C}}}}-SO_3Na$$

Aldehyde — Sodium bisulphite (Crystalline compound)

$$H_3C-\overset{O}{\overset{\|}{C}}-H \xrightarrow{Na^+HSO_3^-} H_3C-\overset{OH}{\overset{|}{\underset{SO_3Na}{\underset{|}{C}}}}-H$$

Acetaldehyde — Bisulphite of acetaldehyde

Mechanism:

$$R-\overset{O}{\overset{\|}{C}}-H \xrightarrow{HSO_3^-} R-\overset{O^-}{\overset{|}{\underset{SO_2OH}{\underset{|}{C}}}}-H \underset{}{\overset{H^+}{\rightleftharpoons}} R-\overset{OH}{\overset{|}{\underset{SO_2O^-}{\underset{|}{C}}}}-H \overset{Na^+}{\rightleftharpoons} R-\overset{OH}{\overset{|}{\underset{SO_3Na}{\underset{|}{C}}}}-H$$

Aldehyde

For Ketones:

$$H_3C-\overset{O}{\overset{\|}{C}}-CH_3 \xrightarrow[H_2O]{Na^+HSO_3^-} H_3C-\overset{CH_3}{\overset{|}{\underset{OH}{\underset{|}{C}}}}-SO_3Na$$

Acetone — Bisulphite of acetone

The product obtained is heated with dilute acid or sodium carbonate to reproduce the original aldehyde or ketone.

II. Addition–Elimination reactions: Some types of reagents react with aldehydes and ketones to yield an addition product which further undergoes elimination reaction *via* loss of water and leads to the formation of double bond in the final product. These kinds of reactions are known as addition-elimination reactions.

$$\underset{\text{Aldehyde/Ketone}}{R-\overset{\overset{\displaystyle O}{\|}}{C}-R'} \xrightarrow[\text{Addition}]{\text{H-NuH}} \underset{\text{Addition product}}{R-\overset{\overset{\displaystyle OH}{|}}{\underset{\underset{\displaystyle NuH}{|}}{C}}-R'} \xrightarrow[\text{Elimination}]{-H_2O} \underset{\text{Product}}{R-\underset{\underset{\displaystyle Nu}{\|}}{C}-R'}$$

Addition of nitrogen nucleophiles or Formation of imines:

Aldehydes and ketones react with primary amines ($R-NH_2$) and other derivatives of ammonia to give imines (Imine: Organic compound with carbon–nitrogen double bond).

$$\underset{\text{Aldehyde or Ketone}}{R-\overset{\overset{\displaystyle O}{\|}}{C}-R'} + \underset{1^\circ\text{Amine}}{R''NH_2} \underset{\Delta}{\overset{H^+}{\rightleftharpoons}} \underset{\text{Imine}}{(R)(R')C=NR''}$$

Like amines, imines are basic. Substituted imine is known as Schiff base.

The following compounds are ammonia derivatives in which one of the hydrogen of ammonia is substituted.

NH_2OH (Hydroxylamine).

NH_2-NH_2(Hydrazine).

NH_2-$NHCONH_2$ (Semicarbazide).

NH_2 - R Primaryamines.

(i) Reaction with primary amines: Aldehydes and ketones react with primary amines to give an imine.

$$\underset{\text{Aldehyde or Ketone}}{R-\overset{\overset{\displaystyle O}{\|}}{C}-R'} + \underset{1^\circ\text{amine}}{R''NH_2} \underset{\Delta}{\overset{H^+}{\rightleftharpoons}} \underset{\text{Imine}}{(R)(R')C=NR''}$$

$$\underset{\text{Acetone}}{H_3C-\overset{\overset{\displaystyle O}{\|}}{C}-CH_3} + \underset{\text{Ethyl amine}}{CH_3CH_2NH_2} \underset{\Delta}{\overset{H^+}{\rightleftharpoons}} \underset{\text{Imine}}{H_3C-\overset{\overset{\displaystyle CH_3}{|}}{C}=NCH_2CH_3} + H_2O$$

The important condition of imine formation reaction is maintaining proper pH. The second half of the reaction is acid catalyzed hence the solution must be acidic.

(ii) Reaction with secondary amines: Aldehydes and ketones react with secondary amines to gives enamines or vinylamines *via* the formation of iminium ions.

$$\underset{\text{Acetaldehyde}}{H_3C-\overset{\overset{\displaystyle O}{\|}}{C}-H} + \underset{\text{Dimethylamine (2° Amine)}}{HN(CH_3)_2} \underset{-H_2O}{\overset{H^+}{\rightleftharpoons}} \underset{\text{Iminium ion}}{\left[H_2\overset{\overset{\displaystyle H}{|}}{C}-\underset{H}{C}=\overset{+}{N}(CH_3)_2\right]} \overset{-H^+}{\rightleftharpoons} \underset{\text{Enamine}}{H_2C=\overset{H}{C}-\ddot{N}(CH_3)_2}$$

An enamine is an unsaturated tertiary amine. The double bond present in the final product enamine comes from aldehyde and ketones.

The mechanism of formation of enamine is similar to that imine formation except the last step. In imine formation, loss of the proton from the nitrogen atom produces the neutral imine. But in the formation of enamine, the hydrogen atom is not bonded to the positively charged nitrogen atom so a proton must be removed from α-carbon of an aldehyde or ketone to form a stable neutral molecule.

$$H_3C-\overset{O}{\overset{\|}{C}}-CH_3 + R-\overset{R'}{\overset{|}{N}}H \longrightarrow H_3C-\overset{:\ddot{O}:^-}{\underset{+HN-R}{C}}-CH_3 \underset{-H^+}{\overset{H^+}{\rightleftharpoons}} H_3C-\overset{:\ddot{O}H}{\underset{+HN-R}{C}}-CH_3$$

Acetone

$$\underset{-H^+}{\overset{H^+}{\rightleftharpoons}} H_3C-\overset{OH}{\underset{:N-R}{C}}-CH_3 \underset{-H_2O}{\rightleftharpoons} H_2\overset{H}{C}-\underset{+N-R'}{\overset{\|}{C}}-CH_3 \xrightarrow{-H^+} CH_2=\underset{:N-R'}{C}-CH_3$$

Neutral tetra hedral intermediate

Enamine

In aqueous acidic solution, an enamine is hydrolysed to give carbonyl compounds.

(iii) Addition of hydrazine and its related compounds: Aldehydes and ketones react with hydrazine to gives hydrazones along with the elimination of water molecule.

$$-\overset{O}{\overset{\|}{C}}- + NH_2\text{-}NH_2 \longrightarrow -\overset{|}{C}=N\text{-}NH_2 + H_2O$$

Aldehyde/Ketone Hydrazine Hydrazone

$$H_3C-\overset{O}{\overset{\|}{C}}-H + NH_2\text{-}NH_2 \longrightarrow H_3C-\underset{H}{C}=N\text{-}NH_2 + H_2O$$

Acetaldehyde Acetaldehyde hydrazone

$$H_3C-\overset{O}{\overset{\|}{C}}-CH_3 + NH_2\text{-}NH_2 \longrightarrow H_3C-\overset{CH_3}{\overset{|}{C}}=N\text{-}NH_2 + H_2O$$

Acetone Hydrazine Acetone hydrazone

$$C_6H_5-\overset{O}{\overset{\|}{C}}-CH_3 + NH_2\text{-}NH_2 \longrightarrow C_6H_5-\overset{N\text{-}NH_2}{\overset{\|}{C}}-CH_3 + H_2O$$

Acetophenone Acetophenone hydrazone

$$C_6H_5-\overset{O}{\overset{\|}{C}}-H + NH_2\text{-}NH_2 \longrightarrow C_6H_5-\overset{N\text{-}NH_2}{\overset{\|}{C}}-H + H_2O$$

Benzaldehyde Benzaldehyde hydrazone

(iv) Addition of hydroxylamine (-NH_2OH): Aldehydes and ketones react with hydroxylamines to give oximes which are called aldoximes and ketoximes respectively. These are crystalline solids and may be used for the identification of aldehydes and ketones.

$$>C=O + NH_2\text{-}OH \xrightarrow{-H_2O} >C=N\text{-}OH$$

Aldehyde/Ketone Hydroxylamine Oxime

$$H_3C-CH=O + NH_2\text{-}OH \longrightarrow H_3C-CH=NOH + H_2O$$

Acetaldehyde Hydroxylamine Acetaldoxime

$$(H_3C)_2C=O + NH_2\text{-}OH \longrightarrow (H_3C)_2C=NOH + H_2O$$

Acetone Hydroxylamine Acetoxime

(v) Addition with phenyl hydrazines: Aldehydes and ketones condense with phenyl hydrazines to yield phenyl hydrazones.

$$>C=O + NH_2NHC_6H_5 \longrightarrow >C=NNHC_6H_5 + H_2O$$

Aldehyde/Ketone Phenyl hydrazine Phenyl hydrazone

$$(H)(H_3C)C=O + NH_2NHC_6H_5 \longrightarrow (H)(H_3C)C=NNHC_6H_5 + H_2O$$

Acetaldehyde Phenyl hydrazine Acetaldehyde phenyl hydrazone

$$(H_3C)_2C=O + NH_2NHC_6H_5 \longrightarrow (H_3C)_2C=NNHC_6H_5 + H_2O$$

Acetone Phenyl hydrazine Acetone Phenyl hydrazone

(vi) Condensation or Addition with semicarbazides: Semicarbazides condenses with aldehydes and ketones to give semicarbazones (well defined crystals).

$$>C=O + NH_2NH-\underset{\underset{O}{\|}}{C}-NH_2 \longrightarrow >C=N-NH\underset{\underset{O}{\|}}{C}NH_2 + H_2O$$

Aldehyde/Ketone Semicarbazide Semicarbazone

$$(H_3C)_2C=O + NH_2NH-\underset{\underset{O}{\|}}{C}-NH_2 \longrightarrow (H_3C)_2C=N-NH\underset{\underset{O}{\|}}{C}NH_2 + H_2O$$

Acetone Semicarbazide Acetone semicarbazone

(vii) **Reaction with 2,4-dinitrophenyl hydrazine (2,4-DNP):** 2,4-Dinitrophenyl hydrazine condenses with aldehydes and ketones to yields 2,4-dinitrophenyl hydrazones which is orange color crystalline compound.

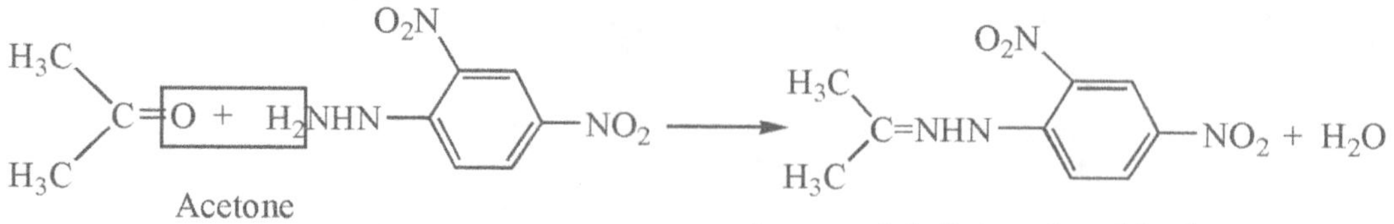

$$(H_3C)_2C{=}O + H_2NHN{-}C_6H_3(NO_2)_2 \longrightarrow (H_3C)_2C{=}NHN{-}C_6H_3(NO_2)_2 + H_2O$$

Acetone + 2,4-DNP → Acetone-2,4-dinitrophenyl hydrazone

This test is used as an identification test for aldehydes and ketones.

(viii) **Reaction with thioalcohols (RSH or Mercaptans):** Aldehydes and ketones react with thioalcohols to give mercaptals and mercaptols respectively.

$$>C{=}O + 2RSH \longrightarrow >C(SR)_2 + H_2O$$

Aldehyde/ ketone + Thiol → Mercaptals/ mercaptols

$$(H_3C)(H)C{=}O + 2RSH \longrightarrow CH_3CH{=}(SR)_2 + H_2O$$

Acetaldehyde → Mercaptals

$$(H_3C)_2C{=}O + 2RSH \longrightarrow (CH_3)_2C{=}(SR)_2 + H_2O$$

Acetone → Mercaptols

Table 15.4 Some N-substituted derivatives of Aldehydes and ketones (> C = N – Z).

Z	Reagent name	Carbonyl derivative	Product name
-H	Ammonia	$>C{=}NH$	Imine
-R	Amine	$>C{=}NR$	Substituted imine (Schiff's base)
-OH	Hydroxylamine	$>C{=}N{-}OH$	Oxime
$-NH_2$	Hydrazine	$>C{=}N{-}NH_2$	Hydrazone

Table 15.4 *Contd...*

$-NH-C_6H_5$	Phenylhydrazine	$>C{=}N-NH-C_6H_5$	Phenylhydrazone
$-NH-C_6H_3(NO_2)_2$ (2-O_2N, 4-NO_2)	2,4-Dinitrophenyl-hydrazine	$>C{=}N-NH-C_6H_3(NO_2)_2$ (2-O_2N, 4-NO_2)	2,4-dinitrophenyl hydrazone
$-NH-\overset{O}{\overset{\Vert}{C}}-NH_2$	Semicarbazide	$>C{=}N-NH-\overset{O}{\overset{\Vert}{C}}-NH_2$	Semicarbazone

III. Oxidation of aldehydes and ketones: Aldehydes and ketones react with oxidising agents quite differently. Aldehydes easily react but ketones need some vigorous reaction conditions.

(i) Oxidation of aldehydes: Aldehydes are oxidized in the presence of sodium dichromate or potassium dichromate in acidic solution or $KMnO_4$ to give carboxylic acids.

$$\underset{\text{Aldehyde}}{R-\overset{O}{\overset{\Vert}{C}}-H} \xrightarrow[H_2SO_4]{K_2Cr_2O_7} \underset{\text{Carboxylic acid}}{R-\overset{O}{\overset{\Vert}{C}}-OH}$$

$$\underset{\text{Acetaldehyde}}{CH_3CHO} \xrightarrow[H_2SO_4]{K_2Cr_2O_7} \underset{\text{Acetic acid}}{CH_3COOH}$$

Reaction with Tollen's reagent: The composition of Tollen's reagent is a dilute solution of silver oxide in ammonical solution. The silver oxide acts as a mild oxidizing agent. It oxidizes aldehyde but it is not used as oxidizing agent to oxidize other functional groups such as alcohol *etc.*

$$\underset{\text{Aldehyde}}{R-\overset{O}{\overset{\Vert}{C}}-H} + \underset{\text{Tollens reagent}}{2Ag(NH_3)_2OH} \longrightarrow R-\overset{O}{\overset{\Vert}{C}}-\overset{-}{O}\overset{+}{N}H_4 + \underset{\text{Metallic silver}}{2Ag\downarrow} + H_2O + 3NH_3$$

The oxidizing agent in Tollen's reagent is Ag^+ ion. During oxidation it reduces to metallic silver. This is the basic principle for Tollen's test. If Tollen's reagent is added to the aldehyde in a test tube, a shiny mirror is coated inside the test tube which indicates the presence of aldehydes.

Reaction with Fehling's solution: Aldehydes are oxidized with Fehling's solution to give a copious red or brown color precipitate. Cupric ions get reduced to cuprous. The formation of copious red or brown color indicates the presence of aldehyde molecule.

$$\underset{\text{Aldehyde}}{RCHO} + 2\,Cu(OH)_2 + NaOH \longrightarrow RCO\overset{-}{O}\overset{+}{Na} + \underset{\substack{\text{Cuprous oxide}\\ \text{(Red/Brown ppt)}}}{Cu_2O\downarrow} + 3H_2O$$

Fehling's solution composition: It is an alkaline solution of cupric ion complexed with sodium, potassium tartarate ions.

Reaction with Benedict's solution: Aldehydes react with Benedict's solution in the same way as that of Fehling's solution. Benedict's solution is an alkaline solution of cupric ion complexed with citrate ions.

(ii) Oxidation of ketones: Ketones are oxidized by strong oxidizing agents such as hot concentrated HNO_3 or alkaline $KMnO_4$ to give two carboxylic acids with fewer carbon atoms less than that of the original ketone. The oxidizing agent attacks both sides of the carbonyl group to give two acids.

$$\underset{\text{Ketone}}{R{-}\overset{O}{\overset{\|}{C}}{-}R'} \xrightarrow[\text{Alk. } KMnO_4,\ \Delta]{\text{Conc.} HNO_3} \underset{\text{Carboxylic acid mixture}}{R{-}\overset{O}{\overset{\|}{C}}{-}OH + R'{-}\overset{O}{\overset{\|}{C}}{-}OH}$$

$$\underset{\text{Acetone}}{H_3C{-}\overset{O}{\overset{\|}{C}}{-}CH_3} \xrightarrow[\text{Alk. } KMnO_4]{\text{Conc. } HNO_3} \underset{\text{Acetic acid}}{CH_3COOH} + \underset{\text{Formic acid}}{HCOOH}$$

$$\underset{\text{Methyl propyl ketone}}{H_3C{-}\overset{O}{\overset{\|}{C}}{-}CH_2CH_2CH_3} \xrightarrow[\text{Alk. } KMnO_4]{\text{Conc. } HNO_3} \underset{\text{Acetic acid}}{CH_3COOH} + \underset{\text{Propionic acid}}{CH_3CH_2COOH}$$

Oxidation by peroxy acid (Baeyer-Villiger oxidation):

Both aldehydes and ketones are oxidized by peroxy acid. Aldehydes give carboxylic acids and ketones give esters. A peroxyacid (also known as acyl hydroperoxide or percarboxylic acid) contains one more oxygen atom than that of carboxylic acid and this atom is inserted between the carbonyl carbon and the hydrogen of an aldehyde or the "R" group of a ketone. This reaction is known as Baeyer - Villiger oxidation.

$$\underset{\text{Butanal}}{CH_3CH_2CH_2\overset{O}{\overset{\|}{C}}H} + \underset{\text{Peroxy acid}}{R{-}\overset{O}{\overset{\|}{C}}{-}OOH} \longrightarrow \underset{\text{Butanoic acid}}{CH_3CH_2CH_2COOH} + \underset{\text{Carboxylic acid}}{RCOOH}$$

$$\underset{\text{Methyl propyl ketone}}{CH_3CH_2CH_2\overset{O}{\overset{\|}{C}}CH_3} + \underset{\text{Peroxy acid}}{R{-}\overset{O}{\overset{\|}{C}}{-}OOH} \longrightarrow \underset{\text{Methyl butanoate}}{CH_3CH_2CH_2COOCH_3} + \underset{\text{Carboxylic acid}}{RCOOH}$$

IV. Reduction reactions of aldehydes and ketones: Aldehydes and ketones undergo reduction reactions in the presence of variety of reducing agents to give an alcohol or amine or a hydrocarbon. The product obtained depends upon the type of reagents used.

$$\underset{\text{Aldehyde/ketone}}{R{-}\overset{O}{\overset{\|}{C}}{-}R'} \longrightarrow \underset{\text{Alcohol}}{R{-}\overset{OH}{\overset{|}{\underset{H}{C}}}{-}R'} \Big/ \underset{\text{Alkane}}{RCH_2R} \text{ (or) } \underset{\text{Amine}}{R\overset{NR_2''}{\overset{|}{C}}HR'}$$

(i) Catalytic hydrogenation: Aldehydes when reduced with catalyst like platinum, palladium or Raney nickel yield respective alcohols.

$$\underset{\text{Acetaldehyde}}{H_3C{-}\overset{O}{\overset{\|}{C}}{-}H} + H_2 \xrightarrow{Pt} \underset{\text{Ethanol (1° Alcohol)}}{CH_3CH_2OH}$$

The conditions of the reaction depend upon the nature of the compound being reduced and the type of catalyst used. Structurally unhindered ketones are used as a staring material. They are reduced at room temperature under 4 atm pressure with Pt or Pd as catalyst but in vigorous conditions Raney nickel, copper chromite *etc* are used as catalyst.

(ii) Reduction with $LiAlH_4$ (or) sodium borohydride: Aldehydes and ketones are reduced with $LiAlH_4$ (Lithium aluminium hydride) or sodium borohydride to yield corresponding alcohols.

$$H_3CH_2CH_2C-\overset{\overset{O}{\|}}{C}-CH_3 \xrightarrow[2)\ H^+, H_2O]{1)\ LiAlH_4} H_3CH_2CH_2C-\overset{\overset{OH}{|}}{CH}-CH_3$$

2-Pentanone → 2-Pentanol

$$H_3CH_2CH_2C-\overset{\overset{O}{\|}}{C}-CH_3 \xrightarrow[2)\ H^+, H_2O]{1)\ NaBH_4} H_3CH_2CH_2C-\overset{\overset{OH}{|}}{CH}-CH_3$$

2-Pentanone → 2-Pentanol

(iii) Clemmensen reduction: Aldehydes and ketones are reduced with Zn(Hg)/HCl (Zinc amalgam and HCl) to yield alkanes. This catalyst or method is suitable for compounds unstable in base but stable in acid.

$$H_3CH_2C-\overset{\overset{O}{\|}}{C}-CH_3 \xrightarrow[HCl]{Zn/Hg} CH_3CH_2CH_2CH_3$$

2-Butanone → *n*-Butane

(iv) Wolf-Kishner reduction: Aldehydes and ketones are first converted into their derivatives such as hydrazones, semicarbazones, azines (by treatment with hydrazines, semicarbazides *etc*) which upon further reaction with a strong base such as potassium hydroxide or potassium *tert*- butoxide in dimethyl sulphoxide (DMSO) solvent yield alkanes. This reaction is suitable for the carbonyl compounds that are stable in base.

$$C_6H_{11}-\overset{\overset{O}{\|}}{C}-CH_3 \xrightarrow[-H_2O]{NH_2NH_2} C_6H_{11}-\overset{\overset{NNH_2}{\|}}{C}-CH_3 \xrightarrow{KOH} C_6H_{11}-CH_2CH_3$$

Cyclohexyl methyl ketone → Cyclohexyl methyl ketone hydrazone → Ethyl cylohexane

(v) Meerwein Pondorf-Verley Reduction: Aldehydes and ketones are reduced to the corresponding alcohols by treating with aluminium isopropoxide in isopropanol solution.

$$\underset{R^1}{\overset{R}{>}}C{=}O + (CH_3)_2CHOH \xrightleftharpoons{Al[OCH(CH_3)_2]_3} \underset{R^1}{\overset{R}{>}}CHOH + (CH_3)_2C{=}O$$

Ketone + Isopropanol → 2° Alcohol + Acetone

The acetone so produced is removed by slow distillation, so the reaction occurs in forward direction only.

Mechanism:

$R(R^1)C{=}O + Al[OCH(CH_3)_2]_2\text{–}O\text{–}CH(CH_3)_2 \underset{-\ CH_3COCH_3}{\rightleftharpoons} R(R^1)CH\text{–}O(H)\cdots Al[OCH(CH_3)_2]_2 \xrightarrow{(CH_3)_2CHOH} R(R^1)CHOH + Al[OCH(CH_3)_2]_3$

2° Alcohol; Aluminium *iso*-propoxide

Note:

1. Aluminium isopropoxide is mostly used because it gives superior yield (60%-70%), when compared with other alkoxides.
2. Reaction is generally carried out in the presence of isopropanol and the acetone formed is easily removed from the reaction mixture.

Synthetic uses:

(a) Synthesis of Chlormycetin or Chloramphenicol:

$O_2N\text{–}C_6H_4\text{–}CO\text{–}CH(NHCOCH_3)CH_2OH \xrightarrow[(CH_3)_2CHOH]{i)\ Al[OCH(CH_3)_2]_3} O_2N\text{–}C_6H_4\text{–}CH(OH)\text{–}CH(NHCOCH_3)CH_2OH + CH_3COCH_3$ (Acetone)

$\xrightarrow{HCl} O_2N\text{–}C_6H_4\text{–}CH(OH)\text{–}CH(NH_3)CH_2OH \xrightarrow[-\ CH_3OH]{Cl_2CHCOOCH_3} O_2N\text{–}C_6H_4\text{–}CH(OH)\text{–}CH(NHCOCHCl_2)CH_2OH$

(b) Synthesis of oestradiol:

Oestrone $\xrightarrow[(CH_3)_2CHOH]{Al[OCH(CH_3)_2]_3}$ Oestradiol + $CH_3\overset{OH2}{\overset{\|}{C}}CH_3$ (Acetone)

(vi) Willgerodt reaction: The conversion of aryl ketones into an amide by using yellow ammonium polysulphide (sulphur is dissolved in ammonium sulphide solution) is called Willgerodt reaction. Sometimes 40% dioxan or pyridine is used as a solvent to improve the yield. The amide obtained is having same number of carbon atom as that of the starting ketone.

$C_6H_5\overset{O}{\overset{\|}{C}}CH_3 \xrightarrow{(NH_4)_2Sx} C_6H_5CH_2CONH_2 + C_6H_5CH_2COONH_4 + C_6H_5CH_2CH_3$

Acetophenone; Phenyl acetamide (50% yield); Ammonium phenyl acetate (13.5% yield); Ethyl benzene

(vii) **Darzein condensation:** Condensation of an aldehyde or a ketone with an α-halogen ester in the presence of sodium ethoxide or sodamide to yield an α, β-epoxy ester (glycidic ester).

$$\underset{\text{Ketone}}{R_2C=O} + \underset{\text{Chloroethyl acetate}}{ClCH_2COOC_2H_5} \xrightarrow{C_2H_5ONa} \underset{\alpha,\beta\text{-Epoxy ester}}{R_2C\overset{O}{\frown}CH.COOC_2H_5} + NaCl + C_2H_5OH$$

Mechanism:

$$ClCH_2COOC_2H_5 \underset{}{\overset{C_2H_5\bar{O}}{\rightleftharpoons}} Cl\bar{C}HCOOC_2H_5 \overset{R_2C=O}{\rightleftharpoons} R_2C(\bar{O})-CH(Cl).COOC_2H_5 \overset{-\bar{Cl}}{\rightleftharpoons} \underset{\alpha,\beta\text{-Epoxy ester}}{R_2C\overset{O}{\frown}CH.COOC_2H_5}$$

(viii) **Reductive amination:** Aldehydes and ketones are treated with ammonia or 1° amine in the presence of hydrogen to yield imines. The C=N group present in the imines undergoes catalytic hydrogenation to give amines.

$$\underset{\text{2-Butanone}}{H_3C-\overset{O}{\overset{\|}{C}}-CH_2CH_3} \xrightarrow[-H_2O]{CH_3NH_2} \underset{\text{An imine}}{H_3C-\overset{NCH_3}{\overset{\|}{C}}-CH_2CH_3} \xrightarrow{H_2,\ Pt} \underset{\text{N-Methyl-}sec\text{-butylamine}}{H_3C-\overset{NHCH_3}{\overset{|}{C}H}-CH_2CH_3}$$

Note: It is a best method for the preparation of amines with a secondary alkyl group.

V. Condensation Reactions:

(i) **Aldol condensation:** A condensation of two or more molecules of same or different aldehydes and ketones with or without the elimination of small molecules such as water, ammonia, methanol *etc.* is known as aldol condensation. Aldehydes possessing α – hydrogen atoms undergo this reaction in the presence of base to yield the product known as **aldols**. The product obtained is β – hydroxy aldehyde or ketone as it contains both aldehyde/ketone and alcohol.

$$\underset{\text{Aldehyde/ketone}}{R-\overset{O}{\overset{\|}{C}}-CH_2-R' + R-\underset{O}{\underset{\|}{C}}-CH_2-R'} \overset{H^+/OH^-}{\rightleftharpoons} \underset{\text{Aldol product}}{R-\overset{OH}{\overset{|\beta}{C}}(CH_2-R')-\overset{\alpha}{C}H(R')-\underset{O}{\underset{\|}{C}}-R} \overset{H^+/OH^-}{\rightleftharpoons} \underset{\alpha,\ \beta\text{-Unsaturated aldehyde/ketone}}{R-\overset{\beta}{C}(CH_2-R')=\overset{\alpha}{C}(R')-\underset{O}{\underset{\|}{C}}-R} + H_2O$$

Mechanism:

(a) Base catalysed aldol condensation.

(b) Acid catalysed aldol condensation.

(a) **Base Catalysed aldol condensation:** Two molecules of acetaldehyde condense in the presence of dilute alkali like potassium carbonate to give a syrupy liquid aldol.

$$H_3C-\overset{O}{\overset{\|}{C}}-H + H_3C-\overset{O}{\overset{\|}{C}}-H \xrightarrow{-OH^-} H_3C-\underset{H}{\overset{OH}{\overset{|}{C}}}-\overset{H_2}{C}-\underset{O}{\underset{\|}{C}}-H$$

3-Hydroxybutanal (aldol)

Mechanism: The aldol condensation is reversible one.

Step 1: Formation of the enolate carbanion from one molecule of acetaldehyde.

$$\underset{\text{Acetaldehyde}}{H-\underset{O}{\underset{\|}{C}}-\overset{H}{\overset{|}{C}}-H + \overset{-}{O}H} \rightleftharpoons \left[H-\underset{O}{\underset{\|}{C}}-\overset{H}{\ddot{C}}{}^{-}-H \longleftrightarrow \overset{H}{\underset{:\ddot{O}^{\ominus}}{C}}=C\begin{matrix}H\\H\end{matrix} \right] + H_2O$$

Enolate ion of acetaldehyde
(Acts as a strong nucleophile)

Step 2: Attack of nucleophile to the carbonyl carbon of the 2nd molecule of acetaldehyde.

$$\underset{\text{Enolate ion (carbanion)}}{H-\underset{O}{\underset{\|}{C}}-\ddot{C}H_2^{-}} + \underset{\text{Acetaldehyde}}{H-\underset{:O:}{\underset{\|}{C}}-CH_3} \rightleftharpoons H-\underset{\alpha}{\underset{|}{\overset{\overset{:\ddot{O}:^{\ominus}}{|\beta}}{C}}}-CH_3 \text{ (attached to } H-\underset{O}{\underset{\|}{C}}-CH_2\text{)}$$

$$\xrightleftharpoons{H-OH} H-\overset{:\ddot{O}H}{\overset{|\beta}{C}}-CH_3 + OH^- \text{ (attached to } H-\underset{O}{\underset{\|}{C}}-\underset{\alpha}{CH_2}\text{)}$$

Aldol (50 %)

In aldol, the α-hydrogen atom is still present, it can be removed as a water molecule on heating in the presence of base.

$$\underset{\text{Aldol (50 \%)}}{H-\overset{OH}{\overset{|\beta}{C}}-CH_3 \text{ on } H-\underset{O}{\underset{\|}{C}}-\underset{\alpha}{CH_2}} \xrightarrow[-H_2O]{\Delta} \underset{\text{Crotonaldehyde}}{CH_3\text{-}CH\text{=}CH\text{-}CHO}$$

For Ketones: Ketones with α–hydrogen atom undergoes aldol condensation in the presence of barium hydroxide to give diacetone alcohol.

$$\underset{\text{Acetone}}{\begin{matrix}H_3C\\H_3C\end{matrix}\!\!>C=O} + \underset{\text{Acetone}}{H-\underset{H_2}{C}-\underset{O}{\underset{\|}{C}}-CH_3} \xrightarrow{Ba(OH)_2} \underset{\text{Diacetone alcohol}}{\begin{matrix}H_3C\\H_3C\end{matrix}\!\!>\underset{OH}{\underset{|}{C}}-CH_2-\underset{O}{\underset{\|}{C}}-CH_3}$$

(b) **Acid catalysed aldol condensation:** Aldol condensation also occurs in the presence of acid. In this condition, enol acts as a weak nucleophile and attacks the carbonyl carbon.

Step 1: Condensation of enol to the protonated carbonyl.

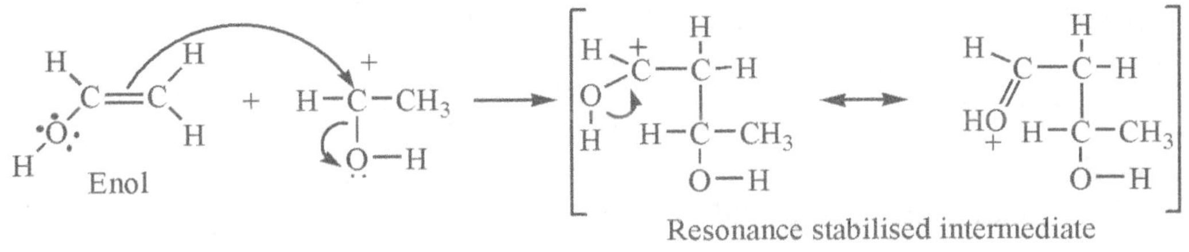

Enol

Resonance stabilised intermediate

INTERESTING FACT

Aldases are enzymes obtained from aldols and are commonly used in the metabolism of sugars. These are also used in organic synthesis for key transformation.

Step 2: Deprotonation

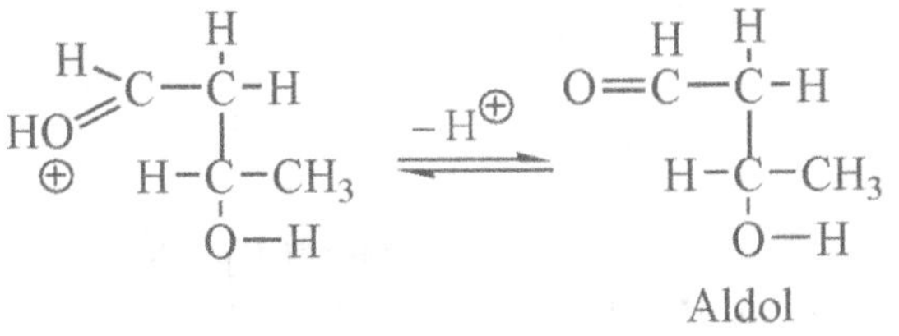

Aldol

Mixed/Crossed Aldol Condensation: Two different kinds of aldehydes or ketones are involved in the aldol condensation which leads to the formation of four products is known as mixed / cross aldol condensation. Since each carbonyl compound can react with itself as well as with other carbonyl compound. The four products obtained possess similar physical properties so they are difficult to separate. This reaction is not synthetically important. In mixed aldol condensation, the main condition is any one of the aldehyde must have α-hydrogen. For example: Cross aldol condensation between acetaldehyde and propanaldehyde. Depending upon the reaction condition, the four possible aldols are obtained as given below.

Step 1:

(i) Enolate of acetaldehyde adds to propanaldehyde.

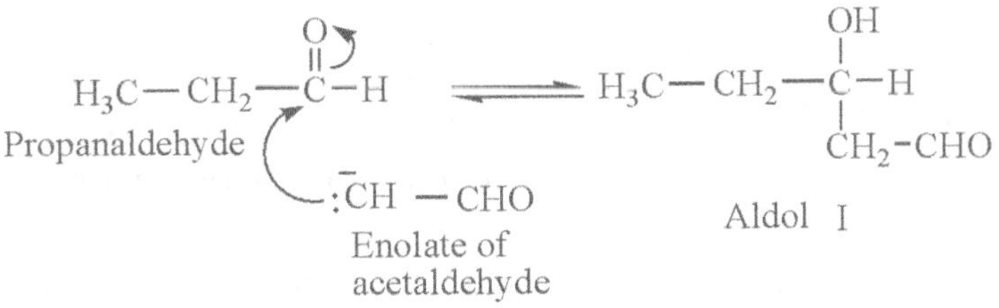

$$H_3C-CH_2-\overset{O}{\overset{\|}{C}}-H + {}^{-}:CH-CHO \rightleftharpoons H_3C-CH_2-\overset{OH}{\overset{|}{\underset{CH_2-CHO}{\underset{|}{C}}}}-H$$

Propanaldehyde

Enolate of acetaldehyde

Aldol I

$$CH_3-\overset{O}{\overset{\|}{C}}-H \;\text{(Acetaldehyde)} + CH_3-\ddot{C}H-CHO \;\text{(Enolate of propanaldehyde)} \rightleftharpoons H_3C-\underset{HC-CHO}{\overset{OH}{C}}-H \;\text{(Aldol II)}$$

Step 2:

(i) Self condensation of acetaldehyde.

$$CH_3-\overset{O}{\overset{\|}{C}}-H + \bar{:}CH_2-CHO \rightleftharpoons H_3C-\underset{CH_2-CHO}{\overset{OH}{C}}-H \;\text{(Aldol III)}$$

(ii) Self-condensation of propanaldehyde.

$$H_3C-CH_2-\overset{O}{\overset{\|}{C}}-H + CH_3-\underset{H}{\ddot{C}}-CHO \rightleftharpoons H_3C-CH_2-\underset{H_3C-CH-CHO}{\overset{OH}{C}}-H \;\text{(Aldol IV)}$$

A cross aldol condensation is always effective one if one reactant produce enolate ion. If any one of reactant has only one α-hydrogen atom, then one enolate ion is formed.

Common Features of Aldol Condensation

1. The reaction is reversible one and the equilibrium is always present on the left hand side of the reaction. By using catalyst or special experimental conditions, the product yield is increased upto 70 %.
2. The aldol condensation is facilitated by – I groups on the aldehyde or ketone and decreased by + I groups on the aldehyde or ketone.
3. The condensed products formed are not isolated easily. Some of the products are readily converted into hemiacetal.

$$H_3C-\underset{OH}{\overset{H}{C}}-CH_2-CHO \;\text{(Acetaldol)} \xrightarrow{CH_3CHO} \text{Hemiacetal (cyclic, with } H_3C, OH, CH_3 \text{ substituents)}$$

4. The aldol product is easily dehydrated to give α, β-unsaturated aldehyde by heating alone with mineral acid or base.
5. When an aryl group is attached to the β-carbon atom, then only isolated product comes from that reaction sequence.
6. If the high concentration of alkali is used for the dehydration of aldol, the α, β–unsaturated aldehyde further undergoes condensation to yield resins.

INTERESTING FACT

Biological aldol condensation: The most abundant sugar present in nature is D-glucose which is synthesized by biological system from two molecules of pyruvate by gluconeogensis. The reverse reaction (breakdown) is called glycolysis. One of the steps in the biosynthesis of D-Glucose is aldol condensation.

The major protein which constitutes about one fourth of the total protein in mammals is collagen. It is the main fibrous component of bones, teeth, skin, cartilages and tendons. It is responsible for holding group of cells together. Tropocollagen, an individual collagen molecule is only isolated from tissues of young animals. As animal ages, the individual collagen molecules become cross linked. This is the reason why meat from older animals is tougher than meat from younger animals. Cross linking of collagen is an example of aldol condensation. Before the cross linking, the primary amino acid group of lysine residues of collagen molecule is reduced to aldehyde groups which is catalysed by the enzyme *lysyl oxidase*. An aldol condensation between the two-aldehyde residue leads to formation of cross-linked protein. The sequence of the reaction is as follows.

$$O{=}C(-NH-)-CH(CH_2)_2CH_2CH_2\overset{+}{N}H_3 \quad + \quad H_3\overset{+}{N}CH_2CH_2(CH_2)_2HC(-NH-)-C{=}O$$

(Lysine Residue)

$$\xrightarrow{\text{Lysyl oxidase}}$$

$$O{=}C(-NH-)-CH(CH_2)_2CH_2\overset{O}{\overset{\|}{C}}H \quad + \quad H\overset{O}{\overset{\|}{C}}CH_2(CH_2)_2HC(-NH-)-C{=}O$$

$$\xrightarrow{\text{Aldol condensation}}$$

$$O{=}C(-NH-)-CH(CH_2)_2CH_2CH{=}C(HC{=}O)(H_2C)_2CH(-NH-)-HC{=}O$$

Cross-linked collagen

(ii) ***Reformatsky reaction:*** It involves the reaction of aldehydes and ketones with α- bromoester in the presence of zinc metal to give β-hydroxy ester after hydrolysis.

$$R-\underset{\|}{\overset{}{C}}(=O)-R' + Zn + BrCH_2COOC_2H_5 \longrightarrow R-\underset{OZnBr}{\overset{R'}{C}}-CH_2COOC_2H_5 \xrightarrow[H_2O]{\text{Hydrolysis}} R-\underset{OH}{\overset{R'}{C}}-CH_2COOC_2H_5 + ZnBr(OH)$$

β-hydroxyester

The reaction is carried out by adding a mixture of carbonyl compound and α-haloester in ether or benzene to zinc metal or add zinc metal to the reaction mixture.

(iii) Cannizaro Reaction: Aldehydes which lack of α-hydrogen atom gives this reaction. Aldehydes when heated with Concentrated NaOH undergoes self redox reactions to form primary alcohol and a salt of an acid.

$$\underset{\text{Formaldehyde}}{2HCHO} + NaOH \longrightarrow \underset{\text{Methanol}}{CH_3OH} + \underset{\text{Sodium formate}}{HCOONa}$$

Mechanism:

Step 1: Nucleophilic attack on the carbonyl carbon of the formaldehyde.

$$H-\overset{\overset{O^{\delta-}}{\parallel}}{\underset{\delta^+}{C}}-H + \bar{O}H \longrightarrow \underset{\text{(I)}}{H-\overset{\overset{\bar{O}}{|}}{\underset{\underset{OH}{|}}{C}}-H}$$

Step 2: Hydride ion transfer between (I) and the second molecule of formaldehyde.

$$H-\overset{\overset{\bar{O}}{|}}{\underset{\underset{OH}{|}}{C}}-H + \begin{matrix}H\\H\end{matrix}\!\!>\overset{\delta+}{C}=\overset{\delta-}{O} \longrightarrow \left[H-\overset{\overset{O}{\parallel}}{\underset{\underset{OH}{|}}{C}} + CH_3\bar{O}\right] \longrightarrow H-\overset{\overset{O}{\parallel}}{\underset{\underset{O^-}{|}}{C}} + CH_3OH$$

Step 3: Acid – base reaction.

$$H-\overset{\overset{O}{\parallel}}{\underset{\underset{O^-}{|}}{C}} + Na^+ \longrightarrow \underset{\text{Sodium formate}}{H-\overset{\overset{O}{\parallel}}{C}-ONa}$$

In the above said reaction, one molecule of aldehyde is oxidized to the salt of carboxylic acid, while the second molecule of aldehyde is reduced to the corresponding alcohol. This type of self oxidation reduction (Redox) reactions is known as **disproportionation or dismutation reaction.**

Crossed Cannizaro Reaction: Cannizaro reaction occuring between two different aldehydes is known as crossed Cannizaro reaction. For Example: Reaction between formaldehyde and benzaldehyde.

$$\underset{\text{Formaldehyde}}{HCHO} + \underset{\text{Benzaldehyde}}{C_6H_5CHO} \xrightarrow{NaOH} \underset{\text{Sodium formate}}{HCOONa} + \underset{\text{Benzyl alcohol}}{C_6H_5CH_2OH}$$

In the above reaction, formaldehyde is oxidized to formate ion and benzaldehyde is reduced to benzylalcohol. The reason is the formaldehyde is more reactive than other aldehydes towards nucleophiles and add a hydroxide ion to the less reactive aldehyde (Benzaldehyde).

Sometimes formaldehyde reacts with another aldehyde containing α- hydrogen atom or atoms when heated with NaOH. Formaldehyde first undergoes aldol condensation to form a hydroxyaldehyde with no α-hydrogen atom which further undergoes Cannizaro reaction with formaldehyde.

$$\underset{\text{Formaldehyde}}{HCHO} + \underset{\text{Dialkyl acetaldehyde}}{\begin{matrix}R\\R'\end{matrix}\!\!>CH-CHO} \xrightarrow[\text{Aldol condensation}]{NaOH} \underset{\text{Hydroxyaldehyde with no }\alpha\text{-hydrogen atom}}{\begin{matrix}R\\R'\end{matrix}\!\!>C<\!\!\begin{matrix}CH_2OH\\CHO\end{matrix}} \xrightarrow[NaOH]{HCHO} \begin{matrix}R\\R'\end{matrix}\!\!>C<\!\!\begin{matrix}CH_2OH\\CH_2OH\end{matrix} + HCOONa$$

Heterocyclic compounds containing aldehyde functional group also gives Cannizaro reaction. For Example:

2 Thiophen-2-aldehyde (CHO) $\xrightarrow{NaOH}$ Thiophen-2-alcohol (CH_2OH) + Sodium salt of thiophene carboxylic acid (COONa)

2 Furfural (CHO) $\xrightarrow{NaOH}$ Furfuryl alcohol (CH_2OH) + Sodium salt of furoic acid (COONa)

Intermolecular (or) internal Cannizaro Reaction

1, 2-dialdehydes and α-ketoaldehydes when treated with a strong base undergoes internal Cannizaro reaction. For Example:

Phenyl glyoxal when treated with NaOH yield mandelic acid.

$$C_6H_5\text{-}C(=O)\text{-}C(H)=O \;\underset{}{\overset{OH^-}{\rightleftharpoons}}\; C_6H_5\text{-}C(=O)\text{-}C(H)(OH)\text{-}O^- \longrightarrow C_6H_5\text{-}C(O^-)\text{-}C(H)(OH)=O \longrightarrow C_6H_5\text{-}C(H)(OH)\text{-}C(=O)O^-$$

Phenyl glyoxal → Mandelate ion

$$\xrightarrow{Na^+} C_6H_5\text{-}C(H)(OH)\text{-}C(=O)O^-Na^+$$

Sodium mandelate

Applications of Cannizaro Reactions

1. ***Synthesis of Pentaerythritol:*** This is an example for the industrial application of Cannizaro reaction coupled with aldol condensation.

   HCHO and acetaldehyde (contains α-hydrogen atom) first undergoes aldol condensation to give trihydroxy pivaldehyde (no α-hydrogen atom) which further undergoes Cannizaro reaction with formaldehyde to give penta erythritol.

$$H\text{-}CH_2\text{-}CHO + 3HCHO \xrightarrow[\text{Aldol Condensation}]{Ca(OH)_2} HOH_2C\text{-}C(CH_2OH)_2\text{-}CHO + HCHO \xrightarrow[\text{Crossed Cannizaro reaction}]{NaOH} HOH_2C\text{-}C(CH_2OH)_2\text{-}CH_2OH + HCOONa$$

Acetaldehyde; Trihydroxypivaldehyde (No α–hydrogen atom); Pentaerythritol; Sodium formate

Esterification of pentaerythritol with polybasic acids give resin polymers. Its nitrate esters are also used as explosives and medicinally used as vasodilators.

$$C(CH_2OH)_4 + 3HNO_3 \xrightarrow{H_2SO_4} OHCH_2C(CH_2ONO_2)_3$$

Pentaerythritol → Pentaerythritol trinitrate (PETRIN)

$$C(CH_2OH)_4 + 4HNO_3 \xrightarrow{H_2SO_4} C(CH_2ONO_2)_4$$

Pentaerythritol → Pentaerythritol tetranitrate (PETN)

Limitations: Substituted aldehydes because of steric hinderance do not undergoes Cannizaro reaction. For Example: Di-ortho-substituted benzaldehyde does not give Cannizaro reaction.

(iv) **α-Halogenation:** Ketones readily undergoes halogenation at α- carbon atom in the presence of either an acidic catalyst or basic conditions. Here base is a reactant and acid acts as catalyst.

Base catalysed α-halogenation:

$$R-\overset{O}{\overset{\|}{C}}-CH_3 + X_2 + O\bar{H} \longrightarrow R-\overset{O}{\overset{\|}{C}}-\underset{H}{\overset{X}{C}}-H + X^- + H_2O$$

Ketone

$$H_3C-\overset{O}{\overset{\|}{C}}-CH_3 + Br_2 + O\bar{H} \longrightarrow H_3C-\overset{O}{\overset{\|}{C}}-CH_2-Br + B\bar{r} + H_2O$$

Acetone → Bromoacetone

X_2 = Cl_2, Br_2 *etc*

Haloform reaction: Acetone or Methyl ketone reacts with halogens in the presence of excess base (i.e in hypohalite solutions) to yield trihalo substituted ketone wherein three hydrogen atoms of the methyl group are replaced by halogen atoms. This reaction is known as haloform reaction.

$$H_3C-\overset{O}{\overset{\|}{C}}-CH_3 + 3Br_2 + 4NaOH \longrightarrow H_3C-\overset{O}{\overset{\|}{C}}-\bar{O}Na^+ + CHBr_3 + 3H_2O + 3NaBr$$

Acetone → Sodium acetate, Bromoform

Mechanism: This reaction is an example for base catalyzed halogenation. It involves following reaction.

Step 1: Abstraction of a proton by a base to give enolate carbanion.

$$H_3C-\overset{O}{\overset{\|}{C}}-\underset{H}{CH_2} + O\bar{H} \longrightarrow H_3C-\overset{O}{\overset{\|}{C}}-\ddot{C}H_2^- + H_2O$$

Acetone → Enolate carbanion

Step 2: Electrophilic attack by the halogen to the enolate carbanion.

$$H_3C-\overset{O}{\overset{\|}{C}}-\ddot{C}H_2^- + X-X \longrightarrow H_3C-\overset{O}{\overset{\|}{C}}-CH_2X + \bar{X}:$$

Step 3: The above two steps are repeated until all the three hydrogen atoms of the methyl group are replaced by three halogen atoms.

$$H_3C-\overset{O}{\overset{\|}{C}}-\underset{H}{\overset{X}{C}}-H \underset{-H_2O}{\overset{:\bar{O}H}{\rightleftharpoons}} H_3C-\overset{O}{\overset{\|}{C}}-\underset{H}{\overset{X}{C}}:^- \xrightarrow[-X^-]{X-X} H_3C-\overset{O}{\overset{\|}{C}}-\underset{H}{\overset{X}{C}}-X$$

$$\xrightarrow{-H_2O \,|\, O\bar{H}} H_3C-\overset{O}{\overset{\|}{C}}-\underset{X}{\overset{X}{C}}:^- \xrightarrow[-X^-]{X-X} H_3C-\overset{O}{\overset{\|}{C}}-\underset{X}{\overset{X}{C}}-X$$

Trihalo acetone

The trihaloacetone obtained is unstable one and further reacts with aqueous alkali. The base (OH^-) attacks the carbon atom of the trihalo acetone and leads to the cleavage of carbon bond between the carbonyl group and the trihalomethyl group. The products obtained are acetate ion or sodium acetate and haloform.

$$H_3C-\overset{O}{\overset{\|}{C}}-CX_3 + \bar{O}H \xrightleftharpoons{(1)} H_3C-\underset{OH}{\underset{|}{\overset{\bar{O}}{\overset{|}{C}}}}-CX_3 \xrightleftharpoons{(2)} H_3C-\underset{OH}{\underset{|}{\overset{O}{\overset{\|}{C}}}} + \bar{C}X_3 \xrightarrow[\text{Proton exchange}]{(3)} H_3C-\overset{O}{\overset{\|}{C}}-\bar{O} + CHX_3$$

Acetate ion Haloform

So the above reaction comprise three steps of EX_3 to form the adduct.

Step 1: Abstraction of proton (or) nucleophillic attack by OH^- (Base).

Step 2: Formation of acetic acid due to loss of CX_3 form the adduct.

Step 3: Proton exchange leads to the formation of a more stable acetate ion and haloform pair.

The reaction carried out by using iodine and ketone is known as **Iodoform reaction.**

Applications:

1. This iodoform test is used to find out acetaldehyde and methyl ketone. This reaction is given by the compounds containing the following groups.

$$-\underset{O}{\underset{\|}{C}}-CH_3 \quad \text{and} \quad -\underset{OH}{\underset{|}{CH}}-CH_3$$

The product of iodoform is the formation of yellow color precipitate.

2. This test is used to differentiate methyl ketones from other ketones. For Example:

$$\underset{\text{2-Pentanone}}{CH_3\text{-}CH_2\text{-}CH_2\text{-}\overset{O}{\overset{\|}{C}}\text{-}CH_3} \xrightarrow[4NaOH]{3I_2} \underset{\substack{\text{Iodoform}\\\text{(Yellow ppt)}}}{CHI_3} + CH_3CH_2CH_2\,COONa + 3H_2O + 3NaI$$

$$\underset{\text{3-Pentanone}}{CH_3\text{-}CH_2\text{-}\overset{O}{\overset{\|}{C}}\text{-}CH_2CH_3} \xrightarrow[NaOH]{I_2} \text{No formation of iodoform (No yellow ppt)}$$

Note: This test is not only given by methyl ketone but primary and secondary alcohols also give positive test.

Some of the Reactions of Aldehydes which are not given by Ketones

Aldehydes give some of the following reactions but ketones do not. They are as follows.

1. **Polymerization.**
2. **Acetal formation.**
3. **Addition reaction with ammonia.**
4. **Cannizaro reaction (action of caustic soda).**
5. **Tischenko Reaction.**
6. **Schiff's test.**

Polymerization: Generally, aldehydes easily undergo polymerization but ketones do not. Reaction between two or more molecules of aldehydes without elimination of small molecules as such as water (H_2O) or CH_3OH or NH_2 *etc.* leads to the formation of a new compound is known as polymerization.

For example.

$$\underset{\text{Formaldehyde}}{\text{(n)HCHO}} \xrightarrow{\text{Evoporation}} \underset{\textit{para}\text{-formaldehyde}}{(CH_2O)_n}$$

$$\underset{\text{(Aqueous solution)}}{\text{3HCHO}} \xrightarrow{\text{On standing}} \underset{\textit{meta}\text{ formaldehyde}}{(HCHO)_3}$$

$$\underset{\text{Acetaldehyde}}{3CH_3CHO} \xrightarrow{\text{Conc. } H_2SO_4} \text{Paraldehyde}$$

CH3 / O O / CH3 O CH3

Paraldehyde

Tischenko reaction (Modified Cannizaro Reaction): All aldehydes can be made to undergo Cannizaro reaction by treatment with aluminium ethoxide. The alcohol and acid molecules so obtained react further to give an ester. For example:

Acetaldehyde and propionaldehyde gives ethyl acetate and propyl propionate when treated with aluminium ethoxide.

$$\underset{\text{Acetaldehyde}}{2CH_3CHO} \xrightarrow{Al(OC_2H_5)_2} \underset{\text{Ethyl acetate}}{CH_3COOC_2H_5}$$

$$\underset{\text{Propionaldehyde}}{2C_2H_5CHO} \xrightarrow{Al(OC_2H_5)_2} \underset{\text{Ethyl propionate}}{C_2H_5COOC_2H_5}$$

Schiff's test: Schiff's reagent is a dilute solution of rosaniline hydrochloride whose red or pink colour disappears by passing SO_2. Aldehydes when treated with Schiff's reagent, restore its pink colour.

Summary of Chemical Reactions

Chemical Reactions of Aldehydes

Nucleophilic Addition Reactions

(i) Addition of Carbon Nucleophiles

R–C(=O)–H (Aldehyde):

- $\xrightarrow[\text{Aldehyde}]{HCN}$ R–C(OH)(CN)–H (Cyanohydrin)
- $\xrightarrow{RMgX}$ [R–C(OMgX)(H)–H] $\xrightarrow{H^+/H_2O}$ R–C(OH)(H)–H + Mg(X)(OH) (1° Alcohol)
- $\xrightarrow{\ddot{H}^-}$ R-C(O^-)(H)-H (Alkoxide ion) $\overset{H^+}{\rightleftharpoons}$ R-C(OH)(H)-H (1° Alcohol)

(ii) Addition of Oxygen Nucleophiles

$$R-\overset{O}{\overset{\|}{C}}-H \ (\text{Aldehyde}) \longrightarrow \begin{cases} \xrightleftharpoons[H^+]{H_2O} R-\underset{OH}{\overset{OH}{C}}-H \ (\text{Gem-diol}) \\ \xrightleftharpoons[H^+]{R'OH} R-\underset{OH}{\overset{OR'}{C}}-H \ (\text{Hemiacetal}) \xrightleftharpoons{R''OH,\ H^+} R-\underset{OR''}{\overset{OR'}{C}}-H + H_2O \ (\text{Acetal}) \\ \xrightarrow[H_2O]{Na^+HSO_3^-} R-\underset{H}{\overset{OH}{C}}-SO_3Na \ (\text{Addition product}) \end{cases}$$

Chemical Reactions of Ketones

Nucleophilic Addition Reactions

(i) Addition of Carbon Nucleophiles

$$R-\overset{O}{\overset{\|}{C}}-R' \ (\text{Ketone}) \longrightarrow \begin{cases} \xrightarrow{HCN} R-\underset{CN}{\overset{OH}{C}}-R' \ (\text{Ketone cyanohydrin}) \\ \xrightarrow[H^+/H_2O]{RMgX} R-\underset{R}{\overset{OH}{C}}-R' + Mg\begin{smallmatrix} X \\ OH \end{smallmatrix} \ (3^o \text{ Alcohol}) \\ \xrightarrow{2\ddot{H}} R-\underset{H}{\overset{OH}{C}}-R' \ (2^o \text{ Alcohol}) \end{cases}$$

(ii) Addition of Oxygen Nucleophiles

$$R-\overset{O}{\overset{\|}{C}}-R \ (\text{Ketone}) \longrightarrow \begin{cases} \xrightleftharpoons[H^+]{H_2O} R-\underset{OH}{\overset{OH}{C}}-R' \ (\text{Diol}) \\ \xrightleftharpoons[R''OH,H^+]{R'OH,\ H^+} R-\underset{OR''}{\overset{OR'}{C}}-R' + H_2O \ (\text{Ketal}) \\ \xrightarrow[H_2O]{Na^+HSO_3^-} R-\underset{R'}{\overset{OH}{C}}-SO_3Na \ (\text{Addition product}) \end{cases}$$

Addition Elimination Reactions of Aldehyde and Ketones

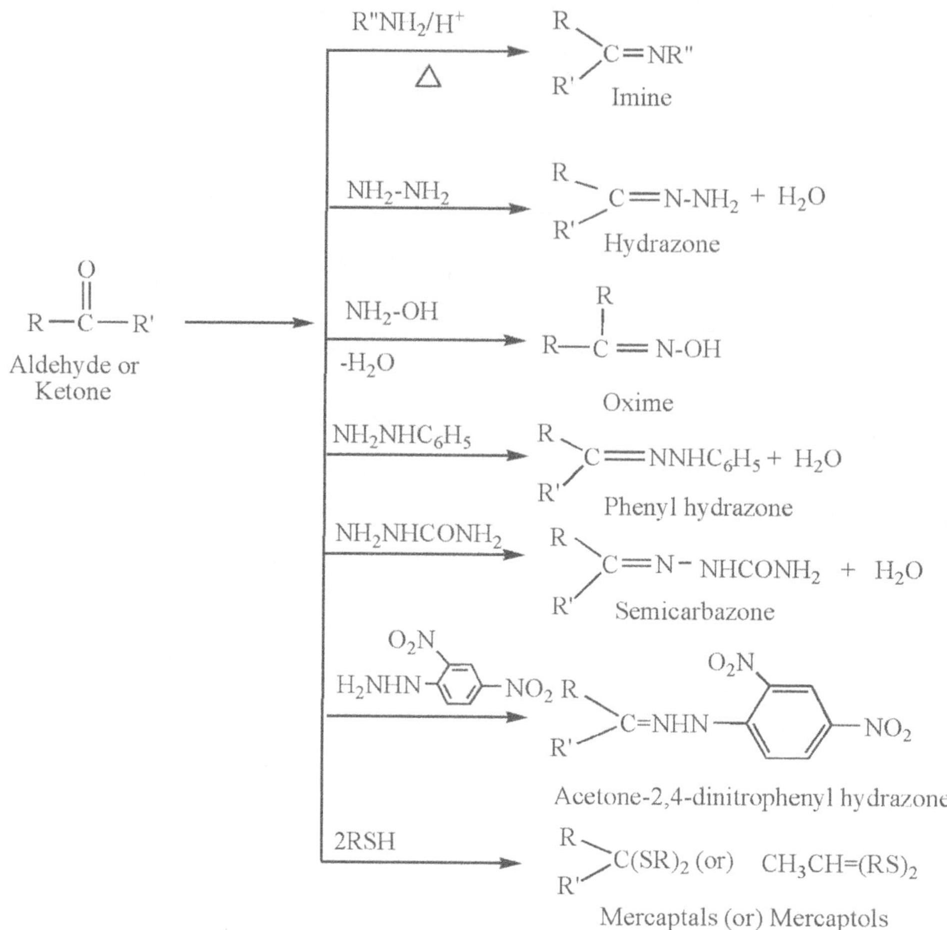

Oxidation Reactions of Ketones

$R-\overset{O}{\overset{\|}{C}}-R'$ (Ketone) ⟶

- $\xrightarrow[\text{Alk.KMnO}_4,\ \Delta]{\text{Conc.HNO}_3}$ $R-\overset{O}{\overset{\|}{C}}-OH + R'-\overset{O}{\overset{\|}{C}}-OH$ — Carboxylic acid mixture
- $\xrightarrow{2Ag(NH_3)_2OH}$ $R-\overset{O}{\overset{\|}{C}}-\overset{-}{O}\overset{+}{N}H_4 + 2Ag\downarrow + H_2O + 3NH_3$ — Metallic silver
- $\xrightarrow{2Cu(OH)_2 + NaOH}$ $RCOO\overset{+}{Na} + Cu_2O\downarrow + 3H_2O$ — Cuprous oxide (Red/Brown ppt)

Oxidation Reactions of Aldehydes

$$CH_3CH_2CH_2\overset{O}{\overset{\|}{C}}H + R-\overset{O}{\overset{\|}{C}}-OOH \longrightarrow CH_3CH_2CH_2COOH + RCOOH$$

Butanal + Peroxy acid ⟶ Butanoic acid + Carboxylic acid

$R-\overset{O}{\overset{\|}{C}}-H$ (Aldehyde) ⟶

- $\xrightarrow[H_2SO_4]{K_2Cr_2O_7}$ $R-\overset{O}{\overset{\|}{C}}-OH$ — Carboxylic acid
- $\xrightarrow{2Ag(NH_3)_2OH}$ $R-\overset{O}{\overset{\|}{C}}-\overset{-}{O}\overset{+}{N}H_4 + 2Ag\downarrow + H_2O + 3NH_3$ — Metallic silver
- $\xrightarrow{2Cu(OH)_2 + NaOH}$ $RCOO\overset{+}{Na} + Cu_2O\downarrow + 3H_2O$ — Cuprous oxide (Red/Brown ppt)

Table 15.5 Carbonyl compounds of Pharmaceutical importance.

Compound name & Structure	Uses	Qualitative test
Formaldehyde HCHO	Aqueous solution (40 %) of formaldehyde (known as formalin) is mainly used. It is used as Antiseptic and germicide. It is used in preservation of biological and anatomical specimens. In leather industries, it is used as a substitute for tannings as it hardens glue or gelatin and renders them insoluble in water. It is also used for the preparation of synthetic resins such as Bakelite and Melmac.	Dilute small amount of sample solution with 10 ml of water in a test tube. To this add 1 ml of silver ammonium nitrate solution. Metallic silver is produced either in the form of a finely divided, grey precipitate or as a bright metallic mirror is formed on the sides of the test tube. Add small quantity of sample solution to the acidified solution of salicylic acid and warm the liquid very gently, a permanent deep red colour appears.
Chloral hydrate $CCl_3CH(OH)_2$	It is used to treat alcohol withdrawal symptoms and to relieve anxiety due to withdrawal of some drugs such as narcotics and barbiturates. It is also used as sedative.	To the small amount of sample solution add 1ml of solution of sodium hydroxide. Acidify the supernatant aqueous solution with acetic acid and boil the solution with mercuric chloride. A precipitate is produced. Heat the small amount of sample with a few drops each of aniline and solution of sodium hydroxide. Phenyl isocyanide is produced which is identified by its disagreeable odour.
Paraldehyde CH_3, O, O, H_3C, O, CH_3	It is used as an anticonvulsant and sedative.	Heat small amount of sample solution with 0.1 ml of sulphuric acid. Acetaldehyde is evolved which is recognizable by its odour. To the small amount of sample solution add 5 ml of ammonical silver nitrate solution in a test tube and heat it on a water bath. Metallic silver is deposited as a mirror on the sides of the tube.
Hexamine H_2 C, N, N, H_2C, H_2C, N, CH_2, H_2C, CH_2, N	It is used as an urinary antiseptic. It is also used in industry as an accelerator in vulcanizing rubber, plastics and in the manufacture of cyclonite (high explosive).	Mix pinch of sample with an equal weight of salicylic acid and heat with 1 ml of sulphuric acid, a carmine red colour is produced. Heat sample solution with dilute sulphuric acid, decomposition takes place and formaldehyde is produced. Add excess solution of sodium hydroxide, ammonia is evolved.

Table 15.5 *Contd...*

Compound name & Structure	Uses	Qualitative test
Benzaldehyde C_6H_5CHO	It is used for the synthesis of α,β-unsaturated acids and anhydrides. It is also used as a flavouring agent, in perfume preparations and in the manufacturing of triphenyl methane dyes.	**Schiff's reagent test:** To the small amount of sample solution add 3 ml of Schiff's reagent and shake well. Pink or purple color is formed immediately. **Indicates the presence of aromatic aldehyde**.
Cinnamaldehyde $C_6H_5CH{=}CHCHO$	It is used as cinnamon essence and flavouring agent.	**Schiff's reagent test:** To the small amount of sample solution add 3 ml of Schiff's reagent and shake well. Pink or purple color is formed on standing.
Acetone or Dimethyl ketone 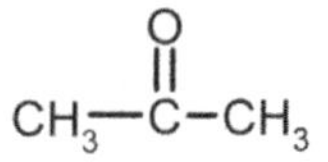 $CH_3{-}C({=}O){-}CH_3$	It is used for the production of chloroform and diacetone alcohol and used as a solvent for varnishes, lacquers, resins etc. It is also used for the manufacturing of thermo softening plastic such as Perspex.	**Iodoform test:** To the small amount of sample solution in appropriate solvent (water/ methanol/ dioxane), add 1 ml of 10% NaOH solution and I_2-KI reagent dropwise. Allow it to stand for 2-3 minutes. Yellow precipitate is appeared. Indicates the presence of ketones.
Vanillin 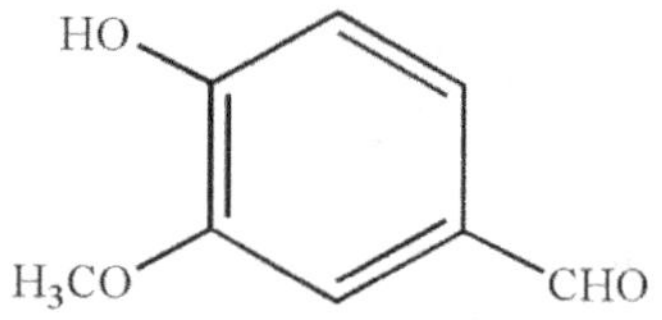 HO, H_3CO, CHO	It is mainly used as a flavouring agent.	**Identification tests:** 1. To the sample solution add few drops of ferric chloride solution, it forms blue color. 2. It is also identified by spectrophotometric method.

Table 15.6 Identification tests for aldehyde & ketones.

S. No.	Procedure	Observation	Inference
Tests for aldehydes:			
1.	**Fehling's test:** To the small amount of sample solution, add 3 ml of both Fehling's solution I and II. Warm the solution in the water bath.	Copious red precipitate is obtained.	Indicates the presence of aldehyde.
	$\underset{\text{Aldehyde}}{R{-}\overset{O}{\overset{\|\|}{C}}{-}H} + 2Cu(OH)_2 \xrightarrow{\Delta} \underset{\text{(copious red precipitate)}}{Cu_2O} + RCOOH + 2H_2O$		

Table 15.6 Contd...

S. No.	Procedure	Observation	Inference
2.	**Benedict's test:** To the small amount of sample solution, add 2 ml of Benedict's reagent and boil for 2-3 minutes.	Copious red precipitate is obtained.	Indicates the presence of aldehyde.
	$R-\overset{O}{\overset{\Vert}{C}}-H + 2Cu(OH)_2 \xrightarrow{\Delta} Cu_2O + RCOOH + 2H_2O$ Aldehyde (copious red precipitate)		
3.	**Tollen's test:** To the small amount of sample solution add 2 ml of Tollen's reagent and boil for 5-6 minutes.	Silver mirror is deposited on the sides of the test tube.	Indicates the presence of aldehyde.
	$R-\overset{O}{\overset{\Vert}{C}}-H + 2Ag(NH_3)_2NO_3 \xrightarrow{\Delta} 2Ag\downarrow + RCOONH_4 + 3NH_3 + 2H_2O$ Aldehyde; Ammonical silver nitrate; Silver mirror		
4.	**Schiff's reagent test:** To the small amount of sample solution add 3 ml of Schiff's reagent and shake well.	(a) Pink or purple color is formed immediately.	Indicates the presence of aliphatic aldehyde.
		(b) Pink or purple color is formed on standing.	Indicates the presence of aromatic aldehyde.
Tests for ketones:			
1.	**Sodium nitro prusside test:** To the sample solution add excess amount of 30% sodium nitro prusside solution.	Orange color changes to pink color.	Indicates the presence of methyl ketones.
2.	**Zimmerman test:** To the sample add *m*-dinitro benzene powder and excess of sodium hydroxide solution and mix it well.	(a) Red color appears immediately.	Indicates the presence of methyl ketones.
		(b) Light red color is formed on standing.	Indicates the presence of benzophenone.
		(c) Yellow to red on standing.	Indicates the presence of salicylaldehyde.
3.	**Iodoform test:** To the small amount of sample solution in appropriate solvent (water/ methanol/dioxane), add 1 ml of 10% NaOH solution and I_2-KI reagent drop wise. Allow it to stand for 2-3 minutes.	Yellow precipitate is appeared.	Indicates the presence of ketones.
	$NaOH + I_2/KI \longrightarrow NaOI + HI$ $R-\overset{O}{\overset{\Vert}{C}}-CH_3 + 3\,NaOI \longrightarrow R-\overset{O}{\overset{\Vert}{C}}-CI_3 + 3NaOH$ $R-\overset{O}{\overset{\Vert}{C}}-CI_3 + NaOH \longrightarrow R-\overset{O}{\overset{\Vert}{C}}-ONa + CHI_3$ (Iodoform, yellow precipitate) $R-\overset{O}{\overset{\Vert}{C}}-ONa + H_2O \longrightarrow R-\overset{O}{\overset{\Vert}{C}}-OH + NaOH$		

Probable Questions

1. What are carbonyl compounds? Explain their importance.
2. Write a detailed note on nomenclature of aldehydes & ketones.
3. List out general methods of preparation of aldehydes & ketones and explain any five methods in detail.
4. Explain Rosenmund's reduction.
5. Write a note on Stephen's method of synthesizing aldehydes.
6. Explain about Wacker process of carbonyl compound preparation.
7. How are carbonyl compounds synthesized from alcohols?
8. Write a note on ozonolysis of alkenes.
9. How are carbonyl compounds synthesized from alkynes?
10. Write about general structure & various physical properties of aldehydes & ketones.
11. Discuss in detail about various chemical properties of carbonyl compounds.
12. Explain nucleophilic addition reactions of carbonyl compounds.
13. Write short notes on addition – elimination reactions of aldehydes & ketones.
14. Write a note on oxidation reactions of aldehydes & ketones.
15. Write a note on reduction reactions of aldehydes & ketones.
16. Explain the relative reactivity of carbonyl compounds in nucleophilic addition reactions.
17. Write the general mechanism involved in nucleophilic addition reactions.
18. Describe in detail about two types of nucleophilic addition reactions of aldehydes & ketones.
19. How do aldehydes / ketones react with various amine derivatives?
20. Write the reactions of aldehyde with Tollens reagent, Benedict's reagent & Fehling's solution.
21. Explain in detail about Baeyer – Villiger oxidation.
22. What is Clemmensen reduction & Wolf-Kishner reduction?
23. Describe in detail about Meerwin-Pondorf-Verley reduction with its reaction mechanism.
24. Explain Willgerodt reaction & Darzein condensation reactions.
25. Write in detail about Aldol condensation reactions.
26. Write short notes on mixed/crossed Aldol condensation.
27. Explain the applications of Aldol condensation reactions.
28. Write the reaction mechanism, drawbacks & synthetic uses of Reformatsky reaction.
29. Write the reaction mechanism & applications of Cannizaro reaction.
30. Explain crossed Cannizaro reaction with suitable example.
31. What is intermolecular or internal Cannizaro reaction?
32. Explain the limitations of Cannizaro reaction.
33. Explain in detail some of the reactions of aldehydes which are not given by ketones.
34. What is Tischenko reaction?
35. Write the reaction mechanism, & synthetic applications of haloform reaction.
36. Write the chemical structure and uses of formaldehyde, paraldehyde, acetone, chloralhydrate, hexamine, benzaldehyde, vanilin, and cinnamaldehyde.

16

Carboxylic Acids

Introduction

These are the carboxyl derivatives of hydrocarbons in which one or more hydrogen atoms are replaced with one or more carboxyl (–COOH) groups as in the following examples.

$$\underset{\text{Methane}}{CH_4} \xrightarrow[\text{by one - COOH group}]{\text{Replacement of one H}} \underset{\text{Acetic acid}}{CH_3COOH}$$

$$\underset{\text{Methane}}{CH_4} \xrightarrow[\text{by two - COOH group}]{\text{Replacement of two H}} CH_2(COOH)_2$$

Malonic acid

Carboxylic acid containing one – COOH group is known as mono carboxylic acids while others having two or three – COOH groups are known as di and tri-carboxylic acids respectively. The carboxyl group consists of a **carbo**nyl group attached to a hydro**xyl** group and therefore named as **carboxyl.**

Importance of Carboxylic Acids

Carboxylic acids of lower members (C_1-C_6) along with –OH, -NH_2, -CO *etc* are important intermediates in the biosynthesis of fats, proteins *etc*. Fatty acids (*i.e,* higher members of carboxylic acids possessing more than ten carbons exist in nature as ester with glycerol as fats). Fatty acids with biological significance are called as prostaglandins which cause inflammation and rise in body temperature.

Carboxyl groups when attached with various chemical structures elicit different kind of pharmacological responses based on the core structure to which it is attached. Some of the drugs which possess carboxyl groups and their therapeutic uses in the treatment of various diseases are shown in Table 16.1.

Table 16.1 Drugs containing carboxyl groups and their therapeutic uses.

Category	Drug	Structure
Antiparkinsonism drug	Levodopa	3,4-(HO)$_2$C$_6$H$_3$–H_2C–C(H)(NH_2)–COOH

Table 16.1 *Contd...*

Antiparkinsonism drug	Carbidopa	HO–C₆H₃(HO)–H_2C–C(CH_3)($NHNH_2$)–COOH
Antihyperlipidemic drug	Nicotinic acid (Niacin)	Pyridine-3-COOH
Antihypertensive drug	Ethacrynic acid	H_3CH_2C–C(=CH_2)–C(=O)–C₆H₂(Cl)(Cl)–OCH_2COOH
	α-Methyl dopa	HO, HO–C₆H₃–CH_2–C(NH_2)(CH_3)–COOH
	Captopril	CH_3; CO–CH–CH_2SH; N (pyrrolidine), H, COOH

Nomenclature of Carboxylic Acids

They can be named by three ways:

1. **Common name system:** Based on the sources of their origin their name is derived from the Greek or Latin words. They follow the suffix –ic acid only.
   In this system, Greek letters α, β and γ *etc* are used to indicate the position of substituents. The carbon atom adjacent to –COOH group is named as α and the next carbon atom is β and so on.

$$CH_3-\underset{\beta\;H}{\overset{CH_3}{C}}-\underset{\alpha\;H}{\overset{CH_3}{C}}-COOH$$

α,β-Dimethylbutyric acid

2. **Derived name system:** In this method, the carboxylic acids are named as derivatives of acetic acid.

$$H_3C-\underset{CH_3}{\overset{CH_3}{C}}-COOH \quad \text{Trimethyl acetic acid}$$

$$C_6H_5-CH_2-COOH \quad \text{Phenylacetic acid}$$

3. **IUPAC name system :** In this system, the name of individual acid is derived by replacing the end **- e** of alkane with **-oic** acid. For example the name pentanoic acid is derived from pentane.

$$\underset{\text{Pentane}}{CH_3-CH_2-CH_2-CH_2-CH_3} \xrightarrow[\text{+ oic acid}]{\text{-e}} \underset{\text{Pentanoic acid}}{CH_3-CH_2-CH_2-CH_2-COOH}$$

While naming the higher acids, the longest chain with -COOH group is selected and numbered by starting with 1 (one) for the carbon of -COOH group and the position of substituents are indicated. For example. The common and IUPAC name of some carboxylic acids are shown in Table 16.2.

$$\underset{6}{CH_3}-\underset{5}{CH_2}-\underset{4}{CH_2}-\overset{CH_3}{\underset{H}{\underset{3}{C}}}-\underset{2}{CH_2}-\underset{1}{COOH}$$

3-Methylhexanoic acid

Table 16.2 Common name and IUPAC name of some carboxylic acids.

Structure	Common name	IUPAC name
HCOOH	Formic acid	Methanoic acid
CH_3COOH	Acetic acid	Ethanoic acid
CH_3CH_2COOH	Propionic acid	Propanoic acid
$CH_3CH_2CH_2COOH$	Butyric acid	Butanoic acid
$(CH_3)_2CHCOOH$	*iso*-Butyric acid	2-Methylpropanoic acid
HOOC—COOH	Oxalic acid	Ethanedioic acid
$HOOC-CH_2-COOH$	Malonic acid	Propanedioic acid
$HOOC-(CH_2)_2-COOH$	Succinic acid	Butanedioic acid
$HOOC-(CH_2)_3-COOH$	Glutaric acid	Pentanedioic acid
$HOOC-(CH_2)_4-COOH$	Adipic acid	Hexanedioic acid
$HOOC-CH_2-\underset{COOH}{CH}-CH_2-COOH$	–	Propane-1,2,3-tricarboxylic acid
C_6H_5—COOH	Benzoic acid	Benzenecarboxylic acid (Benz +oic acid)
C_6H_5—CH_2COOH	Phenylacetic acid	2-Phenylethanoic acid
C_6H_4(COOH)(COOH) (1,2-)	Phthalic acid	Benzene-1,2-dicarboxylic acid

Mono Carboxylic Acids or Fatty Acids

The general formula for the carboxyl homologues series is $C_nH_{2n+1}COOH$. The general formula of monocarboxylic acid is R-COOH. Monocarboxylic acids are also called fatty acid, because they can be obtained from fats. Some of the higher members of monocarboxylic acids are obtained from fats by hydrolysis, example, Palmitic acid ($C_{15}H_{31}COOH$) and stearic acid ($C_{17}H_{35}COOH$).

General Methods of Preparation

1. **By the oxidation of alcohols, aldehydes or ketones with dichromate solution:** Ethanol (1° alcohol) undergoes oxidation with potassium dichromate to give acetic acid *via* the formation of acetaldehyde.

$$\underset{\substack{\text{Ethanol} \\ (1^\circ\ \text{Alcohol})}}{CH_3CH_2OH} \xrightarrow[K_2Cr_2O_7]{(O)} \underset{\text{Acetaldehyde}}{CH_3CHO} \xrightarrow[K_2Cr_2O_7]{(O)} \underset{\text{Acetic acid}}{CH_3COOH}$$

Similarly, 2° alcohols undergo oxidation with potassium dichromate to give mixture of acids *via* the formation of ketones.

$$\underset{2^\circ\ \text{Alcohol}}{RR'CHOH} \xrightarrow[K_2Cr_2O_7]{(O)} \underset{\text{Ketone}}{RR'C=O} \xrightarrow[K_2Cr_2O_7]{(O)} \underbrace{RCOOH + R'COOH}_{\text{Mixture of fatty acids}}$$

$$\underset{\substack{\text{Butan-2-ol} \\ 2^\circ\ \text{Alcohol}}}{(H_3C)(H_3CH_2C)CHOH} \xrightarrow[K_2Cr_2O_7]{(O)} \underset{\text{Butan-2-one}}{(H_3C)(H_3CH_2C)C=O} \xrightarrow[K_2Cr_2O_7]{(O)} \underset{\text{Acetic acid}}{CH_3COOH} + \underset{\text{Propanoic acid}}{CH_3CH_2COOH}$$

Note: Sometimes an ester is obtained as a product because the part of alcohol is oxidized to the acids which further reacts and form ester with remaining unreacted alcohol.

2. **Hydrolysis of cyanides with acid or alkali:** This method is a very good synthetic method for the synthesis of carboxylic acids.

$$\underset{\text{Alkyl nitrile}}{R-C\equiv N} \xrightarrow{H_2O} \underset{\substack{\text{Amide} \\ \text{Intermediate} \\ \text{product (not isolated)}}}{R-\overset{O}{\overset{\|}{C}}-NH_2} \xrightarrow{H_2O} \underset{\text{Acid}}{RCOOH} + NH_3\uparrow$$

The alkyl nitrile needed for this reaction is easily obtained from alkyl halides.

Mechanism:

Acid hydrolysis:

Step 1: Protonation of nitrogen of alkyl nitrile.

$$\underset{\text{Nitrile}}{R-C\equiv N:} \overset{H^+}{\rightleftharpoons} \left[R-C\equiv N^+H \longleftrightarrow RC^+=\ddot{N}H\right]$$

Step 2: Attack of water molecule to the positively charged carbon atom to give an amide intermediate.

$$\left[R-C\equiv N^{+}H \longleftrightarrow R-\overset{+}{C}=\ddot{N}H\right] \xrightleftharpoons{H\ddot{O}H} \left[\begin{matrix}R-C=\ddot{N}H\\ |\\ H\overset{+}{\ddot{O}}H\end{matrix}\right] \underset{H^+}{\overset{-H^+}{\rightleftharpoons}} \left[\begin{matrix}RC=\ddot{N}H\\ |\\ :\ddot{O}H\end{matrix}\right]$$

$$\left[\begin{matrix}RC=\ddot{N}H\\ |\\ :\ddot{O}H\end{matrix}\right] \overset{H^+}{\rightleftharpoons} \left[\begin{matrix}RC=\overset{+}{N}H_2\\ |\\ :\ddot{O}H\end{matrix}\right] \longleftrightarrow \begin{matrix}R-C-\ddot{N}H_2\\ \|\\ \ddot{O}:\end{matrix}$$

(Amide intermediate)

Step 3 : Hydrolysis of amide intermediate gives carboxylic acid.

$$\underset{\text{Amide intermediate}}{\begin{matrix}R-C-\ddot{N}H_2\\ \|\\ \ddot{O}:\end{matrix}} \xrightarrow{H_2O/H^+} \underset{\text{Carboxylic acid}}{\begin{matrix}OH\\ |\\ R-C\\ \|\\ \ddot{O}:\end{matrix}} + NH_4^+$$

Note: Ammonium ion is formed by the reaction between ammonia and H^+, so excess amount of acid is used in this reaction.

Alkaline hydrolysis:

Step 1: Nucleophile (-OH) attacks the Carbonium ion.

$$R-C\equiv N + :\ddot{O}H^- \overset{\Delta}{\rightleftharpoons} \left[\begin{matrix}R-C=\ddot{N}:\\ |\\ :\ddot{O}H\end{matrix}\right] \xrightarrow[-OH^-]{H-OH} \left[\begin{matrix}R-C=\ddot{N}H\\ |\\ :\ddot{O}-H\end{matrix}\right] \xrightarrow[-H_2O]{H-\ddot{O}H} \underset{\text{Amide intermediate}}{\begin{matrix}R-C-\ddot{N}H_2\\ \|\\ :\ddot{O}\end{matrix}}$$

Step 2: Hydrolysis of an amide intermediate to give carboxylic acid.

$$\underset{\text{Amide intermediate}}{\begin{matrix}R-C-\ddot{N}H_2\\ \|\\ \ddot{O}:\end{matrix}} \xrightarrow[-NH_3]{OH^-} \underset{\text{Carboxylate ion}}{\begin{matrix}O^-\\ |\\ R-C\\ \|\\ \ddot{O}:\end{matrix}} \xrightarrow{H^+} \underset{\text{Carboxylic acid}}{\begin{matrix}R-C-OH\\ \|\\ O\end{matrix}}$$

3. **From Grignard reagent:** Grignard reagent on treatment with CO_2 followed by hydrolysis gives carboxylic acids.

$$\underset{\text{Methyl magnesium iodide}}{CH_3MgI} + \underset{\text{Carbondi oxide}}{\begin{matrix}O\\ \|\\ C=O\end{matrix}} \longrightarrow \underset{\text{Addition product}}{CH_3-\overset{\overset{O}{\|}}{C}-OMgI} \xrightarrow{H_2O} \underset{\text{Acetic acid}}{CH_3COOH} + Mg\begin{matrix}\diagup OH\\ \diagdown I\end{matrix}$$

Mechanism:

$$CH_3-MgI + \overset{O}{\overset{\|}{C}}=O \longrightarrow \underset{\text{Addition product}}{CH_3-\overset{O}{\overset{\|}{C}}-OMgI} \xrightarrow{H_2O} \underset{\text{Acetic acid}}{CH_3-\overset{O}{\overset{\|}{C}}-OH} + Mg\begin{matrix}OH\\I\end{matrix}$$

4. **From malonic or aceto acetic ester:** Several kind of fatty acids are conveniently synthesized from malonic or acetoacetic ester. For example, alkyl substituted acetoacetic ester on hydrolysis with conc. KOH yields the respective alkyl acetic acid.

$$\underset{\text{Alkyl aceto acetic ester}}{CH_3-\overset{OH}{\overset{O}{\overset{\|}{C}}}\,\vdots\,\overset{H}{\overset{R}{\overset{|}{C}H}}-\overset{OH}{\overset{O}{\overset{\|}{C}}}\,\vdots\,\overset{H}{}OC_2H_5} \xrightarrow[\text{with KOH}]{\text{Hydrolysis}} \underset{\text{Acetic acid}}{CH_3COOH} + \underset{\text{Alkyl acetic acid}}{RCH_2COOH} + \underset{\text{Ethanol}}{C_2H_5OH}$$

5. **Hydrolysis of esters:** Esters are boiled with aqueous NaOH solution to give sodium salt of an acid which upon further hydrolysis yields respective carboxylic acid.

$$\underset{\text{Ester}}{RCOOR'} + NaOH \xrightarrow{\Delta} \underset{\text{Sodium carboxylate}}{RCOONa} + \underset{\text{Alcohol}}{R'-OH}$$

$$\underset{\text{Sodium carboxylate}}{RCOONa} \xrightarrow{\text{Dil. HCl}} \underset{\text{Carboxylic acid}}{RCOOH} + NaCl$$

$$\underset{\text{Ethyl acetate}}{CH_3-\overset{O}{\overset{\|}{C}}-O-C_2H_5} + NaOH \xrightarrow{\Delta} \underset{\text{Sodium acetate}}{CH_3\overset{O}{\overset{\|}{C}}-ONa} + \underset{\text{Ethanol}}{C_2H_5OH}$$

$$\underset{\text{Sodium acetate}}{CH_3-\overset{O}{\overset{\|}{C}}-ONa} \xrightarrow{\text{Dil. HCl}} \underset{\text{Acetic acid}}{CH_3COOH} + NaCl$$

6. **By heating dicarboxylic acids:** On heating dicarboxylic acids, one molecule of CO_2 gets eliminated to yield respective carboxylic acids.

$$\underset{\text{Malonic acid}}{H_2C\begin{matrix}COOH\\COOH\end{matrix}} \xrightarrow{\Delta} \underset{\text{Acetic acid}}{H_3C-COOH} + CO_2\uparrow$$

$$\underset{\text{Oxalic acid}}{\begin{matrix}COOH\\|\\COOH\end{matrix}} \xrightarrow{\Delta} \underset{\text{Formic acid}}{HCOOH} + CO_2\uparrow$$

Mechanism:

$$R-\underset{\underset{COOH}{|}}{CH}-\overset{\overset{O}{\|}}{C}-O-H \xrightarrow{-CO_2} R-CH_2-COOH$$

Alkyl malonic acid → Alkyl aceticacid

7. **From sodium alkoxide:** On heating sodium alkoxide with carbon monoxide under pressure, the sodium salt of carboxylic acid is obtained which upon further acid hydrolysis gives carboxylic acid.

$$\underset{\text{Sodium alkoxide}}{R-ONa} + CO \xrightarrow[\text{Pressure}]{\Delta} \underset{\text{Sodium carboxylate}}{RCOONa} \xrightarrow[-NaCl]{HCl} \underset{\text{Carboxylic acid}}{RCOOH}$$

$$\underset{\text{Sodium methoxide}}{CH_3-ONa} + CO \xrightarrow[\text{Pressure}]{\Delta} \underset{\text{Sodium acetate}}{CH_3COONa} \xrightarrow[-NaCl]{HCl} \underset{\text{Acetic acid}}{CH_3COOH}$$

8. **Koche reaction or carboxylation of alkenes:** Alkenes when heated with carbon monoxide and steam under pressure with phosphoric acid at 400 °C yields carboxylic acids.

$$\underset{\text{Ethylene}}{CH_2=CH_2} + CO + H_2O \xrightarrow[400\ ^{o}C]{H_3PO_4} \underset{\text{Propionic acid}}{CH_3CH_2COOH}$$

$$\underset{\text{Propylene}}{CH_3-CH=CH_2} + CO + H_2O \xrightarrow[400\ ^{o}C]{H_3PO_4} \underset{\textit{iso}\text{-Butyric acid}}{CH_3-\underset{\underset{CH_3}{|}}{CH}-COOH}$$

9. **Hydrolysis of natural fats:** Higher fatty acids are obtained by the hydrolysis of natural fats, example hydrolysis of tristearin glyceride gives stearic acid.

$$\underset{\text{Tristearin glyceride}}{\begin{array}{l}CH_2-O-COC_{17}H_{35}\\ |\\ HC-O-COC_{17}H_{35}\\ |\\ CH_2-O-COC_{17}H_{35}\end{array}} + 3NaOH \longrightarrow \underset{\text{Glycerol}}{\begin{array}{l}CH_2OH\\ |\\ CHOH\\ |\\ CH_2OH\end{array}} + \underset{\text{Sodium stearate}}{3C_{17}H_{35}COONa}$$

$$3C_{17}H_{35}COONa \xrightarrow[-3NaCl]{3HCl} \underset{\text{Stearic acid}}{3C_{17}H_{35}COOH}$$

Summary of Methods of Preparation

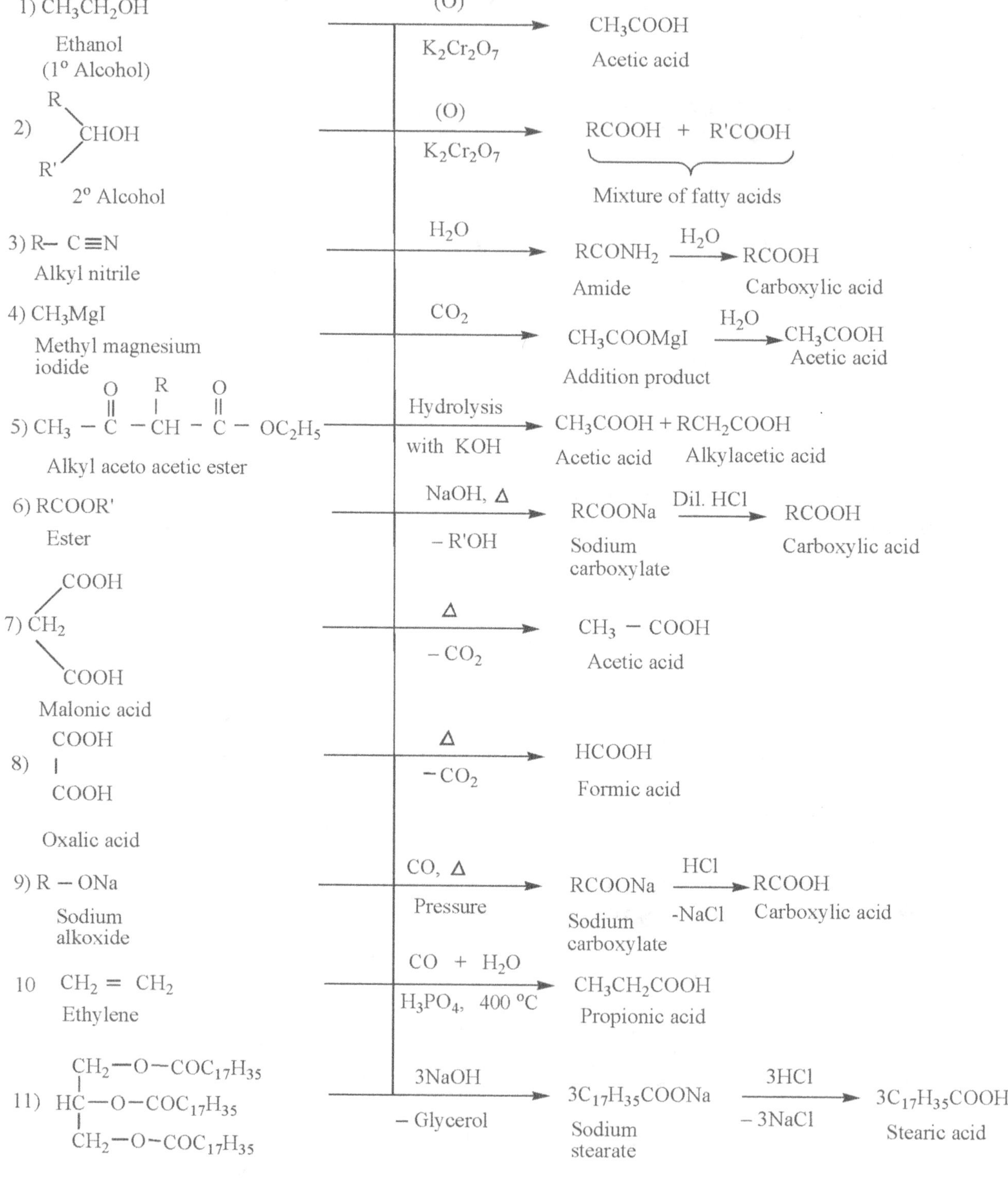

Structure

Carboxylic group is represented by two resonating structures with different stability. Structure (1) is more stable than (2). Actual structure (3) is average of these two.

$$\underset{(1)}{-\overset{\overset{O}{\|}}{C}-\ddot{O}H} \rightleftharpoons \underset{(2)}{-\overset{\overset{O^-}{|}}{C}=\overset{+}{O}-H} \qquad \underset{(3)}{-\overset{\overset{\delta^-}{O}}{\vdots}{C}\,\text{---}\,\overset{\delta^+}{O}-H}$$

Based on the molecular orbital theory, structure (3) is represented as follows. From the bond length value C-O bond length in -COOH group is found to be 1.22A⁰, which is very shorter than normal C-O σ bond (1.30A°).

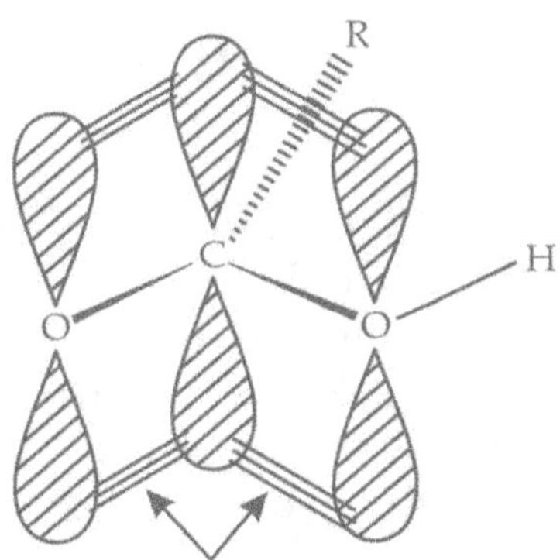

Figure 16.1 Structure of carboxylic acid.

Physical Properties

1. The first three members of carboxylic acid are colorless, corrosive and pungent smelling liquids. From $C_4H_8O_2$ (butyric acid) to $C_9H_{18}O_2$ (nonanoic acid) are oily liquid with an odour of goat butter and higher members are colorless solids.
2. First two members of acids are heavier than water. So the specific gravity gradually decreases from lower members to higher members. For example: Specific gravity of formic acid is 1.22 but specific gravity of stearic acid is 0.845.
3. The melting point of carboxylic acids shows fluctuation effect *i.e.*, carboxylic acids having even number of carbon atoms possess higher boiling point than that of acids with odd number of carbon atoms immediately below and above it.
4. Boiling points of carboxylic acids increase at regular intervals as their molecular weight increases.

Table 16.3 Melting points and boiling points of some carboxylic acids.

IUPAC Name	Common Name	Molecular Formula	M.P (°C)	B.P. (°C)
Methanoic acid	Formic acid	$HCOOH$	8	101
Ethanoic acid	Acetic acid	CH_3COOH	16	118
Propanoic acid	Propionic acid	CH_3CH_2COOH	– 21	141
Butanoic acid	Butyric acid	$CH_3(CH_2)_2COOH$	– 7	163
Pentanoic acid	Valeric acid	$CH_3(CH_2)_3COOH$	-35	186

5. When molecular weight increases, acidity decreases however acidity persists among higher molecules and they give salts and esters.

6. **Water solubility:** The first four members of acids are very freely soluble in water due to formation of hydrogen bonding. Solubility decreases when molecular weight increases. Generally, all fatty acids dissolve readily in alcohol or ether. Water solubility of carboxylic acid is explained as follows.
   (i) Carboxylic acid possesses both carbonyl and hydroxyl groups. Due to these polar groups, carboxylic acids forms hydrogen bonding with water molecule. So these are more soluble in water than that of corresponding aldehydes, ketones, alcohols *etc.* with comparable molecular weight.
   (ii) Let us consider the following example of pentanoic acid.

$$\underbrace{CH_3 - CH_2 - CH_2 - CH_2}_{\text{Hydrophobic alkyl chain}} - \underbrace{\overset{O}{\overset{\|}{C}} - OH}_{\text{Hydrophilic chain}}$$

This molecule contains two regions of different polarity. It contains a polar carboxylic acid group (increase water solubility) and non-polar alkyl chain (decrease water solubility). As, the size of alkyl side chain increases, water solubility decreases.

7. **Dimer formation:** Carboxylic acids have higher boiling point than that of alcohol and aldehydes of comparable molecular weight, because in solid or liquid state, carboxylic acids form dimeric structure (dimers) due to polarity and intermolecular hydrogen bonding.

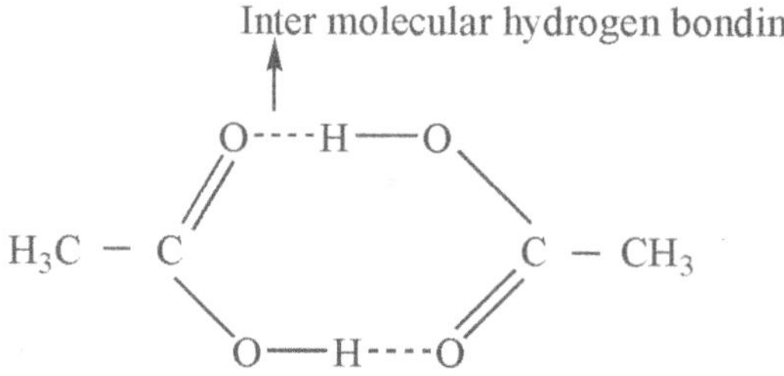

Acetic acid dimer

Chemical Properties

The molecule of carboxylic acid (R-COOH) is made up of two groups.
(1) An alkyl group (-R) and
(2) The carboxylic acid (-COOH) group.
The carboxylic carbon is less electrophilic than carbonyl carbon because of resonance. So the chemical properties of carboxylic acid are the properties of these two groups. They are described below.

$$-C(=O)-\ddot{O}-H \longleftrightarrow -C^{+}(-O^{-})-\ddot{O}-H \longleftrightarrow -C(-O^{-})=\overset{+}{O}-H$$

(1) **Reactions of alkyl groups:**
   (i) **Halogenation or Hell-Volhard-Zelinsky (HVZ) reaction:** Carboxylic acids are readily halogenated in α-position (Hydrogen atom attached to the carbon adjacent to the carboxyl group). Carboxylic acid reacts with Cl_2 or Br_2 in the presence of diffused sunlight or in the presence of halogen carrier (Iodine, red phosphorous) to yield mono halogenated derivatives.

$$R-CH_2COOH + Br_2 \xrightarrow[-HBr]{P} R-\underset{}{\overset{Br}{\overset{|}{C}}}H-COOH + Br_2 \xrightarrow[-HBr]{P} R-\underset{\underset{Br}{|}}{\overset{\overset{Br}{|}}{C}}-COOH$$

Carboxylic acid (Alkyl acetic acid) — α-Halogenated acid (Alkyl α-bromoacetic acid) — α - α' - Dihalogenated acid (Alkyl α, α'-Dibromoacetic acid)

$$CH_3-CH_2COOH + Br_2 \xrightarrow[(or)\ PBr_3\ -HBr]{P} CH_3-\overset{\overset{Br}{|}}{C}H-COOH + Br_2 \xrightarrow[-HBr]{P} CH_3-\underset{\underset{Br}{|}}{\overset{\overset{Br}{|}}{C}}-COOH$$

Propanoic acid — α-Bromo propanoic acid — α, α' - Dibromo propanoic acid

Similarly, in acetic acid, the three hydrogen atoms of the alkyl group are completely replaced by three halogen atoms.

$$CH_3COOH + Cl_2 \xrightarrow{P} ClCH_2COOH + HCl$$

Acetic acid — Monochloro acetic acid

$$ClCH_2COOH + Cl_2 \xrightarrow{P} Cl_2CHCOOH + HCl$$

Monochloro acetic acid — Dichloro acetic acid

$$Cl_2CHCOOH + Cl_2 \xrightarrow{P} Cl_3CCOOH + HCl$$

Dichloro acetic acid — Trichloro acetic acid

Mechanism:

Step 1: Attacking of PBr_3 to a carboxylic acid gives acyl bromide.

$$CH_3CH_2-\overset{\overset{\ddot{O}:}{||}}{C}-OH \xrightarrow[\text{Substitution reaction}\ -OH^-]{PBr_3} CH_3CH_2-\overset{\overset{\ddot{O}:}{||}}{C}-Br \rightleftharpoons CH_3CH=\overset{\overset{:\ddot{O}H}{|}}{C}-Br$$

Propanoic acid — An acyl bromide (Keto form) — Enol form

Step 2: Bromination of the enol form of the brominated acyl bromide.

$$CH_3CH=\overset{\overset{:\ddot{O}H}{|}}{C}-Br + Br-Br \longrightarrow CH_3\underset{\underset{Br}{|}}{C}H-\overset{\overset{+\ddot{O}H}{||}}{C}-Br + Br^- \quad (I)$$

Enol form

Step 3: Deprotonation of the above product (I) to give α-brominated acyl bromide which upon further hydrolysis gives an α-brominated carboxylic acid.

$$CH_3\underset{\underset{Br}{|}}{C}H-\overset{\overset{+\ddot{O}H}{||}}{C}-Br \underset{H^+}{\overset{-H^+}{\rightleftharpoons}} CH_3\underset{\underset{Br}{|}}{C}H-\overset{\overset{\ddot{O}:}{||}}{C}-Br \xrightarrow[-HBr]{H_2O} CH_3-\underset{\underset{Br}{|}}{C}H-COOH$$

α–Bromo acetic acid

(ii) Oxidation: The alkyl group at β-position of the carboxylic acids gets oxidized when treated with mild oxidizing agents such as hydrogen peroxide. Oxidation of butyric acid yields β-hydroxy butyric acid.

$$\underset{\text{Butyric acid}}{CH_3CH_2CH_2COOH} \xrightarrow[H_2O_2]{(O)} \underset{\beta\text{–Hydroxy butyric acid}}{CH_3\overset{\overset{OH}{|}}{C}HCH_2COOH}$$

(2) Reactions of carboxyl group: It involves three types of reactions.

(a) Reactions involving replaceable hydrogen atom.

(b) Reactions involving hydroxyl group.

(c) Reactions involving carboxyl group as a whole.

(a) Reactions involving replaceable hydrogen atom: Carboxylic acids ionize in polar solvents to give hydrogen ion and carboxylate ions.

$$\underset{\text{Carboxylic acid}}{R-COOH} \xrightleftharpoons{\text{Polar solvents}} \underset{\text{Carboxylate ion}}{RCOO^-} + H^+$$

$$\underset{\text{Acetic acid}}{CH_3-COOH} \xrightleftharpoons{\text{Polar solvents}} \underset{\text{Acetate ion}}{CH_3COO^-} + H^+$$

The above reactions indicate that carboxylic acid reacts with metals, alkali metals and alkalis to give salts. This reaction is described as follows.

(i) With alkalis and carbonates: Carboxylic acid reacts with alkalis to give salts. It decomposes carbonates and bicarbonates by evolving carbon dioxide gas with effervescence.

$$\underset{\text{Carboxylic acid}}{RCOOH} + NaOH \longrightarrow \underset{\text{Sodium carboxylate}}{RCOONa + H_2O}$$

$$\underset{\text{Acetic acid}}{CH_3COOH} + \underset{\text{Sodium hydroxide}}{NaOH} \longrightarrow \underset{\text{Sodium acetate}}{CH_3COONa} + H_2O$$

$$\underset{\text{Acetic acid}}{2CH_3COOH} + \underset{\text{Sodium carbonate}}{Na_2CO_3} \longrightarrow \underset{\text{Sodium acetate}}{2CH_3COONa} + CO_2\uparrow + H_2O$$

$$\underset{\text{Acetic acid}}{CH_3COOH} + \underset{\text{Sodium bicarbonate}}{NaHCO_3} \longrightarrow \underset{\text{Sodium acetate}}{CH_3COONa} + CO_2\uparrow + H_2O$$

This property of carboxylic acid is used to convert water insoluble acid into its water soluble salt.

(ii) Reaction with metals: Carboxylic acid reacts with strongly electropositive metals like Na, Zn *etc* and form salts with the liberation of hydrogen gas.

$$2RCOOH + Zn \longrightarrow (RCOO)_2Zn + H_2\uparrow$$
Carboxylic acid — Zinc carboxylate

$$2CH_3COOH + Zn \longrightarrow (CH_3COO)_2Zn + H_2\uparrow$$
Acetic acid — Zinc acetate

$$2RCOOH + 2Na \longrightarrow 2RCOONa + H_2\uparrow$$
Carboxylic acid — Sodium carboxylate

$$2CH_3COOH + 2Na \longrightarrow 2CH_3COONa + H_2\uparrow$$
Acetic acid — Sodium acetate

The salt of carboxylic acid is converted to the original acid by treating with dilute mineral acid.

$$(RCOO)_2Zn + 2HCl \longrightarrow 2RCOOH + ZnCl_2$$
Zinc carboxylate — Carboxylic acid

$$CH_3COONa + HCl \longrightarrow CH_3COOH + NaCl$$
Sodium acetate — Acetic acid

Acidity of Carboxylic Acids

Acidity of carboxylic acid is proportional to the relative ease with which it losses a proton leaving behind the anion. Acid strength of an acid is determined by difference in the stability of the acid and its anion. Both acid and alcohols release a proton but the former does it much more readily than later. Let us account for this fact.

Alcohols and alkoxide ions, each are represented by a single structure *i.e.,*

$$R-OH \rightleftharpoons R-O^- + H^\oplus$$
Alcohol — Alkoxide ion (No resonancehybrid)

But the carboxylic acid and the carboxylate ion are represented by two resonance structures *i.e,* resonance hybrids.

$$\left[R-C(=O)-\ddot{O}-H \longleftrightarrow R-C(-O^-)=\overset{+}{O}H\right] \rightleftharpoons H^+ + \left[R-C(=\ddot{O})-\ddot{O}:^- \longleftrightarrow R-C(-:\ddot{O}:^-)=\ddot{O}\right]$$

Carboxylic acid
Resonance hybrid of non-equivalent structures

Carboxylate anion
Resonance hybrid of equivalent structures

Both the carboxylic acid (R-COOH) and the carboxylate anions ($RCOO^-$) are stabilized by resonance but the stabilization is far greater for the anion than that for the acid due to the contribution of two equivalent resonance structures of carboxylate anion. So, the acidity or acid strength of carboxylic acid is due to the powerful resonance stabilization of its carboxylate anion.

Alternatively, the alkoxide ion produced by an alcohol does not show resonance and it is not stabilized. This lack of stabilization is responsible for the very weak acidity of alcohols than carboxylic acids.

Hence greater the stability of carboxylate anion, greater is its acid strength. The order of acid strength of various groups is as follows

$$RCOOH > HOH > ROH > HC{\equiv}CH > NH_3 > RH$$

The order of basic strength is

$$RCO\bar{O} > O\bar{H} > R\bar{O} > HC{\equiv}\bar{C} > \bar{N}H_2 > \bar{R}$$

Effect of substituents on acidity:

The above facts indicate clearly that any factor that stabilizes the anion more than it stabilizes the acid which increase the acidity. Any factor that makes the anion less stable decreases the acidity of carboxylic acid.

Effect of electron withdrawing substituents:

G ← C(O)(O)⊖ — An electron withdrawing substituent 'G' withdraws electrons and stabilises the anion (Acidity increases)

An electron withdrawing substituents stabilizes the carboxylate anion by dispersing the negative charge and therefore increases the acidity.

Effect of electron donating or electron releasing substituents:

G → C(O)(O)⊖ — An electron releasing substituent 'G' releases electrons and destabilises the anion (Acidity decreases)

Electron releasing substituents intensify the negative charge on the anion which leads to decrease in the stability of the carboxylic acids and consequently a decrease in the acidity of the acid.

For example, the electron withdrawing substituents such as halogen atom increases the acidity, while the electron releasing substituents such as alkyl groups decreases the acidity or acid strength. Acetic acid which contains alkyl group is about one-tenth as strong as formic acid and propionic acid which contains a large alkyl group is still weaker.

Due to the presence of chlorine atom, chloroacetic is 100 times stronger acid than acetic acid. Similarly dichloroacetic acid is still stronger and tri chloro acetic acid is 10,000 times very stronger than acetic acid.

So the order of relative acidity of halogen substituted acids is as follows, because electron withdrawing property of various halogens is in the same order.

$$F > Cl > Br > I$$

$$FCH_2COOH > ClCH_2COOH > BrCH_2COOH > ICH_2COOH$$

Similarly, α-chloro propionic acid is stronger acid than β-chloro propionic acid because inductive effect of chlorine in α-position is stronger than β-position.

$CH_3—CH(Cl)—COOH$
α–Chloropropionic acid
(Strong acid)

$H_2C(Cl)—CH_2—COOH$
β–Chloropropionic acid
(Weak acid)

Direct attachment of groups like phenyl or vinyl to the carboxylic acid increases acidity contrary to the expected decrease due to resonance effect as shown below.

$$H_2C{=}CH{-}C(=O)OH \longleftrightarrow H_2\overset{+}{C}{-}CH{=}C(O^-)OH$$

This is due to greater electronegativity of sp^2 hybridised carbon to which -COOH group is attached. Presence of electron withdrawing groups on the phenyl ring increases acidity due to more conjugation while the electron donating groups decreases acidity. Hence the order of acidity is 4-methoxy benzoic acid < benzoic acid < 4-nitro benzoic acid. More pK_a value means weak acid.

COOH / OCH_3 — 4-Methoxy benzoic acid (pka = 4.46)

COOH — Benzoic acid (pka = 4.19)

COOH / NO_2 — 4-Nitro benzoic acid (pka = 3.41)

Determination of acidity constant or acidity of carboxylic acids:

The ionization of carboxylic acid in water is as follows.

$$RCOOH + H_2O \rightleftharpoons RCO\bar{O} + H_3O^+$$

The equilibrium constant for the system is known as ionization constant (K_a) which indicates the measure of the acidity of the acids.

$$K_a = \frac{[RCO\bar{O}]\,[H_3O^+]}{[RCOOH]}$$

The acid strength is indicated by pKa value rather than Ka value.

$$pK_a = -\log Ka$$

The pKa of HCl, CF_3COOH, C_6H_5COOH and CH_3COOH are -7.0, 0.23, 4.19 and 4.76 respectively.

An acid dissociation constant, K_a, (also known as acidity constant, or acid-ionization constant) is a quantitative measure of the strength of an acid in solution.

Table 16.4 pK_a values of some phenols and ethanol.

Compound	Mol. Formula	pKa
o-Nitrophenol	O—NO_2—C_6H_4—OH	7.2
m-Nitrophenol	m —NO_2—C_6H_4—OH	8.3
p-Nitrophenol	p —NO_2—C_6H_4—OH	7.1
Phenol	C_6H_5—OH	10.0
o-Cresol	O—CH_3—C_6H_4—OH	10.2
m-Cresol	m —CH_3—C_6H_4—OH	10.1
p-Cresol	P—CH_3 —C_6H_4—OH	10.2
Ethanol	C_2H_5OH	15.9

(b) Reactions involving the hydroxyl group:

(i) Reaction with alcohols or esterification reaction: Carboxylic acid reacts with alcohols in the presence of dehydrating agents like conc. H_2SO_4 or anhydrous zinc chloride to give esters. This reaction is also called **Fischer esterification.**

$$\underset{\text{Carboxylic acid}}{RCOOH} + \underset{\text{Alcohol}}{OH - R'} \xrightleftharpoons{\text{Conc. } H_2SO_4} \underset{\text{Ester}}{RCOOR'} + H_2O$$

$$\underset{\text{Acetic acid}}{CH_3COOH} + \underset{\text{Ethanol}}{HOC_2H_5} \xrightleftharpoons{\text{Conc. } H_2SO_4} \underset{\text{Ethyl acetate}}{CH_3COOC_2H_5} + H_2O$$

The reaction is favoured to the right side of the equation by the use of excess alcohol or removal of water by using dehydrating agent like conc. H_2SO_4.

Mechanism:

Step 1: Protonation of the carbonyl portion of carboxylic acid.

$$R - \overset{O}{\overset{\|}{C}} - OH + H^+ \rightleftharpoons R - \underset{+}{\overset{OH}{\overset{|}{C}}} - OH$$

Step 2: Attack of nucleophile.

$$R - \underset{+}{\overset{OH}{\overset{|}{C}}} - OH + H\ddot{O} - R' \rightleftharpoons R - \underset{\underset{OH}{|}}{\overset{OH}{\overset{|}{C}}} - \underset{+}{\overset{H}{\overset{|}{O}}} - R'$$

Step 3: Transfer of hydrogen ion (Hydride ion transfer).

$$R - \underset{\underset{OH}{|}}{\overset{:\ddot{O}H}{\overset{|}{C}}} - \underset{+}{\overset{H}{\overset{|}{O}}} - R' \rightleftharpoons R - \underset{\underset{OH}{|}}{\overset{\overset{+}{O}H_2}{\overset{|}{C}}} - O - R'$$

Step 4: Removal of water molecule and proton to give an ester.

$$R - \underset{\underset{O-H}{|}}{\overset{\overset{+}{O}H_2}{\overset{|}{C}}} - O - R' \rightleftharpoons \underset{\text{Ester}}{R - \underset{\underset{O}{\|}}{C} - OR'} + H_2O + H^+$$

(ii) Reaction with thionyl chloride or phosphorous halides: Carboxylic acids react with thionyl chloride or phosphorous trichloride or phosphorous pentachloride to give acid chlorides.

$$RCOOH + SOCl_2 \longrightarrow RCOCl + SO_2 + HCl$$

Carboxylic acid, Thionyl chloride, Acid chloride

$$CH_3COOH + SOCl_2 \longrightarrow CH_3COCl + SO_2 + HCl$$

Acetic acid, Thionyl chloride, Acetyl chloride

$$CH_3COOH + PCl_5 \longrightarrow CH_3COCl + POCl_3 + HCl$$

Acetic acid, Phosphorous penta chloride, Acetyl chloride, Phosphoryl chloride

$$3CH_3COOH + PCl_3 \longrightarrow 3CH_3COCl + H_3PO_3$$

Acetic acid, Phosphorous trichloride, Acetyl chloride, Phosphoric acid

Mechanism:

Step 1: Formation of acyl chlorosulphite intermediate by reaction of carboxylic acid and thionyl chloride.

$$R-\overset{O}{\overset{\|}{C}}-OH + SOCl_2 \xrightarrow{-HCl} R-\overset{O}{\overset{\|}{C}}-O-\overset{O}{\overset{\|}{S}}-Cl$$

Carboxylic acid → Acylchloro sulphite

Step 2: Formation of tetrahedral intermediate

The chloride nucleophile attacks the carbonyl carbon of the acyl chlorosulphite intermediate to yield tetrahedral intermediate.

$$R-\overset{O}{\overset{\|}{C}}-O-\overset{O}{\overset{\|}{S}}-Cl + Cl^- \longrightarrow R-\underset{Cl}{\underset{|}{\overset{O^-}{\overset{|}{C}}}}-O-\overset{O}{\overset{\|}{S}}-Cl$$

Acyl chlorosulphite → Tetrahedral intermediate

Step 3: Elimination of chlorosulphite ion leads to the formation of acyl chloride. The chloro sulphite group further breaks down into SO_2 and chloride ion.

$$R-\underset{Cl}{\underset{|}{\overset{O^-}{\overset{|}{C}}}}-O-\overset{O}{\overset{\|}{S}}-Cl \longrightarrow R-\overset{O}{\overset{\|}{C}}-Cl + SO_2 + Cl^-$$

Tetrahedral intermediate → Acyl chloride

(iii) Dehydration or formation of anhydrides: Acetic acid vapours when passed over a heated catalyst ($NaNH_4 . HPO_4 + BPO_4$) at 875-895 K give acid anhydride.

$$2CH_3COOH \xrightarrow[877\text{-}895\ K]{NaNH_4HPO_4/BPO_4} (CH_3CO)_2O + H_2O$$

Acetic acid — Acetic anhydride

(OR)

$$CH_3-\overset{O}{\overset{||}{C}}-O-H + HO-\underset{O}{\underset{||}{C}}-CH_3 \xrightarrow[877\text{-}895\ K]{NaNH_4HPO_4/BPO_4} (CH_3-\overset{O}{\overset{||}{C}})_2O + H_2O$$

Acetic acid — Acetic anhydride

Higher acid anhydrides are prepared by the reaction between acids and acetic anhydride.

$$2RCOOH + (CH_3CO)_2O \rightleftharpoons (RCO)_2O + 2CH_3COOH$$

Carboxylic acid — Acid anhydride

Acid anhydrides are also obtained by the reaction between acid chlorides and sodium salts of carboxylic acids.

$$R-\overset{O}{\overset{||}{C}}-ONa + Cl-\overset{O}{\overset{||}{C}}-R' \longrightarrow R-\overset{O}{\overset{||}{C}}-O-\overset{O}{\overset{||}{C}}-R' + NaCl$$

Sodium carboxylate — Acid chloride — Acid anhydride

$$CH_3-\overset{O}{\overset{||}{C}}-ONa + Cl-\overset{O}{\overset{||}{C}}-CH_3 \longrightarrow (CH_3-\overset{O}{\overset{||}{C}})_2O + NaCl$$

Sodium acetate — Acetyl chloride — Acetic anhydride

(c) Reactions involving carboxyl group as whole:

(i) Reduction: The reduction of carboxylic acid under normal condition is a difficult process because -COOH group is very resistant to reduction. Prolonged heating of carboxylic acid under pressure with concentrated hydroiodic acid and small quantity of red phosphorous gives paraffins or alkanes. Reduction can also be achieved by heating carboxylic acid with hydrogen at high temperature and pressure in presence of metal catalyst such as Ni.

$$RCOOH + 3H_2 \xrightarrow[\Delta]{Ni} RCH_3 + 2H_2O$$

Carboxylic acid — Alkane

Reduction with lithium aluminium hydride:

Carboxylic acids are reduced with $LiAlH_4$ to give primary alcohols with better yield.

$$RCOOH \xrightarrow{LiAlH_4,\ Ether\ or\ THF} RCH_2OH + LiOH + Al(OH)_3$$

Carboxylic acid — 1° alcohol

This reaction is carried out in the presence of tetrahydrofuran (THF) or ether or some aprotic solvents because in the presence of water $LiAlH_4$ violently reacts with hydrogen of the water molecule to form salts of lithium, aluminium and hydrogen gas.

(ii) Dry distillation: The anhydrous alkali salts of fatty acids undergo dry distillation with soda lime to give alkanes or paraffins. This reaction is called as **decarboxylation.**

$$\underset{\text{Sodium Carboxylate}}{R-COONa} + NaOH \xrightarrow{CaO} \underset{\text{Alkane}}{R-H} + Na_2CO_3$$

$$\underset{\text{Sodium acetate}}{CH_3-COONa} + NaOH \xrightarrow{CaO} \underset{\text{Methane}}{CH_4} + Na_2CO_3$$

(iii) Kolbes reaction: Electrolysis of concentrated aqueous solution of alkali metal salts of carboxylic acids yields paraffins.

$$\underset{\text{Sodium acetate}}{CH_3COONa} \rightleftharpoons \underset{\text{Acetate ion}}{CH_3COO^-} + Na^+$$

At anode:

$$2CH_3COO^- \longrightarrow \underset{\text{Ethane}}{CH_3-CH_3} + 2CO_2 + 2e^-$$

At cathode:

$$2H_2O + 2e^- \longrightarrow 2OH^- + H_2$$

(iv) Formation of acid amides: Strong heating of ammonium salts of carboxylic acids yields acid amides.

$$\underset{\text{Ammonium salt of carboxylic acid}}{RCOONH_4} \xrightarrow{\Delta} \underset{\text{Acid amide}}{RCONH_2} + H_2O$$

$$\underset{\text{Ammonium acetate}}{CH_3COONH_4} \xrightarrow{\Delta} \underset{\text{Acetamide}}{CH_3CONH_2} + H_2O$$

(v) Formation of cyanide: Dehydration of ammonium salts of carboxylic acid with phosphorous pentoxide gives alkyl cyanide or nitrile.

$$\underset{\text{Ammonium acetate}}{CH_3COONH_4} \xrightarrow{P_2O_5} \underset{\text{Methyl cyanide}}{CH_3-C\equiv N} + 2H_2O$$

(vi) Formation of aldehyde: Calcium salt of fatty acid other than formic acid when strongly heated yields ketone.

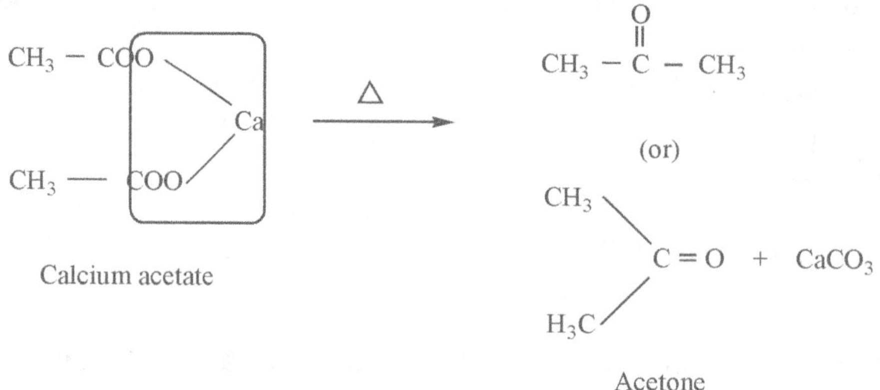

When calcium acetate is heated with calcium formate aldehyde is formed

$$\underset{\text{Calcium acetate}}{(CH_3-COO)_2Ca} + \underset{\text{Calcium formate}}{Ca(OOC-H)_2} \xrightarrow{\Delta} \underset{\text{Acetaldehyde}}{2CH_3CHO} + 2CaCO_3$$

(vii) Hunsdiecker reaction: Silver salt of carboxylic acid on heating with halogen gives an alkyl halide.

$$\underset{\text{Silver carboxylate}}{RCOOAg} + Br_2 \xrightarrow{\Delta} \underset{\text{Alkyl bromide}}{R-Br} + AgBr + CO_2\uparrow$$

$$\underset{\text{Silver acetate}}{CH_3-\overset{O}{\overset{\|}{C}}-O^-\,Ag^+} + Br_2 \xrightarrow{\Delta} \underset{\text{Methyl bromide}}{CH_3Br} + AgBr + CO_2\uparrow$$

(viii) Oxidation: All the monocarboxylic acids except formic acid are extremely resistant to oxidation. These are oxidized to carbon dioxide and water on prolonged heating with strong oxidizing agents. Mild oxidizing agents oxidises carboxylic acid into β-hydroxy acids.

Summary of Chemical Reactions

$R-COOH$ (Carboxylic acid) →

- Polar solvents ⇌ $RCOO^- + H^+$ (Carboxylate ion)
- $NaOH$, Na_2CO_3 / $NaHCO_3$ → $RCOONa + H_2O + CO_2\uparrow$ (Sodium carboxylate)
- Zn → $(RCOO)_2Zn + H_2\uparrow$ (Zinc carboxylate)
- $R'OH$ / Conc.H_2SO_4 ⇌ $RCOOR' + H_2O$ (Ester)
- $SOCl_2 / PCl_5$ (or) PCl_3 → $RCOCl + SO_2 + HCl$ (Acid chloride)
- $-H_2O$ (2Moles) → $(RCO)_2O$ (Acid anhydride)
- $(CH_3CO)_2O$ (2Moles) ⇌ $(RCO)_2O + 2CH_3COOH$ (Acid anhydride, Acetic acid)
- $Ni/3H_2$, Δ → $RCH_3 + 2H_2O$ (Alkane)
- $LiAlH_4$, Ether (or) THF, H_2O → $RCH_2OH + LiOH + Al(OH)_3$ (1° Alcohol)
- $NaOH + CaO$ → $R-H + Na_2CO_3$ (Alkane)
- Electrolysis, Kolb's reaction ⇌ R-R (Alkane)
- Δ, NH_3 → $RCONH_2 + H_2O$ (Acid amide)
- Br_2, Ag, Δ → $R-Br + AgBr + CO_2\uparrow$ (Alkyl bromide)

$R-CH_2COOH$ (Alkyl acetic acid) → $2Br_2$, $-2HBr$ → $R-CBr_2-COOH$

α -α' - Dihalogenated acid (Alkyl α,α'-Dibromoacetic acid)

Higher Fatty Acids

Palmitic Acid

$C_{15}H_{31}COOH$ (Hexadecanoic acid)

It occurs as glycerol ester in palm oil and it is a colorless, waxy solid with a M.P of 337 K.

Stearic Acid

$C_{17}H_{35}COOH$ (Octadecanoic acid)

Its industrial preparation involves hydrolysis of tristearin. It is a solid fat so called as stearic acid. It is a colorless waxy solid used in the manufacture of candle.

Unsaturated Monocarboxylic Acids

Oleic Acid

$$CH_3(CH_2)_7CH = CH(CH_2)_7COOH$$

It occurs as triolein in oils and fats. It is a colorless oil, insoluble in water but soluble in organic solvents.

Uses:

1. It is used in soap preparation.
2. Nutritional value: It reduces blood pressure and prevents ulcerative colitis, type-2 diabetes and supercharges muscles.

Table 16.5 Carboxylic acids of Pharmaceutical importance.

$CH_3-C(=O)-OH$ Acetic acid **Uses**: 1. It is used in the manufacture of white vinegar cellulose acetate, polyvinyl acetate and acetone ester. 2. Used in perfumes, dyes, plastics and pharmaceuticals.	$CH_3-CH(OH)-COOH$ Lactic acid **Uses**: 1. It is used in dairy products, as acidulant in beverages & candies, as a mordant. 2. Calcium and iron lactates are used in medicines.
$HO-C(CH_2COOH)_2-COOH$ Citric acid **Uses**: 1. It is used as medicine in the form of effervescent salts. 2. Magnesium citrate used as a laxative. 3. Used as acidulant in carbonated soft drinks, jams, jellies & candies. 4. Used as as a mordant, as esters (tributyl citrate) that are good plasticizers for lacqers and varnishes.	$CH_2COOH-CH_2COOH$ Succinic acid **Uses**: 1. It is used in the manufacture of lacquers and dyes.

Table 16.5 *Contd...*

COOH | COOH Oxalic acid **Uses:** 1. It is used in the manufacture of inks and metal polishes. 2. For preparing allyl alcohol and formic acid. 3. Used as mordant in dyeing calcico printing. 4. In the volumetric analysis used as primary standard for acid-base titrations.	COOH OH Salicylic acid **Uses:** It is used as an analgesic and anti-inflammatory agent.
COOH $OCOCH_3$ Aspirin **Uses:** 1. It is used as analgesic and anti-inflammatory agent. 2. It is also used to treat Kawasaki disease, pericarditis. 3. Long term use of aspirin prevents ischaemic strokes and blood clots. 4. It may decrease certain type of cancer, mainly colorectal cancer.	$COOCH_3$ OH Methyl salicylate **Uses:** 1. It is used as analgesic and anti-inflammatory agent. 2. Low concentrations (0.04%) solution is used as a flavouring agent in chewing and mints. 3. It is also used as immune histochemistry.
COOH Benzoic acid **Uses:** 1. It is mainly used as food preservative (0.05-0.1%). 2. It is a constituent of whitfield's ointment which is used for the treatment of an athelet's foot and other fungal diseases.	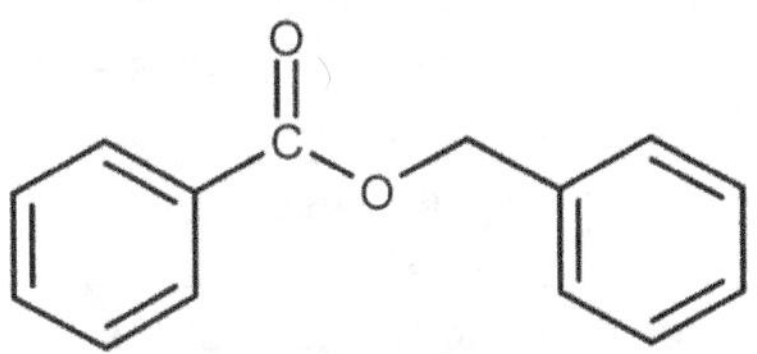 Benzyl benzoate **Uses:** 1. It is used as a topical scabicide, pediculicide in veterinary hospitals. 2. It has vasodilating and spasmolytic effect. 3. It is the one of the ingredient in some testosterone replacement medications to treat hypogonadism. 4. It is also used in perfume industries.

Table 16.5 *Contd...*

```
        COOH
         |
    H ── C ── OH
         |
   HO ── C ── H
         |
        COOH
```

Tartaric acid

Use:

1. It is used for the preparation of Rochelle salt.
2. It is used as laxative.
3. It is also used in silvering mirrors and tanning leather.

Identification tests:

1. To the sample solution add pinch of resorcinol and 2ml of Conc. H_2SO_4, mix it well. Heat the mixture. A violet colour is formed.
2. To the aqueous solution add few crystals of ferrous sulphate and shake. To this add 2 drops of 2% H_2O_2 solution and 2 ml of 5% NaOH solution. A deep violet color is appeared.

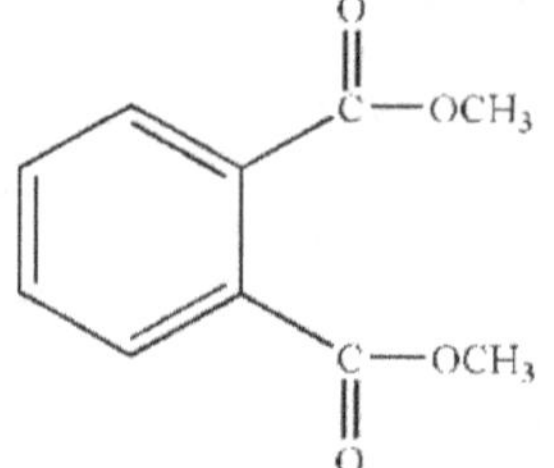

Dimethyl phthalate

Use:

1. It is used as an insect (mosquitoes and flies) repellant.
2. It is also used as an ectoparasiticide, solid rock propellant etc.

Identification tests:

To the sample add little amount of methanolic KOH solution, boil it and evaporate. Add 1ml of HCl filter and collect the precipitate. To the above precipitate add resorcinol, 2 to 3 drops of $CHCl_3$. Heat the mixture for 3 minutes at 180 °C, cool and add 1 ml of 5M NaOH, pour into water. An intense yellowish green fluorescence is produced.

Table 16.6 Qualitative tests for carboxylic acids.

S. No.	Procedure	Observation	Inference
Test for carboxylic acid			
1.	To the sample solution add 1 ml of 5% aqueous sodium bicarbonate solution.	Brisk effervescence is evolved.	Indicates the presence of acid. (Phenol does not gives effervescence)

$$RCOOH + NaHCO_3 \longrightarrow RCOONa + CO_2\uparrow + H_2O$$

S. No.	Procedure	Observation	Inference
2.	To the sample solution, add 2 ml of 5% KI and KIO_3. Boil for 2 minutes and add 1 drop of starch solution.	Appearance of blue color.	Indicates the presence of acid.
3.	**Ferric chloride test:** To the sample solution, add 3 drops of neutral ferric chloride solution. Boil if necessary.	(a) Reddish brown color is formed.	Presence of acetic acid.
		(b) Appearance of violet yellow color.	Presence of oxalic acid.
		(c) Buff color precipitate is formed.	Presence of benzoic acid.
4.	**Sodalime test:** To the sample add sodalime and mix it well in a dry test tube.	Smoky flame appears.	Indicates the presence of aromatic acid.
5.	**Esterification test:** To the 1 portion of sample, add 2 parts of ethanol and 1 part of concentrated H_2SO_4. Warm the mixture for 2 minutes. Cool and pour it into beaker containing aqueous NaOH solution. Observe the smell.	(a) Sweet fruity odour.	Presence of carboxylic acid.
		(b) Banana and apple odour.	Presence of acetic acid.
		(c) Gum like odour.	Presence of formic acid
		(d) Mint odour.	Presence of benzoic acid.
		(e) Balmy odour.	Presence of salicylic acid.

$$R\text{-}COOH + HO\text{-}R' \xrightarrow[\Delta]{Conc.\ H_2SO_4} R\text{-}COOR' + H_2O$$

Table 16.6 *Contd...*

Test for Ester			
1.	**Hydroxamic acid test:** To the sample, add 2 gm of hydroxylamine hydrochloride in ethanol. To this add saturated solution of KOH in alcohol until the solution becomes basic to litmus. Boil, cool and acidify with HCl and then add 2 drops of $FeCl_3$ solution.	Appearance of violet or wine color.	Indicates the presence of ester.
$RC(=O)OR + NH_2OH \longrightarrow R-C(=O)-NHOH + R'OH$			
2.	To the sample solution, add 1 drop of phenolphthalein and add NaOH solution drop by drop, then heat it.	Disappearance of color and odour.	Indicates the presence of ester.
Test for amides			
1.	**Biuret test:** Melt small amount of sample, cool and add 10% sodium hydroxide solution. To this add 1% of copper sulphate solution.	Appearance of violet and purple color.	Indicates the presence of diamide.
$2\ NH_2C(=O)NH_2 \longrightarrow NH_2\text{-}C(=O)\text{-}NH\text{-}C(=O)\text{-}NH_2 + NH_3$			
2.	To the sample solution add 10% NaOH solution.	White fumes are formed.	Indicates the presence of amide.
$RC(=O)NH_2 + NaOH \longrightarrow RCOONa + NH_3\uparrow$ (White fumes)			
3.	To the above solution add 5 drops of concentrated HCl.	White precipitate is formed.	Indicates the presence of amide.
4.	To the ice cold solution of sample add 1 ml of acetic acid and ice cold sodium nitrite solution.	N_2 is evolved as brisk effervescence.	Indicates the presence of amide.
Test for Amines			
1.	**Diazotisation test for aromatic amines** To the sample solution, add 3 ml of conc. HCl, cool it thoroughly to 0-5 °C in ice bath. To this add ice cold solution of sodium nitrite from a dropper. To this solution add ice cold solution of β-naphthol in sodium hydroxide. Stir it well and keep the test tube in ice cold condition.	Formation of (a) An orange red azo dye.	Indicates the presence of (a) 1° Aromatic amine.
		(b) Green solid.	(b) 2° romatic amine.
		(c) No color, but forms blue color with starch iodide paper.	(c) 3° Aromatic amine.
2.	**Carbylamine test:** Dissolve small amount of sample in alcoholic NaOH solution, add 4 drops of $CHCl_3$ and warm gently.	Foul odour is observed.	Indicates the presence of 1° amine.

Probable Questions

1. What are carboxylic acids? & explain their importance.
2. Write a detailed note on nomenclature of carboxylic acid.
3. Enlist various general methods of preparation of mono carboxylic acids and explain any five methods in detail.
4. Explain Koche reaction of alkene.
5. Write a note on carboxylic acid synthesis from alcohol.
6. How carboxylic acids are synthesized from cyanides? Explain its reaction mechanism.
7. How monocarboxylic acids are synthesized from dicarboxylic acids?
8. Write about general structure & various physical properties of carboxylic acids.
9. Discuss in detail about various chemical properties of carboxylic acids.
10. Explain various reactions of alkyl group of carboxylic acids.
11. Write short notes on reactions of carboxylic group of carboxylic acids.
12. What is HVZ reaction? Explain with suitable example & reactions mechanism.
13. Write a note on acidity of carboxylic acids.
14. Explain the effects of various substituents on acidity of carboxylic acids.
15. How acidity constant or acidity of carboxylic acids is determined?
16. Describe in detail about esterification reaction with mechanism of reaction.
17. Explain in detail about Kolbe's reaction.
18. Discuss in detail about reduction reactions of carboxylic acid with its reaction mechanism.
19. Write a note on reactions of salts of carboxylic acids.
20. What is Hunsdiecker reaction?
21. Explain the pharmaceutical importance of carboxylic acid.
22. Write a note on qualitative tests used for identification of carboxylic acid.
23. Write a note on qualitative tests used for identification of amides.
24. Write a note on qualitative tests used for identification of esters.
25. Write the chemical structure and uses of acetic acid, lactic acid, tartaric acid, citric acid, succinic acid, oxalic acid, benzoic acid, benzyl benzoate, dimethyl phthalate, methyl salicylate and acetyl salicylic acid.

17

Fats and Oils

Introduction

Fats and oils are the simple lipids. Lipids are the group of organic substances of animal or plant origin soluble in non polar organic solvents and are comprised of fixed oils, fats and waxes. Although they have several uses in medicine and industries, the function of oils and fats is mainly storage of energy.

The polar groups present in lipids are much smaller than their nonpolar portion, hence they are insoluble in water. The nonpolar portions of lipids supply the water repellent or hydrophobic property. They are extracted from cells by organic solvents like ether or chloroform which have low polarity.

Occurence of Lipids

Throughout the plant and animal kingdom lipids are distributed widely. In plants they occur in the seeds nuts and fruits. In animals, lipids are stored in adipose tissues, nervous-tissues and bone marrow.

Biological Functions of Lipids

The various biological functions of lipids are as follows:

1. Dietary fat is important for the sufficient absorption of the essential fatty acid and fat soluble vitamins from the food.
2. The most important role of lipids is as a concentrated fuel reserve of the body.
3. It serve as important building blocks of some biologically active materials, example acetic acid is derived from lipid.

Classification of Lipids

Lipids are mainly divided into three types, based on the products obtained upon hydrolysis.

1. **Simple lipids:** These lipids upon hydrolysis yield one or more fatty acids and an alcohol.
2. **Compound lipids:** These lipids upon hydrolysis yield fatty acids and alcohol. In addition to these, compounds like phosphoric acid, sugars, proteins *etc* are also obtained.
3. **Derived lipids:** Inspite of not having any ester linkage they are considered as lipids, due to the fact that they have been derived from naturally occurring esterified materials. Majority of them are alcohols of complex nature. example sterols, certain vitamins and plant pigments.

   Schematic representation of the classification of the lipids is shown below.

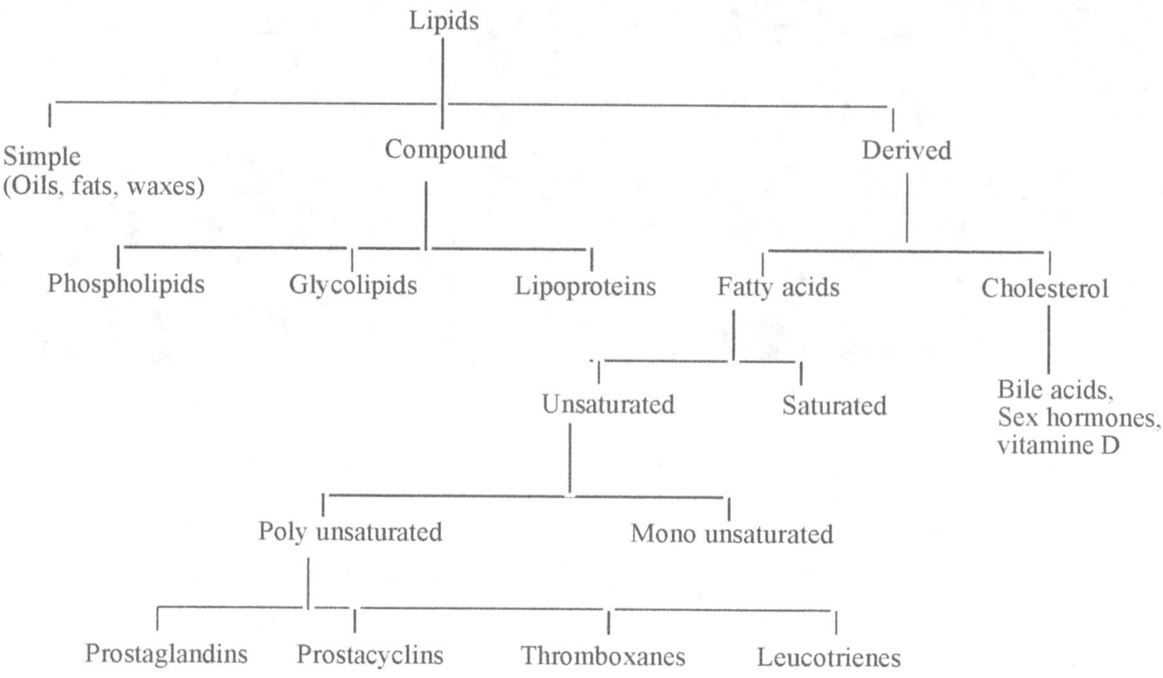

Simple Lipids (Fats and Oils)

These simple lipids are esters of fatty acids with alcohols. Example. neutral fats and waxes.

(a) Fats and oils: These are esters of fatty acids and alcohols.

Fats and oils are glyceryl esters of higher fatty acids like palmitic, stearic, oleic, linoleic and linolenic acid.

The fats and oils (triglyceride) consists of three molecules of fatty acids with one molecule of glycerol. The three molecules of fatty acids are esterified with the three hydroxyl groups of glycerol.

$$\begin{matrix} CH_2OH & & R^1\,COOH & & CH_2OCOR^1 \\ | & & & & | \\ CHOH & + & R^2\,COOH & \longrightarrow & CHOCOR^2 \\ | & & & & | \\ CH_2OH & & R^3\,COOH & & CH_2OCOR^3 \\ \text{Glycerol} & & \text{Fatty acid} & & \text{Triglyceride (Triacylglycerol)} \end{matrix}$$

If the fatty acids of a triglyceride are same, the triglyceride is called as simple triglyceride. Whereas, if the fatty acids are not identical, the triglyceride is called as mixed triglyceride.

Natural fats are mainly comprised of mixed glycerides. As these glycerides have no free acid or basic groups they are often termed as neutral fats.

Waxes: These are esters of long chain fatty acids and long chain monohydric alcohols or sterols.

Differences between fats and oils:

1. They are distinguished based on their melting ranges, the oils are liquids at ordinary temperature, while the fats are solids or semisolids (This distinction is not sharp and depends upon the climate and seasonal variation. Example coconut oil, sesame oil and ghee are liquids in summer and solids in winter in our country).
2. Fats contain mostly saturated fatty acids like stearic and palmitic acids and melts at higher temperature, while the oils contain mostly unsaturated fatty acids like oleic acid and melts at lower temperature.

Some of the common examples, of naturally occurring fats and oils in a classified manner is shown below with examples.

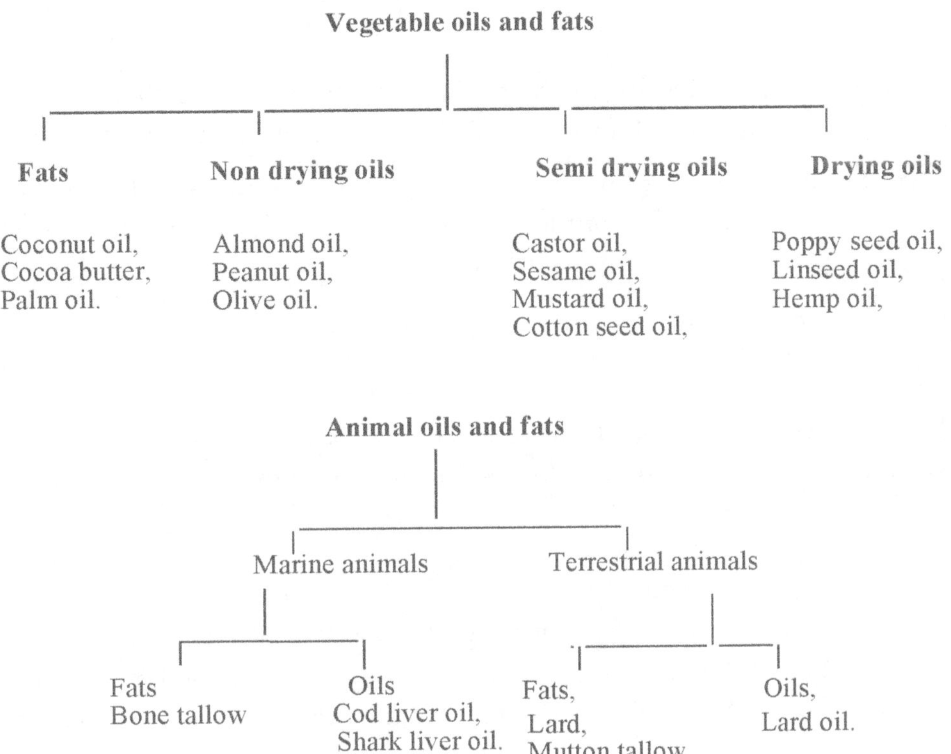

Nomenclature of Fats

If the three acid radicals are same the glyceride is known as **simple**, and if they are different the glyceride is **mixed**.

For simple glycerides the name is made up of the alcohol (glycerol) or its radical (glyceryl) and the name of the acid or the name of the acid concerned is changed to suffix 'in'.

$$\begin{array}{l} CH_2OCOC_3H_7 \\ | \\ CHOCOC_3H_7 \\ | \\ CH_2OCOC_3H_7 \end{array}$$

Glycerol (or Glyceryl) tributyrate [or Tributyrin]

$$\begin{array}{l} CH_2OCOC_{17}H_{35} \\ | \\ CHOCOC_{17}H_{35} \\ | \\ CH_2OCOC_{17}H_{35} \end{array}$$

Glycerol (or Glyceryl) tristearate [or Tristearin]

For mixed glycerides, the positions and the names of the acid groups are specified as 1, 2, 3, or α, β, γ.

$$\begin{array}{lll} 1 & CH_2OCOC_{17}H_{35} & \alpha \\ & | & \\ 2 & CHOCOC_{15}H_{31} & \beta \\ & | & \\ 3 & CH_2OCOC_{17}H_{35} & \gamma \end{array}$$

β-Palmito-α,γ-distearin

Extraction of Fats

By the following general procedure fats and oils are extracted from the parent material.

Rendering

Usually animal fats are extracted by this process. The animal tissues are chopped off and liver, kidney *etc* are boiled with water or steam. The fat melts and floats at the surface is then withdrawn.

Pressing: Seeds containing vegetable fats or oils are crushed by rollers and pressed in hydraulic press or continuous expellers. The oil separates out leaving behind the oil cake which is used as cattle feed.

The pressed cake produced above, is heated and pressed further to get oil of second grade which has undesirable components and odour.

Solvent Extraction

The pressed cake, after hot pressing has to be crushed and extracted with organic solvents such as carbon tetrachloride, benzene, petroleum ether *etc.* to obtain remaining oil.

Refining

To neutralise free acid and also to coagulate any colloidal impurities present, the crude oil or fat is treated with a little alkali. It is then bleached by warming at 70-80 °C with animal charcoal. After half an hour the decolorised oil is filtered through filter press and then decolourised by passing superheated steam through it. The oil is then quickly cooled and withdrawn from the apparatus so that its odour may not be impaired.

Physical and Chemical Properties of Fats and Oils

Physical Properties

Fats and oils are colourless or pale yellow, insoluble in water and polar solvents, but soluble in non polar solvents like ether and carbon tetrachloride.

Chemical Properties of Fats

1. **Hydrolysis:** The ester linkages of the fats are cleaved by sodium hydroxide to yield glycerol and sodium salt of long chain fatty acids (soaps) (alcoholic solution of alkali hydrolysis the fats more rapidly than the aqueous solution). Hydrolysis of fats may also be achieved by heating with water under pressure.

$$\underset{\text{Fat}}{\begin{array}{l}CH_2OCOR\\ |\\ CHOCOR\\ |\\ CH_2OCOR\end{array}} \xrightarrow{3\,NaOH} \underset{\text{Glycerol}}{\begin{array}{l}CH_2OH\\ |\\ CHOH\\ |\\ CH_2OH\end{array}} + \underset{\text{Soap}}{3\,RCOONa}$$

   If the fats are hydrolysed with enzyme *lipase*, cleavage of fats occurs stepwise from triglyceride to diglyceride to mono glyceride and finally to glycerol and fatty acids. The most important lipase of the digestive system is *pancreatic lipase* or *steapsin* found in the pancreatic juice.

$$\begin{array}{c} CH_2OCOR^1 \\ | \\ CHOCOR^2 \\ | \\ CH_2OCOR^3 \end{array} \xrightarrow[-\ R^3COOH]{H_2O,\ Lipase} \begin{array}{c} CH_2OCOR^1 \\ | \\ CHOCOR^2 \\ | \\ CH_2OH \end{array} \xrightarrow[-\ R^2COOH]{H_2O,\ Lipase} \begin{array}{c} CH_2OCOR^1 \\ | \\ CHOH \\ | \\ CH_2OH \end{array}$$

Triglyceride (Triacylglycerol) — Diglyceride (Diacylglycerol) — Monoglyceride (Monoacylglycerol)

$$\xrightarrow[-\ R^1COOH]{H_2O,\ Lipase} \begin{array}{c} CH_2OH \\ | \\ CHOH \\ | \\ CH_2OH \end{array}$$

Glycerol

2. **Rancidity:** Fats and oils when exposed to atmospheric moisture or oxidation by the air or to the action of aerobic bacteria they are oxidised or hydrolysed into aldehydes, ketones, acids *etc*. Leads to the formation of bad taste and odour, we call this oil or fat become rancid. Two types of rancidity are hydrolytic and oxidative.

   Hydrolytic rancidity: Hydrolysis is brought about by certain enzymes like *lipase* which are mainly present in naturally occurring fats, especially, animal fats. The hydrolysis of triglycerides may produce short chain fatty acids, which have bad odours.

   Oxidative rancidity: Oxidative rancidity takes place in presence of oxygen and catalyzed by light, heat, moisture and traces of certain metal like copper. It occurs due to the oxidation of unsaturated fatty acid at their double bonds with the formation of shorter-chained acids, aldehydes and ketones. The time required for the onset of rapid oxidation is known as the induction period. Due to the presence of traces of antioxidants, some of the fats and oils have longer induction period.

   Fats and oils should be kept refrigerator, to prevent rancidity. These reactions are slower at low temperatures and in dark bottles. Antioxidants are often added to fats and oils to prevent rancidity. Vitamin E, Vitamin C, phenols, hydroquinone, tannins *etc* which are antioxidants present in vegetables prevent development of rancidity. Hence vegetable fats can be preserved for a longer period than animal fats. Commonly used artificial antioxidants in food stuffs are gallic acid and its esters, nordihydroguaiaretic acid (NDGA), dilaurylthiodipropionate and 3-tertiary butyl-4-hydroxy anisole.

   Test for oxidative rancidity (kreis test): The fat to be tested is treated with ether, phloroglucinol and hydrochloric acid appearance of red color indicates the oxidative rancidity.

3. **Drying:** Certain oils, when exposed to air become thick and finally harden to resin like solid substances, this effect is known as drying. Drying takes place due to the addition of oxygen to the unsatured linkages present in the esters of the oil, and also due to some amount of polymerization. According to the behaviour of oils when exposed to air, they are divided into three types, drying, semi drying and non-drying oils.

   Non-drying oils: They consist of triolein largely. Under the influence of micro-organisms, the oil gets decomposed into glycerol and saturated and unsaturated fatty acids. The saturated acids get decomposed under the influence of micro-organisms *via* the β-oxidation pattern to form ketones leads to unpleasant rancid odor. The unsaturated acids get oxidised into aldehydes and acids with fewer carbon atoms in the molecule. Example for non drying oils are palm oil, coconut oil, olive oil and almond oil.

$$\underset{\text{Fatty acid}}{CH_3(CH_2)_4CH_2CH_2COOH} \xrightarrow{\text{Oxidation}} \underset{\beta\text{-Keto acid}}{CH_3(CH_2)_4-\overset{\displaystyle O}{\overset{\|}{C}}-CH_2COOH} \xrightarrow{-CO_2} \underset{\text{Methyl amyl ketone}}{CH_3(CH_2)_4-\overset{\displaystyle O}{\overset{\|}{C}}-CH_3}$$

Drying oils: They are glycerides of linoleic acid and linolenic acids. Upon exposure to light and air they become dry and form a solid elastic film. A good drying oil dries within 4-5 hours. Example for drying oils are linseed oil and hemp oil.

Semi-drying oils: They possess a high content of linoleic acid than non drying oils and low content of linolenic acid than that of the drying oils. Example for semi-drying oils are sesame oil, sunflower oil and cotton seed oil.

4. **Hydrogenolysis:** When hydrogen under pressure, is passed through oil or fat in the presence of copper-chromium catalyst it is cleaved into glycerol and higher aliphatic alcohols.

$$\underset{\text{Stearin}}{\begin{array}{l} CH_2OCOC_{17}H_{35} \\ | \\ CHOCOC_{17}H_{35} \\ | \\ CH_2OCOC_{17}H_{35} \end{array}} \xrightarrow[200\ atm]{Cu\text{-}Cr,\ 6\ H_2} \underset{\text{Glycerol}}{\begin{array}{l} CH_2OH \\ | \\ CHOH \\ | \\ CH_2OH \end{array}} + \underset{\text{Octadecyl alcohol}}{3\ C_{17}H_{35}CH_2OH}$$

5. **Acrolein formation:** When fats are heated while cooking, some of the glycerol may be released by the hydrolysis, which undergoes decomposition to form acrolein, it may be responsible for headache or severe smarting of the eyes.

Analysis of Oils and Fats

According to the degree of unsaturation, molecular weight, and also acidity from hydrolysis the properties of oils and fats vary. For the analysis of fats and oils following physical constants and chemical constants are measured.

Physical Constants

The important physical constants of fats and oils are

1. Specific gravity
2. Refractive index

3. Viscosity and
4. Melting point.

The fats are mixtures of chemical compounds, the melting point is not sharp so solidification point is generally determined.

Chemical Constants

1. **Acid value:** It is the number of milligrams of potassium hydroxide required to neutralise the free fatty acids in one gram of the fat or oil.

   Procedure: Weigh about 10 g of substance being examined in an iodine flask or round bottom flask. Prepare 50 ml of mixture of equal volumes of ethanol (95%) and ether, to this add 0.5ml of phenophthalein solution and titrate it against 0.1 N aqueous KOH solution to make it neutral. Dissolve the weighed quantity of substance in above neutralised solution if the sample does not dissolve in the cold solvent, connect the flask with a reflux condenser and warm slowly with frequent shaking until the sample dissolves. Add 1ml of phenolphthalein solution and titrate with 0.1N aqueous KOH solution until the appearance of permanent faintly pink colour.

   Acid value $= 5.61 \times n/w$

   Where n = the number of ml of 0.1 N KOH solution W = the weight of substance in gm.

   Significance: Acid value give indication about the degree of rancidity of the given fat. Rancid oils and fats give high acid values.

2. **Saponification value:** It is defined as the number of milligrams of KOH required to completely saponify 1 g of fat or oil. i.e. To neutralise the fatty acids resulting from complete hydrolysis of 1g of the oil or fat.

   Procedure: Weigh about 2 g of substance being examined in an iodine flask fitted with reflux condenser. Add 25 ml of 0.5 M ethanolic KOH solution and boil under reflux on water bath for 30 min. Remove the condenser and add 1ml of phenolphthalein solution and tittrate immediately with 0.5 M HCl. Note the reading as 'a'. Repeat the above procedure without the substance being examined (called as blank titration) note the reading as'b'. Calculate the saponification value from the following equation.

   Saponification value $= 28.05 \times (b-a)/W$

   where, W = weight of the substance taken in gm.

   Significance:

   - To identity the given fatty oil.
   - To distinguish between fatty oils and mineral oils.
   - It also helps to calculate the amount of alkali required for converting a definite amount of fat or oil into soap and in detecting the adulteration of a fat or oil.

3. **Ester value:** It is the number of milligrams of potassium hydroxide required to saponify the esters present in 1 gm of the substance.

   It is determined by substracting acid number from the saponification number.

4. **Acetyl value:** It is defined as the number of milligrams of KOH required to neutralize the acetic acid obtained by saponifying 1 gm of acetylated fat.

   Method: It is determined by saponifying a weighed sample of acetylated fat with alcoholic KOH, neutralising the so formed potassium salt of fatty acids with mineral acid equivalent to the KOH used in saponification, separating the acetic acid (soluble) from the rest of fatty acid (insoluble) by filtration, and finally titrating the acetic acid against standard alkali.

   Significance: Acetyl number is a measure of the number of -OH group present in a fat or oil.

   Most of the fats and oils have a low acetyl number ranging between 3 and 15. Acetyl castor oil which is high in ricinoleic acid $C_{17}H_{32}(OH)COOH$, has a high acetyl number of above 150.

5. **Iodine value:** It is defined as the number of grams of iodine that combine with 100 gm of an oil or fat.
   Method: It is determined by treating a known amount of the fat or oil dissolved in carbontetrachloride solution with a known excess of Wij's reagent (solution of iodine monochloride in glacial acetic acid). The stopper of the flask has been moistened with potassium iodide solution and the flask is kept in dark for 30-60 min. The solution is then mixed with 15 ml of 10% KI and titrated with standard 0.1 M sodium thiosulphate solution using starch mucilage as indicator. The blank determination is also performed. The difference between the volume in ml of sodium thiosulphate solution indicate the equivalent of iodine absorbed.
   Thus Iodine value = (a-b) × 1.27/W
   Where, a = reading for the blank experiment
   b = reading for actual experiment
   W = weight of fat or oil taken.
   Significance: Iodine value gives indication about the proportion of unsaturated fatty acids present both in free and combined forms of esters.
6. **Reichert Meissl value (or) RM value:** It is defined as number of ml of 0.1 N KOH solution required to neutralise the volatile water soluble fatty acids obtained by 5 gm of fat.
   Method: The sample is heated with a solution of KOH in glycerol until completely saponified. The mixture is then acidified with sulphuric acid. It is then distilled (fatty acids possessing upto 10 carbon atoms are volatile with steam and thus appear in the distillate), the distillate is filtered and the water soluble acids in the filtrate are neutralised with 0.1 N KOH solution.
   Significance:
   - It is a measure of volatile water soluble acid contents of the fat.
   - It is used to determined the purity of the butter and ghee.

   The Reichert-Meissl value of pure butter and ghee lies in between 20-30 because hydrolysis of the butter fat gives C_4, C_6, C_8 acids which are volatile in steam. Whereas the cotton seed oil value is less than 1.
7. **Polenski value:** It is the number of ml of 0.1 N KOH required to neutralize the water-insoluble, steam-distillable acid liberated by the hydrolysis of 5 gm of the fat.
   Method: The water insoluble fatty acids obtained by filtration of the distillate in Reichert Meissl value of determination, are dissoloved in alcohol and tiltrated against 0.1 N KOH.
   Significance:
   - It is a measure of the steam distillable water insoluble acid constituents of the fat.
   - The Reichert Meissl and polenski numbers are of greater importance in testing for the adulteration of butter

Fatty Acids

Fatty acids occur mainly as esters in neutral fats and oils but do occur in the unesterified form as free fatty acids.

Fatty acids of natural fats are usually straight chain derivatives, contains even number of carbon atoms, as they are synthesized from 2-carbon units. The chain may be saturated or unsaturated.

Nomenclature of Fatty Acids

Generally used systematic nomenclature is based on naming the fatty acid on the basis of hydrocarbon with the same number of carbon atoms, -oic being substituted for the final 'e' in the name of the hydrocarbon.

Thus saturated acids ends with 'anoic' e.g. Octanoic acid, and unsaturated acids with double bonds ends with 'enoic' e.g. octadecenoic acid (Oleic acid).

Carbon atoms are numbered from the carboxy carbon which is carbon no. 1. The carbon atom adjacent to the carboxyl carbon no.2, is also known as the α-carbon. Carbon atom no.3 is the β-carbon and the next carbon is known as the δ carbon and so on.

$$\underset{\omega}{\overset{6}{H_3C}}-\underset{\delta}{\overset{5}{H_2C}}-\underset{\gamma}{\overset{4}{CH_2}}-\underset{\beta}{\overset{3}{CH_2}}-\underset{\alpha}{\overset{2}{CH_2}}-\overset{1}{COOH}$$

Various conventions are used for indicating the number and position of the double bonds e.g. E.g. Δ^9 indicates a double bond between carbon atoms 9 and 10 of the fatty acid. It is the widely used conventions to indicate the number of carbon atoms, the number of double bonds and the positions of the double bonds.

Linoleic acid [C_{18}, $\Delta^{9,12}$]

$$\overset{18}{H_3C}-(H_2C)_4-\overset{13}{CH}=\overset{12}{HC}-\overset{11}{H_2C}-\underset{H}{\overset{10}{C}}=\overset{9}{CH}-(CH_2)_7-\overset{1}{COOH}$$

Arachidonic acid [C_{20}, $\Delta^{5,8,11,14}$]

$$\overset{20}{H_3C}-(H_2C)_4-\underset{H}{\overset{15}{C}}=\underset{H}{\overset{14}{C}}-\overset{13}{H_2C}-\underset{H}{\overset{12}{C}}=\overset{11}{CH}-\overset{10}{H_2C}-\underset{H}{\overset{9}{C}}=\underset{H}{\overset{8}{C}}-\overset{7}{H_2C}-\underset{H}{\overset{6}{C}}=\overset{5}{CH}-(CH_2)_3-\overset{1}{COOH}$$

Linolenic acid [C_{18}, $\Delta^{9,12,15}$]

$$\overset{18}{H_3C}-\overset{17}{H_2C}-\underset{H}{\overset{16}{C}}=\overset{15}{CH}-\overset{14}{H_2C}-\underset{H}{\overset{13}{C}}=\underset{H}{\overset{12}{C}}-\overset{11}{H_2C}-\underset{H}{\overset{10}{C}}=\overset{9}{CH}-(CH_2)_7-\overset{1}{COOH}$$

Classification of fatty acids:

1. **Depending on the total number of carbon atoms they are classified as follows.**
   (a) **Even chain:** Consists of even number of carbon atoms (2, 4, 6 and similar series). Most of the naturally occurring lipids contain even chain fatty acids.
   (b) **Odd chain:** Consists of odd number of carbon atoms (3, 5, 7 *etc*). Odd numbered fatty acids are present in milk and also found in microbial cell walls.
2. **Depending on the length of hydrocarbon chain they are classified as follows.**
   (a) **Short chain:** Consists of 2 to 6 carbon atoms
   (b) **Medium chain:** Consists of 8 to 14 carbon atoms
   (c) **Long chain:** Consists of 16 and above usually upto 24 carbon atoms.
3. **Depending on the nature of hydrocarbon chain they are classified as**
   (a) **Saturated fatty acids:** Contains no double bonds:
   They have the general structural formula CH_3-$(CH_2)_n$-COOH. They are numbered by adding the suffix 'enoic' after the hydrocarbon with same number of carbon atoms. Eg stearic acid, palmitic acid *etc.*
   (b) **Unsaturated fatty acids:** Contains one or more double bonds
   They are named by adding the suffix -enoic after the systematic name. They may be further subdivided according to degree of unsaturation.
   (i) Monounsaturated (Monoethenoid, monoenoic) acids: Containing one double bond.
   (ii) Polyunsaturated (polyethenoid, polyenoic) acid: Containing two or more double bonds.

Eg: Linoleic and linolenic acids. These are called essential fatty acids because they cannot be synthesised by the body and have to be supplied in the diet.

(c) **Eicosanoids:** Eicosanoids are derived from 20 C arachidonic acid, consists of prastanoids (prostaglandins (PG) prastacyclins (PGI), and thromboxanes (TX) and leukotrienes (LT).

Table 17.1 Commonly occurring Fatty acids, their systemic name and occurrence.

Common name	Systemic name	Occurrence
(i) Even chain saturated fatty acid		
Caproic acid	n-Hexanoic acid	Butter
Caprylic acid	n-Octanoic acid	Butter
Capric acid	n-Decanoic acid	Butter and Coconut oil
Lauric acid	n-Dodecanoic acid	Coconut oil
Myristic acid	n-Tetradecanoic acid	Coconut oil
Palmitic acid	n-Hexadecanoic acid	Adipose tissue
Stearic acid	n-Octadecanoic acid	Adipose tissue
Arachidic acid	n-Eicosanoic acid	Peanut oil
(ii) Odd chain saturated fatty acid		
Propionoic acid	n-Propionoic acid	Metabolic intermediate
Valeric acid	n-Pentanoic acid	Metabolic intermediate
(iii) Even chain unsaturated fatty acids		
Palmitoleic acid	9-Hexadecenoic acid	Adipose tissue
Oleic acid	9-Octadecenoic acid	Adipose tissue
Linoleic acid	9,12-Octadeca dienoic acid	Vegetable oils
Linolenic acid	9,12,15-Octadeca trienoic acid	Vegetable oils
Arachidonic acid	5,8,11,14-Eicosa tetraenoic acid	Vegetable oils and phospholipids
(iv) Branched fatty acids		
Iso valeric acid	3-Methyl butanoic acid	Metabolic intermediate
Iso caproic acid	4-Methyl pentanoic acid	Metabolic intermediate
(v) Hydroxy fatty acids		
Cerebronic acid	2-Hydroxy tetracosanoic acid	Brain lipids
Ricinoleic acid	12-Hydroxy-9-octadecenoic acid	Brain lipids

Isomerism in unsaturated fatty acids: Due to different configuration around the double bond unsaturated fatty acids show cis-trans (geometrical) isomerism. All the naturally occurring fatty acids have the ***cis*** – configuration. However in the body during metabolism trans fatty acids are produced.

cis form (Oleic acid)

$$\begin{array}{ccc} H_3C-(H_2C)_7 & - & CH \\ & & \| \\ COOH-(CH_2)_7 & - & CH \end{array}$$

trans form (Elaidic acid)

$$\begin{array}{ccc} & HC & -(CH_2)_7-CH_3 \\ & \| & \\ COOH-(CH_2)_7 & -CH & \end{array}$$

In *cis* configuration the molecules being bent 120 degrees at the double bond thus oleic acid has an **L-shape** whereas elaidic acid remains straight at its *trans* double bond.

Increase in the number of cis double bonds in a fatty acid leads to a variety of possible spatial configurations of the molecule eg arachidonic acid with 4 double cis bonds, may have **kinks or a U-shape.** The presence of trans double bonds will alter these spatial relationships.

During the saturation of fatty acid in the process of hydrogenation or hardening of natural oils in the manufacture of margarine, trans fatty acids are present in certain foods as a by product.

Reactions of Fatty Acids

1. **Formation of esters or esterification reaction:** Fatty acid reacts with alcohols in the presence of concentrated sulphuric acid and give esters.

$$\underset{\text{Fatty acid}}{RCOOH} + \underset{\text{Ethanol}}{OHC_2H_5} \longrightarrow \underset{\text{Ethylester of fatty acid}}{RCOOC_2H_5} + H_2O$$

2. **Soap formation or reaction with alkalis:** Soaps obtained by the reaction between fatty acids and alkalis are important in day today life. Fatty acids reacts with sodium hydroxide or potassium hydroxide to give sodium and potassium soaps respectively. Sodium soaps are hard but potassium soaps are soft and high cost. Small amount of silicate or sodium carbonate is added to sodium soaps to make them soft and usable at toilet soaps. Calcium and magnesium soaps of fatty acids are insoluble in water and do not lather, hence not suitable for washing purposes.

$$\underset{\text{Fatty acid}}{RCOOH} + \underset{\text{Sodium hydroxide}}{NaOH} \longrightarrow \underset{\text{Soap (Sodium salt of fatty acid)}}{RCOONa} + H_2O$$

Detergents: These are non soap cleansing agents contain sodium salt of lauryl sulphate, sulphated lauryl monoglyceride or salts of long chain fatty acids with quaternary ammonium as ingredients. Similar to soaps, they are also good wetting agents and emulsifiers. They give lather in hard water also.

1. **Reactions due to the presence of double bonds in unsaturated fatty acids:**
   (a) **Hydrogenation:** Unsaturated fatty acids undergo hydrogenation under suitable temperature, pressure and in the presence of catalysts to yields corresponding saturated fatty acid.

$$\underset{\text{Oleic acid}}{RCH{=}CHCOOH} + 2H \longrightarrow \underset{\text{Stearic acid}}{RCH_2CH_2COOH} \quad R — R = C_{15}H_{30}$$

   (b) **Halogenation:** Unsaturated fatty acid undergo halogenation under suitable condition to yield corresponding halogenated saturated fatty acid.

$$\underset{\text{Oleic acid}}{RCH{=}CHCOOH} + Br_2 \longrightarrow \underset{\text{1,2- Dibromo stearic acid}}{RCHBrCHBr\,COOH}$$

   (c) **Oxidation:** Unsaturated fatty acid undergo oxidation and the double bond gets broken to yield lower fatty acid and dicarboxylic acid.

$$\underset{\text{Oleic acid}}{RCH{=}CHCOOH} + [O] \longrightarrow \underset{\text{Fatty acid}}{RCOOH} + \underset{\text{Oxalic acid}}{\begin{array}{c}COOH\\|\\COOH\end{array}}$$

Probable Questions

1. What are fats & oils? List out various biological functions of lipids in detail.
2. Define & classify lipids.
3. Write a short note on differences between fats and oils.
4. Explain the nomenclature of fats.
5. Describe in detail about extraction of fats & oils.
6. Explain the various physical properties & chemical properties of fats & oils.
7. Write a note on rancidity.
8. Write short notes on hydrolysis reactions of fats.
9. What is drying oil, non drying oil & semi-drying oil? Give example for each class.
10. How fats & oils are analysed?
11. Write a note on various physical constants of fats and oils.
12. Write a note on various chemical constants of fats and oils.
13. What is acid value? How it is determined? & mention its significance.
14. What is saponification value? How is it determined? Mention its significance.
15. What is ester value? How it is determined?Mention its significance.
16. What is acetyl value? How it is determined? & Mention its significance.
17. What is iodine value? How it is determined? & Mention its significance.
18. What is Reichert meissl value? How it is determine? Mention its significance.
19. What is Polenski value? How it is determined? Mention its significance.
20. What are fatty acids? Eplain its nomenclature in detail.
21. Define & classify fatty acids.
22. Explain the isomerism in unsaturated fatty acids.
23. Write a detailed note on various reactions of fatty acids.

18

Amines, Alkyl Nitrites and Alkyl Nitrates

Introduction

Amines are alkyl or aryl derivatives of ammonia (NH_3) in which one or more of the hydrogen atoms of the NH_3 are replaced by alkyl group. According to the number of alkyl groups present, they are classified as primary amines (one alkyl group directly linked with nitrogen), secondary amines (two alkyl groups directly linked with nitrogen) and tertiary amines (three alkyl groups directly linked with nitrogen).

NH_3	$R\text{-}NH_2$	R_2NH	R_3N	$R_4N^{\oplus}X^{\ominus}$
Ammonia	Alkylamine (Primary)	Dialkylamine (Secondary)	Trialkylamine (Tertiary)	Quaternary ammonium salts

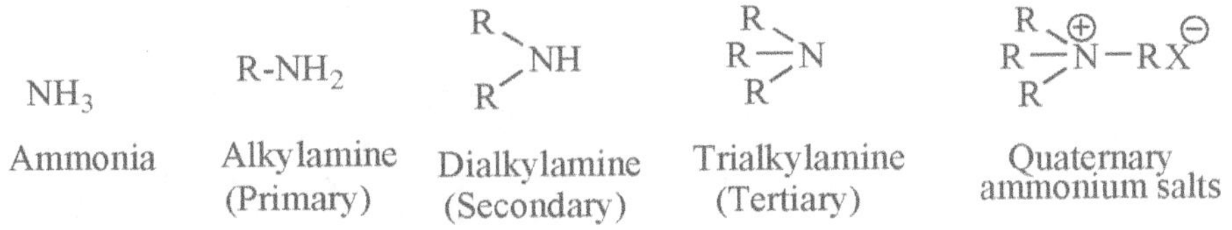

As the amines are basic in nature due to the presence of lone pair of electrons on nitrogen, they form salts with acids. The salts are considered related to ammonium (NH_4^+) in which all the four hydrogen atoms are replaced by alkyl groups and are called as quaternary ammonium salts.

$CH_3\text{-}NH_2$	$(CH_3)_2NH$	$(CH_3)_3N$	$(CH_3)_4N^+\ \bar{O}H$
Methylamine (1° Amine)	Dimethylamine (2° Amine)	Trimethylamine (3° Amine)	Tetramethyl ammonium hydroxide (4° Amine)

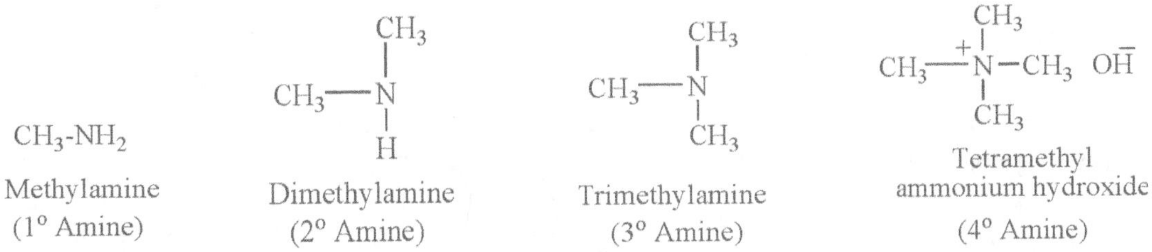

Importance of Amines

In nature they occur as proteins, vitamins, alkaloids, hormones *etc.* The bioactive compounds adrenaline and noradrenaline which are used to increase blood pressure contains amino group. Quaternary ammonium salts are used as an intermediate in the preparation of surfactants.

Amino groups when attached with various chemical structures elicit different kind of pharmacological activities based on the core structure to which it is attached. Some of the drugs which possess amino groups and their therapeutic uses in the treatment of various diseases are shown in Table 18.1.

Table 18.1 Drugs containing amino groups and their therapeutic uses.

Category	Drug	Structure
Antihistamine	Doxylamine	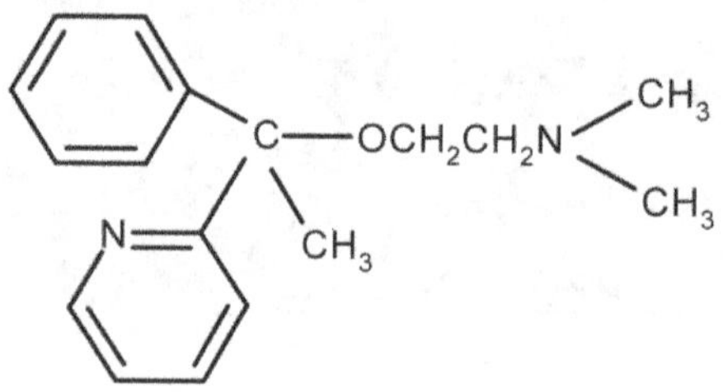 $C—OCH_2CH_2N$ CH_3 CH_3 CH_3 N
	Diphenhydramine	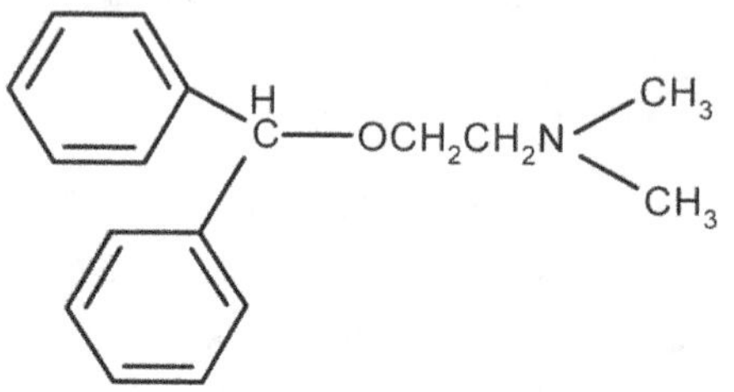 H $C—OCH_2CH_2N$ CH_3 CH_3
General anaesthetic	Ketamine	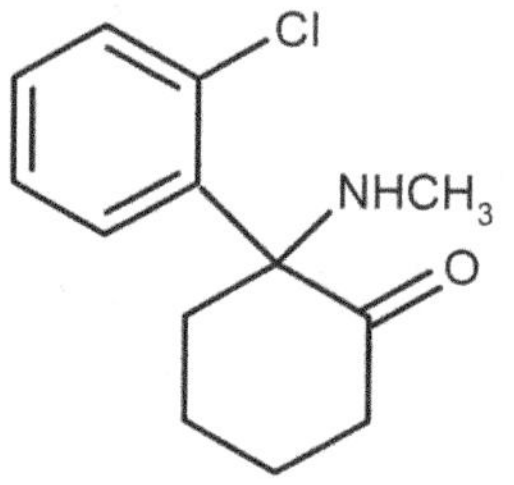 Cl $NHCH_3$ O
Tranquilizer	Triflupromazine	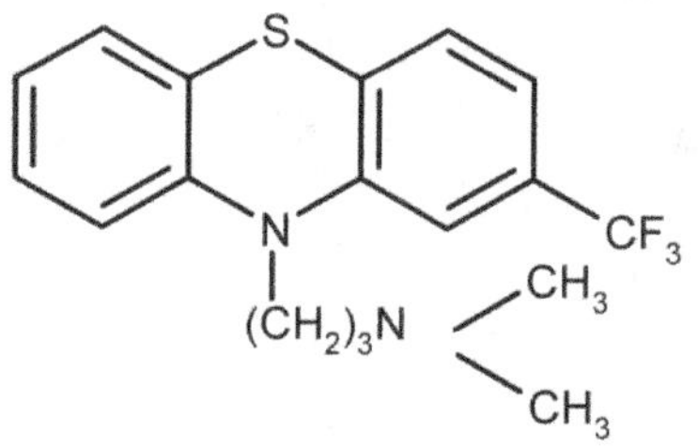 S N CF_3 $(CH_2)_3N$ CH_3 CH_3
Antidepressant	Amitryptiline	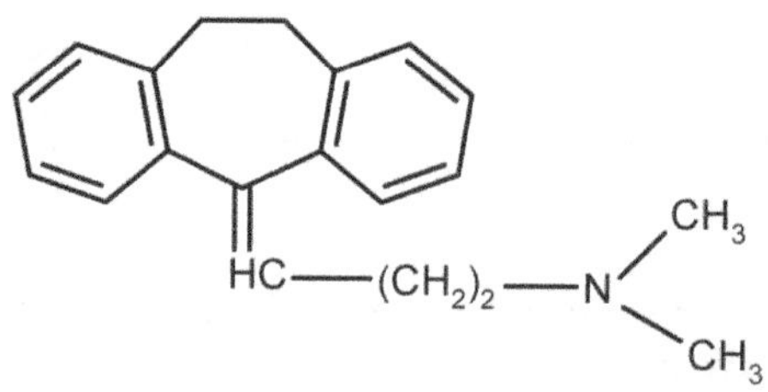 $HC—(CH_2)_2—N$ CH_3 CH_3
	Doxepin	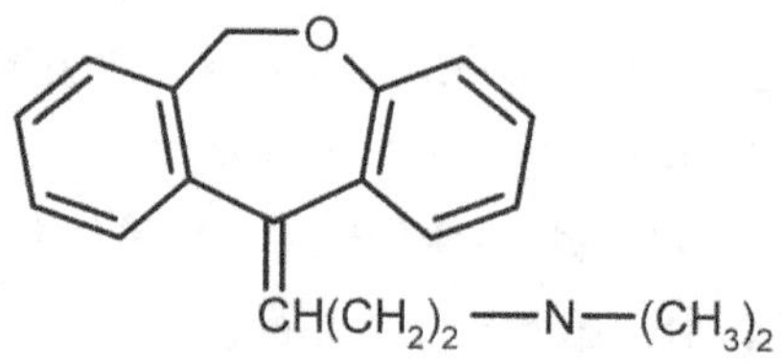 O $CH(CH_2)_2—N—(CH_3)_2$

Nomenclature

Individual amines derive their names from the alkyl groups present in it. If the alkyl groups present are **same then it is called simple amines**, if the alkyl groups present are different then it is called **mixed amines.**

The rules for naming amines as per IUPAC nomenclature are described below.

1. Name the alkyl groups attached to nitrogen atom followed by the suffix- amine.

$CH_3\text{-}NH_2$ — Methylamine

$CH_3CH_2NH_2$ — Ethylamine

2. If two or more identical alkyl groups are attached to the nitrogen the prefix di or tri *etc* is added to the amine name.

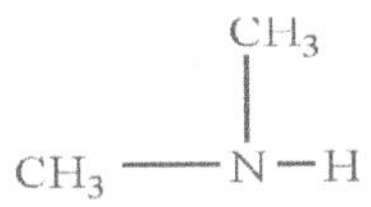

$CH_3-N(CH_3)-H$ — Dimethylamine

$CH_3-N(CH_3)-CH_3$ — Trimethylamine

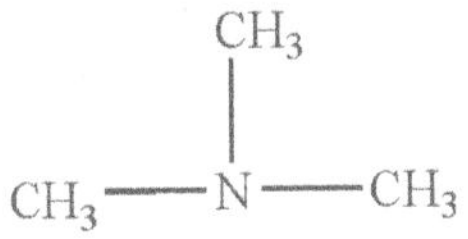

3. If two or more different alkyl groups are present, then they are named alphabetically.

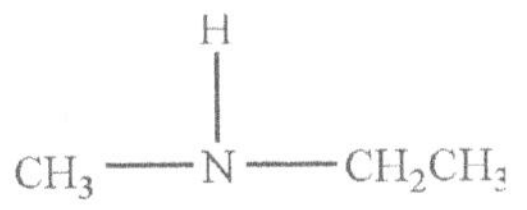

$CH_3-N(H)-CH_2CH_3$ — Ethyl methylamine

$CH_3-N(CH_2CH_3)-CH_2CH_2CH_3$ — Ethyl methyl propylamine

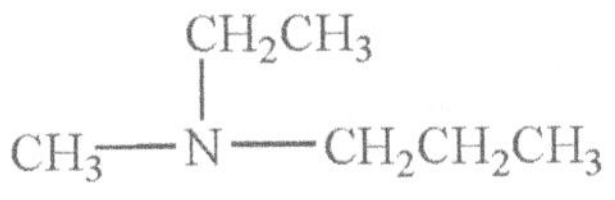

4. If the amine is complex, then the amino group is considered as substituent and lowest possible number is assigned to it in the carbon chain.

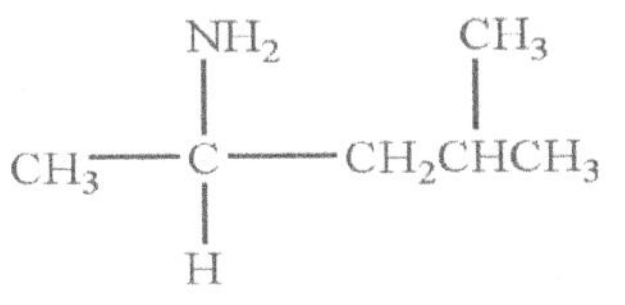

$CH_3-CH(NH_2)-CH_2CH(CH_3)CH_3$

2-Amino-4-methyl pentane

5. The amine salts are considered as substituted ammonium salts. The alkyl groups are named first and word ammonium is added as a suffix. The name of anion present is written as second word.

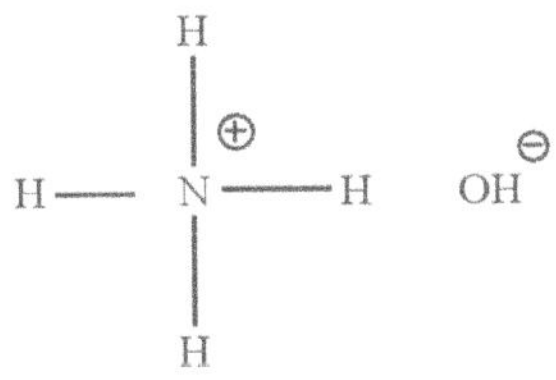

$[H-N^{\oplus}(H)(H)-H]\ OH^{\ominus}$ — Ammonium hydroxide

$[H_3C-N^{\oplus}(CH_3)(CH_3)-CH_3]\ OH^{\ominus}$ — Tetramethyl ammonium hydroxide

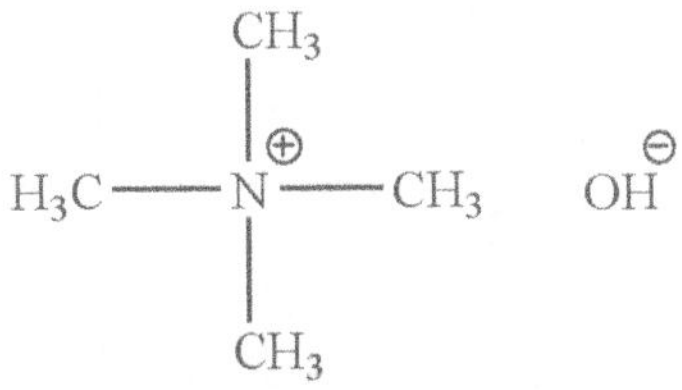

Some of the amines, their common and IUPAC name are depicted in the Table 18.2.

Table 18.2 Common name and IUPAC name of some Alkylamines and Arylamines.

Amine	Common name	IUPAC name
$CH_3-CH_2-NH_2$	Ethylamine	Ethanamine
$CH_3-CH_2-CH_2-NH_2$	*n*-Propylamine	Propan-1-amine
$CH_3-CH(NH_2)-CH_3$	*iso*-Propylamine	Propan-2-amine
$CH_3-N(H)-CH_2-CH_3$	Ethyl methylamine	N-Methylethanamine
$CH_3-N(CH_3)-CH_3$	Trimethylamine	N, N-Dimethyl methanamine
$C_2H_5-N(C_2H_5)-\overset{1}{C}H_2-\overset{2}{C}H_2-\overset{3}{C}H_2-\overset{4}{C}H_3$	N,N-Diethylbutylamine	N, N-Diethylbutan-1-amine
$NH_2-\overset{1}{C}H_2-\overset{2}{C}H=\overset{3}{C}H_2$	Allylamine	Prop-2-en-1-amine
$H_2N-(CH_2)_6-NH_2$	Hexamethylene diamine	Hexane-1,6-diamine
$C_6H_5NH_2$ (benzene ring with NH_2)	Aniline	Aniline or Benzenamine
benzene ring with NH_2 and ortho CH_3	*o*-Toluidine	2-Methylaniline
benzene ring with NH_2 and para Br	*p*-Bromoaniline	4-Bromo benzenamine or 4-Bromoaniline
benzene ring with $N(CH_3)_2$	N,N-Dimethylaniline	N,N-Dimethyl benzenamine

General Methods of Preparation

1. **Methods yielding mixture of amines:**
   (a) **By reacting alkyl halide with ammonia:** Heating of alkyl halide with alcoholic ammonia in a sealed tube yields mixture of 1°, 2°, 3° amines and 4° ammonium compounds which have to be separated.

$$\underset{\text{Alkyl halide}}{RX} + \underset{\text{Ammonia}}{NH_3} \xrightarrow[\text{Sealed tube}]{\text{Alcohol}} \underset{\text{Methylamine}}{RNH_2} + HX$$

$$\underset{\text{Methyl iodide}}{CH_3I} + \underset{\text{Ammonia}}{NH_3} \xrightarrow[\text{Sealed tube}]{\text{Alcohol}} \underset{\text{Methylamine}}{CH_3NH_2} + HI$$

$$\underset{\text{Methyl amine}}{CH_3NH_2} + \underset{\text{Methyl iodide}}{CH_3I} \longrightarrow \underset{\text{Dimethylamine}}{(CH_3)_2NH} + HI$$

$$\underset{\text{Dimethyl amine}}{(CH_3)_2NH} + \underset{\text{Methyl iodide}}{CH_3I} \longrightarrow \underset{\text{Trimethylamine}}{(CH_3)_3N} + HI$$

$$\underset{\text{Trimethyl amine}}{(CH_3)_3N} + \underset{\text{Methyl iodide}}{CH_3I} \longrightarrow \underset{\text{Tetramethyl ammonium iodide}}{(CH_3)_4\overset{\oplus}{N}\overset{\ominus}{I}}$$

   1° Amines are major product which further reacts with little excess of alkyl halide to form secondary, tertiary amines and quaternary compounds.
   Amines on any given stage forms quaternary ammonium salt by reacting with hydrogen halide. For example with hydrogen iodide it forms

$$\underset{\text{Methyl ammonium iodide}}{CH_3\overset{\oplus}{N}H_3\overset{\ominus}{I}} \qquad \underset{\text{Dimethyl ammonium iodide}}{(CH_3)_2\overset{\oplus}{N}H_2\overset{\ominus}{I}} \qquad \underset{\text{Trimethyl ammonium iodide}}{(CH_3)_3\overset{\oplus}{N}H\overset{\ominus}{I}}$$

   From the quaternary ammonium salts, amines can be liberated by reacting with alkali.

   (b) **By ammonolysis of alcohols:** By passing the mixture of vapours of alcohol and ammonia through overheated alumina at 623 K, a mixture of amines is obtained which is separated by fractional distillation.

$$\underset{\text{Methanol}}{CH_3OH} + \underset{\text{Ammonia}}{NH_3} \longrightarrow \underset{\text{Methyl amine}}{CH_3NH_2} + H_2O$$

$$\underset{\text{Methyl amine}}{CH_3NH_2} + \underset{\text{Methanol}}{CH_3OH} \longrightarrow \underset{\text{Dimethyl amine}}{(CH_3)_2NH} + H_2O$$

$$\underset{\text{Dimethyl amine}}{(CH_3)_2NH} + \underset{\text{Methanol}}{CH_3OH} \longrightarrow \underset{\text{Trimethyl amine}}{(CH_3)_3N} + H_2O$$

   Primary amine can be obtained as a major product if excess of ammonia is used.

2. **Methods yielding primary amines:**
   (a) **Hoffmann's bromamide reaction or Hoffmann's degradation of amides:** By the action of bromine and caustic potash on amide, primary amine is obtained. This is the most convenient method for preparing primary amine.

$$CH_3CONH_2 + Br_2 + 4KOH \longrightarrow CH_3NH_2 + 2KBr + K_2CO_3 + 2H_2O$$

Acetamide → Methyl amine

Mechanism:

Step 1: Abstraction of proton by base leads to the formation of amide anion.

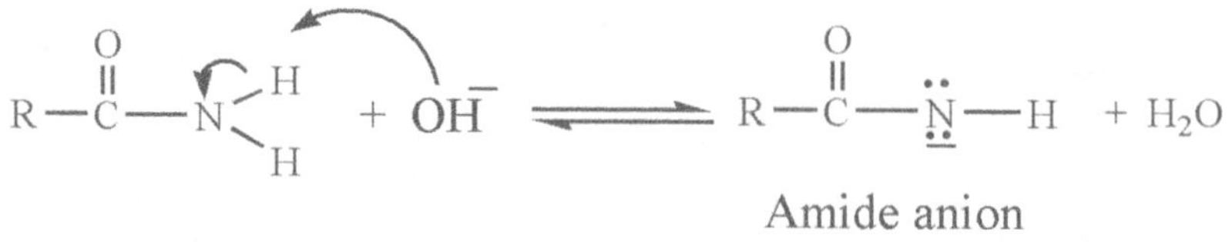

$$R-C(=O)-NH_2 + \bar{O}H \rightleftharpoons R-C(=O)-\ddot{N}^{-}-H + H_2O$$

Amide anion

Step 2: Attack of amide anion by Br_2 molecule to give N-bromoamide.

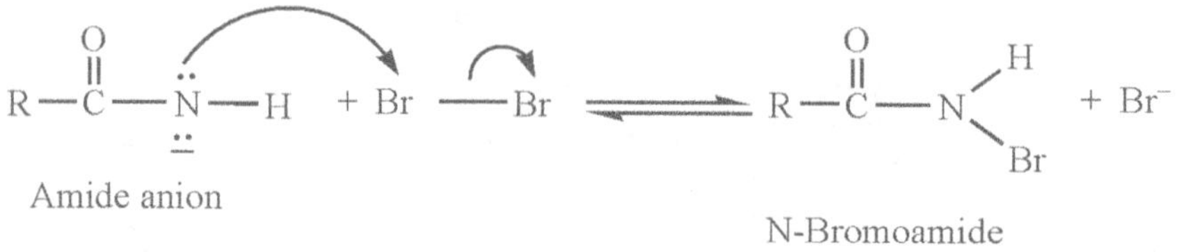

$$R-C(=O)-\ddot{N}^{-}-H + Br-Br \rightleftharpoons R-C(=O)-N(H)(Br) + Br^-$$

Amide anion → N-Bromoamide

Step 3: Abstraction of second proton from N-bromoamide by base to yield N-bromoamide anion.

$$R-C(=O)-N(H)(Br) + \bar{O}H \rightleftharpoons R-C(=O)-\ddot{N}^{-}-Br + H_2O$$

N-Bromoamide anion

Step 4: Removal of bromide ion from N-bromoamide anion leads to the formation of alkyl isocyanate.

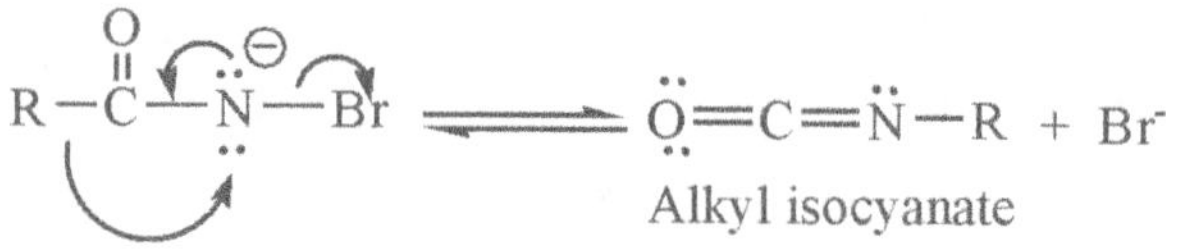

$$R-C(=O)-\ddot{N}^{\ominus}-Br \rightleftharpoons \ddot{O}=C=\ddot{N}-R + Br^-$$

Alkyl isocyanate

Step 5: Attack of OH^- to double bonded carbon to yield 1° amine and CO_2 *via* the formation of N-alkyl carbamic acid.

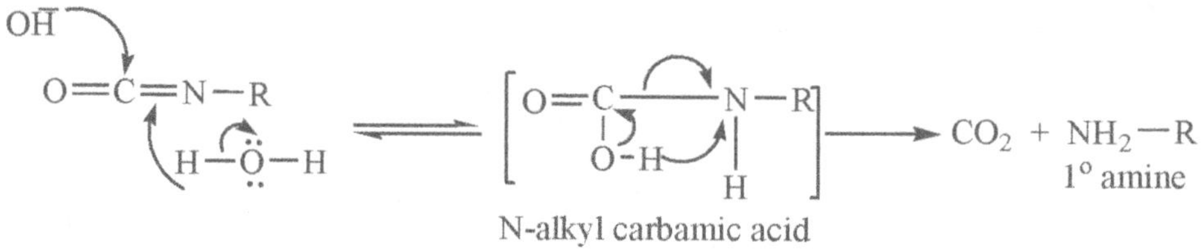

$$\bar{O}H + O=C=N-R + H-\ddot{O}-H \rightleftharpoons [O=C(OH)-N(H)-R] \longrightarrow CO_2 + NH_2-R$$

N-alkyl carbamic acid → 1° amine

Note:

(i) The obtained amine having one carbon atom less than the starting amide.

(ii) This reaction involves the complete removal of carbonyl group from the amide.

(b) Reduction of nitro alkanes: By the reduction of nitro paraffin using H_2 / Pt or nickel or $LiAlH_4$, 1° amines are obtained.

$$CH_3NO_2 + 3H_2 \xrightarrow{Pt} CH_3NH_2 + 2H_2O$$

Nitro methane → Methyl amine

(c) Mendius reaction or reduction of nitriles: By the reduction of alkyl cyanides with hydrogen / nickel or $LiAlH_4$, 1° amines are obtained.

$$\underset{\text{Methyl cyanide}}{CH_3C{\equiv}N} + 4[H] \xrightarrow{LiAlH_4} \underset{\text{Ethyl amine}}{CH_3CH_2NH_2}$$

(d) Reduction of amides: By the reduction of amides using $LiAlH_4$, 1° amines are obtained.

$$\underset{\text{Acetamide}}{CH_3-\overset{\overset{\displaystyle O}{\|}}{C}-NH_2} + 4[H] \xrightarrow{LiAlH_4} \underset{\text{Ethyl amine}}{CH_3CH_2NH_2} + H_2O$$

(e) Reduction of oximes: Oximes of aldehydes (aldoximes) or oximes of ketones (ketoximes) upon reduction by using $LiA1H_4$ or hydrogen/nickel or sodium/ ethanol yields 1° amines.

$$\underset{\substack{\text{Actaldoxime}\\\text{(Aldoxime)}}}{CH_3CH{=}NOH} + 4[H] \xrightarrow{LiAlH_4} \underset{\text{Ethyl amine}}{CH_3CH_2NH_2} + H_2O$$

(f) Gabriel phthalimide method: Phthalimide upon reaction with alcoholic KOH yields potassium phthalimide which upon further reaction with alkyl halide yields N-alkyl phthalimide and further refluxing with KOH yields 1° amine and potassium phthalate.

Phthalimide $\xrightarrow[-H_2O]{KOH}$ Potassium phthalimide (N-K) $\xrightarrow[-KX]{RX}$ N-Alkyl phthalimide (N–R) $\xrightarrow{2KOH}$ Potassium phthalate (C(=O)OK, C(=O)OK) + $R-NH_2$ (1° Amine)

Mechanism:

Step 1: Abstraction of proton from the nitrogen of phthalimide by base to yield imide anion.

Phthalimide (N–H) + $\overline{OH}$ $\xrightarrow{-H_2O}$ Imide anion ($:\overset{-}{N}$)

Step 2: Alkyl halide reacts with imide anion to yield alkyl amide.

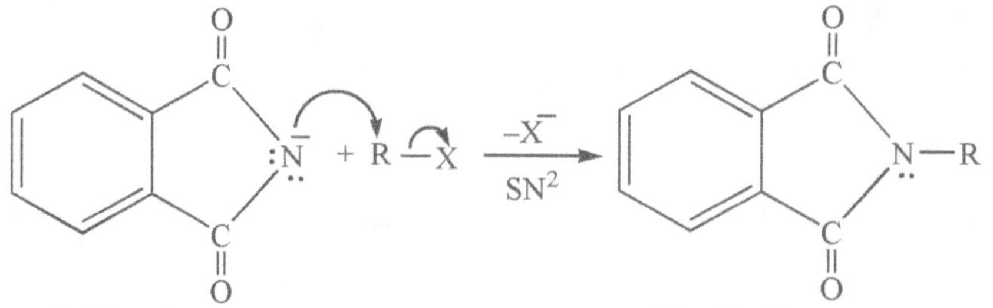

$$\text{Imide anion} + R\text{—}X \xrightarrow[SN^2]{-X^-} \text{Alkyl imide}$$

Imide anion

Alkyl imide

Step 3: Hydrolysis of alkyl imide by base to yield 1° amine.

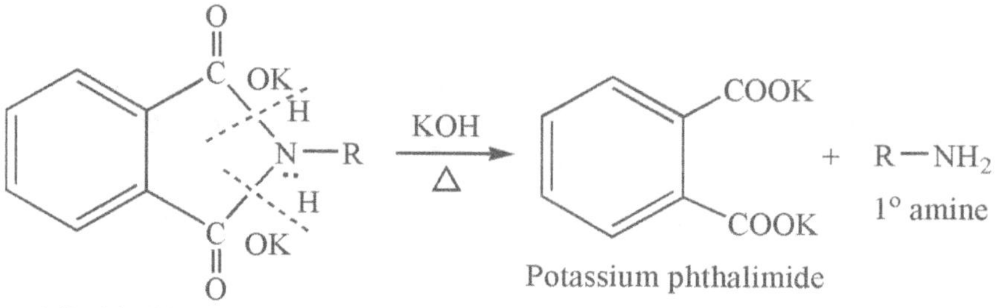

$$\text{Alkyl imide} \xrightarrow[\Delta]{KOH} \text{Potassium phthalimide} + R\text{—}NH_2$$

Alkyl imide

Potassium phthalimide

1° amine

(g) By reductive amination of aldehydes and ketones:

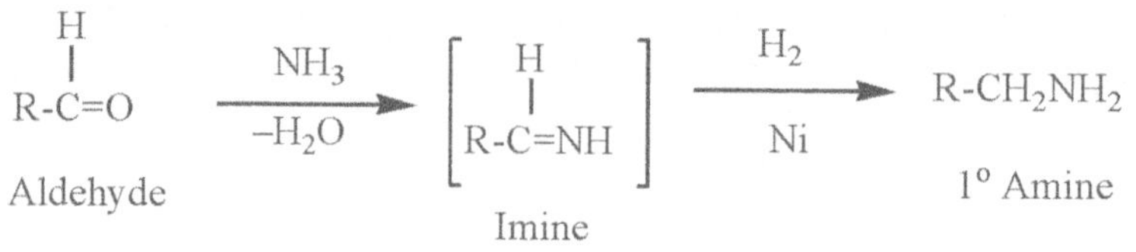

$$\underset{\text{Aldehyde}}{R\text{-}C(H)\text{=}O} \xrightarrow[-H_2O]{NH_3} \underset{\text{Imine}}{\left[R\text{-}C(H)\text{=}NH\right]} \xrightarrow[Ni]{H_2} \underset{1^\circ\ \text{Amine}}{R\text{-}CH_2NH_2}$$

By passing aldehyde or ketone, hydrogen and ammonia on a nickel catalyst at high temperature yields 1° amine as the product.

Mechanism (Reductive amination)

Step 1: Attack of ammonia on the carbonyl carbon of a ketone or aldehyde.

$$R\text{—}C(=O)\text{—}R + \ddot{N}H_3 \rightleftharpoons R\text{—}C(O^-)(R)\text{—}N^+H_3$$

Step 2: Transfer of proton from the nitrogen to the oxygen atom.

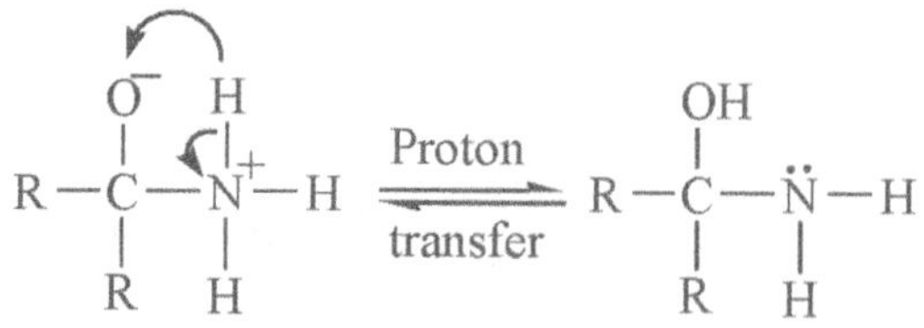

$$R\text{—}C(O^-)(R)\text{—}N^+H_2\text{—}H \xrightleftharpoons[\text{transfer}]{\text{Proton}} R\text{—}C(OH)(R)\text{—}\ddot{N}H\text{—}H$$

Step 3: Loss of hydroxide ion to from iminium ion intermediate.

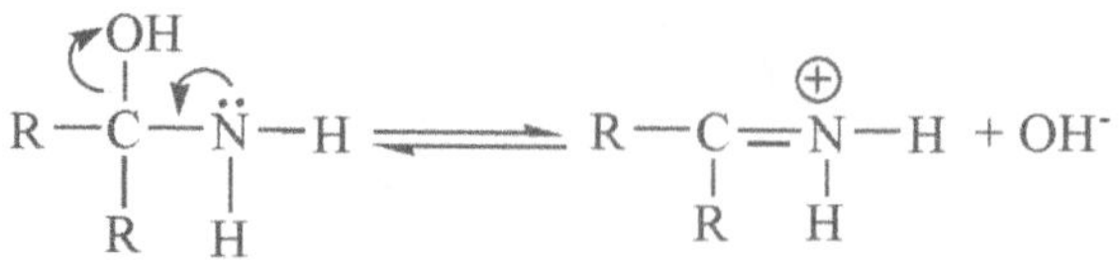

$$R\text{—}C(OH)(R)\text{—}\ddot{N}H\text{—}H \rightleftharpoons R\text{—}C(R)\text{=}\overset{\oplus}{N}H\text{—}H + OH^-$$

Iminium ion intermediate

Step 4: Loss of water molecule fron iminium ion to yield imine which upon reduction gives 1⁰ amine.

$$R-\underset{R}{\underset{|}{C}}=\overset{+}{N}\underset{H}{\underset{|}{-}}H \xrightleftharpoons{-H_2O} R-\underset{R}{\underset{|}{C}}=N-H \xrightarrow{H_2/Ni} R-\underset{R}{\underset{|}{\overset{H}{\overset{|}{C}}}}-NH_2$$

$\bar{O}H$ Imine 1° Amine

(h) By decarboxylation of amino acids: Amino acids upon decarboxylation with barium hydroxide or on treatment with putrifying bacteria forms 1°amines.

$$\underset{\text{Glycine}}{NH_2CH_2COOH} \xrightarrow{Ba(OH)_2} \underset{\text{Methyl amine}}{CH_3NH_2} + CO_2 \uparrow$$

(i) Curtius rearrangement: Acyl azides undergo hydrolysis to give 1° amines which in turn are prepared by the reaction between acid halide and NaN_3.

$$\underset{\text{Acid halide}}{R-\overset{O}{\overset{||}{C}}-Cl} \xrightarrow{NaN_3} \underset{\text{Acyl azide}}{R-\overset{O}{\overset{||}{C}}-\ddot{\bar{N}}-\overset{+}{N}\equiv N} + NaCl$$

$$\underset{\text{Acyl azide}}{R-\overset{O}{\overset{||}{C}}-\ddot{\bar{N}}-\overset{+}{N}\equiv N} \xrightarrow[\Delta]{H_2O} \underset{1^\circ\text{ Amine}}{RNH_2} + CO_2 + N_2$$

3. **Methods yielding secondary amines:**

(a) **Reaction of primary amine with alkyl halides:** By heating primary amines with calculated quantity of alkyl halides, secondary amines are obtained.

$$\underset{\text{Methyl amine}}{RNH_2} + \underset{\text{Alkyl halide}}{RI} \xrightarrow{\Delta} \underset{2^\circ\text{ Amine}}{R_2NH} + HI$$

$$\underset{\text{Methyl amine}}{CH_3NH_2} + \underset{\text{Methyl iodide}}{CH_3I} \xrightarrow{\Delta} \underset{\text{Dimethyl amine}}{(CH_3)_2NH} + HI$$

(b) **From *p*- nitroso dialkyl aniline:** By heating *p*-nitroso dialkyl aniline with a strong solution of NaOH, pure 2° amine is obtained. This is one of the best method for preparing 2° amine.

$$\underset{p\text{-Nitroso diethyl aniline}}{ON-C_6H_4-N(C_2H_5)_2} \xrightarrow{NaOH} \underset{p\text{-Nitrosophenol}}{ON-C_6H_4-OH} + \underset{2^\circ\text{ Amine}}{(C_2H_5)_2NH}$$

(c) **Reduction of alkyl iso cyanides:** Reduction of alkyl iso cyanides (iso nitriles) using hydrogen and platinum yields 2° amines. In this method the 2° amine obtained always have methyl group in it.

$$\underset{\text{Alkyl iso cyanide}}{R\text{-}\overset{+}{N}\equiv\overset{-}{C}} + 2H_2 \xrightarrow{Pt} \underset{\text{2° Amine}}{R-\overset{\overset{H}{|}}{N}-CH_3}$$

(d) **Reduction of N-substituted amides:** N-substituted amides upon reduction with $LiAlH_4$ forms 2° amines (The carbonyl group is reduced to $-CH_2$).

$$\underset{\text{N- methyl acetamide}}{CH_3\text{-}\overset{\overset{O}{\|}}{C}\text{-}NH\text{-}CH_3} + 4[H] \xrightarrow[-H_2O]{LiAlH_4} \underset{\text{Ethyl methyl amine}}{CH_3CH_2NH\text{-}CH_3}$$

(e) **By hydrolysis of dialkyl cyanamine:** Dialkyl cyanamine upon hydrolysis with acid or alkali yields 2° amine.

$$\underset{\text{Dialkylcyanamine}}{R_2NCN} \xrightarrow[H^+]{2H_2O} \underset{\text{2° Amine}}{R_2NH} + CO_2 + NH_3$$

(f) **Reductive amination of aldehydes & ketones:** Upon passing aldehydes or ketones, hydrogen and 1° amine on nickel at high temperature, 2° amines are obtained.

$$\underset{\text{Aldehyde}}{R-\overset{\overset{H}{|}}{C}=O} \xrightarrow[-H_2O]{R'NH_2} \underset{\text{Imine}}{\left[R-\overset{\overset{H}{|}}{C}=NR'\right]} \xrightarrow[\text{Ni, High temperature}]{H_2} \underset{\text{2° Amine}}{R\text{-}CH_2\text{-}NHR'}$$

4. Methods yielding tertiary amines:

(a) **Reduction of alkyl halides with ammonia:** Excess of alcoholic ammonia solution when heated with alkyl halide, trialkyl ammonium halide is formed which upon reaction with NaOH yields free 3° amine.

$$\underset{\text{Alkyl halide}}{3RX} + NH_3 \longrightarrow \underset{\text{Trialkyl ammonium halide}}{R_3\overset{\oplus}{N}H\overset{\ominus}{X}} + 2HX$$

$$\underset{\text{Trialkyl ammonium halide}}{R_3\overset{\oplus}{N}H\overset{\ominus}{X}} + NaOH \longrightarrow \underset{\text{3° Amine}}{R_3N} + H_2O + NaX$$

(b) **Reduction of N,N-disubstituted amides:** 3° amines can be obtained by the reduction of N,N-di substituted amides with $LiAlH_4$.

$$\underset{\text{N,N-Dimethyl acetamide}}{CH_3\text{-}\overset{\overset{O}{\|}}{C}\text{-}N(CH_3)_2} + 4[H] \xrightarrow[-H_2O]{LiAlH_4} \underset{\text{Ethyl dimethyl amine}}{CH_3CH_2N(CH_3)_2}$$

(c) **By decomposition of tetra alkyl ammonium hydroxide:** 3° amines can be obtained by strongly heating tetra alkyl ammonium hydroxides, which in turn are obtained by treating tetra alkyl ammonium halides with moist silver oxide (AgOH). This method is best adopted for preparing only tri methylamine.

$$\underset{\text{Tetra methyl ammonium iodide}}{(CH_3)_4\overset{\oplus\ominus}{N}I} + AgOH \longrightarrow \underset{\text{Tetramethyl ammonium hydroxide}}{(CH_3)_4\overset{\oplus\ominus}{N}OH} + AgI$$

$$\underset{\text{Tetra methyl ammonium hydroxide}}{(CH_3)_4\overset{\oplus\ \ominus}{N}OH} \xrightarrow{\Delta} \underset{\text{Trimethylamine}}{(CH_3)_3N} + CH_3OH$$

Separation of Mixture of Amines

The mixture of amines containing 1°, 2°, 3° amines and 4°ammonium salt is distilled with KOH. The 1°, 2°, 3° amines distill over (because it is volatile), the 4°ammonium salt is non-volatile and hence it is left in the solution. The distillate containing 3 amines (1°, 2°, 3°) are separated by any of the following methods.

1. **Fractional distillation:** As the boiling point of 1°, 2°, 3° amines are different, so they are separated by fractional distillation. This method is used widely in the industry.

   The mixture of amines is treated with **Hinsberg reagent (benzene sulphonyl chloride)** followed by NaOH and ether.

   (a) Primary amine yields N-alkyl benzene sulphonamide, this forms salt with NaOH which is soluble in water.

$$\underset{1^\circ\text{ Amine}}{R\text{-}NH_2} + \underset{\text{Benzene sulphonyl chloride}}{C_6H_5\text{-}SO_2Cl} \xrightarrow{-HCl} \underset{\substack{\textit{N}\text{-Alkylbenzene sulphonamide}\\ \text{(Insoluble salts)}}}{C_6H_5\text{-}SO_2\overset{H}{\overset{|}{N}}-R} \xrightarrow[-H_2O]{NaOH} \underset{\substack{\text{(Soluble salts)}\\ \text{Sodium-N-alkyl benzene sulphonamide}}}{C_6H_5\text{-}SO_2\overset{Na}{\overset{|}{N}}-R}$$

   (b) Secondary amine gives N,N-dialkyl benzene sulphonamide which does not form salt with NaOH (since no acidic proton present) and it is insoluble in alkali solution.

$$\underset{2^\circ\text{ Amine}}{R_2NH} + \underset{\text{Benzene sulphonyl chloride}}{C_6H_5\text{-}SO_2Cl} \xrightarrow{-HCl} \underset{\substack{\textit{N,N}\text{-Dialkylbenzene sulphonamide}\\ \text{(Insoluble salts)}}}{C_6H_5\text{-}SO_2\overset{R}{\overset{|}{N}}-R} \xrightarrow{NaOH} \text{No reaction}$$

   (c) Tertiary amine does not react with Hinsberg reagent. Upon distillation, 3° amine passes over and the remaining mixture is filtered. The filtrate on treatment with HCl gives sulphonamide of 1° amine and the insoluble solid is sulphonamide of 2° amine. Thus, two sulphonamides isolated can be hydrolysed with conc. HCl and distilled over NaOH to yield the respective amines.

$$C_6H_5\text{-}SO_2\overset{H}{N}\text{-}R + HCl \xrightarrow{H_2O} C_6H_5\text{-}SO_2OH + R\text{-}\overset{\oplus}{N}H_3\overset{\ominus}{Cl} \xrightarrow[\Delta]{NaOH} RNH_2 + NaCl + H_2O$$

N-Alkylbenzene sulphonamide — 1° Amine

$$C_6H_5\text{-}SO_2\overset{R}{N}\text{-}R + HCl \xrightarrow{H_2O} C_6H_5\text{-}SO_2OH + R_2\text{-}\overset{\oplus}{N}H\overset{\ominus}{Cl} \xrightarrow[\Delta]{NaOH} R_2NH + NaCl + H_2O$$

N,N- dialkyl benzene sulphonamide — 2° Amine

Now a days benzene sulphonyl chloride is replaced by *p*-toluene sulphonyl chloride because these substituted sulphonamides are stable solids and can be recrystallized easily.

2. **Hoffmann's method:** This method involves the reaction of **diethyl oxalate** with primary, secondary, tertiary amines followed by reaction with alkali.

Primary amine: Reacts with diethyl oxalate and a solid oxamide is formed.

$$\begin{matrix} CO\text{-}OC_2H_5 \\ | \\ CO\text{-}OC_2H_5 \end{matrix} + \begin{matrix} HHNR \\ \\ HHNR \end{matrix} \longrightarrow \begin{matrix} CO\text{-}NHR \\ | \\ CO\text{-}NHR \end{matrix} + 2C_2H_5OH$$

Diethyl oxalate — 1° Amine — Oxamide (solid)

Secondary amine: Reacts with diethyl oxalate and gives liquid oxamic ester.

$$\begin{matrix} CO\text{-}OC_2H_5 \\ | \\ CO\text{-}OC_2H_5 \end{matrix} + HNR_2 \longrightarrow \begin{matrix} CO\text{-}NR_2 \\ | \\ COOC_2H_5 \end{matrix} + C_2H_5OH$$

Diethyl oxalate — 2° Amine — Oxamic ester (Liquid)

Tertiary amine: Does not react

$$\begin{matrix} CO\text{—}OC_2H_5 \\ | \\ CO\text{—}OC_2H_5 \end{matrix} + NR_3 \longrightarrow \text{No reaction}$$

Diethyl oxalate — 3° Amine

The solid oxamide is separated by filtration and reacted with KOH to yield primary amine.

$$\begin{matrix} CO\text{-}NHR \\ | \\ CO\text{-}NHR \end{matrix} + \begin{matrix} KOH \\ \\ KOH \end{matrix} \longrightarrow \begin{matrix} COOK \\ | \\ COOK \end{matrix} + 2RNH_2$$

Oxamide (Solid) — Potassium oxalate — 1° Amine

From the filtrate the tertiary amine is distilled off and the liquid oxamic ester is reacted with KOH to liberate secondary amine.

$$\begin{matrix} CO\text{-}NR_2 \\ | \\ COOC_2H_5 \end{matrix} + 2KOH \longrightarrow \begin{matrix} COOK \\ | \\ COOK \end{matrix} + R_2NH + C_2H_5OH$$

Oxamide (Liquid) — Potassium oxalate — 2° Amine

Summary of Methods of Preparation

Preparation of Primary amines

1) RX (Alkyl halide) $\xrightarrow[\text{Sealed tube, } -HX]{NH_3,\ \text{alcohol}}$ RNH_2 (1°Amine) $\xrightarrow[-HX]{RX}$ R_2NH (2°Amine) $\xrightarrow[-HX]{RX}$ R_3N (3°Amine) $\xrightarrow[-HX]{RX}$ $(R_4)N^+X^-$ (Quaternary ammonium compounds)

2) ROH (Alcohol) $\xrightarrow[-H_2O]{NH_3}$ RNH_2 (1°Amine) $\xrightarrow[-H_2O]{ROH}$ R_2NH (2°Amine) $\xrightarrow[-H_2O]{ROH}$ R_3N (3°Amine)

3) CH_3CONH_2 (Acetamide) $\xrightarrow[3H_2/Pt]{Br_2 + 4KOH}$ $CH_3NH_2 + 2KBr + K_2CO_3 + 2H_2O$ (Methyl amine)

4) CH_3NO_2 (Nitro methane) $\longrightarrow$ $CH_3NH_2 + 2H_2O$ (Methyl amine)

5) $CH_3C \equiv N$ (Methyl cyanide) $\xrightarrow[LiAlH_4]{4[H]}$ $CH_3CH_2NH_2$ (Ethyl amine)

6) CH_3CONH_2 (Acetamide) $\xrightarrow[LiAlH_4]{4[H]}$ $CH_3CH_2NH_2 + H_2O$ (Ethyl amine)

7) $CH_3CH{=}NOH$ (Actaldoxime (Aldoxime)) $\xrightarrow[LiAlH_4]{4[H]}$ $CH_3CH_2NH_2 + H_2O$ (Ethyl amine)

8) Phthalimide $\xrightarrow[-H_2O]{KOH}$ Potassium phthalimide $\xrightarrow[-KX]{RX}$ N-Alkyl phthalimide $\xrightarrow{2KOH}$ $R\text{-}NH_2$ (1° Amine) + Potassium phthalate ($C_6H_4(COOK)_2$)

9) $R\text{-}CH{=}O$ (Aldehyde) $\xrightarrow[-H_2O]{NH_3}$ $[R\text{-}CH{=}NH]$ (Imine) $\xrightarrow[Ni]{H_2}$ $R\text{-}CH_2NH_2$ (1° Amine)

10) NH_2CH_2COOH (Glycine) $\xrightarrow{Ba(OH)_2}$ $CH_3NH_2 + CO_2\uparrow$ (Methyl amine)

11) $R\text{-}\overset{O}{\overset{\|}{C}}\text{-}Cl$ (Acid chloride) $\xrightarrow[-NaCl]{NaN_3}$ $R\text{-}\overset{O}{\overset{\|}{C}}\text{-}\bar{N}\text{-}\overset{+}{N}{\equiv}N$ (Acyl azide) $\xrightarrow[\Delta]{H_2O}$ $R\text{-}NH_2 + CO_2 + N_2$ (1° Amine)

Preparation of Secondary amines

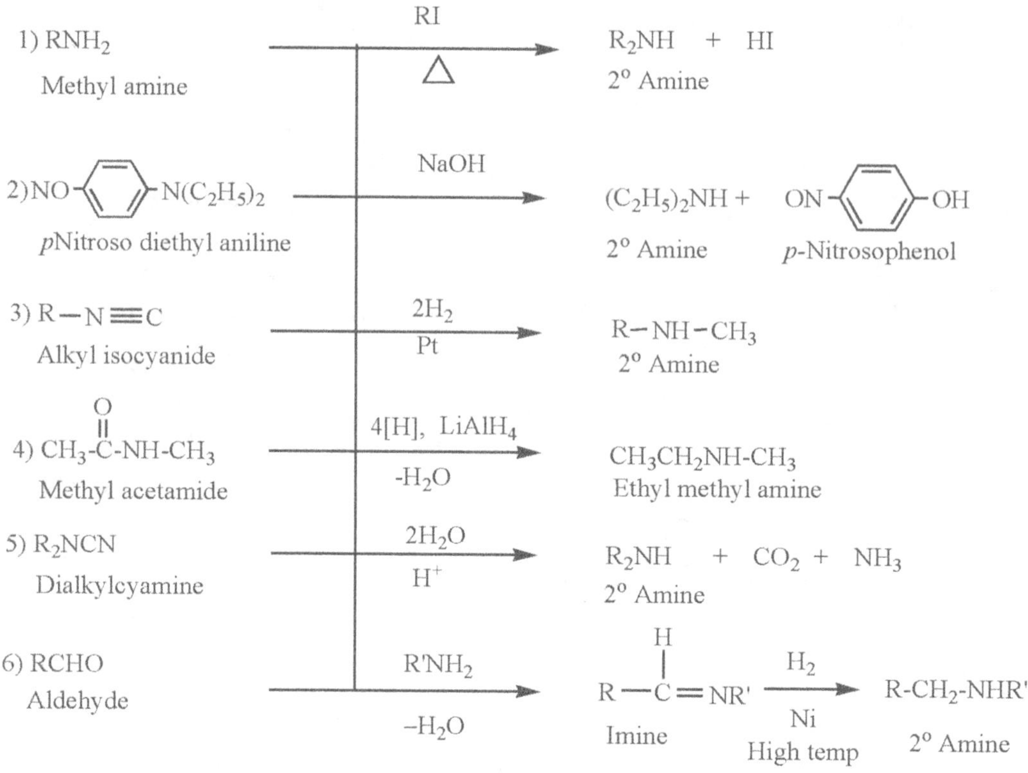

Preparation of Tertiary amines

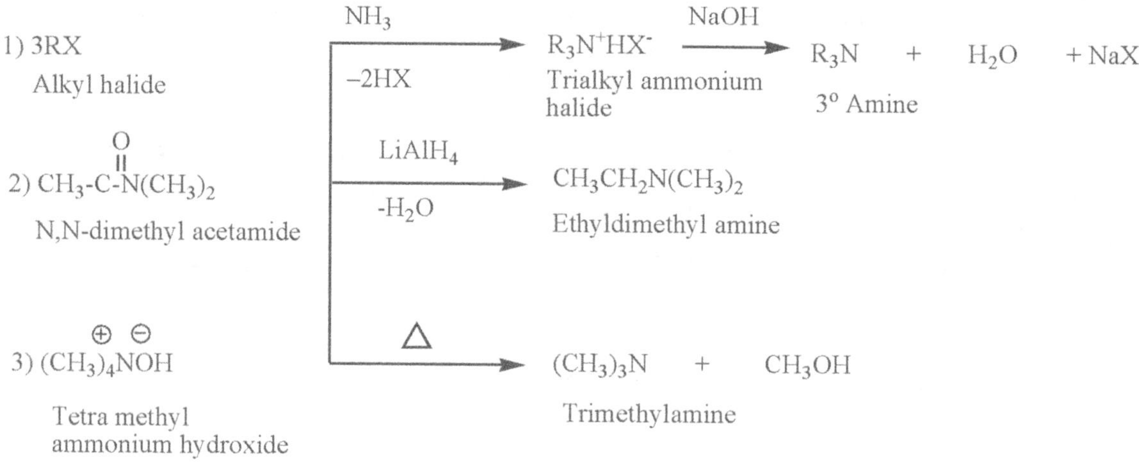

Structure

Amines have structure similar to that of ammonia. Nitrogen of amines is sp^3 hybridised, one of the sp^3 hybridized orbitals are filled up with two electrons, so it cannot participate in bond formation. Remaining sp^3 hybridised orbitals are directed towards the corners of a tetrahedran. The bond of angle C-N-C in trimethylamine is 108^0 and the structure of amines is shown below.

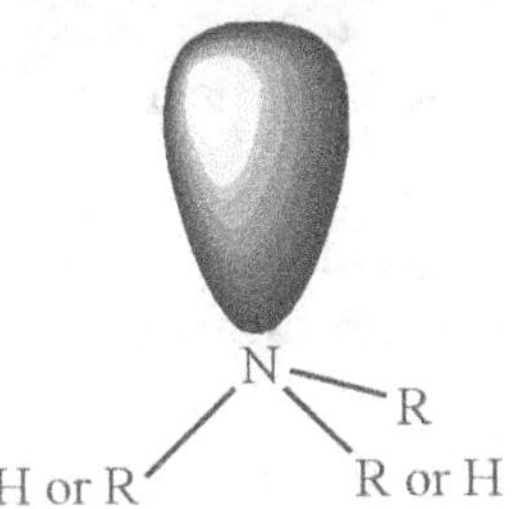

Physical Properties

1. Lower members of the amines are gases, the members with 3 or more carbon atoms are liquids and still higher amines are solids.
2. Low molecular weight amines are soluble in water due to hydrogen bonding while higher members are insoluble in water due to increase in hydrophobic group of amines.

```
      H              H              H
      |              |              |
H — N: ········ H — N: ········ H — N:
      |              |              |
      R              R              R
```

Hydrogen bond
Hydrogen bonding in amines

```
                     H
                     |
H — O: ········ H — N: ········ H — O:
      |              |              |
      H              R              H
```

Hydrogen bond
Hydrogen bonding between amine and water

3. The lower members are having fishy smell and are combustible; the odour fades with the increase in molecular weight.
4. The boiling point increases with increase in molecular weight.

Basicity of Amines

As the lone pair of electrons can be donated to a proton, the amines are basic in nature and react with acid to form salt. The lower members are stronger bases than ammonia but weaker bases than hydroxide ion.

$$\underset{\text{Amine (Base)}}{RNH_2} + \underset{\text{(Acid)}}{HCl} \longrightarrow \underset{\text{Alkyl ammonium chloride (Salt)}}{R\overset{+}{N}H_3\overset{\ominus}{Cl}}$$

Strong bases like NaOH completely ionize in aqueous solution but amines are weak bases, hence they are partially ionized. The extent of ionization is known as basicity constant.

$$\underset{\text{Amine}}{RNH_2} + H_2O \rightleftharpoons R\overset{+}{N}H_3\overset{\ominus}{O}H \qquad K_b = \frac{[R\overset{\oplus}{N}H_3\overset{\ominus}{O}H]}{[RNH_2]}$$

Aliphatic amines are stronger base than ammonia, because of fact that alkyl groups are electron releasing, they increase the electron density around the nitrogen, hence increase in the availability of lone pair of electrons.

As the number of electron releasing groups increase, the greater the availability of lone pair of electrons of nitrogen and hence aliphatic amines are stronger bases than ammonia

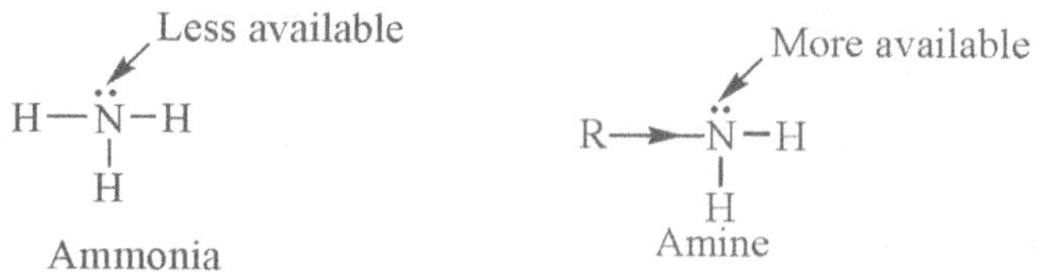

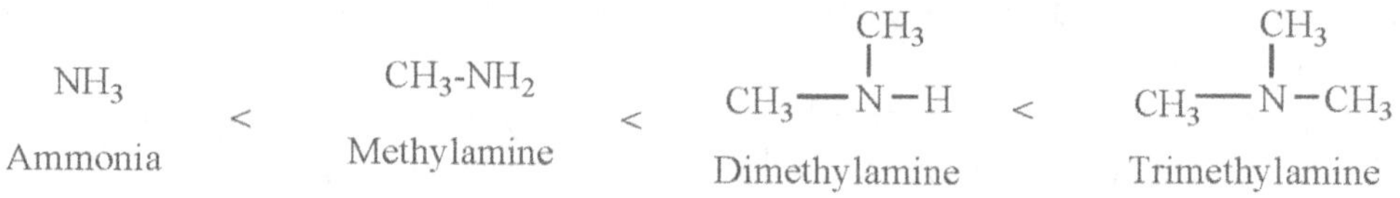

Less available: H—N̈—H (with H below) — Ammonia

More available: R→N̈—H (with H below) — Amine

NH_3 (Ammonia) < CH_3-NH_2 (Methylamine) < CH_3—N(CH$_3$)—H (Dimethylamine) < CH_3—N(CH$_3$)—CH_3 (Trimethylamine)

Dimethylamine is stronger base than methyl amine, which in turn is more basic than ammonia. But trimethyl amine is not more basic than dimethyl amine even though electron density of nitrogen is increased more, it is due to steric hindrance of the methyl groups, the path towards nitrogen for the proton to form bond is blocked. As a result, tertiary amines are least basic.

Alkylamines versus Ammonia

As per above discussion, due to +I effect of alkyl amines and crowding of alkyl groups increases from primary to tertiary amines, the order of basicity is 3° > 2° >1°.

Protonation of alkylamine and ammonia

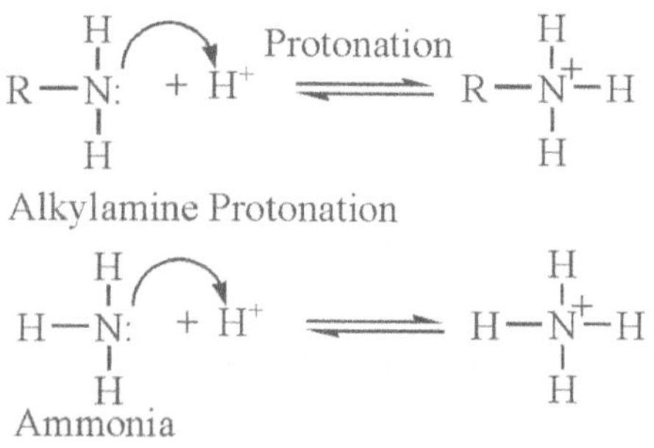

$$R\text{—}NH_2: + H^+ \xrightleftharpoons{\text{Protonation}} R\text{—}\overset{+}{N}H_3$$

Alkylamine Protonation

$$H\text{—}NH_2: + H^+ \rightleftharpoons H\text{—}\overset{+}{N}H_3$$

Ammonia

The ammonium cation is also stabilized by solvation and hydrogen bonding. As the size of the ion increases, solvation will decrease and stability of ion also decreases. The more is the stability of substituted ammonium cation, stronger will be the corresponding amine as base. Solvation of 'N' atom of ammonium cation of the primary amine is favorable for hydrogen bonding whereas the secondary or tertiary 'N' atom are not easily accessible for solvation and hydrogen bonding due to steric hindrance, hence according to this concept the order of basicity of amines is 1° > 2° > 3°. Thus, the basicity of alkyl amines is decided not only by +I effect but also steric hindrance. Based on the all these facts the order of basicity of methylamines and ethylamines in aqueous solution is as follows

$$(C_2H_5)_2NH > (C_2H_5)_3N > C_2H_5NH_2 > NH_3$$

$$(CH_3)_2NH_2 > CH_3NH_2 > (CH_3)_3N > NH_3$$

Aryl amines versus ammonia:

In aniline the 'N' atom of $-NH_2$ group is attached directly to benzene nucleus, hence the lone pair of electrons of N atom enters into conjugation with benzene ring which makes the lone pair of electrons less available for protonation and hence aniline is less basic than ammonia.

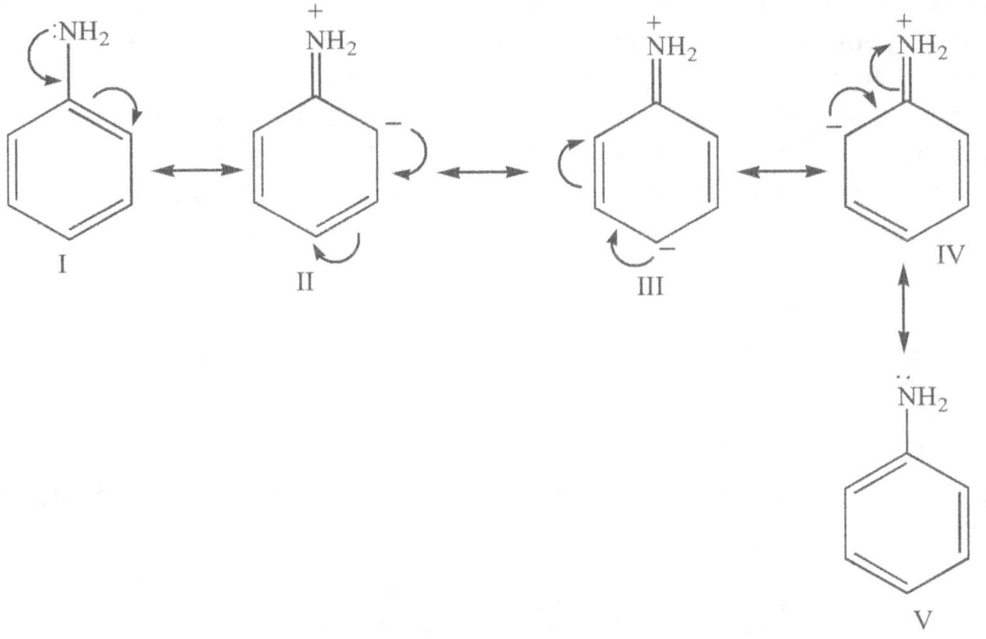

:NH_2 $\overset{+}{N}H_2$ $\overset{+}{N}H_2$ $\overset{+}{N}H_2$

I II III IV

$\ddot{N}H_2$

V

The anilinium ion obtained upon protonation of aniline can have two resonating structures. The more the number of resonating structures, more is the stability. Hence the proton acceptability or basic nature of aromatic amines is less than ammonia.

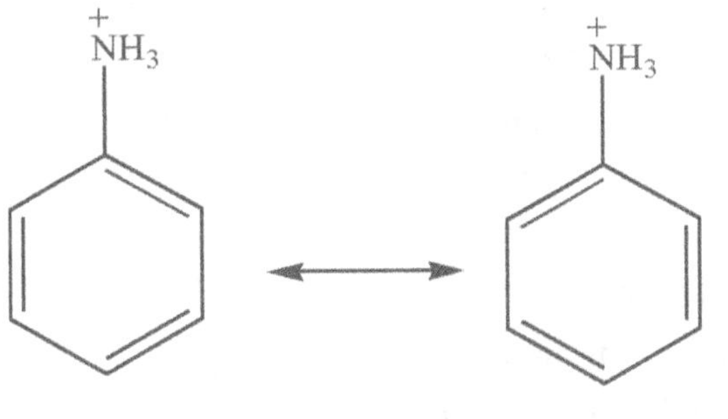

$\overset{+}{N}H_3$ $\overset{+}{N}H_3$

Hence aniline and all aryl amines are less basic than ammonia. The electron releasing groups increases the basicity while the electron withdrawing groups decreases the basicity.

Chemical Properties

The chemistry of amines can be compared as follows.

Primary and secondary amines are compared with alcohols, tertiary amines resembles ethers and the quaternary ammonium salts are compared with alkyloxonium ions .

CH_3-NH_2	$(CH_3)_2NH$	$(CH_3)_3N$	$(CH_3)_4\overset{+}{N}$ $(CH_3)_3\overset{+}{N}H$
1° Amine	2° Amine	3° Amine	Alkyl ammonium ions
CH_3OH	$(CH_3)_2O$		$(CH_3)_3\overset{+}{O}$ $(CH_3)_2\overset{+}{O}H$
Alcohol	Ether		Alkyloxonium ions

The reactions of amines can be described as follows:

1. **Reaction of the lone pair of electrons:** All amines have lone pair of electrons and the reactions involving these lone pair of electrons are important.
   (a) **Basic nature of amines:** Described in detail under **Basicity of amines.**
   (b) **Alkylation:** Amines are nucleophilic because of the presence of lone pair of electrons, hence they form bond with carbon. For example: Amines react with alkyl halides and form a new N-C bond.

$$\underset{\text{Methyl amine}}{CH_3\ddot{N}H_2} + \underset{\text{Methyl bromide}}{CH_3Br} \longrightarrow CH_3-\overset{\overset{H}{|}}{\underset{\underset{H}{|}}{\overset{\oplus}{N}}}-CH_3\ \overset{\ominus}{Br} \xrightarrow{-HBr} \underset{\text{2° Amine}}{CH_3-\overset{\overset{H}{|}}{N}-CH_3}$$

Thus, this reaction can be used for successful replacement of hydrogen atoms of amines with alkyl groups.

$$\underset{\text{1° Amine}}{CH_3NH_2} \xrightarrow[-HBr]{CH_3Br} \underset{\text{2° Amine}}{(CH_3)_2NH} \xrightarrow[-HBr]{CH_3Br} \underset{\text{3° Amine}}{(CH_3)_3N} \xrightarrow[-HBr]{CH_3Br} \underset{\text{4° Ammonium salts}}{(CH_3)_4\overset{\oplus}{N}\overset{\ominus}{Br}}$$

   (c) **Acetylation:** Acid chlorides react with primary and secondary amines and replaces hydrogen atom forming acyl derivatives called amides.

$$\underset{\text{Methylamine (1° Amine)}}{CH_3NH_2} + \underset{\text{Acetyl chloride}}{CH_3COCl} \xrightarrow{-HCl} \underset{\text{Acylmethylamine (or) } N\text{-Methylacetamide}}{CH_3NHCOCH_3}$$

$$\underset{\text{Dimethylamine (2° Amine)}}{(CH_3)_2NH} + \underset{\text{Acetyl chloride}}{CH_3COCl} \xrightarrow{-HCl} \underset{\text{Acyldimethylamine (or) } N,N\text{-Dimethylacetamide}}{(CH_3)_2NCOCH_3}$$

2. **Reactions of N-H bond:**
   (a) **Deprotonization:** By using strong base amines can be deprotonated.

$$\underset{\text{2° Amine}}{(C_2H_5)_2NH} + \underset{\text{Methyl lithium}}{CH_3Li} \longrightarrow \underset{\text{Diethyl lithium}}{(C_2H_5)_2\overset{\ominus\oplus}{NLi}} + \underset{\text{Methane}}{CH_4}$$

   (b) **Action with sodium:** When heated with sodium, primary and secondary amines form sodium salts with the liberation of hydrogen gas.

$$\underset{\text{1° Amine}}{2RNH_2} + 2Na \longrightarrow 2[R\bar{N}H]\overset{+}{Na} + H_2\uparrow$$

$$\underset{\text{2° Amine}}{2R_2NH} + 2Na \longrightarrow 2[R_2\bar{N}H_2]\overset{+}{Na} + H_2\uparrow$$

(c) **Reaction with halogens:** Primary and secondary amines react with halogens in the presence of alkali to give halogen amines.

$$\underset{1^\circ\text{ Amine}}{RNH_2} \xrightarrow[-HX]{X_2/NaOH} \underset{\text{Monohalogen derivative}}{RNHX} \xrightarrow[-HX]{X_2/NaOH} \underset{\text{Dihalogen derivative}}{RNX_2}$$

$$\underset{2^\circ\text{ Amine}}{R_2NH} \xrightarrow{X_2/NaOH} \underset{\text{Monohalogen derivative}}{R_2NX} + HX$$

(d) **Reaction with nitrous acid:** Amines reacts with nitrous acid as follows.

(i) **Primary amine:** It forms alcohol with the liberation of nitrogen gas as bubbles.

$$\underset{1^\circ\text{ Amine}}{RNH_2} \xrightarrow{HONO} \underset{\text{Alcohol}}{R\text{-}OH} + N_2\uparrow + H_2O$$

(ii) **Secondary amines:** Forms N-nitrosoamines (Yellow oily liquid).

$$\underset{2^\circ\text{ Amine}}{R_2NH} + HONO \longrightarrow \underset{\text{Nitrosoamine (Yellow oil)}}{R_2N\text{-}NO} + H_2O$$

Nitrosoamines when reacted with phenol and concentrated H_2SO_4 gives brown red colour which soon changes to blue green to red upon the dilution with water, further it changes to blue or violet on reaction with alkali. This colour change is called as **Libermann's nitroso reaction,** and hence used for identification of secondary amines.

(iii) Tertiary amines on cold condition form salt $R_3N.HNO_2$. These salts upon warming decompose to give nitrosoamine and alcohol.

$$\underset{\text{Salt}}{R_3N.HNO_2} \longrightarrow \underset{\text{Nitrosoamine}}{R_2N\text{-}NO} + \underset{\text{Alcohol}}{ROH}$$

(e) **Carbylamine reaction:** Primary amines react with chloroform and KOH and gives isocyanides with the offensive smell of carbylamines. Secondary, tertiary amines do not react.

$$\underset{\text{Methylamine}}{CH_3NH_2} + \underset{\text{Chloroform}}{CHCl_3} + 3KOH \longrightarrow \underset{\text{Methylisocyanide}}{CH_3N{\equiv}C} + 3KCl + 3H_2O$$

(f) **Reaction with Grignard reagent:** Primary, secondary amines react with Grignard reagent to give hydrocarbons and the tertiary amine do not react.

$$\underset{\text{Methyl magnesium iodide}}{Mg\begin{smallmatrix} I \\ CH_3 \end{smallmatrix}} + \underset{1^\circ\text{ Amine}}{RNH_2} \longrightarrow \underset{\text{Methane}}{CH_4} + Mg\begin{smallmatrix} I \\ NHR \end{smallmatrix}$$

$$\underset{\text{Ethyl magnesium iodide}}{Mg\begin{matrix} I \\ C_2H_5 \end{matrix}} + \underset{2^\circ \text{ Amine}}{R_2NH} \longrightarrow \underset{\text{Ethane}}{C_2H_6} + Mg\begin{matrix} I \\ NR_2 \end{matrix}$$

(g) **Reaction with carbon disulphide:** When amines are warmed with carbon disulphide, it behaves as follows.

(i) Primary amine forms alkyl dithiocarbamic acid which is decomposed with mercuric chloride to yield alkyl isothiocyanate, this is called **Hoffman's mustard oil reaction** (Used for identification of primary amines).

$$\underset{1^\circ \text{ Amine}}{RNH_2} \xrightarrow{S=C=S} \underset{\text{Dithiocarbamic acid}}{RNH\text{-}C(=S)SH} \xrightarrow[-HgS,\ -2HCl]{HgCl_2} \underset{\text{Alkyl iso thio cyanate}}{RNCS}$$

(ii) Secondary amine forms dithiocarbamic acid but not decomposed by mercuric chloride

$$\underset{2^\circ \text{ Amine}}{R_2NH} + S=C=S \longrightarrow \underset{\text{Dithiocarbamic acid}}{R_2N\text{-}C(=S)SH} \xrightarrow{HgCl_2} \text{No reaction}$$

(iii) Tertiary amine will not react with carbon disulphide.

$$\underset{3^\circ \text{ Amine}}{R_3N} + S=C=S \longrightarrow \text{No reaction}$$

(h) **Oxidation:** Amines undergo oxidation and the oxidation products formed depend upon the oxidizing agent used.

(i) **Primary amine:**

$$\underset{1^\circ \text{ Amine}}{RCH_2NH_2} \xrightarrow[{[O]}]{KMnO_4} \underset{\text{Aldimine}}{RCH=NH} \xrightarrow{H_2O} \underset{\text{Aldehyde}}{RCHO + NH_3}$$

$$\underset{1^\circ \text{ Amine}}{R_2CHNH_2} \xrightarrow[{[O]}]{KMnO_4} \underset{\text{Ketimine}}{R_2C=NH} \xrightarrow{H_2O} \underset{\text{Ketone}}{R_2CO} + NH_3$$

(ii) **Secondary amine:**

$$\underset{\text{Dialkyl hydroxyl amine}}{R_2NOH} \xleftarrow[\text{Caros acid}]{H_2SO_4} \underset{2^\circ \text{ Amine}}{R_2NH} \xrightarrow[(O)]{KMnO_4} \underset{\text{Tetra alkyl hydrazine}}{R_2N\text{-}NR_2}$$

Tertiary amines are not affected by $KMnO_4$ but can oxidized by Caro's acid or Fenton's reagent to amine oxide.

$$\text{No reaction} \xleftarrow{KMnO_4} \underset{3^\circ \text{ Amine}}{R_3N} \xrightarrow[{[O]}]{\text{Caros acid}} \underset{\text{Amine oxide}}{R_3N \rightarrow O}$$

(i) **Reaction with Benzene sulphonyl chloride:** Primary, secondary amines react with benzene sulphonyl chloride (Hinsberg reagent) and form sulphonamides. Tertiary amines do not react. From the sulphonamides, amines are liberated when treated with NaOH. This method is used for the separation of primary, secondary, tertiary amines (Refer Hinsberg method for separation of amines).

3. Reactions involving N-C bonds:

Cleavage of N-C bonds of alkyl ammonium ions occur when it undergoes nucleophillic attack.

$$(CH_3)_4N^{\oplus}\ \overset{\ominus}{O}H \longrightarrow (CH_3)_3N + CH_3OH$$

Tetra methyl ammoniumhydroxide → Trimethyl amine + Methanol

By the following tests we can distinguish primary, secondary and tertiary amines.

Table 18.3 Various tests employed to distinguish three types of amines.

Test	Primary amines	Secondary amines	Tertiary amines
1. Treat with nitrous acid.	Formation of alcohols with the evolution of nitrogen.	Formation of nitroso amines which give Lieberman's nitroso reaction.	Formation of nitrites in cold.
2. Treat with chloroform and alcoholic potash.	Obnoxious smell of carbyl-amine.	No action.	No action.
3. Treat with Hinsberg's reagent- (benzenesulphonylchloride)	Form sulphonamides soluble in caustic soda.	Form sulphonamides insoluble in caustic soda.	No reaction.
4. Treat with methyl iodide under ordinary pressure.	No reaction.	No reaction.	Form quaternary compounds.
5. Treat with acetyl chloride.	Forms amides.	Forms amides.	No reaction.

Summary of Chemical Reactions of Primary Amines

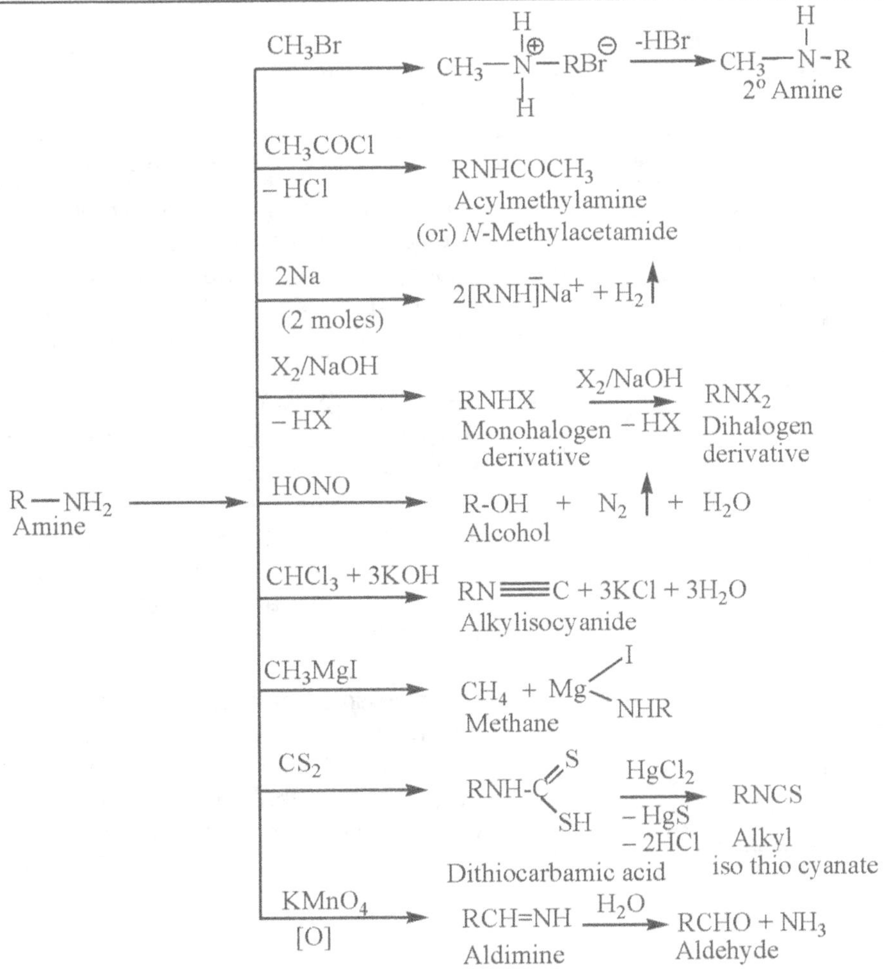

Ascent and Descent of Series

Hoffmann's bromamide reaction is used for the ascent and descent of homologues series of amines as described below.

1. **Ascent of series:** Conversion of methylamine to ethylamine.

$$\underset{\text{Methyl amine}}{CH_3NH_2} \xrightarrow[-N_2,\ -H_2O]{HONO} \underset{\text{Methanol}}{CH_3OH} \xrightarrow{P + I_2} \underset{\text{Methyl iodide}}{CH_3I} \xrightarrow[-KI]{KCN} \underset{\text{Methyl cyanide}}{CH_3CN} \xrightarrow{4(H)} \underset{\text{Ethyl amine}}{CH_3CH_2NH_2}$$

2. **Descent of series:** Conversion of ethylamine to methylamine.

$$\underset{\text{Ethyl amine}}{C_2H_5NH_2} \xrightarrow[-N_2,\ -H_2O]{HONO} \underset{\text{Ethanol}}{C_2H_5OH} \xrightarrow{(O)} \underset{\text{Acetaldehyde}}{CH_3CHO} \xrightarrow{(O)} \underset{\text{Acetic acid}}{CH_3COOH} \xrightarrow{NH_3} \underset{\text{Ammonium acetate}}{CH_3COONH_4}$$

$$\underset{\text{Ammonium acetate}}{CH_3COONH_4} \xrightarrow[-H_2O]{\Delta} \underset{\text{Acetamide}}{CH_3CONH_2} \xrightarrow[-CO_2,\ -KBr,\ -HBr]{Br_2 + KOH} \underset{\text{Methyl amine}}{CH_3NH_2}$$

Table 18.4 Amines of Pharmaceutical Importance.

Compound name	Use	Qualitative test
Amphetamine $C_6H_5-CH_2-CH(NH_2)-CH_3$	It is used as a CNS stimulant. It is also used for weight loss.	To the aqueous solution of sample add 10 ml of sodium hydroxide solution and one drop of benzoyl chloride, shake, add more benzoyl chloride until no further precipitate is produced. Determine the melting range of the product. It should be within the limit. To the sample solution add solution of barium chloride. A white precipitate is obtained which is insoluble in hydrochloric acid. To the sample solution add a solution of lead acetate. A white precipitate is obtained which is soluble in solution of ammonium acetate and in solution of sodium hydroxide. Dissolve small amount of sample in water, add 1 ml of 1N hydrochloric acid, 2 ml of solution of diazotized p-nitroaniline, 4 ml of 1N sodium hydroxide and 2 ml of n-butyl alcohol. Shake it well and allow to separate, a red colour develops in the butyl alcohol layer.

Table 18.4 Contd...

Compound name	Use	Qualitative test
Ethanolamine $OHCH_2CH_2NH_2$	In pharmaceutical formulations, ethanolamine is used for buffering. Used in the preparation of emulsions. It is used as a pH regulator in cosmetics. It is used as a sclerosing agent.	Neutralise the aqueous solution of sample with trinitrophenol, and evaporate to dryness on a water bath. Determine the melting range of the residue (159 – 161 °C).
Ethylene diamine $C_2H_4(NH_2)_2$	It is used as a pharmaceutical aid.	Dilute small amount sample solution with water . To this add 2ml of copper sulphate solution and shake well. A purple blue colour is produced.

Probable Questions

1. What are amines? Mention their importance.
2. Write a detailed note on nomenclature of amines.
3. Write a detailed note on general methods of preparation of amines.
4. Explain Hoffmann's bromamide reaction with reaction mechanism.
5. What is Mendius reaction?
6. Explain Gabriel phthalimide method with reaction mechanism.
7. Write a note on reductive amination.
8. Write short notes on Curtius rearrangement.
9. How are primary amines synthesized? Explain any three methods with suitable example.
10. How are secondary amines synthesized? Explain any three methods with suitable example.
11. How are tertiary amines synthesized? Explain any three methods with suitable example.
12. Explain the various methods used for separation of primary, secondary & tertiary amines.
13. Write a note on general structure and physical properties of amines.
14. Write a short note on basicity of amines.
15. Compare the basicity of ammonia with alkylamines & aryl amines.
16. Write the various chemical properties of amines in detail.
17. Write about chemical reactions involving N-H bond of amine.
18. Explain about reactions involving N-C bonds of amines.
19. Write a note on various tests employed to distinguish three types of amines.
20. Write a note on ascent and descent series of amines with its conversion.
21. Write the chemical structure, qualitative test & uses of ethanolamine, ethyenediamine & amphetamine.
22. Explain the pharmaceutical importance of amines.

19

Introduction to Aromatic Compounds

Introduction

A huge number of organic compounds are obtained from nature. For example: resins, balsams and drugs. They are pleasant smelling hence called as aromatic compounds. They have high amount of carbon *i.e.* minimum 6 carbon atoms when compared with aliphatic compounds. They are ring or closed chain compounds and their properties are entirely different from aliphatic compounds. These days the word aromatic is used for benzene and its derivatives.

Reasons for Separate Classification of Aromatic Compounds

- Aromatic compounds are closed chain or ring compounds whereas the aliphatic compounds are open chain compounds. The ring structure confers on the aromatic compounds extreme stability despite of their unsaturated nature.
- Aromatic compounds easily undergo nitration when heated with conc. HNO_3 to yield nitro compounds (does not takes place easily in aliphatic compounds).
- Aromatic compounds forms sulphonic acid derivatives when treated with fuming or concentrated sulphuric acid but this reaction is not possible for aliphatic compounds.
- Aromatic amines are less basic than aliphatic amines.
- Aromatic hydroxyl compounds are acidic in nature but aliphatic hydroxyl compounds are neutral.
- Aryl halides are chemically less reactive than alkyl halides.
- Aromatic compounds have large amount of carbon atoms than aliphatic compounds.
- A number of chemical reactions shown by aromatic compounds are not common in aliphatic compounds. For example: Friedel-Craft's reaction, Perkin reaction*etc.*
- Alkyl hydrocarbons undergo addition reaction at unsaturation and free radical substitution reaction at other points along the chain but aryl hydrocarbons have a tendency to undergo ionic substitution.

Nomenclature of Aromatic Compounds

Aromatic compounds can be named by IUPAC system using the rules framed. However, even now some common names are also used to mention some of the aromatic compounds. All aromatic compounds we study in basic organic chemistry have benzene structure in it, and hence aromatic compounds are considered as derivatives of benzene.

By replacing one or more hydrogen atoms of benzene, various aromatic compounds are obtained. The group obtained by removal of one hydrogen atom from benzene is called phenyl [As in alkane, removal of one hydrogen from alkane yield alkyl group –R].

$$\underset{\text{(Alkane)}}{R-H} \xrightarrow[-H]{} \underset{\text{(Alkyl)}}{R-}$$

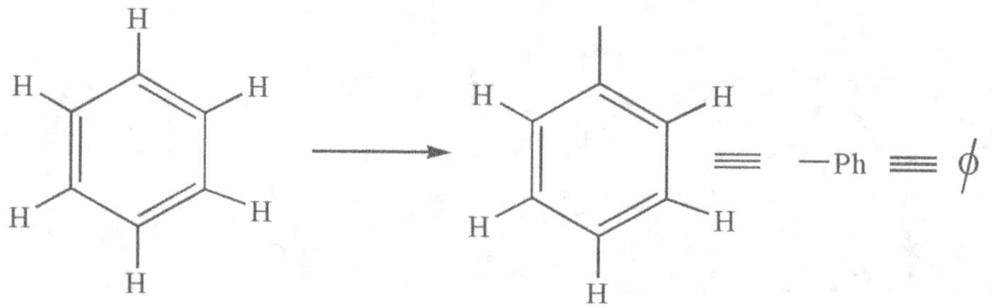

C_6H_6 (Benzene) → $-C_6H_5$ (Phenyl) ≡ —Ph ≡ ϕ

1. **Naming of mono substituted benzene:** Benzene is a regular hexagon with a conjugate double bond or denoted by circle. Hence the mono substituted benzene means any one of six-hydrogen atom is replaced by the substituent.
   (a) Some aromatic compounds are named as substituted benzene. Example: Bromobenzene, chlorobenzene *etc.* In this, benzene is the base name and the substituents are indicated by prefix.

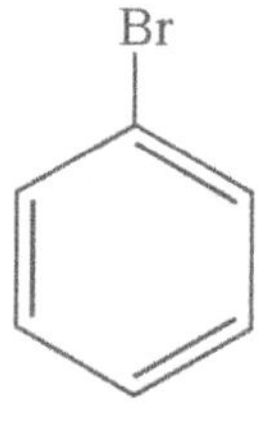

Bromo benzene 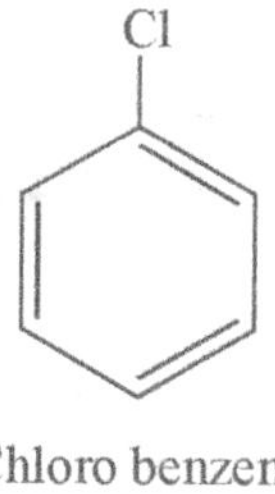 Chloro benzene

   (b) In some other compounds benzene and substituents are included in base name as shown in following examples.

Compound	IUPAC Name	Common name
$C_6H_5CH_3$	Methyl benzene	Toluene
C_6H_5OH	Hydroxy benzene	Phenol
$C_6H_5NH_2$	Amino benzene	Aniline

Sometimes the benzene ring is named as substituent (phenyl group) as shown below.

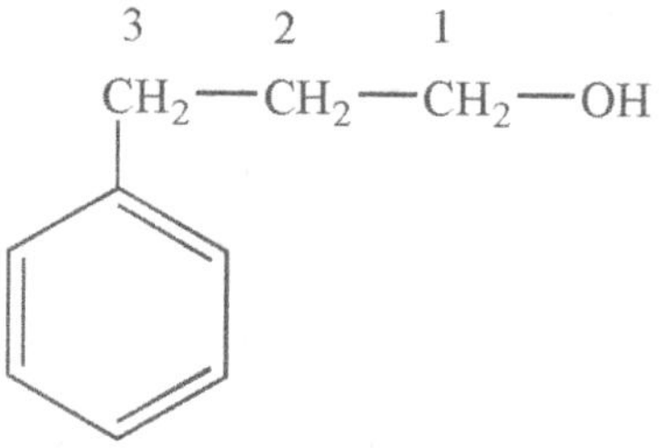

$\overset{3}{CH_2}-\overset{2}{CH_2}-\overset{1}{CH_2}-OH$ (with phenyl on C-3)

3-Phenyl propanol

2. **Naming of disubstituted benzene:**
   (a) When more than one substituent is present in the ring, the principal functional group should be given number 1 and should be oriented with position 1 up and the other positions are numbered clockwise direction. If the prefixes *ortho (o), meta (m) and para (p)* are used to number then the principal functional (in this method position 2, 6 and 3, 5 are same) group should be 1. In general, if the functional group is named as a suffix, this group is given number 1.

(b) When common name is used, the substituent is responsible for the name (Example –OH for phenol) is considered on carbon 1 (C-1).

(c) When the substituents present are different, it should be mentioned in alphabetical order.

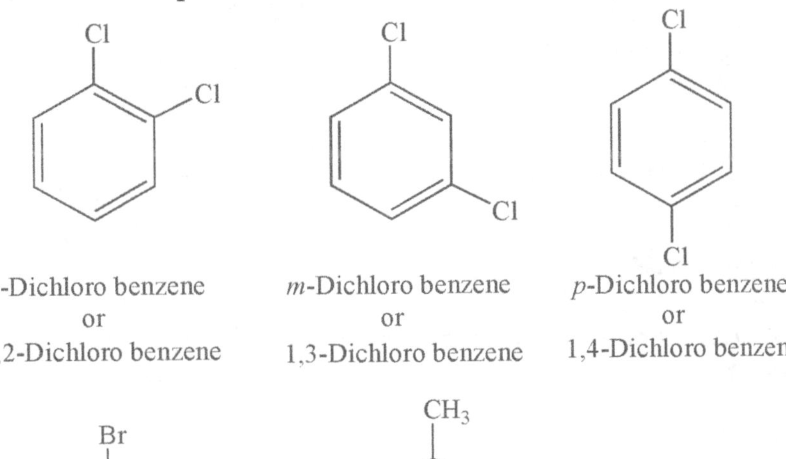

o-Dichloro benzene or 1,2-Dichloro benzene

m-Dichloro benzene or 1,3-Dichloro benzene

p-Dichloro benzene or 1,4-Dichloro benzene

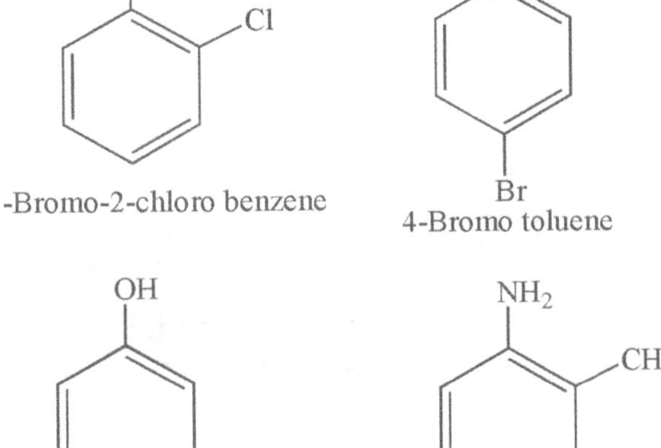

1-Bromo-2-chloro benzene

4-Bromo toluene

3-Methyl phenol

2-Methyl aniline

3. Naming of poly substituted benzene:

(a) All the substituents must be numbered in such a way to give minimum number to the substituents.

(b) Among the substituents, the preference to one group and the numbering of next groups are done adopting the above rules described under disubstituents.

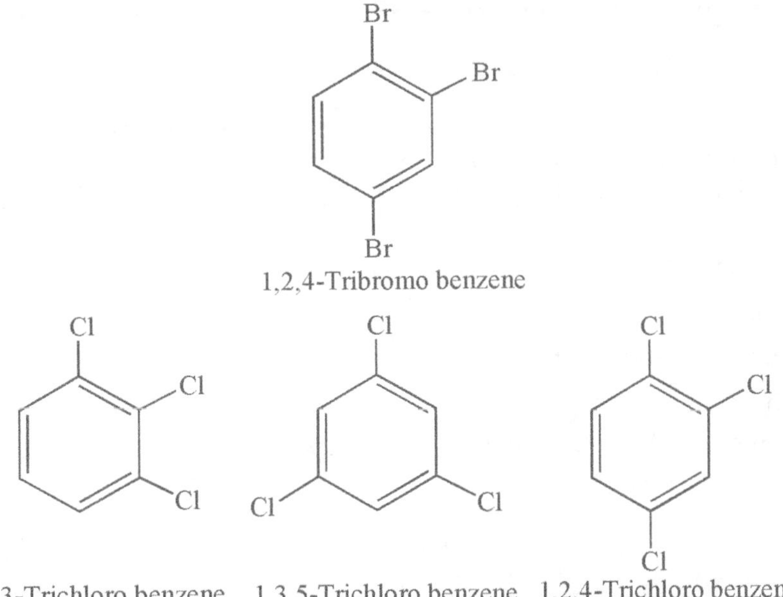

1,2,4-Tribromo benzene

1,2,3-Trichloro benzene

1,3,5-Trichloro benzene

1,2,4-Trichloro benzene

(c) When a group present in the ring yields special name, then the remaining substituents along with the special name is mentioned.

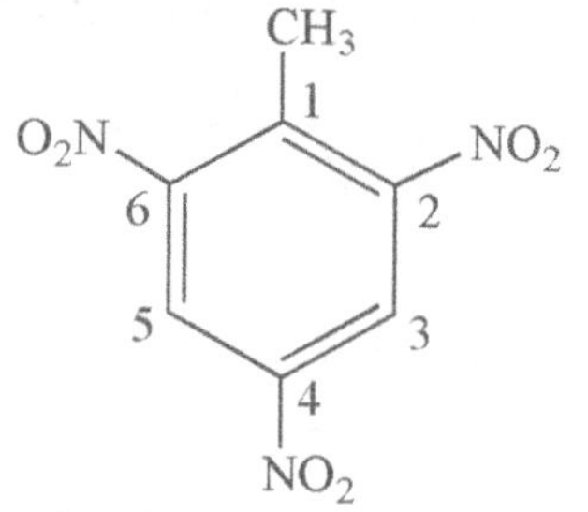

CH_3 1 O_2N NO_2 6 2 5 3 4 NO_2

2,4,6-Trinitro toluene

(d) Arrangement of substituents in alphabetical order is also to be kept in mind, when different groups are present.

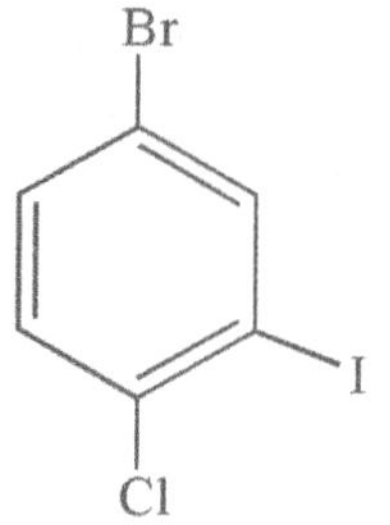

Br I Cl

4-Bromo-1-chloro-2-iodo benzene

4. **Naming of fused benzene ring system:** More than one aryl rings joined together and form poly aromatic ring compounds. The substituents in these rings are indicated as per IUPAC nomenclature.

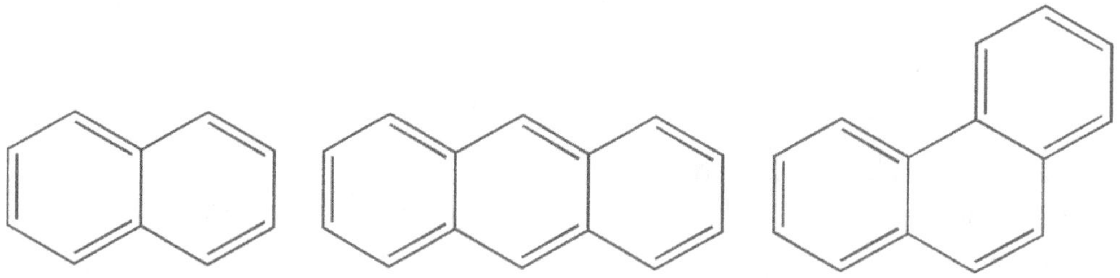

Naphthalene Anthracene Phenanthrene

Orientation

It is defined as the process of finding the relative position of various groups attached to the nucleus in an unknown derivative. For example, consider the sample of xylene, then find out whether it is *o*, *p* or *m*-xylene. Three types of methods are used for finding out the orientation of unknown compounds. They are as follows.

1. Korner's absolute method.
2. The relative method.
3. Dipole moment measurement method.

Let us discuss in detail.

1. **Korner's absolute method:** When a disubstituted compound is converted into a trisubstituted product then *o*-derivative yields two products, the *m*-derivative yields 3 products, while the *p*-derivative yields one trisubstituted product.

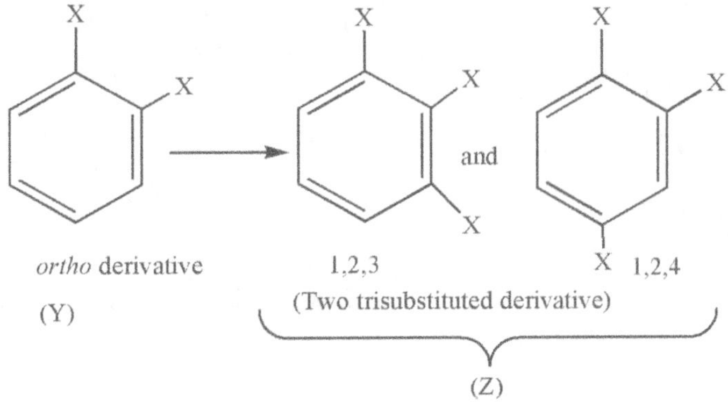

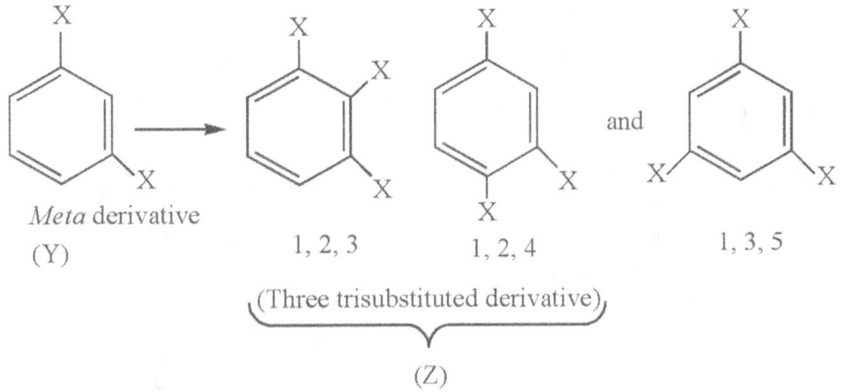

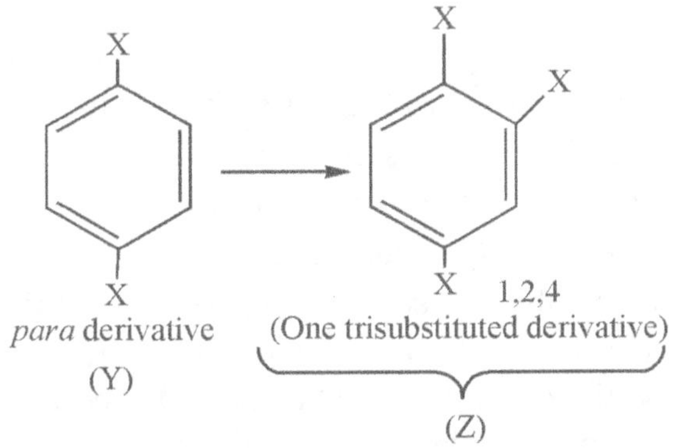

ortho derivative (Y) → 1,2,3 and 1,2,4 (Two trisubstituted derivative) (Z)

Meta derivative (Y) → 1, 2, 3, 1, 2, 4 and 1, 3, 5 (Three trisubstituted derivative) (Z)

para derivative (Y) → 1,2,4 (One trisubstituted derivative) (Z)

Hence for finding out orientation,

(i) Convert the disubstituted compound Y into a trisubstituted product Z and analyse the product Z obtained.

(ii) If Z is a single substance, Y is a *para* derivative. If Z is a mixture of two substances then Y is an *ortho* derivative and if Z is a mixture of three substances Y is a *meta* derivative.

Similarly, when a trisubstituted compound is converted in to a tetra substituted product, the number of derivatives obtained are shown as follows.

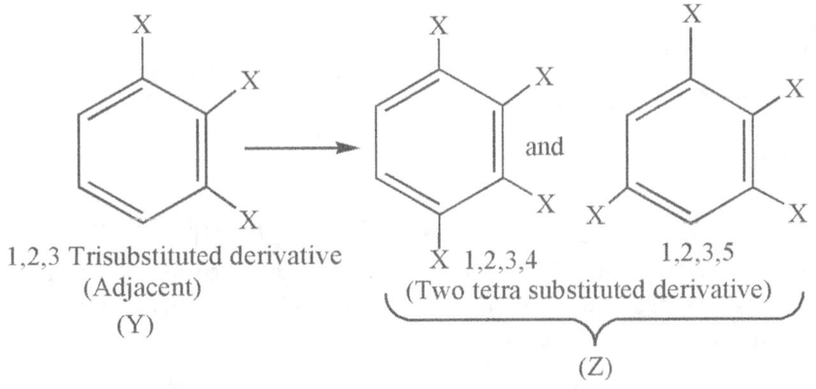

1,2,3 Trisubstituted derivative (Adjacent) (Y) → 1,2,3,4 and 1,2,3,5 (Two tetra substituted derivative) (Z)

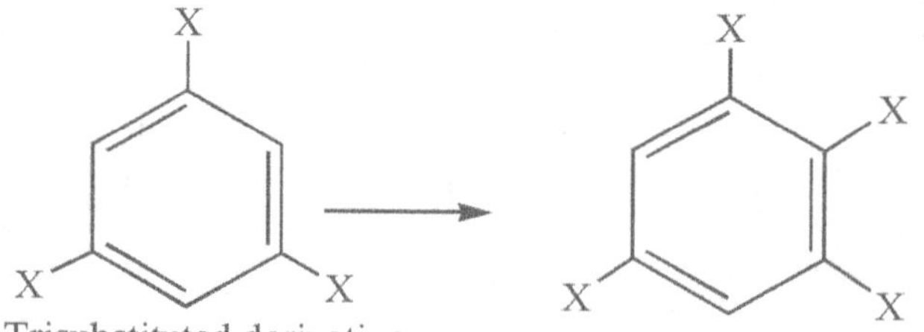

1,3,5 Trisubstituted derivative (Symmetric) (Y) → 1,2,3,5 (Only one tetra substituted derivative) (Z)

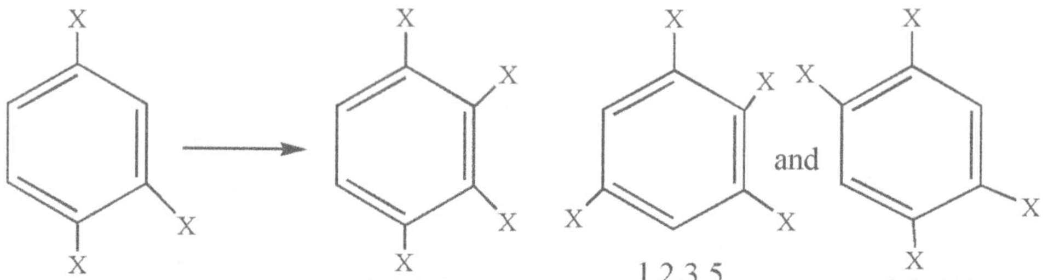

1,2,4 Trisubsytituted derivative (Unsymmetric) (Y) → 1,2,3,4, 1,2,3,5 and 1,2,4,5 (Three tetra substituted derivative) (Z)

Hence for orientation,

(i) Convert the tri substituted derivative Y into tetra derivative Z and analyse the product Z obtained.

(ii) If Z is a single substance, Y is 1,3,5-derivative (symmetric).
If Z is a mixture of two substance Y is 1,2,3-derivative (adjacent).
If Z is a mixture of three substance, Y is 1,2,4-derivative (unsymmetrical).

Disadvantage: All the products cannot be always isolated. So, it is too difficult to perform.

2. **The Relative method:** The relative position of the unknown compound is determined directly by studying its properties. In another way, the compound to be studied is converted into or synthesized from another compound of previously known orientation. For Example:

(i) **By studying its reactions:** Three types of benzene dicarboxylic acids are present. They are phthalic acid, iso phthalic acid, terephthalic acid. Out of these, first one gives anhydride upon heating. It indicates that it must be *ortho*-isomer in which the two-COOH groups, being in close proximity can eliminate a water molecule.

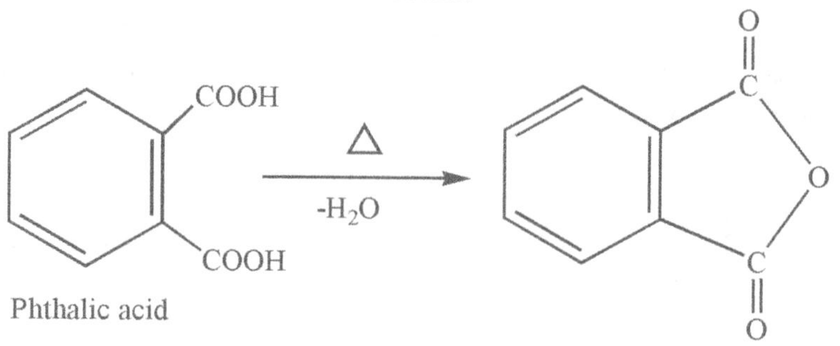

Phthalic acid $\xrightarrow[-H_2O]{\Delta}$ Phthalic anhydride

(ii) **By synthesizing it from another compound of previously known orientation:** When three isomeric xylenes are present. Out of these one has the boiling point 412 K and can be obtained from mesitylene in following way.

$$\text{Mesitylene} \xrightarrow[\text{oxidation}]{\text{Partial}} \text{Mesitylenic acid} \xrightarrow[\text{Sodalime}]{\text{Heat}} \text{Xylene}$$

Hence three methyl groups in mesitylene are in *meta* position to each other. Removal of one of them must have left the remaining two in *meta* position. Thus, xylene with boiling point 412 K is obtained.

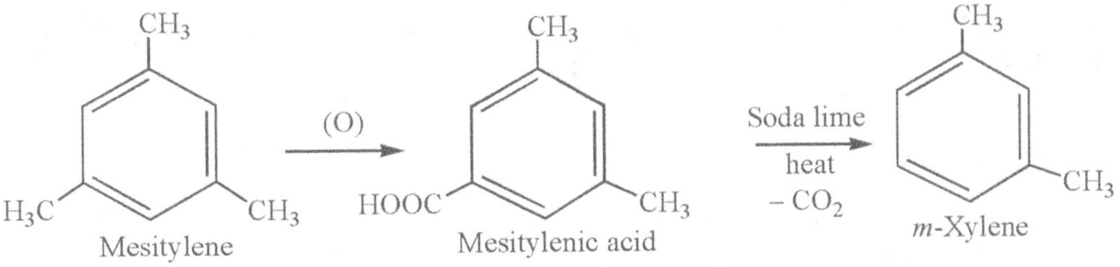

Mesitylene $\xrightarrow{(O)}$ Mesitylenic acid $\xrightarrow[\text{heat}, -CO_2]{\text{Soda lime}}$ *m*-Xylene

(iii) **By converting it in to another compound of previously known orientation:** One of the xylene isomer boils at 417 K and undergoes oxidation to yield phthalic acid which upon further heating yields phtahlic anhydride. Obviously in phthalic acid two -COOH groups are in *ortho* to each other, hence the two $-CH_3$ groups in xylene are present *ortho* to each other. Thus out of the three unknown xylene molecule one is *meta* isomer (B.P. 412 K), the second one is *ortho* isomer (B.P. 417 K) and the third one is *para* isomer (B.P. 411 K). This method is based on the assumption that atoms or groups remain in same position or exchange the position with respect to the incoming substituents. Sometimes this assumption may be incorrect also.

3. **Dipole moment measurement method:** In some cases, measurement of dipole moment of the compound is used to determine the orientation of the groups. For example, in dichlorobenzene, the observed D value of C-Cl bond is 1.55 D. By using these values, we can find out three dichloro benzene. By comparison of the calculated value with the observed one, we can decide the orientation of the given dichloro benzene.

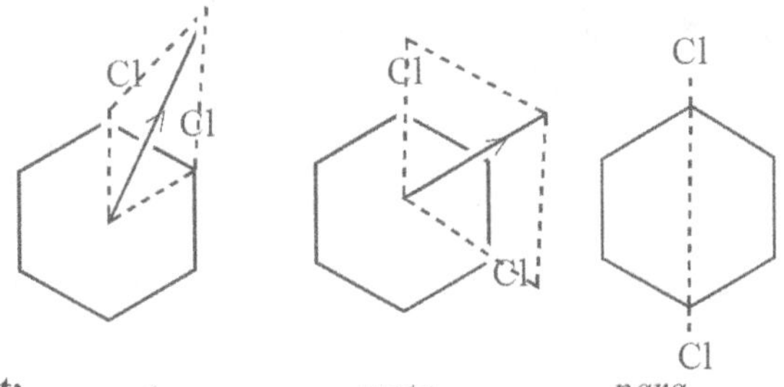

Dipole moment:	*ortho*	*meta*	*para*
Calculated :	2.68 D	1.55 D	0 (zero)
Observed :	2.30 D	1.48 D	0 (zero)

Reactions of Substituted Benzene

Reactions of substituted benzenes are influenced by the nature of substituents attached to it and they are described as follows.

Effect of Substituents on Reactivity

Like benzene, substituted benzenes also undergo electrophilic aromatic substitution reactions such as chlorination, nitration, sulphonation, Friedel-craft's alkylation and acylation. In this chapter we are discussing about the reactivity of substituted benzene. When compared to benzene substituted benzene is more reactive or less reactive than benzene? The answer is "Reactivity depends on the substituent". Some of the substituents make the ring more reactive and some make the ring less reactive than benzene towards electrophilic aromatic substitution. The rate determining step (slow step) for the electrophilic aromatic substitution is the attack of electrophile on the nucleophilic aromatic ring. Increasing the electron density of the benzene ring increases the attractiveness to an electrophile. So, the above fact indicates that "substituents capable of donating electrons to the benzene ring increases the rate of electrophilic aromatic substitution where as substituents capable of withdrawing the electrons from the benzene ring decreases the rate of electrophilic aromatic substitution reaction". Relative rates of electrophilic aromatic substitution reactions is as follows.

Z > (benzene) > Y

Z = Donates electrons to the benzene ring

Y = Withdraws electrons from the benzene ring

Substituents can donate the electrons into the benzene ring by two ways.

1. Electron donation by inductive effect or inductive electron donation.
2. Electron donation by resonance.

Similarly, substituents can withdraw the electrons from benzene ring by two ways.

1. Electron withdrawal by inductive effect or inductive electron withdrawal.
2. Electron withdrawal by resonance.

1. **Electron donation and withdrawal by inductive effect or Inductive electron donation and withdrawal:** If a substituent attached to the benzene ring has less electron withdrawing capacity than hydrogen, then the electrons of the σ bond which connects the benzene ring and the substituents readily move towards the benzene ring. Donation of electrons *via* a σ bond is called as "Inductive electron donation". For example: toluene

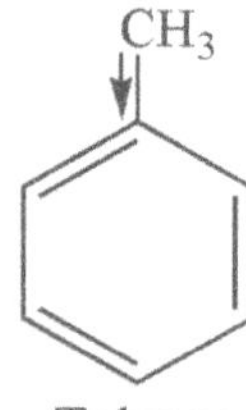

Toluene

In toluene structure, the electron donating capacity of $-CH_3$ group (alkyl) is compared with hydrogen atom. Actually carbon atom is slightly more electronegative than hydrogen, but an alkyl group is more electron donating than hydrogen. So it increases the rate of electrophilic aromatic substitution reaction.

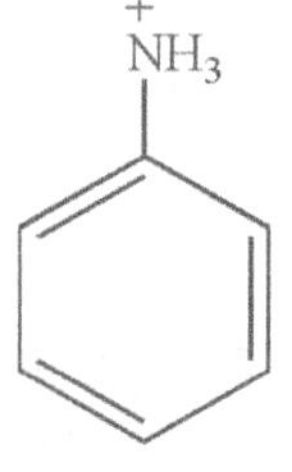

Anilinium ion

In the above structure, the substituent (NH_3^+) is more electronegative than hydrogen so it attracts or withdraws the electron from the ring and decreases the rate of electrophilic aromatic substitution reaction.

2. **Electron donation and withdrawal by resonance:** If a substituent attached to a benzene ring contains a lone pair of electrons, then these electrons are delocalized into the ring by *p*-orbital overlap. These are said to donate their electrons by resonance. For example: NH_2OH, :OR and Cl donate electrons by resonance as well as withdraw electrons by inductive effect due to their more electronegativity than hydrogen.

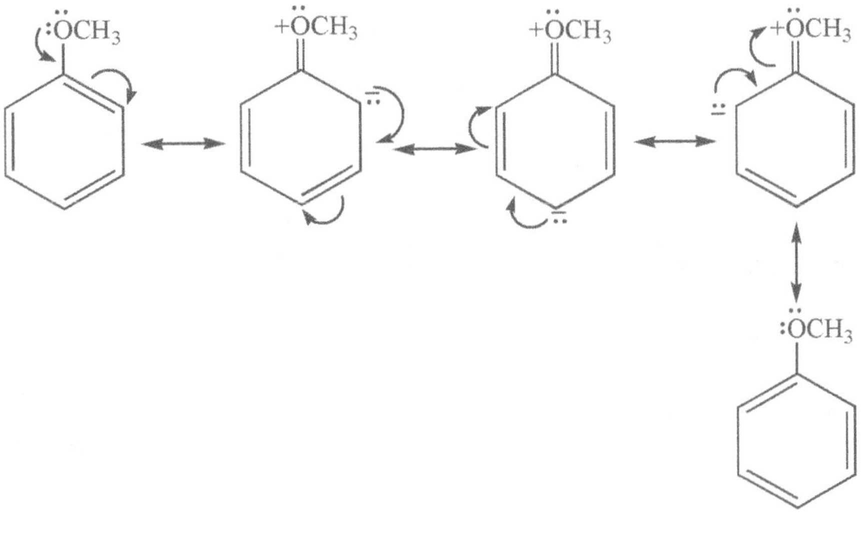

Figure 19.1 Donation of electrons into a benzene ring by resonance.

If the substituent attached to a benzene ring by an atom which is attached to a more electronegative atom (C=O, C≡N,NO_2), it withdraws the electrons from the benzene ring by resonance as well as withdraw the electrons inductively from benzene ring because the atom attached to the benzene ring is more electronegative than hydrogen.

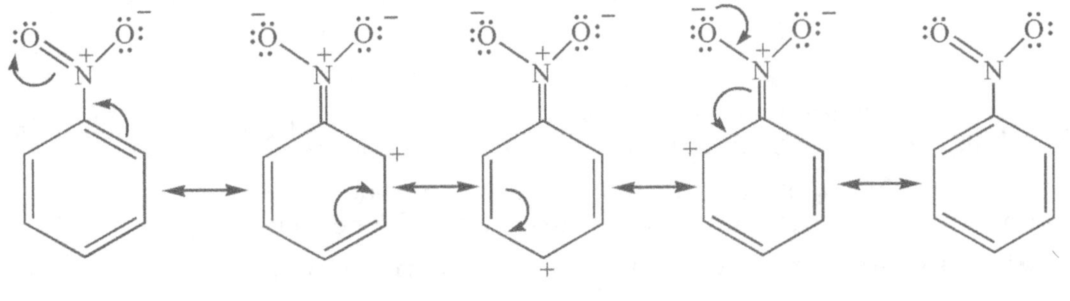

Relative Reactivity of Substituted Benzene

The following table shows the "effect of substituents on the reactivity of substituted benzene towards electrophilic substitution".

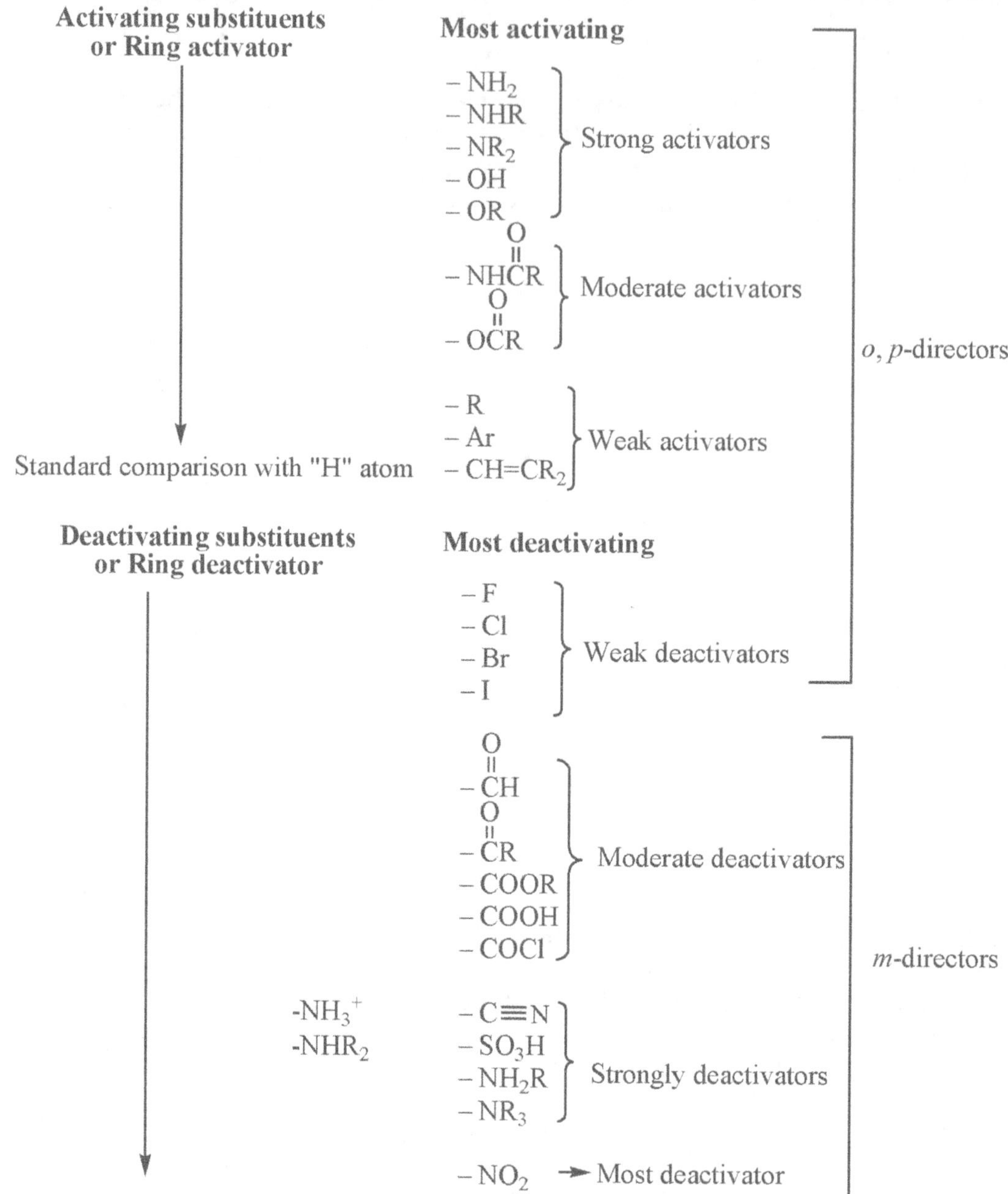

Activators or activating substituents: These groups or substituents make the benzene ring more reactive towards electrophilic aromatic substitution and they donate electrons to the benzene ring.

Deactivators or deactivating substituents: They make the benzene ring less reactive towards electrophilic aromatic substitution and they withdraw the electrons from the benzene ring.

Strongly activating substituents: They activate the benzene ring strongly.

:NH_2 — Aniline

:ÖH — Phenol

:Ö—R — Aryl ether

Moderately activating substituents: They also donate electrons into the benzene ring by resonance and withdraw electrons from the benzene ring by inductive effect. But their electron donation by resonance is less effective than the strongly activating substituents.

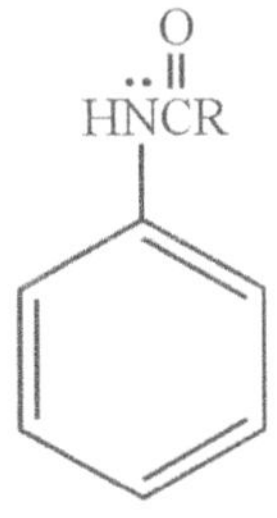

Arylamide 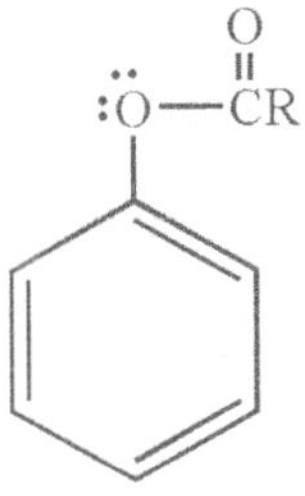 Arylester

These substituents are less effectively donating the electrons to the ring when compared with strong activators. Because the strong ring activators donate the electrons to the ring by resonance only. But moderate activators donate the electrons to the ring by resonance in two competing directions *i.e.,* into the ring and away from the ring. The fact that these substituents are activators, that even though their resonance electron donation is diminished into the ring, overall they donate electrons by resonance more strongly than they withdraw electrons inductively.

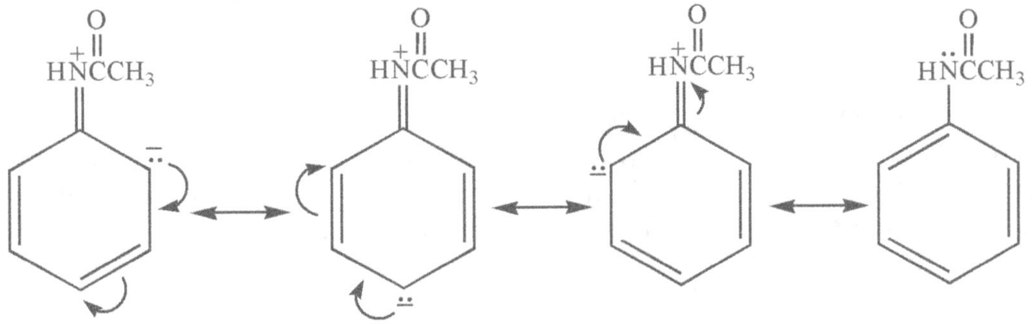

Substituent donate electrons by resonance into the benzene ring

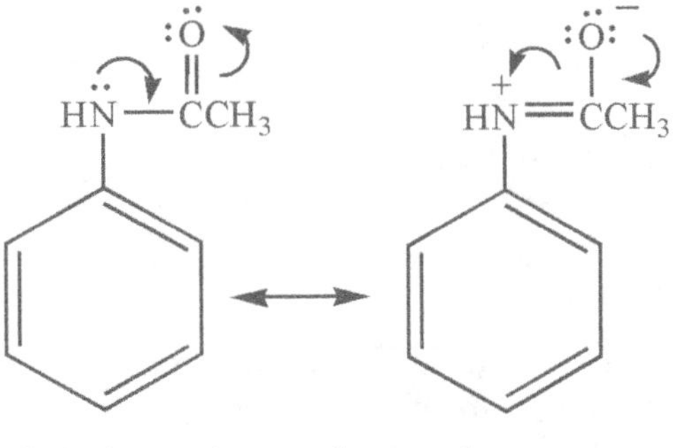

Substituent donate electrons by resonance away from the benzene ring

Weak activators: Alkyl, aryl, -CH=CHR groups are weak activators. Alkyl substituents donate their electrons by inductive effect. But aryl and -CH=CHR substituents donate electrons by resonance and also withdraw electrons from ring by resonance. Due to this, they are weak activators (slightly donate the electrons and more withdraws the electrons).

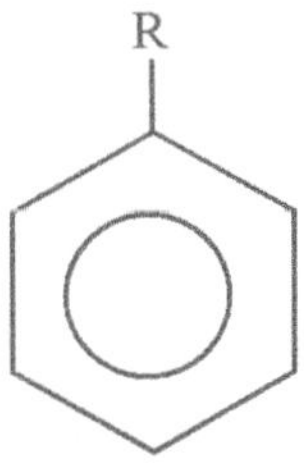

Alkyl group 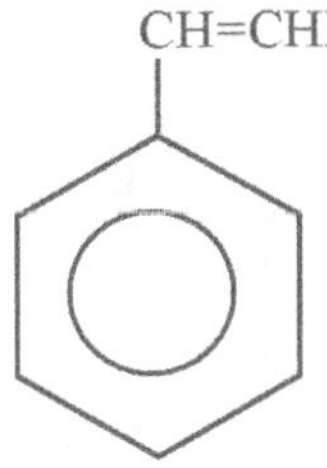 -CH=CH-R group 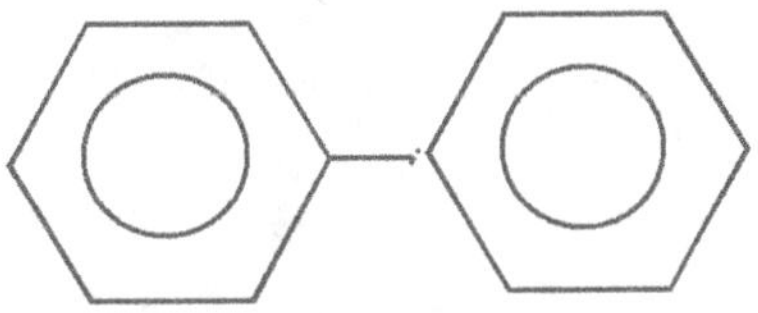 Aryl group

Weak deactivators: Halogen atoms are weak deactivating substituents. They donate electrons into the ring by resonance and withdraw the electrons from the ring by inductive effects. They withdraw the electrons more strongly by inductive effect than donating the electrons by resonance. So they are weak deactivators.

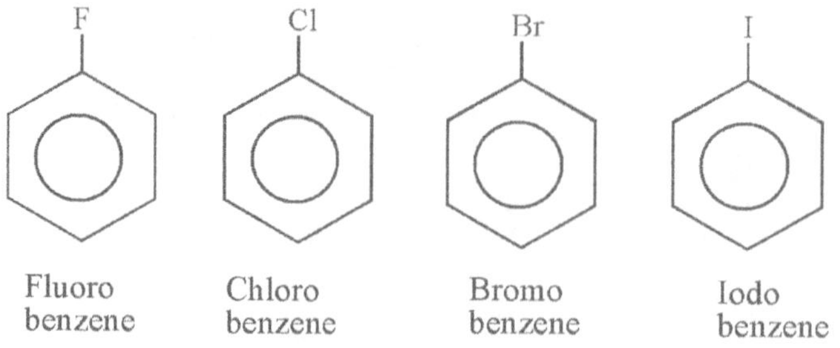

F Cl Br I

Fluoro benzene Chloro benzene Bromo benzene Iodo benzene

Moderate deactivators: Carbonyl group directly attached to the benzene ring is moderate deactivator, because it withdraws the electrons by both inductive effect and resonance.

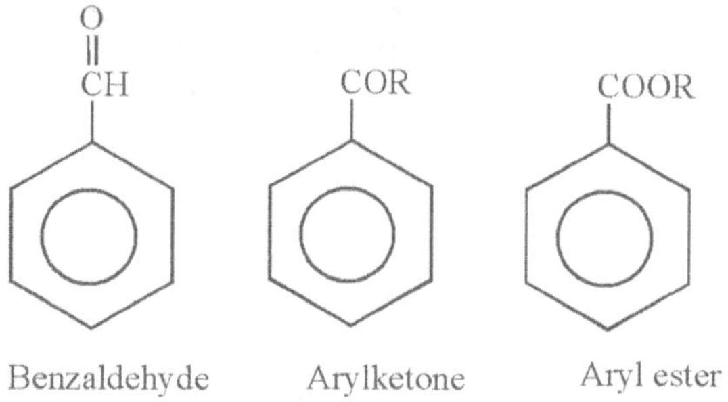

CHO COR COOR

Benzaldehyde Arylketone Aryl ester

Strong deactivators: These are powerful electron withdrawing substituents except ($^{+}NH_3$, $^{+}NH_2R$, $^{+}NHR_2$ and $^{+}NR_3$) because they withdraw electrons by both resonance and inductive effect. Actually, the ammonium ion has no resonance effect but the positive charge on the nitrogen atom causes them to strongly withdraw the electrons by inductive effect.

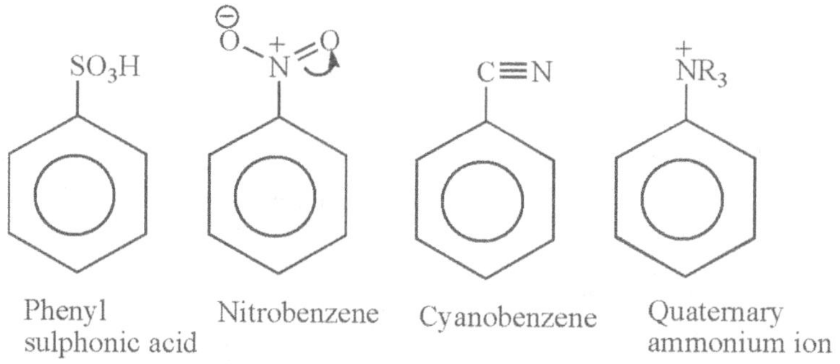

SO_3H NO_2 C≡N $\overset{+}{N}R_3$

Phenyl sulphonic acid Nitrobenzene Cyanobenzene Quaternary ammonium ion

The *ortho/para* ratio: When benzene ring with an *ortho/para* directing substituents undergoes electrophilic aromatic substitution reaction, what percentage of *ortho* product is formed and what percentage of *para* product is formed? Generally, we should expect, *ortho* product is formed more than *para* product because two *ortho* positions are available but only one "*p*" positions. (But the ortho position is sterically hindered and *p*-isomer is formed preferentially than *o*-isomer if either the substituent on the ring or large amount of electrophile is present). For example: Consider the following example, Nitration of toluene with conc. HNO_3/ H_2SO_4 yields two types of products, *o*-nitro toluene and *p*-nitro toluene.

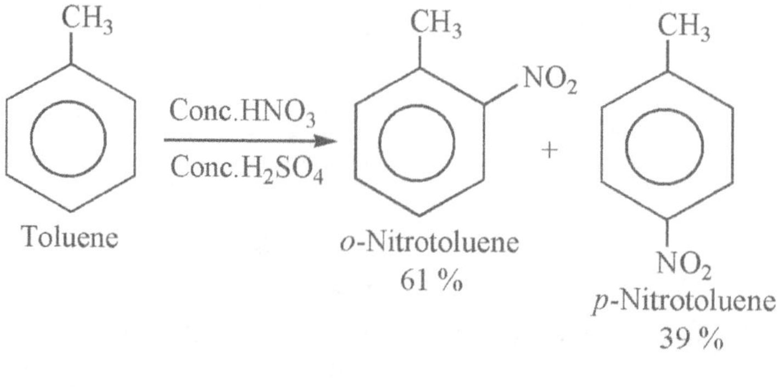

Toluene $\xrightarrow[\text{Conc.}H_2SO_4]{\text{Conc.}HNO_3}$ *o*-Nitrotoluene 61 % + *p*-Nitrotoluene 39 %

CH_2CH_3 (Ethyl benzene) + HNO_3 $\xrightarrow{H_2SO_4}$ CH_2CH_3, NO_2 (50 %, *o*-Nitroethyl benzene) + CH_2CH_3, NO_2 (50 %, *p*-Nitroethyl benzene)

$C(CH_3)_3$ (*tert*-Butyl benzene) + HNO_3 $\xrightarrow{H_2SO_4}$ $C(CH_3)_3$, NO_2 (*o-tert*-Butyl nitro benzene) + $C(CH_3)_3$, NO_2 (*p-tert*-Butyl nitro benzene)

Peculiar difference in the physical properties of *ortho* and *para*-isomers is that they can be separated easily. The electrophilic aromatic substitution reactions that form both *ortho* and *para*-isomers are useful synthetic tools in organic chemistry since the desired product can be readily separated from the reaction mixture.

Extra or additional considerations regarding substituent effects

When performing a chemical reaction, it is important to know whether already existing substituents is an activator or deactivator. For example, sulphonation of benzene needs fuming sulphuric acid, but sulphonation of toluene needs only conc. sulphuric acid. This is because the activating methyl group present in the toluene molecule allows the reaction to take place at a reasonable rate under less vigorous conditions.

Benzene $\xrightarrow{\text{Fuming } H_2SO_4}$ SO_3H (Benzene-*o*-sulphonic acid)

CH_3 (Toluene) + Conc.H_2SO_4 $\longrightarrow$ CH_3, SO_3H (Toluene-*o*-sulphonic acid) + CH_3, SO_3H (Toluene-*p*-sulphonic acid)

1. The five most common electrophilic substitution reactions of benzene and substituted benzene are halogenation, sulphonation, nitration, acylation and alkylation. In the below reaction of substituted

benzene, halogenation takes place under the mildest conditions. For example, bromination of anisole takes place without Lewis acid ($FeBr_3$ or $FeCl_3$) catalyst. This is due to the presence of strongly activating $-OCH_3$ substituent.

OCH_3 + Br_2 ⟶ OCH_3 / Br + OCH_3 / Br

Anisole — *o*-Bromo anisole — *p*-Bromo anisole

2. Among all the five electrophilic aromatic substitution reactions of substituted benzene, Friedel-craft's reactions are the most sluggish. So, if a *meta* director is present on the ring it will deactivate the ring towards Friedel-Craft's acylation and alkylation. For example,

SO_3H + CH_3CH_2Cl $\xrightarrow{AlCl_3}$ No reaction

Benzene sulphonicacid — Ethyl chloride

NO_2 + CH_3COCl $\xrightarrow{AlCl_3}$ No reaction

Nitrobenzene — Acetyl chloride

3. Aniline can't be nitrated because nitric acid is an oxidizing agent, so it produces explosive combination. But tertiary amines are easily nitrated, because the amino group present is strong activator, so the reaction is carried out using HNO_3 in acetic acid.

NH_2 + Conc.HNO_3 $\xrightarrow{H_2SO_4}$ No reaction

Aniline

$N(CH_3)_2$ $\xrightarrow[CH_3COOH]{HNO_3}$ $N(CH_3)_2$ / NO_2 + $N(CH_3)_2$ / NO_2

N,N-Dimethyl aniline — *o*-Nitro-N,N-dimethyl aniline — *p*-Nitro-N,N-dimethyl aniline

4. Aniline and N-substituted aniline can't undergo Friedel-Crafts reaction. For catalyzing the reaction, the $-NH_2$ group must be forming a complex with $AlCl_3$ but this complexation converts the $-NH_2$ substituent into a deactivating *m*-director. So the reaction does not take place.

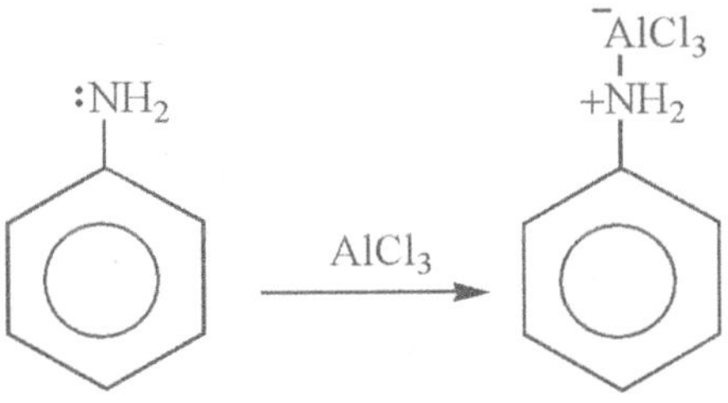

$:NH_2$ $\xrightarrow{AlCl_3}$ $^{-}AlCl_3$–$^{+}NH_2$

Substituent is a *meta* director

The Effect of Substituents on Orientation

When substituted benzene undergoes electrophilic aromatic substitution reaction, where the new incoming substituents will attach and the product formed is *ortho* or *para* or *meta* isomer?

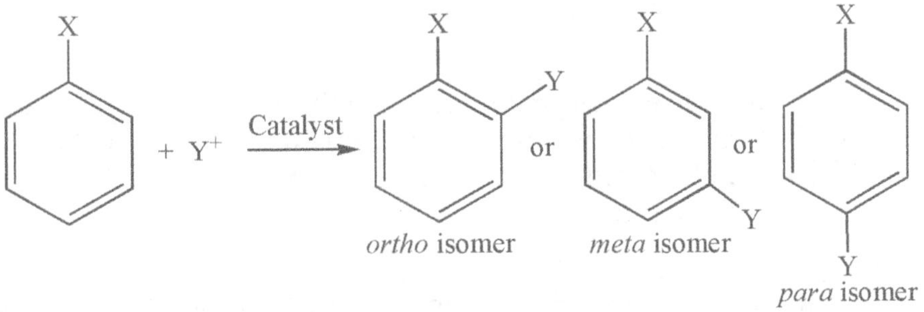

X + Y^+ $\xrightarrow{\text{Catalyst}}$ X, Y or X, Y or X, Y

ortho isomer *meta* isomer *para* isomer

The position of the incoming substituents is determined by the nature of existing substituent in the benzene ring. Substitution takes place as follows.

1. All the ring activators or ring activating substituents direct an incoming substituents or electrophile to the *ortho* and *para* positions.

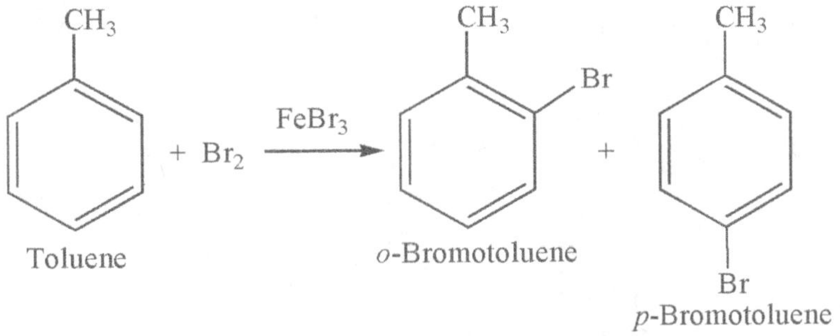

CH_3 + Br_2 $\xrightarrow{FeBr_3}$ CH_3, Br + CH_3, Br

Toluene *o*-Bromotoluene *p*-Bromotoluene

2. The weak deactivators such as halogens also direct the incoming substituents or electrophile to the *ortho* or *para* positions.

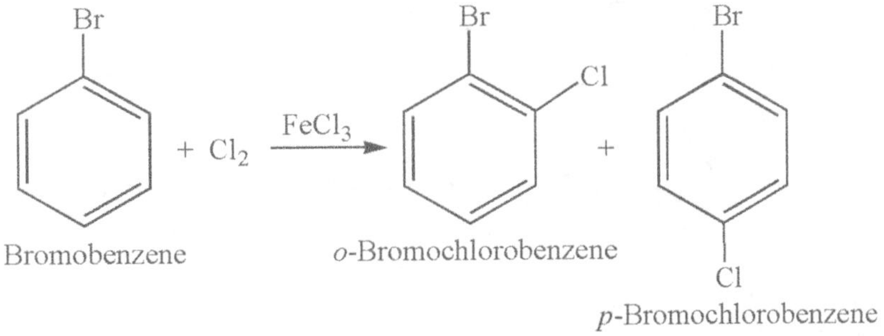

Br + Cl_2 $\xrightarrow{FeCl_3}$ Br, Cl + Br, Cl

Bromobenzene *o*-Bromochlorobenzene *p*-Bromochlorobenzene

3. All strong and moderate deactivating substituents direct an incoming substituent or an electrophile to *meta* position.

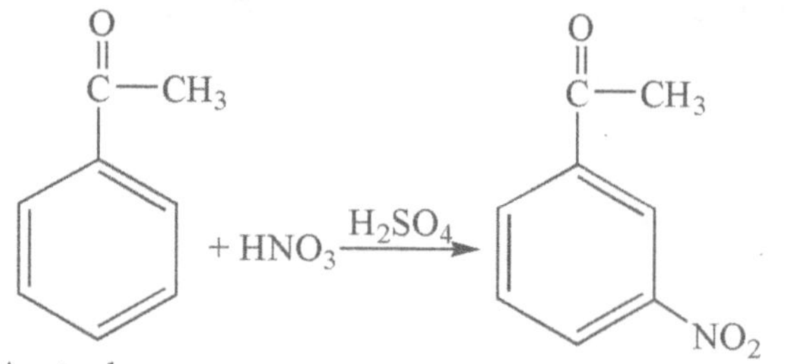

Acetophenone + HNO_3 $\xrightarrow{H_2SO_4}$ *m*-Nitroacetophenone

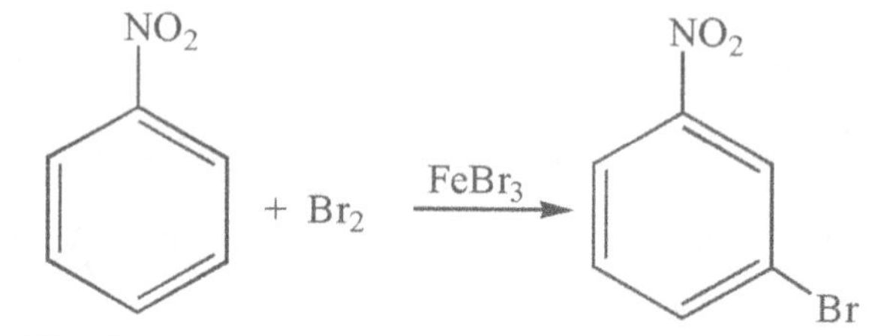

Nitrobenzene + Br_2 $\xrightarrow{FeBr_3}$ *m*-Bromonitrobenzene

Why the existing substituents will direct the incoming substituents to a particular position? This is explained by the stability of carbocation intermediates formed during the reaction. The relative stabilities of the carbocations formed is used to determine the preferred pathway of their action, since more stable carbocation is formed less energy is needed to make it.

This is explained as follows.

1. If an existing substituent donates the electron inductively (-CH_3 group), three types of carbocation intermediates are formed. They are depicted as follows.

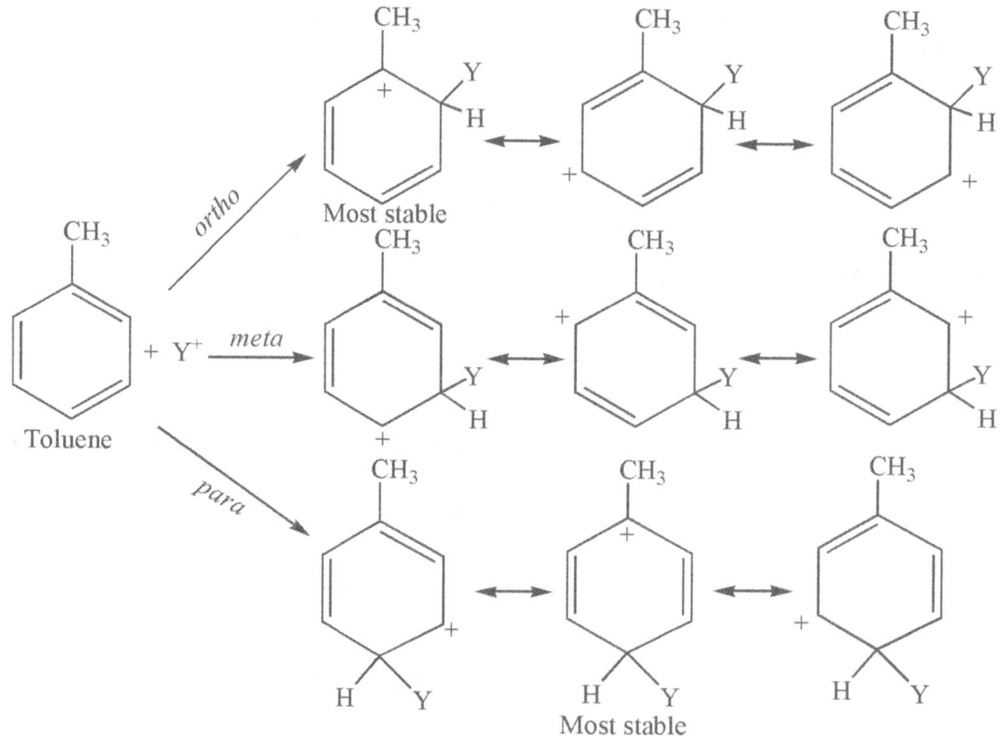

Toluene + Y^+ → *ortho* / *meta* / *para* intermediates

ortho: Most stable

para: Most stable

All of the above carbocation intermediate, *ortho* and *para* substituted carbocations are most stable because the substituents is directly attached to the positively charged carbon atom. In other words the *ortho* or *para* attack make the positive charge in the carbon atom with the CH_3 group, so the electron donating $-CH_3$ group stabilizes the positive charge. If the electrophile attacks the *m*-position to the $-CH_3$ group, it does not produce any carbocation resonance structures which is stabilized by electron donation. Hence, the most stable carbocation is formed by directing the incoming substituents to the *ortho* or *para* positions. It indicates that any substituent donates electrons inductively is an *ortho/ para* director.

If a substituent donates electrons to the ring by resonance, the carbocation intermediates formed are depicted in the following figure:

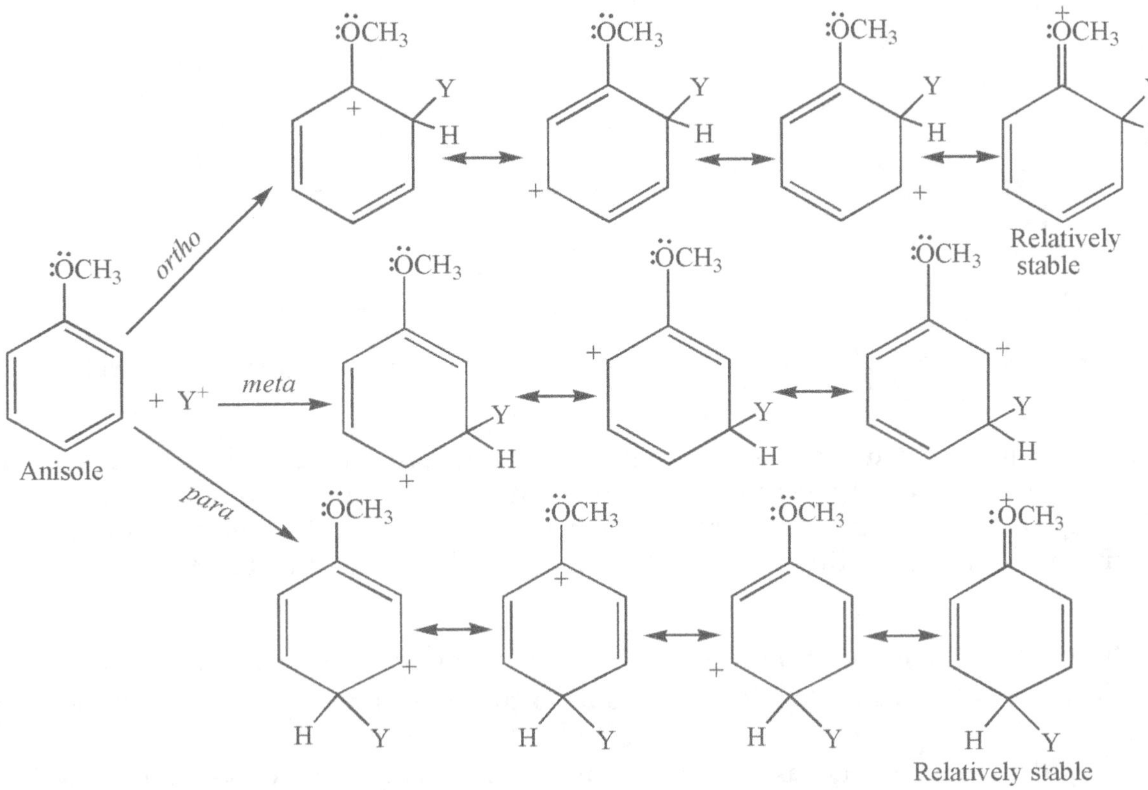

The structure of the carbocation intermediates formed from the reaction of an electrophile with anisole at the *ortho, meta* and *para* positions

In this the fourth resonance contributor is formed which is most stable one because it possess complete octet. It indicates that all substituents that donate electrons by resonance are *ortho/para-* directors.

2. Substituents with a positive charge or a partially positive charge on the atom attached to the benzene ring withdraws electrons from the ring by either inductively or resonance.

   For example: Electrophilic aromatic substituents or substitution reaction of protonated aniline. The three resonance contributors of carbocation intermediates are depicted below. In this, the resonance structures obtained by the *ortho* and *para* attack are least stable because they have the positive charge on two adjacent atoms. Hence the most stable carbocation is formed when the incoming substituent is directed to the *meta* position.

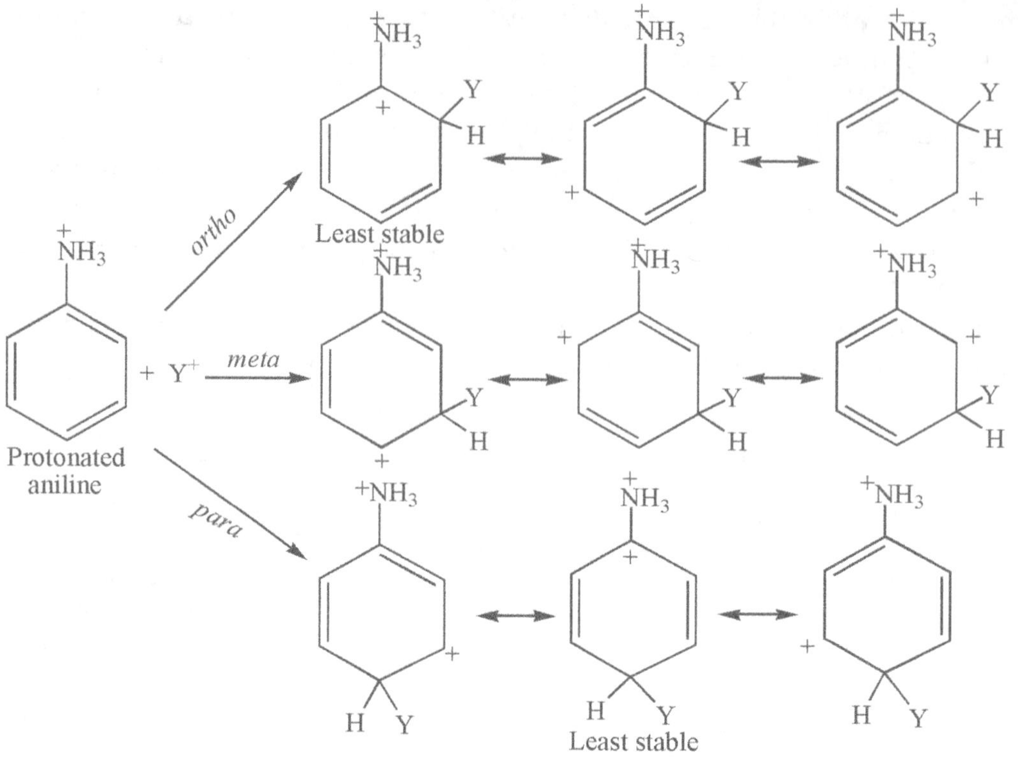

The structure of the carbocation intermediates formed from the reaction of an electrophile with protonated aniline at the *ortho*, *meta* and *para* positions

But the deactivating substituents which are *ortho/para* director are halogens. We have already discussed that halogens are deactivators because they withdraw electron more strongly by inductive effect from the ring than they donate the electron by resonance. Their electron donation by resonance make them to be *ortho/para* directors (stabilizes the carbocation intermediate).

Summary

1. All substituents that donate electrons in to the ring inductively or by resonance are *ortho/para* directors.
2. All the substituents that can't donate electrons in to the ring inductively or by resonance are *meta* directors.
3. All *ortho/para* directors have atleast one lone pair of electrons on the atom directly attached to the benzene ring (Except CH_3, aryl, CH = CHR groups).
4. All *meta* directors have a positive charge or a partial positive charge on the atom attached to the ring.

Electrophilic Aromatic Substitution of Disubstituted Benzene

When disubstituted benzene undergoes electrophilic aromatic substitution reaction, what type of products is obtained? The resulting product is whether *ortho* or *para* or *meta* isomer. For determining this, check the directive effect of the existing substituents by using the following 3 guidelines.

Guide line 1: If the direct effects of two groups strengthen, the new incoming substituent or an electrophile is placed on the position directed by both groups.

Note: The *p*-position to $-CH_3$ group is blocked with $-NO_2$ group and hence only one product is obtained.

Guide line 2: If the directive effects of two groups oppose each other, the more powerful activator predominates.

Guide line 3: Due to crowding or steric hindrance no substitution takes place between two *meta* substituents.

Probable Questions

1. Define aromatic compounds.
2. Write a short note on the characteristics of aromatic compounds.
3. Write any two examples illustrating the directive influence effect of functional group in aromatic compounds.
4. Classify aromatic compounds.
5. Write in detail about nomenclature of benzene and substituted benzenes.
6. Write a note on
   (i) Theory of orientation
   (ii) Theory of reactivity of aromatic compounds
7. Explain about the effect of substituents, reactivity and orientation of benzene.
8. When mono substituted benzene is converted in to disubstituted compound. How does the group already present in the ring exerts a direct influence on the incoming group - Explain with examples.
9. How do you identify whether a certain group present in a ring is an activating or deactivating one? State the rule related to orientation in terms of electron releasing and electron withdrawing groups.
10. Predict the major product obtained on mono nitration of each of the following.

CH_3 / NO_2 OH / CF_3 Br / CN CN / SO_3H

11. Write the products obtained when the following compounds undergo substitution in the benzene ring.
    (a) Nitration of benzonitrile, benzoic acid or benzene sulphonic acid.
    (b) Bromination of nitrobenzene, aniline or phenol.
    (c) Chlorination of phenol, aniline or bromo benzene.
12. Electron donating groups are *ortho* and *para* directors whereas electron withdrawing groups are meta directors. Justify the fact with two examples.

20

Benzene and its Analogues (Arenes)

Introduction

Benzene is one of the great star in organic chemistry. It has lot of applications in organic chemistry and is the basis for manufacturing of many useful organic compounds.

Importance of Benzene

Removal of a hydrogen from benzene yields phenyl group, which is attached with various chemical structures and elicit different kind of pharmacological activities based on the core structure to which it is attached. Some of the drugs which possess phenyl group and their therapeutic uses in the treatment of various diseases are shown in Table 20.1.

Table 20.1 Drugs containing phenyl group and their therapeutic uses.

Category	Drug	Structure
Sedative and Hypnotic	Diazepam	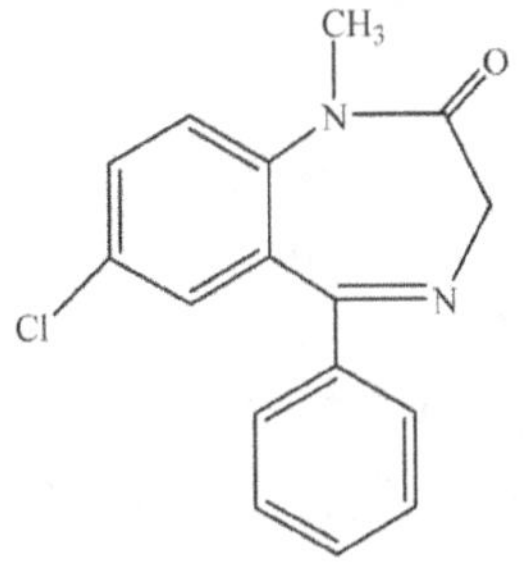
	Prazepam	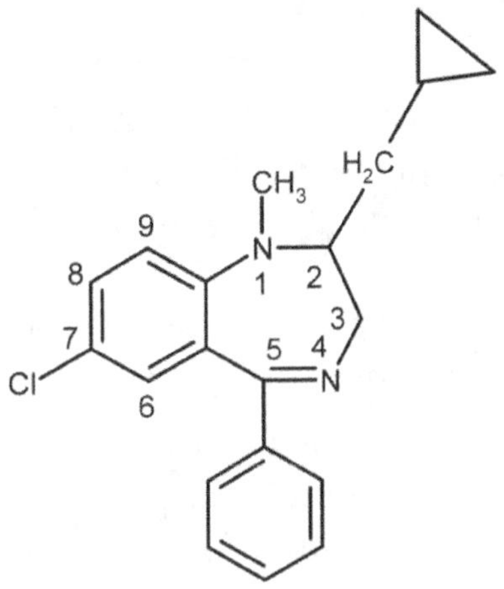

Table 20.1 *Contd...*

Category	Drug	Structure
Local anesthetic	Lidocaine	CH_3, $-N(H)-CO-CH_2N(C_2H_5)(C_2H_5)$, CH_3
Antidepressant	Isocarboxazide	$-CH_2NHNHC(=O)-$, N, O, $-CH_3$

The Great Discovery of Benzene (A Shining Molecule)

The compound benzene was first isolated by Michael Faraday in 1825. He had extracted benzene from a liquid residue left after heating whale oil under pressure to liberate a gas which is used to illuminate buildings in London. Because of its origin, the chemists suggested the name "Pheno" from the Greek word "Phainein" (to shine). The elemental analysis showed that the ratio of carbon and hydrogen in this compound is 1:1 (small ratio) with an empirical formula CH. So, Faraday named it as Carbureted hydrogen or Biocarbureted hydrogen'.

In 1834, the chemist Eilhard Mitscherlich synthesized benzene by heating benzene acid which is isolated from gum benzoin in the presence of lime. He also proposed that the empirical formula of benzene was CH. By using vapor phase density measurement, he found that the molecular weight of the compound having molecular formula C_6H_6 is 78 . So, the new compound derived from gum benzoin was named as benzin, but now known as benzene.

In 19th century, many of the compounds are discovered with relatively little hydrogen compared to the number of carbons which are particularly present in oils obtained from trees and other plants. Earlier chemists suggested that the compounds are aromatic due to their pleasant fragrances. By this way, they distinguish from the aliphatic compounds, which possess higher hydrogen to carbon ratios, that were obtained from the chemical degradation of fats (Aliphatic means fat like).

 **MICHAEL FARADAY (1791 – 1867)**   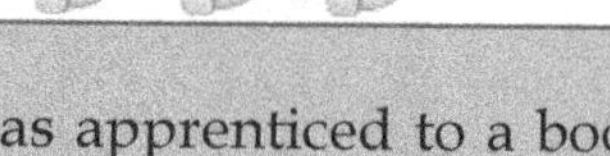

He was born in England, a son of Blacksmith. At the age of 14 years he was apprenticed to a book binder and educated himself reading the books that he bounded. He becomes a professor of chemistry at the Royal institution in 1833. He is best known for his work on electricity.

With the discovery of chemical stability of aromatic compounds, the term "aromatic" is related to stability of compounds not to their odours.

In 1845, Benzene was obtained from coal tar by August W. Von Hofmann. For many years coal tar is the primary source for the industrial preparation of benzene. A substantial production of benzene is converted to styrene which is used for the preparation of polystyrene plastics and films.

Criteria for Aromaticity

A compound can be categorized as aromatic if it fulfills both of the following criteria.

1. **It must possess an uninterrupted cyclic π electrons cloud above and below the plane of the molecule.**
   (a) For the π cloud to be cyclic, the molecule must be cyclic.
   (b) For the π cloud to be uninterrupted, every atom in the ring must have a "*p*" orbital.
   (c) For the π cloud to be formed, each "*p*" orbital must be able to overlap with the "*p*" orbital on either side of it, and therefore the molecule must be planar.
2. **The π cloud must contain an odd number of π electrons pair.**
   In 1866, the German scientist Friedrich August Kekule proposed a cyclic structure for benzene with alternate three double bonds and three single bonds.

EILHARDT MITSCHERLICH (1794 – 1863)

He was born in Germany. He studied oriental languages at the University of Heidelberg and the Sorbonne. He received doctorate in Persian studies. He was a professor of chemistry at the University of Berlin.

Structure of Benzene (Kekule's Structure of Benzene)

Early organic chemists found that benzene had a molecular formula of C_6H_6 and was an unusually stable compound that did not undergo addition reactions which are the characteristic reactions of alkenes. They also suggest the following facts.

1. When a different atom is substituted for one of hydrogen atoms of benzene, only one product is obtained.
2. When the mono substituted product undergoes a second substitution, three products are formed such as A, B and C.

$$\underset{\text{Benzene}}{C_6H_6} \xrightarrow[\text{Replaced one hydrogen by an X}]{-HX} \underset{\text{Monosubstituted benzene (One product)}}{C_6H_5X} \xrightarrow[\text{Replaced one hydrogen by an X}]{-HX} \underset{A}{C_6H_4X_2} + \underset{B}{C_6H_4X_2} + \underset{C}{C_6H_4X_2}$$

Disubstituted benzene (3 products)

According to the early chemist's suggestions, what kind of a structure we would get for benzene? We can know from its molecular formula that benzene has eight hydrogens less than an acyclic alkane with six carbon atoms ($C_nH_{2n+2} = C_6H_{14}$). Therefore, benzene is either an acyclic compound with four π bonds or a cyclic compound with three π bonds and further all the hydrogen atoms in benzene must be identical since only one product is obtained when any one of the hydrogen is replaced with another atom or group. The following two structures which fit these criteria are shown below.

$$H_3C-C\equiv C-C\equiv C-CH_3$$

I

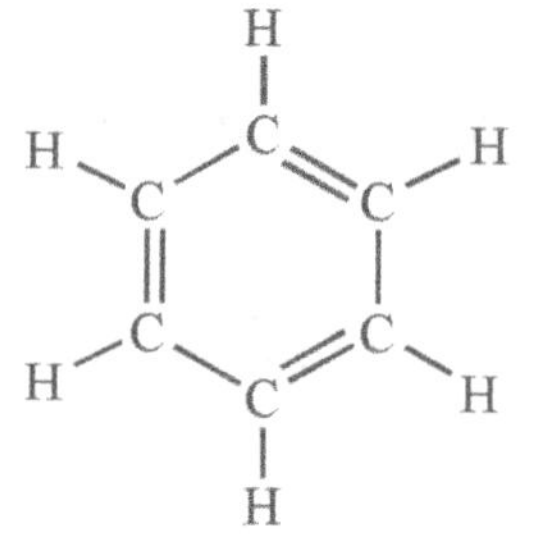

II

According to them first (I) structure, benzene did not behave like alkenes or alkynes, *i.e.*, it does not decolorize $KMnO_4$ or Br_2 water and does not undergo acid catalyzed addition of water.

Neither of these structures is consistent with the observation that three compounds are obtained if a second hydrogen atom is replaced. The structure I yield two disubstituted products.

$$CH_3-C\equiv C-C\equiv C-CH_3 \xrightarrow{\text{Replaced two hydrogen by Br}} CH_3-C\equiv C-C\equiv C-CH(Br)-Br + BrH_2C-C\equiv C-C\equiv C-CH_2Br$$

I

Disubstituted products
(Two products)

But the cyclic structure (II) yields the following four disubstituted products such as 1,3- disubstituted product, 1,4-disubstituted product and two 1,2-disubstituted products – since the two substituents can be placed either on two adjacent carbons joined by a single bond or on two adjacent carbons joined by a double bond.

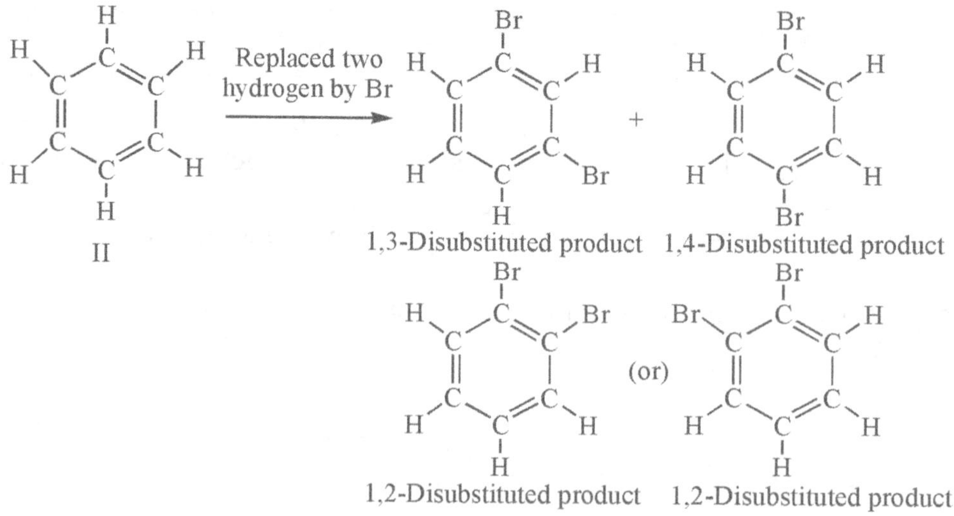

II — Replaced two hydrogen by Br →

1,3-Disubstituted product + 1,4-Disubstituted product

1,2-Disubstituted product (or) 1,2-Disubstituted product

In 1865, the German chemist Friedrich August Kekule Von Stradonitz suggested an answer to this dilemma. He suggested that benzene was not a single compound but it was a mixture of two compounds in rapid equilibrium.

Rapid Equilibrium

So, the above suggestion explains why only three disubstituted products are obtained when benzene undergoes second substitution reaction.

[Diagram: o-dibromobenzene (Br, Br) ⇌ (Rapid Equilibrium) o-dibromobenzene (Br, Br)]

The Kekule's structures of benzene explain the molecular formula of benzene and the number of isomers obtained as a result of substitution. But it fails to account for the following factors.

1. Unusual stability of benzene.
2. The double bonds present in the benzene ring do not undergo addition reactions which is the characteristic reactions of alkenes.

Benzene —$KMnO_4$→ No decolorisation

Benzene —Br_2 water→ No decolorisation

Benzene —H_2O Addition→ No reaction

In 1901, Paul Sabatie found that hydrogenation of benzene gives cyclohexane but it did not solve the puzzle of benzene's structure.

Benzene —H_2/Ni, 150-250 °C→ Cyclohexane

DREAM OF KEKULE

Friedrich August Kekule Von Stradonitz (1829-1896) was born in Germany. He was a professor of chemistry at the University of Heidelberg. In 1892 he gave an excellent speech at the 25th anniversary celebration of his first paper on the cyclic structure of benzene. In this speech he claimed that the Kekule's structure of benzene is a result of dozing off in front of a fire while working on a text book.

He dreamed of chains of carbon atoms twisting and turning in a snake-like motion, when suddenly the head of one snake seized and hold its own tail and formed a spinning ring. Recently, the variety of this snake story has been questioned by those who pointed out there is no written records of the dream, other counters that dreams are not the kinds of evidence one publishes in scientific papers. But it is not uncommon for scientists to report moments of creativity through the unconscious mind when they were not thinking about science. Kekule said **"Let us learn to dream we shall learn the truth". "But let us also beware not to publish our dreams until they have been examined by the wakened mind"**. In 1895, emperor William II made Kekule as a noble man. This is allowed him to add "Von Stradonitz" to his name. Three members of Kekule's students received three of the first noble prizes in chemistry.

Van't hoff in 1901, Fischer in 1902, Beyer in 1905.

Until 1930's the controversy over the structure of benzene is continued. The new techniques of X-ray and electron diffraction studies produced a surprising result. They showed that

1. Benzene is a planar molecule.
2. All carbon-carbon single bonds have the same length *i.e.* 1.39 A°.
3. The C-C single bond length in benzene is shorter (the carbon - carbon single bond actual length is 1.54 A°) and little longer than a carbon-carbon double bond (1.33 A°).

If all the bond lengths (single bonds) present in the benzene are equal or of same length, then they must possess same electron density. If it is true then

(a) The π electrons of benzene are delocalized around the ring rather than having each of π electrons localized between two carbon atoms. Hence benzene is actually a resonance hybrid of two structures.

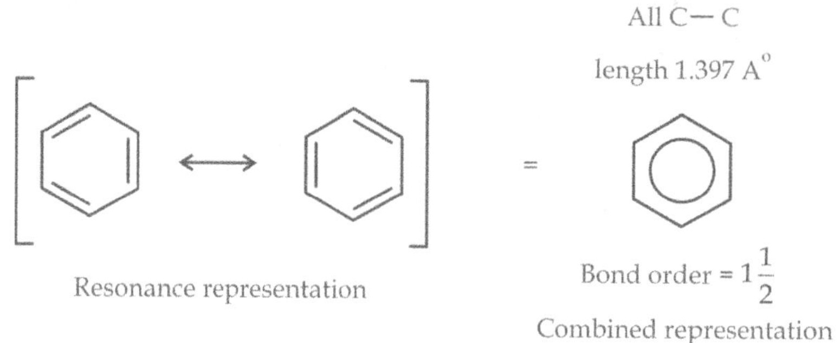

Figure 20.1 Resonance representation of benzene.

Bonding in Benzene

For better understanding of delocalized electrons, let's take a close exposure at the bonding in benzene.

1. Benzene is a planar molecule. Each of six carbon atoms is sp^2 hybridized with a bond angle of 120⁰, which is identical to the planar hexagon bond angle.
2. Each carbon in benzene use two sp^2 orbitals to form a bond to two other carbons.
3. The third sp^2 orbital overlaps the "*s*" orbital of hydrogen to form C-H bonds.
4. Benzene is a planar compound, so each carbon has a "*p*" orbital which is at right angle to the sp^2 orbitals, so the six "*p*" orbitals are parallel.
5. All the *p* orbitals are close enough for side to side overlap or sidewise overlap, so each one *p* orbital overlap the "*p*" orbitals on both adjacent carbons.
6. The overlaps of "*p*" orbitals yield a continuous dough-nut shaped cloud of π electrons above and dough–nut shaped cloud of π electrons below the plane of benzene ring. All the carbon - carbon bonds possess the same electron density.
7. Since, each π electron is neither localised on a single carbon nor in a bond between two carbons. Each π electron is shared by all six carbon atoms.
8. So these delocalized π electrons freely roam within the dough-nut shaped clouds that lie over and under the ring of the carbon atoms.
9. Accordingly, benzene also can be represented by a hexagon containing a circle that indicates the six delocalized π electrons.

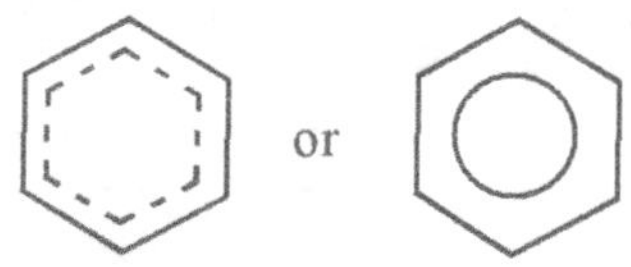

The above type of representation clearly indicates that benzene does not contain any double bond. Now the Kekule's structure of benzene was pretty close to the correct structure. The actual structure of benzene is a Kekule's structure with delocalized electrons Figure 20.2.

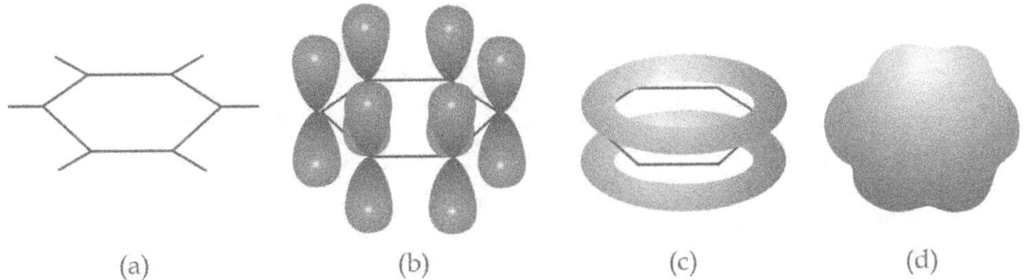

Figure 20.2 Kekule's structure of benzene.

(a) The carbon- carbon and carbon-hydrogen σ bond in benzene.
(b) The *p* orbital of each carbon atom of the benzene ring overlap with adjacent *p* orbitals.
(c) The π electrons cloud present above and below the plane of the benzene ring.
(d) The electrostatic potential map of benzene.

Unusual Stability of Benzene or Resonance Energy of Benzene

Benzene's reluctance to undergo typical alkenes type reactions indicates that it should be stable. For understanding the stability of benzene, we can compare the molar heats of hydrogenation of benzene. The following equations show the heat of hydrogenation (The amount of energy liberated in Kcal/mole when hydrogenation or addition of hydrogen) of benzene, cyclohexene and the cyclohexadienes. All the above compounds are hydrogenated in the presence of catalyst at high pressure to yield cyclohexane.

$$\text{Benzene} + 3H_2 \xrightarrow[\text{High pressure}]{\text{Catalyst}} \text{Cyclohexane} + 49.8 \text{ kcal}$$

$$\text{Cyclohexene} + H_2 \xrightarrow[\text{High pressure}]{\text{Catalyst}} \text{Cyclohexane} + 28.6 \text{ kcal}$$

$$\text{Cyclohexadiene} + 2H_2 \xrightarrow[\text{High pressure}]{\text{Catalyst}} \text{Cyclohexane} + 55.4 \text{ kcal}$$

The Figure 20.3 shows the experimentally determined heat of hydrogenation of cyclohexene, cyclohexadiene and benzene. This graphical representation is used to compare the resonance energies of the above said compounds based on the following reasons.

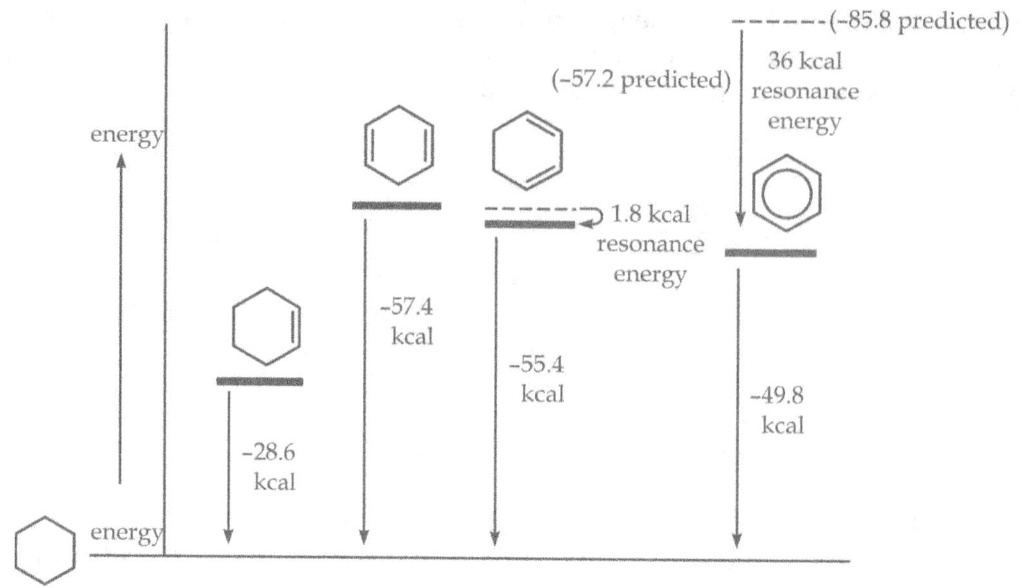

Figure 20.3 Graphical representation of heat of hydrogenation of benzene, cyclohexene and cyclohexa diene.

1. Hydrogenation of cyclohexene is an exothermic reaction with the evolution of 28.6 Kcal/mole (120 kJ/mole).
2. Hydrogenation of 1,4-cyclohexadiene is an exothermic reaction with the evolution of 57.4 kcal/mole (240 kJ/mole). This value is approximately twice the value of heat of hydrogenation of cyclohexene. The resonance energy of the isolated double bonds in 1,4-cyclohexadiene is about zero.
3. Hydrogenation of 1,3- cyclohexadiene is exothermic with the evolution of 55.4 kcal/mole (232 kJ/mole) which is about 1.8 kcal (8 kJ/mole) less than twice the value of cyclohexene. Resonance energy of 1.8 kJ/mole is typical for conjugated diene.
4. Hydrogenation of benzene needs higher pressure and highly active catalyst. Like cyclohexene hydrogenation of benzene is exothermic process with the evolution of 49.8 kcal/mole (208 kJ/mole) which is about 36.0 kcal/mole (152 kJ/mole) i.e. three times less than that of the value for cyclohexene.

Benzene + $3H_2$ —Catalyst / High pressure→ Cyclohexane

ΔH^o = – 49.8 kcal/mole
3 x Cyclohexene = – 85.8 kcal/mole
Resonance energy = 36.0 kcal/mole

The high value of 36 kcal/mole resonance energy of benzene not only depends upon conjugation effects alone. The ΔH^o value of benzene is actually smaller than that of 1,3- cyclohexadiene. The hydrogenation of the first double bond of benzene is endothermic, that should be encountered. In practice it is difficult to stop the reaction after the addition of 1 mole of H_2 to the product such as 1,3-cyclohexadiene which is more easily hydrogenated than benzene itself. It indicates that the benzene ring is exceptionally unreactive.

Benzene + H_2 —Catalyst / High pressure→ 1,3-Cyclohexadiene

ΔH^o
Benzene = – 49.8 kcal/mole
1,3-Cyclohexadiene = – 55.4 kcal/mole
ΔH^o = + 5.6 kcal/mole

Resonance Energy

It is a measure of how much more stable a resonance hybrid structure is than its extreme resonance structures.

(OR)

The extra stability of a compound which gains from delocalized electrons are knows as delocalization energy or resonance energy.

Long term exposure of benzene leads to Leukemia, which is characterized by decreased number of RBC and increase in the number of malfunctioning WBC. It takes place by the binding of metabolite of benzene with proteins and DNA in the bone marrow which alters the production of RBC and WBC.

The Molecular Orbital Structure of Benzene

Benzene is a planar molecule with six sp^2 hybrid carbon atoms. Each carbon atom possesses an unhybridised "*p*" orbital which overlaps with neighboring "*p*" orbitals to form continuous ring of orbitals above and below the plane of carbon atoms.

These overloaded "*p*" orbitals have six π electrons. The six overlapping "*p*" orbitals form a cyclic system of molecular orbitals.

Representation of Molecular Orbital Structure of Benzene

1. Benzene has six atomic "*p*" orbital's that overlap to form the π system of benzene. Therefore, benzene has six molecular orbitals.
2. The lowest energy molecular orbital entirely bonds with constructive overlap between all pairs of adjacent "*p*" orbitals. Because of entire bonding, no nodes are present.
3. When the number of bonds increases, the molecular orbital energy increases.
4. The molecular orbitals (MOs) should be evenly divided between bonding and anti-bonding molecular orbitals with the possibility of nonbonding molecular orbitals in some cases.

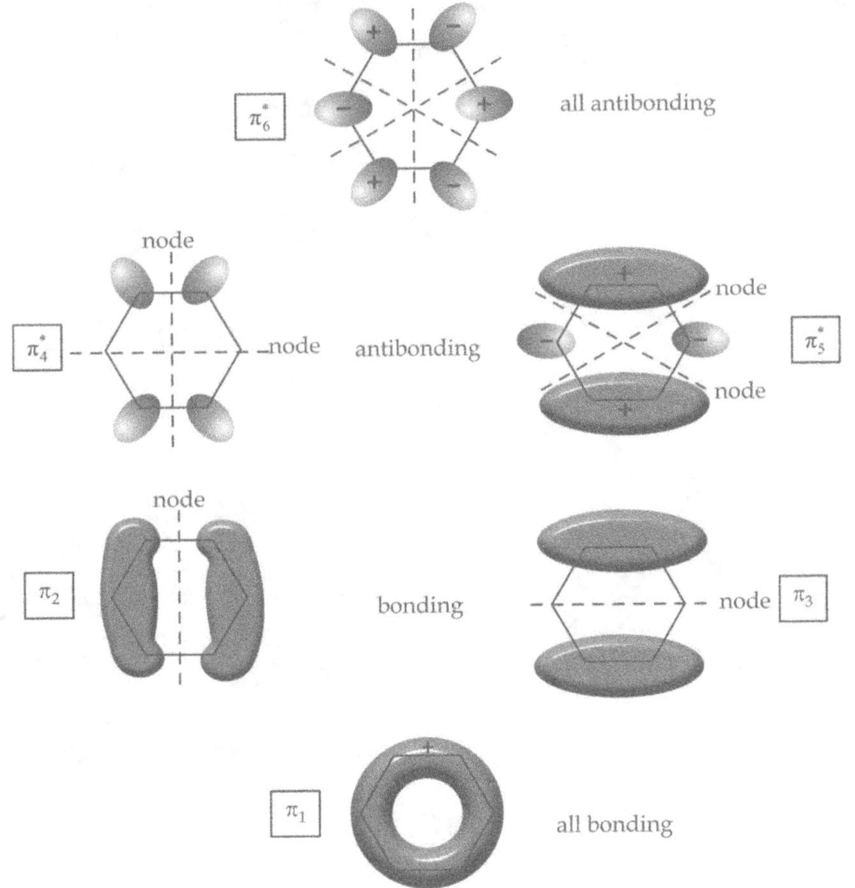

Figure 20.4 The six π MOs of benzene (bonding and antibonding orbitals).

The above figure 20.4 shows the π molecular orbitals of benzene viewed from above. The first molecular orbital (π) is entirely bonded, so no nodes and with very low energy since it has six bonding interactions and the electrons are delocalized over all six carbon atoms. The top lobes of "*p*" orbitals have the same sign similar to the bottom lobes. The six "*p*" orbitals overlaps each other to form a continuous bonding of ring with electron density.

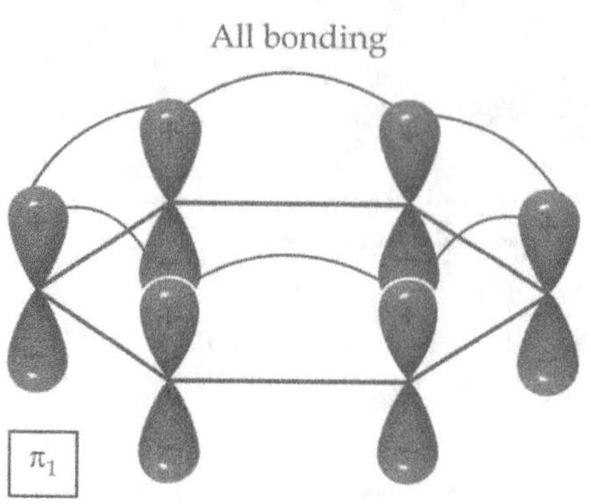

Figure 20.5 All bonding orbitals of benzene.

In a cyclic system of overlapping "*p*" orbitals, the intermediate energy levels degenerate (equal in energy) with two orbitals at each energy level. Both π_2 and π_3 have one nodal plane, as we expect at the second energy level.

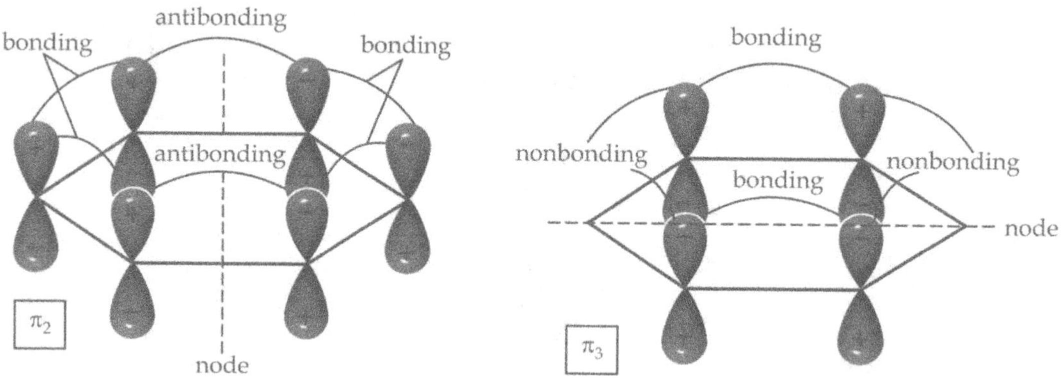

Figure 20.6 Bonding and antibonding and non bonding orbitals of benzene.

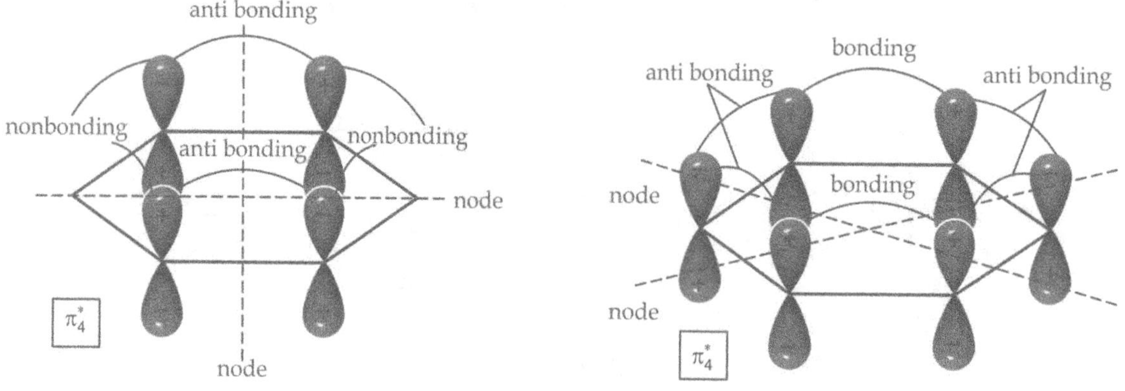

Figure 20.7 Bonding and antibonding and non bonding orbitals of benzene.

Similarly, π_2 orbital has four bonding interactions and two anti bonding interactions. Similarly π_3 orbital's has two bonding interactions and four non bonding interactions, in total two net bonding interactions. The next orbitals π_4^* and π_5^* are also degenerate with two nodal planes in each. The π_4^* orbital has two antibonding interactions and four nonbonding interactions. It is an anti bonding (*) orbital which degenerates partner. π_5^* has four antibonding interactions and two bonding interactions for a total of two antibonding interactions. This degenerate pair of molecular orbitals, π_4^* and π_5^* are about as strong antibonding as π_2 and π_3 are bonding. The all antibonding π_6^* has three nodal planes. Each pair of adjacent "*p*" orbital is out of phase and interacts destructively (Refer Fig. 20.6 & 7).

The Energy Diagram of Benzene

The energy diagram of molecular orbitals (Fig. 20.8) shows that they are symmetrically distributed above and below the non - bonding line (the energy of an isolated *p* orbital). The all bonding and all antibonding orbitals (π_1 and π_6^*) have lowest and highest energy levels respectively. The degenerated bonding orbitals (π_2 and π_3) are higher in energy than π_1, but still bonding. The degenerated pair π_4^* and π_5^* are antibonding, yet not as high in energy as the all anti bonding π_6^* orbital.

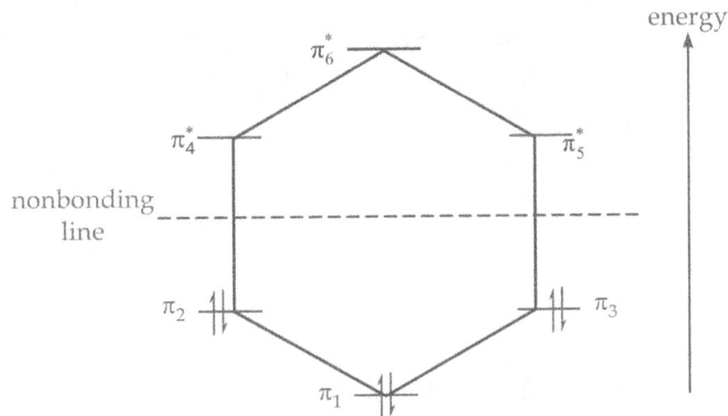

Figure 20.8 Energy diagram of benzene.

The Kekule structure of benzene shows three "π" bonds which represent six electrons or three electron pairs are involved in π bonding. These six electrons fill the three bonding molecular orbitals of the benzene system. This electronic configuration explains the unusual stability of benzene ring. The first molecular orbital are all bonding and particularly stable. The second and third (degenerate) molecular orbitals are still strongly bonding, and all three of these bonding molecular orbitals delocalize the electrons over several nuclei. This configuration, with all the bonding molecular orbitals filled (a closed bonding shell) is energetically very favourable.

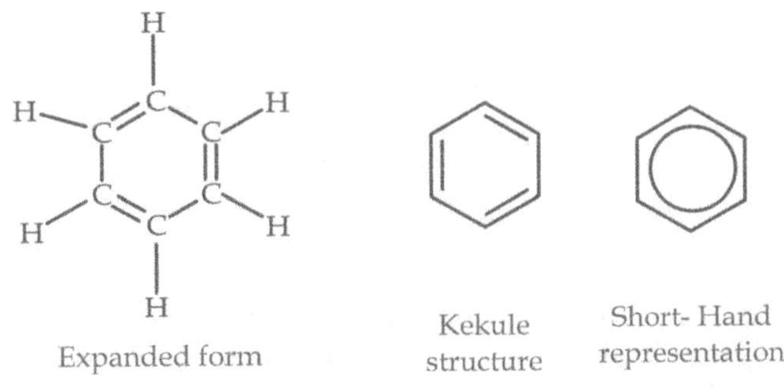

Figure 20.9 Structure of benzene.

Huckle's Rule or Concept of Aromatic Character (Aromaticity)

Benzene and other organic compounds which resemble benzene in some characteristic properties are known as aromatic compounds. These characteristic properties are commonly known as aromatic character or aromaticity. Some of the important properties are as follows:

1. **Unusual stability:** Aromatic compounds are highly stable due to their low heat of hydrogenation and low heat of combustion.
2. **Undergoes substitution reactions rather than addition reactions:** Aromatic compounds have double bonds but do not undergo addition reactions. They undergo electrophilic aromatic substitution reactions like halogenations, nitration, sulphonation, Friedel-Craft's acylation and alkylation *etc.*
3. **Cyclic flat structures:** Generally aromatic compounds contain five, six or seven membered ring and are found to be flat structure.

Theoretical Criteria for Aromaticity

If a compound has an aromatic character, then it must fulfill the following four structural criteria.

Criteria 1: A molecule must be cyclic and planar, so the π cloud must be cyclic. To be aromatic each "*p*" orbital must overlap with the orbitals of two adjacent atoms.

The *p* orbitals on all 6 carbon atoms of benzene continuously overlap, so benzene is an aromatic compound. But 1,3,5-hexatriene has six *p* orbitals, but the two of the terminal carbon atoms can't overlap so 1,3,5-hexatriene is not an aromatic compound.

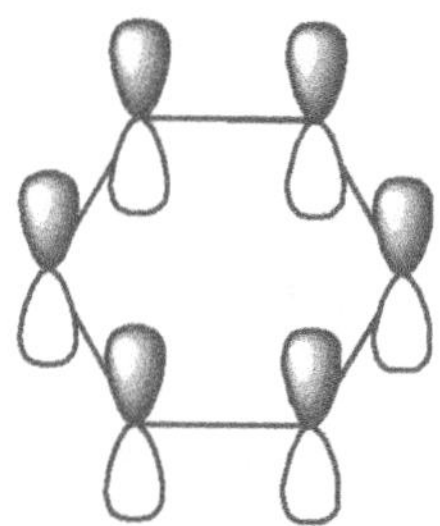

Benzene
Every *p* orbital overlaps with each other, so aromatic

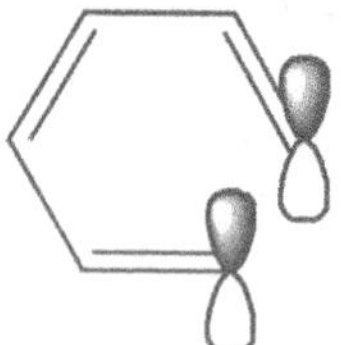

1,3,5-Hexatriene
No overlap between the *p* orbital of terminal carbon, so non-aromatic

Criteria 2: All adjacent "*p*" orbitals must be aligned so that the π electrons density can be delocalized (Must possess the uninterrupted π electrons cloud above and below the cyclic ring). For example; cyclooctatetraene resembles like benzene due to its cyclic ring with alternating single and double bonds. It has tub shaped eight membered rings but not planar molecule, so overlap between adjacent π bonds is impossible. Therefore, cyclooctatetraene is not an aromatic compound, so it undergoes addition reaction like those of alkenes.

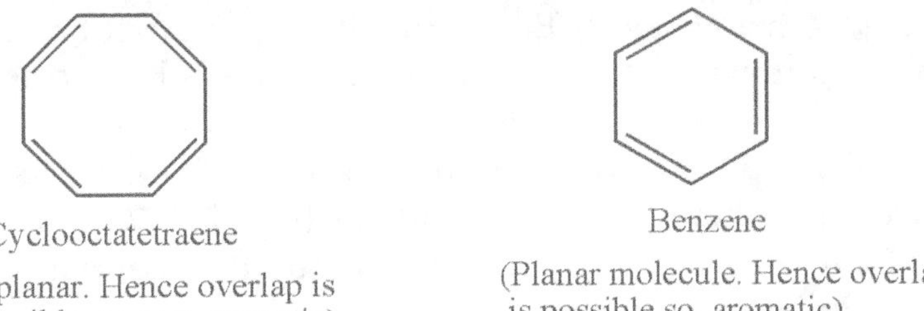

Cyclooctatetraene
(Not planar. Hence overlap is impossible, so non-aromatic)

Benzene
(Planar molecule. Hence overlap is possible so, aromatic)

Criteria 3: If a molecule is to be completely conjugate, they must have a "*p*" orbital on every atom. Let us see the following examples.

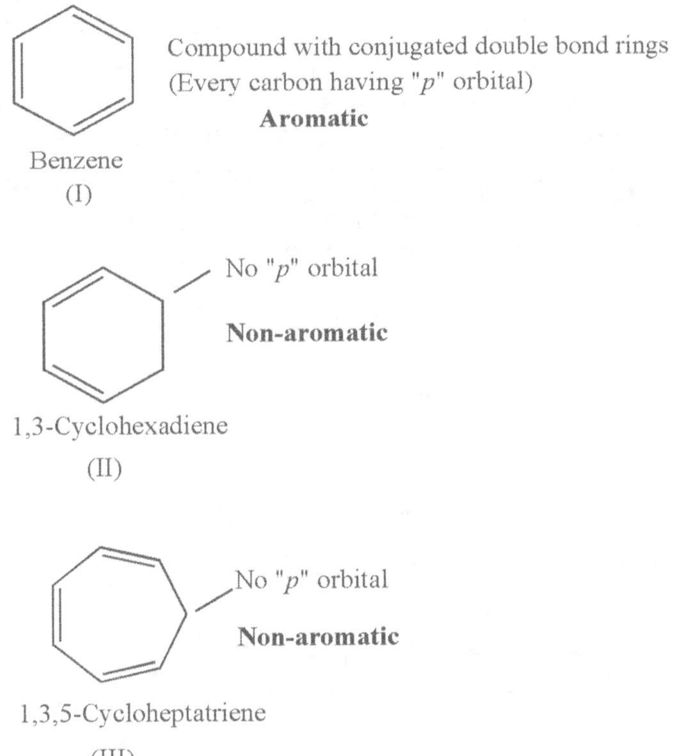

In the above examples the compounds II & III contain at least one carbon atom that does not contain *p* orbital and hence they are not completely conjugated and are non-aromatic.

Criteria 4: A molecule must satisfy **Huckle's rule of (4n+2) π electrons** and contain a particular number of electrons.

Some of the compounds satisfy first three criteria, but still they do not show the stability like aromatic compounds. For example

Cyclobutadiene:

It is highly reactive so it can be prepared at an extremely low temperature. It is a planar, cyclic conjugated molecule without an aromatic character. The above facts indicate that a compound being cyclic, conjugated, planar also needs a particular number of π electrons to be aromatic.

Huckel's Rule: In 1931, the German chemist Erich Huckle was the first one to recognize that an aromatic compound must have an odd number of π electron pairs. This rule is known as Huckel's rule and it includes two parts.

1. The rule states that "For a planar cyclic compound to be aromatic, its uninterrupted π cloud must contain (4n+2) π electrons where "n" is any number (n = 0, 1, 2, *etc*).
2. Cyclic, planar and completely conjugated compounds that contain 4n π electrons are especially unstable and are said to be nonaromatic.

Huckel's rule is just a mathematical way of expressing that an aromatic compound must have an odd number of pair of electrons. It refers to the numbers of π electrons, not the number of atoms in a particular ring.

So, the cyclic compounds with 2, 6, 10, 14, 18 *etc* π electrons are aromatic. So benzene is aromatic and especially stable because it contains 6 π electrons. But cyclobutadiene is nonaromatic and especially unstable because it contains 4 π electrons.

Erich was born in Germany. He was the professor of chemistry in University of Stuttgart and the university of Marbura.

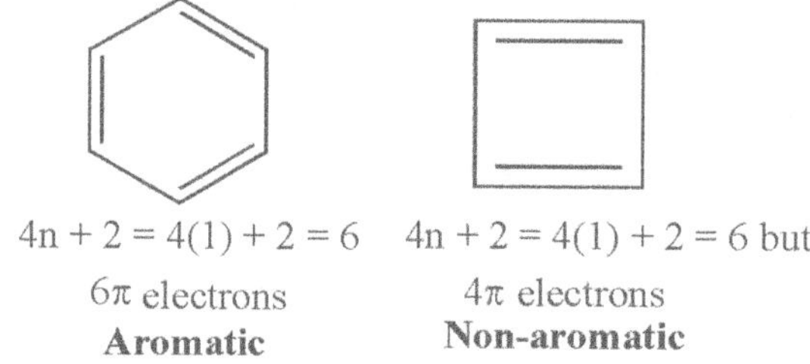

Table 20.2 Number of π electrons and the size of ring.

n (No. of ring system)	(4n+2)
0	2
1	6
2	10
3	14
4	18

Relationship between some aromatic compounds and their similar open chained molecule:

1. An aromatic compound is more stable than a similar acyclic compound containing the same number of π electrons as shown below.

   Benzene is more stable than 1,3,5-hexatriene.

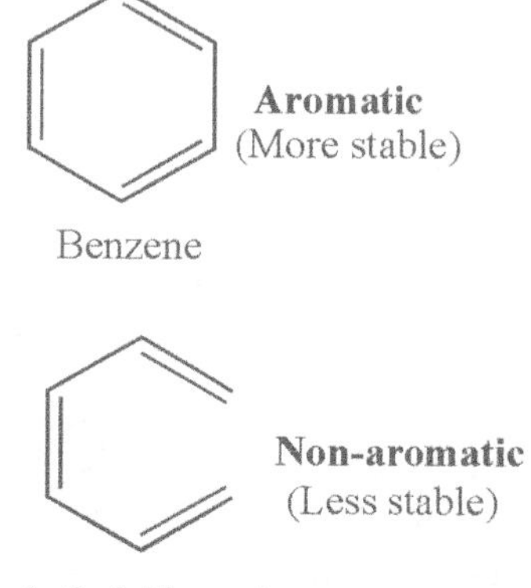

2. A nonaromatic compound is less stable than an acyclic compound having the same number of π electrons. For example, Cyclobutadiene is less stable than 1,3- butadiene.

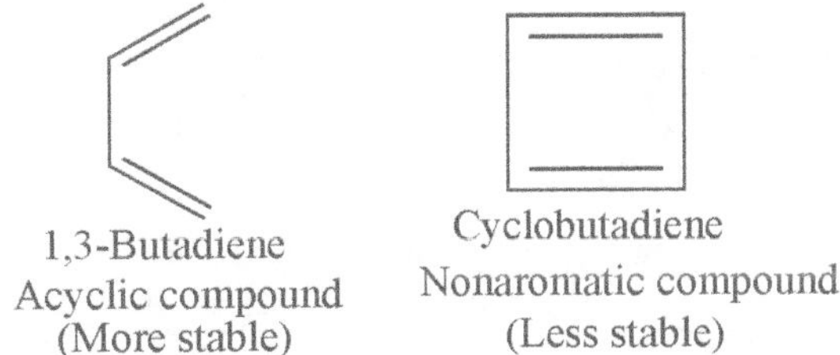

1,3-Butadiene
Acyclic compound
(More stable)

Cyclobutadiene
Nonaromatic compound
(Less stable)

3. A compound that is not aromatic and is similar in stability to an acyclic compound possessing the same number of π electrons. For example. 1,3-cyclohexadiene is similar in stability to *cis*- 2,4-hexadiene and is not aromatic.

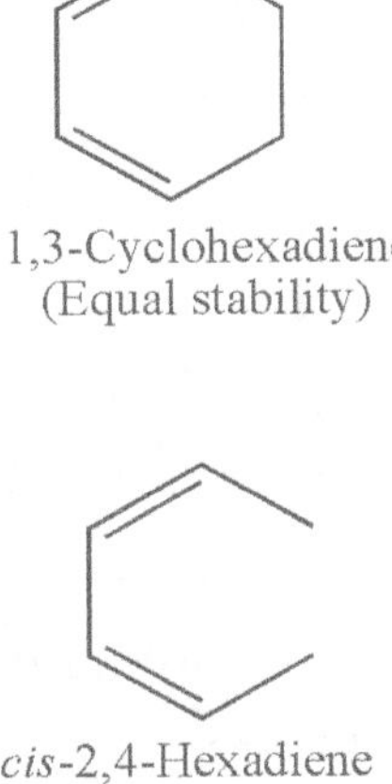

1,3-Cyclohexadiene
(Equal stability)

cis-2,4-Hexadiene
(Equal stability)

On the basis of the above consideration, it indicates that an aromatic compound need not necessary to contain a benzene ring. Actually, several compounds without benzene are found to be aromatic in nature. In other words, an aromatic compound may be benzenoid or non- benzenoid. Infact all that is required for aromaticity is the presence of flat ring, with (4n+2) π electrons in the form of continuous electron cloud above and below the plane of the ring.

The following are the cyclic compounds which fulfill the above two conditions of aromaticity and are found to be aromatic in nature.

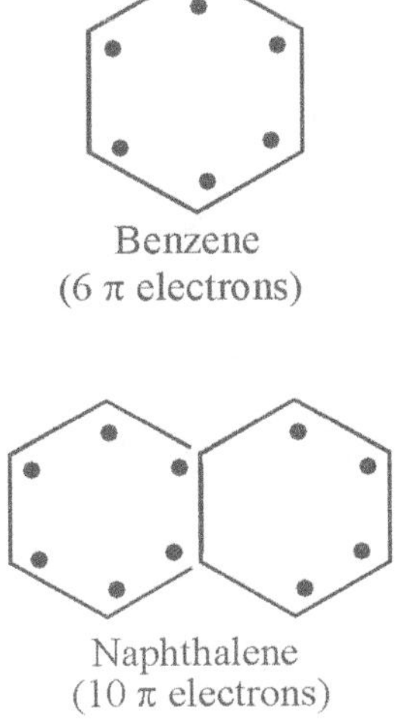

Benzene
(6 π electrons)

Naphthalene
(10 π electrons)

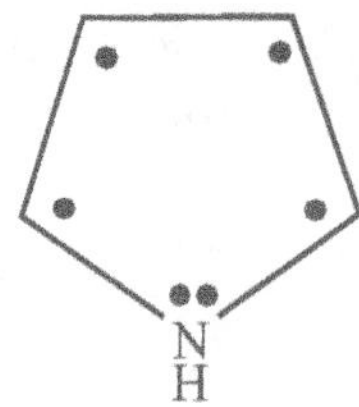

Pyrrole
(6 π electrons)

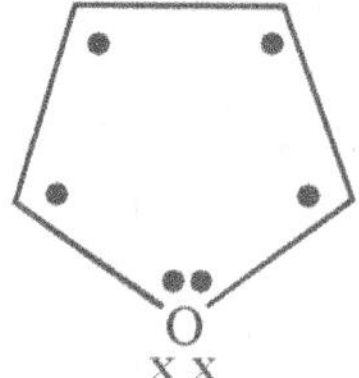

Furan
(6 π electrons)

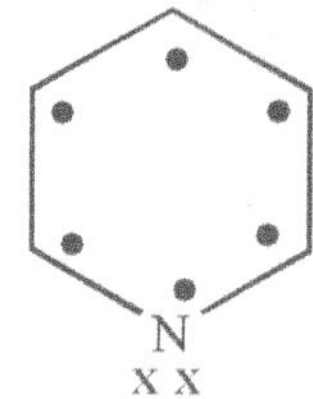

Pyridine
(6 π electrons)

Note: Dot or a pair of dots indicate the electrons present in the "*p*" orbital lying parallel to each other and hence capable of overlapping and ultimately produce electron clouds.

In addition to the above neutral molecule, there are number of cyclic ions which are aromatic according to Huckel's rule as shown below.

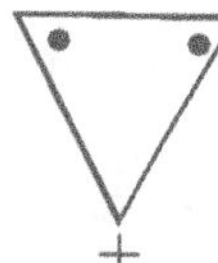

Cyclopropenyl cation
(2 π electrons)
(n= 0)

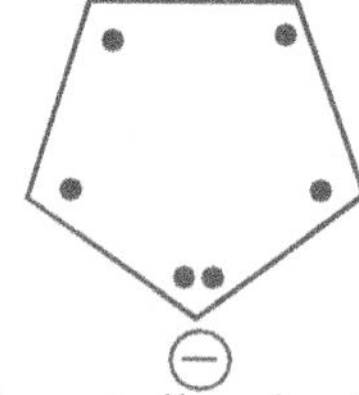

Cyclopentadienyl cation
(6 π electrons)
(n = 1)

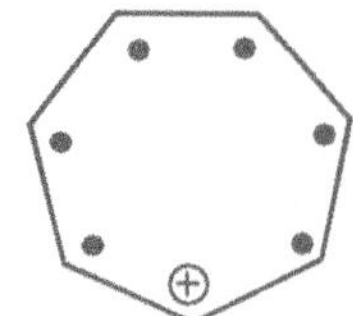

Cycloheptatrienyl cation
(6 π electrons)
(n = 1)

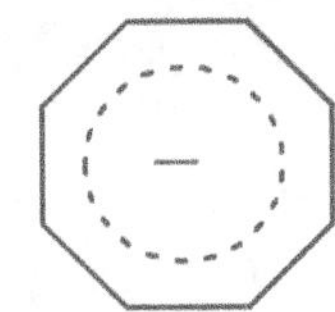

Cyclooctatetraenyl dianion
(10 π electrons)
(n = 2)

It is important to note that the Huckel's rule is strictly valid only for peripheral π-electrons and its application to condensed polycyclic system gives misleading results. So it works for naphthalene and phenanthrene but not for the aromatic compounds such as diphenyl and acenaphthylene.

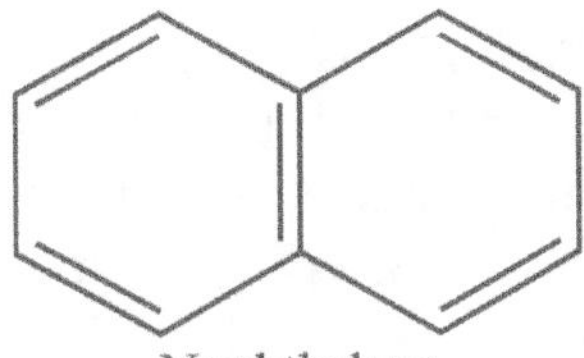

Naphthalene
(n = 2, 10 π electrons)

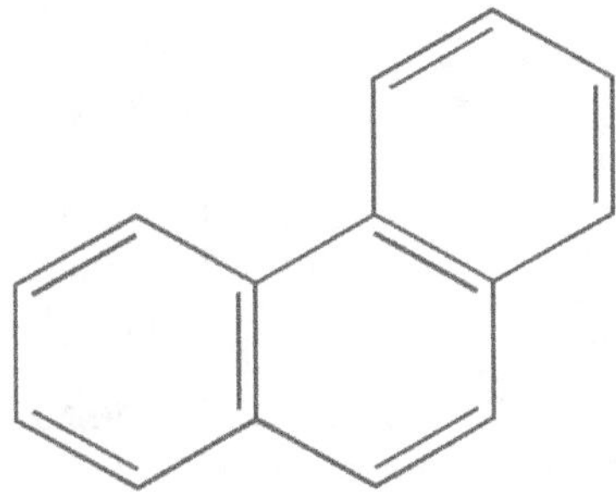

Phenanthrene
(n = 3, 14 π electrons)

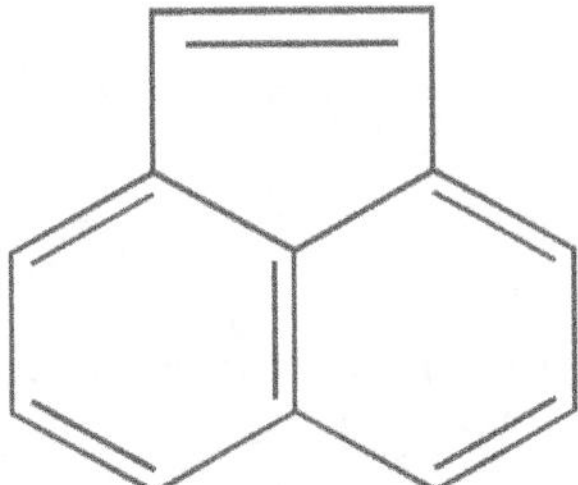

Acenaphthylene
(n = 2.5, 12 π electrons)

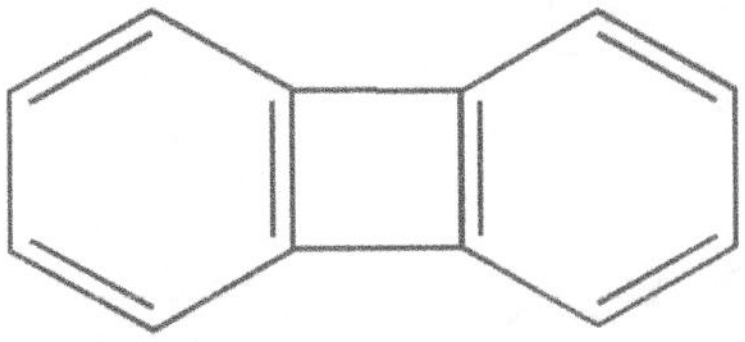

Diphenylene
(n = 2.5, 12 π electrons)

It may be noted that the Huckle's number "6" is the most common in aromatic system and aromatic sextet is commonly used for these 6π-electrons.

Finally, it must be noted that "Benzene is a perfect aromatic compound with π electrons over a flat hexagonal ring. So, the regular hexagonal angle (120^0) is exactly same as that of sp^3 hybridised orbitals, there is no angle strain at all in the molecule.

General Methods of Preparation of Benzene and its Analogues

(A) Small Scale Preparation: The general methods of preparation of benzene and its analogues include the following two methods.

***Method* I:** Removal of the group already present in the parent nucleus.

***Method* II:** Substitution or replacement of hydrogen atom of the benzene nucleus by one or more alkyl groups **(Preparation of benzene homologues)**.

***Method* I:** Removal of the group already present in the parent nucleus.

1. ***From aromatic carboxylic acids*:** Aromatic carboxylic acid (– COOH) and substituted acid or its sodium salt when heated with soda lime gives benzene or its analogues.

$$C_6H_5COONa + NaOH \xrightarrow[\Delta]{CaO} C_6H_6 + Na_2CO_3$$

Sodium benzoate → Benzene

$$H_3C{-}C_6H_4{-}COONa + NaOH \xrightarrow[\Delta]{CaO} C_6H_5CH_3 + Na_2CO_3$$

4-Methylsodium benzoate → Toluene

2. ***Removal of –OH group*:** Phenol and *o*/*m*/*p* – methyl phenol vapours are passed through heated zinc dust to yield benzene and toluene respectively.

$$C_6H_5OH + Zn \longrightarrow C_6H_6 + ZnO$$

Phenol → Benzene

$$H_3C{-}C_6H_4{-}OH + Zn \longrightarrow H_3C{-}C_6H_5 + ZnO$$

p-Methyl phenol → Toluene

3. ***Removal of –SO_3H group*:** Benzene sulphonic acid or toluene sulphonic acid (*p*-methyl benzene sulphonic acid) is hydrolyzed with super heated steam or boiling with dilute HCl under pressure at 400-447 K to yield benzene and toluene respectively.

$$C_6H_5SO_3H \xrightarrow[\text{or Dil.HCl}]{H_2O \text{ Super heated steam}} C_6H_6 + H_2SO_4$$

Benzene sulphonic acid → Benzene

$$H_3C{-}C_6H_4{-}SO_3H \xrightarrow[\text{or Dil.HCl}]{H_2O \text{ Super heated steam}} H_3C{-}C_6H_5 + H_2SO_4$$

Toluene sulphonic acid (*p*-Methylbenzene sulphonic acid) → Toluene

4. ***Removal of $N_2^+Cl^-$ group***: Benzene diazonium salts or arene diazonium salts or toluene diazonium salts when reduced with hypophosphorous acid yields benzene and toluene respectively.

$$C_6H_5-\overset{+}{N}\equiv N-Cl + H_3PO_2 + H_2O \longrightarrow C_6H_6 + N_2 + HCl + H_3PO_3$$

Benzene diazonium chloride; Hypophosphorous acid; Benzene

$$H_3C-C_6H_4-\overset{+}{N}\equiv N-Cl + H_3PO_2 + H_2O \longrightarrow H_3C-C_6H_5 + N_2 + HCl + H_3PO_3$$

Toluene diazonium chloride; Hypophosphorous acid; Toluene

5. ***From chlorobenzene***: Aryl halides when treated with magnesium yield aryl magnesium halide which upon treatment with HCl yields benzene.

$$C_6H_5-Cl \xrightarrow{Mg} C_6H_5-MgCl \xrightarrow{H^+/H_2O} C_6H_6 + Mg(OH)Cl$$

Chlorobenzene; Phenyl magnesium chloride; Benzene

Method II: It involves substitution or replacement of hydrogen atom of the benzene nucleus by one or more alkyl groups (Preparation of benzene homologues).

1. ***Friedel - Craft's Alkylation***: Benzene reacts with alkyl halides in the presence of anhydrous aluminium chloride (catalyst) to yield substituted benzene. This reaction is known as Friedel-Craft's alkylation.

$$C_6H_6 + CH_3Cl \xrightarrow{AlCl_3} C_6H_5-CH_3 + HCl$$

Benzene; Methyl chloride; Toluene

$$C_6H_5-CH_3 + CH_3Cl \xrightarrow{AlCl_3} C_6H_4(CH_3)_2 + HCl$$

Toluene; Methyl Chloride; *o*-Xylene

2. ***Wurtz-Fittig Reaction***: Aryl halides or any one of its halo derivatives and an alkyl halide reacts with metallic sodium in dry ether to give aryl homologues.

$$C_6H_5-Br + 2Na + CH_3I \xrightarrow{\text{Dry ether}} C_6H_5-CH_3 + NaBr + NaI$$

Phenyl bromide; Methyl iodide; Toluene

$$o\text{-}Br-C_6H_4-CH_3 + 2Na + CH_3I \xrightarrow{\text{Dry ether}} C_6H_4(CH_3)_2 + NaBr + NaI$$

o-Bromo toluene; Methyl iodide; *o*-Xylene

3. ***By Using Grignard Reagents***: Aryl magnesium bromide when treated with alkyl halide in the presence of dry ether yields benzene homologues.

$$C_6H_5MgBr + CH_3I \xrightarrow{\text{Dry ether}} C_6H_5CH_3 + Mg\langle^{Br}_{I}$$

Phenyl magnesium bromide; Methyl iodide; Toluene

4. ***Clemmensen Reduction***: Ketones are reduced with zinc amalgam and HCl to give alkyl benzene.

$$C_6H_5COCH_3 + 4(H) \xrightarrow[HCl]{Zn/Hg} C_6H_5CH_2CH_3 + H_2O$$

Acetophenone; Ethyl benzene

The ketone *i.e.*, acetophenone needed for this reaction is prepared by Friedel-Craft's acylation.

5. ***From Alkanes***: The vapours of long chain alkanes obtained from petroleum are passed over chromium catalyst (Cr_2O_3 supported on alumina) under pressure at 775 K to yield benzene.

$$CH_3(CH_2)_4CH_3 \xrightarrow[\text{775 K, Under pressure}]{Cr_2O_3} C_6H_6 + 4H_2$$

n-Hexane; Benzene

Similarly, toluene and xylene are prepared from *n*-heptane and *n*-octane respectively.

$$CH_3\text{-}(CH_2)_5CH_3 \xrightarrow[\text{500 °C, 15 atm}]{Pt/Al_2O_3} C_6H_5CH_3 + 4H_2$$

n-Heptane; Toluene

$$CH_3\text{-}(CH_2)_6CH_3 \xrightarrow[\text{500 °C, 15 atm}]{Pt/Al_2O_3} C_6H_4(CH_3)_2 + H_3CH_2C\text{-}C_6H_5 + 4H_2$$

n-Octane; *o*-Xylene; Ethyl benzene

This reaction is otherwise called as catalytic reforming or hydroforming.

(B) Large Scale Production of benzene and its homologues: Berthelot in 1870 first synthesized benzene from acetylene. Acetylene is passed through a red-hot tube to give benzene.

$$3CH\equiv CH \xrightarrow[\text{tube}]{\text{Red hot}} C_6H_6$$

Acetylene; Benzene

Fractional Distillation of Coal-tar: Different type of compounds present in coal–tar is separated by fractional distillation which is carried out as follows.

Coal–tar is taken in a pre-heater, and then it is heated in a pipe through which hot vapours are passed from the still. Here most of the water present is removed (if it presents, it produces violent frothing) along with some low boiling hydrocarbons. From the preheater it flows to the iron still which is heated in a furnace. Different fractions are distilled over at different temperatures and are collected separately.

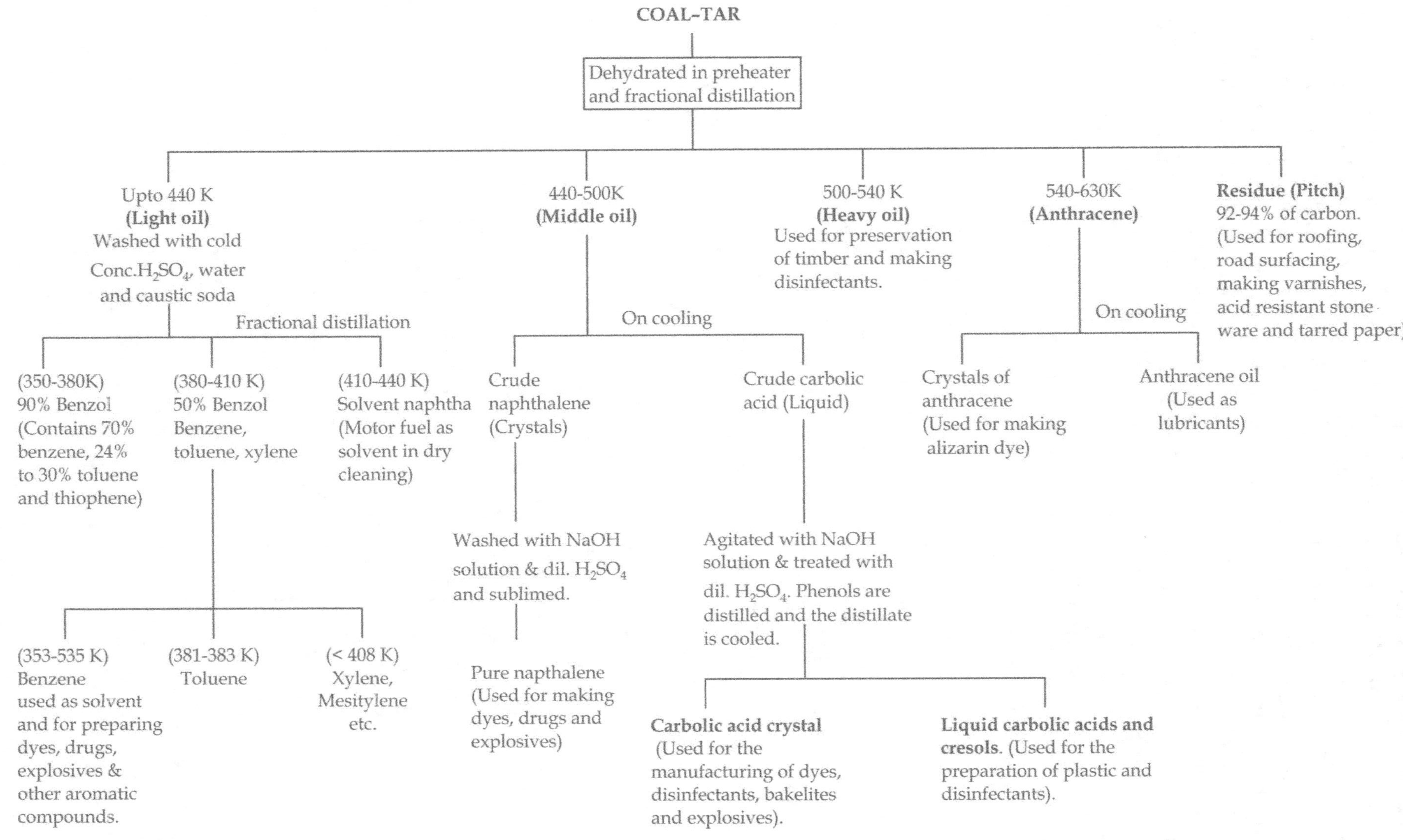
COAL–TAR
Dehydrated in preheater and fractional distillation
Upto 440 K (Light oil)
Washed with cold Conc.H_2SO_4, water and caustic soda
Fractional distillation
(350-380K) 90% Benzol (Contains 70% benzene, 24% to 30% toluene and thiophene)
(380-410 K) 50% Benzol Benzene, toluene, xylene
(410-440 K) Solvent naphtha (Motor fuel as solvent in dry cleaning)
(353-535 K) Benzene used as solvent and for preparing dyes, drugs, explosives & other aromatic compounds.
(381-383 K) Toluene
(< 408 K) Xylene, Mesitylene etc.
440-500K (Middle oil)
On cooling
Crude naphthalene (Crystals)
Washed with NaOH solution & dil. H_2SO_4 and sublimed.
Pure napthalene (Used for making dyes, drugs and explosives)
Crude carbolic acid (Liquid)
Agitated with NaOH solution & treated with dil. H_2SO_4. Phenols are distilled and the distillate is cooled.
Carbolic acid crystal (Used for the manufacturing of dyes, disinfectants, bakelites and explosives).
Liquid carbolic acids and cresols. (Used for the preparation of plastic and disinfectants).
500-540 K (Heavy oil)
Used for preservation of timber and making disinfectants.
540-630K (Anthracene)
On cooling
Crystals of anthracene (Used for making alizarin dye)
Anthracene oil (Used as lubricants)
Residue (Pitch)
92-94% of carbon. (Used for roofing, road surfacing, making varnishes, acid resistant stone ware and tarred paper)

Different Products of Coal - tar distribution and their uses:

1. ***Light Oil or Crude Naphtha fraction (up to 440 K):*** It is the first fraction and lighter than water, so it is named as light oil fraction. Benzene is mainly obtained from this fraction. First, this is treated with cold conc. H_2SO_4 to remove basic substances example, Pyridine and then washed with water. The acidic substances are removed by treatment with caustic soda example, Phenol and the excess H_2SO_4 is removed by washing with water which further undergoes distillation to give the following fractions.
   (a) *90 % Benzol (350-380 K)*: It contains 70% benzene and 30% of toluene. It is known as 90% benzol since 100 ml of this fraction after distillation yields 90% of distillate below 373 K.
   (b) *50 % Benzol (380-410 K)*: It contains 46% of benzene. It also contains toluene and xylene.
   (c) *Benzene or Solvent Naptha (410-440 K)*: It contains xylene, mesitylene *etc.*

Summary of Methods of Preparation

1) C_6H_5COONa (Sodium benzoate) $\xrightarrow[\Delta]{NaOH/CaO}$ Benzene

2) C_6H_5OH (Phenol) $\xrightarrow[-ZnO]{Zn}$ Benzene

3) $C_6H_5SO_3H$ (Benzene sulphonic acid) $\xrightarrow[\text{Super heated steam or Dil.HCl}]{H_2O, -H_2SO_4}$ Benzene

4) $C_6H_5-\overset{+}{N}\equiv N-\overset{-}{Cl}$ (Benzene diazonium chloride) $\xrightarrow[H_2O]{H_3PO_2}$ Benzene

5) C_6H_5Cl (Chlorobenzene) $\xrightarrow[H^+/H_2O,\ -Mg(OH)Cl]{Mg}$ Benzene

6) n-Hexane ($CH_3-CH_2-CH_2-CH_2-CH_2-CH_3$) $\xrightarrow[\text{775 K, Under pressure}]{Cr_2O_3, -4H_2}$ Benzene

7) $HC\equiv CH$ Acetylene (3 moles) $\xrightarrow[\text{tube}]{\text{Red hot}}$ Benzene

Summary of Methods of Preparation of Homologues of Benzene

1) $H_3C-C_6H_4-COONa$ (4-Methylsodium benzoate) $\xrightarrow[\Delta]{NaOH/CaO}$ $C_6H_5-CH_3$ (Toluene)

2) $H_3C-C_6H_4-OH$ (*p*-Methyl phenol) $\xrightarrow{Zn}$ $C_6H_5-CH_3$ (Toluene)

3) $H_3C-C_6H_4-SO_3H$ (Toluene sulphonic acid) $\xrightarrow[\text{or Dil.HCl}]{H_2O \text{ Super heated steam}}$ $C_6H_5-CH_3$ (Toluene)

4) $H_3C-C_6H_4-\overset{+}{N}\equiv N-Cl^-$ (Toluene diazonium chloride) $\xrightarrow{H_3PO_2,\ H_2O}$ $C_6H_5-CH_3$ (Toluene)

5) C_6H_6 (Benzene) $\xrightarrow[AlCl_3]{CH_3Cl}$ $C_6H_5-CH_3$ (Toluene)

6) C_6H_5-Br (Phenyl bromide) $\xrightarrow[\text{Dry ether}]{2Na + CH_3I}$ $C_6H_5-CH_3$ (Toluene)

7) C_6H_5-MgBr (Phenyl magnesium bromide) $\xrightarrow[\text{Dry ether}]{CH_3I}$ $C_6H_5-CH_3$ (Toluene)

8) $C_6H_5-\overset{O}{\overset{\|}{C}}-CH_3$ (Acetophenone) $\xrightarrow[Zn/Hg/\ HCl]{4(H)}$ $C_6H_5-CH_2CH_3$ (Ethyl benzene)

9) $CH_3\text{-}(CH_2)_5CH_3$ (n-Heptane) $\xrightarrow[15\ atm,\ 500\ ^\circ C]{Pt/Al_2O_3}$ $C_6H_5-CH_3$ (Toluene)

Physical Properties of Benzene: Benzene and its homologues are colorless, refractive liquids with peculiar smell. Boiling point is 80 °C. It is insoluble in water but miscible with organic solvents in various proportions. It is inflammable with a smoky flame which is a characteristic reaction of aromatic compounds. It acts as a good solvent for fats, resins, iodine and sulphur *etc.*

The boiling points, melting points and densities of benzene and its derivatives are indicated in the Table 20.3. Benzene derivatives tend to be more symmetrical than similar aliphatic compounds, so they are packed better into crystals and have higher boiling points. For example, melting point of benzene is 6 °C but hexane is -95 °C. Likewise, *p*-disubstituted benzene is more symmetrical than *o* and *m*-isomers with higher melting points. The relative boiling points of many benzene derivatives are related to their dipole moments. For example, the boiling points of dichlorobenzene follow their dipole moments.

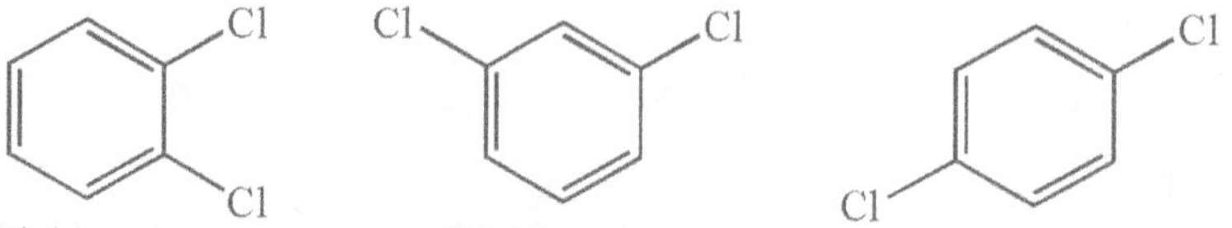

o-Dichloro benzene	*m*-Dichloro benzene	*p*-Dichloro benzene
B.P 181 °C	B.P 173 °C	B.P 170 °C
M.P -17 °C	M.P -25 °C	M.P 54 °C

o-Dichloro benzene has large dipole moment, so it has higher boiling point, *m*-dichloro benzene has little or small dipole moment, so it has less B.P. than *o*-dichlorobenzene. But *p*- dichlorobenzene has zero dipole moment, so very less B.P. when compared with *o* & *m*–dichlorobenzene. In contrast, *p*-dichloro benzene packs best in to a crystal structure so it has highest melting point than other two (*o*, *m*) dichlorobenzene derivative.

Table 20.3 Physical properties of benzene and its analogues.

Compound name	Melting point (°C)	Boiling point (°C)	Density (wt/ml)
Benzene	6	80	0.88
Ethyl benzene	95	136	-
Ethynyl benzene	-45	142	0.93
Toluene	-95	111	0.87
Phenol	43	182	1.07
Anisole	37	156	0.98
Aniline	-6	186	1.02
Benzoic acid	122	249	1.31
Benzyl alcohol	-15	205	1.04
o-Xylene	-26	144	0.88
o-Cresol	30	192	1.03
o-Dicholro benzene	-17	181	1.31

Chemical reactions of benzene: The important characteristic reactions of benzene involve five types.

(I) Electrophilic aromatic substitution reactions.
(II) Addition reactions of benzene.
(III) Oxidation reaction.
(IV) Nucleophilic aromatic substitution reactions.
(V) Free radical substitution reactions.

(I) Electrophilic Aromatic Substitution Reactions: Benzene contains π electrons which are present above and below the plane of its cyclic ring. An electrophile (Y^+) attacks the π electrons, attaches to the benzene nucleus itself and forms a carbocation (carbonium ion) intermediate.

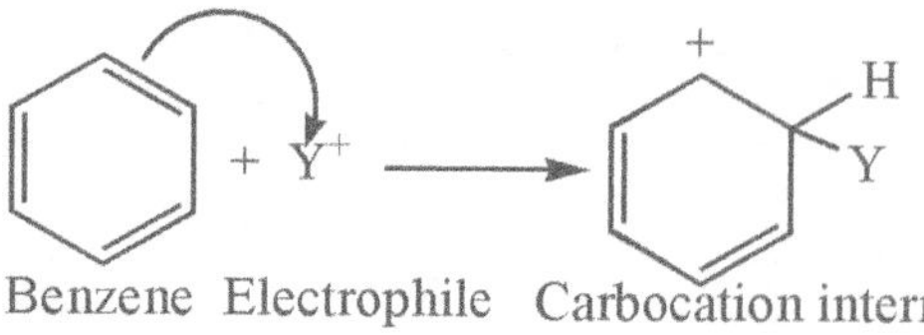

Benzene + Y^+ (Electrophile) ⟶ Carbocation intermediate

This step is similar to that of the electrophilic addition reaction of an alkene - *i.e.,* Alkene reacts with electrophile to form a carbocation intermediate. The second step in electrophilic addition reaction of alkene is the reaction of nucleophile (Z) with carbocation intermediate to form an addition product.

RCH=CHR + Y^+ ⟶ RCH(+)-CH(Y)R $\xrightarrow{Z^-}$ RCH(Z)-CH(Y)R

Alkene Electrophile Carbocation intermediate Addition product

In alkenes, the nucleophile adds to the carbocation intermediate to give addition product of alkene. In electrophilic aromatic substitution reaction if the aromatic substitution takes place in a similar way like electrophilic addition of alkene, the product obtained is not an aromatic compound (in electrophilic aromatic substitution reaction, the carbocation losses a proton from the site when electrophile attacks instead of adding a nucleophile). By this way, the aromaticity of benzene ring is restored. Hence the overall reaction involves the substitution of an electrophile for one of the hydrogen atom attached to the benzene ring.

Benzene + Y^+ ⟶ Carbocation intermediate

Benzene Electrophile Carbocation intermediate

Carbocation intermediate + :Z^- —a→ An aromatic compound + HZ (Product of electrophillic substitution)

—b→ A non aromatic compound

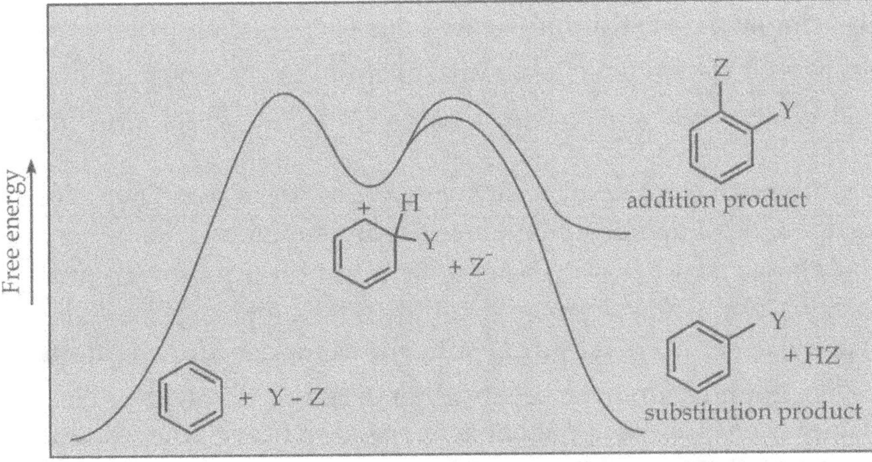

Figure 20.10 Reaction coordination diagram of electrophilic substitution and electrophilic addition reactions of benzene molecule.

Figure 20.10 indicates the reaction coordination diagram of benzene to from a substituted product and for electrophilic addition product of benzene.

It shows that the ΔG° value of substituted product of benzene is near or close to zero. But, benzene forms a much less stable nonaromatic addition product (ΔG° value is high),then it would have been endergonic reaction. It indicates that benzene undergoes an electrophilic substitution reaction that retains aromaticity.

General Mechanism: In an electrophilic aromatic substitution reaction, an electrophile substitutes hydrogen of an aromatic compound.

***Step 1:* Formation of electrophile:** Formation of electrophile takes place by heterolysis of unsymmetrical or symmetrical reagent.

$$\text{Y-Nu} + \text{Catalyst} \longrightarrow \text{Y}^{+} + \overline{\text{Nu:}}$$

Where, Y = Electrophile, Nu: Nucleophile

***Step 2:* Formation of carbocation (carbonium ion) intermediate:**

Formation of carbocation (carbonium ion) intermediate takes place by the attack of electrophile to the benzene. The carbocation intermediate can be stabilized by three resonance structures or contributors. This step is slow and the aromatic benzene is converted into non-aromatic intermediate.

Benzene + Y^+ (Electrophile) —Slow→ [Carbocation intermediate]

***Step* 3:** Loss of proton leads to the formation of substituted product: A nucleophile in the reaction mixture pulls of a proton from the carbocation intermediate and the electrons which held the proton move in to the ring to restabilise its aromaticity.

Carbocation —$\overline{\text{Nu:}}$, Fast→ Substituted product (Y) + Nu:H

The third step is fast and strongly exergonic because this step restores the stability enhancing aromaticity.

Types of Electrophilic Aromatic Substitution Reactions:

The most common five type of electrophilic aromatic substitution reactions are mentioned below:

In an electrophilic substitution reaction, an electrophile (Y^+) is put on the ring and H^+ removed from the ring.

1. **Halogenations:** A bromine, chlorine or iodine substitutes for a hydrogen atom.
2. **Nitration:** A nitro ($-NO_2$) group substitutes for a hydrogen atom.
3. **Sulfonation:** A sulphonic acid ($-SO_3H$) group substitutes for a hydrogen atom.
4. **Friedel - Craft's acylation:** An acyl group substitutes for a hydrogen atom.
5. **Friedel - Craft's alkylation:** An alkyl group substitutes for a hydrogen atom.

All of the above five electrophilic aromatic substitution reactions follow the same three step mechanism, differ only in the formation of electrophile (Y^+) that is required to initiate the reaction. Once Y^+ is formed all five reactions follow the same three steps mechanism.

Note: In an electrophilic substitution reaction, an electrophile (Y^+) is put on the ring and H^+ removed from the ring.

1. **Halogenation of Benzene:** Benzene reacts with halogens (Br_2 or Cl_2 or I_2) in the presence of Lewis acid catalyst ($FeBr_3$ or $FeCl_3$) to give bromobenzene or chlorobenzene or iodobenzene respectively. Lewis acid is a compound that accepts a pair of electrons.

Bromination:

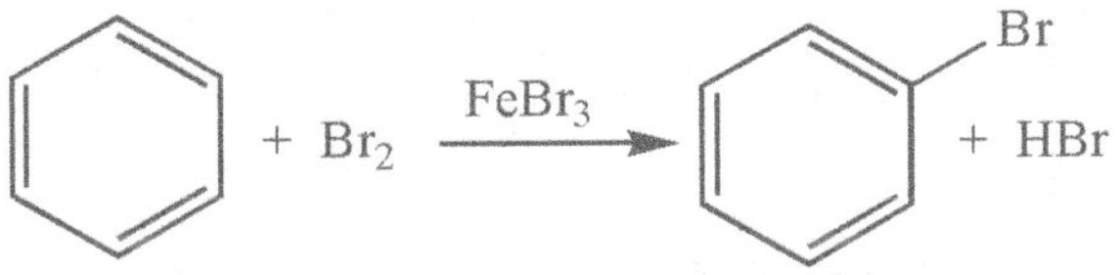

$$\text{Benzene} + Br_2 \xrightarrow{FeBr_3} \text{Bromobenzene} + HBr$$

Benzene Bromobenzene

Chlorination:

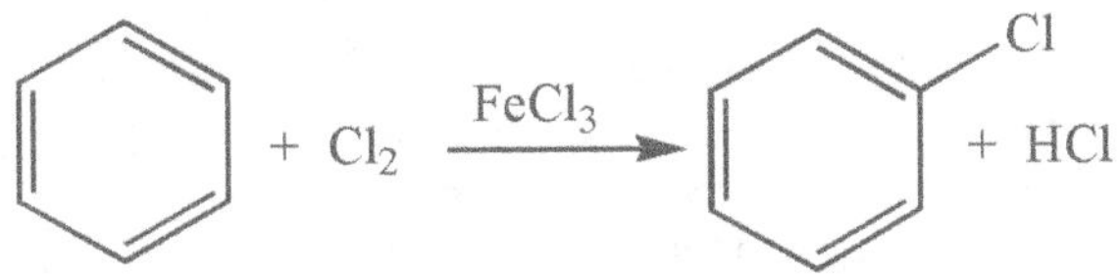

$$\text{Benzene} + Cl_2 \xrightarrow{FeCl_3} \text{Chlorobenzene} + HCl$$

Benzene Chlorobenzene

Iodination:

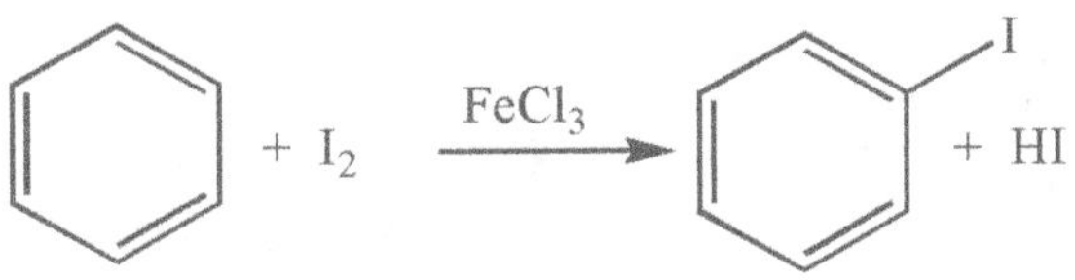

$$\text{Benzene} + I_2 \xrightarrow{FeCl_3} \text{Iodobenzene} + HI$$

Benzene Iodobenzene

Mechanism of bromination:

***Step* 1: Formation of electrophile:** Formation of electrophile (Br^+) takes place by heterolysis of bromine molecule.

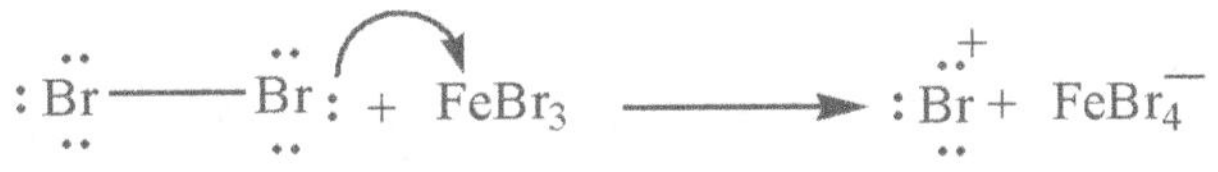

$$:\ddot{Br}-\ddot{Br}: + FeBr_3 \longrightarrow :\ddot{Br}^+ + FeBr_4^-$$

***Step* 2: Formation of carbocation intermediate:** Electrophile attacks the benzene nuclei and forms resonance stabilized carbocation intermediate (carbocation intermediate with two more resonance contributors). For easy understanding of this mechanism only one of the three resonance contributors of the carbocation intermediate is taken.

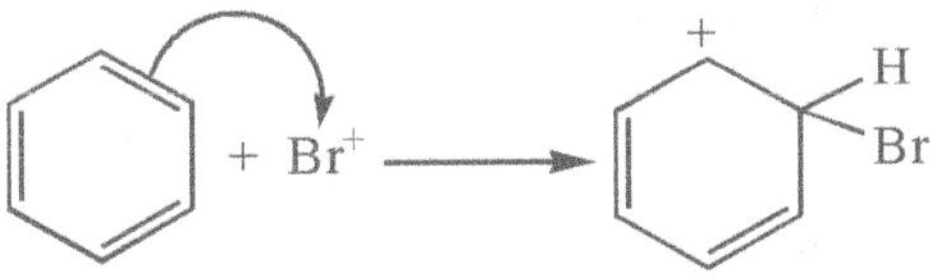

$$\text{Benzene} + Br^+ \longrightarrow \text{Resonance stabilized carbocation}$$

Benzene Resonance stabilized carbocation

***Step* 3: Loss of proton leads to the formation of substituted product:** A base from the reaction mixture ($FeBr_4^-$ / Br^- or the solvent) removes the proton from the carbocation to yield the product.

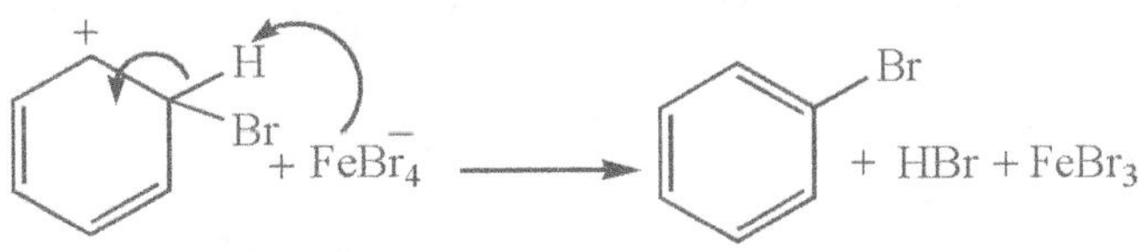

$$\text{Resonance stabilised carbocation} + FeBr_4^- \longrightarrow \text{Bromobenzene} + HBr + FeBr_3$$

Resonance stabilised carbocation Bromobenzene

Mechanism of Chlorination:

It follows the similar mechanism of bromination.

***Step* 1: Formation of electrophile:** Formation of electrophile (Cl^+) takesplace by heterolysis of chlorine molecule.

$$:\ddot{Cl}—\ddot{Cl}: + FeCl_3 \longrightarrow \overset{+}{Cl} + FeCl_4^-$$

Chlorine

***Step* 2: Formation of carbocation intermediate:** Electrophile (Cl^+) attacks the benzene nuclei and forms resonance stabilized carbocation intermediate.

Benzene + Cl^+ ⟶ Resonance stabilized carbocation

***Step* 3: Loss of proton leads to the formation of substituted product:** A base from the reaction mixture ($FeCl_4^-$ / Cl^- or the solvent) removes the proton from the carbocation to yield the product.

Resonance stabilised carbocation + $FeCl_4^-$ ⟶ Chlorobenzene + HCl + $FeCl_3$

In bromination and chlorination, the catalysts $FeBr_3$ and $FeCl_3$ easily react with moisture in the air during handling, which inactivates them as catalysts. So the $FeBr_3$ or $FeCl_3$ is generated "*insitu*" by adding iron filling and Br_2 or Cl_2 to the reaction mixture. Therefore, the reagent halogens match the halogen in the catalyst (Cl_2 + $FeCl_3$, Br_2 + $FeBr_3$)

Mechanism of iodination:

***Step* 1: Formation of electrophile:** The electrophile (I^+) is formed by oxidizing I_2 with an oxidizing agent such HNO_3.

$$I_2 \xrightarrow{HNO_3} 2I^+ + 2e^-$$

***Step* 2: Formation of carbocation intermediate:** Electrophile (I^+) attacks the benzene nuclei and forms resonance stabilized carbocation intermediate.

Benzene + I^+ ⟶ Resonance stabilized carbocation

***Step* 3: Loss of proton leads to the formation of substituted product:** A base from the reaction mixture removes the proton from the carbocation to yield the product.

Carbocation + I^+ ⟶ Iodobenzene + HI

2. **Nitration:** Benzene reacts with Conc. HNO_3 acid and conc. H_2SO_4 (nitrating mixture) at 60 °C to give nitrobenzene

$$\text{Benzene} + HNO_3 \xrightarrow{\text{Conc. } H_2SO_4} \text{Nitrobenzene } (C_6H_5NO_2) + H_2O$$

Nitration of benzene with HNO_3 required H_2SO_4 as a catalyst. Here sulphuric acid protonates HNO_3 which produces nitronium ion by the loss of water molecule. Nitronium ion acts as an electrophile which is needed for nitration.

Note: Any base such as H_2O, HSO_4^- or solvent present in the reaction mixture can remove the proton from the carbocation intermediately.

Mechanism:

***Step* 1: Formation of electrophile:** The electrophile (NO_2^+) is formed by the reaction between sulphuric acid and nitric acid. Here sulphuric acid protonates HNO_3 which produces nitronium ion by the loss of water molecule.

$$\underset{\text{Nitric acid}}{HO{-}NO_2} + \underset{\text{Sulphuric acid}}{H{-}SO_4H} \xrightleftharpoons{-HSO_4^-} H\overset{+}{O}(H){-}NO_2 \rightleftharpoons \underset{\text{Nitronium ion}}{\overset{+}{N}O_2} + H_2O$$

***Step* 2: Formation of carbocation intermediate:** Electrophile (NO_2^+) attacks the benzene nuclei and forms resonance stabilized carbocation intermediate.

$$\text{Benzene} + \underset{\text{Nitronium ion}}{\overset{+}{N}O_2} \longrightarrow \text{Resonance stabilized carbocation}$$

***Step* 3: Loss of proton leads to the formation of substituted product:** A base from the reaction mixture removes the proton from the carbocation to yield the product.

$$\text{Carbocation (H, } NO_2) + HSO_4^- \longrightarrow \text{Nitrobenzene} + H_2SO_4$$

3. **Sulphonation:** Benzene reacts with fuming sulphuric acid (7 % solution of SO_3 in H_2SO_4) or excess concentrated sulphuric acid to give benzene sulphonic acid.

$$\text{Benzene} + \underset{\text{(Excess)}}{H_2SO_4} \xrightleftharpoons{\Delta} \text{Benzene sulphonic acid } (C_6H_5SO_3H) + H_2O$$

$$\text{Benzene} + SO_3 \rightleftharpoons \text{Benzene sulphonic acid } (C_6H_5SO_3H)$$

***Step* 1: Formation of electrophile:** A substantial amount of electrophile (SO_3H^+) is formed on heating with conc. H_2SO_4 (loss of water molecule and a proton from H_2SO_4).

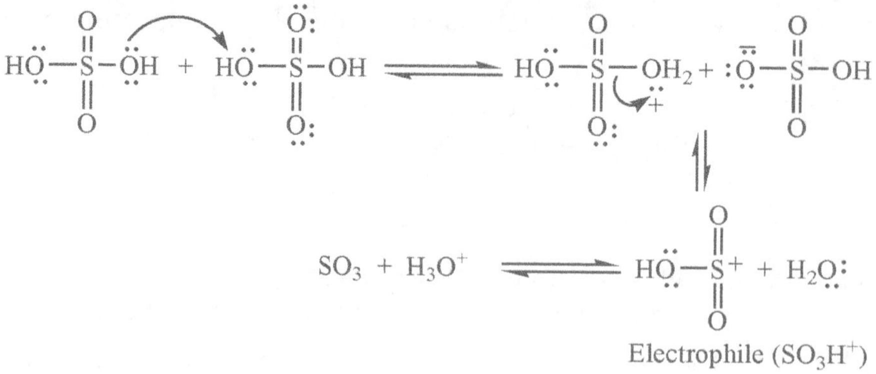

$$HO-SO_2-OH + HO-SO_2-OH \rightleftharpoons HO-SO_2-OH_2^+ + {}^-O-SO_2-OH$$

$$SO_3 + H_3O^+ \rightleftharpoons HO-\overset{+}{S}O_2 + H_2O$$

Electrophile (SO_3H^+)

***Step* 2: Formation of carbocation intermediate:** Electrophile attacks the benzene nuclei and forms resonance stabilized carbocation intermediate.

Benzene + $\overset{+}{S}O_3H$ ⇌ Resonance stabilized carbocation (H, SO_3H)

Benzene

Resonance stabilized carbocation

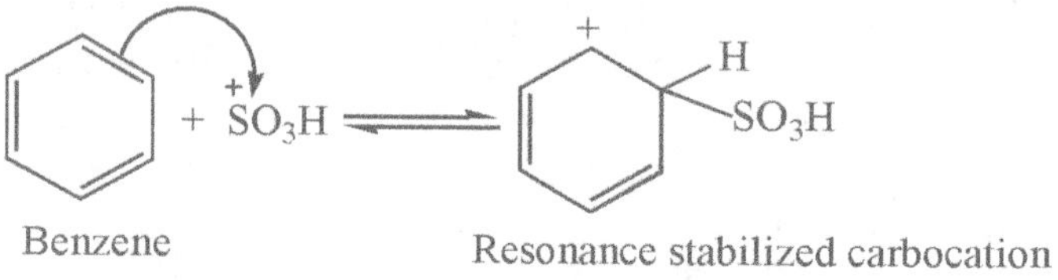

INTERESTING FACT

Substantial quantities of sulphonated aromatic compounds are released in to the environment as a result of industrial and domestic use. The major source is laundry detergent, which contains alkyl benzene sulphonates.

***Step* 3: Loss of proton leads to the formation of substituted product:** A base from the reaction mixture removes the proton from the carbocation to yield the product.

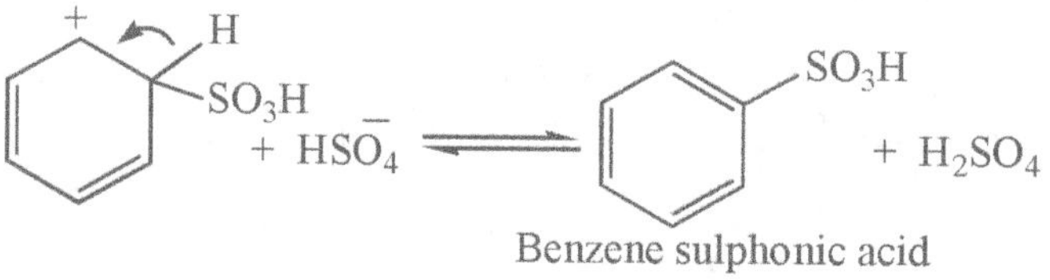

Carbocation (H, SO_3H) + HSO_4^- ⇌ Benzene sulphonic acid (SO_3H) + H_2SO_4

Benzene sulphonic acid

Sulphonic acids are strong acids because of the presence of three electron withdrawing oxygen atoms and the electron delocalization in the sulfonate ion.

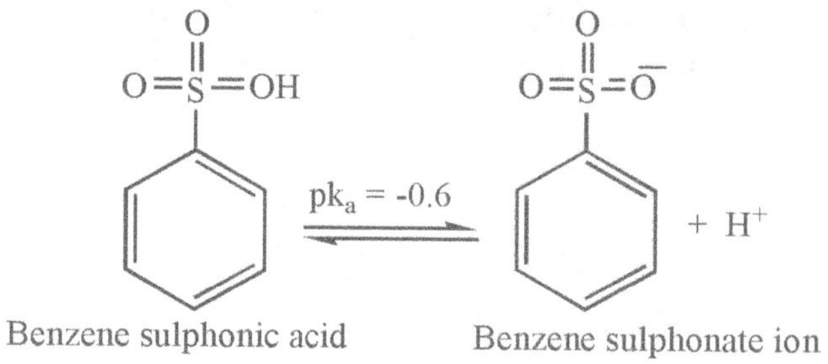

$$C_6H_5-SO_2-OH \underset{}{\overset{pk_a = -0.6}{\rightleftharpoons}} C_6H_5-SO_2-O^- + H^+$$

Benzene sulphonic acid

Benzene sulphonate ion

Sulphonation of benzene is a reversible reaction. If the product of sulphonation is heated in dilute acid, the reaction proceeds in the reverse direction.

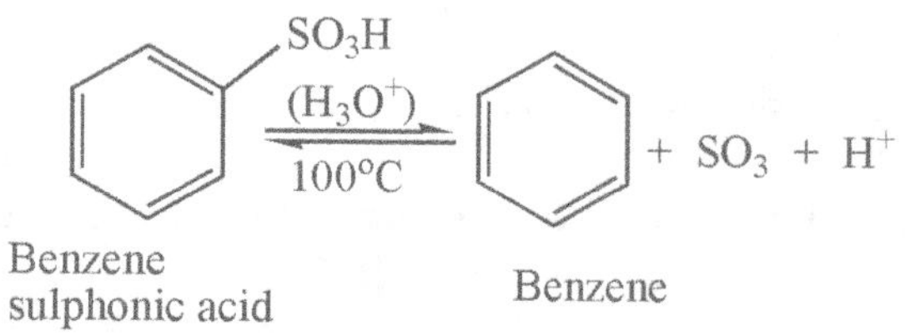

$$C_6H_5SO_3H \underset{100°C}{\overset{(H_3O^+)}{\rightleftharpoons}} C_6H_6 + SO_3 + H^+$$

Benzene sulphonic acid

Benzene

The above reaction is explained by the principle of microscopic reversibility *i.e.,* mechanism of a reaction in the reverse direction must retrace each step of the mechanism in the forward direction in microscopic detail. That means the intermediate of the both forward and backward reaction is same and the rate determining "energy hill" is similar in both reactions.

For example, the energy coordination diagram of sulphonation and desulphonation of benzene is shown in figure 20.11. Sulphonation is going from left to right. So, the desulphonation is right to left. The rate determining step in sulphonation is the attack of benzene by SO_3H^+. In desulphonation the rate determining step is loss of SO_3H^+ from the benzene ring.

Mechanism of desulphonation:

$C_6H_5SO_3H + H^+ \rightleftharpoons$ (arenium ion with H and SO_3H) $\rightleftharpoons C_6H_6 + \overset{+}{S}O_3H$

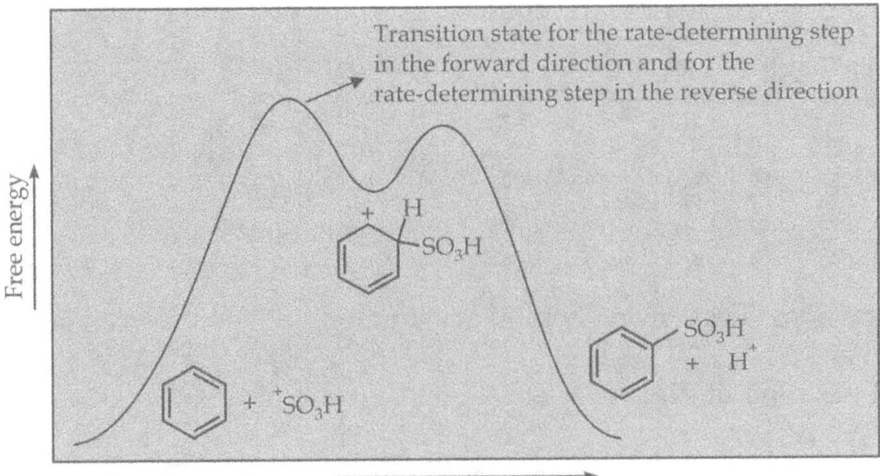

Figure 20.11 Energy coordination diagram of desulphonation.

4. **Friedel - Crafts Acylation and Alkylation of Benzene:** Two electrophilic aromatic substitution reactions of benzene bear the names of two chemists Charles Friedel and James Crafts. The Friedel – Crafts acylation adds an acyl group on a benzene ring and the Friedel - Crafts alkylation adds an alkyl group to a benzene ring.

R—C(=O)— An acyl group   R— Alkyl group

(a) **Friedel - Crafts Acylation:** Benzene reacts with acid chlorides or acid anhydrides in the presence of $AlCl_3$ to give aromatic ketones.

$$\underset{\text{Benzene}}{C_6H_6} + \underset{\text{Acyl chloride}}{R\text{-}C(=O)\text{-}Cl} \xrightarrow[H_2O]{AlCl_3} \underset{\text{Acyl benzene}}{C_6H_5\text{-}C(=O)\text{-}R} + HCl$$

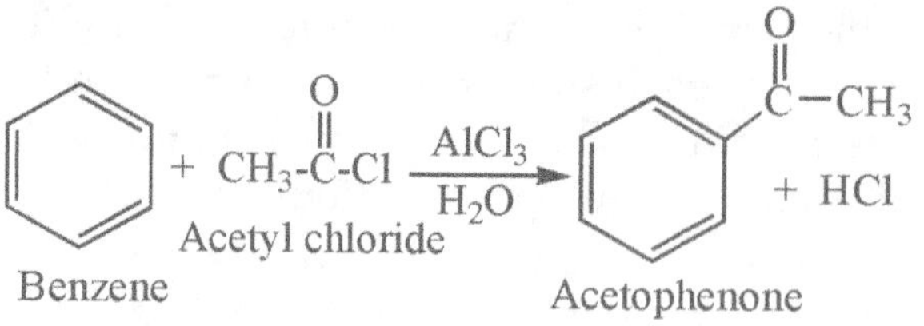

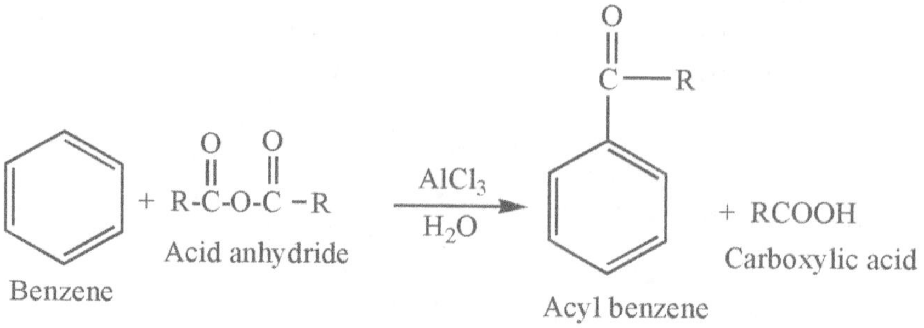

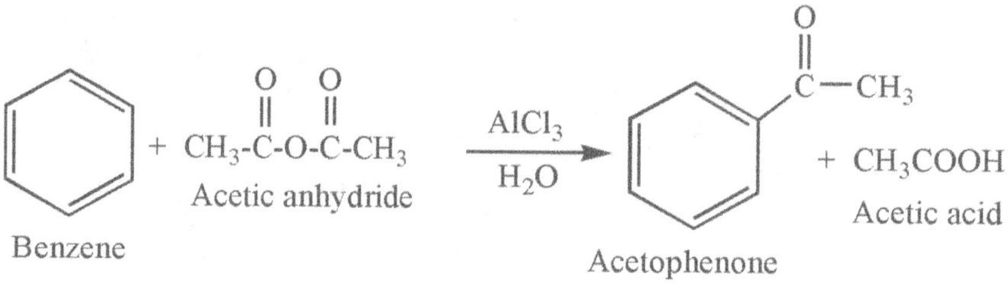

Benzene + CH_3-C(=O)-Cl (Acetyl chloride) $\xrightarrow[H_2O]{AlCl_3}$ Acetophenone (C_6H_5-C(=O)-CH_3) + HCl

Benzene + R-C(=O)-O-C(=O)-R (Acid anhydride) $\xrightarrow[H_2O]{AlCl_3}$ Acyl benzene (C_6H_5-C(=O)-R) + RCOOH (Carboxylic acid)

Benzene + CH_3-C(=O)-O-C(=O)-CH_3 (Acetic anhydride) $\xrightarrow[H_2O]{AlCl_3}$ Acetophenone (C_6H_5-C(=O)-CH_3) + CH_3COOH (Acetic acid)

In this reaction, the electrophile is an acylium ion.

Mechanism:

***Step* 1: Formation of electrophile:**

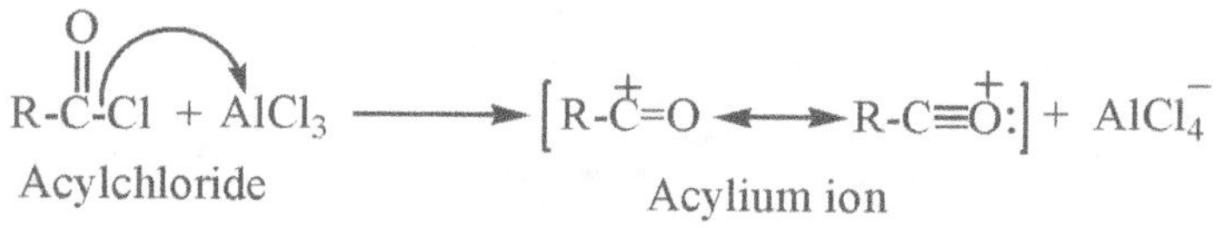

R-C(=O)-Cl (Acylchloride) + $AlCl_3$ ⟶ [R-$\overset{+}{C}$=O ⟷ R-C≡$\overset{+}{O}$:] (Acylium ion) + $AlCl_4^-$

***Step* 2: Formation of carbocation intermediate:** Electrophile (acylium ion) attacks the benzene nuclei and forms resonance stabilized carbocation intermediate.

Benzene + R-$\overset{+}{C}$=O (Acylium ion) ⇌ Resonance stabilised carbocation

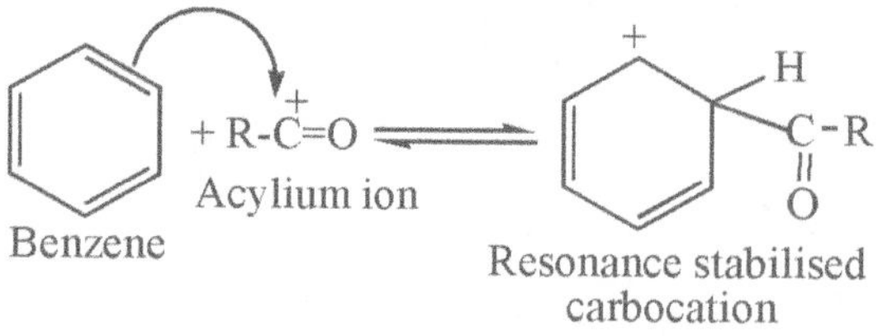

Step 3: Loss of proton leads to the formation of substituted product: A base from the reaction mixture removes the proton from the carbocation to yield the product.

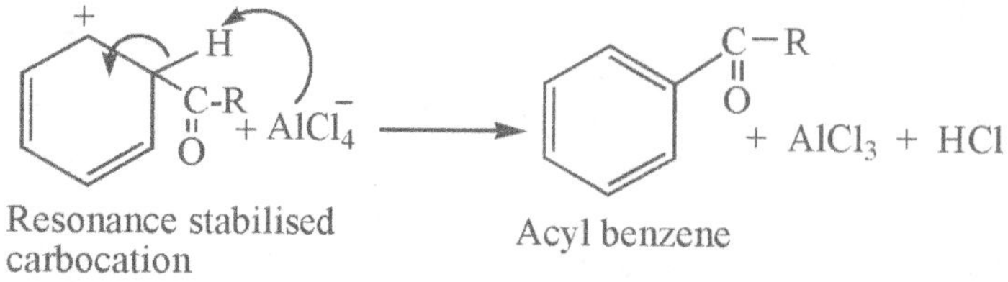

Resonance stabilised carbocation + $AlCl_4^-$ ⟶ Acyl benzene + $AlCl_3$ + HCl

Note: Friedel–Crafts acylation reactions must be carried out with more than equivalent of $AlCl_3$, because the product formed contains a carbonyl group which can complex with $AlCl_3$. When the reaction is over, water is added to the reaction mixture to liberate the product from the complex by hydrolyzing the aluminium salts.

$$C_6H_5COR + AlCl_3 \rightleftharpoons C_6H_5C(R)=\overset{+}{O}-\overset{-}{A}lCl_3 \xrightarrow{3H_2O} C_6H_5COR + Al(OH)_3 + 3HCl$$

Acyl benzene — Complex — Acyl benzene

Aromatic nitro compounds are components of many drugs. In our body, the nitro compounds are reduced to arylamines e,g. Chloramphenicol – Antibiotic.Charles – Friedel (1832-1899) was born in Strasbourg, France. He was a professor of chemistry and director of research at the Sorbonne. His interest in mineralogy led him to make synthetic diamonds. He met Crafts when they both were doing research in Medicine at Paris. They collaborated scientifically for most of their lives. They discover Friedel – Crafts reactions in Friedel's Laboratory in 1877.

Limitations:

1. The acylium ion is resonance stabilized so no rearrangement takes place.
2. The product obtained is deactivated so no possibility of further reaction.
3. Acylation fails with strongly deactivated aromatic rings.

Intramolecular Friedel-Craft's Acylation

The starting material containing both benzene and electrophile undergoes intra molecular reaction to form a ring. For example: 4-phenyl butanoyl chloride undergoes intramolecular Friedel Craft's acylation in the presence of $AlCl_3$ to form α–tetralone.

$$C_6H_5CH_2CH_2CH_2COCl \xrightarrow{AlCl_3} \alpha\text{-Tetralone} + HCl$$

4-Phenylbutanoylchloride — α-Tetralone

Friedel - Crafts Alkylation

Benzene reacts with alkyl halides in the presence of $AlCl_3$ or $FeCl_3$ to give alkyl benzene.

$$C_6H_6 + RCl \xrightarrow{AlCl_3} C_6H_5R + HCl$$

Benzene — Alkyl chloride — Alkyl benzene

$$C_6H_6 + CH_3Cl \xrightarrow{AlCl_3} C_6H_5CH_3 + HCl$$

Benzene — Methyl chloride — Toluene

Mechanism:

***Step* 1: Formation of electrophile (R^+):**

$$R\text{-}Cl + AlCl_3 \longrightarrow R^+ + AlCl_4^-$$

***Step* 2: Formation of carbocation intermediate:** Electrophile (R^+) attacks the benzene nuclei and forms resonance stabilized carbocation intermediate.

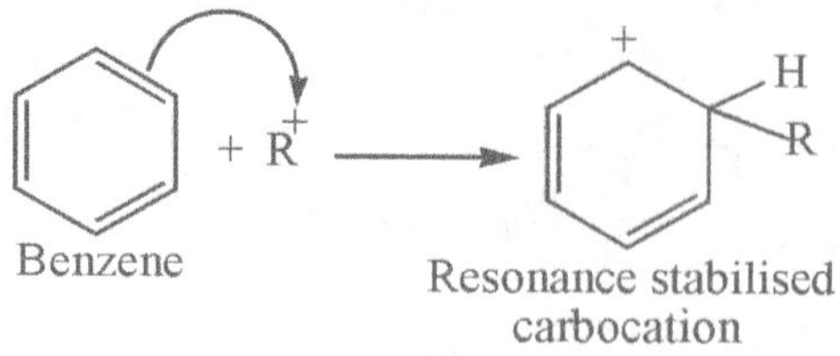

Benzene + $\overset{+}{R}$ ⟶ Resonance stabilised carbocation

***Step* 3: Loss of proton leads to the formation of substituted product:** A base from the reaction mixture removes the proton from the carbocation to yield the product.

Resonance stabilised carbocation + $AlCl_4^-$ ⟶ Alkyl benzene + $AlCl_3$ + HCl

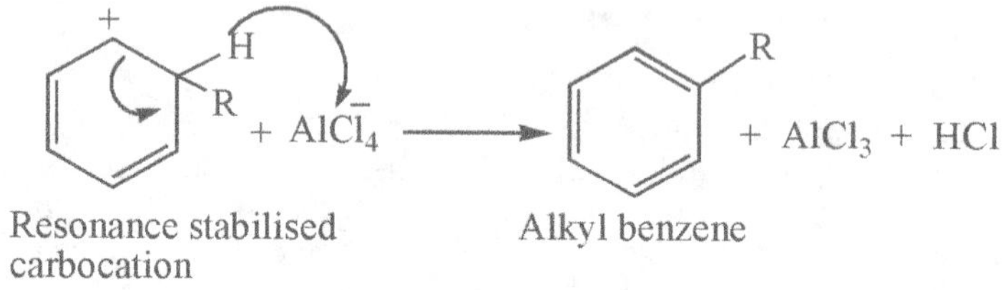

Example: Benzene reacts with methyl chloride in the presence of aluminium chloride to yield toluene.

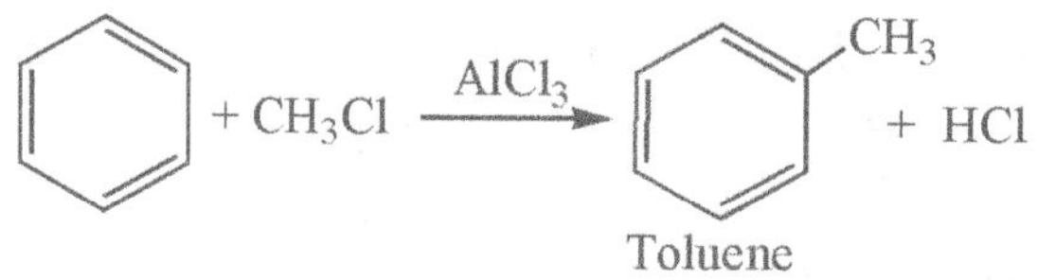

Benzene + CH_3Cl $\xrightarrow{AlCl_3}$ Toluene + HCl

Mechanism:

***Step* 1: Formation of electrophile (CH_3^+):**

$$CH_3\text{-}Cl + AlCl_3 \longrightarrow CH_3^+ + AlCl_4^-$$

***Step* 2: Formation of carbocation intermediate:** Electrophile (CH_3^+) attacks the benzene nuclei and forms resonance stabilized carbocation intermediate.

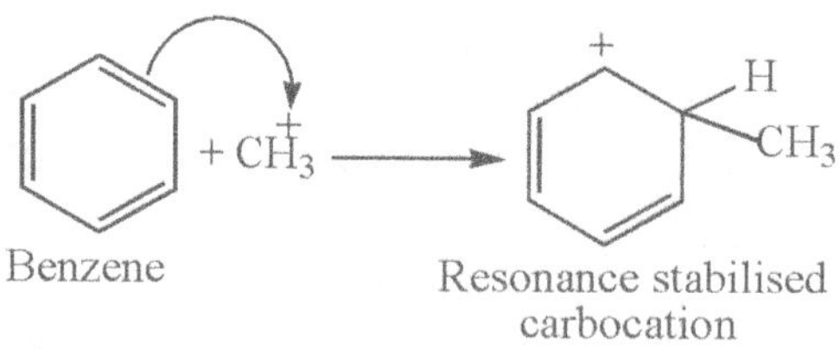

Benzene + $\overset{+}{C}H_3$ ⟶ Resonance stabilised carbocation

***Step* 3: Loss of proton leads to the formation of substituted product:** A base from the reaction mixture removes the proton from the carbocation to yield the product.

Resonance stabilised carbocation + $AlCl_4^-$ ⟶ Toluene + $AlCl_3$ + HCl

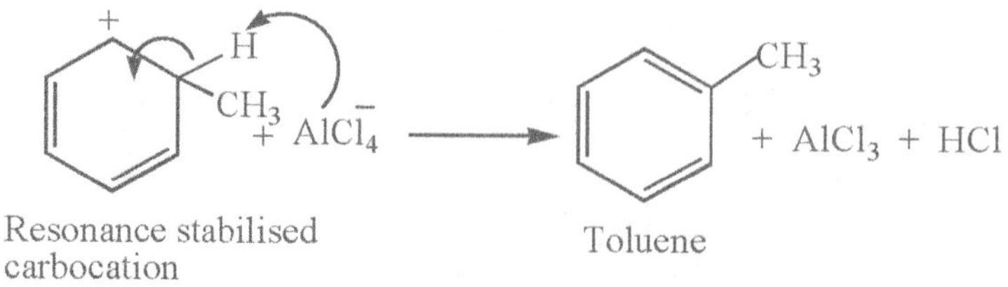

The alkyl substituted benzene is more reactive than benzene. To prevent further reaction or further alkylation of alkylated benzene, a large excess amount of benzene is used in Friedel - Crafts alkylation to ensure that the electrophile is more likely to encounter a molecule of benzene than a molecule of alkyl substituted benzene.

Other examples of Friedel – Crafts alkylation are shown below:

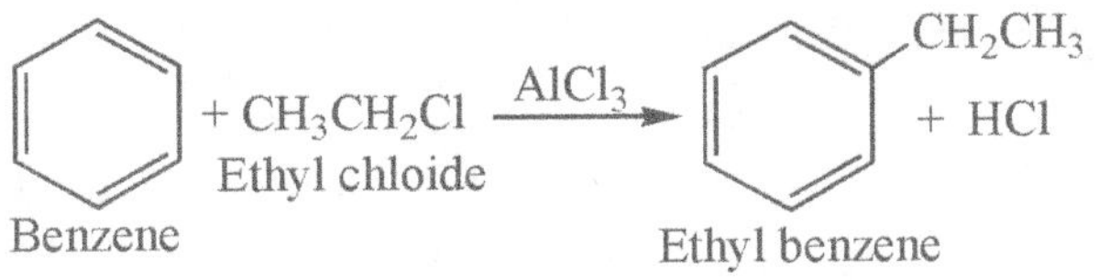

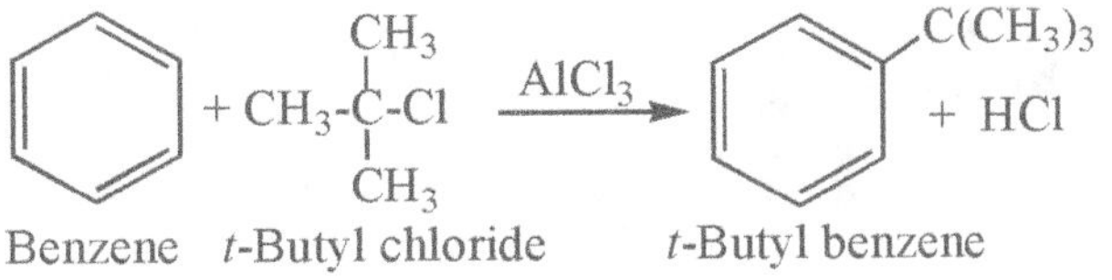

$$C_6H_6 + CH_3CH_2Cl \xrightarrow{AlCl_3} C_6H_5CH_2CH_3 + HCl$$

Benzene + Ethyl chloide → Ethyl benzene

$$C_6H_6 + (CH_3)_3C\text{-}Cl \xrightarrow{AlCl_3} C_6H_5C(CH_3)_3 + HCl$$

Benzene + *t*-Butyl chloride → *t*-Butyl benzene

Benzene reacts with 1° alkyl halides: The primary carbocation formed is unstable one. Here the actual electrophile is a complex of $AlCl_3$ with the alkyl halide. In this complex, the C-X bond is weak bond and carbon atom gets the considerable positive charge.

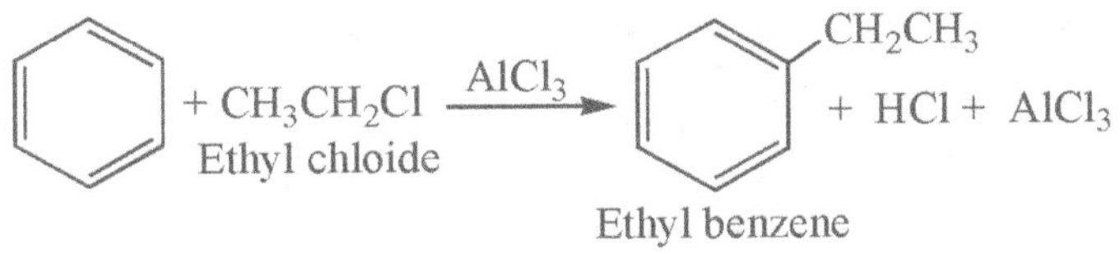

$$C_6H_6 + CH_3CH_2Cl \xrightarrow{AlCl_3} C_6H_5CH_2CH_3 + HCl + AlCl_3$$

Ethyl chloide → Ethyl benzene

Mechanism:

$$CH_3\text{-}CH_2Cl + AlCl_3 \rightleftharpoons CH_3\text{-}\overset{\delta+}{CH_2}\cdots Cl\cdots \overset{\delta-}{AlCl_3} \text{ or } CH_3CH_2^+ + \overline{AlCl_4}$$

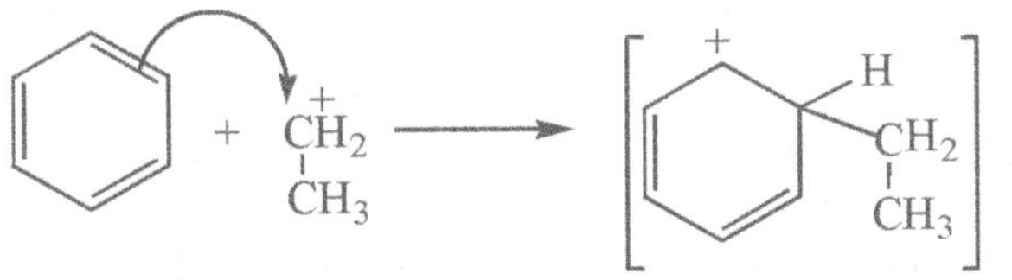

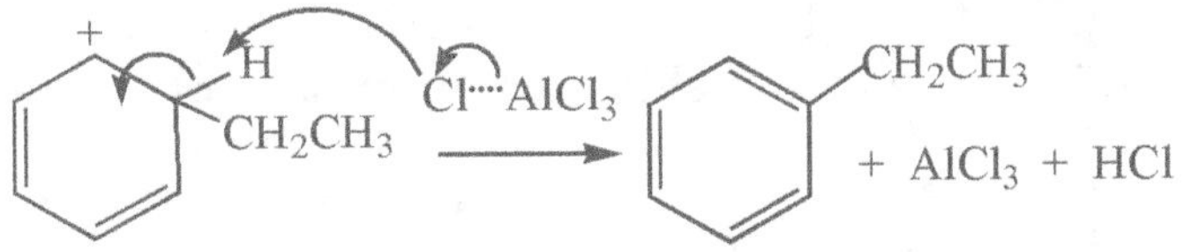

$$C_6H_6 + \overset{+}{CH_2}\text{-}CH_3 \longrightarrow \left[C_6H_6^+(H)(CH_2CH_3) \right]$$

$$C_6H_6^+(H)(CH_2CH_3) + \bar{Cl}\cdots AlCl_3 \longrightarrow C_6H_5CH_2CH_3 + AlCl_3 + HCl$$

Limitations: Although the Friedel – Crafts alkylation looks good in principle, it has major three limitations that severely restrict its use.

***Limitation* 1:** Friedel – Crafts alkylation is only given by benzene, halo benzene and other activated benzene derivatives. Strongly deactivated systems like nitrobenzene, phenylketones and benzene sulphonic acid do not give Friedel – Crafts alkylation.

***Limitation* 2:** Like other carbocation reactions, Friedel – Crafts alkylation also involves carbocation rearrangements. Then the major product obtained will be the product with the rearranged alkyl group on the benzene ring. The relative amount of rearranged and unrearranged product obtained depends upon the increase in the carbocation stability which is achieved by rearrangement.

For example: Benzene reacts with 1-chlorobutane to give rearranged and unrearranged products. Here a primary carbocation obtained is less stable which rearranges into more stable secondary carbocation.

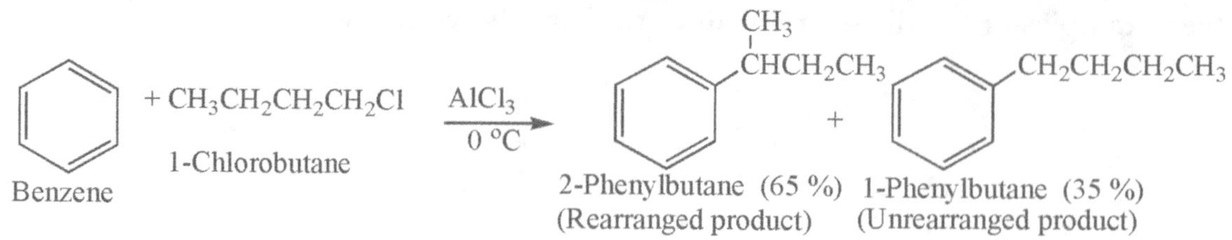

Benzene + $CH_3CH_2CH_2CH_2Cl$ (1-Chlorobutane) $\xrightarrow[0\ °C]{AlCl_3}$ 2-Phenylbutane (65 %) (Rearranged product) + 1-Phenylbutane (35 %) (Unrearranged product)

Mechanism:

***Step* 1: Formation of electrophile.**

$CH_3CH_2CH_2CH_2Cl + AlCl_3$ (1-Chloro butane) $\xrightarrow[AlCl_4^-]{}$ $CH_3CH_2CH_2\overset{+}{C}H_2$ (1° Carbocation (Less stable)) $\xrightarrow{\text{1,2-Hydride shift}}$ $CH_3CH_2\overset{+}{C}HCH_3$ (2° Carbocation (More stable))

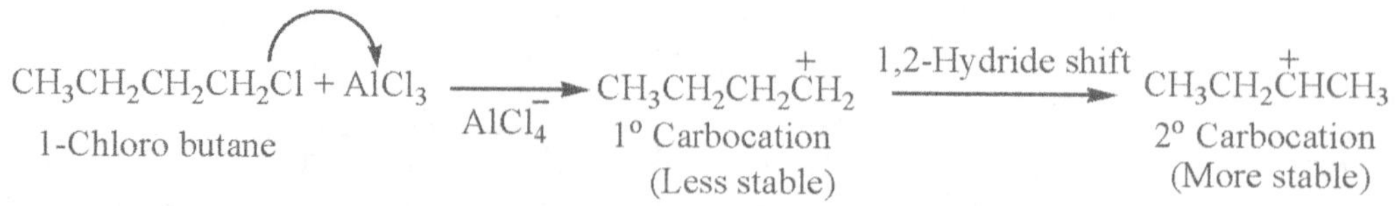

***Step* 2: Formation of carbocation.**

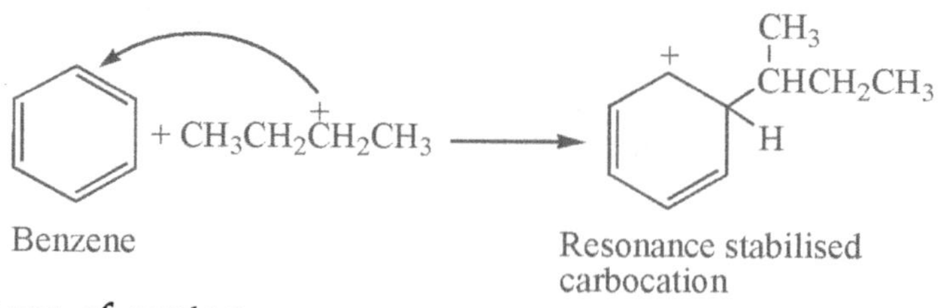

Benzene + $CH_3CH_2\overset{+}{C}H_2CH_3$ $\longrightarrow$ Resonance stabilised carbocation

***Step* 3: Loss of proton**

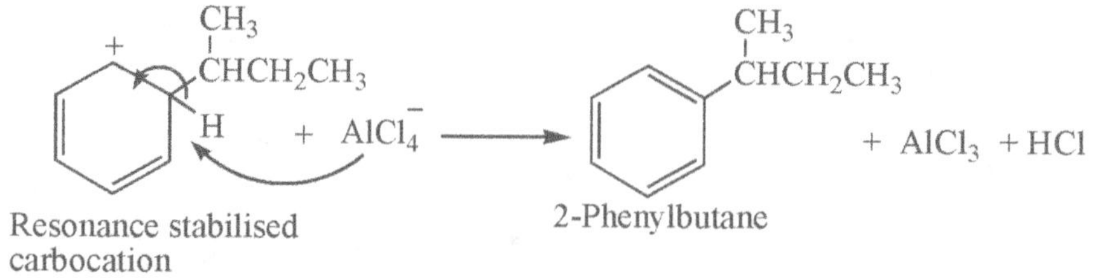

Resonance stabilised carbocation + $AlCl_4^-$ $\longrightarrow$ 2-Phenylbutane + $AlCl_3$ + HCl

Example No. 2: Benzene reacts with 1-chloro-2,2-dimethyl propane to give 100 % rearranged product and 0 % unarranged products.

Here, the primary carbocation formed is rearranged into highly stable tertiary carbocation. , The greater increase in the stability of which yields greater amount of rearranged product.

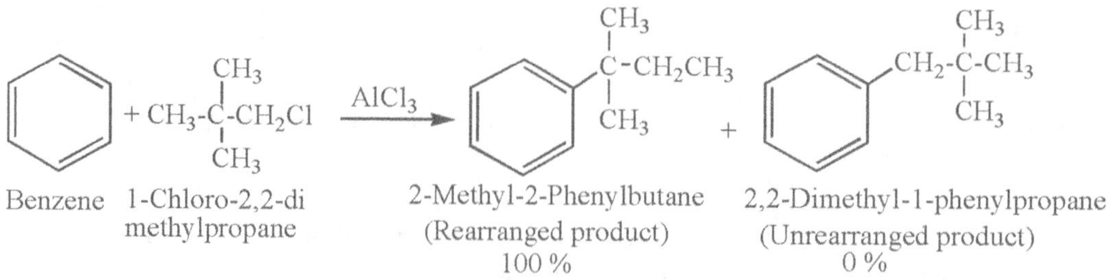

Benzene + $CH_3\text{-}C(CH_3)_2\text{-}CH_2Cl$ (1-Chloro-2,2-dimethylpropane) $\xrightarrow{AlCl_3}$ 2-Methyl-2-Phenylbutane (Rearranged product) 100 % + 2,2-Dimethyl-1-phenylpropane (Unrearranged product) 0 %

Mechanism:

***Step* 1:** Formation of electrophile and carbocation intermediate.

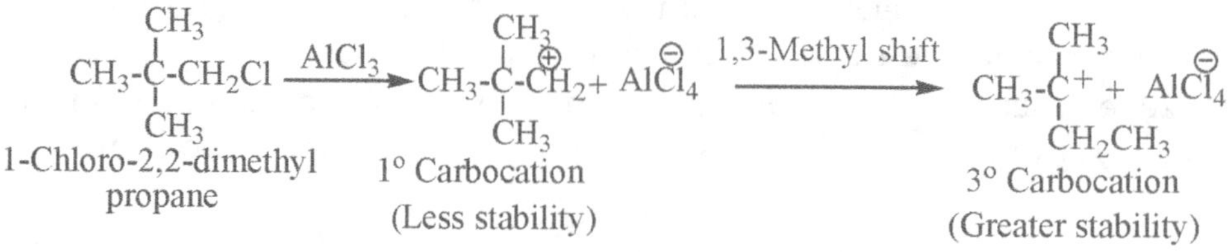

$CH_3\text{-}C(CH_3)_2\text{-}CH_2Cl$ (1-Chloro-2,2-dimethyl propane) $\xrightarrow{AlCl_3}$ $CH_3\text{-}C(CH_3)_2\text{-}\overset{\oplus}{C}H_2 + AlCl_4^{\ominus}$ (1° Carbocation (Less stability)) $\xrightarrow{\text{1,3-Methyl shift}}$ $CH_3\text{-}\overset{+}{C}(CH_3)\text{-}CH_2CH_3 + AlCl_4^{\ominus}$ (3° Carbocation (Greater stability))

$$C_6H_6 + CH_3\text{-}\overset{+}{C}(CH_3)\text{-}CH_2CH_3 \longrightarrow C_6H_6^{+}(H)\text{-}C(CH_3)_2CH_2CH_3$$

Benzene — Resonance stabilised carbocation

***Step* 2:** Loss of proton.

$$C_6H_6^{+}(H)\text{-}C(CH_3)_2CH_2CH_3 \xrightarrow{AlCl_4^-} C_6H_5\text{-}C(CH_3)_2CH_2CH_3 + AlCl_3 + HCl$$

Resonance stabilised carbocation

James Mason Crafts (1839-1917) was born in Boston, the son of a woolen goods manufacturer. He was a professor of Chemistry at Cornell University and at the Massachusetts Institute of Technology. He was the president of MIT from 1897 to 1900, when he was forced to retire because of chronic poor health.

Unstable to be formed in the solution. It is concluded that the 1° carbocation is never formed in Friedel - Crafts alkylation reaction. The carbocation remains complexed with the catalyst, as an incipient carbocation. Carbocation rearrangement still occurs which means that the incipient carbocation has sufficient carbocation character to permit the rearrangement.

$$\underset{\text{Butyl chloride}}{CH_3CH_2CH_2CH_2Cl} + \underset{\text{Aluminium chloride}}{AlCl_3} \longrightarrow CH_3CH_2CH_2\overset{\delta+}{C}H_2....\overset{\delta-}{Cl}....AlCl_3 \xrightarrow{\text{1,2-Hydride shift}} CH_3CH_2\overset{\delta+}{C}HCH_3\,\cdots Cl\cdots\overset{\delta-}{AlCl_3}$$

***Limitation* 3:** The product of Friedel - Crafts alkylation is more active than the starting material due to the presence of alkyl groups which are active substituents. So it is difficult to stop or avoid multiple alkylations. For example, if we want to synthesize ethyl benzene, we use 1 mole of ethyl chloride and 1 mole of benzene with a small amount of $AlCl_3$. During this reaction, some amount of ethyl benzene is formed first, once the reaction is activated, reaction is even faster than benzene itself. So, the product is a mixture of some *ortho* and *para* diethyl benzenes, some triethyl benzenes, a small amount of ethyl benzene and some un reacted benzene.

$$\underset{\substack{\text{Benzene}\\(1\text{ mole})}}{C_6H_6} + \underset{\substack{\text{Ethyl chloride}\\(1\text{ mole})}}{CH_3CH_2Cl} \xrightarrow{AlCl_3} \underset{p\text{-Diethyl benzene}}{p\text{-}C_6H_4(CH_2CH_3)_2} + \underset{o\text{-Diethyl benzene}}{o\text{-}C_6H_4(CH_2CH_3)_2} + \underset{\text{Benzene}}{C_6H_6} + \underset{\text{Ethyl benzene}}{C_6H_5CH_2CH_3} + \text{Triethyl benzene}$$

This multiple alkylation is avoided by using large excess quantity of benzene such as one mole of ethyl chloride is mixed with 50 moles of benzene. [The concentration of ethyl benzene is always low; the electrophile is more likely to react with benzene rather than with ethyl benzene]. The products are separated from benzene by distillation. This is common industrial procedure since a continuous distillation can recycle the unreacted benzene.

Finally, from the above conclusion, the Friedel Craft's alkylation is a useful method for introducing just one group to the benzene without rearrangement or without formation of polyalkylated products.

Difference between Friedel-Craft's acylation and Friedel-Craft's-alkylation.

S. No.	Friedel-Craft's acylation	Friedel-Craft's-alkylation
1.	Only benzene, halo benzene and activated benzene derivatives undergo this reaction.	Strongly deactivated benzene derivatives do not give alkylation.
2.	No rearrangement is possible (formation of resonance stabilized acylium ions).	Rearrangement of carbocation intermediate takes place (less stable carbocation is converted into more stable carbocation).
3.	The acylation leads to the formation of deactivated acyl benzene which does not react further.	Polyalkylation is a common problem.

Friedel-Craft's Alkylation - Uses Other Carbocation Sources

Carbocations are obtained generally by several ways and most of these can be used for Friedel Crafts alkylation.

Alkenes are used as the source: Alkenes are protonated by HF (hydro fluoric acid) to form carbocations in the presence of BF_3. Fluoride ion is weak nucleophile so it does not attack the benzene immediately; if activated benzene or its derivatives are present, then electrophilic substitution takes place. It follows Markovnikov's rule, forms the more stable carbocation, which alkylate the aromatic ring. For example: Alkylation of benzene by alkene.

Benzene + $CH_3CH{=}CHCH_3$ (2-Butene) $\xrightarrow{HF}$ *sec*-Butyl benzene ($C_6H_5CH(CH_3)CH_2CH_3$)

Mechanism:

***Step* 1: Formation of electrophile.**

$CH_3CH{=}CHCH_3$ (2-Butene) + H—F $\rightleftharpoons$ $CH_3\overset{+}{C}HCH_2CH_3$ (Electrophile) + F^-

***Step* 2: Formation of carbocation intermediate:** Electrophile attacks the benzene nuclei and forms resonance stabilized carbocation intermediate.

Benzene + $CH_3\overset{+}{C}HCH_2CH_3$ $\longrightarrow$ Resonance stabilised carbocation (H, $CH(CH_3)CH_2CH_3$)

***Step* 3: Loss of proton leads to the formation of substituted product:** A base from the reaction mixture removes the proton from the carbocation to yield the product.

Resonance stabilised carbocation + F^- $\longrightarrow$ *sec*-Butyl benzene ($C_6H_5CH(CH_3)CH_2CH_3$) + HF

Alcohols are used as the source: Alcohols are used as another source of carbocation. Alcohols produce carbocations in the presence of Lewis acid such as BF_3. If activated benzene or its derivatives are present, electrophilic substitution takes place. A full equivalent of the Lewis acid is utilized during the reaction which means that the reaction is promoted by BF_3 rather than catalyzed by BF_3

$$\text{C}_6\text{H}_6 + CH_3CH(OH)CH_3 \xrightarrow[\Delta]{H_2SO_4} C_6H_5CH(CH_3)CH_3 + H_2O$$

Benzene 2-Propanol Cumene (Isopropyl benzene)

Mechanism:

***Step* 1: Formation of electrophile.**

$$CH_3CH(OH)CH_3 \xrightarrow{H_2SO_4} CH_3\overset{+}{C}HCH_3 + H\overline{SO_4} + H_2O$$

2-Propanol 2° Carbocation

***Step* 2: Formation of carbocation intermediate:** Electrophile attacks the benzene nuclei and forms resonance stabilized carbocation intermediate.

$$\text{C}_6\text{H}_6 + CH_3\overset{+}{C}HCH_3 \longrightarrow \text{C}_6\text{H}_6^{+}(H)CH(CH_3)_2$$

Benzene Resonance stabilized carbocation

***Step* 3: Loss of proton leads to the formation of substituted product:** A base from the reaction mixture removes the proton from the carbocation to yield the product.

$$\text{C}_6\text{H}_6^{+}(H)CH(CH_3)_2 + H\overline{SO_4} \longrightarrow C_6H_5CH(CH_3)_2 + H_2SO_4$$

Resonance stabilised carbocation Cumene

Alkylation of Benzene by Acylation – Reduction: Alkyl benzene containing straight chain alkyl group can't be prepared in better yield by Friedel-Crafts – alkylation or in other words alkyl benzene other than methyl, ethyl or tertiary alkyl benzene cannot be prepared by Friedel-Crafts – alkylation, because the incipient primary carbocation will rearrange into a more stable carbocation. For example, In the reaction between benzene and *n*-butyl chloride, the butyl benzene is obtained as a minor product and *sec*-butyl benzene obtained as major product.

$$\text{C}_6\text{H}_6 + CH_3CH_2CH_2CH_2Cl \xrightarrow{AlCl_3} C_6H_5CH(CH_3)CH_2CH_3 + C_6H_5CH_2CH_2CH_2CH_3$$

Benzene, *n*-Butyl chloride, *Sec*-Butyl benzene Major product (More stable 2° carbocation), *n*-Butyl benzene Minor product (Less stable 1° carbocation)

To solve this problem, the straight chain alkyl benzenes are not prepared by direct alkylation of benzene but by acylation followed by reduction sequence.

Acylium ions do not undergo rearrangement consequently; hence a straight chain alkyl group is attached to the benzene ring by using Friedel Crafts – acylation reaction, followed by the reduction of carbonyl groups to a methylene group by Clemmensen reduction or Wolf Kishner reduction. The main advantage of this method is to avoid the excess usage of benzene, because acylated benzene is less reactive than alkylated benzene, so it will not easily undergo polyacylation.

Benzene + $CH_3CH_2CH_2\overset{O}{\overset{\|}{C}}Cl$ (n-Butanoyl chloride) $\xrightarrow[-HCl]{AlCl_3}$ $C_6H_5\overset{O}{\overset{\|}{C}}CH_2CH_2CH_3$ (Acyl substituted benzene) $\xrightarrow{\text{Reduction}}$ $C_6H_5CH_2CH_2CH_2CH_3$ (n-Butyl benzene)

Two methods are commonly used to reduce the carbonyl group in to alkane. They are Clemmenson reduction and Wolf - Kishner reduction.

Clemmensen reduction uses acidic solution of zinc in mercury as the reducing agent. (Zn (Hg)/HCl)

Butanoyl benzene ($C_6H_5\overset{O}{\overset{\|}{C}}CH_2CH_2CH_3$) $\xrightarrow[\text{Clemmensen reduction}]{Zn(Hg)/HCl}$ $C_6H_5CH_2CH_2CH_2CH_3$ (n-Butyl benzene)

Butanoyl benzene $\xrightarrow[\text{Wolf-Kishner reduction}]{NH_2\text{-}NH_2/OH^-}$ $C_6H_5CH_2CH_2CH_2CH_3$ (n-Butyl benzene)

Note: The Wolf - Kishner reduction uses hydrazine in basic conditions.

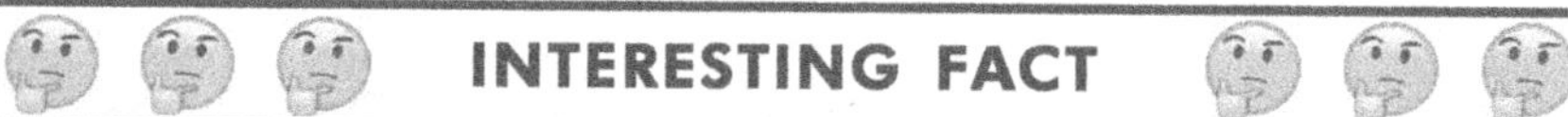

INTERESTING FACT

E.C - Clemmensen (1876 - 1941) was born in Denmark. He worked as a scientist at Clemmensen corp. in Newyork. Ludwig Wolf (1857-1919) was born in Germany; He was a professor at the University of Jena in Germany.

N.M. Kishner (1867-1935) was born in Moscow. He received Ph.D. from the Moscow University under the guidance of Markovnikovs. He was a professor in Tomsk and Moscow University.

All the above Friedel Craft's reactions follow the intermolecular reaction of benzene with an electrophile.

(II) Addition Reactions of Benzene: Benzene more commonly undergoes substitution reaction, but it may undergo addition reaction under drastic conditions. Following reactions are the examples of addition reactions of benzene.

1. **Halogenation:** When benzene molecule is treated with an excess amount of chlorine in the presence of heat under pressure or irradiation with light gives benzene hexa chloride (BHC). During this reaction, three π bonds of the benzene ring is substituted with 6 chlorine atoms (1,2,3,4,5,6–hexa chloro cyclohexane).

$C_6H_6 + 3Cl_2 \xrightarrow[\text{Pressure}]{\text{Heat /light}} C_6H_6Cl_6$

Benzene hexachloride (BHC)

This addition reaction of chlorine to benzene follows a free radical mechanism. So it is impossible to stop the reaction at intermediate stage. The first addition destroys the aromaticity and the next addition of 2 moles of Cl_2 takes place very rapidly. All eight possible isomers are produced or formed in various amounts. The most important isomer that is commercially used as an insecticide is lindane. It is used in shampoos to kill head lice.

H Cl H Cl Cl H Cl H Cl H H H Cl

Lindane (Insecticides)

Mechanism: The reaction takes place *via* the formation of chlorine free radical.

$$Cl_2 \xrightarrow{UV} Cl\cdot + Cl\cdot$$

Chlorine — Chlorine free radical

Benzene + Cl· → (H, ·) → Cl_2 → (H, Cl; H, Cl) + Cl· $\xrightarrow[\text{propagation}]{\text{Chain}}$ $C_6H_6Cl_6$ BHC

2. **Catalytic Hydrogenation or Addition of Hydrogen to the Benzene:** Benzene undergoes reduction or hydrogenation at elevated temperature and pressure or in the presence of Pt or Pd or Ni, Ru, or Rhodium (Rh). Substituted benzene gives substituted cyclohexane. Similarly, disubstituted benzenes give mixtures of *cis* and *trans* isomers. This reaction is the commercial method for the preparation of cyclohexane and substituted cyclohexane derivatives.

Benzene $\xrightarrow[\text{Pt, Pd, Ru, Rh, Ni}]{3H_2\text{,1000 psi}}$ Cyclohexane (100 %)

Toluene (CH_3) $\xrightarrow[\text{Pt, Pd, Ru, Rh, Ni}]{3H_2\text{,1000 psi, 100 °C}}$ Methylcyclohexane (CH_3)

m-Xylene (CH_3, CH_3) $\xrightarrow[\text{Ru/Rh, 1000 °C}]{3H_2\text{,1000 psi}}$ 1,3-Dimethylcyclohexane (cis and trans isomeric mixture) (CH_3, CH_3)

3. **Birch reduction:** A.J. Birch, an Australian chemist in 1944 found that the reduction of benzene or its derivatives with sodium or lithium in a mixture of liquid ammonia and alcohol gives 1,4-cyclohexadiene. This reaction is known as Birch reduction. It is a convenient method for synthesizing a wide variety of useful cyclic dienes.

Benzene $\xrightarrow[\text{Liquid ammonia, ROH}]{\text{Na/Li}}$ Cyclohexadiene

Mechanism: The mechanism of Birch reduction is similar to that of the mechanism of reduction of alkynes to *trans* alkynes in the presence of sodium in liquid ammonia.

***Step* 1: Formation of solvated electrons:** Sodium reacts with liquid ammonia to produce solvated electrons.

$$Na + NH_3 \rightleftharpoons N\dot{H}_3 e + \overset{+}{Na}$$

(Dark blue solution contain solvated electrons)

***Step* 2: Formation of cyclohexadienyl radical anion:**

The radical anion produced by the addition of solvated electrons to the benzene upon further reaction with alcohol abstracts a proton (due to its high basicity) from an alcohol to yield cyclohexadienyl radical.

Benzene —e⁻ (electron)→ Radical anion —H–Ö–R→ Free radical + R–Ö:⁻

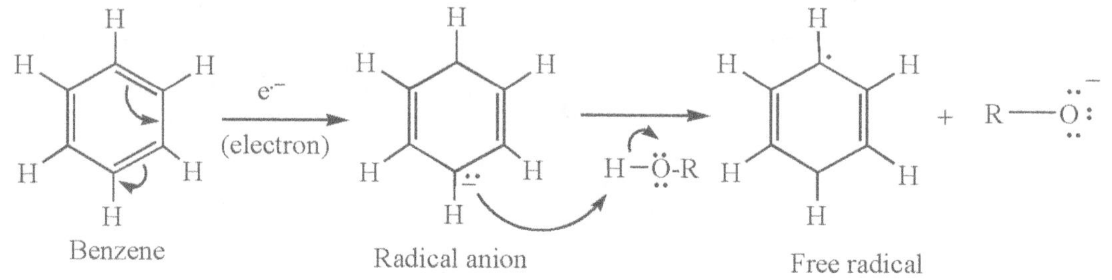

***Step* 3: Formation of 1, 4 - cyclohexadiene *via* the carbanion intermediate:**

The cyclohexadienyl radical fastly abstract another solvated electron to form cyclohexadienyl carbanion. Protonation of this carbanion leads to the formation of 1,4-cyclohexadiene.

Freeradical —Solvated e-→ Cyclohexadienyl carbanion —H–Ö–R→ 1,4-Cyclohexadiene + R-Ö:⁻

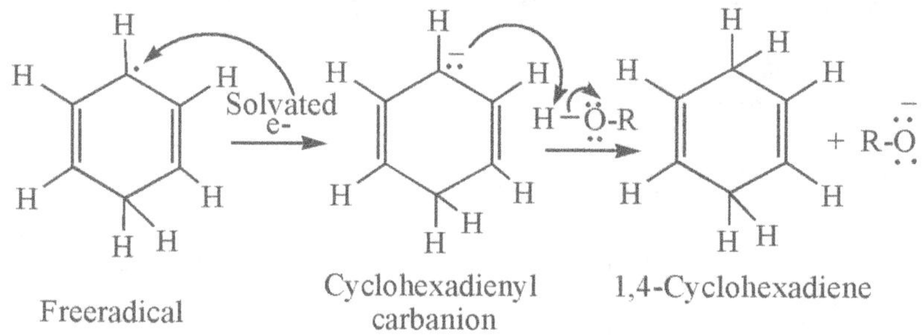

Effect of substituents on Birch reduction: Generally, electron withdrawing substituents stabilize the carbanion while electron donating substituents destabilizes. So, the Birch reduction takes place on carbon atoms bearing electron withdrawing substituents (carbon containing - C=O group) and not carbon atoms bearing electron releasing substituents (such as alkyl and alkoxyl groups). Generally a carbon atom containing electron withdrawing carbonyl group is reduced.

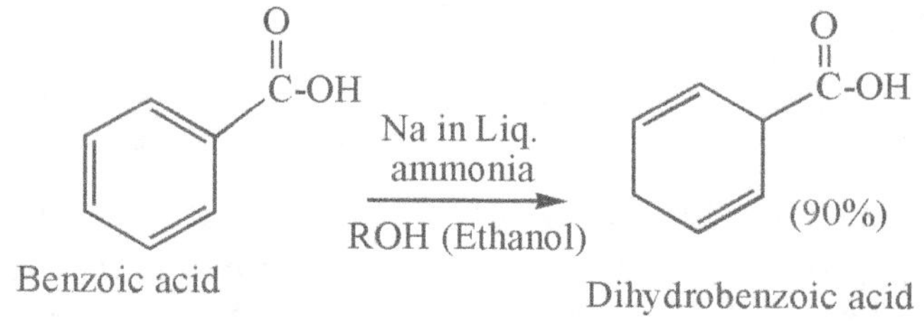

Benzoic acid —Na in Liq. ammonia, ROH (Ethanol)→ Dihydrobenzoic acid (90%)

A carbon containing an electron - releasing group such as alkoxyl group is not reduced.

Electron - releasing substituents deactivates the aromatic ring towards Birch reduction. Here lithium is used instead of sodium together with a co solvent (THF, Tetra hydro furan) and a weaker proton source (*t*-butyl alcohol). The weaker proton source combined with the stronger reducing agent enhances the reduction.

$$C_6H_5OCH_3 \xrightarrow[NH_3/THF]{Li,\ (CH_3)_3COH} \text{1-methoxycyclohexa-1,4-diene}$$

(III) Oxidation reaction:

1. **Ozonolysis:** Benzene reacts with ozone to yield a triozonide which upon hydrolysis with zinc yield glyoxal.

$$\underset{\text{Benzene}}{C_6H_6} + 3O_3 \longrightarrow \text{Triozonide} \xrightarrow[H_2O]{Zn,} 3\ \underset{\text{Glyoxal}}{\begin{matrix} CH{=}O \\ | \\ CH{=}O \end{matrix}}$$

2. **Vapour phase oxidation:** Benzene oxidizes with air or oxygen in the presence of vanadium pentoxide at 450 °C to yield maleic anhydride.

$$2\ \underset{\text{Benzene}}{C_6H_6} + 9O_2 \xrightarrow[450^{\circ}C]{V_2O_5} 2\ \text{Maleic anhydride} + 4H_2O + 4CO_2$$

3. **Mercuration:** When benzene is heated with mercuric acetate, one of the hydrogen atom of benzene ring is replaced by acetoxy mercuric group ($HgCOOCH_3$). (Since the mercury atom directly bonded can be readily replaced by other atom or group, this reaction is now increasingly used as a synthetic tool for preparation of valuable mercurial drugs).

$$\underset{\text{Benzene}}{C_6H_6} + \underset{\text{Mercuric acetate}}{(CH_3COO)_2Hg} \longrightarrow \underset{\text{Acetoxy mercuric benzene (Phenylmercuric acetate)}}{C_6H_5HgOOCCH_3} + CH_3COOH$$

4. **Thallation:** Benzene reacts with thallium trifluoroacetate (TTFA), (TI $(OOCCF_3)_3$) to give organo or benzene thallium reagent. This reaction is known as thallation.

$$\underset{\text{Benzene}}{C_6H_6} + \underset{\text{TTFA}}{TI(OCOCF_3)_3} \longrightarrow \underset{\text{Benzene thallium (or) Organo thallium reagent}}{C_6H_5\text{-}TI(OCOCF_3)_2} + \underset{\text{Trifluoroacetic acid}}{CF_3COOH}$$

TTFA needed for this reaction is prepared by the reaction between thallium oxide and trifluoro acetic acid.

$$Tl_2O_3 + 6CF_3COOH \longrightarrow 2Tl(O\overset{O}{\overset{||}{C}}CF_3)_3 + 3H_2O$$

Thallium oxide — Trifluoro acetic acid — Thallium trifluoro acetate

Synthetic uses: The organo thallium reagent obtained by this method is used for the preparation of substituted aromatic compounds in higher yield.

Benzene thallium ($H_3C\text{-}\overset{O}{\overset{||}{C}}\text{-}O\text{-}Tl\text{-}O\text{-}\overset{O}{\overset{||}{C}}\text{-}CH_3$ attached to benzene ring):

- $\xrightarrow{NH_3}$ Aniline ($C_6H_5NH_2$)
- $\xrightarrow{CuCN}$ Benzonitrile ($C_6H_5C\equiv N$)
- $\xrightarrow{KF}$ Fluorobenzene (C_6H_5F)
- $\xrightarrow{KI}$ Iodobenzene (C_6H_5I)

Note: Thallium was discovered in 1861. Previously, due to the toxicity of thallium ions, thallium reagent was not used routinely. But in recent years many types of thallium reagents are designed and used for organic synthesis.

5. **The Gattermann - Koche Formylation (Synthesis of Benzaldehyde):** By using Friedel-Craft's acylation reaction, a formyl group cannot be added to the benzene ring in usual manner (reagent formyl chloride which is unstable and it cannot be bought or stored).

$$\left[H\text{-}\underset{O}{\underset{||}{C}}\text{-}Cl\right] \qquad H\text{-}\overset{O}{\overset{||}{C}}\text{-}$$

Unstable Formyl chloride — Formyl group

Since, the formyl chloride needed for the preparation of benzaldehyde is prepared by following way. Carbon monoxide is treated with HCl in the presence of a mixture of cuprous chloride (CuCl) and aluminium chloride as catalyst. The reaction mixture generates the formyl cation, possibly through a small concentration of formylchloride, which reacts with benzene to give formyl benzene or benzaldehyde. This reaction is known as Gattermann - Koche formylation which is used for the synthesis of arylaldehydes.

$$CO + HCl \rightleftharpoons \left[H\text{-}\underset{O}{\underset{||}{C}}\text{-}Cl\right] \xrightarrow[CuCl]{AlCl_3} \left[H\text{-}\overset{+}{C}=\ddot{O}:\right]AlCl_4^-$$

Carbon monoxide — Formyl chloride — Formyl cation

$$C_6H_6 + H\text{-}\overset{+}{C}{=}\ddot{O}: \xrightarrow{+AlCl_4^-} C_6H_5\text{-}\overset{O}{\overset{\|}{C}}\text{-}H + HCl + AlCl_3$$

Benzene Formyl cation Benzaldehyde

Summary of Chemical Reactions

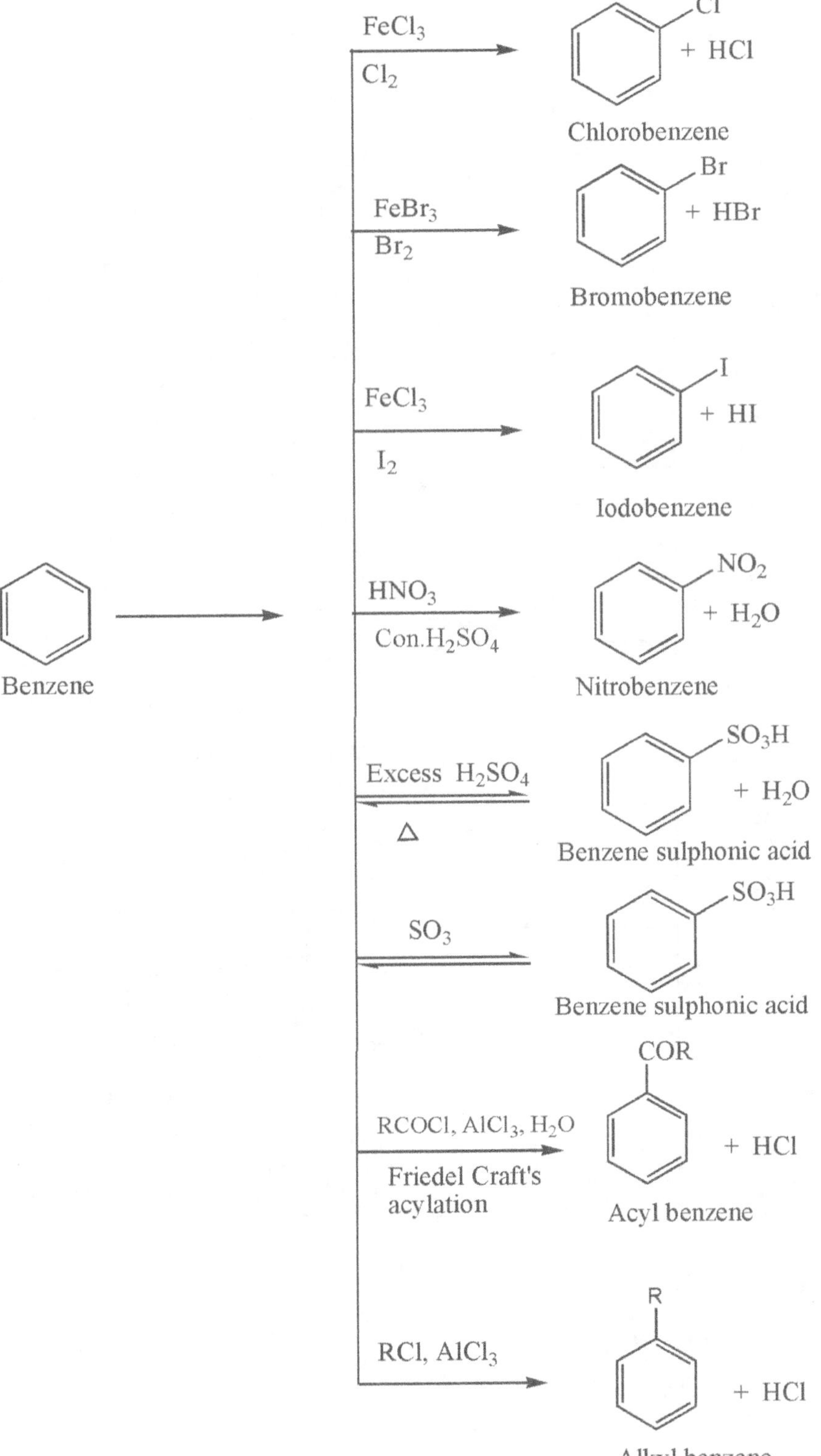

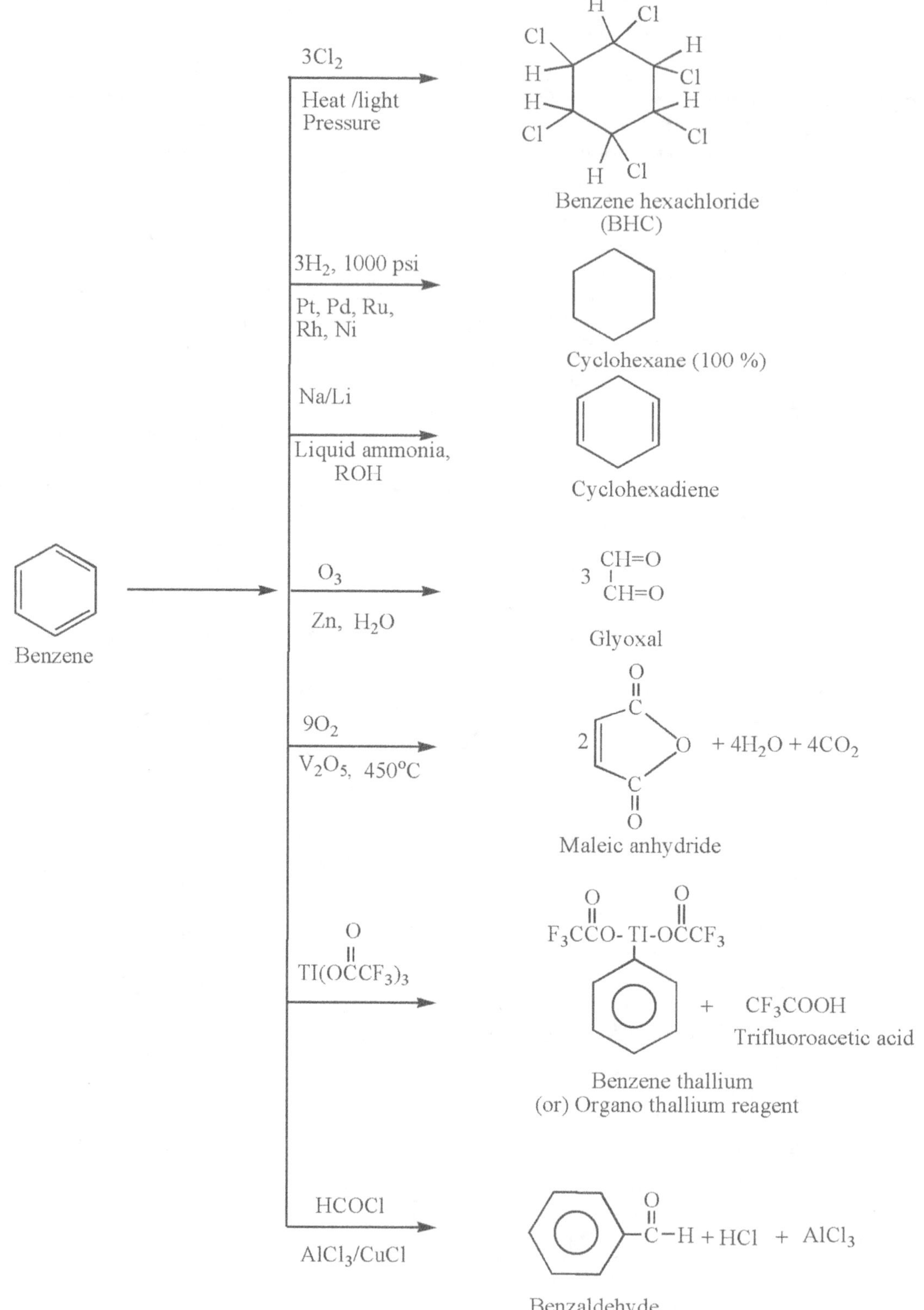

Benzene
3Cl$_2$
Heat /light
Pressure
Benzene hexachloride
(BHC)
3H$_2$, 1000 psi
Pt, Pd, Ru,
Rh, Ni
Cyclohexane (100 %)
Na/Li
Liquid ammonia,
ROH
Cyclohexadiene
O$_3$
Zn, H$_2$O
3 CH=O | CH=O
Glyoxal
9O$_2$
V$_2$O$_5$, 450°C
+ 4H$_2$O + 4CO$_2$
Maleic anhydride
TI(OCCF$_3$)$_3$
F$_3$CCO- TI-OCCF$_3$
+ CF$_3$COOH
Trifluoroacetic acid
Benzene thallium
(or) Organo thallium reagent
HCOCl
AlCl$_3$/CuCl
C−H + HCl + AlCl$_3$
Benzaldehyde

   **INTERESTING FACT**   

Bucky Balls and AIDS: Apart from the discovery of diamond and graphite, a third form of pure carbon was discovered while scientists were conducting experiments designed to understand how long chain molecules are formed in outer space. In 1996, R.E. Smalley, R.F. Curl, Jr. and H. W. Kroto, shared three Nobel Prize in chemistry for their discovery of the third form of pure carbon. They named it Buckminster fullerence (Fullerene) because it reminded them of the geodesic domes popularized by R. Buckminster Fuller, an American Architect and Philosopher. Its nick named Bucky Ball.

Fullerene is the most symmetrical large molecule known, consisting of a hollow cluster of 60 carbons. Each molecule has 32 interlocking rings (20 hexagons and 12 pentagons) which fit together like the seams of soccer Ball.

Due to its benzene like structure, it appears to be aromatic at first glance. However it does not undergoes electrophilic substitution reactions, but undergoes electrophilic addition. This lack of aromaticity apparently caused by the curvature of the ball prevents the molecule from fulfilling the first criterion for aromaticity.

Bucky balls have extraordinary chemical and physical properties. They are exceedingly rugged and are capable of surviving the extreme temperatures of outer space. Because they are essentially hollow cages, they can be manipulated to make materials never before known. For example, hen Buck ball is doped by inserting caesium or potassium in to its cavity, it becomes the best organic super conductor known.

These molecules are presently being studied for use in many other applications such as new polymers and catalysts and new drug delivery system.

The discovery of buckyballs is a strong reminder of the technological advances that can be achieved as a result of conducting basic research.

Scientists have even turned their attention to buckyballs in their quest for a cure for AIDS. An enzyme required by the HIV to reproduce exhibits a non – polar pocket in its three-dimensional structure. If this is blocked, the virus production is blocked. Because buckyballs are non-polar and have approximately same diameter as the pocket of enzyme, they are being considered as possible blockers.

The first step in pursuing this possibility was to equip the Bucky ball with polar side chains to make it water soluble so that it flows through the blood stream. Scientists have now modified the arms so that they can bind to the enzyme. It's still a long way from a cure for AIDS, but this represents one example of the many and varied approaches that scientists are taking to find a cure for this disease.

Richard T. Smalley was born in 1943 in Akron. He was a professor of chemistry at Rice University.

Robert F. Curl Jr. was born in Texas in 1933. He was a professor of chemistry in Rice University.

Sir Harold W. Kroto was born in England in 1939. He was the professor of chemistry at the University of Sussex.

A few words of warning about Benzene (One of the great stars in organic chemistry):

Benzene is widely used in chemical synthesis and is frequently used as a solvent, but it is toxic. It produces major toxic effect on CNS and bone marrow. Chronic exposure to benzene causes aplastic anemia and leukemia. A higher than average incidence of leukemia has been found in industrial workers with long term exposure of little amount (1 ppm) benzene in the atmosphere.

Toluene has replaced the benzene as a solvent because even though toluene is a CNS depressant, it does not cause aplastic anaemia or leukemia.

"Glue sniffers" seek the narcotic CNS effects of solvents like toluene. This is a highly dangerous activity.

Annulenes

Annulenes are conjugated monocyclic hydrocarbons. The size of ring is indicated by a number of carbons in brackets.

Example: Benzene is [6] annulene and cyclooctatetraene is [8] annulene.

Huckel's rule predicts that annulene is an aromatic compound and satisfy (4n+2) electrons and possess planar carbon ring. For example: [6] annulene is aromatic whereas [8] annulene is not aromatic.

Benzene
[6]
Aromatic

Cyclooctatetraene
[8] annulene
Non-aromatic

Structure and uses of medicinal important compounds

DDT **(Dichloro diphenyl trichloroethane)**	**Sachharin/saccharin sodium**
Uses: 1. It is used as an effective insecticide and pesticide especially for mosquitoes. 2. It is used to completely destroy the malarial parasite but it is genotoxic to human beings.	**Uses:** 1. The soluble salt of saccharin is known as saccharin sodium. 2. It is used as sweetening agent in toothpaste and aerated drinks etc. 3. It is also used as a calorie free sugar substitute for diabetic patients.
Chloramine-T **(Sodiumsalt of Tosylchloramide)**	**Benzene hexachloride (BHC)** **(Hexa chlorobenzene)**
Uses: 1. It is used as an algicide, bactericide, virucide & fungicide. 2. It is also effective against myco-bacterium tuber-culosis, foot & mouth diseases and avian influenza. 3. It is also used for drinking water disinfection and used to destroy parasite.	**Use:** 1. It is used as an insecticide.

Probable Questions

1. What are aromatic hydrocarbons? Mention the examples.
2. Write the importance of benzene and its homologues.
3. Describe in detail about the discovery of benzene.
4. Explain about the various criteria of aromaticity.
5. Explain in detail about the Kekule's structure of benzene with evidences.

6. Define resonance energy of benzene.
7. Write a short note on type of bonding in benzene.
8. How will you explain the stability of benzene by considering heat of hydrogenation of benzene as supporting evidence?
9. Define aromaticity. Mention the examples of non benzenoid aromatic compounds. Explain the conditions for the compound to be aromatic.
10. Explain various theoretical criteria of aromaticity.
11. Define Huckel's rule and explain it with examples.
12. Explain in detail about the small scale preparation of benzene and its homologues.
13. How is benzene and its analogues (toluene) prepared form distillation of coal tar?
14. Explain the general methods of preparation of benzene and its homologues.
15. Enumerate the products of destructive distillation of coal tar and their uses.
16. Define the following terms with examples.
    (a) Friedel Craft's alkylation
    (b) Wurtz – Fittig reaction
    (c) Clemmensen reduction
17. Write a note on physical properties of benzene and its analogues.
18. Describe the mechanism of electrophilic aromatic substitution reaction with examples.
19. Explain the following substitution reaction,
    (a) Halogenation
    (b) Nitration
    (c) Sulphonation
    (d) Fridel- Crafts Alkylation
    (e) Acylation
20. How will you differentiate the following isomers?

    CH_3 CH_3 (para-xylene) CH_3 CH_3 (ortho-xylene)

21. Write the structure and uses of a. DDT, b. Saccharin, c. BHC, d. Chloramine.
22. Does cyclo octa tetraene possess aromatic character? Justify your answer with suitable evidence.
23. Start the following synthesis of the compounds by using benzene, toluene or aniline.
    (a) *o*-Acetyl toluene
    (b) *o*-Toluic acid
    (c) *n*-Butylbenzene
    (d) *m*-Nitrobenzyl chloride
24. What is the action of
    (a) Ozone
    (b) Nitric acid and Sulphuric acid on benzene.
25. Write the structure and uses of DDT, saccharin, BHC & Chloramine.

21

Aromatic Amines

Introduction

These are aromatic hydrocarbons in which a hydrogen atom of the benzene ring is replaced by an amino group (-NH_2). In aromatic compounds, the amino group may be attached directly to the benzene nucleus or it may be attached to the side chain. Those compounds in which the amino group is directly attached to a nuclear carbon atom are called as aromatic amino compounds or aromatic amines. For example: aniline.

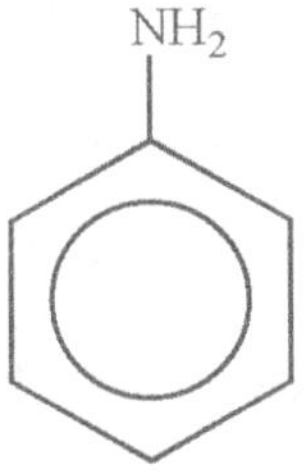

NH_2

Aniline

Like aliphatic amines, aromatic amines are classified into primary, secondary and tertiary amines according to the number of aryl groups replacing the hydrogen atoms of the ammonia.

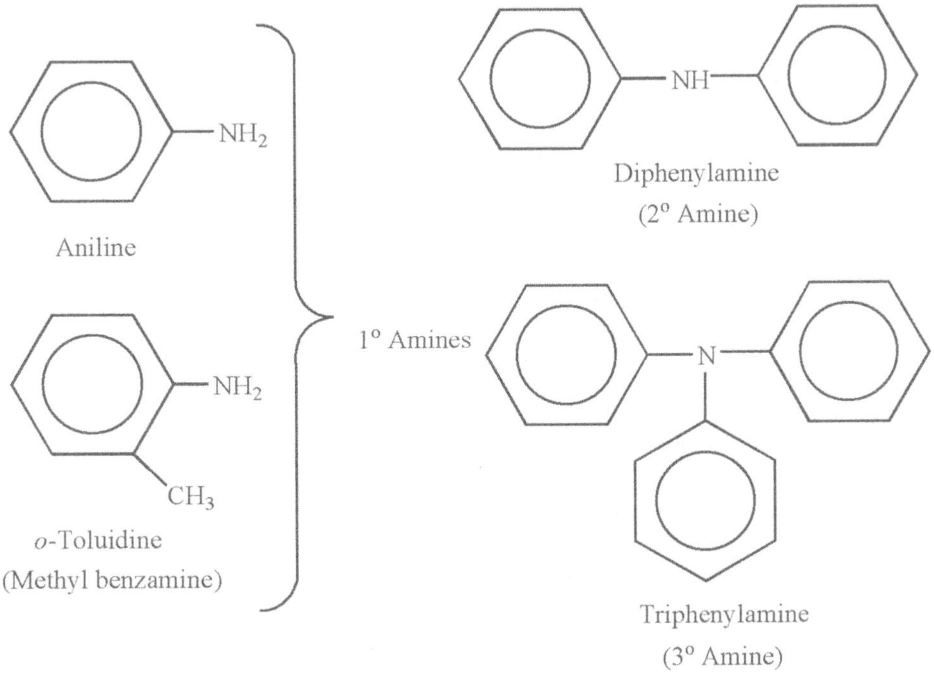

NH_2

Aniline

NH_2

CH_3

o-Toluidine
(Methyl benzamine)

1° Amines

NH

Diphenylamine
(2° Amine)

N

Triphenylamine
(3° Amine)

Simple amines: If the hydrogen atoms of ammonia are replaced by either alkyl or aryl groups then they are called as simple amines.

Mixed amines: Some times, in secondary and tertiary amines, all the hydrogen atoms of ammonia molecule may not be only replaced by either alkyl or aryl group but may be replaced by both alkyl and aryl groups. Such types of amines are known as mixed amines as shown in example:

$C_6H_5 - NH - CH_3$
N-Methyl aniline
(or) Methyl phenyl amine
(2° Amine)

$C_6H_5 - N - (CH_3)_2$
N,N-Dimethyl aniline (or)
Dimethylphenyl amine
(3° Amine)

Amino group attached to the side chain of the aromatic ring is known as **araliphatic amines.** For example:

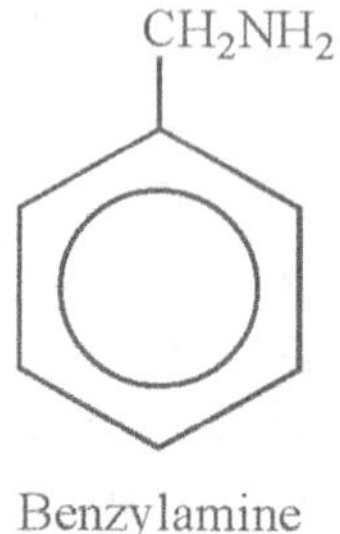

Benzylamine

Importance of Aryl Amines

Aryl amino group when attached with various chemical structures elicit different kind of pharmacological activities based on the core structure to which it is attached. Some of the drugs which possess aryl amino groups and their therapeutic uses in the treatment of various diseases are shown in Table 21.1.

Table 21.1 Drugs containing aryl amino group and their therapeutic uses.

Category	Drug	Structure
Antitubercular drug	*p*-Aminosalicylic acid	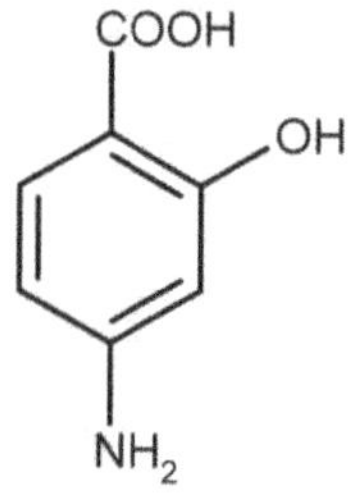 COOH, OH, NH_2
Anticancer drug	Amino glutethemide	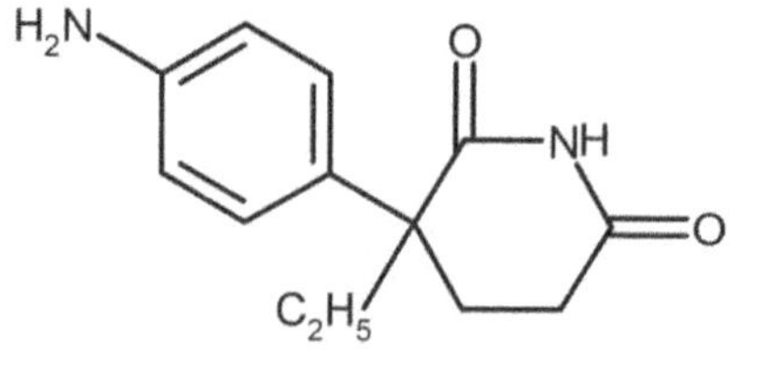 H_2N, O, NH, =O, C_2H_5
Antithyroid drug	Sulphanilamide	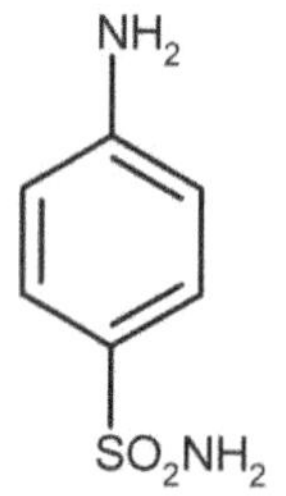 NH_2, SO_2NH_2

Table 21.1 *Contd...*

Category	Drug	Structure
Antithyroid drug	Sulfaguanidine	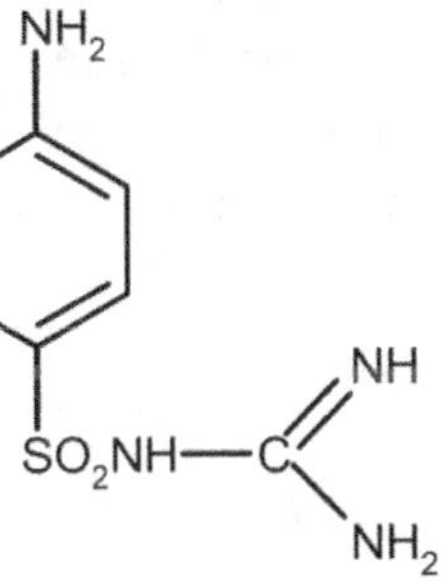
Antihypertensive drug	Triamterene	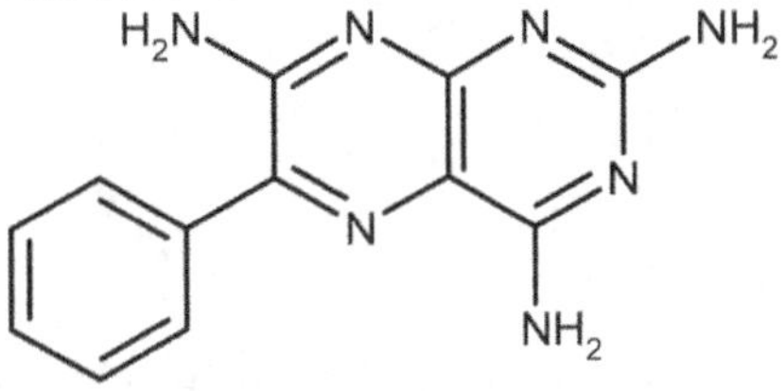
Local anaesthetic	Benzocaine	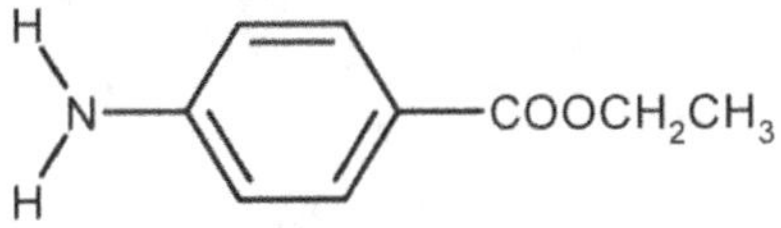
Sedative and hypnotic	Aprobarbital	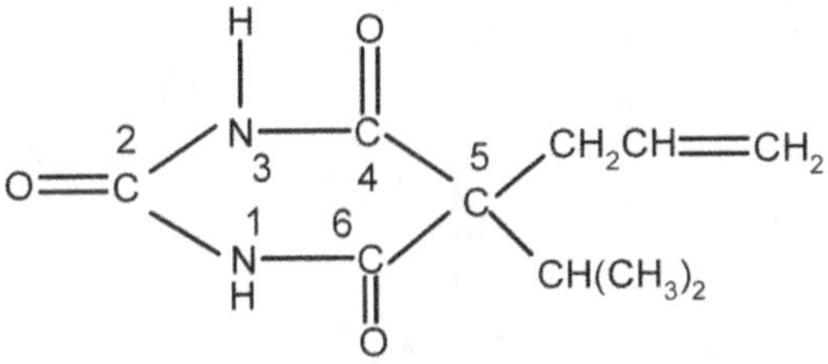
Diuretic	Amiloride	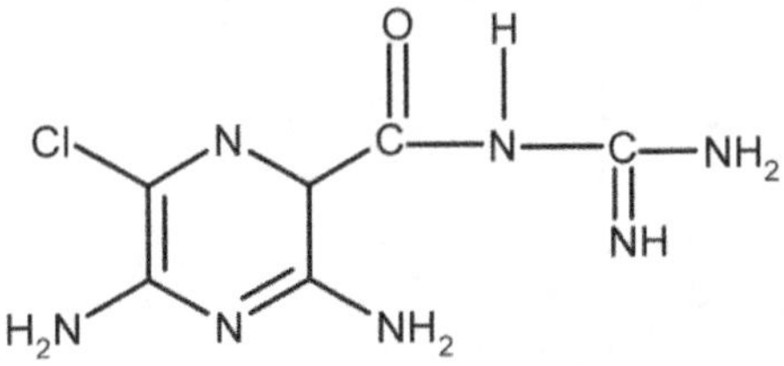
Adrenergic blocker	Prazosin	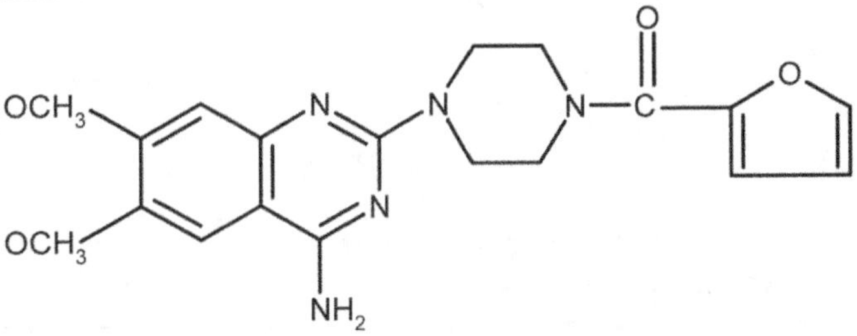
	Terazosin	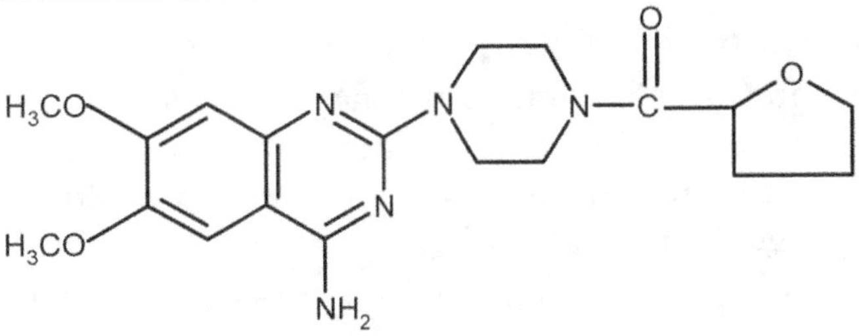

Nomenclature

Simple aromatic amines are named as derivatives of aniline. Substitution in the benzene ring is indicated by numerical or *o, m* or *p* as prefix. Substitution in amino group is indicated by the prefix N for monosubstitution, N,N for disubstitution and so on as shown in the following examples.

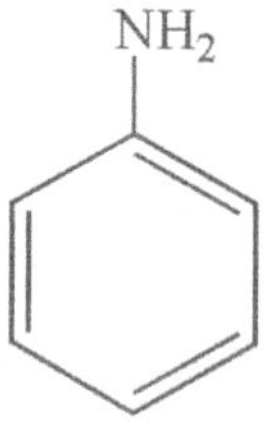

NH_2 NH_2 NO_2 $NHCH_3$

Aniline 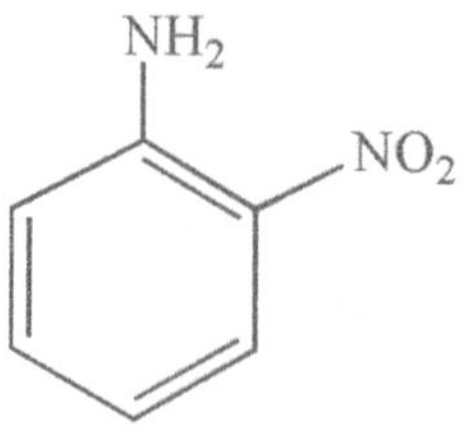 2-Nitroaniline or *o*-Nitroaniline *N*-Methylaniline

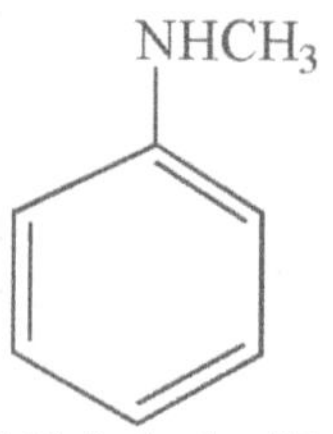

Methyl and **methoxy** derivatives of aniline are given special names as toluidines and anisidines respectively.

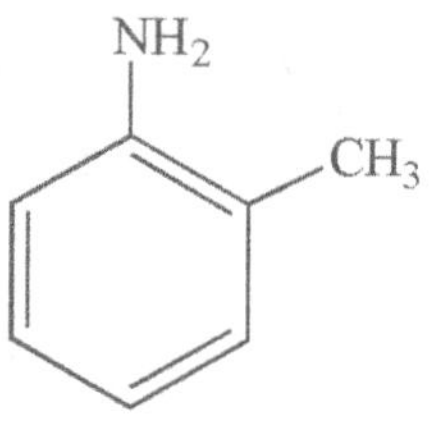

NH_2 CH_3 NH_2 CH_3 NH_2 CH_3

o-Toluidine 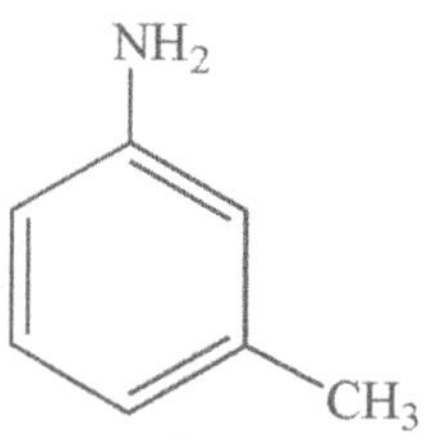 *m*-Toluidine  *p*-Toluidine

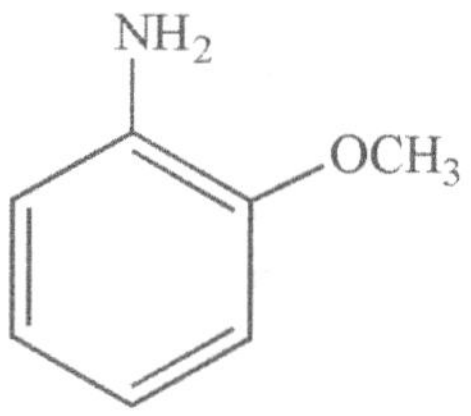

NH_2 OCH_3 NH_2 OCH_3 NH_2 H_3CO

o-Anisidine 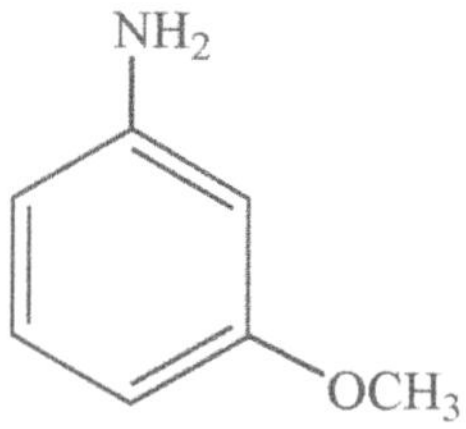 *m*-Anisidine 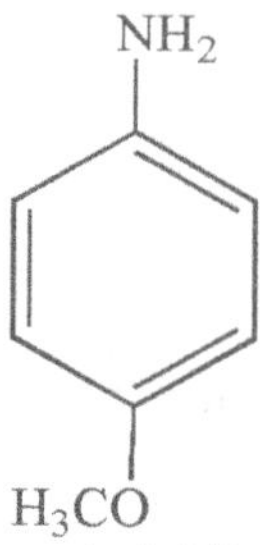 *p*-Anisidine

Primary Amino Compounds

General Methods of Preparation

1. **Preparation of aromatic primary amines.**
   (i) **By reduction of aromatic nitro compounds:** Aromatic nitro compounds are reduced with different kinds of reducing agents to yield aromatic 1° amines. The various types of reducing agents used are as follows.
   (1) Hydrogen (H_2) in the presence of metal catalysts like Ni, Pd, Pt.
   (2) Sn or Fe and HCl.
   (3) Lithium aluminium hydride ($LiAlH_4$).

$$Ar - NO_2 \xrightarrow{3H_2/\ Ni} Ar - NH_2 + 2H_2O$$

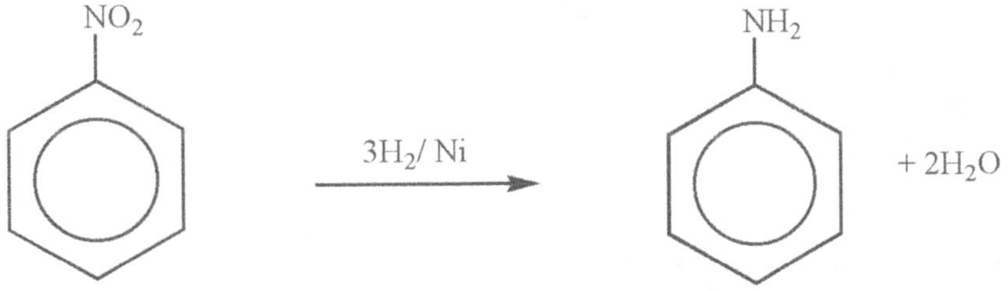

Nitroarene → Arylamine

Nitrobenzene $\xrightarrow{3H_2/\ Ni}$ Aniline $+ 2H_2O$

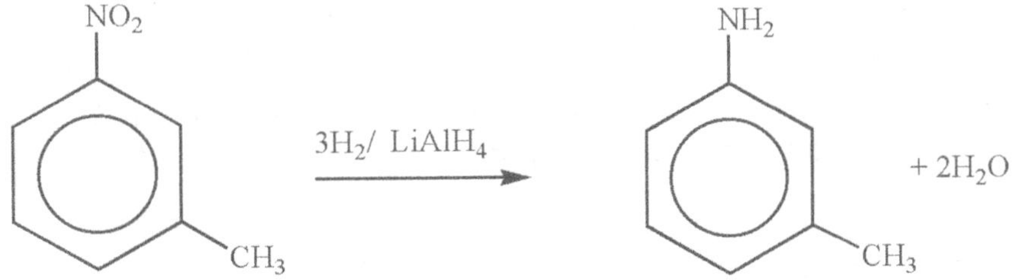

m-Nitrotoluene $\xrightarrow{3H_2/\ LiAlH_4}$ *m*-Toluidine $+ 2H_2O$

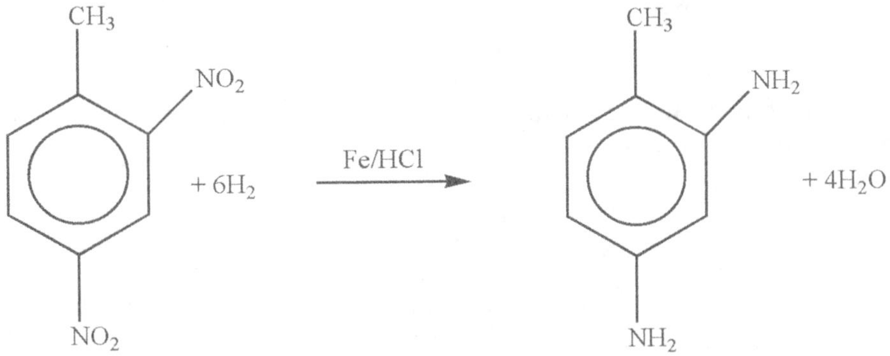

2,4-Dinitrotoluene $+ 6H_2 \xrightarrow{Fe/HCl}$ 2,4-Diaminotoluene $+ 4H_2O$

Sometimes ammonium sulphide is used as a reducing agent to reduce only one nitro group. Ammonium sulphide needed for this reaction is prepared by the reaction between ammonia and hydrogen sulphide.

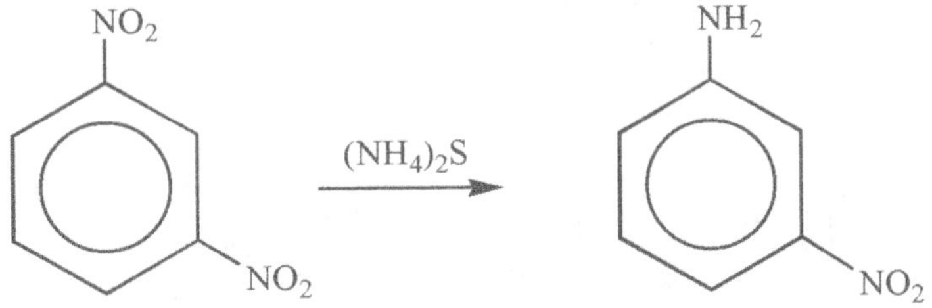

m-Dinitrobenzene $\xrightarrow{(NH_4)_2S}$ *m*-Nitroaniline

(ii) **Hoffmann's hypobromide method or Hofmann's degradation or Hofmann's rearrangement (Conversion of amides to amines):** Amides on treating with bromine or chlorine in alkaline solution undergo degradation or rearrangement to yield aromatic primary amines.

$$\underset{\text{Amide}}{Ar-\overset{\overset{O}{\|}}{C}-NH_2} \xrightarrow[(Br_2 + NaOH)]{NaOBr} \underset{1^\circ \text{ Amine}}{Ar-NH_2} + CO_2 + HBr + NaBr$$

$$\underset{o\text{-Carbamoyl benzoic acid}}{C_6H_4(COOH)(CONH_2)} \xrightarrow[Br_2 + NaOH]{NaOBr} \underset{\text{Aniline-2-Carboxylic acid}}{C_6H_4(COOH)(NH_2)} + CO_2 + HBr + NaBr$$

(iii) **Ammonolysis of phenols:** Phenols are reduced with ammonia in the presence of zinc chloride at 570 K to give aromatic primary amines.

$$\underset{\text{Phenol}}{C_6H_5-OH} + NH_3 \xrightarrow[570\text{ K}]{ZnCl_2} \underset{\text{Aniline}}{C_6H_5-NH_2} + H_2O$$

(iv) **Reduction of nitroso compounds:** Nitroso compounds are reduced with Sn/HCl to yield aromatic primary amines.

$$\underset{\text{Nitrosobenzene}}{C_6H_5-NO} \xrightarrow[Sn/\ HCl]{4(H)} \underset{\text{Aniline}}{C_6H_5-NH_2} + H_2O$$

(v) **Ammonolysis of aryl halogen compounds:** Aryl halides when treated with liquid ammonia under pressure and high temperature in the presence of copper catalyst (Cu_2O) yields corresponding aromatic primary amines.

$$2\underset{\text{Chlorobenzene}}{C_6H_5Cl} + 2NH_3 + Cu_2O \xrightarrow{470K} 2\underset{\text{Aniline}}{C_6H_5NH_2} + 2CuCl + H_2O$$

Mechanism: This is not a simple displacement reaction. It occurs by special type of mechanism known as "elimination and then addition". The intermediate formed during the reaction is called as benzyne.

(a) **Elimination Stage:** It involves two steps.

Step 1: Formation of carbanion: Abstraction of hydrogen atom by $-NH_2$ ion to yield NH_3 and carbanion I.

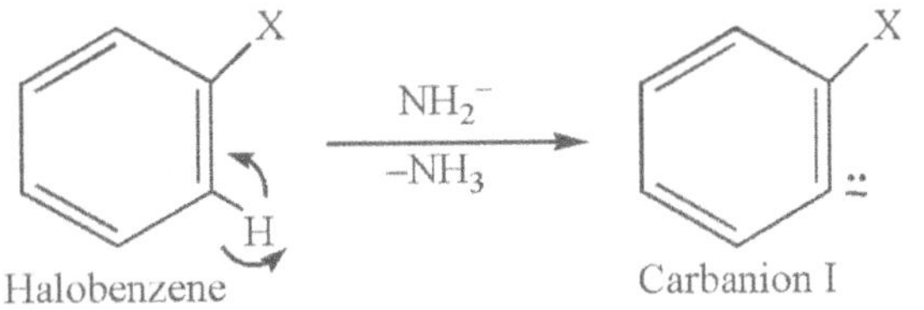

Halobenzene $\xrightarrow[-NH_3]{NH_2^-}$ Carbanion I

Step 2: Formation of benzyne intermediate by the loss of halide ion from the carbanion I.

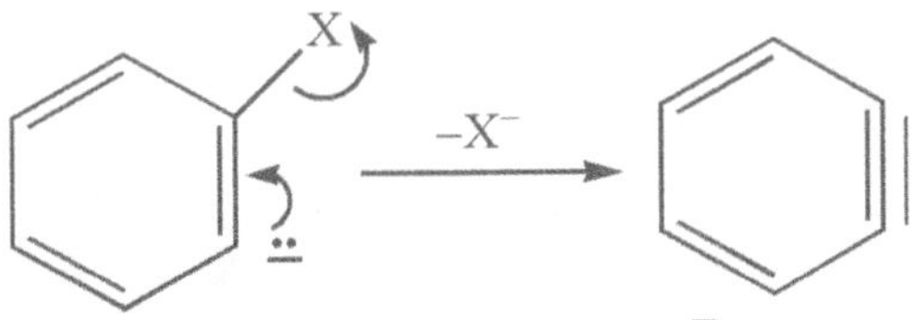

Carbanion I $\xrightarrow{-X^-}$ Benzyne

(b) **Addition stage:** It involves two steps.

Step 3: Formation of carbanion II by the addition of amine ion ($-NH_2$) to the benzyne.

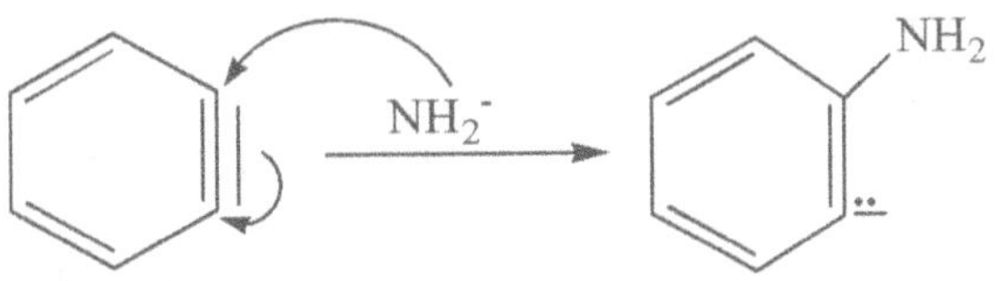

Benzyne $\xrightarrow{NH_2^-}$ Carbanion II

Step 4: Formation of primary amine: Carbanion II reacts with NH_3 to form the product by the abstraction of hydrogen.

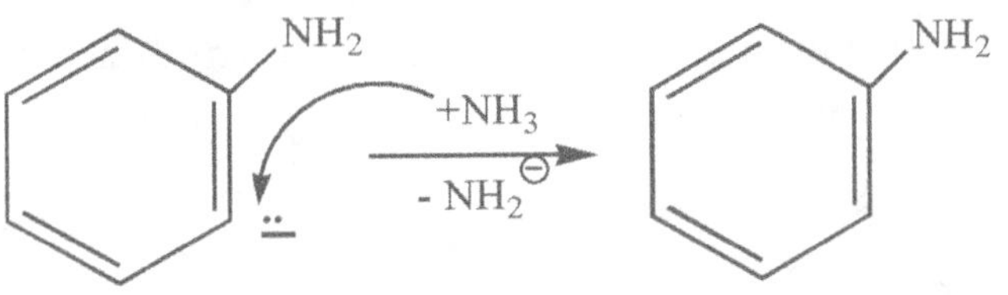

$\xrightarrow[-NH_2^{\ominus}]{+NH_3}$ Aniline

(vi) **Reductive amination of aldehydes and ketones:** Aldehydes and ketones are reduced by catalysis or by chemical reduction in the presence of ammonia to yield primary amines. This process is known as reductive amination *via* the formation of an imine as an intermediate product.

$$C_6H_5\text{-}CH{=}O + NH_3 \underset{}{\overset{-H_2O}{\rightleftharpoons}} C_6H_5\text{-}CH{=}NH \xrightarrow[\text{90 atm, 310-340 K}]{H_2/Ni} C_6H_5\text{-}CH_2NH_2$$

Benzaldehyde → Imine → Benzylamine

Secondary amines are obtained by the reaction between benzaldehyde and ethylamine.

$$C_6H_5\text{-}CH{=}O + H_2NC_2H_5 \overset{-H_2O}{\rightleftharpoons} C_6H_5\text{-}CH{=}NC_2H_5 \xrightarrow{LiBH_3CN/CH_3OH} C_6H_5\text{-}CH_2NHC_2H_5$$

Benzaldehyde + Ethylamine → Imine → Benzylethylamine (2° Amine)

Note: $LiBH_3CN$, acts as a reducing agent similar to that of $LiBH_4$ (lithium boro hydride). It reduces imine group effectively than that of carbonyl group.

(vii) Reduction of Azo compounds: Azo compounds are reduced with H_2 in the presence of metal catalyst to give the aryl amines *via* the formation of hydrazo compound.

$$Ar-N=N-Ar \xrightarrow[\Delta]{H_2/Ni} Ar-NH-NH-Ar \xrightarrow[\Delta]{H_2/Ni} 2Ar-NH_2$$

Diazo compound → Hydrazo compound → Amines

Summary of Methods of Preparation

Nitrobenzene ($C_6H_5-NO_2$) $\xrightarrow[-2H_2O]{3H_2/Ni}$ Aniline ($C_6H_5-NH_2$)

o-Carbamoyl benzoic acid ($C_6H_4(COOH)(CONH_2)$) $\xrightarrow[(Br_2 + NaOH),\ -CO_2,\ -HBr,\ -NaBr]{NaOBr}$ Aniline ($C_6H_5-NH_2$)

Phenol (C_6H_5-OH) $\xrightarrow[570\ K,\ -H_2O]{ZnCl_2, NH_3}$ Aniline ($C_6H_5-NH_2$)

Nitrosobenzene (C_6H_5-NO) $\xrightarrow[Sn/HCl,\ -H_2O]{4(H)}$ Aniline ($C_6H_5-NH_2$)

Chlorobenzene (C_6H_5-Cl) $\xrightarrow[470K]{2NH_3 + Cu_2O}$ $2\ C_6H_5-NH_2 + 2\ CuCl + H_2O$ (Aniline)

* 2 moles of chlorobenzene

Benzaldehyde ($C_6H_5-CH=O$) $\xrightarrow[H_2/Ni,\ 90\ atm,\ 310\text{-}340\ K]{NH_3, -H_2O}$ Benzylamine ($C_6H_5-CH_2NH_2$)

Diazo compound ($Ar-N=N-Ar$) $\xrightarrow[\Delta]{2H_2/Ni}$ $2Ar-NH_2$ (Amines)

Physical Properties

1. Aromatic amines are colorless solids or liquids with characteristic unpleasant odour. They change to brown color when exposed to air due to oxidation.
2. They have high melting point and boiling point than that of respective alkane analogues and high solubility in aqueous media. It is due to the formation of intermolecular hydrogen bonds with other amine molecules.

$$C_6H_5-NH_2 \cdots H-NH-C_6H_5$$

3. These are soluble in benzene and other organic solvents.
4. These are purified by steam distillation method due to their steam volatile property.

Chemical Properties

The chemical properties of aromatic amines are similar to that of aliphatic amines. They show the difference from the aliphatic amines in the following aspects.

1. Oxidation reactions.
2. Reaction with nitrous acid.
3. Rearrangement reactions.
4. Electrophilic aromatic substitution reactions of aromatic ring.

(I) Common chemical properties of aromatic amines and aliphatic amines:

(i) Reaction with acids: Aromatic amines are basic in nature, so they react with HCl or H_2SO_4 to yield salts.

$$\underset{\text{Arylamine}}{Ar-NH_2} + HCl \longrightarrow \underset{\text{Arylammoniumchloride}}{Ar-NH_3^+Cl^-}$$

$$\underset{\text{Aniline}}{C_6H_5-NH_2} + HCl \longrightarrow \underset{\text{Anilinium chloride}}{C_6H_5-NH_3^+Cl^-}$$

(ii) Acylation (Formation of anilides): They react with acid chloride or halides or acid anhydrides, the hydrogen atom of the amino group is replaced by acyl group to yield anilides.

$$\underset{\text{Aromatic amine}}{Ar-NH_2} + \underset{\text{Acetyl chloride}}{CH_3-\overset{O}{\overset{\|}{C}}-Cl} \xrightarrow{-HCl} \underset{\text{Anilides (or) N-Arylacetamide}}{Ar-NH-\overset{O}{\overset{\|}{C}}-CH_3}$$

$$\underset{\text{Aniline}}{C_6H_5-NH_2} + \underset{\text{Acetyl chloride}}{CH_3-\overset{O}{\overset{\|}{C}}-Cl} \xrightarrow{-HCl} \underset{\text{Acetanilide}}{C_6H_5-NH-\overset{O}{\overset{\|}{C}}-CH_3}$$

$$\underset{\text{Aromatic amine}}{Ar-NH_2} + \underset{\text{Acetic anhydride}}{(CH_3-\overset{O}{\overset{\|}{C}})_2O} \longrightarrow \underset{\text{Anilides (or) N-arylacetamide}}{Ar-NH-\overset{O}{\overset{\|}{C}}-CH_3} + \underset{\text{Acetic acid}}{CH_3COOH}$$

$$C_6H_5-NH_2 + (CH_3CO)_2O \longrightarrow C_6H_5-NH-CO-CH_3 + CH_3COOH$$

Aniline + Acetic anhydride → Acetanilide

$$C_6H_5-NH_2 + C_6H_5-CO-Cl \xrightarrow{-HCl} C_6H_5-NH-CO-C_6H_5$$

Aniline + Benzoyl chloride → Benzanilide

(iii) Alkylation: Aromatic amines when treated with alkyl halides, the hydrogen atoms of the amino group are displaced by alkyl groups to give mixed secondary and tertiary amino compounds which are further treated with aqueous NaOH to yield free secondary or teritiary amines.

Reaction for 1° amines

$$\underset{\text{Primary amine}}{Ar-NH_2} + \underset{\text{Alkyl halide}}{RX} \longrightarrow \underset{2^\circ\text{ Amine}}{Ar-NH(R)} + HX \longrightarrow \underset{3^\circ\text{ Amine}}{Ar-NH_2^+(R)X^-} \xrightarrow[-NaX,\ -H_2O]{NaOH} \underset{2^\circ\text{ Amine}}{Ar-NH(R)}$$

$$\underset{\text{Aniline}}{C_6H_5-NH_2} + CH_3-I \longrightarrow \underset{\text{N-Methyl aniline}}{C_6H_5-NH(CH_3)} + HI \longrightarrow \underset{\text{N-Methyl anilinium iodide}}{C_6H_5-NH_2^+(CH_3)I^-} \xrightarrow{NaOH} \underset{\text{N-Methyl aniline}}{C_6H_5-NH(CH_3)} + NaI + H_2O$$

Reaction for 2° amines:

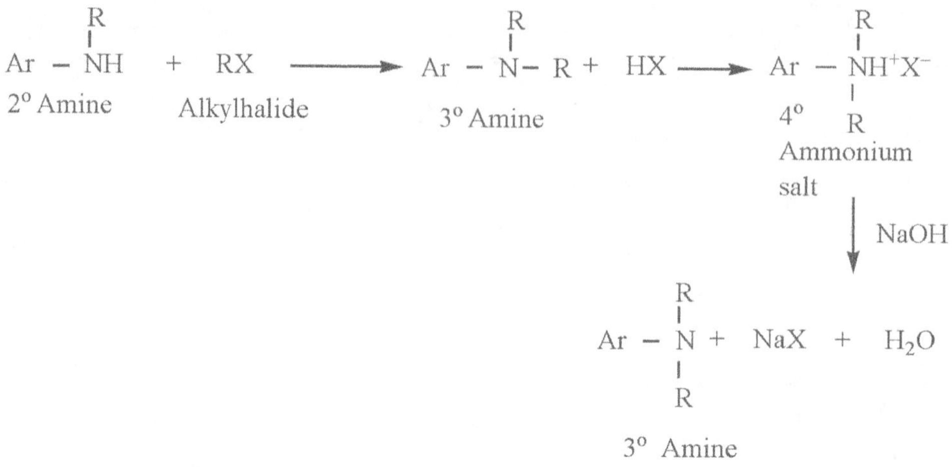

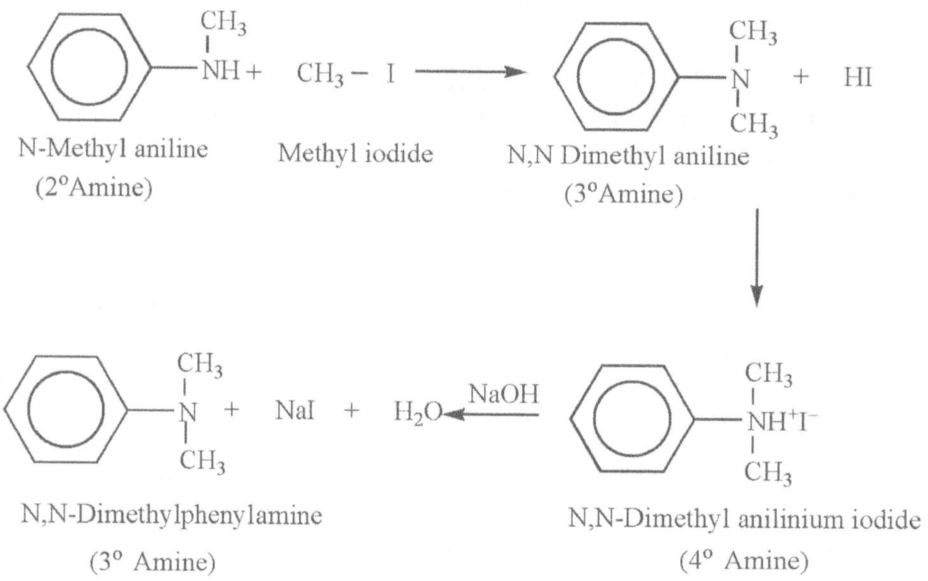

$$Ar-NH(R) + RX \longrightarrow Ar-N(R)-R + HX \longrightarrow Ar-NH^{+}X^{-}(R)(R)$$

2° Amine, Alkylhalide, 3° Amine, 4° Ammonium salt

$$\xrightarrow{NaOH} Ar-N(R)(R) + NaX + H_2O$$

3° Amine

$$C_6H_5-NH(CH_3) + CH_3-I \longrightarrow C_6H_5-N(CH_3)_2 + HI$$

N-Methyl aniline (2°Amine), Methyl iodide, N,N Dimethyl aniline (3°Amine)

$$C_6H_5-NH^{+}I^{-}(CH_3)_2 \xrightarrow{NaOH} C_6H_5-N(CH_3)_2 + NaI + H_2O$$

N,N-Dimethyl anilinium iodide (4° Amine), N,N-Dimethylphenylamine (3° Amine)

(iv) Reaction with Grignard reagent: Aromatic amines react with Grignard reagent to form respective hydrocarbons.

$$Ar-NH_2 + RMg-X \longrightarrow RH + ArNH-MgX$$

Primary amine, Grignard reagent, Alkane

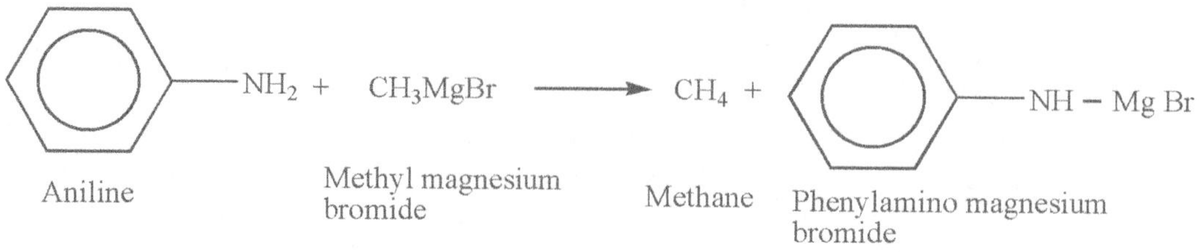

$$C_6H_5-NH_2 + CH_3MgBr \longrightarrow CH_4 + C_6H_5-NH-MgBr$$

Aniline, Methyl magnesium bromide, Methane, Phenylamino magnesium bromide

(v) Reaction with aldehydes or condensation reaction (formation of Schiff's bases): Aromatic amines react or condense with aldehydes to yield imines or Schiff's bases. Reduction of Schiff base gives secondary amines.

$$\underset{\text{Primary amine}}{Ar-NH_2} + \underset{\text{Aldehydes}}{O{=}HC-R} \longrightarrow \underset{\text{Imines}}{Ar-N=CH-R} + H_2O$$

$$\underset{\text{Aniline}}{C_6H_5-NH_2} + \underset{\text{Benzaldehyde}}{O{=}CH-C_6H_5} \longrightarrow \underset{\text{Benzylidine aniline}}{C_6H_5-N=CH-C_6H_5} + H_2O$$

(vi) Carbylamine reaction: Primary aromatic amines when heated with alcoholic KOH solution and chloroform yield carbylamine or isocyanides with an obnoxious smell. It is the identification test for primary amines.

$$\underset{\text{Primary amine}}{Ar-NH_2} + \underset{\text{Chloroform}}{CHCl_3} + 3KOH \xrightarrow{\Delta} \underset{\text{Carbylamine}}{ArNC} + 3KCl + 3H_2O$$

$$\underset{\text{Aniline}}{C_6H_5-NH_2} + \underset{\text{Chloroform}}{CHCl_3} + 3KOH \xrightarrow{\Delta} \underset{\text{Phenyl isocyanide (Carbylamine)}}{C_6H_5-NC} + 3KCl + 3H_2O$$

Mechanism:

$$CHCl_3 + \ddot{O}H \rightleftharpoons :CCl_3^- + H_2O$$

$$:CCl_3^- \longrightarrow \underset{\text{Dichlorocarbene}}{:CCl_2} + Cl^-$$

$$\underset{\text{Aniline}}{C_6H_5-\ddot{N}H_2} + \underset{\text{Dichloro carbene}}{:CCl_2} \longrightarrow C_6H_5-NH_2^+-{}^-CCl_2$$

$$C_6H_5-NH_2^+-{}^-CCl_2 \xrightarrow{-H^+} C_6H_5-NHCCl_2$$

$$C_6H_5-NHCCl_2 \xrightarrow[-H^+]{-Cl^-} C_6H_5-N=CCl^-$$

$$C_6H_5-N=CCl^- \xrightarrow{-Cl^-} C_6H_5-N\equiv C$$

Carbylaminobenzene/Phenylisocyanide (Carbylamine)

(vii) Reaction with carbonyl chloride: Aromatic amines react with carbonyl chloride to give carbonyl amino benzene or aryl isocyanate.

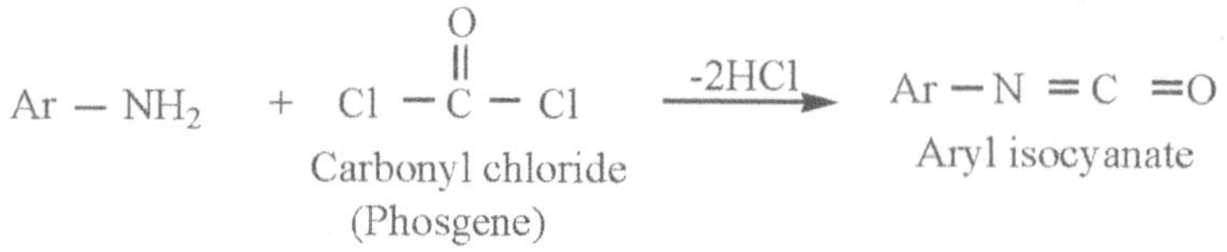

$$Ar-NH_2 + Cl-\overset{\overset{O}{\|}}{C}-Cl \xrightarrow{-2HCl} Ar-N=C=O$$

Carbonyl chloride (Phosgene) → Aryl isocyanate

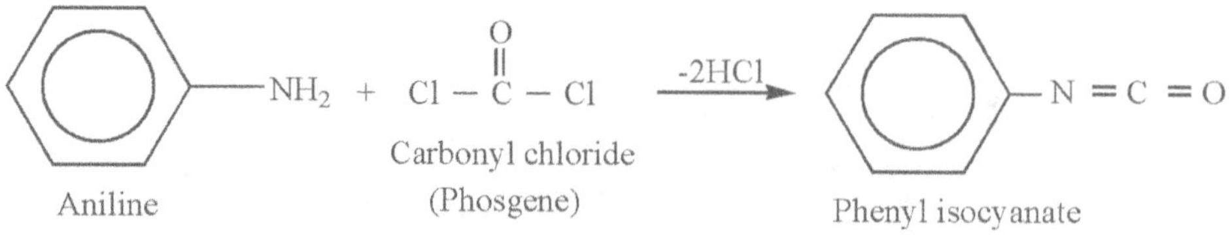

$$C_6H_5-NH_2 + Cl-\overset{\overset{O}{\|}}{C}-Cl \xrightarrow{-2HCl} C_6H_5-N=C=O$$

Aniline + Carbonyl chloride (Phosgene) → Phenyl isocyanate

(viii) Reaction with carbon disulphide: Two molecules of aromatic amines react with alcoholic carbon disulphide solution and solid KOH to give *s*- diphenyl thiourea or thiocarbanilide which is used for vulcanization of rubber.

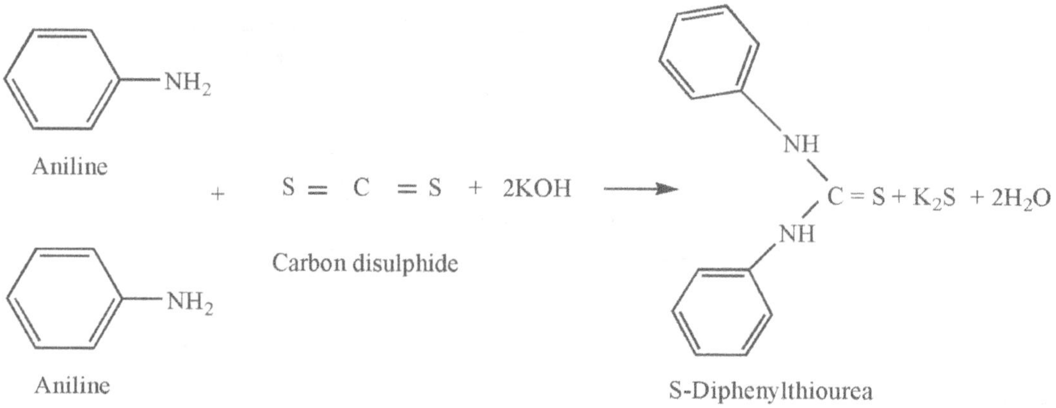

$$2\,C_6H_5-NH_2 + S=C=S + 2KOH \longrightarrow (C_6H_5NH)_2C=S + K_2S + 2H_2O$$

Aniline, Aniline, Carbon disulphide → S-Diphenylthiourea

(ix) Diazotization reaction: The process of conversion of aromatic amines into diazonium salt is known as diazotization reaction. When an ice cold solution of aniline in HCl (0-5 °C) is treated with ice cold solution of sodium nitrite, gives benzene diazonium chloride, which decomposes on heating to yield phenols.

The reaction takes place *via* the formation of nitrosonium ion intermediate. The nitrous acid required for this reaction is produced "*in situ*" by the reaction between sodium nitrite and HCl.

$$NaNO_2 + HCl \longrightarrow HNO_2 + NaCl$$

Sodium nitrite → Nitrous acid

Mechanism:

$$C_6H_5-NH_2 + HNO_2 + HCl \xrightarrow{0-5\ ^\circ C} C_6H_5-N_2^+Cl^- + 2H_2O$$

Aniline + Nitrous acid → Benzene diazonium chloride

$$C_6H_5-N_2^+Cl^- \xrightarrow[-N_2,\ -HCl,\ \Delta]{H_2O} C_6H_5-OH$$

Phenol

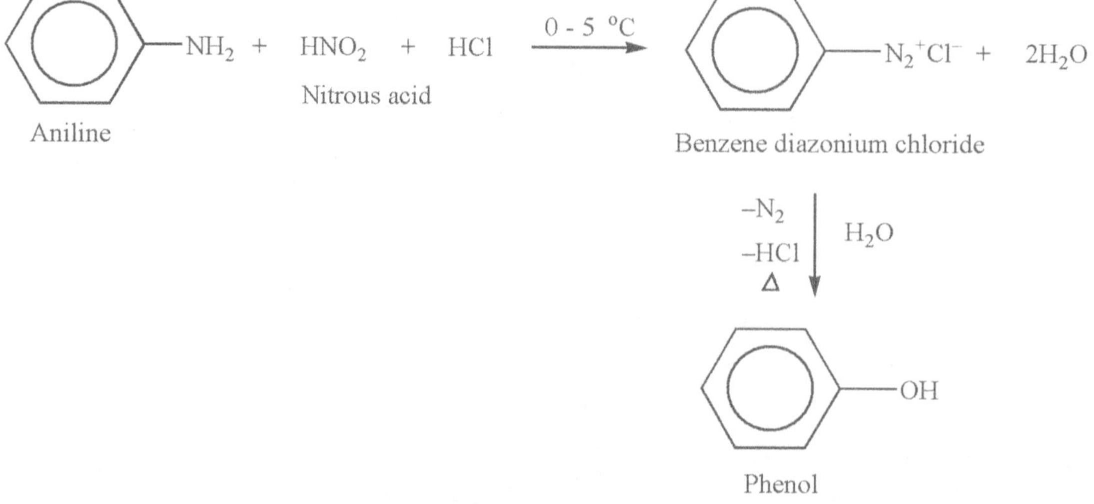

(II) Chemical properties or Reactions of aromatic amines which are different from aliphatic amines are described as follows:

(i) Reaction with nitrous acid:

(a) Primary aromatic amines react with nitrous acid in ice cold condition to yield benzene diazonium salts.

$$\underset{1^\circ \text{ Amine}}{Ar-NH_2} + HCl + HNO_2 \xrightarrow{0-5\,^\circ C} \underset{\text{Ary diazonium salts}}{ArN_2^+Cl^-} + 2H_2O$$

But aliphatic 1° amines do not give the diazonium salts (they produce N_2 gas and bubbles are evolved)

(b) 2° amines react with nitrous acid to yield N-nitrosamine (N-nitrosation)

$$\underset{2^\circ \text{ Amine}}{Ar-N(H)-R} + \underset{\text{Nitrous acid}}{HONO} \xrightarrow{HCl} \underset{\text{N-Nitrosoamine}}{Ar-N(N=O)-R} + H_2O$$

(c) 3° amines react with nitrous acid to yield N,N- dialkyl nitrosoarylamine.

$$\underset{3^\circ \text{ Amines}}{Ar-N(R)-R} + \underset{\text{Nitrous acid}}{HONO} \xrightarrow{HCl} \underset{\text{N,N-Dialkyl nitrosoarylamine}}{Ar-N(N=O)(R)-R} + H_2O$$

(OR)

$$\underset{\text{N,N-Dialkyl aniline}}{C_6H_5NR_2} + \underset{\text{Nitrous acid}}{HONO} \xrightarrow{HCl} \underset{\text{N,N - dialkyl } p\text{-nitrosoaniline}}{p\text{-}O{=}N{-}C_6H_4{-}NR_2} + H_2O$$

This reaction is used to distinguish 1°, 2° and 3° aromatic amines.

(ii) Oxidation: The high electron density in the aromatic ring of aromatic amines makes the compound easily prone to oxidation. But the product formed depends upon the reaction condition. For example, the color of aniline is changed to dark red color when exposed to air due to oxidation. More dark color is obtained when oxidized by stronger oxidizing agents.

Compound	Oxidizing agent	Color formed
Aniline ($C_6H_5-NH_2$)	Bleaching powder	Deep violet
	$Na_2Cr_2O_7$ + Conc. H_2SO_4.	Intense blue
	$Na_2Cr_2O_7$ + $CuSO_4$ + dilute acid.	Aniline black dye

Controlled oxidation of aniline in the presence of sodium dichromate and sulphuric acid gives *p*-benzoquinone.

$$C_6H_5NH_2 \xrightarrow[H_2SO_4]{Na_2Cr_2O_7} O{=}C_6H_4{=}O$$

Aniline p-Benzoquinone

(iii) Hofmann-Martius Rearrangement: N,N-dialkylaniline undergo intermolecular rearrangement in the presence of acid at 400 °C to yield 2,4-dialkyl aniline.

$$C_6H_5N(CH_3)_2\cdot HCl \xrightarrow{400\ ^\circ C} p\text{-}CH_3C_6H_4NH(CH_3)\cdot HCl \longrightarrow 2,4\text{-}(CH_3)_2C_6H_3NH_2\cdot HCl \xrightarrow[-NaCl,\ -H_2O]{NaOH} 2,4\text{-}(CH_3)_2C_6H_3NH_2$$

N,N-dimethyl aniline hydrochloride; N-Methyl p-Toludine hydrochloride; 2,4-Dimethyl aniline hydrochloride; 2,4-Dimethyl aniline

(iv) Reaction with alkali metals or hyphohalous acid: Aromatic amine when treated with hyphohalous acid, the hydrogen atoms of the amino group are replaced by halogen atoms. For example, aniline reacts with hyphohalous acid to yield N,N- dichloroaniline.

$$C_6H_5NH_2 + 2HOCl \longrightarrow C_6H_5NCl_2 + 2H_2O$$

Aniline; Hypochlorous acid; N,N-Dichloroaniline

(v) Reaction with Sodium: Aniline when heated with sodium or potassium, it releases hydrogen gas.

$$2C_6H_5NH_2 + 2Na \xrightarrow{\Delta} 2C_6H_5NH\cdot Na + H_2\uparrow$$

Aniline; Sodium salt of aniline

(vi) Reaction of Benzene Nucleus (Electrophilic aromatic substitution reactions): Aromatic amines easily react with electrophiles to yield substituted product. The amino group present in the ring directs the incoming substituents to the *ortho* and *para* positions.

(a) Halogenation: Aniline reacts with bromine to yield 2,4,6-tribromo aniline (light yellow precipitate).

$$C_6H_5NH_2 + 3Br_2 \longrightarrow 2,4,6\text{-}Br_3C_6H_2NH_2 + 3HBr$$

Aniline; 2,4,6-Tribromoaniline

For chlorination, the reaction should be carried out in the presence of solvent such as water free chloroform, otherwise oxidation takes place.

$$C_6H_5NH_2 \; (\text{Aniline}) + 3Cl_2 \xrightarrow{CHCl_3} C_6H_2Cl_3NH_2 \; (\text{2,4,6-Trichloroaniline}) + 3HCl$$

NH2 | Aniline | Cl | Cl | Cl | 2,4,6-Trichloroaniline

(b) Nitration: Aniline is easily oxidized by nitric acid, so direct nitration of aniline by HNO_3 is not possible. Hence the amino group present in aniline is protected by alkyl group or acyl group before nitration.

$$C_6H_5NH_2 \; (\text{Aniline}) + ClCOCH_3 \; (\text{Acetyl chloride}) \xrightarrow{-HCl} C_6H_5NHCOCH_3 \; (\text{Acetanilide})$$

$$\xrightarrow[\text{Nitration}]{HNO_3 \; H_2SO_4} o\text{-Nitro acetanilide} + p\text{-Nitro acetanilide}$$

$$\xrightarrow[-CH_3COOH]{H_2O \; \text{Hydrolysis}} o\text{-Nitroaniline (Minor product)} + p\text{-Nitroaniline (Major product)}$$

NHCOCH3 | NO2 | o-Nitro acetanilide | NHCOCH3 | NO2 | p-Nitro acetanilide

NH2 | NO2 | o-Nitroaniline (Minor product) | NH2 | NO2 | p-Nitroaniline (Major product)

(c) **Sulphonation:** Aniline when heated with fuming H_2SO_4 at 450-470 K gives *p*-amino benzene sulphonic acid (sulphanilic acid). Sulphonation is a reversible reaction and the *p*-isomer is the most stable one.

$$C_6H_5-NH_2 \xrightarrow[450-470\ K]{\text{Fuming } H_2SO_4} SO_3H-C_6H_4-NH_2$$

Aniline → Sulphanilic acid

(d) **Coupling reaction:** Aniline reacts with diazonium salts and gives azodyes.

$$C_6H_5-NH_2 + C_6H_5-N \equiv N-Cl \xrightarrow{-HCl} C_6H_5-N=N-NH-C_6H_5$$

Aniline + Benzene diazonium chloride → Diazo aminobenzene (Yellow dye)

$$C_6H_5-N=N-NH-C_6H_5 \xrightarrow[\text{(Left in mother liquor)}]{\text{Warm at } 300-310\ K} C_6H_5-N=N-C_6H_4-NH_2$$

p-amino azobenzene
(Brilliant orange red dye)

Note: Basicity of Amines-Refer the Chapter Aliphatic amines.

Summary of Chemical Reactions

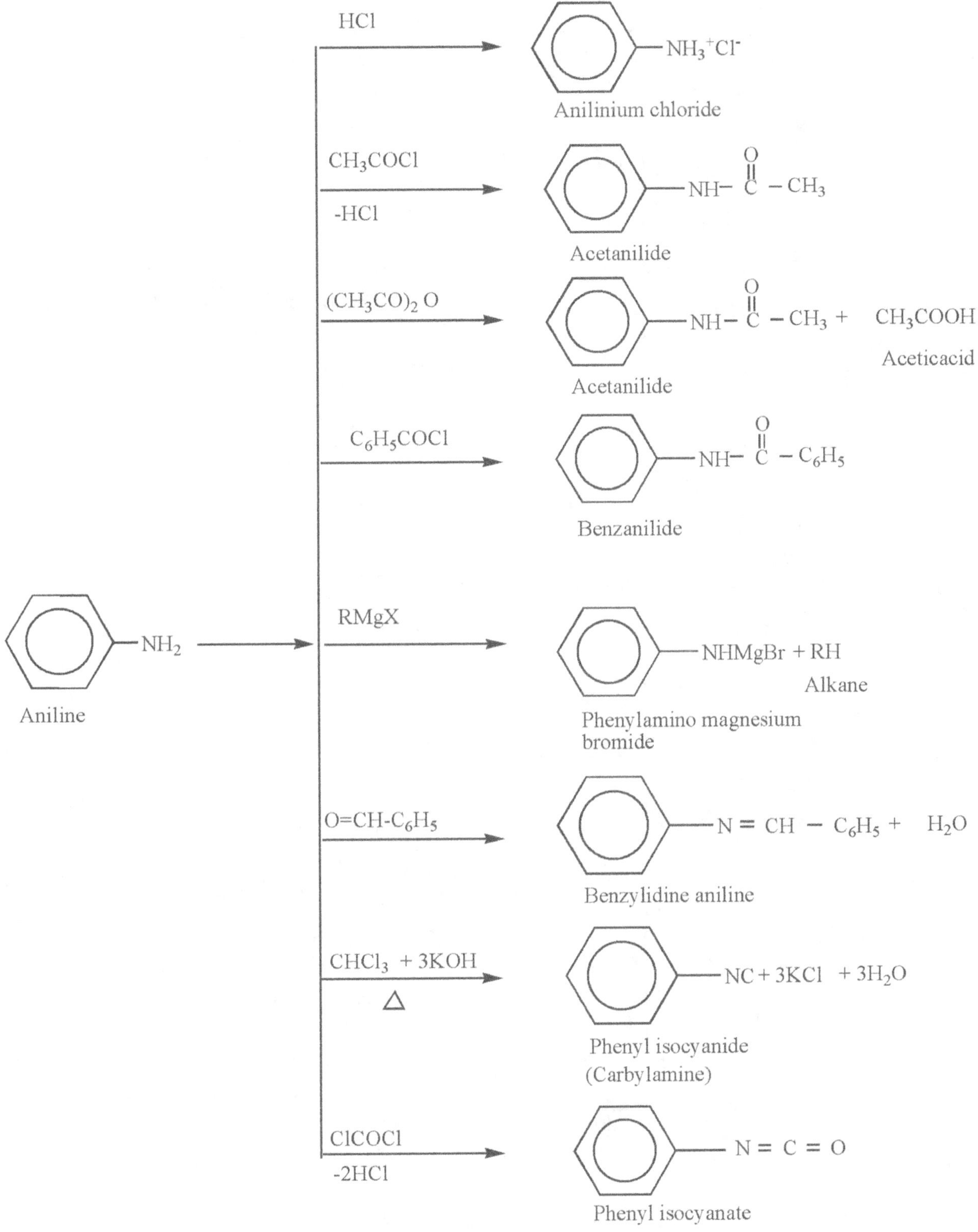

Aniline $\xrightarrow{}$

- CS_2, 2KOH (2 Moles) → S-Diphenylthiourea (PhNH–C(=S)–NHPh) $+ K_2S + 2H_2O$
- $HNO_2 + HCl$, 0 - 5 °C → Benzene diazonium chloride ($C_6H_5N_2^+Cl^-$) $\xrightarrow[-HCl,\ H_2O]{-N_2,\ \Delta}$ Phenol
- $Na_2Cr_2O_7$, H_2SO_4 → *p*-Benzoquinone
- 2HOCl → N,N-Dichloroaniline ($C_6H_5NCl_2$) + $2H_2O$
- Na, Δ → $C_6H_5NH\cdot Na + 1/_2 H_2\uparrow$
- $3Br_2$ → 2,4,6-Tribromoaniline $+ 3HBr$
- $3Cl_2$, $CHCl_3$ → 2,4,6-Trichloroaniline $+ 3HCl$

Tests for distinguishing primary, secondary and tertiary amines

Treatment with	Primary amine			Secondary amine		Tertiary amine	
	Aromatic amine Ar-NH_2	Araliphatic amine $ArCH_2NH_2$	Aliphatic amine RNH_2	$(Ar)_2$- NH_2	R-NH_2	Ar-NH_2	R-NH_2
$CHCl_3$ + Alcoholic KOH	Carbylamine is formed	Carbylamine is formed	Carbylamine is formed	No reaction	No reaction	No reaction	No reaction
HNO_2 ($NaNO_2$+HCl)	Forms diazonium compounds in cold and phenol in hot.	Forms aromatic alcohols and N_2.	Forms aromatic alcohols and N_2.	Gives nitroso amines (Yellow)	Gives nitroso amines (Yellow)	Gives green nitroso compound.	Gives nitroso amine and alcohol on warming.
Alkylation (Heat under pressure)	Reacts with 3 molecules	Reacts with 3 molecules	Reacts with 3 molecules	Reacts with 2 molecule	Reacts with 2 molecule	Combine with one molecule under ordinary pressure	Combine with one molecule under ordinary pressure
Hinsberg's reagent	The product obtained is soluble in water	The product obtained is soluble in water	The product obtained is soluble in water	The product is insoluble in NaOH. Soluble in ether	The product is insoluble in NaOH. Soluble in ether	No reaction	No reaction
Con.HNO_3/ H_2SO_4 or Cl_2	Easily nitrated or chlorinated	Nitrated	Nitrated	Reactions are possible	Nitrated	Reactions are possible	No reaction

Probable Questions

1. Define and classify aromatic amines with examples.
2. Write the importance and nomenclature of aromatic amines.
3. Write the general methods of preparation of primary aromatic amines.
4. Define Hofmann's degradation of amides and explain its mechanism.
5. Explain the reductive amination of aldehydes and ketones.
6. Explain the structure and general physical properties of aromatic amines.
7. Enumerate the common reactions of aromatic amines as well as aliphatic amines.
8. Write the reaction and mechanism of Schiff's reaction.
9. Describe the reaction of carbylamine reaction.
10. Enumerate the reactions of aromatic amines and how they differ from aliphatic amines.
11. Explain Hofmann-Martius rearrangement.
12. Describe the electrophilic aromatic substitution reactions of aromatic amines.
13. Explain about the basicity of amines.
14. Explain about the methods of preparation, chemical properties and uses of aniline.
15. What is meant by Ulmann's reaction?
16. Explain why aniline is less basic than *n*-methyl aniline.

22

Aryl Diazonium Salts

Introduction

These are important class of aromatic compounds which possess the functional group $-\overset{+}{N}\equiv N$ (Diazonium ion), directly attached to a benzene ring. Diazonium compounds are the salts of strong bases, they form salts with anions such Cl^-, Br^-, NO_2 *etc.* which are known as aryl diazonium salts. The general formula of aryl diazonium salts is shown below.

$$Ar - N \equiv N^+ X^-$$

Example $C_6H_5-N\equiv N^+Cl^-$ Benzene diazonium chloride

Preparation: Aryl diazonium salts are mainly prepared by diazotization reaction. The formation of a diazonium salt by the reaction between primary aromatic amine and sodium nitrite solution in the presence of inorganic acid in ice cold condition is known as **diazotization reaction.**

$$\underset{\text{Aniline}}{Ar - NH_2} + \underset{\text{Sodium nitrite}}{NaNO_2} + 2HCl \xrightarrow{0\text{-}5\ ^\circ C} \underset{\text{Aryl diazonium chloride}}{Ar - N \equiv N^+ Cl^-} + 2H_2O + NaCl$$

Mechanism: Described under chapter 21 - aromatic amines.

Note: In some cases, the diazo compounds may also be prepared from aliphatic compounds containing primary amino group ($-NH_2$) which is attached to a carbon atom which is adjacent to electron withdrawing group like -CN, -COR, -COOR.

Example,

$$\underset{\text{Ethyl glycine}}{NH_2CH_2COOC_2H_5} \xrightarrow[0\text{-}5\ ^\circ C]{HONO/HCl} \underset{\text{Ethyl diazoacetate}}{N_2CH - COOC_2H_5} + 2H_2O$$

Physical Properties

1. Aryl diazonium salts are colorless, crystalline solids.
2. Completely soluble in water, but less soluble in alcohol.
3. Changes to brown color on exposure to air.
4. Highly explosive in dry condition and extremely sensitive to shock. But some of the diazonium salts such as aromatic diazonium fluoroborates (ArN_2BF_4) are insoluble in water, stable in dry air.

Chemical Properties: General chemical reactions of aryl diazonium salts are divided into two categories.

(I) Reactions involving replacement of the diazo group $\left(\overset{+}{N_2}\overset{-}{X}\right)$ by another univalent group with the liberation of nitrogen gas.

(II) Reactions in which the two nitrogen atoms are retained.

(I) Replacement reactions: These reactions are the examples of unimolecular nucleophilic substitution reaction (S_N1) of the diazo group. For example,

Benzene diazonium ion —Slow→ Phenyl cation + $N \equiv N$ (Nitrogen gas)

Phenyl cation —Fast→
- H_2O → $C_6H_5\overset{+}{O}H_2$ —$-H^+$→ Phenol (C_6H_5OH)
- Cl^- → Chlorobenzene (C_6H_5Cl)
- CH_3OH → $C_6H_5\overset{+}{O}(H)CH_3$ —$-H^+$→ Anisole ($C_6H_5OCH_3$)

1. **Synthesis of benzene (-N_2X group is replaced by hydrogen):** Aryl diazonium chloride when reduced with sodium stannite (alkaline stannous chloride) or hypo phosphorous acid gives benzene.

$$C_6H_5N_2^+Cl^- + H_3PO_2 + H_2O \longrightarrow C_6H_6 + N_2 + H_3PO_3 + HCl$$

Benzene diazonium chloride; Phosphorous acid; Benzene; Phosphoric acid

The simplest way of carrying out the reaction is to use hypophosphorous acid along with the diazotising agent as the co-agent. The diazonium salt formed is reduced immediately. This reaction is takes place by free radical mechanism.

Applications: This reaction is used for the removal of -NH_2 or NO_2 group from benzene ring.
For example: Synthesis of 2,4,6-tribromo benzoic acid.

3-Aminobenzoic acid $\xrightarrow[H_2O,\ -3HBr]{3Br_2}$ 3-Amino-2,4,6-tribromo benzoic acid $\xrightarrow[\text{Diazotisation},\ -\text{Nacl},\ -H_2O]{NaNO_2/2HCl,\ 0-5\ ^\circ C}$ (2,4,6-tribromo-3-carboxy benzene diazonium chloride, $N_2^+Cl^-$) $\xrightarrow[-N_2,\ -H_3PO_3,\ -HCl]{H_3PO_2,\ H_2O}$ 2,4,6-Tribromo benzoic acid

2. **Synthesis of phenols (–N_2X group is replaced by -OH group):** Aryl diazonium chloride solution when steam distilled or boiled yields phenol.

$$\underset{\text{Benzene diazonium chloride}}{C_6H_5\overset{+}{N_2}Cl^-} + HOH \xrightarrow{\Delta} \underset{\text{Phenol}}{C_6H_5OH} + N_2\uparrow + HCl$$

Applications: This reaction is used for the synthesis of substituted phenols which are not prepared by electrophilic aromatic substitution reaction of phenol. For example, Synthesis of *m*-nitro phenol.

m-Nitroaniline $\xrightarrow[-NaCl,\ -2H_2O]{NaNO_2/2HCl,\ 0\text{-}5\ ^\circ C}$ 2-Nitro benzene diazonium chloride ($N_2^+Cl^-$, NO_2) $\xrightarrow[\Delta]{H_2O}$ *m*-Nitrophenol + HCl + N_2

3. **Replacement by halogens:**

(a) **Sand Mayer's reaction (–N_2X group is replaced by a halogen atom):** Aryl diazonium chloride solution reacts with cuprous halide (Cuprous chloride) solution which is dissolved in a corresponding halogen acid to give corresponding phenyl halide.

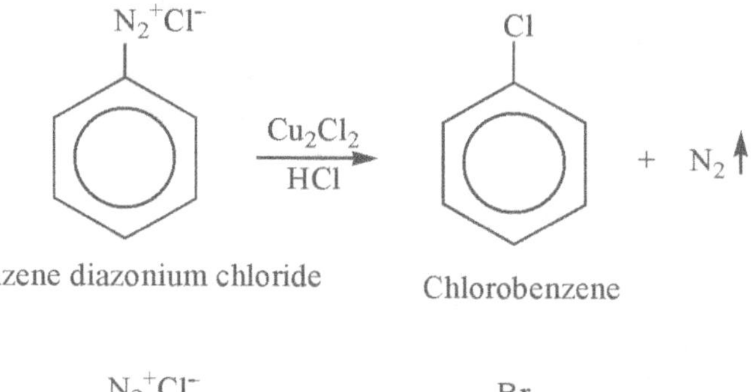

Benzene diazonium chloride → Chlorobenzene

$N_2^+Cl^-$ (benzene ring) $\xrightarrow[HBr]{Cu_2Br_2}$ Br (benzene ring) + N_2↑

Benzene diazonium chloride → Bromobenzene

Mechanism: This reaction takes place *via* free radical mechanism.

Step 1: Cuprous chloride or bromide gives a complex ion (cupric chloride).

$$Cu_2Cl_2 + 2Cl^- \longrightarrow 2CuCl_2$$

Cuprous chloride → Cupric chloride

Step 2: Diazonium salt reacts with the above complex ion to yield a phenyl free radical that is formed by an electron transfer coupled with loss of nitrogen.

N_2^+ (benzene ring) + $Cu-Cl_2^-$ $\xrightarrow{Slow}$ (benzene ring)$^\cdot$ + N_2 + $Cu-Cl_2^{(+\cdot)}$

Benzene diazonium ion → Phenyl free radical

Step 3: Phenyl free radical abstracts a chlorine atom from cupric chloride to yield chloro benzene.

2 (benzene ring)$^\cdot$ + $2CuCl_2$ ⟶ 2 Cl (benzene ring) + Cu_2Cl_2

Phenyl free radical → Chlorobenzene

(b) Gattermann reaction: The cuprous salt used in the Sandmeyer reaction is replaced by finely divided copper is known as Gattermann reaction (Cuprous is replaced with finely divided copper).

$N_2^+Cl^-$ (Benzene diazoniumchloride) $\xrightarrow[\text{Copper}]{\text{Cu/HCl Finely divided}}$ Cl (Chlorobenzene) + $N_2\uparrow$

Benzene diazoniumchloride — Chlorobenzene

$N_2^+Cl^-$ (Benzene diazoniumchloride) $\xrightarrow[\text{Copper}]{\text{Cu/HBr Finely divided}}$ Br (Bromobenzene) + $N_2\uparrow$

Benzene diazoniumchloride — Bromobenzene

(c) Synthesis of aryl iodo compounds (–N_2X group is replaced by iodine atom): Aryl diazonium chloride when boiled with aqueous potassium iodide solution, yields iodo benzene.

$N_2^+Cl^-$ + KI ⟶ I + $N_2\uparrow$ + KCl

Benzene diazoniumchloride — Iodobenzene

Applications: This method is used for introducing an iodine atom in to the benzene ring.

(d) Synthesis of aryl fluoro compounds (Schliemann reaction): Aryl diazonium chloride when treated with fluoroboric acid (HBF_4) gives arene diazonium borides which is blistered, washed and dried. The product obtained is heated and the dry diazonium borate decomposes to yield the corresponding aryl fluoride. This reaction is called as Schliemann reaction.

$N_2^+Cl^-$ + HBF_4 ⟶ F + BF_3 + $N_2\uparrow$ + HCl

Benzene diazoniumchloride — Fluorobenzene

4. **Synthesis of cyano compounds ($-N_2X$ group is replaced by -CN):** The solution of aryl diazonium chloride when treated with cuprous cyanide dissolved in aqueous potassium cyanide solution with copper powder gives benzonitrile.

$$C_6H_5N_2^+Cl^- + KCN \xrightarrow{K_3Cu(CN)_4} C_6H_5CN + N_2\uparrow + KCl$$

Benzene diazoniumchloride → Benzonitrile

$$C_6H_5N_2^+Cl^- + KCN \xrightarrow{\text{Cu Powder}} C_6H_5CN + N_2\uparrow + KCl$$

Benzene diazoniumchloride → Benzonitrile

Applications: The benzonitrile obtained from the above reaction is easily converted in to benzoic acid and benzylamine by hydrolysis and reduction respectively.

$$C_6H_5CN \xrightarrow{HOH} C_6H_5CONH_2 \xrightarrow[-NH_3]{HOH} C_6H_5COOH$$

Benzonitrile → Benzamide → Benzoic acid

$$C_6H_5CN \xrightarrow{LiAlH_4\ 4(H)} C_6H_5CH_2NH_2$$

Benzonitrile → Benzylamine

5. **Replacement by aryl group (Gomberg reaction):** Aryl diazonium chloride when treated with ethanol and copper powder gives diphenyl.

$$2\ C_6H_5N_2^+Cl^- \xrightarrow[C_2H_5OH]{Cu} C_6H_5-C_6H_5 + 2N_2\uparrow + CuCl_2$$

Benzene diazoniumchloride → Diphenyl or Biphenyl

It is also produced by the reaction between the alkaline solution of aryl diazonium chloride and benzene.

$$C_6H_5N_2^+Cl^- + C_6H_6 \xrightarrow{NaOH} C_6H_5{-}C_6H_5 + N_2\uparrow + NaCl + H_2O$$

Benzene diazoniumchloride; C_6H_6 Benzene; Diphenyl or Biphenyl

Mechanism (Free radical mechanism): This reaction takes place *via* the formation of free radicals.

$$Ar-\overset{\oplus}{N}\equiv N-\bar{Cl} \xrightarrow{\bar{OH}} Ar-N=N-OH + Cl^-$$

$$2Ar-N=N-OH \xrightarrow[-H_2]{} Ar-N=N-O-O-N=N-Ar$$

$$\xrightarrow{\text{Homolysis}} Ar-N=N-\dot{O}$$

$$C_6H_6 + \dot{Ar} \longrightarrow [\text{Ar, H-substituted cyclohexadienyl radical}] \xrightarrow{Ar-N=N-\dot{O}} C_6H_5{-}Ar + Ar-N=N-OH$$

6. **Synthesis of nitro compounds (-N_2X group is replaced by -NO_2):** Aryl diazonium chloride when heated with $NaNO_2$ in the presence of copper powder yields nitrobenzene.

$$C_6H_5N_2^+Cl^- + NaNO_2 \xrightarrow[\Delta]{Cu} C_6H_5NO_2 + N_2\uparrow + NaCl$$

Benzene diazoniumchloride; Nitrobenzene

Applications: This reaction is used for the synthesis of *p*-dinitro benzene from aniline.

NH_2 Aniline —[$(CH_3CO)_2O$, $-CH_3COOH$]→ $NHCOCH_3$ Acetanilide —[Nitration, HNO_3/H_2SO_4, H_2O/H^+, 0-10 °C]→ NH_2, NO_2 *p*-Nitroaniline —[$NaNO_2/HCl$, 0 - 5 °C, Diazotisation]→ $N_2^+Cl^-$, NO_2 *p*-Nitroaryl diazonium chloride —[$NaNO_2/Cu$, $-N_2$, $-NaCl$]→ NO_2, NO_2 *p*-Dinitrobenzene

7. **Synthesis of thiophenol (-N_2X group is replaced by -SH):** Aryl diazonium chloride reacts with potassium hydrogen sulphide to give thiophenol.

$N_2^+Cl^-$ Benzene diazonium chloride + KSH Potassium hydrogen sulphide → SH Thiophenol + $N_2\uparrow$ + KCl

8. **Meerwin Reaction:** Aryl diazonium salts react with compounds containing activated ethylene bond (═) in acetone solution in the presence of cupric salt and yields aryl halides.

$N_2^+Cl^-$ Benzene diazoniumchloride + $CH_2 = CHCN$ Acrylo nitrile —[Cu_2^+]→ $CH_2CH(Cl)CN$ 2-Chloro-3-phenyl propio nitrile + N_2

$N_2^+Cl^-$ Benzene diazoniumchloride + $C_6H_5-CH=CH-COOH$ Cinnamic acid → $CH=CH-C_6H_5$ Stillbene + $CO_2\uparrow$ + N_2 + HCl

(II) Reactions in which the diazo group is retained:

1. **Reduction:** Aryl diazonium chloride is reduced with stannous chloride & HCl or with sodium sulphide to yield phenyl hydrazine.

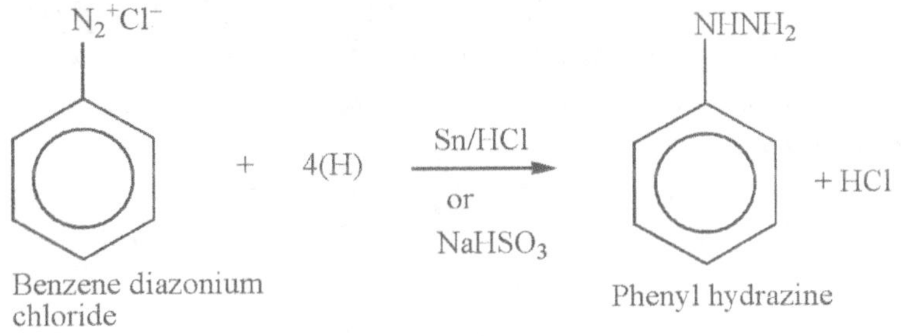

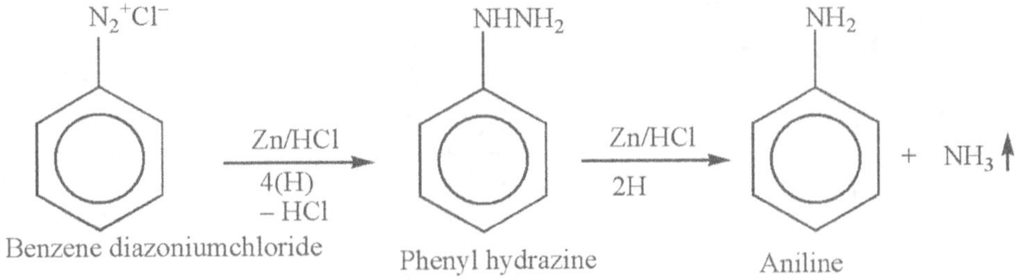

$$C_6H_5N_2^+Cl^- + 4(H) \xrightarrow[\text{or } NaHSO_3]{Sn/HCl} C_6H_5NHNH_2 + HCl$$

Benzene diazonium chloride — Phenyl hydrazine

$$C_6H_5N_2^+Cl^- \xrightarrow[4(H),\ -HCl]{Zn/HCl} C_6H_5NHNH_2 \xrightarrow[2H]{Zn/HCl} C_6H_5NH_2 + NH_3\uparrow$$

Benzene diazoniumchloride — Phenyl hydrazine — Aniline

2. **Diazo coupling reaction:** Aryl diazonium salts when treated with aromatic compound containing strong electron donor group (phenol, aniline *etc.*) the two aromatic rings are joined together to form an azo compound that contains a double bond between two nitrogen atoms.

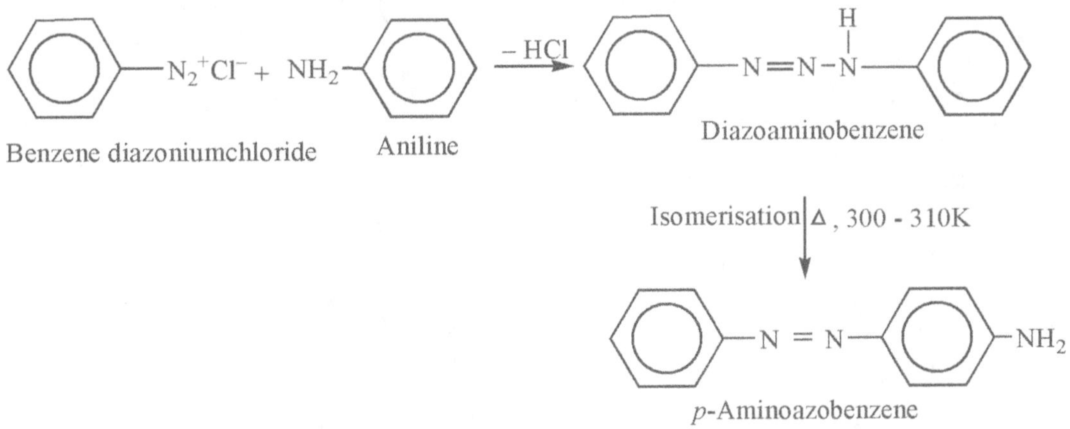

$$C_6H_5N_2^+Cl^- + NH_2C_6H_5 \xrightarrow{-HCl} C_6H_5-N{=}N-NH-C_6H_5$$

Benzene diazoniumchloride — Aniline — Diazoaminobenzene

$$\downarrow \text{Isomerisation}\ \Delta,\ 300 - 310K$$

$$C_6H_5-N = N-C_6H_4-NH_2$$

p-Aminoazobenzene

Mechanism:

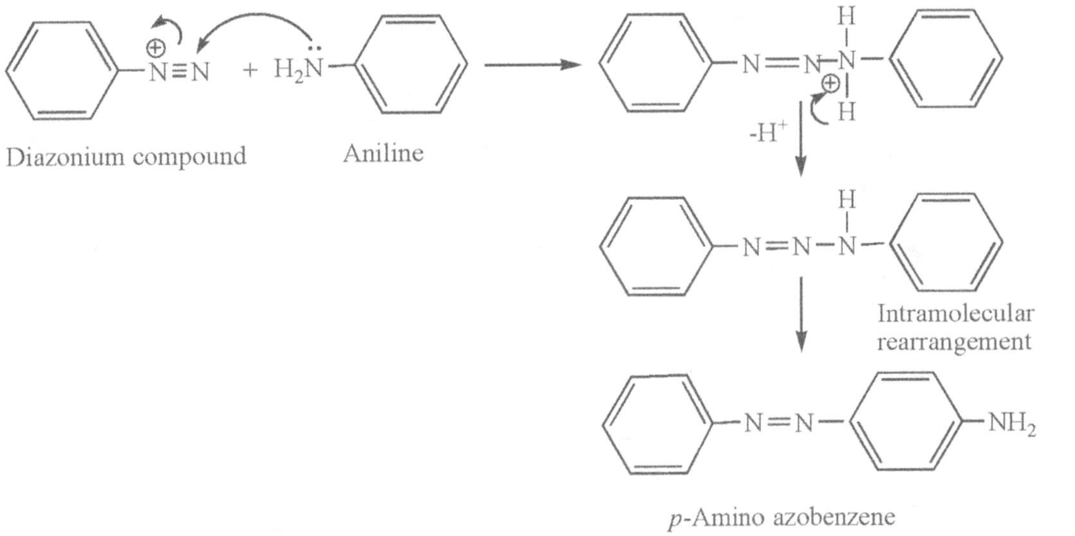

$$C_6H_5-\overset{\oplus}{N}{\equiv}N + H_2\ddot{N}-C_6H_5 \longrightarrow C_6H_5-N{=}N-\overset{\oplus}{N}H_2-C_6H_5$$

Diazonium compound — Aniline

$$\downarrow -H^+$$

$$C_6H_5-N{=}N-NH-C_6H_5$$

$$\downarrow \text{Intramolecular rearrangement}$$

$$C_6H_5-N{=}N-C_6H_4-NH_2$$

p-Amino azobenzene

Note:

1. Coupling with phenols, best occurs in a faintly alkaline solution and coupling with amines occurs in faintly acidic solution.
2. The azo group is mainly attached or enters into the *p*-position to -OH group or $-NH_2$ group, but in case this position is not vacant, coupling occurs in *o*-position but never in the *m*-position.

Summary of Chemical Reactions

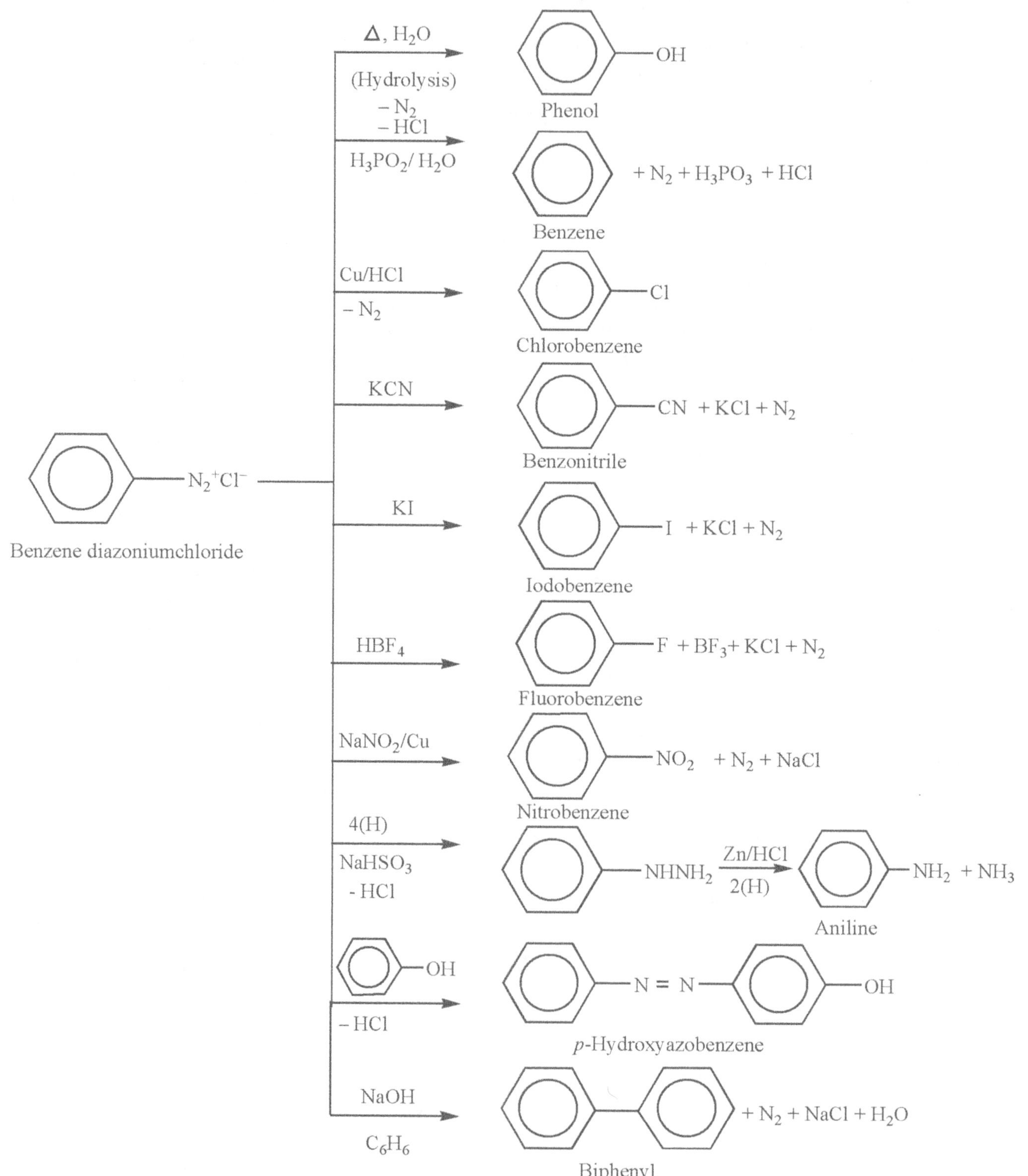

Probable Questions

1. Define aromatic diazonium compounds and write their general formula.
2. Define diazotization reaction and write its mechanism.
3. Write down the importance of aryl diazonium compounds and explain its nomenclature.
4. Explain the general methods of preparation of aryl diazonium compounds.
5. Write the general physical properties of aryl diazonium compounds.
6. Enumerate the reactions of diazonium salts involving the replacement of diazo group.
7. Explain Schliemann reaction.
8. Write a note on Gomberg reaction.
9. Explain Sandmeyer's reaction and diazo coupling reaction.
10. Describe Meerwin reaction.
11. Explain the reactions of diazonium salts wherein the diazo group is retained.
12. Write the structure and uses of diazonium salts in organic synthesis.

23

Phenols

Introduction

Phenols are compounds containing hydroxy (-OH) group attached directly into an aromatic ring. According to the number of –OH group present in it, phenols are called as monohydric, dihydric, trihydric or polyhydric phenols. Some of the commonly known phenols are shown below.

1. **Monohydric phenols**

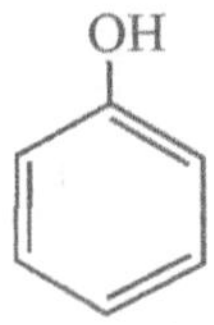

Phenol

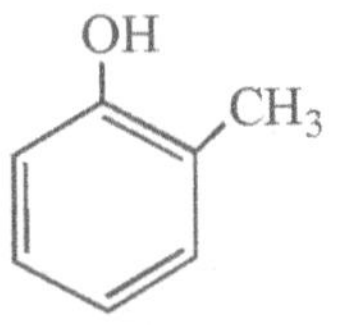

o-Cresol (2-Methyl phenol)

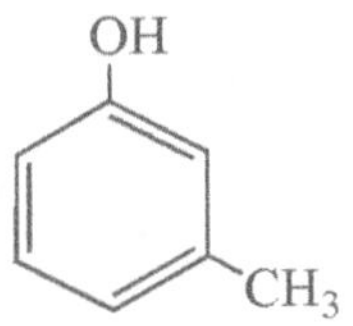

m-Cresol (3-Methyl phenol)

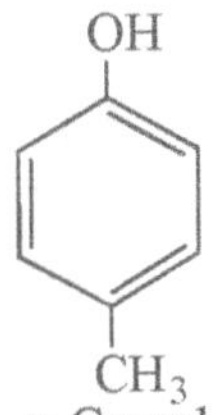

p-Cresol (4-Methyl phenol)

p-Nitrophenol (4-Nitro phenol)

2. **Dihydric phenols**

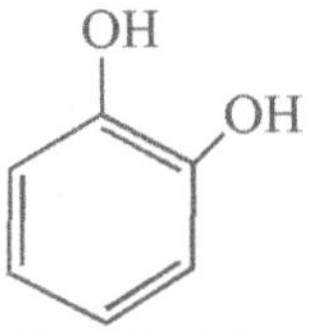

Catechol (1,2-Dihydroxy benzene)

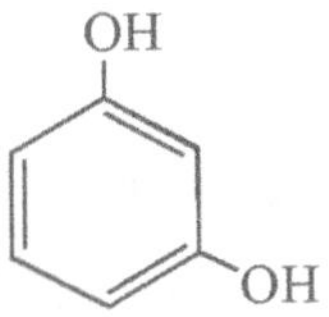

Resorcinol (1,3-Dihydroxy benzene)

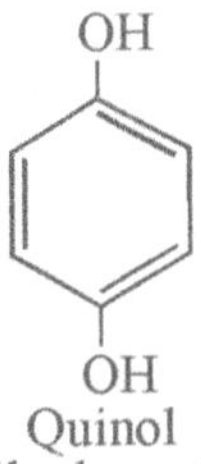

Quinol (1,4-Dihydroxy benzene)

3. **Trihydric phenols**

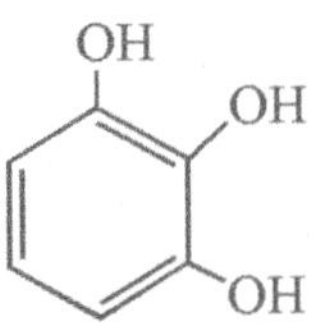

Pyrogallol (1,2,3-Trihydroxy benzene)

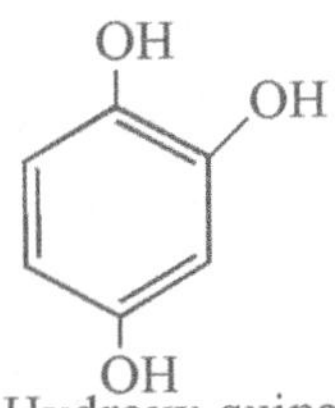

Hydroxy quinol (1,2,4-Trihydroxy benzene)

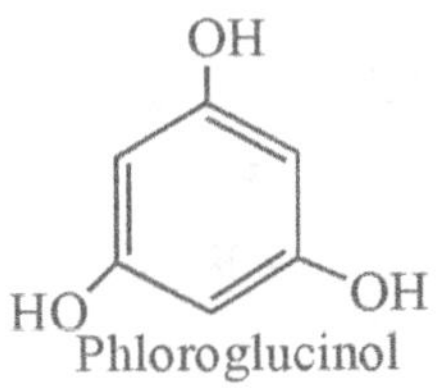

Phloroglucinol (1,3,5-Trihydroxy benzene)

Some of the aromatic hydroxy compounds contain –OH group in a side chain which is attached to the aromatic ring (-OH group is not directly linked to aryl ring). They are regarded as aryl derivatives of aliphatic alcohols. Some of the examples are given below.

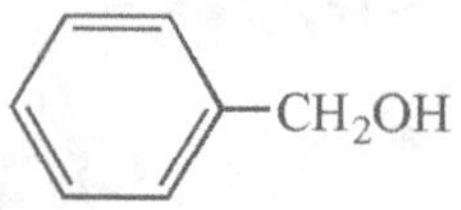

$C_6H_5-CH_2OH$

Benzylalcohol
(Phenyl methanol)

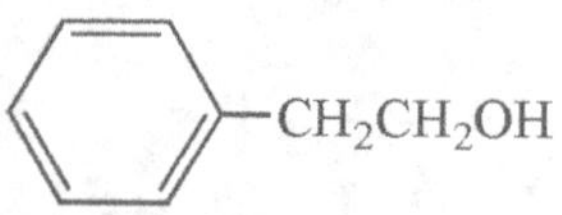

$C_6H_5-CH_2CH_2OH$

2-Phenyl ethanol
(Benzyl methanol)

Importance of Phenols

Phenols when attached with various chemical structures elicit different kind of pharmacological activities based on the core structure to which it is attached. Some of the drugs which possess phenolic groups and their therapeutic uses in the treatment of various diseases are shown in Table 23.1.

Table 23.1 Drugs containing phenolic groups and their therapeutic uses.

Category	Drug	Structure
Adrenergic drug	Dobutamine	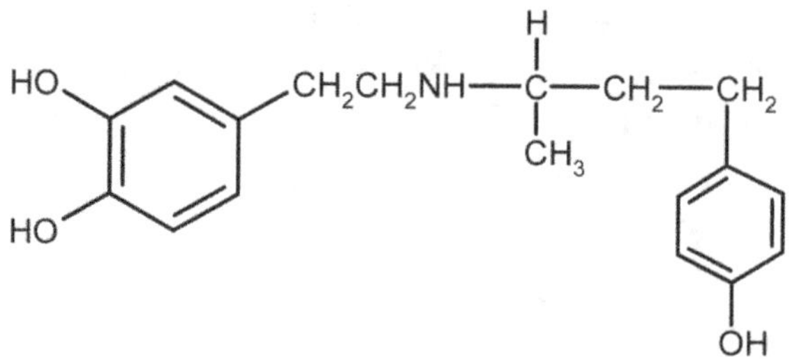 HO, HO (3,4-dihydroxyphenyl)–$CH_2CH_2NH-C(H)(CH_3)-CH_2-CH_2$–(4-hydroxyphenyl, OH)
	Salbutamol	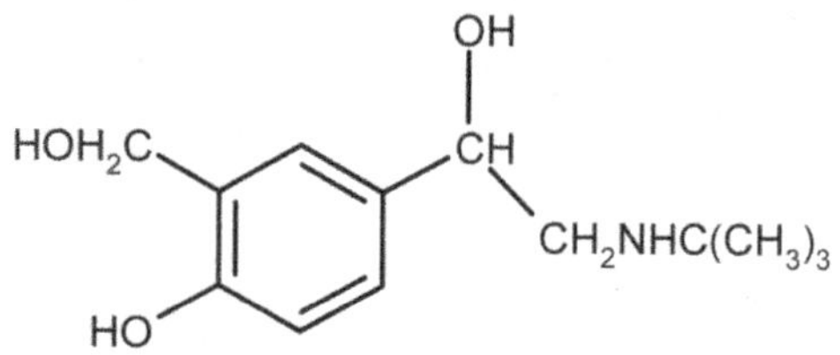 HOH_2C, HO (phenyl ring)–CH(OH)–$CH_2NHC(CH_3)_3$
	Phenylephrine	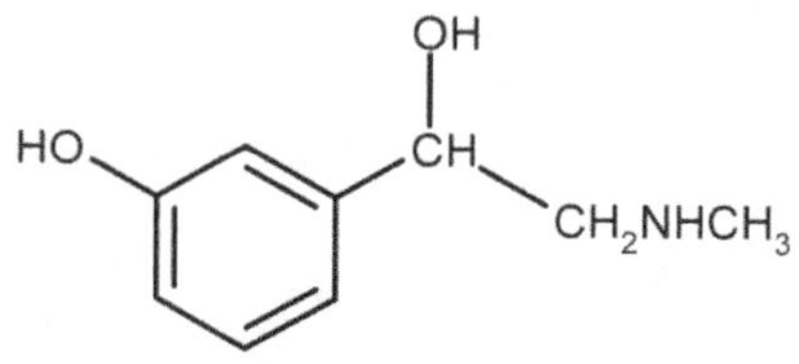 HO (phenyl ring)–CH(OH)–CH_2NHCH_3
Antihyperlipidemic drug	Ezetimibe	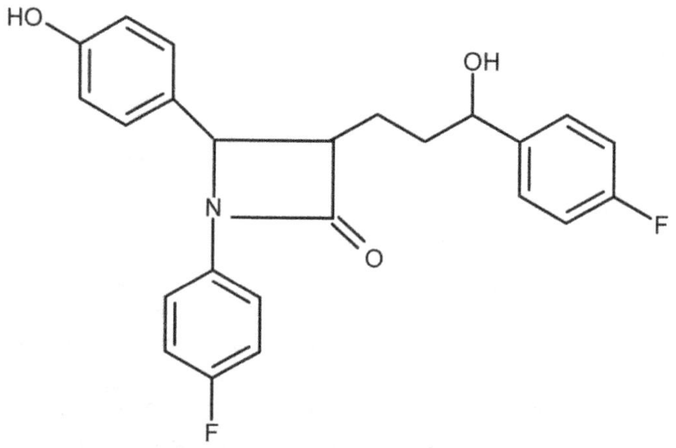 HO; OH; N; O; F; F

Phenol

Phenol is the simplest and important member of the phenolic compounds.

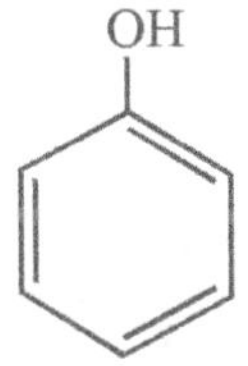

OH

Phenol

Preparation

1. **From sodium benzenesulphonate:** By fusing sodium benzene sulphonate with sodium hydroxide followed by treatment with dilute hydrochloric acid, phenol is obtained.

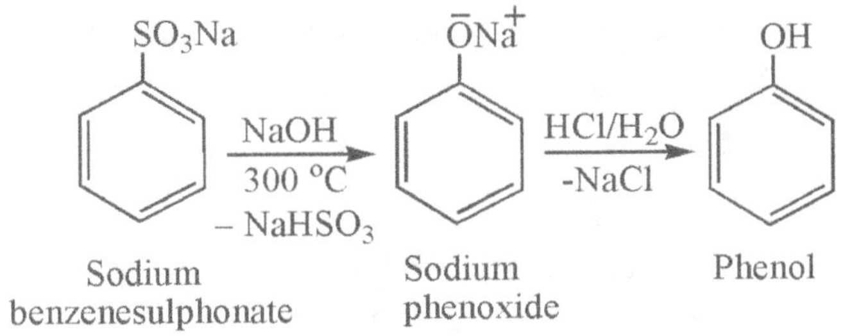

$$\text{Sodium benzenesulphonate } (C_6H_5SO_3Na) \xrightarrow[300\,^{\circ}C,\ -NaHSO_3]{NaOH} \text{Sodium phenoxide } (C_6H_5\bar{O}\overset{+}{Na}) \xrightarrow[-NaCl]{HCl/H_2O} \text{Phenol } (C_6H_5OH)$$

2. **From chlorobenzene (Dow process):** Hydrolysis of chlorobenzene with NaOH at high temperature and pressure gives sodium phenoxide which further reacts with HCl to yield phenol.

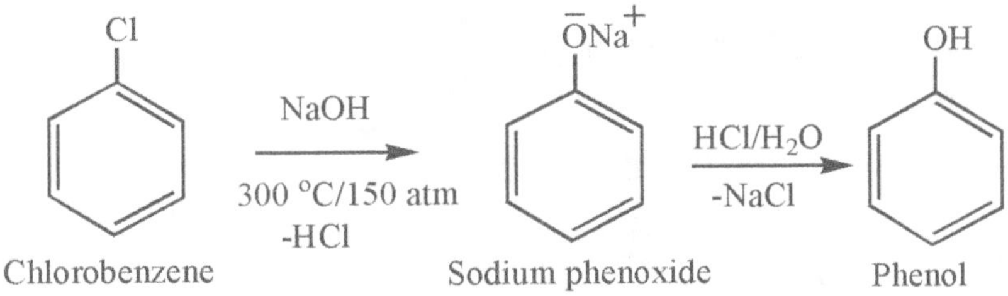

$$\text{Chlorobenzene } (C_6H_5Cl) \xrightarrow[300\,^{\circ}C/150\ atm,\ -HCl]{NaOH} \text{Sodium phenoxide } (C_6H_5\bar{O}\overset{+}{Na}) \xrightarrow[-NaCl]{HCl/H_2O} \text{Phenol } (C_6H_5OH)$$

Mechanism of Dow process

It follows addition-elimination mechanism *via* formation of benzyne intermediate.

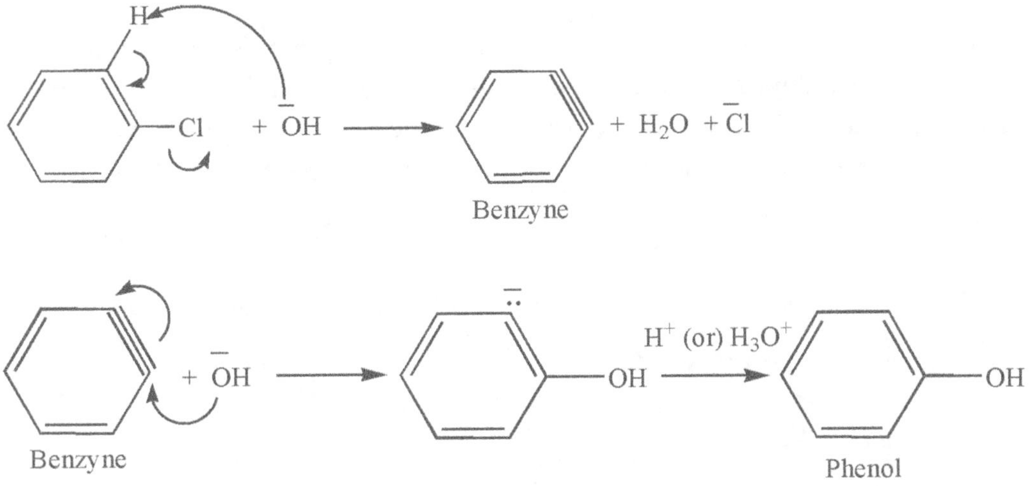

$$C_6H_5Cl + \bar{O}H \longrightarrow \text{Benzyne } (C_6H_4) + H_2O + \bar{Cl}$$

$$\text{Benzyne} + \bar{O}H \longrightarrow \bar{C}_6H_4OH \xrightarrow{H^+ \text{ (or) } H_3O^+} \text{Phenol } (C_6H_5OH)$$

3. **From cumene:** Oxidation of cumene followed by treatment with HCl yields phenol.

Cumene (benzene ring bearing $H_3C-CH(CH_3)-$) $\xrightarrow[\Delta]{100\ °C,\ O_2}$ Cumenehydroperoxide (benzene ring bearing $H_3C-C(CH_3)-OOH$) $\xrightarrow{HCl/H_2O}$ Phenol (benzene ring bearing OH) + CH_3COCH_3 (Acetone)

Advantage: 80 % of phenol is synthesized by cumene process only. The success of this method is due to easy availability of benzene and propene from petroleum (starting materials for the synthesis of cumene) and the formation of valuable by product acetone.

4. **From benzene diazonium salts:** By warming benzene diazonium chloride at 50 °C, phenol is obtained.

Benzenediazonium chloride ($C_6H_5\overset{+}{N_2}\overset{-}{Cl}$) $\xrightarrow[H_2O,\ 50\ °C]{\Delta}$ Phenol (OH) + N_2 + HCl

5. **By oxidation of benzene:** By passing a mixture of benzene and air over vanadium pentoxide at high temperature, phenol is formed.

2 Benzene + O_2 $\xrightarrow[\Delta]{V_2O_5}$ 2 Phenol (OH)

6. **From coal tar:** The middle oil fraction of coal tar contains phenols, cresols and naphthalene. Upon cooling naphthalene is deposited as solid and removed by centrifuging the mixture. The fraction of oil left, upon agitation with sodium hydroxide solution dissolves phenols and cresols as sodium salts. Phenols are liberated by passing CO_2 into the solution and isolated by fractional distillation.

2 Sodium phenoxide ($\bar{O}\overset{+}{Na}$) + CO_2 + H_2O $\longrightarrow$ 2 Phenol (OH) + Na_2CO_3

Summary of Methods of Preparation

1) Sodium benzene sulphonate ($C_6H_5SO_3Na$) $\xrightarrow[\text{-NaCl}]{HCl/H_2O}$ Phenol

2) Chlorobenzene $\xrightarrow[\text{HCl/}H_2O\text{, -NaCl}]{\text{NaOH/ -HCl}}$ Phenol

3) Cumene ($H_3C\text{-}CH(CH_3)\text{-}C_6H_5$) $\xrightarrow[HCl/H_2O,\ \Delta]{O_2,\ 100^\circ C}$ Phenol + CH_3COCH_3

4) Benzenediazonium chloride ($C_6H_5\overset{+}{N_2}\overset{-}{Cl}$) $\xrightarrow[\Delta,\ -N2,\ -HCl]{H_2O,\ 50^\circ C}$ Phenol

5) Benzene (2moles) $\xrightarrow[\Delta]{O_2,\ V_2O_5}$ Phenol

6) Sodium phenoxide ($C_6H_5\overset{-}{O}\overset{+}{Na}$) (2 moles) $\xrightarrow{CO_2 + H_2O}$ 2 Phenol + Na_2CO_3

Physical Properties: Phenol is colorless, needle shaped crystalline substance. Phenol is partially miscible at temperature below 65.8 °C with water but above this temperature it is miscible in all proportions, more soluble in ethanol, ether and other organic solvents. Liquefied phenol contains 5 % water at room temperature, known as carbolic acid. The needle shaped crystals are corrosive, hygroscopic and poisonous, turn pink on exposure to air. Phenols are more polar than cycloalkanols which have similar skeleton hence they have higher melting point and boiling point. Further, the *ortho* haloisomers have lower melting point and solubilities are more than their *para* and *meta* analogues due to the intramolecular hydrogen bonding in *ortho* isomers.

o- Fluoro phenol (Intramolecular hydrogen bonding) Low melting point

p- Fluoro phenol (Intermolecular hydrogen bonding) High melting point

Phenol is poisonous and produces painful blisters when comes in contact with skin. Phenol have characteristics carbolic odour. The boiling point of phenol is higher than that of aliphatic alcohols due to stronger intermolecular hydrogen bonding.

Chemical Properties: In phenols, the-OH group is attached to sp^2 hybridised carbon of an aromatic ring. The C-O bond length (136 pm) in phenol is slightly less than methanol (142 pm). This is due to the following facts.

1. Partial double bond character because of conjugation of unshared pair of electrons of oxygen with aromatic ring.
2. sp^2 hybridised state of carbon to which oxygen is attached.

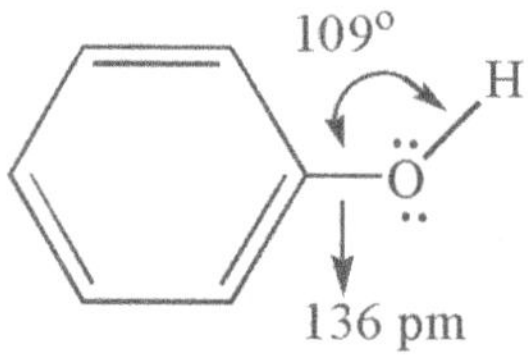

109°
H
Ö
136 pm
Phenol

The chemical reactions of phenols are described as follows.

(I) Reactions of -OH group.

(II) Reactions of benzene ring.

As alcohols also contain -OH group, some of the reactions of phenols are similar to alcohols.

(I) Reactions of -OH group:

1. **Formation of salts:** Phenols are acidic in nature and hence react with NaOH or Na metal to form sodium salts. But the acidity is weak; hence it cannot react with $NaHCO_3$ and Na_2CO_3.

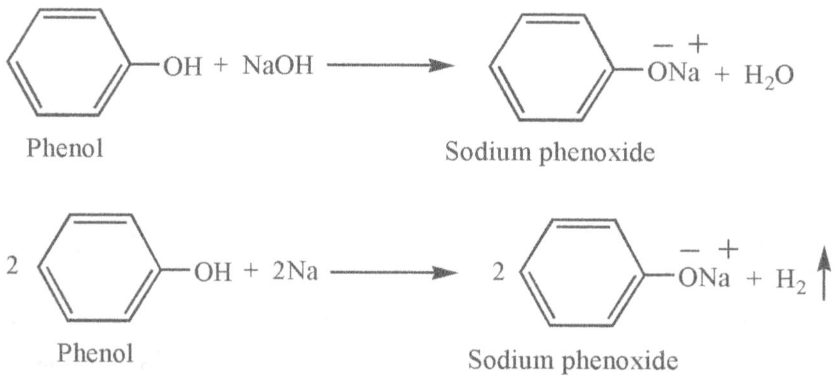

$$C_6H_5OH + NaOH \longrightarrow C_6H_5\overset{-}{O}\overset{+}{Na} + H_2O$$

Phenol → Sodium phenoxide

$$2C_6H_5OH + 2Na \longrightarrow 2C_6H_5\overset{-}{O}\overset{+}{Na} + H_2\uparrow$$

Phenol → Sodium phenoxide

But alcohols react only with sodium metal not with NaOH

$$2CH_3CH_2OH + 2Na \longrightarrow 2CH_3CH_2ONa + H_2\uparrow$$

Sodium ethoxide

Acidity of phenols: The hydroxyl (-OH) group in phenol is acidic in nature. The acidity of phenol is much more as compared to alcohols but less than carboxylic acids.

The acidity of phenol is due to the formation of stable phenoxide ions in aqueous solutions.

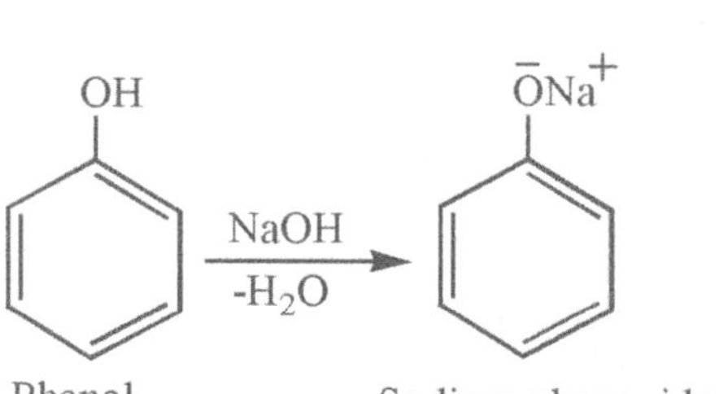

$$C_6H_5OH \xrightarrow[-H_2O]{NaOH} C_6H_5\overset{-}{O}\overset{+}{Na}$$

Phenol → Sodium phenoxide

The stronger acidity of phenols than alcohols is due to the fact that phenol exists as resonance hybrid of the following structures depicted below.

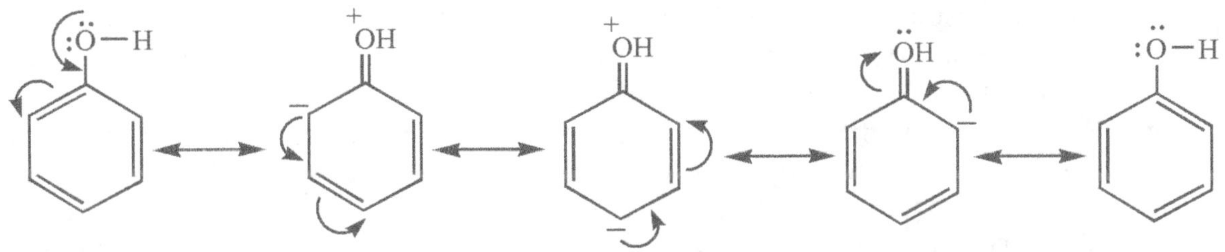

(a) The oxygen atom gets positive charge due to resonance and attracts the electron pair of the O-H bond and hence facilitates the release of *proton*.

(b) Carbon atom of the -C-OH group in phenol being sp^2 hybridised is more electron attracting than sp^3 carbon atom in alcohols. Thus the –I effect in phenol is greater and facilitates proton release. However the inductive effect is less significant as compared to resonance effect.

(c) The phenoxide ion formed after proton release is stabilized by resonance, where as in alcohols the resonance is not possible and hence proton is not released and held tightly with oxygen.

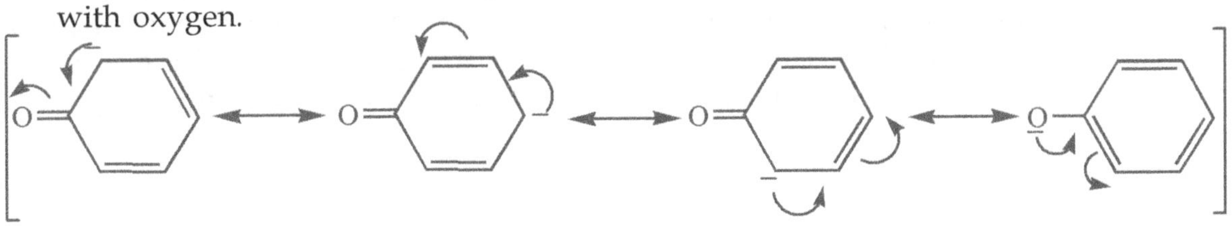

Effect of substituents

(a) Nature of substituents: Electron withdrawing groups like $-NO_2$, -Cl, -CN, -CHO, -COOH substituents disperse the negative charge of the phenoxide ion and help to stabilize the ion and increase acidity of phenol (Example: *p*-nitrophenol is more acidic than phenol). Electron releasing groups like $-CH_3$, $-OCH_3$, $-NH_2$ on the aromatic ring tend to intensify the negative charge on phenoxide ion and inhibit the delocalization which leads to destabilization of phenoxide ion and results in decreasing its acidity. Example, *p*-cresol is less acidic than phenol.

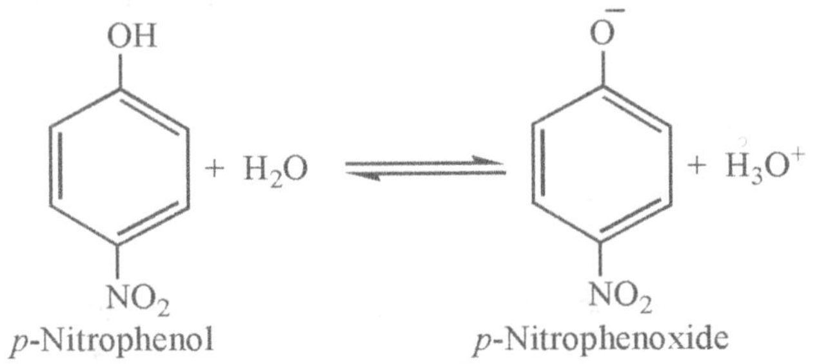

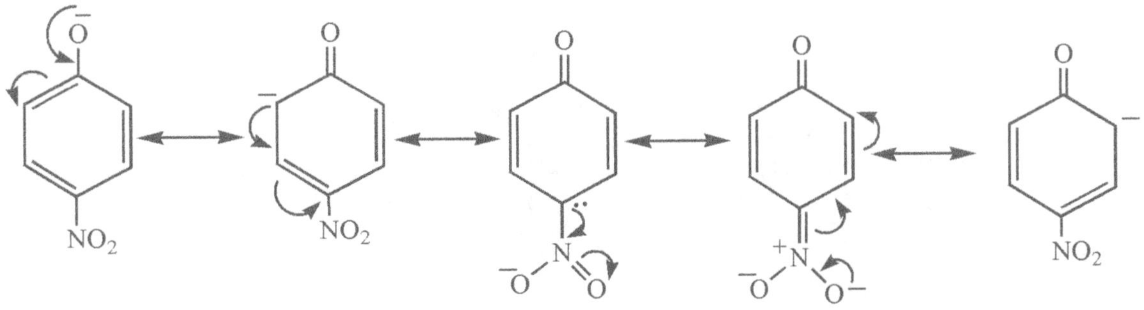

p-Nitrophenol + H_2O ⇌ *p*-Nitrophenoxide + H_3O^+

(b) **Ability of the substituent to enter into resonance with -OH group:** Electron releasing substituents inhibit the release of lone pair of electrons of oxygen of -OH group or phenoxide ion and hence resonance will be blocked that leads to destabilization of phenol or phenoxide ion and results in decrease in acidity. Hence electron releasing groups reduce the acidity by decreasing the resonance. On the other hand electron withdrawing groups enhances the resonance and increases acidity.

(c) **Position of substituents:** As nitro group can enter into resonance with -OH group only from *o*- and *p*-position, these isomers are more acidic than phenol, where as *m*-isomer cannot resonate with -OH group of phenol hence *m*-nitro phenol is less acidic. Picric acid (2,4,6-tri nitro phenol) is strongly acidic due to cumulative effect of three nitro groups. Thus *o*-and-*p* nitro phenols are more acidic than *m*-nitro phenol, which in turn is more acidic than phenol. The increased acidity in *o* and *p* nitro phenols than *m*-nitro phenol is due to both –R and –I effect. But in *m* isomer it has only –I effect.

Methoxy (-OCH_3) and amino (-NH_2) groups when present in *p*-position exert +R effect and –I effect. As +R effect is stronger than the –I effect, they reduce acidity where as when they are present in *m*-position, –I effect is stronger than +R effect and hence *m*-isomers with -OCH_3, –NH_2 are more acidic than phenol.

(d) **Steric hindrance:** This also plays a role in acidity. For example, consider the following substituted phenols: a decrease in acid strength from I to II is due to + I effect of two *ortho* methyl groups in II. Drastic decrease in acid strength from I to III is due to steric hindrance of resonance of NO_2 group with -OH by the two *ortho* methyl groups.

OH, NO_2 — I — 4-Nitrophenol — pKa 7.15

OH, H_3C, CH_3, NO_2 — II — 2,6-Dimethyl-4-nitrophenol — 7.22

OH, H_3C, CH_3, NO_2 — III — 3,5-Dimethyl-4-nitrophenol — 8.25

2. **Formation of benzene:** Upon distillation of phenol with zinc dust, benzene is obtained.

Phenol–OH + Zn $\xrightarrow{\Delta}$ Benzene + ZnO

But aliphatic alcohols are reduced to hydrocarbons on treatment with HI acid.
Phenol is also reduced to benzene with hydrogen at atmospheric pressure in the presence of molybdenum oxide as a catalyst.

Phenol (OH) + H_2 $\xrightarrow{MoO_2}$ Benzene + H_2O

3. **Formation of ethers:** Phenol upon reaction with alkali forms sodium phenoxide, which upon further reaction with alkyl halide yields ethers. Alcohols undergo similar reaction. This is called **Williamson's ether synthesis.**

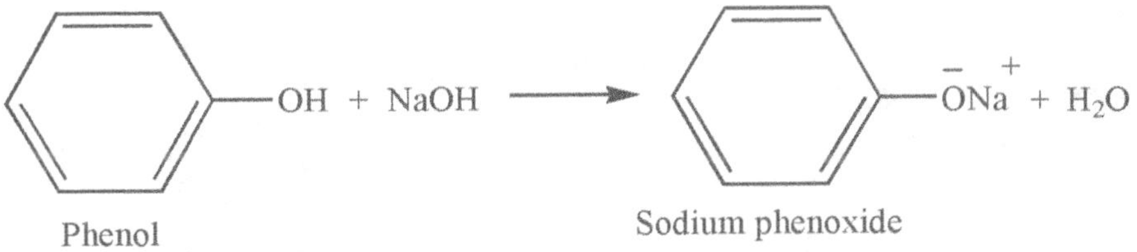

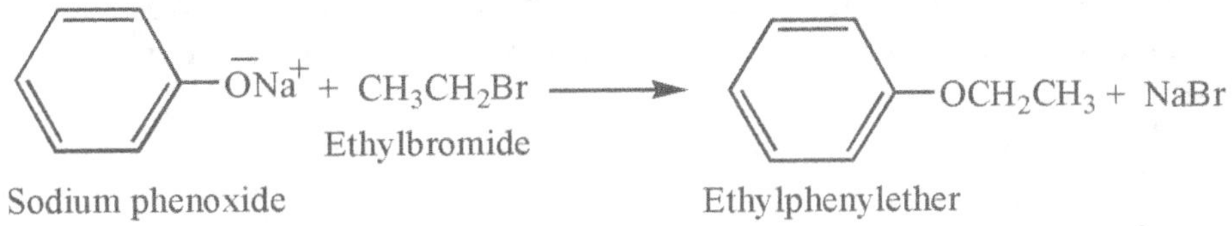

$$C_6H_5\text{-}OH + NaOH \longrightarrow C_6H_5\text{-}\overset{-}{O}\overset{+}{Na} + H_2O$$

Phenol — Sodium phenoxide

$$C_6H_5\text{-}\overset{-}{O}\overset{+}{Na} + CH_3CH_2Br \longrightarrow C_6H_5\text{-}OCH_2CH_3 + NaBr$$

Sodium phenoxide — Ethylbromide — Ethylphenylether

$$CH_3CH_2ONa + CH_3CH_2Br \longrightarrow CH_3CH_2OCH_2CH_3 + NaBr$$

Sodium ethoxide — Ethylbromide — Diethylether

When phenol is treated with dimethyl sulphate, anisole (methoxy benzene) is obtained

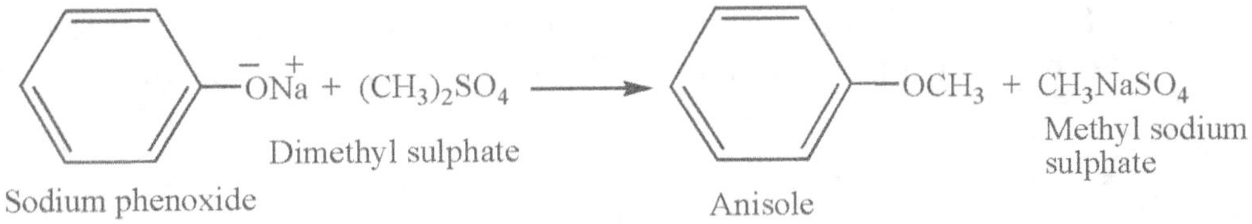

$$C_6H_5\text{-}\overset{-}{O}\overset{+}{Na} + (CH_3)_2SO_4 \longrightarrow C_6H_5\text{-}OCH_3 + CH_3NaSO_4$$

Sodium phenoxide — Dimethyl sulphate — Anisole — Methyl sodium sulphate

4. **Formation of esters:** Phenol in alkali solution reacts with acid chlorides or acid anhydride and forms ester. This reaction is called esterification. Alcohols also undergo same reaction. The reaction of phenol with benzoyl chloride is known as **Schotten- Baumann reaction.**

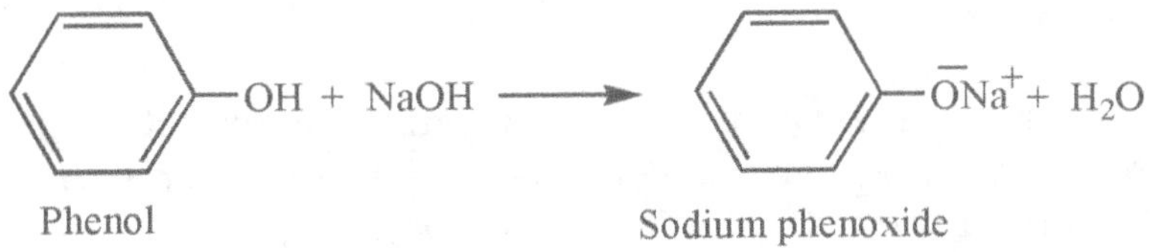

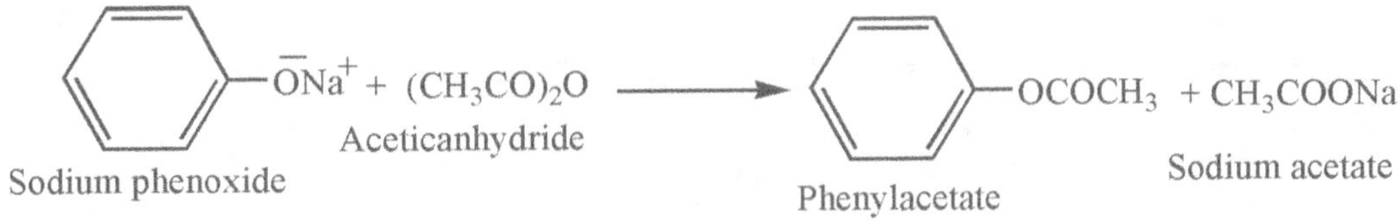

$$C_6H_5\text{-}OH + NaOH \longrightarrow C_6H_5\text{-}\overset{-}{O}\overset{+}{Na} + H_2O$$

Phenol — Sodium phenoxide

$$C_6H_5\text{-}\overset{-}{O}\overset{+}{Na} + (CH_3CO)_2O \longrightarrow C_6H_5\text{-}OCOCH_3 + CH_3COONa$$

Sodium phenoxide — Aceticanhydride — Phenylacetate — Sodium acetate

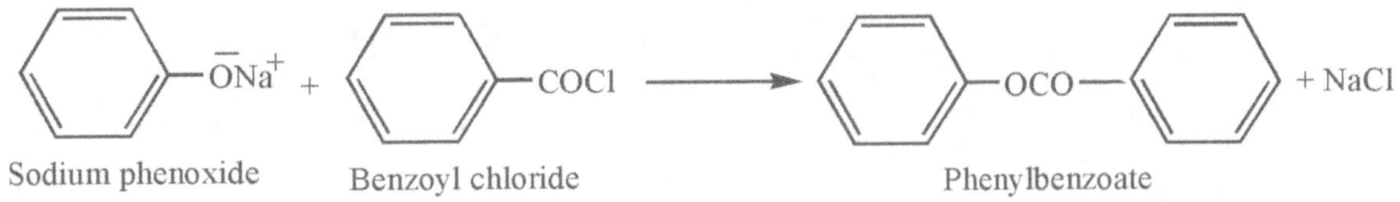

$$C_6H_5\text{-}\overset{-}{O}\overset{+}{Na} + C_6H_5\text{-}COCl \longrightarrow C_6H_5\text{-}OCO\text{-}C_6H_5 + NaCl$$

Sodium phenoxide — Benzoyl chloride — Phenylbenzoate

5. **Reaction with neutral ferric chloride:** Most of the phenols react with neutral ferric chloride solution and gives purple colour complex. This is one of the important identification tests for phenols.

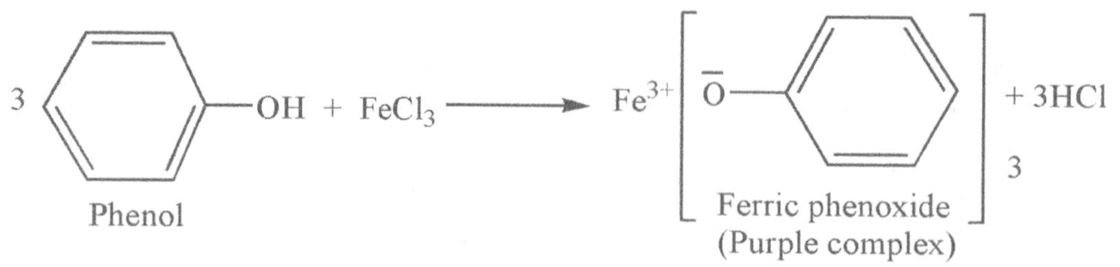

$$3\ C_6H_5OH + FeCl_3 \longrightarrow Fe^{3+}[\bar{O}-C_6H_5]_3 + 3HCl$$

Phenol; Ferric phenoxide (Purple complex)

6. **Formation of halides:** The hydroxyl group of phenol reacts with phosphorus halides but it is difficult to form aryl halides, where as alcohols react with phosphorus halides easily to form alkyl halides.

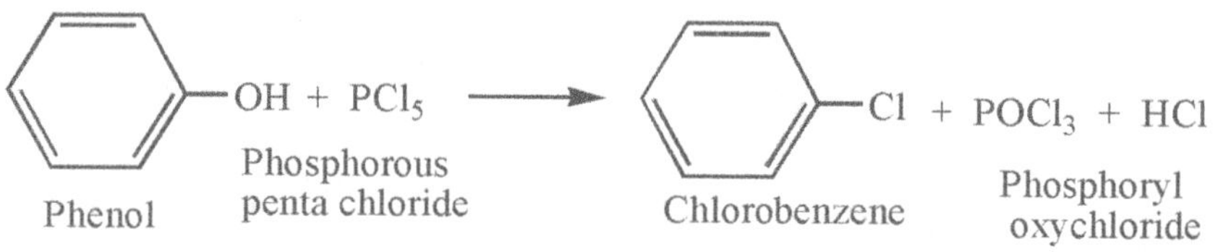

$$C_6H_5OH + PCl_5 \longrightarrow C_6H_5Cl + POCl_3 + HCl$$

Phenol; Phosphorous penta chloride; Chlorobenzene; Phosphoryl oxychloride

$$CH_3CH_2OH + PCl_5 \longrightarrow CH_3CH_2Cl + POCl_3 + HCl$$

Ethanol; Ethyl chloride

7. **Formation of amines:** Phenol upon reaction with ammonia in presence of anhydrous zinc chloride or calcium chloride yields aniline. Alcohols react with amines and yield aliphatic amines.

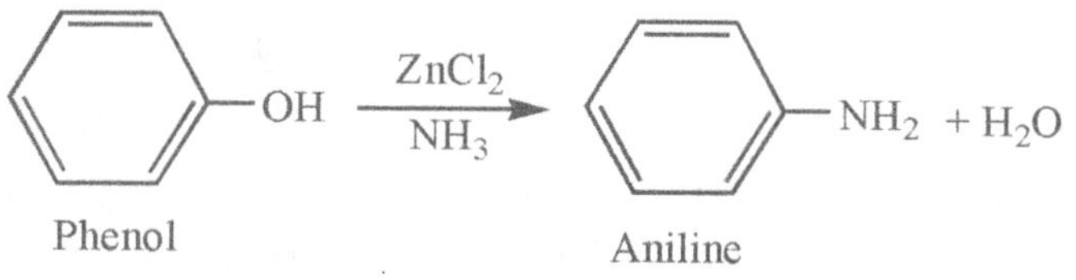

$$C_6H_5OH \xrightarrow[NH_3]{ZnCl_2} C_6H_5NH_2 + H_2O$$

Phenol; Aniline

$$CH_3CH_2OH \xrightarrow[\Delta]{NH_3} CH_3CH_2NH_2 + H_2O$$

Ethanol; Ethylamine

8. **Oxidation:** Phenol on oxidation in air and light turns pink. The oxidation product is quinone, which usually forms complex with phenol and yields brilliant red phenoquinone (in which two phenol molecules link with quinone through hydrogen bonding.

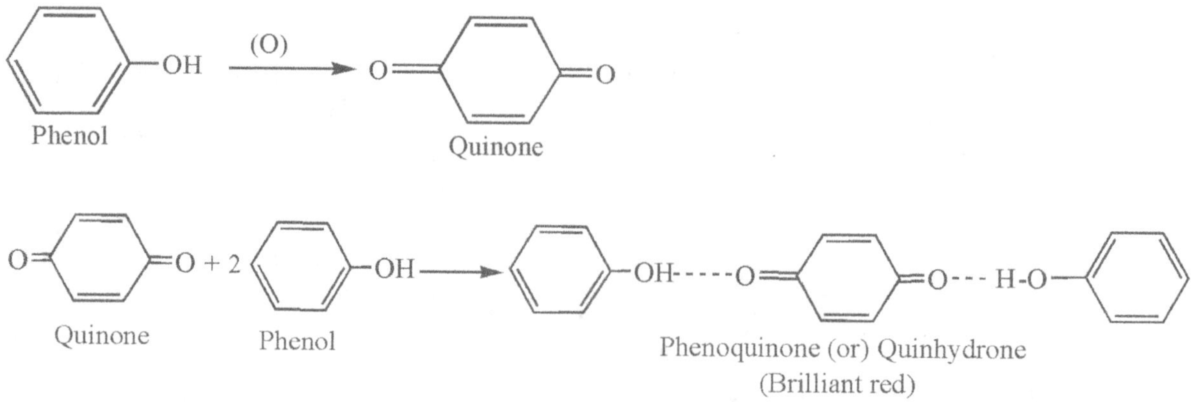

$$C_6H_5OH \xrightarrow{(O)} O{=}C_6H_4{=}O$$

Phenol; Quinone

$$O{=}C_6H_4{=}O + 2\ C_6H_5OH \longrightarrow C_6H_5-OH\text{----}O{=}C_6H_4{=}O\text{---}H\text{-}O-C_6H_5$$

Quinone; Phenol; Phenoquinone (or) Quinhydrone (Brilliant red)

Phenol on oxidation with potassium persulphate in alkaline solution yields dihydric phenol. The new -OH group enters into *p*-position of phenolic -OH group. If the *p*-position is not free then it enters into *o*-position. This is known as **Elbs persulphate oxidation.**

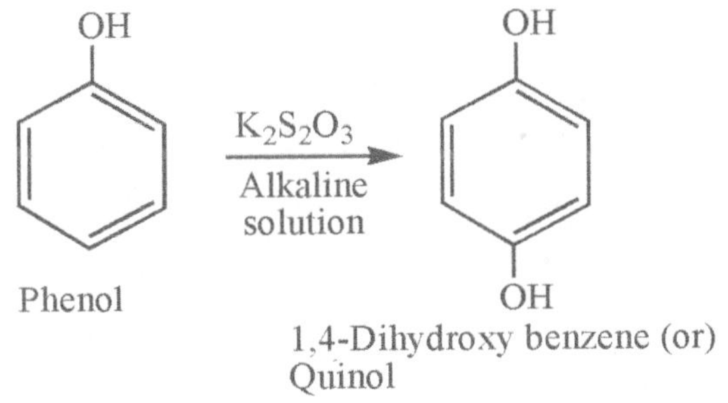

OH

$K_2S_2O_3$
Alkaline solution

Phenol

OH

OH

1,4-Dihydroxy benzene (or) Quinol

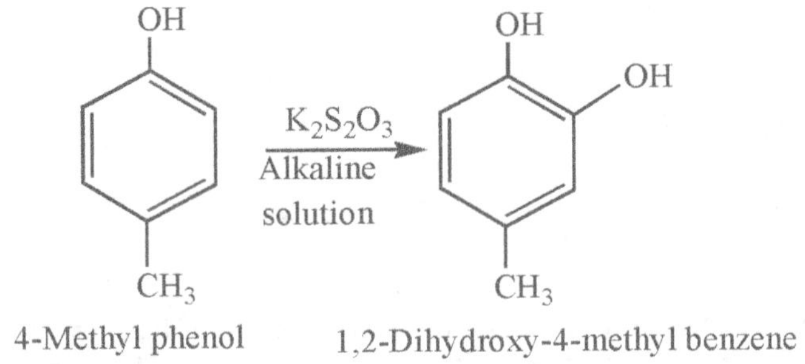

OH

$K_2S_2O_3$
Alkaline solution

CH_3

4-Methyl phenol

OH
OH

CH_3

1,2-Dihydroxy-4-methyl benzene

Phenol or its derivatives are oxidised by potassium permanganate, ring cleavage takes place. But if phenolic group is protected by acylation or alkylation, phenols gets oxidised by $KMnO_4$ yielding phenolic acids.

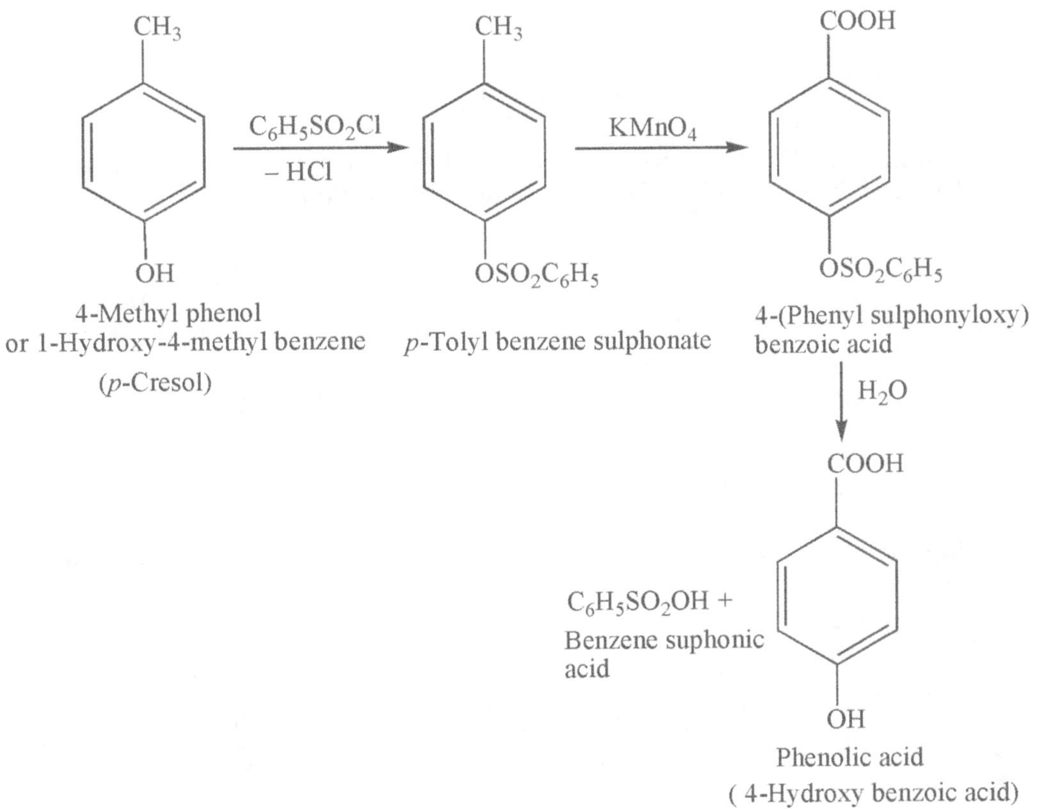

CH_3

OH

4-Methyl phenol
or 1-Hydroxy-4-methyl benzene
(*p*-Cresol)

$C_6H_5SO_2Cl$
$-HCl$

CH_3

$OSO_2C_6H_5$

p-Tolyl benzene sulphonate

$KMnO_4$

COOH

$OSO_2C_6H_5$

4-(Phenyl sulphonyloxy) benzoic acid

H_2O

$C_6H_5SO_2OH$ +
Benzene suphonic acid

COOH

OH

Phenolic acid
(4-Hydroxy benzoic acid)

(II) Reactions of benzene ring: Phenol undergoes electrophilic aromatic substitution reaction. The -OH group in phenol is *o, p*-director and ring activator.

1. **Nitration:** Phenol reacts with dilute nitric acid and gives *o*- and *p*-nitro phenol. With conc. HNO_3, picric acid is obtained.

Phenol + Dil. HNO_3 → *o*-Nitrophenol + *p*-Nitrophenol

Phenol + Conc. HNO_3 → 2,4,6-Trinitrophenol (Picric acid)

2. **Nitrosation:** Phenol reacts with nitrous acid (obtained from $NaNO_2$/HCl at 5 °C) to form p-nitrosophenol exclusively, which on reaction with HNO_3 gives *p*-nitro phenol.

Phenol —($NaNO_2$/HCl, 5 °C, – NaCl, – H_2O)→ *p*-Nitrosophenol —(HNO_3, – HNO_2)→ *p*-Nitrophenol

3. **Sulphonation:** Phenol upon reaction with sulphuric acid at 20 °C gives *o*-phenol sulphonic acid whereas at 100 °C it yields *p*-phenol sulphonic acid.

p-Phenolsulphonic acid ←(H_2SO_4, 100 °C, – H_2O)— Phenol —(H_2SO_4, 20 °C, – H_2O)→ *o*-Phenolsulphonic acid

4. **Halogenation:** Bromination of phenol using bromine in CS_2 or CCl_4 forms a mixture of *o*-and *p*-bromophenol.

Phenol + $2Br_2$ —(CS_2/CCl_4)→ *o*-Bromophenol + *p*-Bromophenol + 2HBr

Aqueous solution of phenol reacts with bromine water in presence of $NaHSO_3$ and yield 2,4,6-tribromo phenol. Chlorine also reacts in same manner.

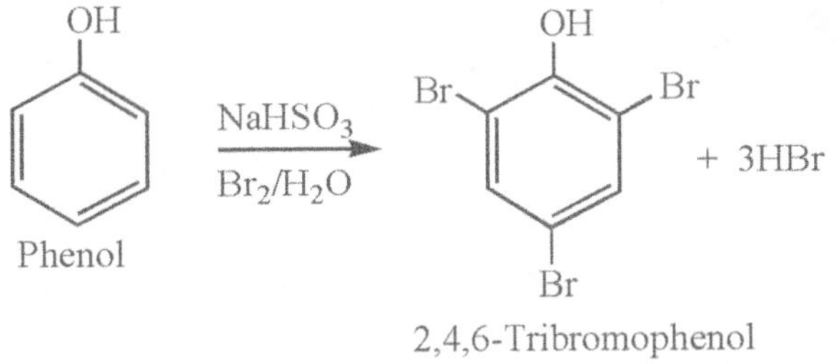

$$\text{Phenol} \xrightarrow[Br_2/H_2O]{NaHSO_3} \text{2,4,6-Tribromophenol} + 3HBr$$

5. **Reimer - Tiemann reaction:** Reaction of phenol with chloroform in aqueous NaOH solution, followed by acid hydrolysis yields salicylaldehyde. If CCl_4 is used in place of chloroform, salicylic acid is obtained. Reimer -Tiemann reaction is used to introduce -CHO or -COOH group *ortho* to -OH group of phenol.

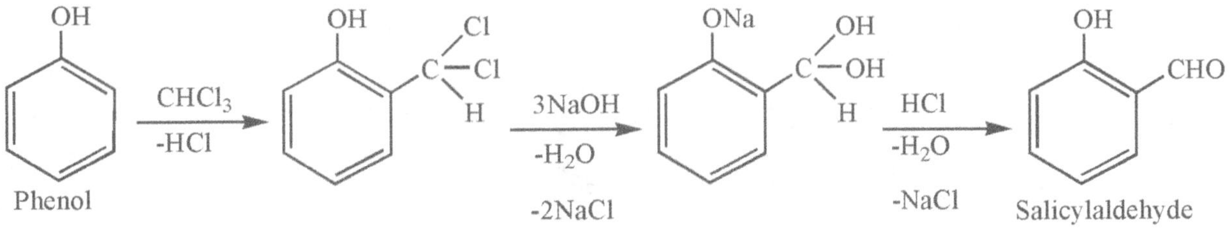

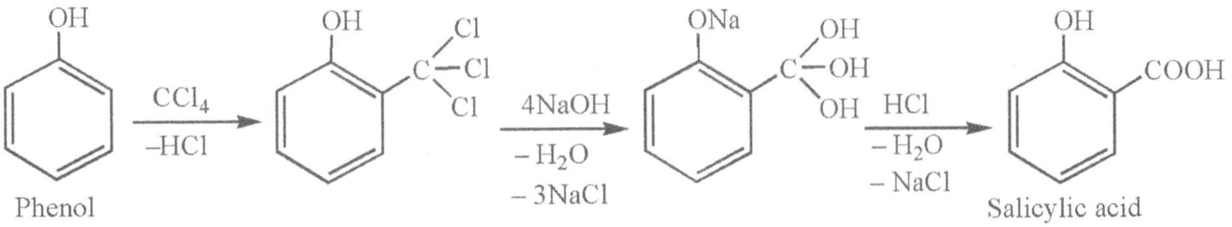

$$\text{Phenol} \xrightarrow[-HCl]{CHCl_3} C_6H_4(OH)(CHCl_2) \xrightarrow[-H_2O,\ -2NaCl]{3NaOH} C_6H_4(ONa)(CH(OH)_2) \xrightarrow[-H_2O,\ -NaCl]{HCl} \text{Salicylaldehyde}$$

$$\text{Phenol} \xrightarrow[-HCl]{CCl_4} C_6H_4(OH)(CCl_3) \xrightarrow[-H_2O,\ -3NaCl]{4NaOH} C_6H_4(ONa)(C(OH)_3) \xrightarrow[-H_2O,\ -NaCl]{HCl} \text{Salicylic acid}$$

Mechanism:

Step 1: Abstraction of proton from chloroform by base to give trichloro methyl anion (reactive species with electron deficient)

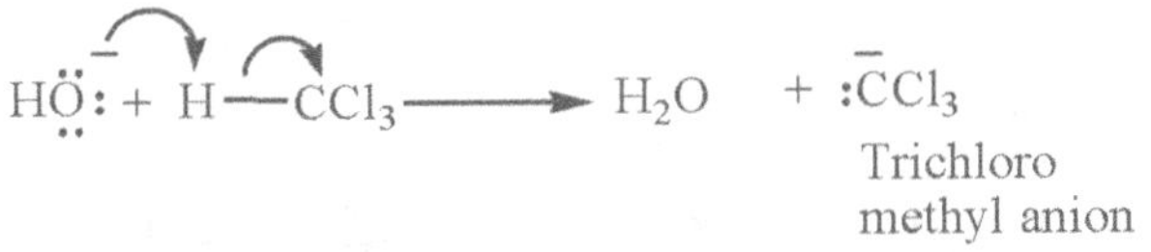

$$H\ddot{O}:^- + H-CCl_3 \longrightarrow H_2O + :\bar{C}Cl_3$$

Trichloro methyl anion

Step 2: Dissociation of trichloro methyl anion to give dichloro carbene.

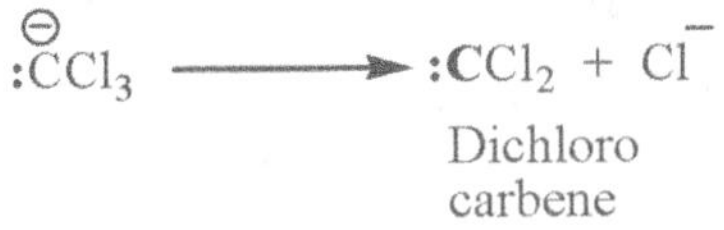

$$:\overset{\ominus}{C}Cl_3 \longrightarrow :CCl_2 + Cl^-$$

Dichloro carbene

Step 3: Addition of dichlorocarbene to phenoxide ion.

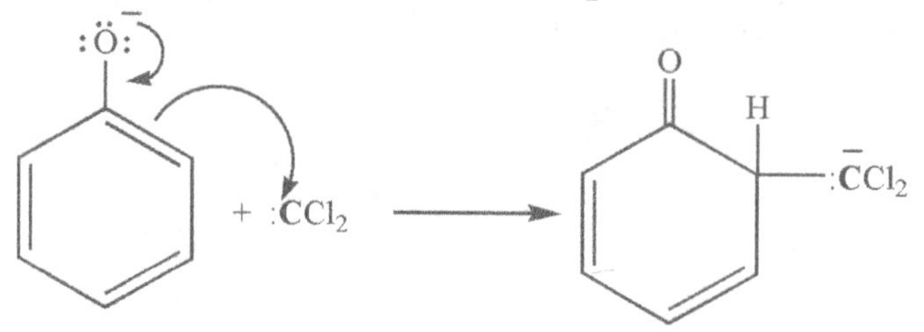

Step 4: Abstraction of proton from water.

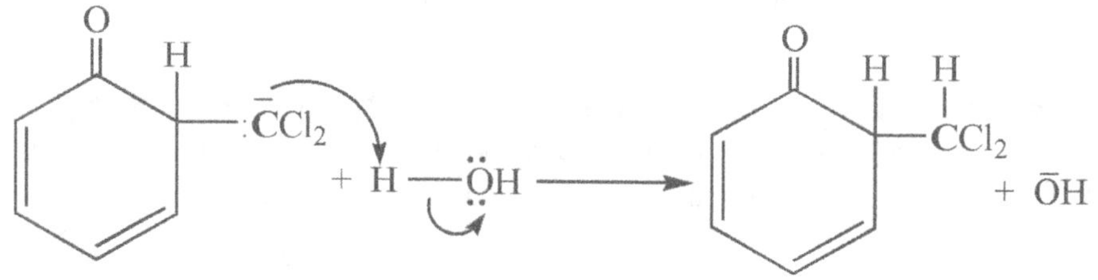

Step 5: Tautomerisation to yield a substituted phenol.

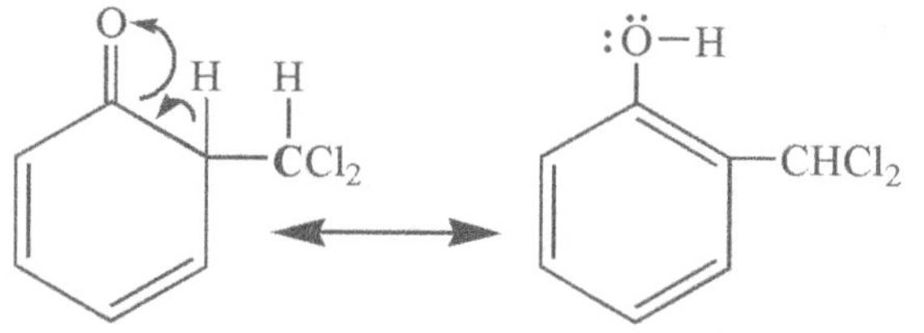

Step 6: Displacement of halide ion (chloride ion) by hydroxide ion.

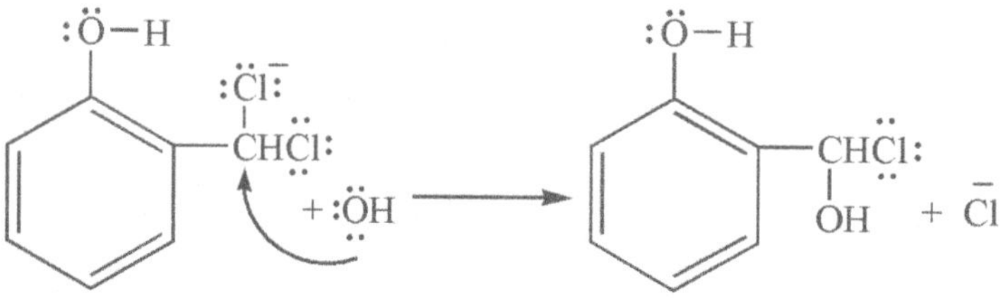

Step 7: Elimination of HCl to form salicylaldehyde.

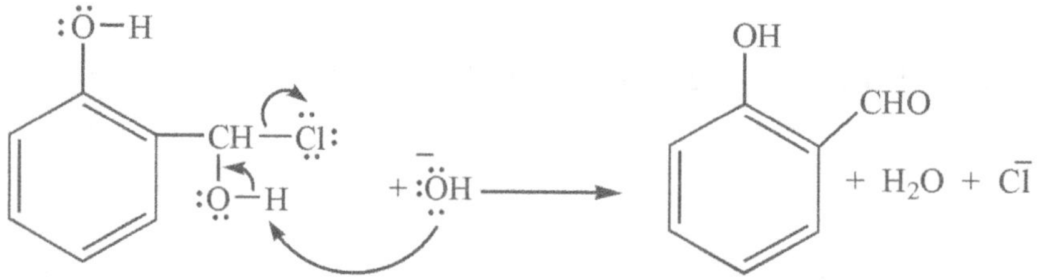

Salicylaldehyde

6. **Kolbe's reaction:** Phenol upon reaction with CO_2 in presence of NaOH at high atmospheric pressure followed by hydrolysis yields salicylic acid. This reaction is used to introduce -COOH group in *ortho* position. It is otherwise known as Kolbe's-Schmidt reaction.

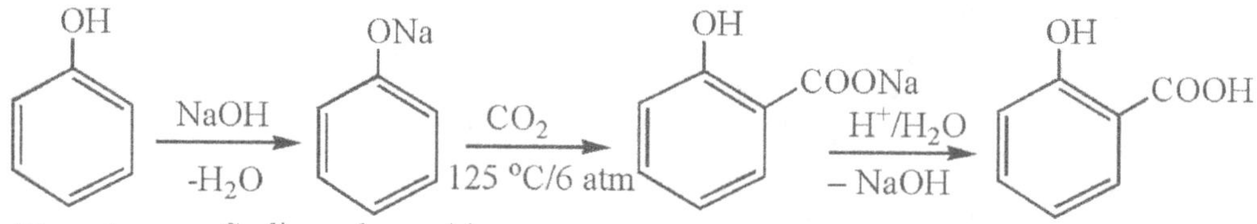

Phenol $\xrightarrow[-H_2O]{NaOH}$ Sodium phenoxide $\xrightarrow[125\,^\circ C/6\ atm]{CO_2}$ Sodium salicylate $\xrightarrow[-NaOH]{H^+/H_2O}$ Salicylic acid

Mechanism:

Step 1: Formation of electrophilic carbon center.

Phenoxid ion

Step 2: Addition of electrophilic carbon to CO_2.

Salicylate anion

Step 3: Abstraction of H^+ leads to formation of salicylic acid.

Salicylate anion $\xrightarrow{H^+}$ Salicylic acid

7. **Gattermann reaction:** Phenol upon reaction with HCN and HCl in presence of $AlCl_3$ catalyst, salicylaldehyde is formed.

Phenol $\xrightarrow[AlCl_3]{HCN/HCl}$ [intermediate with $-CH{=}NH$] $\xrightarrow[H_2O]{\Delta}$ Salicylaldehyde + NH_3

8. **Fries rearrangement :** Phenol when reacts with acetic anhydride in NaOH yields phenyl acetate. Upon reaction of ester with $AlCl_3$ catalyst, the acyl group migrates into *o*- or *p*- position to yield *o*-and *p*-hydroxy acetophenone. This method is used to introduce -COR in *o*- and *p*- positions.

Phenol $\xrightarrow[NaOH, -CH_3COOH]{(CH_3CO)_2O}$ Phenylacetate ($OCOCH_3$) $\xrightarrow[\Delta]{AlCl_3}$ *o*-Hydroxyacetophenone + *p*-Hydroxyacetophenone

9. **Libermanns nitroso reaction:** When phenol reacts with conc. H_2SO_4 and $NaNO_2$, brown or red colour is formed, which changes into blue-green soon. The colour changes to red upon dilution with water and deep blue when treated with NaOH.

(a) Phenol reacts with nitrous acid ($NaNO_2$ + H_2SO_4) to yield *p*-nitrosophenol.

Phenol —HONO, $-H_2O$→ *p*-Nitrosophenol ⇌ Monoxime of quinone

(b) In presence of conc. H_2SO_4, monoxime of quinone condense with phenol to yield blue-green solution of indophenol monosulphate which on dilution gives indophenol (red color). With NaOH, sodium salt of indophenol (deep blue colour) is produced.

Monoxime of quinone + Phenol —H_2SO_4, $-H_2O$→ Indophenol monosulphate (Bluegreen) —$-H_2SO_4$→ Indophenol (Red) —NaOH, $-H_2O$→ Sodium salt of indophenol (Deep blue color)

10. Hydrogenation: Phenol upon hydrogenation in presence of nickel catalyst gives cyclohexanol.

Phenol + $3H_2$ —Ni→ Cyclohexanol

11. Friedel -Crafts reaction: Phenol upon reaction with alkyl halide in presence of $AlCl_3$ gives *p*-alkyl phenol with small quantity of *ortho* alkyl phenol (*o*-methyl phenol).

Phenol + CH_3Cl (Methyl chloride) —$AlCl_3$, Δ, $-HCl$→ *o*-Methylphenol + *p*-Methylphenol

12. **Reaction with benzene diazonium chloride:** Phenol couples with benzene diazonium chloride in alkaline solution to form azo dye called *p*-hydroxydiazo benzene.

$$C_6H_5OH + C_6H_5N_2Cl \xrightarrow[-HCl]{NaOH} C_6H_5-N{=}N-C_6H_4-OH$$

Phenol, Benzene diazonium chloride, p-Hydroxydiazobenzene

13. **Reaction with formaldehyde:** When phenol reacts with formaldehyde (in alkali) a mixture of *ortho* and *para* hydroxy benzyl alcohol is obtained.

$$C_6H_5OH + HCHO \xrightarrow{Aq.NaOH} o\text{-}HOC_6H_4CH_2OH + p\text{-}HOC_6H_4CH_2OH$$

Phenol, *o*-Hydroxybenzyl alcohol, *p*-Hydroxybenzyl alcohol

If reaction is carried out at high temperature and excess of formaldehyde, polymerization takes place and hard thermosetting plastic called **bakelite** is formed.

$$n\,C_6H_5OH + n\,HCHO \longrightarrow \text{Bakelite}$$

Phenol, Bakelite

14. **Phthalein reaction:** When phenol reacts with pthalic anhydride in presence of conc.H_2SO_4 phenolphthalein is obtained.

$$2\,C_6H_5OH + \text{Phthalic anhydride} \xrightarrow[\Delta]{Conc.\ H_2SO_4} \text{Phenolphthalein} + H_2O$$

Phenol, Phthalic anhydride, Phenolphthalein

Summary of Chemical Reactions

Phenol (C_6H_5OH) →

- $\xrightarrow{NaOH}$ $C_6H_5O^-Na^+ + H_2O$ — Sodium phenoxide
- $\xrightarrow{2Na}$ $2\ C_6H_5O^-Na^+ + H_2\uparrow$ — Sodium phenoxide
- $\xrightarrow[\Delta]{Zn}$ $C_6H_6 + ZnO$ — Benzene
- $\xrightarrow{H_2,\ MoO_2}$ $C_6H_6 + H_2O$ — Benzene
- $\xrightarrow[\text{Sodium phenoxide}]{CH_3CH_2Br}$ $C_6H_5OCH_2CH_3 + NaBr$ — Ethylphenylether
- $\xrightarrow[\text{Sodium phenoxide}]{(CH_3)_2SO_4}$ $C_6H_5OCH_3 + CH_3NaSO_4$ — Anisole
- $\xrightarrow[\text{Sodium phenoxide}]{(CH_3CO)_2O}$ $C_6H_5OCOCH_3 + CH_3COONa$ — Phenylacetate
- $\xrightarrow[\text{Sodium phenoxide}]{C_6H_5COCl}$ $C_6H_5OCO{-}C_6H_5 + NaCl$ — Phenylbenzoate
- $\xrightarrow{PCl_5}$ $C_6H_5Cl + POCl_3 + HCl$ — Chlorobenzene

Phenol →

$FeCl_3$, *3 Moles → Fe^{3+} [$\bar{O}$–C_6H_5]$_3$ + 3HCl

Ferric phenoxide, purple complex

$ZnCl_2$, NH_3 → C_6H_5–NH_2 + H_2O

Aniline

(O) → O=C_6H_4=O

Quinone

$K_2S_2O_3$, Alkaline solution → OH–C_6H_4–OH

Quinol

Dil. HNO_3 → o-(NO_2)C_6H_4–OH + HO–C_6H_4–NO_2

o-Nitrophenol *p*-Nitrophenol

Conc. HNO_3 → (O_2N)$_3C_6H_2$–OH

2,4,6-Trinitrophenol (Picric acid)

$NaNO_2$/HCl, HNO_3, 5°C → o-(NO)C_6H_4–OH + HO–C_6H_4–NO

o-Nitrosophenol *p*-Nitrosophenol

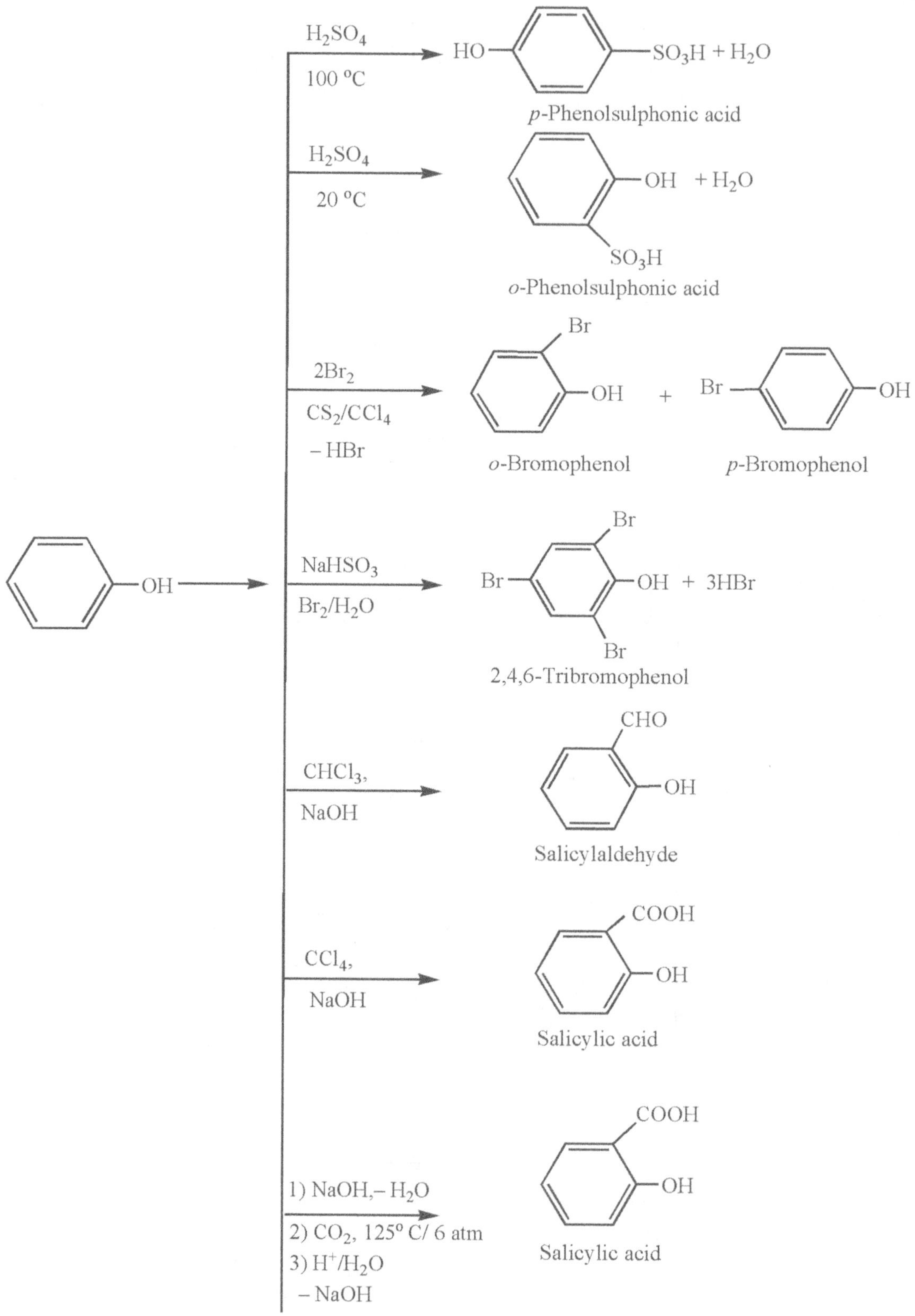

Phenol-OH →

H_2SO_4, 100 °C → HO–C_6H_4–SO_3H + H_2O
p-Phenolsulphonic acid

H_2SO_4, 20 °C → (2-OH)$C_6H_4SO_3H$ + H_2O
o-Phenolsulphonic acid

$2Br_2$, CS_2/CCl_4, – HBr → *o*-Bromophenol + *p*-Bromophenol (Br–C_6H_4–OH)

$NaHSO_3$, Br_2/H_2O → Br, Br, Br–C_6H_2–OH + 3HBr
2,4,6-Tribromophenol

$CHCl_3$, NaOH → (CHO)(OH)C_6H_4
Salicylaldehyde

CCl_4, NaOH → (COOH)(OH)C_6H_4
Salicylic acid

1) NaOH,– H_2O
2) CO_2, 125° C/ 6 atm
3) H^+/H_2O
– NaOH
→ (COOH)(OH)C_6H_4
Salicylic acid

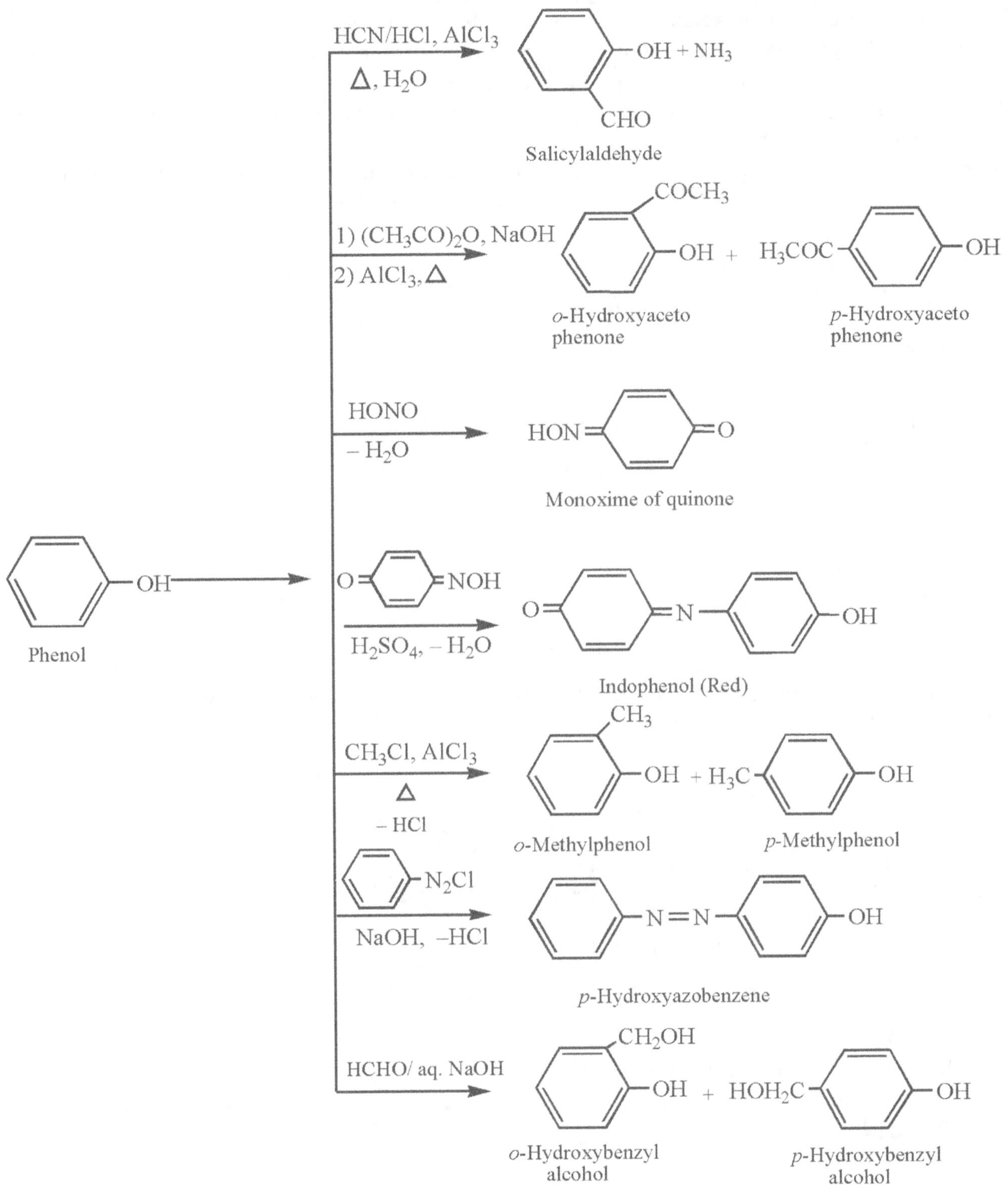

Phenol

- $\xrightarrow[\Delta,\ H_2O]{HCN/HCl,\ AlCl_3}$ Salicylaldehyde (o-CHO, OH) + NH_3
- $\xrightarrow[2)\ AlCl_3,\ \Delta]{1)\ (CH_3CO)_2O,\ NaOH}$ o-Hydroxyacetophenone + p-Hydroxyacetophenone (H_3COC–C_6H_4–OH)
- $\xrightarrow[-H_2O]{HONO}$ HON=C_6H_4=O — Monoxime of quinone
- $\xrightarrow[H_2SO_4,\ -H_2O]{O=C_6H_4=NOH}$ O=C_6H_4=N–C_6H_4–OH — Indophenol (Red)
- $\xrightarrow[\Delta,\ -HCl]{CH_3Cl,\ AlCl_3}$ o-Methylphenol + p-Methylphenol (H_3C–C_6H_4–OH)
- $\xrightarrow[NaOH,\ -HCl]{C_6H_5N_2Cl}$ C_6H_5–N=N–C_6H_4–OH — p-Hydroxyazobenzene
- $\xrightarrow{HCHO/\ aq.\ NaOH}$ o-Hydroxybenzyl alcohol (CH_2OH, OH) + p-Hydroxybenzyl alcohol (HOH_2C–C_6H_4–OH)

Uses of Phenol:

1. Used in the preparation of drugs.
2. Used in the preparation of several dyes.
3. Used as an antiseptic.

Tests for identification

1. Phenol has a characteristic smell.
2. **Ferric chloride test:** Phenol gives violet colour with neutral ferric chloride solution.
3. It gives positive test for Libermanns nitroso reaction.
4. **With bromine water:** An aqueous solution of phenol gives white precipitate of tribromo phenol when treated with bromine water.
5. **Phenolphthalein test:** Upon heating phenol and phthalic anhydride, phenolphthalein is formed which gives pink colour with alkali.

Homologues of Phenol

Methyl Phenols (Cresols)

$CH_3C_6H_4OH$

Methyl phenols are also called a cresols. They are of 3 types.

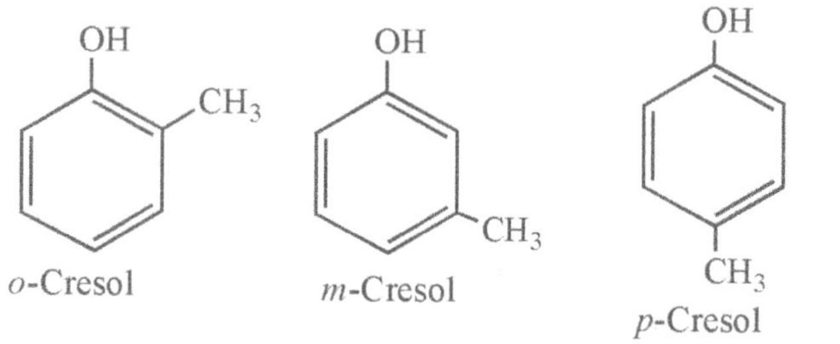

OH, CH_3 — *o*-Cresol; OH, CH_3 — *m*-Cresol; OH, CH_3 — *p*-Cresol

They occur in middle oil fraction of coal tar. Three isomers are separated by fractional distillation. They are synthesized from toludine as follows.

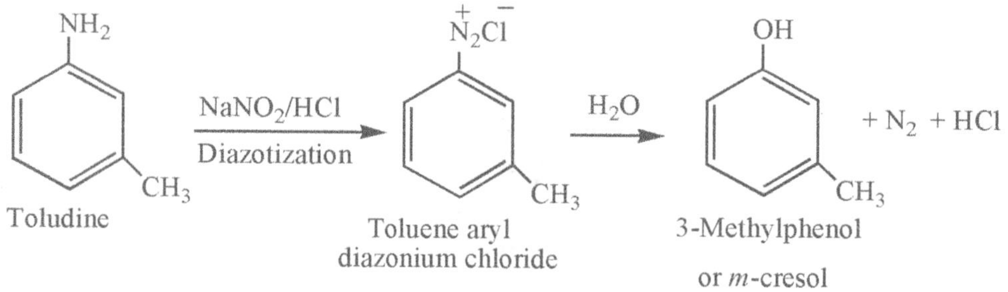

$$\text{Toludine } (NH_2, CH_3) \xrightarrow[\text{Diazotization}]{NaNO_2/HCl} \text{Toluene aryl diazonium chloride } (\overset{+}{N_2}\overset{-}{Cl}, CH_3) \xrightarrow{H_2O} \text{3-Methylphenol or } m\text{-cresol } (OH, CH_3) + N_2 + HCl$$

Physical Properties: They are colourless substances having phenolic odour.

Uses:

A soap solution of methyl phenol is known as Lysol which is used as a disinfectant.

Dihydric Phenols

They are having three isomeric forms.

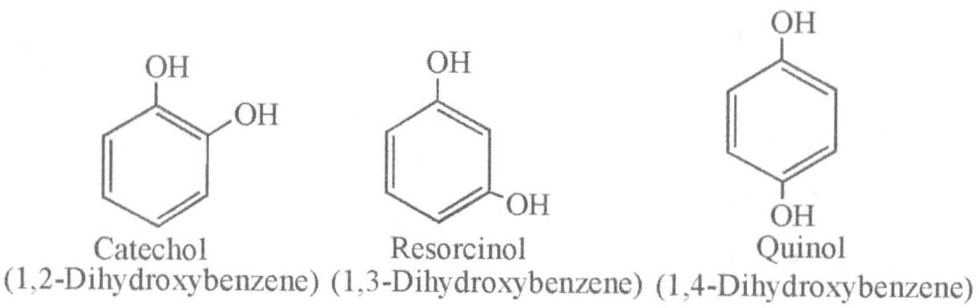

OH, OH — Catechol (1,2-Dihydroxybenzene); OH, OH — Resorcinol (1,3-Dihydroxybenzene); OH, OH — Quinol (1,4-Dihydroxybenzene)

Catechol

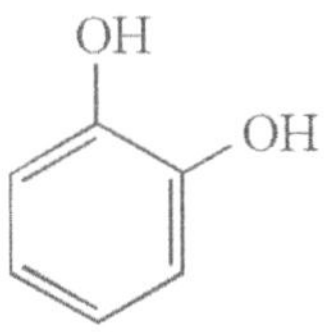

It was first obtained by distillation of Indian catechu.

Physical properties: It is a colourless solid, melting point 104 °C, soluble in water, alcohol and ether. It gives green colour with $FeCl_3$ solution which changes to red upon reaction with Na_2CO_3. It is used in the preparation of an important derivative of catechol called **adrenaline** (a hormone secreted from adrenal glands).

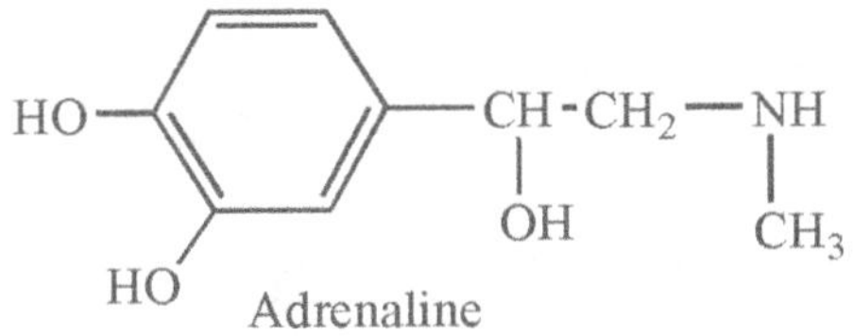

Adrenaline

Resorcinol

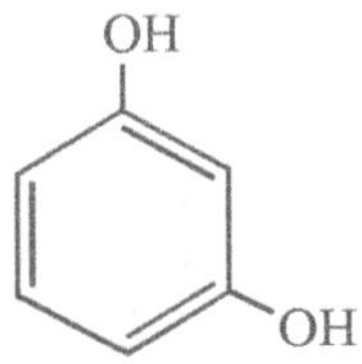

Preparation: It is prepared by alkaline fusion of 1,3-benzene disulphonic acid.

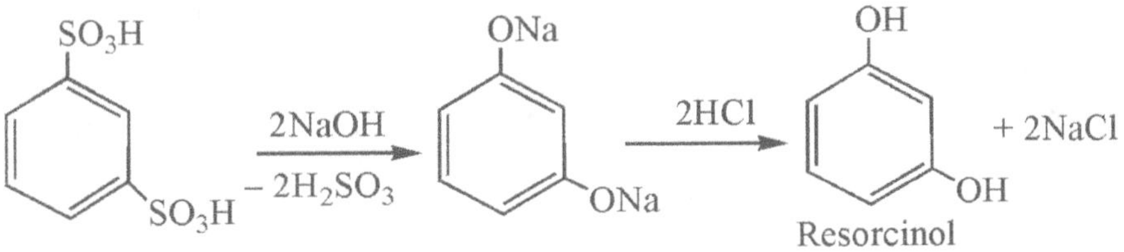

$$\text{1,3-Benzene disulphonic acid} \xrightarrow[-\,2H_2SO_3]{2NaOH} C_6H_4(ONa)_2 \xrightarrow{2HCl} \text{Resorcinol} + 2NaCl$$

Physical properties: It is a colourless crystalline solid, M.P. 110 °C, very soluble in water, alcohol and ether. It gives violet colour with $FeCl_3$ solution.

Uses:

It is used in the preparation of dyes like fluorescein and as an antiseptic.

Quinol

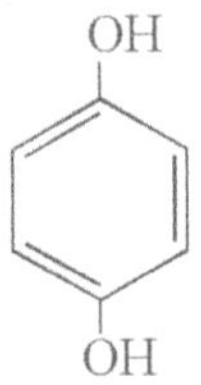

Preparation: It is prepared by reducing *p*-benzoquinone with sulphurous acid (H_2SO_3).

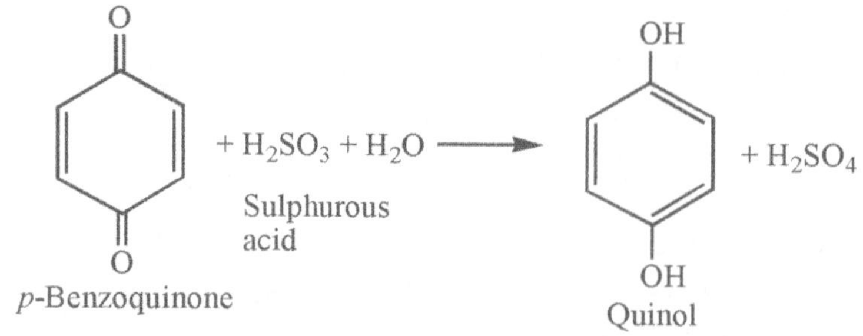

$$p\text{-Benzoquinone} + \underset{\text{Sulphurous acid}}{H_2SO_3} + H_2O \longrightarrow \text{Quinol} + H_2SO_4$$

Physical Properties: It is a colourless crystalline solid, M.P. 173 °C, very soluble in water, alcohol and ether. It gives transient blue colour with $FeCl_3$ solution.

Uses: It is used as an antiseptic.

Trihydric Phenols

They exist as three isomeric forms as shown below.

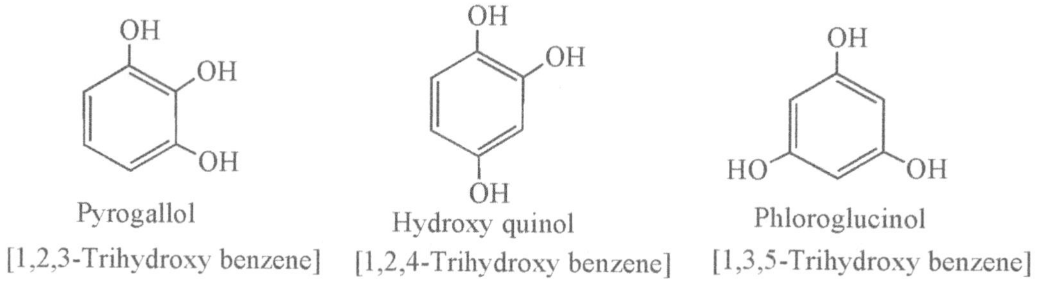

Pyrogallol [1,2,3-Trihydroxy benzene]   Hydroxy quinol [1,2,4-Trihydroxy benzene]   Phloroglucinol [1,3,5-Trihydroxy benzene]

Pyrogallol

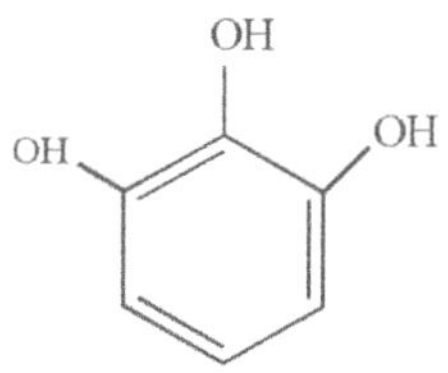

Preparation: It is prepared by reacting gallic acid in a current of NO.

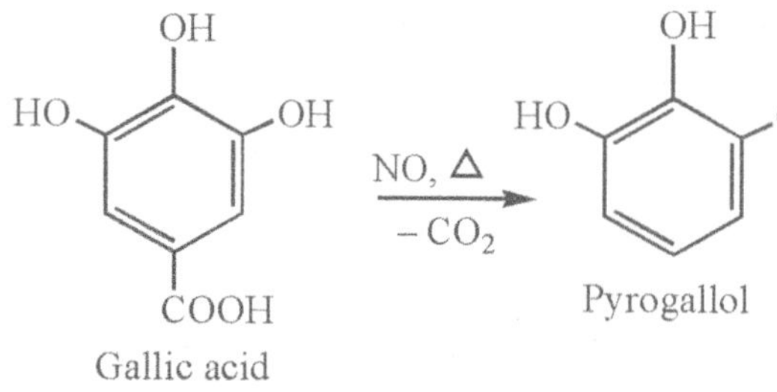

$$\text{Gallic acid} \xrightarrow[-CO_2]{NO,\ \Delta} \text{Pyrogallol}$$

Physical Properties: It is a colourless crystalline solid, M.P. 485 K, very soluble in water, alcohol and ether. It gives red colour with $FeCl_3$.

Uses: It is used as an antiseptic.

Hydroxy Quinol

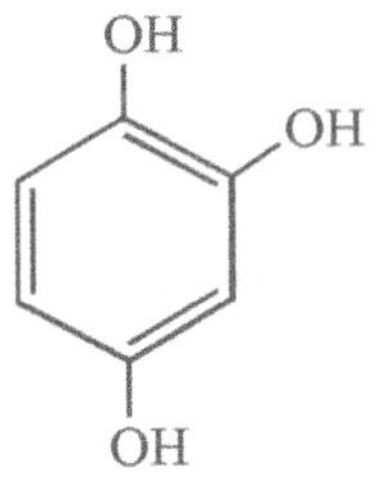

OH
OH
OH

Preparation: It is prepared by alkaline fusion of 1,4-dihydroxy benzene in air.

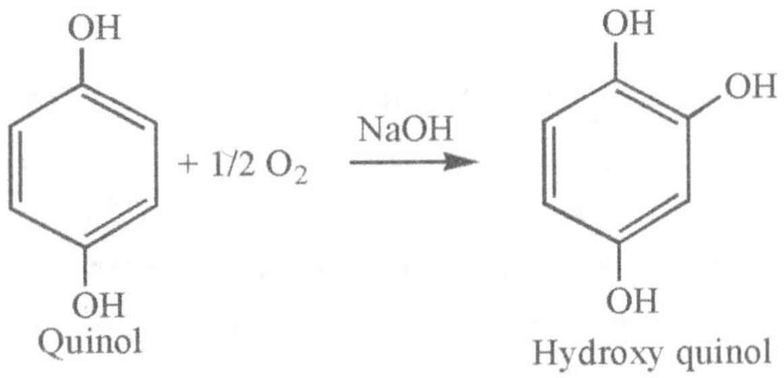

$$\text{Quinol} + 1/2\,O_2 \xrightarrow{NaOH} \text{Hydroxy quinol}$$

Quinol

Hydroxy quinol

Physical Properties: It is a colourless, crystalline solid, very soluble in water. It gives greenish-brown colour with $FeCl_3$ solution.

Phloroglucinol

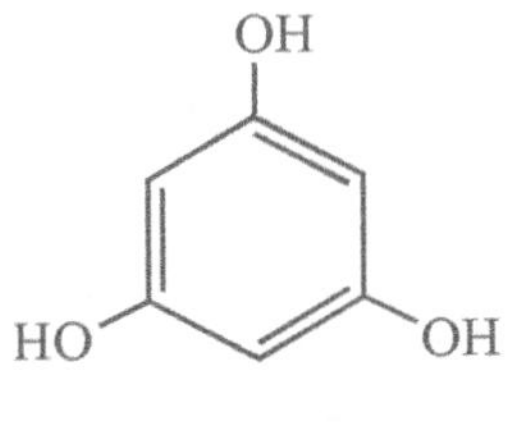

OH
HO OH

Preparation: It is prepared by fusing resorcinol (1,3-dihydroxyl benzene) with NaOH in air.

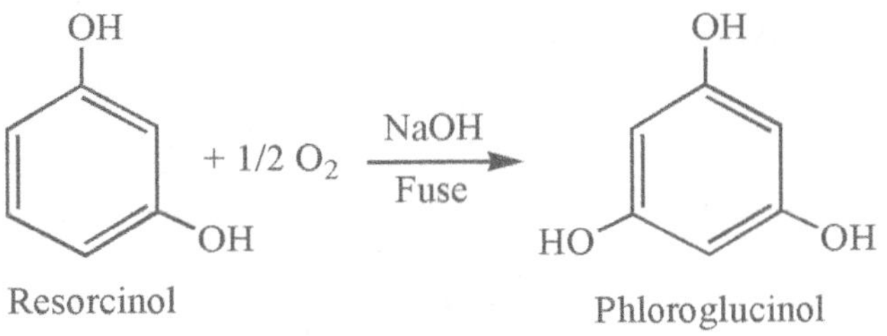

$$\text{Resorcinol} + 1/2\,O_2 \xrightarrow[\text{Fuse}]{NaOH} \text{Phloroglucinol}$$

Resorcinol

Phloroglucinol

Physical Properties:

It is a colourless solid, M.P. 491 K, soluble in water. It gives bluish-violet colour with $FeCl_3$ solution. It gives red colour with carbohydrates, hence used in the detection of carbohydrates.

Aromatic Alcohols

They are aryl derivatives of aliphatic alcohols.

Benzyl Alcohol (Phenyl Methanol)

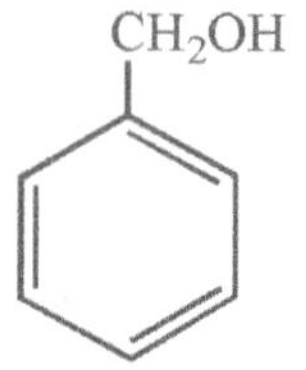

CH_2OH

It occurs in nature as free and ester in jasmine oil. The benzyl alcohol is prepared by following methods:

Uses:

1. Used as an antiseptic.
2. Used in the preparation of benzyl benzoate that is used to treat asthma and whooping cough.

Amino Phenols

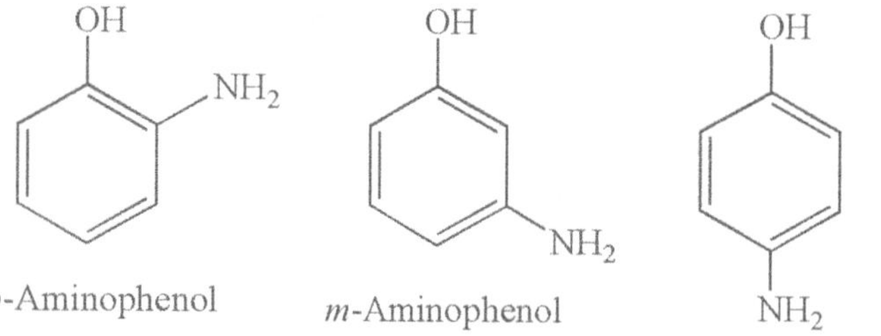

OH, NH_2 — *o*-Aminophenol

OH, NH_2 — *m*-Aminophenol

OH, NH_2 — *p*-Aminophenol

Aromatic Ethers

Two important phenolic ethers are methoxy benzene and ethoxy benzene.

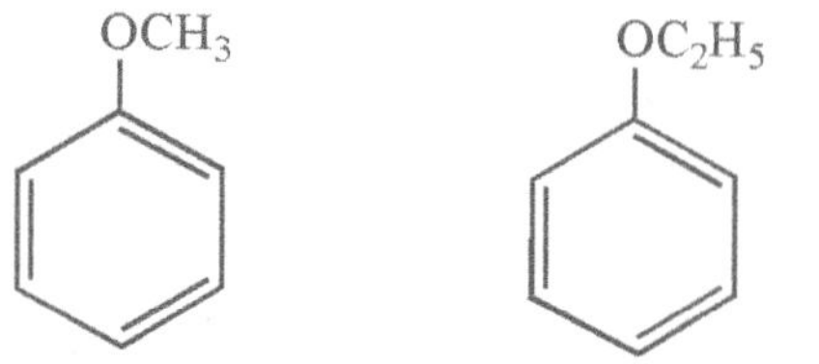

OCH_3

Anisole
(Methoxy benzene)

OC_2H_5

Phenetole
(Ethoxy benzene)

Structure and Uses of Medicinal important compounds

Phenol

OH

Uses:

1. Used in the preparation of drugs.
2. Used in the preparation of several dyes.
3. Used as an antiseptic and disinfectant.

Cresols

OH

CH_3

o-Cresol

Uses:

1. Used is for the synthesis of Mephenesin which used as muscle relaxant.
2. O-cresol and its derivatives used as herbicides.

Cresols

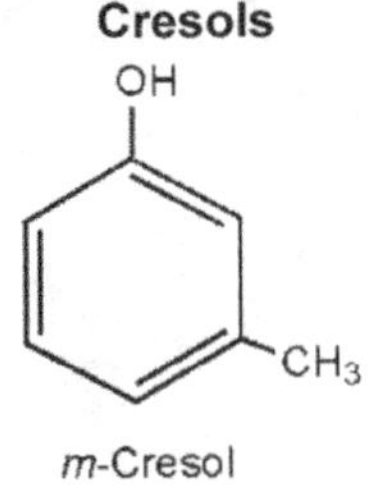

m-Cresol

Uses:

1. It is used as a precursor for tolamolol and Vitamin E.
2. It is also used for the synthesis of pesticides (Penthione).
3. It is used as preservative in some insulin.

Cresols

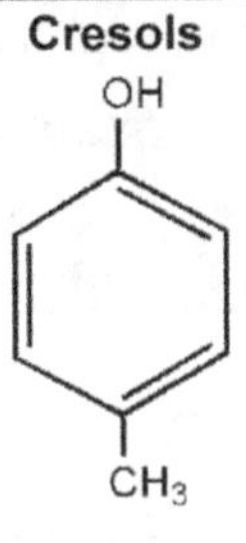

p-Cresol

Uses:

1. Used for the Production of Butylated hydroxy toluidine (BHT) this is used as an antioxidant.
2. It is also used for the synthesis of Buproponolol (Beta Blocker).

Resorcinol

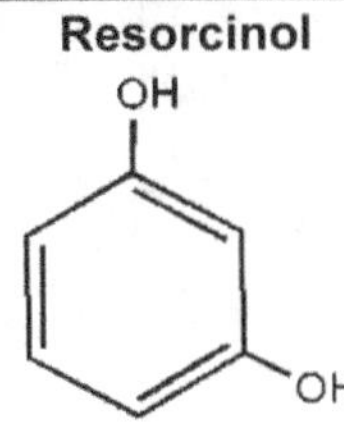

Uses:

It is used in the preparation of dyes like fluorescein and as an antiseptic.

Naphthols

1-Naphthol

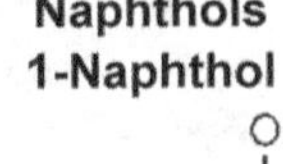

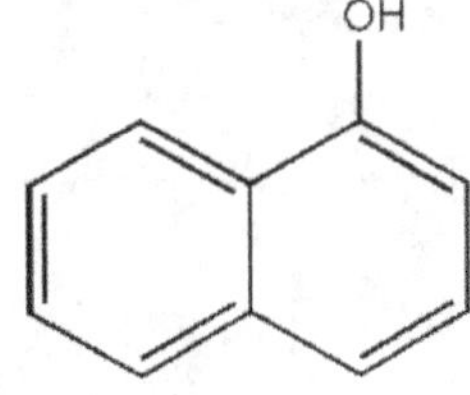

α-Naphthol

Uses:

1. It is used as a precursor for drug synthesis.
2. It is the main ingredient of Molisch reagent.
3. It is also used for the synthesis of dyes.

Naphthols

2-Naphthol

OH

β-Naphthol

Uses:

1. It is used as an antiseptic for skin diseases.
2. Methyl & ethyl ethers of 2-naphthol such as nerolins are used in perfume industries.
3. It is also used for synthesizing dyes.

Probable Questions

1. What are phenols? How will you differentiate phenols from alcohols?
2. Write in detail about the nomenclature and importance of phenols and aromatic alcohols.
3. Mention any two important ways of introducing hydroxyl group (- OH) in benzene nucleus. Compare and contrast the properties of phenol with those of benzyl alcohol.
4. Write the general methods of preparation of monohydric phenols.
5. Arrange the following according to their increasing order of acidity.
   p-Nitrophenol, *p*-Methyl phenol and *p*-Cresol.
6. Explain the followings.
   (a) Reimer-Tiemann reaction.
   (b) Kolbe's reaction.
7. Explain the reaction and mechanism of Dow process.
8. How can you convert the followings in to phenol?
   (a) Benzene diazonium chloride
   (b) Cumene
   (c) Chlorobenzene
9. How phenol is recovered from coal-tar?
10. Explain in detail about the acidity of phenols.
11. Write the effect of substituents on acidity of phenol.
12. Boiling point of phenol is much higher than the hydrocarbons of roughly same molecular weight. Justify it with suitable examples.
13. Among the three isomers of phenols, the *ortho* halo isomers possess lower melting point than *meta* and *para* - Explain.
14. Write a role on steric hindrance in the acidity of phenol.
15. Enumerate the various chemical reactions of monohydric phenols.
16. Write a short note on a) Williamson's ether synthesis b) Schotten- Baumann reaction c) Elbs persulphate oxidation.
17. Enumerate the various electrophilic aromatic substitution reactions of phenol.
18. Explain the following.
    (a) Gattermann reaction
    (b) Fries rearrangement
    (c) Libermanns nitroso reaction
    (d) Friedel-Craft's reaction.
19. Mention the products obtained when phenol reacts with following and write the appropriate chemical equations.
    (a) Methyl chloride
    (b) Diazonium salts
    (c) HCHO
    (d) Phthalic anhydride
20. Write the identification tests for phenols.
21. Mention the reactions of phenols which are similar to those of alcohols.
22. Mention the reactions of phenols which are different from alcohols.
23. Enumerate the various uses of phenols.

24. Write the structure, synthesis, physical properties, chemical properties and uses of picric acid.
25. Define methyl phenols and mention the structure of three types of methyl phenols along with their uses.
26. What is meant by dihydric phenols and mention its three isomeric forms?
27. Describe the structure, physical properties and uses of the following.
    (a) Catechol
    (b) Resorcinol
    (c) Quinol
28. What is meant by trihydric phenols? and mention its three isomeric forms.
29. Write the structure, synthesis, physical properties and uses of the following.
    (a) Pyrogallol
    (b) Hydroxy quinol
    (c) Phloroglucinol
30. Write the general methods of preparation and chemical properties of dihydric phenols.
31. Write a short note on phenolic ethers.
32. What is meant by aromatic ethers and write the structure, the synthesis, chemical properties and uses of anisole?
33. Write the structure and uses of Phenols, Cresols, Resorcinol and Naphthols.

24

Aromatic Carboxylic Acids and their Derivatives

Introduction

These are obtained from aromatic hydrocarbons by the replacement of one or more hydrogen atoms by carboxylic acid groups (-COOH). The carboxylic acids groups may be directly attached to the benzene ring or may be present in the side chain of the benzene ring. Some of the typical aromatic acids are mentioned below.

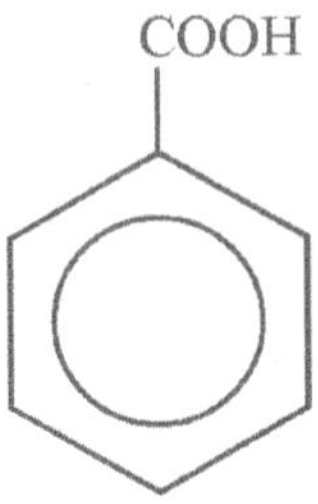

COOH

Benzoic acid

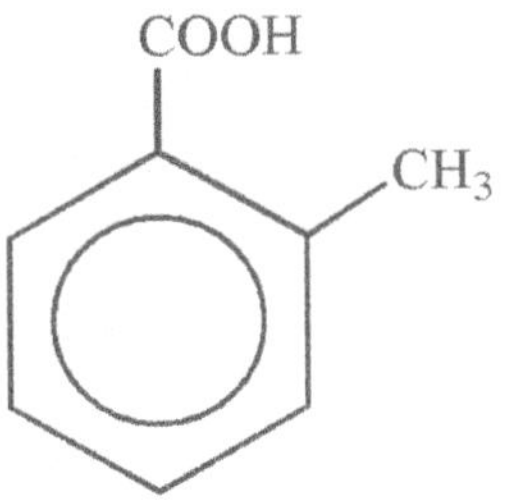

COOH, CH_3

o-Toluic acid

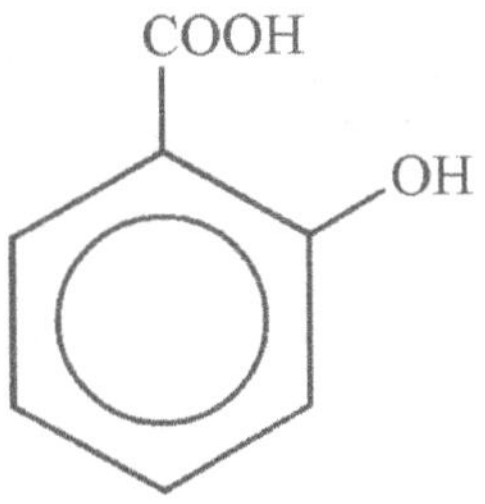

COOH, OH

Salicylic acid

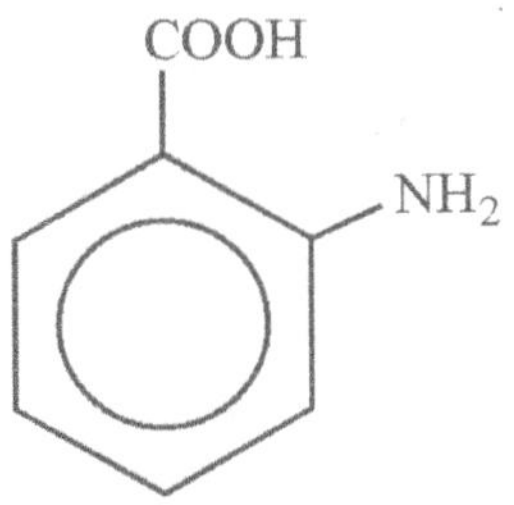

COOH, NH_2

Anthranilic acid

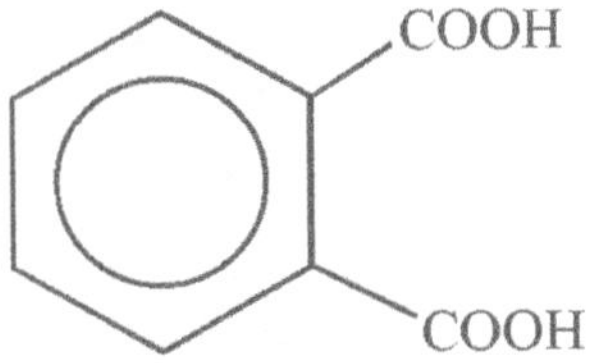

COOH, COOH

Phthalic acid

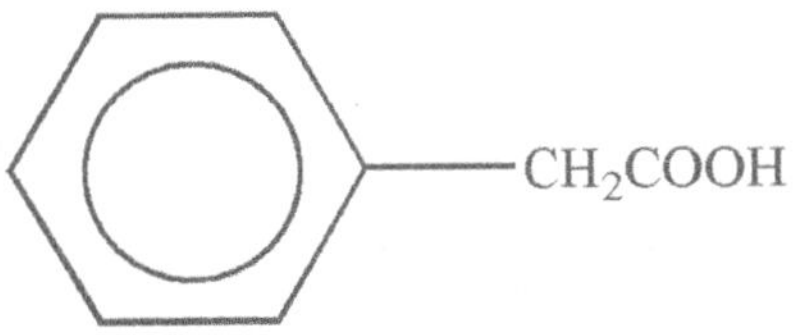

CH_2COOH

Phenyl acetic acid

In the above examples, the acids containing carboxylic group in the side chain are known as araliphatic carboxylic acids but they are classified as aromatic acids due to the presence of an aromatic ring.

Importance of Aryl Carboxylic Acids

Aryl carboxylic acid groups when attached with various chemical structures elicit different kind of pharmacological activities based on the core structure to which it is attached. Some of the drugs which possess aryl carboxylic acid groups and their therapeutic uses in the treatment of various diseases are shown in Table 24.1.

Table 24.1 Drugs containing aryl carboxylic acids and their therapeutic uses.

Category	Drug	Structure
Antihypertensive drug	Telmisartan	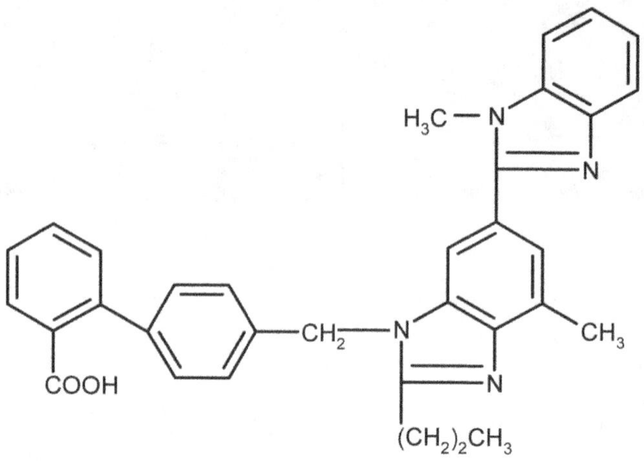 H_3C—N, N, COOH, CH_2—N, N, CH_3, $(CH_2)_2CH_3$
Anticoagulant & Analgesic	Aspirin	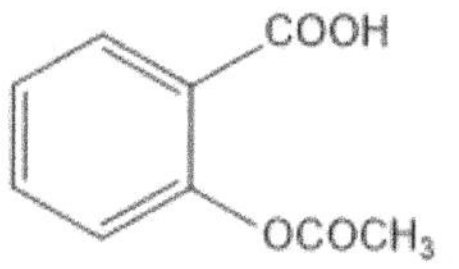 COOH, $OCOCH_3$
Diuretics	Bumetanide	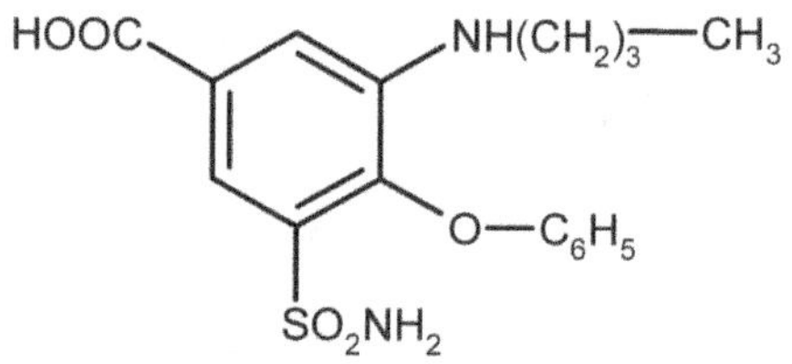 HOOC, $NH(CH_2)_3$—CH_3, O—C_6H_5, SO_2NH_2
NSAIDs	Probenecid	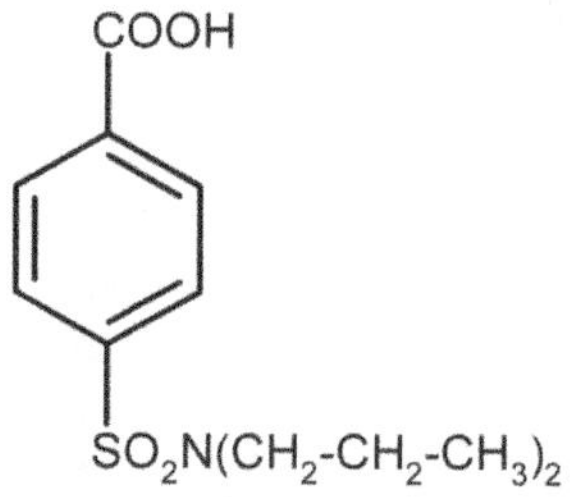 COOH, $SO_2N(CH_2\text{-}CH_2\text{-}CH_3)_2$
	Diflunisal	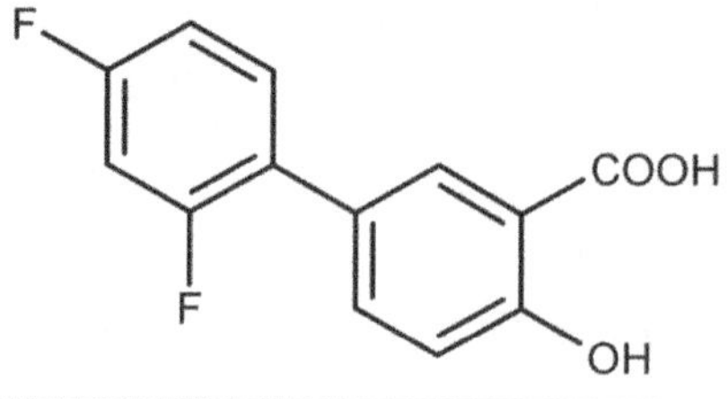 F, F, COOH, OH
	Mefenamic acid	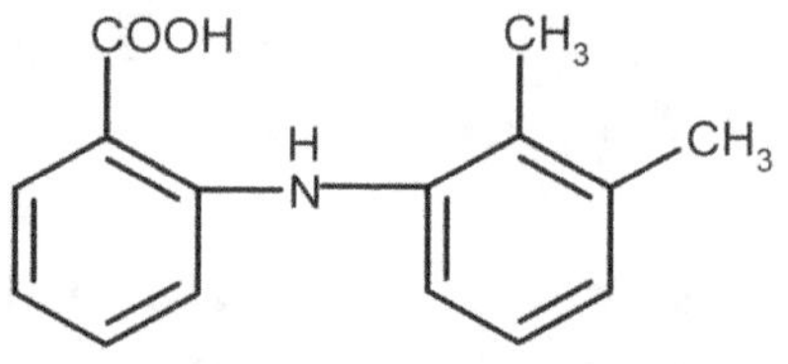 COOH, H N, CH_3, CH_3

Table 24.1 Contd...

Local anaesthetic	Hexylcaine	$H-C_6H_4-COO-CH(CH_3)-CH_2-NH-C_6H_{11}$
	Meprylcaine	$H-C_6H_4-COO-CH_2-C(CH_3)_2-NH-CH_2CH_2CH_3$
	Benzocaine	$H_2N-C_6H_4-COO-CH_2CH_3$

Nomenclature

Carboxyl (-COOH) group attached directly to aryl ring are called as aryl carboxylic acids. They are named generally by common name or as derivatives of benzoic acid. When the substituents are present on the benzoic acid, then they are numbered consecutively after - COOH group. If the -COOH group is linked to alkyl group then they are called as aryl substituted aliphatic carboxylic acids. Some of the examples are shown below.

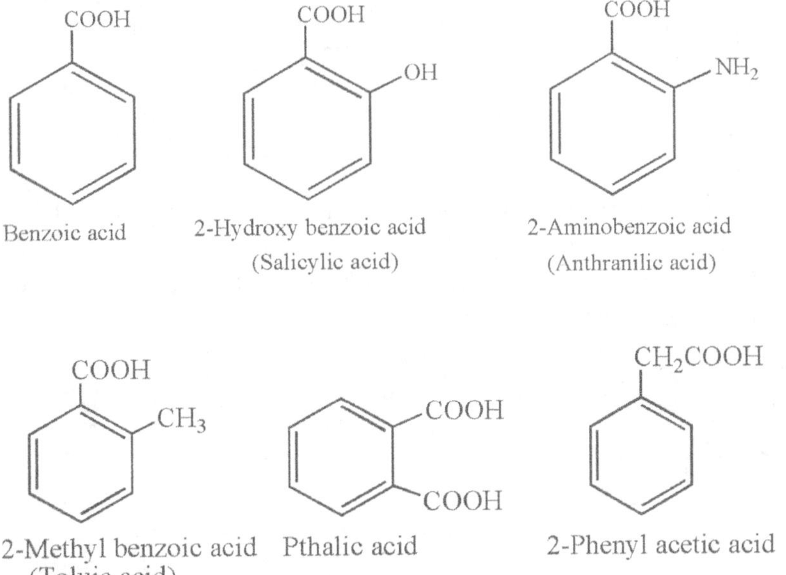

Benzoic acid; 2-Hydroxy benzoic acid (Salicylic acid); 2-Aminobenzoic acid (Anthranilic acid)

2-Methyl benzoic acid (Toluic acid); Pthalic acid; 2-Phenyl acetic acid

General Methods of Preparation

Aromatic acids are prepared by the methods similar to that of general methods of preparation of aliphatic carboxylic acid. All the preparations, physical properties and chemical properties are explained by taking a simple aryl carboxylic acid (benzoic acid) as an example.

Monobasic Acids with Carboxyl Group Directly Attached to the Nucleus

General Methods of Preparation

1. **Oxidation of corresponding aromatic aldehydes or alcohols:** Benzaldehyde or benzyl alcohol is oxidized by strong oxidizing agents such as $KMnO_4$ or $K_2Cr_2O_7$ gives benzoic acid.

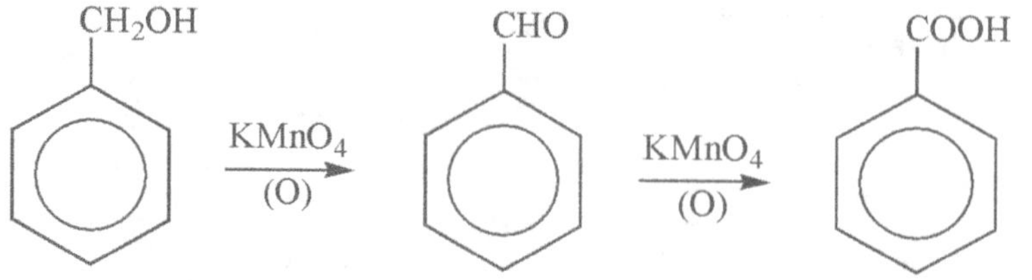

CH_2OH CHO $COOH$

$KMnO_4$ (O) $KMnO_4$ (O)

Benzyl alcohol Benzaldehyde Benzoic acid

2. **Hydrolysis of aromatic cyanides or aryl nitriles:** Benzonitrile or substituted benzonitrile undergoes hydrolysis to yield aromatic acids.

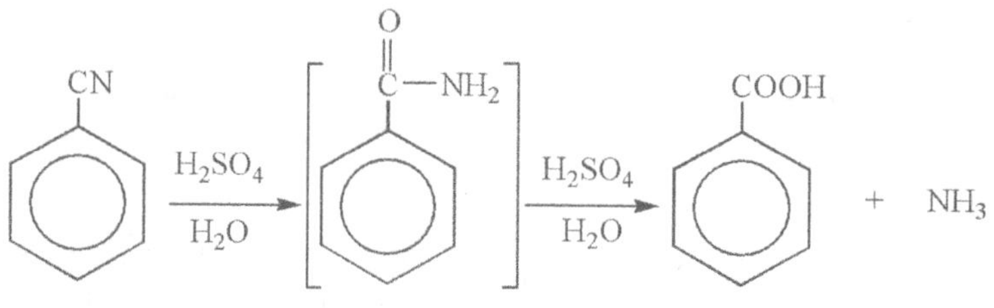

CN O C—NH_2 COOH

H_2SO_4 H_2O H_2SO_4 H_2O + NH_3

Benzonitrile Benzamide Benzoic acid

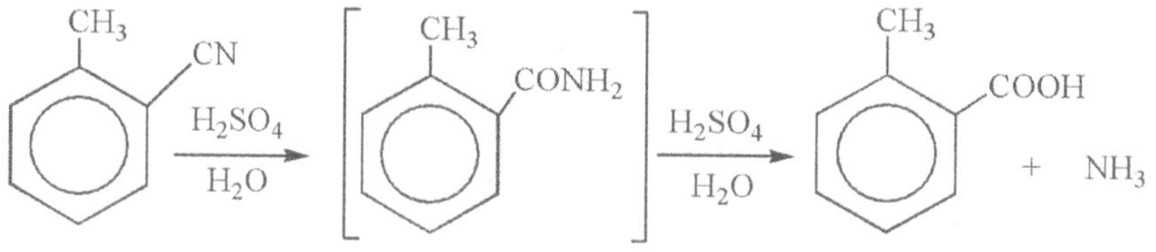

CH_3 CN CH_3 $CONH_2$ CH_3 COOH

H_2SO_4 H_2O H_2SO_4 H_2O + NH_3

o-Toluene carbonitrile *o*-Methyl benzamide *o*-Toluic acid

The aryl nitrile needed for the reaction is obtained by Sandmeyer's reaction from salts of sulphonic acid and KCN.

Sandmeyer's reaction - Benzene diazonium chloride when reacts with cuprous cyanide dissolved in aqueous potassium cyanide yields benzonitrile.

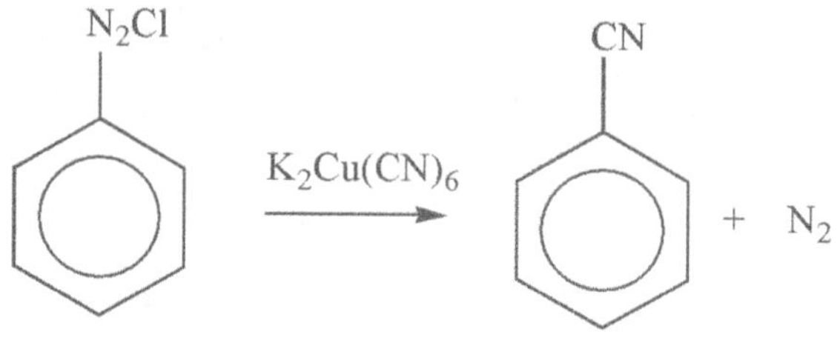

N_2Cl CN

$K_2Cu(CN)_6$ + N_2

Benzene diazonium chloride Benzonitrile

The benzonitrile is also obtained from the alkali salts of sulphonic acid with potassium cyanide.

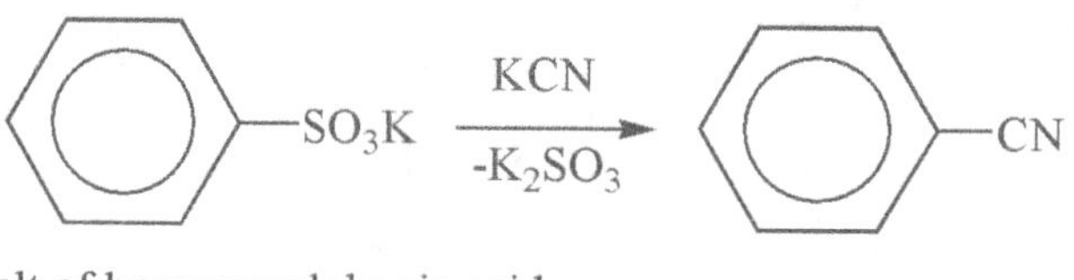

SO_3K KCN $-K_2SO_3$ CN

Potassium salt of benzenesulphonic acid Benzonitrile

3. **Friedel-Craft's acylation reaction:** Benzene when treated with carbonyl chloride in the presence of aluminium chloride gives benzoyl chloride which upon hydrolysis yields benzoic acid. Here excess amount of carbonyl chloride is necessary otherwise benzophenone may be obtained as a main product. The yield of benzoic acid obtained is 55-58 %.

$$C_6H_6 + COCl_2 \xrightarrow[-HCl]{AlCl_3} C_6H_5COCl \xrightarrow[-HCl]{H_2O} C_6H_5COOH$$

Benzene Carbonyl chloride Benzoyl chloride Benzoic acid

4. **By the oxidation of homologues of benzene:** Toluene is oxidized with dilute nitric acid or potassium dichromate / sulphuric acid or alkaline potassium permanganate to yield benzoic acid.

$$C_6H_5CH_3 \xrightarrow[KMnO_4]{(O)} C_6H_5CH_2OH \xrightarrow{(O)} C_6H_5COOH$$

Toluene Benzyl alcohol (Intermediate product) Benzoic acid

5. **From Grignard's reagent:** Aryl magnesium halides when treated with CO_2 gives addition compound which upon further hydrolysis yield benzoic acid.

$$C_6H_5MgBr + O=C=O \longrightarrow C_6H_5-\overset{O}{\overset{\|}{C}}-OMgBr$$

Phenyl magnesium bromide Addition product

$$\downarrow HCl$$

$$C_6H_5-\overset{O}{\overset{\|}{C}}-OH + Mg(Br)(Cl)$$

Benzoic acid

6. **From benzene trichloride:** Alkaline hydrolysis of benzene trichloride yields benzoic acid. Benzene trichloride is obtained by chlorination of toluene with Cl_2 in the presence of UV light.

$$C_6H_5CH_3 + 3Cl_2 \xrightarrow[-3HCl]{UV} C_6H_5CCl_3 \xrightarrow[H^+,\ -3NaCl]{3NaOH} [C_6H_5C(OH)_3]$$

Toluene Benzene trichloride/ Trichlorotoluene

$$[C_6H_5C(OH)_3] \xrightarrow{NaOH,\ -2H_2O} C_6H_5COONa \xrightarrow[H_2O,\ -NaOH]{H^+} C_6H_5COOH$$

Sodium benzoate Benzoic acid

Summary of Methods of Preparation

1) Benzyl alcohol ($C_6H_5CH_2OH$) $\xrightarrow[(O)]{KMnO_4}$ Benzoic acid (C_6H_5COOH)

2) Benzonitrile (C_6H_5CN) $\xrightarrow[H_2O]{H_2SO_4}$ Benzoic acid (C_6H_5COOH) + NH_3

3) Benzene $\xrightarrow[AlCl_3,\ H_2O,\ -2HCl]{COCl_2}$ Benzoic acid (C_6H_5COOH)

4) Toluene ($C_6H_5CH_3$) $\xrightarrow[(O)]{KMnO_4}$ Benzoic acid (C_6H_5COOH)

5) Phenyl magnesium bromide (C_6H_5MgBr) $\xrightarrow[-\ Mg(Br)Cl]{CO_2,\ HCl}$ Benzoic acid (C_6H_5COOH)

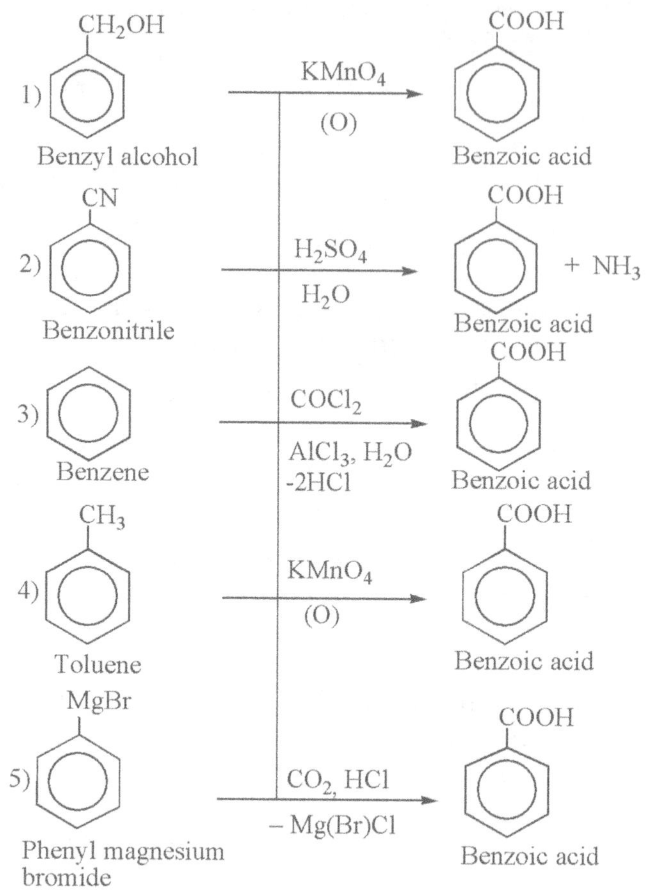

General Properties: Aromatic acids are less volatile, slightly soluble in water and are slightly stronger acids. This is due to the presence of phenyl group. Aromatic acids get decarboxylated immediately when treated with soda lime.

Individual Members

Benzoic Acid

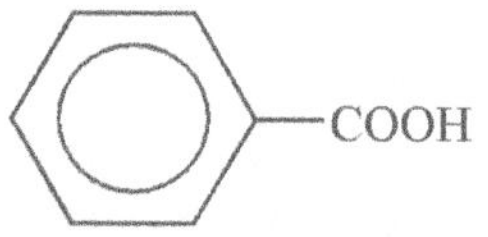

C_6H_5—COOH

Benzoic acid

It is mainly found in resins and balsams particularly gum benzoin. It is also present in the urine of horses as hippuric acid.

Preparation: Described above under general methods of preparations of carboxylic acids.

Physical Properties: It occurs as white pearly flakes, melting point 395 K, sparingly soluble in cool water but readily soluble in hot water, alcohol, ether and benzene. It is volatile in steam, slightly acidic than acetic acid. Its vapours possess irritating smell, provoke sneezing and coughing.

Chemical Properties: Chemical properties of aromatic acids are similar to that of aliphatic carboxylic acids. The chemical reactions of benzoic acid are classified into two types.

(I) Reactions of carboxyl group (– COOH).

(II) Reactions of aromatic ring or phenyl group.

(I) Reactions of carboxyl group (-COOH)

(a) Reaction with sodium hydroxide or sodium carbonate ($NaOH/Na_2CO_3$).
(b) Reaction with ammonia (NH_3).
(c) Reaction with phosphorous pentachloride or thionylchloride (PCl_5 or $SOCl_2$).
(d) Reduction with $LiAlH_4$.
(e) Reaction with sodalime or Decarboxylation.
(f) Esterification.

(a) Reaction with NaOH or Na_2CO_3 (salt formation): Benzoic acid reacts with alkali metal hydroxide or carbonate to yield corresponding salt.

$$C_6H_5\text{-}COOH + NaOH \longrightarrow C_6H_5\text{-}COONa + H_2O$$

Benzoic acid, Sodium hydroxide, Sodium benzoate

$$2\,C_6H_5\text{-}COOH + Na_2CO_3 \longrightarrow 2\,C_6H_5\text{-}COONa + CO_2\uparrow + H_2O$$

Benzoic acid, Sodium carbonate, Sodium benzoate

(b) Reaction with ammonia (NH_3): Benzoic acid reacts with ammonia to give ammonium benzoate which upon further heating yields benzamide.

$$C_6H_5\text{-}COOH + NH_3 \longrightarrow C_6H_5\text{-}COO^-NH_4^+$$

Benzoic acid, Ammonia, Ammonium benzoate

$$\downarrow \Delta,\ -H_2O$$

$$C_6H_5\text{-}\overset{O}{\overset{\|}{C}}\text{-}NH_2 + H_2O$$

Benzamide

(c) Reaction with PCl_5 or $SOCl_2$ (Formation of acyl halide): Benzoic acid reacts with PCl_5 or $SOCl_2$ and gives benzoyl chloride.

$$C_6H_5\text{-}COOH + PCl_5 \longrightarrow C_6H_5\text{-}\overset{O}{\overset{\|}{C}}\text{-}Cl + POCl_3 + HCl$$

Benzoic acid, Phosphorous penta chloride, Benzoyl chloride

$$C_6H_5\text{-}COOH + SOCl_2 \longrightarrow C_6H_5\text{-}\overset{O}{\overset{\|}{C}}\text{-}Cl + SO_2 + HCl$$

Benzoic acid, Benzoyl chloride

Mechanism:

***Step* 1:** Formation of chlorosulphite: Benzoic acid reacts with thionyl chloride to give chlorosulphite or benzene chlorosulphite.

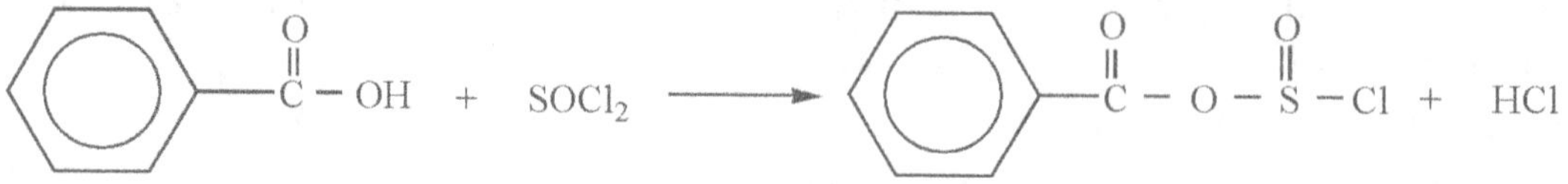

$$C_6H_5-C(=O)-OH + SOCl_2 \longrightarrow C_6H_5-C(=O)-O-S(=O)-Cl + HCl$$

Benzoic acid Chlorosulphite

***Step* 2:** Nucleophile (Cl^-) attacks the chlorosulphite to give tetrahedral intermediate.

$$C_6H_5-C(=O)-O-S(=O)-Cl + :Cl^- \longrightarrow \left[C_6H_5-C(O^-)(Cl)-O-S(=O)-Cl\right]$$

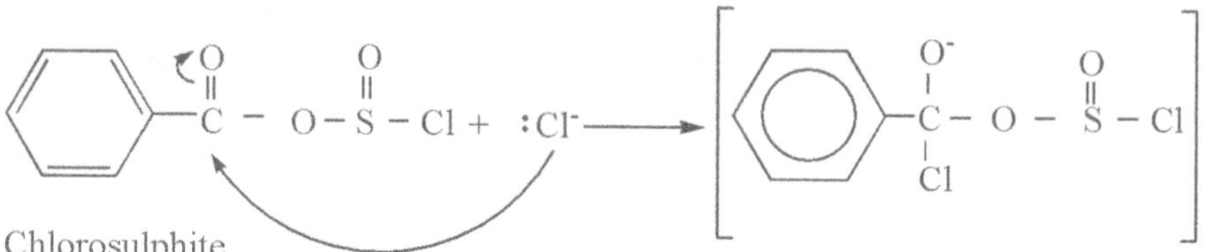

Chlorosulphite Tetrahedral intermediate

***Step* 3:** Elimination of SO_2 and Cl^- from the tetrahedral intermediate gives benzoyl chloride.

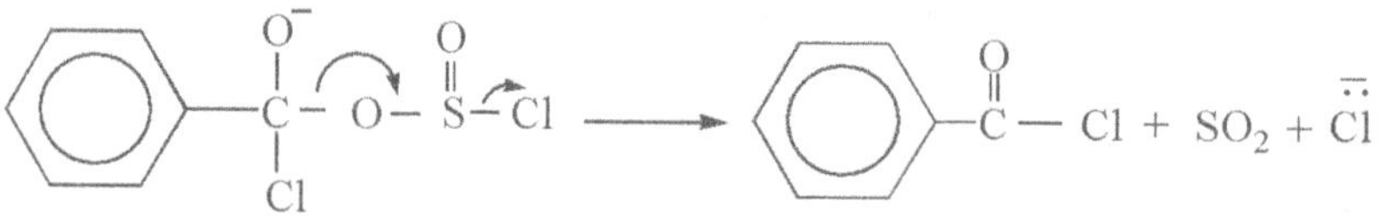

$$C_6H_5-C(O^-)(Cl)-O-S(=O)-Cl \longrightarrow C_6H_5-C(=O)-Cl + SO_2 + \ddot{C}l$$

Tetrahedral intermediate Benzoyl chloride

(d) Reduction with $LiAlH_4$: Benzoic acid when reduced with $LiAlH_4$ gives benzyl alcohol.

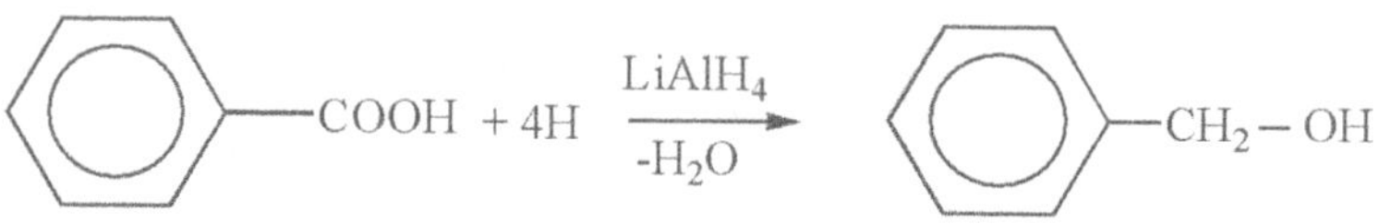

$$C_6H_5-COOH + 4H \xrightarrow[-H_2O]{LiAlH_4} C_6H_5-CH_2-OH$$

Benzoic acid Benzyl alcohol

Mechanism: The mechanism of this reaction follows acyl nucleophilic substitution reaction.

***Step* 1:** Attack of carbonyl carbon by hydride ion leads to the formation of tetrahedral intermediate. The hydride ion needed for this reaction is obtained from $LiAlH_4$.

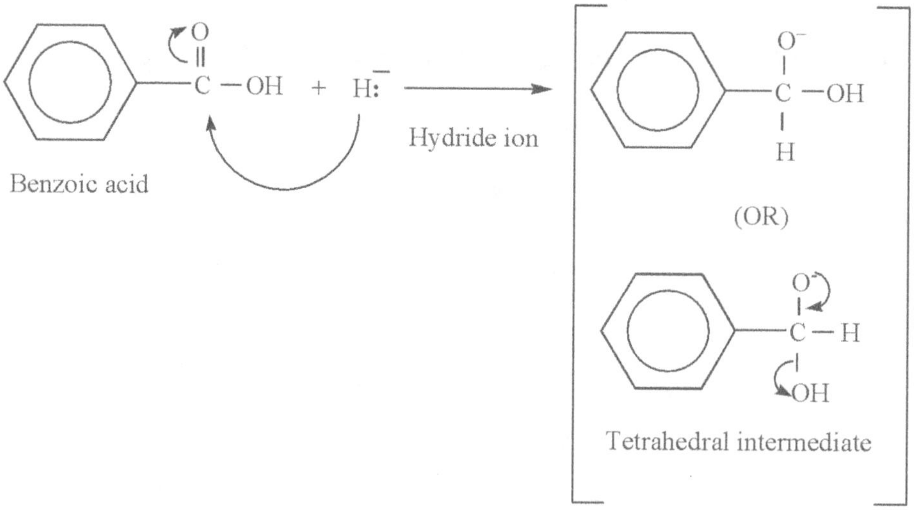

$$C_6H_5-C(=O)-OH + H:^- \longrightarrow \left[C_6H_5-C(O^-)(H)-OH \text{ (OR) } C_6H_5-C(O^-)(OH)-H\right]$$

Benzoic acid Hydride ion Tetrahedral intermediate

***Step* 2:** Removal of- OH^- ion.

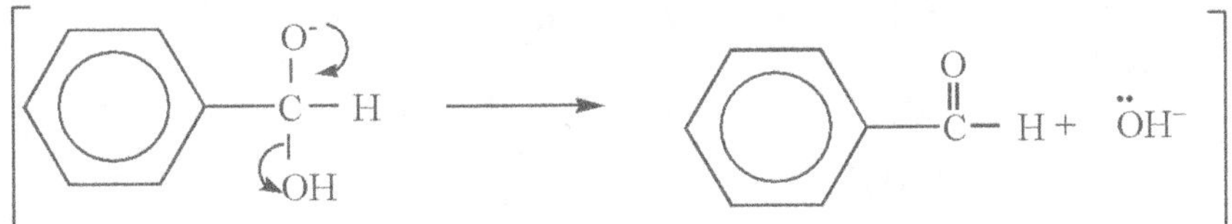

Tetrahedral intermediate → Benzaldehyde (Not isolated) + OH^-

***Step* 3:** Hydride ion transfer and hydrolysis

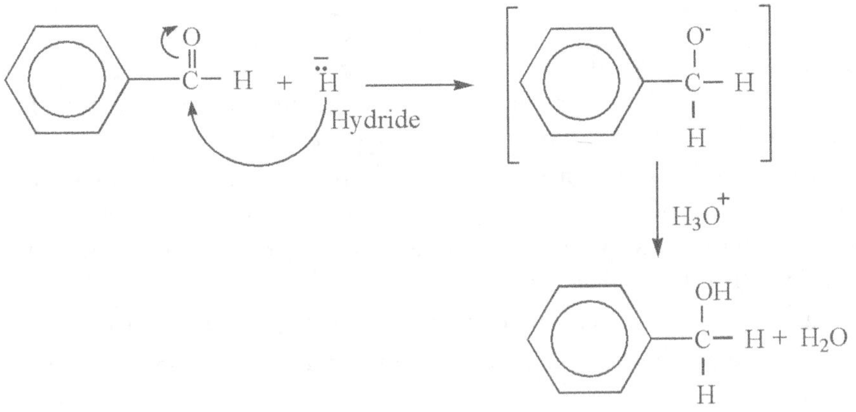

$C_6H_5-CHO + \ddot{H}^-$ (Hydride) → $[C_6H_5-CH_2-O^-]$ $\xrightarrow{H_3O^+}$ $C_6H_5-CH_2OH + H_2O$

Benzyl alcohol

Other reducing agents used for reduction is borane in tetrahydrofuran (THF). The reaction takes place at RT (room temperature) and this agent is preferred than $LiAlH_4$ because of the relative ease and safety with BH_3.

(e) Reaction with sodalime: Benzoic acid when heated with soda lime gives benzene by the loss of CO_2 molecule.

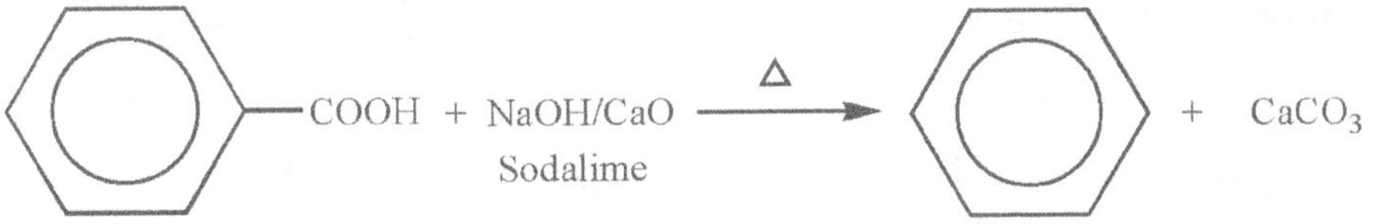

$$C_6H_5-COOH + NaOH/CaO \text{ (Sodalime)} \xrightarrow{\Delta} C_6H_6 + CaCO_3$$

Benzoic acid → Benzene

(f) Esterification: Benzoic acid readily forms esters when refluxed with alcohol in the presence of conc. H_2SO_4.

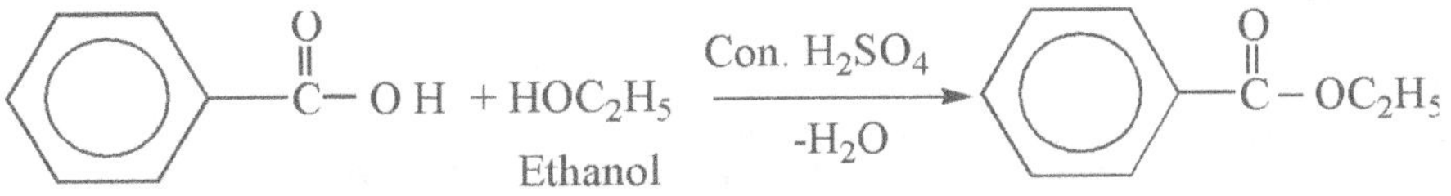

$$C_6H_5-COOH + HOC_2H_5 \text{ (Ethanol)} \xrightarrow[-H_2O]{\text{Con. } H_2SO_4} C_6H_5-COOC_2H_5$$

Benzoic acid → Ethyl benzoate

(II) Reactions of aromatic ring or phenyl group

Benzoic acid gives the usual electrophilic aromatic substitution reactions of benzene nucleus such as chlorination, sulphonation and halogenation to yield *meta* substituted products in each case.

The nature of carboxylic acid group present in the benzene nucleus is

(a) meta director and

(b) Ring deactivator.

(a) ***meta* director:** The -COOH group attached to the benzene ring directs the incoming substituent to the *m*-position which is explained as follows:

Benzoic acid can be represented by the following resonance structures.

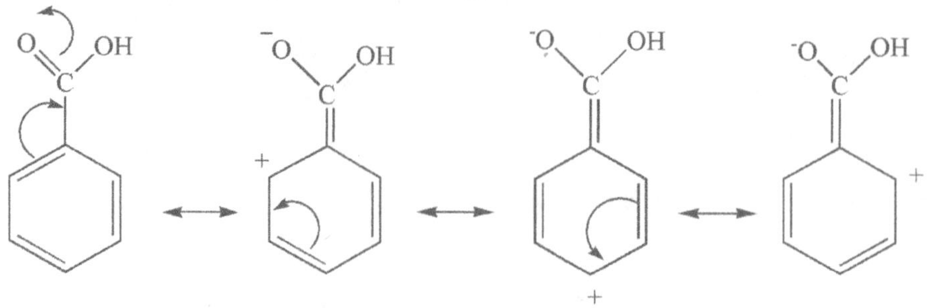

In the above structures, the *ortho* and the *para* positions carry the positive charge so the electrophile cannot attack these positions (similar charges repel). It will attack the electron rich *m*-position, hence the -COOH group directs all incoming group to the *m*-position.

(b) **Ring deactivator:** Electrophilic aromatic substitution of benzoic acid takes place slowly with vigorous reaction conditions than that of benzene. Because the -COOH group withdraws the electrons from the ring by resonance that decreases the electron density of the ring and makes it less attractive to an incoming electrophile (deactivates the benzene ring).

Example for electrophilic aromatic substitution reaction of benzoic acid

Nitration: Benzoic acid reacts with a mixture of conc. HNO_3 and conc. H_2SO_4 to yield *m*-nitrobenzoic acid.

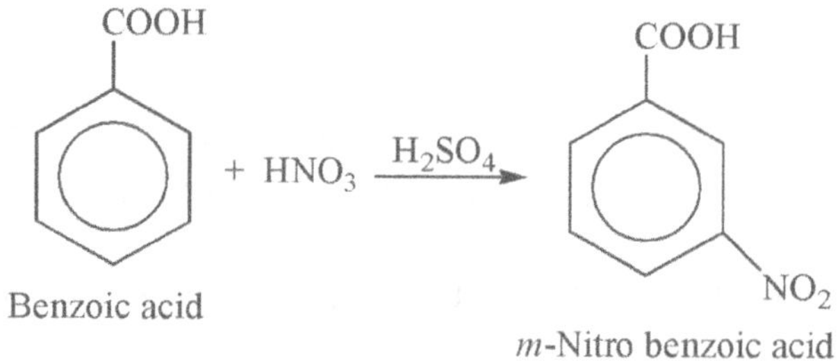

Benzoic acid + HNO_3 $\xrightarrow{H_2SO_4}$ *m*-Nitro benzoic acid

Mechanism: The reaction takes place *via* the formation of carbonium ion intermediate.

***Step* 1:** Formation of electrophile.

$$HNO_3 + 2H_2SO_4 \longrightarrow NO_2^+ + 2HSO_4^- + H_3O^+$$

***Step* 2:** Attack of electrophile to the benzene ring in *m*-position to yield resonance stabilised carbonium ion.

Benzoic acid + NO_2^+ ⟶ Resonance stabilised carbonium ion

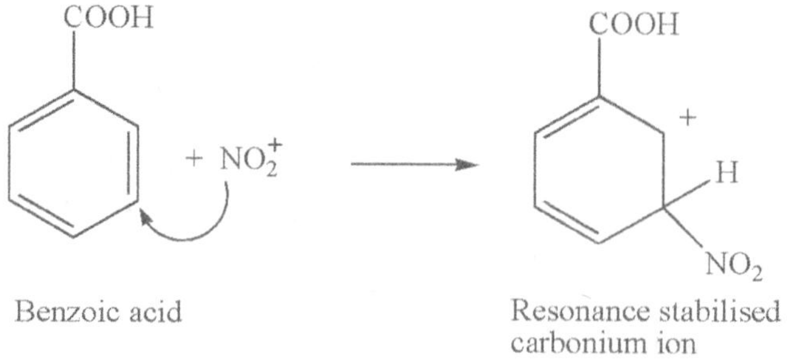

***Step* 3:** Removal of proton to give *m*-nitro benzoic acid.

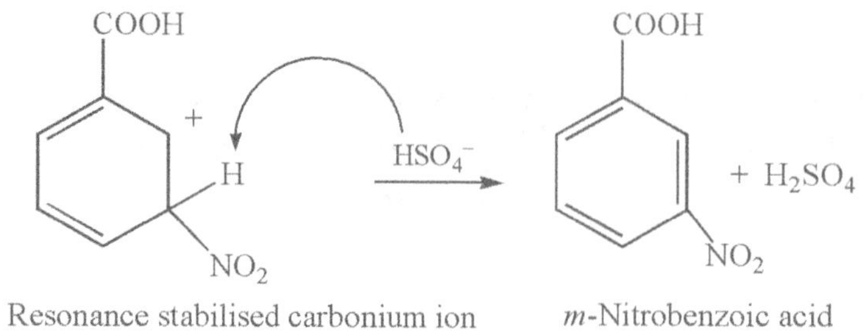

Resonance stabilised carbonium ion $\xrightarrow{HSO_4^-}$ *m*-Nitrobenzoic acid + H_2SO_4

Summary of Chemical Reactions

Benzoic acid (C_6H_5–COOH) →

- NaOH → C_6H_5–COONa + H_2O (Sodium benzoate)
- Na_2CO_3, * 2 moles → 2 C_6H_5–COONa + H_2O + $CO_2\uparrow$ (Sodium benzoate)
- NH_3, Δ → C_6H_5–C(=O)–NH_2 + H_2O (Benzamide)
- PCl_5 → C_6H_5–C(=O)–Cl + $POCl_3$ + HCl (Benzoyl chloride)
- $SOCl_2$ → C_6H_5–C(=O)–Cl + SO_2 + HCl (Benzoyl chloride)
- 4H, $LiAlH_4$, -H_2O → C_6H_5–CH_2–OH (Benzyl alcohol)
- NaOH/CaO → C_6H_6 + $CaCO_3$ (Benzene)
- C_2H_5OH, -H_2O, Con. H_2SO_4, Δ → C_6H_5–C(=O)–O–CH_2CH_3 (Ethyl Benzoate)
- HNO_3, H_2SO_4 → O_2N–C_6H_4–COOH (*m*-Nitro benzoic acid)

Chemical tests for benzoic acid

(a) Benzoic acid dissolves in hot water but separates again on cooling as white shining flakes.

(b) It liberates CO_2 gas from a warm solution of Na_2CO_3.

(c) It reacts with sodalime and liberates the inflammable vapours of benzene.

(d) The neutral solution of benzoic acid reacts with the ferric chloride solution to give buff colored precipitate.

(e) It gives ester when warmed with ethanol and Conc. H_2SO_4.

Comparison of Benzoic acid and Phenol

(a) Both liberates hydrogen gas when react with sodium metal.
(b) Both are acidic in nature, hence turn blue litmus to red and neutralize the alkalis.
(c) Both reacts with PCl_5. Phenol gives chlorobenzene and benzoic acid forms benzoyl chloride.
(d) Both give electrophilic aromatic substitution reactions like nitration, bromination, *etc.*

The properties in which both differ from each other are shown in Table 24.2.

Table 24.2 Distinguishing properties of benzoic acid and phenol.

S. No.	Benzoic acid	Phenol
1.	It reacts with bicarbonates to liberate CO_2 gas.	It is a weaker acid, no reaction with $NaHCO_3$.
2.	It reacts with $FeCl_3$ solution to give buff colored precipitate.	It reacts with $FeCl_3$ solution to give violet colour solution.
3.	When heated with **soda lime** yields benzene.	When reacts with **zinc dust** phenol reduces to benzene.
4.	It does not give Lieberman's nitroso reaction.	It gives a positive Lieberman's nitroso reaction.
5.	No reaction with aryl diazonium salts.	It gives diazocoupling reaction with aryl diazonium salts.
6.	It reacts with CH_3COCl.	No reaction with CH_3COCl.
7.	Esterification reaction occurs and gives esters when reacts with alcohol and conc. sulphuric acid.	No esterification reaction takes place.
8.	It possesses highly irritating smell.	It possess phenolic odour.

Note: Acidity of Aromatic Carboxylic acid: Refer the Chapter Aliphatic Carboxylic acids.

Probable Questions

1. Define aromatic carboxylic acids with example.
2. Write the nomenclature and importance of aromatic carboxylic acids.
3. Explain the nomenclature of aromatic carboxylic acids.
4. Mention different methods of introduction of carboxylic acid groups in to benzene ring and explain the term benzoylation.
5. How will you prepare benzoic acid from Grignard reagent?
6. Write the general methods of preparation of aromatic carboxylic acids.
7. Explain the industrial or technical scale of preparation of benzoic acid.
8. Write a note on general physical properties of aromatic carboxylic acids.
9. Write the chemical properties of aromatic carboxylic acids.
10. Benzoic acid is *meta* director and ring deactivator. Explain.
11. How will you convert benzoic acid into the following?
    (a) Benzene
    (b) Benzoyl chloride
    (c) Ethyl benzoate.
12. Write the chemical tests for benzoic acid.
13. Explain about the acidity of substituted benzoic acid.
14. Mention any one method of preparation of the following.
    (a) Benzoic acid
    (b) Cinnamic acid
    (c) Phthalic acid

15. Arrange the following in order of increasing acidity and justify it.
    p-Nitro benzoic acid, *p*-Nitro phenol, *p*-Cresol, *p*-Hydroxy benzoic acid.
16. Carboxylic acid is *m*-directing while carboxylate anion is *o* & *p* - directing in electrophilic aromatic substitution reaction - Explain.
17. How will you differentiate benzoic acid and phenol?
18. Explain in detail about acidity of aromatic carboxylic acides.

25

Polynuclear Hydrocarbons

Introduction

Polynuclear hydrocarbons or condensed nuclear hydrocarbons are compounds in which two or more carbon atoms are shared commonly by two or more aromatic rings. Simple examples of these type of compounds are naphthalene, anthracene, phenanthrene and their derivatives. In naphthalene two carbon atoms are shared by two benzene rings as shown below.

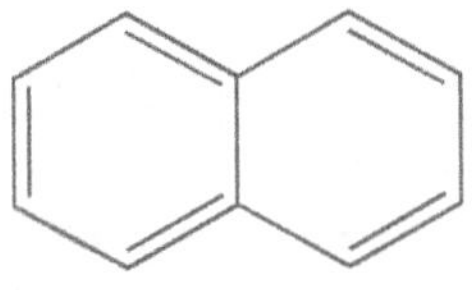

Naphthalene Or Naphthalene

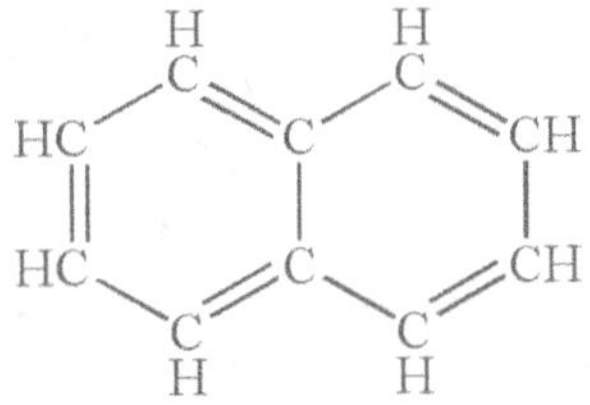

Types: Polynuclear hydrocarbons are classified into two types.

1. **Compounds in which the rings are isolated,** example: Biphenyl, diphenyl methane *etc.*

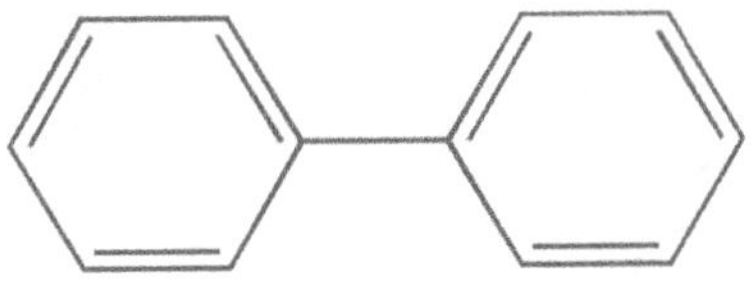

Biphenyl 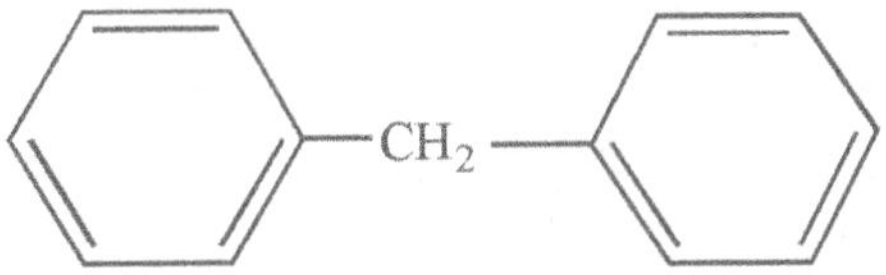   Diphenyl methane

2. **Compounds in which the two or more rings are fused in *o*-positions:** Naphthalene, anthracene and phenanthrene.

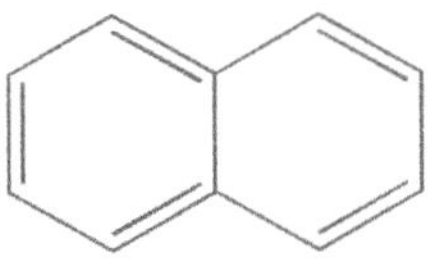

Naphthalene 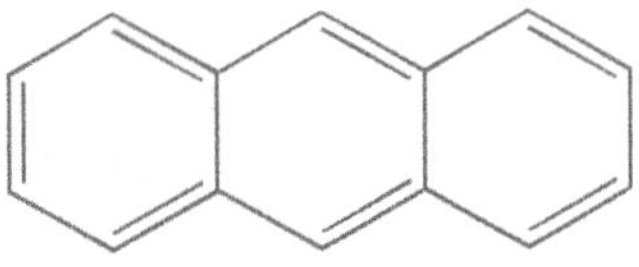   Anthracene 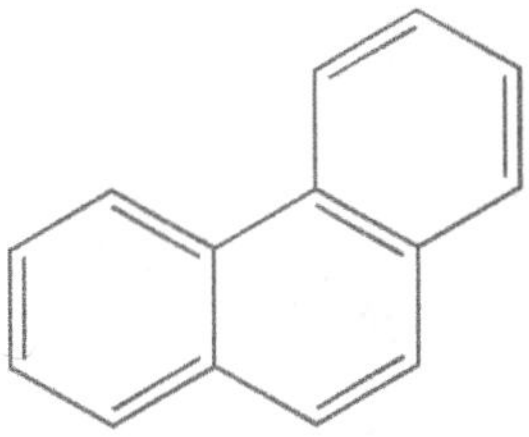   Phenanthrene

Naphthalene

Manufacture of naphthalene: Major source of naphthalene is coal tar. Coal tar contains 6-10% of naphthalene and it is present in the middle oil fraction of coal tar and obtained by distillation process. Allow the middle oil to cool, maximum amount of naphthalene is crystallized out and is collected by centrifugation or pressing out the oil in a hydraulic press. The crystals obtained are washed with water and with sodium hydroxide solution in a centrifuger to remove adhered oil and phenols. Then it is treated with conc. H_2SO_4 to remove the alkaline impurities, the crude naphthalene obtained is purified by sublimation and further purified by recrystallisation with petroleum ether. Nowadays the "hot processing process" is replaced by continuous washing or distillation.

Synthesis

1. **From petroleum fractions:** Petroleum fractions are passed over a heated copper catalyst at 680 °C at atmospheric pressure to give naphthalene and methyl naphthalene and the later undergoes hydro dealkylation to give naphthalene.

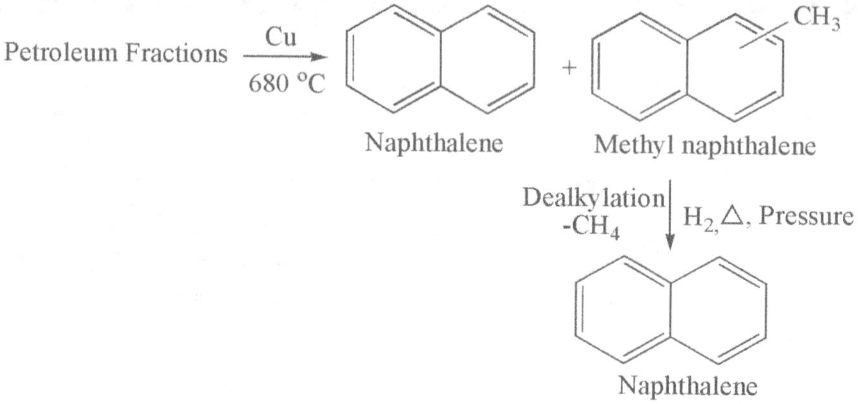

Petroleum Fractions —Cu, 680 °C→ Naphthalene + Methyl naphthalene (CH_3)

Dealkylation $-CH_4$ ↓ H_2, Δ, Pressure

Naphthalene

2. **Haworth synthesis:** Benzene is treated with succinic anhydride followed by reduction to yield γ-phenyl butyric acid. The later one undergoes ring closure reaction in the presence of conc. H_2SO_4 to yield α-tetralone. Reduction of α-tetralone with Zn(Hg)/HCl yields tetrahydronaphthalene (tetralin) which upon further dehydrogenation yields naphthalene by heating with selenium or palladised charcoal. The sequence of the chemical reactions are as follows.

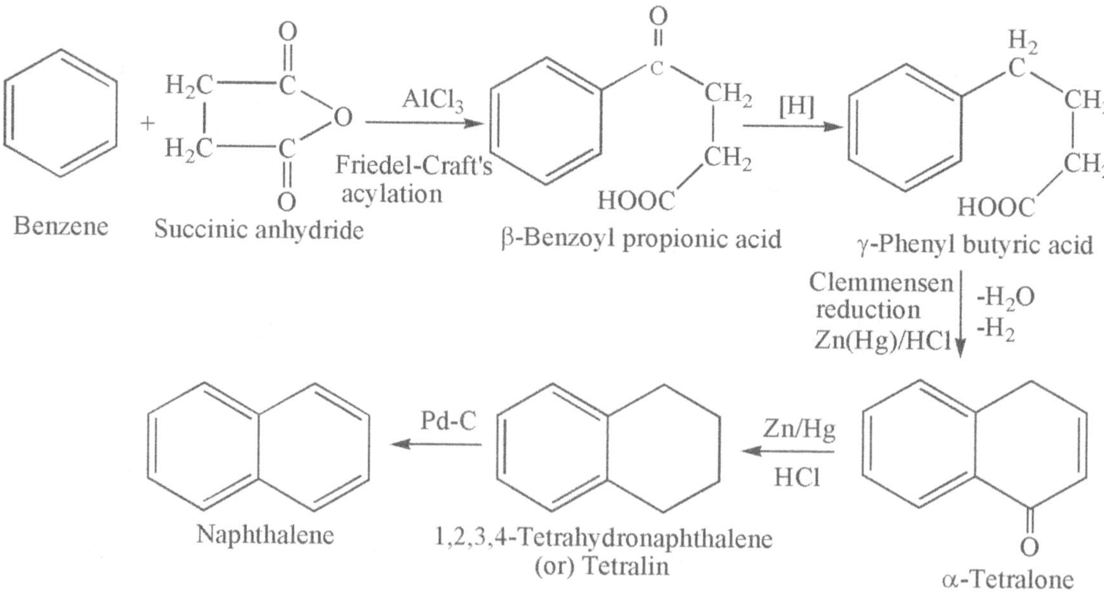

Benzene + Succinic anhydride —$AlCl_3$, Friedel-Craft's acylation→ β-Benzoyl propionic acid —[H]→ γ-Phenyl butyric acid

γ-Phenyl butyric acid —Clemmensen reduction Zn(Hg)/HCl, $-H_2O$, $-H_2$→ α-Tetralone —Zn/Hg, HCl→ 1,2,3,4-Tetrahydronaphthalene (or) Tetralin —Pd-C→ Naphthalene

Ring closure is also effected by Friedel-Crafts reaction on acid chlorides as shown below.

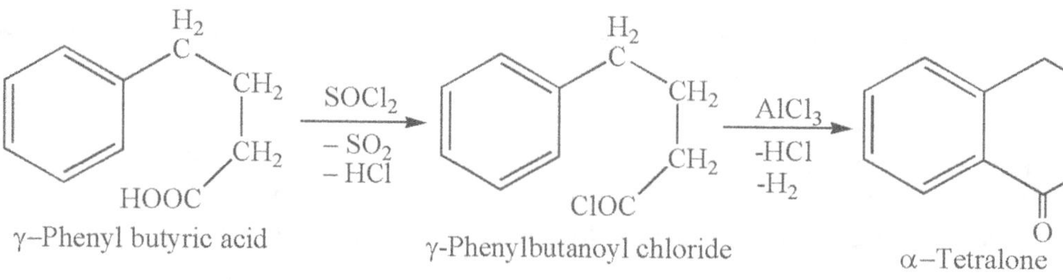

γ-Phenyl butyric acid —$SOCl_2$, $-SO_2$, $-HCl$→ γ-Phenylbutanoyl chloride —$AlCl_3$, $-HCl$, $-H_2$→ α-Tetralone

3. **From 4-phenyl-1-butene:** When 4-phenyl-1-butene is passed over red-hot calcium oxide naphthalene is obtained.

$CH_2CH_2CH{=}CH_2$ — Red hot CaO, Δ → Naphthalene $+ 2H_2$

4-Phenyl-1-butene   Naphthalene

4. **From 4-phenyl-3-butenoic acid:** When 4-phenyl-3-butenoic acid is warmed with conc. H_2SO_4 1-napthol is formed which upon further distillation with Zn dust yields naphthalene.

4-Phenyl-3-butenoic acid —H_2SO_4, $-H_2O$→ [1-Naphthol (Keto form)] —Isomerisation→ 1-Naphthol (Enol form) —Zn, $-ZnO$→ Naphthalene

Physical Properties: It is a colorless crystalline substance with characteristic odour (moth ball odour). It is very volatile and readily sublimes on heating. It is insoluble in water but soluble in organic solvents.

Chemical Properties: Chemical properties of naphthalene resembles to that benzene. It is less aromatic than benzene and forms substitution products more readily than benzene. Like alkenes it forms addition products readily than benzene. But as soon as one of the rings is saturated or destroyed by oxidation, the second ring is stable as the benzene ring. The important characteristic reactions of naphthalene are as follows.

(I) Addition reactions.

(II) Electrophilic aromatic substitution reactions.

(I) Addition reactions

1. **Addition of hydrogens:** Naphthalene gives different kind of products depending upon the type of reducing agents used.

   (a) With catalytic reduction using nickel yields decalin or decahydro naphthalene

   Naphthalene —H_2/Ni→ Tetralin —H_2/Ni→ Decalin

   (b) With sodium and alcohol, naphthalene gives 1,4-dialin (dihydronaphthalene).

   Naphthalene —Na/C_2H_5OH→ 1,4-Dialin

(c) With sodium and isopentanol naphthalene gives 1,2,3,4- tetrahydro naphthalene or tetralin.

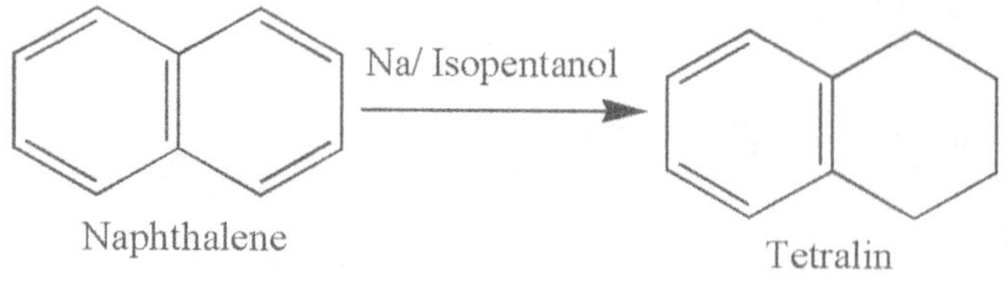

Na/ Isopentanol

Naphthalene

Tetralin

2. **Addition of chlorine:** Solid naphthalene reacts with dry chlorine to give naphthalene di and tetra chlorides. Both of them undergo oxidation to yield phthalic acid. It indicates that the halogen atoms are present in the same ring.

The naphthalene dichloride when heated at 310 K, loses one molecule of hydrogen chloride and gives 1-chloro naphthalene. Naphthalene tetrachloride on treatment with alkali gives dichloro naphthalene.

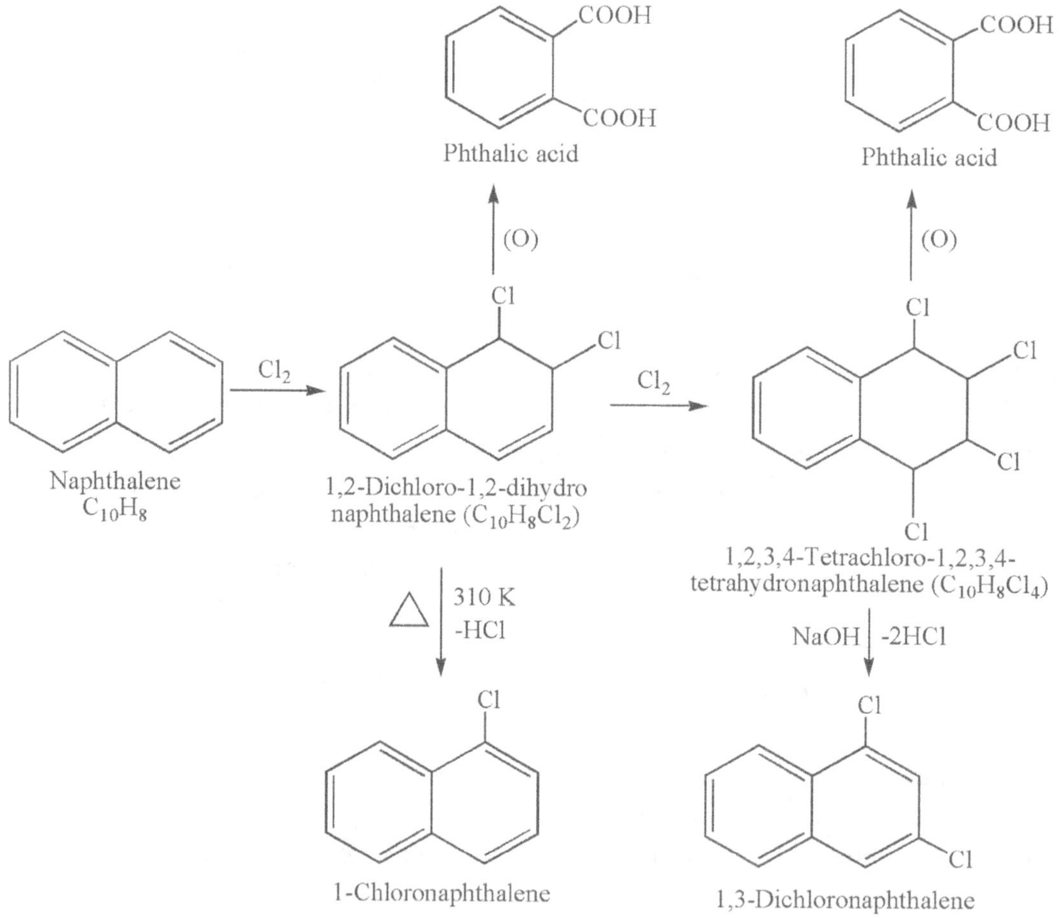

COOH COOH Phthalic acid

COOH COOH Phthalic acid

(O)

(O)

Cl_2

Cl_2

Naphthalene $C_{10}H_8$

1,2-Dichloro-1,2-dihydro naphthalene ($C_{10}H_8Cl_2$)

1,2,3,4-Tetrachloro-1,2,3,4-tetrahydronaphthalene ($C_{10}H_8Cl_4$)

△ 310 K -HCl

NaOH -2HCl

1-Chloronaphthalene

1,3-Dichloronaphthalene

3. **Addition of sodium:** Naphthalene upon reaction with sodium gives 1,4-disodio naphthalene which reacts further with carbon dioxide with the formation of sodium salt 1,4-dihydro naphthalene-1,4-di carboxylic acid.

2Na

$2CO_2$

Naphthalene

1,4-Disodio-1,4-dihydro naphthalene

Sodium salt of 1,4-dihydro naphthalene-1,4-dicarboxylic acid

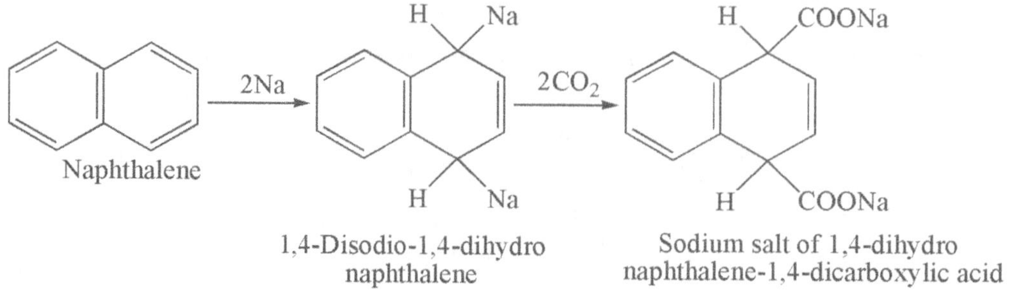

(II) Electrophilic aromatic substitution reactions: Naphthalene undergoes electrophilic aromatic substitution reactions and the major substitution occurs at C_1 (α-position) and C_2 positions.

Naphthalene + E^+ (Electrophile) → [arenium ion intermediates] → $-H^+$ → 1-Substituted naphthalene

Naphthalene + E^+ (Electrophile) → [arenium ion intermediate] → $-H^+$ → 2-Substituted naphthalene

The product by C-1 attack is always predominates because the carbonation intermediate obtained by C-1 attack is more stable (due to resonance), the C-2 attack is possible only when the reaction occurs at high temperature or when bulkier solvents are used.

1. **Halogenation (Bromination):** Naphthalene reacts with bromine in boiling carbon tetrachloride solution to give 1-bromonapthalene. Further bromination gives mainly the 1,4-dibromo naphthalene with small amount of 1,2-dibromo naphthalene.

Naphthalene $\xrightarrow[-HBr]{Br_2/CCl_4}$ 1-Bromonaphthalene $\xrightarrow[-HBr]{Br_2/CCl_4}$ 1,4-Dibromonaphthalene (Major) + 1,2-Dibromonaphthalene (Minor)

Chlorination: Naphthalene reacts with sulphuryl chloride in the presence of aluminium chloride. Naphthalene reacts with one equivalent of sulphuryl chloride at 298 K to give 1-chloro naphthalene and with two equivalents of sulphuryl chloride at 370-410 K to gives 1,4-dichloronaphthalene.

Naphthalene $\xrightarrow[298\ K,\ -SO_2,\ -HCl]{SO_2Cl_2}$ 1-Chloronaphthalene

Naphthalene $\xrightarrow[370\text{-}410\ K,\ -2SO_2,\ -2HCl]{2SO_2Cl_2}$ 1,4-Dichloronaphthalene

Naphthalene reacts with chlorine in the presence of iodine above 610 K and gives a mixture of 1 and 2-chloro naphthalene.

Naphthalene $\xrightarrow[610K]{Cl_2/I_2}$ 1-Chloronaphthalene + 2-Chloronaphthalene

2. **Nitration:** Naphthalene reacts with nitric acid at room temperature to yield 1-nitro naphthalene and at high temperature yield 1,5 & 1,8-dinitro naphthalene.

Naphthalene $\xrightarrow[RT,\ -H_2O]{HNO_3}$ 1-Nitronaphthalene

Naphthalene $\xrightarrow[\text{High temperature},\ -2H_2O]{HNO_3}$ 1,5-Dinitronaphthalene + 1,8-Dinitronaphthalene

3. **Sulphonation:** Naphthalene reacts with conc. H_2SO_4 at 350 K and yields 1-naphthalene sulphonic acid, but at 430 K it gives 2-naphthalene sulphonic acid.

Naphthalene $\xrightarrow[350\ K]{H_2SO_4}$ 1-Naphthalene sulphonic acid

Naphthalene $\xrightarrow[430\ K]{H_2SO_4}$ 2-Naphthalene sulphonic acid

4. **Fridel-Craft's reaction:** Carry out this reaction at low temperature in the presence of anhydrous aluminium chloride. Because at high temperature, one of the naphthalene ring opened.
   (a) **With methyl iodide:** Naphthalene reacts with methyl iodide in the presence of anhydrous $AlCl_3$ and gives 1 & 2-methyl naphthalene.

Naphthalene $\xrightarrow[AlCl_3,\ -HI]{CH_3I}$ 1-Methyl naphthalene + 2-Methyl naphthalene

But naphthalene reacts with ethyl bromide to yield 2-ethyl naphthalene.

C_2H_5Br, $AlCl_3$, $-HBr$

Naphthalene → 2-Ethyl naphthalene (C_2H_5)

(b) **With acetyl chloride:** Naphthalene reacts with acetyl chloride and $AlCl_3$ to yield a mixture of 1-naphthyl methyl ketone & 2-napthyl methyl ketone. The composition of mixture obtained depends upon the nature of the solvent and the temperature used. For example, in the presence of CS_2 at 260 K, the 1 and 2 - derivatives are obtained in 3:1 ratio while in the presence of nitrobenzene at 298 K the product obtained is in 1:9 ratios.

CH_3COCl, $AlCl_3$, $-HCl$

Naphthalene → 1-Naphthyl methyl ketone ($COCH_3$) + 2-Naphthyl methyl ketone ($COCH_3$)

5. **Chloro methylation:** Naphthalene reacts with formaldehyde/HCl and glacial acetic acid, gives 1-chloro methyl naphthalene with a small amount of 1,5-bis chloromethyl naphthalene.

Naphthalene $+ HCHO + HCl \xrightarrow{-H_2O}$ 1-Chloromethyl naphthalene (CH_2Cl) + 1,5-Bischloromethyl naphthalene (CH_2Cl, CH_2Cl)

6. **Oxidation:** Naphthalene gives different kind of products on oxidation. The products formed depend upon the nature of the oxidizing agents used.

The various kind of oxidation reactions are as follows.

(i) **With potassium permanganate:** Naphthalene reacts with acidic $KMnO_4$ and alkaline $KMnO_4$ to give phthalic acid and phthalonic acid respectively.

Naphthalene — Acid $KMnO_4$ → Phthalic acid (COOH, COOH)

Naphthalene — Alkaline $KMnO_4$ → Phthalonic acid (C(=O)-COOH, COOH)

(ii) **With chromic acid:** Naphthalene when oxidised with chromic acid yields 1,4-naphthaquinone.

Naphthalene $\xrightarrow{H_2CrO_4}$ 1,4-Naphthaquinone

(iii) Naphthalene reacts with conc. H_2SO_4 and mercuric sulphate or air in the presence of vanadium pentoxide and yields phthalic anhydride.

Naphthalene $\xrightarrow[\text{or Air + }V_2O_5]{\text{Conc.}H_2SO_4/HgSO_4}$ Phthalic anhydride

(iv) With ozone: Naphthalene is oxidized with ozone to give diozonide which upon hydrolysis gives phthalaldehyde.

Naphthalene $\xrightarrow{2O_3}$ Diozonide $\xrightarrow{\text{Hydrolysis}}$ Phthalaldehyde

Uses:

1. It is used as an insecticide and for preventing moths in clothes.
2. Large amount of naphthalene is used in industry for the manufacture of various dye stuffs such as indigo, azo dye and eosin.
3. It is also used for manufacturing of phthalic anhydride, phthalic acid and phthalimide *etc.*
4. Local gas is carbureting with naphthalene.

Medicinally useful compounds containing naphthalene

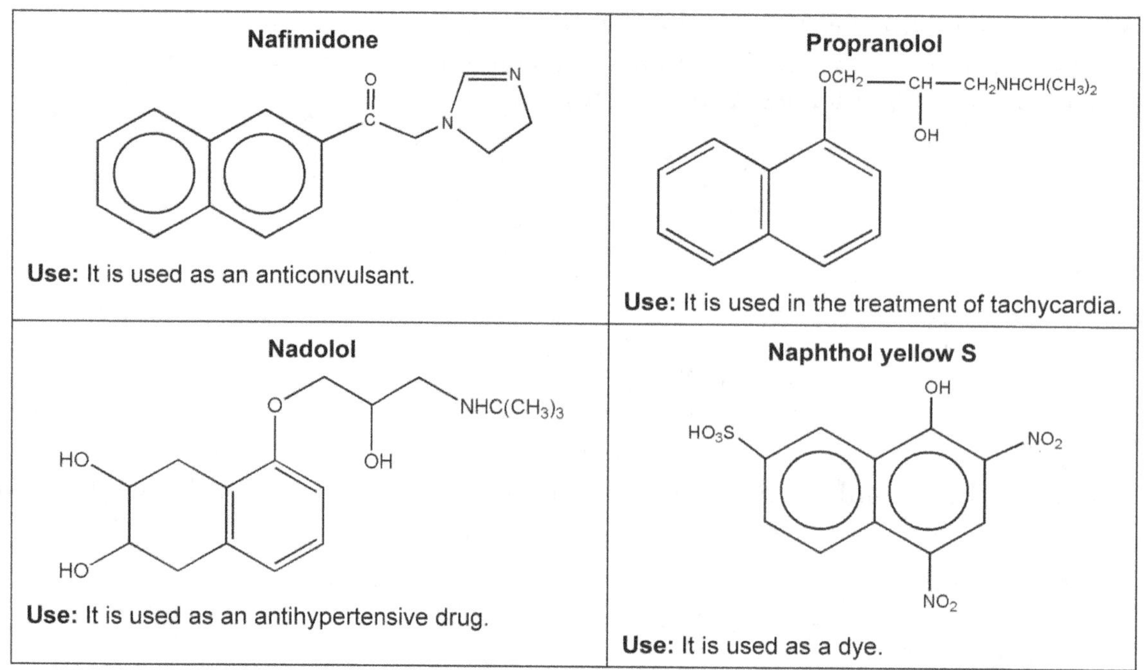

Nafimidone	Propranolol
Use: It is used as an anticonvulsant.	**Use:** It is used in the treatment of tachycardia.
Nadolol	**Naphthol yellow S**
Use: It is used as an antihypertensive drug.	**Use:** It is used as a dye.

Table Contd...

Mordant green 4	Para red azo dye
NO, OH Use: It is used as a dye.	HO, O_2N, N=N Use: It is used as a dye.

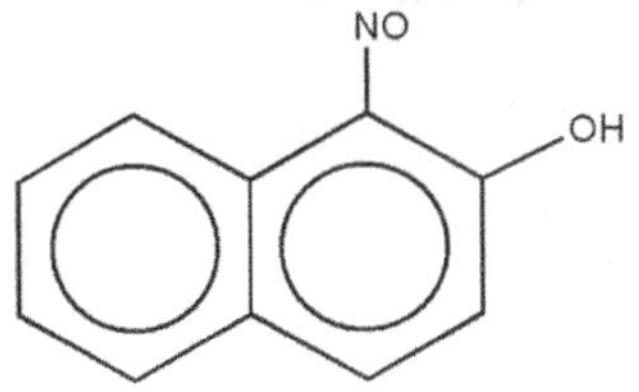

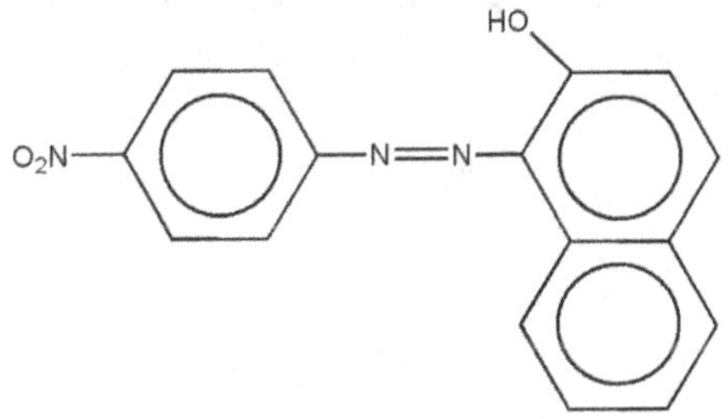

Anthracene

Anthracene is another example of fused aromatic hydrocarbons in which three benzene rings fused together in the *o*-positions.

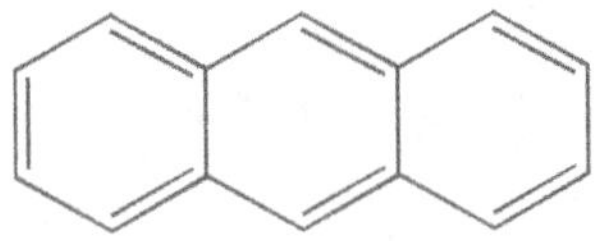

Anthracene is present in coal tar less than 1%. The name derives from the Greek word anthraz (means coal). High boiling point fraction of coal tar also contains anthracene, hence called as anthracene oil.

Preparation of anthracene: Coal tar is a chief source of anthracene and it contains 0.25-0.45% of anthracene. It is present in anthracene oil or green oil along with phenanthrene, carbazole and other substances.

Anthracene oil is allowed to stand in shallow tanks where by a viscous mass separates out and the crude anthracene (20%) is removed by filtration. During this, the anthracene content increases up to 35% and some amount of oil is removed by centrifuge when 50% anthracene is obtained. The resulting product is powdered and washed with solvent naphtha which dissolves out phenanthrene and then treated with pyridine which removes carbazole. Anthracene is finally crystallized out from benzene.

Synthesis

1. **Haworth synthesis:** Benzene reacts with phthalic anhydride in the presence of $AlCl_3$to yield o-benzoyl benzoic acid, which upon reaction with conc. H_2SO_4 to yield 9,10-anthraquinone. The later obtained is distilled with Zn, and yields anthracene.

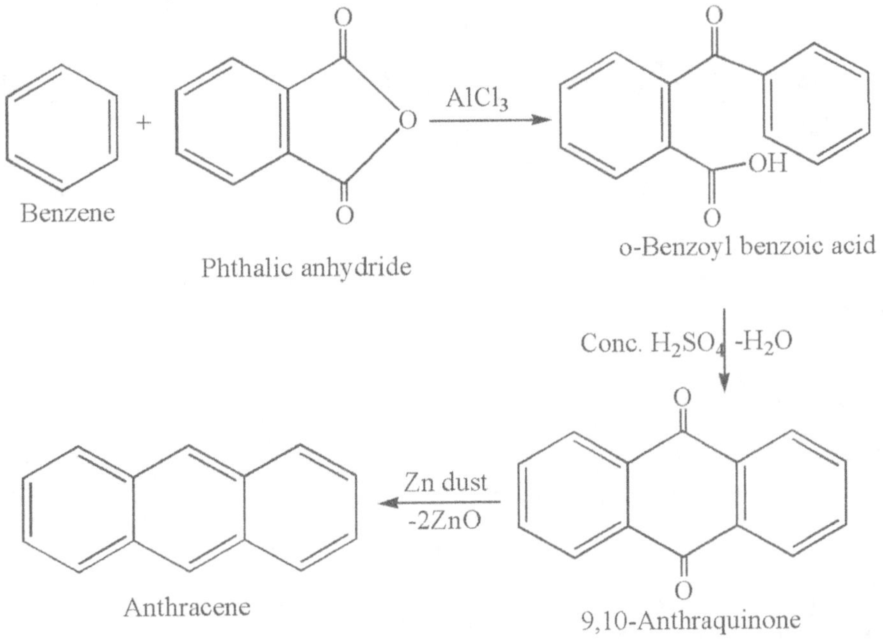

Benzene + Phthalic anhydride —$AlCl_3$→ o-Benzoyl benzoic acid (OH)

o-Benzoyl benzoic acid —Conc. H_2SO_4 -H_2O→ 9,10-Anthraquinone

9,10-Anthraquinone —Zn dust, -2ZnO→ Anthracene

2. **Diel's-Alder reaction:** 1,4-Napthaquinone reacts with 1,3-butadiene to yield Diel's -Alder product which upon oxidation with chromic acid in the presence of glacial acetic acid yields 9,10-anthraquione which upon distilled with Zn dust yields anthracene.

1,4-Napthaquinone + 1,3-Butadiene → Diel's-Alder adduct

Diel's-Alder adduct $\xrightarrow{H_2CrO_4/CH_3COOH\ (O)}$ 9,10-Anthraquinone

9,10-Anthraquinone $\xrightarrow[-2ZnO]{Zn\ dust}$ Anthracene

Physical properties: It is a colorless solid, melting point is 218 °C and boiling point 340 °C. It produces blue fluorescence when exposed to UV light. This property of anthracene is used in criminal detection work.

Chemical properties: Anthracene undergoes two type of reactions.

1. **Addition reactions.**
2. **Electrophilic aromatic substitution reactions.**

1. **Reaction with sodium:** Anthracene reacts with sodium in liquid ammonia to yield deep - blue colour 9,10-disodium anthracene which when heated with an alkyl halide yields 9,10-dialkyl anthracene.

Anthracene + 2Na $\xrightarrow{Liquid\ NH_3}$ 9,10-Disodioanthracene (Deep blue)

9,10-Disodioanthracene $\xrightarrow[-2NaCl]{\Delta,\ 2CH_3CH_2Cl}$ 9,10-Diethylanthracene

2. **Halogenation:** When chlorine is passed in to the solution of anthracene in CS_2, anthracene dichloride is formed. On the further heating with alkali, loses HCl molecule to give 9-chloro anthracene.

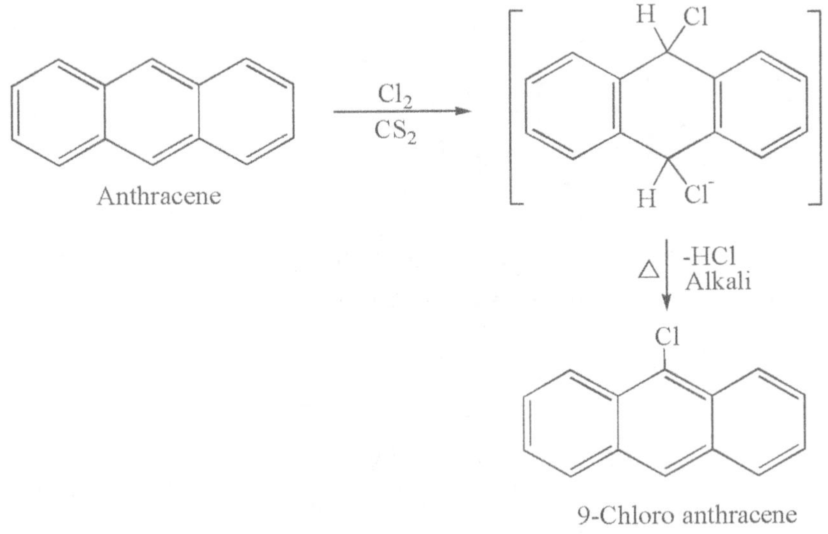

H Cl

Cl_2
CS_2

Anthracene

H Cl

△ -HCl
Alkali

Cl

9-Chloro anthracene

3. **Nitration:** Anthracene reacts with conc. HNO_3 in acetic anhydride at 288-293 K to give 9-nitro anthracene and 9,10-dinitro anthracene. Aqueous nitric acid oxides anthracene to anthraquinone, hence nitric acid in glacial acetic acid is used as a nitrating mixture.

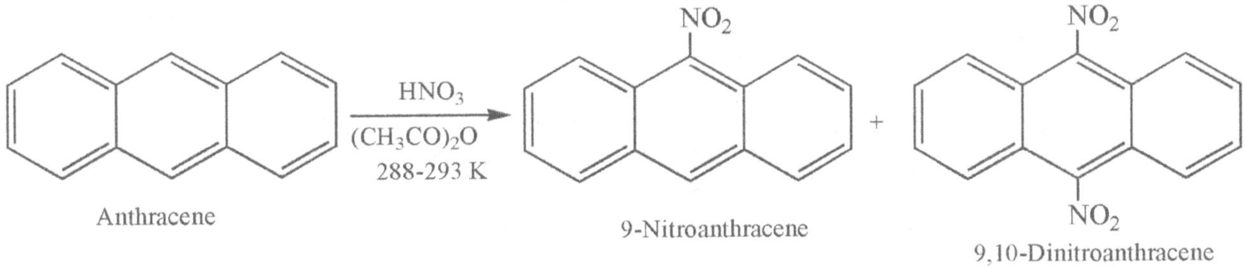

NO_2 NO_2

HNO_3
$(CH_3CO)_2O$
288-293 K

Anthracene 9-Nitroanthracene NO_2
9,10-Dinitroanthracene

4. **Sulphonation:** Anthracene reacts with conc. H_2SO_4 to yield 1-and 2-anthracene sulphonic acid. At low temperature 1-anthracene sulphonic acid is obtained as major product. At higher temperature 2-anthracene suphonic acid is obtained as major product.

SO_3H SO_3H

Dil H_2SO_4

Anthracene 1-Anthracene sulphonic acid 2-Anthracene sulphonic acid

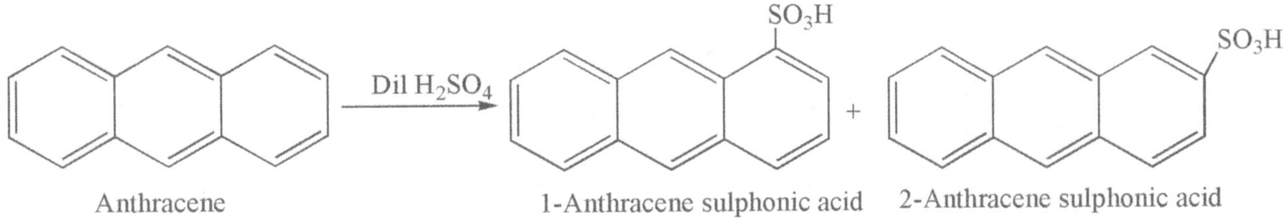

With excess amount of conc. H_2SO_4 anthracene-1,8-disulphonic acid is obtained a major product at low temperature but at high temperature anthracene-2,7-disulphonic acid obtained as a major product.

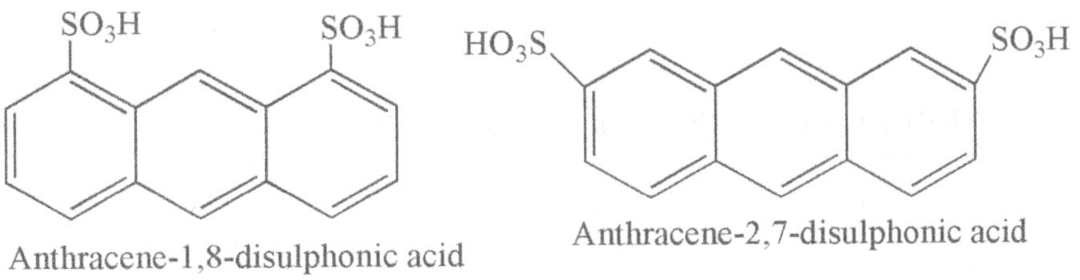

SO_3H SO_3H HO_3S SO_3H

Anthracene-1,8-disulphonic acid Anthracene-2,7-disulphonic acid

5. **Reduction:** Anthracene when reduced with sodium and isopentanol give 9,10-dihydro anthracene. Upon further treatment with conc. H_2SO_4 it gives back anthracene.

Anthracene ⇌ 9,10-Dihydroanthracene (Na/isopentanol; Conc. H_2SO_4 or heat)

Catalytic reduction of anthracene with nickel at 470-520 K gives tetra, hexa, octa and finally gives perhydroanthracene depending upon the amount of hydrogen used.

$$\underset{\text{Anthracene}}{C_{14}H_{10}} \xrightarrow[\text{Ni}]{2H_2} \underset{\text{Tetrahydro anthracene}}{C_{14}H_{14}} \xrightarrow[\text{Ni}]{H_2} \underset{\text{Hexahydro anthracene}}{C_{14}H_{16}} \xrightarrow[\text{Ni}]{H_2} \underset{\text{Octahydro anthracene}}{C_{14}H_{18}} \xrightarrow[\text{Ni}]{H_2} \underset{\text{Perhydro anthracene}}{C_{14}H_{24}}$$

6. **Oxidation:** Anthracene undergoes oxidation with sodium dichromate and sulphuric acid to yield 9,10-anthraquinone.

Anthracene $\xrightarrow[H_2SO_4]{Na_2Cr_2O_7}$ 9,10-Anthraquinone

7. **Diels-Alder Reaction:** Anthracene reacts with maleic anhydride to yield respective Diel's-Alder adduct.

Anthracene + Maleic anhydride ⟶ Diel's-Alder adduct

Uses: Anthracene is used in dye stuff industry for the manufacture of alizarine and anthraquinone dyes.

Medicinally useful compounds containing anthracene

Apomorphine 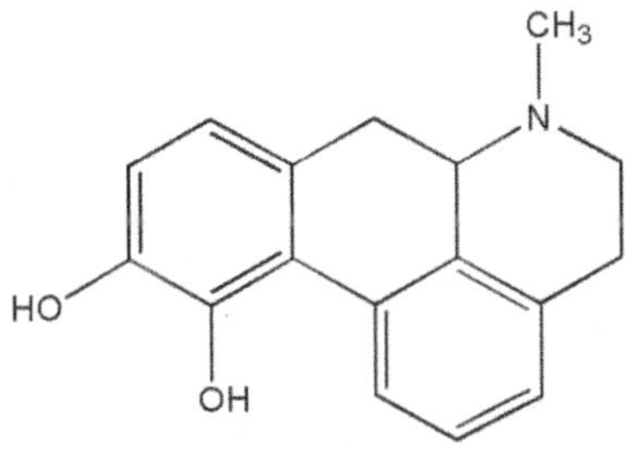 **Use:** It is used as an antiparkinsonism agent.	**Alizarin** 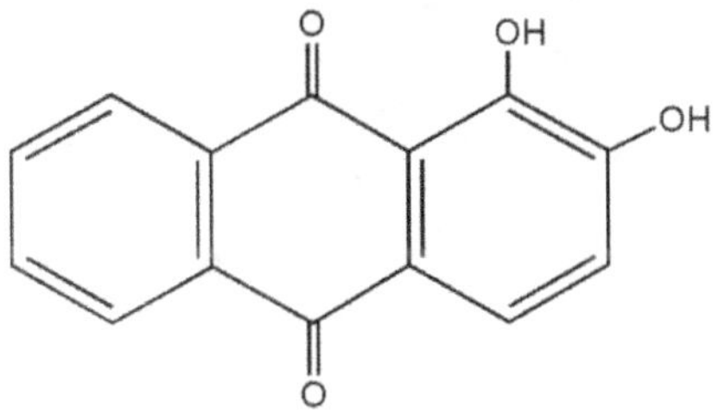 **Use:** It is used as a dye.
Sennoside A 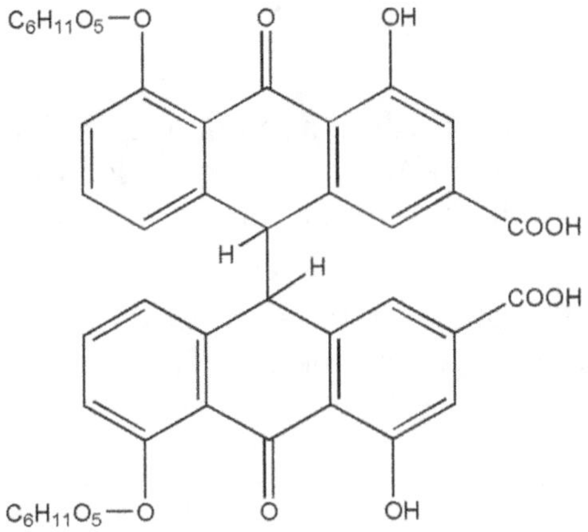 **Use:** It is used to treat constipation.	**Sennoside B** 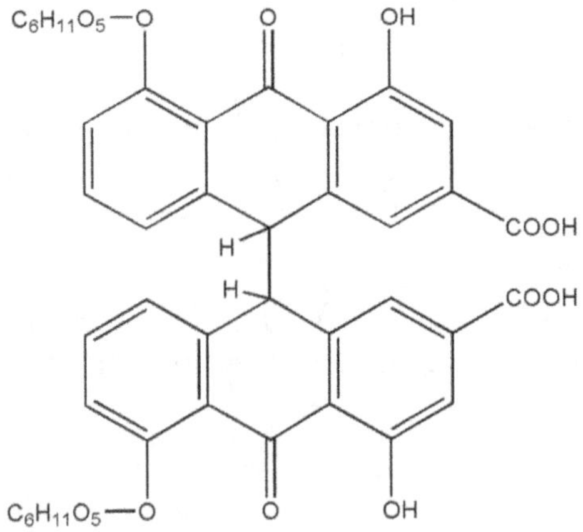 **Use:** It is used to treat constipation.
Sennoside C 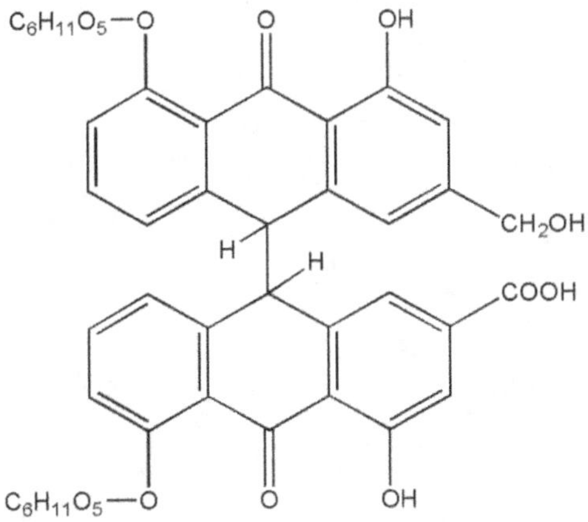 **Use:** It is used to treat constipation.	**Sennoside D** 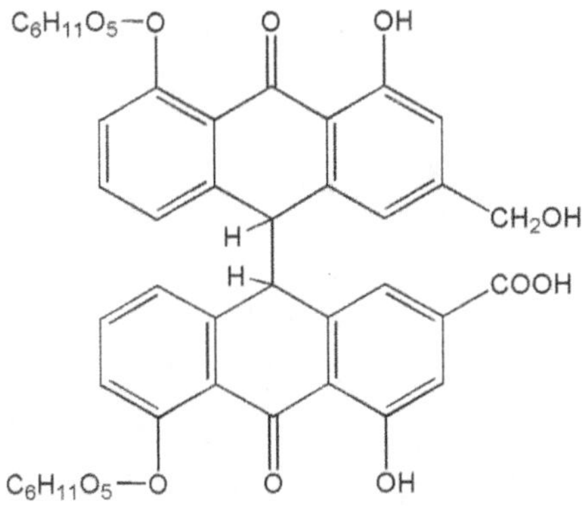 **Use:** It is used to treat constipation.
Aloe emodin 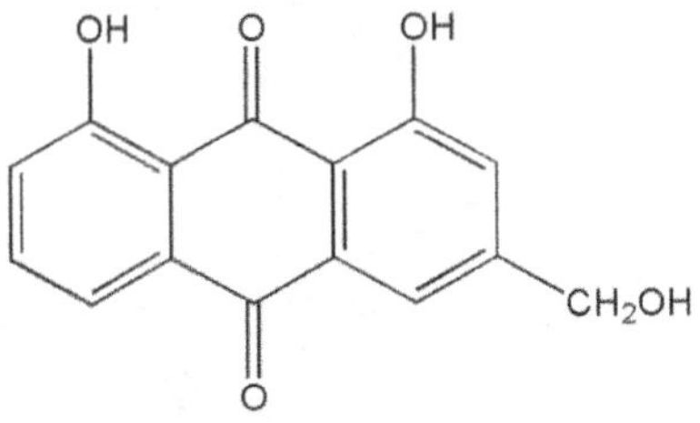 **Use:** It is used as a laxative.	

Phenanthrene

Phenanthrene is isomeric of anthracene. It is present in coal tar up to 4%. It is present in green oil fraction of coal tar along with anthracene.

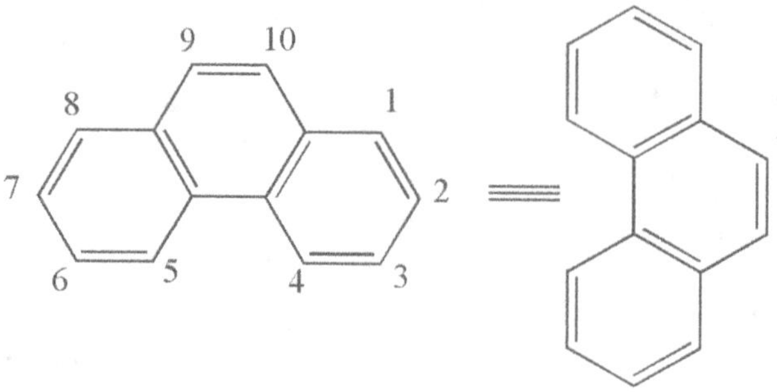

9 10 8 1 7 2 6 5 4 3

Preparation

Isolation: Phenanthrene is isolated from anthracene oil at 540-630 K. Upon cooling the anthracene oil, solid cake is obtained which contains anthracene, phenanthrene and carbazole. This is powdered and phenanthrene present in it is extracted with solvent naphtha. The solution obtained on evaporation yields phenanthrene.

Synthesis: Naphthalene reacts with succinic anhydride in the presence of $AlCl_3$ to yield an intermediate which upon Clemmensen reduction and ring closure reaction yields phenanthrene.

Naphthalene + Succinic anhydride $\xrightarrow[AlCl_3]{\text{Friedel-Craft's acylation}}$

$\xrightarrow{Zn(Hg)/HCl}$

$\xleftarrow[\Delta\ (-H_2O)]{H_2SO_4}$

$\xrightarrow{Zn(Hg)/HCl}$

1,2,3,4-Tetrahydrophenanthrene $\xrightarrow[\Delta,\ -2H_2]{Pd}$ Phenanthrene

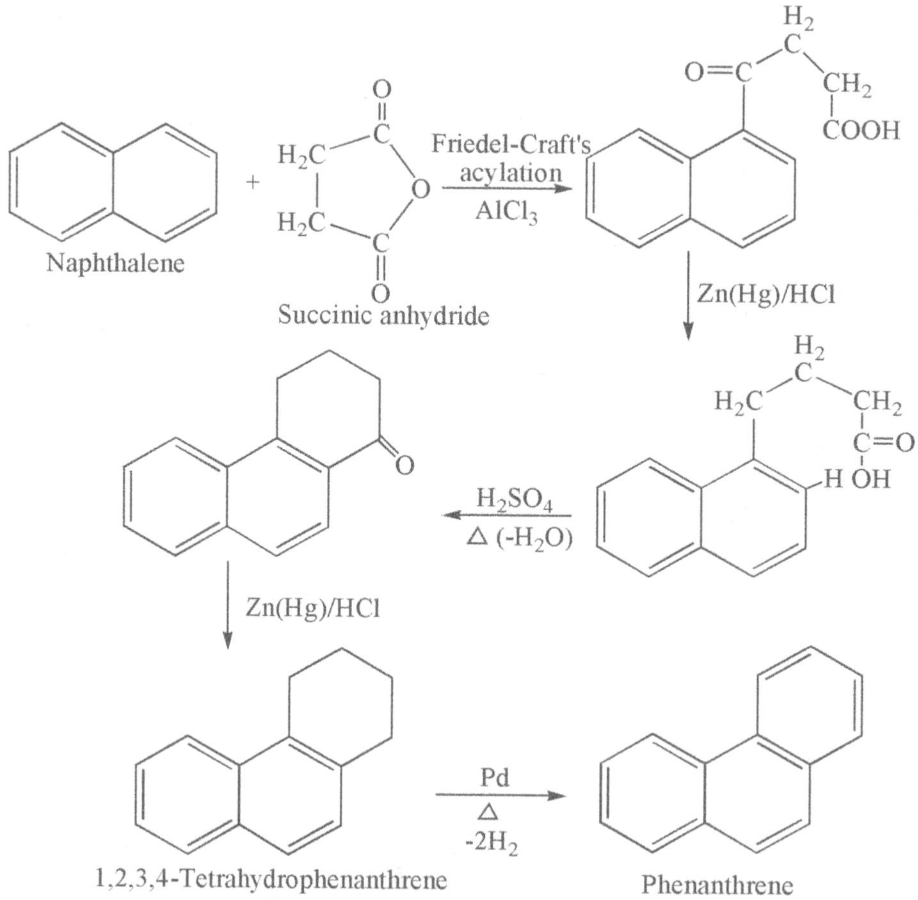

Physical Properties: It is a colourless solid, insoluble in water but soluble in organic solvents. The solution of phenanthrene in benzene gives a blue fluorescence.

Chemical Properties: Phenanthrene easily undergoes oxidation, addition and electrophilic substitution reactions. Like anthracene, this reaction preferentially occurs at C-9 and C-10 positions. Some of the important reactions are as follows.

1. **Reduction:** Phenanthrene on reduction with H_2 using Cuo+Cr_2O_2 as a catalyst gives 9,10-dihydrophenanthrene.

Phenanthrene + H_2 $\xrightarrow{CuO + Cr_2O_2}$ 9,10-Dihydro phenanthrene

2. **Halogenations:** Phenanthrene reacts with bromine in the presence of halogen carrier (iron) to yield 9-bromophenanthrene.

Phenanthrene + Br_2 $\xrightarrow[-HBr]{Fe}$ 9-Bromophenanthrene

Chlorination: Phenanthrene reacts with chlorine in the presence of CCl_4 gives 9,10-dichlorophenanthrene which upon heating a molecule HCl is eliminated to yield 9-chlorophenanthrene

Phenanthrene $\xrightarrow[CCl_4]{Cl_2}$ 9,10-Dichlorophenanthrene $\xrightarrow[-HCl]{\Delta}$ 9-Chlorophenanthrene

3. **Oxidation:**

(a) Phenanthrene undergoes oxidation with sodium dichromate/glacial CH_3COOH to yield phenanthroquinone which upon future oxidation with dichromate and sulphuric acid gives diphenic acid.

Phenanthrene $\xrightarrow[\text{in } CH_3COOH]{Na_2Cr_2O_7}$ Phenanthroquinone $\xrightarrow{Na_2Cr_2O_7/H_2SO_4}$ Diphenic acid

(b) **Oxidation with ozone:** Phenanthrene gets oxidised with ozone to yield mono-ozonide which upon further oxidation gives diphenic acid.

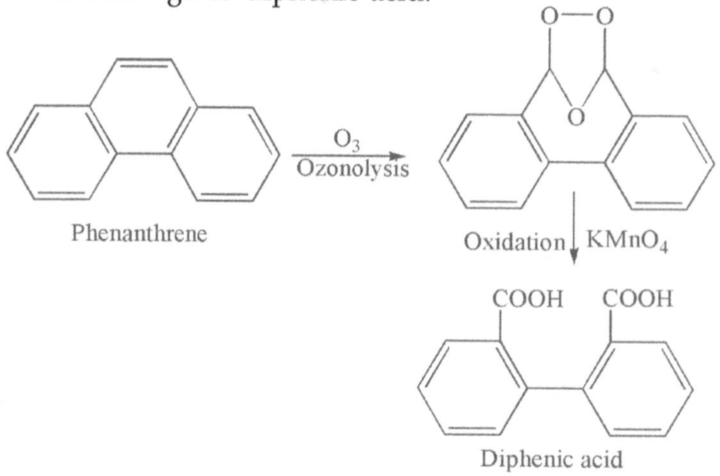

O_3 Ozonolysis

Phenanthrene

Oxidation, $KMnO_4$

COOH COOH

Diphenic acid

(c) **Nitration:** Phenanthrene undergoes nitration with nitric acid/H_2SO_4 to gives 9- nitrophenanthrene.

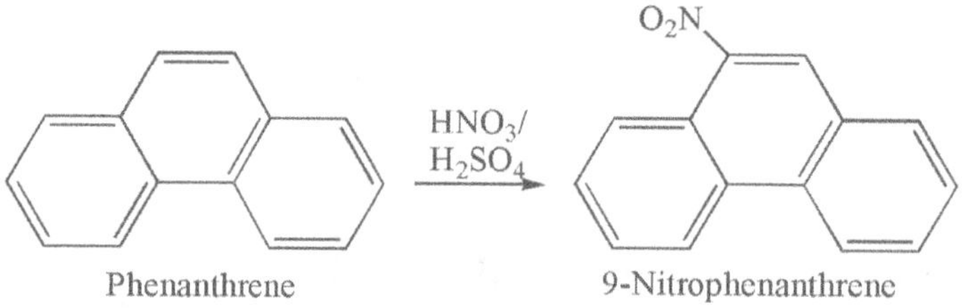

O_2N

HNO_3/ H_2SO_4

Phenanthrene

9-Nitrophenanthrene

4. **Sulphonation:** Phenanthrene reacts with H_2SO_4 at 120 °C to yield a mixture of 2-phenanthrene sulphonic acid and 3-phenanthrene sulphonic acid.

2

Phenanthrene

H_2SO_4 120 °C

$-SO_3H$ +

2-Phenanthrene sulphonic acid

SO_3H

3-Phenanthrene sulphonic acid

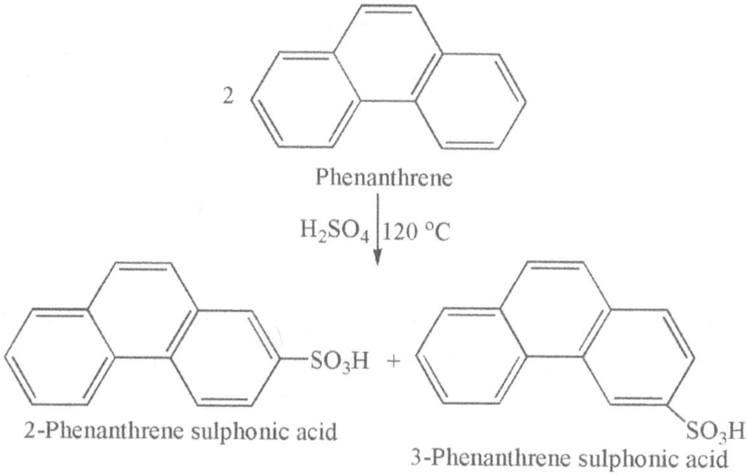

5. **Friedel-Craft's reaction:** Phenanthrene undergoes acylation with acetyl chloride in the presence of aluminium chloride at 0 °C to yield 9-acetyl phenanthrene.

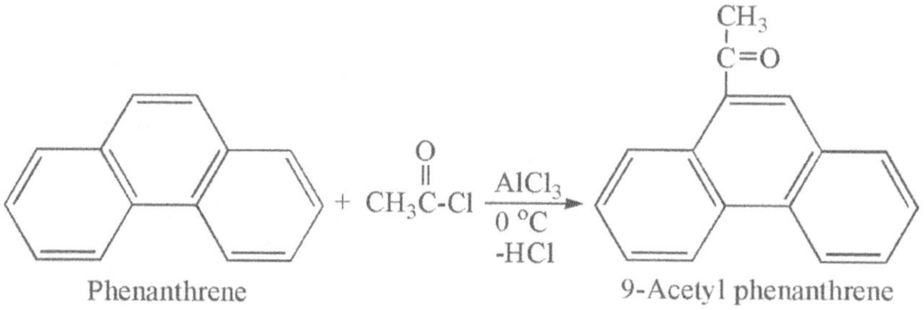

CH_3 C=O

+ $CH_3C(=O)$-Cl $\xrightarrow[0\ °C,\ -HCl]{AlCl_3}$

Phenanthrene

9-Acetyl phenanthrene

Uses: Phenanthrene is widely distributed in natural compounds such as bile acids, hormones, cholesterol and morphine alkaloids.

Medicinally useful compounds containing phenanthrene

Morphine

Use: It is used as a narcotic analgesic.

Heroin

Use: It is used as a narcotic analgesic.

Codeine Phosphate

Use: It is used as an expectorant and antitussive.

Dextromethorphan HBr

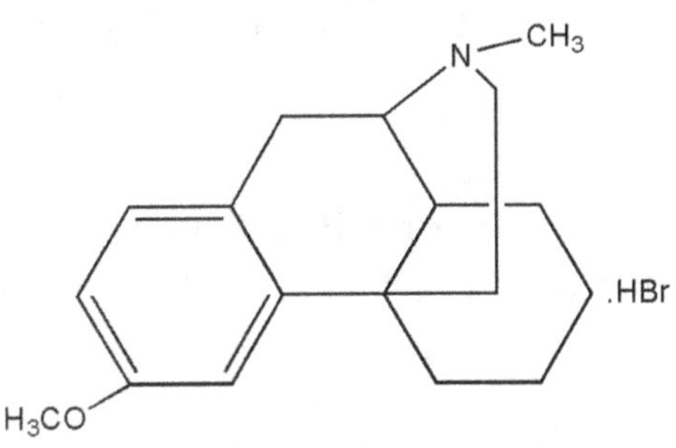

Use: It is used as an expectorant and antitussive.

Halofantrine (Hafan)

Use: It is used as an antimalarial agent.

Hydromorphine

Use: It is used as a narcotic analgesic.

Diphenyl Methane

General Methods of Preparation

1. **Friedel-Craft's reaction:** Benzyl chloride reacts with benzene in the presence of $AlCl_3$ to yield diphenyl methane.

$$C_6H_5-CH_2Cl + C_6H_6 \xrightarrow[AlCl_3,\ -HCl]{\text{Friedel-Craft's alkylation}} C_6H_5-CH_2-C_6H_5$$

Benzyl chloride, Benzene, Diphenyl methane

2. **From formaldehyde:** One molecule of benzaldehyde reacts with two molecules of benzene in the presence of conc. H_2SO_4 to yield diphenyl methane.

$$2\,C_6H_6 + HCHO \xrightarrow{H_2SO_4} C_6H_5-CH_2-C_6H_5 + H_2O$$

Benzene, Formaldehyde, Diphenyl methane

3. **From benzophenone:** Benzophenone reacts with HI and red phosphorous at 160 °C under pressure by Wolf-Kishner reduction or by LAH-$AlCl_3$ reduction to yields diphenyl methane.

$$C_6H_5-CO-C_6H_5 \xrightarrow[HI/P,\ 160\ °C]{\text{Wolf-Kishner reduction}} C_6H_5-CH_2-C_6H_5$$

Benzo phenone, Diphenyl methane

4. **From Grignard reaction:** Phenyl magnesium bromide reacts with benzyl chloride to yield diphenyl methane.

$$C_6H_5MgBr + C_6H_5CH_2Cl \xrightarrow{60\text{-}70\ °C} C_6H_5\text{-}CH_2\text{-}C_6H_5 + Mg(Br)Cl$$

Phenyl magnesium bromide, Benzyl chloride, Diphenyl methane

Chemical Properties

1. **Nitration:** Diphenyl methane reacts with nitrating mixture whereby first nitro group enter in to the 4th position and the second nitro group enters into the 4'-position.

$$C_6H_5-CH_2-C_6H_5 \xrightarrow[H_2SO_4]{HNO_3/} O_2N-C_6H_4-CH_2-C_6H_4-NO_2$$

Diphenyl methane, 4,4'-Dinitrodiphenyl methane

2. **Bromination:** Diphenyl methane reacts with bromine to yield diphenyl methyl bromide. Bromination occurs in the methylene group and which is not in the phenyl ring.

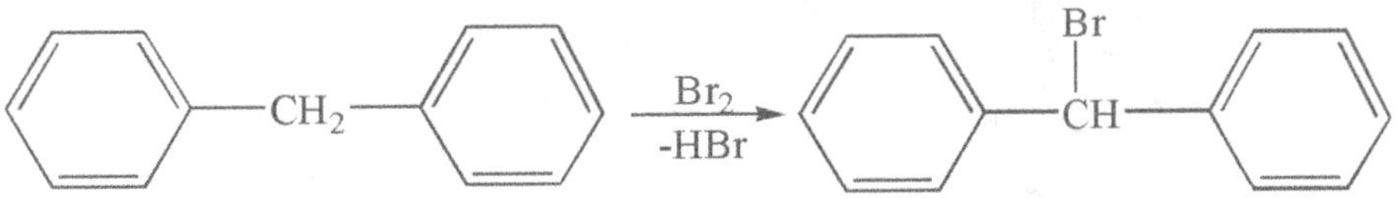

$$C_6H_5-CH_2-C_6H_5 \xrightarrow[-HBr]{Br_2} C_6H_5-CH(Br)-C_6H_5$$

Diphenyl methyl bromide

Medicinally useful compounds containing diphenylmethane

Diphenhydramine	**Clidinium**
$(C_6H_5)_2CH-O-CH_2-CH_2-N(CH_3)_2$ **Use:** It is used as an antihistamine.	CH_3, $N^{\oplus}$, OH, C, C, O, O, $Br^{\ominus}$ **Use:** It is used as an anticholinergic agent.
Clemastine	**Diphenylpyraline**
Cl, CH_3, $C-O-CH_2CH_2$, N, CH_3 **Use:** It is used as an antihistamine.	H, C, O, $N-CH_3$ **Use:** It is used as an antihistamine.
Chlophedianol hydrochloride	**Diphenoxylate HCl**
$(H_3C)_2N-H_2C-H_2C-C-OH$, Cl **Use:** It is used as an expectorant and antitussive.	CN, $C-CH_2-CH_2-N$, O, $C-OC_2H_5$, C_6H_5 **Use:** It is used as an antidiarrhoeal drug.

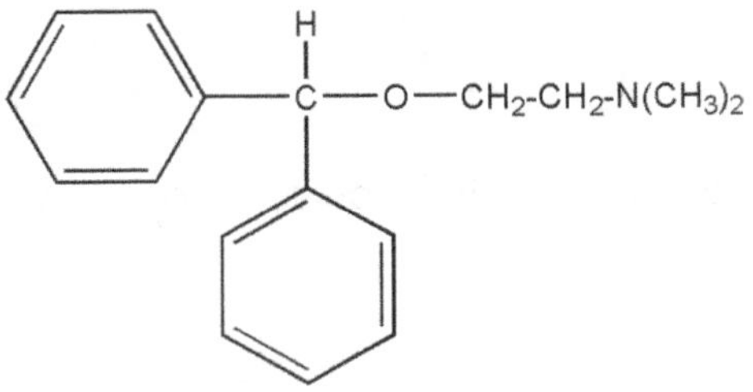

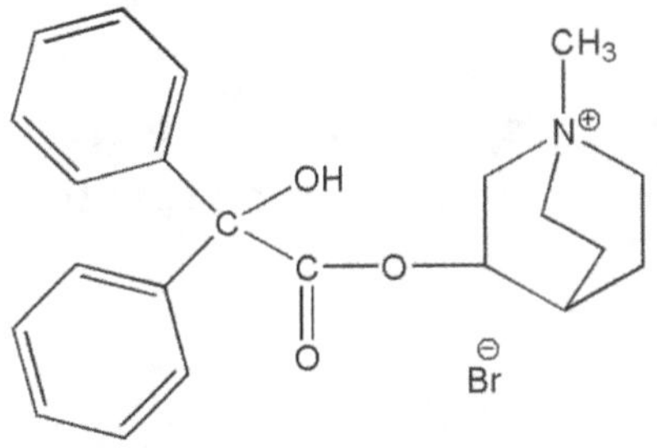

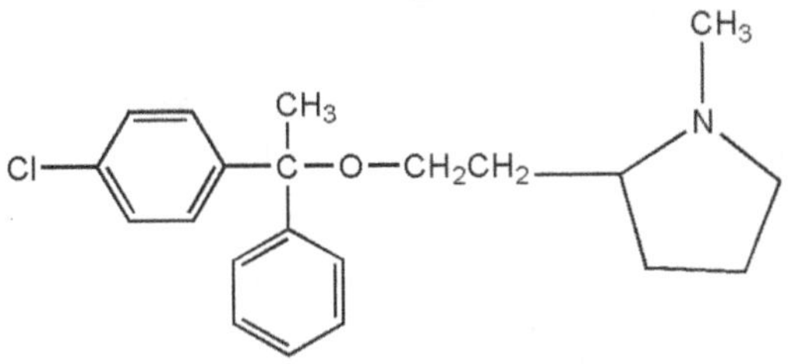

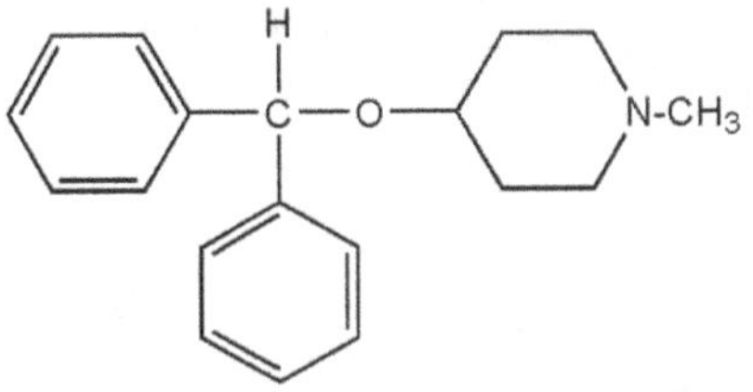

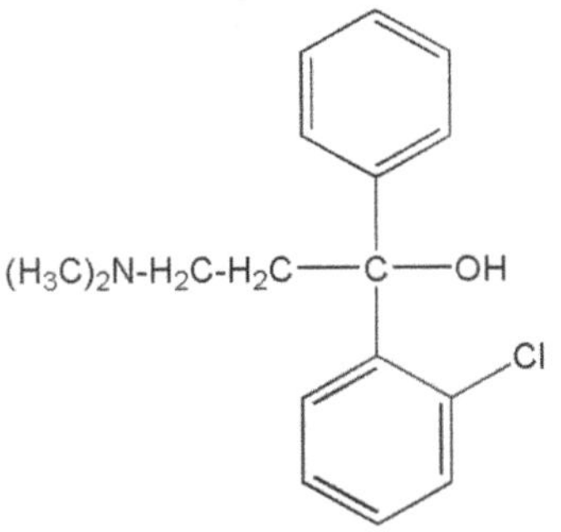

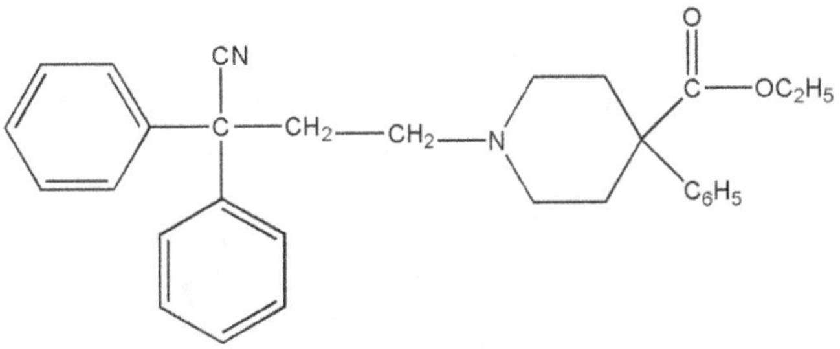

Triphenyl Methane or Tritane

General Methods of Preparation

1. **From benzene:** Benzene reacts with chloroform in the presence of $AlCl_3$ to yield triphenyl methane.

$$3\,C_6H_6 + CHCl_3 \xrightarrow{AlCl_3} (C_6H_5)_3CH + 3\,HCl$$

Benzene Choroform Triphenyl methane

2. **From benzaldehyde:** Benzaldehyde reacts with benzene in the presence of $ZnCl_2$ to yield triphenyl methane.

$$C_6H_5CHO + 2\,C_6H_6 \xrightarrow{ZnCl_2} (C_6H_5)_3CH + H_2O$$

Benzaldehyde Benzene Triphenyl methane

3. **From biphenyl:** Biphenyl reacts with dichloro toluene to yield triphenyl methane.

Biphenyl + Dichloro toluene ($C_6H_5CHCl_2$) $\xrightarrow{AlCl_3}$ Triphenyl methane ($(C_6H_5)_3CH$) + 2HCl

Physical Properties: Melting point of triphenyl methane is 93 °C.

Chemical Properties

1. **Bromination:** Triphenyl methane reacts with bromine to yield triphenyl methyl bromide.

$$(C_6H_5)_3CH + Br_2 \xrightarrow{-HBr} (C_6H_5)_3C{-}Br$$

Triphenyl methane — Triphenyl methyl bromide

2. **Reaction with sodium:** Triphenyl methane is more acidic than diphenyl methane and the conjugate base obtained is stabilized by resonance. It reacts with sodium in ethereal solution or in liquid ammonia to yield triphenyl methyl sodium.

$$(C_6H_5)_3CH + Na \xrightarrow{Ether} (C_6H_5)_3\overset{-}{C}{-}\overset{+}{Na} + 1/2\ H_2$$

Triphenyl methane — Triphenyl methyl sodium

Uses: Used in the preparation of drugs and drug intermediates.

Medicinally useful compounds containing triphenyl methane

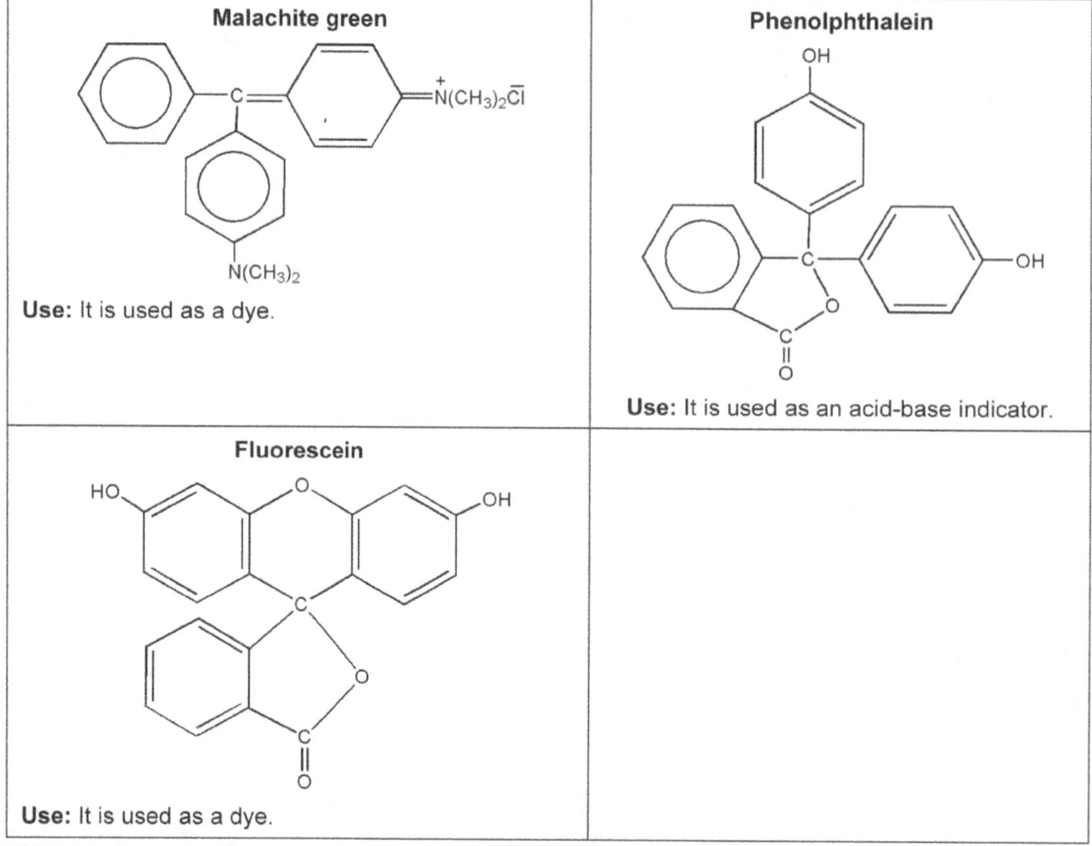

Probable Questions

1. Define polynuclear aromatic hydrocarbons.Give examples.
2. Explain the type of polynuclear hydrocarbons.
3. Describe the isolation and preparation of naphthalene from coal tar.
4. How will you prepare naphthol from naphthalene? Write the action of chlorine on naphthalene.
5. Explain the methods of preparation of two naphthols and write the distinguishing tests between them.
6. Explain about the electrophilic substitution reactions of naphthalene.
7. Mention the sequence of steps involved in the conversion of naphthalene to naphthylamine and indicates the reaction conditions and reagents used.
8. Write a note on Haworth synthesis.
9. Describe the isolation of anthracene from coal tar.
10. Explain about the electrophilic substitution reactions of anthracene.
11. Explain chemical reactions of anthracene.
12. Oxidation of anthracene with chromic acid gives 9,10- anthraquinone but not 1,4- anthra quinine - Explain.
13. How will you obtain alizarin from anthracene?
14. Write the synthesis of anthraquinone and anthrone.
15. Mention the methods of synthesis of phenanthrene.
16. Write any two important reactions of phenanthrene
17. Write the preparation, structure, chemical properties and uses of diphenyl methane.
18. Write the preparation, structure, chemical properties and uses of triphenyl methane.
19. What is the major source of naphthalene, anthracene and phenanthrene?
20. Write the oxidation and reduction reactions of anthracene.

26

Isomerism

Introduction

Number of entirely different variety of toys can be made with the same set of parts in "Mechano" for children; similarly, same set of atoms can be arranged in a number of ways to produce more than one organic compound. For example: the molecular formula of butane is C_4H_{10}. The 4 carbon atoms can be arranged in two different ways such as a straight or branched chain as given below.

So, C_4H_{10} is the molecular formula of two different kind of compounds.

Various type of organic compounds represented by the same molecular formula are known as **isomers** or **isomerides** (compounds with same molecular formula but different physical and chemical properties). This phenomenon is known as **isomerism**. Hence, isomers possess same molecular formula but they differ in properties due to the different arrangement of atoms within the molecule.

$$H_3C - CH_2 - CH_2 - CH_3$$

n-Butane

$$CH_3 - \underset{\underset{CH_3}{|}}{CH} - CH_3$$

iso-Butane

(or)

$$H - \underset{\underset{H}{|}}{\overset{\overset{H}{|}}{C}} - \underset{\underset{H}{|}}{\overset{\overset{H}{|}}{C}} - \underset{\underset{H}{|}}{\overset{\overset{H}{|}}{C}} - \underset{\underset{H}{|}}{\overset{\overset{H}{|}}{C}} - H \qquad H - \underset{\underset{H}{|}}{\overset{\overset{H}{|}}{C}} - \underset{\underset{H-\underset{\underset{H}{|}}{C}-H}{|}}{\overset{\overset{H}{|}}{C}} - \underset{\underset{H}{|}}{\overset{\overset{H}{|}}{C}} - H$$

Types of Isomerism

The two types of isomerism are as follows.

I. Structural isomerism.

II. Stereoisomerism.

I. Structural isomerism: Two or more compounds with same molecular formula but different structural formulae are known as structural isomers and this phenomenon is called as structural isomerism. Structural isomerism is of five types.

(i) Chain or nuclear isomerism.

(ii) Positional isomerism.

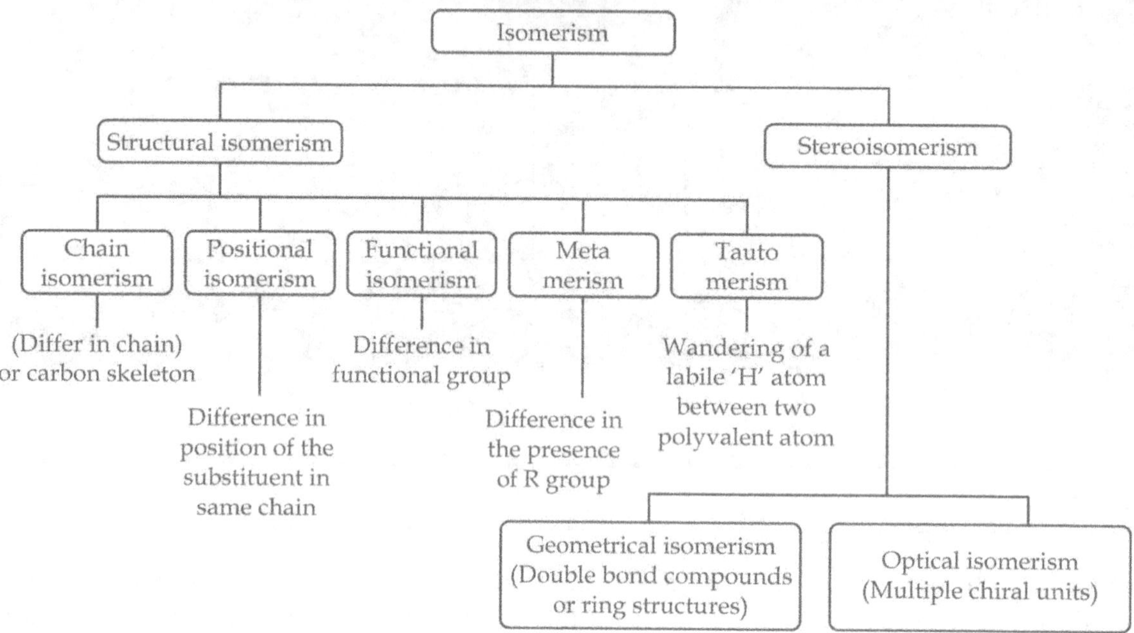

(iii) Functional isomerism.

(iv) Metamerism.

(v) Tautomerism.

II. Stereoisomerism: Some of the compounds having identical molecular formula and structural formula but they exhibits isomerism. For example, lactic acid and 2-butene, but it can't be explained easily. In those cases, the phenomenon can be explained on the assumption that the various atoms in the molecule occupy different position in space. Space models can be used to present such type of isomerism.

"The phenomenon of isomerism exhibited by two or more compounds with the identical molecular and structural formula, but different spatial arrangement of atoms or groups is known as stereoisomerism. The term stereoisomerism was coined by *Victor Meyer* in 1888. In Greek, "stereo means solid".

Let us discuss the above type of isomerisms in detail.

Structural Isomerism or Constitutional Isomerism

Structural isomers are compounds which have same molecular formula but differ in their arrangement of atoms. It is classified in to five types.

1. **Chain isomerism or nuclear isomerism:** Isomers with similar molecular formula but differ in their nature of carbon chain. For example, *n*-butane and *iso*-butane. Both of the compounds are having same molecular formula, but *n*-butane is straight chain hydrocarbon and *iso*-butane is branched chain hydrocarbon.

$$H_3C - CH_2 - CH_2 - CH_3$$

n-Butane

$$CH_3 - \underset{\displaystyle CH_3}{\underset{|}{CH}} - CH_3$$

iso-Butane

Similarly, pentane also exhibits chain isomerism. Three isomers of pentane are as follows.

$H_3C-CH_2-CH_2-CH_2-CH_3$

n-Pentane

$H_3C-CH_2-CH(CH_3)-CH_3$

iso-Pentane

(Branched chain with 2° carbon atom)

$H_3C-C(CH_3)_2-CH_3$

(Branched chain with one 3° carbon atom)

neo-Pentane

2. **Positional isomerism:** Isomerism produced by difference in the position of the same substituent in the main chain is known as positional isomerism. For example, *n*- propyl iodide and *iso*-propyl iodide.

$CH_3CH_2CH_2I$

n-Propyl iodide

(1-Iodopropane)

$CH_3-CH(I)-CH_3$

iso-Propyl iodide

(2-Iodopropane)

Both of the above two structures possess, straight chain formula but differ in the position of the halogen atom. In the first structure, the iodine atom is attached to first carbon atom and in second structure iodine is attached to the middle carbon atom.

Similarly, the alkene, butene exhibits two structures.

$\underset{1}{CH_2}=\underset{2}{CH}-\underset{3}{CH_2}-\underset{4}{CH_3}$

1-Butene

$\underset{1}{CH_3}-\underset{2}{CH}=\underset{3}{CH}-\underset{4}{CH_3}$

2-Butene

Both of the above structures have straight chain formula but differ in their position of the double bond. Some more examples are also shown below.

$CH_3-CH_2-CH_2OH$

n-Propanol

1-Propanol

$CH_3-CH(OH)-CH_3$

2-Propanol

iso-Propanol

$CH_3\text{-}C(=O)\text{-}CH_2\text{-}CH_2\text{-}CH_3$

2-Pentanone

$CH_3-CH_2-C(=O)-CH_2-CH_3$

3-Pentanone

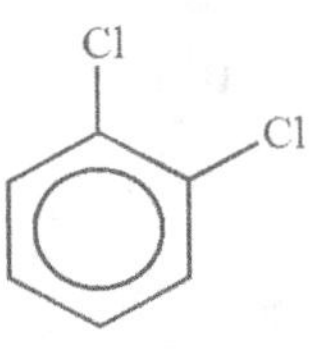

o-Dichlorobenzene

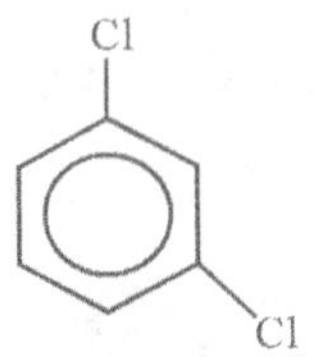

m-Dichlorobenzene

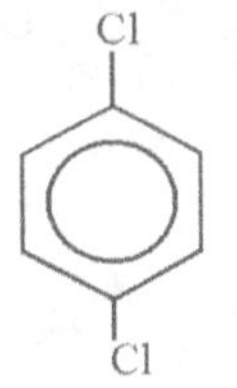

P-Dichloro benzene

3. **Functional isomerism:** When different functional groups are exhibited by same molecular formula it is known as functional isomerism and this kind of compounds are termed as functional isomers. For example, monohydric alcohols are isomeric with ethers. The general formula for both of these compounds is $C_n H_{2n+2}O$.

Molecular formula		
C_2H_6O	CH_3CH_2OH Ethanol	$CH_3 - O - CH_3$ Dimethyl ether
C_3H_8O	$CH_3 - CH_2 - CH_2OH$ *n*-Propanol	$C_2H_5 - O - CH_3$ Ethyl methyl ether

Some other examples are shown below.

(i) **Common formula: C_3H_6O**

$CH_3 - CH_2 - CHO$ Propionaldehyde

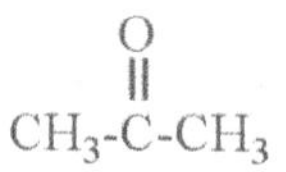

$CH_3-C(=O)-CH_3$ Acetone

(ii) **Common formula: C_3H_6**

$CH_3 - CH = CH_2$ 1-Propene

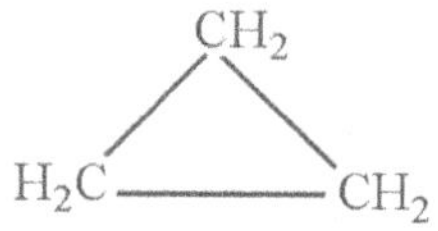

Cyclopropane (CH_2, H_2C—CH_2 ring)

(iii) **Common formula: $C_2H_4O_2$**

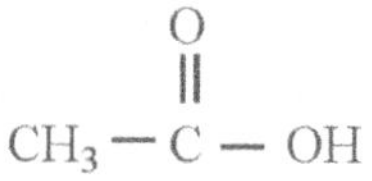

$CH_3-C(=O)-OH$ Acetic acid

$H-C(=O)-OCH_3$ Methyl formate

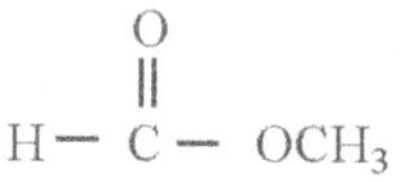

4. **Metamerism:** This type of isomerism is exhibited by the compounds of same homologues series. This isomerism is due to the difference in the nature of the alkyl groups attached to the polyvalent atom of the functional groups.

For example, three isomeric ketones are possible for the molecular formula of $C_5H_{10}O$ *etc.*

$C_2H_5 - C(=O) - C_2H_5$ Diethyl ketone

$(H_3C)(CH_3CH_2CH_2)C = O$ Methyl-*n*-propyl ketone

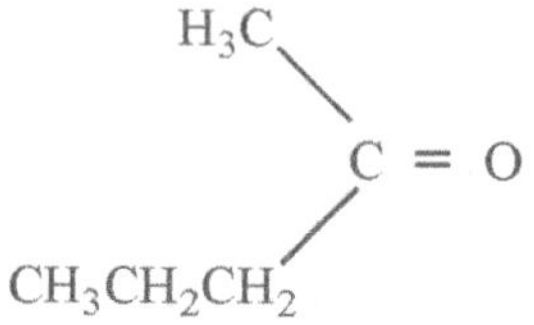

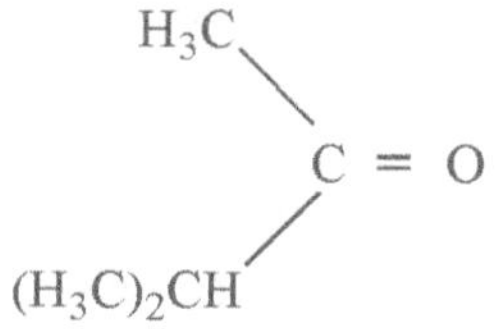

$(H_3C)((H_3C)_2CH)C = O$ Methyl *iso*-propyl ketone

Similarly, three isomeric ethers are represented by the molecular formula $C_4H_{10}O$ as follows.

$C_2H_5-O-C_2H_5$ Diethyl ether

$CH_3-O-CH_2-CH_2-CH_3$ Methyl-*n*-propyl ether

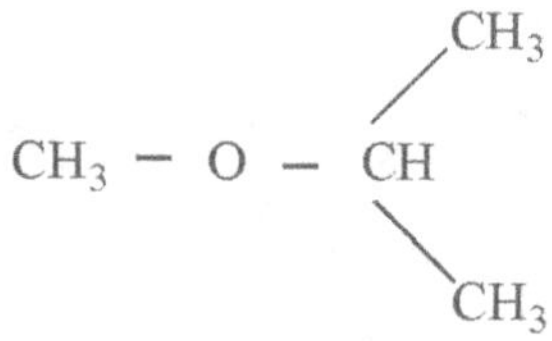

$CH_3 - O - CH(CH_3)_2$ Methyl *iso*-propyl ether

5. **Tautomerism:** It is a type of dynamic isomerism (functional isomerism) in which one isomer is constantly changing into the other and *vice versa*. Tautomerism occurs by wandering of labile hydrogen atom between two polyhedral atoms, carbon and nitrogen. For example, alkyl cyanides (RCN) and alkyl isocyanides (RNC) are possessing tautomerism. For explaining this concept, it is assumed that hydrocyanic acid exists in two isomeric forms.

$$H-C\equiv N \rightleftharpoons H-\overset{\oplus}{N}\equiv\overset{\ominus}{C}$$

Hydrocyanic acid — Iso hydrocyanic acid

And the alkyl derivatives of HCN also exhibit tautomerism.

$$R-C\equiv N \rightleftharpoons R-\overset{\oplus}{N}\equiv\overset{\ominus}{C}$$

Other examples:

(i) Nitro alkanes and isonitro alkanes.

$$R-CH_2-N(=O)\rightarrow O \rightleftharpoons R-CH=N(-OH)\rightarrow O$$

Nitro alkanes — *iso*-nitroalkanes

(ii) In room temperature, acetoacetic ester (AAE) present in two isomeric forms. One is keto form and another one is unsaturated hydroxy compound (enol form).

$$CH_3-\overset{O}{\overset{\|}{C}}-CH_2-COOC_2H_5 \rightleftharpoons CH_3-\overset{OH}{\overset{|}{C}}=CH-COOC_2H_5$$

I (Keto form) — II (Enol form)

Form I is constantly converted into form II; hence it shows the reactions of keto compound as well as an unsaturated hydroxy compound. This type of dynamic equilibrium is due to the wandering of labile hydrogen atom, which oscillates like a pendulum between two polyvalent atoms such as carbon and oxygen.

In the above two said structures, the form I possess keto group and form II possess an enol group. Hence this isomerism is known as keto - enol tautomerism.

Enols: Compounds having a double bond as in alkenes and an -OH group as in an alcohol.

(iii) Cyanic acid and isocyanic acid.

$$H-O-C\equiv N \rightleftharpoons O=C=N-H$$

Cyanic acid gives rise to cyanates — isocyanic acid gives rise to isocyanates

Tautomerism is discussed in detail below.

Tautomerism

Two structural isomers which are mutually interchangeable and exist in dynamic equilibrium are called **tautomers** and the phenomenon is **tautomerism**. Tautomers are constitutional isomers of organic compounds which readily interconvert due to relocation of a proton. In the following example of acetone, we can observe that the keto form changes into enol form by the migration of a proton to carbonyl oxygen and then the electrons of C-H bond shifted to C-C bond.

$$CH_3-\overset{O}{\overset{\|}{C}}-\overset{H}{\overset{|}{C}}H_2 \rightleftharpoons CH_3-\overset{OH}{\overset{|}{C}}=CH_2$$

Acetone (Keto form) Acetone (Enol form)

Tautomerism differs from resonance as follows. In case of resonance, resonating structures do not have real independent existence of structure and in resonance only electrons shifting takes place (without shifting of atoms).

Some important tautomerism occur in organic chemistry are as follows.

1. Keto-enol tautomerism.
2. Lactam-lactim tautomerism.
3. Amide-imide tautomerism.
4. Amine-imine tautomerism.

Let us discuss these tautomerism briefly.

1. **Keto-enol tautomerism:** Compounds containing carbonyl groups like aldehydes, ketones and esters can exist in two tautomeric forms, called keto form (contains C=O group) and enol form (contains double bonds and -OH group) by the migration of a proton from α-carbon to carbonyl oxygen. This property of interconverting keto form to enol form is called as **keto-enol tautomerism.**

$$CH_3-\overset{O}{\overset{\|}{C}}-\overset{H}{\overset{|}{C}}H_2 \rightleftharpoons CH_3-\overset{OH}{\overset{|}{C}}=CH_2$$

Acetone (Keto form) Acetone (Enol form)

Mechanism:

1. Due to electron withdrawing property of oxygen of carbonyl group, α-hydrogen of keto compound polarized and forms carbanion (enolate ion) and it is stabilized by resonance.

$$CH_3-\overset{O}{\overset{\|}{C}}-\overset{H}{\overset{|}{C}}H_2 \xleftrightarrow{-H^+} CH_3-\overset{:\ddot{O}:^-}{\overset{|}{C}}=CH_2 \equiv CH_3-\overset{\overset{\delta^-}{O}}{\overset{\vdots}{C}}\cdots\underset{\delta^-}{CH_2}$$

2. Upon acidification, protonation at oxygen produces enol form while protonation at carbon yields keto form.

$$CH_2=\overset{OH}{\overset{|}{C}}-CH_3 \underset{-H^+}{\overset{+H^+}{\rightleftharpoons}} \underset{\delta^-}{H_2C}\cdots\overset{\overset{\delta^-}{O}}{\overset{\vdots}{C}}-CH_3 \underset{-H^+}{\overset{+H^+}{\rightleftharpoons}} H_3C-\overset{O}{\overset{\|}{C}}-CH_3$$

Enol form Intermediate Keto form

Tautomerism

(As both reactions are reversible, both forms are interconvertible)

The amount of keto or enol form present can be determined by NMR spectroscopy.

In acetone the enol form exists less than 0.1 %, when the carbonyl group, separated by methylene group, the existence of enol will be more amount as shown in ethyl acetoacetate.

Keto-enol tautomerism in ethyl acetoacetate (Aceto aceticester)

As significant amount of ethyl acetoacetate exist in keto and enol form. Both form exhibits its characteristic reactions of keto form and enol form.

$$CH_3-\overset{O}{\overset{\|}{C}}-CH_2-\overset{O}{\overset{\|}{C}}-OC_2H_5 \rightleftharpoons CH_3-\overset{OH}{\overset{|}{C}}=CH_2-\overset{O}{\overset{\|}{C}}-OC_2H_5$$

Keto form (90 %) Enol form (10 %)

Both forms can be isolated and shows distinctive properties. Hence the reactions of ethyl acetoacetate can be described in two heads as

(i) Reactions of keto form

(ii) Reactions of enol form.

2. **Lactam -lactim tautomerism:** This type of tautomerism occurs due to the shifting of proton between Lactam-Lactim form of organic compounds as shown below.

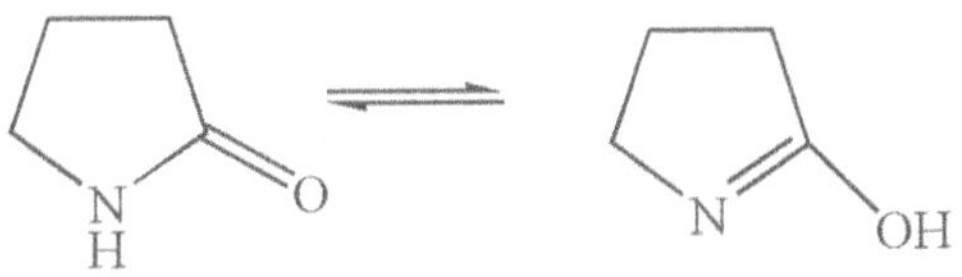

Lactam form ⇌ Lactim form

3. **Amide-imide tautomerism:** It occurs due to the interconversion of proton between amide and imide form of organic compound.

$$R-\overset{O}{\overset{\|}{C}}-\underset{H}{N}-R' \rightleftharpoons R-\overset{OH}{\overset{|}{C}}=N-R'$$

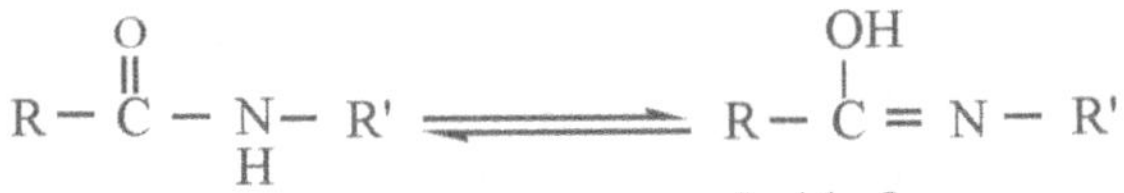

Amide form Imide form

4. **Amine-imine tautomerism:** This type of tautomerism is due to the exchange of proton between amine-imine forms as shown below.

NH_2 ⇌ NH

Amine form Imine form

Stereoisomerism

Isomerism exhibited by two or more compounds with the identical molecular and structural formula but differ in their spatial arrangement of atoms or groups is termed as stereo isomerism. It is of two types. They are geometrical isomerism and optical isomerism. **It is discussed in detail under chapter No. 27.**

Probable Questions

1. Define isomerism and isomers.
2. Enumerate various types of isomerism with examples.
3. What is meant by structural isomerism and explain its types with examples.
4. Define stereoisomerism and write its types with examples.
5. Explain in detail about metamerism.
6. What is meant by tautomerism?
7. Define geometrical isomerism with example.
8. What is the type of isomerism exhibited in alkanes?
9. Differentiate *cis* and *trans* isomerism and write isomers with minimum two structural examples.
10. What is meant by *syn* and *anti* isomer?

27

Stereoisomerism

Introduction

Isomerism exhibited by two or more compounds with the identical molecular and structural formula but differs in their spatial arrangement of atoms or groups is termed as stereoisomerism. It is of two types. They are geometrical and optical isomerism.

Optical Isomerism

Optical rotation: Enantiomers possess many of the identical properties such as m.p., b.p. and solubility. But some of the properties are not same and depends upon the nature of attachment *i.e.* spatial arrangement of atoms or groups of that molecule to the chiral center. One of the important properties of enantiomers is interaction with plane polarized light.

Plane polarized light: Normal light consists of electromagnetic waves that oscillate in all planes passing through the direction of the light travels. Plane polarized light oscillates only in a single plane passing through the direction of the light travel. It is obtained by passing normal light through a polarizer *i.e.,* Nickel prism or lens (Fig. 27.1).

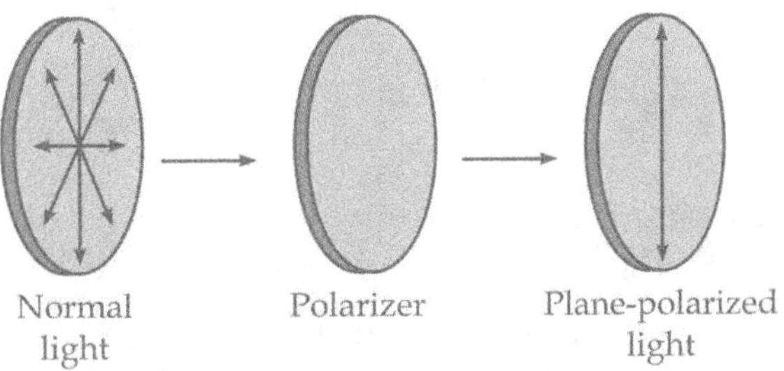

Figure 27.1 Plane polarized light.

INTERESTING FACT

Polarized sunglasses pass or allows only light oscillating in a single plane to pass *via* them, so they block glare (Reflections) more effectively than non-polarized glasses. *William Nicol* (1768-1851) was the professor at the University of Edinburgh. He developed the first prism that produced plane polarized light. He also developed methods to make thin slices of materials for use in microscopical studies.

In 1815, the physicist *Jean-Baptiste Biot* discovered some natural organic substances such as camphor and turpentine oil rotates polarized light. He noticed that some compounds rotate the plane polarized light in clockwise direction and some of them in anticlockwise direction while others did not rotate the plane-polarized light. He predicted that "the ability of a particular compound to rotate a plane-polarized light is due to the presence of asymmetrical centre or some asymmetry in the molecule. *Van't Hoff* and *LeBel* later determined that the molecular asymmetry was the result of possessing one or more chiral centers (Fig. 27.2).

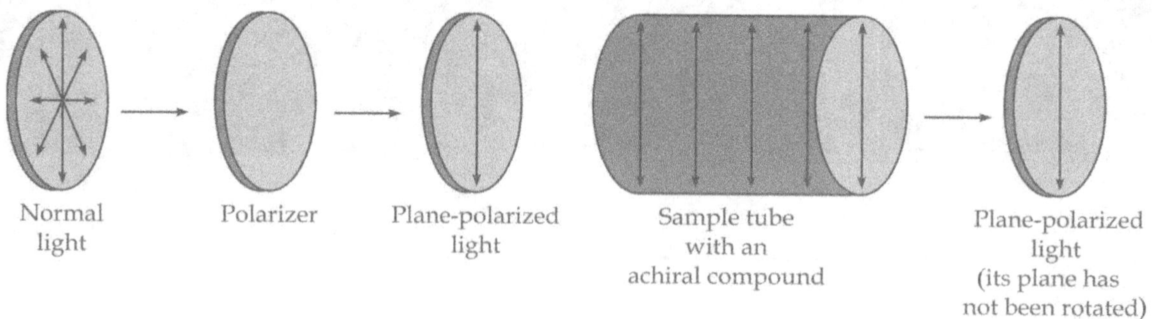

Figure 27.2 Direction of light propagation in achiral molecule.

When plane of polarized light passes through a solution of chiral compound, plane of polarization changes because the molecules are asymmetric. Hence, a chiral compound rotates plane-polarized light either clockwise or anticlockwise direction. If one enantiomer rotates the plane polarized light in clockwise direction and its mirror image (another enantiomer) will rotate the plane polarized light exactly same amount in anticlockwise direction (Fig. 27.3).

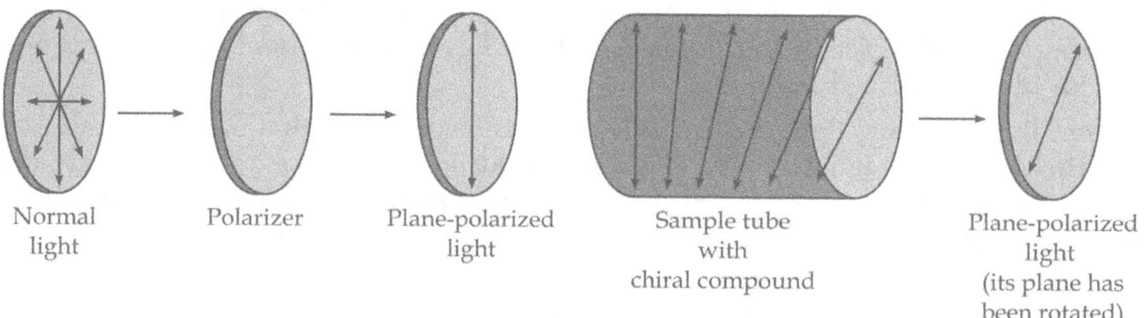

Figure 27.3 Direction of light propagation in chiral molecule.

A compound which rotates a plane polarized light is said to be optically active. Otherwise, chiral compounds are optically active and achiral compounds are optically inactive.

Dextrorotatory: If an optically active compound or chiral compound rotates the plane polarized light in a clockwise direction, it is known as dextrorotatory and indicated by the symbol (+) or "d".

Levorotatory: If the optically active compound or chiral compound rotates the plane polarized light in anticlockwise direction, it is known as levorotatory and indicated by the symbol (-) or "*l*". "Dextro and levo" are Latin prefixes for "to the right" and "to the left" respectively.

While naming stereochemical compounds, the D, L, R and S indicates stereochemical relation (structural features) and the (+) or (-) indicates actual direction of rotation of plane polarized light.

Note: Don't get confuse (+) and (-) with R and S or D and L. The (+) and (-) symbol indicates the direction in which a chiral compound rotates a plane polarized light. But R and S or D and L indicates the arrangement of the groups around or about a chiral centre.

Specification of configuration: The arrangement of atoms that characterises a particular stereoisomer is known as configuration.

Chiral carbon: A carbon atom attached with four different atoms or groups is called as chiral carbon or chiral center. The chiral center is denoted by an asterisk.

For example

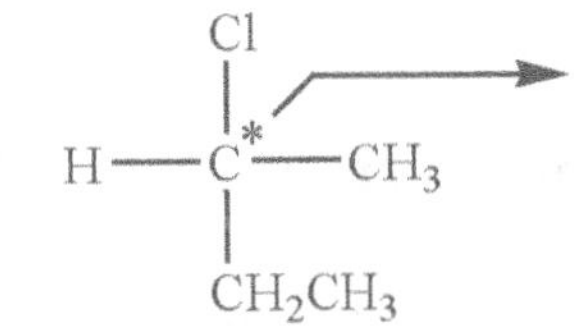

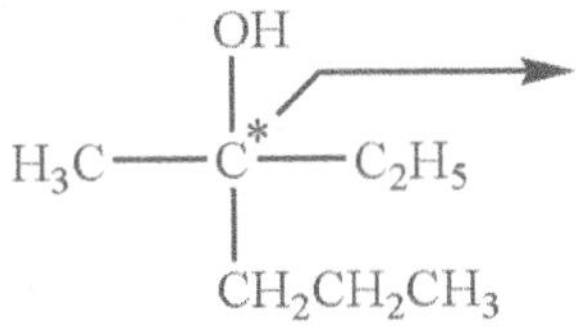

Cl
H—C*—CH_3 → Chiral carbon/center
CH_2CH_3

OH
H_3C—C*—C_2H_5 → Chiral carbon/center
$CH_2CH_2CH_3$

Enantiomers: Any compound with chiral carbon exists in two different forms of isomers. They cannot be superimposable and they are the analogous of left and right hand *i.e.*, they are mirror images of each other. The non-superimposable mirror images of a chiral compound are known as enantiomers. In Greek *enantion* means opposite, *morphos* means form.

Example 1:

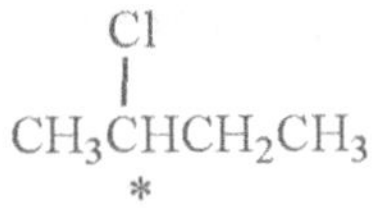

Cl
$CH_3\overset{*}{C}HCH_2CH_3$
2-Chloro butane

Cl, C, H, H_3CH_2C, CH_3 | Cl, H, C, H_3C, CH_2CH_3

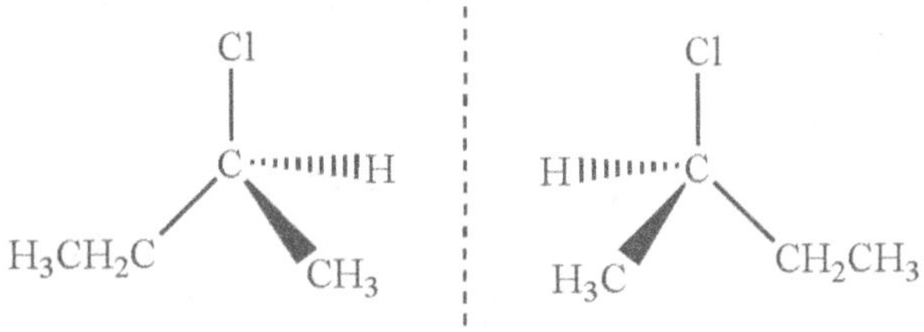

Enantiomers of 2-chloro butane

Example 2:

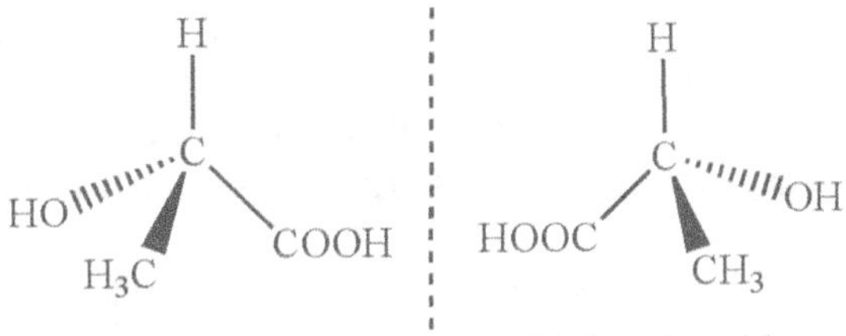

H, HO, C, H_3C, COOH | H, C, OH, HOOC, CH_3

(+)-Lactic acid (−)-Lactic acid

Enantiomers of lactic acid

Enantiomers have identical physical properties and one enantiomer is easily converted in to another one by interchanging of any two groups.

Example 3:

```
     Cl OH
     |  |
  CH3CHCHCH3
      *  *
3-Chloro-2-butanol
```

It contains two chiral centers. Hence it possess four stereoisomers. They are as follows.

```
      CH3              CH3                 CH3               CH3
      |                |                   |                 |
  H — C — OH      HO — C — H          H — C — OH       HO — C — H
      |                |                   |                 |
  H — C — Cl      Cl — C — H         Cl — C — H         H — C — Cl
      |                |                   |                 |
      CH3              CH3                 CH3               CH3

      1                2                   3                 4
      Erythro enantiomers                  Threo enantiomers
```

In the above 4 structures, the 1 and 2 are non-superimposable mirror images, hence enantiomers. The stereoisomers 3 and 4 are also enantiomers. Hence the four stereoisomers of 3-chloro-2-butanol contain two sets of enantiomers.

Erythro enantiomers: The set or pair of enantiomers with similar groups on the same side of the carbon chain in Fischer projection is known as erythro enantiomers.

Threo enantiomers: The set or pair of enantiomers with similar groups on the opposite side of the carbon chain in Fischer projection is known as threo enantiomers.

Diastereomers: Diastereomers are configurational isomers which are not enantiomers. They possess non-identical mirror images. For example.

```
      CH3              CH3                 CH3               CH3
      |                |                   |                 |
  H — C — OH      HO — C — H          H — C — OH       HO — C — H
      |                |                   |                 |
  H — C — Cl      Cl — C — H         Cl — C — H         H — C — Cl
      |                |                   |                 |
      CH3              CH3                 CH3               CH3

      1                2                   3                 4
      Erythroenantiomers                   Threoenantiomers
```

In the above example stereoisomers 1 & 3 are not identical and not mirror images hence they are called diastereoisomers. 1 & 4, 2 & 3 and 2 & 4 are also diastereomers. Example: Tartaric acid

```
      COOH               COOH              COOH
      |                  |                 |
  H — C — OH         H — C — OH       HO — C — H
      |                  |                 |
  H — C — OH        HO — C — H         H — C — OH
      |                  |                 |
      COOH               COOH              COOH

      1                  2                 3
  Meso tartaric acid          Enantiomers
                            (Mirror images)
```

In the above structures 1 & 3, 1 & 2 are diastereomers but 2 & 3 are enantiomers. Diastereomers have more than one chiral center (Minimum two). They possess different physical properties. Once we know the number of chiral centers of a compound we can calculate the possible number of stereoisomers for that compound.

If any compound have 'n' chiral centers, then the maximum number of stereoisomers for that compound = 2^n.

S. No.	Enantiomers	Diastereomers
1	Non- identical mirror images with identical physical properties. Similar mp, bp and solubilities.	Non- identical mirror images with different physical properties. Different mp, bp and solubilities.
2	They possess identical chemical properties.	They differ in their chemical properties.
3	They react with same achiral reagent with similar rate.	They react with same achiral reagent with different rate.
4	Separation is a difficult one.	Easily separated by crystallization or chromatographic technique.

Meso Compounds

A meso compound is an achiral molecule which has two or more chiral centers. In Greek word *'Meso'* means middle. It has plane of symmetry. It is optically inactive and it does not rotate the plane polarized light. They have stereogenic carbons.

It is a set of stereoisomers in which atleast two of them are optically active. A meso compound is not superimposable on its mirror image (Diasteromer). For example, tartaric acid

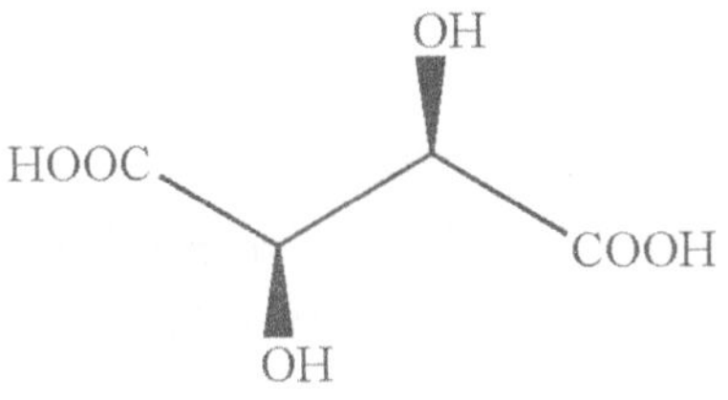

OH HOOC COOH OH
(+)-Tartaric acid

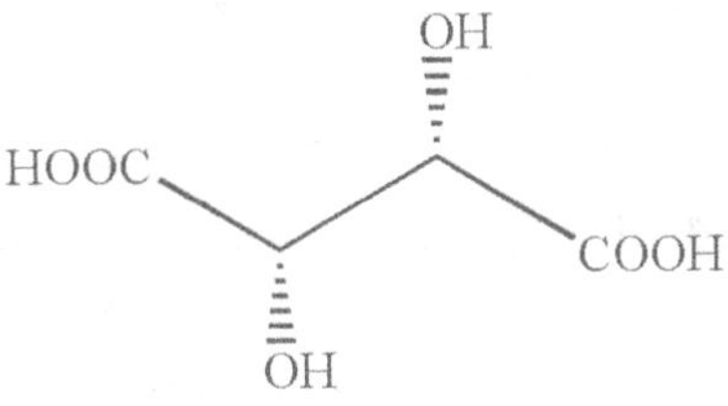

OH HOOC COOH OH
(-)-Tartaric acid

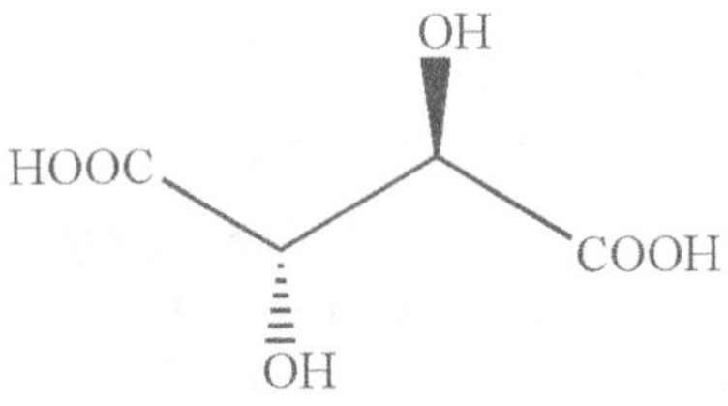

OH HOOC COOH OH
meso-Tartaric acid

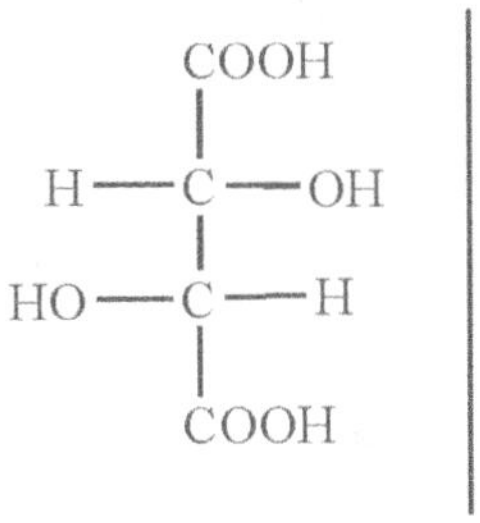

COOH
H—C—OH
HO—C—H
COOH
(+)-Tartaric acid

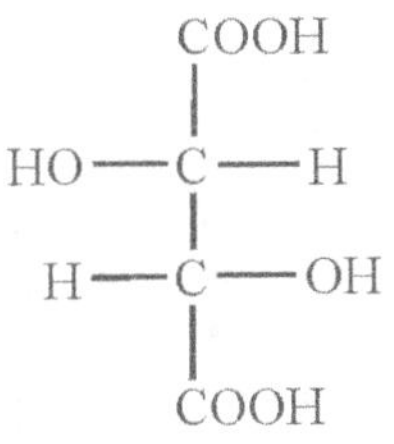

COOH
HO—C—H
H—C—OH
COOH
(-)-Tartaric acid

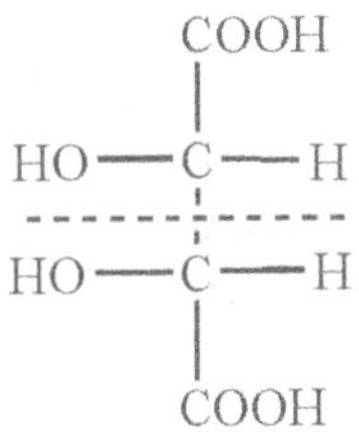

COOH
HO—C—H
HO—C—H
COOH
meso-Tartaric acid

If a compound with two chiral centers has the same four groups bonded to each chiral center, one of its stereoisomers will be a meso compound and other two will be a pair of enantiomers.

For example

A meso compound A pair of enantiomers

Cyclic meso compounds: In the following example, the *cis*-isomer is the meso compound and the *trans* isomer will exist as pair of enantiomers.

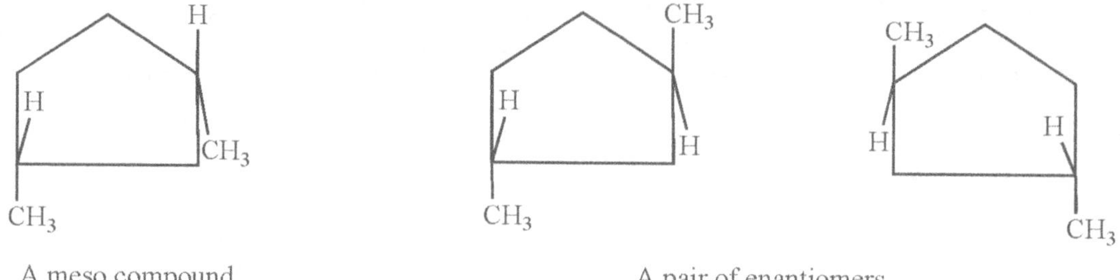

A meso compound A pair of enantiomers

Chirality and Symmetry of elements: A molecule is said to have symmetry when it is having a single chiral centre. Another method used to recognize them is to look for the presence of certain symmetry of elements.

For example, If the molecule will not be chiral if it contains,

(i) Plane of symmetry indicated by the symbol σ (sigma).

(ii) Centre of symmetry indicated by the symbol i.

(iii) Any n - fold (n = even number) alternative axis of symmetry indicated by C_n or S_n.

Plane of symmetry indicated by the symbol σ (sigma): A plane which divides an object into two identical halves is known as plane of symmetry. For example, a ball can be divided by a plane into two identical halves, so the ball is achiral or non-chiral molecule (Fig. 27.4).

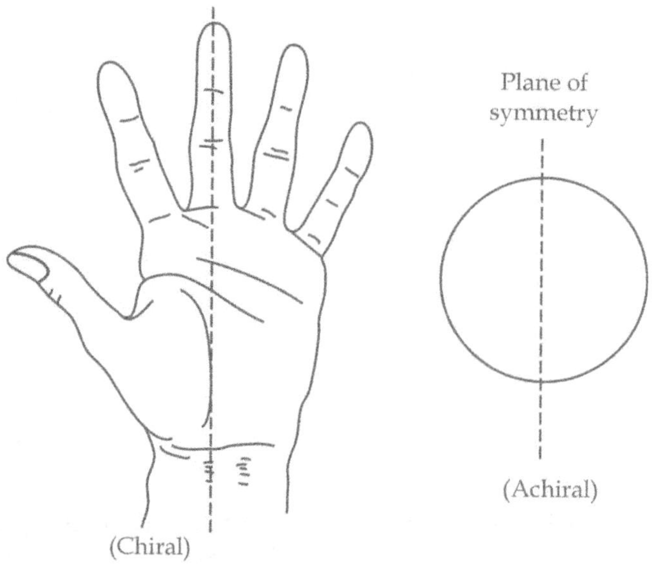

Figure 27.4 Plane of symmetry in chiral and achiral molecule.

From this example it is clearly denoted that a plane of symmetry is an imaginary plane which cuts the molecule in to two parts so that each part is the mirror image of the other. So, these are optically inactive.

Both of the above cases, there are two asymmetric carbon atoms present even then they are optically inactive. Because an imaginary plane can divide both the molecules in to two parts (mirror image of each other). Eg: meso-tartaric acid (structure shown earlier).

Centre of Symmetry: An imaginary point in the molecule such that if a line is drawn from the any group of the molecule to this point and then extended to an equal distance beyond the point, it meets the original group. For example, the substituted diketo piperazine having two asymmetric carbon atoms even then it is optically inactive because it possesses centre of symmetry.

trans-1,4-dimethyl diketopiperazine

When a line is drawn from one of the methyl groups on C-1 to the centre of symmetry and extended beyond this centre by an equal distance, if it meets the methyl group at C-4, it is optically inactive.

Alternating axis of symmetry: It is an imaginary axis such that when the structure possessing such an axis is rotated *via* the angle of $2\pi/n$ about this axis and then reflected across a plane perpendicular to this axis leads to identical structures. For example, if the molecule rotated through an imaginary axis by an angle of 60° leads to the formation of mirror images or identical structures. It is said to be possess a sixfold axis (360°/60°) of symmetry. Only less number of compounds are inactive due to this elements of symmetry.

cis-1-Bromo-3-chloro cyclobutane *trans* -1-Bromo-3-chloro cyclobutane

Chiral molecules

Chirality is a geometric property of some compounds. It is defined as **"a carbon atom attached with four different groups or atoms is known as chiral carbon and the molecule is known as chiral molecule.** The term chirality is derived from the Greek word *"chirality* mean handed". Chiral molecules rotate plane polarised light. A chiral molecule does not contain plan of symmetry and inversion center. Many of the aminoacids, sugars and enzymes are chiral molecules. Chiral carbon is also known as chiral center and denoted by asterisk. The term chirality was coined by the scientist Lord Kelvin in 1894.

Lactic acid Alanine Glyceraldehyde

The two forms of chiral molecules (mirror images) are called as enantiomers. Many of the drugs also possess chiral centers. Some of the examples are as follows.

D-penicillamine is used as chelating agent for iron poisoning. L-penicillamine inhibits the pharmacological action of vitamin B hence it is toxic. S-Citalopram used as an antidepressant drug. L-aspartame has sweet taste hence used as an artificial sweetner. D-aspartame is tasteless.

If a compound is achiral it is superimposable on its mirror image. The stereo centre present in the achiral molecule is known as meso compound. It possess plane of symmetry and inversion centre.

Achiral molecules: A carbon atom is known as achiral when any two or more of the four atoms or groups attached to it are identical. In this condition, the carbon atom does not rotate the plane polarised light.

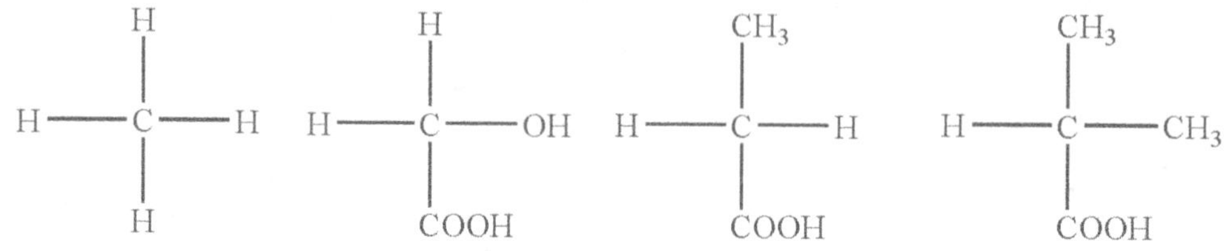

H—C(H)(H)—H

Methane

H—C(H)(COOH)—OH

Glycollic acid

H—C(CH_3)(COOH)—H

Propionic acid

H—C(CH_3)(COOH)—CH_3

iso-butyric acid

HOOC- - - - -C(OH)—H

HOOC- - - -C(OH)—H

Plane of symmetry

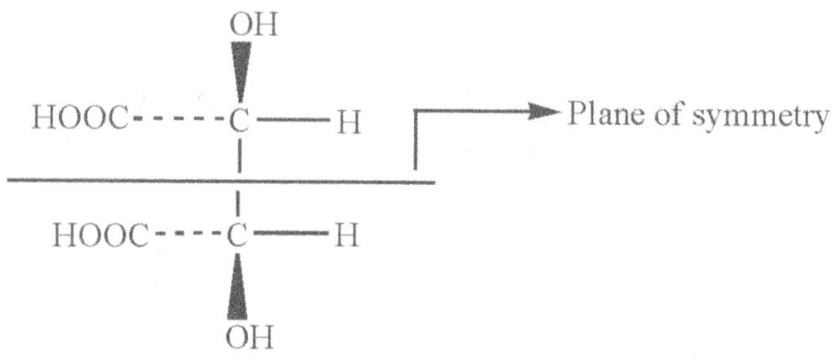

Tartaric acid

DL System of Nomenclature: Arrangement of atoms which describes a particular stereoisomer is known as configuration. For example, Lactic acid has two types of stereoisomers. One rotates plane polarized light to the right (+) sign and another rotates left side (-) sign. How can one assign configuration and how to indicate a specific or particular configuration in a simple and convenient manner without drawing orbital structure? It is described as follows.

There are two types of stereochemical notations are commonly used. They are **D (Dee)** and **L (Ell)** notation.

Symbolic representation of a specific configuration of a molecule is an easiest method which involves letters D and L for adopting names to them. This is explained by considering the structure of a carbohydrate glyceraldehyde as an example.

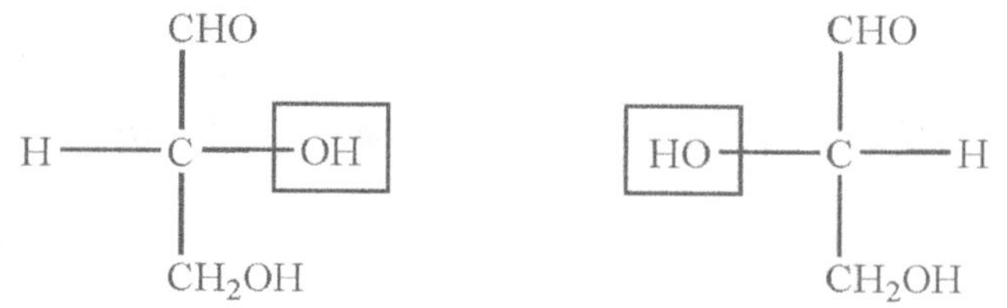

CHO
H—C—OH
CH_2OH

D-Glyceraldehyde
I

CHO
HO—C—H
CH_2OH

L-Glyceraldehyde
II

I & II are two configurations of glyceraldehyde, (I) with –OH group on the right hand side of the molecule is designated as D-glyceraldehyde while the other glyceraldehyde (II) with –OH group on the left hand side of the molecule is designated as L-glyceraldehyde.

For hydroxy acids such as (-) glyceric acid which possess the similar configuration as D-glyceraldehyde, hence it can be designated as D- family.

$$\begin{array}{c} COOH \\ | \\ H—C—OH \\ | \\ CH_2OH \end{array}$$

D(–)-Glyceric acid

Due to the presence of $-NH_2$ group in the structure of all naturally occurring α-amino acids they belongs to the L-family irrespective of the direction in which they rotate the plane polarised light. For example,

$$\begin{array}{c} COOH \\ | \\ H_2N—C—H \\ | \\ CH_2C_6H_5 \end{array}$$

L(–)-Phenyl alanine

For assigning D & L configuration of any molecule the following points are to be considered.

1. Hang it from the most oxidised carbon atom, For example, in glyceraldehyde keep –CHO group at the top.
2. The group with carbon atom forming a part of the chain is kept away from you. For example, in glyceraldehyde keep the - CH_2OH group at the bottom.
3. The remaining two groups should be projecting towards you.

   Now if the –OH group is towards your right hand side it has the D- configuration while the –OH group is towards your left hand side it has the L- configuration.

   Note: Always remember that

   (a) In olden days, d (dextro) and *l* (levo) were also used to indicate the sign of the rotation of plane polarised light instead of (+) and (-) signs.

   (b) D & L-notation denotes to the absolute configuration of a molecule (3D- structure).

   (c) Signs (+) and (-) denotes to the direction of the rotation of plane polarised light.

   (d) Reader should not confuse with the D & L notation (which is based on structural formula) and the (+) & (-) sign which denotes the actual direction of rotation of plane polarised light. Hence following isomers are possible to exist D(+), D(-), L(+) and L(-).

$$\begin{array}{c} CHO \\ | \\ H—C—OH \\ | \\ HO—C—H \\ | \\ H—C—OH \\ | \\ H—C—OH \\ | \\ CH_2OH \end{array} \qquad \begin{array}{c} CH_2OH \\ | \\ C{=}O \\ | \\ HO—C—H \\ | \\ H—C—OH \\ | \\ H—C—OH \\ | \\ CH_2OH \end{array}$$

D(+)-Glucose D(–)-Fructose

In the above examples, the D indicates both have same configuration at 5th carbon atom (penaltimate carbon) but they differ in direction of rotation of plane polarised light i.e., glucose turns in to right hence (+) sign while fructose turns in to left, hence (-) sign.

Nomenclature of Enantiomers or the R, S - System of Nomenclature of R and S Notation of Optical Isomers [*Cahn – Ingold – Prelog* System of Nomenclature]: Let us consider the following example of 2-bromobutane which has two stereoisomers. We need to name the two individual isomers to differentiate them. For this purpose we need a systematic naming system. *Cahn, Ingold* and *Prelog* proposed a schematic representation of enantiomers by using the prefix *R* (derived from Greek word *Rectus* means right handed) and the prefix *S* (derived from the Greek word *Sinister* means left handed). The assignment of the configuration to an enantiomer is based upon the type or nature of groups attached to the chiral centre.

Br Br

H_3CH_2C C H CH_3 H C CH_2CH_3 H_3C

The enantiomers of 2-bromobutane

The nature of groups attached to the chiral carbon is determined by determining the priority of groups which is done on the basis of sequence rules indicated below.

Sequence Rules

Rule 1: When four different atoms or groups are attached to the chiral carbon, priority depends upon their atomic number *i.e.,* the atom with high atomic number gets highest priority. Some of the important atoms and their atomic numbers given below.

Groups: **I > Br > Cl > SH > OH > NH_2 > CH_3 > H**

Atomic Number: 53 > 35 > 17 > 16 > 8 > 7 > 6 > 1

Cl

H—C—SO_3H

I

Chloro iodomethane sulphonic acid

In the above example the order of priority is I, Cl, S, H

Rule 2: If the two atoms are isotopes of the same element, atom of higher mass numbers gets higher priority. Examples of isotopes of some atoms and their mass numbers are given below.

Isotopes: **O > O > O > C > C > C > H > H > H**

Mass number: 18 >17 > 16 > 14 > 13 > 12 > 3 > 2 > 1

H

CH_3—C—Br

D

α-Deuterio methyl bromide

In the above example the order of priority sequence is Br, CH_3, D, H.

Rule 3: If the two atoms attached to the chiral carbon are same, rule 1 cannot used to decide the priority. In such cases, consider the first atom of a group, then find the next set of atoms and continue until the priority can be allotted or assigned.

For Example: In secondary butyl chloride

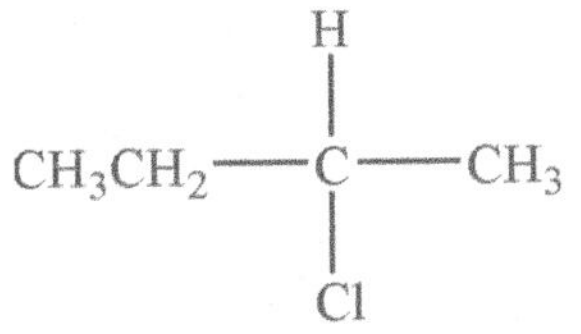

$$CH_3CH_2-\overset{\displaystyle H}{\underset{\displaystyle Cl}{C}}-CH_3$$

2° – Butyl chloride

The relative priority of CH_3 & C_2H_5 is fixed as follows.

In CH_3, the second atoms are three hydrogen (H, H, H), but in C_2H_5 they are C, H, H. As carbon atoms has higher atomic number than the hydrogen, C_2H_5 has higher priority than the methyl ($-CH_3$) group. Hence the sequence of priority is Cl, C_2H_5, CH_3, H. Some of the example of groups is given below.

Groups: **$-CH_2Cl > -CH_2OH > -CH_2NH_2 > -CH_2CH_3 > CH_2D > -CH_2H$**

Order of priority: 17 > 8 > 7 > 6 > 2 > 1

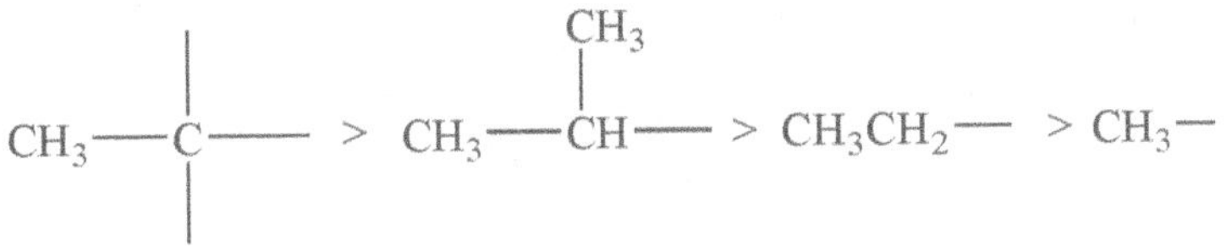

$$CH_3-\overset{|}{\underset{|}{C}}- \;>\; CH_3-\overset{CH_3}{\overset{|}{CH}}- \;>\; CH_3CH_2- \;>\; CH_3-$$

Rule 4: In case of double and triple bonds, atoms participating in the double or triple bond are considered to be bonded to an equivalent number of single bonds, that is atoms of double and triple bonds are replicated as shown in the following examples.

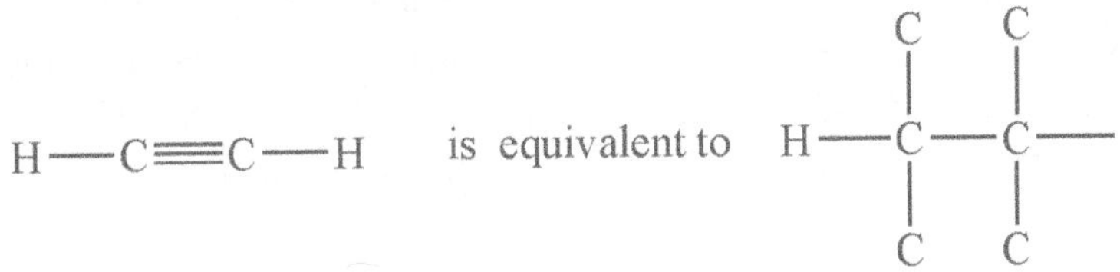

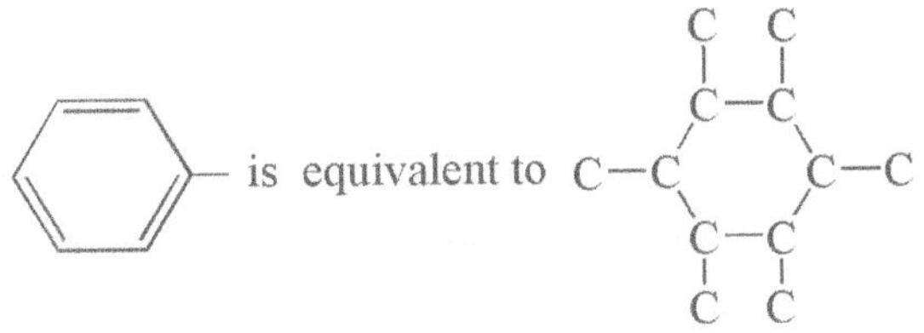

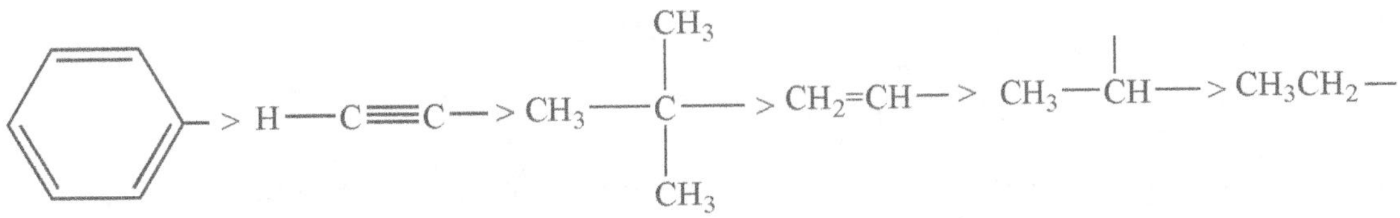

$$H_2C{=}C\langle^{H} \quad \text{is equivalent to} \quad H-\overset{H}{\underset{C}{C}}-\overset{H}{\underset{H}{C}}-$$

$$H-C{\equiv}C-H \quad \text{is equivalent to} \quad H-\overset{C}{\underset{C}{C}}-\overset{C}{\underset{C}{C}}-$$

(benzene ring) is equivalent to (ring of six carbons, each bearing a duplicated C: C–C, C–C, C–C, C–C, C–C, C–C)

$$C_6H_5- \;>\; H-C{\equiv}C- \;>\; CH_3-\overset{CH_3}{\underset{CH_3}{C}}- \;>\; CH_2{=}CH- \;>\; CH_3-\overset{|}{CH}- \;>\; CH_3CH_2-$$

Example 1: In glyceraldehyde molecule, the four groups attached to the chiral carbon are

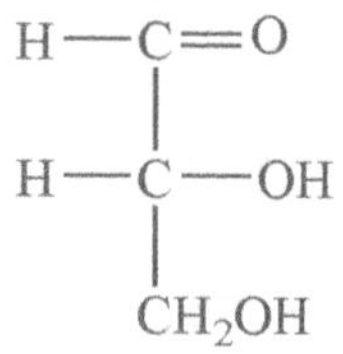

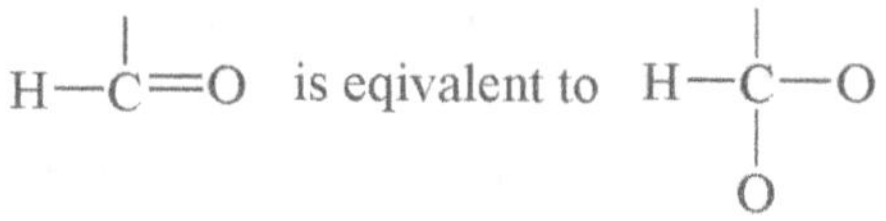

H—C=O
|
H—C—OH
|
CH_2OH

H—C=O is eqivalent to H—C—O with O

Out of these four groups, the -OH gets higher priority. The O, O, H takes the second priority than -CH_2OH. Hence the sequence is -OH, -CHO, CH_2OH and -H.

Example 2: 2–Methyl–1-phenyl propanamine

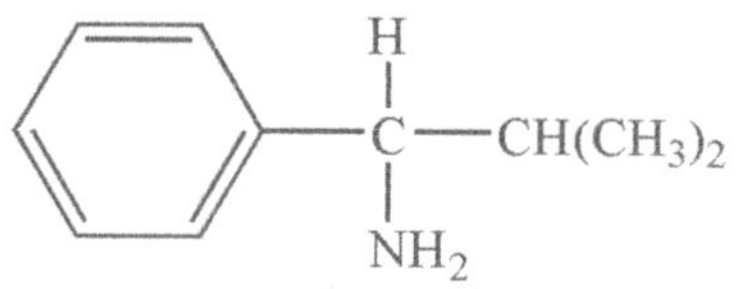

H
|
C_6H_5—C—$CH(CH_3)_2$
|
NH_2

The different groups attached to the chiral carbon atom are -C_6H_5, -NH_2, -$CH(CH_3)_2$. Out of these, "N" atom of -NH_2 gets highest priority, next C, C, C of phenyl gets priority over C, C, H of isopropyl group. Hence the sequence is -NH_2 > -C_6H_5 > -C_3H_7 > -H

Determination of Configuration: Following steps are adopted to assign the configurational system to chiral compounds.

Let's first look at how we would determine the configuration of the compound with three dimensional molecular models.

Step 1: Rank the various groups attached to chiral carbon according to the priority which depends upon the atomic number of the atoms.

Step 2: Orient the molecule so that the atom grouped with the lowest priority is directed away from us. Then draw an imaginary arrow starts from the highest priority group to the next highest priority group (Descending order). If the arrow points in a clockwise direction the chiral carbon is said to have R configuration (Rectus – R). If the arrow points in anticlockwise direction, the chiral carbon or centre is said to S configuration (Sinister – S).

For Example:

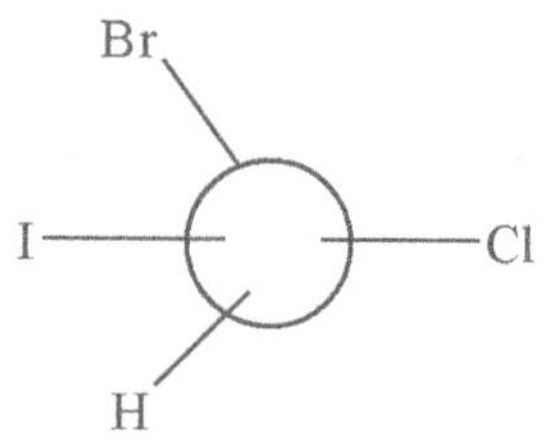

Bromochloro iodomethane

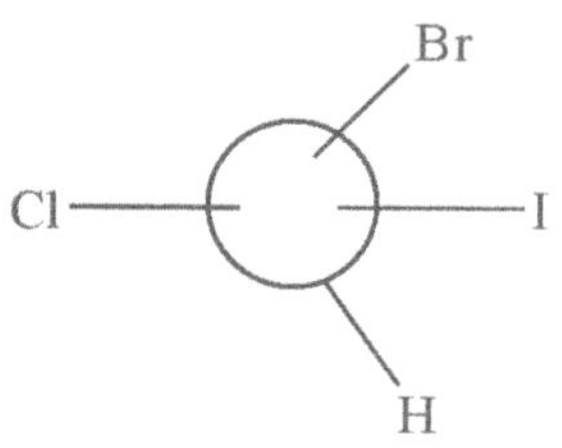

Step 1: Follow the sequence rules and assign the sequence of priority to the four different groups attached to the chiral carbon atom. According to this the sequence of priority for the above example is I, Br, Cl and H.

Step 2: Then orient the molecule in such a way that the lowest priority group directs away from us. Draw an imaginary plane which starts from higher priority group to the next higher priority group. Accordingly, viewed the structure I and II of bromo chloro iodo methane the configuration is represented as follows.

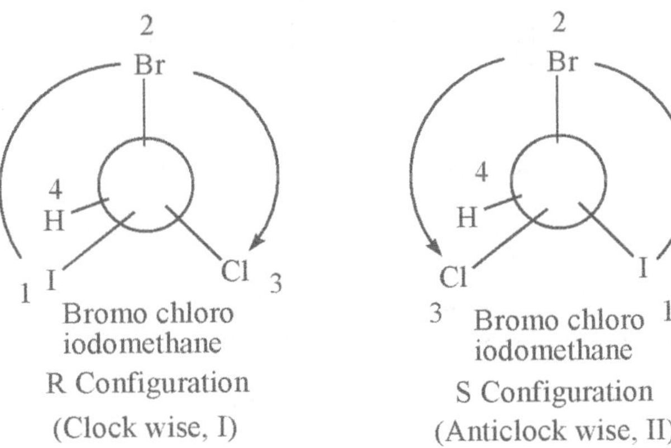

Bromo chloro iodomethane
R Configuration
(Clock wise, I)

Bromo chloro iodomethane
S Configuration
(Anticlock wise, II)

Example for compound showing optical isomerism:

1. **Lactic acid or 2-hydroxy propanoic acid**

   Lactic acid contains one chiral carbon atom.

$$CH_3 - \overset{\overset{OH}{|}*}{\underset{\underset{H}{|}}{C}} - COOH$$

Lactic acid

In lactic acid, the central carbon atom is attached with 4 different groups, so it exhibits two 3D structures. These structures are not identical because they cannot be superimposable on their mirror image. This type of non-superimposable mirror image of optical isomers are known as enantiomers. Three forms of lactic acid exist. They are as follows.

(i) **(+) Lactic acid:** It rotates the plane polarized light to the clockwise (right hand side) direction (dextrorotatory)

(ii) **(-) Lactic acid:** It rotates the plane polarized light to the anticlockwise (left hand side) direction (levorotatory)

(iii) **Racemic mixture:** It does not rotate plane polarised light (optically inactive). It contains equimolar mixture of (+) and (-) forms.

(I)	(II)	(III) Both
OH above C; HOOC, H, CH_3 (wedge) below	OH above C; H, COOH, CH_3 (wedge) below	± Lactic acid
3D-Structures		M.P 18 °C
$$\begin{array}{c} COOH \\ | \\ HO-C-H \\ | \\ CH_3 \end{array}$$	$$\begin{array}{c} COOH \\ | \\ H-C-OH \\ | \\ CH_3 \end{array}$$	(Optically inactive)
Fischer projection structures		
(+) – Lactic acid M.P 26 °C (Optically active)	(–) – Lactic acid M.P 26 °C (Optically active)	

2. **Tartaric acid or 2,3-dihyroxy butanedioic acid:** Tartaric acid possesses two chiral centers.

```
        OH
        |
H  -  C* -  COOH
        |
H  -  C* -  COOH
        |
        OH
```

Tartaric acid

Four forms of tartaric acid exist in which two of them are optically active (enantiomers) and other two are optically inactive.

```
      COOH          :          COOH
        |           :            |
H  -  C  -  OH      :    HO  -  C  -  H
        |           :            |
HO -  C  -  H       :    H   -  C  -  OH
        |           :            |
      COOH          :          COOH
```

(+) - Tartaric acid

M.P = 170 °C

(-) - Tartaric acid

M.P = 170 °C

(Optically active)

```
      COOH
        |
H  -  C  -  OH
--------:---------
H  -  C  -  OH
        |
      COOH
```

meso tartaric acid

M.P = 143 °C

(Optically inactive)

(i) **(+) Tartaric acid:** It rotates the plane polarized light in the clockwise direction (right hand side) dextrorotatory.

(ii) **(-) Tartaric acid:** It rotates the plane polarized light in the anticlockwise direction (left hand side) levorotatory.

(iii) **Mesotartaric acid:** It is having plane of symmetry hence the optical rotation of one chiral carbon is compensated by other chiral carbon (internally cancelled or compensated).

(iv) **Racemic mixture:** It does not rotate the plane polarized light. It is an equimolar mixture of (+) and (-) forms.

Measurement of optical activity: The amount of plane polarized light rotated by an optically active compound is measured by an instrument called as polarimeter (Fig. 27.5). In a polarimeter, monochromatic light passes through a polarizer and converted into plane polarized light. Then it passes through the empty sample cell or optically inactive solvent (standard cell) and emerging out unchanged. Then finally it passes through analyzer that is a second polarizer mounted on an eyepiece with a dial marked in degrees. The user rotates the analyzer until he or she finds out total darkness. At this point no light passes through the analyzer (analyzer is right angle to the first polarizer). This setting of analyzer corresponds to zero rotation. Then the sample to be analyzed is placed in the sample tube. If the sample is optically active and rotates the plane polarized light, then the analyzer no longer blocks all the light and the light reaches the users eye. The amount of light rotated by the analyzer can be directly read from the dial and indicates the difference between active sample and inactive sample. This is known as optical rotation (α) and the unit is degrees. The amount of rotation caused or produced by the sample depends upon the number of optically active molecules when the light encounters in the sample, which in turn depend on the concentration of the sample and the length of the sample tube. The observed rotation depends upon the wavelength and temperature. Each and every optically active compound possesses a characteristic specific rotation. It is defined as "the number of degree of rotation caused by a solution of 1 gm of the compound per ml of solution in a sample tube of 1.0 cm long at a particular temperature and wavelength".

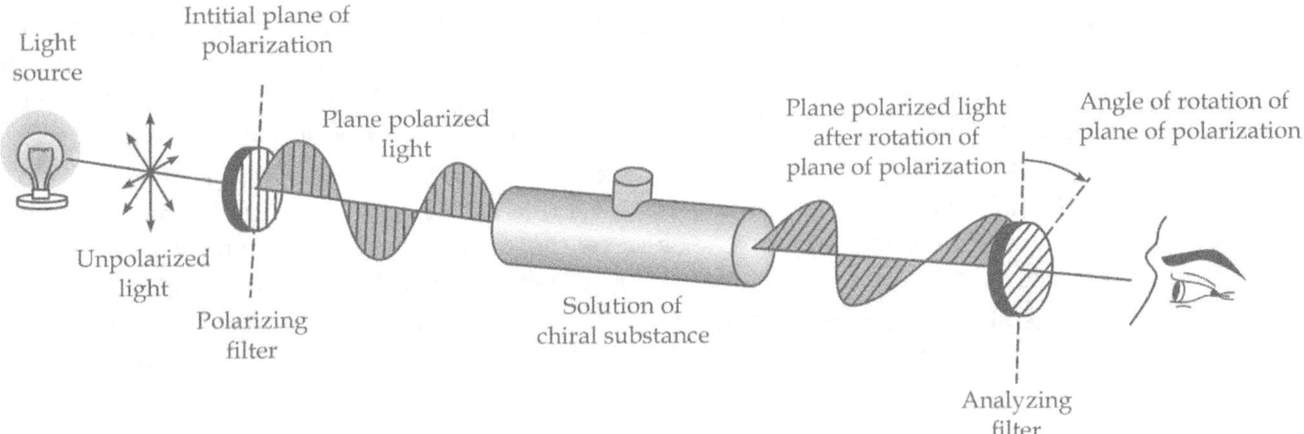

Figure 27.5 Measurement of optical activity by polarimeter.

$$\boxed{[\alpha]_{\lambda}^{T} = \frac{\alpha}{l \times C}}$$

Where, α = Specific rotation

T = Temperature (°C)

λ = Wavelength of the incident light

l = Length of sample tube

C = Concentration of sample in g/ml

Optical purity: Whether a sample consists of a single enantiomer or mixture of enantiomers can be determined by using observed specific rotation. For example, if the sample is enantiomerically pure then one enantiomer is present.

Optical purity of the particular sample can be calculated by using the following formula,

$$\text{Optical purity} = \frac{\text{Observed specific rotation}}{\text{Specific rotation of the pure enantiomer}}$$

Reactions of Chiral Molecules

Introduction: We have already discussed about stereoisomers and their products *etc.* But this area covers the formation of stereoisomers in chemical reaction and various reactions of stereoisomers in which they are utilized or consumed and the conditions of the reagents used as such as optically active or inactive reagents.

The main aspects of this topic are as follows:

1. Conversion of achiral compound into chiral compound with the formation of chiral center.
2. Reactions of chiral compounds in which the bonds to chiral center are not broken and the reactions in which second chiral centers are formed.
3. Determination of relative configuration of optically active compounds and their products (Bonds to the chiral centers are not broken).
4. Formation of second chiral center.
5. Reactions of chiral compounds with optically active reagents (Resolution of racemic modification).
6. Mechanism of free radical chlorination:

1. Conversion of achiral compound into chiral compound with the formation of chiral center:

Consider the following example. Free radical chlorination of *n*-butane to give secondary butyl chloride. *n*-Butane is an achiral compound but *sec*-butyl chloride is a chiral compound which exists in two forms (R & S forms) of enantiomers. The product obtained does not rotate the plane polarized light, because it involves racemic modification. It is explained as follows.

The first step of this reaction involves the attack of chlorine to *n*-butane and forms HCl and *sec*-butyl free radical. The carbon atom in the *sec*-butyl free radical is sp^2 hybridised and partially flat in structure. Hence in the second step the chlorine molecule attacks the free radical on both sides in equal amount which leads to the formation of enantiomers of products in equal amounts.

Hence, from the above fact it is concluded that "The conversion of achiral to chiral compounds lead to the formation of enantiomers of products (R and S forms) in equal amounts."

$$CH_3CH_2CH_2CH_3 + Cl_2 \xrightarrow{UV/\Delta} CH_3\overset{*}{C}H(Cl)CH_2CH_3 + HCl$$

n-Butane (Achiral compound) → 2-Chloro butane (Chiral compound)

2. Reactions of chiral compounds in which the bonds to chiral center are not broken and the reactions in which second chiral centers are formed [Bond breaking: A chemical does not involve the breaking of a bond to chiral center proceeds with retention of configuration]:

Let us consider the chlorination of optically active (*S*)-*sec*-butyl chloride. During this reaction, the each hydrogen atom of the $-CH_3$ group is replaced by chlorine atom. Hence, the bond cleavage between chiral carbon and other atoms does not takes place and form the product with retention of configuration (The product formed contains similar spatial arrangement of atoms with respect to starting materials). In the product the spatial arrangement of atoms about the chiral center is not changed such as $-CH_2Cl$ occupies the same relative position that was previously occupied by $-CH_3$.

$$H-C(CH_3)(C_2H_5)-Cl \xrightarrow[-HCl]{Cl\bullet} H-C(\dot{C}H_2)(C_2H_5)-Cl \xrightarrow{Cl_2} H-C(CH_2Cl)(C_2H_5)-Cl + Cl\bullet$$

(*S*)-*sec*-Butyl chloride → (*R*)-1,2-Dichloro butane

The above mechanism indicates that, any chemical reaction involves the optically active substrate and the reaction does not involve the breaking bond between the chiral carbon and other atoms leads to the formation of product with retention of configuration about the chiral center.

3. **Reactions of chiral molecules: Determination of relative configuration:**

   Generally X-ray diffraction studies are used to determine the configuration of enantiomers. But it is time consuming as well as complex procedure. Hence it can be determined by another one way *i.e.*, the configurational relationship between two optically active compounds can be determined by converting one into another by reactions that do not involve breaking of a bond to the chiral center.

   For example: (*S*)-(-)-2-methyl-1-butanol (optically active) reacts with HCl and yields (*S*)-(+)-1-chloro-2-methyl butane (optically active product).

   CH_3 / H—C*—CH_2OH / C_2H_5 (A) $\xrightarrow[-H_2O]{HCl}$ CH_3 / H—C—CH_2Cl / C_2H_5 (B)

   (*S*)-(−)-2-Methyl-1-butanol (Optically active) (*S*)-(+)-1-Chloro-2-methyl butane (Optically active product)

   In the above equation without the knowledge of mechanism we can conclude that the bond between carbon and oxygen atom is broken but no bond breaking occurs in chiral carbon or chiral center. Hence in the product $-CH_2Cl$ retains its position in the same as that of $-CH_2OH$ position in the starting material. When we check the optical rotating property of the product by polarimeter shows that it rotate the plane polarized light on the right hand side. Hence the product obtained is (+)-1-chloro-2-methyl butane indicates that the optically active substrate and the product must possess the configuration of A and B respectively. Hence, the reactions of chiral molecules provide the knowledge about the relative configurations of two optically active compounds.

4. **Formation of second chiral center**: The reactions of chiral compound sometimes leads to the formation of second chiral center in the product. It is explained as follows.

   Chlorination of *sec*-butyl chloride follows free radical mechanism and forms more than one product. During this reaction, the compound 2,3-dichlorobutane is obtained as one of the product that contains second chiral center. It exists as one meso compound and one set of enantiomers.

   $CH_3CH_2\overset{Cl}{C}HCH_3$ (*) $\xrightarrow[(or)\ h\nu]{Cl_2,\ \Delta}$ $CH_3\overset{Cl}{C}H\overset{Cl}{C}HCH_3$ (* *) + Remaining possible products

   New chiral centre

   sec-Butyl chloride 2,3-Dichloro butane

If the starting material is an optically active isomer such as (*S*)-isomer of *sec*-butyl chloride, what types of stereoisomers are obtained as products? It is illustrated by the following mechanism.

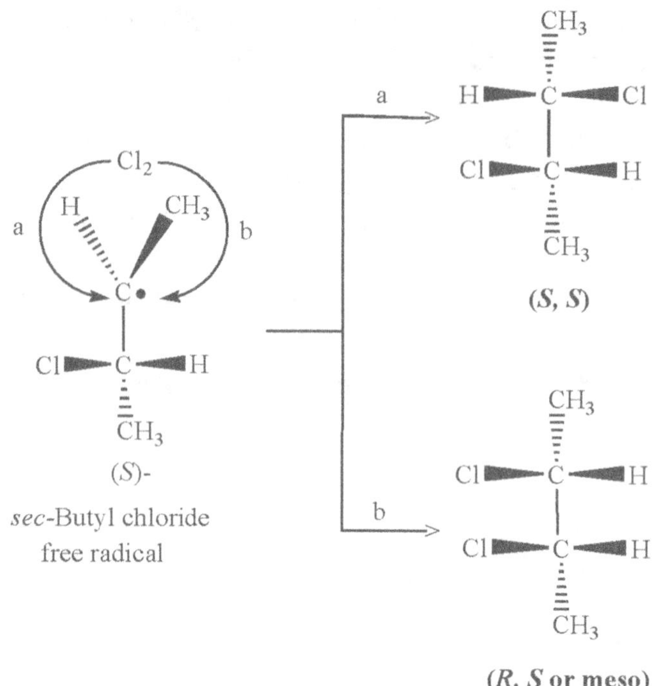

Unequal amount of diastereomers

If the chlorine atoms attacks on both sides (front and back) then it gives unequal amount of diastereomers.

Conclusion:

(i) The above reaction stated that there is no bond breaking between original chiral center, hence the configuration of the starting optically active compound is retained in all the final products. It must be applied in all cases where second chiral center is formed.

(ii) Due to the presence of two new chiral centers, two possible configurations of products are formed [Diastereomers (*S, S*) and *R, S* or meso products]

(iii) Unequal amounts of diastereomers are formed because of the unequal attack of chlorine molecule to the both sides of the flat *sec-butyl free radical.*

The above facts concluded that "Formation of new chiral center in an optically active substrate yields the optically active products containing unequal amounts of diastereomers".

5. **Reactions of chiral compounds with optically active reagents (Resolution of racemic modification):** Described under last part of this chapter.

6. **Mechanism of free radical chlorination:**

Stereochemistry of one chemical reaction plays a major role for establishing the reaction mechanism. [What about the stereochemistry of reactions in which the bonds to the chiral center are broken?]. It depends upon the mechanism of that reaction. It is explained as follows.

Let us consider the halogenation of alkanes. The main step of reaction is chain propagation step. It occurs by two ways or two alternative mechanisms.

1] $X\bullet + R{-}H \longrightarrow H{-}X + R\bullet \longrightarrow$ (2a)

$R\bullet + X{-}X \longrightarrow R{-}X + X\bullet \longrightarrow$ (3a)

(or)

2] $X\bullet + R{-}H \longrightarrow R{-}X + H\bullet \longrightarrow$ (2b)

$H\bullet + X{-}X \longrightarrow H{-}X + X\bullet \longrightarrow$ (3b)

The above two mechanisms are explained by considering the following reaction.

Example: (S)-(+)-1-chloro-2-methyl butane, an optically active compound undergoes photochemical halogenation to yield a number of isomeric products depending upon the attack at different positions in the molecule. We are considering only one product which is formed by the attack at C-2 chiral center of the molecule.

$$CH_3CH_2\overset{CH_3}{\underset{*}{C}}HCH_2Cl \xrightarrow[-HCl]{Cl_2/light} CH_3CH_2\overset{CH_3}{\underset{Cl}{C}}CH_2Cl$$

(*S*)-(+)-1-Chloro-2-methyl butane (Optically active) → (±)-1,2-Dichloro-2-methyl butane (Optically inactive)

This reaction takes place *via* the formation of alkyl free radical intermediate (It is indicated in 2a and 3a) The structure of the radical is flat (no chirality), hence it is attacked on both sides to yields an equal amounts of enantiomers [racemic mixtures].

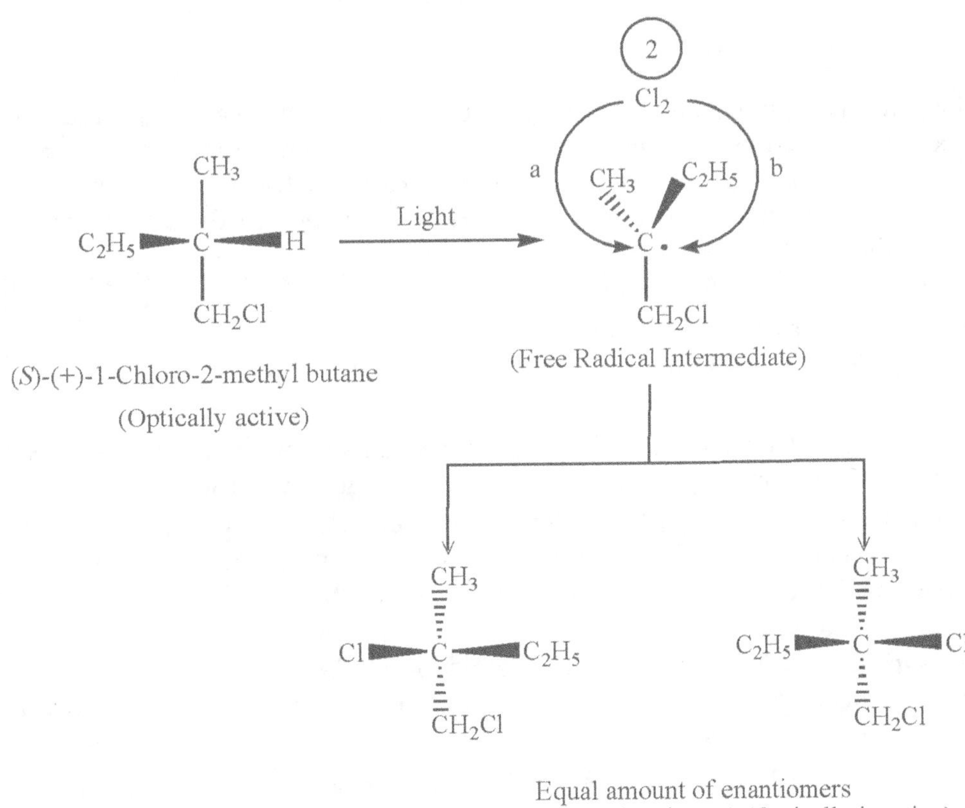

The alternate mechanism such as 2b, 3b the chlorine atom is attached and the hydrogen free radical is formed. In this reaction there is no evidence for the formation of enantiomers. It indicates that stereochemistry is an important tool for finding out a mechanism of reaction.

The other examples of reactions of chiral compounds are stereochemistry of S_N^1 and S_N^2 reactions as well as E_1 and E_2 reactions – For detailed information refer Chapter No 14 .

Racemic Modification

A racemic modification is defined as an equimolecular mixture of pair of enantiomers. *i.e.*, (+) and (-) forms and is denoted by ±. It can be prepared by different ways as mentioned below.

1. Mixing of enantiomers.
2. By synthesis.
3. By racemisation.

1. **Mixing of enantiomers:** It is most useful and trivial methods of producing racemic modification. This can be done by exactly mixing equal amounts of the two enantiomers.
2. **By synthesis:** It involves the synthesis of asymmetrical compound from symmetrical compound in the absence of optically active reagents. For example, Synthesis of lactonitrile from acetaldehyde.

$$\underset{\text{L-Lactonitrile}}{HO-\overset{CH_3}{\underset{CN}{C}}-H} \longleftarrow \underset{\text{Acetaldehyde}}{H_3C-\overset{O}{\overset{\|}{C}}-H} \xrightarrow{\overset{\ominus}{CN}} \underset{\text{D-Lactonitrile}}{H-\overset{CN}{\underset{CH_3}{C}}-OH}$$

3. **By racemisation:** The phenomenon of converting optically active compound into optically inactive racemic mixture is called as racemisation. For example, when (+) tartaric acid in water is heated, it is completely transformed into optically inactive mixture of the racemic and mesotartaric acid. The phenomenon of spontaneous conversion of the *levo* variety to the *dextro* variety and *vice versa* is called as racemisation.

Racemisation can be done by the following ways.

(i) The action of heat or thermal racemisation.
(ii) Treatment with chemical reagents.
(iii) Autoracemisation.

(i) The action of heat or thermal racemisation: Increase in temperature always leads to the formation of racemic mixture. For example, lactic acid, tartaric acid gives racemic mixture when heating.

(ii) Treatment with chemical reagents: Generally, presence of foreign substances increases the chance of formation of racemisation. Such kind of substances undergoes racemisation when treated with chemical reagents. For example, mandelic acid on reaction with HBr forms respective (±) bromo acid.

(iii) Auto racemisation: In some compounds, racemisation takes place at room temperature. For example, dimethyl bromosuccinate isomerizes on standing at room temperature. This kind of process is called as autoracemisation.

Mechanism of racemisation: The compounds undergo racemisation in which the chiral carbon is attached to the hydrogen atom and an electron withdrawing group. It will readily undergo tautomeric change and racemisation takes place *via* enolisation. For example,

$$\underset{\text{Dextro acid}}{R-\overset{H}{\underset{R}{C}}-\overset{O}{\overset{\|}{C}}-OH} \rightleftharpoons \underset{\text{Enol form with achiral carbon}}{\left\{R-\underset{R}{C}=\overset{OH}{C}-OH\right\}} \rightleftharpoons (\pm)\text{-acid or racemic mixture}$$

The intermediate enol form obtained is achiral in nature that reverses to the stable form. There are chances to form *dextro* and *levo* form and hence it gives a racemic mixture.

Epimerization: An optically active compound contains two or more chiral carbon atoms and the configuration of only one of the carbon atom changes in a reaction, this process is called as epimerization.

Walden Inversion

It was discovered by a chemist *Paul Walden* in 1896. Walden inversion is the inversion of a chiral centre of chemical compound in a chemical reaction. So, a molecule can produce two enantiomers around a chiral centre. The Walden inversion converts the configuration of the molecule from one enantiomerically forms to other. For example, in S_N2 reaction Walden inversion occurs at tetrahedral carbon atom. It can be visualized by imagining an umbrella turned inside out in a gale.

For detailed information refer S_N2 reaction mechanism-chapter-14-Nucleophillic substitution and elimination reactions.

To convert an optically active *levo* malic acid into its enantiomers takes place by series of substitution reactions.

$$\underset{\text{(I)}}{\underset{\text{(-)-Malic acid}}{\begin{array}{l}CH(OH)COOH\\|\\CH_2COOH\end{array}}} \underset{\substack{KOH\\-POCl_3\\-HCl}}{\overset{PCl_5}{\rightleftharpoons}} \underset{\text{(II)}}{\underset{\text{(+)-Chloro succinicacid}}{\begin{array}{l}CH(Cl)COOH\\|\\CH_2COOH\end{array}}} \xrightarrow[-AgCl]{AgOH} \underset{\text{(III)}}{\underset{\text{(+)-Malic acid}}{\begin{array}{l}CH(OH)COOH\\|\\CH_2COOH\end{array}}}$$

The interchanging of the position between the two groups takes place in way as given above. If the configuration of II corresponds to that of (I), then the inversion occurs in the second step *i.e.,* between II & III.

Mechanism: There is a no exact way to explain in which step of the reaction inversion occurs. According to *Kenyon* (1925), inversion takes place only when the bond in chiral carbon atom is ruptured.

For example:

1. **Formation of acid chloride:** Carboxylic acid is treated with thionyl chloride ($SOCl_2$) to yield acid chloride without change of configuration because the chiral centre remains unaffected.

$$\underset{\text{Acid (D-isomer)}}{H_3CH_2C-\overset{\overset{\displaystyle H}{|}}{\underset{\underset{\displaystyle CH_3}{|}}{C}}-COOH} \xrightarrow{SOCl_2} \underset{\substack{\text{Acid chloride}\\\text{(D-isomer, no change in configuration))}}}{H_3CH_2C-\overset{\overset{\displaystyle H}{|}}{\underset{\underset{\displaystyle CH_3}{|}}{C}}-COCl} + SO_2 + HCl$$

2. 2-Bromobutane when treated with NaI solution in acetone results in the change of configuration. This reaction takes place by S_N2 mechanism.

$$\underset{\substack{\text{2-Bromo butane}\\\text{(D-isomer)}}}{H_3CH_2C-\overset{\overset{\displaystyle H}{|}}{\underset{\underset{\displaystyle Br}{|}}{C}}-CH_3} \xrightarrow[\text{acetone}]{\text{NaI in}} \underset{\substack{\text{2-Iodobutane}\\\text{(L-isomer)}}}{H_3CH_2C-\overset{\overset{\displaystyle H}{|}}{\underset{\underset{\displaystyle I}{|}}{C}}-CH_3} + NaBr$$

Resolution or Separation of Racemic Mixture

In synthetic organic chemistry, some of the steps in the synthetic process leads to a racemic modification and the next step generally needed a specific enantiomer in a pure form, so it is necessary to separate the racemic modification in to its components for further reactions. So the process of separation of racemic modification into its two pure form of enantiomers is called as resolution. It is a very important step in synthetic organic chemistry, especially for natural products.

Partial Resolution: Whenever the isolated form is not optically pure, one is contaminated with other then the process is known as partial resolution.

Generally, enantiomers with similar physical properties can be separated by physical methods such as, distillation or fractional crystallisation. Hence several methods are used for the separation. They are described as follows

1. Mechanical separation or spontaneous resolution.
2. Preferential crystallisation by inoculation.
3. By diastereomers.
4. Biochemical separation.
5. Chromatography.
6. Complex formation.
7. Kinetic method.
8. Asymmetric transformation.

1. **Mechanical separation:** It was discovered by *Pasteur* in 1884 for the separation of sodium ammonium tartarate which crystallizes out in racemic form below 27 °C. The crystals of the two forms have different shapes and separated by magnifying lens and forceps. This is applicable only to racemic mixture where the crystal shapes of the optical isomers are enantiomorphs.
2. **Preferential crystallisation by inoculation:** This method was discovered by *Gernez* in 1866. In this method, the saturated solution of the racemic mixture to be separated is seeded with the pure sample of any one of the enantiomers and now the solution is supersaturated and after some time one enantiomer crystallises out. (For example: Total optical resolution of free α- amino acids by using one odd isomers of corresponding amino acid.)
3. **Biochemical separation:** In this method, some of the microorganisms are grown in the dilute solution of the racemic mixture and they are rapidly accumulated in one enantiomers rather than other. For examples: *penicillium glaucum* preferentially destroys the (+) form of ammonium tartarate and leaves the (-) form.
4. **By diastereomers**: It is the best method for the resolution of racemic mixture. According to this method the enantiomers of the racemic modification is converted into diastereomers by the addition of enantiomers of any other compounds. The diastereomers obtained are non-identical and possess different physical properties and separated by fractional distillation. Then the diastereomers obtained is converted into original chemical compounds by different ways. The groups which are resolved by this method are acids, bases, alcohols, amino acids, aldehydes and ketones.
5. **Chromatography:** The use of chromatography for the resolution of racemic mixture is limited only. In this, the solution of racemic mixture is allowed to pass through the optically active adsorbents. The adsorbate formed contains enantiomers and adsorbent are diastereomers. The two enantiomers are separated from the diastereomers by careful elution.

   For example: Resolution of (±) – phenylene bis iminocamphor.
6. **Complex formation:** Some of the racemic modification forms complex with a dissymmetric reagents. If the two diastereomeric complexes have different solubilities, one will precipitate out faster than other. The two original enantiomers present in the two complexes are obtained or separated by decomposition process with the aid of heat, dissolution, chromatography or chemical reactions. For example, resolution of 1-naphthyl-*s*-butylether.
7. **Kinetic method**: This method is based on the principle that one of the enantiomers of racemic modification reacts faster than other with an optically active compound. For example resolution of (±) mandelic acid with menthol.

8. **Asymmetric transformation:** This is exhibited by the optically unstable compound, *i.e.* when the two enantiomers are readily interconvertible.

$$(+) \rightleftharpoons (-)$$

For example: Resloving of *N*-benzene sulphonyl-8-nitro-1-napthylglycine with brucine as a base and acetone or methanol as a solvent.

Asymmetric Synthesis

Generally, when an optically active compound is synthesized in the laboratory the product obtained may be a racemic mixture which has to be resolved to get optically active constituents. There is no need of resolution process when one optically active isomer is synthesized alone in the laboratory.

The production or the synthesis of an optically active compound from symmetric compounds (optically inactive) with or without the use of optically active reagent is called as **asymmetric synthesis.**

Types of asymmetric synthesis: Two types of asymmetric synthesis are follows.

(a) Partial asymmetric synthesis
(b) Absolute asymmetric synthesis.

Let us discuss in detail.

Methods of asymmetric synthesis: Various methods used for asymmetric synthesis are described as follows.

(a) **With optically active reagents or partial asymmetric synthesis**: This method was introduced by *Markwald* in 1904. He synthesized *iso*-valeric acid by heating the half brucine salt of ethyl methyl malonic acid at 440 K (Brucine is naturally occurring optically active base).

H_3C, COOH, C_2H_5, COOH on C — Ethyl methyl malonic acid

$\xrightarrow[\text{Brucine}]{\text{Levo}}$

H_3C, COOH, C_2H_5, COOH on C — (–)-Brucine + H_3C, COOH, C_2H_5, COOH on C — (–)-Brucine

$\downarrow$ 440K, $-CO_2$, HCl

H_3C, COOH, C_2H_5, H on C — (–)-Brucine + H_3C, H, C_2H_5, COOH on C — (–)-Brucine

$\downarrow$ HCl

H_3C, COOH, C_2H_5, H on C — *iso*-Valeric acid (Dextro) + H_3C, H, C_2H_5, COOH on C — Levo — *iso*-Valeric acid (Levo)

(b) Absolute asymmetric synthesis using circularly polarized light: Optically active compounds can be synthesized by carrying out asymmetric synthesis under the influence of circularly polarized light. This is a special type of asymmetric synthesis in which no optically active reagent is used hence known as absolute asymmetric synthesis.

1. Synthesis of slightly *levo* rotatory lactic acid from pyruvic acid esters in which alcohol is optically active.

$$\underset{\text{Pyruvic acid ester}}{CH_3\text{-}CO\text{-}COOR} + 2(H) \longrightarrow \underset{\text{Lactic ester}}{CH_3\text{-}CHOH\text{-}COOR} \xrightarrow[-ROH]{H_2O} \underset{(-)\text{-Lactic acid}}{CH_3\text{-}CHOH\text{-}COOH}$$

2. With enzymes: (-) Mandelic acid is obtained by treating benzaldehyde with HCN in the presence of enzyme "emulsin" with continued stirring.

$$\underset{\text{Benzaldehyde}}{C_6H_5CHO} + HCN \longrightarrow \underset{\text{Mandelo nitrile}}{C_6H_5CH(OH)CN} \xrightarrow[-NH_3]{2HOH} \underset{(-)\text{-Mandelic acid}}{C_6H_5CH(OH)COOH}$$

Probable Questions

1. Define stereoisomerism.
2. Define chiral carbon.
3. Write about enantiomers.
4. What are meso compounds.
5. Write in detail about chiral and achiral molecules.
6. Discuss about sequence rules with examples.
7. Explain about optical isomerism in detail with examples.
8. Write a note on various reactions of chiral molecules.
9. Define racemic modification.
10. Write in detail about different methods of racemisation.
11. Explain the mechanism of racemisation with example.
12. Explain in detail about Walden inversion with example.
13. Describe methods of separation of racemic mixture.
14. Write in detail about asymmetric synthesis and various methods used for it.
15. Write the examples for D and L isomers.

28

Geometrical Isomerism

Introduction

Let us consider the example of 2-butene. It exhibits in the following two structures.

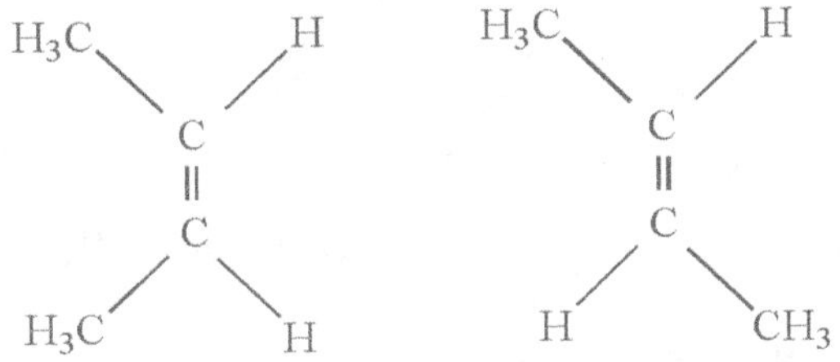

cis-2-butene	trans-2-butene
$H_3C(H)C{=}C(H)CH_3$ with both CH_3 on the same side	$H_3C(H)C{=}C(H)CH_3$ with CH_3 on opposite sides
cis-2-butene	*trans*-2-butene
(B.P - 227 K)	(B.P - 274 K)
(I)	(II)

We can easily observe that, atoms can be arranged in two different ways. In form I, the two methyl groups are present on same side, while in form II these lie on opposite sides of the molecule.

The conversion of form I into form II involves a rotation about carbon-carbon double bond and that is made up of one sigma bond and one π bond. The π bond is formed by the sidewise overlap of the "*p*" orbitals. To convert the form I to II, the 2-butene molecule gets twisted so that the "*p*" orbitals no longer overlaps and the π bond is broken.

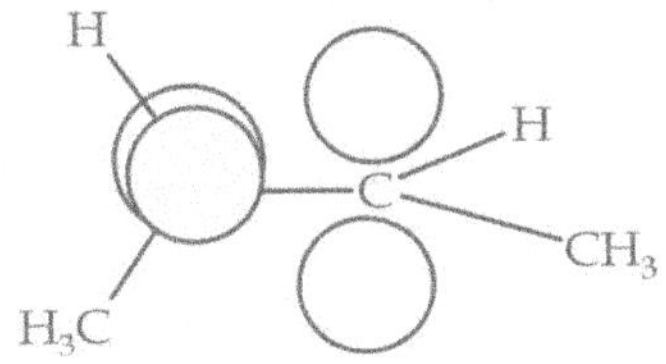

H
H
C
CH_3
H_3C

Figure 28.1 Hindered rotation about carbon-carbon double bond which prevents the overlap of *p* orbitals leads to breaking of π bond.

For breaking of this π bond approximately 167.2 KJ of energy is needed and an insufficient number of collisions possess this much energy. So interconversion of I and II is extremely small. *i.e.,* rotation about the carbon-carbon double bond is hindered on account of this 167.2 KJ energy barrier. Due to this hindered rotation, the positions of different groups attached to the two carbon atoms are fixed in space. **"The isomerism caused by the difference in spatial arrangement of groups about the double bonded carbon atoms is known as geometrical isomerism.** For example: 2-butene (refer the above structures I & II).

The two different compounds are distinguished from each other by the terms *cis* and *trans*.

***Cis* isomer**: Isomer in which similar groups lie on the same side is known as *cis* isomer (In Latin word *cis* means "same side").

***Trans* isomer:** Isomer in which similar groups lie on opposite side is known as trans isomer (In Latin term *trans* means "across").

Due to *cis* and *trans* isomers, geometrical isomerism is otherwise known as *cis-trans* isomerism.

Some other examples of compounds showing geometrical isomerism:

1. **Isomerism of maleic and fumaric acid:** Maleic acid and fumaric acid possess same molecular formula of $C_4H_4O_4$, but they differ in most of their physical properties and some of their chemical properties. This is due to the assumption that "there is no free rotation about the double bond, so two spatial arrangements of atoms are possible for the above molecular formula". Maleic acid is readily converted into maleic anhydride on heating. Maelic acid exist as *cis*-form and fumaric acid as *trans*. Hence maleic and fumaric acid exhibits geometrical isomerism.

```
H       COOH          H       COOH
 \     /               \     /
   C                     C
   ||                    ||
   C                     C
 /     \               /     \
H       COOH       HOOC       H
```

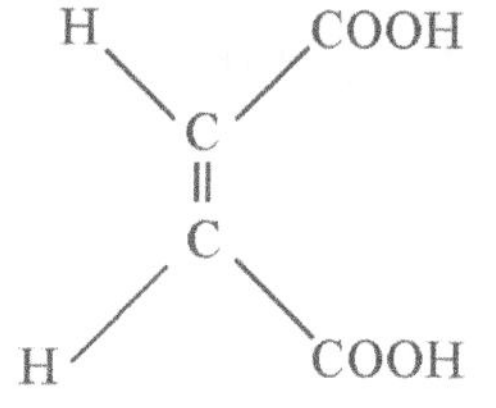

cis form (Maleic acid) 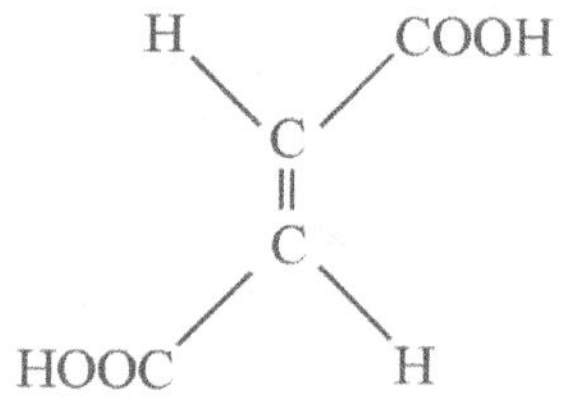 *trans* form (Fumaric acid)

2. **Geometrical or *cis-trans* isomerism of oximes:** Carbon atoms connected by double bond cannot rotate freely. Similarly a carbon atom and a nitrogen atom connected by a double bond also cannot rotate freely. So, the positions of various groups attached to C=N, are fixed in space and such type of compounds exhibit geometrical isomerism. Hence, the aldoxime from benzaldehyde shows geometrical isomerism and the two isomers have been isolated.

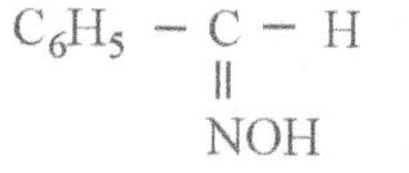

```
C6H5 - C - H        C6H5 - C - H
       ||                  ||
       NOH              HO-N
```

α–Benzaldoxime (*syn* form-m.p 308 K) β–Benzaldoxime (*anti* form-m.p 303 K)

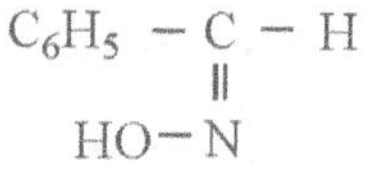

***syn* or *cis*-isomer:** The isomer in which H and OH groups are on the same side is known as *syn* or *cis* isomer.

***anti* or *trans* isomer:** The isomer in which H and OH groups on the opposite side is known as *anti* or *trans* isomer.

The *anti* form of benzaldoxime loses its water molecule more readily and forms phenyl cyanide than its *syn* form when heated with acetic anhydride.

Which type of compounds show geometrical isomerism? Answer for this question is "Theoretically all double bonded compounds exhibit geometrical isomerism and converts into one another, so the double bond has to be broken". But not all double bonded compounds exhibit geometrical isomerism. Let us examine which compounds will be expected to show geometrical isomerism. In order to predict the existence of geometrical isomerism, we must draw the existence of possible isomers after breaking the double bond and then find out the resulting structures which are identical in nature. For example, the following compounds should not show geometrical isomerism.

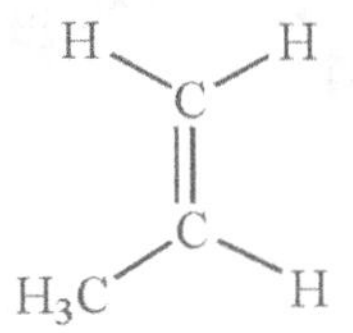

Propene 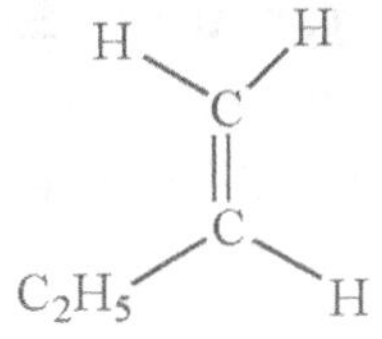 1-Butene 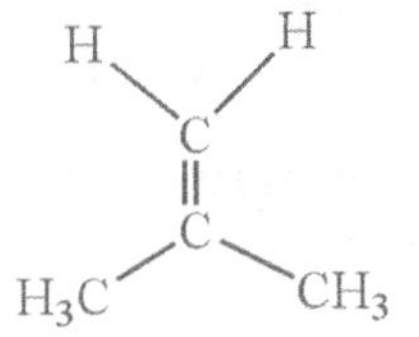 *iso*-Butene

But **1,2-dichloroethylene** shows geometrical isomerism in the following way.

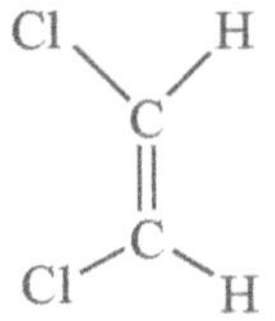

cis-Dichloroethylene 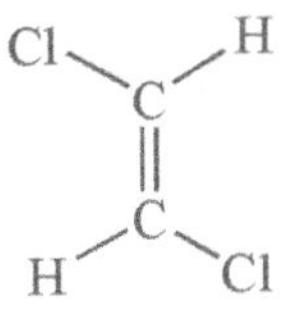 *trans*-Dichloroethylene

It is also concluded that geometrical isomerism cannot exist in the compounds if either of the carbons joined by a double bond contains two identical groups. Hence any one can predict the geometrical isomerism for the following series of compounds.

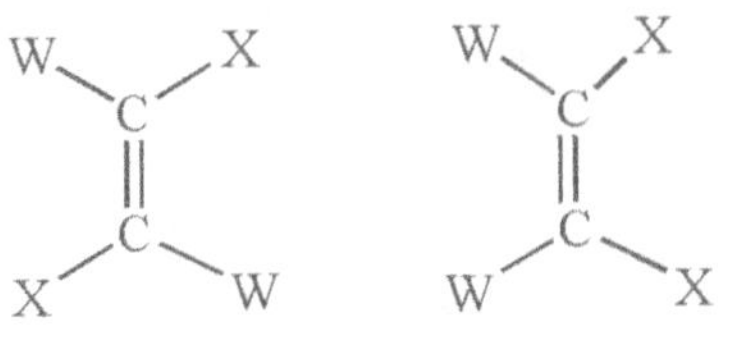

Shows isomerism 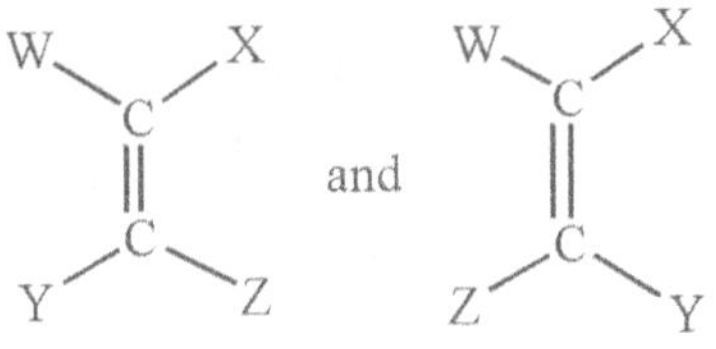 Shows isomerism

While the followings

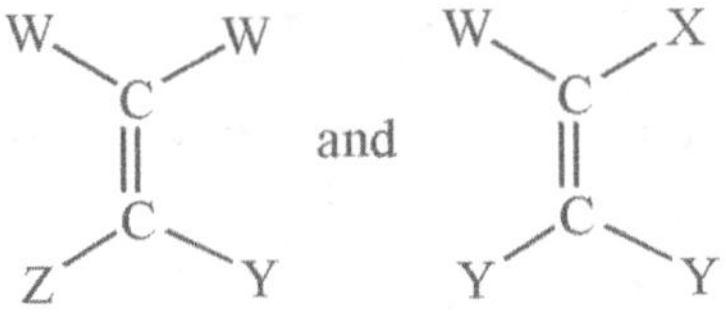

Do not show geometrical isomerism

Nomenclature of Geometrical isomers (The E, Z system): The terms *cis/syn* or *trans/anti* are being used for naming the geometrical isomerism even today also. But in some cases, these terms are inadequate in describing the isomers without ambiguity. For example, consider the following example, the structure I is denoted by *cis* but the structure 2 is not exactly *cis or trans*.

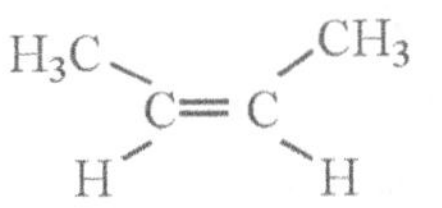

I 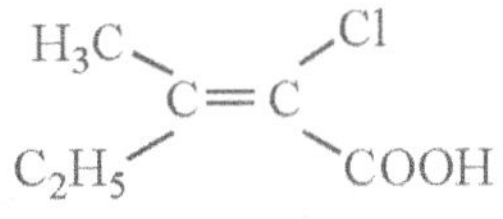 II

Hence a new system of nomenclature of geometrical isomers (E, Z system) has been introduced. The E and Z are derived from the German words *Entgegen* (opposite) and *Zusammen* (following together). The following steps are involved in the proposed nomenclature.

Step 1: Based on sequence rules (For details, refer under nomenclature of optical isomers) a sequence of priority (a and b) is assigned to the two groups at one end of the double bond. Similarly, a sequence of priority (a and b) is assigned independently to the two groups at the other end of the double bond.

Step 2: The configuration in which the two "a" groups are present on same side, labelled as (Z) and another in which the two "a" groups are on opposite sides, is labelled as (E).

After these two steps of priority assignments and labelling the configuration, the names of the above compounds (I) and (II) are as follows,

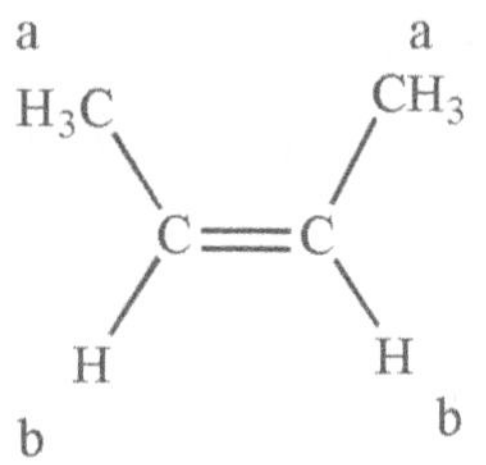

(Z)-2-Butene

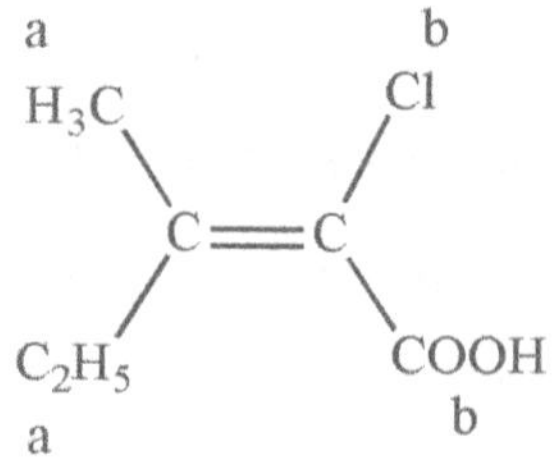

(E)-2-Chloro-3-methyl-2-pentenoic acid

Example

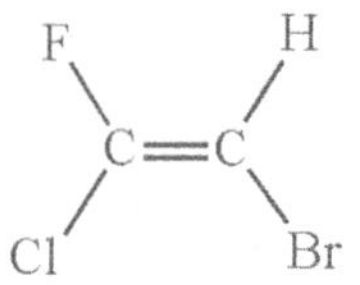

(i)

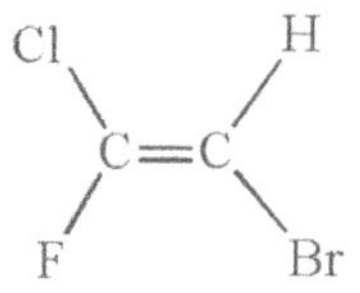

(ii)

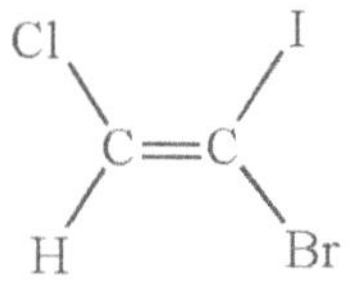

(iii)

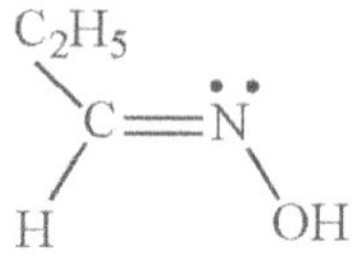

(iv)

On priority assignment and labeling the E, Z nomenclature of the above compounds are given below.

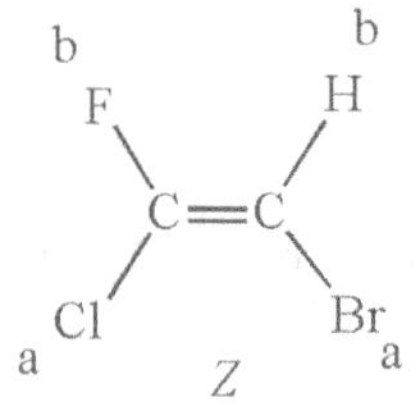

Z (Two 'a' groups are on the same side) (i)

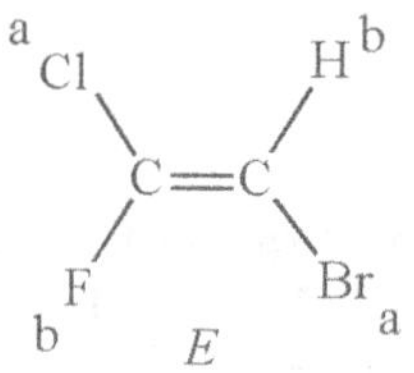

E (Two 'b' groups are on the opposite side) (ii)

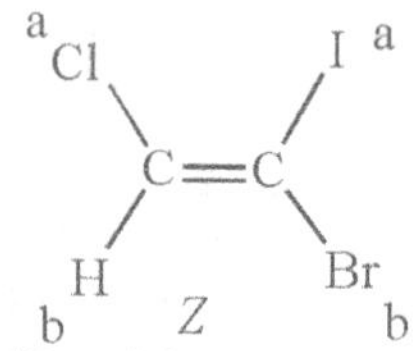

Z (Two 'a' groups are on the same side) (iii)

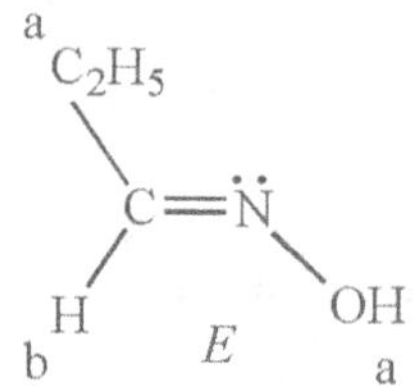

E (Two 'a' groups are on the opposite side) (iv)

Nomenclature of geometrical isomers is difficult when the geometrical isomers possess more than one double bond. In such cases, the compound is considered to be a derivative of longest chain which contains the maximum number of double bonds and the prefix *cis* and *trans* are placed before the numbers indicating relative positions of the carbon atoms in the main chain.

Determination of configuration of geometrical isomers: Various methods are useful for the determination of the configuration of geometrical isomers. They are as follows.

1. Geometrical isomers differ in their physical and chemical properties (dipole moment, melting point, boiling point, refractive index *etc*). So, these differences are used to determine the configuration.
2. By converting the geometrical isomers into compounds of known configuration.
3. The two isomers can be determined by X-ray analysis.
4. *cis* and *trans* isomers of geometrical isomers differ in their UV, IR and NMR spectra. These differences are used for determination of configuration of geometrical isomers.

Geometrical isomers differ from each other in their many physical and chemical properties due to the different arrangement of group in space. Hence the various properties of these *cis* & *trans* isomers can be used to determine the configuration of geometrical isomers. Like optical isomers, configuration of geometrical isomers cannot be determined by a single method. Various methods have been developed and used for the determination of configuration of geometrical isomers. Some of the methods are described in detail below:

1. Method of cyclisation.
2. By converting the compound into known configuration.
3. Method based on physical properties.
   (i) Dipole moment.
   (ii) Acid strength.
   (iii) Melting point, Boiling Point, Density, Solubility and Refractive Index.
   (iv) X-ray and Electron diffraction.
   (v) Spectroscopy method.
4. Methods of optical activity.
5. By stereoselective addition and elimination reactions.

Let us discuss each methods in detail.

1. ***Method of cyclization:*** The basic principle behind this method is that the intramolecular reactions take place easily when the reacting groups are close proximity together.

   For Example,

   Example 1: Consider fumaric acid and maleic acid. If both acids are gently heated, maleic acid yield maleic anhydride. It indicates that two carboxylic acid groups (-COOH) are close together, hence maleic acid is the *cis*-isomer and fumaric acid is the *trans*-isomer.

   Fumaric acid is converted into maleic acid by strong heating *via* isomerisation.

   Note: This method is used only for the geometrical isomers (Both *cis* & *trans*) in which either of the form is easily capable of forming ring.

H—C—COOH ‖ HOOC—C—H (Fumaric acid) $\xrightarrow[\text{Strongly}]{\text{Heat}}$ H—C—COOH ‖ H—C—COOH (Maleicacid) $\xrightarrow[-H_2O]{\text{Gentle heat}}$ Maleic anhydride

Fumaric acid $\xrightarrow{\text{Gentle heat}}$ No anhydride formation

Example 2: Citraconic & Mesaconic acids. Among these two acids, citraconic acid is present in *cis*-form, hence it easily undergoes cyclization and forms anhydride.

2. **By converting into compounds of known configuration:** By this method, the pairs of geometrical isomers are converted into a compound of known configuration.

$H_3C-C-COOH$ / $HOOC-C-H$ (Mesoconic acid, *trans* form) $\xrightarrow{\Delta}$ No anhydride formation

$H_3C-C-COOH$ / $H-C-COOH$ (Citraconic acid, *cis* form) $\xrightarrow[-H_2O]{\Delta}$ Citraconic anhydride

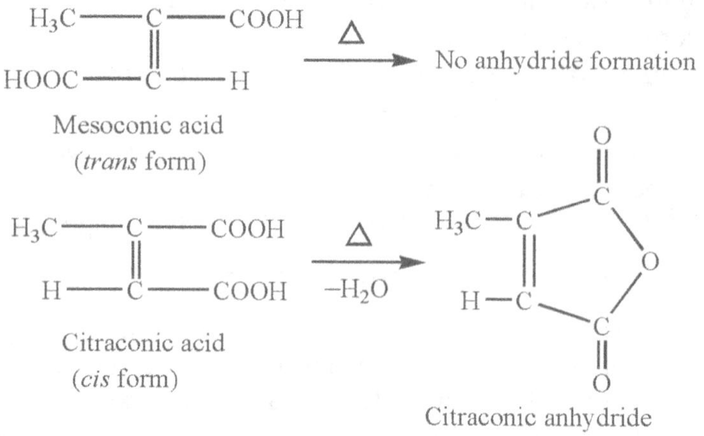

For example, consider the geometrical isomers of trichlorocrotonic acid. The one form of this compound produces fumaric acid by hydrolysis which indicates that, the above said trichlorocrotonic acid must be in *trans* form. The same trichlorocrotonic acid gives crotonic acid on reduction.

Another isomer of trichlorocrotonic acid does not provide fumaric acid on hydrolysis & it produces isocrotonic acid by reduction.

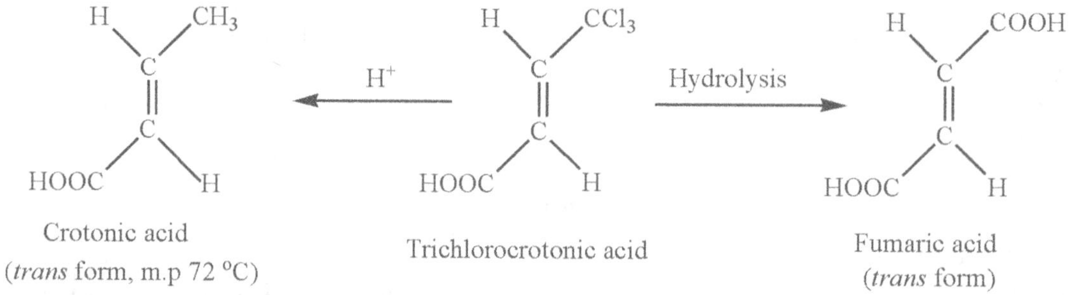

Crotonic acid (*trans* form, m.p 72 °C) $\xleftarrow{H^+}$ Trichlorocrotonic acid $\xrightarrow{\text{Hydrolysis}}$ Fumaric acid (*trans* form)

The above reaction indicated that, isocrotonic acid and the respective trichlorocrotonic acid are *cis* isomers

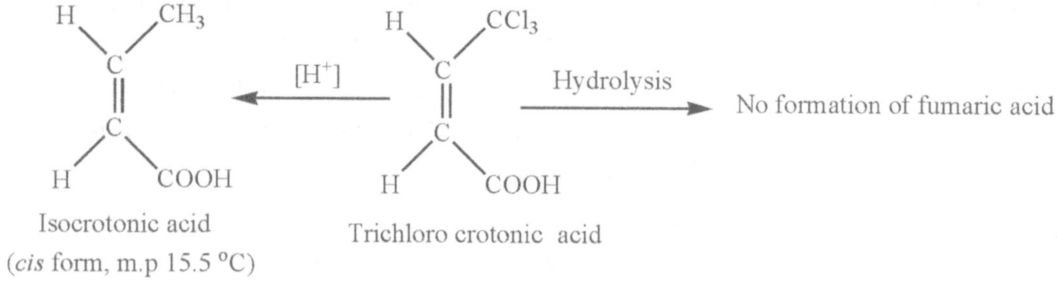

Isocrotonic acid (*cis* form, m.p 15.5 °C) $\xleftarrow{[H^+]}$ Trichloro crotonic acid $\xrightarrow{\text{Hydrolysis}}$ No formation of fumaric acid

3. **Methods based on physical properties:** Sometimes, the physical properties of the geometrical isomers can be used to determine their configurations. But care should be taken to before assigning a particular configuration to a geometrical isomer, it must be confirmed by as much as informations as possible. Various physical properties used in this method are described as follows.

(i) **Dipole Moment:** It is used to determine the configuration of the following kind of compounds

(a) **Ethylenic carbon containing compounds:** When each of the isomers possess similar groups, the *trans* isomer has centre of symmetry, hence the dipole moment is zero. For example: *trans* and *cis* isomers of 1,2-dibromo ethylene.

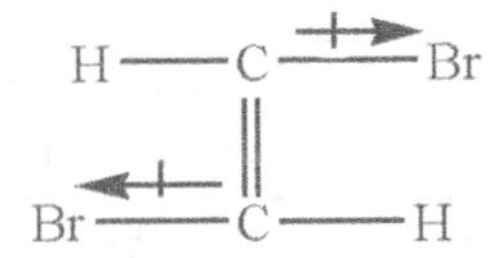

trans-1,2-Dibromoethylene

(μ = 0.0 D)

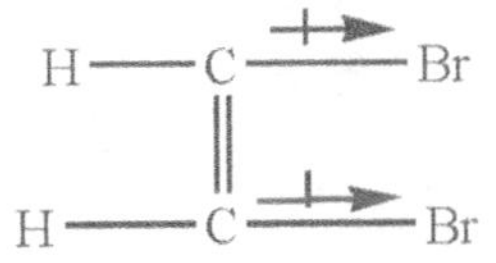

cis-1,2-Dibromoethylene

(μ = 1.89 D)

In the above example, the trans 1,2-dibromo ethylene has centre of symmetry, hence dipole moment is zero, but in *cis* isomer, no centre of symmetry is present, The D Value is 1.89.

The two groups attached to the ethylenic carbon atom are different from each other, the *trans* isomer has low dipole moment (but not zero) than the *cis* isomer.

(b) When a compound contains electron donating and electron withdrawing group, the *trans* isomer possess the higher dipole moment than the *cis* isomer.

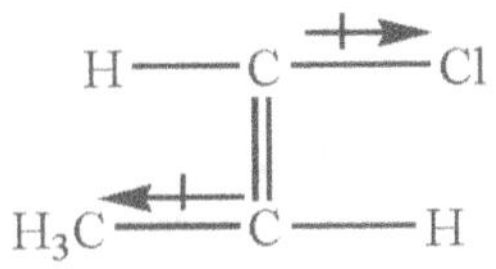

trans-isomer

(μ = 1.97 D)

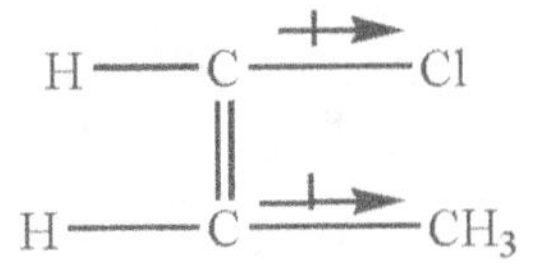

cis-isomer

(μ = 1.71 D)

Note: This method is satisfactory one as long as the groups attached to the ethylenic carbon atoms have linear moments. When the groups are non-linear, the difference in the dipole moments of the two isomers are very small. For example, the diethyl maleate & diethyl fumarate.

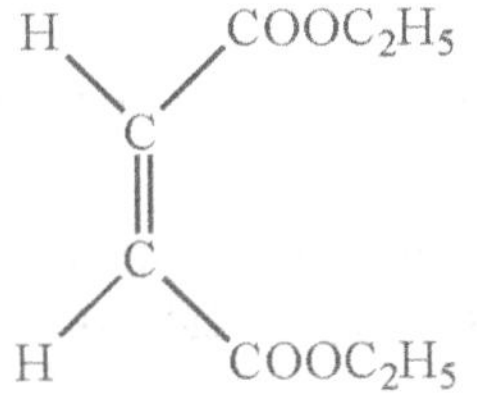

Diethyl maleate

μ = 2.54 D

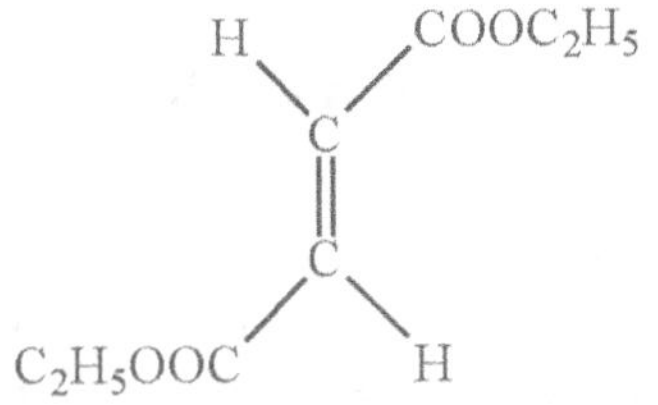

Diethyl fumarate

μ = 2.38 D

(ii) **Acid Strength:** The acidic strength of the *cis* isomer of monobasic acid is more than the corresponding *trans* isomer. For example, the pKa value of *cis* and *trans* isomers of crotonic acid is 4.44 and 4.70 respectively.

(iii) **Melting Point, Boiling Point, Solubility, Density & Refractive Index:** The melting point, solubility of the *cis* isomer are always lesser than the respective *trans* isomer.

Isomers	Melting point (°C)	Solubility in water at 25 °C
Maleic acid (*cis* form)	130	78.8 g/100 ml
Fumaric acid (*trans* form)	300	0.78 g/100 ml

For Example, the boiling points, densities and refractive indxes also used to determine the configurations of geometrical isomers. An isomer with higher values of either of the above three constant will also possess the higher values of other two constants.

According to the **dipole rule** (A.E.Van Arkel) any isomer with higher dipole moment also possess the higher values of the above physical constants. For example: ClCH=CHCl

Isomers	Boiling point (°C)	Density (g/ml)	nD^{20}
cis	60.3	1.2835	1.4486
trans	48.4	1.2565	1.4454

The Table 28.1 shows the relative physical constants of *cis* & *trans* isomer.

Table 28.1 Relative physical constants of *cis* & *trans* isomers.

S. No.	Physical constants	*cis* isomer	*trans* isomer
1	Solubility	High	Low
2	Melting point	Low	High
3	Boiling point	High	Low
4	Density	High	Low
5	Refractive index	High	Low
6	Heat of combustion	High	Low
7	Dipole moment	High	Low
8	Thermal stability	Low	High

(iv) X-ray and electron diffraction method: X-ray crystallographic analysis is the best and easily available method for determining the configuration of the geometrical isomers.

For Example, X-ray crystallographic analysis of sorbic acid shows the *trans-trans* structure.

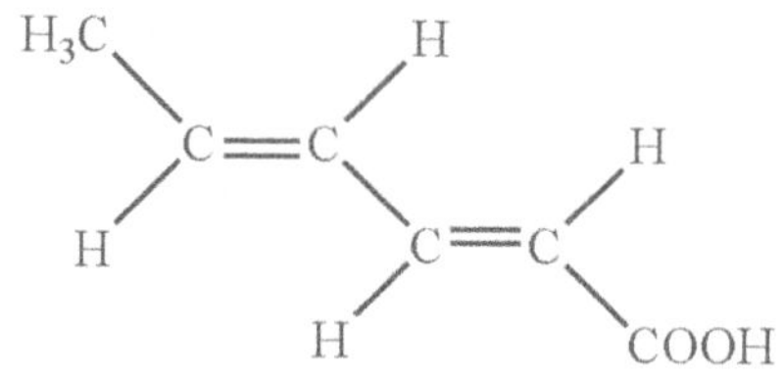

H_3C H C=C H H C=C H COOH

In some of the cases, electron diffraction measurements are used to identify the configuration of the geometrical isomers.

(v) Spectroscopy method: The configuration of geometrical isomers can be determined by ultra-violet spectra. The *trans*-isomers shows l_{max} value higher than the respective *cis*-isomer. The steric factor present in the cis isomer hinders the resonance, reduces the l_{max} and *trans*–isomer vice versa.

For example, the l_{max} value of *cis* & *trans* stilbene is 278 mμ and 294 mμ respectively.

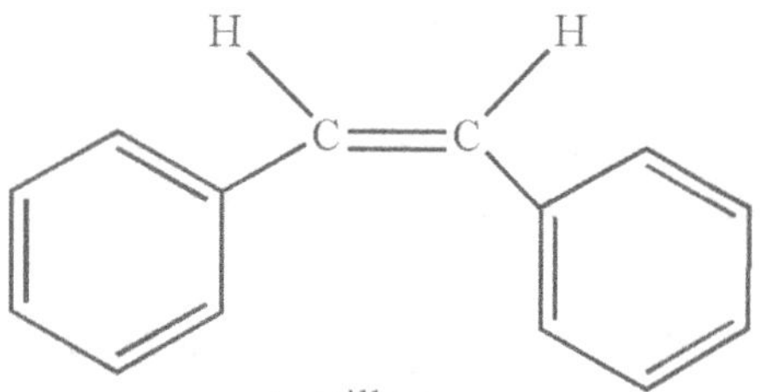

H H C=C

cis-stilbene

(λ_{max} = 278 mμ)

(Resonance decreased due to the steric hindrance of two benzene rings same side)

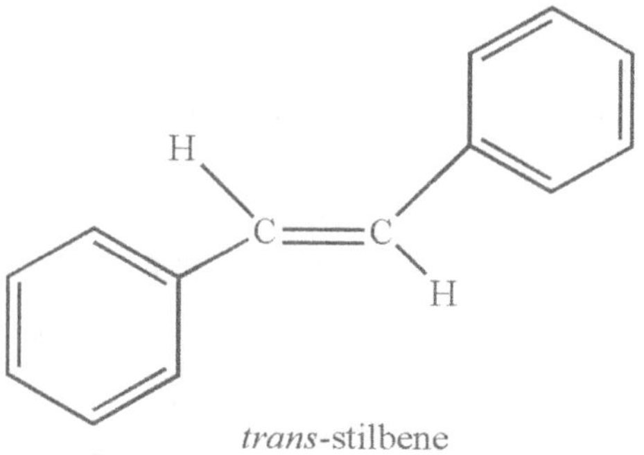

H C=C H

trans-stilbene

(λ_{max} = 294 mμ)

(No steric hindrance because both benzene rings are in opposite side)

IR Spectroscopy is also used to determine the configuration of geometrical isomers. For example, the alkenes of the following type such as RCH=CHR[1] shows a characteristic intense peak at 965 cm^{-1}.

Similarly NMR spectroscopy also used to distinguish geometrical isomers, because *trans*-vinyl protons are strongly coupled to each other than *cis*-vinyl protons.

For example: Geometrical isomers of stilbenes.

trans-stilbene

J_{HH} = 15.8 Hz

cis-stilbene

J_{HH} = 12.3 Hz

Hence the coupling constant value (J_{HH}) of *trans* is more than that of *cis*-isomer.

4. **Method of optical activity:** In some case, anyone of the geometrical isomer shows optical activity, whereas another isomer is optically inactive due to the presence of an element of symmetry. So, the optically active isomer is resolved and may used to determine its configuration. For example, hexahydrophthalic acid, the *trans* form of which has been resolved.

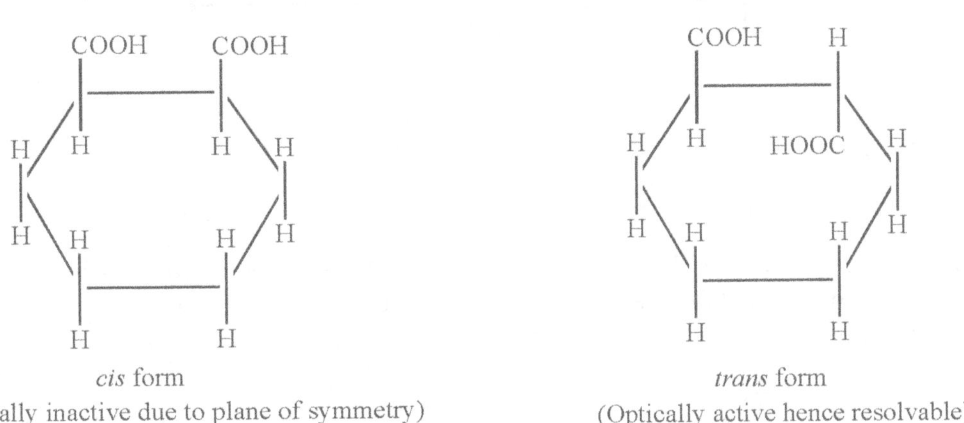

cis form
(Optically inactive due to plane of symmetry)

trans form
(Optically active hence resolvable)

5. **By stereo selective addition & elimination reactions:**

Reactions in which two isomers are obtained as products can be classified into two types according to the stereochemistry of the reaction. They are

(a) Stereoselective reactions.

(b) Stereospecific reactions.

(a) **Stereo selective reactions:** In these type of reactions, the product obtained is selective with respective to the stereoisomeric starting material *i.e.*, one isomer is formed in more amount than the other. For example: Hydrogenation of alkyne in the presence of Na or lithium in liquid ammonia yields a mixture of *cis* & *trans* isomers in which *trans* isomer predominates.

$$R-C\equiv C-R + H_2 \xrightarrow[\text{Liquid } NH_3]{Na/Li} (H)(R)C=C(H)(R) + (H)(R)C=C(R)(H)$$

Alkyne

cis-form

trans form
(High yield)

(b) Stereospecific reactions: It is defined as a reaction in which the reactant exist as stereoisomers and each reactants form different stereoisomeric product or a different set of stereoisomeric products. Example: Hydrogenation of alkenes. Those know the stereochemical course of addition or elimination reactions or the given reaction takes place in *cis* or *trans* manner, can determine the configuration of the starting material by knowing the nature of the product formed. Some of the reactions are described as follows.

(i) Reduction: Catalytic reduction of alkene or alkyne gives the *cis* addition product, hence the reaction will give *meso*-product with the *cis*-isomer and (±) product with the trans-isomer. For example: Hydrogenation of *cis*-2,3-diphenyl butane yields *meso* compound and a mixture of (±) 2,3-diphenyl butane.

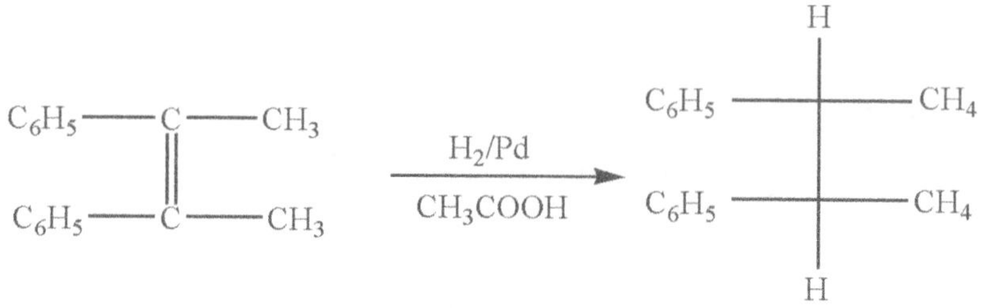

C_6H_5—C—CH_3 ‖ C_6H_5—C—CH_3 $\xrightarrow[CH_3COOH]{H_2/Pd}$ H; C_6H_5—+—CH_4; C_6H_5—+—CH_4; H

cis-2,3-Diphenyl butene   *meso*-2,3-Diphenyl butene

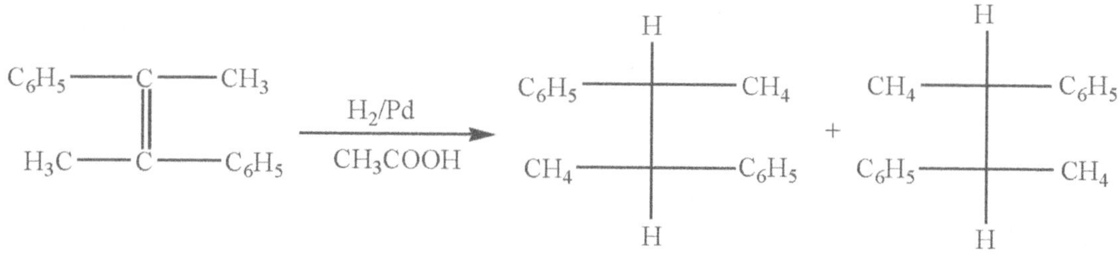

C_6H_5—C—CH_3 ‖ H_3C—C—C_6H_5 $\xrightarrow[CH_3COOH]{H_2/Pd}$ H; C_6H_5—+—CH_4; CH_4—+—C_6H_5; H + H; CH_4—+—C_6H_5; C_6H_5—+—CH_4; H

trans-2,3-Diphenyl butene   (±)Diphenyl butane

(ii) Free radical addition: Free radical addition of double bonds in cyclic rings takes place *via* trans-manner. For example, heat or light promoted addition of HBr to 1-bromocyclohexene gives cis 1,2-dibromocyclohexane.

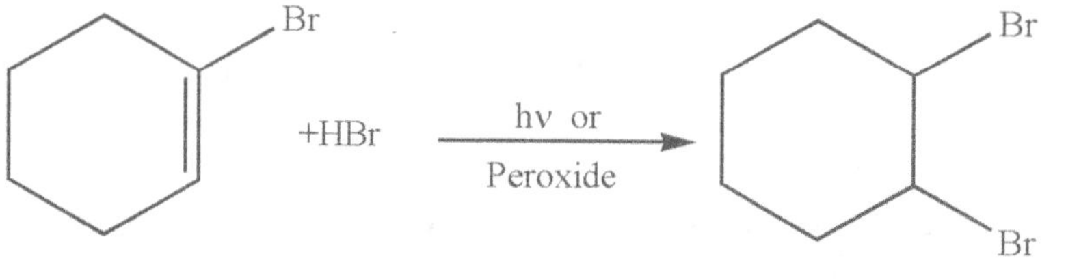

1-Bromocyclo hexene (Br) +HBr $\xrightarrow[\text{Peroxide}]{h\nu \text{ or}}$ 1,2-Dibromocyclo hexene (Br, Br)

trans-addition

(iii) Hydroxylation: Catalytic hydroxylation of olefins in the presence of $KMNO_4$ / H_2O_2 or osmium tetroxide takes place in *cis*-manner.

Example: Oxidation of maleic acid (*cis*-isomer) and fumaric acid (*trans*-isomer) with either $KMnO_4$ or H_2O_2 yields *meso-* and (±)-tartaric acids respectively.

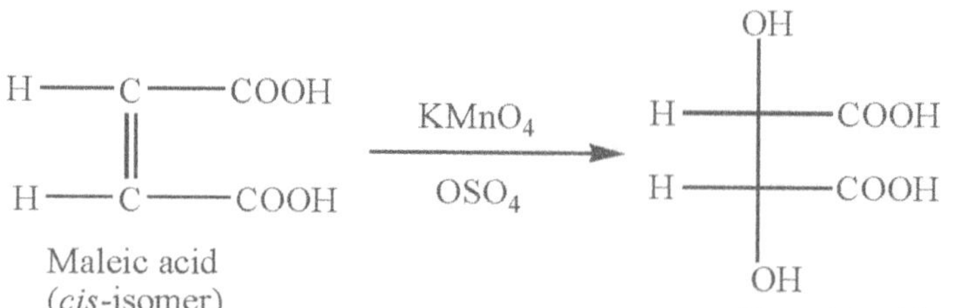

H—C—COOH ‖ H—C—COOH $\xrightarrow[OSO_4]{KMnO_4}$ OH; H—+—COOH; H—+—COOH; OH

Maleic acid (*cis*-isomer)   *meso*-Tartaric acid

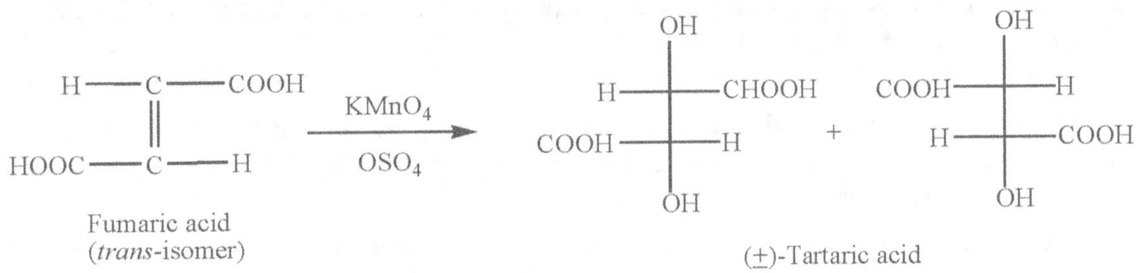

H—C—COOH ‖ HOOC—C—H $\xrightarrow[OSO_4]{KMnO_4}$ H—C(OH)—CHOOH / COOH—C(OH)—H + COOH—C(OH)—H / H—C(OH)—COOH

Fumaric acid (*trans*-isomer) (±)-Tartaric acid

On the other way hydroxylation of olefins in the presence of H_2O_2 or peracids, SeO_2 gives *trans*-addition products. For example: Maleic acid and fumaric acid leads to (±) and *meso* tartaric acids, respectively on oxidation with peracids.

(iv) Diel's-Alder reaction: Since the configuration of a cyclic adduct can be easily determined, the Diel's-Alder reaction provides a means of determining the configuration of the dienophile.

For example: Determination of the two isomers of cinnamic acid configuration by the determination of the configuration of their stereomeric adducts with butadiene.

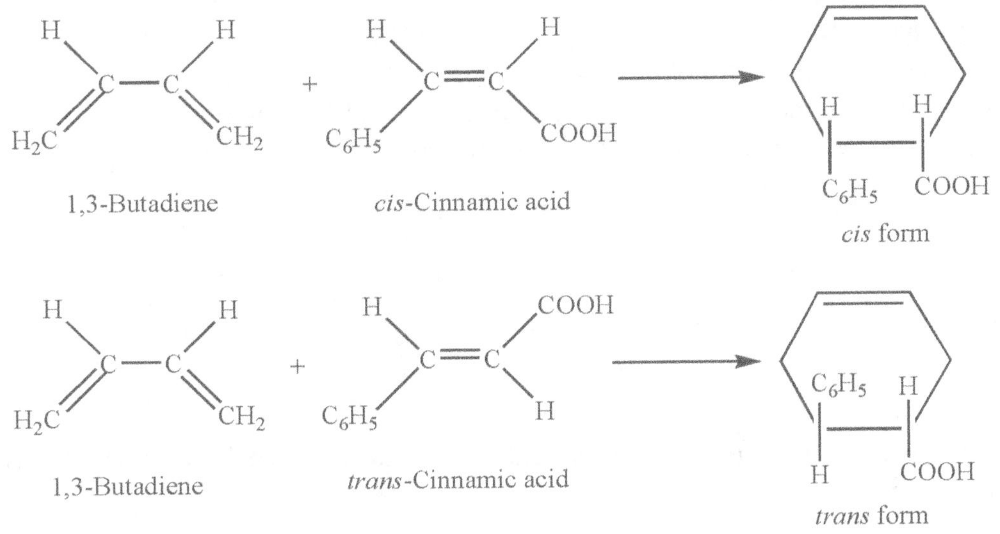

1,3-Butadiene + *cis*-Cinnamic acid ⟶ *cis* form

1,3-Butadiene + *trans*-Cinnamic acid ⟶ *trans* form

(v) Elimination reactions: Like addition reaction some of the elimination reactions also stereospecific. The formation of alkenes/acetylene follows the stereoselective elimination.

(a) Formation of acetylenes: The elimination reaction takes place rapidly when the removal of atoms or groups are in the *trans*-position with respect to each other than the *cis*-position.

For example,

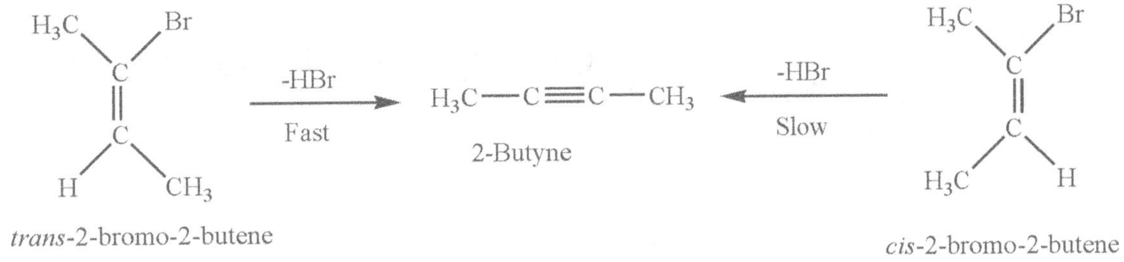

trans-2-bromo-2-butene $\xrightarrow[\text{Fast}]{\text{-HBr}}$ $H_3C—C≡C—CH_3$ (2-Butyne) $\xleftarrow[\text{Slow}]{\text{-HBr}}$ *cis*-2-bromo-2-butene

(b) Formation of alkenes: The good example for this reaction is E_2 elimination. It occurs place occurs when the two eliminated groups are in the *trans*-position to each other and the two carbon atoms bearing such groups lie in one plane.

Conformations of Alkanes

The carbon-carbon single bond is formed by the overlap of sp^3 orbital of first carbon atom with the sp^3 orbital of the second carbon atom, so rotation about a carbon-carbon single bond is possible without altering the amount of orbital overlap.

The various kinds of spatial arrangements of the atoms that can be converted into one another by rotation about C-C single bond are called **conformations.** The rotation of C-C single bond in ethane molecule leads to the formation of two extreme conformations. *i.e.*, staggered conformation and eclipsed conformation. Along with these, an infinite number of conformations are possible between these two extremes.

Generally, compounds having 3D-structures, but 2D structures are used for showing the structure on paper. Commonly, three types of methods are used to represent the 3D structure of molecule or 3D spatial arrangements of atoms on paper that results from the rotation about C-C single bond. These methods are described as follows:

1. Perspective formulae.
2. Sawhorse projections.
3. Newman projections.

Perspective formulae: In perspective formula, solid lines are used for bonds that lie in the plane of paper, solid wedges are used for bonds protruding out (standing out) from the plane of paper and hatched wedges are used for bonds protruding into the paper as shown below.

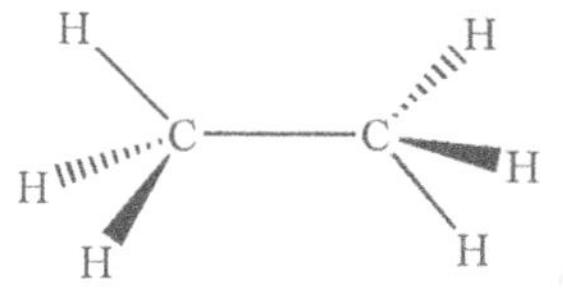

Ethane (Perspective formula)

Sawhorse projections: In this formula the molecule is viewed at the carbon - carbon bond from an oblique angle, as shown below.

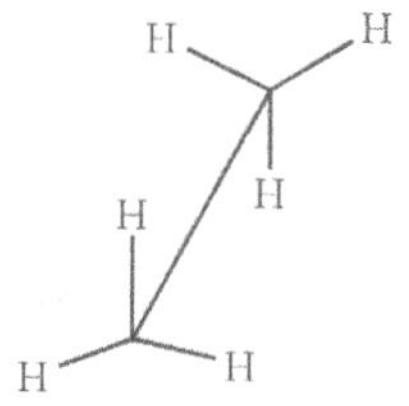

Ethane (Sawhorse projection)

Newman projections: In this formula molecule is viewed from the length of a particular carbon - carbon bond. The carbon atom in front is indicated by the point at which three bonds intersect, and the carbon atom in back is indicated by a circle.

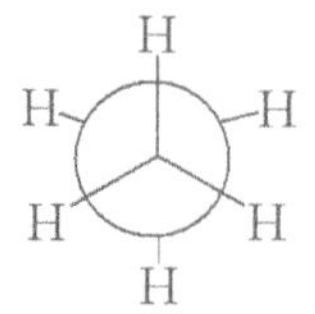

Ethane (Newman Projection)

The three lines emanating from each of the carbon atom indicates the carbon's other three bonds. Newman projection is easy to draw as it clearly represents the spatial relationship of the substituents on the two carbon atoms. The electrons present in C-H bond will repel the electrons in another C-H bond if the bonds come too close to each other.

Staggered conformation is the most stable one because the C-H bonds are as far away from each other. The eclipsed conformation is the least stable conformation because in no other conformation the C-H bonds are as close as to one another.

Conformational analysis: The investigation of the various conformations of a compound and their relative stabilities is known as conformational analysis.

Conformational Isomerism in Ethane

1. The type of hybridization in ethane molecule is sp^3 hybridisation (C-C is formed by the overlap of sp^3 orbital of one carbon atom overlaps with the sp^3 orbital of another carbon atom).
2. Hence the rotation about carbon-carbon single bond is taking place without change in the amount of orbital overlap that leads to different spatial arrangements of atoms (Fig. 28.2).

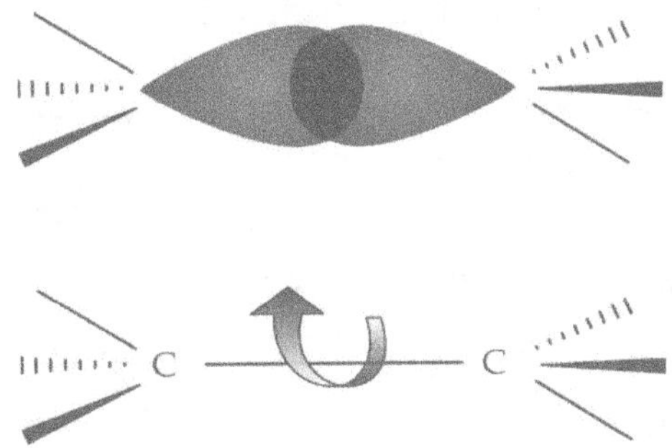

Figure 28.2 Formation of carbon-carbon bond in ethane by the overlap of cylinderically symmetrical sp^3 orbitals.

3. **Conformations:** Rotation about carbon-carbon single bonds leads to different spatial arrangements of atoms in a molecule is known as conformations. A particular conformation is known as conformer or conformational isomer.

The energy difference between the eclipsed and staggered conformers are rotation about carbon-carbon single bonds not completely free. The eclipsed conformer always possesses energy maxima and this energy must be overcome by the occurrence of carbon-carbon single bond rotation. But the enegy barrier is less in ethane molecule, it allows the conformers to interconvert millions of times per second RT. The Fig. no. 28.3. shows the various potential energy of ethane and its various conformers due to the rotation about carbon-carbon single bond (360° rotation).

In this graph, all the eclipsed conformers are present in energy maxima and all the staggered conformers are present in energy minima.

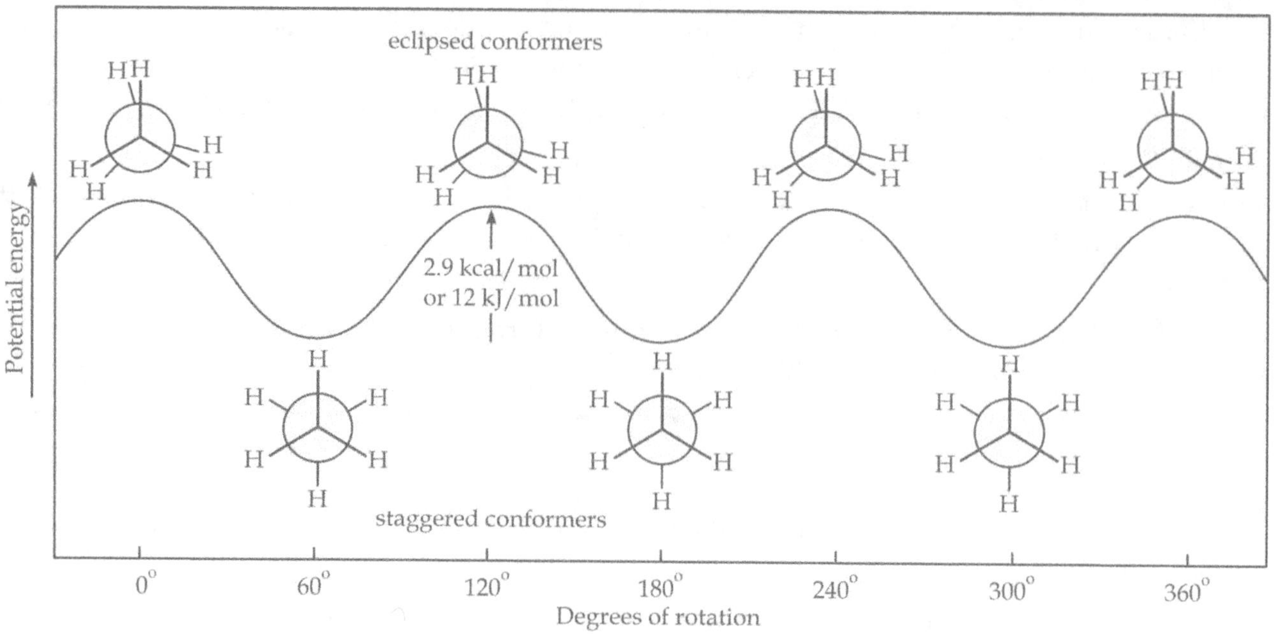

Figure 28.3 Potential energy of ethane molecule due to the angle of rotation about the carbon-carbon bond.

Conformational isomerism in *n*-Butane

The structural formula of *n*-butane is as follows. It contains three carbon-carbon single bonds.

The carbon-carbon single bond between C-1 and C-2 gives to the staggered and eclipsed conformers.

The energy of staggered conformers obtained by C-1 and C-2 bond has same energy while those are obtained by C-2 and C-3 bond. The energy of staggered conformers obtained from the C-1 and C-2 rotation of butane differs from those obtained from the C-2 and C-3 rotation of butane. The various staggered conformers obtained by the rotation of C-2 and C-3 bond in butane is described below.

The conformer (4) is a more stable conformer because the two methyl groups are present as far apart as possible, which is more stable than other two staggered conformations 2 and 5.

Anticonformer: The more stable staggered conformer is known as anticonformer (4). Bulky substituents are present opposite to each other.

Gauche conformer: The less stable staggered conformers are known as Gauche conformers. In this, the bulky substituents are adjacent to each other (2 and 6).

Steric strain: The strain is produced due to the presence of atoms or groups too close to each other (electron clouds of too closer atoms or groups repel to each other).

Gauche interaction: Gauche conformer has more steric strain than anticonformer because the two methyl groups are closer to each other in the former one. This phenomenon is called as Gauche interaction: Similarly, the eclipsed conformers of butane obtained by the C-2 and C-3 rotation also not have equal energy. The eclipsed conformer with two methyl groups are closest to each other (1) is less than the eclipsed conformer with two methyl groups are present far apart 3 and 5. The Fig No. 28.4 shows the various potential energy about C-2 and C-3 bond rotation of butane. The energy of various conformers is indicated by their respective numbers in the graph.

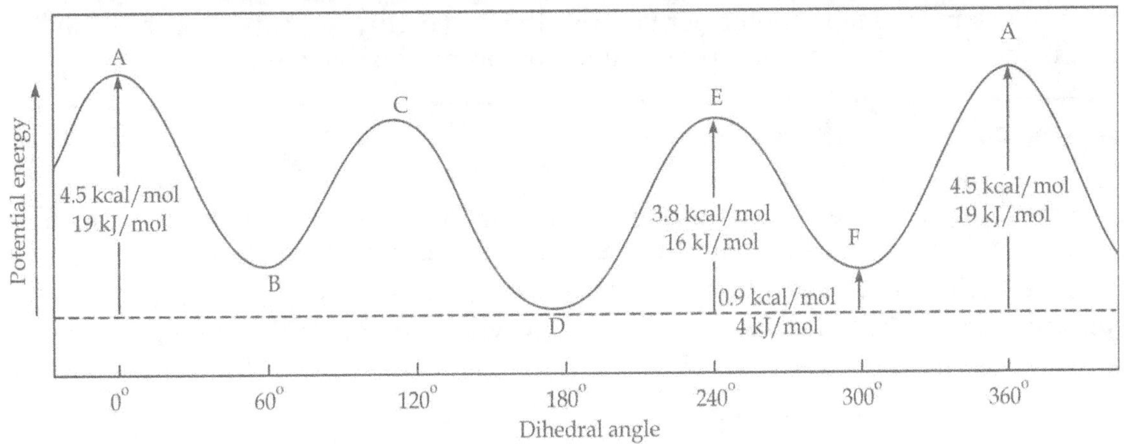

Figure 28.4 Potential energy of butane molecule due to the rotation of C-2 & C-3 bond.

The relative number of molecules in a particular conformation at any one time depends on the stability of the conformation - the more stable the conformation, the greater the fraction of molecules that will be in that conformation. Hence most of the molecules are present in staggered conformation and more molecules are in an anticonformation than in a Gauche conformation.

Carbon 1
Carbon 2

Staggered conformation obtained by the rotation of C-1 and C-2 bond in butane molecule

Eclipsed conformation obtained by the rotation of C-1 and C-2 bond in butane molecule

The energy of staggered conformers obtained by C-1 and C-2 bond has same energy while those are obtained by C-2 and C-3 bond. The energy of staggered conformers obtained from the C-1 and C-2 rotation of butane differs from those obtained from the C-2 and C-3 rotation of butane. The various staggered conformers obtained by the rotation of C-2 and C-3 bond in butane is described below.

1 2 3

4 5 6

The conformer 4 is a more stable conformer because the two methyl groups are present as far as apart as possible, which is more stable than other two staggered conformations 2 and 5.

Conformations of Cycloalkanes

Most common naturally available cyclic compounds contain six member rings because such rings exist in almost completely strain free conformation. This is called as chair conformation.

In the chair conformation, the bond angle is 111° which are near or closer to tetrahedral bond angle 109°28′ and all the adjacent carbon atoms are staggered which increases the stability.

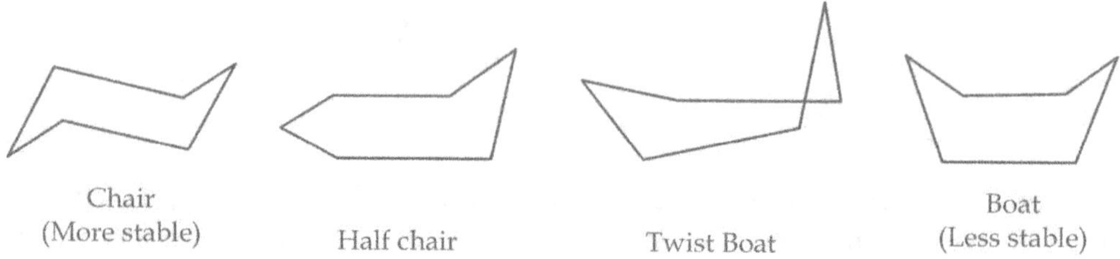

Figure 28.5 Different conformations of cyclohexane.

Drawing Chair Form of Cyclohexane Ring

The chair conformer of cyclohexane is important and more stable one. It should be drawn by as follows.

1. Draw two parallel lines of the same length, slanted upward position. Both lines should start at the same height.

2. Connect the tops of lines with a V; the left hand of the V should be slightly longer than the right hand side. Connect the bottoms of lines with an inverted V; the lines of the both V should be parallel. It forms the complete frame work of six membered ring system.

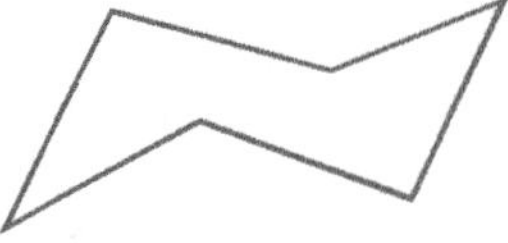

3. Each carbon possesses axial and equatorial hydrogens. The axial bonds are vertical and alternate above and below the ring. The one on the uppermost carbon is up and the next are down, the next is up and so on.

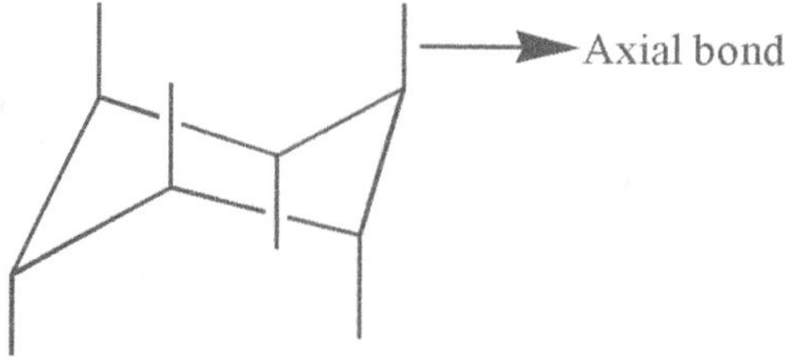

4. The equatorial bonds point towards out from the ring. If the axial bond is up, the equatorial bond on the same carbon is below the plane perpendicular to the bottom of the axial bond. If the axial bond is down, the equatorial bond on the same carbon is above the plane perpendicular to the bottom of the axial bond.

Equatorial bond

5. Observe that each equatorial bond is parallel to a ring bond two carbons over. The cyclohexane ring is always viewed on edge. The lower bonds of the ring are in front and the upper bonds of the ring are in back.

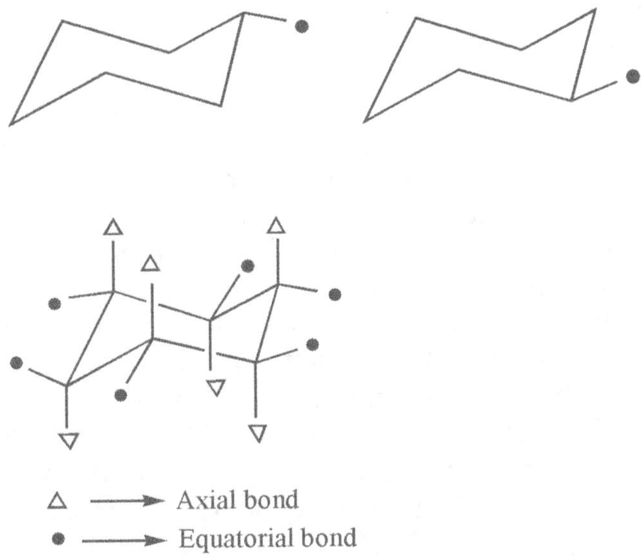

Due to the presence of rotation about carbon-carbon single covalent bond, cyclohexane interconverts between two stable chair conformations. This interconversion is called as ring flip (Fig. 28.6).

pull this carbon down

push this carbon up

Ring-flip

Figure 28.6 Ring flip in cyclohexane.

It is also free from angle strain, but it is less stable since some of the bonds are eclipsed which leads to torsional strain. It is also destabilized by the hydrogen atoms which are positioned in bow and stem position of the boat (flagpole hydrogen) which induces steric strain.

Interconversion of Conformations of Cyclohexane

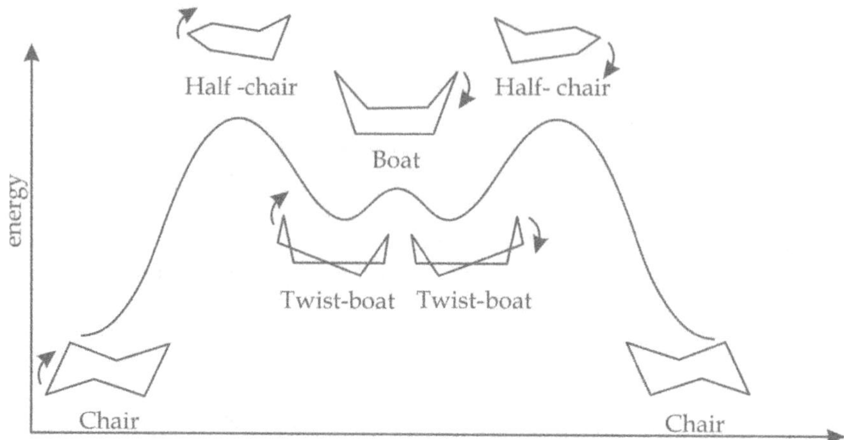

Figure 28.7 Various conformations of cyclohexane and their relative energies.

The inter conversion of one conformer to another conformer of cyclohexane is shown in graph.

1. To convert from the boat conformer to one of chair conformer, the top most carbon atoms of the boat conformer must be pulled down, it is converted to the bottom most carbon.
2. **Twist board** or **skew boat** conformer: It is obtained by just slightly pulling down the top most carbon. This conformer is more stable than that of boat conformer since the flag pole hydrogens in the boat conformer is far apart which leads to slight reduction in the steric strain.
3. **Half chair conformer:** When the top most carbon is pulled down to the point from where it is in the same plane as the sides of the boat then it leads to unstable half chair conformer.
4. Pulling the top most carbon down further produces chair conformer.
5. The energy barrier of cyclohexane molecule when it undergoes interconversion is shown in the Figure 28.7.

The energy barrier value of cyclohexane during the interconversion is calculated as 10.8 kcal/mol which shows that cyclohexane undergoes 10^5 ring flips per second at room temperature (the two chair forms are in equilibrium). The chair conformers of cyclohexane are more stable and more number of chair conformers is exhibited than that of other conformers.

INTERESTING FACT

Some of the highly strained cycloalkanes are synthesized by researchers. They are cyclo (1.1.0) butane, cubane, prismane. Octanitro cyclooctane is most powerful explosive.

Conformations of Mono Substituted Cyclohexanes

In previous section we discussed about the two equivalent chair conformers of cyclohexane. Unlike cyclohexane, the two chair conformers of mono substituted cyclohexane are not equivalent. In the substituted cyclohexane, the methyl group is present in an equatorial position in one conformer and in an axial position in another conformer.

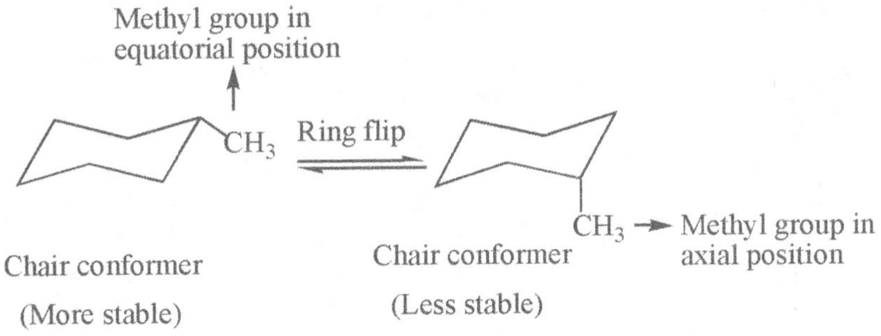

Methyl group in equatorial position

CH_3 Ring flip

CH_3 → Methyl group in axial position

Chair conformer

(More stable)

Chair conformer

(Less stable)

The chair conformer with methyl group in equatorial position is more stable because, a substitution has more space, therefore, lesser steric interactions when it is in an equatorial position. When the methyl group is in an equatorial position, it is anti to the C-3 and C-5 carbons. So, the methyl group extends into space away from the rest of the molecules.

In contrast, the methyl group is at axial position, it is gauche to the C-3 and C-5 carbons. These two positions are depicted in the following figure.

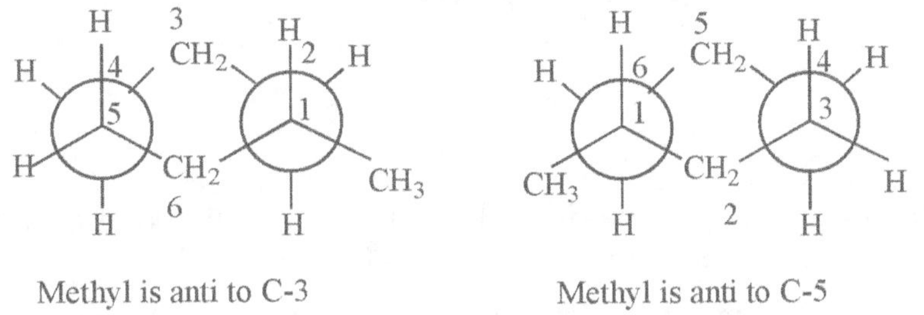

Methyl is anti to C-3

Methyl is anti to C-5

An equatorial substituent on the C-1 carbon is anti to the C-3 and C-5 carbons.

Methyl is gauche to C-3

Methyl is gauche to C-5

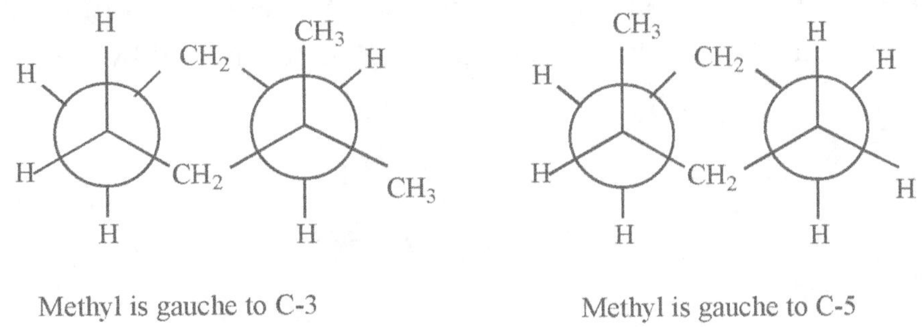

Due to this, the three axial bonds on the same side of the ring are parallel to each other; therefore, any axial substituent will be closer to axial substituents on the other two carbon atoms. Because the interacting substituents are on 1,3-positions relative to each other. Unfavourable steric interactions are known as 1,3-diaxial interactions.

Conformations of Disubstituted Cyclohexanes

If the cyclohexane possesses two substituents, then both substituents are to be considered while determining which of the two chair conformers is a more stable one. Let us consider 1,4-dimethyl cyclohexane. It exhibits geometrical isomerism (*cis-trans* isomers). One isomer is *cis* –isomer (both methyl groups are present in the same side of the cyclohexane ring) and the second isomer is the *trans*-isomer in which the two methyl groups are present in the opposite side of the cyclohexane ring.

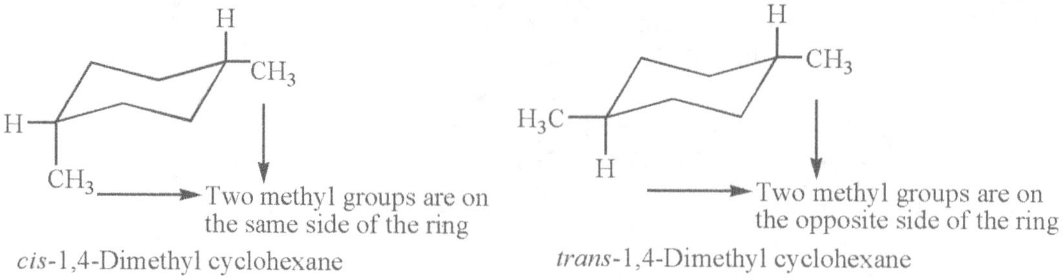

First of all we shall determine which of the two chair conformers of *cis*-1,4-dimethyl cyclohexane is more stable. For finding out this the following two structures are considered.

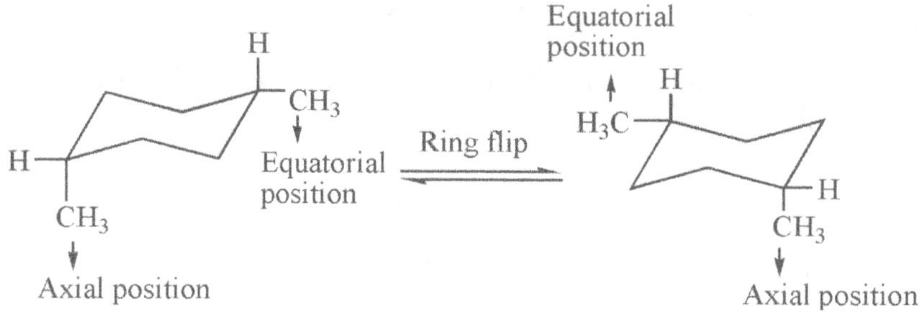

In the above structures, both conformers are having one methyl group in equatorial position and other methyl group in axial position. So, both chair conformers of *cis*-1,4-dimethyl cyclohexane are more stable.

Stereoisomerism in Biphenyl Compounds

(Atropisomerism or optical isomerism occurs due to restricted rotation around single bond).

Biphenyl or diphenyl compounds in which two benzene rings are connected by single bond.

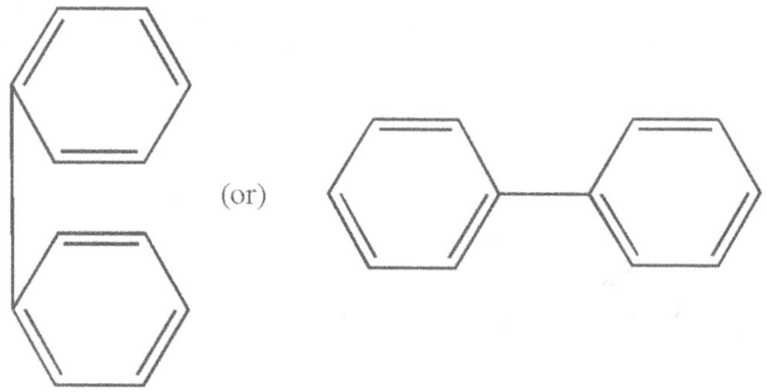

The substituted biphenyls exhibit chirality with individual chiral carbon. This chirality is due to the restricted rotation of single bond between two benzene rings. Therefore they exists in two forms which are non-superimposable mirror images of each other. This kind of isomerism is called as atropisomerism and the stereoisomers are known as atropisomers.

The word atropisomer (in greek atrops means 'without turn') was first introduced by a German biochemist Richard Kuhn.

The compound 6,6′-diphenic acid was the first diphenyl molecule to show optical isomerism.

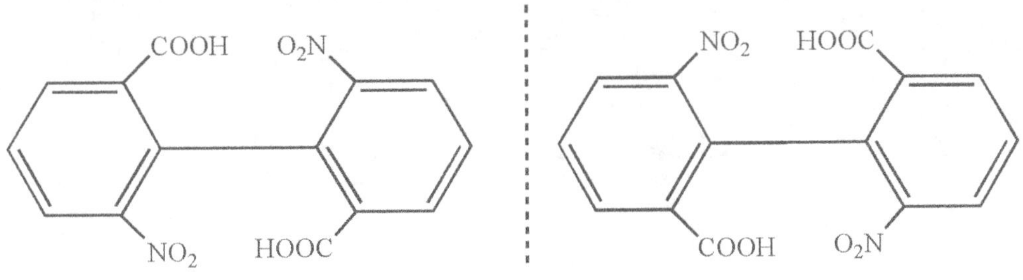

COOH O$_2$N | NO$_2$ HOOC

NO$_2$ HOOC | COOH O$_2$N

Enantiomers of 6,6'-diphenic acid

In the above diphenyl compound, the two benzene nuclei are in collinear. The introductions of bulky groups in the two *o*-positions prevent the free rotation about the co-axis. So the two *o*-substituted benzene ring cannot be coplanar (the compound is dissymmetric).

Conditions for Biphenyls to Show Optical Isomerism

1. Each ring of the biphenyl should be unsymmetrically substituted [Neither of the two ring must have a vertical plane of symmetry].

   For example:

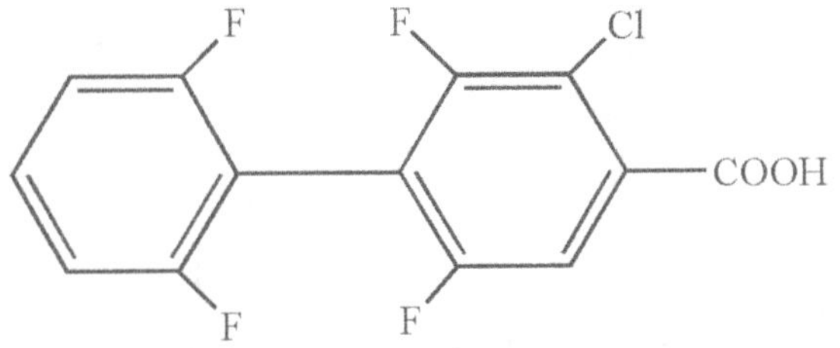

   F F Cl

   COOH

   F F

2. The *o*-positions (minimum two) of the rings must be substituted by bulky groups to prevent the free rotation about single bond. If the *o*-positions are substituted by small groups, the two benzene ring becomes planar and free rotation about single bond is possible and the compound can't exhibit optical isomerism.

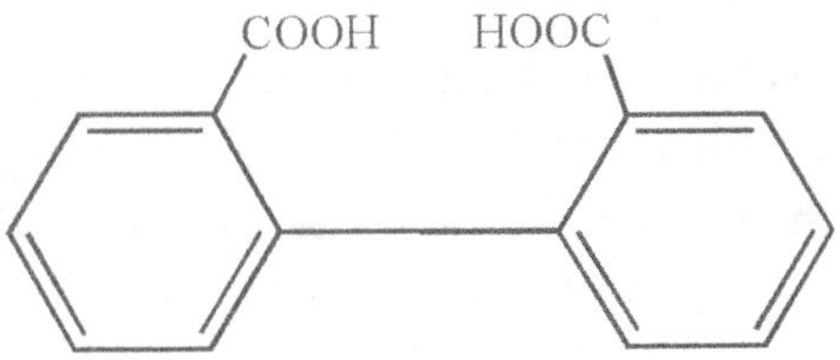

   COOH HOOC

   Diphenic acid
   (optically inactive due to plane of symmetry)

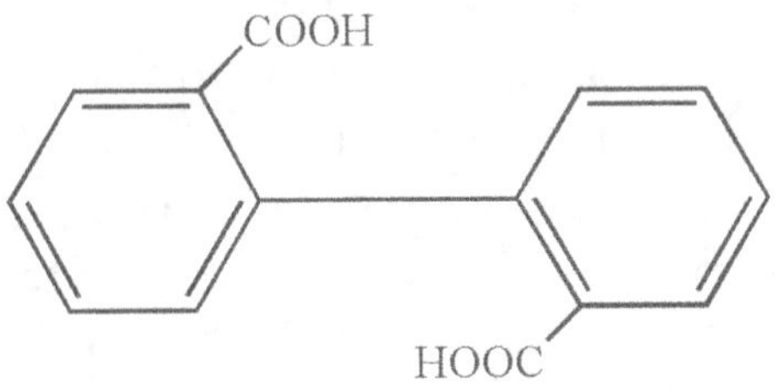

   COOH

   HOOC

   Diphenic acid
   (optically inactive due to centre of symmetry)

3. Size of the '*o*' substituent also important for optical isomerism in biphenyls. Compare the following two compounds.

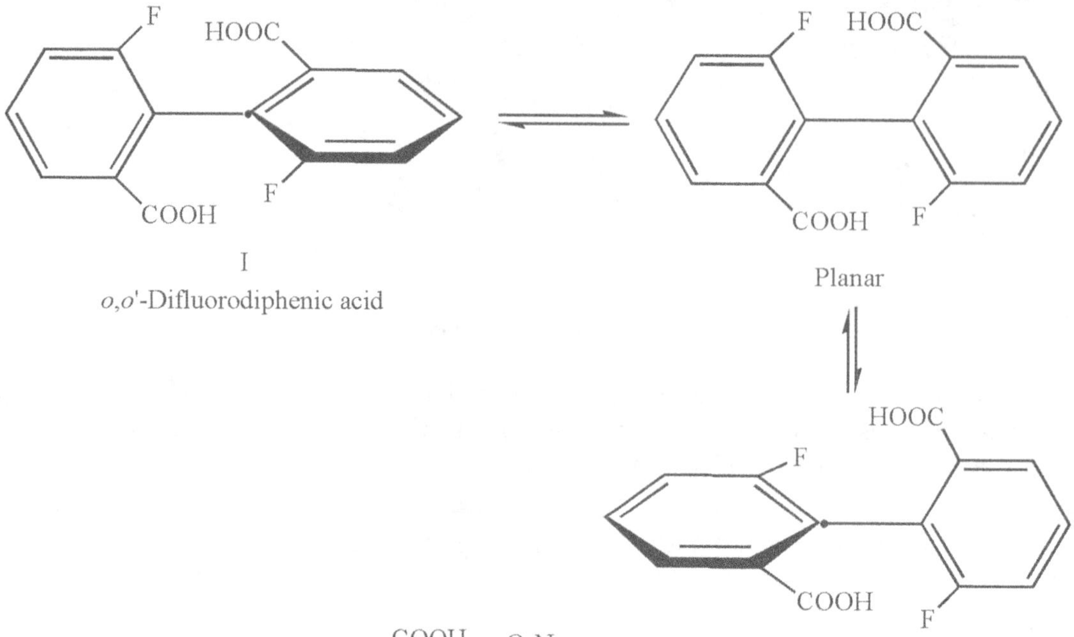

I

o,o'-Difluorodiphenic acid

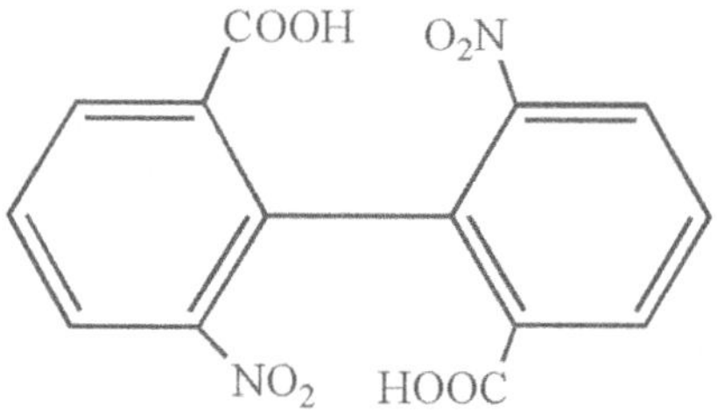

Planar

II

o,o'-Dinitro diphenic acid

The compound I is easily resolvable and racemizes more than the compound II. It is because the fluorine atoms are smaller in size which forms less interference for the rotation about single bond; hence the two benzene rings possess planar conformation at one time which leads to molecular symmetry and racemization.

Conditions for optical activity:

1. The principle condition for one compound to exhibit optical activity or optical isomerism is that it must be asymmetric or chiral.
2. The second one is the compound or the geometrical structure of the compound should not be super imposable on its mirror image (They should be in enantiomeric form).
3. But some of the compounds have asymmetric center, but does not show optical activity. Example: meso-tartaric acid as well as some of the compounds don't have asymmetric centre but dissymmetric and show optical isomerism. Example: Biphenyl compounds.

Stereoselective and stereospecific reactions: In all the previous chapters, we learned about the organic reactions, their mechanism, uses *etc* especially in alkenes, we studied about the electrophilic addition reactions and their step by step mechanism also. We also determined the type of end products formed. But we did not examine the kind of stereoisomers formed.

Stereochemistry: It is the branch of organic chemistry, deals about the three dimensional structures of organic compounds. For studying and understanding the stereochemistry of organic reactions, the following points should be considered.

1. If the product of the organic reaction is stereoisomers or a chemical reaction produces stereoisomers as products, the reaction forms a single stereoisomer or a set of specific stereoisomers or all possible stereoisomers?

2. If the reactant molecule in a chemical reaction is stereoisomer, all stereoisomers react to form same stereoisomeric product or form different stereoisomer or a set of different stereoisomers.

For determining the stereochemistry of organic reaction, we should know about some of the important terms. They are as follows.

Regioselective reaction: Regioselective reaction is an organic reaction in which two constitutional isomers are obtained as products but one of the isomer is obtained more than the other. It selects for a particular constitutional isomer.

$$\underset{\text{Isobutene}}{CH_3-\overset{\overset{CH_3}{|}}{C}=CH_2} + HBr \longrightarrow \underset{\substack{\text{Major product}\\\text{(Isobutyl bromide)}}}{CH_3-\underset{\underset{Br}{|}}{\overset{\overset{CH_3}{|}}{C}}-CH_3} + \underset{\substack{\text{Minor product}\\\text{(2-Methyl propyl bromide)}}}{CH_3-\overset{\overset{CH_3}{|}}{CH}-CH_2Br}$$

$$\underset{\text{2-Methyl-2-butene}}{CH_3-CH=\overset{\overset{CH_3}{|}}{C}-CH_3} + HI \longrightarrow \underset{\substack{\text{2-Iodo-2-methyl butane}\\\text{(Major product)}}}{CH_3CH_2-\underset{\underset{I}{|}}{\overset{\overset{CH_3}{|}}{C}}-CH_3} + \underset{\substack{\text{2-Iodo-3-methyl butane}\\\text{(Minor product)}}}{CH_3\underset{\underset{I}{|}}{CH}-\overset{\overset{CH_3}{|}}{CH}-CH_3}$$

Stereoselective reaction: Stereoselective reaction is a reaction in which one stereoisomer is formed more than other rather than constitutional isomer.

Depending upon the preference of the formation of any one of stereoisomer, it is classified into moderately stereoselective, highly stereoselective and completely stereoselective. If the products formed are enantiomers, then known as enantioselective reactions. If the products formed are diastereomers, known as diastereoselective reactions.

Stereospecific reactions: Stereospecific reaction is defined as "a reaction in which the reactant exist as stereoisomers and each reactants form different stereoisomeric product or a different set of stereoisomeric products".

The above two reactions and their mechanism are explained by considering the addition reaction of hydrogen atom to alkene molecule.

Alkene reacts with hydrogen, the two hydrogen atoms are added to the same side of the double bond, known as syn-addition.

$$>C=C< + H_2 \xrightarrow{Pd/Pt} \overset{H}{C}-\overset{H}{C}$$

Consider the following example. (Syn-addition of H_2 to cis-alkene)

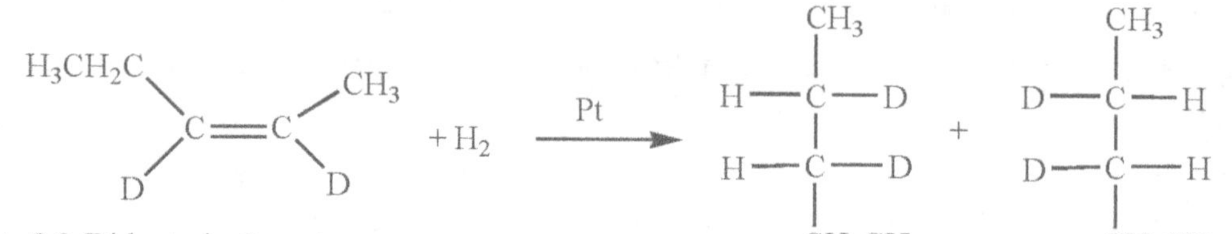

cis-2,3-Dideuterio-2-pentene

Fischer projections of erythro enantiomers

(or)

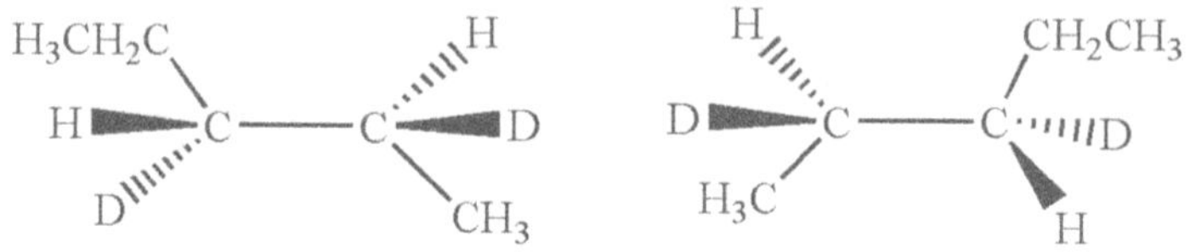

Perspective formulas of products

1. The addition of hydrogen to *syn*-alkene leads to the formation of a product with two chiral centers, only two of the four possible isomers are formed.
2. One of the stereoisomer is obtained by the addition of hydrogen atoms from the plane above the double bond.
3. Another stereoisomer is obtained by the addition of the hydrogen atoms from the plane below the double bond.
4. Hence, the type of stereoisomeric product formed in the reaction depends on whether the reactant is *cis* or *trans* isomer.
5. Syn addition of hydrogen to *cis* alkene produces only the erythro pair of enantiomers. [Erythro pair of enantiomers: It is the pair of enantiomers in which identical groups are present in the same side of the carbon chain in Fischer projections].

Consider the synaddition of H_2 to trans-alkene, the product obtained is the threo pair of enantiomers. Hence, the addition of hydrogen to alkene is a stereo specific reaction because the product obtained from the addition to *cis*-isomer is different from the addition to the *trans*-isomer.

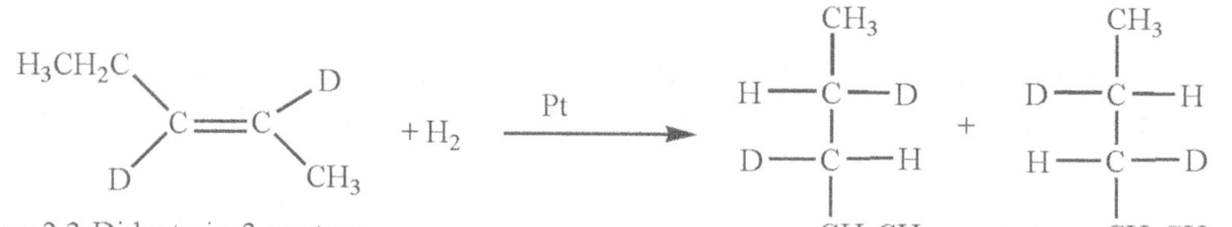

trans-2,3-Dideuterio-2-pentene

Fischer projections of threo enantiomers

(or)

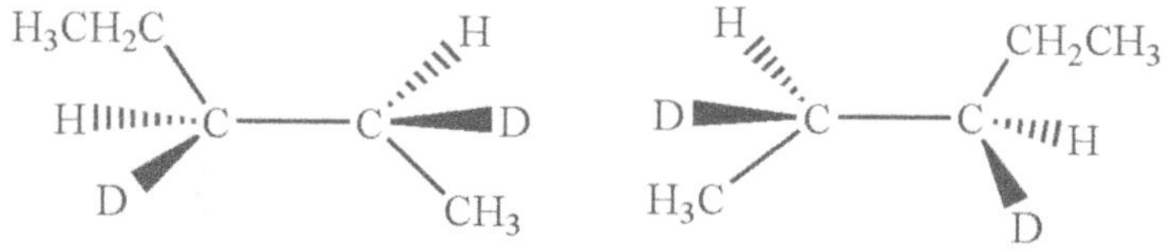

Perspective formulas of products

Note: All stereo specific reactions are stereo selective but all stereo selective reactions are not stereo specific because there are stereo selective reactions in which the reactant does not have a carbon-carbon double bond or a chiral center, so it cannot exist as stereoisomers.

Probable Questions

1. Define Geometrical isomerism.
2. What are *cis* and *trans* isomers, explain with examples.
3. Write about conformations of cycloalkanes.
4. Write a note on Newman projection formula.
5. Write in detail about conformational isomerism in n-Butane.
6. Discuss inter conversion of conformations of cyclohexane.
7. Explain about optical isomerism in detail with examples.
8. Write a note on conformations of mono substituted cyclohexanes.
9. Write a note on conformations of disubstituted cyclohexanes.
10. Explain stereoisomerism of biphenyl compounds with examples.
11. What are the conditions for optical activity.
12. Define stereochemistry.
13. Write in detail about stereospecific reactions.
14. Write the examples for E and Z isomers.
15. Write a note on methods for determination of configuration of geometrical isomers.

29

Heterocyclic Compounds (Part 1)

Introduction to Heterocyclic compounds

Heterocyclic compounds are cyclic compounds in which the ring carbon atoms are displaced or substituted with one or more polyvalent atoms. The major substituted polyvalent atoms are nitrogen, oxygen and sulphur as shown in the following examples.

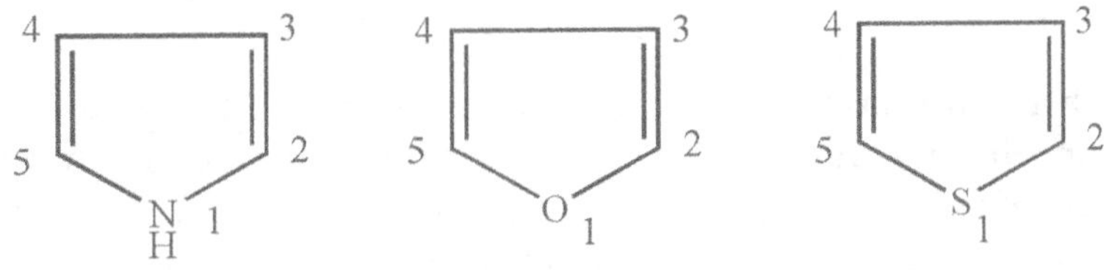

Pyrrole Furan Thiophene

Heterocycles possess wide range of applications in day to day life, such as drugs, veterinary products and agrochemicals. They are also used as antioxidants, dyestuffs, pigments and additives *etc.*

Why heterocyclic compounds are very important in organic chemistry?

1. "Easy manipulation of ring system is possible to attain a required modification leads to the formation of different structural variations with different properties. This phenomenon is an useful strategy for developing new drugs. For example, consider the following compound I, which acts as a fungicide and it is highly lipophilic in nature; the water solubility of the compound is increased by the replacement of the benzene ring with suitable heterocycles.

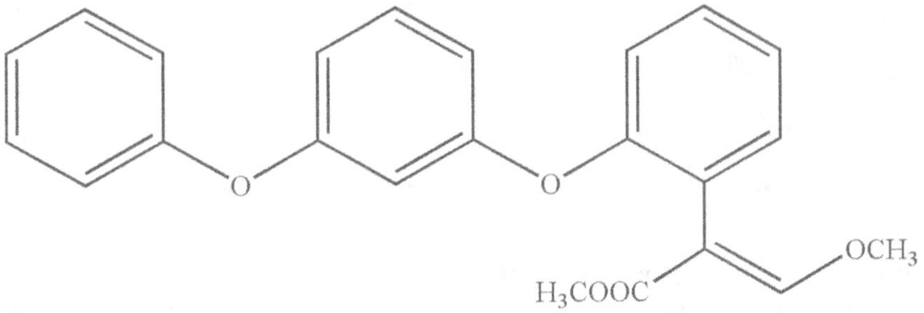

OCH_3 H_3COOC

(I)

2. Another important fact is we can easily accommodate any functional groups into the heterocycles framework as a substituent or as a part of the ring system itself. It makes heterocyclic chemistry as a very peculiar part in organic chemistry.
3. Heterocycles or heterocyclic compounds are widely distributed in nature as important fundamental units of living systems as shown below with examples.
   (i) Nucleic acid contains pyrimidine and purine rings, which are responsible for cell replication.

(ii) Chlorophyll and heme contains porphyrin ring system which is needed for photosynthesis and for respiration (oxygen transport) in higher plants and animals respectively.

(iii) Vitamins: Thiamine (vitamin B_1), riboflavin (vitamin B_2), pyridoxol (vitamin B_6), ascorbic acid (vitamin C) - All are heterocyclic compounds.

(iv) Amino acids: Histidine, prolines, and tryptophan are possesing heterocyclic rings.

Pharmaceutical applications of heterocyclic compounds: Many of the drugs are heterocyclic in nature and possess heterocyclic ring skeleton. They are not extracted from the nature due to the difficulties in extraction and purification process. Hence they are manufactured or synthesized in the laboratories.

The origin of organic chemistry is based on the natural products and many of the drug candidates are developed subsequently. Some of the examples are described below.

Development of Histamines and Antihistamines

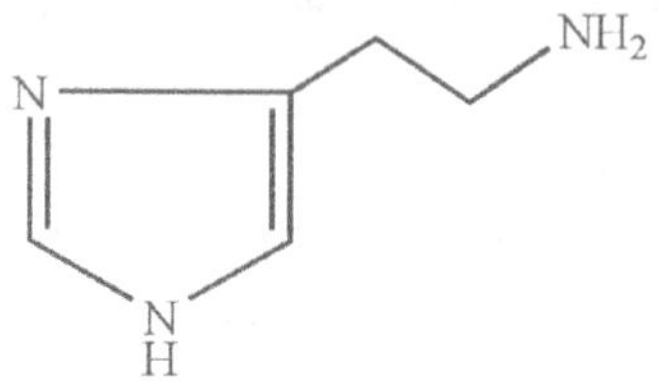

NH_2 N N H

(1)
Histamine

Histamine (1), a monosubstituted imidazole ring system released in our body from the amino acid histidine by decarboxylation. The main pharmacological actions of histamine includes contraction of smooth muscle, fall in BP (hypotension), producing allergic reactions and regulation of gastric acid secretion. For antagonising the action of histamine several kind of drugs are synthesized from 1940. One of the important drug is pyrilamine, a pyridine derivative (2).

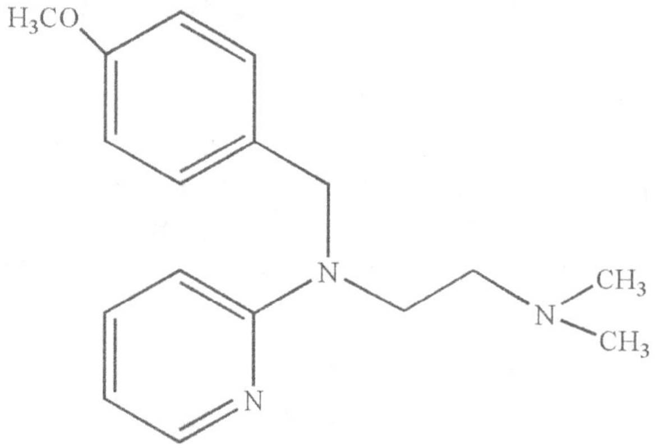

H_3CO N N CH_3 CH_3 N

Pyrilamine (2)

Pyrilamine antagonizes or inhibits the several actions of histamine but it does not block the gastric acid secretion.

In 1976, the lacunae of pyrilamine is encountered by chemists with modification of the structure of histamine and discovered an active drug known as cimetidine (3), used for the treatment of peptic ulcer.

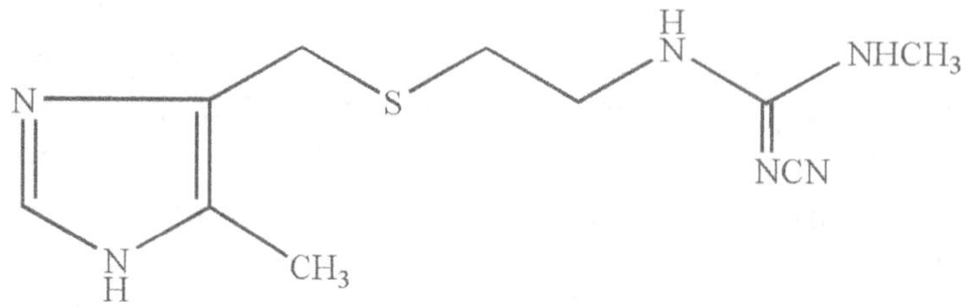

H N $NHCH_3$ N S NCN N H CH_3

Cimetidine (3)

Cimetidine is one of the major leading drug in 1970s and 1980s being the first non-surgical treatment of peptic ulcer. Further development of cimetidine analogues are promoted by changing the heteroyclic ring in the cimetidine molecule.

Ranitidine, pyrrole ring of the cimetidine is replaced by furan, is also a successful drug for the treatment of peptic ulcer.

H_3C N H_3C O S N H NO_2 $NH-CH_3$

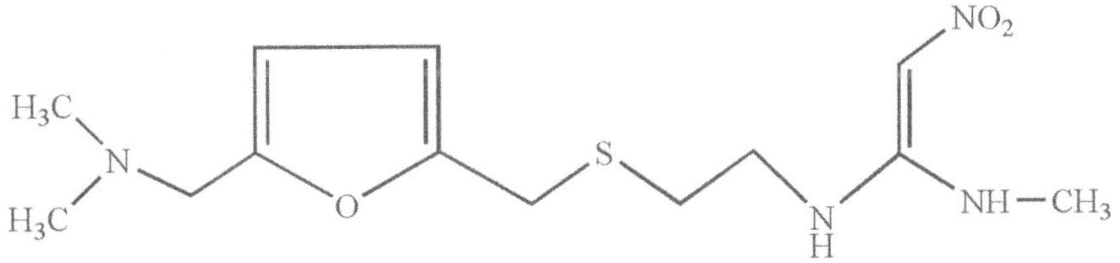

Ranitidine

Development of Nucleoside Analogues

In the search for drugs to combat cancer and virus modification in the structure of DNA is the best approach by synthesizing the nucleosides.

These analogues consists of pyrimidine and purine nucleus which is attached to a sugar moiety as shown in the following examples.

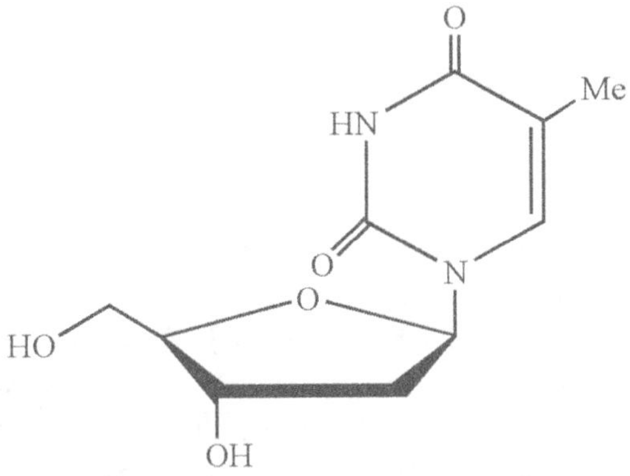

O HN Me O N O HO OH

2-Deoxythymidine

Various nucleic acid analogues are developed by the structural modifications of heterocycles or sugar or both and yielded the important drugs as mentioned below.

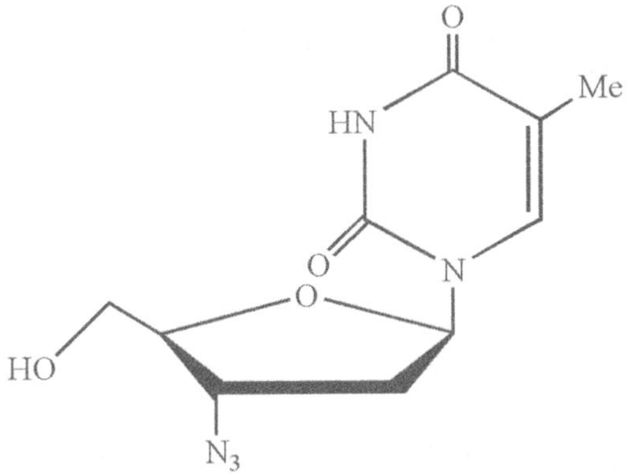

O HN Me O N O HO N_3

Zidovudine

(Used in the treatment of AIDS,
Close analogue of 2'-deoxythymidine)

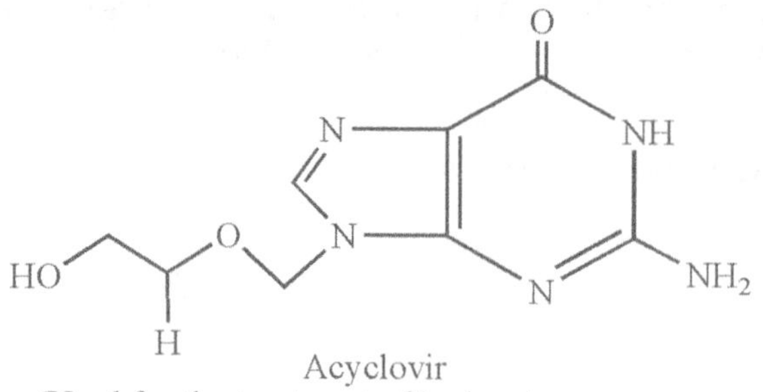

O
N NH
N
HO O N NH2
H

Acyclovir
(Used for the treatment of herbs virus infections, Analogue of 2'-deoxyguanosine)

Development of Alkaloidal Drugs

Development of Serotonin Analogues

Serotonin: Vasoconstrictor drug obtained from natural source is widely distributed in nature but present only in very low concentration. It was first extracted from natural source in 1948.

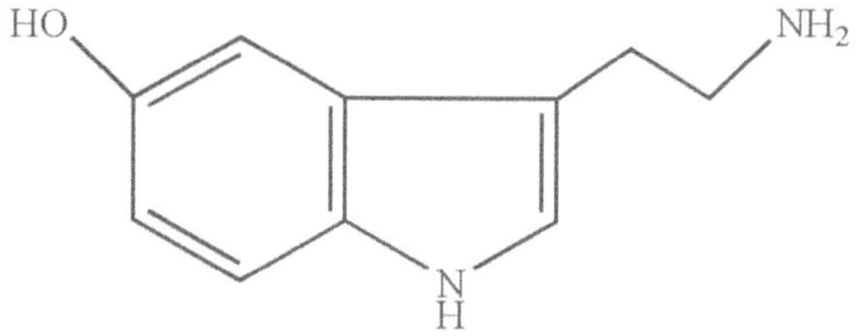

HO NH2
N
H

Serotonin

Synthesis of this drug was done in laboratories after few years and it is used for investigating the mechanism of action. The main pharmacological actions of serotonin includes constriction of brain arteries, behavourial changes in the body *etc*,. But the main drawback of serotonin is that it is too rapidly metabolised in the brain. This was overcome by designing the serotonin analogues and the simple indole derivative of serotonin such as sumatriptan was developed by researchers which acts as a selective agonist at serotonin receptor sites in the brain and used as a drug of choice for the treatment of migraine.

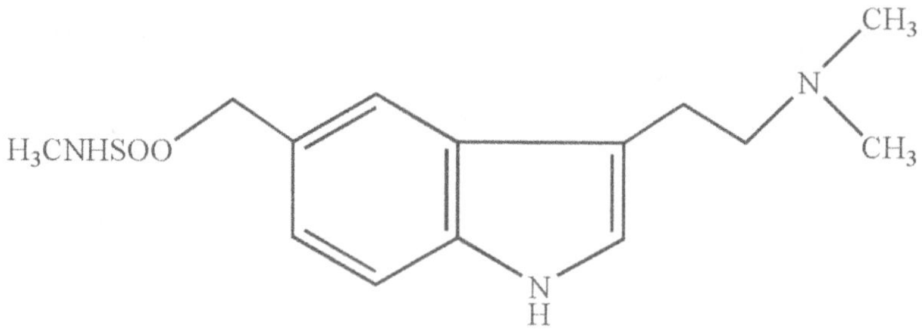

CH3
N
H3CNHSOO CH3
N
H

Sumatriptan
(Indole derivative of serotonin used for migraine)

Ergot alkaloids:

Ergotamine: An indole alkaloid possesing aminoethyl side chain at 3rd position, is used in migraine at low doses. But the drug is highly toxic and not in use. But its synthetic analogue lysergic acid diethylamide (LCD) is discovered and it is now notorious as a hallucinogen.

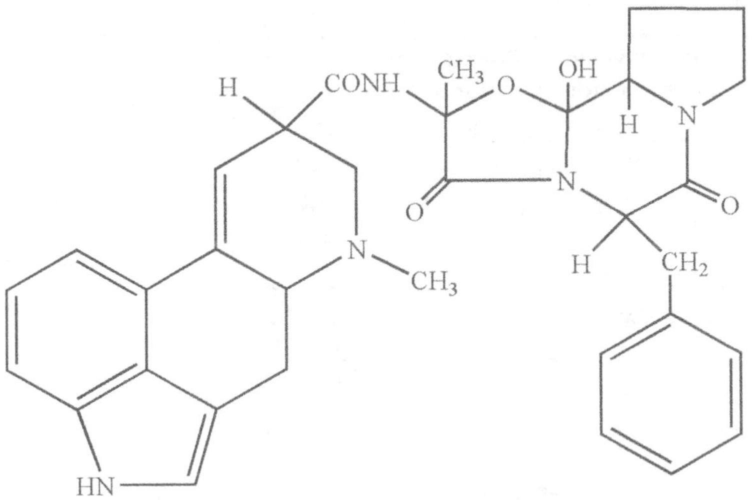

H CONH CH_3 O OH H N O N O N CH_3 H CH_2 HN

Ergotamine

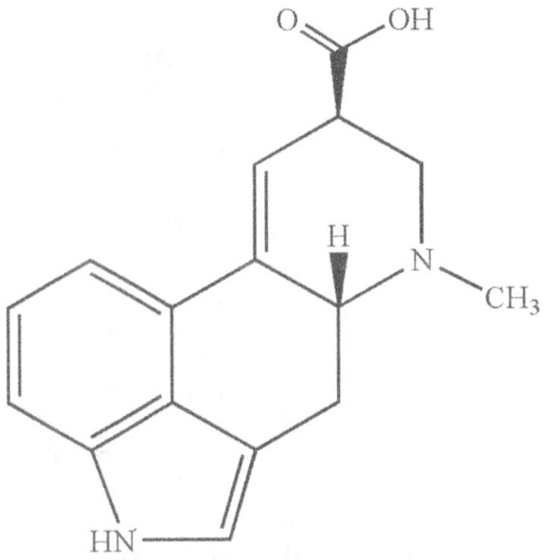

O OH H N CH_3 HN

Lysergic acid

Nomenclature of Heterocyclic Compounds

Heterocyclic compounds are named by using trivial names as well as systematic names. Trivial name does not provide the detailed information about the structure but still it is used (In recent years the IUPAC has made efforts to systematize the nomenclature of heterocyclic compounds).

According to the IUPAC system, following guidelines are used for naming the heterocycles.

1. The monocyclic compounds are named by a prefix which is derived from the nature of the hetero atom present in it (Eliding "a" where necessary). Some of the rings and their prefixes are indicated in the following Table 29.1.

Table 29.1 Prefix for Hetero Atoms.

Nature of hetero atom	Symbol	Respective prefix
Oxygen	O	Oxa
Nitrogen	N	Aza
Sulphur	S	Thia
Phosphorous	P	Phospha
Selenium	Se	Selena
Silicon	Si	Sila
Germanium	Ge	Germa

2. If the same hetero atom is present more than one time, the prefixes di, tri *etc* are used.

   For example: dioxa, triaza *etc.*

3. If different heteroatoms are present in the ring, the naming starts from the atom which is high group in periodic table and as low in atomic number in that group. Hence the order of naming is as follows O, S, N, P, Si *etc.*

4. When the size of the monocyclic ring is from 3 to 10, they are indicated with suffixes which are mentioned in the Table 29.2.

Table 29.2 Common name endings for heterocycles.

Nature of the ring size	Suffixes for completely unsaturated compounds		Suffixes for completely saturated compounds	
	With N	Without N	With N	Without N
3	-irine	-irene	-iridine	-irane
4	-ete	-ete	-etidine	-etan
5	-ole	-ole	-olidine	-olane
6	-ine	-in	-	-ane
7	-epine	-epin	-	-epane
8	-ocine	-	-ocin	-ocane
9	-onine	-onin	-	-onan
10	-ecine	-ecin	-	-ecan

5. The nature of hydrogenation is indicated by the suffixes as mentioned in the above table or prefixes dihydro, tetrahydro *etc.* or by prefixing the parent unsaturated compound with the symbol H preceded by a number indicating the position of saturation. For example:

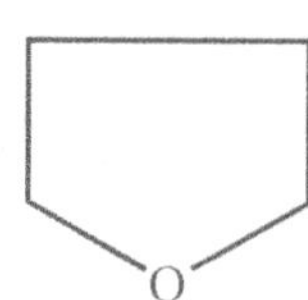

Tetrahydrofuran

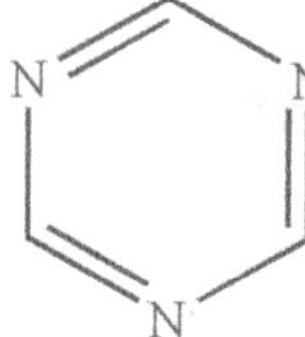

1,3,5-Triazine

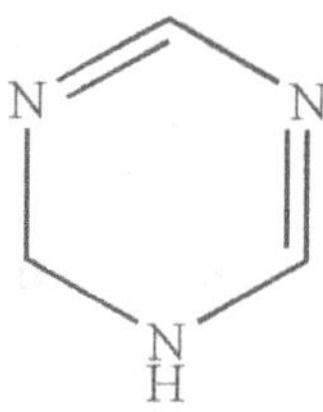

1,2-Dihydro-1,3,5-triazine

6. In monocyclic ring system with one hetero atom, numbering start from that atom only.

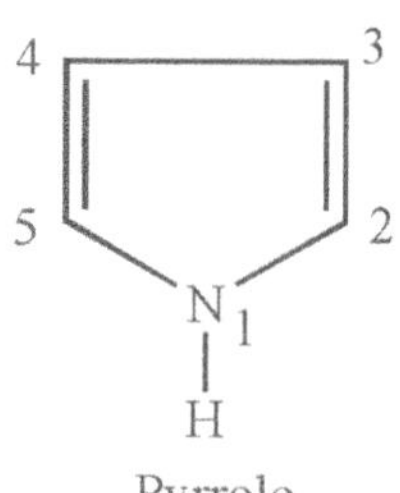

Pyrrole

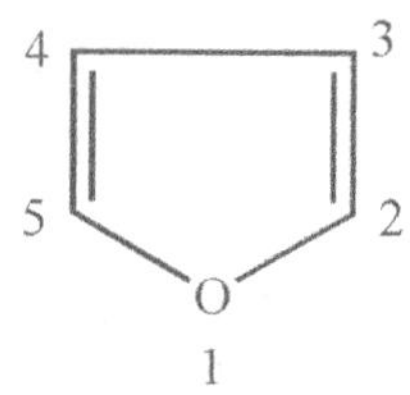

Furan

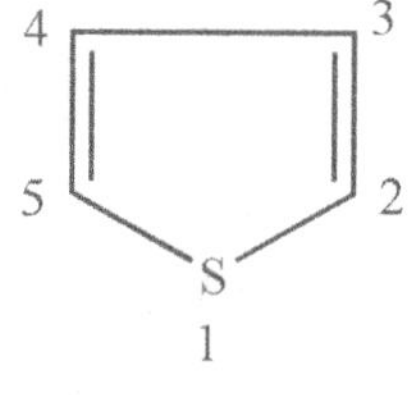

Thiophene

According to the above system, some of the heterocycles and their nomenclature are mentioned below.

Example 1:

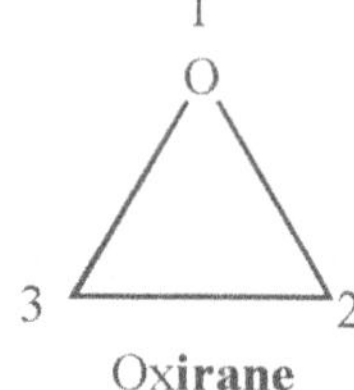

Ox**irane**

[Three membered ring without N and full saturation]

Ox**irene**

[Three membered ring without N and unsaturation]

Example 2:

Aziridine

[Three membered ring with N and saturation]

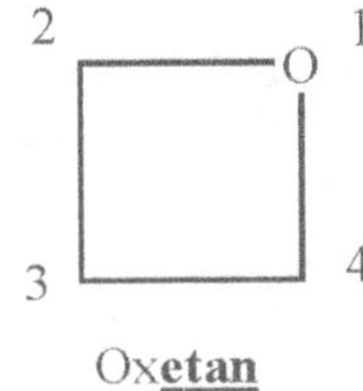

Oxetan

[Four membered ring with O and saturation]

Example 3:

If two or more different atoms are present in the ring, naming is given by combining the prefixes of the heteroatoms.

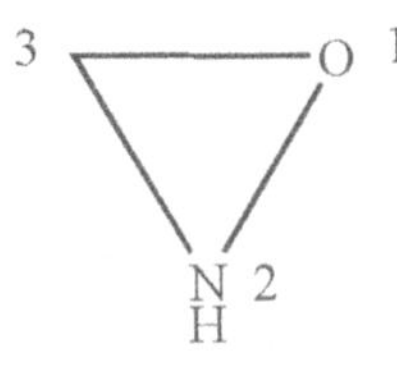

Oxa aziridine

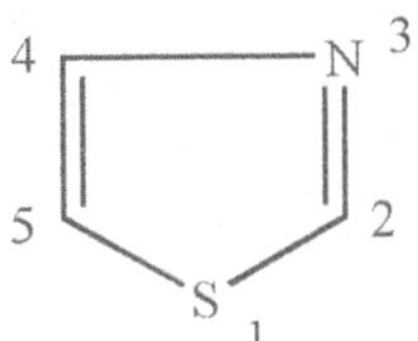

Thiazole

Nature of the substituents: In substituted heterocycles, the numbering starts from the hetero atom (assigned position 1) and the substituents are numbered with a lowest possible numbers. The name of substituents should be mentioned in alphabetical order.

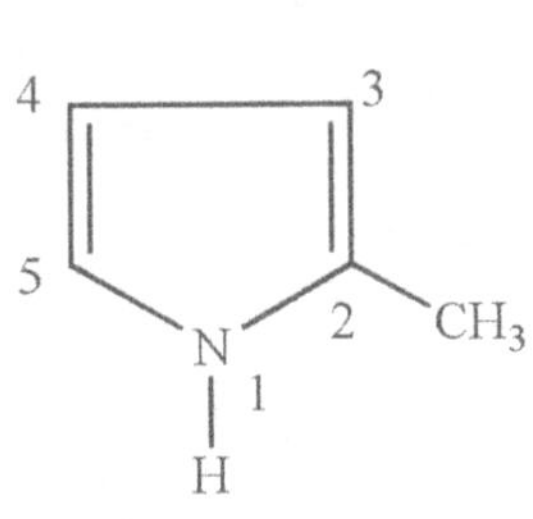

2-Methyl pyrrole

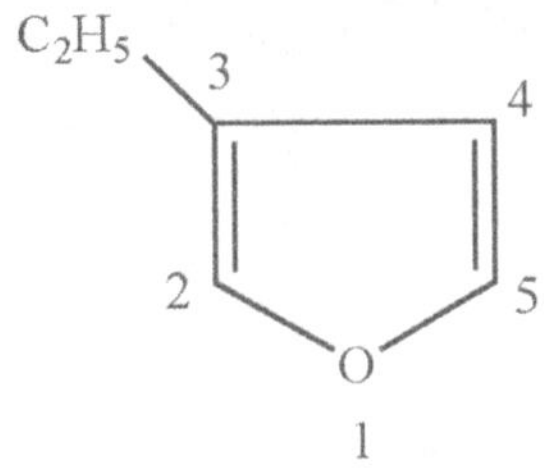

3-Ethyl furan

If the heterocycles contain more than one hetero atom, the order of preference is O, S, N. The ring is numbered from the atom of preference and preceeded in such a way so as to give the smallest possible number to the other hetero atoms in the ring. In this case, the substituents numbering need not be bothered.

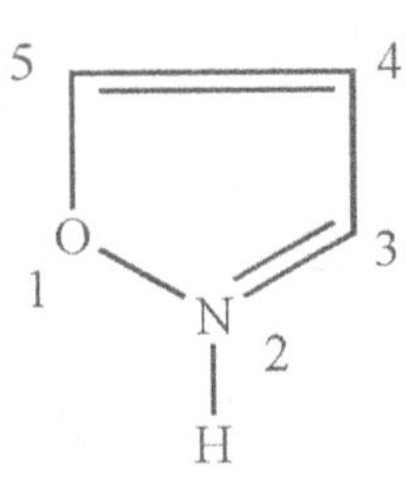

[The preference of number starts from O and the next heteroatom attains lowest possible number]

Isoxazole

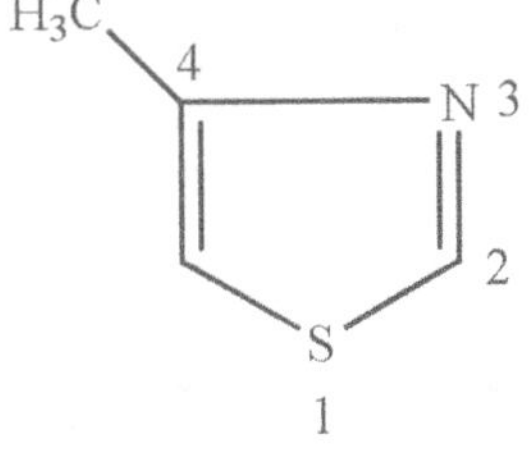

[The preference of numbering of heteroatom starts from S and the next heteroatom attains lowest possible number, no preference for subtituents]

4-Methyl thiazole

7. Apart from the systematic method, many number of heterocycles are also named by common names or non-systematic names which are widely used. Few examples are shown below.

Five membered rings:

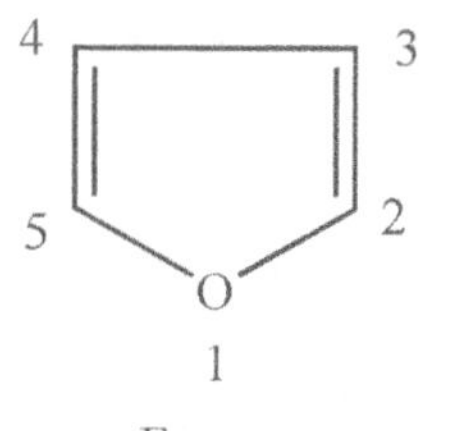
Furan

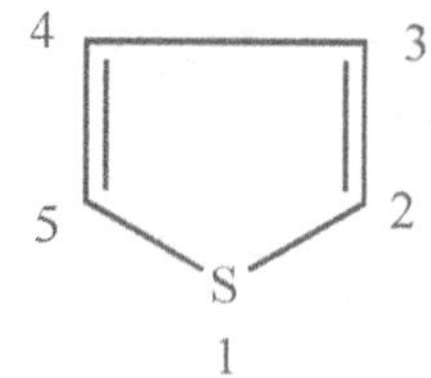
Thiophene

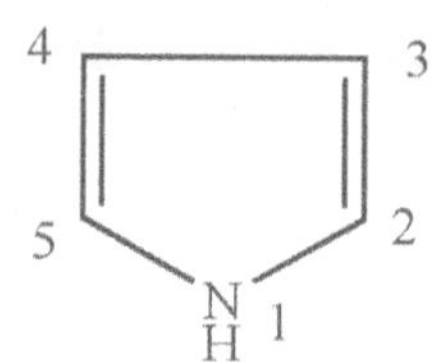
Pyrrole

Six membered rings:

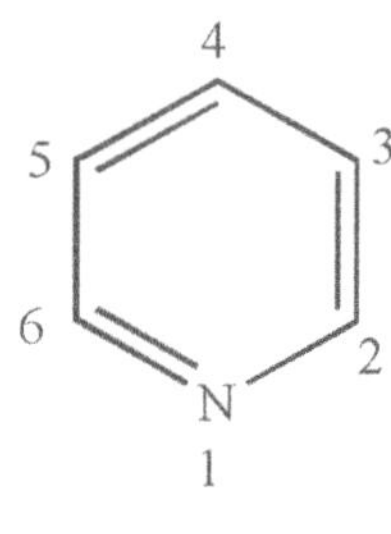

Pyridine

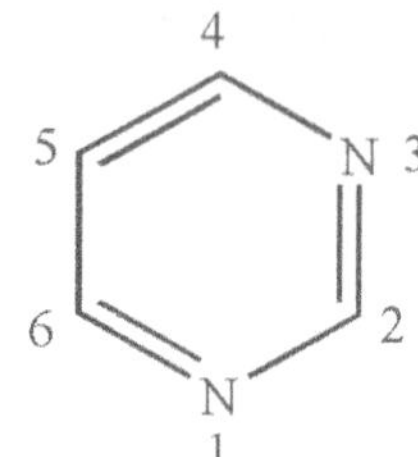
Pyrimidine

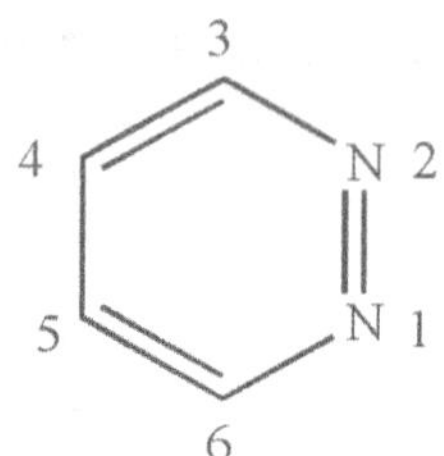
Pyridazine

Condensed heterocycles:

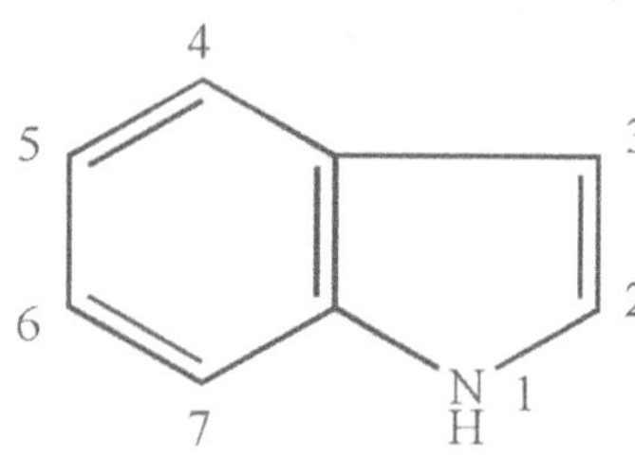
Indole

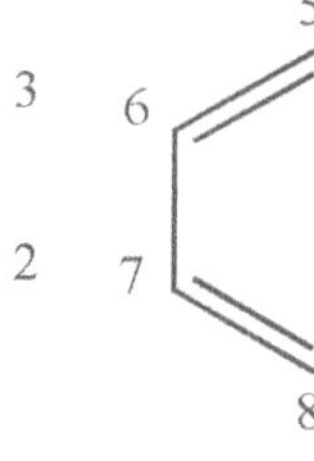
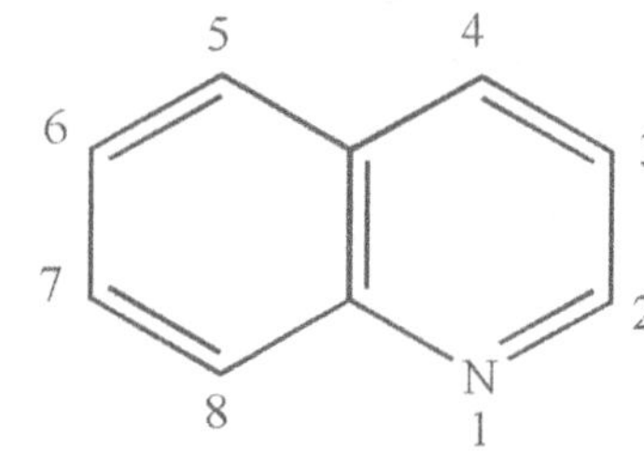
Quinoline

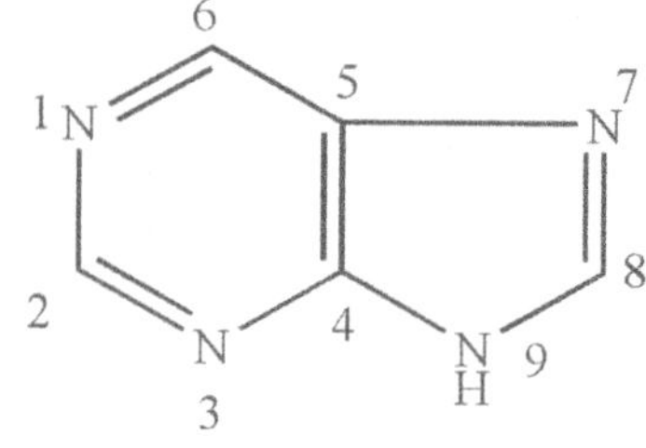
Purine

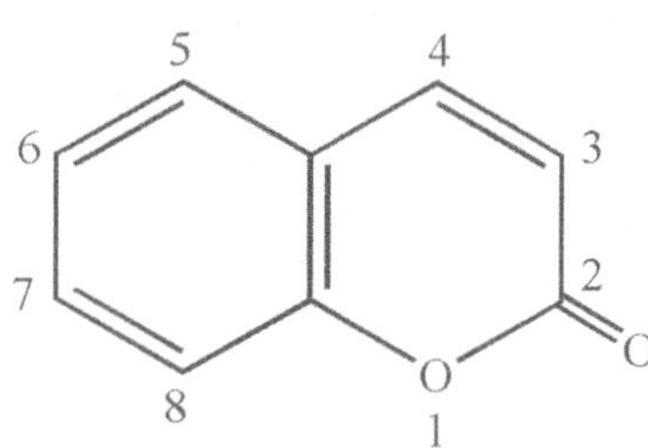
Coumarin

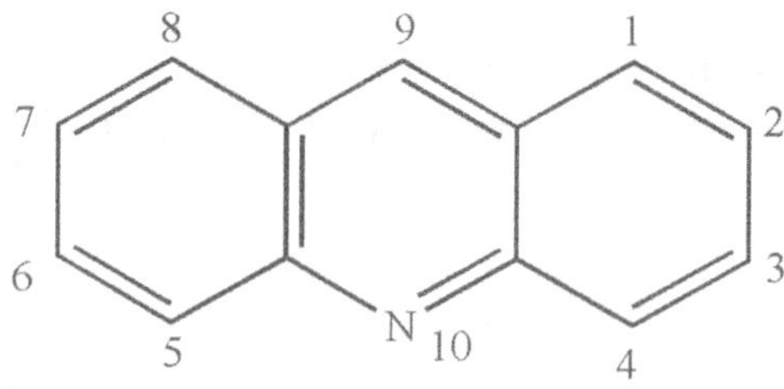
Acridine

4 3
5
N
2
6
N1
7 H

Indazole

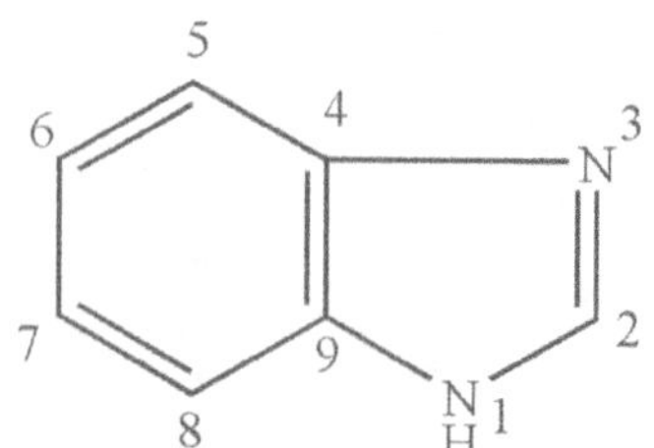
Benzimidazole

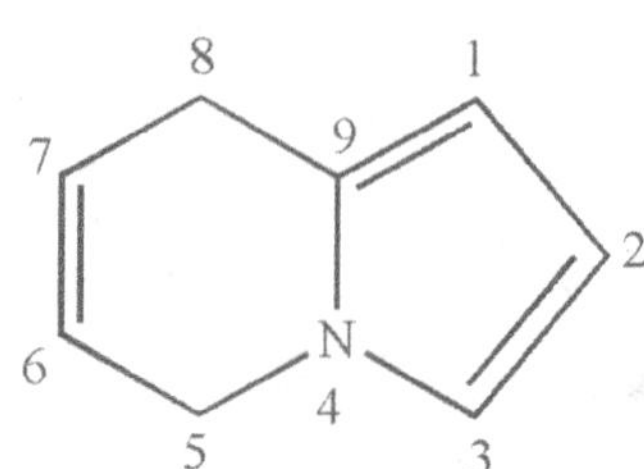
Indolizine

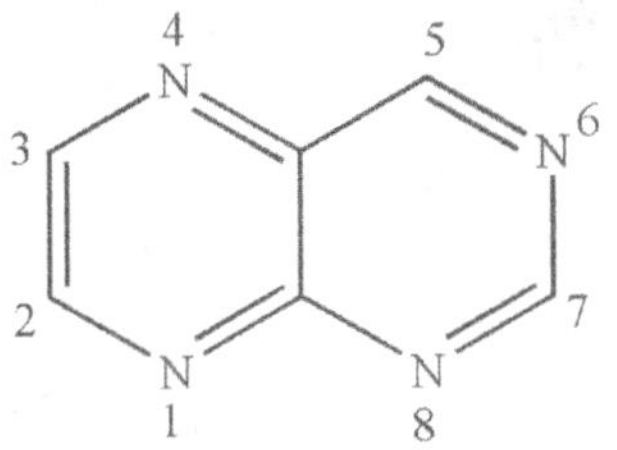

Pteridine

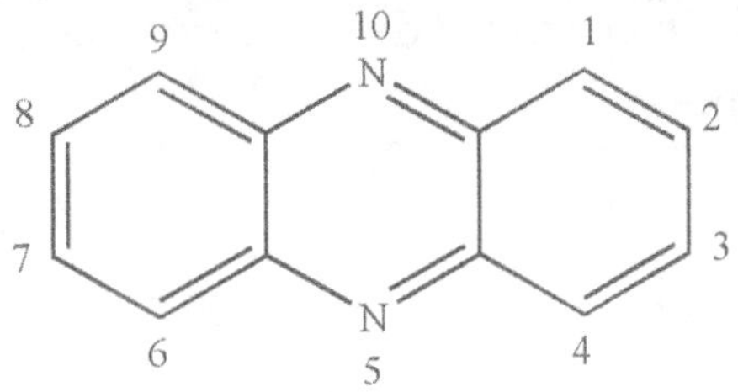

Phenazine

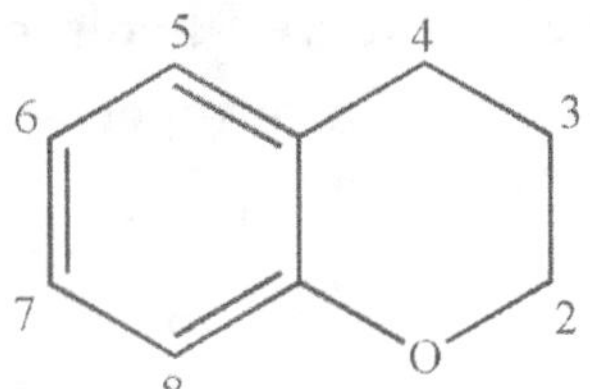

Chroman

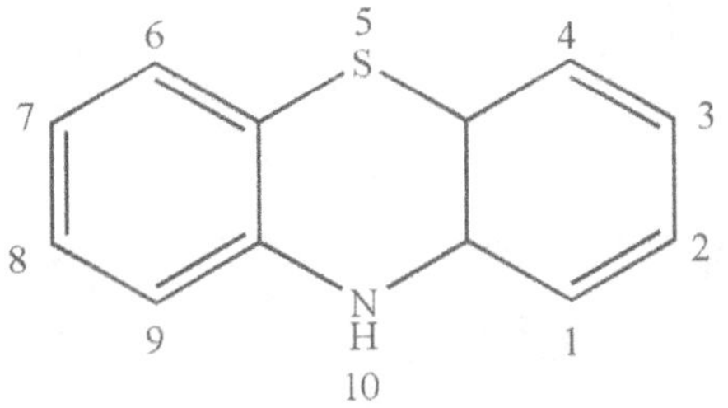

Phenothiazine

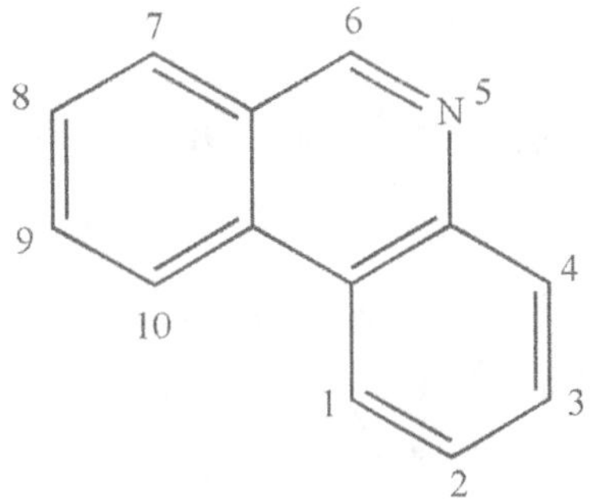

Phenanthridine

Classification of Heterocyclic Compounds

I. Three membered heterocyclic compounds with hetero atom.

(a) With nitrogen

Saturated Compound

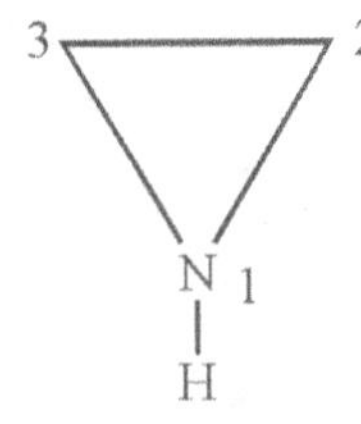

Aziridine

Unsaturated Compound

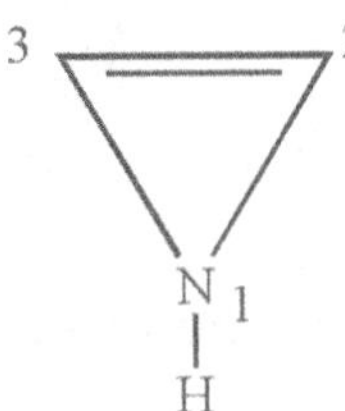

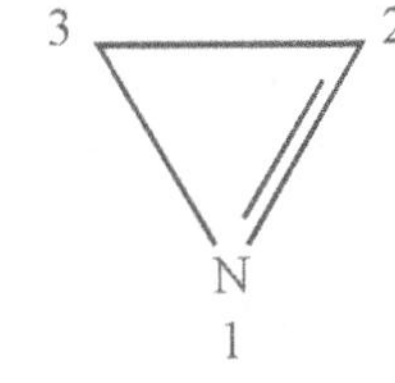

Azirine

(b) With oxygen

Saturated Compound

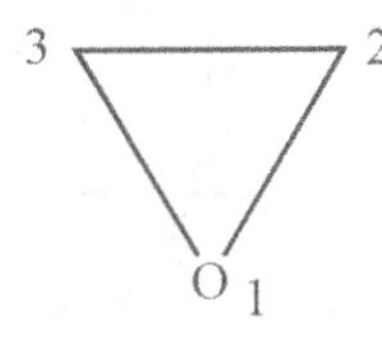

Oxirane

Unsaturated Compound

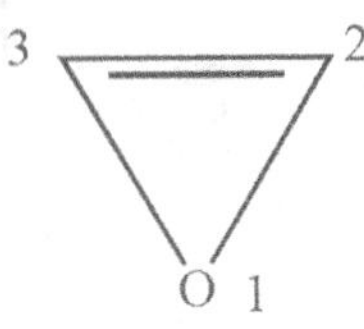

Oxirene

(c) With sulphur

Saturated Compound

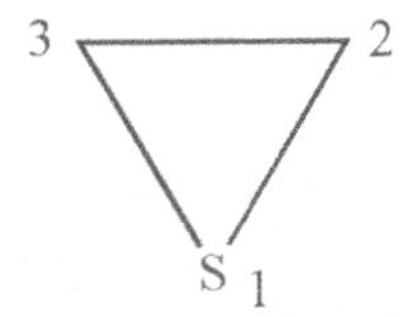

Thiirane

Unsaturated Compound

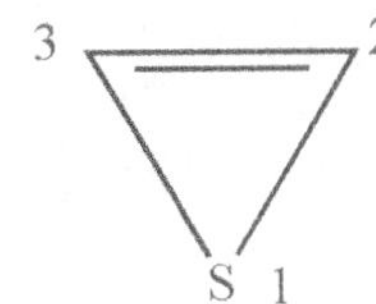

Thiirene

II. Three membered heterocyclic compounds with two hetero atoms.

(a) With two nitrogen atoms

Saturated compound

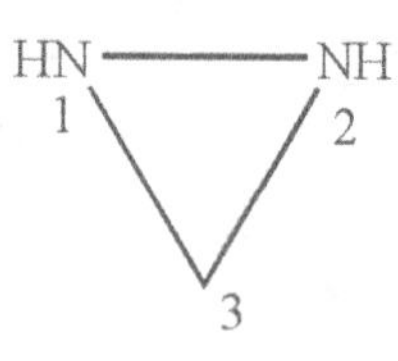

Diaziridine

Unsaturated compounds

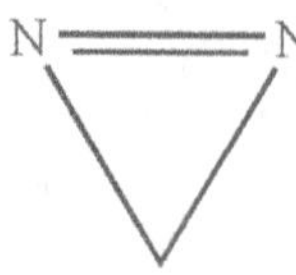

Diazirine

(b) With one nitrogen and one sulphur atoms:

Saturated compound

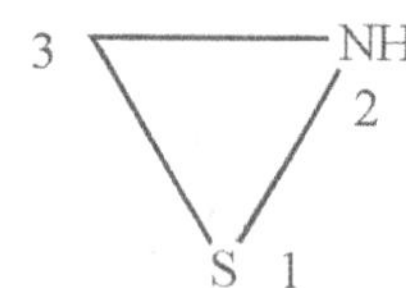

Thiaziridine

Unsaturated compound

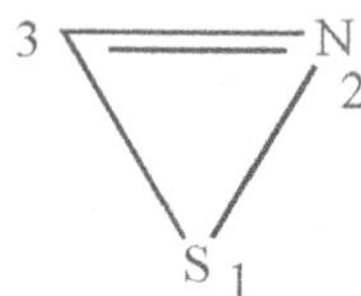

Thiazirene

(c) With one nitrogen and one oxygen atoms:

Saturated compound

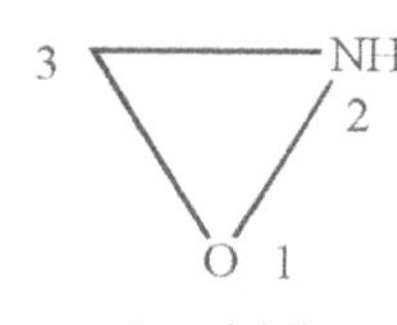

Oxaziridine

Unsaturated compound

Oxaziren
e

III. Four membered heterocyclic compounds with one hetero atom.

(a) With nitrogen

Saturated Compound

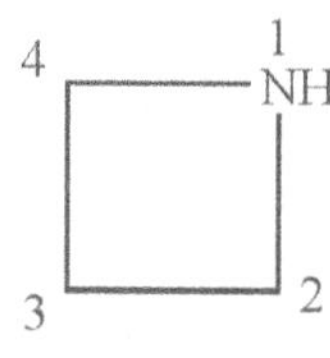

Azetidine

Partially Unsaturated Compounds

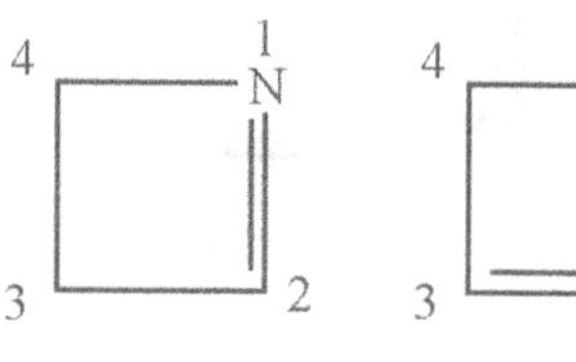

1-Azetine

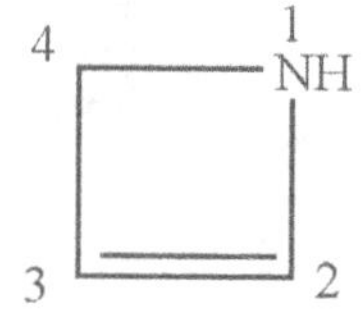

2-Azetine

Completely Unsaturated Compound

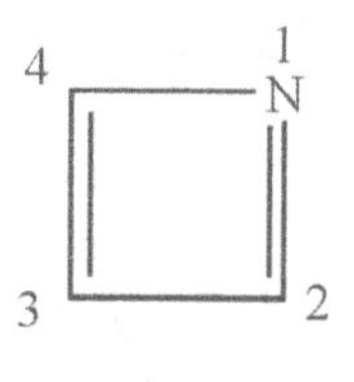

Azete
(Antiaromatic)

(b) With oxygen atom

Saturated Compound

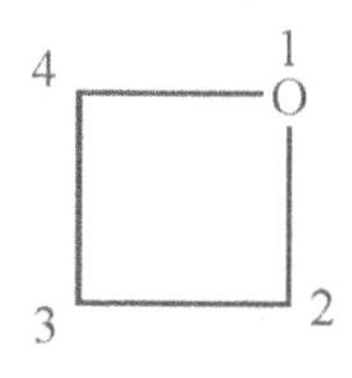

Oxetane

Unsaturated Compound

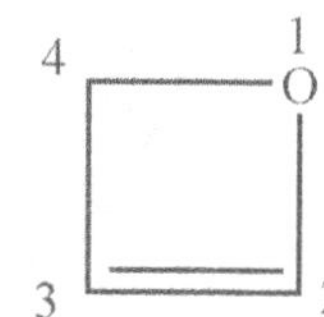

2-Oxetene

(c) With sulphur atom

Saturated Compound

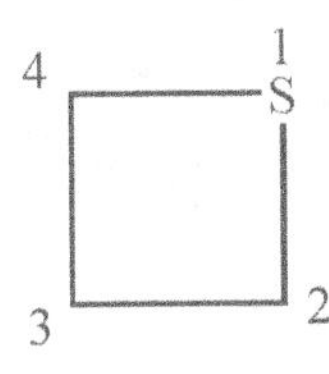

Thietane

Unsaturated Compound

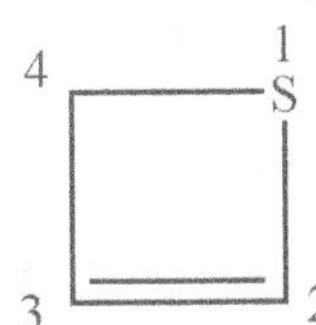

Thietene or Thiete

IV. Five membered heterocyclic compounds with one heteroatom.

(a) With nitrogen

Unsaturated Compound

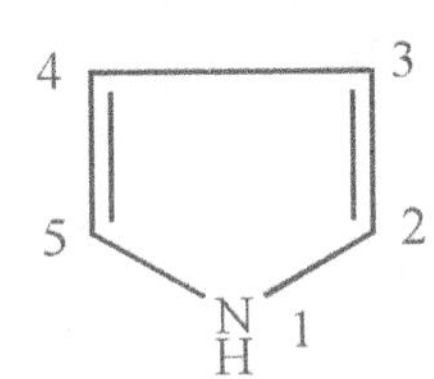

Pyrrole

Saturated Compound

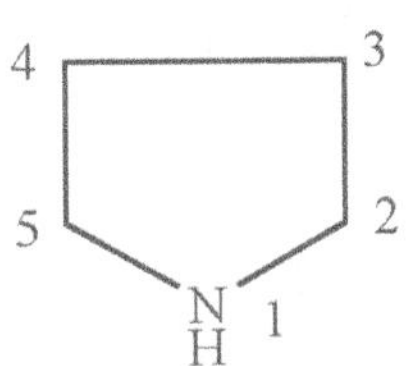

Pyrrolidine

Partially Saturated Compounds

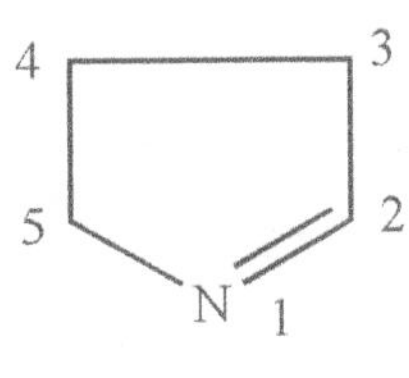

Δ^1-Pyrroline

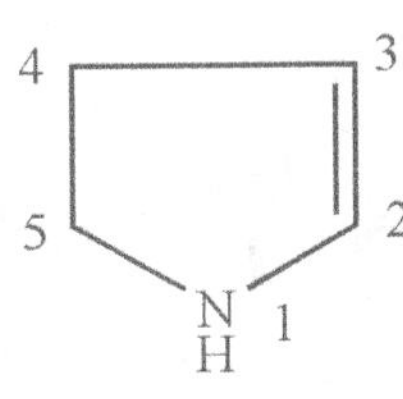

Δ^2-Pyrroline

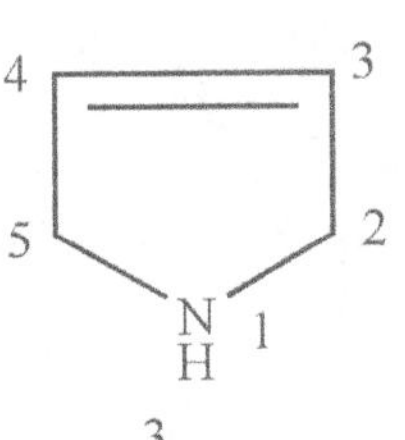

Δ^3-Pyrroline

(b) With oxygen atom

Unsaturated Compound

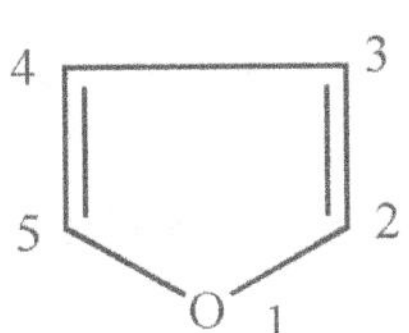

Furan

Saturated Compound

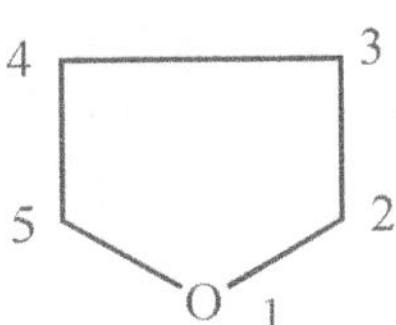

Tetrahydro furan

Partially Saturated Compounds

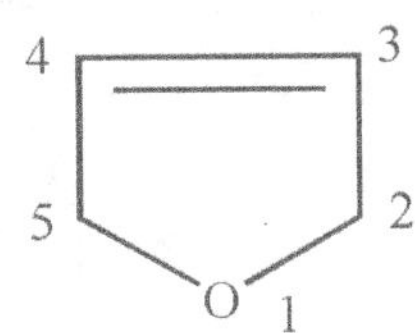

2,5-Dihydrofuran

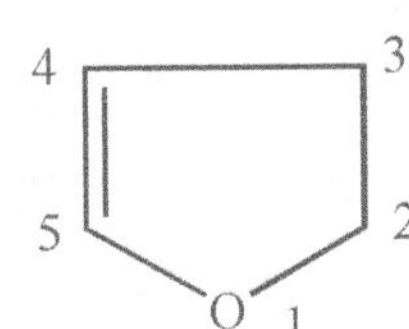

2,3-Dihydrofuran

(c) With sulphur atom

Unsaturated Compound

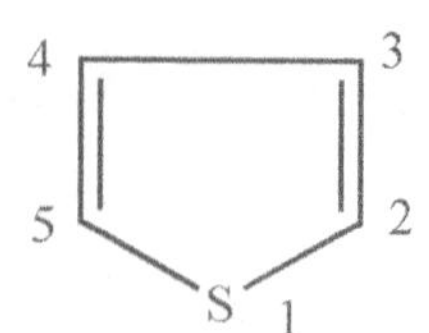

Thiophene

Fully Saturated Compound

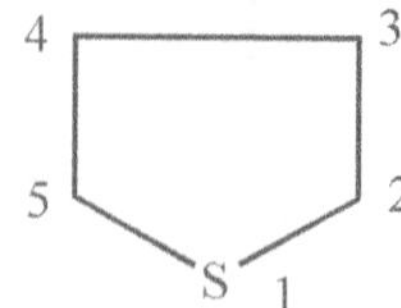

Tetrahydro thiophene (Thiolene)

Partially Saturated Compound

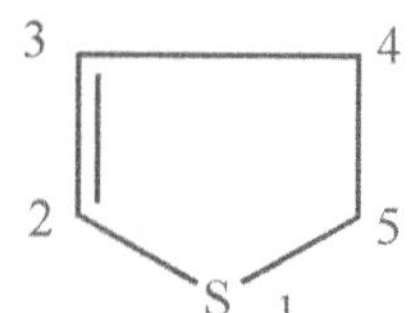

2-Thiolene
(2,3-Dihydrothiophene)

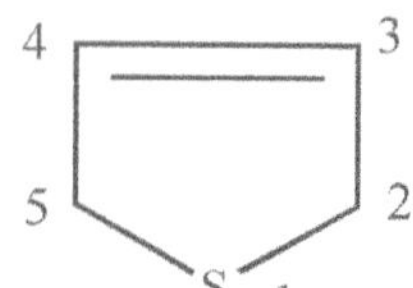

3-Thiolene
(3,4-Dihydrothiophene)

V. Five membered heterocyclic compounds with two hetero atoms.

(a) With two nitrogen atoms

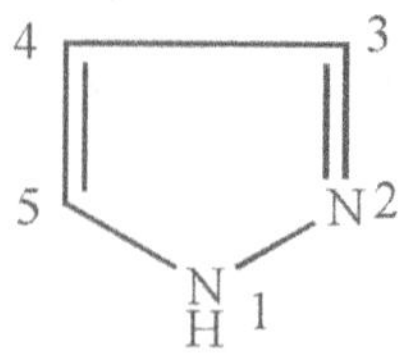

Pyrazole

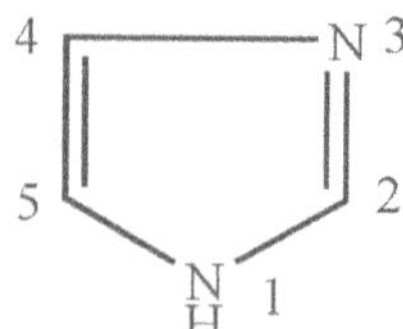

Imidazole

(b) With one nitrogen and one oxygen atom

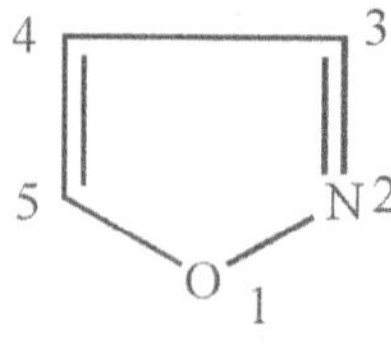

Isooxazole

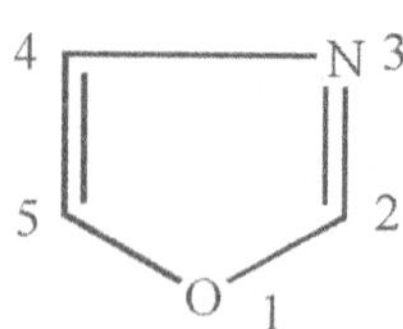

Oxazole

(c) With one nitrogen and one sulphur atom

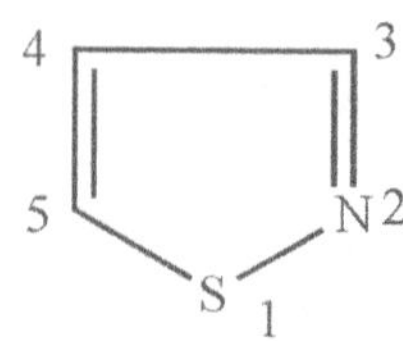

Isothiazole

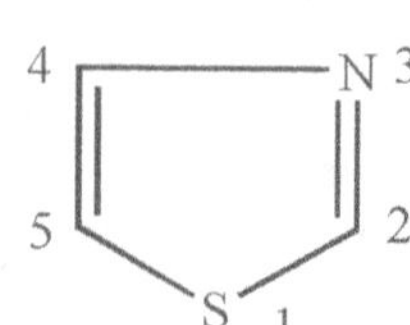

Thiazole

(d) With two oxygen atoms

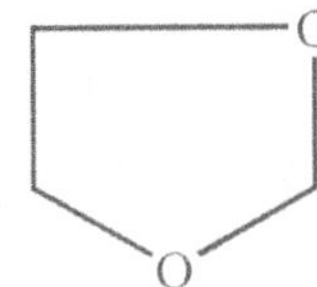

Dioxolane

VI. Six membered heterocyclic compounds with one hetero atom.

(a) With nitrogen atom

Unsaturated Compound

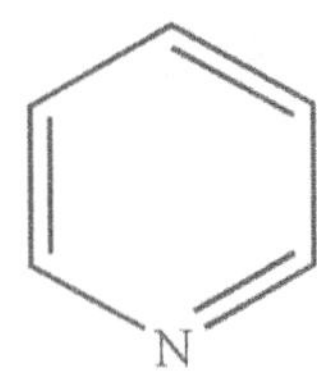

Pyridine

Fully Saturated Compound

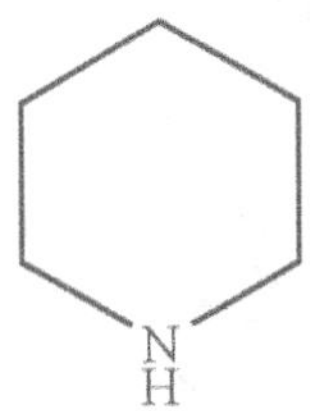

Piperidine

Partially Saturated Compounds

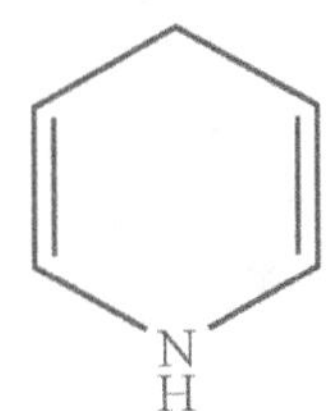

Dihydropyridine

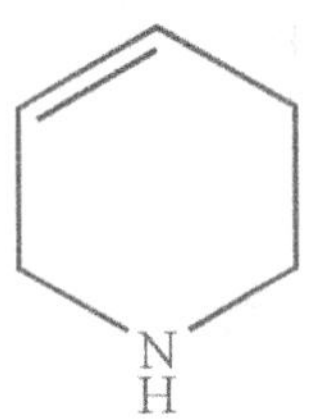

Tetrahydropyridine

(b) With oxygen

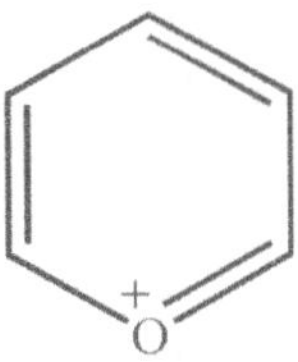

Pyrylium salt

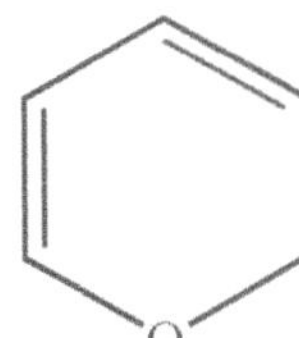

α-Pyrrones

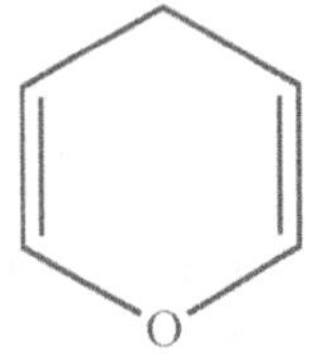

γ-Pyrrones

(Unstable and less aromatic)

(c) With sulphur

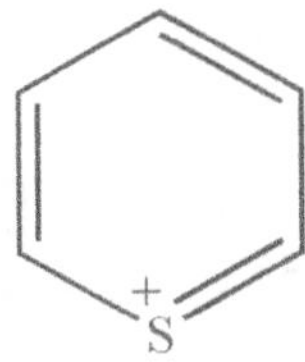

Thiopyrilium salt

Thiopyran

VII. Six membered heterocyclic compounds with two hetero atoms.

(a) With two nitrogen atoms

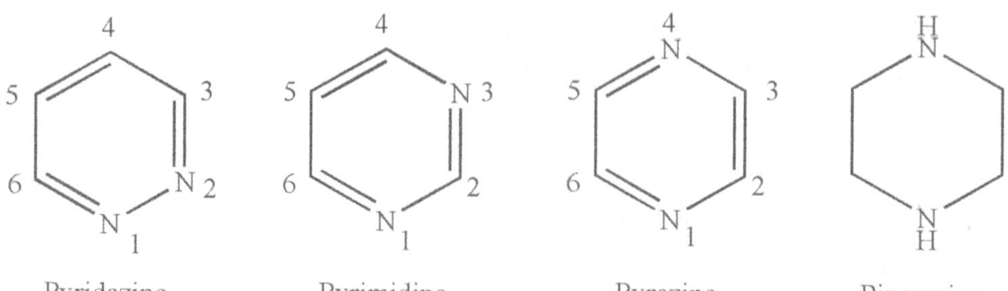

Pyridazine Pyrimidine Pyrazine Piperazine

(b) With oxygen and nitrogen atoms

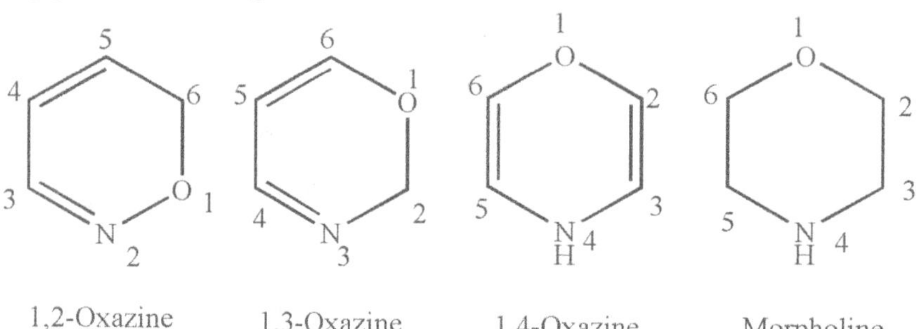

1,2-Oxazine 1,3-Oxazine 1,4-Oxazine Morpholine

(c) With nitrogen and sulphur atoms

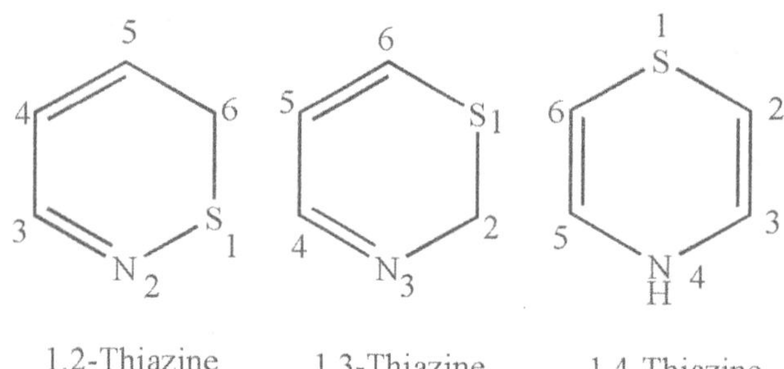

1,2-Thiazine 1,3-Thiazine 1,4-Thiazine

(d) With three nitrogen atoms

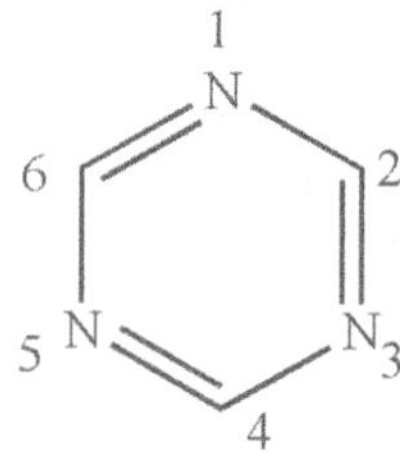

1,3,5-Triazine

Fused rings

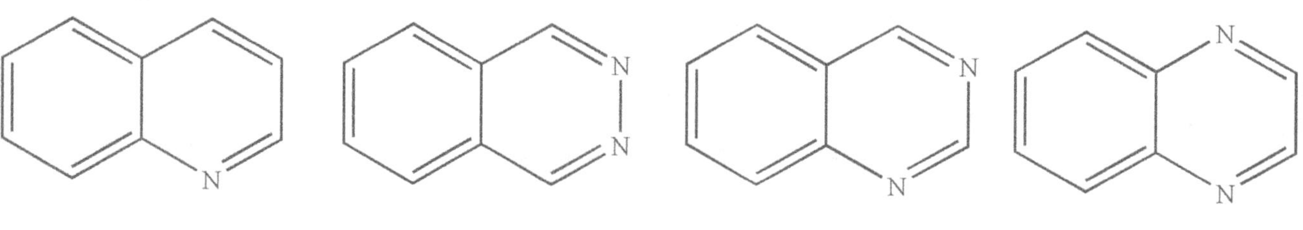

Quinoline Pthalazine Quinazoline Quinoxaline

VIII. Seven membered heterocyclic compounds.

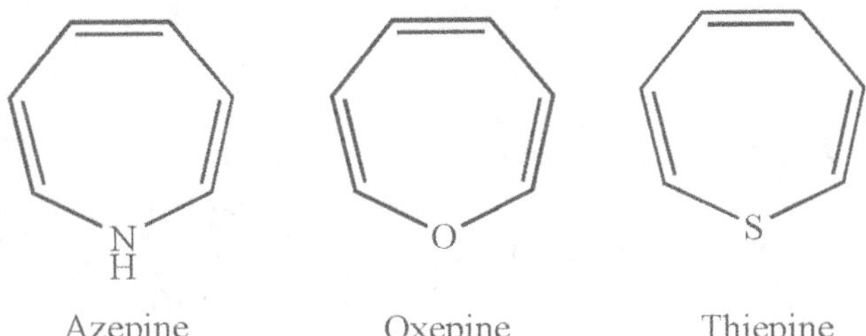

IX. Bicyclic ring systems from pyrrole, furan and thiophene (fused five membered ring system).

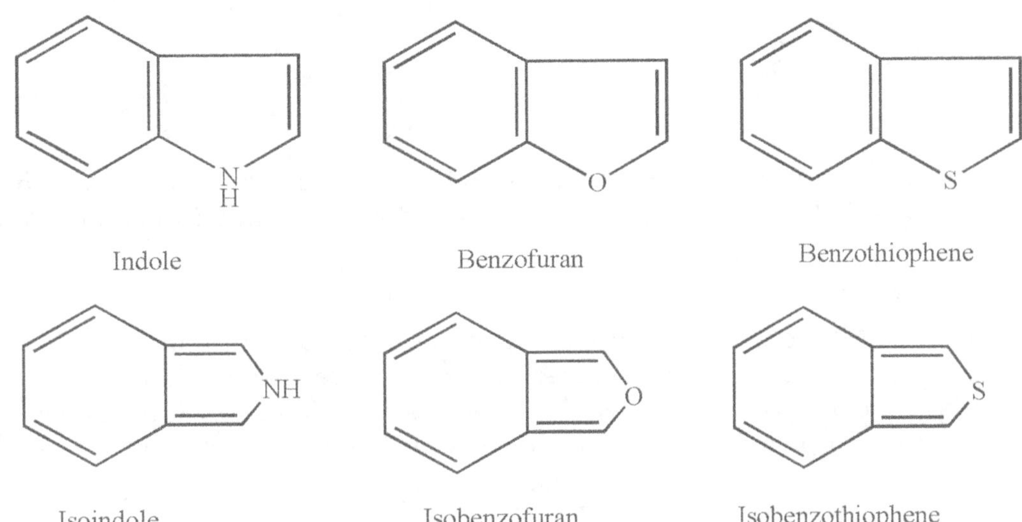

X. Tricyclic heterocyclic compounds.

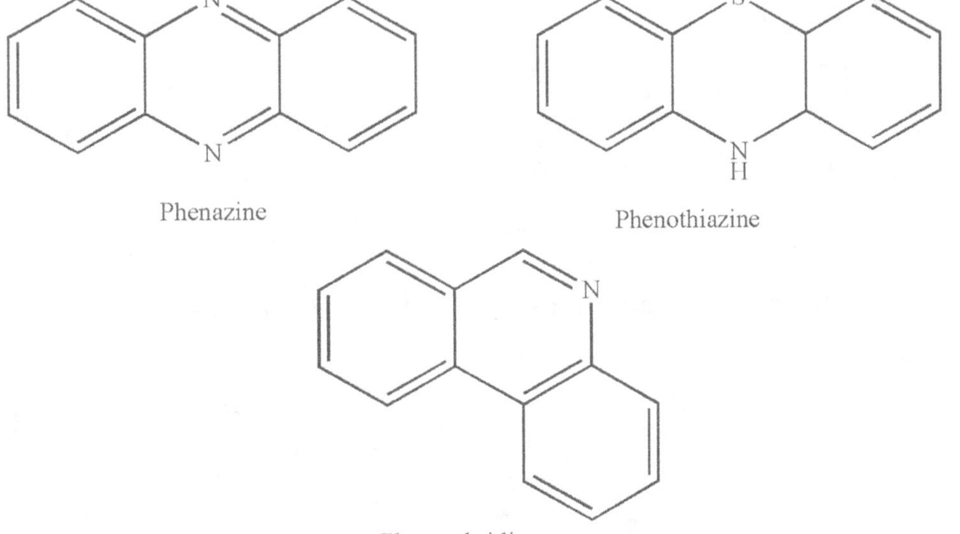

Heterocycles with Trivial names.

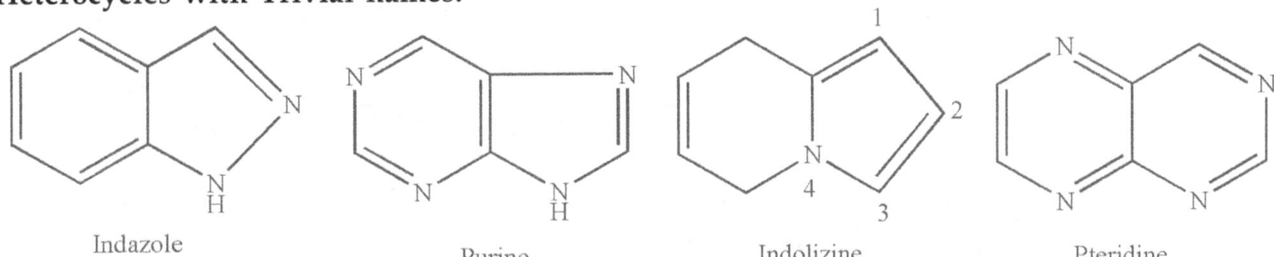

Pyrrole (Azacyclopenta-2,4-diene)

Introduction

It is a five membered heterocyclic compound having a nitrogen atom. Pyrrole is a very important one among the other five membered heterocycles because it occurs in various important natural products. It is widely present throughout the animal as well as plant kingdom. For example, it is present in haemoglobin as haem, chlorophyll (colouring matter of leaves) and vitamin B_{12}, bilirubin (coloring matter of biles) and alkaloids *etc.*

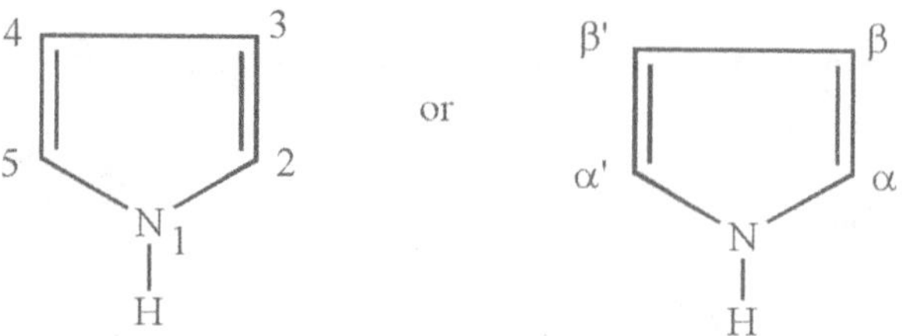

Isolation: Pyrrole mainly occurs in coal tar, bone oil and in products that are derived from proteins or protein derivatives. It was first isolated by Runge in 1834 and subsequently isolated by Anderson in 1857 from coal tar.

General methods of preparation:

1. ***Isolation from bone oil:*** Bone oil is first treated with dilute alkali to remove acidic substances and further washed with mild acid to remove the basic substances such as pyridine *etc,.* Then fractional distillation of bone oil, pyrrole distills over between the temperature of 370 and 420 K. This is fused with KOH and the solid potassio pyrrole obtained is further steam distilled to give pure pyrrole.

$$\underset{\text{Potassio pyrrole}}{C_4H_4NK} + H_2O \xrightarrow[\text{distillation}]{\text{Steam}} \underset{\text{Pyrrole (Pure)}}{C_4H_4NH} + KOH$$

2. ***From Furan:*** Furan is passed over ammonia, steam and heated at 400 °C in the presence of aluminium oxide catalyst to give pyrrole.

$$\text{Furan} \xrightarrow[400\ ^\circ C,\ \Delta]{NH_3,\ Al_2O_3} \text{Pyrrole} + H_2O$$

3. ***From Succinimide:*** Distillation of succinimide with zinc dust yields pyrrole.

Succinimide (Keto form) ⇌ (Enol form) $\xrightarrow[\text{Distillation} \\ -\text{Zno}]{\text{Zn dust}}$ Pyrrole (or)

(or)

$$\text{Succinimide} \xrightarrow[-\text{Zno}]{\text{Zn}} \text{Pyrrole}$$

4. ***From ammonium mucate:*** This method was discovered in 1860 by Schwartz. When the ammonium salt of mucic acid is heated, it is dissociated to give free acid, which upon further dehydration, decarboxylation and cyclization in the presence of ammonia yields pyrrole.

$$\begin{array}{c} (CHOH)_2COONH_4 \\ | \\ (CHOH)_2COONH_4 \end{array} \xrightarrow[-2NH_3]{\Delta} \begin{array}{c} (CHOH)_2COOH \\ | \\ (CHOH)_2COOH \end{array} \xrightarrow[-2H_2O,\ -2CO_2]{NH_3} \text{Pyrrole}$$

Ammonium mucate — Mucic acid — Pyrrole

5. ***From Acetylene:*** The mixture of acetylene and ammonia are passed through a red hot tube to yield pyrrole.

$$\underset{\text{Acetylene}}{CH{\equiv}CH} \quad \underset{\text{Acetylene}}{CH{\equiv}CH} + \underset{\text{Ammonia}}{H{-}\underset{H}{\underset{|}{N}}{-}H} \xrightarrow{\text{Red hot tube}} \text{Pyrrole}$$

6. ***Paal-Knorr synthesis:*** This is the general method for the synthesis of pyrrole. In this method, succinaldehyde is treated with ammonia to give pyrrole.

$$\underset{\text{Succinaldehyde}}{\begin{array}{cc} CH_2 & - CH_2 \\ | & | \\ CHO & CHO \end{array}} \xrightarrow{NH_3} \begin{array}{cc} CH_2 & - CH_2 \\ | & | \\ HO-\underset{H}{C}-NH_2 & CHO \end{array} \xrightarrow{-H_2O} \begin{array}{cc} CH & - CH_2 \\ \| & | \\ HC-\ddot{N}H_2 & CHO \end{array}$$

$$\text{Pyrrole} \xleftarrow{-H_2O} \begin{array}{ccc} CH & - & CH_2 \\ \| & & | \\ HC & - \underset{H}{N} - & C(OH)H \end{array}$$

This method is widely used for the preparation of substituted pyrroles.

Structure: Pyrrole is a planar molecule. The molecular orbital picture of pyrrole shows that the four carbon atoms are sp^2 hybridised. Each of the carbon atoms has one electron in p_z orbital and the nitrogen atom has two electrons in the *p*-orbital. These *p*-orbitals overlap leads to the formation of π electron clouds (one above and one below the plane of the ring) which contain six electrons in the π system (The aromatic sextet). The stability of the ring system is due to the delocalization of π electrons. It shows aromaticity because the 6 electrons satisfies the Huckel's rule.

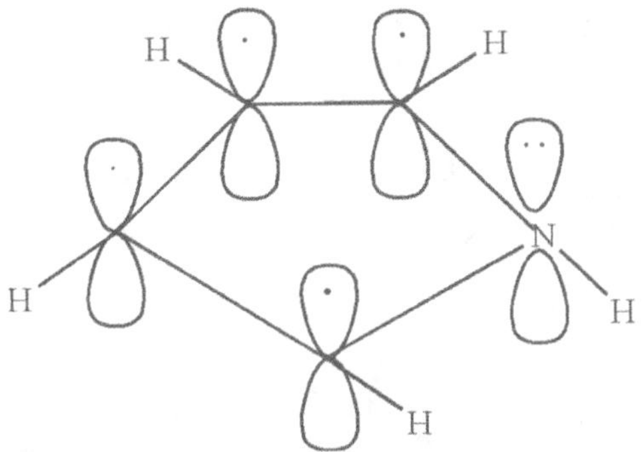

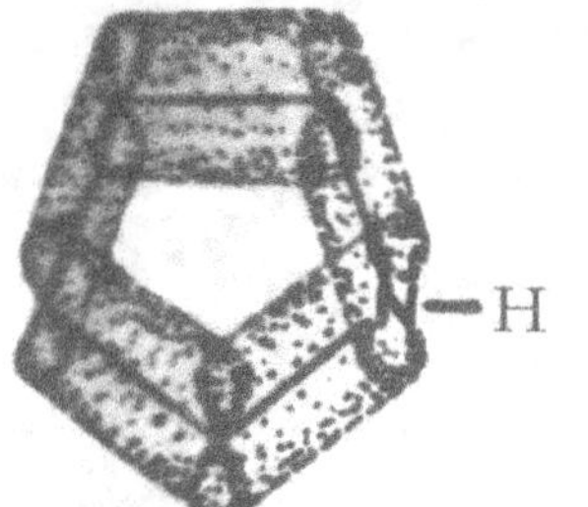

Nonbonding pair on nitrogen part of aromatic system

Figure 29.1 Orbital structure of pyrrole with electron clouds above and below the plane of the pyrrole ring.

Resonance in pyrrole: Pyrrole possess the following resonance structures.

The main resonance contributing structures are I, II and III.

Physical properties: Pyrrole is a colorless liquid, B.P. 404 K. It is sparingly soluble in water, completely soluble in alcohol and ether. On exposure to air, red colour turns to dark color and finally produces a resinous mass. It has odour similar to chloroform. The boiling point of pyrrole is higher than furan and thiophene because the formation of intermolecular hydrogen bond between the N-H group of one pyrrole molecule with the π-system of the another one.

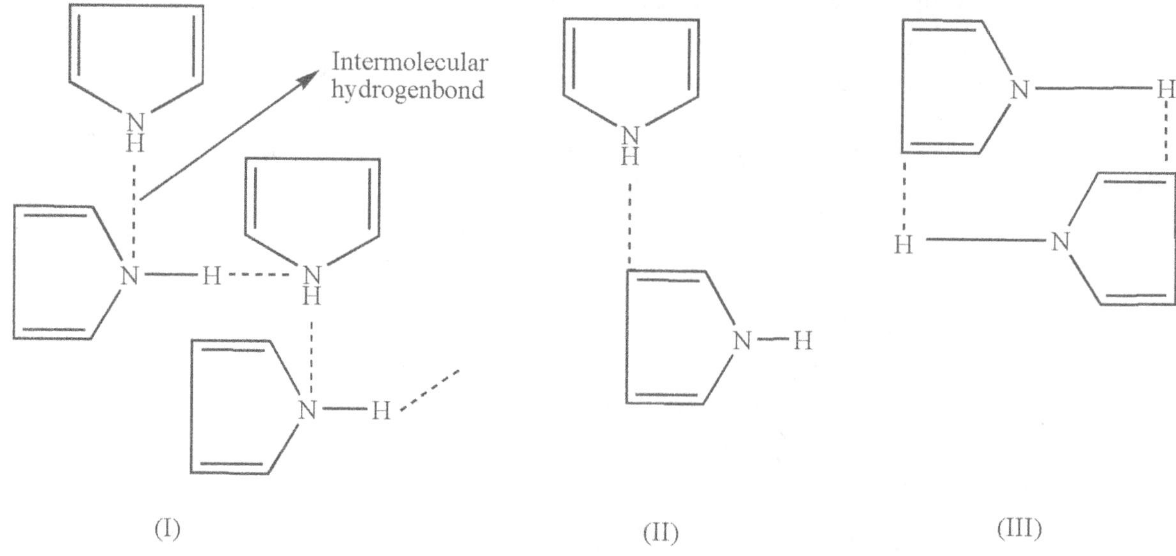

Pyrrole vapors turns a pine splint moistened HCl to red color.

Chemical properties: Pyrrole is less aromatic than thiophene but more aromatic than furan. It gives the following type of chemical reactions.

1. Electrophilic substitution reactions.
2. Condensation reactions.
3. Ring opening reactions.

Basic Nature of Pyrrole: Pyrrole is a weak base because the lone pair of electrons present in the "N" atom involving in the formation of (4n+2) π electron molecule, hence the lone pair of electrons are not readily available for protonation. However in acidic solution protonation takes place at 2nd and 3rd position, but in concentrated solution, pyrrole polymerises to yield pyrrole-red polymer.

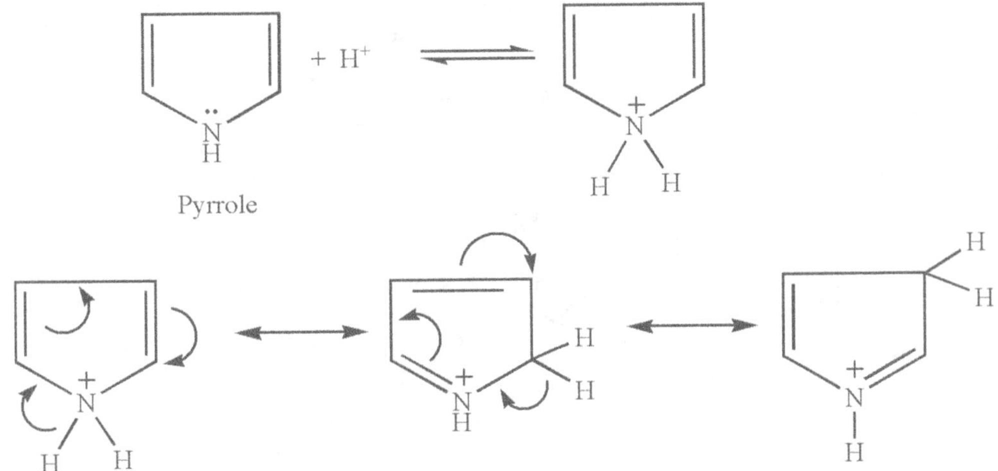

Acidity of Pyrrole: Pyrrole is weakly acidic in nature due to the presence of imino hydrogen atom. It is explained as follows.

(a) Higher s-character of N-H bond in pyrrole, the bonding electrons are highly held to the pyrryl nitrogen and hence hydrogen is eliminated as a proton.

(b) The pyrrole anion obtained is stabilized by resonance.

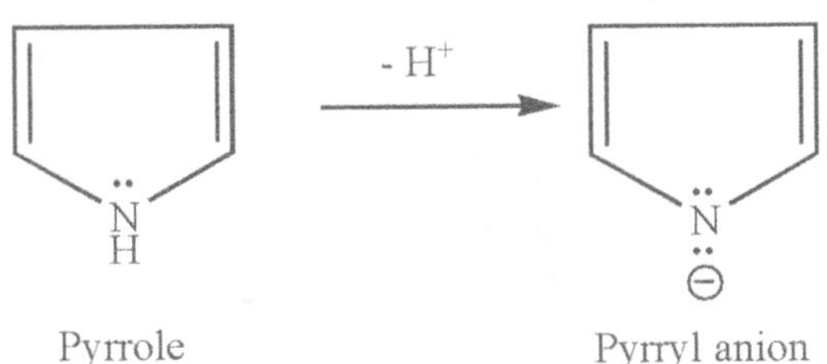

Hence, the imino-hydrogen of pyrrole is easily replaced by sodium, potassium and alkyl or aryl radicals. For example: When pyrrole is heated with solid potassium hydroxide to give potassiopyrrole.

Pyrrole + KOH $\xrightarrow{\Delta}$ Potassio pyrrole + H_2O

1. **Electrophilic substitution reactions:** Pyrrole is more reactive when compared with thiophene, benzene and furan. This high reactivity is due to the presence of excess π electron density of pyrrole *i.e.*, the π electron density of pyrrole is greater than benzene. Hence the electrophilic substitution of pyrrole takes place in a manner similar to that of benzenoid molecules.

   Pyrrole undergoes electrophilic substitution reactions and substitution chiefly at C-2 position. If the two C-2 positions are blocked, then substitution occurs at C-3 position. This is explained as follows.

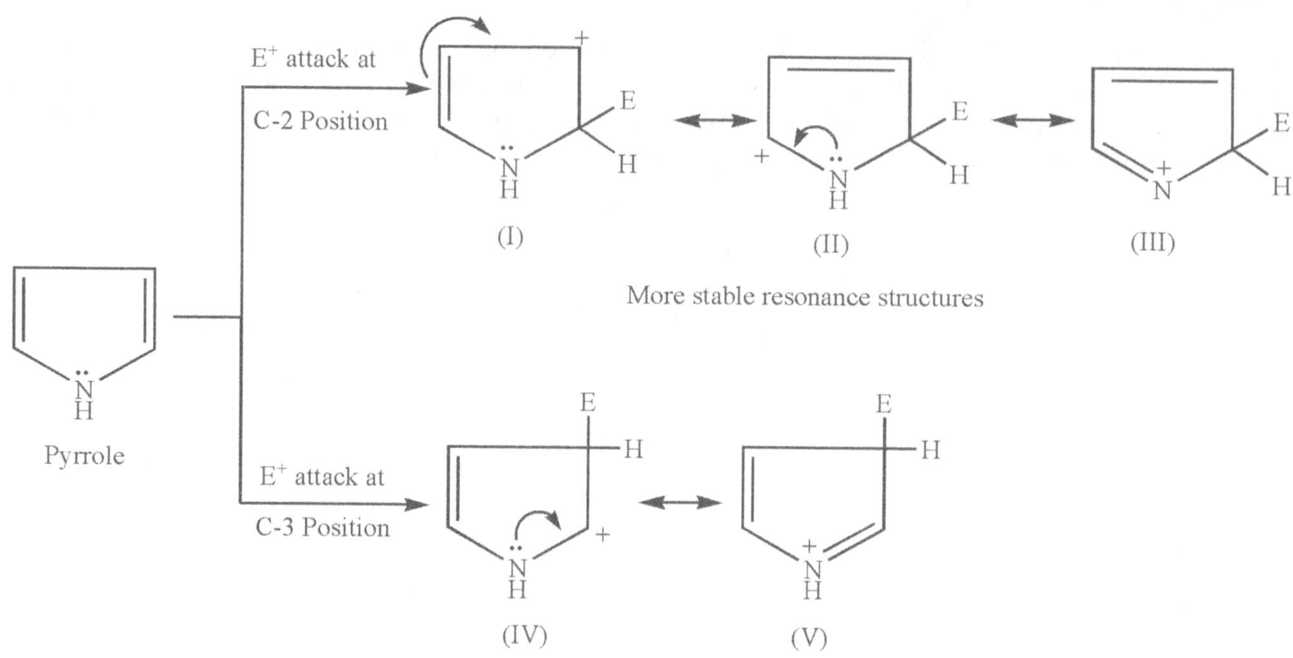

The first step of electrophilic substitution reaction is the formation of resonance stabilized carbocation intermediate. If the electrophile attacks the 2nd position, produces three resonance forms whereas there are only two resonance forms are obtained when attacks at C-3 position. The extra stabilization of (III) makes this carbocation more stable, hence C-2 substitution predominates.

(i) ***Halogenation:***

(a) ***Chlorination:*** Pyrrole vigorously reacts with SO_2Cl_2 in ether at 0 °C and gives tetrachloro pyrrole.

SO_2Cl_2 (4 moles), Ether, 0 °C, – 4 HCl, – 4 SO_2

Pyrrole → Tetrachloropyrrole

(b) ***Bromination:*** Pyrrole reacts with Br_2/CH_3COOH and gives tetrabromopyrrole.

Br_2/CH_3COOH

Pyrrole → Tetrabromopyrrole

(c) ***Iodination:*** Pyrrole reacts with I_2 /KI_3 (I_3/KI_3) and gives tetraiodopyrrole (Iodole) which is used as a substitute for iodoform.

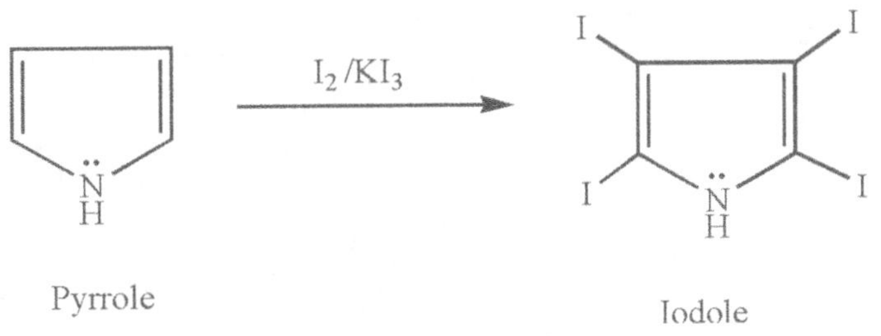

I_2/KI_3

Pyrrole

Iodole

The halopyrroles obtained are highly unstable and easily decomposes in light and air.

(ii) ***Nitration:*** Pyrrole reacts with nitric acid in the presence of acetic anhydride at 260 K and yields 2-nitro pyrrole.

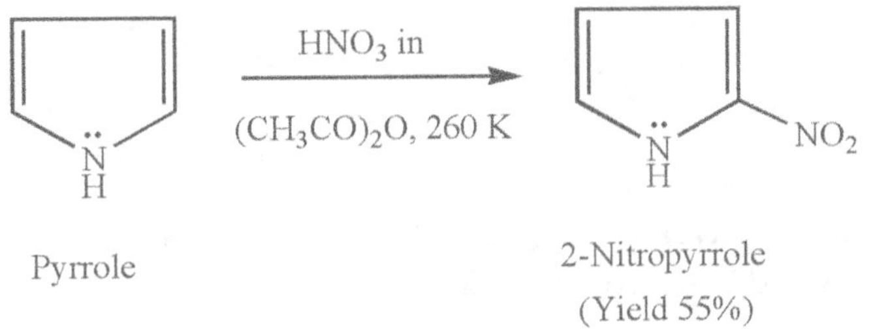

HNO_3 in $(CH_3CO)_2O$, 260 K

NO_2

Pyrrole

2-Nitropyrrole (Yield 55%)

Note:

Nitration is also carried out with general nitrating mixture (Conc. HNO_3 + Conc. H_2SO_4), but it leads to extensive decomposition of pyrrole ring and forms tar.

(iii) ***Sulphonation:*** Pyrrole reacts with H_2SO_4 at room temperature and gives resinous material but in the presence of a mild sulphonating agent such as pyridine-sulphur trioxide it yields 2-pyrrolesulphonic acid.

Pyridine-SO_3 in $C_2H_4Cl_2$

SO_3H

Pyrrole

2-Pyrrole sulphonic acid

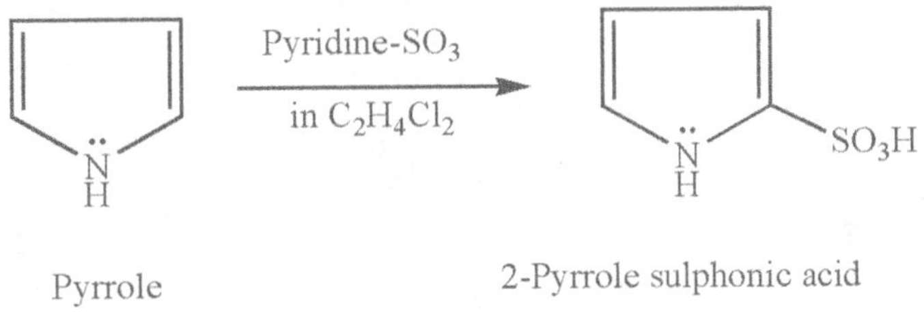

(iv) ***Friedel-Craft's Acylation:*** Pyrrole reacts with acetic anhydride in the presence of an acid scavenger such as triethylamine and gives 2-acetylpyrrole. It also reacts with acid chlorides in the absence of catalyst.

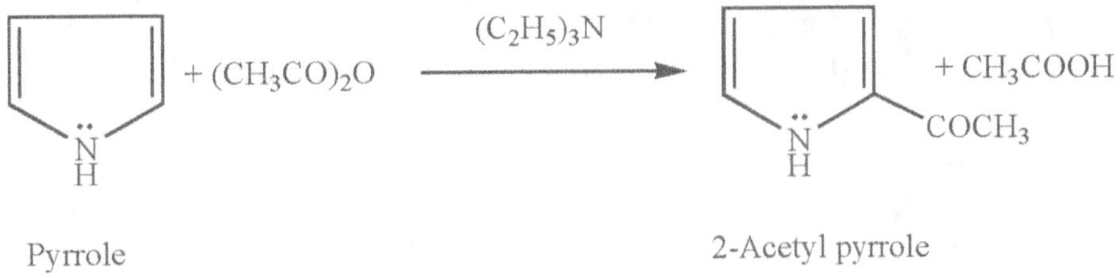

+ $(CH_3CO)_2O$ $\xrightarrow{(C_2H_5)_3N}$ + CH_3COOH

$COCH_3$

Pyrrole

2-Acetyl pyrrole

(v) ***Gatterman-Koche Reaction or Formylation:*** When pyrrole is heated with phosphorous oxychloride and dimethyl formamide it gives 2-pyrrolecarbaldehyde.

Pyrrole + Dimethyl formamide $\xrightarrow{POCl_3}$ 2-pyrrolyl–$\overset{+}{CH}=N(CH_3)_2$ Cl^- $\xrightarrow[\text{Vilsmeier-Haack reaction}]{CH_3COONa,\ H_2O}$ 2-Pyrrolecarbaldehyde + $(CH_3)_2\overset{+}{N}H_2$

(vi) ***Reaction with chloroform (Reimer-Tiemann Reaction):*** Pyrrole reacts with chloroform in the presence of NaOH to yield 2-pyrrolecarbaldehyde.

Pyrrole $+ CHCl_3 + NaOH \longrightarrow$ 2-Pyrrolecarbaldehyde $+ NaCl + H_2O + Cl_2$

(vii) ***Kolbe-Schmidt Reaction:*** Potassium salt of pyrrole reacts with CO_2 to give 2- and 3- pyrrole carboxylic acid.

Potassio pyrrole $+ CO_2 \longrightarrow$ 2-Pyrrole carboxylic acid + 3-Pyrrole carboxylic acid

(viii) ***Coupling Reaction:*** Like phenol, pyrrole undergoes coupling reaction with diazonium salt in acidic/basic condition. In acidic solution coupling occurs at C-2 position and in basic solution it occurs in C-2 and C-5 position to yield bisazo compound.

Pyrrole $+ 2C_6H_5N_2Cl \xrightarrow[\text{Condition } -2HCl]{\text{Alkaline}}$ C_6H_5–N=N–(pyrrole)–N=N–C_6H_5

Bisazo-Compound

Pyrrole $+ C_6H_5N_2Cl \xrightarrow[\text{Condition}]{\text{Acidic}}$ 2-(Diazo benzene) pyrrole

(ix) ***Ring expansion:*** Pyrrole reacts with sodium methoxide and methylene iodide and yields pyridine.

Pyrrole $+ 2CH_3ONa + CH_2I_2 \longrightarrow$ Pyridine $+ 2NaI + 2CH_3OH$

(x) ***Houben-Hoesch Synthesis:*** Pyrrole reacts with methyl cyanide in the presence of $ZnCl_2$ and gives 2-acetyl pyrrole.

Pyrrole $+ CH_3CN \xrightarrow{ZnCl_2}$ 2-$C(CH_3)=NH$ pyrrole $\xrightarrow[-NH_3]{H_2O}$ 2-Acetyl pyrrole ($COCH_3$)

(xi) ***Reaction with Grignard Reagent:*** Pyrrole reacts with methyl magnesium iodide and gives *N*-pyrrole magnesium iodide, which further reacts with CH_3I, CO_2 & dil.HCl and gives 2-substituted products.

Pyrrole $+ CH_3MgI \xrightarrow[-CH_4]{ZnCl_2}$ N-MgI pyrrole $\rightleftharpoons$ ($-MgI^+$) pyrrole anion ($N^{\ominus}$)

Pyrrole anion $\xrightarrow{CH_3I}$ 2-Methyl Pyrrole (CH_3)

Pyrrole anion $\xrightarrow{CO_2,\ \text{dil.HCl}}$ 2-Pyrrole carboxylic acid ($COOH$)

2. **Reduction:** Pyrrole is reduced with Pt, Pd and Raney nickel at 470 K to yield pyrrolidine. It is also reduced with zinc/CH_3COOH and gives pyrroline.

Zn / CH_3COOH ← Pyrrole → Pt/Ni, 470 K

Pyrroline (2,5-Dihydropyrrole) — Pyrrole — Pyrrolidine (Tetrahydropyrrole)

3. **Oxidation:** Pyrrole gets oxidized when exposed to air and light to yield reddish brown color. When oxidized with ozone or CrO_3/H_2SO_4 it gives maleimide.

Pyrrole —CrO_3/H_2SO_4→ Maleimide

4. **Hofmann Exhaustive Methylation (Ring opening reaction):**

Pyrrole —Pt/H_2→ pyrrolidine —$2CH_3I$, -HI→ $[N,N\text{-dimethylpyrrolidinium}]^+ I^-$ —AgOH, -AgI→ $[N,N\text{-dimethylpyrrolidinium}]^+ OH^-$ —Δ, 180-200 °C ($-H_2O$)→ $H_2C{=}CH{-}CH_2{-}CH_2{-}N(CH_3)_2$ —i. CH_3I, ii. AgOH, iii. Δ→ $CH_2{=}CH{-}CH{=}CH_2$

1,3-butadiene

Pyrrole Derivatives

Pyrrole-2-aldehyde

(Structure: pyrrole ring with CHO at position 2)

Preparation:

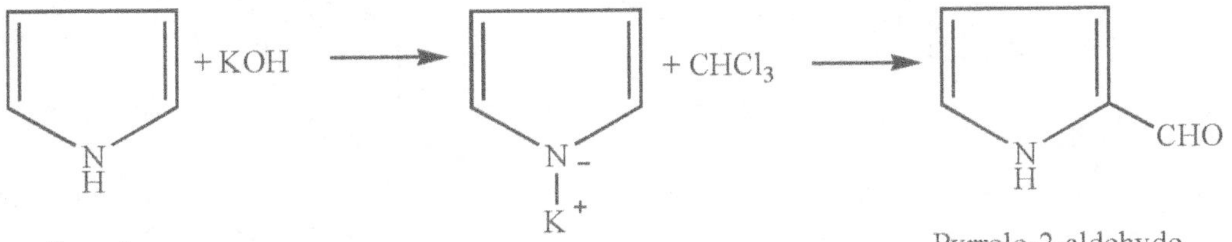

Pyrrole + KOH ⟶ (pyrrole N⁻ K⁺) + $CHCl_3$ ⟶ Pyrrole-2-aldehyde

Pyrrole carboxylic acid

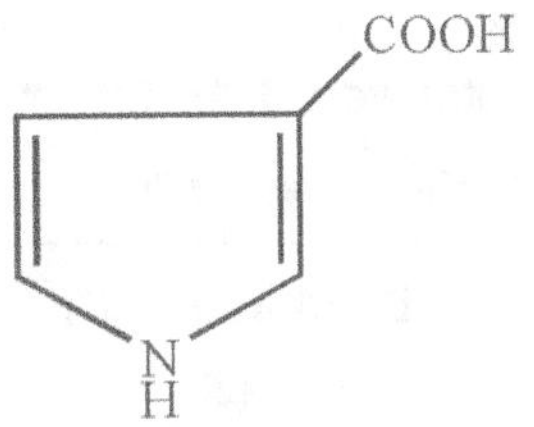

Medicinal Compounds Containing Pyrrole (Or) Pyrrole Derivatives Used in Medicine

Ethosuximide 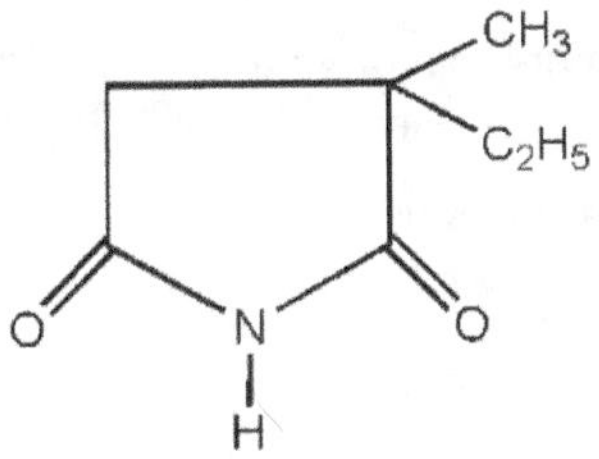 3-Ehyl-3-methylpyrrolidine-2,5-dione **Use:** It is used as an anti convulsant	**Xilobam** 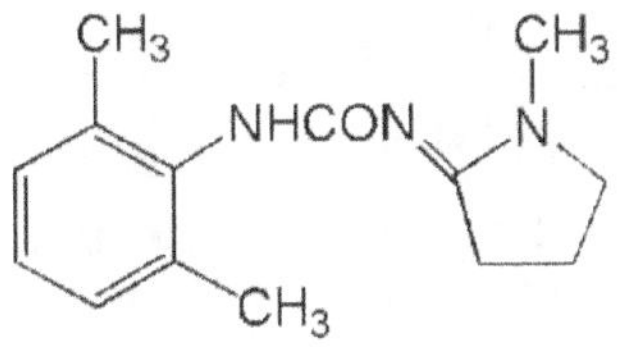 1-(2,6-Dimethylphenyl)-3-(1-methylpyrrolidin-2-ylidene)urea **Use:** It is used skeletal muscle relaxant
Pindalol 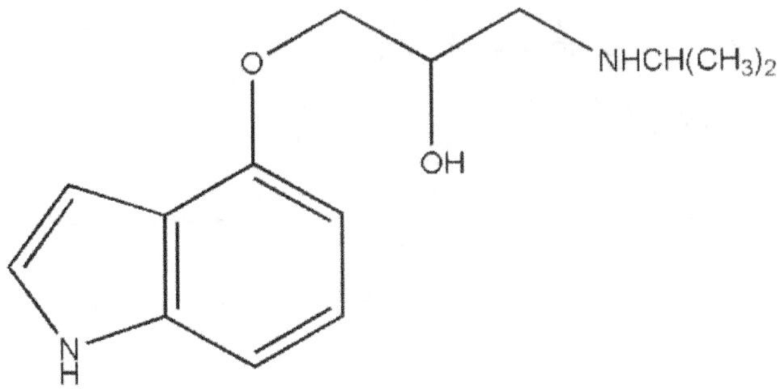 1-(1*H*-Indol-4-yloxy)-3-(propan-2-ylamino)propan-2-ol **Use:** It is used as an antihistaminic agent	**Clemastine** 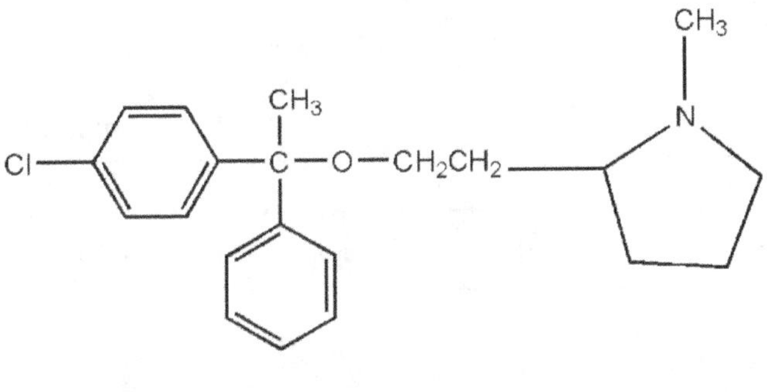 (2R)-2-{2-[(1R)-1-(4-Chlorophenyl)-1-phenylethoxy]ethyl}-1-methylpyrrolidine **Use:** It is used as an antihistaminic agent

Table *Contd...*

Gliclazide

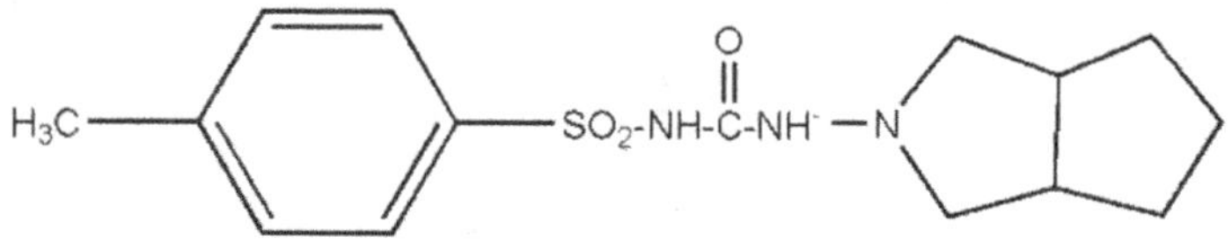

1-(3,3a,4,5,6,6a-hexahydro-1*H*-cyclopenta[c]pyrrol-2-yl)-3-(4-methylphenyl)sulfonylurea

Use: It is used as an oral hypoglycaemic drug

Malindone

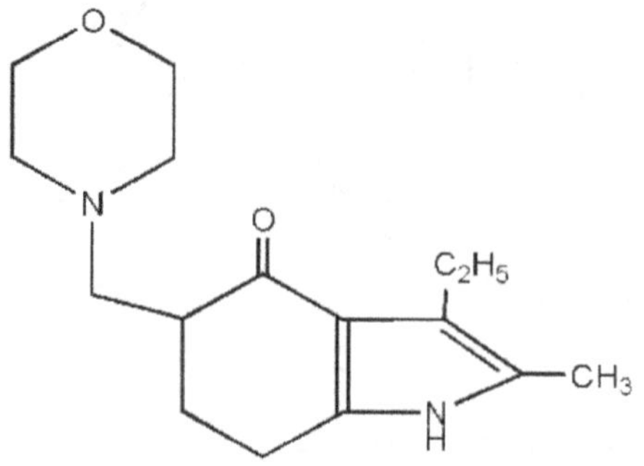

3-Ethyl-2-methyl-5-(morpholin-4-ylmethyl)-1,5,6,7-tetrahydro-4H-indol-4-one

Use: It is used as tranquillizer

Physostigmine (Eserine)

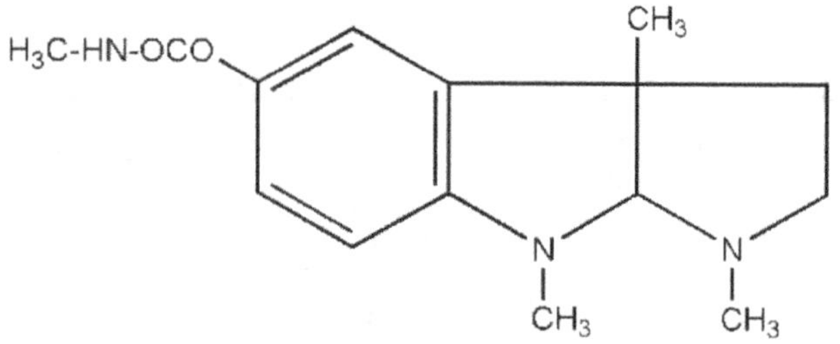

3,4,8b-Trimethyl-2,3a-dihydro-1*H*-pyrrolo[2,3-b]indol-7-yl] N-methylcarbamate

Use: It is used as cholinergic drug

Doxapram. HCl

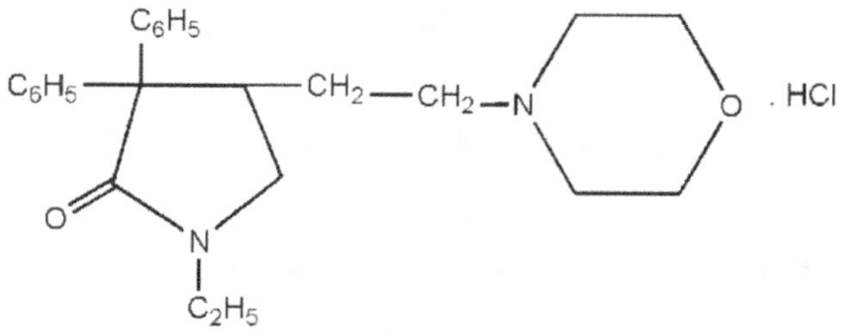

1-Ethyl-4-(2-morpholin-4-ylethyl)-3,3-diphenylpyrrolidin-2-one; hydrochloride

Use: It is used as CNS stimulant

Tolmetin sodium (Tolectin)

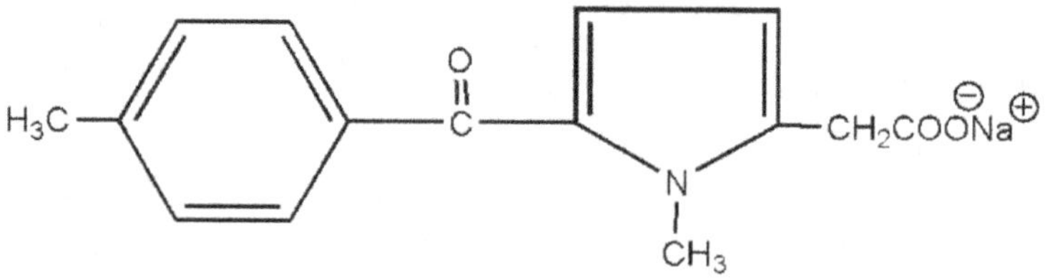

Sodium 2-(1-methyl-5-(4-methylbenzoyl)-1*H*-pyrrol-2-yl)acetate

Use: It is used as a NSAID

Furan (Furfuran) (Oxacyclopenta-2,4-diene)

Introduction

Furan is a five membered organic heterocyclic compound containing four carbon atoms and one oxygen atom. The first derivative of furan is 2-furoic acid which was reported by Carl Wilhem Scheele. In 1870, the chemist Heinrich limpricht first synthesized furan and he had given the name tetraphenol. The name furan derived from the latin word furfur (Latin furfur means bran) because the aldehyde of furan i.e., furfural is prepared by distilling bran with HCl. It occurs mainly in secondary plant metabolites especially in terpenoids.

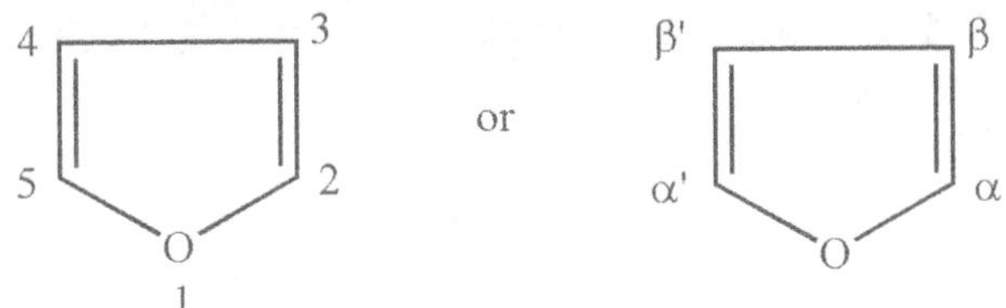

The positions of the substituents are indicated by numbers or Greek letters. The numbering starts from the oxygen atom which is assigned to position 1.

General Methods of Preparation:

1. ***By isolation:*** Furan is mainly isolated by distillation of pine wood.
2. ***From Furfural:*** Oxidation of furfural gives furoic acid which upon decarboxylation gives furan.

Furfural (CHO) $\xrightarrow{(O)}$ Furoic acid (COOH) $\xrightarrow[\Delta]{-CO_2}$ Furan

3. ***From Mucic acid:*** Furan is obtained by the dry distillation of mucic acid which the further decarboxylation yields furan.

Mucic acid (COOH–$(CHOH)_4$–COOH) $\xrightarrow[-CO_2,\ -3H_2O]{\text{Dry distillation}}$ Furoic acid (COOH) $\xrightarrow[\Delta]{-CO_2}$ Furan

4 ***From carbohydrates:*** The major source of furan is furfural. The furfural is obtained from polysaccharides present in corn cobs, oat and husks by acid hydrolysis. The furfural obtained undergoes oxidation and yields 2-furoic acid which upon decarboxylation yields furan.

Structure: The four carbon atoms and one oxygen atom in the furan molecule is sp^2 hybridized. All the four sp^2 orbitals overlaps with each other and overlaps with *s* orbitals of hydrogen atoms forms the C-C, C-H and C-O bonds which are lie in the same plane. In this stage each carbon atom has one unused *p* orbital with one π electron and the oxygen atom has one lone pair of electrons in a single *p* orbital. These all six electrons overlap with each other to form π-molecular orbital. Hence, furan is a planar molecule.

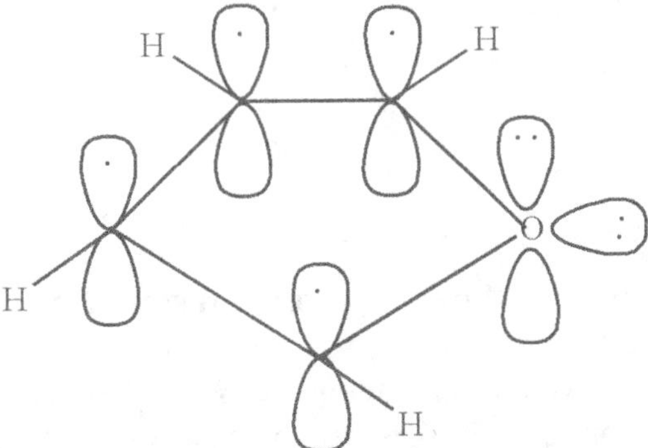

Resonance in Furan: The aromaticity of molecule depends upon the availability of lone pair of electrons for resonance which in turn depends on the electro negativity of heteroatom. More electronegative atoms will have a greater hold on the lone pair which will therefore, be localized. Hence furan is less aromatic than pyrrole or thiophene. The following are the resonance structures of furan, the lone pair of electrons are donated to the ring atoms.

Physical properties: It is a colorless liquid, B.P. is 305 K, insoluble in water but soluble in organic solvents. It turns a pine split moistened with HCl to green color.

Chemical properties: The chemical properties of furan are similar to that of typical diene. Some of the chemical reactions of furan are as follows.

1. **Electrophilic Substitution Reactions:** Furan undergoes electrophilic substitution like benzene due to the presence of excess π-electron density of the ring carbons in furan.

   Furan is less aromatic than thiophene and pyrrole. The C-2 position of the furan is preferentially reactive towards electrophilic attack than C-3 position. It is explained as follows.

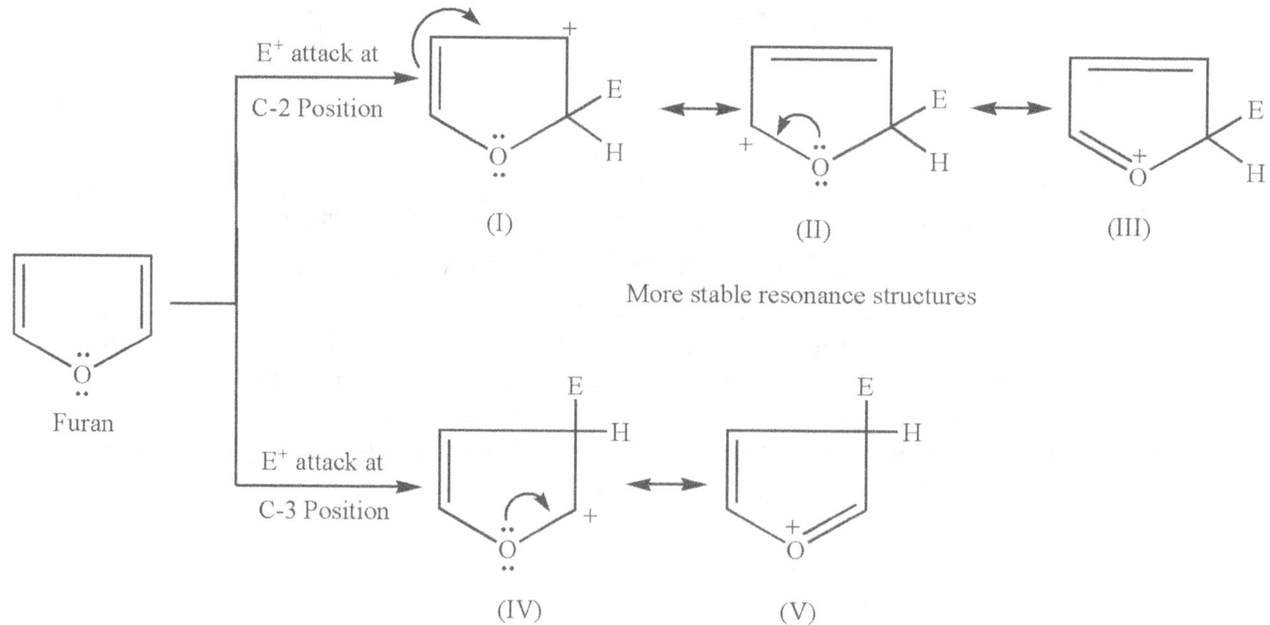

Less stable resonance structures

The first step of electrophilic substitution reaction is the formation of resonance stabilized carbocation intermediate. If the electrophile attacks the 2nd position, it produces three resonance forms whereas there are only two resonance forms are obtained when it attacks at C-3 position. The extra stabilization of (III) makes this carbocation more stable, hence C-2 substitution predominates.

(i) ***Halogenation:*** Furan reacts with halogens readily and gives a mixture of mono and poly substituted products or resins. Hence, mono halogen derivatives are obtained indirectly. For example, bromination of furoic acid gives 5-bromo furoic acid, which upon decarboxylation yields 5-bromo furan.

Furoic acid $\xrightarrow{Br_2}$ 5-Bromo furoic acid $\xrightarrow[-CO_2]{\Delta}$ 5-Bromo furan

Gradual reaction between furan and less amount of Cl_2 in dichloromethane at 40 °C gives a mixture of 2-chloro, 2,5-dichloro and 2,3,5-trichloro furan, while increasing the chlorine concentration yields tetrachloro furan.

(ii) ***Nitration:*** Furan gets nitrated with mild nitrating agents such as acetyl nitrate (a mixture of acetic anhydride and nitric acid) at -5 °C to -30 °C to give 2- nitro furan.

Furan $\xrightarrow[(CH_3CO)_2O,\ -5\text{ to }-30\ ^\circ C]{HNO_3}$ 2-Nitro furan

(iii) ***Sulphonation:*** Furan gets sulphonated with pyridine–SO_3 complex to yield 2-furan sulphonic acid (90 % yield). It reacts with excess amount of pyridine–SO_3 complex and gives furan 2,5-disulphonic acid (80 % yield).

2-Furansulphonic acid $\xleftarrow{Pyridine\text{-}SO_3}$ Furan $\xrightarrow[\text{Pyridine-}SO_3\text{ Complex}]{\text{3 fold excess}}$ Furan-2,5-disulphonic acid

(iv) ***Friedel-Crafts Alkylation:*** Furan does not undergo Friedel-Craft's alkylation because of polymerization and poly alkylation due to catalyst.

(v) ***Friedel-Crafts Acylation:*** Furan reacts with acetic anhydride in the presence of mild Lewis acid catalyst such as $SnCl_4$ or $ZnCl_2$ to give 2-acetyl furan.

Furan $+ (CH_3CO)_2O \xrightarrow{SnCl_4/ZnCl_2}$ 2-Acetyl furan ($COCH_3$) $+ CH_3COOH$

(vi) ***Gatterman – Koche Reaction:*** Furan reacts with a mixture of hydrogen cyanide and hydrogen chloride in the presence of aluminium chloride and gives 2-furan carbaldehyde.

$$HCN + HCl \longrightarrow NH{=}CHCl$$

Furan $+ NH{=}CHCl \longrightarrow$ 2-furyl–$CH{=}\overset{+}{N}H_2Cl^-$ $\xrightarrow{H_2O}$ 2-Furancarbaldehyde (CHO)

(vii) ***Mercurification:*** Furan is heated with mercuric chloride and aqueous sodium acetate to give 2-chloro mercurifuran. The mercurifuran further reacts with halides and acyl halides and gets easily replaced with halides and acyl group.

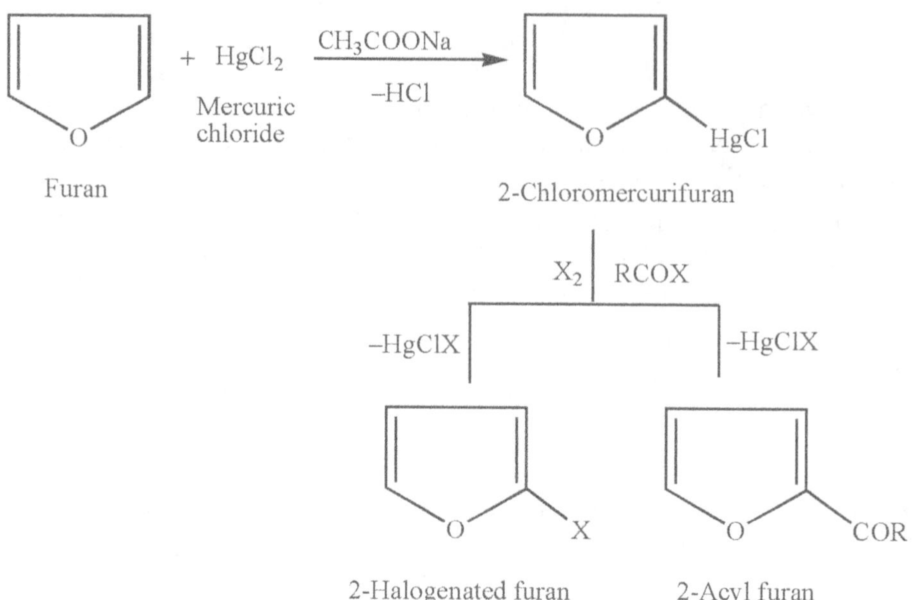

2. Gomber's Reaction: Furan reacts with diazonium salt in alkaline solution and yields aryl furans.

+ $Ar\overset{+}{N_2}\overset{-}{Cl}$ $\xrightarrow[-N_2,\ -HCl]{NaOH}$ 2-Aryl furan

Furan   Aryl diazonium chloride   2-Aryl furan

3. Reaction with *n*-Butyl lithium: Furan reacts with *n*-butyl lithium and gives 2-lithiated furan which further reacts with CO_2 and H^+ and yields furoic acid.

Furan + C_4H_9Li (n-Butyl lithium) $\xrightarrow[-C_4H_{10}]{NaOH}$ 2-Li furan $\xrightarrow[H^+]{CO_2}$ Furoic acid (COOH)

4. Reaction with Carbenes: Carbenes react with furan and adds to the 2,3-carbon double bond.

Furan + CH_2N_2 (Diazomethane) $\xrightarrow[-N_2]{CH_3COONa}$

5. **The Diel's-Alder Reaction:** Furan reacts with maleic anhydride at 25 °C and gives a mixture of exoadduct (90% yield) and an endoadduct (10% yield).

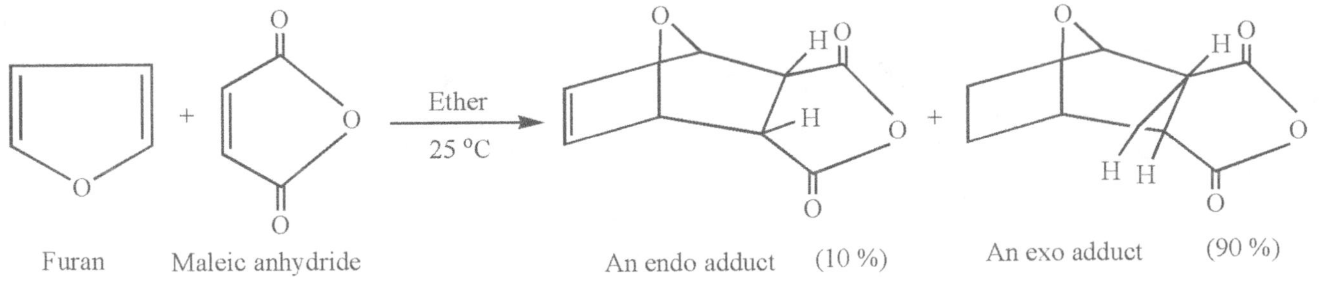

Ether, 25 °C

Furan + Maleic anhydride → An endo adduct (10 %) + An exo adduct (90 %)

6. **Reduction:** Furan is catalytically reduced with Raney nickel at 125 °C and 100 atm pressure to yield tetrahydrofuran (THF) which is used as solvent.

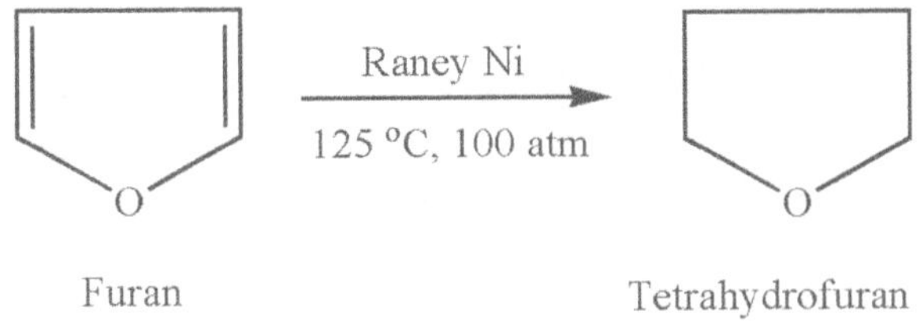

Furan —(Raney Ni, 125 °C, 100 atm)→ Tetrahydrofuran

7. **Oxidation:** Furan is easily oxidised in the presence of air and light.

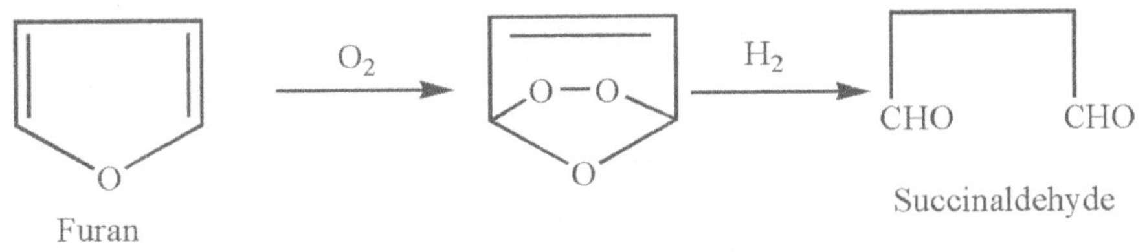

Furan —(O_2)→ —(H_2)→ CHO CHO Succinaldehyde

8. **The Tishchenko Reaction:** Furan 2-aldehyde undergoes dimerisation in the presence of alkoxide to yield respective esters.

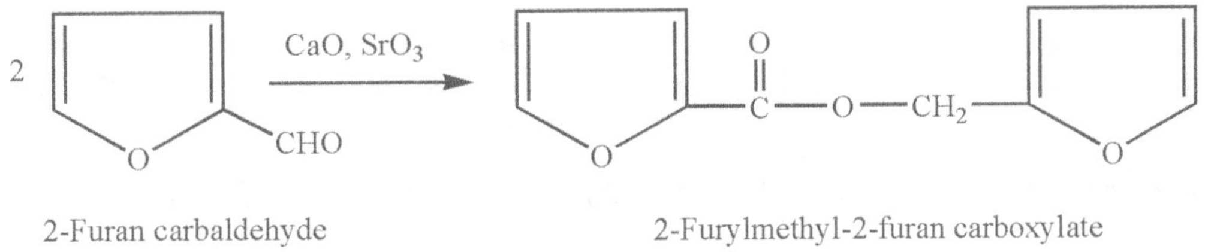

2 2-Furan carbaldehyde (CHO) —(CaO, SrO_3)→ C(=O)—O—CH_2 2-Furylmethyl-2-furan carboxylate

9. **Reactions with acids:** Furan is easily hydrolysed by acids in the presence of mild conditions to yield succinaldehyde *via* the formation of protonated intermediate.

Furan —(H^+)→ [protonated intermediate] —(H_2O)→ OHC CHO Succinaldehyde

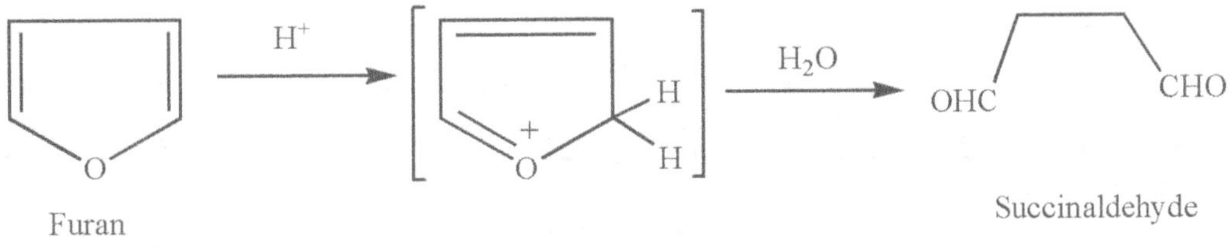

10. **Alkylation:** Furan gets alkylated with isobutylene in the presence of phosphoric acid in Kiesulghur and gives a mixture of 2 and 3-*t*-butyl furan.

$$\text{Furan} + CH_3-\underset{}{\overset{CH_3}{\overset{|}{C}}}=CH_2 \ (\text{Isobutylene}) \xrightarrow{H_3PO_4} \text{2-}t\text{-Butyl furan} + \text{3-}t\text{-Butyl furan}$$

Furan Dervatives

Tetra Hydro Furan(THF)/Oxolane/Oxacyclopentane

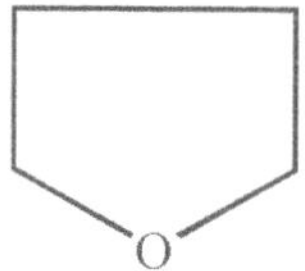

General methods of preparation:

1. **From acetylene:** Acetylene condensed with formaldehyde to yield butyne-1,4-diol which upon reduction and followed by dehydrative cyclisation yields THF.

$$\underset{\text{Acetylene}}{CH \equiv CH} + \underset{\text{Formaldehyde}}{2HCHO} \xrightarrow[\text{Condensation}]{Cu^+/NH_4Cl} \underset{\text{Butyne-1,4-diol}}{OHCH_2-C\equiv C-CH_2OH}$$

$$\xrightarrow{Ni/H_2} CH_2OH-CH_2-CH_2-CH_2OH \xrightarrow[-H_2O]{H_3PO_4,\ 280\ ^\circ C} \text{Tetrahydrofuran}$$

2. **Catalytic reduction of furan:** Furan is catalytically reduced with Ni or Pd-PdO and gives THF.

$$\text{Furan} + 2H_2 \xrightarrow[\text{Pd-PdO}]{Ni} \text{Tetrahydrofuran}$$

Physical properties: It is a colorless volatile and poisonous liquid, miscible with water and organic solvents. It easily forms explosive peroxides.

Chemical properties:

1. **Reaction with HCl:** THF reacts with HCl and yields 4-chlorobutanol.

Tetrahydrofuran + HCl ⟶ $Cl\text{-}CH_2\text{-}CH_2\text{-}CH_2CH_2OH$ (4-Chlorobutanol)

2. **Reaction with H_2S:** THF reacts with H_2S at 300° C in the presence of Al_2O_3 and yields tetrahydrothiophene.

Tetrahydrofuran + H_2S $\xrightarrow[300\,^{\circ}C]{Al_2O_3}$ Tetrahydrothiophene + H_2O

3. **Reaction with ammonia:** THF when heated with NH_3 at 300 ° C yields pyyrolidine.

Tetrahydrofuran + NH_3 $\xrightarrow{300\,^{\circ}C}$ Pyrrolidine + H_2O

4. **Acid hydrolysis:** THF undergoes hydrolysis with H_2SO_4 and yields 1,4-butanediol

Tetrahydrofuran + H_2O $\xrightarrow{H_2SO_4}$ $OH\text{-}CH_2\text{-}CH_2\text{-}CH_2\text{-}CH_2OH$ (1,4-Butanediol)

Uses:

1. THF is used in the manufacturing of raw materials for the preparation of nylon.
2. It is mainly used as a solvent.

Furfural/furfuraldehyde/Furan-2-aldehyde/2-Formyl furan

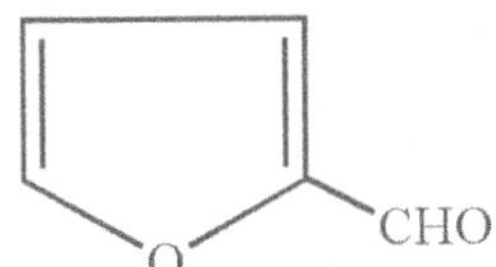

It is the most important derivative of furan. It was produced by boiling bran with dilute acid, hence the name furfural (Latin: Furfur means bran).

General method of preparation:

It is mainly pepared by the distillation of pentoses or oat husks, straw, cobs and brans with dilute mineral acids (Acid catalysed dehydration).

$$\text{Pentose} \xrightarrow[\Delta]{\text{Dil.}H_2SO_4} \text{Furfural} + 3H_2O$$

Physical properties: It is a colorless liqiud, B.P 161.7 °C. It possess pleasant aromatic odour. Miscible with all organic solvents and slightly soluble in water.

Chemical properties: Some of the important chemical reactions of furfural are summarized as follows.

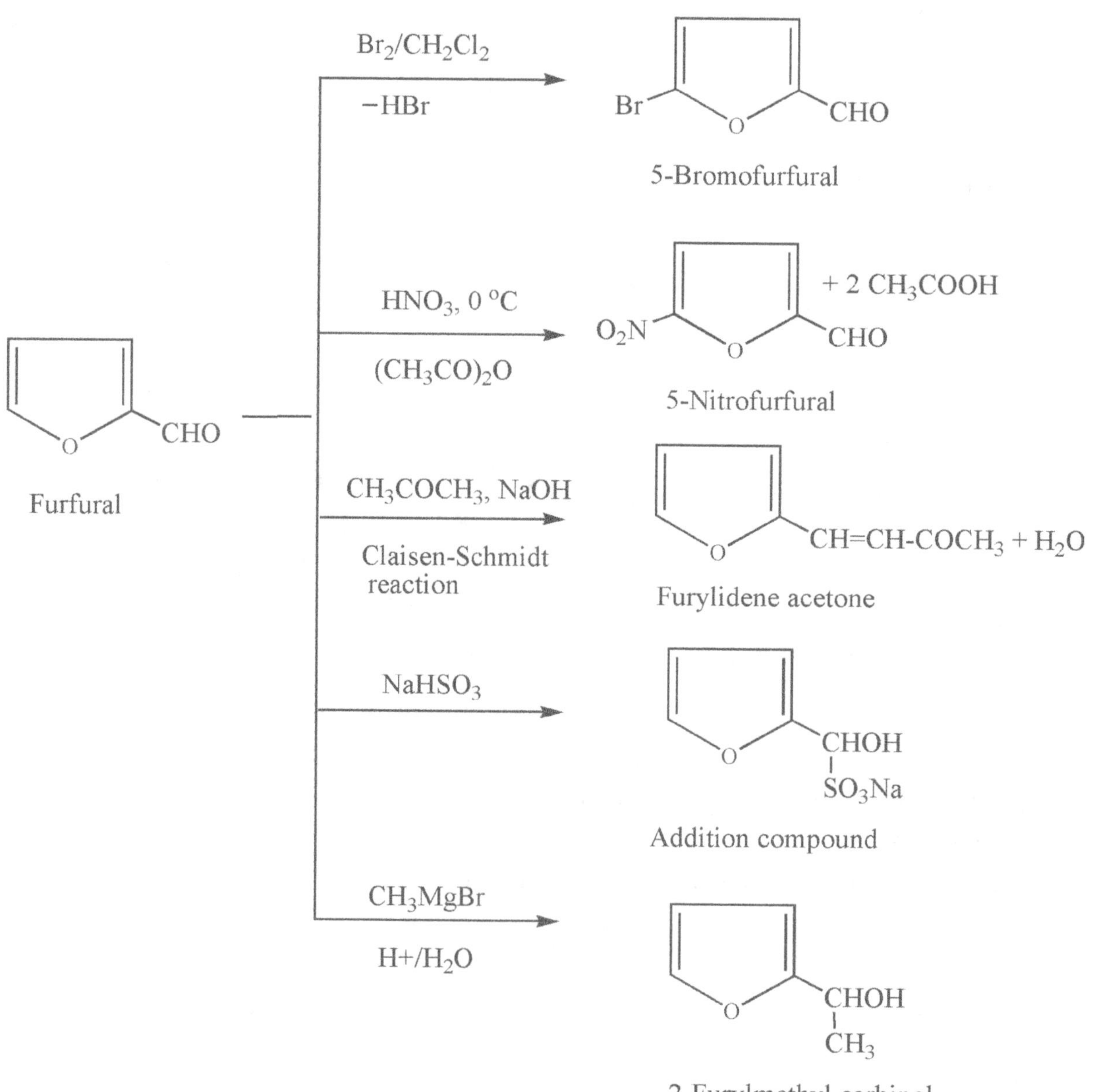

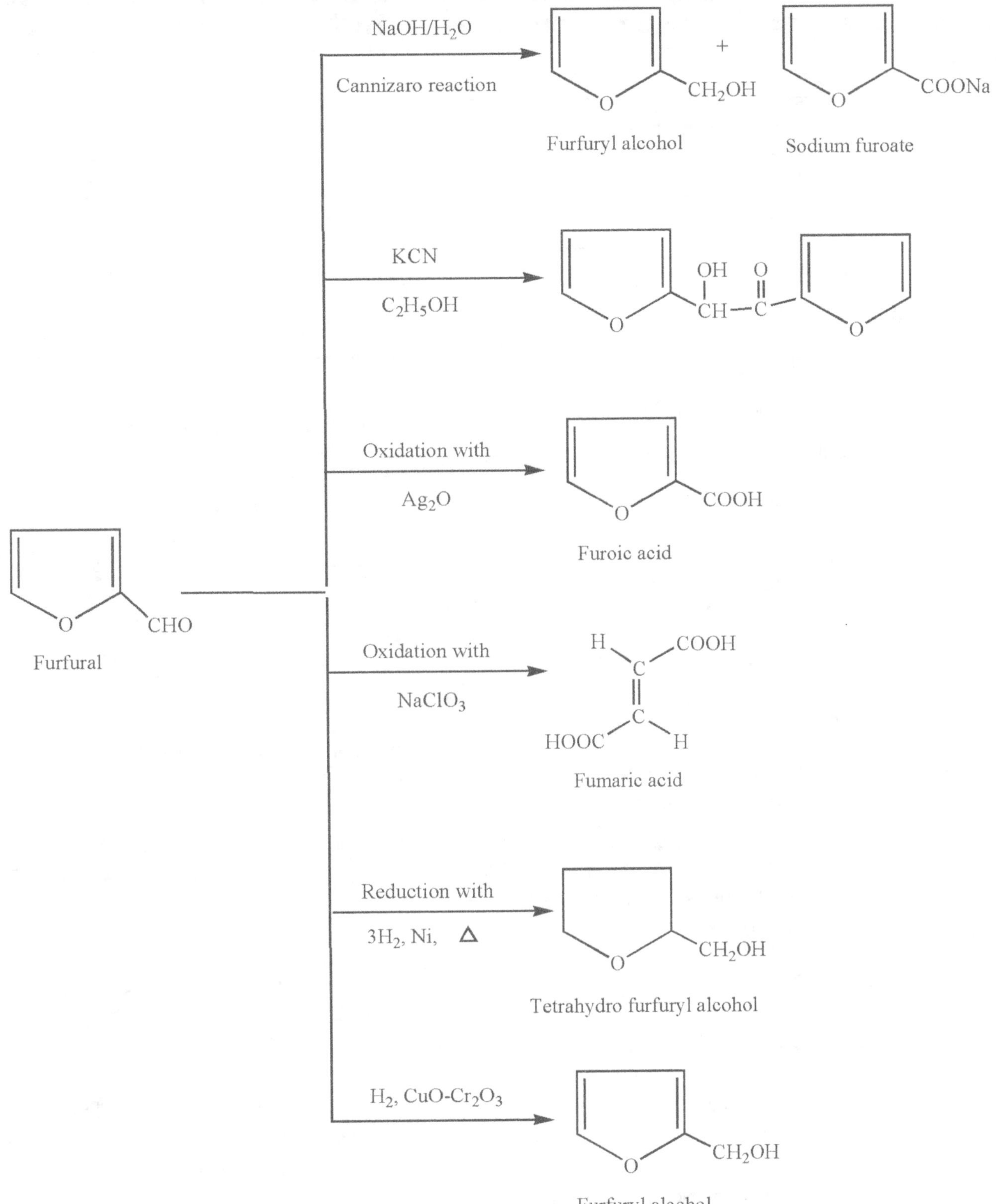

Uses:

1. It is used in the manufacture of resins and also used to remove butadiene from butane and butene mixture.
2. In poultry houses, it is used as a washing liquid to destroy lice.
3. It is also used as a solvent for cellulose esters.

Medicinal Compounds Containing Furan (or) Furan Derivatives Used in Medicine

Furosemide (or) Frusemide	**Pilocarpine**
COOH; NHCH$_2$; O; H$_2$NO$_2$S; Cl 4-Chloro-*N*-furfuryl-5-sulphamoylanthranilic acid **Use:** It is used as loop diuretic	H$_2$C; CH$_3$; N; C$_2$H$_5$; O; O; N 3-Ethyl-4-[(1-methyl-imidazole-5-yl)methyl]tetrahydro-2-furanone **Uses**: It is used as cholinergic agent
Griseofulvin	**Ranitidine**
OCH$_3$; O; OCH$_3$; O; H$_3$CO; O; H$_3$C; Cl 7-Chloro-3',4,6-trimethoxy-5'-methylspiro[1-benzofuran-2,4'-cyclohex-2-ene]-1',3-dione **Uses:** It is used as an antifungal agent	CH$_2$SCH$_2$CH$_2$NHCNHCH$_3$; CHNO$_2$; CH$_3$; H; N; C; O; CH$_3$; H 1-*N*'-[2-[[5-[(Dimethylamino)methyl]furan-2-yl]methylsulfanyl]ethyl]-1-*N*-methyl-2-nitroethene-1,1-diamine **Uses:** It is used as an antiulcer agent
Diloxanide furoate	**Dantrolene**
H$_3$C—N—COCHCl$_2$; O; C; O; O 4-(2,2-Dichloro- *N*-methylacetamido)phenyl furan-2-carboxylate **Uses:** It is used as an antiameobic agent	O; C=N—N; NH; H; O; O$_2$N; O 1-[(5-(4-Nitrophenyl)furan-2-yl)methylamino] imidazolidine-2,4-dione **Uses:** It is used as skeletal muscle relaxant

Table *Contd...*

Terazosin 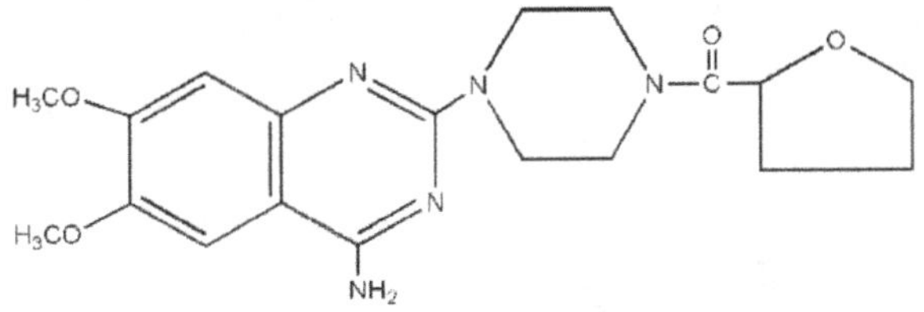 2{[4-(Tetrahydro-2-furanyl)-carbonyl]-1-piperazinyl}-6,7-dimethoxy-4-quinnolinamine **Uses:** It is used as an antihypertensive agent	**Prazosin** 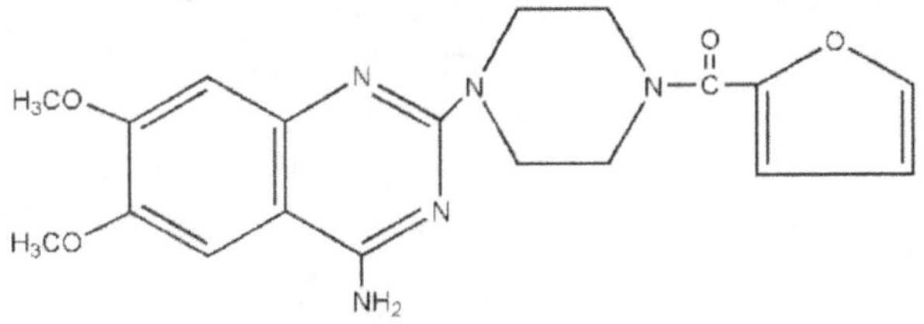 1-(4-Amino-6-7-dimethoxy-2-quinazolinyl) 4-(-2-furanyl carbonyl)-piperazine **Uses:** It is used as adrenergic blocker
Bisucuilline 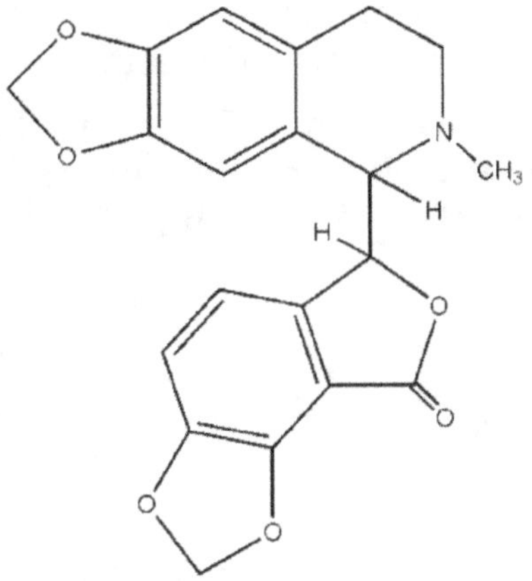 **Uses:** It is used as CNS stimulant	**Canrenone** 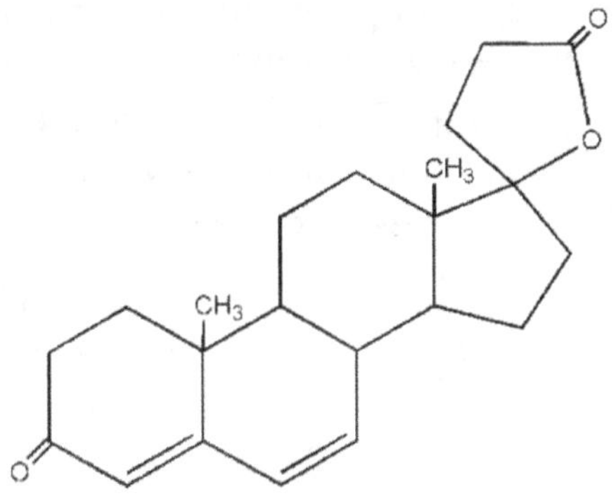 **Uses:** It is used as diuretic
Lupitidine 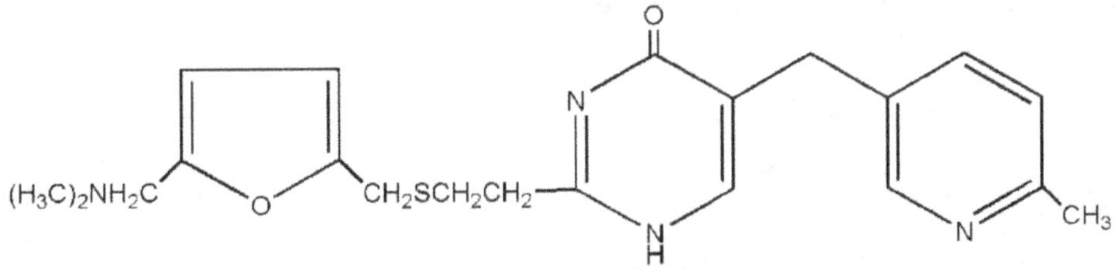 2-[2-[[5-(2-Aminopropan-2-yl)furan-2-yl]methylsulfanyl]ethylamino]-5-[(6-methylpyridin-3-yl)methyl]-1*H*-pyrimidine-6-one **Use:** It is used as an antiulcer drug	

Thiophene (Thiocyclopenta-2,4-diene)

Introduction

It is an organic heterocyclic compound contains four carbon atoms and one sulphur atom in its ring. Thiophene and its derivatives are mainly present in petroleum. It is an important building block for agrochemicals and many drugs. The positions of the substituents are indicated by numbers or Greek letters in which the sulphur atom is assigned to position-1.

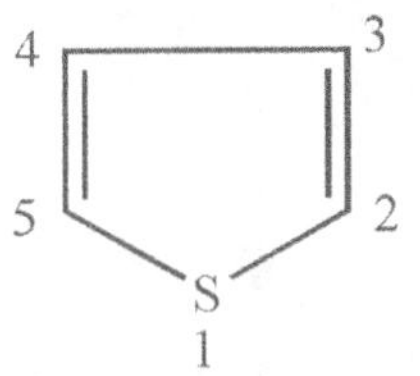

or

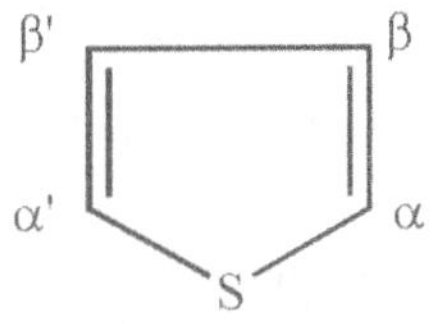

The reduced form of thiophene are indicated or named by trivial names. For example, 2,3-dihydrothiophene is named as 2-thiolene *etc.*

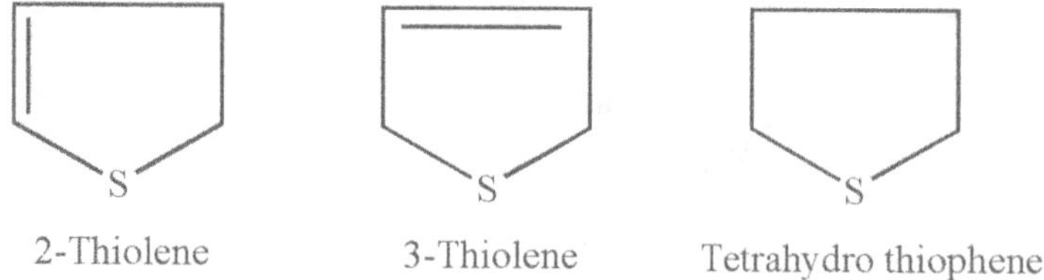

2-Thiolene 3-Thiolene Tetrahydro thiophene

In coaltar thiophene is present along with benzene. Separation of benzene and thiophene by fractional distillation is a difficult one due to their closeness in boiling point; hence thiophene is an unavoidable impurity in commercial benzene.

Isolation of thiophene: The coaltar containing benzene and thiophene mixture is treated with cold Conc. H_2SO_4 and converts thiophene into thiophene sulphonic acid which is dissolved out in water. The pure thiophene is obtained by treating thiophene sulphonic acid with super heated steam.

A good method for separation of thiophene from benzene is as follows. Refluxing the mixture with aqueous mercuric acetate, during this thiophene get mercurated while benzene remains unchanged. The pure form of thiophene is recovered by distilling the mercurated thiophene with HCl.

Thiophene in coaltar + $(CH_3COO)_2Hg$ (Mercuric acetate) → Thiophene mercuric acetate ($HgOCOCH_3$) $\xrightarrow{HCl}$ Thiophene (Pure) + CH_3COOH+HgCl

General Methods of preparation:

1. ***Manufacturing process:*** A mixture of acetylene and hydrogen sulphide is passed through a tube containing Al_2O_3 at 670 K to yield thiophene.

$$2CH{\equiv}CH + H_2S \xrightarrow{Al_2O_3} \text{Thiophene} + H_2$$

Acetylene Thiophene

2. ***From n-Butane:*** Thiophene is also manufactured by the reaction between *n*-butane and sulphur in vapour phase.

$$C_4H_{10} + 4S \xrightarrow{920\ K} \text{Thiophene} + 3H_2S$$

n-Butane Thiophene

3. ***From Sodium succinate (Laboratory method of preparation):*** Sodium succinate is heated with phosphorous trisulphide to give thiophene.

CH_2COONa–CH_2COONa $\xrightarrow{P_2S_3}$ Thiophene

Sodium succinate Thiophene

Structure: According to molecular orbital theory the four carbon atoms in the thiophene molecule is sp^2 hybridised. Each of four carbon atom possess one electron in the p_z orbital and the sulphur atom possess two p electrons to form the π- electron cloud above and below the ring. The sulphur atom in the thiophene ring is less electronegative than oxygen and nitrogen, hence it releases the electrons into the ring to form a π-sextet electrons needed for electricity. So thiophene is aromatic and also possesses excessive π electron density.

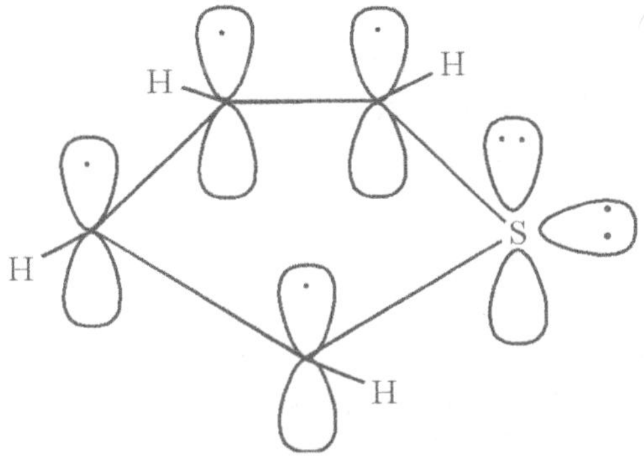

The sulphur atom present in the thiophene ring belongs to the second row elements in the periodic table. During hybridisation it expands its valency shell by utilising the empty '*d*' orbitals. Consequently instead of having six electrons in five orbitals (as in case of furan and pyrrole) thiophene by using pd^2 hybrid orbitals might six electrons in the six orbitals. This effect is responsible for the extreme stability of thiophene and its resemblance to benzene in physical and chemical properties.

Resonance in Thiophene: The resonance structure of thiophene is as follows:

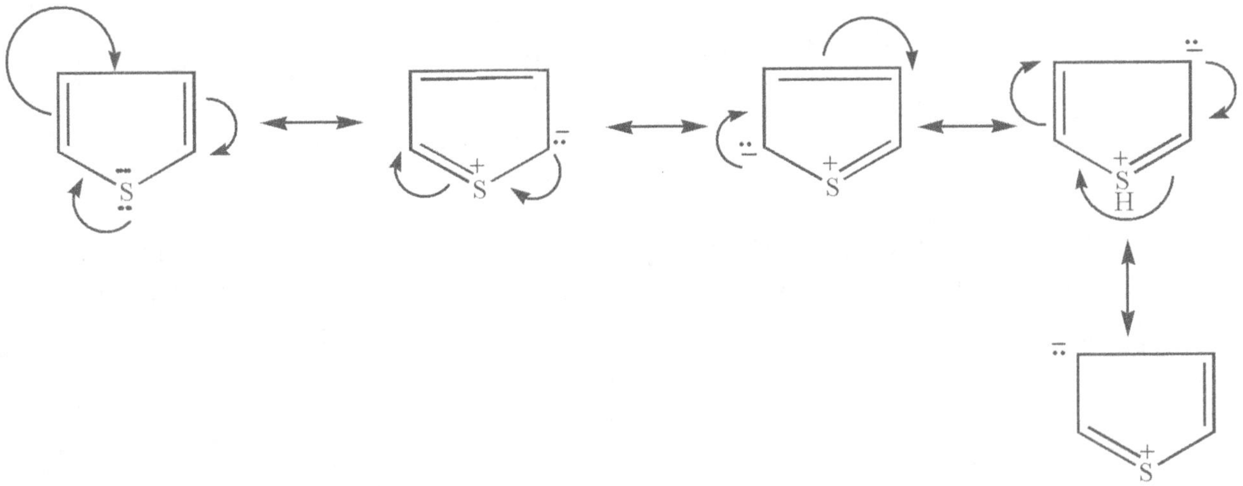

Physical Properties: It is colourless liquid, B.P. 357 K. Benzene like odour, insoluble in water but soluble in organic solvent. It does not show basic character.

Chemical Properties: The chemical properties of thiophene are similar to that of benzene.

1. **Reaction with acids:** Thiophene is more stable to acids, but very strong acids produces resins. For example thiphene reacts with orthophosphoric acid under mild conditions and gives a trimer.

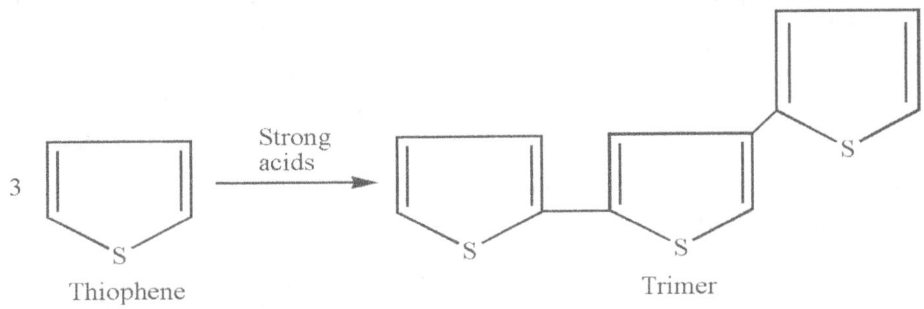

3 Thiophene → (Strong acids) Trimer

2. **Electrophilic Substitution Reactions:** Thiophene is least reactive when compared to pyrrole and furan but more reactive than benzene. The electrophilic attack of thiophene is preferentially occurs in C-2 and C-5 positions. Practically C-2 position predominates.

(i) ***Halogenation:***

(a) ***Chlorination:*** Thiophene reacts with chlorine and gives substituted as well as addition products. But it reacts with sulphurylchloride and yields 2-chloro thiophene.

Thiophene $+ SO_2Cl_2 \longrightarrow$ 2-Chlorothiophene $+ SO_2 + HCl$

Thiophene $+ Cl_2 \xrightarrow{50\,^0C}$ 2-Chlorothiophene (Major product) + 2,5-Dichlorothiophene (Minor product)

(b) ***Bromination:*** Thiophene reacts with Br_2/CH_3COOH and yields 2-bromothiophene.

Thiophene $+ Br_2 \xrightarrow[-HBr]{CH_3COOH}$ 2-Bromothiophene

2-Bromothiophene is also obtained by the reaction with thiophene with *N*-bromosuccinimide.

(c) ***Iodination:*** Thiophene reacts with I_2 and HgO and gives mono and diiodothiophenes.

Thiophene $+ I_2 \xrightarrow{HgO}$ 2-Iodothiophene (Major product) + 2,5-Diiodothiophene (Minor product)

The reaction between 2,5-dibromothiophene with Li, Na, LDA *etc* gives rearranged products. This is known as dance mechanism .

For example: 2,5–dibromothiophene undergoes rearrangement with LDA in THF at -70 °C and yield 3,5-dibromothiophene.

2,5-Dibromothiophene $\xrightarrow[\text{THF, -70 °C}]{\text{LDA}}$ 3,5-Dibromothiophene

(ii) ***Nitration:*** Thiophene is nitrated with a mixture of nitric acid and acetic anhydride and yields a mixture of 2 and 2,5-dinitrothiophenes

Thiophene $+ HNO_3 \xrightarrow{(CH_3CO)_2O}$ 2-Nitrothiophene + 2,5-Dinitrothiophene

(iii) ***Sulphonation:*** Thiophene reacts with 90% H_2SO_4 at room temperature and yields thiophene-2-sulphonic acid.

Thiophene $\xrightarrow{95\ \%\ H_2SO_4}$ Thiophene-2-sulphonic acid (SO_3H)

(iv) ***Friedel-Craft's acylation reaction:*** Thiophene reacts with acetyl chloride in the presence of stannic chloride and gives 2-acetylthiophene.

Thiophene $+ CH_3COCl \longrightarrow$ 2-Acetylthiophene ($COCH_3$)

It also reacts with pthalic anhydride in the presence of Lewis acid and $AlCl_3$ to give 2,2′-thienoyl benzoic acid.

Thiophene + Pthalic anhydride $\xrightarrow{AlCl_3}$ 2,2'-Thienoyl benzoic acid (COOH)

(v) ***Alkylation:*** Thiophene reacts with alkene and H_2SO_4 to yield alkylated thiophene.

Thiophene $+ CH_3-C(CH_3)=CH_2$ (Isobutylene) $\xrightarrow{80\ \%\ H_2SO_4}$ *t*-Butyl thiophene ($C(CH_3)_3$)

(vi) *Vilsemeier Formylation*: Thiophene reacts with DMF and $POCl_3$ to give 2-thiophenecarbaldehyde.

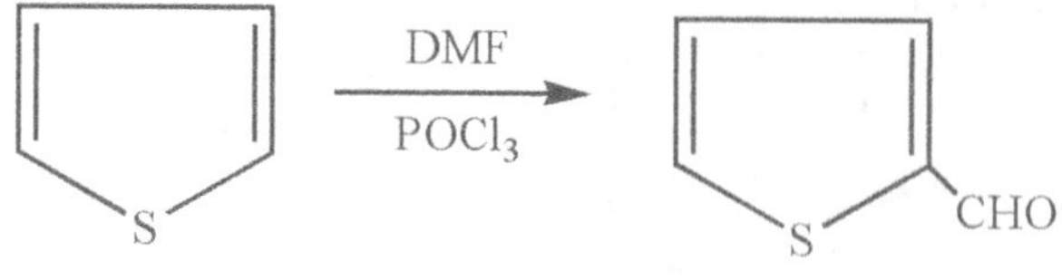

Thiophene 2-Thiophenecarbaldehyde

3. **Reaction with carbene:** Thiophene reacts carbethoxy carbene to yield ethyl thiophene β-acetate *via* the formulation of cyclopropane intermediate.

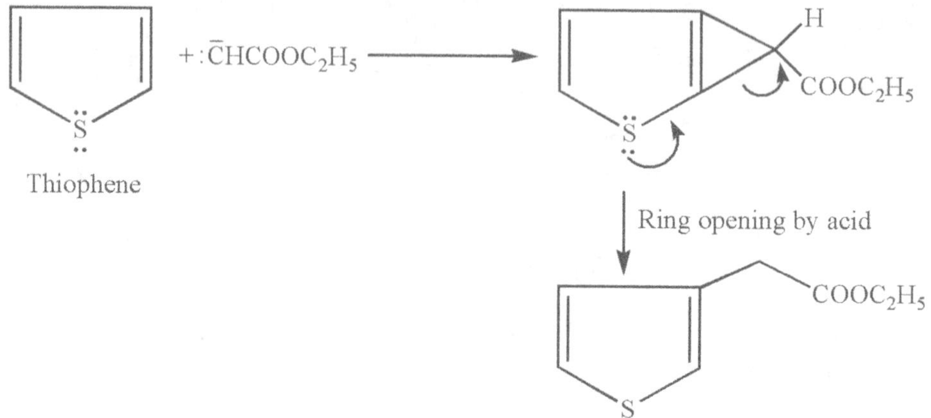

Ethyl thiophene β-acetate

4. **Reaction with Nitrenes:** Thiophene reacts with ethoxycarbonitrene and gives *N*-carboethoxypyrrole.

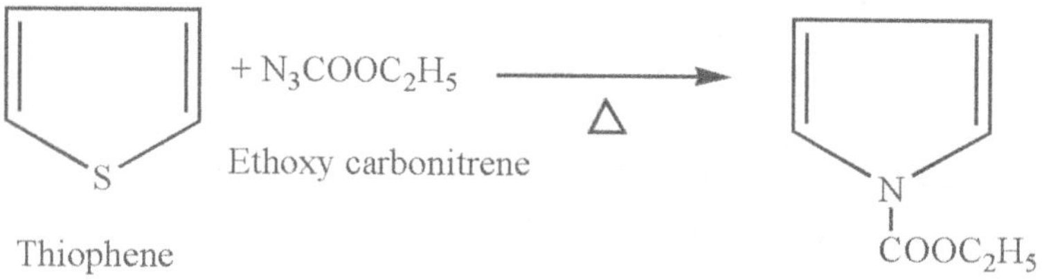

$+ N_3COOC_2H_5$ — Δ →

Ethoxy carbonitrene

Thiophene

$COOC_2H_5$

N-Carboethoxypyrrole

5. **Mercuration:** Thiophene reacts with mercuric chloride in the sodium acetate and gives thiophene-2-mercuric chloride.

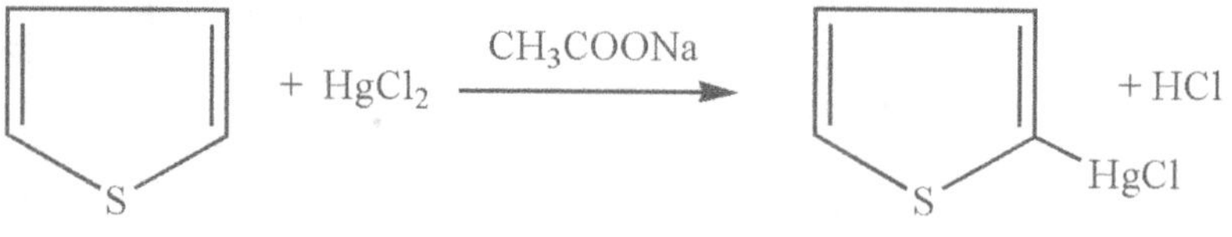

$+ HgCl_2$ —CH_3COONa→ HgCl + HCl

Thiophene Thiophene-2-mercric chloride

6. **Reaction with *n*-alkyl lithium:** Thiophene reacts with *n*-butyl lithium to yield 2-lithium thiophene, which upon reaction with CO_2 yields 2-thiophene carboxylic acid.

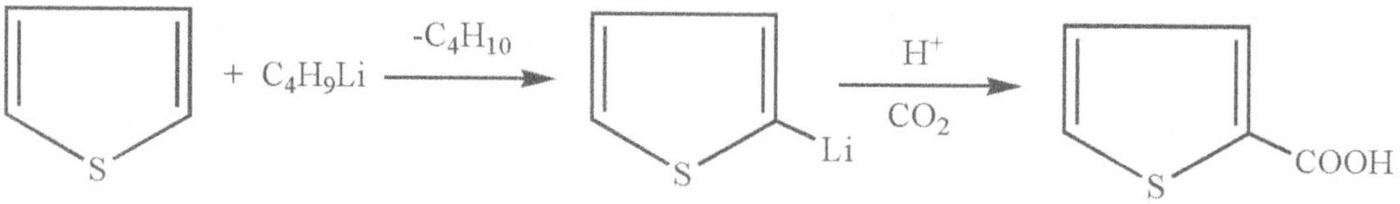

$+ C_4H_9Li$ —$-C_4H_{10}$→ Li —H^+, CO_2→ COOH

Thiophene 2-Thiophene carboxylic acid

7. **Reduction:** Thiophene gets reduced with sodium in methanol and ammonium at -40 °C to yield mixture of products.

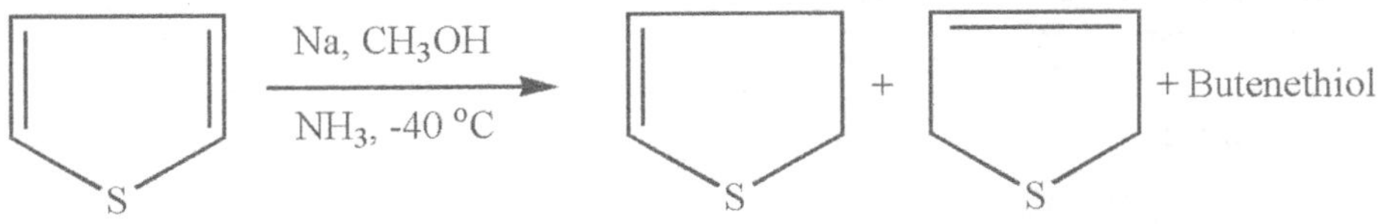

Na, CH_3OH / NH_3, -40 °C

Thiophene → 2-Thiolene + 3-Thiolene + Butenethiol

It also reduced with Palladium-charcoal catalysts and gives tetrahydrothiophene.

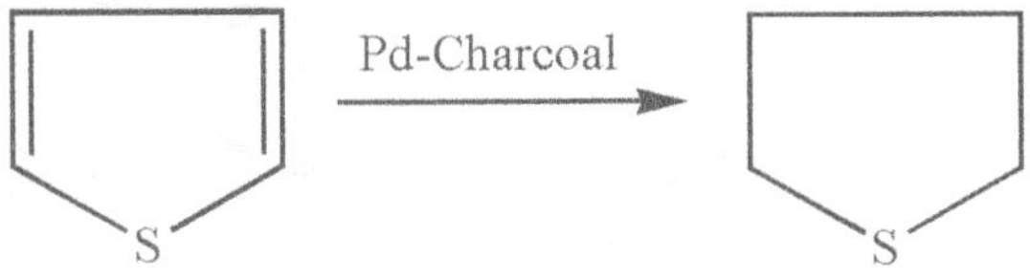

Pd-Charcoal

Thiophene → Tetrahydro thiophene

Thiophene reacts with Raney nickel in acetic anhydride to yield *n*-butane by hydrogenative desulphurization.

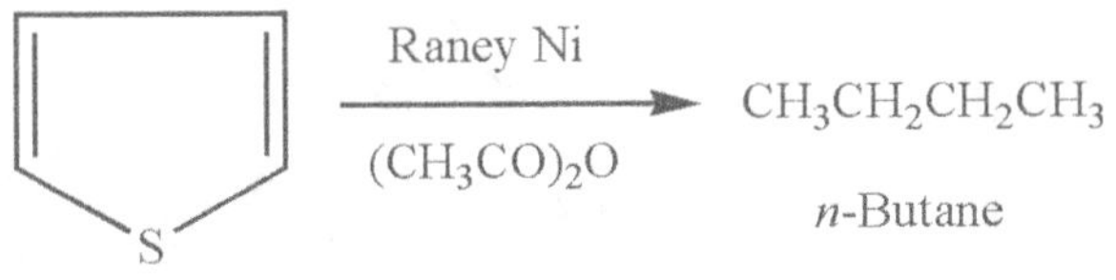

Raney Ni / $(CH_3CO)_2O$

Thiophene → $CH_3CH_2CH_2CH_3$ *n*-Butane

This reaction is most important one for the structural determination of thiophene.

8. **Oxidation:** Thiophene is resistant to mild oxidizing agents. But it is oxidized to maleic acid and oxalic acid by the treatment with HNO_3.

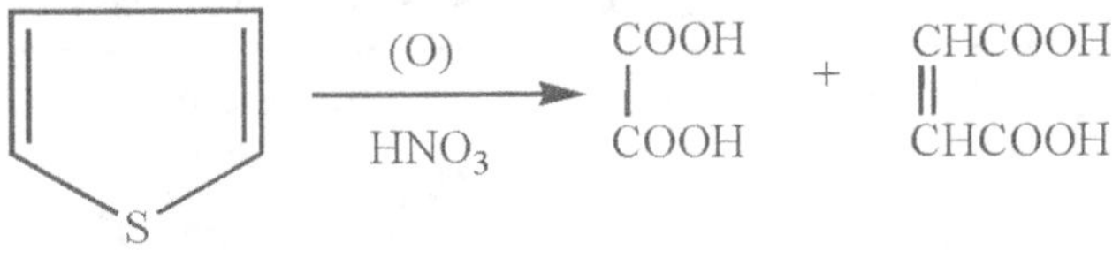

(O) / HNO_3

Thiophene → COOH–COOH (Oxalic acid) + CHCOOH=CHCOOH (Maleic acid)

It also gets oxidised to sulphone.

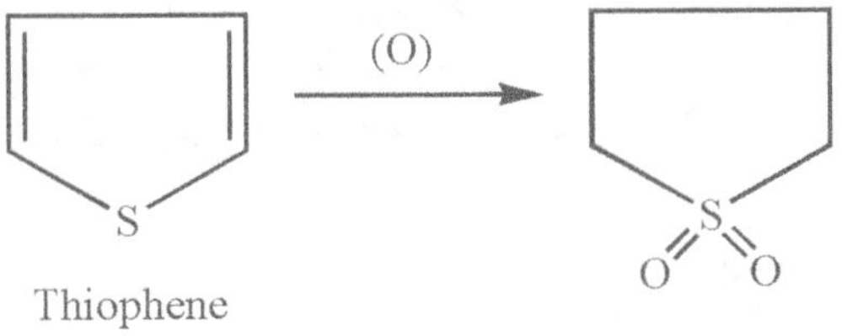

(O)

Thiophene → Sulphone

9. **Diels-Alder Reaction:** Thiophene reacts with acetylenic dienophiles and gives a benzene derivative *via* the formation of unstable intermediate.

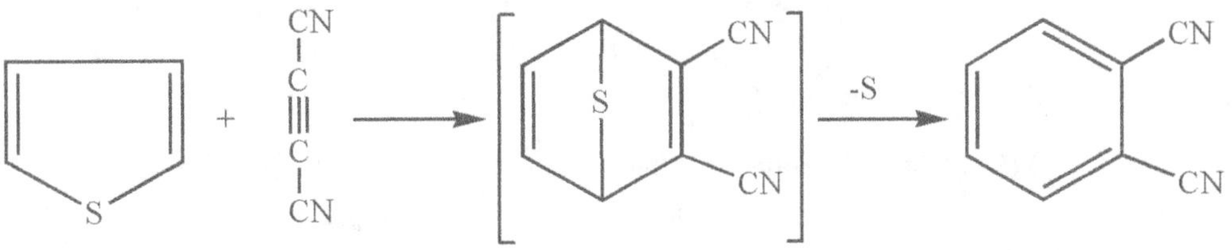

Thiophene + CN–C≡C–CN → [intermediate] —(-S)→ Dicyanobenzene

Medicinal Compounds Containing Thiophene or Thiophene Derivatives Used in Medicine

Ketotifene **Use:** It is used as an antihistaminic agent	**7-Chloro ketotifene** **Use:** It is used as an antihistaminic agent
Pyrantel pamoate 1-Methyl-2-[(E)-2-thiophen-2-ylethenyl]-5,6-dihydro-4*H*-pyrimidine **Use:** It is used as antihelmintic	**Tigabin** (−)-(3*R*)-1-[4,4-Bis(3-methyl-2-thienyl)-3-buten-1-yl]-3-piperidinecarboxylic acid **Use:** It is used as an anti convulsant
Ticrynafen 2-[2,3-Dichloro-4-(thiophene-2-carbonyl)phenoxy]acetic acid **Use:** It is used as diuretic	**Clopidogrel** Methyl (2*S*)-2-(2-chlorophenyl)-2-(6,7-dihydro-4*H*-thieno[3,2-c]pyridin-5-yl)acetate **Use:** It is used as an antiplatelet agent
Methapyrilene (−)-(3*R*)-1-[4,4-Bis(3-methyl-2-thienyl)-3-buten-1-yl]-3-piperidinecarboxylic acid **Use:** It is used as an antihistamine	

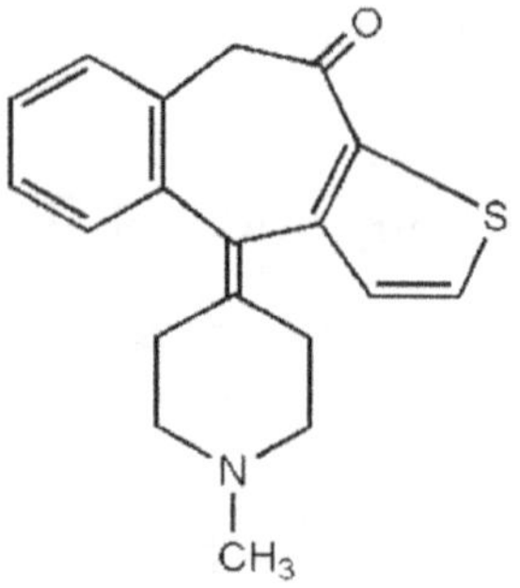

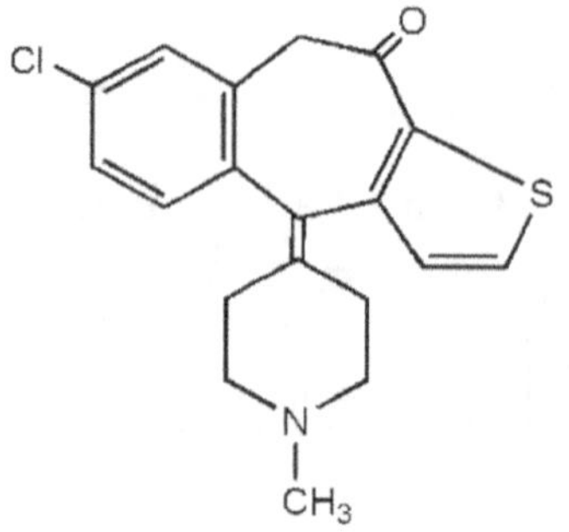

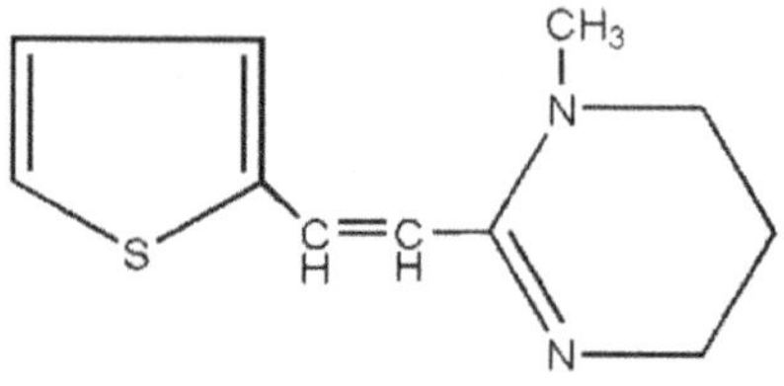

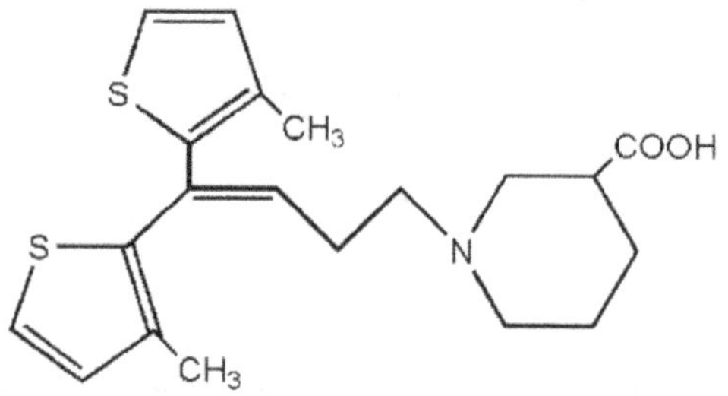

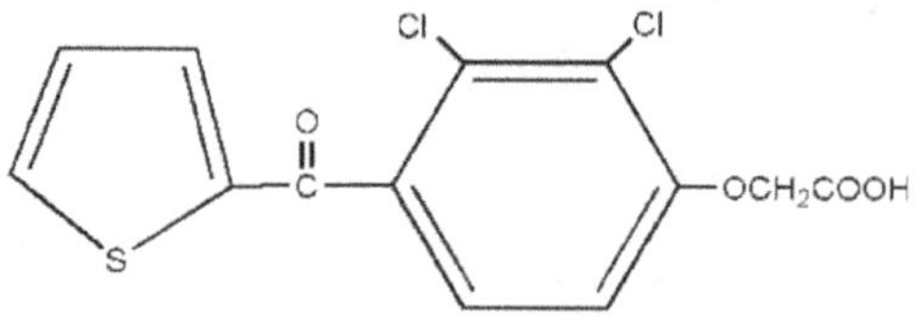

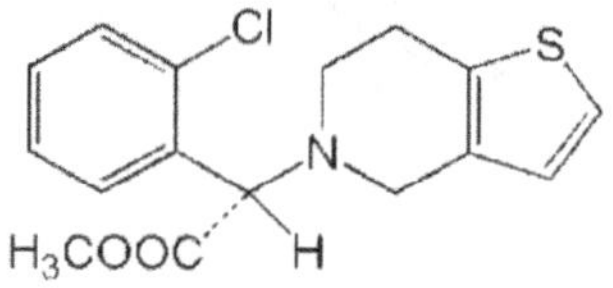

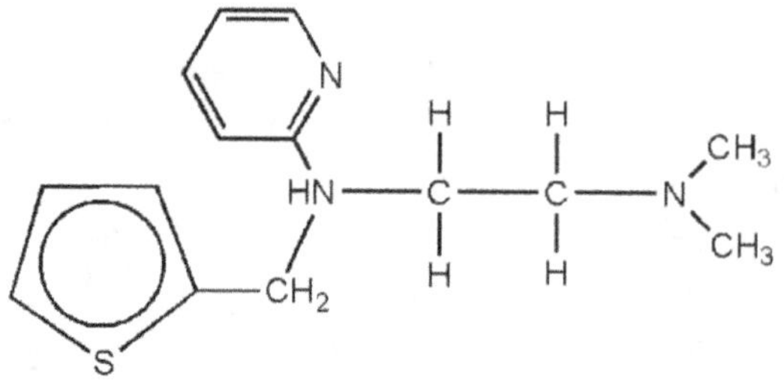

Relative Aromaticity and Reactivity of Pyrrole, Furan and Thiophene

The five membered heterocycles with one hetero atom such as pyrrole, furan and thiophene have four carbon π electrons (one on each carbon) and two / one pair of electrons on heteroatom combines to form a aromatic sextet and satisfies Huckle's rule of aromaticity (4n+2) π electrons. The order of aromaticity or the decreasing order of aromaticity or stability of these compounds is as follows **Thiophene > Pyrrole > Furan.**

The comparative order of these three compounds with benzene is: Furan > Pyrrole > Thiophene > Benzene.

The above fact indicates that thiophene is more aromatic or least reactive when compared with pyrrole and furan which is the least aromatic. The reactivity and aromaticity of these three compounds are explained and evidenced by the following facts.

1. Furan behaves like normal diene and gives Diel's Alder adduct with maleic anhydride (Diels Alder reaction), but pyrrole does not give Diel's Alder adduct. Instead it undergoes Michael type of addition whereas thiophene not at all reacts with maleic anhydride.
2. **Reaction with H_2SO_4:** Furan undergoes polymerisation with H_2SO_4 very easily, pyrrole polymerises *via* the formation salts but Thiophene does not react at all.
3. **Electrophilic Substitution reactions:**
   (a) ***Halogenation:*** Furan violently reacts with halogens leads to the formation of halogenated furan. This reaction is not controlled easily due to the formation of halogen acid, hence furan is halogenated always indirectly. But the direct halogenation of pyrrole and thiophene can be controlled easily.
   (b) ***Nitration and Sulphonation:*** Thiophene easily undergoes nitration and sulphonation with normal nitrating & sulphonating reagents. In contrast nitration & sulphonation of pyrrole and furan take place in the presence of strong acids which leads to polymerisation.
   (c) ***Friedel - Craft's Reaction:*** Pyrrole readily undergoes with Friedel - Craft's reaction & combines with acetic anhydride in presence of any catalyst at hot condition. But thiophene is acetylated in the presence of stannic chloride / H_3PO_4. But furan very easily undergoes Friedel - Craft's acetylation with stannic chloride.
4. **Reduction reaction:** The order of reactivity towards reduction reactions of these three compounds comparing with benzene molecule is as follows,

   Pyrrole > Furan > Thiophene > Benzene.

   Pyrrole easily gets reduced to give dihydro and tetrahydro derivative. Similarly Furan also undergoes reduction easily and gives THF. But reduction of thiophene is resistant because the sulphur atom present in the thiophene ring poisons the catalyst.
5. **Coupling reaction:** The greater reactivity of these heterocyclic rings is compared with benzene which is illustrated by their coupling reaction with diazonium salt.

   On the other hand in the benzene series only amines or phenols with sufficient electron density undergo this reaction.

ArN_2X/ weak acid, $-HX$ ← (2-N_2Ar heterocycle, Z) → ArN_2X in Base, $-HX$ → (2,5-bis(ArN_2) heterocycle)

Z   Z N_2Ar ArN_2 Z N_2Ar

Where Z= O, NH, S

Probable Questions

1. Write in detail about the nomenclature of heterocyclic compounds.
2. Write the classification of heterocyclic compounds.
3. Write the four general methods of preparation of pyrrole.
4. Describe about the structure and basic & acidic nature of pyrrole.
5. Why pyrrole, furan and thiophene undergoes electrophilic substitution reactions chiefly at C-2 position?
6. Explain about the chemical reactions of pyrrole.
7. Write any five medicinal compounds containing pyrrole (or) pyrrole derivatives used in medicine.
8. Write the preparation methods and chemical reactions of furan.
9. Write the medicinal compounds containing furan (or) furan derivatives used in medicine.
10. Write the methods of preparation and reactions of thiophene.
11. Write medicinal compounds containing thiophene or thiophene derivatives used in medicine.
12. Describe the relative aromaticity and reactivity of pyrrole, furan and thiophene.

30

Heterocyclic Compounds (Part 2)

Azoles

Five membered heterocycles with nitrogen and additional hetero atoms are known as azoles.

Types of azoles:

Pyrazole: Azole contains two nitro atoms in 1,2-position.

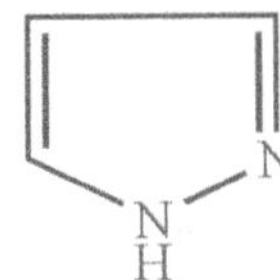

Isooxazole: Azole contains one oxygen and one nitrogen atom in 1,2-position.

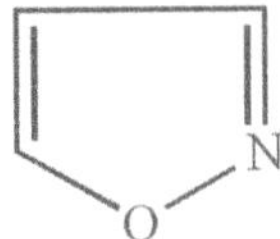

Isothiazole: Azole contains one sulphur and one nitrogen in 1,2-position.

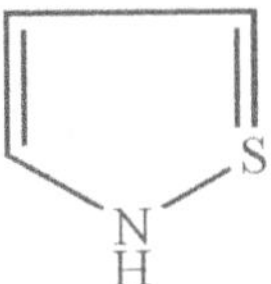

Imidazole: Azole contains two nitrogen atoms in 1,3-position.

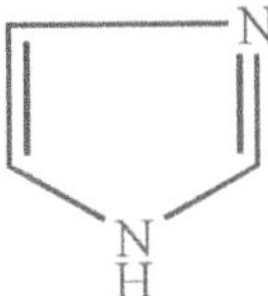

Oxazole: Azole contains one oxygen and one nitrogen atom in 1,3-position.

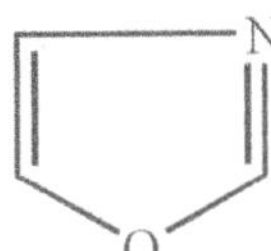

Thiazole: Azole contains one sulphur and one nitrogen atom in 1,3-position.

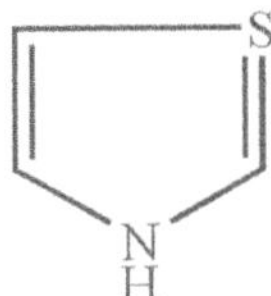

Structure of all Azoles

All the carbon atom of azoles are sp^2 hybridized. Each carbon atom posses one p_z electron, the nitrogen atom of azole posses fourth electron, the second hetero atom provides two electrons to complete the aromatic sextet leads to the formation of molecular orbital.

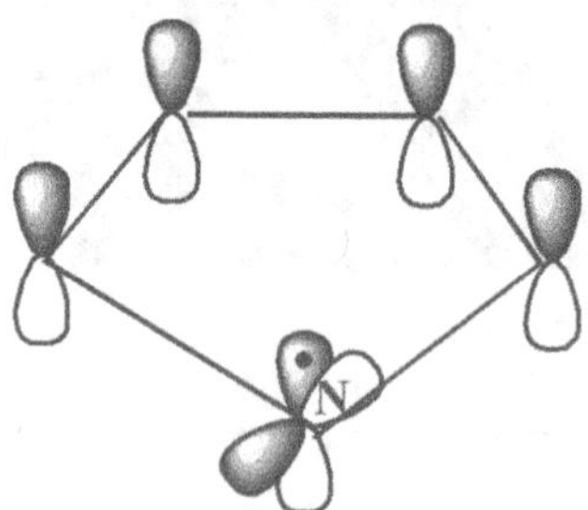

The above diagram shows that the lone pair of electrons of the nitrogen atom is present in orthogonal to the molecular π cloud, hence azoles behave as bases as well as nucleophiles.

The various resonance contributing structures of 1,2 & 1,3-azoles are as follows.

1,2-Azoles:

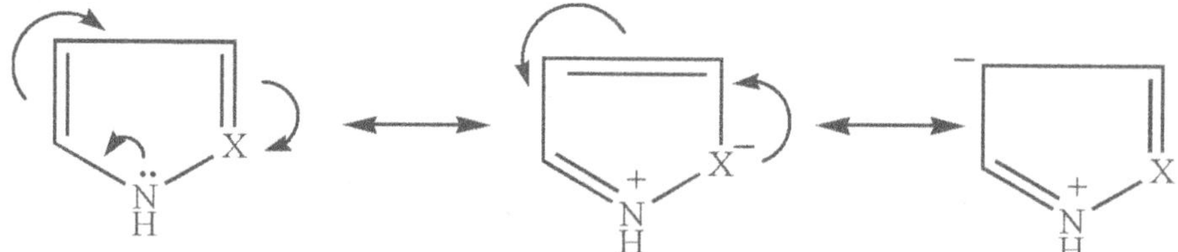

1,3-Azoles:

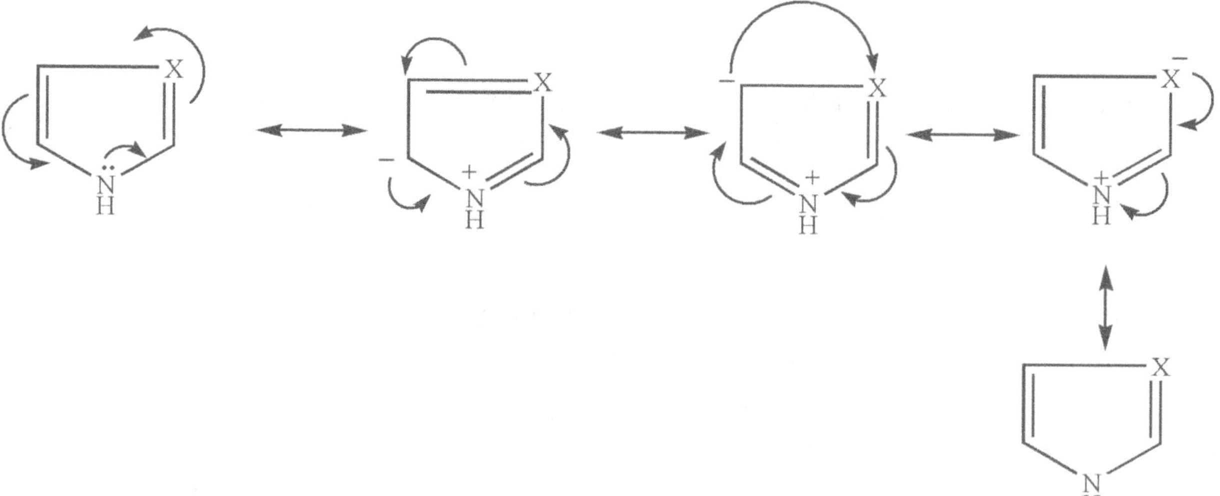

When compared with 1,3-azoles, 1,2-azoles are less reactive. Hence 1,2-azoles does not usually undergo electrophilic substitution reactions and electrophilic attack normally occurs in C-4 & C-5 positions.

Pyrazole

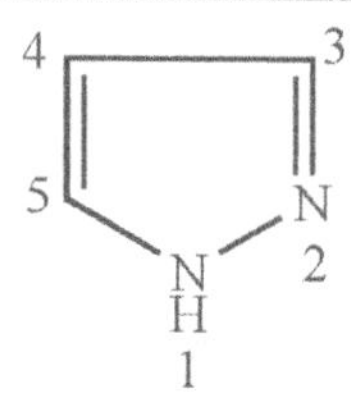

Pyrazole is an organic heterocyclic compound. It is a five membered ring with two nitrogen atoms in adjacent sides. Pyrazole was first discovered in 1889 by Buchner.

General methods of preparation:

1. **From acetylene:** Acetylene is passed through a cold solution of diazomethane to yield pyrazole.

CH≡CH Acetylene; $CH_2=\overset{\oplus}{N}=\overset{\ominus}{N}$ Diazomethane; Cold → [HC=CH, H_2C, N=N] → Pyrazole

2. **From pyrazole carboxylic acids:** Decarboxylation of various pyrazole carboxylic acids gives pyrazole.

Pyrazole-3,4,5-tricarboxylic acid —300 °C, Δ, $-3CO_2$→ Pyrazole

3. **From α,β-unsaturated aldehydes, ketones or acids:** Condensation between α,β-unsaturated aldehydes, ketones or acids with hydrazines yield pyrazolines, which upon oxidation with Br_2 or mercuric oxide yields pyrazole.

Acrolein (α,β-Unsaturated aldehyde) ($CH_2=CH-CHO$) + NH_2-NH_2 Hydrazine —$-H_2O$→ $CH_2=CH-CH=N-NH_2$ —Br_2, $-2HBr$→ Pyrazole

Synthesis of substituted pyrazoles:

1. **From dicarbonyl compound:** 1,3-Dicarbonyl compounds react with hydrazine or its derivative and yield substituted pyrazole.

$CH_3-\overset{O}{\overset{\|}{C}}CH_2\overset{O}{\overset{\|}{C}}-CH_3$ 1,3-Dicarbonyl compond

—NH_2-NH_2, $-2H_2O$→ 3,5-Dimethyl pyrazole

—$C_6H_5NHNH_2$, $-2H_2O$→ 3,5-Dimethyl-1-phenyl pyrazole

2. **From α,β-ethylene carbonyl compounds:** α,β-Ethylene carbonyl derivatives react with hydrazine and its derivatives and yield substituted pyrazoles.

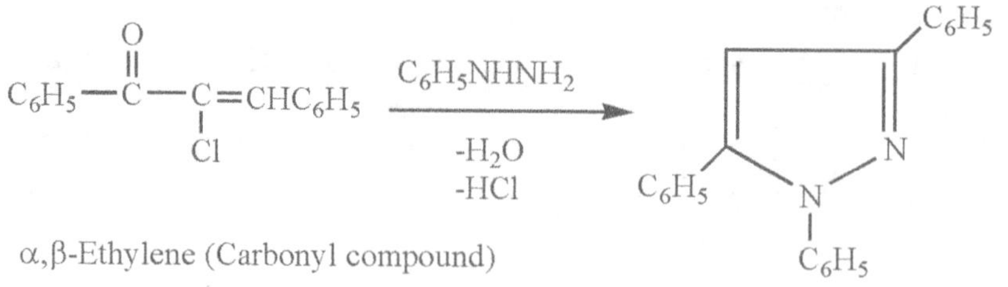

$$C_6H_5-\overset{O}{\overset{\|}{C}}-\underset{Cl}{\underset{|}{C}}=CHC_6H_5 \xrightarrow[-H_2O,\ -HCl]{C_6H_5NHNH_2} \text{1,3,5-Triphenyl pyrazole}$$

α,β-Ethylene (Carbonyl compound)

1,3,5-Triphenyl pyrazole

Physical properties: It is a colorless liqiud, M.P. 70 °C. It exhibits tautomerism.

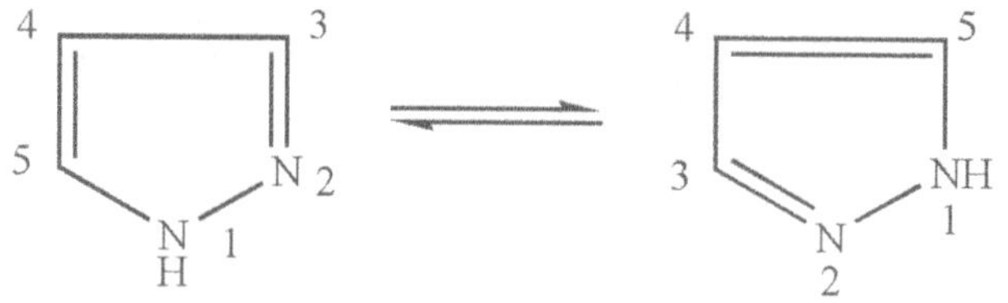

Chemical Properties:

1. **Reaction with acids:** Pyrazole is a basic compound and reacts with HCl, H_2SO_4 *etc* to yield salts.

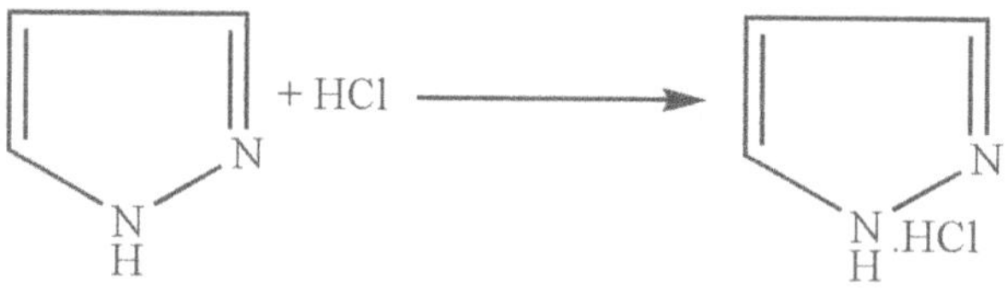

Pyrazole + HCl → Pyrazole hydrochloride

2. **Alkylation:** Pyrazole reacts with methyl iodide and yields N-methyl pyrazole.

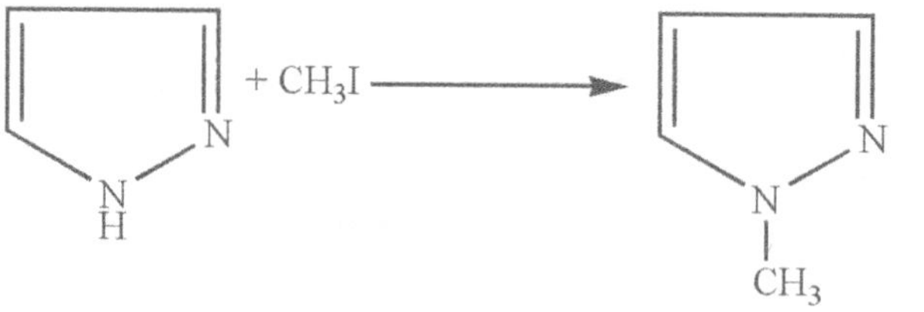

Pyrazole + CH_3I → *N*-Methyl pyrazole

3. **Electrophilic substitution reaction:** Pyrazole undergoes electrophilic substitution reaction and substitution chiefly occurs at position C-4.

 (i) ***Chlorination:*** Pyrazole reacts with Cl_2 or SO_2Cl_2 and yields 4-chloro pyrazole.

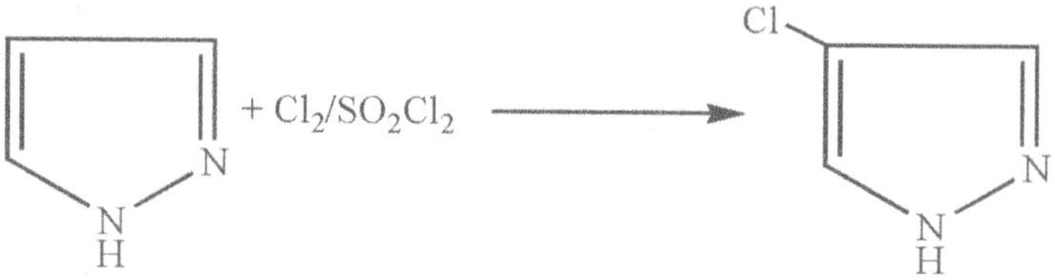

Pyrazole + Cl_2/SO_2Cl_2 → 4-Chloropyrazole

(ii) ***Bromination:*** Pyrazole reacts with Br_2 in dioxane and yields 4-bromo pyrazole.

Pyrazole + Br_2 $\xrightarrow{\text{Dioxane}}$ 4-Bromopyrazole + HBr

(iii) ***Nitration:*** Pyrazole undergoes nitration with nitrating mixture (HNO_3 + H_2SO_4) under severe condition to yield 4-nitro pyrazole.

Pyrazole + HNO_3 $\xrightarrow{H_2SO_4}$ 4-Nitropyrazole

(iv) ***Sulphonation:*** Pyrazole reacts with oleum to yield pyrazole-4-sulphonic acid.

Pyrazole + H_2SO_4 (Oleum) $\xrightarrow{SO_3}$ Pyrazole-4-sulphonic acid + H_2O

4. Oxidation: Pyrazole is resistant to oxidation like benzene but alkylated pyrazole is oxidized with alkaline $KMnO_4$ and yields respective pyrazole carboxylic acid.

N-Phenyl pyrazole $\xrightarrow[\Delta]{\text{Alkaline } KMnO_4}$ *N*-Phenyl pyrazole carboxylic acid

5. Reduction: Catalytic reduction of pyrazole yields pyrazoline and pyrazolidine.

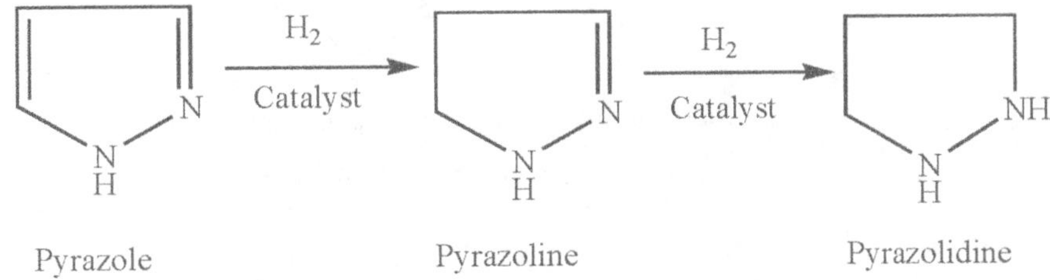

Medicinal Compounds Containing Pyrazole (or) Pyrazole Derivatives Used in Medicine

Celecoxib	Sulphaphenazole
CF_3; H_3C; N; N; SO_2NH_2 4-[5-(4-Methyl phenyl)-3-(trifluoro)-1*H*-pyrazo-lyl]benzene sulphonamide **Use:** It is used as an anti-inflammatory and analgesic agent	H_2N—SO_2NH—; N; N; C_6H_5 4-Amino-*N*-(2-phenylpyrazol-3-yl)benzenesulfonamide **Use:** It is used as an antibacterial agent
Phenylbutazone	**Oxyphenbutazone**
N; N; O; O; $(CH_2)_3CH_3$ 1,2-Diphenylpyrazolidine-3,5-dione **Use:** It is used as an anti-inflammatory and analgesic agent	OH; N; N; O; O; $(CH_2)_3CH_3$ 4-Butyl-1-(4-hydroxyphenyl)-2-phenylpyrazolidine-3,5-dione **Use:** It is used as an anti-inflammatory and analgesic agent
Bifonazole	
H; C—N; N 1-[Biphenyl-4-yl(phenyl)methyl]imidazole **Use:** It is used as an antifungal agent	

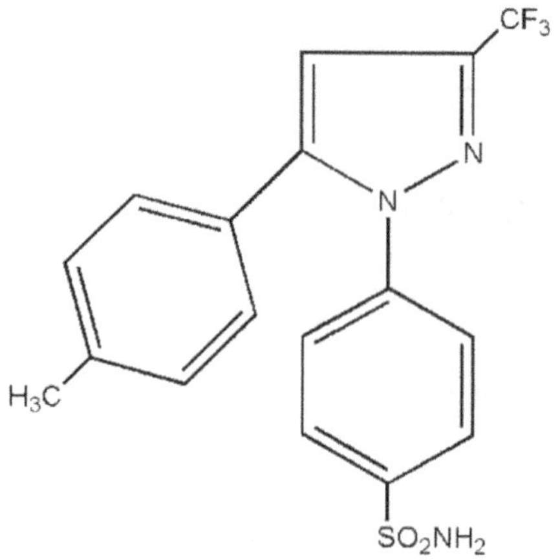

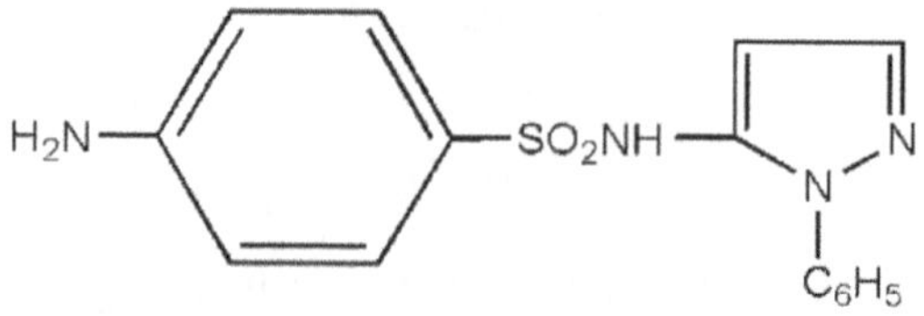

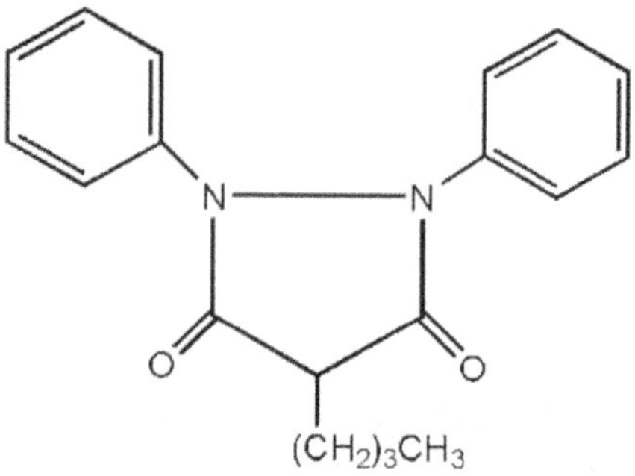

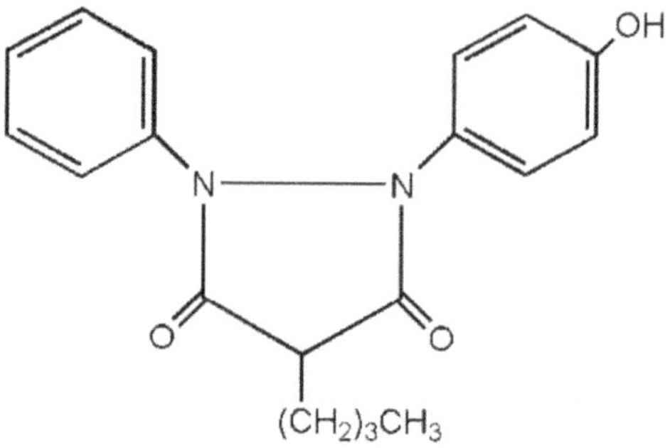

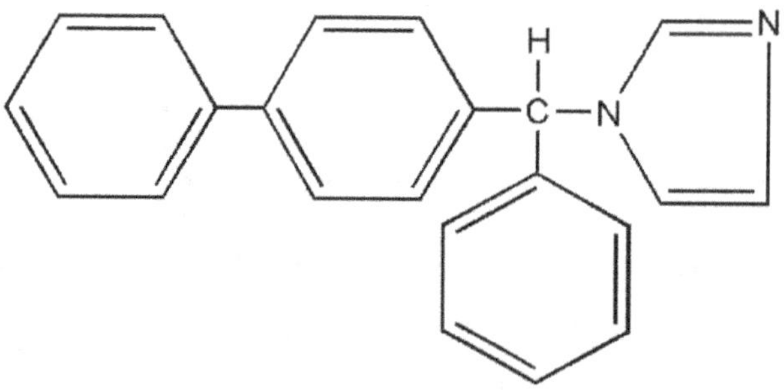

Table *Contd...*

Imidazole (iminazoline)

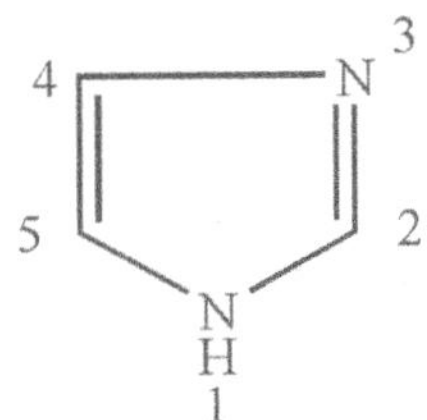

Imidazole is an azapyrrole in which the nitrogen atoms are separated by one carbon atom. It is also known as glyoxazoline and was first synthesized from glyoxal and ammonia. The imino nitrogen is present in position-1-and the tertiary nitrogen atom is present in position-3. It occurs mainly in nature as alkaloids. It is the important building block of histidine and histamine. The name imidazole was given by the chemist Arthur Rudolf Hantzch. When it is fused with pyrimidine it forms the important fused heterocyclic ring purine which is the back bone of DNA and RNA. It is amphoteric in nature. Imidazole derived histidine derivatives play a major role in intracellular buffering system. Imidazole is also used for the purification of Hig-tagged proteins in immobilized metal affinity chromatography.

General Methods of Preparation:

1. **From glyoxal:** Glyoxal reacts with ammonia, formaldehyde and yields imidazole.

CHO–CHO (Glyoxal) + HCHO + $2NH_3$ ⟶ Imidazole + $3H_2O$

2. **From tartaric acid dinitrate:** Tartaric acid dinitrate reacts with ammonia and formaldehyde to give imidazole 4,5-dicarboxylic acid and which upon further decarboxylation gives imidazole.

COOH–$CHONO_2$–$CHONO_2$–COOH (Dinitrotartaric acid) —($-2HNO_2$)⟶ COOH–C=O–C=O–COOH —($2NH_3$, HCHO, $-3H_2O$)⟶ Imidazole-4,5-dicarboxylic acid (HOOC, HOOC)

Imidazole-4,5-dicarboxylic acid —(△, Cu, 300 °C, $-2CO_2$)⟶ Imidazole

Synthesis of substituted imidazoles:

1. **Dehydrogenation of imidazolines:** Imidazolines are dehydrated in the presence of sulphur or barium manganate and yields 2-substituted imidazoles.

$BaMnO_4$

Imidazoline 2-Substituted imidazole

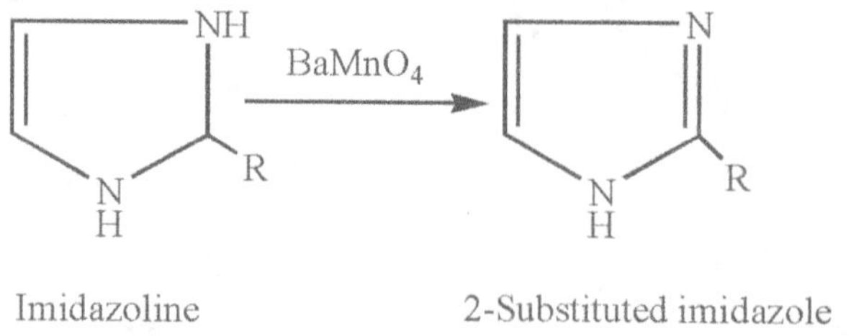

2. **From α-haloketones:** α-Haloketones react with amidines to yield substituted imidazoles.

$BrCH_2—CO—C_6H_5$ + $HN=C(NH_2)—C_6H_5$ → ($-H_2O$, $-HBr$)

α-Halo ketones Amidines 2,4-Diphenyl imidazole

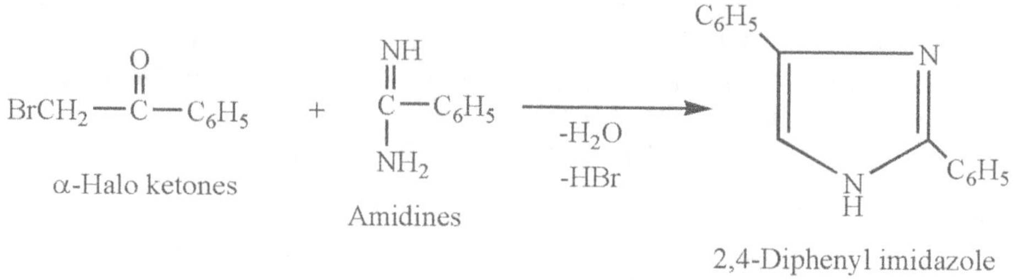

Physical properties: It is weak base but stronger than isomeric pyrazoles. Like pyrazole it also shows tautomerism.

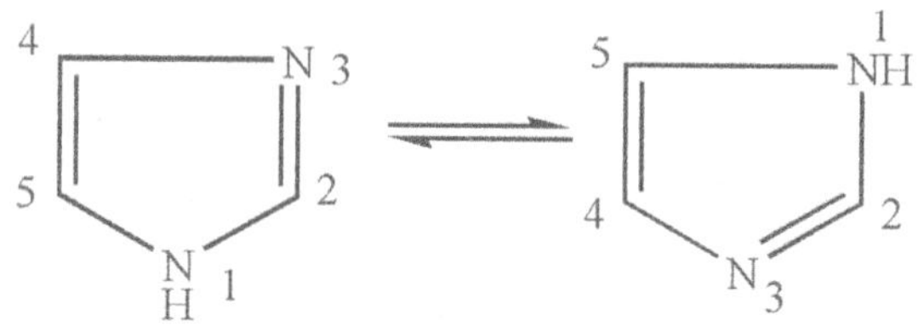

Chemical Properties:

1. **Electrophilic substitution reaction:** The contribution of charge is more in imidazole, hence the reactivity towards electrophiles are more when compared with benzene. It is more susceptible to electrophilic attack than pyrazole and thiazole. Electrophilic attack takes place at C-4 or C-5 position in the imidazole moiety.

   (i) ***Halogenation:*** The halogenation reaction of imidazole is a complex process and depends upon the conditions of substrate and reagents used in the reaction.

   (a) ***Bromination:*** Imidazole reacts with Br_2 in chloroform at -10 °C and yields 2,4,5-tri bromo imidazole.

Imidazole + $3Br_2$ $\xrightarrow{CHCl_3}$ 2,4,5-Tribromoimidazole + 3HBr

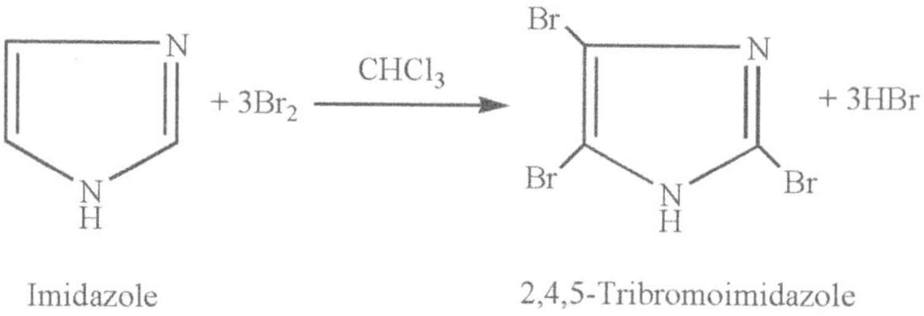

(b) ***Iodination:*** Imidazole reacts with I_2 in alkaline condition and yields 2,4,5-triiodo imidazole.

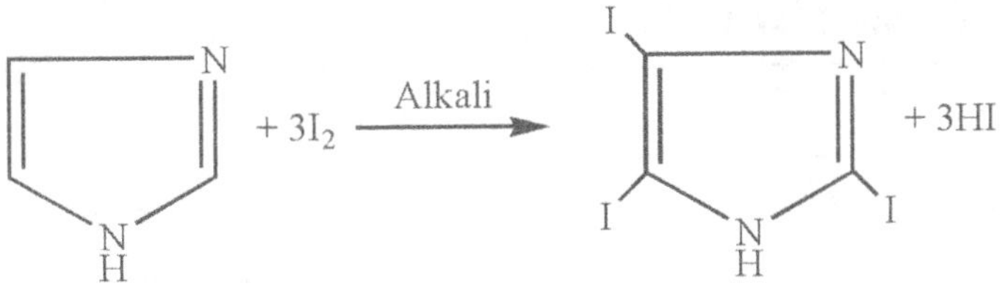

$$\text{Imidazole} + 3I_2 \xrightarrow{\text{Alkali}} \text{2,4,5-Triiodooimidazole} + 3HI$$

Imidazole

2,4,5-Triiodooimidazole

(c) ***Nitration:*** Imidazole reacts with HNO_3/H_2SO_4 and yields 4 or 5-nitro imidazole.

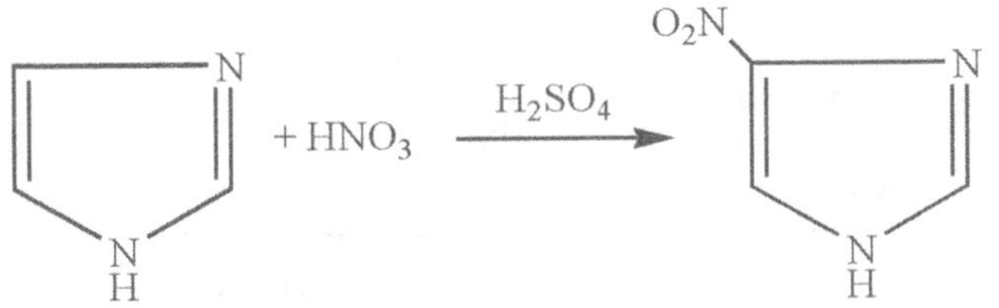

$$\text{Imidazole} + HNO_3 \xrightarrow{H_2SO_4} \text{4-Nitroimidazole}$$

Imidazole

4-Nitroimidazole

(d) ***Sulphonation:*** Imidazole reacts with H_2SO_4 and gives 4 or 5-imidazole sulphonic acid.

$$\text{Imidazole} + H_2SO_4 \longrightarrow \text{4-Imidazole sulphonic acid}$$

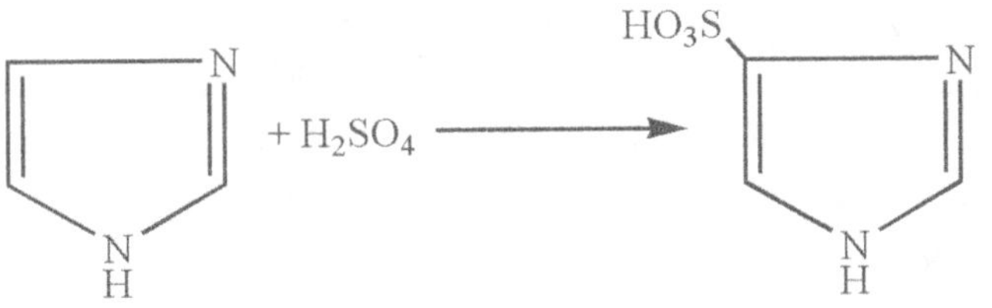

Imidazole

4-Imidazole sulphonic acid

2. Oxidation: Imidazole is stable to oxidation but reacts with H_2O_2 and gives oxamide.

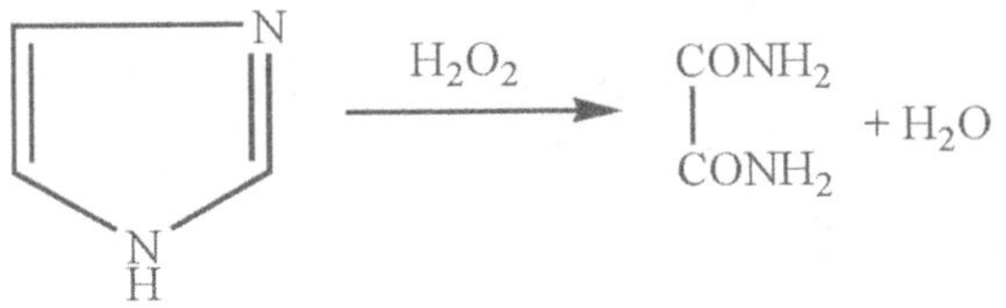

$$\text{Imidazole} \xrightarrow{H_2O_2} \begin{array}{c} CONH_2 \\ | \\ CONH_2 \end{array} + H_2O$$

Imidazole

Oxamide

3. Reaction with benzoyl chloride: Imidazole reacts with benzoyl chloride and NaOH to give di-(benzoyl amino)ethylene.

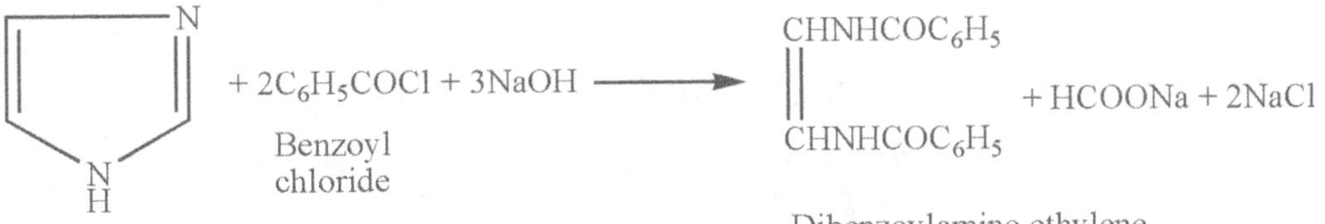

$$\text{Imidazole} + 2C_6H_5COCl + 3NaOH \longrightarrow \begin{array}{c} CHNHCOC_6H_5 \\ \| \\ CHNHCOC_6H_5 \end{array} + HCOONa + 2NaCl$$

Benzoyl chloride

Imidazole

Dibenzoylamino ethylene

Medicinal Compounds Containing Imidazole (or) Imidazole Derivatives Used in Medicine

Carbimazole	**Methimazole**
$COOC_2H_5$, N, S, N, CH_3 2-Methyl -2-thioxo-1,2-dihydroimidazole-3-ethyl carboxylate **Uses:** It is used as an antithyroid agent	NH, S, N, CH_3 3- Methyl-1*H*-imidazole-2(3*H*)-thione **Uses:** It is used as an antithyroid agent
Tolazoline	**Naphazoline**
N, NH, H, C, H 3- Benzylimidazoline **Use:** It is used as an antihypertensive agent	H, C, H, HN, N **Use:** It is used as an adrenergic agent
Clotrimazole	**Xylometazoline**
C, N, N, Cl 1-[(2-Chlorophenyl)diphenyl methyl]-1*H*-imidazole **Use:** It is used as an antifungal agent	CH_3, $(H_3C)_3C$, CH_2, HN, N, . HCl, CH_3 2-[(4-*tert*-butyl-2,6-dimethylphenyl)methyl]-4,5-dihydro-1*H*-imidazole **Use:** It is used as an adrenergic agent

Table *Contd...*

Clonidine	**Moxonidine**
N-(2,6-Dichlorophenyl)-4,5-dihydro-1*H*-imidazol-2-amine **Use:** It is used as an antihypertensive agent	4-Chloro-*N*-(4,5-dihydro-1*H*-imidazol-2-yl)-6-methoxy-2-methylpyrimidin-5-amine **Use:** It is used as an anticancer agent
Etomidate	**Phentolamine**
Ethyl-1-(1-methyl benzyl) imidazole-5-carboxylate **Use:** It is used as a general anaesthetic	3-[(4,5-Dihydro-1-imidazole-2-yl)methyl](4-methyl phenyl)amino phenol **Use:** It is used as an antihypertensive agent
Pantoprazole	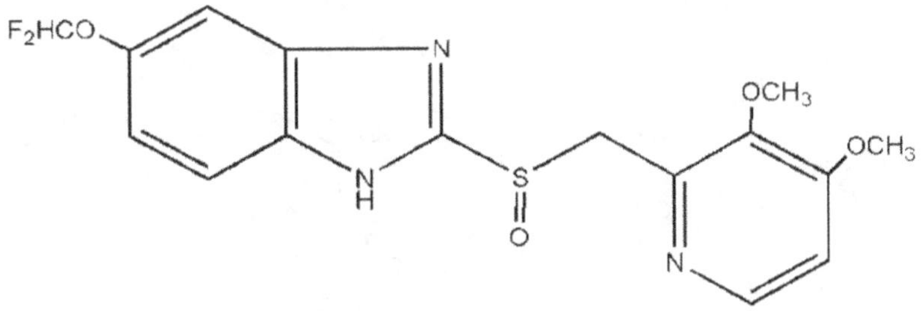
6-(Difluoromethoxy)-2-[(3,4-dimethoxypyridin-2-yl)methylsulfinyl]-1*H*-benzo[*d*]imidazole **Use**: It is used as an antiulcer agent	
Etintidine	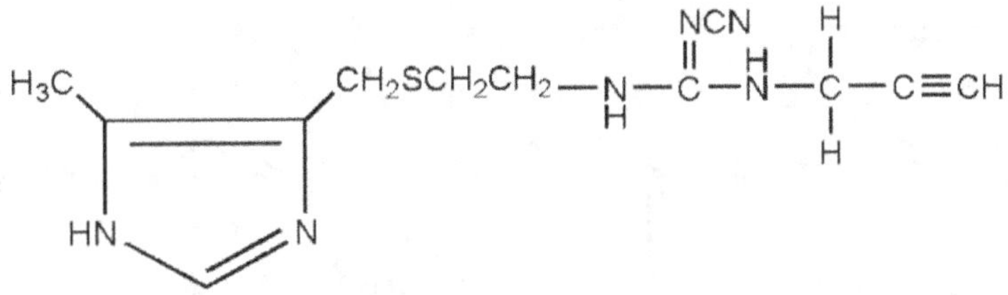
Use: It is used as an antiulcer agent	

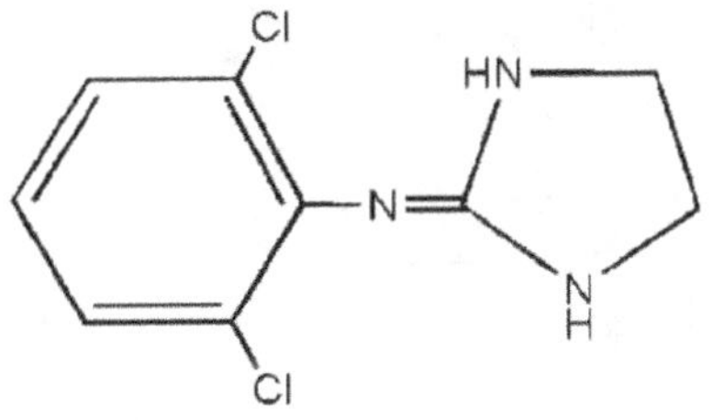

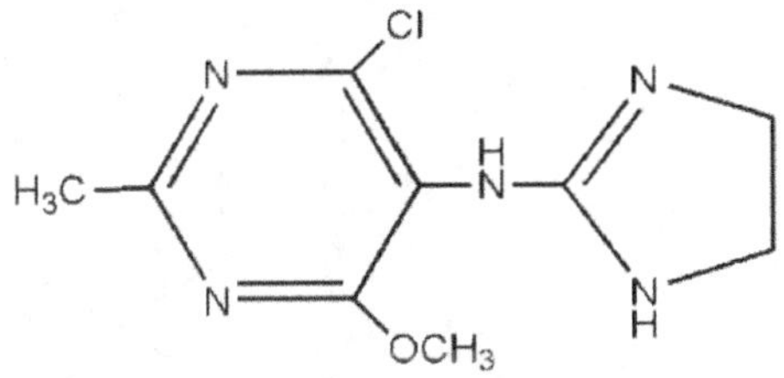

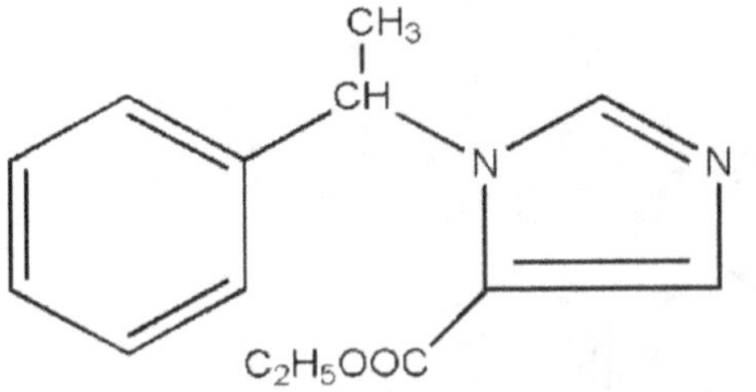

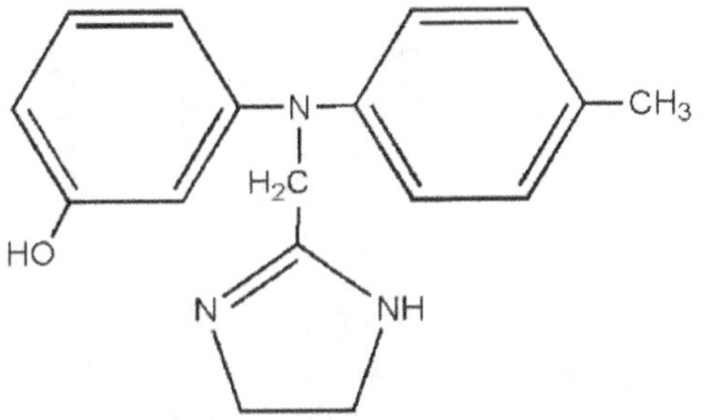

Table *Contd...*

Ketoconazole

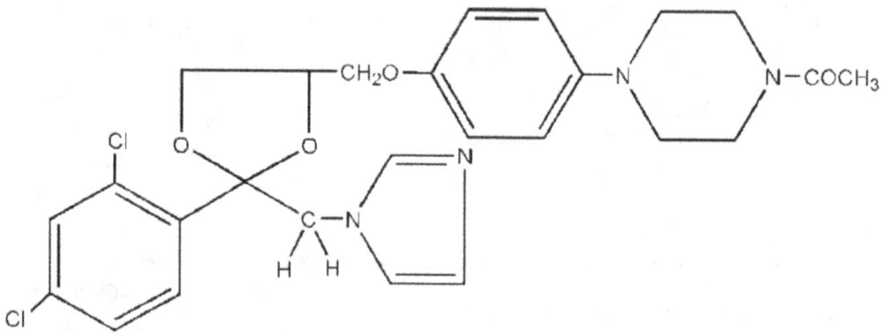

Use: It is used as an antifungal agent

Lansoprazole

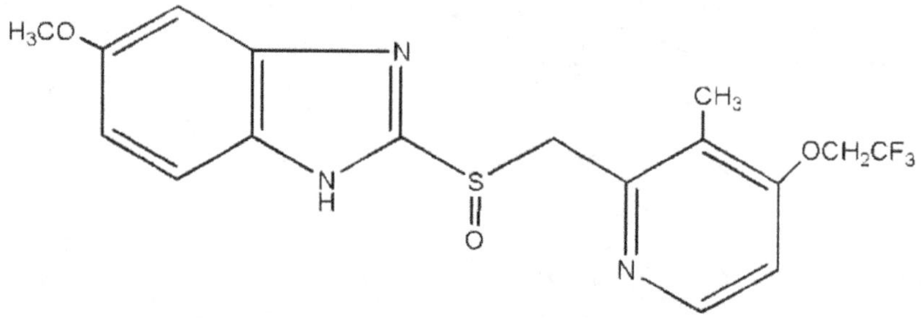

5-Methoxy-2-((3-methyl-4-(2,2,2-trifluoroethoxy)pyridin-2-yl)methylsulfinyl)-1*H*-benzo[*d*]imidazole

Use: It is used as an antiulcer agent

Rabeprazole

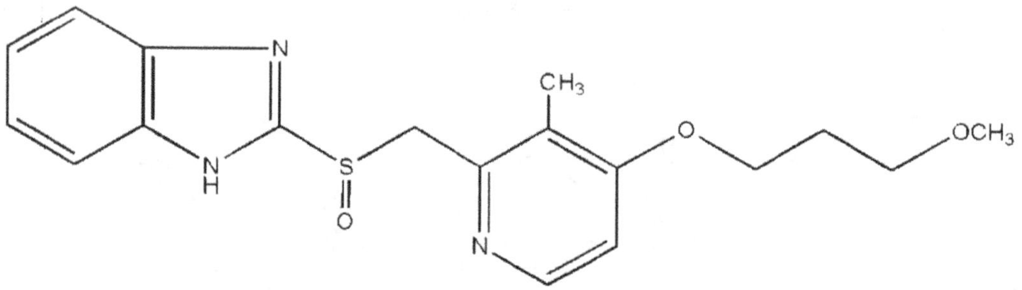

2-((4-Methoxypropoxyl)-3-methyl)pyridin-2-yl)methylsulfinyl)-1*H*-benzo[*d*]imidazole

Use: It is used as an antiulcer agent

Oxmetidine

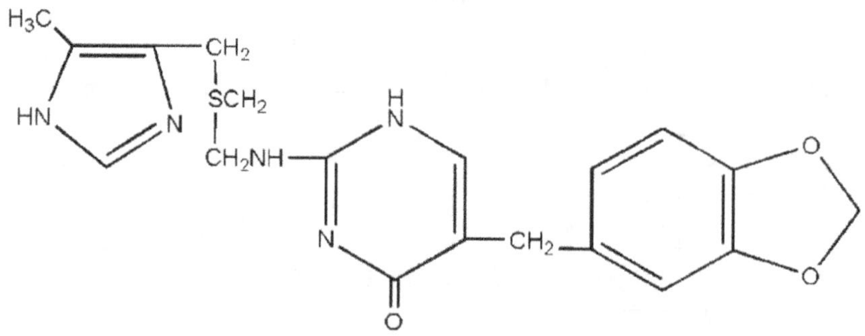

Use: It is used as an antiulcer agent

Oxazole

Five membered ring system contains one oxygen and one nitrogen atom in 1,3-position is called as oxazole. It was first introduced in 1887 by Hantzsch. It does not present in nature. In this the oxygen atom and the nitrogen atom is separated by one carbon atom. The first oxazole was synthesized in the year 1800. It is used for the preparation of many of the important medicinal compounds.

General Methods of Preparation:

1. **From ethyl α-hydroxy ketosuccinate:** Ethyl α-hydroxy keto succinate reacts with formamide to yield diethyl oxazole-4,5-dicarboxylate which upon hydrolysis and decarboxylation yield oxazole.

α-Hydroxy keto succinate + $HCONH_2$ (Formamide) —(100 °C, $-2H_2O$)→ diethyl oxazole-4,5-dicarboxylate —($Ba(OH)_2$, $2H_2O$, $-2C_2H_5OH$)→ oxazole-4,5-dicarboxylic acid —(Ca, Quinoline, Δ, $-2CO_2$)→ Oxazole

Synthesis of substituted oxazole:

1. **Robinson-Gabriel synthesis:** α-acylamino ketones undergo cyclisation and dehydration in the presence of strong mineral acid or P_2O_5 and gives substituted oxazoles.

α-Acylamino Ketone → (hydroxy oxazoline intermediate) —($-H_2O$)→ Substituted Oxazole

2. **From α-aminocarbonyl compounds:** α-Amino carbonyl compounds are refluxed with imino ester and gives substituted oxazole.

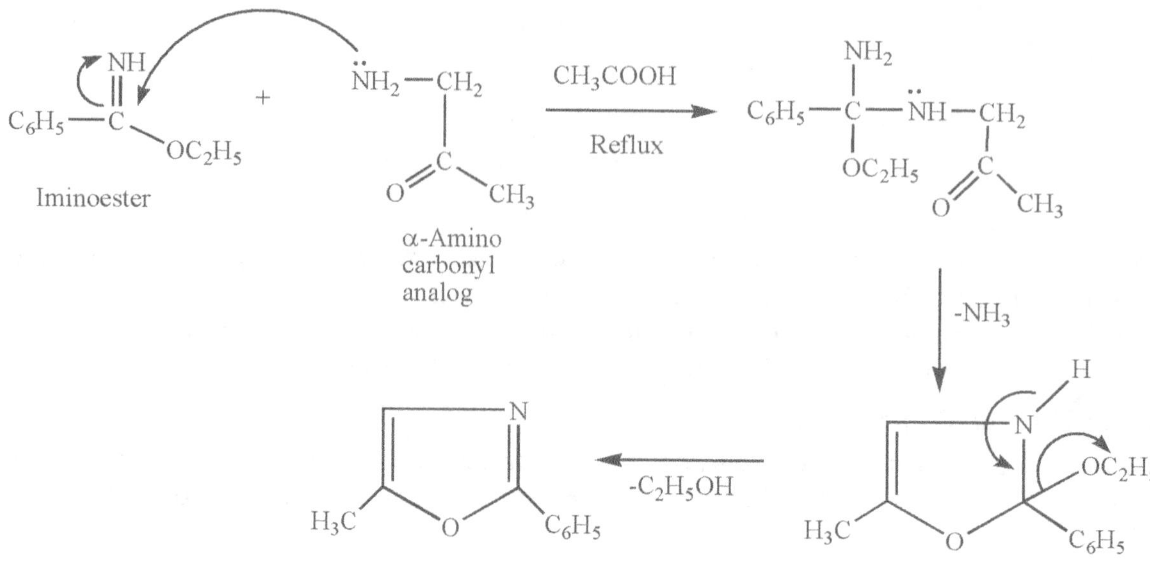

Iminoester

α-Amino carbonyl analog

CH_3COOH Reflux

$-NH_3$

$-C_2H_5OH$

5-Methyl-2-phenyl oxazole

Chemical Properties:

1. **Electrophilic substitution reactions:**

When compared with thiazole, oxazole is more reactive towards electrophilic reagents but less reactive than imidazole. Electrophile is substituted at C-5 position. Electrophilic substitution takes place easily when the ring is attached with electron donating substituents as shown below.

(i) ***Bromination of 2-phenyl oxazole:*** 2-phenyl oxazole reacts with NBS (N-Bromo succinimide) and yields 5-bromo 2-phenyl oxazole

Oxazole → (NBS) → 5-Bromo-2 phenyl oxazole

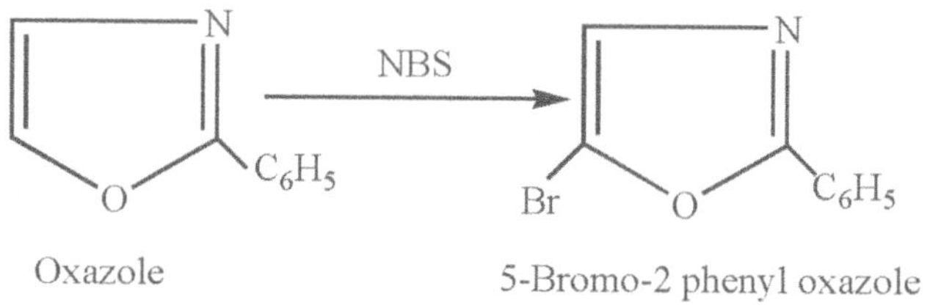

(ii) ***Nitration and sulphonation:*** Due to the presence of pyridine type "N" atom, imidazole does not undergo nitration and sulphonation.

2. **Oxidation:** It is easily oxidised by $KMnO_4$, ozone and chromic acid. During this reaction, the imidazole ring is opened or cleaved.
3. **Reduction:** Oxazole is reduced to oxazolidine with the aid of sodium in ethanol.

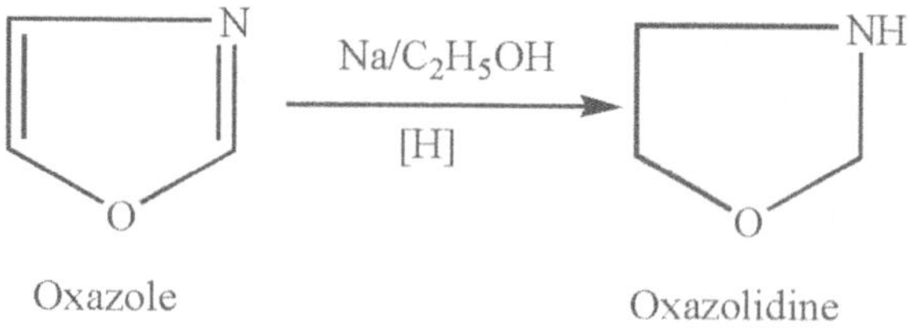

Oxazole → (Na/C_2H_5OH, [H]) → Oxazolidine

Medicinal Compounds Containing Oxazole Derivatives Used in Medicine

Pemoline	Fenozolone
2-Amino-5-phenyloxazol-4(5*H*)-one **Use:** It is used as a CNS stimulant	2-(Ethylamino)-5-phenyl-1,3-oxazol-4(5*H*)-one **Use:** It is used as a CNS stimulant
Tozalinone	**Trimethadione**
2-(Dimethylamino)-5-phenyl-2-oxazolin-4-one **Use:** It is used as a CNS stimulant	3,5,5-Trimethyl oxazolidine-2,4-dione **Use:** It is used as an anticonvulsant
Paramethadone 5-Ethyl-3,5-dimethyl oxazolidine-2,4-dione **Use:** It is used as an anticonvulsant	

Thiazole

It is a five membered heterocycle contains one sulphur and one nitrogen atoms in the ring at 1,3-position. It was first discovered by Hantzsch Weber. Thiazole ring is a part of pencillins structure and vitamin B *etc.*

General Methods of Preparation:

1. **From thiourea:** Reaction between thiourea and α-halocarbonyl compound gives thiazole.

Thiourea + Chloro acetaldehyde $\xrightarrow[-HCl]{-H_2O}$ 2-Amino thiazole $\xrightarrow[NaNO_2,\ HCl]{Diazotisation}$ Thiazole diazonium chloride $\xrightarrow{C_2H_5OH,\ Boil}$ Thiazole

2. **From thioamides:** Thioamides react with α-halo carbonyl compound and gives substituted thiazole.

CH_3CO–CH_2Cl + $HN{=}C(SH)$–CH_3 ⟶ 2,4-Dimethyl thiazole + HCl + H_2O

Chloroacetaldehyde (α-Halecarbonyl compound) Thioamide 2,4-Dimethyl thiazole

Physical properties: Thiazole is a colorless liquid.B.P. 177 °C. It smells like pyridine, miscible in water.

Chemical properties: It is a weak base and behaves as a tertiary base.

1. **Reaction with acids:** The lone pair of electrons of nitrogen atom in thiazole does not involve in aromatic sextet, hence available for protonation. Thiazole reacts with strong acids and forms salts.
2. **Electrophilic Substitution Reactions:** The reactivity of thiazole with electrphilic reagent is intermediate between thiophene and pyridine, but less reactive than imidazole. The substitution preferentially takes place at C-5 position.
   (i) ***Bromination:*** Gaseous phase bromination of thiazole yields 2-bromo thiazole.

Thiazole $\xrightarrow[250\text{-}400\ °C]{Br_2}$ 2-Bromothiazole

   (ii) ***Nitration:*** Thiazole does not undergo direct nitration in presence of oleum at 160 °C.
   (iii) ***Sulphonation:*** Thiazole gets sulphonated under drastic conditions with mercury sulphate catalyst to yield thiazole-5-sulphonic acid.

Thiazole $\xrightarrow[\text{Drastic condition}]{H_2SO_4,\ HgSO_4}$ Thiazole-5-sulphonic acid (HO_3S at C-5)

3. **Oxidation:** Thiazole is resistant to oxidation especially with HNO_3.
4. **Reduction:** Substituted thiazoles are reduced with H_2/Ni in methanol and yields sulphur (desulphurisation reaction).

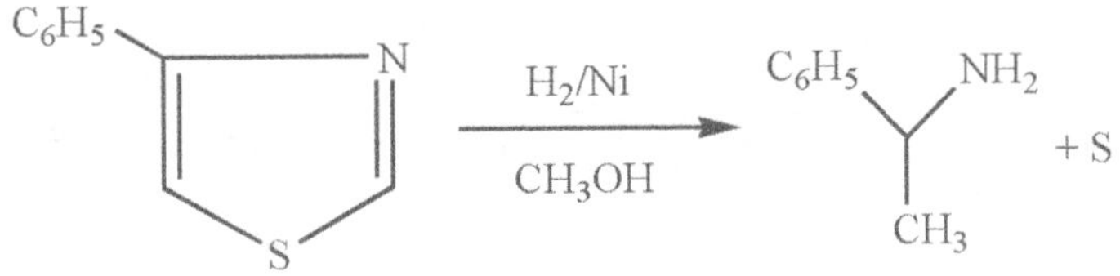

C_6H_5 N S — H_2/Ni, CH_3OH → C_6H_5 NH_2 CH_3 + S

4-Phenyl thiazole — Phenyl ethyl amine

Medicinal Compounds Containing Thiazole or Thiazole Derivatives Used in Medicine

Famotidine 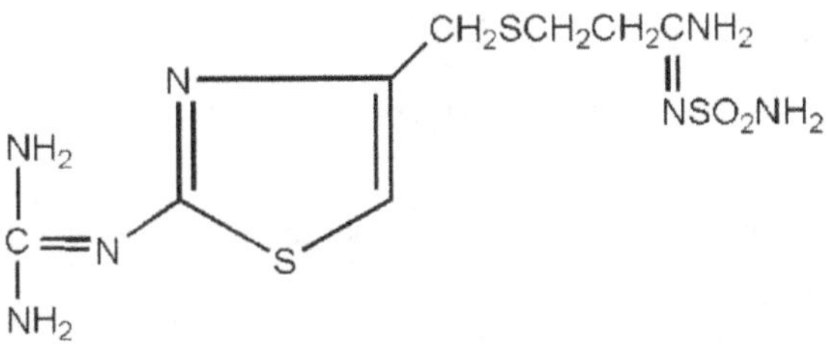 $CH_2SCH_2CH_2CNH_2$, NSO_2NH_2, N, S, NH_2, C=N, NH_2 **Use:** It is used as an antiulcer agent	**Sulphasomizole** 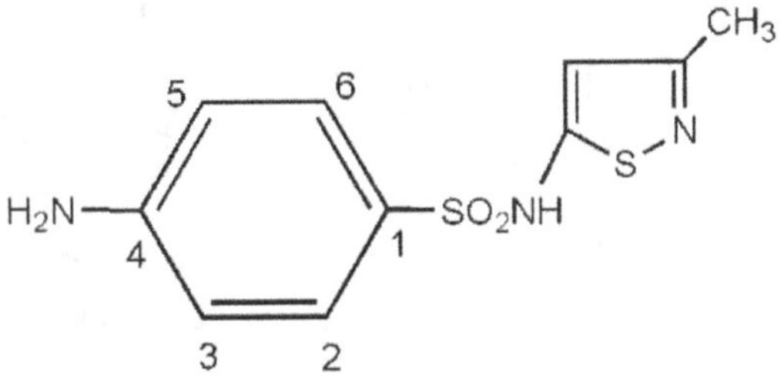 H_2N, SO_2NH, CH_3, N, S, 1, 2, 3, 4, 5, 6 4-Amino-*N*-(3-methyl-1,2-thiazol-5-yl)benzenesulfonamide **Use:** It is used as an antibacterial agent
Ethoxozolamide 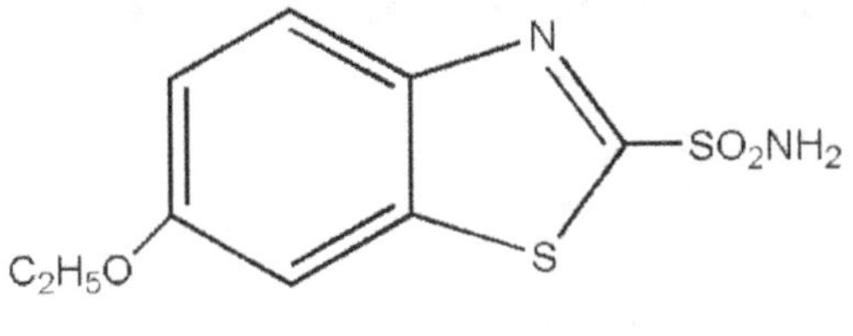 N, S, SO_2NH_2, C_2H_5O 6-Ethoxy-1,3-benzothiazole-2-sulfonamide **Use:** It is used as a diuretic agent	**Sulphathiazole** 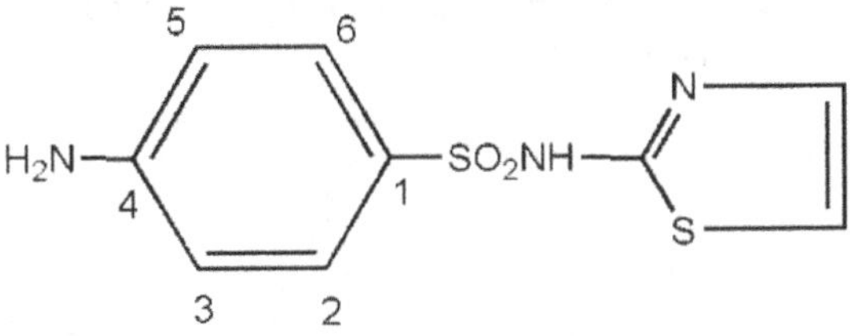 H_2N, SO_2NH, N, S, 1, 2, 3, 4, 5, 6 4-Amino-*N*-(1,3-thiazol-2-yl)benzenesulfonamide **Use:** It is used as an antibacterial agent
Nitazoxanide 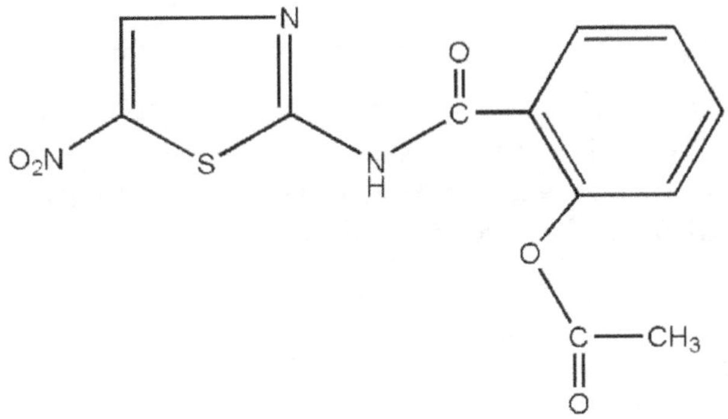 O_2N, N, S, N H, C=O, O, C—CH_3, O 2-((5-Nitrothiazol-2-yl)carbamoyl)phenyl acetate **Use:** It is used as an antiamoebic agent	

Table *Contd...*

Tiotidine

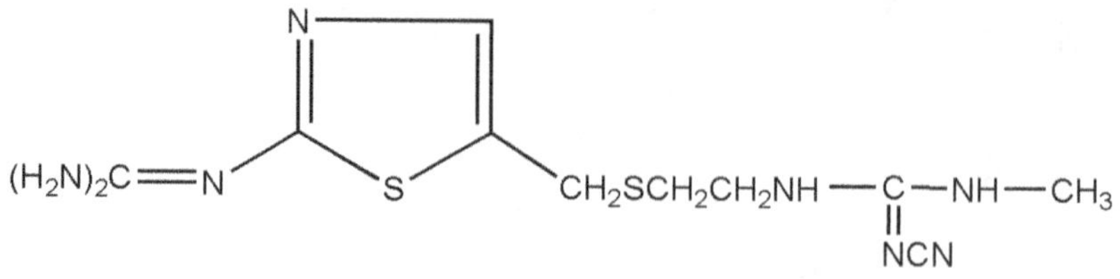

1-Cyano-3-[2-[[2-(diaminomethylideneamino)-1,3-thiazol-4-yl]methylsulfanyl]ethyl]-2-methylguanidine

Use: It is used as an antiulcer agent

Nizatidine

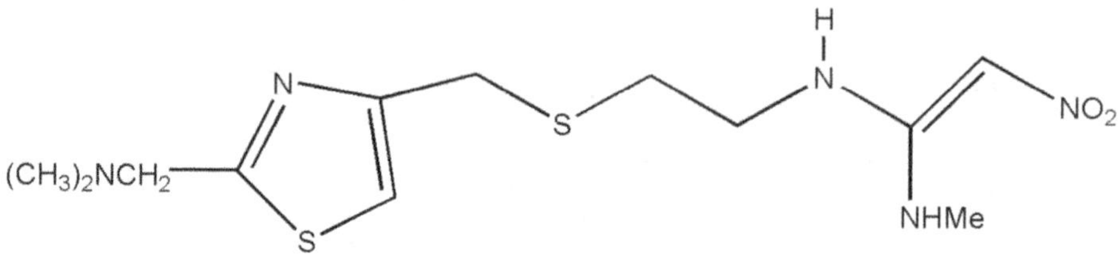

(*E*)-1-*N'*-[2-[[2-[(Dimethylamino)methyl]-1,3-thiazol-4-yl]methylsulfanyl]ethyl]-1-*N*-methyl-2-nitroethene-1,1-diamine

Use: It is used as an antiulcer agent

Six Membered Heterocycles with One Hetero Atom

Replacement of the methine (-CH=) group in six membered benzene ring by hetero atoms such as nitrogen, oxygen or sulphur yields important kind of compounds called six membered heterocyclic compounds. They are as follows.

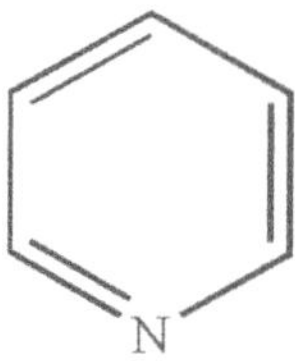

Pyridine

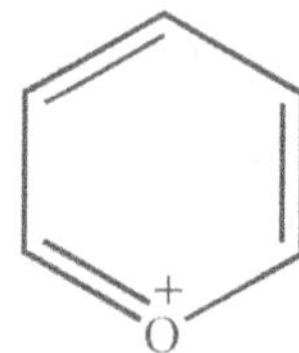

Pyrilium salt (Less stable)

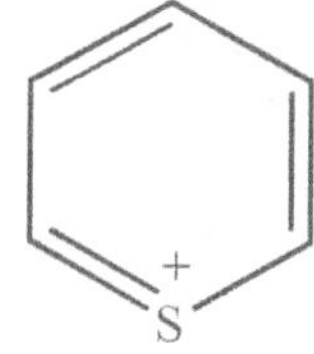

Thiopyrylium salt

Pyridine

The replacement of methine group (-CH=) in benzene ring by a trivalent nitrogen atom yields pyridine. It is the simplest and important and well known six membered heterocyclic ring. It was first discovered by Anderson and isolated from bone oil. It is also present in coal tar. The pyridine ring and its derivatives are highly distributed in nature and mainly present in alkaloids.

Nomenclature:

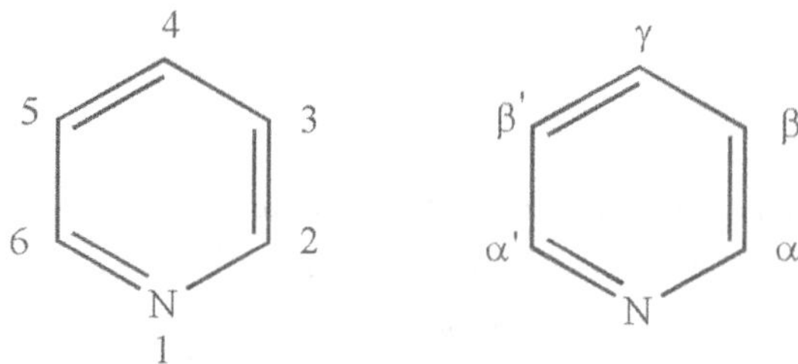

The nitrogen atom in pyridine ring is assigned as position 1. Pyridine possess element of asymmetry, hence there are three mono substituted pyridines are possible. The positions on the mono sunstituted pyridine are indicated by either Greek alphabet or numbering system.

The alkyl derivatives of pyridine are named by both trivial names as well as by systematic nomenclature.

α-Picoline (2-Methyl pyridine) β-Picoline (3-Methyl pyridine) γ-Picoline (4-Methyl pyridine)

Dimethyl pyridines nomenclature: There are six types of dimethyl pyridines available, commonly known as lutidines. Similarly trimethyl pyridines are known as collidines.

2,6-Lutidine (2,6-Dimethyl pyridine) 2,4,6-Collidines (2,4,6-Trimethyl pyridine)

Reduced form of pyridine:

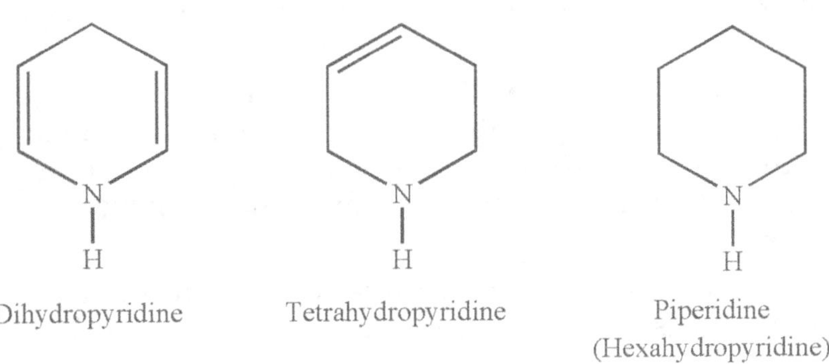

Dihydropyridine Tetrahydropyridine Piperidine (Hexahydropyridine)

Structure of Pyridine

The structure of pyridine is closely similar to that of benzene. The molecular orbital structure of pyridine is shown in the following figure. Each of the five carbon and the nitrogen atom is sp^2 hybridised (the two sp^2 carbon atom are overlap with the two hybrid orbitals of two adjacent carbon atoms and one with *1s* orbital of the hydrogen atom). Hence each of the five carbon atoms of pyridine has three π bonds. Likewise, the two sp^2 hybrid orbitals of nitrogen atom form two π bonds with adjacent carbon atoms, and third are fully occupied by the "N" atom and acts as a lone pair. The remaining fifth electron remains in the *p*-orbital.

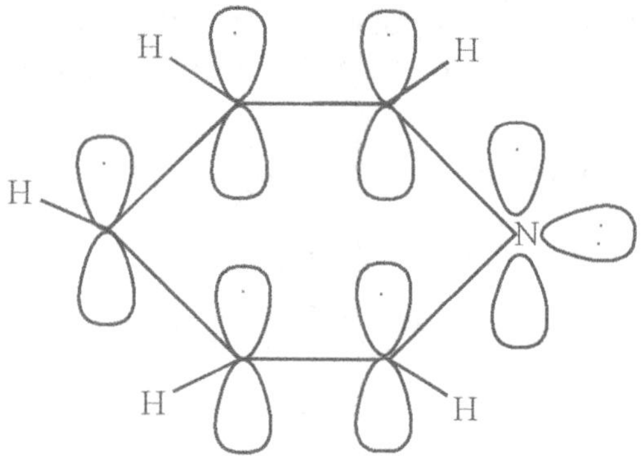

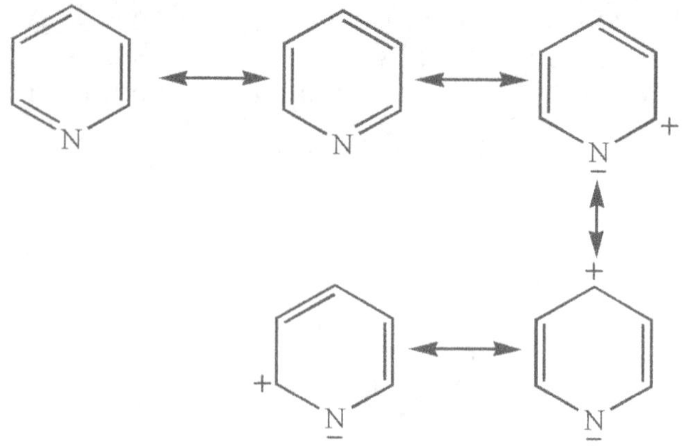

Resonance Structures of Pyridine

The six "p" orbitals (five from 5 carbon atom and one from nitrogen atom) having one electron overlap with each other to form a π molecular orbital which contains six electrons.

Resonance forms of pyridine: Various forms of resonance hybrid structures of pyridine are shown above.

Basicity of Pyridine: Pyridine behaves like a tertiary base and forms salts with acids *i.e.,* it reacts with HCl and forms pyridinium hydrochloride.

Pyridine + HCl → Pyridinium hydrochloride

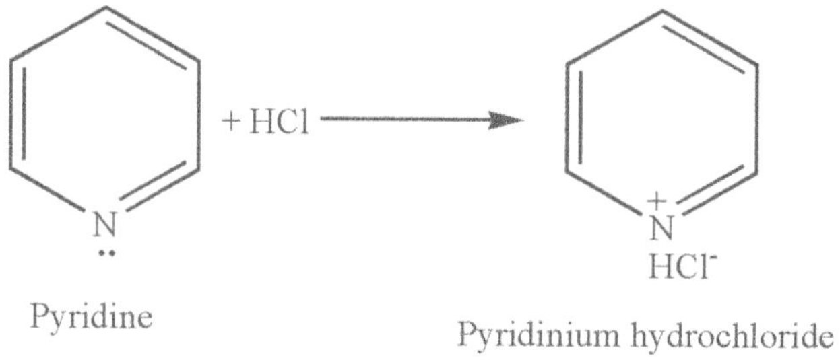

The lone pair of electrons present in the nitrogen atom of the pyridine ring is solemnely responsible for its basic character. This lone pair of electrons are not involving in aromatic character, hence it is readily available for protonation with mineral acids. But pyridine is less basic than alkylamines and piperidine. It is explained by the following facts.

(i) The nitrogen atom in pyridine is sp^2 hybridised (more *s* character) and possess less *p* charecter.

(ii) Hence the nitrogen atom of pyridine is more electrons attracting than sp^3 hybridised nitrogen atom of alkylamines.

(iii) Due to this, the lone pair of electrons in the nitrogen atom of pyridine is tightly attached and not readily available for protonation. So pyridine is less basic than alkylamines and piperidine. Similarly pyridine is stronger base than pyrrole because the lone pair of electrons in nitrogen atom of pyridine does not involve in the formation of aromatic sextet but in pyrrole the lone pair is involved in aromatic sextet.

Like tertiary amine pyridine forms quarternary ammonium salts. This nature of lone pair of electrons of nitrogen of the pyridine ring makes it basic and produces remarkable effect in electrophilic substitution reaction.

General Methods of Preparation:

1. **From acetylene**: A mixture of acetylene and hydrogen cyanide are passed through a red hot tube to yield pyridine.

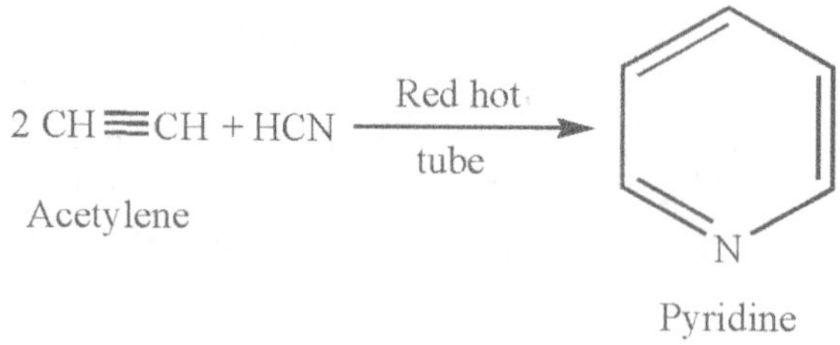

$$2\ CH{\equiv}CH + HCN \xrightarrow[\text{tube}]{\text{Red hot}} \text{Pyridine}$$

Acetylene

2. **From pentamethylene diamine hydrochloride:** Heating of pentamethylene diamine HCl yields piperidine which upon further reaction with H_2SO_4 at 300 °C gives pyridine.

H_2C CH_2 H_2C CH_2 H_2C CH_2 NH_2 $NH_2 \cdot HCl$ —Δ, $-NH_4Cl$, $-HCl$→ Piperidine —H_2SO_4, 300 °C→ Pyridine + 3 H_2

Pentamethylene diamine hydrochloride   Piperidine   Pyridine

3. **From glutaconic aldehyde:** Glutaconic aldehyde reacts with ammonia and yields pyridine.

CHOH —NH_3, $-2H_2O$→ Pyridine

Glutaconic aldehyde   Pyridine

4. **Guareschi method (synthesis of alkylated pyridine):** Cyano acetamide is treated with a diketone to yield substituted pyridine which is converted into pyridine by H_2SO_4 and PCl_5 or $POCl_3$.

Diketone + Cyano acetamide → (R, CN, NH, O substituted pyridone) —H_2SO_4, Δ, $PCl_5/POCl_3$→ (R, Cl substituted pyridine) —$-HCl$, H_2/Pd→ (R substituted pyridine)

Chemical Properties:

1. **Basic properties:** Pyridine is basic in nature, because of the lone pair of electrons present in the pyridine ring which is readily available for extra bonding it shows the ideal characteristics of tertiary amine. Pyridine easily reacts with HCl and forms pyridinium chloride.

Pyridine —HCl→ Pyridinium chloride (Cl^-)

Pyridine also reacts with $AlCl_3$ and forms complex.

Pyridine + $AlCl_3$ → pyridine–$AlCl_3$ complex (N^+–Al^-Cl_3)

Pyridine

2. **Electrophilic substitution reactions:** Due to the presence of electronegative nitrogen atom in pyridine, the five carbon atoms are electron deficient, hence pyridine is less reactive towards electrophiles than benzene. Electerophilic substitution takes place only in extreme severe conditions. Electrophilic substitution occurs at C-3 position of the pyridine (electronegative nitrogen atom does not require to accept a positive charge).
   (i) ***Halogenation:*** Halogenation of pyridine takes place only under vigorous conditions.
      (a) ***Chlorination:*** Pyridine reacts with Cl_2 in the presence of large excess amount of $AlCl_3$ and yields 3-chloropyridine (30-35% yield). Small amount of 3,5-dichloro pyridine also formed.

Pyridine + Cl_2 —(Large excess $AlCl_3$)→ 3-Chloropyridine + HCl

      (b) ***Bromination:*** Pyridine reacts with Br_2 in the presence of oleum at 130 °C and yields 3-bromopyridine.

Pyridine + Br_2/Oleum —(130 °C)→ 3-Bromopyridine + HBr

      (c) ***Iodination:*** Pyridine reacts with iodine in the presence of oleum at 320 °C and yield 3-iodopyridine (yield is 18%).

Pyridine + I_2 —(Oleum, 320 °C)→ 3-Iodopyridine + HI

(ii) ***Nitration:*** Nitration of pyridine takes places in vigorous conditions only. Pyridine reacts with HNO_3 / H_2SO_4 at 300 °C and yield 3-nitropyridine in poor yield.

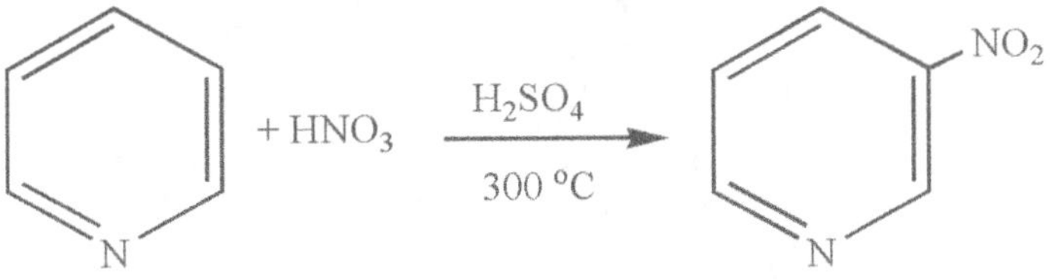

Pyridine + HNO_3 $\xrightarrow[300\ °C]{H_2SO_4}$ 3-Nitropyridine

(iii) ***Sulphonation:*** Pyridine reacts with Conc. H_2SO_4 or oleum in the presence of mercuric sulphate catalyst at 230 °C upon prolonged heating and yields pyridine-3-sulphonic acid (yield 75-85%).

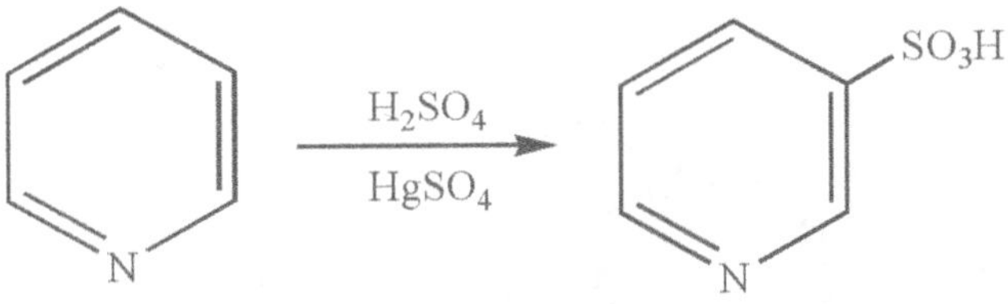

Pyridine $\xrightarrow[HgSO_4]{H_2SO_4}$ Pyridine-3-sulphonic acid

(iv) ***Friedel-craft's reaction:*** Pyridine does not undergo Frieldel-craft's reactions.

(v) ***Mercuration:*** Pyridine is heated with mercuric acetate at 170-180 °C to give a salt which upon rearrangement to yield 3-pyridyl mercuric acetate.

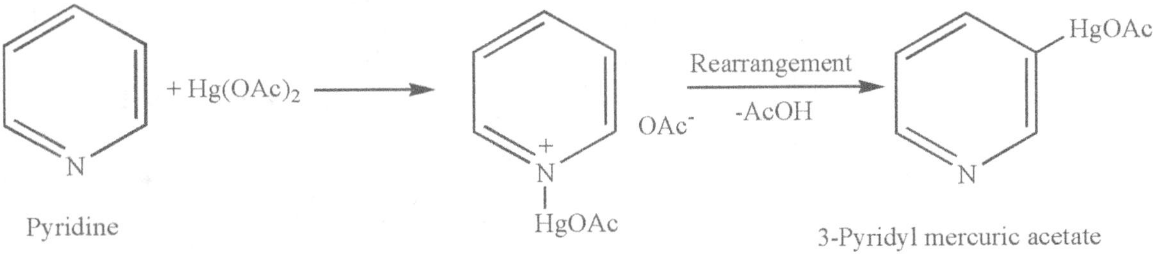

Pyridine + $Hg(OAc)_2$ ⟶ N^+–HgOAc OAc^- $\xrightarrow[-AcOH]{Rearrangement}$ 3-Pyridyl mercuric acetate

3. Nucleophilic substitution reactions: The electron deficient nature of the C-4 & C-5 positions of pyridine makes the ring to easily susceptible for nucleophilic attack. For example,

(i) ***Chichibabin reaction (preparation of amino pyridine):*** Pyridine reacts with sodamide in toluene at 100 °C and yields 2-amino pyridine.

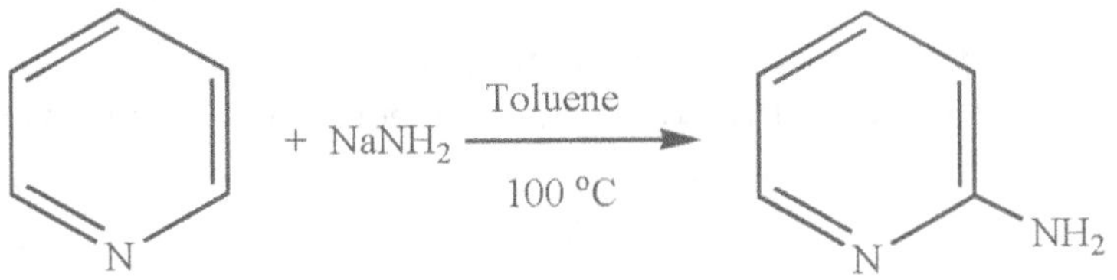

Pyridine + $NaNH_2$ $\xrightarrow[100\ °C]{Toluene}$ 2-Aminopyridine

Mechanism:

i) $NaNH_2$ ⟶ $Na^+ + NH_2^-$

ii) Pyridine + NH_2^- $\xrightarrow[Attack]{Nucleophilic}$ (adduct with NH_2, H at C-2, N^-) $\xrightarrow[Elimination]{Hydride\ ion}$ 2-Aminopyridine

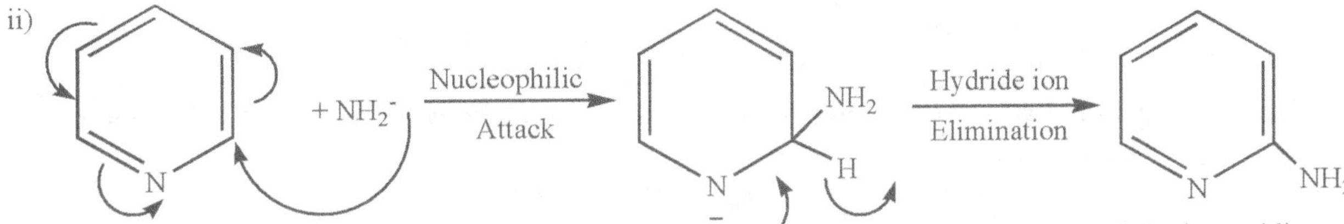

The other nucleophilic substitution reactions of pyridine are summarized as follows.

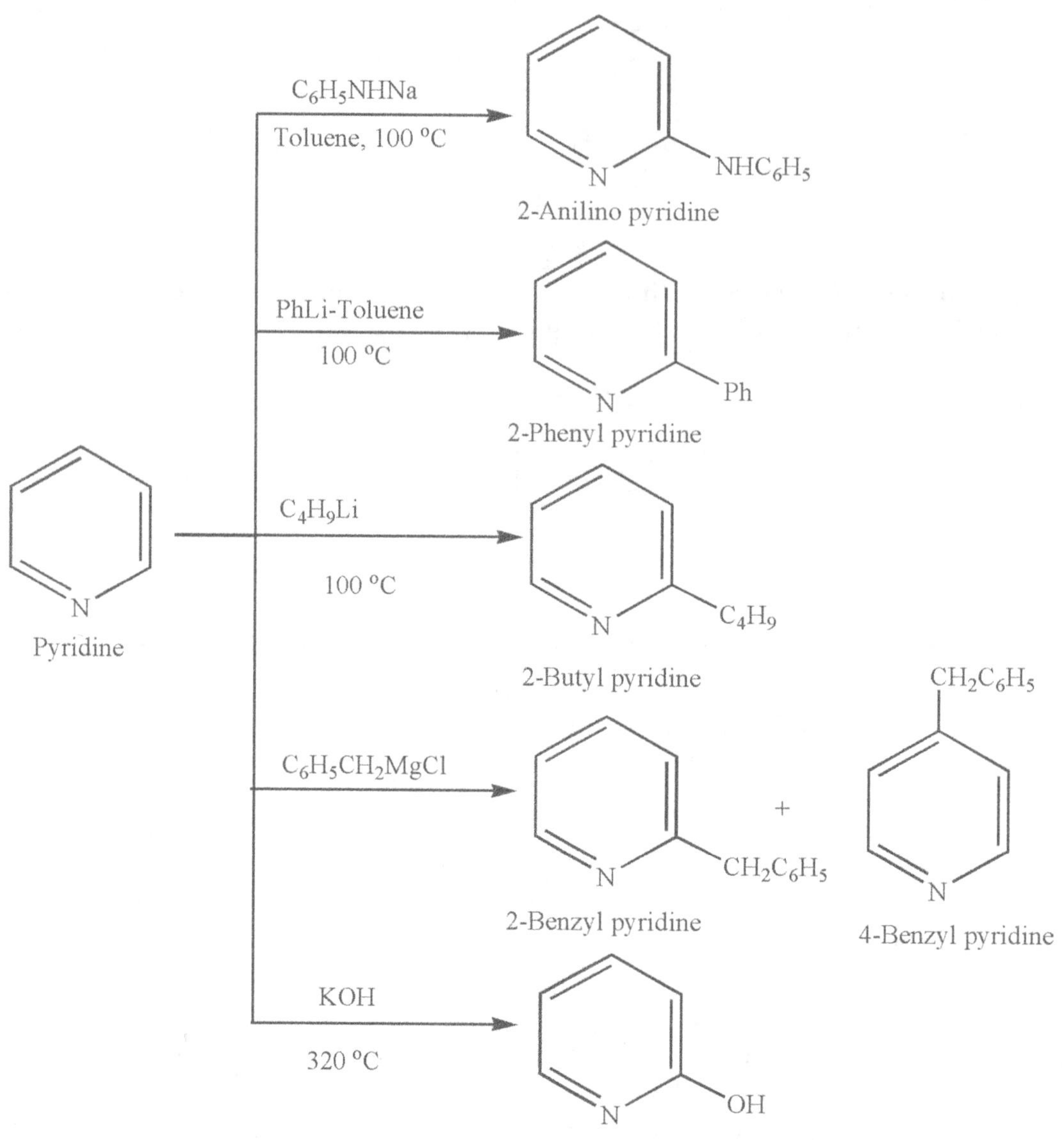

4. **Oxidation:** Pyridine is highly resistant to oxidizing reagents due to the presence of low electron dense carbon atoms. But the nitrogen atom is oxidized by H_2O_2 or peracids to yield pyridine-*N*-oxide.

N

H_2SO_4

$CH_3{\cdot}COOH$

65 °C

$\overset{+}{N}$

O^-

Pyridine

Pyridine-*N*-oxide

5. **Reduction:** Pyridine easily undergoes reduction than benzene due to the presence of low electron density carbon atoms.

Pyridine is easily reduced to piperidine by Raney nickel at 120 °C.

$$\text{Pyridine} + 3H_2 \xrightarrow[120\,°C]{\text{Raney Ni}} \text{Piperidine}$$

Pyridine Piperidine

Pyridine is also reduced by sodium & alcohol or lithium aluminium hydride and yields tetra hydropyridine.

$$\text{Pyridine} \xrightarrow[\text{or Na/ROH}]{LiAlH_4} \text{Tetrahydro pyridine}$$

Pyridine Tetrahydro pyridine

It is also reduced to dihydropyridine by Na in liq. ammonia.

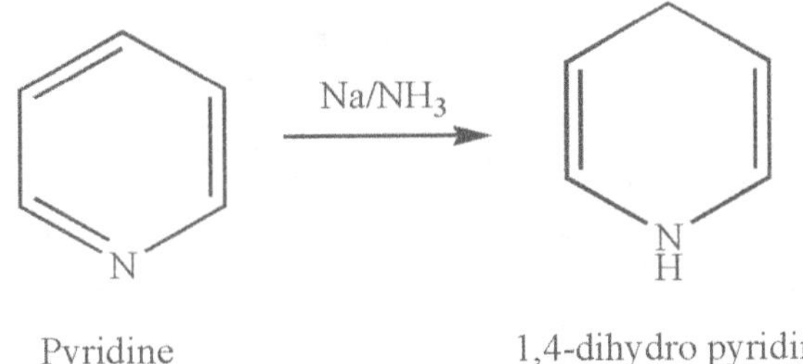

Pyridine 1,4-dihydro pyridine

Medicinal Compounds Containing Pyridine (or) Pyridine Derivatives Used in Medicine

Nicotinic acid (Niacin)	**Methyprylone**
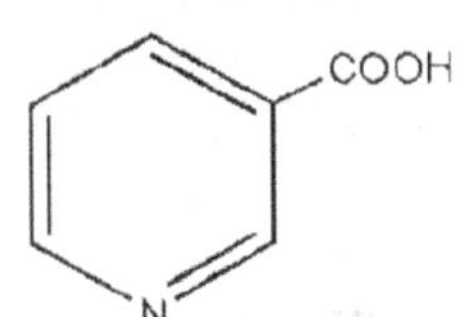 Pyridine-3-carboxylic acid **Use:** It is used as an antihyperlipidaemic agent	(*RS*)-3,3-Diethyl-5-methylpiperidine-2,4-dione **Use:** It is used as a sedative and hypnotic
Glutethimide	**Ethypicone**
3-Ethyl-3-phenylpiperidine-2,6-dione **Use:** It is used as a sedative and hypnotic	**Use:** It is used as a sedative and hypnotic

Table *Contd...*

Dihydroprylone 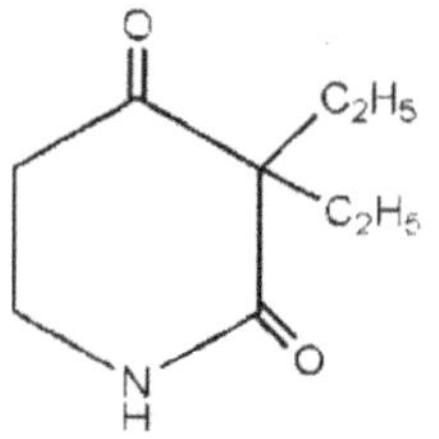 3,3-Diethylpyridine-2,4(1*H*,3*H*)-dione **Use:** It is used as a sedative and hypnotic	**Piperocaine** 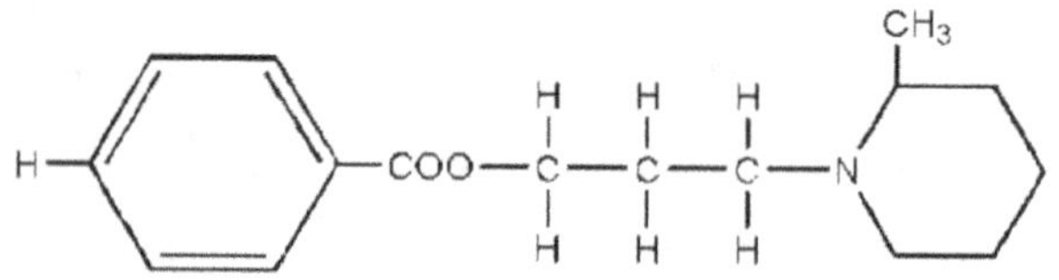 3-(2-Methylpiperidin-1-yl)propyl benzoate **Uses:** It is used as a local anaesthetic
Iproniazid 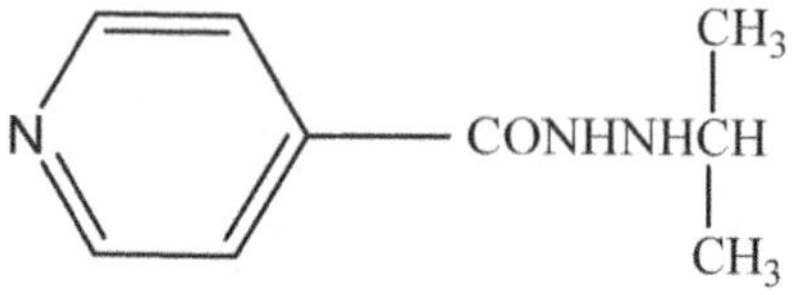 *N'*-Propan-2-ylpyridine-4-carbohydrazide **Use:** It is used as an antidepressants	**Phenindamine** 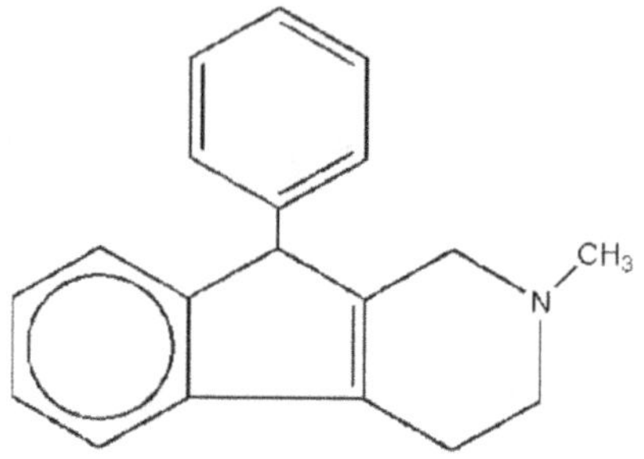 2-Methyl-9-phenyl-1,3,4,9-tetrahydroindeno[2,1-c]pyridine **Use:** It is used as an anti histamines
Metyrapone 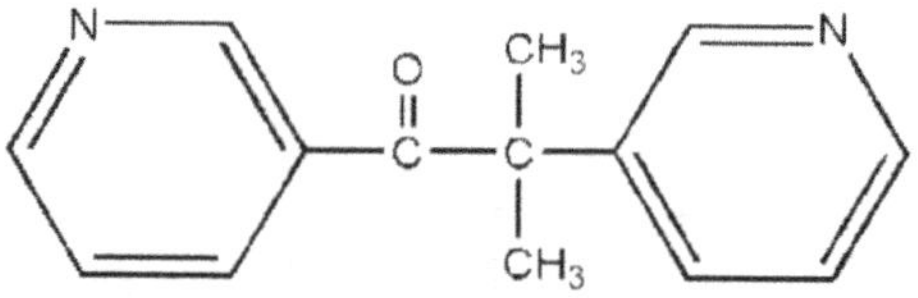 2-Methyl-1,2-di-3-pyridyl-1-propanone **Use:** It is used as a diuretic	**Torsemide** 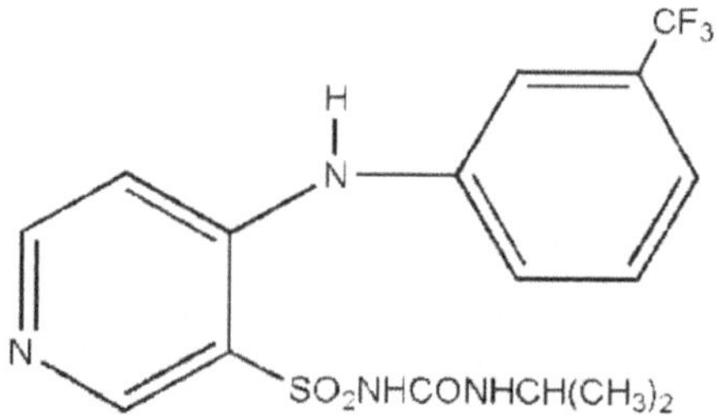 1-[4-(3-Methylanilino)pyridin-3-yl]sulfonyl-3-propan-2-ylurea **Use:** It is used as a diuretic agent
Ketotifene 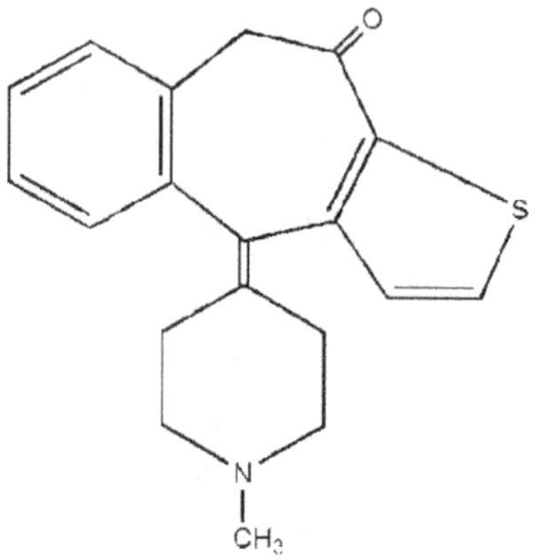 **Use:** It is used as an anti histamines	**Loratadine** 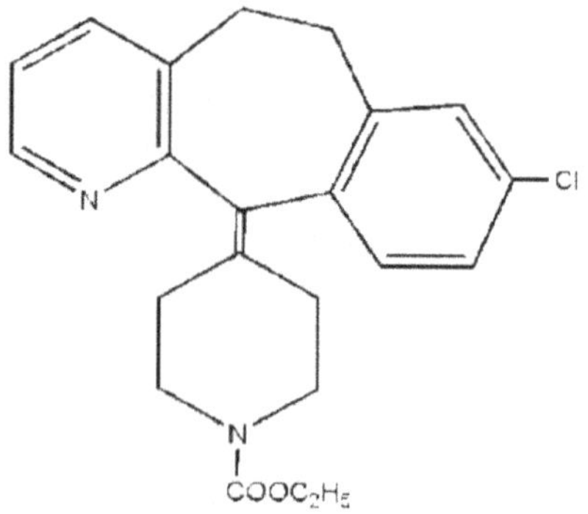 Ethyl 4-(8-chloro-5,6-dihydro-11*H*-benzo[5,6]cyclohepta[1,2-*b*]pyridin-11-ylidene)-1-piperidinecarboxylate **Use:** It is used as an antihistaminic agent

Table Contd...

Nevirapine 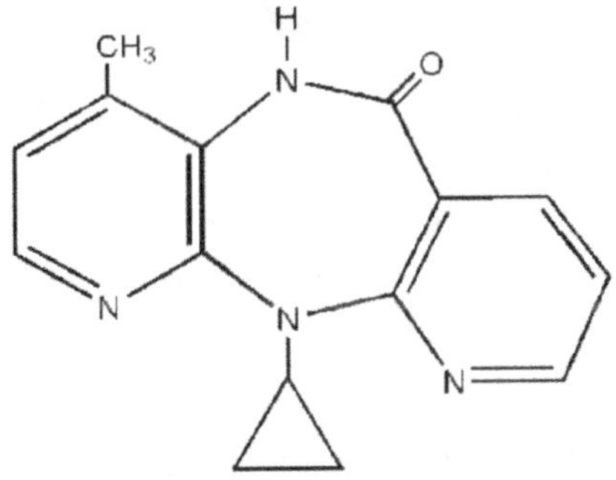 11-Cyclopropyl-4-methyl-5,11-dihydro-6*H*- dipyrido [3,2-*b*:2′,3′-*e*][1,4]diazepin-6-one **Use:** It is used as an anti viral agent	**Rocastine** 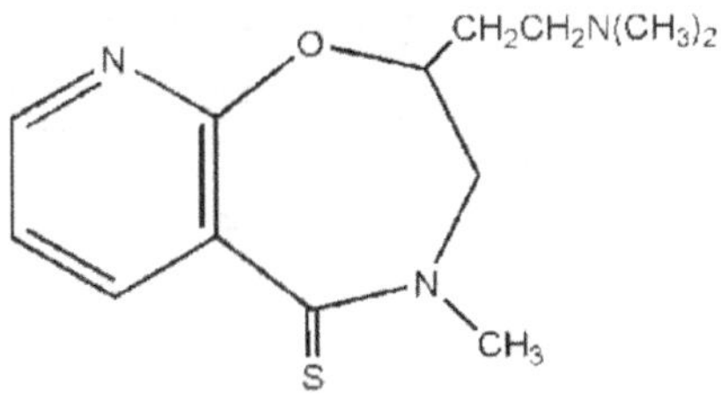 **Use:** It is used as an antihistaminic agent
Roxatidine 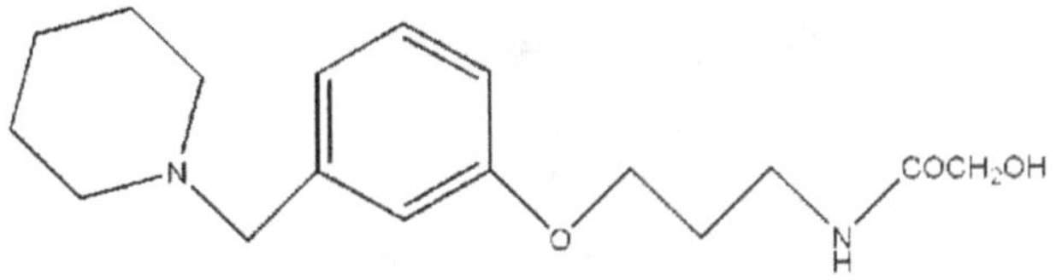 [2-Oxo-2-[3-[3-(piperidin-1-ylmethyl) phenoxy]propylamino]ethyl]acetate **Use:** It is used as an antiulcer agent	**Acrivastine** 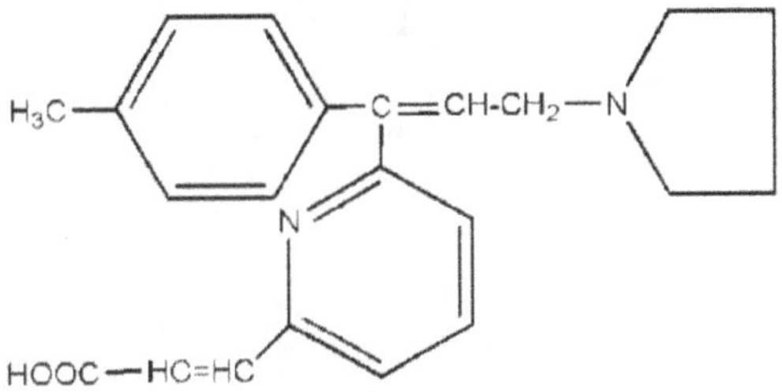 (*E*)-3-{6-[(*E*)-1-(4-methylphenyl)-3-pyrrolidin-1-yl-prop-1-enyl]pyridin-2-yl}prop-2-enoic acid **Use:** It is used as an antihistaminic agent
Repaglinide 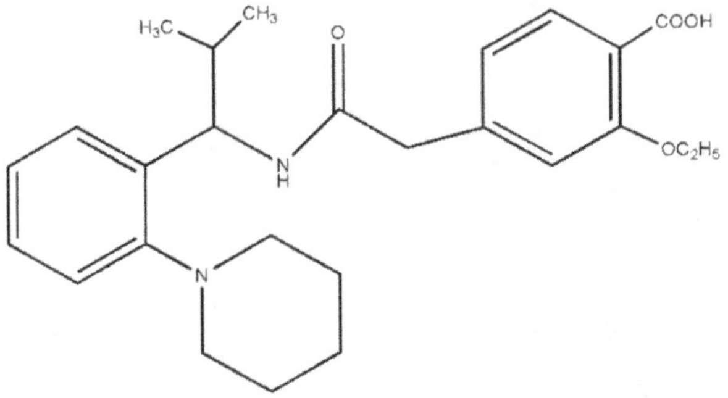 **Use:** It is used as an oral hypoglycemic agent	**Trovirdine** 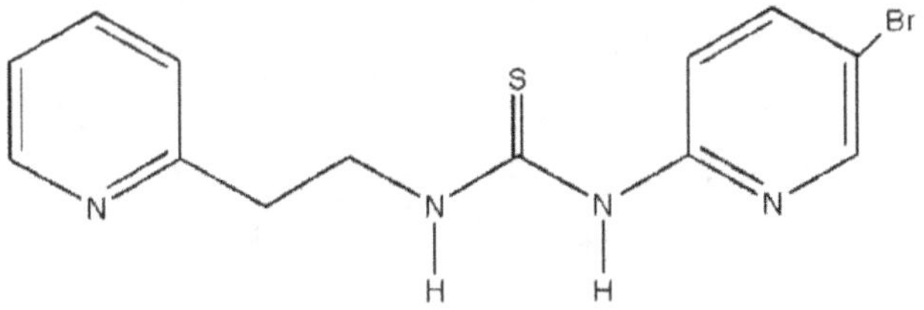 1,5-(Bromo-2-pyridinyl)-3-[2-(2-pyridinyl)ethyl]thiourea **Use:** It is used as an antiviral agent
Indinavir 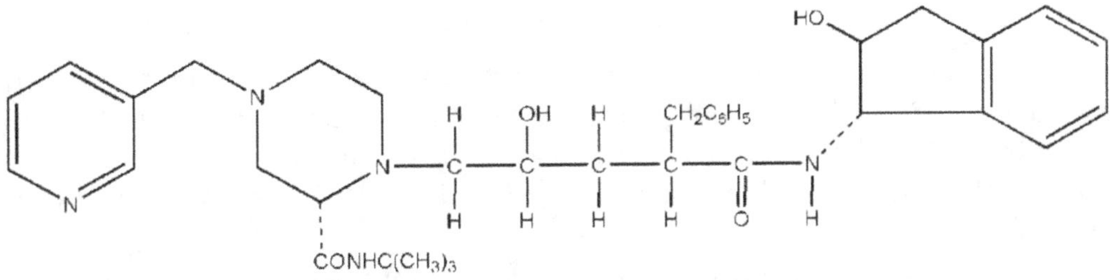 **Use:** It is used as an anti viral agent	

Table *Contd...*

Trifluperidol

1-(4-Fluorophenyl)-4-[4-hydroxy-4-[3-(trifluoromethyl)phenyl]piperidin-1-yl]butan-1-one

Uses: It is used as a tranquillizer

Droperidol

1-(1(-(3-(*p*-Fluorobenzoyl)propyl)-1,2,3,6-tetrahydro-4-pyridyl)-2-benzimidazolinone

Use: It is used as a tranquilizer

Saquinavir

Use: It is used as an antiviral agent

Quinoline (1-Azanaphthalene or Benzopyridine)

Introduction

It was first isolated from coal tar by Friedlieb Ferdin and Runge and given the name leukol (White oil in Greek). It is easily degraded by the microorganism *Rhodococcus* species. It is used for the manufacture of niacin, dyes and hydroxyquinoline. The oxidation of quinoline leads to quinolinic acid which is used as herbicide. Little amount of quinoline is present in crude oil which is removed by the process known as **hydrodentrification.**

Quinoline is a bicyclic ring system derived from pyridine in which benzene is fused with pyridine in α,β-position. The numbering in quinoline starts from nitrogen atom which assigned the position 1.

5, 4, 4a, 3, 6, 7, 2, 8a, N 1, 8 — or — β, α, N

Occurrence: It is present in coal-tar and bone oil. It was first isolated from coal tar by Runge in 1834 and subsequently isolated from the alkaloid cinchonine by alakaline pyrolysis hence known as quinoline.

General Methods of Preparation:

1. **The Skraup synthesis**: It is most popular and important method of preparation of quinoline. In this method aniline is heated with glycerol, Conc.H_2SO_4 and nitrobenzene (oxidizing agent). During this reaction glycerol gets dehydrated with Conc. H_2SO_4 and yield acrolein (formed *insitu*).

$$\underset{\text{Glycerol}}{CH_2(OH)-CH(OH)-CH_2(OH)} \xrightarrow[-2H_2O]{Conc.H_2SO_4} CH_2=C=C(OH)-H \longrightarrow \underset{\text{Acrolein}}{CH_2=CH-CHO}$$

Acrolein further reacts with aniline to yield quinoline. The sequence of reactions are as follows.

Aniline + Acrolein —Michael addition→ Aniline propanal —Ring closure, H^+→ —$-H^+$→ —$-H_2O$→ 1,2-Dihydro quinoline —Oxidation, $C_6H_5NO_2$→ Quinoline

NOTE: Substituted quinolines are obtained when substituted anilines are used as a starting material.

2. **The Friedlander synthesis:** *o*-Aminobenzaldehyde or *o*-aminoacetophenone is condensed with aldehyde or ketone containing an active methylene group in the presence of refluxing alcoholic NaOH solution to give quinoline.

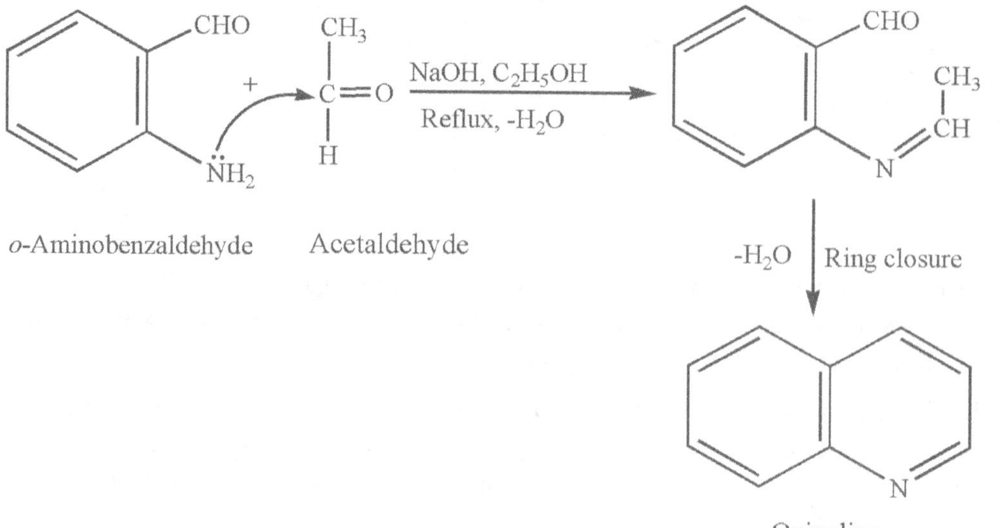

CHO
CH$_3$
NaOH, C_2H_5OH
Reflux, $-H_2O$
NH_2
H
o-Aminobenzaldehyde
Acetaldehyde
CHO
CH$_3$
CH
N
$-H_2O$ Ring closure
N
Quinoline

Physical properties: Quinoline is a colorless liquid, hygroscopic in nature, B.P. 237 °C. It has a characteristic odour like pyridine. On exposure to air produces yellow colour. Insoluble in water (0.7 %), but soluble in all other organic solvents.

Structure: The all carbon atoms and nitrogen atom in quinoline is *sp*^{2} hybridized. The lone pair electrons on the nitrogen atom does not involve in the formation of π molecular orbital. It contains ten π electrons in π molecular orbital satisfies the Huckel's rule, hence aromatic.

Resonance structures of quinoline: Quinoline exhibits the following resonance structures

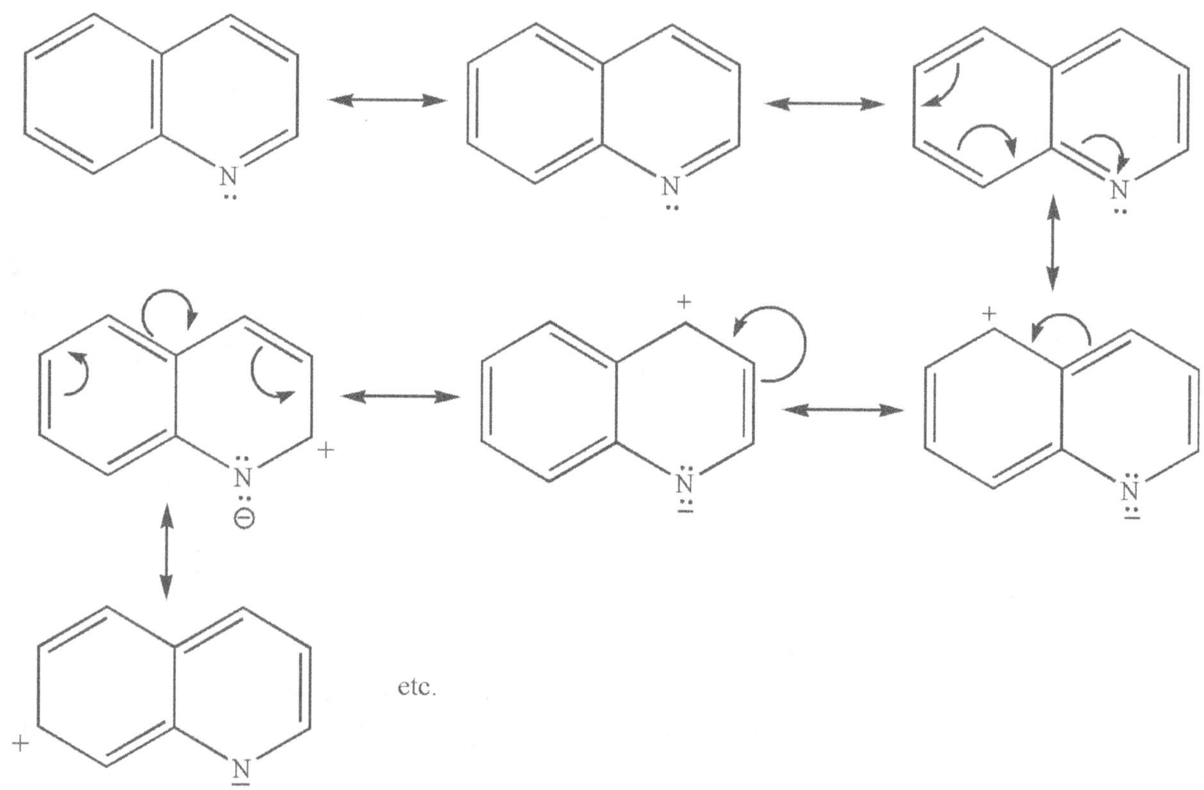

etc.

Chemical Properties:

Quinoline shows the properties of 3° amines, properties of benzenoid and pyridinoid compounds.

1. **Basicity of Quinoline**

   Quinoline is weakly basic in nature (pKa 4.94). The pKa value is intermediate between pyridine (pka 5.17) and aniline (pKa 4.58). It reacts with acids and yield water soluble salts.

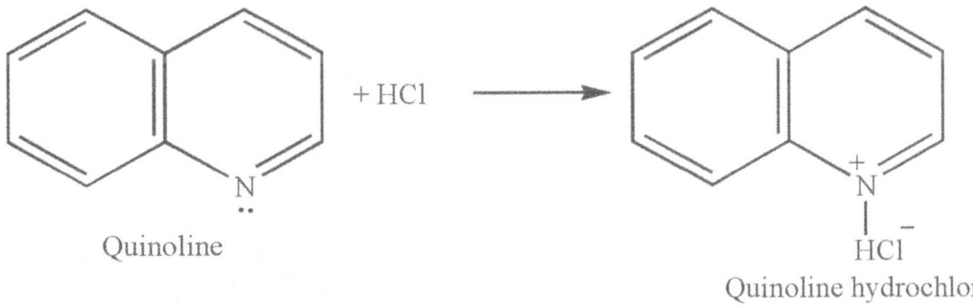

   Quinoline + HCl → Quinoline hydrochloride

   Presence of electron donating substituents at C-2 and C-4 positions of the quinoline ring increases the basicity.

2. **Electrophilic Substitution Reaction**

   The main point of attack of electrophile to the quinoline nucleus is the hetero atom (nitrogen atom). Electrophilic substitution reaction of quinoline occurs at very vigorous conditions only because the nitrogen atom in quinoline is a deactivator. Hence the electrophile attacks the quinoline at C-5 and C-8 position.

   (i) ***Halogenation:*** Halogenation of quinoline depends upon the types of halogenating agent used.

   (a) ***Chlorination:*** Quinoline reacts with sulphuryl chloride to yield 3-chloro quinoline while bromination with bromine and carbontetrachloride yield 3-bromoquinoline.

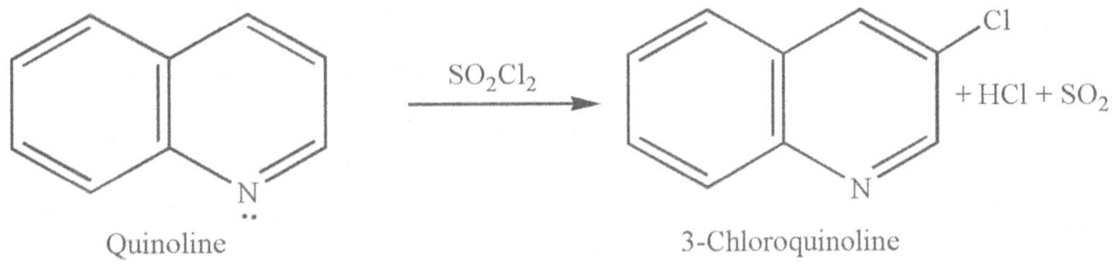

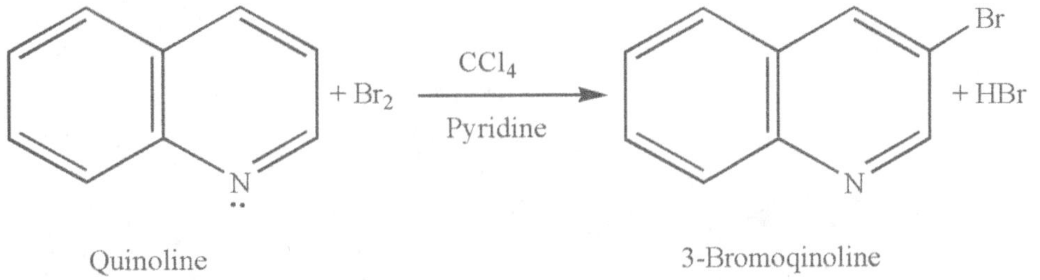

   Quinoline $\xrightarrow{SO_2Cl_2}$ 3-Chloroquinoline + HCl + SO_2

   Quinoline + Br_2 $\xrightarrow[\text{Pyridine}]{CCl_4}$ 3-Bromoqinoline + HBr

   In the presence of strong acids mixture of 5- and 8-bromoquinolines are obtained.

   (b) ***Nitration:*** Quinoline reacts with HNO_3 and H_2SO_4 at 0 °C and gives a mixture of 5- and 8-nitroquinolines.

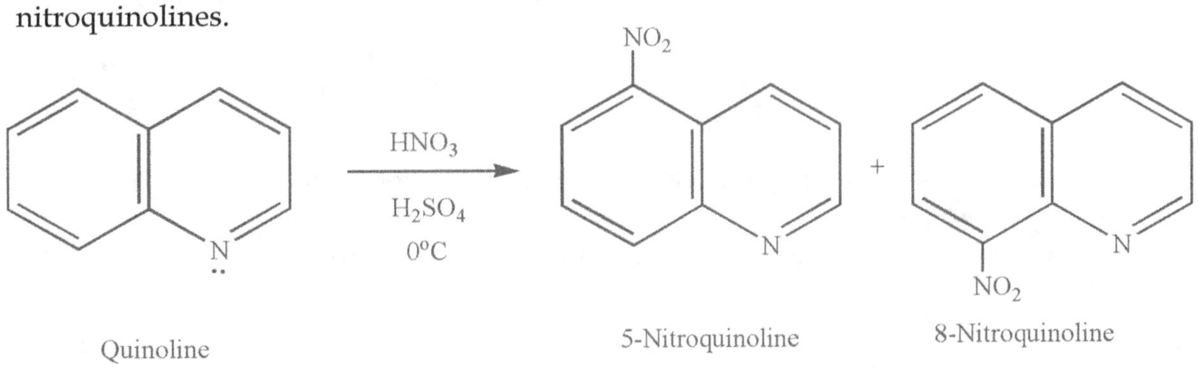

   Quinoline $\xrightarrow[H_2SO_4,\ 0^oC]{HNO_3}$ 5-Nitroquinoline + 8-Nitroquinoline

Note: Dinitrogen tetroxide in the presence of acetic anhydride is also used as nitrating agent.

(c) ***Sulphonation:*** Quinoline reacts with sulphuric acid at 220 °C and gives 8-sulphonic acid. At higher temperature (300 °C) yields 6-quinoline sulphonic acid.

Quinoline + H_2SO_4 —(220 °C)→ Quinoline-8-sulphonic acid (SO_3H at C-8)

Quinoline + H_2SO_4 —(300 °C)→ Quinoline-6-sulphonic acid (HO_3S at C-6)

(d) ***Friedel-Crafts reaction:*** Freidel-Craft's reaction not easily occurs due to deactivation of the benzene ring by the nitrogen atom. But substituted quinolines undergoes acylation and gives 8-methoxy quinoline.

3. **Mercuration:** Quinoline reacts with mercuric acetate and gives quarternary N-mercuric acetate. At 160 °C further substitution occurs and when treated with NaCl and gives a mixture of 3- and 5-mercuric chlorides.

Quinoline —($Hg(OAC)_2$, R.T)→ N-HgOAC quinolinium OAC^- —(NaCl)→ 3-Quinoline mercuric chloride (HgCl) + 5-Quinoline mercuric chloride (HgCl)

4. **Reduction:** Quinoline is easily reduced than benzene. It gets reduced with different kind of reducing agents. For example,

Quinoline gets reduced with Raney nickel or PtO_2, H_2 or ammonium carbonate, Pd/C to yield hydro quinolines.

Quinoline —LiAlH$_4$→ 1,2-Dihydroquinoline + 1,2,3,4-Tetrahydroquinoline

Quinoline —PtO_2, CF_3COOH→ 1,2,3,4-Tetrahydroquinoline

Quinoline gets broken or ruptured at 260-380 °C in the presence of Ni.

5. **Oxidation:** Quinoline is oxidised with $KMnO_4$ and gives quinolinic acid which upon further decarboxyltion gives nicotinic acid or niacin.

Quinoline —$KMnO_4$→ Quinolinic acid (COOH, COOH) —Δ, $-CO_2$→ Niacin (COOH)

6. **Ozonolysis:** Quinoline reacts with ozone and gives pyridine 2,3-dicarbaldehyde and glyoxal.

Quinoline + O_3 → Pyridine-2,3-dicarbaldehyde (CHO, CHO) + Glyoxal (CHO–CHO)

7. **Oxidation with peracids:** Quinoline gets oxidised with peracids and gives quinoline N-oxides.

Quinoline —$CH_3\overset{O}{\overset{\|}{C}}OOH$→ Quinoline-*N*-oxide ($\overset{+}{N}$–$O^-$) ⟷ ($\overset{+}{N}$=O)

8. **Reaction with sodamide:** Quinoline reacts with sodamide in the presence of liq. NH_3 and gives 2-aminoquinoline.

Quinoline → ($NaNH_2$, Liq. NH_3) → 2-Aminoquinoline

9. **Reaction with organolithium reacgent:** Quinoline reacts with *n*-butyl lithium and gives 1,2-addition product.

Quinoline → (*n*-BuLi, Benzene, R.T) → 1,2-Addition product

The resulting 1,2-dihydro quinoline undergo further oxidation and reduction reactions to give 2-alkyl quinoline and 2-substituted 1,2,3,4-tetrahydro quinoline.

→ (H_2O)

Oxidation ($C_6H_5NO_2$, 200 °C) → 2-Butyl quinoline

Reduction (Na/C_2H_5OH) → 2-Butyl-1,2,3,4-tetrahydro quinoline

Medicinal Compounds Containing Quinoline (or) Quinoline Derivatives Used in Medicine

Dibucaine $CONH(CH_2)_2$-N(C_2H_5)(C_2H_5); N; $O(CH_2)_3CH_3$ 2-Butoxy-*N*-[2-(diethylamino)ethyl]quinoline-4-carboxamide **Use:** It is used as an local anasthetic	**Primaquine** H_3CO; N; NH-CH (CH_3)-$(CH_2)_3$-NH_2 (*RS*)-*N*-(6-Methoxyquinolin-8-yl)pentane-1,4-diamine **Use:** It is used as an antimalarial agent
Mefloquine HO; NH; N; CF_3; CF_3 [2,8-bis(Trifluoromethyl)quinolin-4-yl]-piperidin-2-ylmethanol **Use:** It is used as an antimalarial agent	**Oxamniquine** HOH_2C; O_2N; NH; $CH_2NHCH(CH_3)_2$ (*RS*)-1,2,3,4-Tetrahydro-2-isopropylaminomethyl-7-nitro-6-quinolylmethanol **Use:** It is used as an anthelmentic agent
Cholorquine NH-CH$(CH_3)(CH_2)_3$-N$(C_2H_5)_2$; Cl; N (*RS*)-*N'*-(7-Chloroquinolin-4-yl)-*N*,*N*-diethyl-pentane-1,4-diamine **Use:** It is used as an antimalarial agent	**Pentaquine phosphate** H_3CO; N; CH_3; NH -$(CH_2)_5$-NH-CH; CH_3 5-[(6-Methoxyquinolin-8-yl)amino]pentyl-propan-2-ylazanium **Use:** It is used as an antimalarial agent

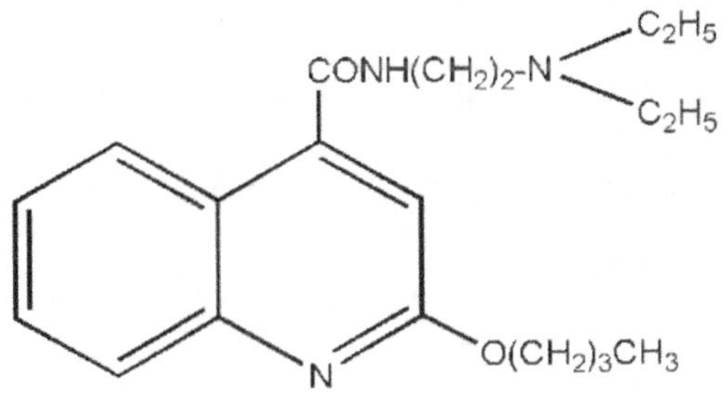

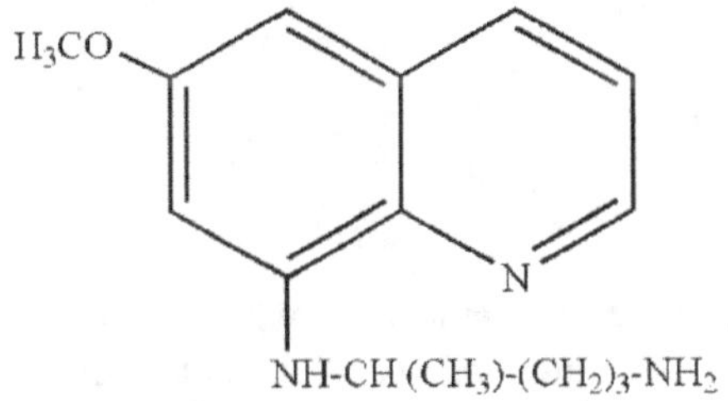

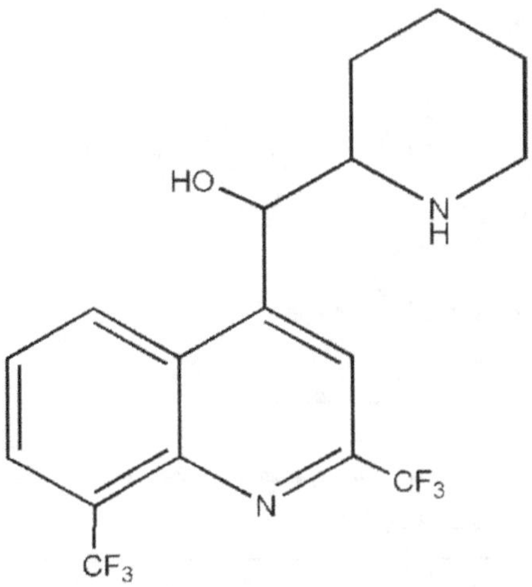

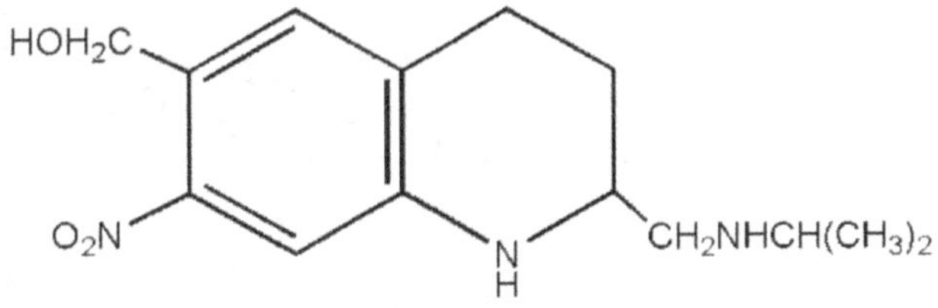

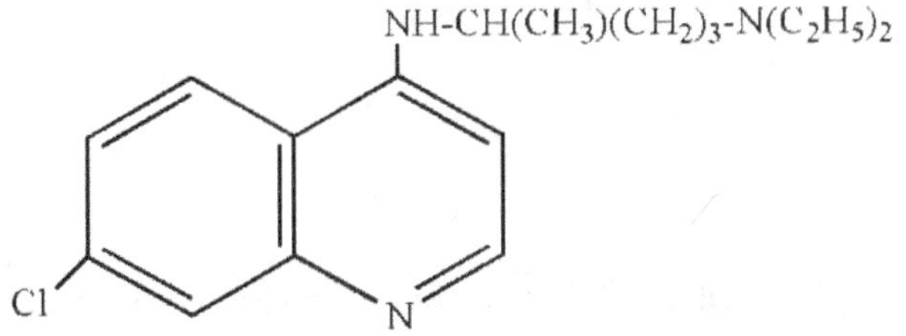

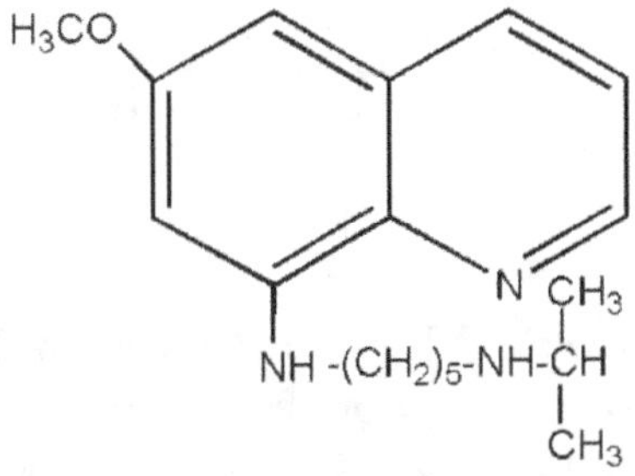

Table *Contd...*

Quinidine	Pamaquine
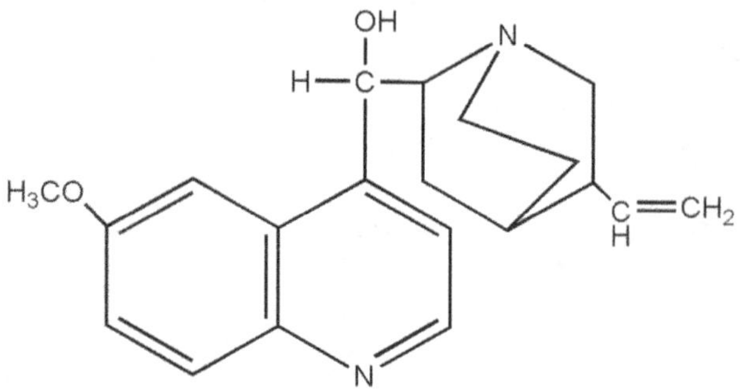 OH, H—C, N, H_3CO, C=CH$_2$, H, N (*S*)-(6-Methoxyquinolin-4-yl)[(1*S*,2*R*,4*S*,5*R*)-5-vinylquinuclidin-2-yl]methanol **Use:** It is used as an antiarrythmic agent	H_3CO, N, NH-CH (CH_3)-(CH_2)$_3$-N(C_2H_5)$_2$ **Use:** It is used as an antimalarial agent

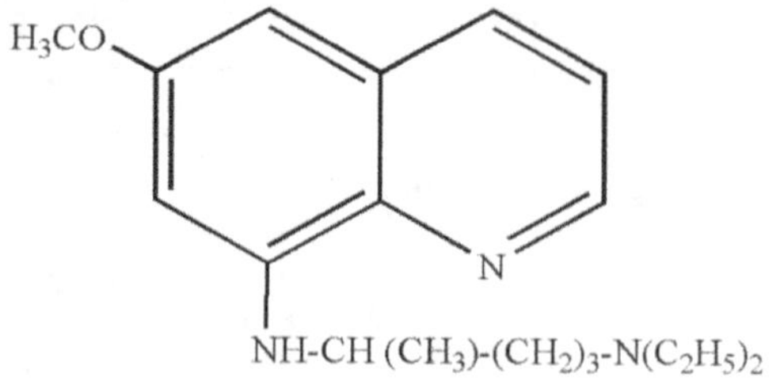

Amodiaquine
$CH_2N(C_2H_5)_2$, NH, OH, Cl, N 4-[(7-Chloroquinolin-4-yl)amino]-2-(diethylaminomethyl)phenol **Use:** It is used as an antimalarial agent

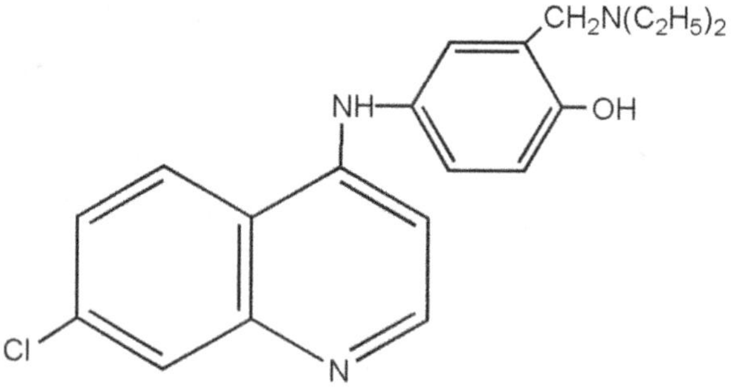

Isoquinoline (2-Azanaphthalene (Or) Benzo[b]Pyridine)

Isoquinoline is similar to the structure of quinoline (Benzene ring is fused with pyridine ring) but the nitrogen atom is present in position-2.

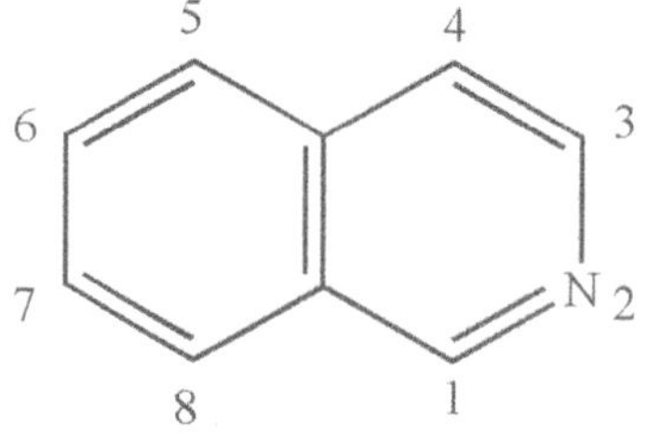

5 4 6 3 7 N 2 8 1

It was first isolated by Hoogewerff and Dorp from fraction of coal tar in 1885. It does not present in nature as free state but more amounts present in alkaloids. The reduced form of isoquinoline such as 1,2-dihydro isoquinoline, 1,2,3,4-tetrahydro isoquinoline *etc,* are important due to their occurrence in alkaloids. 1-Benzyl isoquinoline is the backbone of papaverine alkaloids. Tetrahydroisoquinolines are used as a precursor for neurotoxins.

General Methods of Preparation:

1. **From Quinoline:** Isoquinoline always present in quinoline. It is separated from the mixture by sulphonation. The mixture is first converted into sulphate by treatment with H_2SO_4 then separated

by fractional crystallisation from alcohol in which isoquinoline is sparingly soluble while quinoline is completely soluble.

2. **From Indene:** Indene reacts with ozone at -70 °C in the presence of methanol and gives dialdehyde which upon further reduction with dimethyl sulphide in the presence of ammonium hydroxide yields isoquinoline.

Indene + O_3 $\xrightarrow[-70\ °C]{CH_3OH}$ Dialdehyde (CHO, CH_2CHO)

Dialdehyde $\xrightarrow{NaHCO_3,\ (CH_3)_2S,\ NH_4OH}$ Isoquinoline

3. **From Cinnamaldehyde oxime:** Cinnamaldehyde oxime is heated with phosphorous pentoxide to yields an intermediate which upon further dehydration and ring closure reaction yield isoquinoline.

Cinnamaldehyde oxime $\xrightarrow{\text{Beckmann's rearrangement}}$ [intermediate: HOHC=N] $\longrightarrow$ Isoquinoline

4. **Bischler-Napieralski synthesis / From 2-Phenyl ethylamine:** 2-Phenylethylamine is treated with formyl chloride in the presence of alkali to give *N*-formyl-2-phenylethylamine. The amide intermediate is heated with phosphorous pentoxide in the presence of pyridine to yield 3,4-dihydro isoquinoline which upon further dehydrogenation with palladium yields isoquinoline.

2-Phenyl ethylamine ($CH_2CH_2NH_2$) + HCOCl (Formyl chloride) $\xrightarrow[(-HCl)]{\overline{OH}}$ N-Formyl-2-phenyl ethylamine

N-Formyl-2-phenyl ethylamine $\xrightarrow{P_2O_5,\ \Delta,\ -H_2O}$ 3,4-Dihydro quinoline

3,4-Dihydro quinoline $\xrightarrow[(-H_2)]{Pd,\ \Delta}$ Isoquinoline

Physical properties: It is a colorless solid, M. P. 243 °C. It is steam volatile, sparingly soluble in water but completely soluble in organic solvents. The odour of isoquinoline is similar to that of benzaldehyde.

Chemical properties: Isoquinoline behaves as a weak base due to the presence of nitrogen atom pKa value is 1.54. The electrophilic attack of isoquinoline nucleus chiefly occurs at C-5 and C-8 positions. It is highly aromatic in nature and the various resonance structures of isoquinoline are as follows.

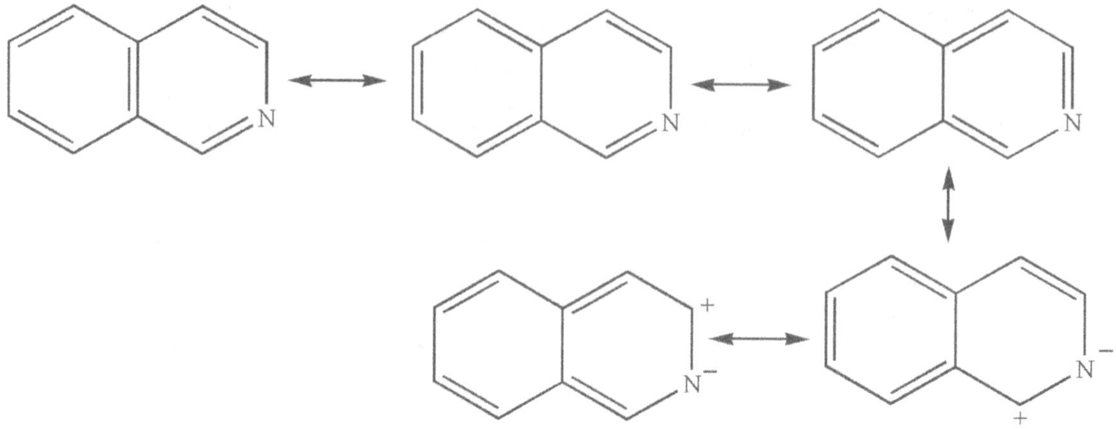

1. **Electrophilic substitution reactions:** Isoquinoline undergoes electrophilic substitution reaction under drastic conditions than pyridine.

   (i) ***Halogenation:***

   (a) ***Chlorination:*** Isoquinoline reacts with Cl_2 in the presence of $AlCl_3$ and yields 5-chloro isoquinoline with lesser yield.

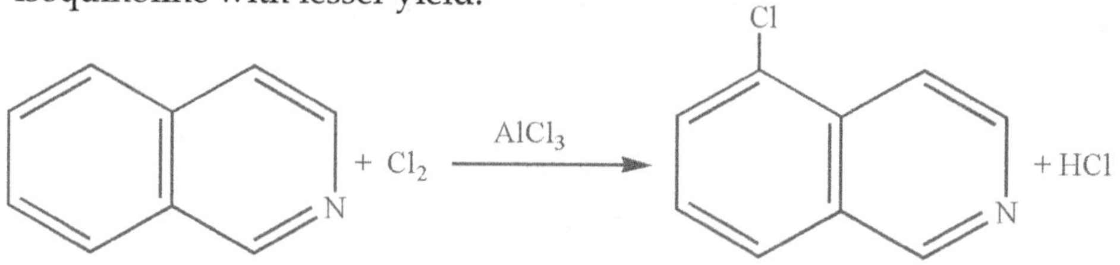

   Isoquinoline + Cl_2 $\xrightarrow{AlCl_3}$ 5-Chloro isoquinoline + HCl

   (b) ***Bromination:*** Isoquinoline reacts with Br_2 and $AlCl_3$ and gives 5-bromo isoquinoline.

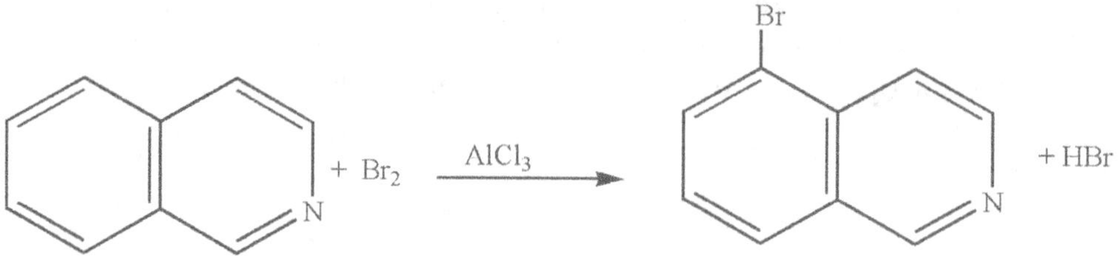

   Isoquinoline + Br_2 $\xrightarrow{AlCl_3}$ 5-Bromoisoquinoline + HBr

   (ii) ***Nitration:*** Isoquinoline reacts with nitrating mixture (HNO_3 and H_2SO_4) at 0 °C and yields 5-nitro isoquinoline (72 %) with little amount of 8-nitro isoquinoline. At higher temperature the amount of 8-isomer is slightly increased.

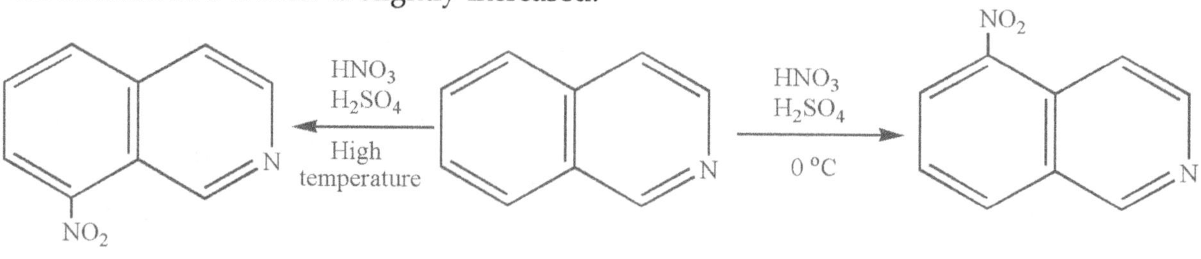

   8-Nitroisoquinoline $\xleftarrow[\text{High temperature}]{HNO_3, H_2SO_4}$ Isoquinoline $\xrightarrow[0\ °C]{HNO_3, H_2SO_4}$ 5-Nitroisoquinoline

(iii) ***Sulphonation:*** Isoquinoline reacts with H_2SO_4 and yields isoquinoline-7-sulphonic acid.

H_2SO_4

HO_3S N

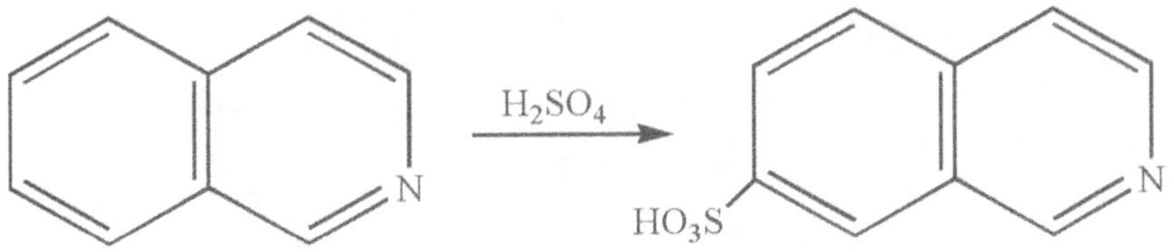

Isoquinoline Isoquinoline-7-sulphonic acid

2. Reaction with sodamide: Isoquinoline reacts with sodamide in the presence of Liq. ammonia and gives 1-amino isoquinoline.

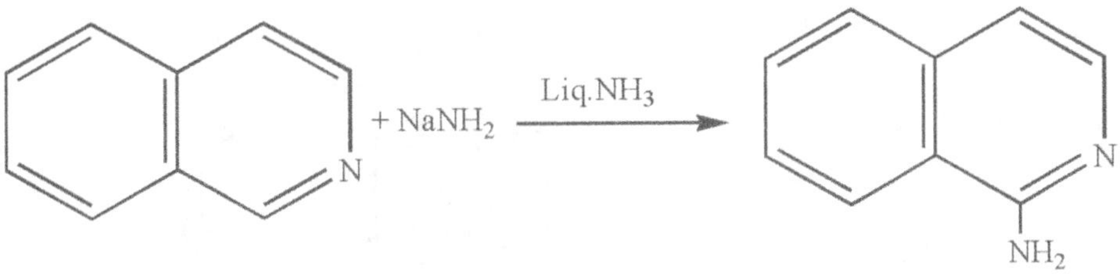

$+ NaNH_2$ Liq.NH_3

N NH_2

Isoquinoline 1-Aminoisoquinoline

3. Reaction with benzoyl chloride: Isoquinoline reacts with benzoyl chloride and gives quaternary salts.

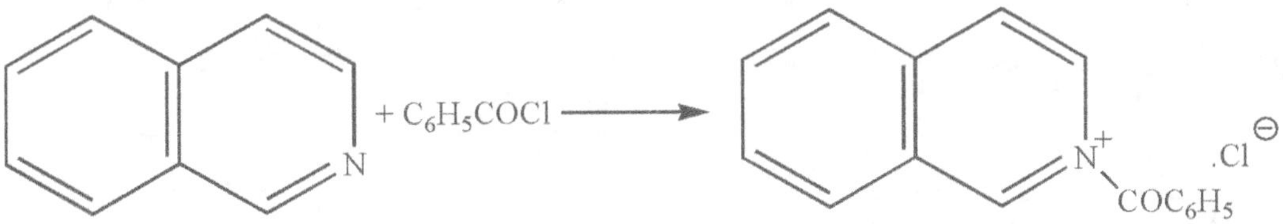

$+ C_6H_5COCl$

N^+ $.Cl^{\ominus}$ COC_6H_5

Isoquinoline 2-Benzoyl isoquinolinium chloride

4. Oxidation: Isoquinoline is oxidised with neutral $KMnO_4$ to yield phthalimide.

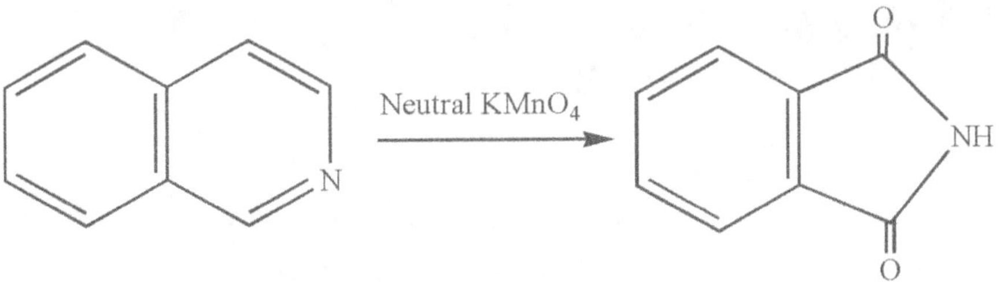

Neutral $KMnO_4$

O NH O

Isoquinoline Pthalimide

It also gets oxidized with alkaline $KMnO_4$ and gives phthalic acid and pyridine-3,4-dicarboxylic acid.

Alkaline $KMnO_4$

COOH COOH + HOOC HOOC N

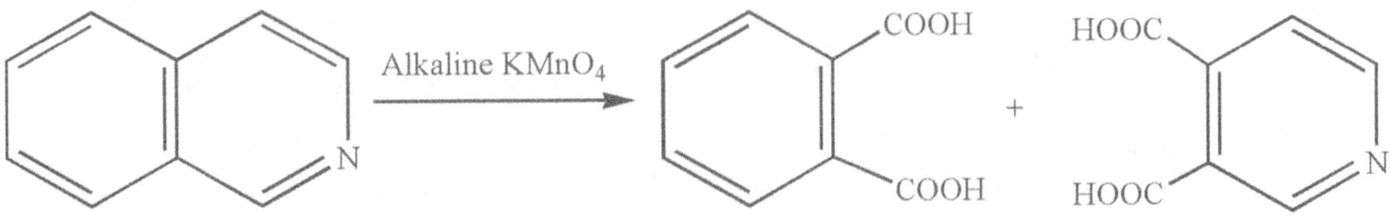

Isoquinoline Pthalic acid Pyridine-3,4-dicarboxylic acid

5. **Reduction:** Isoquinoline reacts with different kinds of reducing agents and yields different products.

Isoquinoline

H_2, Pt / Conc.HCl, RT → Decahydroisoquinoline

Na, C_2H_5OH → 1,2,3,4-Tetrahydroisoquinoline

$(C_2H_5)_2AlH$ / 0 °C → 1,2-Dihydroisoquinoline

6. **Reaction with oxirane:** Isoquinoline reacts with oxirane in acetic acid and yield oxazolidine.

Isoquinoline + Oxirane —CH_3COOH, 25 °C→ Oxazolidine derivative

Medicinal Compounds Containing Isoquinoline or Isoquinoline Derivatives Used in Medicine

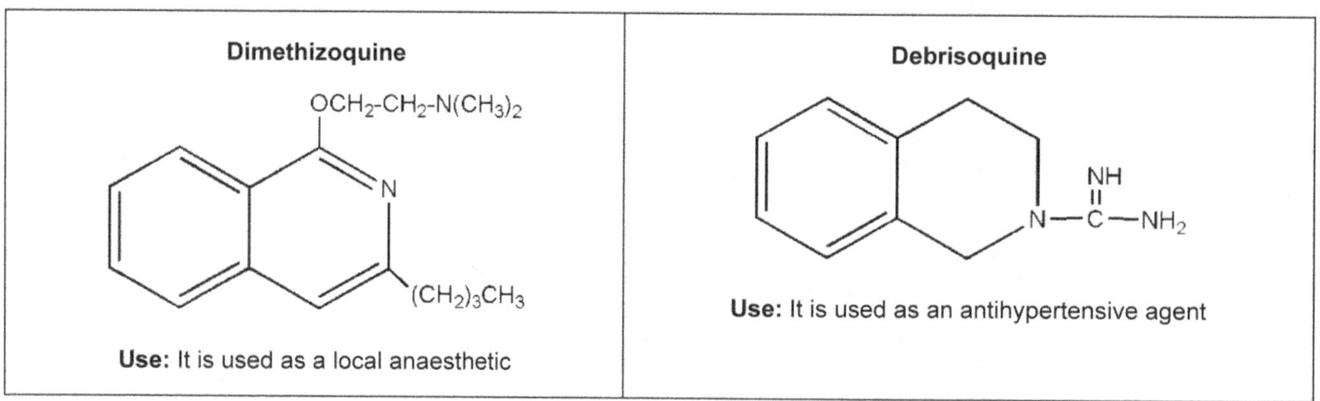

Dimethizoquine	Debrisoquine
OCH_2-CH_2-$N(CH_3)_2$; N ; $(CH_2)_3CH_3$	NH ; N—C—NH_2
Use: It is used as a local anaesthetic	**Use:** It is used as an antihypertensive agent

Table *Contd...*

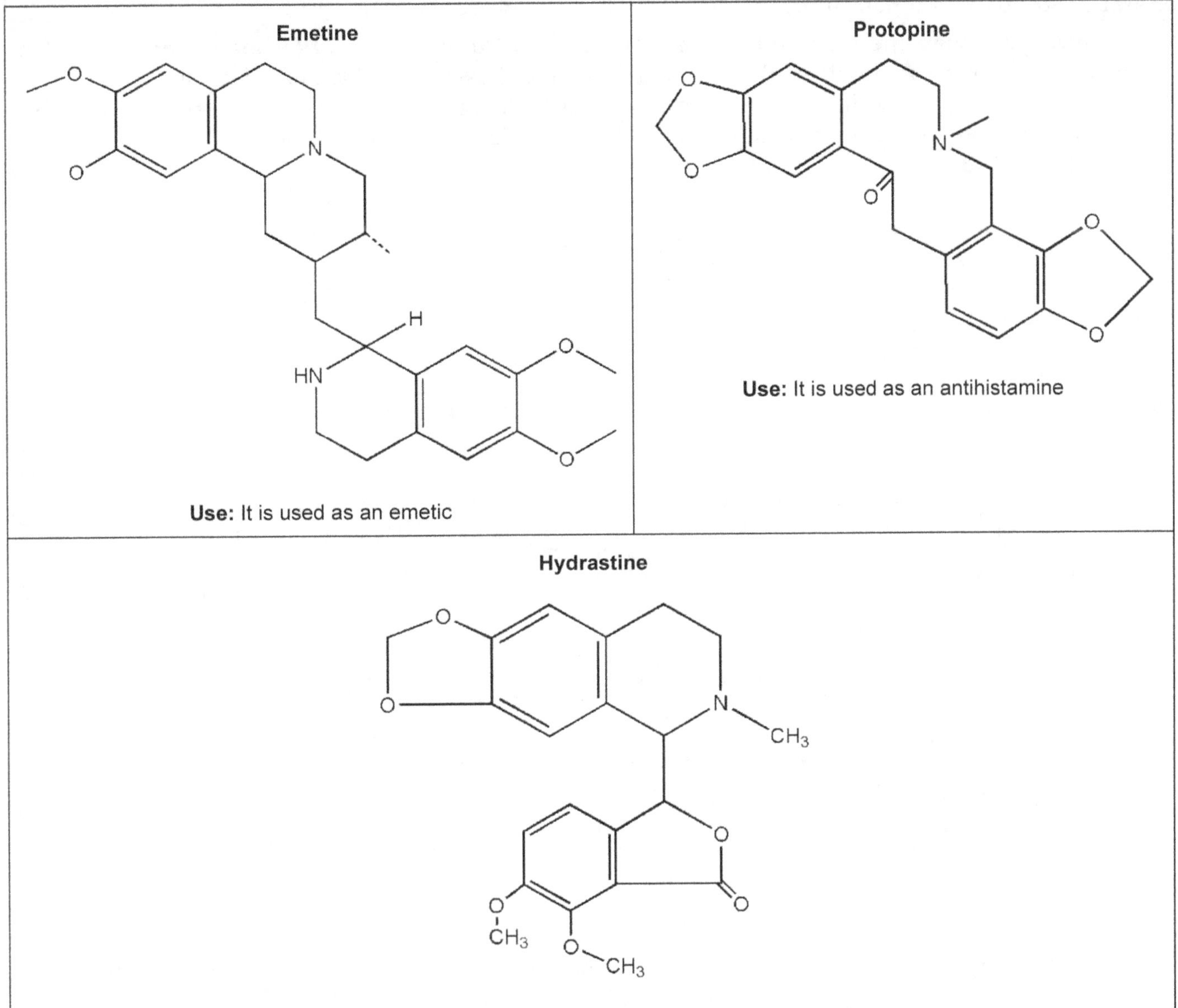

Use: It is used as an emetic

Use: It is used as an antihistamine

Use: It is used as a tyrosine hydroxylase inhibitor

Acridine

Acridine was discovered by Grache and Caro in 1871 from the anthracene fraction of coal tar. This name was given due to its acid and irritating behaviour.

It has the formula of $C_{13}H_9N$. It is structurally resemble to anthracene molecule in which the central -CH group is replaced by N atom. Acridine orange is used for cell cycle determination. Acridine and its derivatives are used as antiseptics. The numbering of acridine skeleton is as follows.

General Methods of Preparation:

1. **From diphenylamine carboxylic acid / from *o*-chlorobenzoic acid/Ullmann method:** *o*-Chloro benzoic acid is treated with aniline in the presence of base to yield diphenylamine-2-carboxylic acid which upon further cyclisation in the presence of H_2SO_4 gives acridinone. The later one is subjected to reduction with suitable reducing agents (H_2/Ni or Na/amyl alcohol) followed by dehydrogenation to yield acridine.

COOH, Cl, H_2N, OH^-, -HCl, COOH, N H

o-Chloro benzoic acid Aniline

Ring closure H_2SO_4 $-H_2O$

O

Reduction

N H N H

Dehydrogenation Acridinone

N

Acridine

2. **From diphenylamine:** Oxidation of methyldiphenylamine with lead oxide yields acridine.

CH_3 PbO Oxidation N H N

Methyl diphenylamine Acridine

3. **From *o*-aminodiphenyl methane:** *o*-Amino diphenyl methane is passed through red hot tube and yields acridine.

H_2 C Δ $-2H_2$ NH_2 N

o-Aminodiphenyl methane Acridine

4. **Goldberg method:** *o*-Aminobenzoic acid is treated with bromobenzene to yields an intermediate which upon cyclisation and yields acridine.

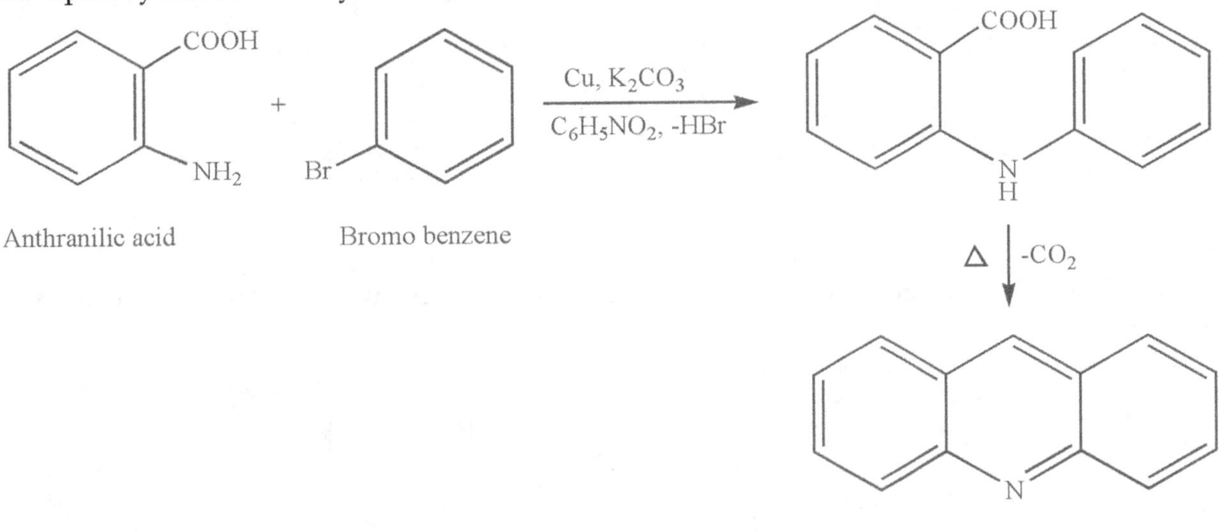

COOH NH$_2$ + Br → (Cu, K_2CO_3 / $C_6H_5NO_2$, -HBr) → COOH N H → (Δ, -CO_2) → N

Anthranilic acid Bromo benzene

Acridine

Physical properties: It is a pale yellow color, crystalline solid, M. P. 114 °C, Acridine and its solutions are highly fluorescent.

Chemical properties: Acridine is an aromatic compound, weakly basic in nature.

1. **Basic character:** Acridine is a weak base and hence it reacts with mineral acids.
2. **Electrophilic substitution reaction:** Electrophilic attack of acridine takes place at C-2 & C-7 positions.
   (i) ***Halogenation:*** Halogenation of acridine yields addition as well as substitution products.
   (a) ***Bromination:*** Acridine reacts with bromine in acetic acid and gives a mixture of 2- and 2,7-dibromoacridine.

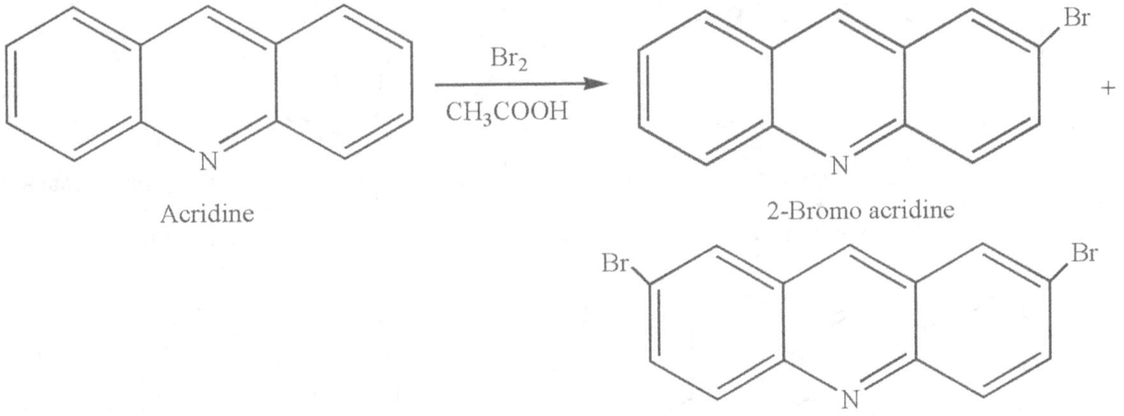

N → (Br_2 / CH_3COOH) → Br N + Br Br N

Acridine 2-Bromo acridine

2,7-Dibromo acridine

   (ii) ***Nitration:*** Acridine N-oxide reacts with Conc. HNO_3 and Conc. H_2SO_4 at 0 °C and yields 9-nitroacridine-*N*-Oxide.

N O$^-$ → (HNO_3 / H_2SO_4) → N O$^-$ NO$_2$

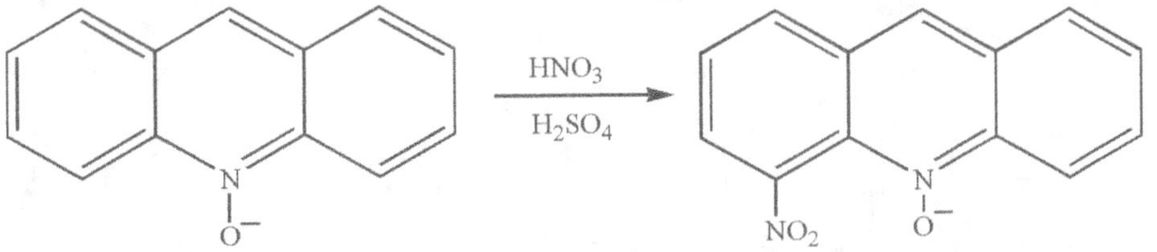

Acridine N-oxide 9-Nitroacridine-*N*-Oxide

3. **Reduction:** Acridine is reduced with Pd/H_2 in HCl, platinum oxide in trifluoroacetic acid to give octahydroacridine.

$$\text{Acridine} \xrightarrow[CF_3COOH]{H_2,\ PIO_3} \text{Octahydroacridine}$$

Acridine Octahydroacridine

Acridine is also reduced with lithium, in liquid ammonia and ethanol and gives 1,4,5,8-tetrahydroacridine.

$$\text{Acridine} \xrightarrow[C_2H_5OH]{Li,\ Liq\ NH_3} \text{1,4,5,8-Tetrahydro acridine}$$

Acridine 1,4,5,8-Tetrahydro acridine

4. **Oxidation:** Acridine is stable one and it does not easily oxidized by normal oxidizing agents. But oxidized with peracids to yield acridine-*N*-oxide.

Medicinal Compounds Containing Acridine (or) Acridine Derivatives Used in Medicine

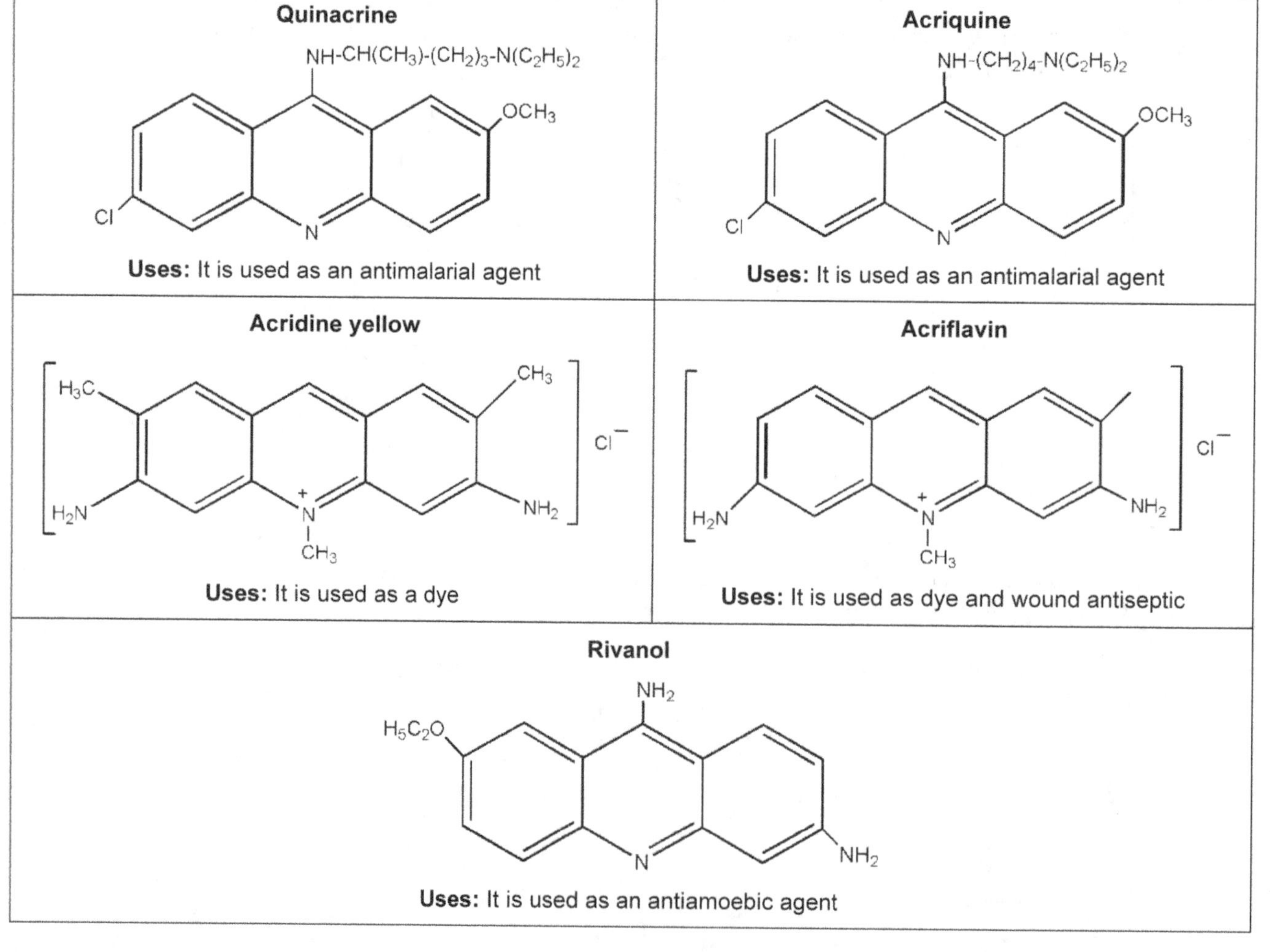

Indole (1*H*-1-Azaindine/Benzopyrrole)

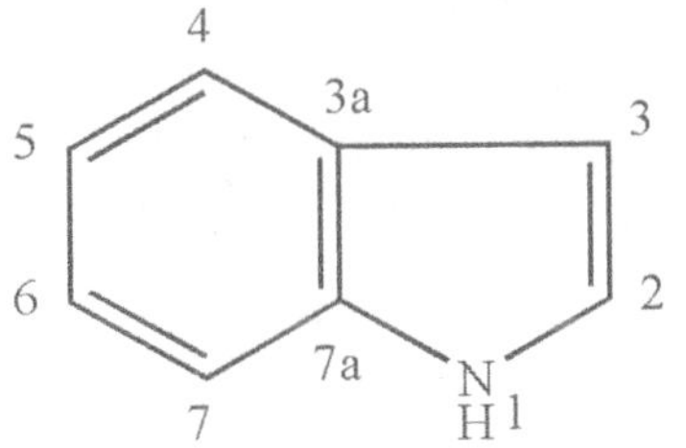

4 3a 3 5 6 2 7a 7 N H 1

Indole molecule is an aromatic heterocyclic compound which is made of a benzene ring fused with 2 and 3 positions of pyrrole ring. The formula of indole is C_8H_7N. It is a benzofused isomer. The IUPAC name is 1*H*-benzo(*b*)pyrrole. It is widely present in nature and it can be produced by various kind of bacteria. It is mainly occur in coal-tar, jasmine flavours and orange blossoms. It was first prepared by Baeyer in 1866 by zinc dust distillation of oxindole. It is also present in tryptophan (amino acids) as a plant growth hormone, in alkaloids and in indigo dye.

Tautomerisation: Indole shows tautomerism and the tautomers are called indolenines.

3*H*-Indolenine ⟷ 2*H*-Indolenine

General Methods of Preparation:

1. **The Fischer indole Synthesis/From pyruvic acid phenylhyrazone:**

   In this reaction, pyruvic acid phenyl hydrazone is heated with polyphoshoric acid or zinc chloride to yield indole-2-carboxylic acid which upon further decarboxylation gives indole. This is one of the most important method for the synthesis of indole.

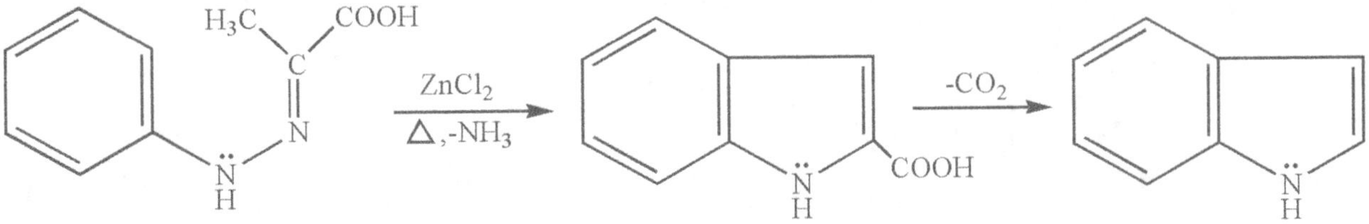

   Phenyl hydrazone of pyruvic acid $\xrightarrow[\Delta, -NH_3]{ZnCl_2}$ Indole-2-carboxylic acid $\xrightarrow{-CO_2}$ Indole

2. **From formyl-*o*-toluidine:** Formyl-*o*-toluidine is heated with sodium or potassium alkoxide to give indole.

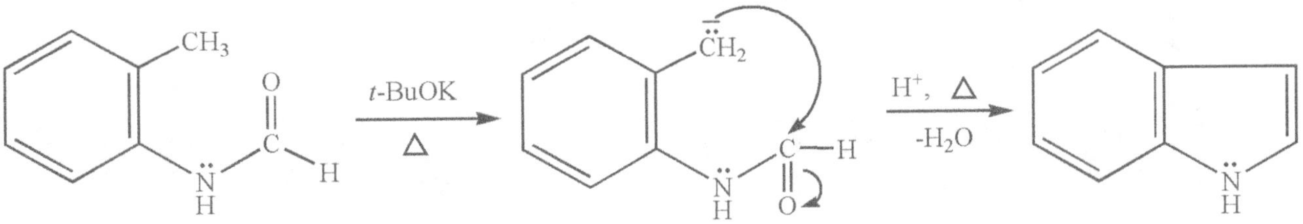

   Formyl-*o*-toluidine $\xrightarrow[\Delta]{t\text{-BuOK}}$ $\xrightarrow[-H_2O]{H^+, \Delta}$ Indole

3. **From *o*-nitrophenyl acetaldehyde:** Reduction of *o*-nitro phenyl acetaldehyde with iron powder and sodium bisulphate solution gives *o*-aminophenyl acetaldehyde which converts into indole by cyclisation.

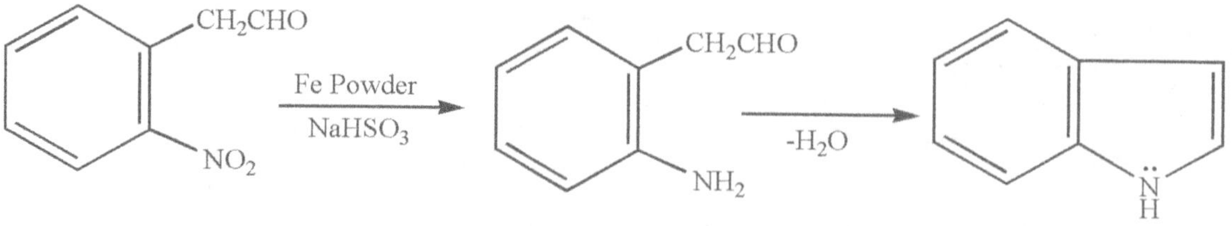

CH_2CHO / NO_2 — Fe Powder, $NaHSO_3$ → CH_2CHO / NH_2 — $-H_2O$ → Indole

o-Nitrophenyl acetaldehyde

Indole

4. **The Reissert Synthesis:** It is a convenient method for the synthesis of indole. It involves the base catalysed condensation of *o*-nitrotoluene with diethyl oxalate in the presence of sodium ethoxide to yield *o*-nitrophenyl pyruvate. Hydrolysis of the resulting ester followed by reduction using Zn/CH_3COOH gives carboxylic acid which gives indole by decarboxylation. The sequence of the reactions are as follows:

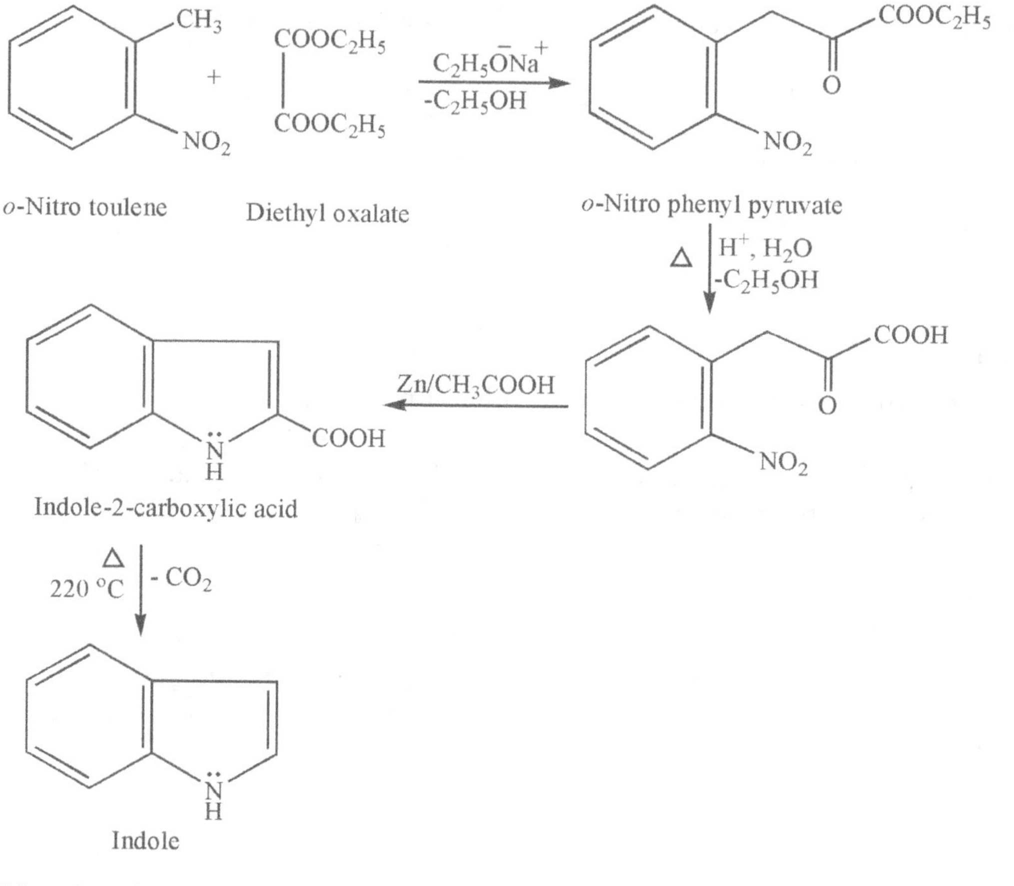

CH_3 / NO_2 + $COOC_2H_5$–$COOC_2H_5$ — $C_2H_5\bar{O}\overset{+}{Na}$, $-C_2H_5OH$ → $COOC_2H_5$ / O / NO_2

o-Nitro toulene Diethyl oxalate *o*-Nitro phenyl pyruvate

Δ H^+, H_2O, $-C_2H_5OH$ ↓

COOH / O / NO_2 — Zn/CH_3COOH → COOH / N H

Indole-2-carboxylic acid

Δ 220 °C, $-CO_2$ ↓

Indole

5. **The Lipp Synthesis/ From *o*-amino-ω-chlorostyrene:** By heating *o*-amino-ω-chlorostyrene with sodium ethoxide gives indole.

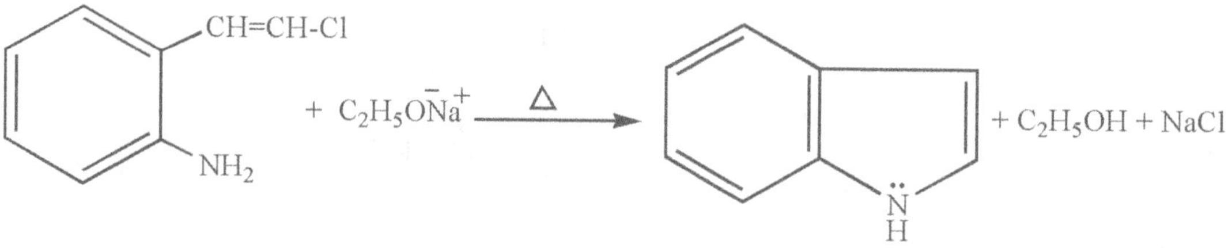

CH=CH-Cl / NH_2 + $C_2H_5\bar{O}\overset{+}{Na}$ — Δ → (Indole) + C_2H_5OH + NaCl

o-Amino-ω-chlorostyrene Indole

Structure: Based on molecular orbital theory, the eight carbon atoms and the one nitrogen atom in indole molecule is sp^2 hybridised. All sp^2 hybridised orbitals overlaps with each other and overlaps with "s" orbital of hydrogen atom to form C-C, C-H, C-N and N-H σ bonds. Each carbon atom possess one electron in p_z orbital and the nitrogen atom posses two electrons, overlapping of these forms a π molecular orbitals with 10 π-electrons, hence it behaves as aromatic compound (satisfies Huckel's rule).

Indole possesses the following resonating structures.

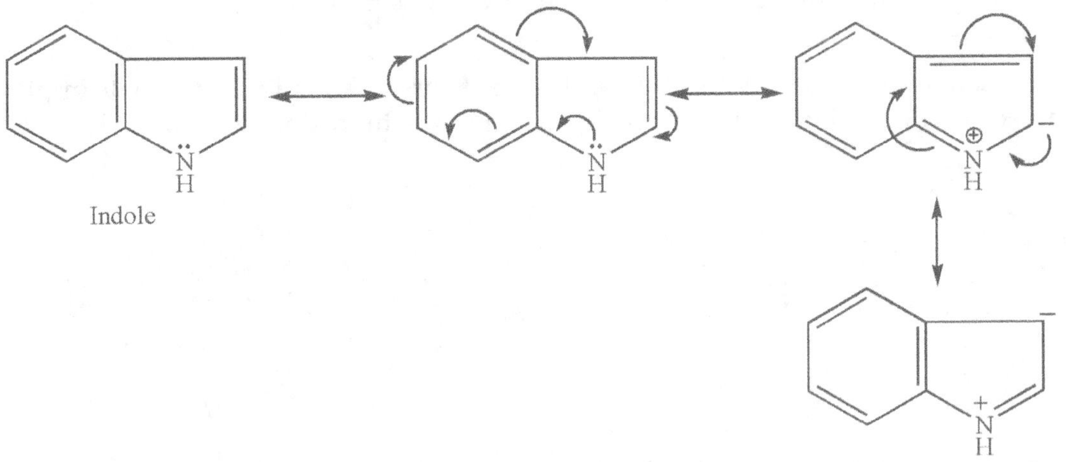

Indole

Physical Properties: Indole is a colorless solid, M.P is 52 °C and soluble in all organic solvents. Pure indole has very pleasant smell and used in perfume preparations and stable in air.

Chemical Properties:

1. **Reaction with Acids or Basicity of Indole:** Indole is a weak base, it get polymerises with acids.
2. **Acidic Character:** Indole is a weak acid and reacts with strong alkalis and Grignard Reagents.
3. **Eletrophilic substitution Reaction:** Similar to pyrrole indole also possess excessive π electrons, hence the density of electron on carbon atoms is greater than benzene. The electrophile chiefly attack at C-3 position than C-2. It is explained as follows. The first step in the electrophilic aromatic substitution reaction is the formation of resonance stabilized carbocation intermediate. Attack of electrophile to the indole at C-2 and C-3 produces following carbocation intermediates.

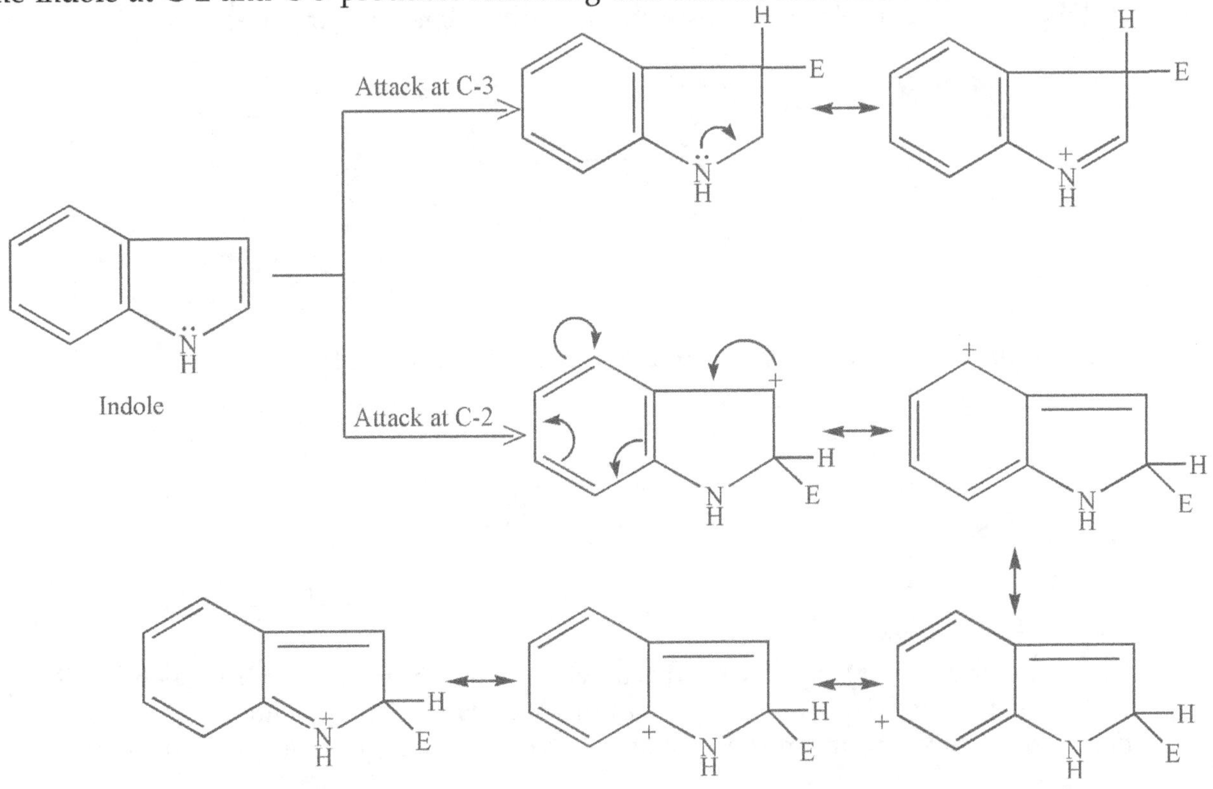

Attack at C-3

Indole

Attack at C-2

In the carbocation obtained by the attack at C-2 position of indole, the charge is distributed on both the benzene nucleus as well as the hetero atom, but the structure with the charge on the hetero atom does not possess aromatic character for longer time. But attack at C-3 position form a carbocation which does not allow the effective distribution of charge throughout the ring but the lone pair of electrons accumulate on the hetero atom of the ring efficiently and retains the aromaticity of the benzene ring. If C-3 position is blocked the substitution occurs at C-2 position. If the both positions are blocked substitution takes place in benzene ring at C-6 position.

(i) ***Halogenation:***

(a) ***Chlorination:*** Indole reacts with sulphuryl chloride or *N,N*-dichloro carbonate or phosphorus pentachloride or *tert*-butyl hypochloride and gives 3-chloro indole.

SO_2Cl_2 ; Cl ; $+ HCl + SO_2$; N̈ H ; N̈ H

Indole   3-Chloro indole

(b) ***Bromination:*** Indole reacts with Br_2/dioxane at 0 °C or *N*-bromo succinimide or Br_2/CH_3COOH and yields 3-bromoindole.

$+ Br_2$; Dioxane ; 0 °C ; Br ; + HBr ; N̈ H ; N̈ H

Indole   3-Bromoindole

If the C-3 position is blocked substitution occurs in benzene ring at C-6 position. For example

$COOC_2H_5$; Br_2/CH_3COOH ; $COOC_2H_5$; Br ; N̈ H ; N̈ H

Ethylindole-3-carboxylate   Ethyl-6-bromoindole-3-carboxylate

(c) ***Iodination:*** Indole reacts with iodine and gives 3-iodo indole.

$+ I_2$; I ; + HI ; N̈ H ; N̈ H

Indole   3-Iodoindole

(ii) ***Nitration:*** Indole very vigorously reacts with normal nitrating mixture such as HNO_3/H_2SO_4 and leads to polymerisation of the final product. Hence it is nitrated with benzoyl nitrate in non - acidic medium in presence of acetonitrile (CH_3CN) at 0 °C to yield 3-nitro indole.

Indole $\xrightarrow[CH_3CN,\ 0\ ^\circ C]{C_6H_5COONO_2}$ 3-Nitroindole

(iii) ***Sulphonation:*** Indole reacts with pyridine–sulphur trioxide complex at 50 °C to yield indole-3-sulphonic acid. The reaction is carried out at mild conditions.

Indole $\xrightarrow[50\ ^\circ C]{C_5H_5\overset{+}{N}-SO_3^-}$ 3-Indolesulphonic acid

(iv) ***Friedel-Craft's acylation:*** Indole reacts with acetyl chloride in the presence of stannous chloride to yield 3-acetylindole.

Indole $+ CH_3COCl \xrightarrow{SnCl_4}$ 3-Acetylindole $+ HCl$

(v) ***Alkylation:*** Indole reacts with CH_3I to yield 3-methylindole.

Indole $+ CH_3I \xrightarrow{DMSO}$ 3-Methylindole $+ HI$

4. Reaction with benzene diazonium chloride: Indole reacts with benzene diazonium chloride in the presence of dilute acids and yield 3-phenylazoindole.

Indole $+$ Benzene diazonium chloride ($C_6H_5N_2Cl$) $\xrightarrow{Dil.Acids}$ 3-Phenylazoindole ($-N{=}N-C_6H_5$) $+ HCl$

5. **Reaction with Aldehydes/ Mannich Reaction**: Indole reacts with formaldehyde and dimethylamine to yield gramine (3-dimethyl aminomethylindole).
6. **Reimer –Tiemann Reaction**: Indole reacts with chloroform and NaOH to yield 3-indole carbaldehyde.

Indole $+ CHCl_3 + NaOH \longrightarrow$ 3-Indolecarbaldehyde (CHO)

7. **Mercuration**: Indole reacts with mercuric acetate and gives 2,3-diacetoxy mercuric indole.
8. **Reaction with Grignard Reagent**: Indole reacts with methyl magnesium bromide to yield indole magnesium bromide which is used for the preparation of substituted indoles.

Indole $+ CH_3MgBr \xrightarrow{-CH_4}$ (N–MgBr) $\xrightarrow[\text{ii) } H^+]{\text{i) } CO_2}$ 3-Indole carboxylic acid (COOH)

9. **Reaction with metals**: Indole reacts with *n*-butyl lithium to give indolelithium. Metallation occurs at *N*-hydrogen which is used for the synthesis of various substituted indoles.

Indole + n-BuLi $\longrightarrow$ (N–Li) $\xrightarrow[\text{ii) } H^+]{\text{i) } CO_2}$ 1-Indole carboxylic acid (N–COOH)

10. **Oxidation**: Indole undergoes autooxidation in the presence of air and light to give indoxyl which reacts further to yield indigo.

Indole $\xrightarrow{O_2}$ Indoxyl (OH)

It is also oxidized with ozone in the presence of formamide to yield *o*-formamino benzaldehyde.

Indole + O_3 $\xrightarrow{HCONH_2}$ *o*-Formaminobenzaldehyde

11. **Reduction**: Indole is treated with zinc and HCl to yield dihydroindole or indoline. If the reaction is carried out with sodium cyanoborohydride in acetic acid indole yields indoline with higher yield (90%).

Indole $\xrightarrow{ZnCl_2}$ Indoline

Indole is reduced with lithium-ammonia-ethanol and yields the product in which benzene nucleus gets reduced.

Indole $\xrightarrow[\text{Ethanol}]{Li, NH_3}$ 4,7-Dihydro indole + 4,5,6,7-Tetrahydro indole

Indole also reduced with powerful reducing agents such as Pt/CH_3COOH to give *cis*-tetrahydro indole.

Indole $\xrightarrow{Pt, CH_3COOH}$ *cis*-Tetrahydro indole

Indole Derivatives

Indoxyl

Ketoform of 3- hydroxy indole is known as indoxyl.

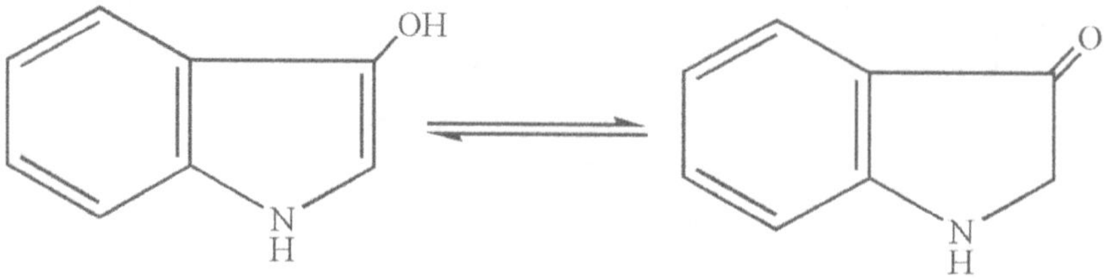

Enol form

Keto form
(Indoxyl)

Oxindole

Three isomeric forms of oxindole are present.

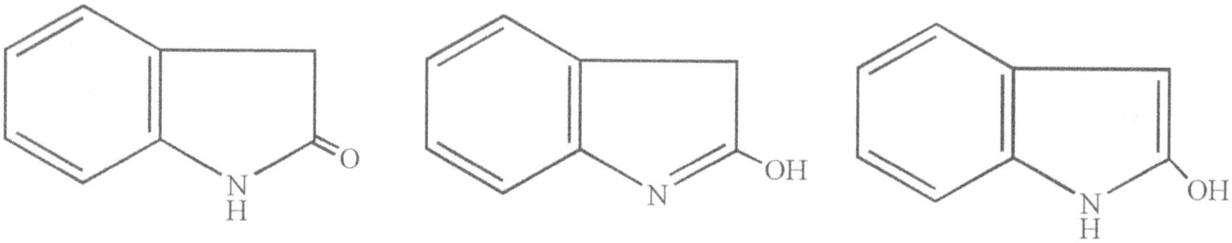

Dioxindole

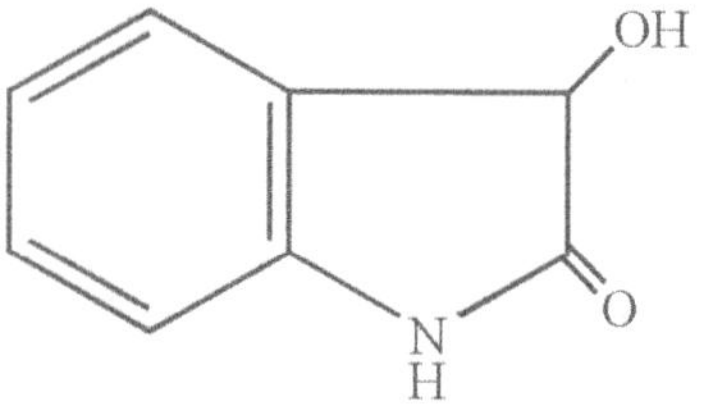

Carbazole or Dibenzopyrole

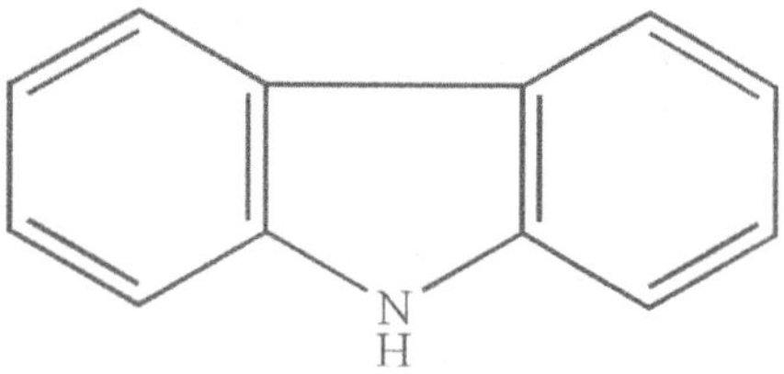

Isatin

It is present in two forms such as lactum and lactim form. It is a derivative of 2,3-dihydroindole.

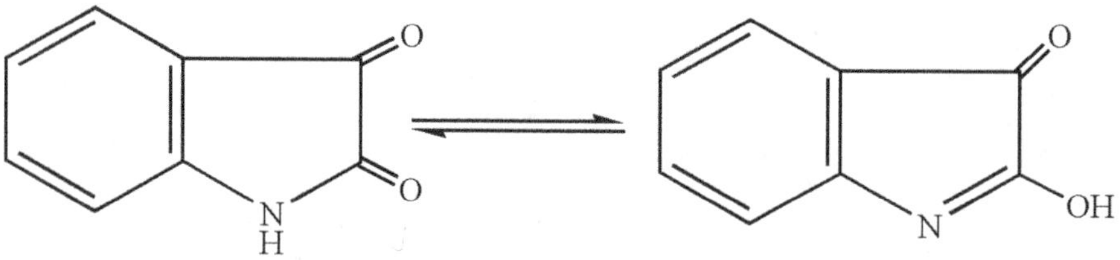

(I)

(II)

Isatin is an example of the amido-imidol tautomeric system,

$$—NH—CH_2— \rightleftharpoons —N{=}C(OH) + —$$

General Methods of Preparation:

1. **From *o*-nitro benzoyl chloride:** *o*-Nitro benzoyl chloride reacts with various kind of reagents (NaCN, $FeSO_4.NH_3$) and yields isatin.
2. **From indigotin:** Isatin was first synthesized from indigotin which was treated with nitric acid.
3. **From aniline:** A mixture of aniline in conc. HCl is heated with chloral hydrate, hydroxyl amine and sodium sulphate and yields isonitrosoacetanilide which upon further treatment with conc.H_2SO_4 gives isatin.

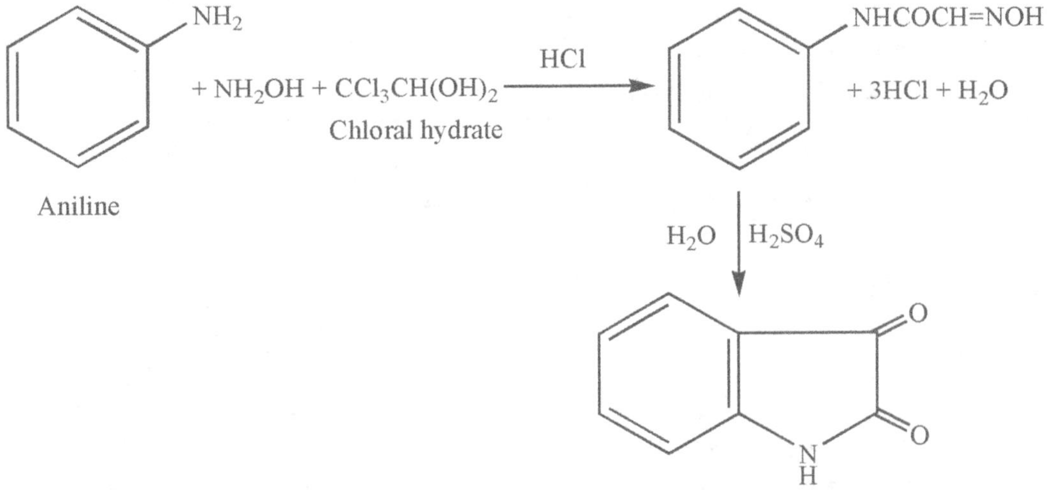

NH₂ + $NH_2OH + CCl_3CH(OH)_2$ (Chloral hydrate) $\xrightarrow{HCl}$ NHCOCH=NOH + $3HCl + H_2O$

Aniline

$\downarrow H_2O \mid H_2SO_4$

Isatin

Physical properties: Isatin is a red colour solid. M.P 200 °C. It is sparingly soluble in water but soluble in ethanol.

Chemical Properties:

1. **Reaction with NaOH:** Isatin when warmed with NaOH yields sodium salt of isatinic acid.

Isatin + NaOH $\xrightarrow{Warm}$ COCOONa, NH₂

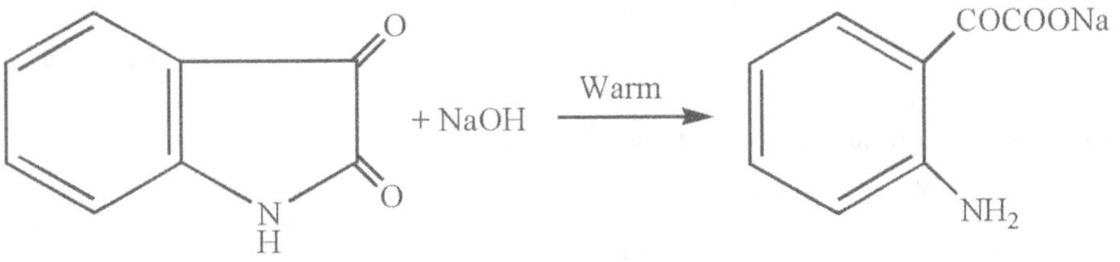

Isatin — Sodium salt of isatinic acid

2. **Reaction with PCl_5:** Isatin reacts with PCl_5 and yields isatin chloride.

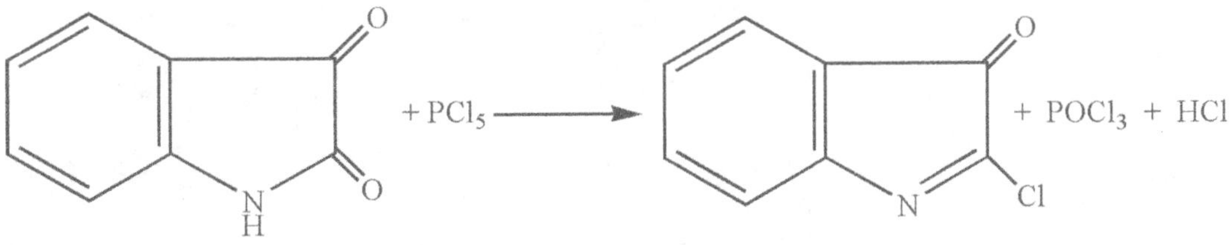

Isatin + PCl_5 $\longrightarrow$ Isatin chloride + $POCl_3$ + HCl

Isatin — Isatin chloride

Medicinal Compounds Containing Indole or Indole Derivatives Used in Medicine

Indapamide	Clorexolone
3-(Aminosulphonyl)-4-chloro-*N*-(2,3-dihydro-2-methyl1*H*-indol-1-yl)- benzamide **Use:** It is used as a diuretic agent	6-Chloro-2-Cyclohexyl-3-oxo-1*H*-isoindole-5-sulfonamide **Use:** It is used as a diuretic agent
Pindolol	**Psilocyn**
1-(1*H*-indol-4-yloxy)-3-(propan-2-ylamino)propan-2-ol **Use:** It is used as an antiadrenergic agent	**Use:** It is used as a CNS stimulant
Methisazone	**Mitomycin C**
(*Z*)-1-(1-Methyl-2-oxoindolin-3-ylidiene)thiosemicarbazide **Use:** It is used as an antiviral agent	**Use:** It is used as an anticancer agent
Reserpine	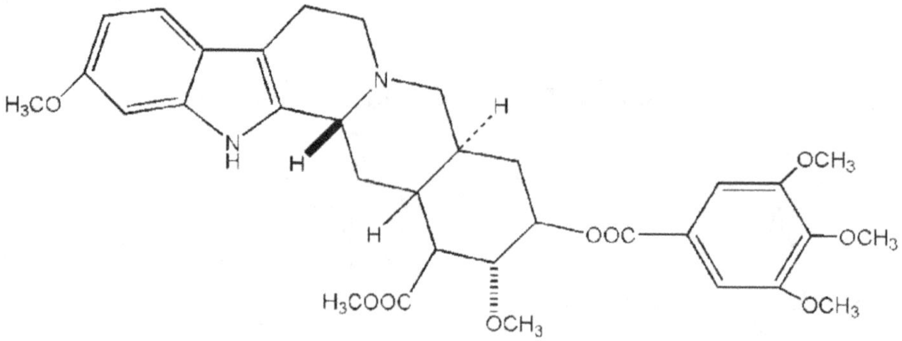
Use: It is used as an antihypertensive agent	

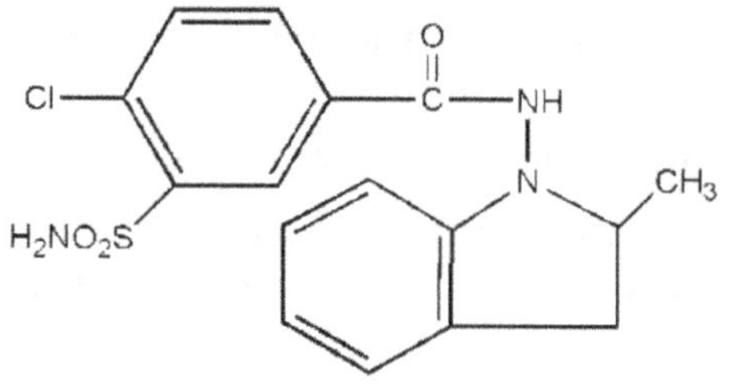

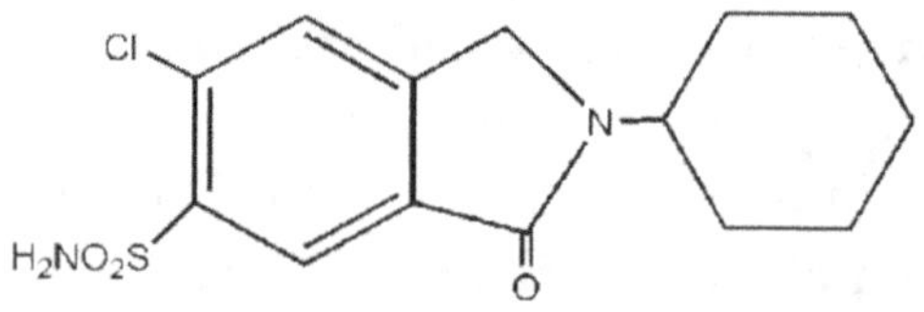

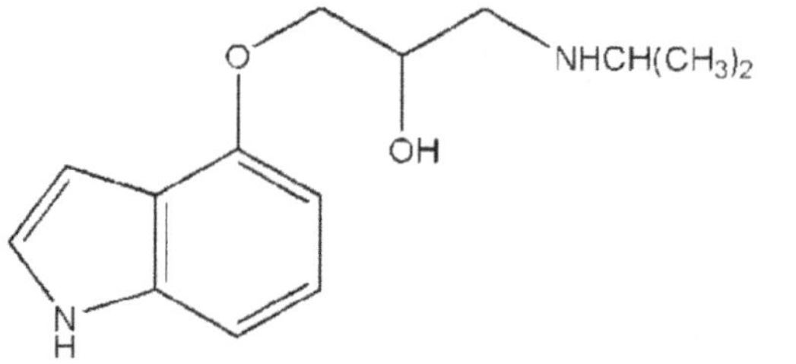

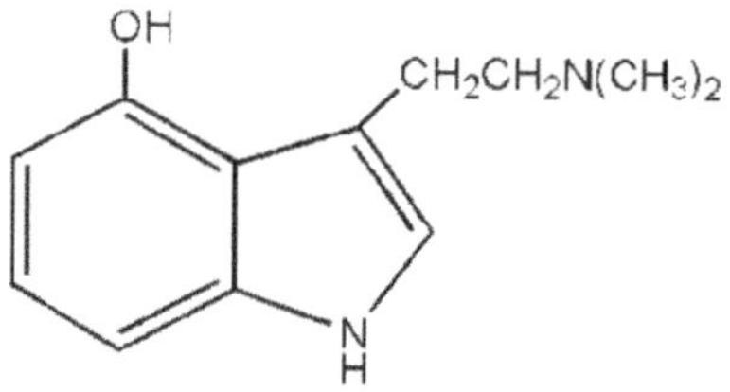

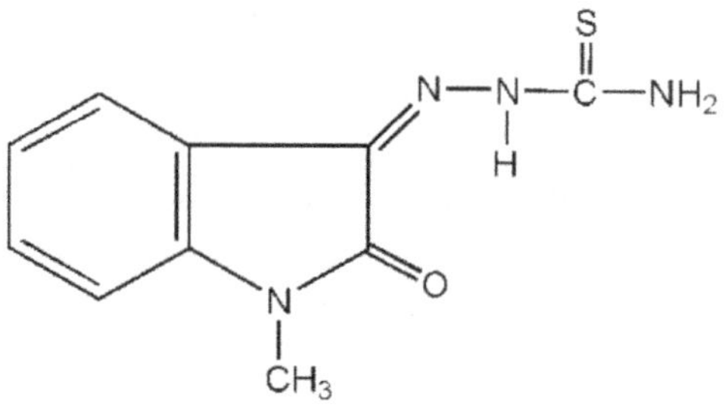

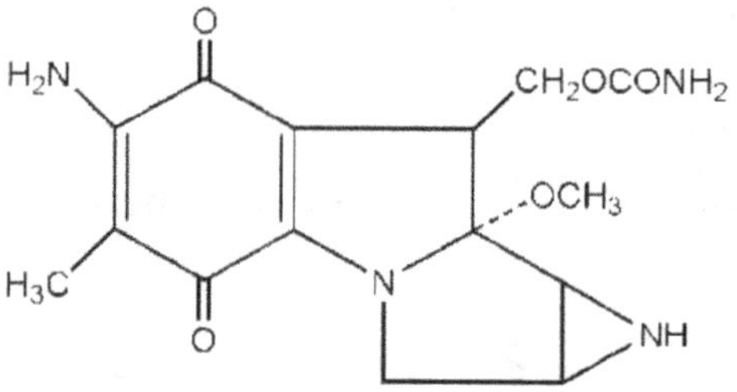

Pyrimidine

Pyrimidine is the most important compounds of diazine series in which the two nitrogen atoms are present in 1st and 3rd positions of the ring. It occurs widely in living organisms. It was first discovered by Gabriel and Colman. In nucleic acids, three types of nucleic acid bases are present viz cytosine, thymine and uracil. The Scientist Pinner was first proposed the name pyrimidine in 1885. The pyrimidine ring system is present in many of the drugs like barbituric acid, antimalarial and some antibacterial drugs. The major disorder of pyrimidine metabolism leads to diseases such as orotic aciduria and Rey's syndrome.

[Structure: pyrimidine ring numbered N1, 2, N3, 4, 5, 6]

General Methods of Preparation:

1. **From 4-dichloropyrimidine:** In this method 2,4-dichloropyrimidine on reductive dechlorination in the presence of catalyst such as Pd-C and MgO gives pyrimidine.

$$\text{2,4-Dichloropyrimidine} \xrightarrow[\text{Pd-C \& MgO}]{2H_2} \text{Pyrimidine} + 2HCl$$

2. **From 1,1,3,3- tetraethoxy propane:** Reaction between 1,1,3,3-tetraethoxy propane and formamide yields pyrimidine.

$$\underset{\text{Formamide}}{2HCONH_2} + \underset{\text{1,1,3,3-Tetraethoxy propane}}{H_2C(CH(OC_2H_5)_2)_2} \xrightarrow[200\,^{\circ}C]{Al_2O_3} \text{Pyrimidine}$$

Medicinal Compounds Containing Pyrimidine (or) Pyrimidine Derivatives Used in Medicine

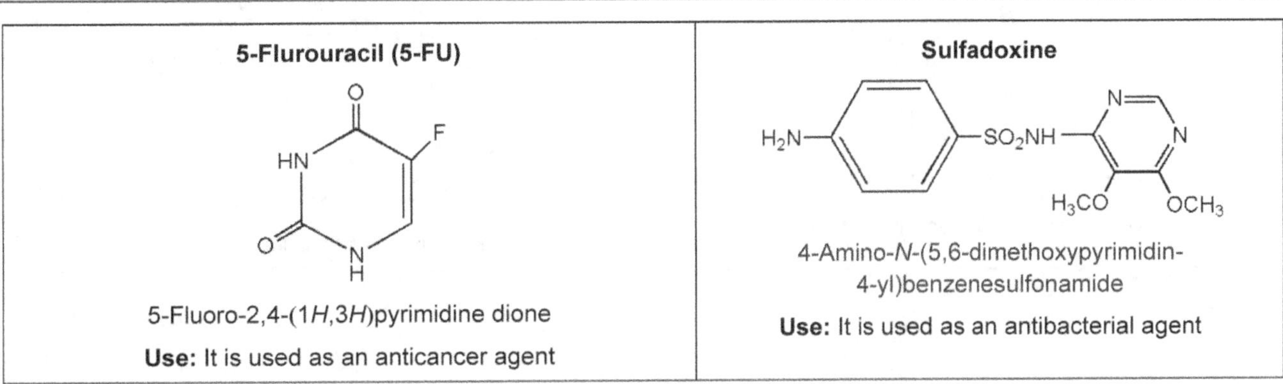

5-Flurouracil (5-FU)	Sulfadoxine
5-Fluoro-2,4-(1*H*,3*H*)pyrimidine dione	4-Amino-*N*-(5,6-dimethoxypyrimidin-4-yl)benzenesulfonamide
Use: It is used as an anticancer agent	**Use:** It is used as an antibacterial agent

Table Contd...

Barbital 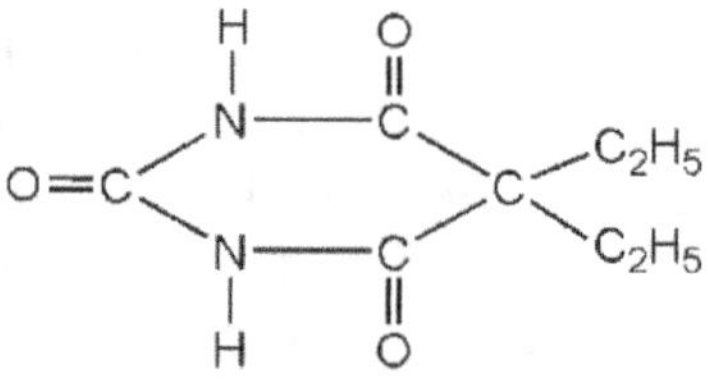 5,5- Diphenyl barbituric acid **Use:** It is used as a sedative and hypnotic	**Phenobarbital** 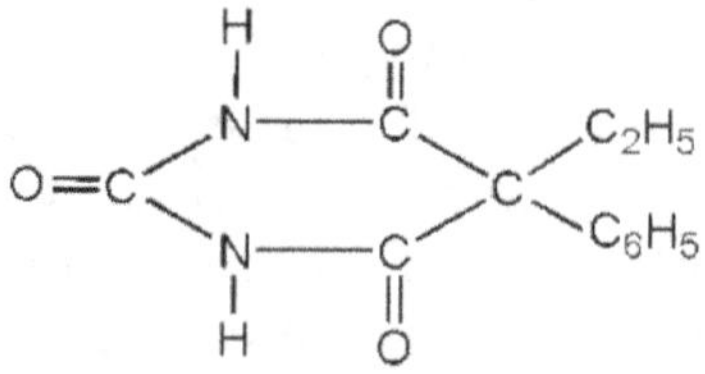 5-Ethyl-5-phenyl barbituric acid **Use:** It is used as a sedative and hypnotic
Pyrimethamine 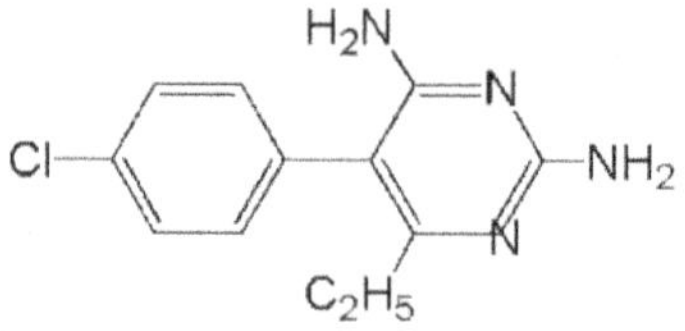 5-(4-Chlorophenyl)-6-ethylpyrimidine-2,4-diamine **Use:** It is used as an antimalarial agent	**Trimethoprim** 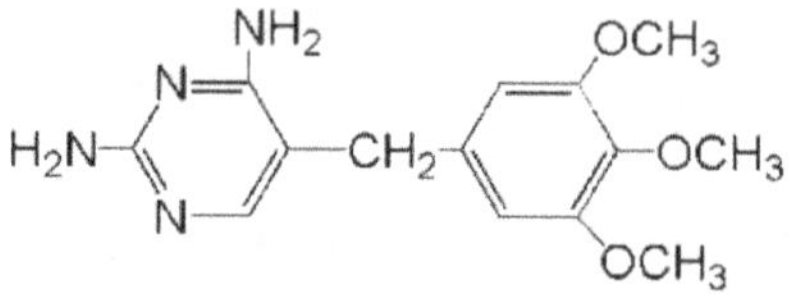 5-(2,3,4-Trimethoxy benzyl)pyrimidine-2,4-diamine **Use:** It is used as an antimalarial agent
Quinethazone 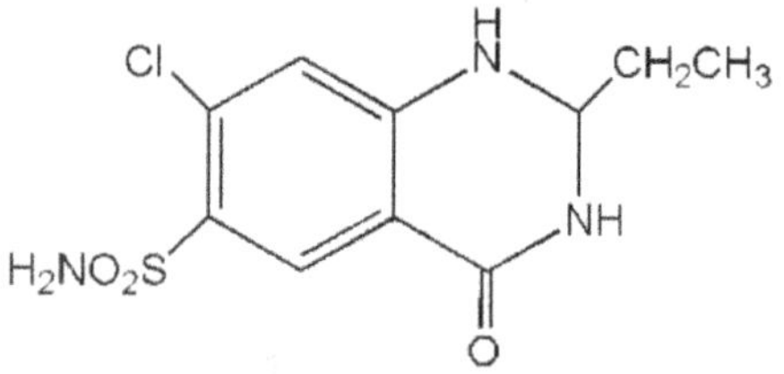 7-chloro-2-ethyl-4-oxo-2,3-dihydro-1*H*-quinazoline-6-sulfonamide **Use:** It is used as a diuretic	**Metolazone** 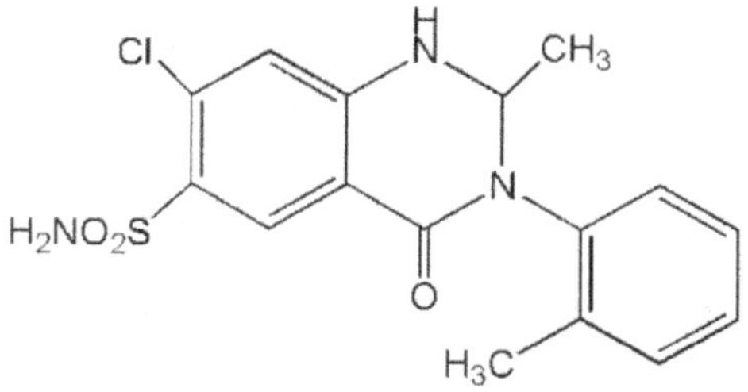 7-Chloro-2-methyl-3-(2-methylphenyl)-4-oxo-1,2-dihydroquinazoline-6-sulfonamide **Use:** It is used as a diuretic
Cidofovir 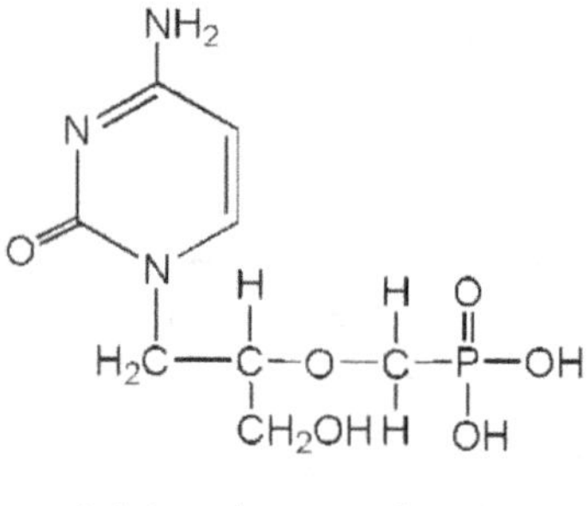 1-[3- Hydroxy-2-(phosphonomethoxy)propyl]cytosine **Use:** It is used as an antiviral agent	**Emivirine** 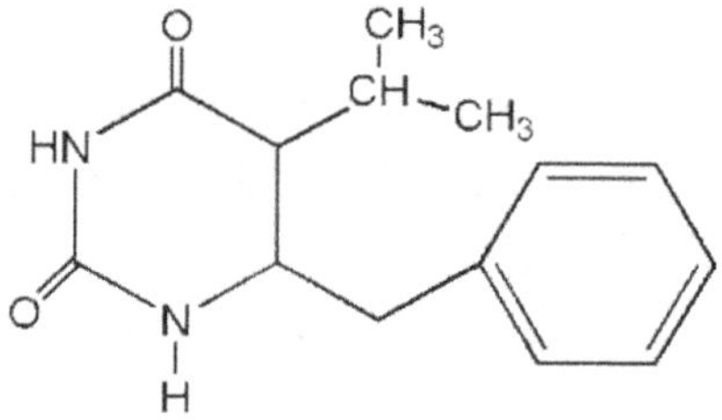 6-Benzyl-5-isopropyl-hexahydropyrimidine-2,4-dione **Use:** It is used as an antiviral agent

Table *Contd...*

Zidovudine 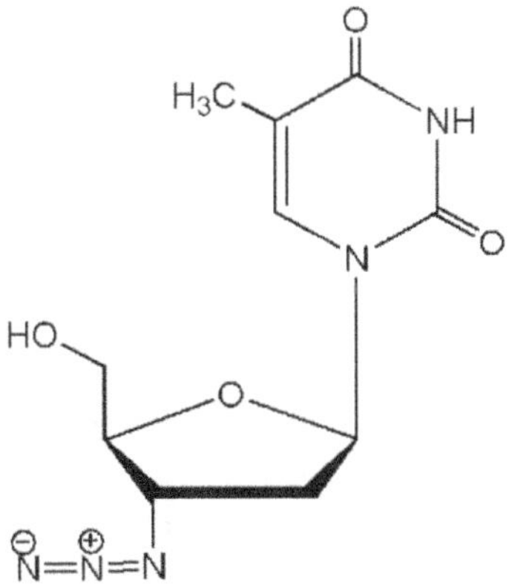 1 [4-Azido-5-(hydroxymethyl)oxolan-2-yl]-5-methylpyrimidine-2,4-dione **Use:** It is used as an antiviral agent	**Zalcitabine** 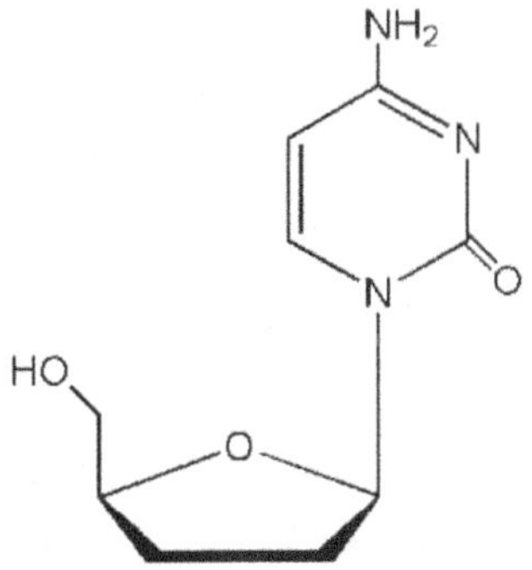 **Use:** It is used as an antiviral agent
Trifluridine 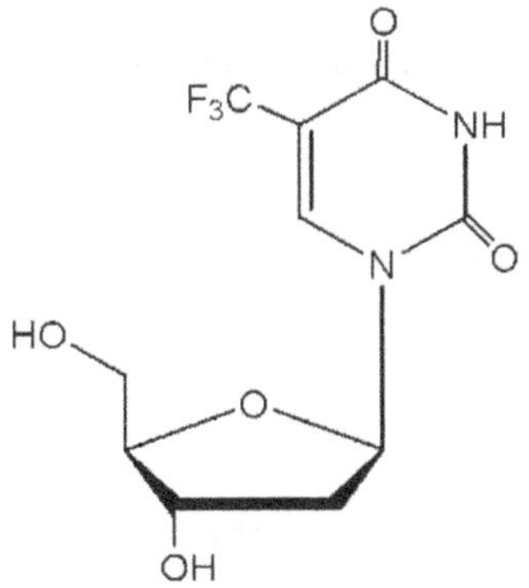 1-[(2*R*,4*S*,5*R*)-4-Hydroxy-5-(hydroxymethyl)oxolan-2-yl]-5-(trifluoromethyl)pyrimidine-2,4-dione **Use:** It is used as an antiviral agent	**Capectitabine** 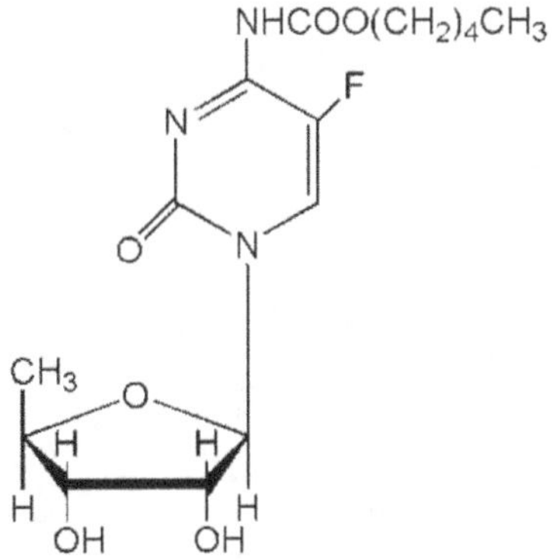 **Use:** It is used as an anticancer agent
Fluoxuridine 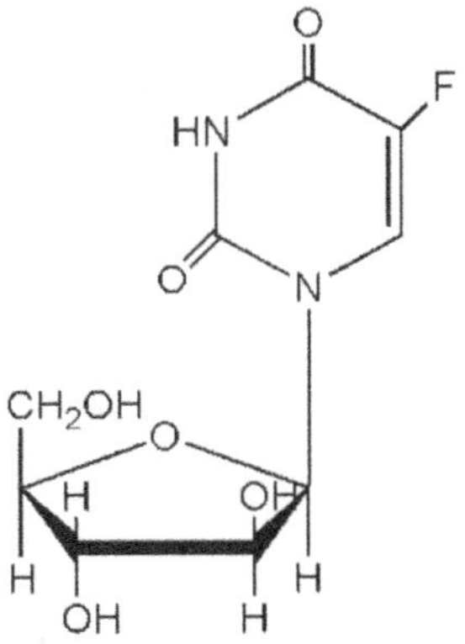 **Use:** It is used as an anticancer agent	**Cytarabine** 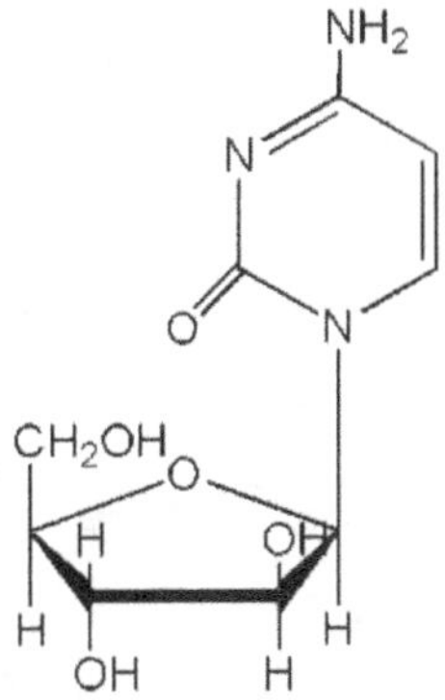 **Use:** It is used as an anticancer agent

Table *Contd...*

Terazocine
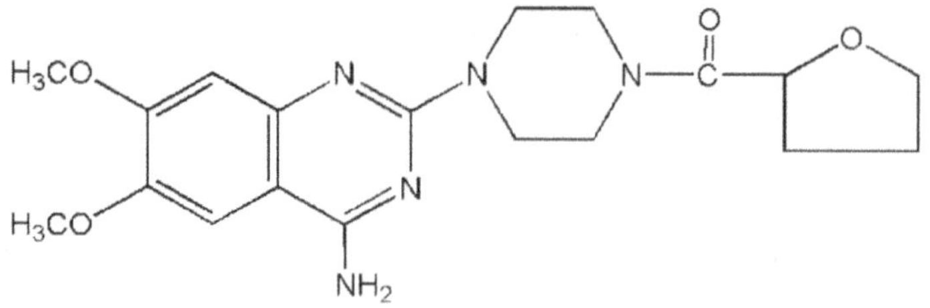 2{[4-(Tetrahydro-2-furanyl)-carbonyl]-1-piperazinyl}-6,7-dimethoxy-4-quinolinamine **Use:** It is used as an antihypertensive agent
Prazosin
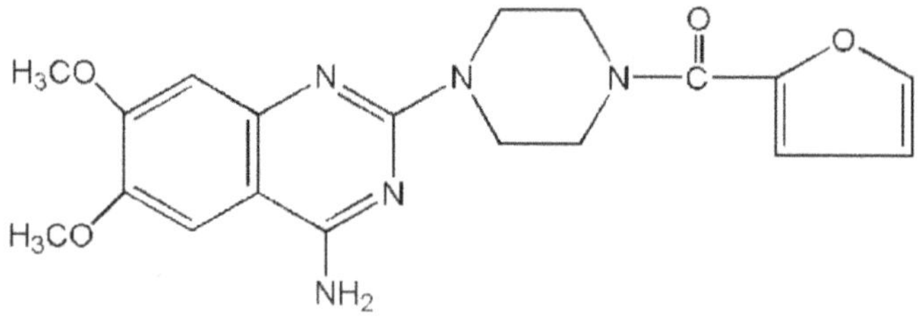 1-(4-Amino-6,7-dimethoxy-2-quinazolinyl)-4-(2-furanyl carbonyl)-pipearzine **Use:** It is used as an antihypertensive agent
Oxmetidine
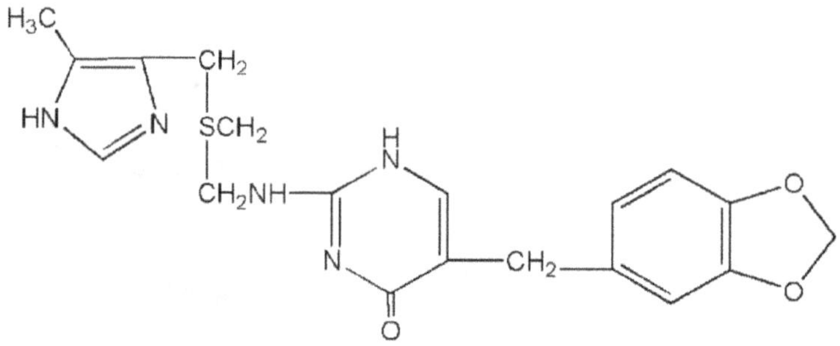 **Use:** It is used as an antiulcer agent

Purine

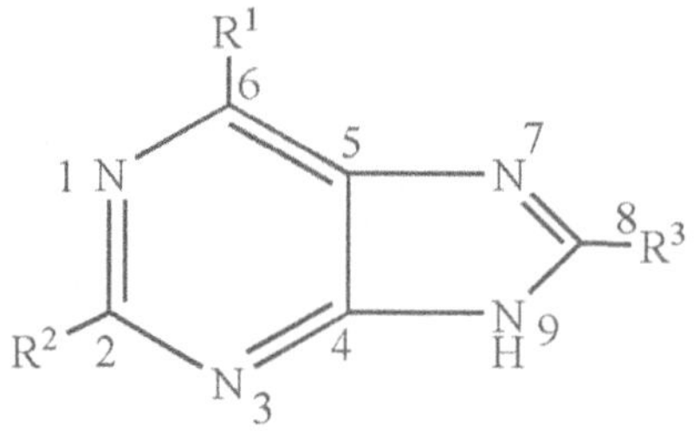

Purine is an organic heterocyclic compound in which pyrimidine ring is fused with imidazole ring. These are the building blocks of DNA and RNA. The two important nucleic acid bases such as adenine and guanine are purine derivatives. High concentration of purine is present in liver and kidney. Plant based products such as spinach, beans and oat meal contains are medium concentration of purines. Some of the important purines are xanthine, hypoxanthine, caffeine, uric acid, theobromine *etc.* In 1884 the German chemist Emil Fischer coined the name purine (pure urine). The metabolic disorder of purine leads to gout. Purine possesses self inhibiting and self stimulating properties. When purines are formed in the cells it inhibits the enzyme responsible for purine synthesis.

General Methods of Preparation:

1. **From 4,5-diaminopyrimidine:** Diamino pyrimidine reacts with formic acid and yields purine.

4,5-Diaminopyrimidine + Formic acid → Purine + H_2O

2. **From Uric acid:**

Uric acid ⇌ Uric acid (Enol form) —$POCl_3$→ 2,6,8-trichloropurine —HI, 0 °C→ 2,6-diiodopurine —Zn dust, H_2O/Boil, –2ZnI→ Purine

3. **Traube synthesis / synthesis of guanine:** 2,5,6-Triaminopyrimidine-4-one reacts with formic acid to yield guanine.

Diamino pyrimidine + NH_2CONH_2 (Urea) —Δ→ (NHCONH$_2$ intermediate) → 8-Oxopurine

4. **Synthesis of 8-oxopurine:** Diamino pyrimidines are heated with urea to yield 8-oxopurines.

Diamino pyrimidine + NH_2CONH_2 (Urea) $\xrightarrow{\Delta}$ → 8-Oxopurine

5. **From imidazole:** Imidazole undergoes cyclisation reaction with formic acid under heat and yield purines.

+ HCOOH $\xrightarrow{\Delta}$ $\xrightarrow{\Delta}$

Medicinal Compounds Containing Purine (or) Purine Derivatives Used in Medicine

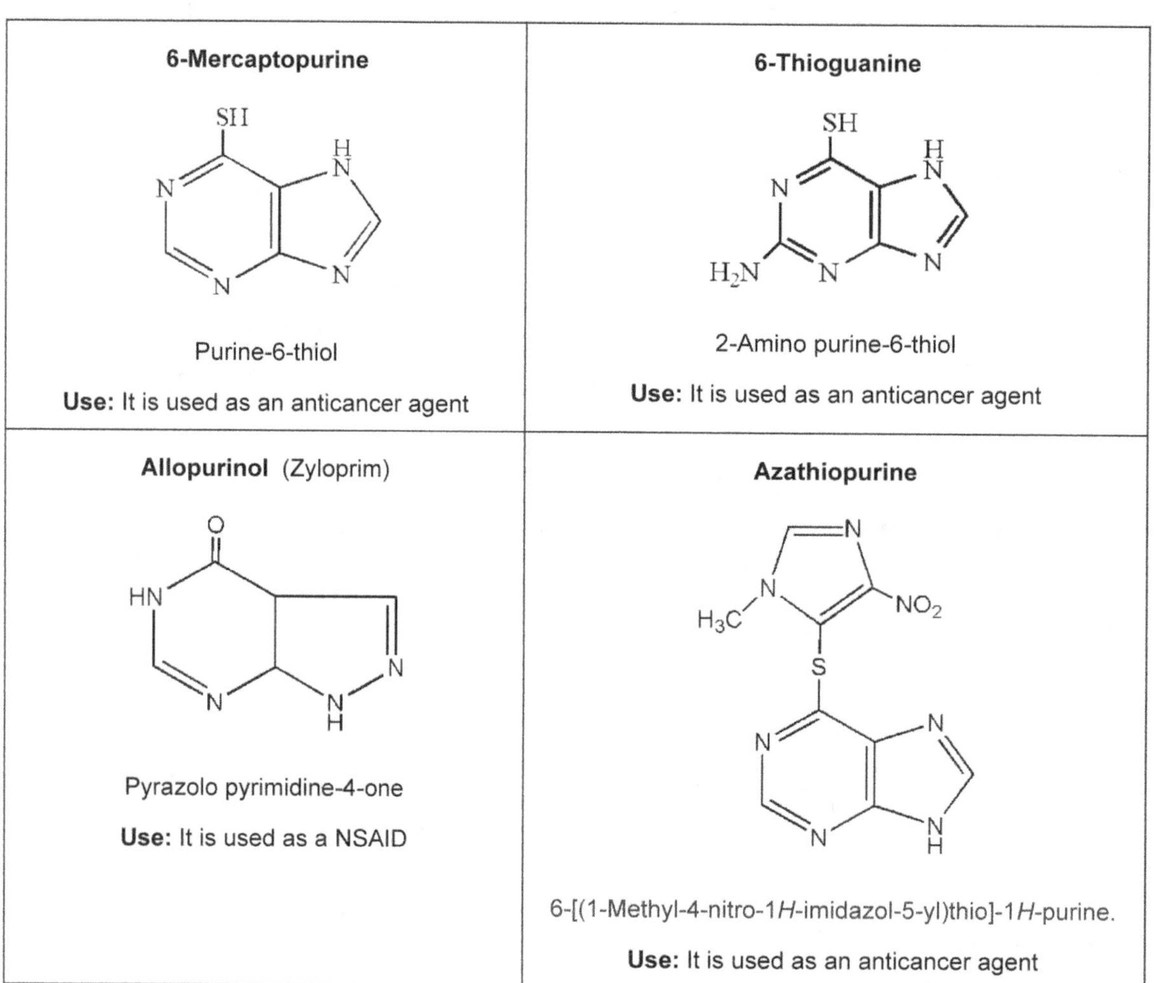

6-Mercaptopurine	**6-Thioguanine**
Purine-6-thiol	2-Amino purine-6-thiol
Use: It is used as an anticancer agent	**Use:** It is used as an anticancer agent
Allopurinol (Zyloprim)	**Azathiopurine**
Pyrazolo pyrimidine-4-one	6-[(1-Methyl-4-nitro-1*H*-imidazol-5-yl)thio]-1*H*-purine.
Use: It is used as a NSAID	**Use:** It is used as an anticancer agent

Table *Contd...*

Fludarabine	Vidarabine
NH2, N, N, N, F, N, HO, O, H, OH, H, OH, H, H	NH2, N, N, N, N, HO, O, HO, OH
Use: It is used as an anticancer agent	**Use:** It is used as an antiviral agent
Acyclovir	**Ganciclovir**
O, HN, N, H_2N, N, N, $CH_2OCH_2CH_2OH$	O, HN, N, H_2N, N, N, CH_2OH, CH_2–O–C–CH_2OH, H
Use: It is used as an antiviral agent	2-Amino-9-((1,3-dihydroxy propan-2yloxy)methyl)-1*H*-purin-one **Use:** It is used as an antiviral agent
Didanosine	**Stavudine**
O, HN, N, N, N, HO, O	O, H_3C, NH, N, O, HO, O
2', 3'-Dideoxyinosine **Use:** It is used as an antiviral agent	5-(Hydroxymethyl)-2,5-dihydrofuran-2-yl]-5-methylpyrimidine-2,4-dione **Use:** It is used as an antiviral agent

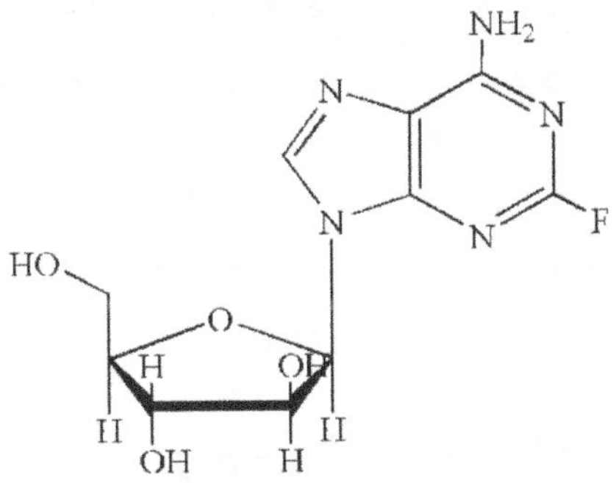

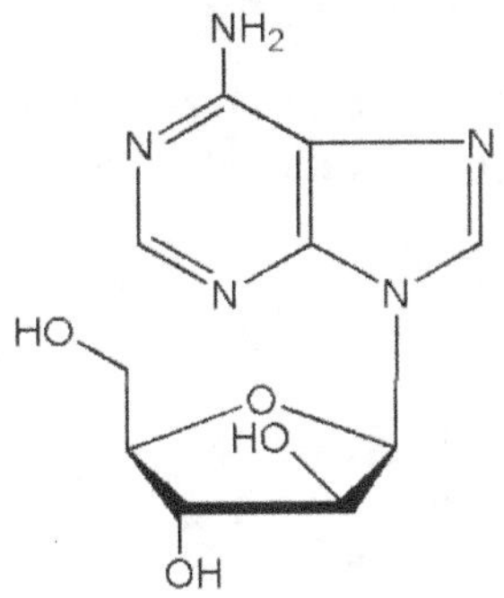

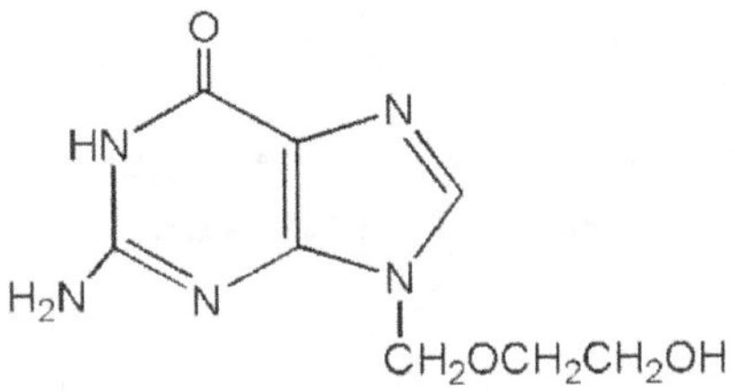

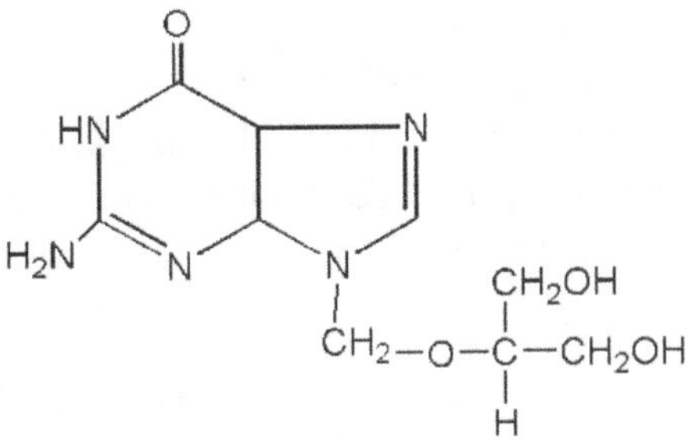

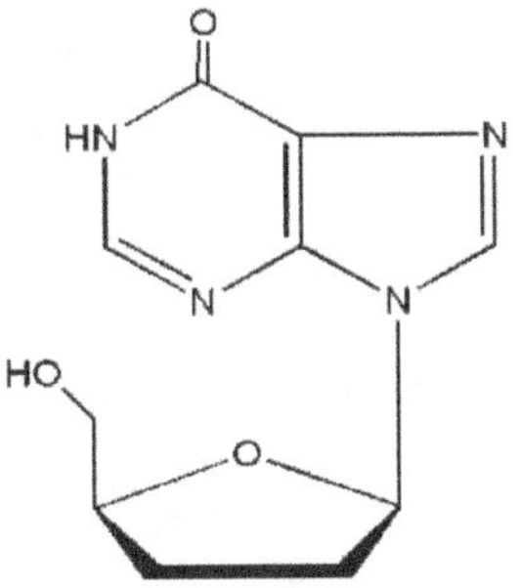

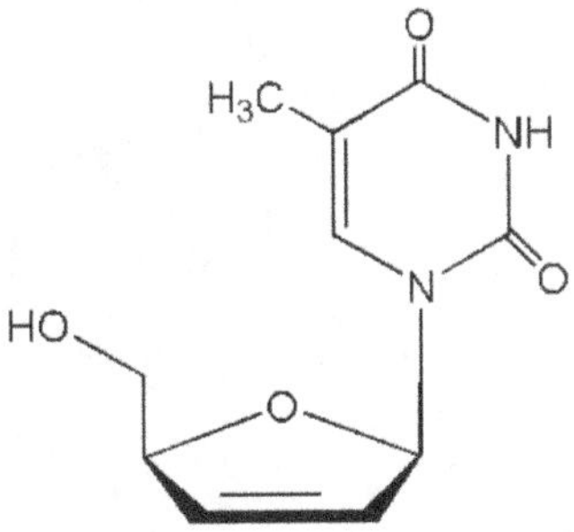

Abacavir

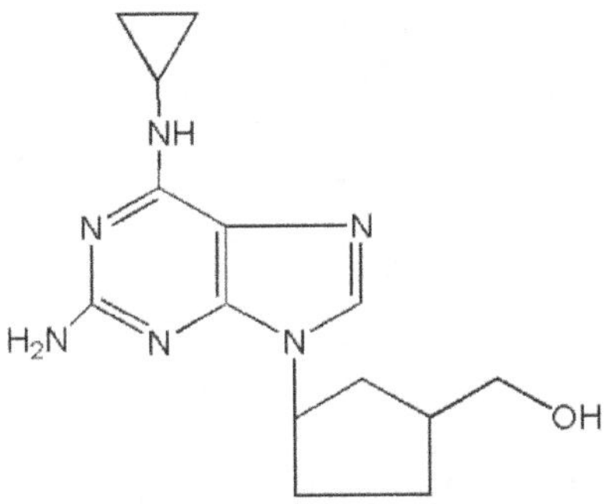

NH, N, N, H_2N, N, N, OH

4-[2-Amino-6-(cyclopropylamino)purin-9-yl]cyclopent-2-en-1-yl]methanol

Use: It is used as an antiviral agent

Azepines

The seven membered ring systems with one nitrogen atom is known as azepines. It doesnot not satisfy Huckel's rule ((4n+2) π electrons) and hence non-aromatic.

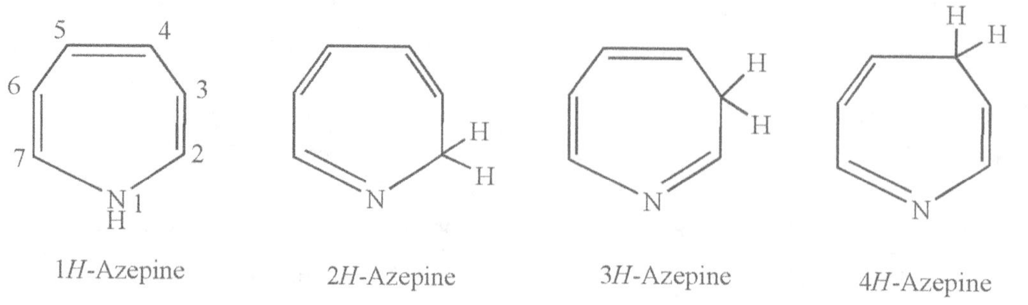

1*H*-Azepine 2*H*-Azepine 3*H*-Azepine 4*H*-Azepine

General Methods of Preparation of Azepine and its Derivatives:

1. **From Benzene or Valence bond Isomerisation:** This method is used in recent years for the synthesis of majority of the organic compounds or molecules. It involves the reorganisation or rearrangement of the σ and π electrons.

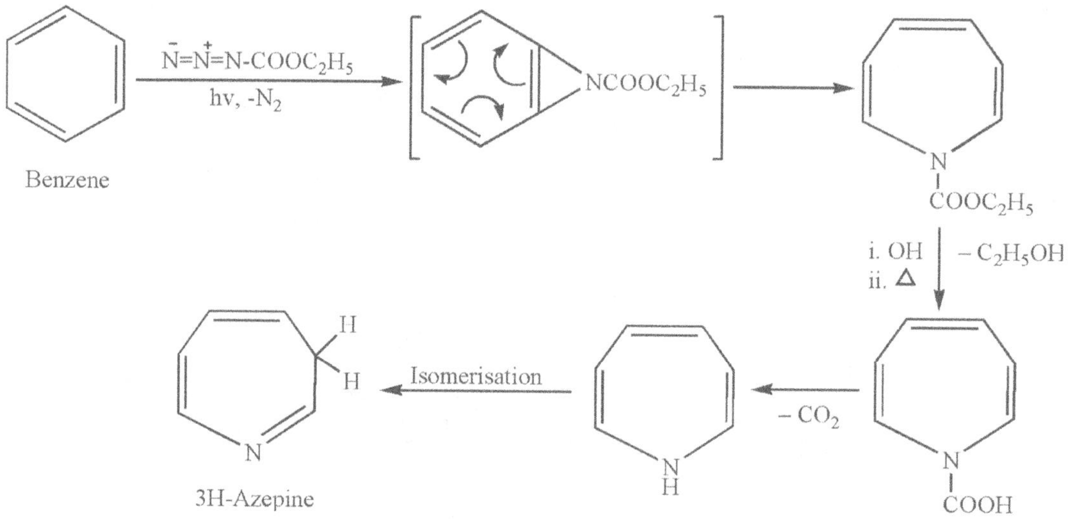

Benzene $\xrightarrow[hv,\ -N_2]{\bar{N}=\overset{+}{N}=N\text{-}COOC_2H_5}$ [$NCOOC_2H_5$] → N-$COOC_2H_5$ azepine $\xrightarrow[ii.\ \Delta]{i.\ OH,\ -C_2H_5OH}$ N-COOH azepine $\xrightarrow{-CO_2}$ NH azepine $\xrightarrow{Isomerisation}$ 3H-Azepine

2. **From Nitrobenzene:** Nitrobenzene reacts with tributyl phosphine and gives arylnitrene which upon further reaction with primary or secondary alcohol yields 2-alkoxy-3H-azepines.

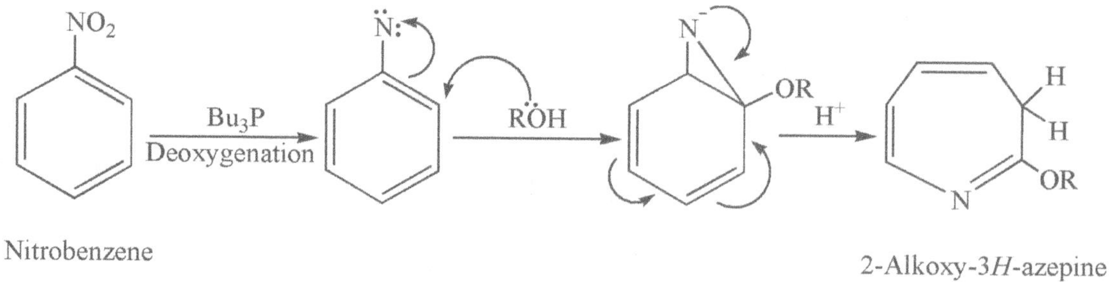

Nitrobenzene (NO_2) $\xrightarrow[Deoxygenation]{Bu_3P}$ aryl nitrene (N:) $\xrightarrow{R\ddot{O}H}$ (N⁻, OR) $\xrightarrow{H^+}$ 2-Alkoxy-3*H*-azepine

3. **From Phenylazide:** Boiling of phenylazide with primary and secondary amine yields substituted azepine *via* the formation of benzazirine intermediate.

$-N_2$ Ring expansion reaction

$(C_2H_5)_2NH$

Isomerisation

Phenylazide

2-Substituted-3*H*-azepine

4. **Synthesis of azepine-2,5-diones:** *p*-Benzoquinone reacts with hydrazoic acid at 0 °C in the presence of H_2SO_4 catalyst and yields azepine-2,5-diones.

NH_3/H^+

HOH $-N_2$

2,3,5-Trimethyl-*p*-benzoquinone

3,6,7-Trimethyl azepine-2,5-dione

Medicinal Compounds Containing Azepine (or) Azepine Derivatives Used in Medicine

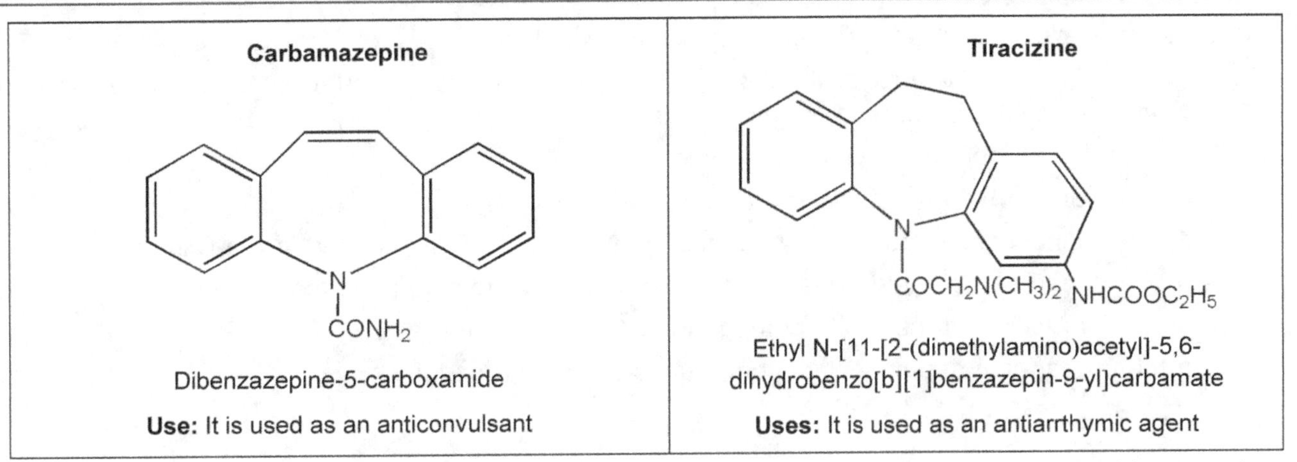

Carbamazepine	Tiracizine
Dibenzazepine-5-carboxamide	Ethyl N-[11-[2-(dimethylamino)acetyl]-5,6-dihydrobenzo[b][1]benzazepin-9-yl]carbamate
Use: It is used as an anticonvulsant	**Uses:** It is used as an antiarrthymic agent

Table *Contd...*

Azelastine

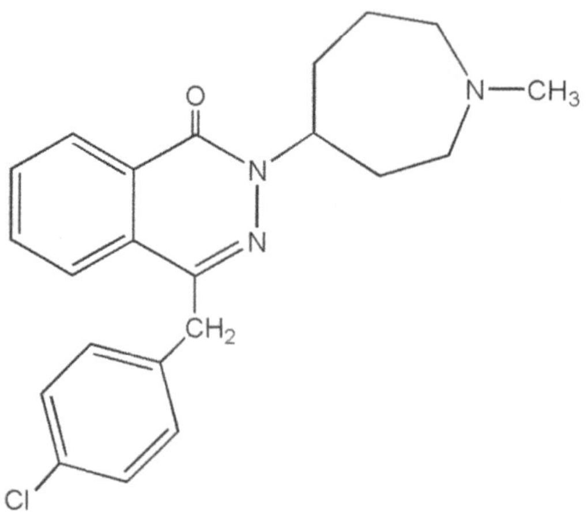

4-[4-Chlorophenyl)methyl] -2-(1-methylazepan-4-yl)-phthalazin-1-one

Use: It is used as an antihistaminic agent

Probable Questions

1. Define azoles and write the different type of azoles. Write about the structure of azoles.
2. Write the general methods of preparation of pyrazole and describe the electrophilic substitution reaction of pyrazole.
3. Write any five medicinal compounds containing pyrazole (or) pyrazole derivatives used in medicine.
4. Explain general methods of preparation of Imidazole and chemical properties of imidazole.
5. Wrie a note on medicinal compounds containing imidazole (or) imidazole derivatives used in medicine.
6. General methods of preparation and reactions of oxazole.
7. Write the medicinal compounds containing oxazole or oxazole derivatives used in medicine.
8. Write the preparation & reactions of thiazole and write its four medicinal uses.
9. Explain about the nomenclature and basicity of pyridine.
10. Write the preparation and reactions of pyridine.
11. Mention medicinal compounds containing pyridine (or) pyridine derivatives used in medicine.
12. Write the preparation and chemical properties of quinoline.
13. Write medicinal compounds containing quinoline and isoquinoline used in medicine.
14. Write the preparation and reactions of isoquinoline.
15. Write general methods of preparation and chemical properties of acridine. Explain the medicinal compounds containing acridine used in medicine.
16. Explain the preparation and reactions of indole.
17. Write the preparation and reactions of isatin.
18. Write the medicinal compounds containing indole or indole derivatives used in medicine.
19. Explain the preparation and medicinal uses of pyrimidine containing used in medicine.
20. Write the preparation and medicinal compounds containing purine.
21. Explain the preparation and medicinal compounds containing azepines.

31

Reactions and Reagents of Synthetic Importance

Sodium Borohydride ($NaBH_4$)

Method of Preparation

It is obtained by the reaction between methyl borate and sodium hydride at elevated temperature.

Advantages

Sodium borohydride is mainly used as an important reducing agent because:

(a) Much milder reducing agent than $LiAlH_4$.

(b) More effective reducing agent and useful for the specific reduction of aldehydes, ketones and acid chlorides without attacking other reducible groups.

(c) $NaBH_4$ is slowly decomposed by water and alcohol, so it can be used for reduction of carbonyl compounds in aqueous solutions. So it is widely used in carbohydrate chemistry (Quantitative reduction of glucose to sorbitol).

$$\underset{\text{Ketone}}{4R_2CO} + NaBH_4 + 3H_2O \longrightarrow \underset{2^o\text{ Alcohol}}{4R_2CHOH} + NaH_2BO_3$$

Some of important applications of sodium borohydride are discussed as follows:

1. **Reduction of aldehydes and ketones:** Carbonyl groups are reduced to respective alcohols by $NaBH_4$.

$$\underset{\text{Acetyl acetone}}{CH_3\text{-}\overset{\overset{O}{\|}}{C}\text{-}CH_2\text{-}\overset{\overset{O}{\|}}{C}\text{-}CH_3} \xrightarrow{NaBH_4} \underset{\text{2,4-Pentanediol}}{CH_3\text{-}\overset{\overset{OH}{|}}{CH}\text{-}CH_2\text{-}\overset{\overset{OH}{|}}{CH}\text{-}CH_3}$$

$$\underset{\text{Glucose}}{CH_2OH(CHOH)_4CHO} \xrightarrow{NaBH_4} \underset{\text{Sorbitol}}{CH_2OH(CHOH)_4CH_2OH}$$

$$\underset{p\text{-Nitrobenzaldehyde}}{O_2N\text{-}C_6H_4\text{-}CHO} \xrightarrow{NaBH_4} \underset{p\text{-Nitrobenzyl alcohol}}{O_2N\text{-}C_6H_4\text{-}CH_2OH}$$

Reduction of ketonic group by $NaBH_4$ into alcoholic group is used for the synthesis of tropine and cholesterol.

2. **Reduction of ozonides:** $NaBH_4$ reduces an ozonide to alcohol.
For example: Cyclohexene ozonides to 1,6-diol.

$$\text{Cyclohexene} \xrightarrow{O_3} \text{Cyclohexene ozonide} \xrightarrow{NaBH_4} \underset{\text{1,6-Diol}}{OH(CH_2)_6OH}$$

3. **Reduction of azides:** If the azide molecule contains sulphur, reduction to amine by hydrogenation is impossible. In that case, the azides are reduced by $NaBH_4$ in isopropanol.
4. **Reduction of tosyl hydrazones:** Tosyl hydrazones are reduced to hydrocarbons with $NaBH_4$. This process is used as a valuable tool for the synthesis of ketosteroids to remove carbonyl groups.

TsNHN $\xrightarrow{NaBH_4}$

5. $NaBH_4$ is an effective catalyst for ester interchange.
6. It is valuable source of diborane.

$$\underset{\text{Boron trifluoride}}{4BF_3} + 3NaBH_4 \longrightarrow \underset{\text{Diborane}}{2B_2H_6} + \underset{\text{Sodium borofluoride}}{3NaBF_4}$$

7. **Preparation of Brown and Brown catalyst:** Brown and Brown catalyst is a finely divided black precipitate of pure platinum and is prepared by the reaction between 10 % solution of chloroplatinic acid and $NaBH_4$. Brown and Brown catalyst is an effective catalyst for hydrogenation of alkenes and alkynes.

Lithium Aluminium Hydride ($LiAlH_4$)

Preparation

It is prepared by slow addition of the limited amount of anhydrous aluminium chloride to a thin paste of lithium hydride in ether.

$$\underset{\text{Lithium hydride}}{4\,LiH} + \underset{\text{Aluminium chloride}}{AlCl_3} \longrightarrow \underset{\text{Lithium aluminium hydride}}{LiAlH_4} + 3LiCl\downarrow$$

The lithium chloride precipitate obtained is removed by filtration and the resulting filtrate is evaporated in the absence of air and carbon dioxide to give a solid $LiAlH_4$.

Properties

It is grey white solid, rapidly hydrolyzed with water with the evolution of hydrogen gas (violent reaction with water). It is an extremely valuable reagent for reducing a wide range of functional groups, *e.g.* carbonyl compounds and their derivatives, nitro compounds, azides, aldoximes, nitriles *etc*. Especially it is a versatile reducing agent for carbonyl compounds.

The reduction reaction with $LiAlH_4$ is usually carried out by adding the compound to a flask containing $LiAlH_4$ dissolved in ether or tetra hydrofuran. Excess amount of reagent is destroyed by adding ethyl acetate.

$$\underset{\text{Ethyl acetate}}{4CH_3COOC_2H_5} + LiAlH_4 \longrightarrow \underset{\text{Ethoxy lithium}}{C_2H_5OLi} + \underset{\text{Tri ethoxy aluminium}}{Al(OC_2H_5)_3} + \underset{\text{Ethanol}}{4C_2H_5OH}$$

The reaction is carried out under mild conditions and room temperature.

Advantages

1. It is a low molecular weight reagent and one molecule of it reduces four molecules of carbonyl compounds, 2 molecules of ester *etc*.
2. It has a special property of not attacking double bonds, so unsaturated aldehydes or ketones can be reduced to unsaturated alcohols.
3. But when phenyl group is attached to the β-carbon atom of a α,β-unsaturated carbonyl compounds, the double bond is reduced as shown in the following example.

$$\underset{\text{Cinnamaldehyde}}{C_6H_5\text{-}CH{=}CH\text{-}CHO} \xrightarrow{LiAlH_4} \underset{\text{Benzylethyl alcohol}}{C_6H_5CH_2CH_2CH_2OH}$$

Applications

1. **Reduction of carbonyl compounds:** Aldehydes and ketones are reduced to alcohol by $LiAlH_4$ *via* the formation of aluminium salt of the alcohol intermediate, which is further treated with aqueous acid to yield alcohol.

$$\underset{\text{Aldehyde or ketone}}{4R\overset{O}{\overset{\|}{C}}R'} + LiAlH_4 \longrightarrow [(RR'CHO)_4Al]^{-}Li^{+} \xrightarrow{H^+ \,|\, 4H_2O} \underset{\text{Alcohol}}{4\,RR'CHOH} + LiOH + Al(OH)_3\downarrow$$

$$\underset{\text{Cinnamaldehyde}}{C_6H_5\text{-}CH{=}CH\text{-}CHO} \xrightarrow{LiAlH_4} \underset{\text{Benzylethyl alcohol}}{C_6H_5CH_2CH_2CH_2OH}$$

2. **Reduction of carboxylic acids and their derivatives:** Carboxylic acids and most of their derivatives are reduced to corresponding alcohols by $LiAlH_4$. But amides are reduced to corresponding amines.

$$\underset{\textit{neo}\text{-Pentanoic acid}}{(H_3C)_3C\text{-}COOH} \xrightarrow{LiAlH_4} \underset{\textit{neo}\text{-Pentyl alcohol}}{(H_3C)_3C\text{-}CH_2OH}$$

$$\underset{\text{Cinnamic acid}}{C_6H_5\text{-}CH{=}CH\text{-}COOH} \xrightarrow{LiAlH_4} \underset{\text{3-Phenyl propanol}}{C_6H_5CH_2CH_2CH_2OH}$$

(a) Reduction of acid chlorides: Acid chlorides are reduced to primary alcohols by $LiAlH_4$.

$$\underset{\text{Acetyl chloride}}{CH_3COCl} \xrightarrow{LiAlH_4} \underset{\text{Ethanol}}{CH_3CH_2OH}$$

(b) Reduction of esters: Esters are reduced to alcohols by $LiAlH_4$.

$$\underset{\text{Ethyl benzoate}}{C_6H_5COOC_2H_5} \xrightarrow{LiAlH_4} \underset{\text{Benzyl alcohol}}{C_6H_5CH_2OH}$$

(c) Reduction of amides: Amides are reduced to amines by $LiAlH_4$.

$$\underset{\text{Amide}}{RCONH_2} \xrightarrow{LiAlH_4} \underset{1^\circ\text{ Amine}}{RNH_2}$$

$$\underset{\text{N- Methyl amide}}{RCONHCH_3} \xrightarrow{LiAlH_4} \underset{2^\circ\text{ Amine}}{RCH_2NHCH_3}$$

3. **Reduction of nitro compounds:** Aliphatic and aromatic nitro compounds are reduced to aliphatic primary amines and azo benzenes respectively.

$$\underset{\text{2-Nitrobutane}}{CH_3\text{-}\underset{NO_2}{\underset{|}{CH}}\text{-}CH_2CH_3} \xrightarrow{LiAlH_4} \underset{\text{2-Aminobutane}}{CH_3\text{-}\underset{NH_2}{\underset{|}{CH}}\text{-}CH_2CH_3}$$

$$\underset{\text{Nitrobenzene}}{C_6H_5NO_2} \xrightarrow{LiAlH_4} \underset{\text{Nitroso benzene}}{C_6H_5NO} + \underset{\text{Phenyl hydroxylamine}}{C_6H_5NHOH} \xrightarrow{-H_2O} \underset{\text{Azoxy benzene}}{\begin{matrix} C_6H_5\text{-}N \rightarrow O \\ \| \\ C_6H_5\text{-}N \end{matrix}} \longrightarrow \underset{\text{Azo benzene}}{\begin{matrix} C_6H_5\text{-}N \\ \| \\ C_6H_5\text{-}N \end{matrix}}$$

4. **Reduction of azides:** Azides are reduced to amines by $LiAlH_4$.

$$\underset{\text{Azides}}{RN_3} \xrightarrow{LiAlH_4} \underset{\text{Amines}}{RNH_2}$$

5. **Reduction of nitriles:** Nitriles are reduced in a specific way. If the reagent is added to nitrile, aldehyde is obtained. If the nitrile is added to the reagent, an amine is obtained.

$$\underset{\text{Nitrile}}{R\text{-}C\equiv N} \xrightarrow{LiAlH_4} [RHC{=}NH] \xrightarrow{H_2O} \underset{\text{Aldehyde}}{RCHO} + NH_3$$

$$\underset{\text{Benzonitrile}}{C_6H_5\text{-}C\equiv N} \xrightarrow{LiAlH_4} \underset{\text{Benzylamine}}{C_6H_5CH_2NH_2}$$

Reduction of nitrile group to amine group is used for the synthesis of strychnine alkaloid. Oximes are also reduced to amines.

$$\underset{\text{Oximes}}{RR'C{=}NOH} \xrightarrow{LiAlH_4} \underset{\text{Amines}}{RR'CHNH_2}$$

6. **Reduction of anilides:** Anilides are reduced to secondary or tertiary amines depending on the nature of anilides.

$$\underset{\text{Acetanilide}}{C_6H_5NHCOCH_3} \xrightarrow{LiAlH_4} \underset{\text{Ethyl aniline}}{C_6H_5NHCH_2CH_3}$$

$$\underset{\text{N-Methyl acetanilide}}{C_6H_5N(CH_3)COCH_3} \xrightarrow{LiAlH_4} \underset{\text{N-Methyl-N-ethyl aniline (3° Amine)}}{C_6H_5N(CH_3)CH_2CH_3}$$

7. **Reduction of sulphonyl chlorides, sulphoxides, sulphones and pyridine N-oxides:** In these reactions, the bond between the S or N atoms and other hetero atoms are replaced with hydrogen or unshared pair of electrons when reduced with $LiAlH_4$.

$$\underset{\text{Sulphonyl chloride}}{RSO_2Cl} \xrightarrow{LiAlH_4} \underset{\text{Thiol}}{RSH}$$

$$\underset{\text{Sulphone}}{RS(=O)_2R'} \xrightarrow{LiAlH_4} \underset{\text{Thioether}}{R{-}S{-}R'}$$

$$R_2S(=O)_2 \xrightarrow{LiAlH_4} \underset{\text{Thioether}}{R_2S}$$

Sulphone (SO_2 ring) $\xrightarrow{LiAlH_4}$ Tetrahydrothiophene (S ring)

Pyridine-N oxide ($N^+{-}O^-$) $\xrightarrow{LiAlH_4}$ Pyridine

8. **Reduction of halides:** Halides are reduced to their respective parent hydrocarbon by $LiAlH_4$.

$$\underset{\text{Alkyl chloride}}{4RCl} \xrightarrow{LiAlH_4} \underset{\text{Alkane}}{4RH} + AlCl_3 + LiCl$$

9. **Reduction of α-hydroxy acetylenes:** Acetylenes with hydroxyl group at α-positions are reduced to alkenes by $LiAlH_4$.

$$R-C\equiv C-C(OH)(R_1)(R_2) \xrightarrow{LiAlH_4} R-CH=CH-C(OH)(R_1)(R_2)$$

10. **Determination of active hydrogen:** $LiAlH_4$ removes molecular hydrogen from the compounds containing active hydrogen, so it is used for the estimation of the active hydrogen.

$$LiAlH_4 + \underset{\text{Alcohol}}{4ROH} \longrightarrow (RO)_4LiAl + 4H_2$$

$$LiAlH_4 + \underset{\text{Amine}}{4RNH_2} \longrightarrow (RNH)_4LiAl + 4H_2$$

11. **Deacetylation and debenzylation:**

(i) $LiAlH_4$ removes acetyl group and benzyl group from O, N, S. Hence $LiAlH_4$ is an important reagent in synthetic organic chemistry.

$$\underset{\text{Ester}}{RO\overset{O}{\overset{||}{C}}CH_3} \xrightarrow{LiAlH_4} \underset{\text{Alcohol}}{ROH} + CH_3\overset{O}{\overset{||}{C}}OH$$

$$\underset{\text{Benzyl alcohol}}{C_6H_5CH_2OH} \xrightarrow{LiAlH_4} \underset{\text{Toluene}}{C_6H_5CH_3} + H_2O$$

(ii) Similarly, acetyl group of an ester is easily cleaved by $LiAlH_4$. Hence it is used for deacetylation process of compounds which are sensitive to acids and bases.

$$\underset{\text{Alkyl benzylamine}}{R-NH-CH_2C_6H_5} \xrightarrow{LiAlH_4} \underset{\text{Amine}}{RNH_2} + \underset{\text{Toluene}}{CH_3-C_6H_5}$$

$$\underset{\text{Alkylamide}}{R-NHCOCH_3} \xrightarrow{LiAlH_4} \underset{\text{Amine}}{RNH_2} + \underset{\text{Ethanol}}{CH_3CH_2OH}$$

Clemmensen Reduction

Aldehydes and ketones are reduced with Zn(Hg)/HCl (Zinc amalgam and HCl) to yield alkanes. This catalyst or method is suitable for compounds unstable in base but stable in acid.

$$\underset{\text{2-Butanone}}{H_3CH_2C-\overset{O}{\overset{||}{C}}-CH_3} \xrightarrow[HCl]{Zn/Hg} \underset{n\text{-Butane}}{CH_3CH_2CH_2CH_3}$$

Mechanism

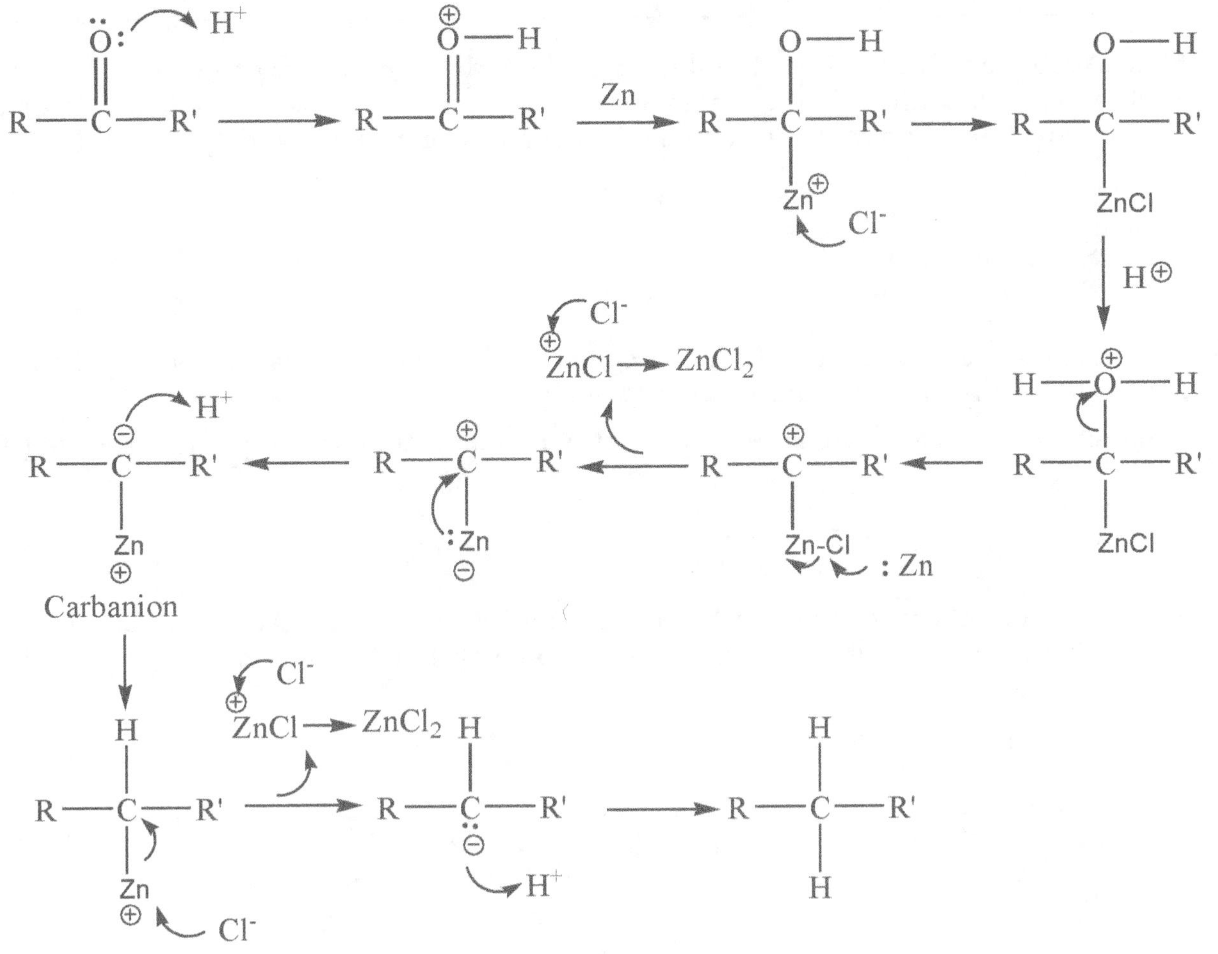

Ö: H⁺ R—C—R' → R—C(=O⊕—H)—R' —Zn→ R—C(OH)(Zn⊕)—R' Cl⁻ → R—C(OH)(ZnCl)—R'

—H⊕→ H—O⊕(H)—C(R)(R')—ZnCl → R—C⊕(R')—Zn-Cl : Zn → R—C⊕(R')—:Zn⊖ (⊕ZnCl → $ZnCl_2$, Cl⁻) → R—C⊖(R')—Zn⊕ H⁺

Carbanion

↓

R—C(H)(R')—Zn⊕ Cl⁻ → (⊕ZnCl → $ZnCl_2$, Cl⁻) R—C(H)(R')⊖ H⁺ → R—C(H)(H)—R'

Application:

1. This reaction is widely used to convert a carbonyl group into a meythylene group.

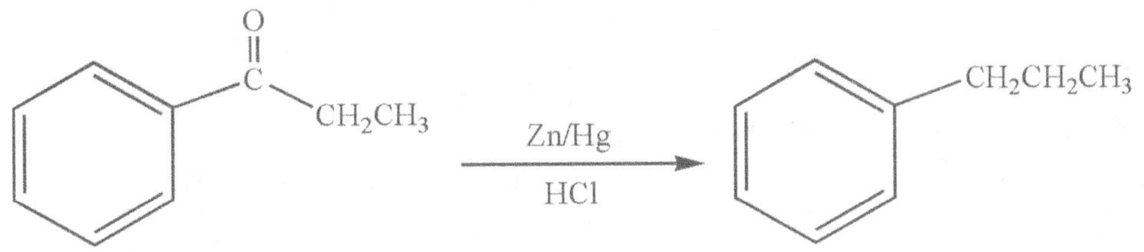

$$C_6H_5COCH_2CH_3 \xrightarrow[HCl]{Zn/Hg} C_6H_5CH_2CH_2CH_3$$

Ethyl phenyl ketone → Propyl benzene

2. In the preparation of polycyclic aromatics and aromatics containing unbranched side hydrocarbon chain.
3. To reduce aliphatic, mixed aliphatic and aromatic carbonyl compounds.

$$CH_3\text{-}(CH_2)_3\text{-}CHO \xrightarrow[HCl]{Zn/Hg} CH_3\text{-}(CH_2)_3\text{-}CH_3$$

Pentanal → Pentane

$$CH_3COCH_3 \xrightarrow[HCl]{Zn/Hg} CH_3CH_2CH_3$$

Acetone → Propane

Birch Reduction

A.J. Birch, an Australian chemist in 1944 found that the reduction of benzene or its derivatives with sodium or lithium in a mixture of liquid ammonia and alcohol gives 1,4-cyclohexadiene. This reaction is known as Birch reduction. It is a convenient method for synthesizing a wide variety of useful cyclic dienes.

Benzene —Na/Li, Liquid ammonia, ROH→ Cyclohexadiene

Mechanism: The mechanism of Birch reduction is similar to that of the mechanism of reduction of alkynes to *trans* alkynes in the presence of sodium in liquid ammonia.

***Step* 1: Formation of solvated electrons:** Sodium reacts with liquid ammonia to produce solvated electrons.

$$Na + NH_3 \rightleftharpoons \dot{NH_3}e + \overset{+}{Na}$$

(Dark blue solution contain solvated electrons)

***Step* 2: Formation of cyclohexadienyl radical anion:** The radical anion produced by the addition of solvated electrons to the benzene upon further reaction with alcohol abstracts a proton (due to its high basicity) from an alcohol to yield cyclohexadienyl radical.

Benzene —e⁻ (electron)→ Radical anion —H–Ö–R→ Free radical + R–Ö:⁻

***Step* 3: Formation of 1,4-cyclohexadiene *via* the carbanion intermediate:** The cyclohexadienyl radical fastly abstract another solvated electron to form cyclohexadienyl carbanion. Protonation of this carbanion leads to the formation of 1,4-cyclohexadiene.

Freeradical —Solvated e-→ Cyclohexadienyl carbanion —H–Ö-R→ 1,4-Cyclohexadiene + R-Ö:⁻

Application:

1) It used to synthesize the cyclohexadiene and substituted cyclohexadienes.
2) It is used in the reduction of aromatic rings and conjugated dienes.
3) It is used in the synthesis of 5,8-dihydro-1-naphthol on reduction of 1-naphthol with Na/Liquid ammonia.
4) It is used in the synthesis of 3,5-diketocyclohexane carboxylic acid by the reduction of 3,4,5-trimethoxybenzoic acid.
5) It is used to reduce the aromatic ring of dehydroabietic acid.

Effect of substituents on Birch reduction: Generally, electron withdrawing substituents stabilize the carbanion while electron donating substituents destabilizes. So, the Birch reduction takes place on carbon atoms bearing electron withdrawing substituents (carbon containing-C=O group) and not carbon atoms bearing electron releasing substituents (such as alkyl and alkoxyl groups). Generally a carbon atom containing electron withdrawing carbonyl group is reduced.

$$\text{Benzoic acid} \xrightarrow[\text{ROH (Ethanol)}]{\text{Na in Liq. ammonia}} \text{Dihydrobenzoic acid (90\%)}$$

A carbon containing an electron releasing group such as alkoxyl group is not reduced.

Electron releasing substituents deactivates the aromatic ring towards Birch reduction. Here lithium is used instead of sodium together with a co solvent (THF, Tetra hydro furan) and a weaker proton source (*t*-butyl alcohol). The weaker proton source combined with the stronger reducing agent enhances the reduction.

$$\text{Anisole} \xrightarrow[NH_3/THF]{Li,\ (CH_3)_3COH} \text{Dihydro anisole}$$

Wolf-Kishner Reduction

Aldehydes and ketones are first converted into their derivatives such as hydrazones, semicarbazones, azines (by treatment with hydrazines, semicarbazides *etc*) which upon further reaction with a strong base such as potassium hydroxide or potassium *tert*-butoxide in dimethyl sulphoxide (DMSO) solvent yield alkanes. This reaction is suitable for the carbonyl compounds that are stable in base.

$$C_6H_{11}-\overset{O}{\overset{\|}{C}}-CH_3 \xrightarrow[-H_2O]{NH_2NH_2} C_6H_{11}-\overset{NNH_2}{\overset{\|}{C}}-CH_3 \xrightarrow{KOH} C_6H_{11}-CH_2CH_3$$

Cyclohexyl methyl ketone → Cyclohexyl methyl ketone hydrazone → Ethyl cylohexane

Mechanism:

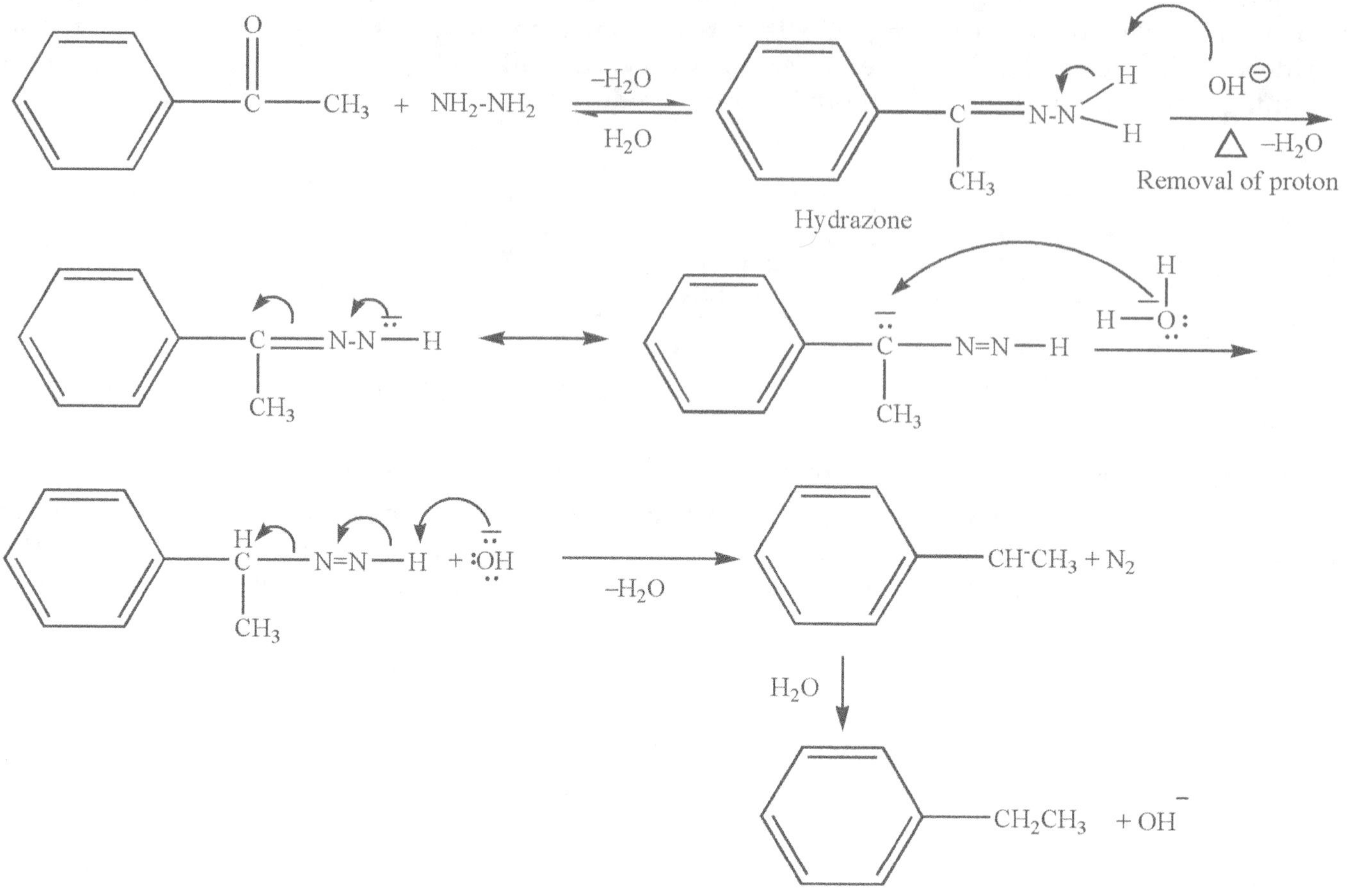

Hydrazone

Removal of proton

$OH^{\ominus}$ Δ $-H_2O$

Application: This reaction used frequently for the reduction of carbonyl group (-C=O) to methylene (-CH_2) during the synthesis of large compounds.

1. Reduction of camphor to camphene

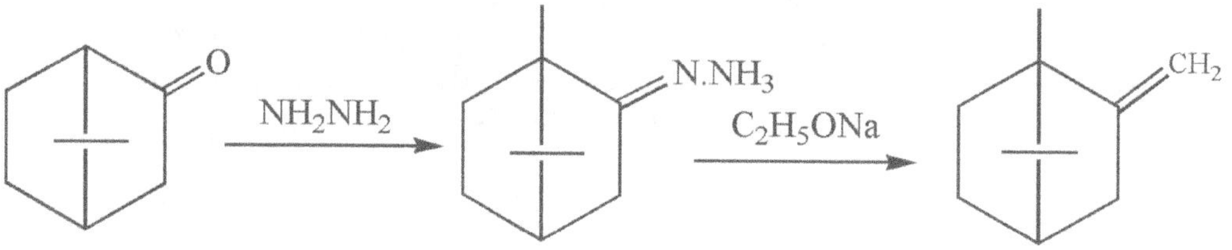

Camphor $\xrightarrow{NH_2NH_2}$ (=N.NH_3) $\xrightarrow{C_2H_5ONa}$ Camphane

2. Synthesis of pyrroles

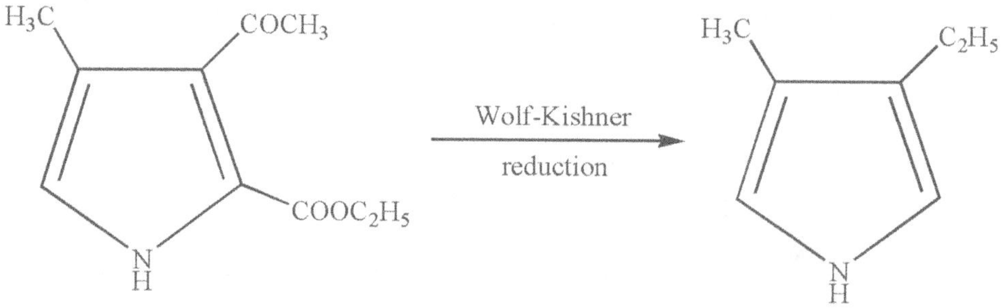

$\xrightarrow{\text{Wolf-Kishner reduction}}$ 3-Ethyl-4-methyl Pyrrole

3. Elucidating the structure of Oestrone

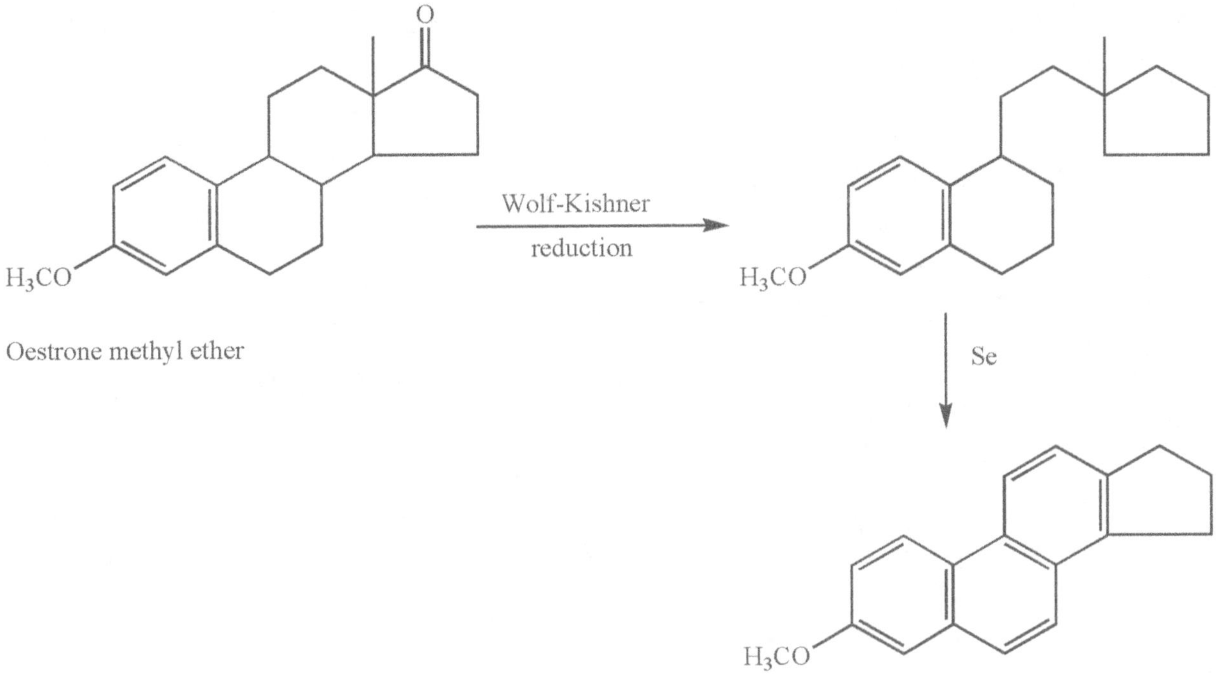

H_3CO

Oestrone methyl ether

Wolf-Kishner reduction

H_3CO

Se

H_3CO

7-Methoxy-1,2-cyclopentenophenanthrene

Oppaneur Oxidation

It is the reverse reaction of Meerwein-Ponndorf-Verley reduction. It involves the oxidation of alcohols (2° alcohols) in to ketones in the presence of aluminium tertiary butoxide in toluene or benzene.

$$\underset{2^\circ \text{ Alcohol}}{R_2CHOH} + \underset{\text{Acetone}}{CH_3-\overset{O}{\overset{\|}{C}}-CH_3} \xrightleftharpoons{Al[OC(CH_3)_3]_3} \underset{\text{Ketone}}{R-\overset{O}{\overset{\|}{C}}-R} + \underset{\text{Isopropanol}}{(CH_3)_2CHOH}$$

The large excess amount of acetone or cyclohexanone is commonly used to proceed the reaction in the desired direction.

Mechanism:

It involves the migration of hydride ion within an acyclic complex.

$$\underset{2^\circ \text{ Alcohol}}{3(R)_2CHOH} + Al[OC(CH_3)_3]_3 \rightleftharpoons [(CH_3)_2CHO]_3Al + \underset{\text{Ketone}}{3RCOR}$$

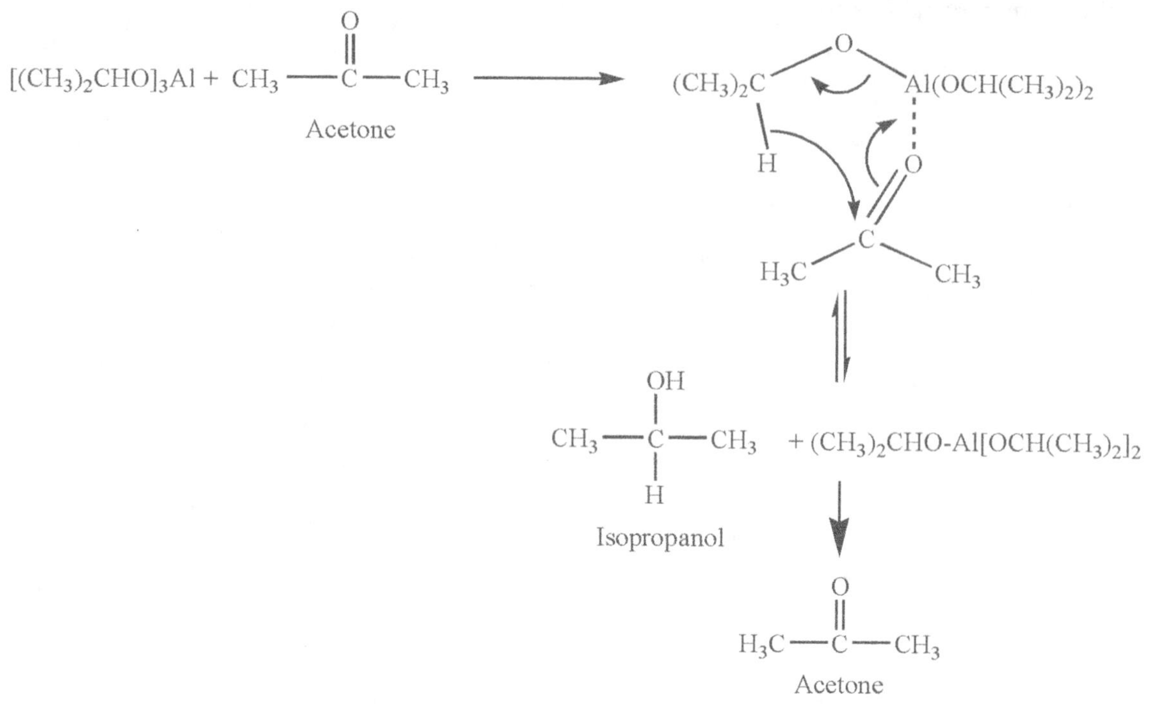

$[(CH_3)_2CHO]_3Al + CH_3-C(=O)-CH_3$ (Acetone) → $(CH_3)_2C$ ... O ... $Al(OCH(CH_3)_2)_2$ / H ... O=C(H_3C)(CH_3)

⇌

$CH_3-CH(OH)-CH_3$ (Isopropanol) $+ (CH_3)_2CHO\text{-}Al[OCH(CH_3)_2]_2$

↓

$H_3C-C(=O)-CH_3$

Acetone

Applications

1. Synthesis of cholestenone

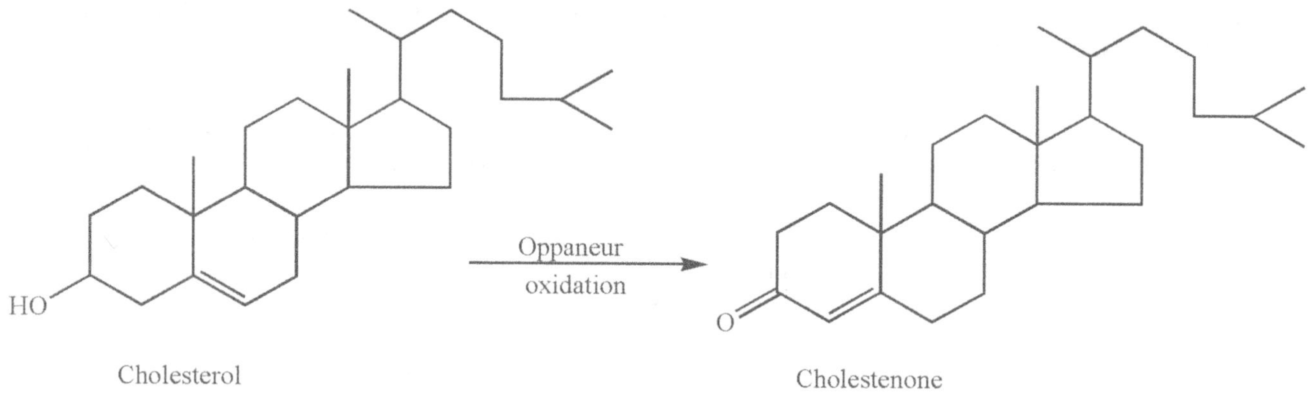

Cholesterol → (Oppaneur oxidation) → Cholestenone

2. Conversion of formates to carbonyl group

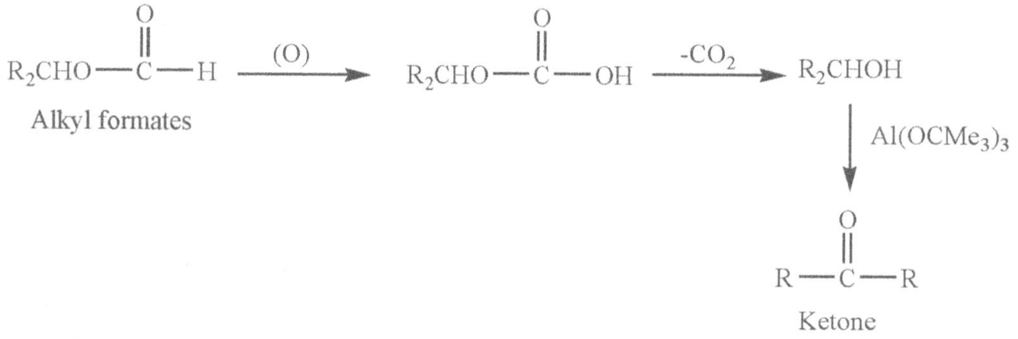

$R_2CHO-C(=O)-H$ (Alkyl formates) $\xrightarrow{(O)}$ $R_2CHO-C(=O)-OH$ $\xrightarrow{-CO_2}$ R_2CHOH $\xrightarrow{Al(OCMe_3)_3}$ $R-C(=O)-R$

Ketone

Dakin's Reaction

The replacement reaction of the *o*-hydroxy, *p*-hydroxy or *o*-amino benzaldehyde or ketone by using hydrogen peroxide is known as Dakin reaction.

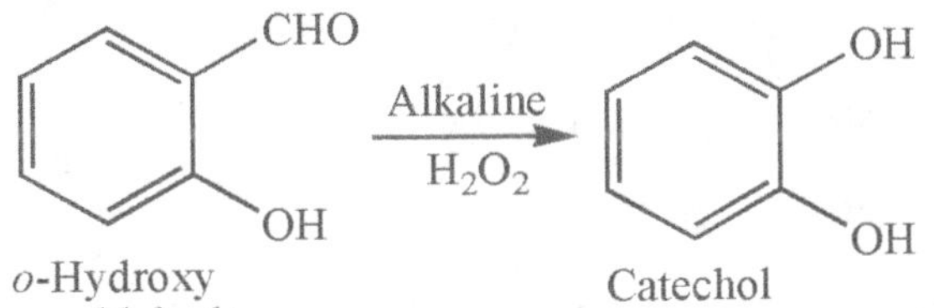

o-Hydroxy benzaldehyde (CHO, OH) —Alkaline H_2O_2→ Catechol (OH, OH)

Mechanism: The mechanism of the reaction is some what similar to that of Bayer-Villiger reaction. In this reaction, the formyl group is removed as formic acid and replaced by hydroxyl group.

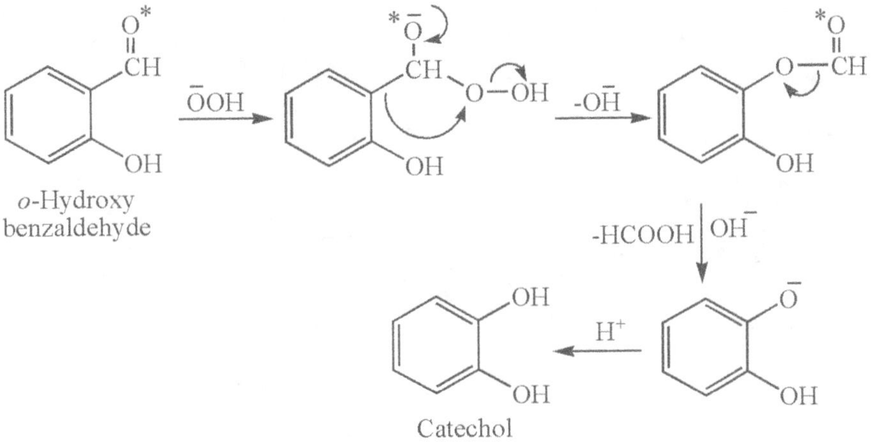

o-Hydroxy benzaldehyde —$\bar{O}OH$→ —$-O\bar{H}$→ —$-HCOOH$ | OH^-→ —H^+→ Catechol

Note: The above mechanism is supported by tracer technique. All of the labelled oxygen is found in formic acid.

Applications: Dakin reaction is used to synthesize polyhydric phenols from naturally occurring hydroxy aldehydes.

Some of the examples are shown below.

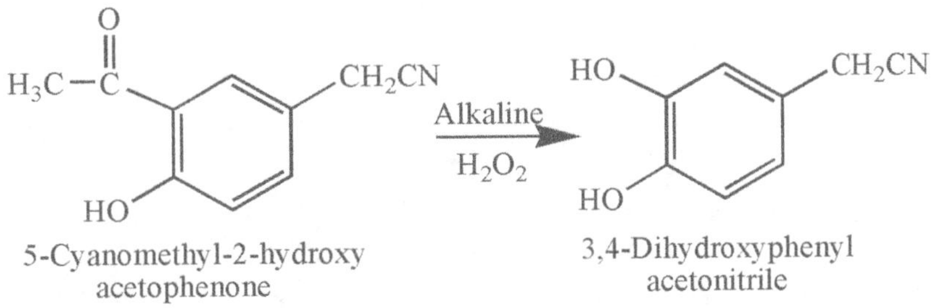

5-Cyanomethyl-2-hydroxy acetophenone (H_3C-CO, CH_2CN, HO) —Alkaline H_2O_2→ 3,4-Dihydroxyphenyl acetonitrile (HO, HO, CH_2CN)

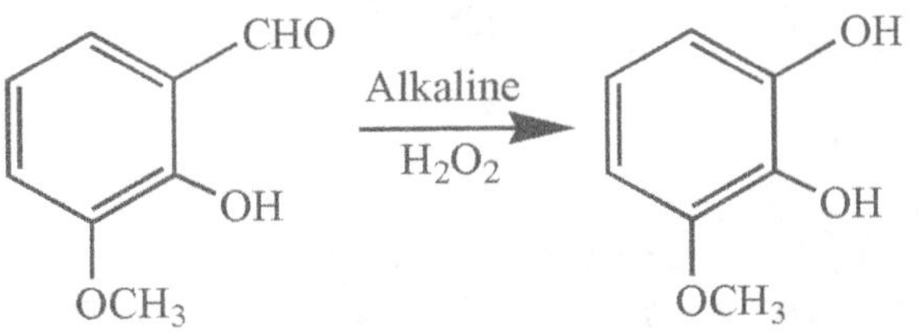

o-Vanillin (CHO, OH, OCH_3) —Alkaline H_2O_2→ Pyrogallol-1-mono methyl ether (OH, OH, OCH_3)

Beckmann Rearrangement

The conversion of ketoximes to N-substituted amides by heating with conc. H_2SO_4 or BF_3 or poly phosphoric acid or PCl_5 or $SOCl_2$ *etc* called as Beckmann rearrangement.

$$\underset{\text{Ketoxime}}{\overset{R}{\underset{R'}{>}}C{=}NOH} \xrightarrow{H_2SO_4} \underset{\text{N-Substituted amide}}{R'CONHR \text{ or } RCONHR'}$$

The two kind of amides (*syn* and *anti*) obtained are identified by their hydrolytic product. In this reaction, the migration of group is always anti (trans) to the leaving group which is denoted as follows.

$$\underset{\text{Ketoxime}}{R{-}\underset{\underset{NOH}{\|}}{C}{-}R'} \xrightarrow{H_2SO_4} \underset{\text{Amide (Enol form)}}{\left[HO{-}\underset{\underset{NR'}{\|}}{C}{-}R\right]} \longrightarrow \underset{\text{Amide (Keto form)}}{R{-}\underset{\underset{NHR'}{|}}{C}{=}O}$$

$$R{-}\underset{\underset{N\text{-}OH}{\|}}{C}{-}R' \xrightarrow{H_2SO_4} \underset{\text{Amide (Enol form)}}{\left[R'{-}\underset{\underset{NR}{\|}}{C}{-}OH\right]} \longrightarrow \underset{\text{Amide (Keto form)}}{R''{-}\underset{\underset{NHR}{|}}{C}{=}O}$$

Mechanism: It takes place by intramolecular mechanism.

$$\overset{R}{\underset{R'}{>}}C{=}NOH \xrightarrow{H^+} \overset{R}{\underset{R'}{>}}C{=}N\text{-}\overset{\oplus}{O}H_2 \xrightarrow[-H_2O]{} \overset{R}{\underset{R'}{>}}C{=}\overset{\oplus}{N}:$$

$$\longrightarrow \left[R{-}\overset{\oplus}{C}{=}\ddot{N}{-}R' \longleftrightarrow R{-}C{\equiv}\overset{\oplus}{N}{-}R'\right] \xrightarrow{H_2O}$$

$$\overset{R}{\underset{H_2\overset{\oplus}{O}}{>}}C{=}NR' \xrightarrow{-H^+} \overset{R}{\underset{H-O}{>}}C{=}N\text{-}R' \longrightarrow \overset{R}{\underset{O}{>}}C{-}N{<}\overset{H}{\underset{R'}{}}$$

This reaction occurs *via* the formation of carbonium ion which accepts hydroxyl ion or a water molecule from a solvent, so the original oxygen atom of the oxime is lost during the reaction.

Application

1. **Synthesis of isoquinoline:** Cinnamaldoxime undergoes Beckmann rearrangement in the presence of P_2O_5 followed by cyclisation yields isoquinoline.

$$\text{Cinnamaldoxime} \xrightarrow{P_2O_5} [\text{intermediate}] \xrightarrow{-H_2O} \text{Isoquinoline}$$

2. It is also used to determine the configuration of ketoximes. Based on the hydrolysis products obtained (the anti and syn product obtained by hydrolysis of Beckmann's rearrangement reaction).

Schmidt Rearrangement (Rearrangement to Electron Deficient Nitrogen)

Carboxylic acids or aldehydes or ketones react with hydrazoic acid in the presence of conc. H_2SO_4 to yield amines or mixture of cyanides and formyl derivatives of primary amines or amides respectively. This reaction is known as Schmidt rearrangement or Schmidt reaction.

$$\underset{\text{Carboxylic acid}}{RCOOH} + \underset{\text{Hydrazoic acid}}{HN_3} \xrightarrow{H_2SO_4} \underset{\text{Amines}}{RNH_2 + CO_2 + N_2}$$

$$\underset{\text{Aldehyde}}{2RCHO} + \underset{\text{Hydrazoic acid}}{2HN_3} \longrightarrow \underset{\text{Cyanide}}{RCN} + \underset{\text{Formyl derivative of amine}}{RNHCHO} + 2N_2 + H_2O$$

$$\underset{\text{Ketone}}{R-\overset{O}{\overset{\|}{C}}-R'} + HN_3 \xrightarrow{H_2SO_4} \underset{\text{Amide}}{R-\overset{O}{\overset{\|}{C}}-NHR'} + N_2$$

Mechanism

Reaction with acids: The reaction takes place *via* the formation of acid azide (it is present as its conjugate acid) in the presence of conc. H_2SO_4. The later one loses nitrogen with or without heating (depending upon the nature of the carboxylic acid used) to yields amine.

1. If the starting material is a simple acid such as benzoic acid, the reaction requires heating for the removal of nitrogen from acid azide. The mechanism of the reaction is as follows.

$$\underset{\text{Acid}}{R-\overset{O}{\overset{\|}{C}}-OH} \xrightarrow{H^+} R-\overset{+}{C}(OH)_2 \xrightarrow{N_3H} R-C(OH)_2-NH-\overset{+}{N}\equiv N \xrightarrow{-H_2O} R-\overset{O}{\overset{\|}{C}}-NH-\overset{+}{N}\equiv N$$

$$\xrightarrow{-N_2} R-\overset{O}{\overset{\|}{C}}-\overset{+}{N}-H \xrightarrow{-H^+} \underset{\text{Isocyanate}}{O=C=N-R} \xrightarrow{H_2O} CO_2 + \underset{\text{Amine}}{NH_2R}$$

Reaction with aldehydes and ketones

For aldehydes

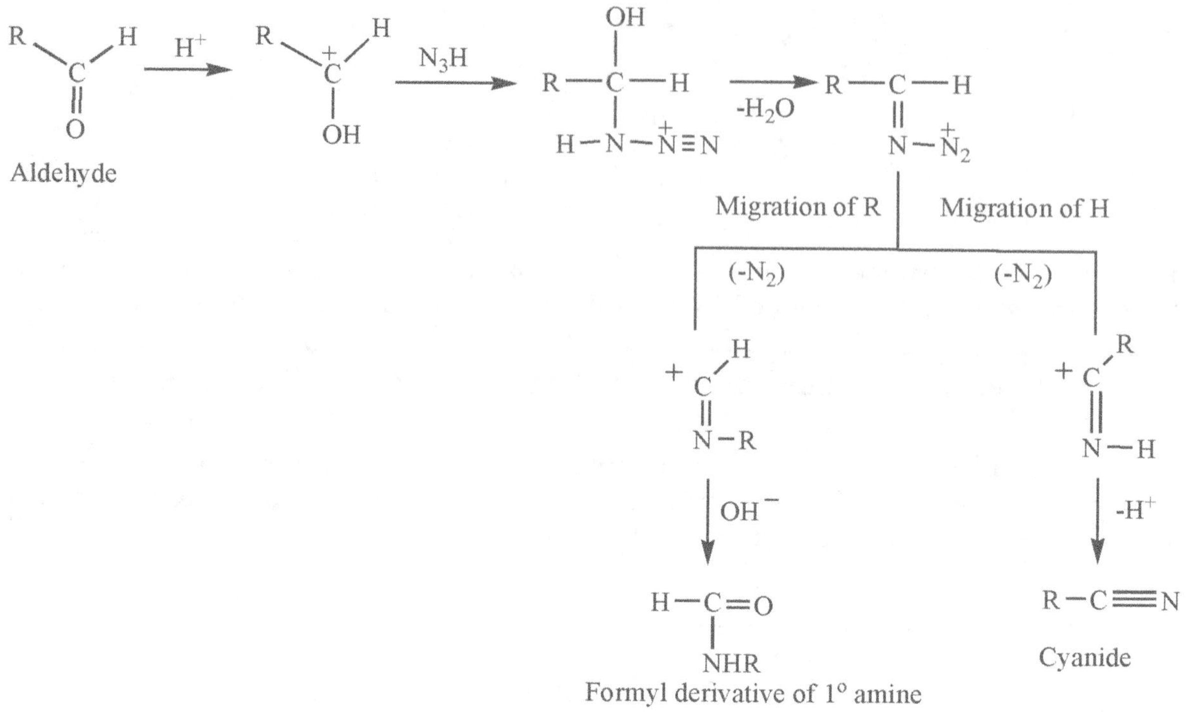

R–CHO (Aldehyde) $\xrightarrow{H^+}$ R–C⁺H–OH $\xrightarrow{N_3H}$ R–CH(OH)–NH–N⁺≡N $\xrightarrow{-H_2O}$ R–CH=N–N⁺₂

Migration of R (−N_2): ⁺C(H)=N–R $\xrightarrow{OH^-}$ H–C(=O)–NHR

Formyl derivative of 1° amine

Migration of H (−N_2): ⁺C(R)=N–H $\xrightarrow{-H^+}$ R–C≡N

Cyanide

For ketones

R–CO–R' $\xrightarrow{H^+}$ R–C⁺(OH)–R' $\xrightarrow{N_3H}$ R–C(OH)(R')–NH–N⁺≡N $\xrightarrow{-H_2O}$ R(R')C=N–N⁺≡N: $\xrightarrow{-N_2}$ R–C⁺=NR' $\xrightarrow{H_2O}$ H_2O^+–C(=NR')–R $\xrightarrow{-H^+}$ HO–C(=NR')–R

Amide (Enol form)

⇅

O=C(NHR')–R

Amide (Keto form)

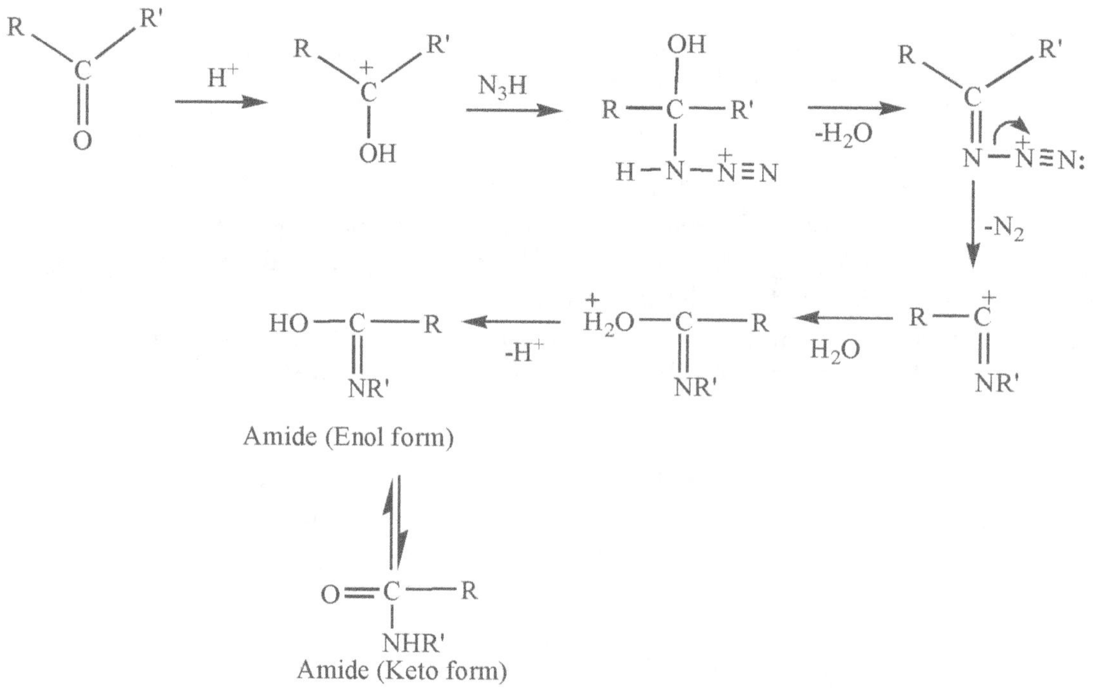

Applications

1. **Synthesis of α-aminoacids:** Schmidt rearrangement is used for the synthesis of α-aminoacids from acetoacetic ester.

$$CH_3CO\text{-}\underset{|}{\overset{R}{C}}H\text{-}COOC_2H_5 \xrightarrow[-N_2]{N_3H/H_2SO_4} CH_3CONH\overset{R}{\overset{|}{C}}HCOOC_2H_5$$

Alkyl acetoacetic ester

$$\downarrow 2H_2O \quad \text{Acid hydrolysis}$$

$$H_2N\text{-}\overset{R}{\overset{|}{C}}H\text{-}COOH + C_2H_5OH + CH_3COOH$$

α-Amino acid

2. Schmidt reaction is the direct method for the preparation of 1° amines from carboxylic acid with good yield. The hydrazoic acid used in this reaction is explosive and poisonous in nature hence it must be used carefully.

Claisen-Schmidt Condensation

The condensation reaction between aromatic aldehydes without α-hydrogen atom and aliphatic aldehydes or ketones or esters with active hydrogen atom in the presence of 10 % alkali solution to give α,β-unsaturated aldehydes or ketones. This reaction is called as Claisen-Schmidt or Claisen reaction.

$$\underset{\text{Benzaldehyde}}{C_6H_5CHO} + \underset{\text{Acetaldehyde}}{CH_3CHO} \xrightarrow{10\ \%\ NaOH} \underset{\text{Cinnamaldehyde}}{C_6H_5CH{=}CH\text{-}CHO} + H_2O$$

$$\underset{\text{Benzaldehyde}}{C_6H_5CHO} + \underset{\text{Acetone}}{CH_3\overset{O}{\overset{||}{C}}CH_3} \xrightarrow{10\ \%\ NaOH} \underset{\text{Benzalacetone}}{C_6H_5CH{=}CH\text{-}COCH_3} + H_2O$$

Note: Dilute alkali should be used for this reaction, because strong alkali leads to Cannizaro reaction.

Mechanism: This reaction occurs *via* the formation of carbanion intermediate.

Step 1:

$$H\text{-}CH_2CHO + OH^- \rightleftharpoons \overset{-}{C}H_2CHO + H_2O$$

Step 2:

$$C_6H_5\text{—}\underset{\underset{H}{|}}{\overset{O}{\overset{||}{C}}} + \overset{-}{C}H_2CHO \rightleftharpoons C_6H_5\text{—}\underset{\underset{H}{|}}{\overset{O^-}{\overset{|}{C}}}\text{— }CH_2CHO \overset{H^+}{\rightleftharpoons} C_6H_5\text{—}\underset{\underset{H}{|}}{\overset{OH}{\overset{|}{C}}}\text{—}CH_2CHO$$

$$\downarrow\uparrow -H^+$$

$$\underset{\text{Cinnamaldehyde}}{C_6H_5CH{=}CH\text{-}CHO} \xleftarrow[-\overset{-}{O}H]{} C_6H_5\text{—}\underset{\underset{H}{|}}{\overset{OH}{\overset{|}{C}}}\text{— }\overset{-}{C}HCHO$$

Applications

1. This reaction is commonly used for the synthesis of α,β-unsaturated carbonyl compounds which are used in perfume preparations.
2. **Synthesis of natural products:** Claisen condensation is successfully used for the synthesis of various kind of natural product such as β-ionone, piperine, flavones *etc*.
   (a) Synthesis of piperine:
   (b) Synthesis of β-ionone.

$$\underset{\text{Benzaldehyde}}{C_6H_5CHO} + \underset{\text{Acetaldehyde}}{CH_3CHO} \xrightarrow{10\%\ NaOH} \underset{\text{Cinnamaldehyde}}{C_6H_5CH{=}CHCHO} + H_2O$$

Piperonal $\xrightarrow[NaOH]{CH_3CHO}$ 3,4-methylenedioxy-C_6H_3–$CH{=}CHCHO$

Perkin reaction: $(CH_3CO)_2O$, CH_3COONa → 3,4-methylenedioxy-C_6H_3–$CH{=}CHCH{=}CHCOOH$ (Piperinic acid)

Piperidine, PCl_5 → 3,4-methylenedioxy-C_6H_3–$CH{=}CH{-}CH{=}CH{-}C(=O){-}N$(piperidine) (Piperine)

Citral + CH_3COCH_3 $\xrightarrow[-H_2O]{Ba(OH)_2}$ $CH{=}CHCOCH_3$ (pseudoionone)

H_2SO_4 → α-Ionone + β-Ionone (each bearing $CH{=}CHCOCH_3$)

Probable Questions

1. Write the method of preparation of lithium aluminium hydride.
2. Explain the properties of lithium aluminium hydride.
3. Write the advantages of lithium aluminium hydride.
4. Write a note on various applications of lithium aluminium hydride.
5. Write the advantages of sodium boro hydride.
6. Write the applications of sodium boro hydride.
7. What is Dakin reaction?
8. Explain the mechanism of Dakin reaction.
9. Write a note on various applications of Dakin reaction.
10. Define Schmidt rearrangement and explain its mechanism in detail.
11. Give the applications of Schmidt rearrangement.
12. What is Beckmann rearrangement?
13. Explain the mechanism of Beckmann rearrangement.
14. Write about applications of Beckmann rearrangement.
15. Define and explain Claisen-Schmidt condensation.

Index

B

C

D

E

F

G

H

I

J

K

L

M

N

O

P

Q

R

S

T

U

V

W

X

Y

Z

www.ingramcontent.com/pod-product-compliance
Lightning Source LLC
Chambersburg PA
CBHW080545110726
47973CB00005B/86

* 9 7 8 9 3 8 9 9 7 4 3 7 9 *